Oxford Resources for IB
Diploma Programme

2023 EDITION

BIOLOGY

COURSE COMPANION

Andrew Allott
David Mindorff

OXFORD
UNIVERSITY PRESS

Great Clarendon Street, Oxford, OX2 6DP, United Kingdom

Oxford University Press is a department of the University of Oxford. It furthers the University's objective of excellence in research, scholarship, and education by publishing worldwide. Oxford is a registered trade mark of Oxford University Press in the UK and in certain other countries.

First published in 2023

British Library Cataloguing in Publication Data
Data available

9781382016339

9781382016377 (ebook)

10 9 8 7 6 5 4 3

Paper used in the production of this book is a natural, recyclable product made from wood grown in sustainable forests.

The manufacturing process conforms to the environmental regulations of the country of origin.

Printed in China by Shanghai Offset Printing Products Ltd

Acknowledgements

The "In cooperation with IB" logo signifies the content in this textbook has been reviewed by the IB to ensure it fully aligns with current IB curriculum and offers high-quality guidance and support for IB teaching and learning.

The Publisher wishes to thank the International Baccalaureate Organization for permission to reproduce their intellectual property.

The Publisher would like to thank the following members of the DP Science 2023 Research Panel for sharing their insights, expertise, and feedback:
B. Jane Taylor, Menna Shawky El Sherbiny, Aura Milena Vera, Jessica K. Hartman, Nilan A. Senaratna, Joanne Walton, Holly O'Donal Allen, Gavin Terry, Arnavaz Kollah, Amanda Lockhart, Lisa Privitera, Ioannis Papatheocharis, Anastasia Denisova, Dr. Regina Katz, Srivalli Singh, Emma Nason, Rosanne Jacobs-Sarkin, Aditya Rao, María Cristina Salvador Crow, Tania M Hodges
The publisher would like to thank the following for permissions to use copyright material:

Cover: Darrell Gulin/Getty Images. **Photos: piv:** International Baccalaureate Organisation; **p2:** Claude Nuridsany & Maria Perennou / Science Photo Library; **p3(t):** Eye of Science / Science Photo Library; **p3(b):** Library of Congress/Science Photo Library; **p4:** Hubble, M. Kornmesser/ESA; **p5:** Andrew Allott; **p6(t):** oliveromg/Shutterstock; **p6(b):** Stephen Barlow; **p8(tl):** Des Callaghan; **p8(tr):** © 2020 Spirina, Voronkova and Ignatov; **p9:** Andrew Allott; **p10(t):** Emilio Ereza/Alamy Stock Photo; **p10(b):** Thekopmylife/Shutterstock; **p12(t):** Markus Varesvuo/Nature Picture Library/Science Photo Library; **p12(b):** Кирилл Уютнов, CC BY-SA 4.0 / Wikimedia Commons; **p13:** John Thomas/Science Photo Library; **p15:** Lebrecht Music & Arts/Alamy Stock Photo; **p16:** Power and Syred/Science Photo Library; **p17:** Science History Images/Alamy Stock Photo; **p23:** James King-Holmes/Science Photo Library; **p27:** PDB ID: 1AOI, DOI Citation: Luger, K., Maeder, A.W., Richmond, R.K., Sargent, D.F., Richmond, T.J. doi: 10.2210/pdb1AOI/pdb; **p28:** LEE D. SIMON / Science Photo Library; **p33:** Madprime, CC0 / Wikimedia Commons; **p34:** Dr Keith Wheeler/Science Photo Library; **p36:** Eduard Muzhevskyi/Science Photo Library/Getty Images; **p37(t):** Corbin17/Alamy Stock Photo; **p37(b):** NOAA; **p38:** M. Kornmesser/ESO; **p39:** Space Science Institute/JPL/NASA; **p40:** Thomas Deerinck, NCMIR/Science Photo Library; **p41:** Matt Heaton/FossilEra; **p42:** Andrew Allott; **p43:** Science Photo Library/Alamy Stock Photo; **p44:** PDB ID: 1FFK, DOI Citation: Ban, N., Nissen, P., Hansen, J., Moore, P.B., Steitz, T.A. doi: 10.2210/pdb1FFK/pdb; **p46:** Jane Gould/Alamy Stock Photo; **p47:** GSFC/METI/ERSDAC/JAROS, and U.S./Japan ASTER Science Team/NASA; **p48:** Noaa Okeanos Explorer Program/Science Photo Library; **p49(t):** Ted Kinsman/Science Source/Science Photo Library; **p49(b):** Dr Yorgos Nikas / Science Photo Library; **p50:** Royal Institution Of Great Britain / Science Photo Library; **p51:** © 2004 MBARI; © 2011 MBARI (Kim Fulton-Bennett) **p52:** Copyright of ZEISS Microscopy; **p54(l):** Andrew Allott; **p54(r):** William Allott; **p55(t):** © Woods Hole Oceanographic Institution; **p55(m):** Professors P. Motta & T. Naguro / Science Photo Library; **p55(b):** Andrew Allott; **p56(tl):** Andrew Allott; **p56(tr):** Peter Andrus, CC BY-SA 4.0 / Wikimedia Commons; **p56(b):** Wim Van Egmond/Science Photo Library; **p57(l):** Jeroen Rouwkema, CC BY-SA 3.0 / Wikimedia Commons; **p57(r):** Copyright of ZEISS Microscopy; **p58(t):** imageBROKER/Alamy Stock Photo; **p58(b):** Volgi archive/Alamy Stock Photo; **p59:** Toyoshi Fujimoto; **pp60-61:** Ge, P., Scholl, D., Prokhorov, N.S. et al. Action of a minimal contractile bactericidal nanomachine. Nature 580, 658–662 (2020). https://doi.org/10.1038/s41586-020-2186-z; **p63(t):** A. Dowsett, National Infection Service /Science Photo Library; **p63(b):** Monkey Business Images/Shutterstock; **p64:** Bénédicte Salin/Université de Bordeaux Cell and Molecular Imaging Service; **p65:** Eric Grave/Science Photo Library; **pp66-67:** Wolfgang Bettighofer, www.protisten.de; **p69:** Prof. Alexandro Bonifaz/ Hospital General de México; **p70(a):** Biophoto Associates / Science Photo Library; **p70(b):** Microscape / Science Photo Library; **p70(c):** Don W. Fawcett / Science Photo Library; **p70(d):** Biophoto Associates / Science Photo Library; **p70(e):** Dr Gopal Murti / Science Photo Library; **p71(a):** Dr Gopal Murti / Science Photo Library; **p71(b):** Science History Images/Alamy Stock Photo; **p71(c):** Dr Kari Lounatmaa / Science Photo Library; **p71(d):** Microscape / Science Photo Library; **p71(e):** Don W. Fawcett / Science Photo Library; **p72(t):** Heiti Paves/Alamy Stock Photo; **P72(m):** Dr Gopal Murti / Science Photo Library; **p72(b):** Dr Gary Gaugler / Science Photo Library; **p73:** Andrew Allott; **p75:** Frank Fox/Science Photo Library; **p76:** Han, X., Zhou, Z., Fei, L. et al. Construction of a human cell landscape at single-cell level. Nature 581, 303–309 (2020). https://doi.org/10.1038/s41586-020-2157-4; **p77(l):** Eye of Science/Science Photo Library; **p77(r):** David Scharf/Science Photo Library; **p78:** Wolfgang Bettighofer, www.protisten.de; **p80(t):** Steve Gschmeissner/Science Photo Library; **p80(b):** Tumor Cell Lab/Phanie/Science Photo Library; **p81:** Xiao C, Kuznetsov YG, Sun S, Hafenstein SL, Kostyuchenko VA, et al. (2009), CC BY 2.5 /Wikimedia Commons; **p86:** Sergio Azenha/Alamy Stock Photo; **p88:** Centers for Disease Control and Prevention; **p92(t):** The Canadian Press / Alamy Stock Photo; **p92(b):** Science Source / Science Photo Library; **p93:** CDC / Science Source / Science Photo Library; **p94(t):** Herve Conge, ISM / Science Photo Library; **p94(b):** Ami Images/Science Photo Library; **p95:** Bob Blaylock, CC BY-SA 3.0 / Wikimedia Commons; **p96:** Kaj R. Svensson/Science Photo Library; **p97(tl):** Cburnett, CC BY-SA 3.0 / Wikimedia Commons; **p97(tr):** Ron Singer, Public domain / Wikimedia Commons; **p97(bl):** Dr. Michael Mares/Sam Noble Museum; **p97(br):** Brian Gratwicke, CC BY 2.5; **p98(t):** Agefotostock/Alamy Stock Photo; **p98(b):** Timothy Hellum/Alamy Stock Photo; **p99:** Andrew Allott; **p100:** RGB Ventures SuperStock/Alamy Stock Photo; **p101(tl):** Mike lane/Alamy Stock Photo; **p101(tr):** EyeLoveBirds from Vancouver, Canada, CC BY-SA 2.0 / Wikimedia Commons; **p101(b):** Corbis; **p103:** Dept. of Clinical Cytogenetics, Addenbrookes Hospital / Science Photo Library; **p104:** The History Collection / Alamy Stock Photo; **p109:** Mauro Fermariello/Science Photo Library; **pp110-112:** Andrew Allott; **p114:** Andrea Izzotti/Shutterstock; **p115(t):** Melissa Lutz Blouin; **p115(b):** Vince F/Alamy Stock Photo; **p116(t):** NOAA Okeanos Explorer Program, 2016 Deepwater Exploration of the Marianas, Leg 3; **p116(b):** Jen Guyton/Nature Picture Library; **p117(t):** Gaertner/Alamy Stock Photo; **p117(b):** World Meteorological Organization; **p118(tl):** Nature Picture Library/Alamy Stock Photo; **p118(tr):** Auscape International Pty Ltd/Alamy Stock Photo; **p118(bl):** Alan Jeffery/Shutterstock; **p118(br):** Vvvita/Shutterstock; **p119(t):** Gabor Csorba, Hungarian Natural History Museum (or G. Csorba, HNHM); **p119(b):** Andrew Allott; **p122(t):** Andrew Allott; **p122(bl):** Basic Local Alignment Search Tool; **p122(bm):** Ken Griffiths/Shutterstock; **p122(br):** Edwin Verin/Shutterstock; **p123:** Andrew Allott; **p127(l):** Matteo Omied / Alamy Stock Photo; **p127(r):** Martin Fowler/Alamy Stock Photo; **p128:** Koepfli, KP., Deere, K.A., Slater, G.J. et al. Multigene phylogeny of the Mustelidae: Resolving relationships, tempo and biogeographic history of a mammalian adaptive radiation . BMC Biol 6, 10 (2008). https://doi.org/10.1186/1741-7007-6-10; **p129:** Ger Bosma/Alamy Stock Photo; **p132(tl):** Didier Descouens/Wikimedia Commons; **p132(tr):** Peter Cairns/Nature Picture Library/ Science Photo Library; **p132(bl):** Quartl, CC BY-SA 3.0 / Wikimedia Commons; **p132(br):** Nature Picture Library / Alamy Stock Photo; **p133(t):** Sandy Rae, CC BY-SA 3.0 / Wikimedia Commons; **p133(b):** M. G. Harasewych/Oxford University Press; **p136:** Lauren Suryanata/Shutterstock; **p137(t):** PAUL D STEWART / Science Photo Library; **p137(b):** Tony Camacho/Science Photo Library; **p138:** Andrew Allott; **p141:** Dora Zett/Shutterstock; **p142(tl):** Steve Huskey, Ph.D.; **p142(tr):** www.opencage photographer, CC BY-SA 2.5/Wikimedia; **p142(bl):** Sklmsta/Wikimedia; **p142(br):** photowind/Shutterstock; **p145(t):** AGAMI Photo Agency/Alamy Stock Photo; **p145(b):** Cyril Ruoso / Nature Picture Library / Science Photo Library; **p146:** Diego Delso, CC BY-SA 4.0 / Wikimedia Commons; **p147:** Bill Coster/Alamy Stock Photo; **p148:** George Turner; **p149:** Alfredo Garcia Saz/Alamy Stock Photo; **p150(t):** Ailurus~frwiki, CC BY-SA 4.0 / Wikimedia Commons; **p150(b):** BrazilPhotos/Alamy Stock Photo; **p151:** Lee Dalton/Alamy Stock Photo; **p152:** AGAMI Photo Agency/Alamy Stock Photo; **p153:** Science History Images / Alamy Stock Photo; **p155(t):** Andrew Allott; **p155(b):** Tanya C Smith/Alamy Stock Photo; **p156(t):** Don Johnston_MA/Alamy Stock Photo; **p156(b):** Nick Upton / Nature Picture Library / Science Photo Library; **p157:** Mannion PD, Upchurch P, Benson RB, Goswami A. The latitudinal biodiversity gradient through deep time. Trends Ecol Evol. 2014 Jan;29(1):42-50. doi: 10.1016/j.tree.2013.09.012. Epub 2013 Oct 17. PMID: 24139126.; **p158(l):** Cephas, CC BY-SA 3.0 / Wikimedia Commons; **p158(r):** Jim Rorabaugh/USFWS, Public domain / Wikimedia Commons; **p160(l):** Frederick William Frohawk, Public domain / Wikimedia Commons; **p160(r):** Natural History Museum, London, CC BY 4.0 / Wikimedia Commons; **p161(t):** Science History Images/Alamy Stock Photo; **p161(b):** © Auscape / ardea.com; **p163(t):** Minden Pictures/Alamy Stock Photo; **p163(b):** NASA. Collage by Producercunningham., Public domain / Wikimedia Commons; **p165:** Stephen Barlow; **p166:** Phil Marsh; **pp169-170:** Andrew Allott; **p171(tl):** Andrew Allott; **p171(bl):** Andrew Allott; **p171(b):** B. Bartel/USFWS & Jitze Couperus, CC BY 2.0 / via Wikimedia Commons; **p174(t):** Mauro Fermariello/Science Photo Library; **p174(b):** United States Department of Agriculture; **p178:** ibreakstock/Shutterstock; **p180:** RM Floral / Alamy Stock Photo; **p185:** BGSmith/Shutterstock; **p188:** Jamie Presland; **p190(l):** m.pilot/Shutterstock; **p190(r):** Jiri Hera/Shutterstock; **p191:** Papa November CC BY-SA 3.0/Wikimedia Commons; **p193:** Fuse/Getty Images; **p195:** RCSB PDB; **p199(t):** Georgios Kollidas/Shutterstock; **p199(b):** February 2001, David Goodsell, doi:10.2210/rcsb_pdb/mom_2001_2; **p200(t):** Glow Images/Getty Images; **p200(b):** Andrew Allott; **p203:** Andrei Lomize, CC BY-SA 3.0/Wikimedia Commons; **p204:** February 2011, David Goodsell, doi:10.2210/rcsb_pdb/mom_2011_2; **p206:** Mark A. Herzik Jr. Cryo-electron microscopy reaches atomic resolution. Nature 587, 39-40 (2020). doi: https://doi.org/10.1038/d41586-020-02924-y; **p207(t):** April 2000, David Goodsell, doi:10.2210/rcsb_pdb/mom_2000_4/; **p207(b):** Science Photo Library / Alamy Stock Photo; **p208(l):** John Birdsall Social Issues Photo Library / Science Photo Library; **p208(r):** Prostock-studio/Shutterstock; **p210:** Biophoto Associates / Science Photo Library; **p211(t):** mikeledray/Shutterstock; **p211(b):** fotofeel/Shutterstock; **p212:** Andrew Allott; **p216:** Science Photo Library / Alamy Stock Photo; **p219:** Andrew Allott, courtesy of Roel Haeren and Hans Vink;

Contents

Introduction

The aim of the International Baccalaureate biology syllabus is to combine a conceptual approach to biology, an understanding of the nature of science and the development of discipline-specific skills. All of these elements are embedded in specific contexts. The syllabus road map is shown in Figure 1.

Topics are organized into four themes and four levels of organization. The theme and level of organization shows possible conceptual lenses through which the topics can be viewed. Students and teachers are encouraged to personalize their approach to the syllabus. This textbook allows you to sequence the course by theme or level of organization. It is structured in the same way as the syllabus, with each chapter corresponding to a topic and divided into numbered understandings. Some understandings will also include reference to the application of skills and the nature of science (NOS).

Theme	Level of organization			
	1. Molecules	2. Cells	3. Organisms	4. Ecosystems
A Unity and diversity	Common ancestry has given living organisms many shared features while evolution has resulted in the rich biodiversity of life on Earth.			
	A1.1 Water **A1.2** Nucleic acids	**A2.1** Origins of cells *[HL only]* **A2.2** Cell structure **A2.3** Viruses *[HL only]*	**A3.1** Diversity of organisms **A3.2** Classification and cladistics *[HL only]*	**A4.1** Evolution and speciation **A4.2** Conservation of biodiversity
B Form and function	Adaptations are forms that correspond to function. These adaptations persist from generation to generation because they increase the chances of survival.			
	B1.1 Carbohydrates and lipids **B1.2** Proteins	**B2.1** Membranes and membrane transport **B2.2** Organelles and compartmentalization **B2.3** Cell specialization	**B3.1** Gas exchange **B3.2** Transport **B3.3** Muscle and motility *[HL only]*	**B4.1** Adaptation to environment **B4.2** Ecological niches
C Interaction and interdependence	Systems are based on interactions, interdependence and integration of components. Systems result in emergence of new properties at each level of biological organization.			
	C1.1 Enzymes and metabolism **C1.2** Cell respiration **C1.3** Photosynthesis	**C2.1** Chemical signalling *[HL only]* **C2.2** Neural signalling	**C3.1** Integration of body systems **C3.2** Defence against disease	**C4.1** Populations and communities **C4.2** Transfers of energy and matter
D Continuity and change	Living things have mechanisms for maintaining equilibrium and for bringing about transformation. Environmental change is a driver of evolution by natural selection.			
	D1.1 DNA replication **D1.2** Protein synthesis **D1.3** Mutations and gene editing	**D2.1** Cell and nuclear division **D2.2** Gene expression *[HL only]* **D2.3** Water potential	**D3.1** Reproduction **D3.2** Inheritance **D3.3** Homeostasis	**D4.1** Natural selection **D4.2** Stability and change **D4.3** Climate change

Figure 1

Nature of science

Science has features that make it different from other pursuits such as the arts, social sciences, mathematics, or the study of language. Science has particular methodologies and purposes.

The effective pursuit of modern scientific work and its theories depends on the **Nature of Science**, which can be summarized in the following eleven aspects:

- **Observations and experiments**
  Sometimes the observations in experiments are unexpected and lead to serendipitous results.
- **Measurements**
  Measurements can be qualitative or quantitative, but all data are prone to error. It is important to know the limitations of your data.
- **Evidence**
  Scientists learn to be sceptical about their observations and they require their knowledge to be fully supported by evidence.
- **Patterns and trends**
  Recognition of a pattern or trend forms an important part of the scientist's work whatever the science.
- **Hypotheses**
  Patterns lead to a possible explanation. The hypothesis is this provisional view and it requires further verification.
- **Falsification**
  Hypotheses can be proved false using other evidence, but it can't be proved definitely true. This has led to paradigm shifts in science throughout history.
- **Models**
  Scientists construct models as simplified explanations of their observations. Models often contain assumptions or unrealistic simplifications, but the aim of science is to increase the complexity of the model, and reduce its limitations.

- **Theories**
  A theory is a broad explanation that takes observed patterns and hypotheses and uses them to generate predictions. These predictions may confirm a theory (within observable limitations) or may falsify it.
- **Science as a shared activity**
  Scientific activities are often carried out in collaboration, such as peer review of work before publication or agreement on a convention for clear communication.
- **Global impact of science**
  Scientists are responsible to society for the consequences of their work, whether ethical, environmental, economic, or social. Scientific knowledge must be shared with the public clearly and fairly.

Course book definition

The IB Diploma Programme course books are resource materials designed to support students throughout their two-year Diploma Programme course of study in a particular subject. They will help students gain an understanding of what is expected from the study of an IB Diploma Programme subject while presenting content in a way that illustrates the purpose and aims of the IB. They reflect the philosophy and approach of the IB and encourage a deep understanding of each subject by making connections to wider issues and providing opportunities for critical thinking.

The books mirror the IB philosophy of viewing the curriculum in terms of a whole-course approach; the use of a wide range of resources, international mindedness, the IB learner profile and the IB Diploma Programme core requirements, theory of knowledge, the extended essay, and creativity, activity, service (CAS).

Each book can be used in conjunction with other materials and, indeed, students of the IB are required and encouraged to draw conclusions from a variety of resources. Suggestions for additional and further reading are given in each book and suggestions for how to extend research are provided.

In addition, the course companions provide advice and guidance on the specific course assessment requirements and on academic honesty protocol. They are distinctive and authoritative without being prescriptive.

IB mission statement

The International Baccalaureate aims to develop inquiring, knowledgeable and caring young people who help to create a better and more peaceful world through intercultural understanding and respect.

To this end, the organization works with schools, governments and international organizations to develop challenging programmes of international education and rigorous assessment.

These programmes encourage students across the world to become active, compassionate and lifelong learners who understand that other people, with their differences, can also be right.

The IB Learner Profile

The aim of all IB programmes to develop internationally minded people who work to create a better and more peaceful world. The aim of the programme is to develop this person through ten learner attributes, as described below.

Inquirers: They develop their natural curiosity. They acquire the skills necessary to conduct inquiry and research and snow independence in learning. They actively enjoy learning and this love of learning will be sustained throughout their lives.

Knowledgeable: They explore concepts, ideas and issues that have local and global significance. In so doing, they acquire in-depth knowledge and develop understanding across a broad and balanced range of disciplines.

Thinkers: They exercise initiative in applying thinking skills critically and creatively to recognize and approach complex problems, and to make reasoned, ethical decisions.

Communicators: They understand and express ideas and information confidently and creatively in more than one language and in a variety of modes of communication. They work effectively and willingly in collaboration with others.

Principled: They act with integrity and honesty, with a strong sense of fairness, justice and respect for the dignity of the individual, groups and communities.

They take responsibility for their own action and the consequences that accompany them.

Open-minded: They understand and appreciate their own cultures and personal histories, and are open to the perspectives, values and traditions of other individuals and communities. They are accustomed to seeking and evaluating a range of points of view, and are willing to grow from the experience.

Caring: They show empathy, compassion and respect towards the needs and feelings of others. They have a personal commitment to service, and to act to make a positive difference to the lives of others and to the environment.

Risk-takers: They approach unfamiliar situations and uncertainty with courage and forethought, and have the independence of spirit to explore new roles, ideas and strategies. They are brave and articulate in defending their beliefs.

Balanced: They understand the importance of intellectual, physical and emotional ballance to achieve personal well-being for themselves and others.

Reflective: They give thoughtful consideration to their own learning and experience. They are able to assess and understand their strengths and limitations in order to support their learning and personal development.

A note on academic integrity

It is of vital importance to acknowledge and appropriately credit the owners of information when that information is used in your work. After all, owners of ideas (intellectual property) have property rights. To have an authentic piece of work, it must be based on your individual and original ideas with the work of others fully acknowledged. Therefore, all assignments, written or oral, completed for assessment must use your own language and expression. Where sources are used or referred to, whether in the form of direct quotation or paraphrase, such sources must be appropriately acknowledged.

How do I acknowledge the work of others?

The way that you acknowledge that you have used the ideas of other people is through the use of footnotes and bibliographies.

Footnotes (placed at the bottom of a page) or endnotes (placed at the end of a document) are to be provided when you quote or paraphrase from another document or closely summarize the information provided in another document. You do not need to provide a footnote for information that is part of a 'body of knowledge'. That is, definitions do not need to be footnoted as they are part of the assumed knowledge.

Bibliographies should include a formal list of the resources that you used in your work. 'Formal' means that you should use one of the several accepted forms of presentation. This usually involves separating the resources that you use into different categories (e.g. books, magazines, newspaper articles, Internet-based resources, CDs and works of art) and providing full information as to how a reader or viewer of your work can find the same information. A bibliography is compulsory in the Extended Essay.

What constitutes malpractice?

Malpractice is behaviour that results in, or may result in, you or any student gaining an unfair advantage in one or more assessment component. Malpractice includes plagiarism and collusion.

Plagiarism is defined as the representation of the ideas or work of another person as your own. The following are some of the ways to avoid plagiarism:

- words and ideas of another person to support one's arguments must be acknowledged
- passages that are quoted verbatim must be enclosed within quotation marks and acknowledged
- email messages, websites on the internet and any other electronic media must be treated in the same way as books and journals
- the sources of all photographs, maps, illustrations, computer programs, data, graphs, audio-visual and similar material must be acknowledged if they are not your own work
- when referring to works of art, whether music, film dance, theatre arts or visual arts and where the creative use of a part of a work takes place, the original artist must be acknowledged.

Collusion is defined as supporting malpractice by another student. This includes:

- allowing your work to be copied or submitted for assessment by another student
- duplicating work for different assessment components and/or diploma requirements.

Other forms of malpractice include any action that gives you an unfair advantage or affects the results of another student. Examples include, taking unauthorized material into an examination room, misconduct during an examination and falsifying a CAS record.

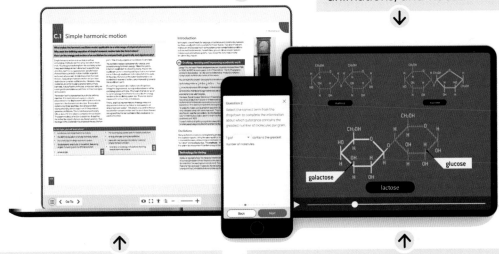

The aim of this book is to develop conceptual understanding, aid in skills development and provide opportunities to cement knowledge and understanding through practice.

Feature boxes and sections throughout the book are designed to support these aims, by signposting content relating to particular ideas and concepts, as well as opportunities for practice. This is an overview of these features:

Developing conceptual understanding

Guiding questions

At the start of every chapter, guiding questions are included to engage you with some of the questions that might arise as they study the material.

Linking questions

At the end of each section, you will find examples of linking questions followed by examples of extended-response questions. The linking questions help you view the course content through a different lens from the themes and levels of organization that guide the syllabus.

Nature of Science

These illustrate NOS using issues from both modern science and science history, and show how the ways of doing science have evolved over the centuries. There is a detailed description of what is meant by NOS and the different aspects of NOS on page iv.

Theory of knowledge

This is an important part of the IB Diploma course. It focuses on critical thinking and understanding how we arrive at our knowledge of the world. The TOK features in this book are modelled on the TOK Exhibition and pose questions for you that highlight these issues.

AHL

Content listed under the SL/HL heading should be learned by all students. Sections marked as additional higher level are required for HL students only.

Developing skills

Approaches to learning

The approaches to learning (ATL) framework seeks to promote skills that will support your learning processes in a way that is useful to all of your IB subjects and in your academic career following your study of the IB. The framework consists of five general skill categories: thinking skills, communication skills, social skills, research skills and self-management skills. Throughout the text, there are a number of examples of how the biology course can support ATL skill development.

Application of skills

Throughout the IB biology syllabus, "Application of skills" items are intended to expose you to a range of experimental and mathematical techniques as well as some suggestions for how technology can support inquiry. A subset of these skills has been designated as "Practising techniques". These are intended to introduce you to a range of possible protocols that can be modified and combined to carry out investigations of your own design. A culminating experience as an IB biology student is an open-ended inquiry called the internal assessment investigation. There is a separate chapter at the end of the book to guide you through this task.

Practicing

Data-based questions

Frequent examples of data-based questions have been included, both embedded within chapters as well as at the end of chapters. Many of these questions come from previous IB exams. Data-based questions teach the skills of data presentation, processing and analysis. In this syllabus, relatively more topic statements focus on presentation and mathematical analysis of data generated from experiments.

End-of-chapter questions

Use these questions at the end of each theme to draw together concepts from that chapter and other parts of the book, and to practise answering exam-style questions. Many of these are past IB biology exam questions.

Activity

These give you an opportunity to apply your biology knowledge and skills.

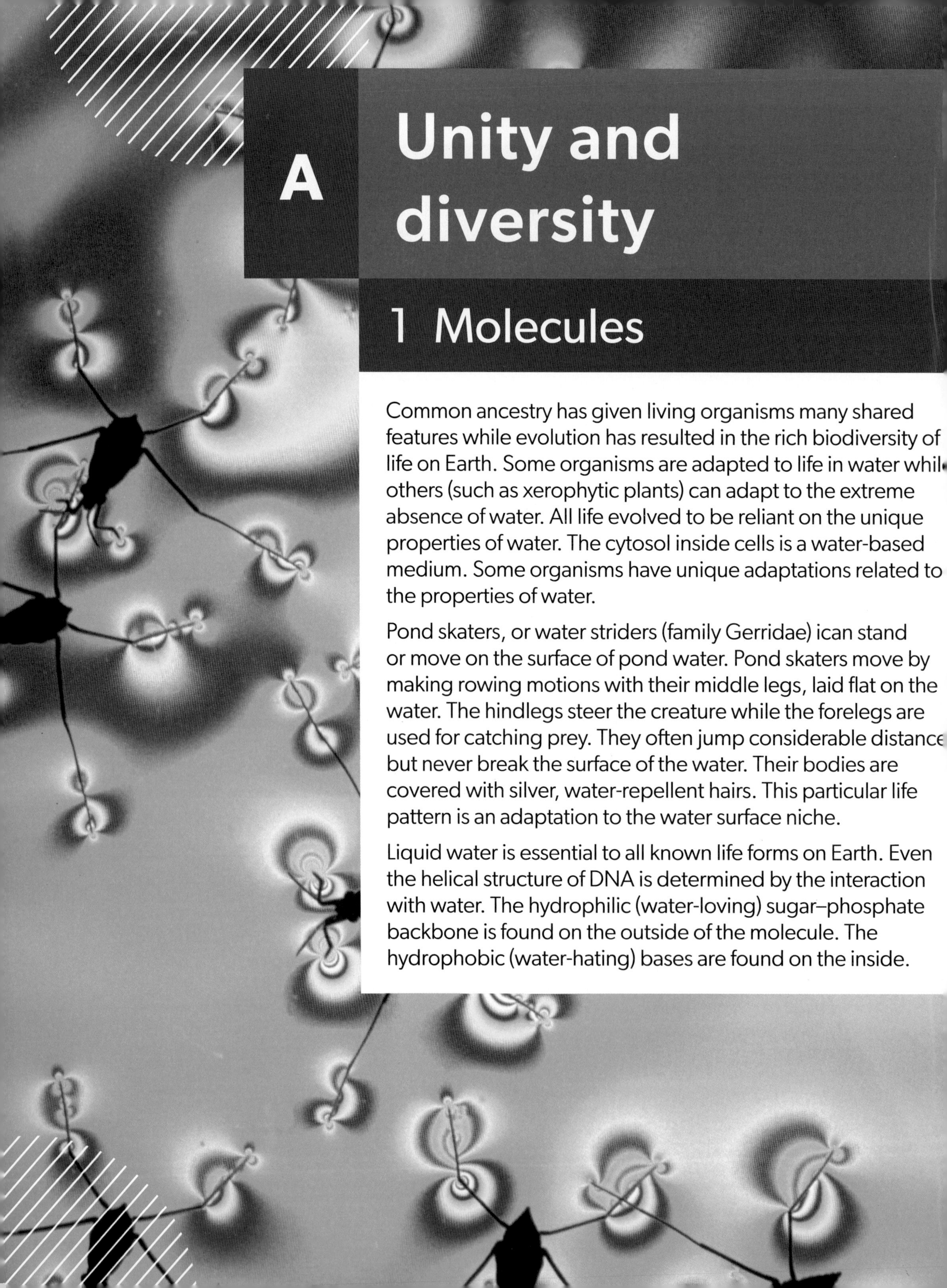

A Unity and diversity

1 Molecules

Common ancestry has given living organisms many shared features while evolution has resulted in the rich biodiversity of life on Earth. Some organisms are adapted to life in water whil• others (such as xerophytic plants) can adapt to the extreme absence of water. All life evolved to be reliant on the unique properties of water. The cytosol inside cells is a water-based medium. Some organisms have unique adaptations related to the properties of water.

Pond skaters, or water striders (family Gerridae) ican stand or move on the surface of pond water. Pond skaters move by making rowing motions with their middle legs, laid flat on the water. The hindlegs steer the creature while the forelegs are used for catching prey. They often jump considerable distanc• but never break the surface of the water. Their bodies are covered with silver, water-repellent hairs. This particular life pattern is an adaptation to the water surface niche.

Liquid water is essential to all known life forms on Earth. Even the helical structure of DNA is determined by the interaction with water. The hydrophilic (water-loving) sugar–phosphate backbone is found on the outside of the molecule. The hydrophobic (water-hating) bases are found on the inside.

A1.1 Water

What physical and chemical properties of water make it essential for life?

Water bears (*Macrobiotus sapiens*) are tiny invertebrates that live in aquatic habitats such as on damp moss. They require water to obtain oxygen by gas exchange. In dry conditions, they can enter a shrivelled dormant state to survive.

What is the longest period of time a water bear can remain dormant? To remain alive, what is the minimum metabolic activity they must perform?

When space scientists look for evidence of life on other planets, why do they begin by searching for the presence of liquid water? What are the physical and chemical properties of water that make it essential for life?

▲ Figure 1 The water bear (*Macrobiotus sapiens*) in its active state

What are the challenges and opportunities of water as a habitat?

After hunting and killing a whale, whalers processed the carcass for oil, blubber and meat. Which orders of mammals have blubber? What role does blubber play in buoyancy? Do birds or whales require more energy to counteract gravity? Do organisms require more energy to move through water than through air? What role does blubber play in thermoregulation? What is the significance of the high thermal conductivity of water for warm-blooded animals? What is the adaptive advantage of the thorough vascularization of the blubber? What other adaptations do whales have for the unique demands of life in an aquatic environment?

▲ Figure 2

SL and HL	AHL only
A1.1.1 Water as the medium for life A1.1.2 Hydrogen bonds as a consequence of the polar covalent bonds within water molecules A1.1.3 Cohesion of water molecules due to hydrogen bonding and consequences for organisms A1.1.4 Adhesion of water to materials that are polar or charged and impacts for organisms A1.1.5 Solvent properties of water linked to its role as a medium for metabolism and for transport in plants and animals A1.1.6 Physical properties of water and the consequences for animals in aquatic habitats	A1.1.7 Extraplanetary origin of water on Earth and reasons for its retention A1.1.8 Relationship between the search for extraterrestrial life and the presence of water

A1.1.1 Water as the medium for life

In 1871, Charles Darwin wrote about the first organisms appearing "in some warm little pond". It is still thought that life began in water; however, most hypotheses today place the first cells in the oceans rather than a pond.

During the formation of the first cells, a small volume of water became enclosed in a membrane. Substances were dissolved in this water and chemical reactions could occur between the solutes. After billions of years of evolution, most molecules of life are still dissolved in water. With water in a liquid state, molecules can move around and interact, allowing the processes of life to happen.

▶ Figure 3 Water vapour has been detected in the atmosphere of K2-18b, a planet 110 million light years away in the constellation Leo. Water in a liquid state is regarded as essential for the evolution of life on any planet

A1.1.2 Hydrogen bonds as a consequence of the polar covalent bonds within water molecules

In a water molecule, there are covalent bonds between oxygen and hydrogen atoms. The sharing of electrons in these bonds is unequal so they are polar covalent bonds. This is because the nucleus of an oxygen atom is more attractive to electrons than the nucleus of a hydrogen atom (Figure 4).

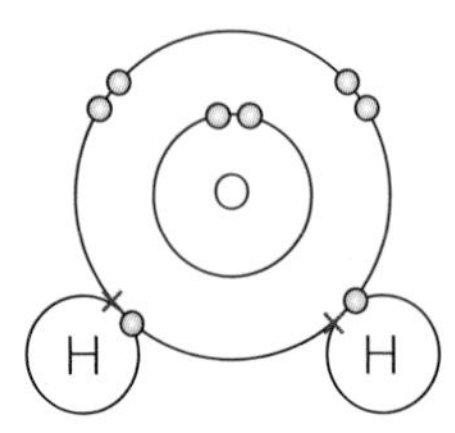

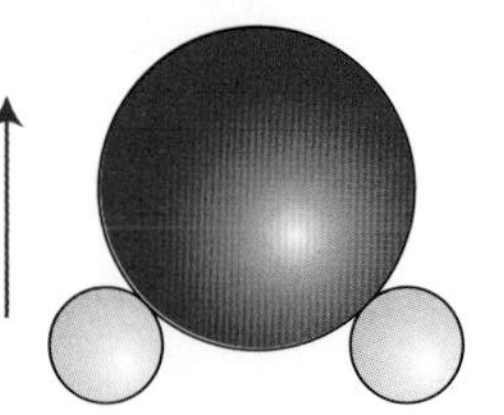

▶ Figure 4 Polarity of water molecules

Unequal sharing of electrons in water molecules gives the hydrogen atoms a partial positive charge and the oxygen atom a partial negative charge. The molecules are bent rather than linear, so the two hydrogen atoms are on the same side of the molecule and form one pole. The oxygen atom forms the opposite pole.

Positively charged particles (positive ions) and negatively charged particles (negative ions) attract each other and form ionic bonds. Water molecules only have partial charges, so the attraction is small—but it is enough to have significant effects. The attraction between two water molecules is called a hydrogen bond although, strictly speaking, it is an intermolecular force rather than a bond. A hydrogen bond is the force that forms when a slightly positive hydrogen atom in one polar molecule is attracted to a slightly negative atom of another polar molecule.

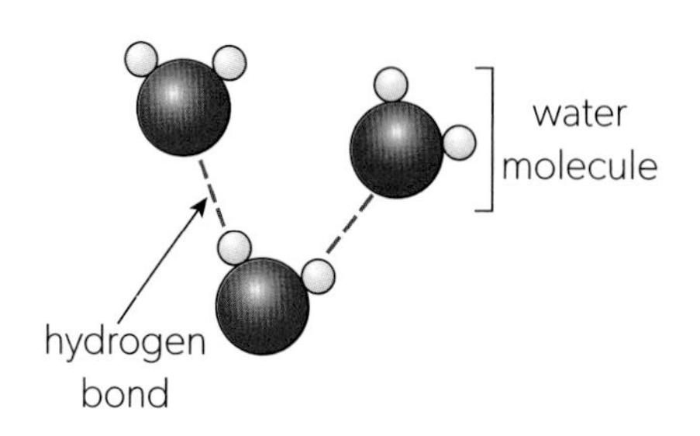

Although a hydrogen bond is a weak intermolecular force, water molecules are small and so there are many of them per unit volume of water. As a result, there are also large numbers of hydrogen bonds (Figure 5) which collectively give water its unique properties. These properties are very important to living things.

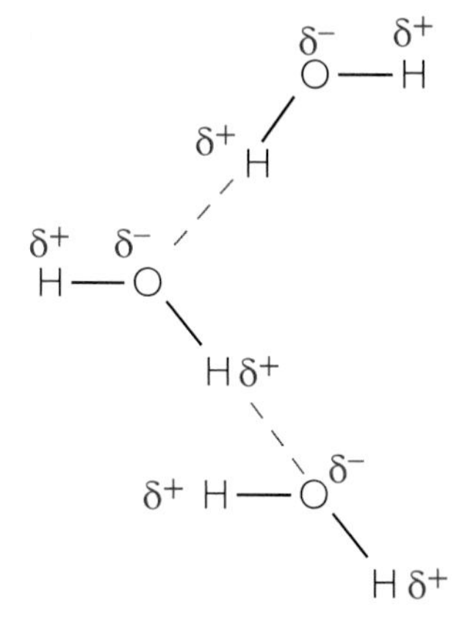

▲ Figure 5 Hydrogen bonds between water molecules can be represented with a dotted or dashed line, whereas covalent bonds within water molecules are represented with a continuous line

Applying techniques: Demonstrating the strength of the hydrogen bond

Assemble the apparatus as in Figure 6, with:

- 10 cm^3 syringe held upside down in a clamp
- weights hanging from the barrel of the syringe
- a tube connected to the nozzle of the syringe
- a gate clip that can be used to close the tube.

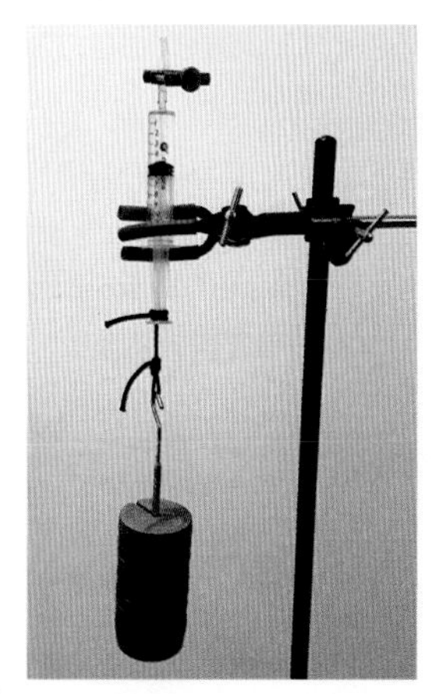

▲ Figure 6

1. Keep the gate clip open and the syringe empty. Begin unweighted and add weights one by one until the syringe plunger is pulled down to the end of the barrel (100 g ≅ 1 N). How much force is needed to overcome the friction between the plunger seal and the inner surface of the barrel?
2. Repeat step 1 with the syringe half-filled with air and the gate clip tightly closed. How much force is needed to increase the volume of air in the syringe to 10 cm^3?
3. Repeat step 1 with the syringe half-filled with water and no air bubbles. Be careful to avoid accidents due to heavy weights falling. How much force is needed to increase the volume of water to 10 cm^3?

A1.1.3 Cohesion of water molecules due to hydrogen bonding and consequences for organisms

There is a mutual attraction between water molecules and hydrogen bonds form between them. Energy is required to break these bonds. In simple terms, water molecules stick together. The scientific term for this property is cohesion. Living organisms use this property. Two examples are the conduction of water in xylem and the use of water surfaces as a habitat.

▲ Figure 7 The rope in a tug of war has to be strong enough to withstand considerable tension forces

Conduction of water under tension in xylem

Cohesion allows the transport of water under tension in plants. Water is sucked upwards from the roots to the leaves along tubular vessels in xylem tissue. There are continuous columns of water in these vessels. Each column of water is under tension (pulling forces), like a rope pulled at both ends in a "tug of war". Tension in the roots is generated by attractions between soil particles and water. Tension in the leaves develops as water is lost by evaporation to the atmosphere; it is also due to attractions between water molecules and the cell walls of leaf cells. Water moves upwards because the pulling forces in the leaves are greater than the forces in the roots.

Water in xylem can withstand tensions because hydrogen bonds make it cohesive. As long as the water column remains continuous, it will be pulled upwards. For a column of water in a xylem vessel to break, many hydrogen bonds must be broken simultaneously at one point along the vessel. This takes more energy than is normally available—hydrogen bonds can withstand surprisingly large tensions. If there were fewer hydrogen bonds, columns of water in xylem would break more easily and trees would not be able to grow so tall.

Use of water surfaces as habitats

The surface of a pond or other body of water acts like an elastic membrane that shrinks to the minimum possible area. This is because water molecules are much more attracted to each other by hydrogen bonding than to air particles. This effect is known as surface tension. All liquids have this property but only a few (such as mercury) have stronger surface tension than water.

Because of surface tension, it is possible to float objects such as steel pins on the surface of water, even though they are denser and we might expect them to sink. This is because cohesion between water molecules due to hydrogen bonds is greater than attractions between water and the floating object. For an object to break through the surface of the water, many hydrogen bonds must be broken simultaneously. The energy to do this may not be available.

Living organisms make use of this property by using water surfaces as a habitat. Water striders (also known as pond skaters) walk on the water surface with their six legs. Mosquito larvae live just below the surface, hanging from it using their siphon.

▲ Figure 8 The raft spider *Dolomedes fimbriatus* hunts prey on the water surface. It detects prey by means of vibrations passing through the water. It has a coat of unwettable hairs that help it to remain on the surface, even though its mass is about 30 g and it is denser than water

Data-based questions: Tall trees

The tallest trees in the world currently are redwoods (*Sequoia sempervirens*) in Humboldt Redwoods State Park, California. Researchers climbed five of these trees, including Hyperion which is 116 m tall (the tallest in the world). They measured the pressure within xylem tissue in small side branches at different heights during the dry season (late September to early October). In Figure 9, the upper group of data points represents measurements taken before dawn. The lower group represents measurements taken at midday.

1. a. In this research, height above ground was the independent variable. Explain what makes it an independent variable. [1]
   b. Xylem pressure was the dependent variable. Explain what makes it the dependent variable. [1]
2. a. State the relationship between height above ground and xylem pressure before dawn. [1]
   b. Suggest reasons for the relationship. [2]
3. a. Compare and contrast the pre-dawn xylem pressures with the pressures at midday. [2]
   b. Suggest a reason for differences. [1]
4. At pressures below −2.0 MPa, columns of water in xylem are prone to breaking. This limits the maximum height of trees. Use the data in the graph to predict a maximum height for redwood trees in Humboldt State Park. Explain your answer. [2]

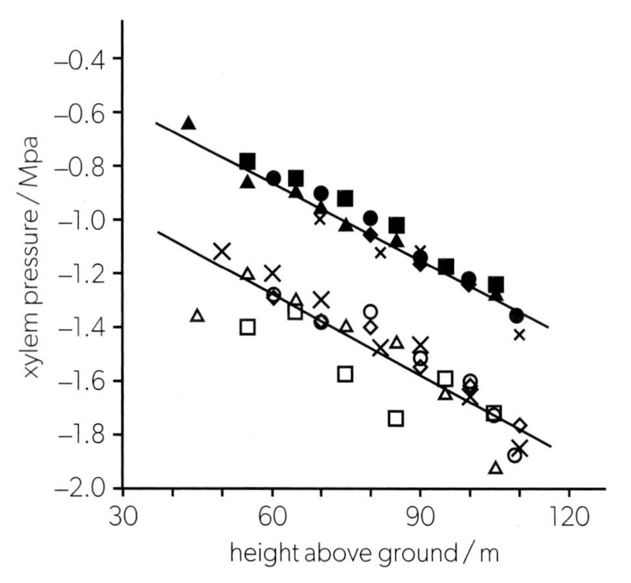

▲ **Figure 9** Source: Koch, G. et al. The limits to tree height. Nature 428, 851–854 (2004). https://doi.org/10.1038/nature02417

Communication skills: Responding to command terms

If a question asks you to "suggest", you need to propose a solution, a hypothesis or another possible answer. You may propose a range of hypotheses but this does not mean that every answer is a good one. The data usually gives clues as to what is possible. Because you are "suggesting", your statements should be possible and testable. However, they do not need to be proven by the available data.

A1.1.4 Adhesion of water to materials that are polar or charged and impacts for organisms

Hydrogen bonds can form between water and the surface of a solid composed of polar molecules. This causes water to stick to the surface of the solid and is called adhesion. It can also cause movement, as when water is drawn through narrow glass tubes. This is called capillary action. The change from air-filled to water-filled results in formation of many hydrogen bonds, so there is a release of energy. As air is replaced by water along the tube, many hydrogen bonds are formed between glass and water, so energy is released. Porous solids such as paper have large amounts of surface area attractive to water. This means they can exert strong suction forces through adhesion. We observe this when water is drawn through the narrow spaces between cellulose molecules in paper towels.

Water is attracted to many chemical substances in soil. If soil is porous, water is drawn by capillary action through dry soil, wetting it. This is how water can rise up from an underground source, even though gravity tends to pull it down.

Capillary action due to adhesion is useful in plants. Water adheres to cellulose molecules in cell walls, so any wall that starts to dry out is automatically rewetted as long as there is a source of water available.

- If water evaporates from the cell walls in leaves and is lost to the atmosphere, adhesive forces cause water to be drawn out of the nearest xylem vessel. This keeps the walls moist so they can absorb carbon dioxide needed for photosynthesis. It also generates the low pressures that draw water up in xylem vessels.
- If a xylem vessel becomes air-filled, adhesion between water and the wall of the vessel can help the vessel to refill with water. For example, the xylem vessels in deciduous trees are air-filled through the winter. In spring, capillary action due to adhesion helps the sap to rise, refilling the vessels.

▲ Figure 10 Some mosses have narrow hair-like structures on their stems, called paraphyllia. The cellulose cell walls of these structures attract water from fog or dew and store it, helping to keep the moss hydrated. The moss on the left is *Dicranum majus*. The moss on the right (at higher magnification) is *Climacium dendroides*, with paraphyllia around a group of developing leaves

Measuring variables: Determining wet and dry mass

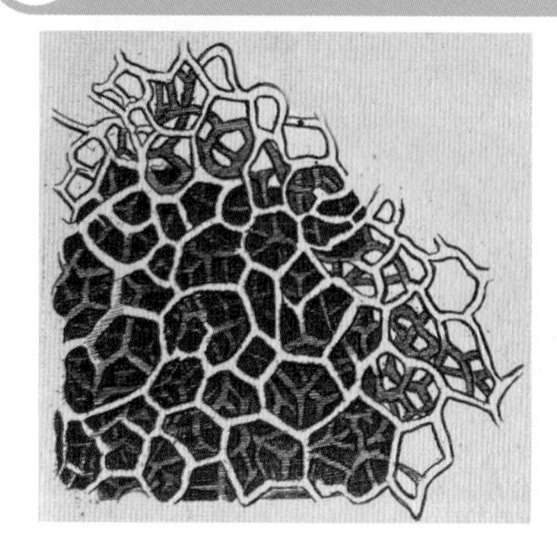

◀ Figure 11 In 1665, Robert Hooke published a drawing of the structure of natural sponge. He described it as "A confus'd heap of the fibrous parts curiously jointed and implicated. The joints are for the most part where three fibres onely meet, for I had very seldom met with any that had four"

Natural sponge is the soft skeleton of animals called sponges (phylum Porifera). The skeleton is composed of the protein spongin, which unusually contains iodine. Spongin is resistant to digestion by most proteases and water adheres to it. Because it has a porous structure and large surface area, natural sponge can absorb and hold large amounts of water. This is why it has been used over thousands of years for washing and other daily tasks.

More recently, artificial sponge has almost entirely replaced natural sponge. This has similar properties but reduces the need for harvesting of wild sponges from marine ecosystems. But 'sponges' made of plastic could cause harm to ocean ecosystems in another way.

➔

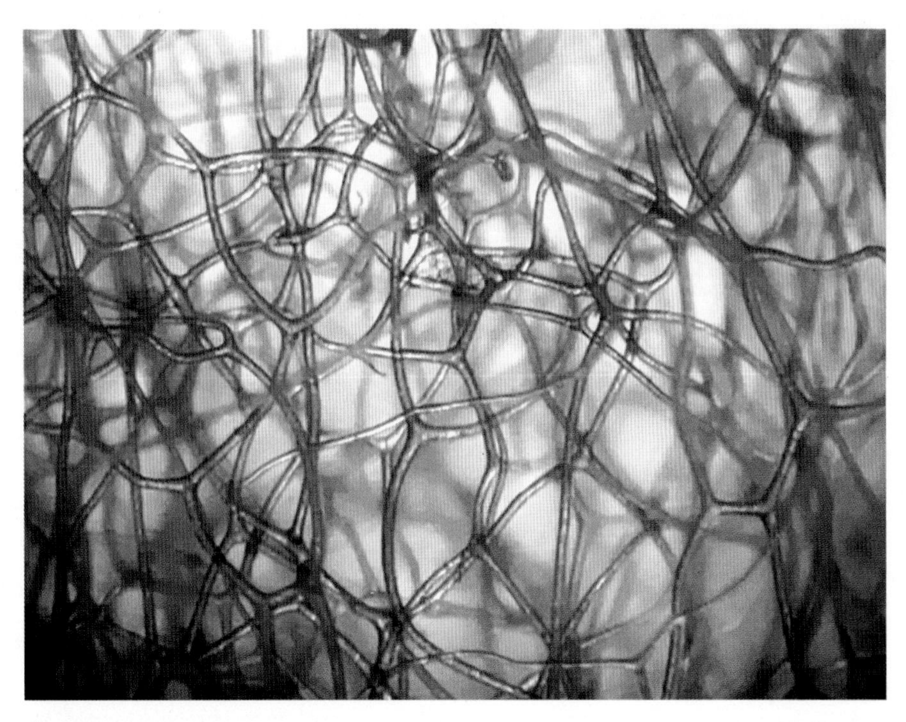

▲ **Figure 12** Image of natural sponge made using a Meiji microscope and an iPhone in 2021 (more than 350 years after Hooke's drawing)

If you can, obtain a sample of natural sponge from a sustainable source and compare it with some synthetic sponge.

1. Examine the structures using a microscope.
2. Dry the sponges—for example, by placing them in an oven at 80°C for 24 hours.
3. Find the mass of each dry sponge.
4. Allow the sponges to soak up as much water as they can.
5. Find the mass of each saturated sponge.
6. Calculate the amount of water retained by each sponge as a percentage of the dry mass.

A1.1.5 Solvent properties of water linked to its role as a medium for metabolism and for transport in plants and animals

When substances such as sugar dissolve, the particles of the substance become separated and dispersed into a liquid. The liquid is the solvent and the separated particles are solutes. The mixture of solvent and solutes is a solution.

Water has important solvent properties. The polar nature of the water molecule means that it forms shells around both charged and polar molecules. This prevents them from clumping together so they remain in solution. Water's partially negative oxygen pole is attracted to positively charged ions and its partially positive hydrogen pole is attracted to negatively charged ions, so both dissolve. Water forms hydrogen bonds with polar molecules.

▲ Figure 13 Several plants that can survive almost complete dehydration are called resurrection plants. Rehydration involves water being drawn rapidly through the desiccated cell walls, by capillary action. *Selaginella lepidophylla* is an example: the dry ball of the plant swells and opens out in a few hours, after which it turns green again and starts to grow

All substances that dissolve in water are hydrophilic (literally, "water-loving"). These substances include both polar molecules such as glucose and particles with positive or negative charges, such as sodium and chloride ions. The term "hydrophilic" is used to describe substances that are chemically attracted to water. Substances that water adheres to but does not dissolve—for example, cellulose—are also hydrophilic.

Other substances are hydrophobic. Although this term literally means "water-fearing", these substances are not repelled by water. They are simply not attracted by it; instead, they are more attracted to other hydrophobic substances. Molecules are hydrophobic if they are non-polar and they do not have negative or positive charges. Hydrophobic substances are insoluble in water although they may dissolve in other solvents such as propanone (acetone). All lipids are hydrophobic, including fats and oils.

In summary, water dissolves many different substances. However, it is not a universal solvent because there are also many substances that do not dissolve in it. Water's solvent properties allow it to be used as a medium for metabolism and for transport.

Metabolism

Cytoplasm is a complex mixture of dissolved substances. It is an aqueous solution, because the solvent is water. The solutes in this aqueous solution can move around and interact. Dissolved enzymes catalyse specific chemical reactions. The many different chemical reactions catalysed in cytoplasm are collectively known as metabolism. Without water, the components of these reactions could not move and come together on the active sites of enzymes. Therefore, water is the medium for metabolism.

Transport

Substances can be transported as an aqueous solution in both plants and animals. Plants have two such transport systems:

- Mineral ions are transported in xylem sap.
- Sucrose and other products of photosynthesis are transported in phloem sap.

Blood transports a diverse range of substances. For example:

- Sodium chloride is an ionic compound that is freely soluble in water; it dissolves to form sodium ions (Na^+) and chloride ions (Cl^-), which are carried in blood plasma.
- Amino acids have both negative and positive charges. Because of this they are soluble in water. Their solubility varies depending on the variable part of the molecule, which is hydrophilic in some amino acids and hydrophobic in others. All amino acids are soluble enough to be carried dissolved in blood plasma.
- Glucose is a polar molecule. It is freely soluble in water so is also carried in plasma.
- Oxygen is a non-polar molecule, composed of two oxygen atoms and sometimes called dioxygen. The small size of this molecule allows it to dissolve in water sparingly. Water becomes saturated with oxygen

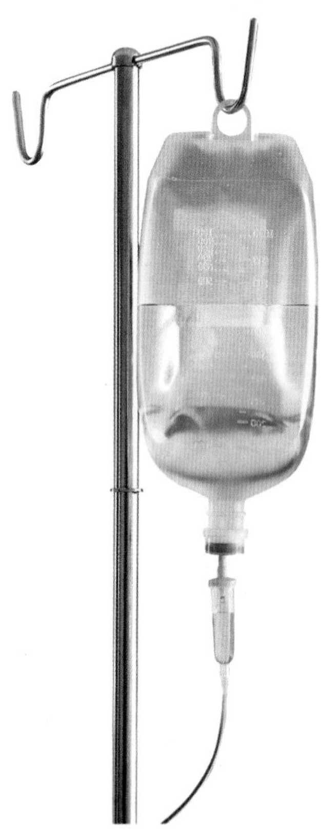

▲ Figure 14 Hospital patients are often given fluid intravenously. Sometimes the fluid is just "saline", which is water with sodium chloride dissolved in it. Other patients are given water containing all the nutrients they need; this is known as total parenteral nutrition (TPN). An emulsifier is required in TPN. Can you explain why?

at relatively low concentrations. As the temperature of water rises, the solubility of oxygen decreases, so blood plasma at 37°C can hold much less dissolved oxygen than plasma at 20°C or lower. Blood plasma cannot transport enough oxygen around the body to provide for aerobic cell respiration. This is why red blood cells contain haemoglobin. Haemoglobin has binding sites for oxygen and greatly increases the capacity of the blood to transport oxygen.

- Fat molecules are entirely non-polar and are larger than oxygen so they are insoluble in water. They tend to coalesce to form large droplets in blood. To prevent this, small fat droplets are coated in a single layer of phospholipids. Phospholipid molecules are hydrophilic at one end and hydrophobic at the other. This means they can prevent contact between water and fat, allowing the small fat droplets to remain suspended in blood plasma while being transported around the body.

A1.1.6 Physical properties of water and the consequences for animals in aquatic habitats

A physical property is a characteristic of a material that can be observed or measured without changing its chemical structure. Water has some distinctive physical properties, with major consequences for living organisms.

Buoyancy

When an object is immersed in a fluid, the fluid exerts an upward force on the object. This force is equal to the weight of the fluid displaced by the object. It is called buoyancy. If the density of the object is lower than the density of the fluid, the force acting on the object due to buoyancy will be greater than the force due to gravity and the object will float. If the density of the object is higher, buoyancy will be less than gravity and the object will sink.

The densities of living tissues are quite variable—for example, bone is denser than water while lung tissue and adipose tissue for fat storage are both less dense than water. However, living organisms have an overall density close to that of water. This makes it easier for them to use water as a habitat, because they do not need to use much energy to float at a particular depth. Bony fish have an air-filled swim bladder which they use to control their overall density. Cyanobacteria have gas vesicles which they use to adjust how close to the surface they float.

Air is much less dense than living organisms and provides negligible amounts of buoyancy. Organisms therefore have to generate lift to stay airborne.

Viscosity

In simple terms, viscosity is the stickiness of a fluid which determines how easily it can flow. Organic solvents such as propanone have low viscosity, whereas treacle has high viscosity. Viscosity is due to internal friction caused when one part of a fluid moves relative to another part. For example, when a fluid flows through a tube, the velocity is greater in the centre of the tube than at the edges, so there is internal friction. The more viscous a fluid, the greater the friction and the resistance to flow.

Pure water has a higher viscosity than organic solvents, because hydrogen bonds cause internal friction. Solutes increase the viscosity even further, so blood does not flow as easily as water. Seawater has a higher viscosity than freshwater because of the dissolved salts, with consequences for organisms that swim in it. The viscosity of air is about 50 times smaller than that of water at the same temperature.

Thermal conductivity

The rate at which heat passes through a material is known as thermal conductivity. Water has a relatively high thermal conductivity. Fats and oils conduct heat about 25% as quickly as water, and air 5%. These materials are therefore useful as heat insulators. On the other hand, aquatic warm-blooded animals are at much greater risk of the loss of body heat than land-based warm-blooded animals. Water is useful when there is a need to absorb and transfer heat. For example, the high water content of blood allows it to carry heat from parts of the body where it is generated (such as contracting muscles) to parts that need more heat or parts that are able to dissipate excess heat to the environment.

Specific heat

The heat required to raise the temperature of 1 g of a material by 1°C (or kelvin, K) is its specific heat capacity. The specific heat capacity of water is $4.18\,J\,g^{-1}\,K^{-1}$. For air, the value is only $1.01\,J\,g^{-1}\,K^{-1}$.

Water has a relatively high specific heat capacity because hydrogen bonds restrict molecular motion. For the temperature of water to increase, hydrogen bonds must be broken and energy is needed to do this. This is why a relatively large amount of heat is needed to raise the temperature of water. To cool down, water must lose an equally large amount of energy. As a result, the temperature of water remains relatively stable compared with air temperatures and aquatic habitats are more thermally stable than terrestrial habitats. The high specific heat capacity of water also helps birds and mammals (which are mostly composed of water) to maintain constant body temperatures.

▲ Figure 15 An adult black-throated loon (*Gavia arctica*) has a length of about 65 cm and a wing span of 120 cm

Physical properties of air and water

Differences between the physical properties of air and water have major consequences for organisms living in different habitats. Consider the ringed seal (a mammal) and the Arctic or black-throated loon (a bird). They are both of moderate size and have overlapping habitats. Both spend time on land rearing their young and in the water foraging for food. However, the Arctic loon flies while the ringed seal spends far more time submerged in the water. The energy requirements for movement in these habitats vary due to differences in buoyancy and viscosity of the medium. Air is less dense, so it provides far less buoyant force. The loon must expend more energy to stay aloft than the ringed seal floating in water. Water is more viscous, so the seal must use more energy to move through it. There is about 800 times more drag on a body moving through water than through air at the same velocity.

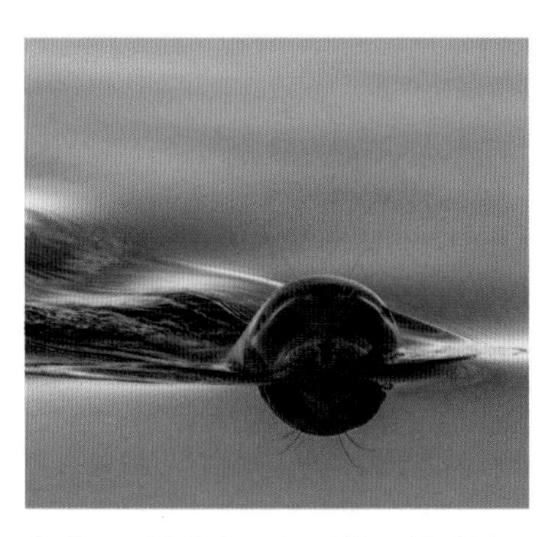

▲ Figure 16 A ringed seal (*Pusa hispida*) peeks its head above water in the Laptev Sea near Russia. A ringed seal rarely grows longer than 150 cm

Water has greater thermal conductivity than air so it conducts heat away from the bodies of submerged animals, while air acts as an insulator. It is therefore easier for a loon in air to maintain a temperature above that of the environment than it is for a ringed seal in water. At the same time, water has a higher specific heat capacity so it resists changes in temperature. Thus it provides a more stable thermal environment for the seal than air does for the loon.

A1.1.7 Extraplanetary origin of water on Earth and reasons for its retention

There are nearly 1.4 billion cubic kilometres of water on Earth and 98.3% of this is in a liquid state. The remainder is solid in ice and snow, or gas as water vapour in the atmosphere. It is unlikely that this water was on Earth when the planet was formed, because temperatures would have been above 100°C so water would have boiled and been lost to space. There are competing hypotheses for the origin of the vast amounts of water on Earth. The most widely supported hypothesis is that water was delivered to the Earth by colliding asteroids.

Currently, large asteroids (with a diameter greater than 5 km) only collide with the Earth about once every 20 million years. This rate of bombardment could not account for all of the water on Earth, especially as a sample of material recently taken from an asteroid and brought back to Earth contained only a small proportion of water. However, there is evidence of much heavier bombardment during the first few hundred million years after Earth's formation. Also, it is likely that asteroids that collided with Earth early in its history contained more water. Asteroids that have been in orbit for billions of years have lost nearly all of their water due to heat from the Sun evaporating the water and gravity being too weak to retain water vapour.

When trying to explain how water was retained on Earth after its delivery by asteroids, two factors are significant.

- The distance of the Earth from the Sun ensures that sunlight never raises temperatures high enough for water to boil. Liquid water is retained much more easily than water vapour due to cohesion from hydrogen bonding.
- Due to its size, the Earth has relatively strong gravity, holding the oceans tightly to its surface and holding gases within the atmosphere. Some hydrogen and helium escape from the atmosphere into space but very little water vapour.

There is evidence for the presence of water on Mars but this seems to have disappeared soon after the planet's formation. It is thought that most of this water was used in hydration reactions with minerals in Martian rock. On Earth, the quantities of these minerals were less so surface water was not used up.

◀ Figure 17 The comet Hyakutake. Comets are mostly formed of ice and dust. They have a diameter of a few kilometres and go around the Sun in highly elongated orbits. When a comet approaches the Sun, the ice vaporizes to form a tail of gas and dust

Data-based questions: Were comets the source of water in Earth's oceans?

Scientists have analysed the ratio of deuterium to hydrogen (D/H) of water in the Earth's oceans (1.56×10^{-4}). They have compared it with the same ratio in:

- meteorites (asteroids that have passed through the Earth's atmosphere)
- comets originating from the Oort Cloud including Halley's comet
- comets of the Jupiter family including 67P/C-G, which was explored by the Rosetta spacecraft.

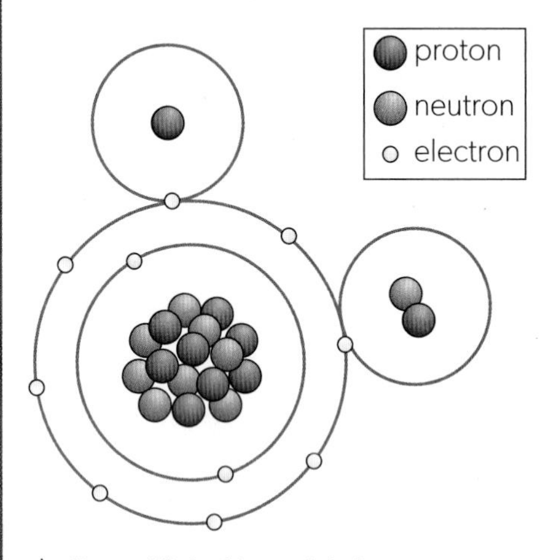

▲ **Figure 18** In this model of a water molecule, one of the hydrogen atoms has a neutron so it is the isotope deuterium. On Earth, 1 in 6,420 atoms of hydrogen are deuterium

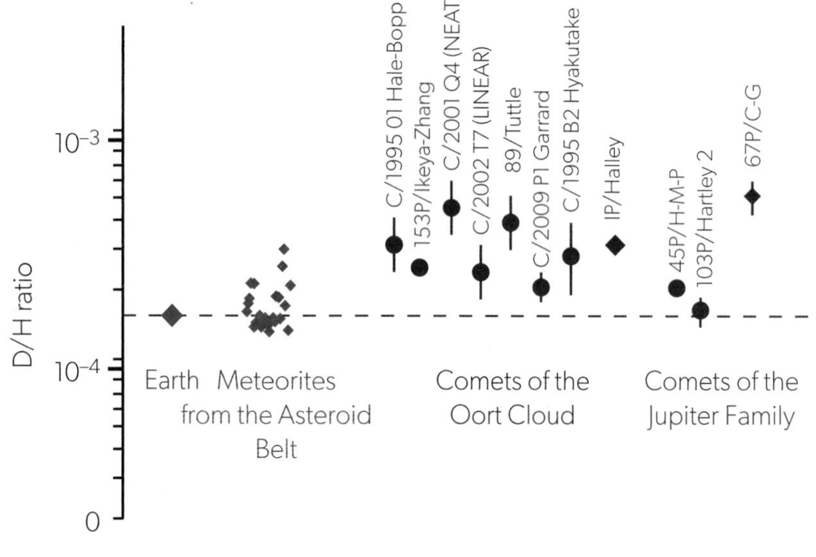

▲ **Figure 19** Source: The European Space Agency

The graph in Figure 19 shows these D/H ratios, arranged on the x-axis according to their distance from the Sun. Diamond-shaped data points represent measurements from *in situ* samples; circles show astronomical data, obtained remotely.

1. Look at Figure 19. The y-axis uses a logarithmic scale. Outline what is meant by a logarithmic scale. [2]
2. Determine the D/H ratio found on Halley's comet. [1]
3. The graph shows data for the 11 comets for which a D/H ratio has been measured. Identify how many of the comets have a D/H ratio that matches that of water on Earth. [2]
4. The D/H ratio for comet 67P/C-G was measured by the Rosetta spacecraft in 2014. Discuss how this ratio changes the likelihood that water on Earth was derived from comets. [2]
5. Using the data in the graph, discuss whether asteroids or comets are more likely to have been the source of the Earth's water. [3]

A1.1.8 Relationship between the search for extraterrestrial life and the presence of water

In the fairy tale, *Goldilocks and the Three Bears*, a young girl tries three bowls of porridge. She finds one of them too hot and another too cold but the third bowl is just the right temperature. This is used as a metaphor for the habitable zone around a star, often called the Goldilocks zone. Liquid water is essential to all known life forms on Earth. If a planet is too close to a star, water will vaporize; too far away and water freezes. However, for planets in the Goldilocks zone, the temperature allows water to exist in a liquid state.

The location of the Goldilocks zone depends on the size of the star and the amount of energy it emits. It also depends on the size of the planet, which determines the strength of gravity and the atmospheric pressure. Within our galaxy alone, it is estimated that there are 40 billion planets within a "Goldilocks zone". The more planets there are in the Goldilocks zone around other stars, the greater the chance that extra-terrestrial life has evolved.

▲ Figure 20 Will the porridge in the large bowl be too hot or too cold?

Linking questions

1. How do the various intermolecular forces of attraction affect biological systems?
   a. Outline how the properties of cohesion and adhesion are important to living things. (A1.1.3)
   b. Describe the role of hydrogen bonding in the structure of DNA. (A1.2.6)
   c. Explain the importance of hydrophobic interactions in the structure of the plasma membrane. (B2.1.2)
2. What biological processes only happen at or near surfaces?
   a. Outline an example of how the surface of water acts as a habitat. (A1.1.3)
   b. Describe the role of cell surface receptors in chemical signalling. (C2.1.6)
   c. Explain the relationship between surface-area-to-volume ratio and materials exchange. (B2.3.6)

A1.2 Nucleic acids

How does the structure of nucleic acids allow hereditary information to be stored?

All of the information encoded on a computer is ultimately based on binary code—a code based on two options, 0 and 1. A computer byte is 8 binary digits. Figure 1 shows the letters converted to binary code. How would the term "DNA" be represented in binary code? Because each digit can have four values instead of two, DNA codons with three symbols have 64 possible values, compared with a binary byte which has 256 possibilities using eight symbols. For this reason, scientists have worked to develop DNA computers.

Character	Binary code	Character	Binary code
A	01000001	N	01001110
B	01000010	O	01001111
C	01000011	P	01010000
D	01000100	Q	01010001
E	01000101	R	01010010
F	01000110	S	01010011
G	01000111	T	01010100
H	01001000	U	01010101
I	01001001	V	01010110
J	01001010	W	01010111
K	01001011	X	01011000
L	01001100	Y	01011001
M	01001101	Z	01011010

▲ Figure 1

How does the structure of DNA facilitate accurate replication?

Cells divide for the purposes of maintenance, repair, growth and reproduction. Why must dividing cells produce new DNA? The structure of DNA is dependent on complementary base pairing—A is always paired with T and C is always paired with G. Complementarity guides accurate replication. Chromosomes are mainly composed of DNA. These chromosomes (Figure 2) are seen during the early stages of cell division. The double structure of each chromosome shows that the DNA has replicated to form two identical strands, known as chromatids. These strands are linked by a region called the centromere.

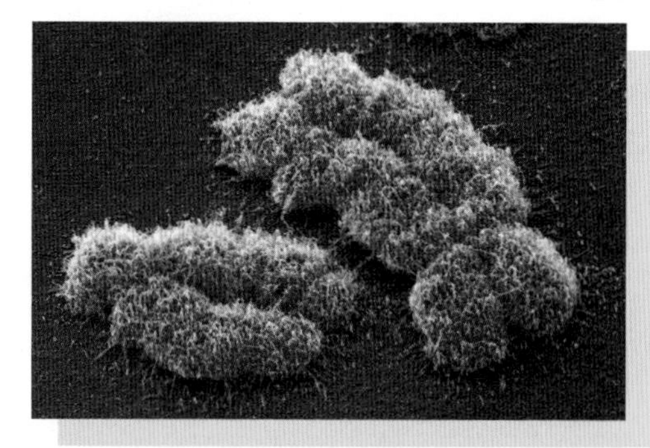

▶ Figure 2 A coloured scanning electron micrograph of two human chromosomes

SL and HL	AHL only
A1.2.1 DNA as the genetic material of all living organisms A1.2.2 Components of a nucleotide A1.2.3 Sugar–phosphate bonding and the sugar–phosphate "backbone" of DNA and RNA A1.2.4 Bases in each nucleic acid that form the basis of a code A1.2.5 RNA as a polymer formed by condensation of nucleotide monomers A1.2.6 DNA as a double helix made of two antiparallel strands of nucleotides with two strands linked by hydrogen bonding between complementary base pairs A1.2.7 Differences between DNA and RNA A1.2.8 Role of complementary base pairing in allowing genetic information to be replicated and expressed A1.2.9 Diversity of possible DNA base sequences and the limitless capacity of DNA for storing information A1.2.10 Conservation of the genetic code across all life forms as evidence of universal common ancestry	A1.2.11 Directionality of RNA and DNA A1.2.12 Purine-to-pyrimidine bonding as a component of DNA helix stability A1.2.13 Structure of a nucleosome A1.2.14 Evidence from the Hershey–Chase experiment for DNA as the genetic material A1.1.15 Chargaff's data on the relative amounts of pyrimidine and purine bases across diverse life forms

A1.2.1 DNA as the genetic material of all living organisms

Genetic material is a store of information. If copied, it can be passed from cell to cell and also from parent to offspring. Because genetic material is inherited it is sometimes called hereditary information. All living organisms use DNA to store hereditary information.

The full name for DNA is deoxyribonucleic acid. The other type of nucleic acid is ribonucleic acid or RNA. Nucleic acids were first discovered in the cell nucleus, hence the name. They are very large molecules, made from subunits called nucleotides which link to form a polymer.

Some viruses use RNA as their genetic material, for example, coronaviruses and HIV. This observation does not seem to fit the theory that genes are made of DNA in all living organisms. However, reproduction is a fundamental property of living organisms and viruses cannot reproduce themselves. Instead, they rely on a host cell for this process so they are not considered to be true living organisms. Therefore, they do not falsify the claim that all living organisms use DNA as their genetic material.

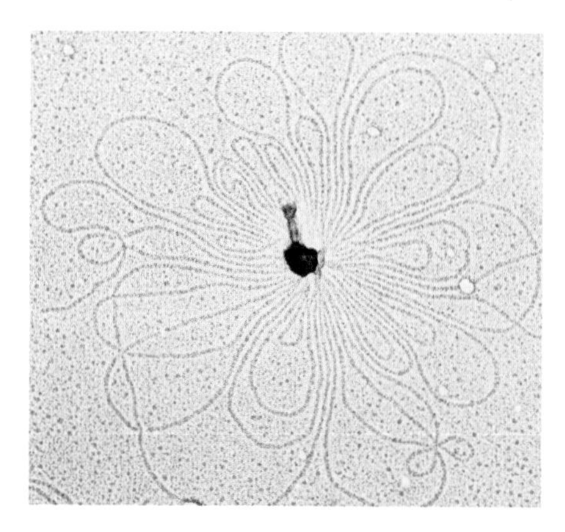

▲ Figure 3 The virus shown in the centre (black structure) uses DNA as its genetic material. The virus has burst open and its DNA has spilled out of the polyhedral head, where it is stored

A1.2.2 Components of a nucleotide

Nucleotides consist of three parts:

- a sugar, which has five carbon atoms so is a pentose sugar
- a phosphate group, which is the acidic and negatively charged part of nucleic acids
- a base that contains nitrogen and has either one or two rings of atoms in its structure.

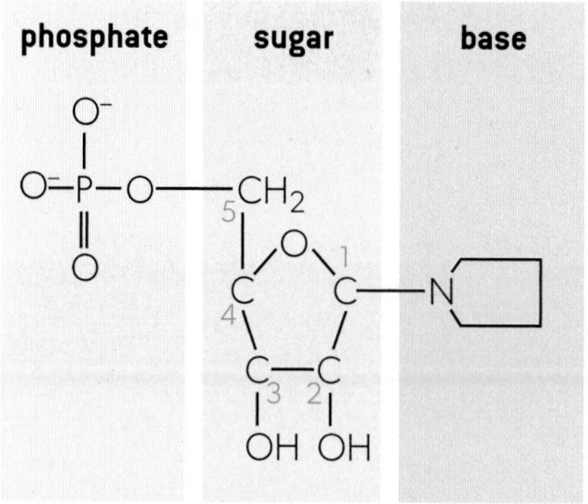

phosphate **sugar** **base**

O^- — P(=O) — O — CH_2 (5) — sugar ring C1–C4 — C1 — N (base); C3 OH, C2 OH

▲ Figure 4 Parts of a nucleotide

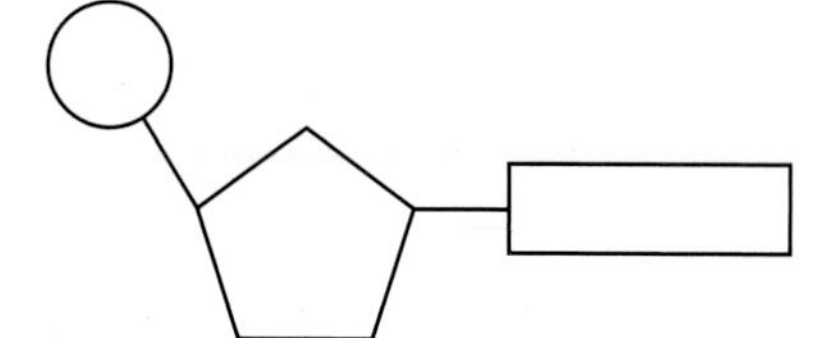

▲ Figure 5 Simple diagram of a nucleotide

Figure 4 shows these parts and how they are linked together to form an RNA nucleotide. The base and the phosphate are both linked by covalent bonds to the pentose sugar. The five carbon atoms in the pentose sugar are numbered, with the base linked to C1 and the phosphate to C5.

Figure 5 shows a nucleotide in symbolic form, with a circle to represent the phosphate, a pentagon for the pentose sugar and a rectangle for the base.

▲ Figure 6 The oxygen atom shown in red forms links between the phosphate of one nucleotide and the pentose sugar of the next nucleotide

bases in DNA	bases in RNA
adenine (A)	adenine (A)
cytosine (C)	cytosine (C)
guanine (G)	guanine (G)
thymine (T)	uracil (U)

▲ Table 1

A1.2.3 Sugar–phosphate bonding and the sugar–phosphate "backbone" of DNA and RNA

To link nucleotides together into a chain or polymer, covalent bonds are formed between the phosphate of one nucleotide and the pentose sugar of the next nucleotide.

Whenever nucleic acids are produced by living organisms, the nucleotides are always added to the growing polypeptide in the same way: the phosphate of the nucleotide being added is linked by a covalent bond to the pentose sugar of the previous nucleotide. Linking together nucleotides in this way creates a series of alternating sugar and phosphate groups, with a chain of carbon, oxygen and phosphorus atoms covalently bonded together. This chain forms a strong sugar–phosphate backbone in DNA and RNA molecules that helps to conserve the sequence of bases.

A1.2.4 Bases in each nucleic acid that form the basis of a code

There are four different bases in DNA and in RNA. Three bases are the same but the fourth one differs. All of the bases contain nitrogen—this is why they are often referred to as nitrogenous bases.

Each nucleotide contains one base so there are four types of nucleotide in DNA and in RNA. Any two nucleotides can be linked to each other, because the phosphate and sugar used to make the bond are the same. Any base sequence is therefore possible along a DNA or RNA molecule and the number of possible sequences is almost infinite.

The sequence of bases is how information is stored. The information is stored in a coded form—this is the universal genetic code that is shared by all organisms.

Data-based questions: Bases in DNA

Look at the molecular models in Figure 7 and answer the following questions.

1. State one difference between adenine and the other bases. [1]
2. Each of the bases has a nitrogen atom bonded to a hydrogen atom in a similar position (shown lower left). Deduce how this nitrogen is used when a nucleotide is being assembled from its subunits. [2]
3. Identify three similarities between adenine and guanine. [3]
4. Compare the structure of cytosine and thymine. [4]
5. Although the bases have some shared features, each one has a distinctive chemical structure and shape. Remembering the function of DNA, explain why it is important for each base to be distinctive. [5]

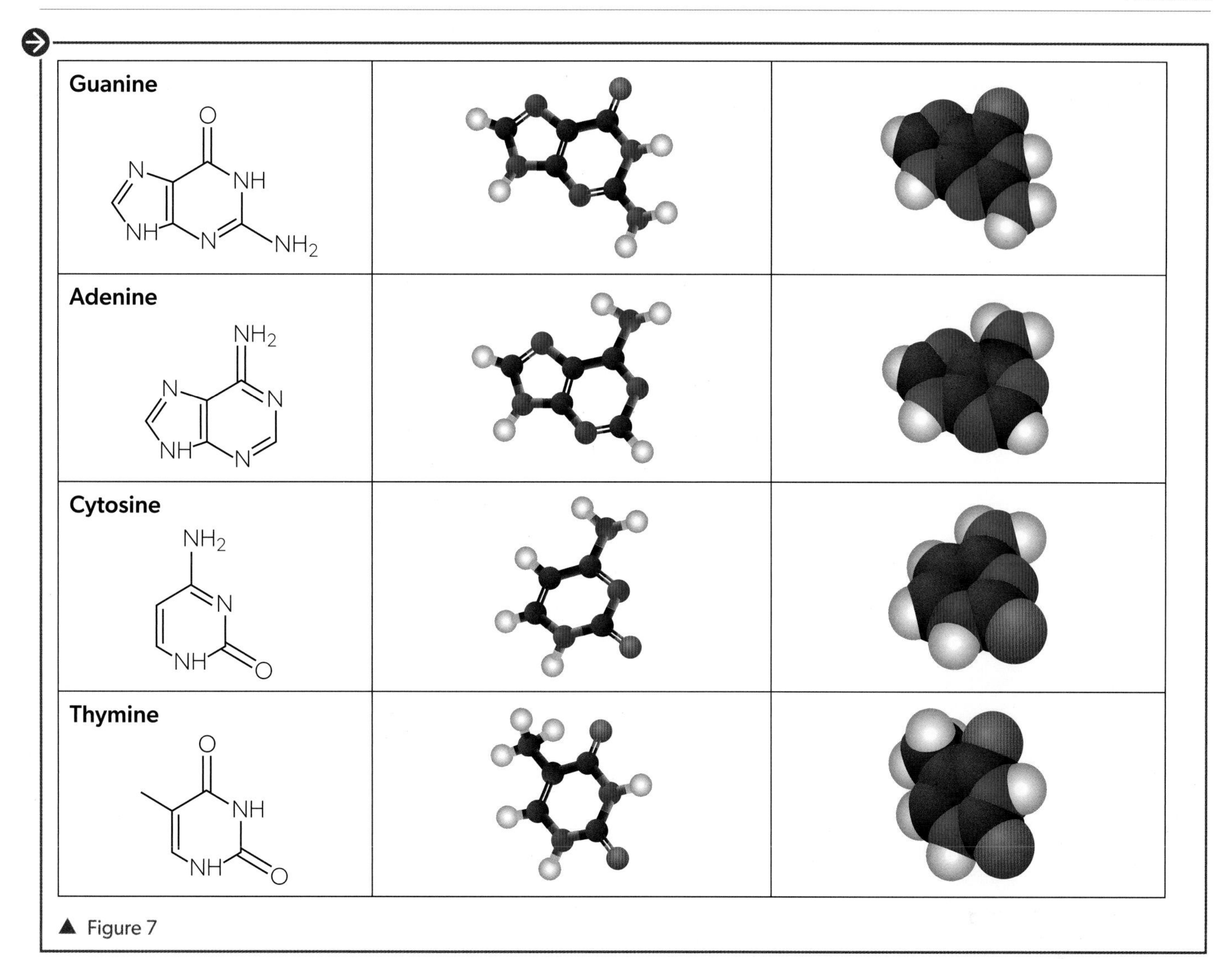

▲ Figure 7

ATL Communication skills: Interpreting and evaluating information presented in different forms

Figure 7 in the data-based questions shows three different representations of each base. The first is a structural formula, the second is a ball and stick model and the third is a space filling model. The command term "evaluate" means to make an appraisal by weighing up strengths and limitations. Evaluate each type of representation. Which was most useful in answering the data-based questions?

A1.2.5 RNA as a polymer formed by condensation of nucleotide monomers

RNA is a single, unbranched polymer of nucleotides. The nucleotides are subunits of a polymer, so they are monomers. The number of nucleotides in a molecule of RNA is unlimited, but they are always linked in the same way, by a condensation reaction.

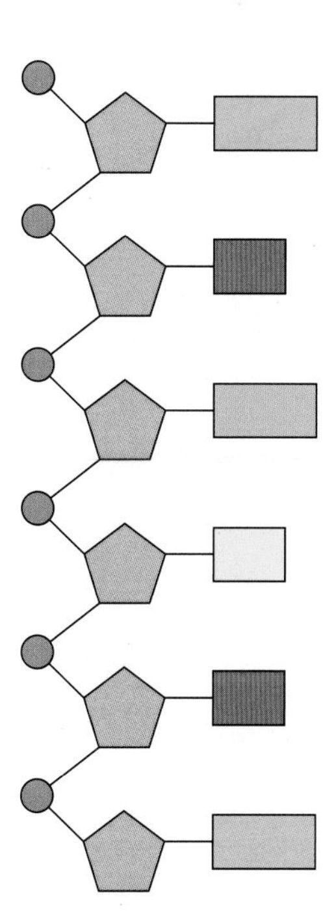

▲ Figure 8 RNA polymers can be represented using circles, pentagons and rectangles

In a condensation reaction, two molecules are combined to form a single molecule and water is eliminated. Hydroxyl groups (OH) on the phosphate of one nucleotide and on the pentose sugar of another nucleotide are used. One of the OH groups is removed entirely. It is combined with the hydrogen from the other OH, producing water. The remaining oxygen forms a new covalent bond, linking the two nucleotides. This is shown in Figure 9.

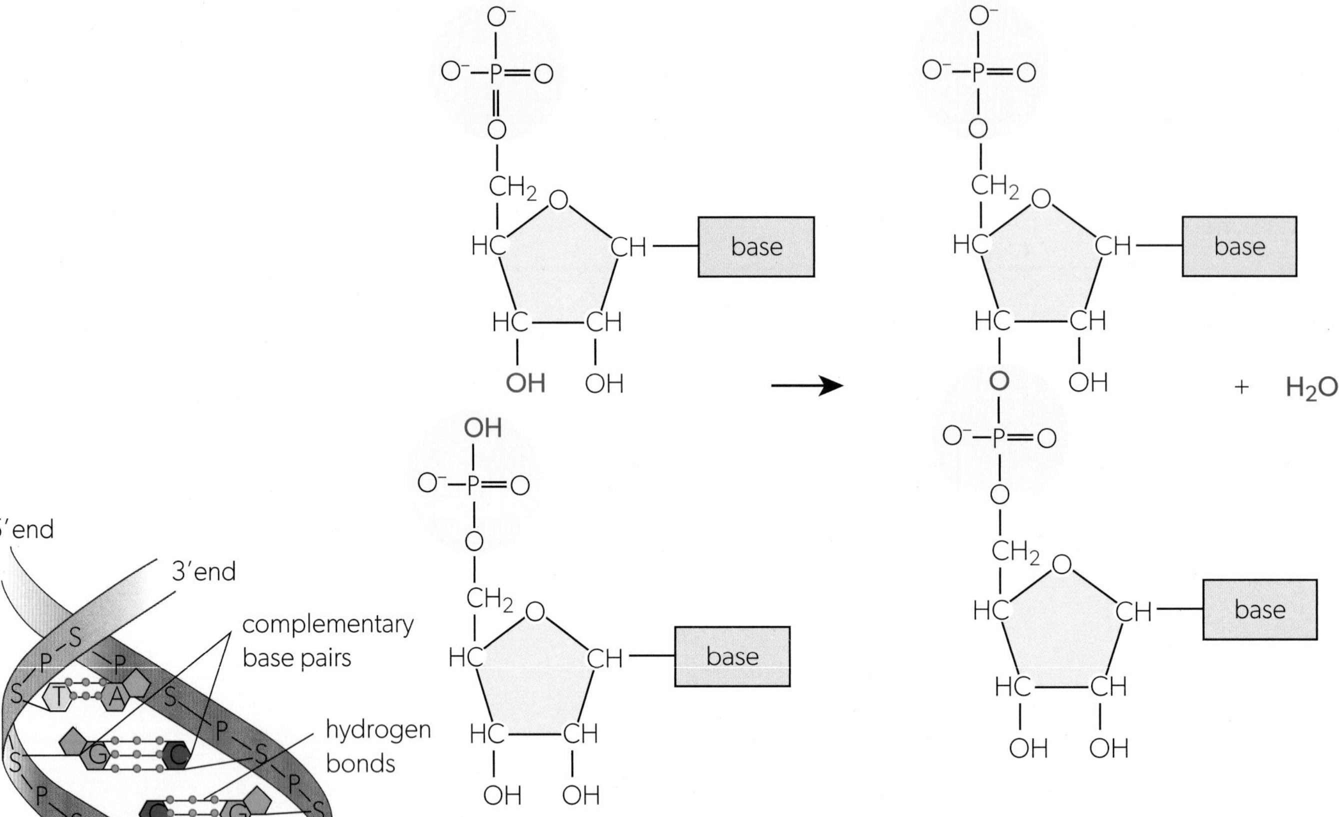

▲ Figure 9 Condensation reaction between two nucleotides

▲ Figure 10 The double helix

A1.2.6 DNA as a double helix made of two antiparallel strands of nucleotides with the two strands linked by hydrogen bonding between complementary base pairs

DNA is composed of strands or polymers of nucleotides. The pentose sugar in each nucleotide is deoxyribose and the bases are adenine, cytosine, guanine and thymine.

A DNA molecule consists of two strands of nucleotides linked to each other by their bases. The links between the bases are hydrogen bonds. Adenine (A) only forms hydrogen bonds with thymine (T). Guanine (G) only forms hydrogen bonds with cytosine (C). This results in complementary base pairing. A and T complement each other by forming base pairs and similarly G and C complement each other by forming pairs.

The two strands of nucleotides are parallel to each other. However, they run in opposite directions so they are said to be antiparallel. For this reason, one strand ends with the phosphate group of the terminal nucleotide while the other strand ends with a deoxyribose. If the two strands were oriented in the same direction, the bases would not be able to form hydrogen bonds with each other.

DNA molecules usually adopt a helical shape. A helix is a coiled structure that has a constant diameter of 2 nanometres (2 nm). Because of the two strands, DNA is a double helix. Figure 10 shows its features.

Drawings of the structure of DNA on paper cannot show all features of the three-dimensional structure of the molecule. Figure 11 shows how the structure of DNA can be represented simply in a diagram.

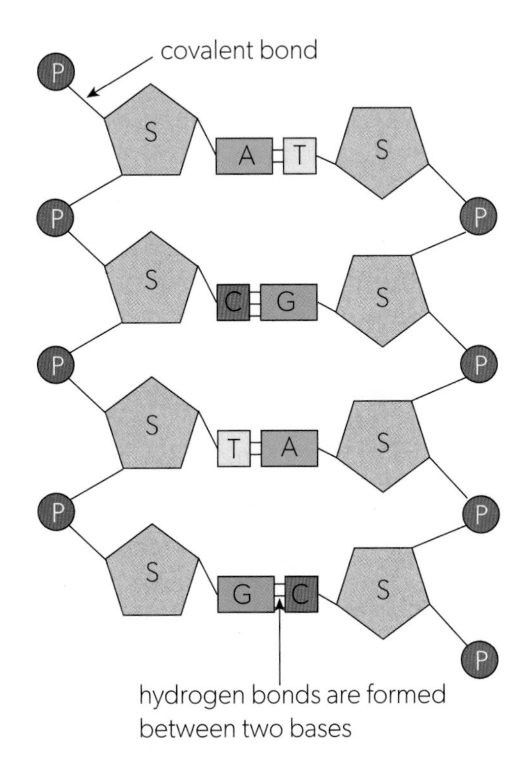

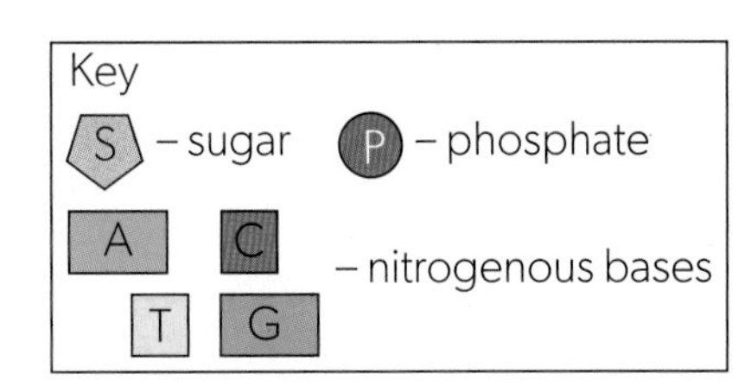

▲ Figure 11 Complementary base pairing between the antiparallel strands of DNA

A1.2.7 Differences between DNA and RNA

There are three important differences between the two types of nucleic acid:

1. There are usually two polymers of nucleotides in DNA, whereas there is only one in RNA. The polymers are often referred to as strands, so DNA is double-stranded and RNA is single-stranded.
2. The four bases in DNA are adenine, cytosine, guanine and thymine. The four bases in RNA are adenine, cytosine, guanine and uracil, so uracil is present instead of thymine in RNA.
3. The pentose sugar within DNA is deoxyribose, whereas the sugar in RNA is ribose. Figure 12 shows that deoxyribose has one fewer oxygen atom than ribose. The full names of DNA and RNA are based on the type of sugar in them—deoxyribonucleic acid and ribonucleic acid.

CH_2OH (5), O, H (1), H, H, OH, OH, OH

ribose

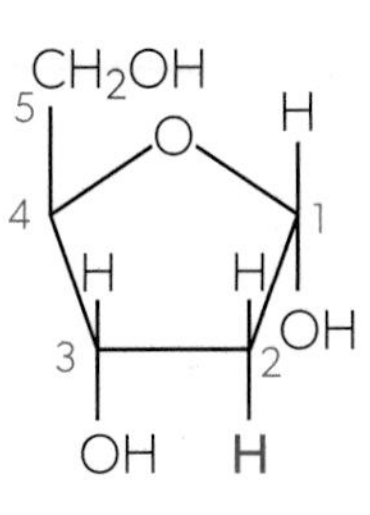

deoxyribose

▲ Figure 12 Ribose has an OH group and an H atom attached to carbon 2, whereas deoxyribose has two H atoms

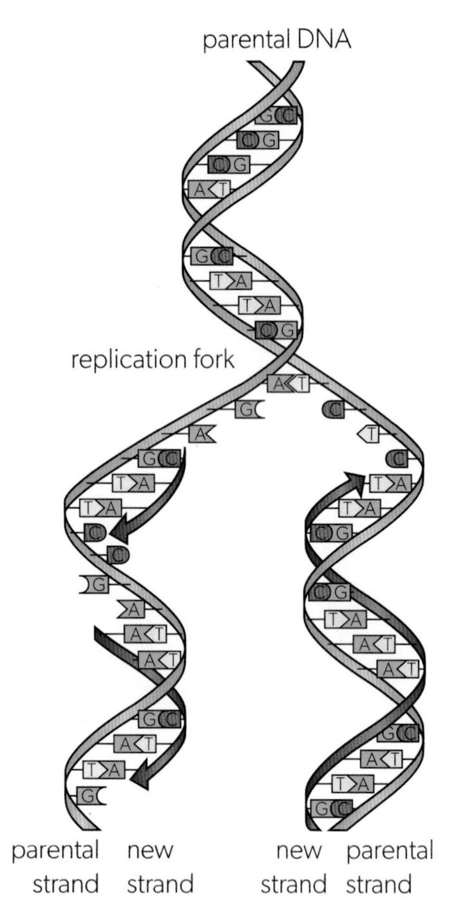

▲ Figure 13 Semi-conservative replication of DNA

A1.2.8 Role of complementary base pairing in allowing genetic information to be replicated and expressed

In DNA, adenine can only pair with thymine and cytosine can only pair with guanine. This is complementary base pairing. It allows an exact copy of a DNA molecule to be made in a process called replication. In DNA replication, the two strands of the double helix separate. Each of the original strands serves as a guide, or template, for the creation of a new strand. The new strands are formed by adding nucleotides one by one and linking them together.

Each nucleotide that is added must be carrying the base that is complementary to the next base on the template strand. This means the newly synthesized strand on each of the two template strands should have exactly the same base sequence as the other template strand. Replication changes one original DNA molecule into two identical DNA molecules, each with one strand from the original molecule and one new strand. This is called semi-conservative replication.

Genetic information consists of sections of DNA called genes. Each gene contains information needed for a particular purpose. When the information in a gene has an effect on the cell, this is called gene expression. The first stage in expressing a gene is the copying of its base sequence, but the copy is made of RNA rather than DNA. Only one of the two DNA strands is used as a template for this. The rules of complementary base pairing are followed but adenine on the template strand pairs with uracil on the new strand of RNA, rather than thymine. This process of making an RNA copy of the base sequence of DNA is called transcription.

RNA that is produced by transcription may have a regulatory or structural role in the cell, or it may be used in protein synthesis. To synthesize a protein, the base sequence of the RNA molecule is translated into the amino acid sequence of a protein. Again, complementary base pairing is involved. Both transcription and translation are more fully described in *Topic D1.2*.

A1.2.9 Diversity of possible DNA base sequences and the limitless capacity of DNA for storing information

Genetic information is stored in the base sequence of one of the two strands of a DNA molecule. Any sequence of bases is possible.

- There are four possibilities for each base in the sequence—A, C, G or T.
- There are 4^2 or 16 possibilities for a sequence of two bases—AA, AC, AG and so on.
- There are 4^3 or 64 possibilities for a sequence of three bases—AAA, AAC, AAG and so on.
- With n bases, there are 4^n possible sequences. As n increases, the number of possibilities becomes immense. With a sequence of just 10 bases, there are over a million possibilities.

DNA molecules can be any length, adding to the potential diversity of base sequences. The range of possible sequences is effectively limitless, which is an ideal feature for an information storage system.

The diameter of a DNA molecule is just 2 nanometres, so immense lengths of DNA can be stored in a very small volume. Compared with data-storage systems devised by humans, DNA is very economical, both in terms of the space it takes up and the amount of material used to make it.

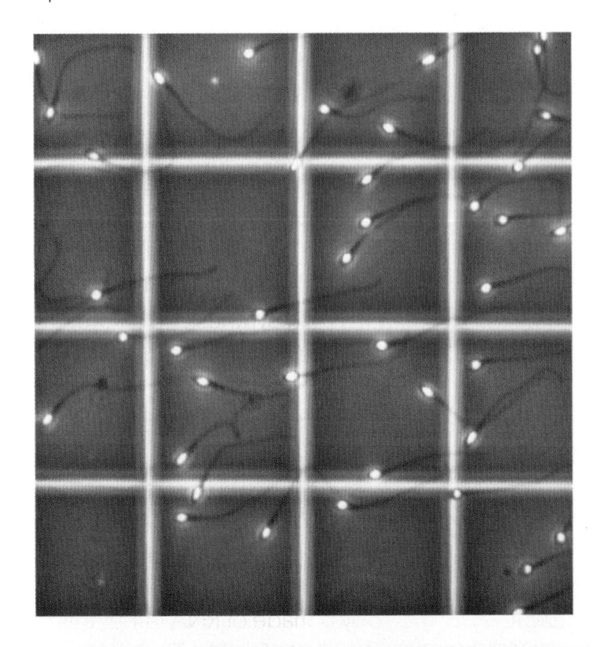

◀ Figure 14 A sperm is a DNA delivery system. These human sperm cells each contain 3.3 picograms of DNA, with a total length of about 2 metres and over 3 billion base pairs in total. The microscope image has a grid of lines 50 micrometres apart. How long is each sperm and how wide is the head where the DNA is stored?

Data-based questions: DNA lengths

1. In *Homo sapiens*, the smallest chromosome (and therefore the shortest DNA molecule) is the Y chromosome which has 57,227,415 base pairs. Assuming that the human genome has 3.08 billion base pairs in total, what percentage of this does the Y chromosome contain? [1]
2. The bacterium *Carsonella ruddii* has just 173,904 base pairs in its genome, with an estimate of 224 genes. Of these, 194 code for proteins. A surprisingly low 7.3% of the bases are guanine. Calculate the percentage of bases that are adenine, cytosine and thymine. [3]
3. Canine circovirus has a genome of 2,063 bases with two protein-coding genes. This type of virus has a protein coat that is only 17 nanometres in diameter. Its genetic material is single-stranded DNA. Suggest one advantage and one disadvantage of this DNA being single-stranded. [2]
4. Bacteria can store genetic information in small circular DNA molecules called plasmids. A plasmid with 1,440 base pairs has been found in the bacterium *Acetobacter pasteurianus*. The main chromosome of this bacterium has 3.155 Mb (Mb = megabase pairs). What is the ratio between the length of the plasmid and the length of the main chromosome? [2]
5. Can you find examples of DNA molecules from animals, bacteria, viruses or plasmids that are shorter than the examples given here? Can you find an example of DNA with less than 7.3% guanine? [2]

A1.2.10 Conservation of the genetic code across all life forms as evidence of universal common ancestry

The sequence of bases in DNA or RNA contains information in a coded form. The information is decoded during protein synthesis. Groups of three bases are called codons and have meanings in the code. There are 64 different codons, because each base in a codon can be any of four, so there are 4 × 4 × 4 combinations. Each of the 64 codons has a meaning:

- most codons specify one particular amino acid
- one codon signals that protein synthesis should start
- three codons signal that protein synthesis should stop.

Details of the genetic code are described in *Topic D1.2*.

It is an extraordinary fact that—with a few minor exceptions—all living organisms and all viruses use the same genetic code. It represents a sort of genetic language. Humans use many different spoken languages, each of which is an effective form of communication. Many different versions of a genetic code could be devised and they would probably function perfectly well, but all life forms use essentially the same version. For this reason, it is called the universal genetic code.

The minor exceptions to the universal genetic code found in some organisms are changes to the meaning of one of the 64 codons. In most cases, one of the three stop codons has changed to code for a specific amino acid instead. Life has been diversifying by evolution over billions of years so it is not surprising that there have been a few very small changes to the genetic code in some organisms. It is noteworthy that the code has changed so little and that all forms of life still speak essentially the same genetic language.

ATL Thinking skills: Evaluating the role of emotions and attitudes in science

The words below were spoken by Marshall Nirenberg, who was awarded the Nobel Prize in Physiology or Medicine in 1968 for his work on the genetic code.

> The finding that the code is universal had a terrific philosophical effect on me. I knew everything about evolution at the time, but these findings were so immediate and so profound, because I understood that most or all forms of life on this planet use the same genetic instructions and so we are all related. We're related to all living things and when I came in the garden and saw the plants, the squirrels and some of the birds, it really had a profound effect on me, which lasts to this day. I think that the feeling of being one with nature is very real and in fact is very true: we all use the same genetic language.

1. Why did the universality of the genetic code have such a profound effect on Marshall Nirenberg and others at the time?
2. What are the implications of the recognition of the unity of life, to scientists and to other people?
3. Are there other examples of scientific discoveries causing a profound change in attitudes?
4. To what extent do emotional responses such as the one described here support or run counter to stereotypical representations of scientists?

ATL Thinking skills: Evaluating the role of languages in science

A language is a code which ascribes agreed meanings to symbols.

1. What are the benefits of sharing a common language?
2. For scientists, why is the standardization of terminology viewed as essential?
3. Esperanto is (see Figure 15) an international language created by Ludwik Zamenhof in 1887. He hoped that a universal second language would promote world peace and understanding. What are the difficulties in creating a new language? Why does Esperanto not persist widely today?

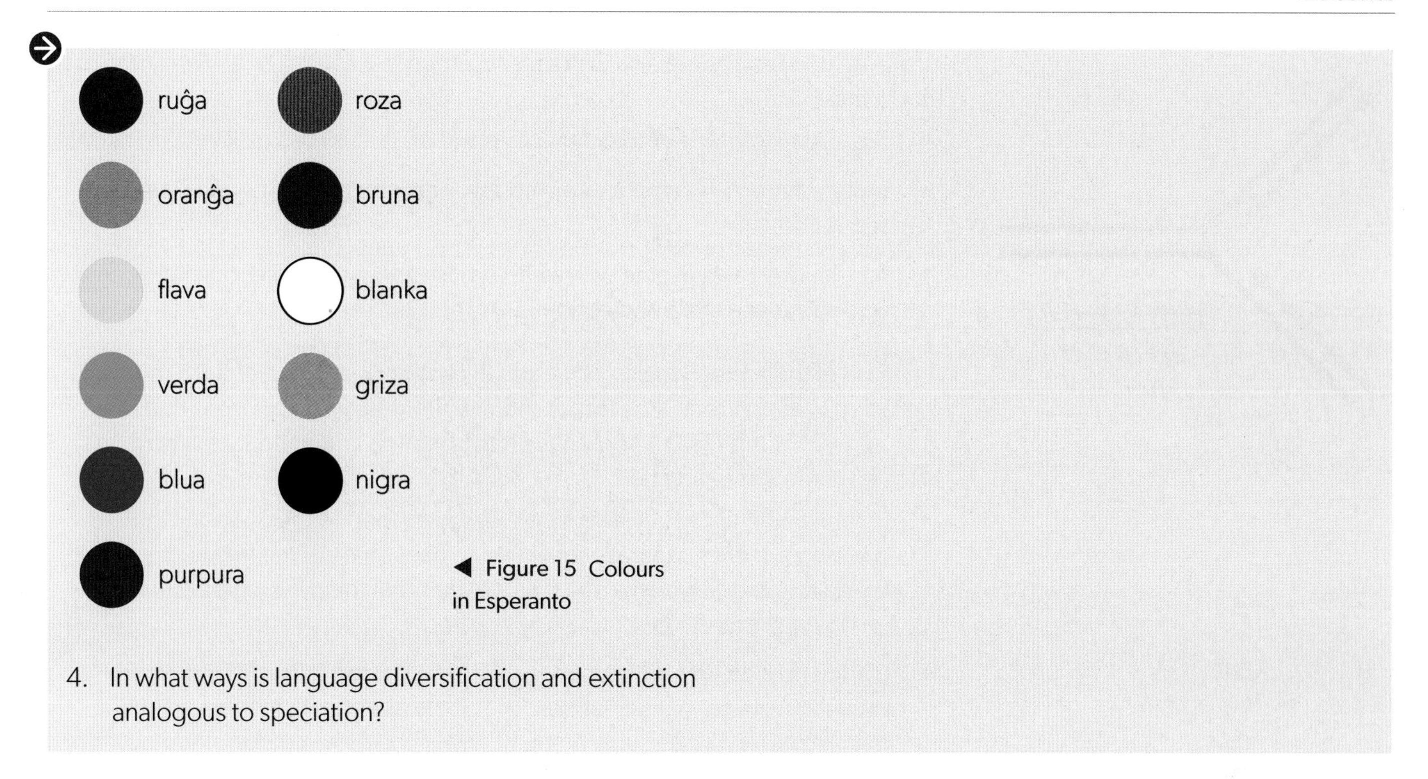

◀ Figure 15 Colours in Esperanto

4. In what ways is language diversification and extinction analogous to speciation?

A1.2.11 Directionality of RNA and DNA

The nucleotides within a strand of DNA or RNA are all linked in the same way: the phosphate group of one nucleotide is linked to the pentose sugar of the next nucleotide. As a result, the nucleotides are all orientated in the same way and the strand as a whole has directionality. The two ends of a strand of DNA or RNA can be distinguished as shown in Figure 16.

- The pentose sugar of the nucleotide at one end of the strand is unlinked. This is the 3′ terminal because C3 (carbon atom number 3) in this sugar is available for linkage to another nucleotide.
- The phosphate group of the nucleotide at the other end of the strand is unlinked. This is called the 5′ terminal, because within a nucleotide the phosphate group is attached to C5 of the pentose sugar.

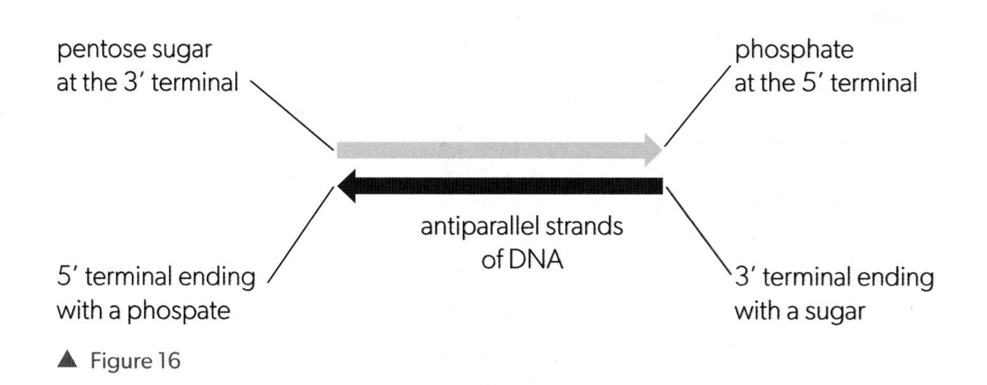

▲ Figure 16

AHL

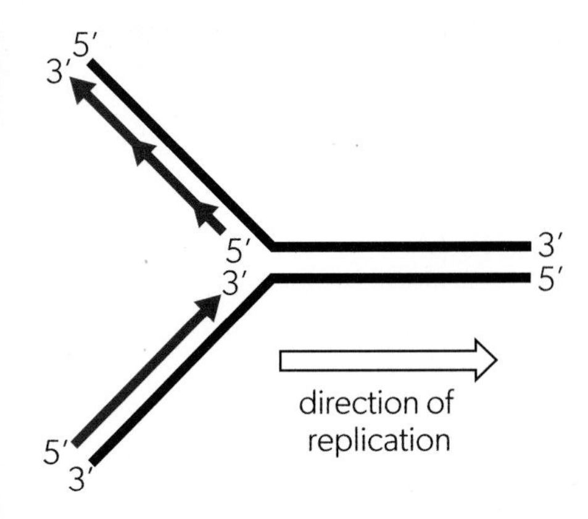

▲ Figure 17 Replication

The directionality of RNA and DNA affects processes carried out by enzymes or ribozymes:

- replication in which DNA polymerases and other enzymes make copies of DNA
- transcription in which RNA polymerase makes an RNA copy of a DNA base sequence
- translation at a ribosome with an RNA base sequence determining the amino acid sequence of a polypeptide.

Because of their directionality, DNA and RNA strands and nucleotides must be facing in the correct direction for them to fit the active sites of enzymes and ribozymes. For this reason, replication, transcription and translation always happen in the same direction.

- In replication, DNA nucleotides are always added to the 3′ end of the growing polymer of nucleotides. The 5′ phosphate of the free nucleotide is linked to the deoxyribose sugar at the 3′ end of the growing polymer. DNA replication is therefore 5′ to 3′.
- In transcription, RNA nucleotides are always added to the 3′ end of the growing polymer of nucleotides. The 5′ phosphate of the free nucleotide is linked to the ribose sugar at the 3′ end of the growing polymer. Transcription, like replication, is therefore 5′ to 3′.
- In translation, a molecule of RNA carries the sequence information for making a polypeptide by linking amino acids together. The ribosome that carries out translation moves along this RNA molecule towards the 3′ end. Translation therefore works in a 5′ to 3′ direction.

Replication

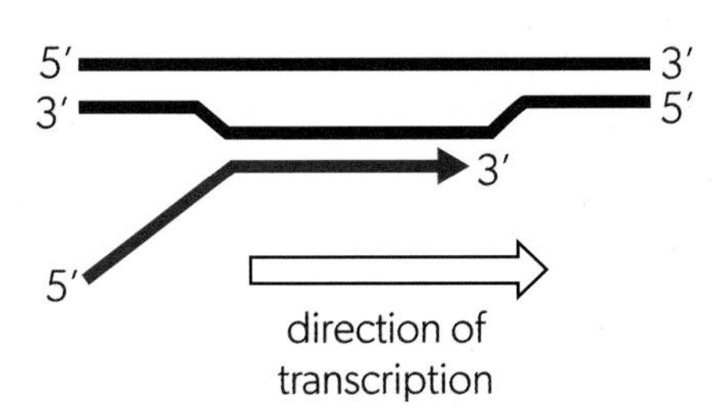

▲ Figure 18 Transcription

DNA nucleotides are always added to the 3′ end of the growing polymer of nucleotides. The 5′ phosphate of the free nucleotide is linked to the deoxyribose sugar at the 3′ end of the growing polymer. DNA replication is therefore 5′ to 3′.

Both of the strands in DNA are used as templates during replication. The two strands are antiparallel. On one strand, 5′ to 3′ assembly of a new strand moves in the same direction as the overall process of replication. On the other strand, it moves in the opposite direction. As a result, there are differences in what happens on the two template strands. These differences are described in *Topic D1.1*.

Transcription

RNA nucleotides are always added to the 3′ end of the growing polymer of nucleotides. The 5′ phosphate of the free nucleotide is linked to the ribose sugar at the 3′ end of the growing polymer. Transcription, like replication, is therefore 5′ to 3′.

Only one of the two strands of DNA is used as a template for making an RNA transcript. This is always the strand that allows the assembly of the RNA strand to move in the same direction as the overall process of transcription.

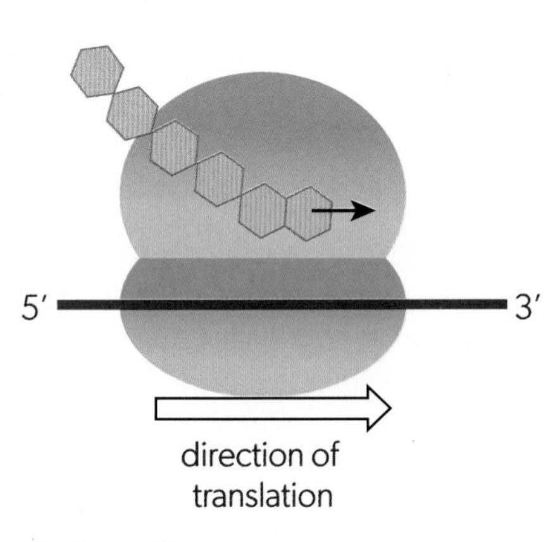

▲ Figure 19 Translation

Translation

A molecule of RNA carries the sequence information for making a polypeptide by linking amino acids together. The ribosome that carries out translation moves along this RNA molecule towards the 3′ end. Translation therefore works in a 5′ to 3′ direction.

A1.2.12 Purine-to-pyrimidine bonding as a component of DNA helix stability

The nitrogenous bases in DNA are in two chemical groups:

- Adenine and guanine are purine bases with molecules that have two rings of atoms.
- Cytosine and thymine are pyrimidine bases with molecules that have only one ring of atoms.

Each base pair in DNA therefore has one purine and one pyrimidine base. As a consequence, the two base pairs are of equal width and require the same distance between the two sugar–phosphate backbones in the double helix. This helps to make the structure of DNA stable and allows any sequence of bases in genes on a DNA molecule.

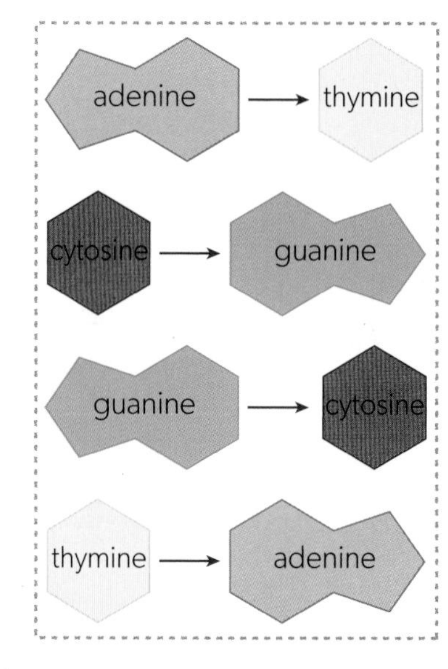

▲ Figure 20 Purine bases have two rings and pyrimidine bases have only one

A1.2.13 Structure of a nucleosome

The DNA of eukaryotes looks like a string of beads when viewed using an electron microscope. Each "bead" is a nucleosome. At the core of a nucleosome are eight histone proteins. Two copies each of four different types of histone together make up a disc-shaped structure. The DNA molecule is wound approximately twice around this protein core.

An additional histone protein molecule called H1 reinforces the binding of the DNA to the nucleosome core. H1 may also help in the packaging of chromosomes when a nucleus is preparing to divide. There is a short section of linker DNA between adjacent nucleosomes.

Plants, animals and other eukaryotes have nucleosomes. Bacteria do not have nuclei and their DNA is "naked" because it is not associated with histones.

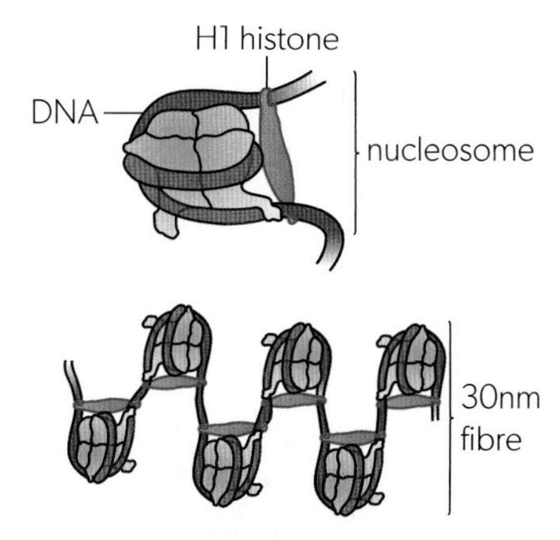

▲ Figure 21 Nucleosomes

Using molecular visualization software: Visualizing nucleosomes

Molecular visualization software can be used to analyse the association between protein and DNA within a nucleosome.

1. Visit the protein data bank at www.rcsb.org and search for "human nucleosome structure".
2. Click "3D View".
3. Rotate the molecule to see the two copies of each histone protein. In Figure 22, they are identified by the tails that extend from the core. Each protein has a tail like this that extends out from the core.
4. Note the approximately 150 bp of DNA wrapped nearly twice around the octamer core.
5. Note the N-terminal tail that projects from the histone core for each protein. Chemical modification of this tail is involved in regulating gene expression.
6. Visualize the positively charged amino acids on the nucleosome core. Suggest how they play a role in the association of the protein core with the negatively charged DNA.

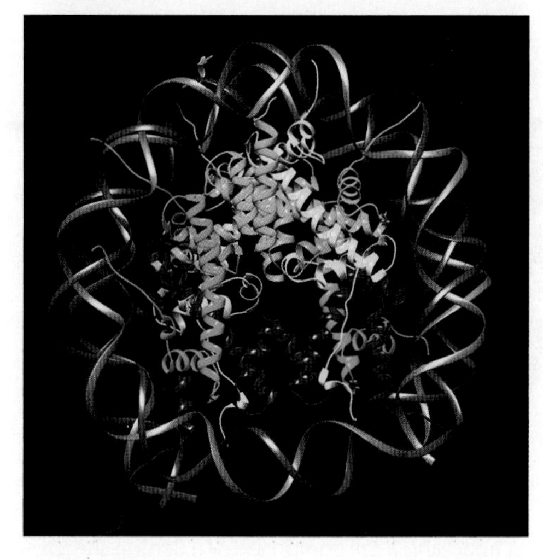

▲ Figure 22

A1.2.14 Evidence from the Hershey–Chase experiment for DNA as the genetic material

From the late 1800s, scientists were convinced that chromosomes played a role in heredity and that the hereditary material had a chemical nature. It was known that chromosomes were composed of protein and DNA but it was not clear which of these molecules was the genetic material. Until the 1940s, most biologists viewed protein as the more likely candidate, because it contains 20 different amino acid subunits, whereas DNA has just four types of nucleotide. In addition, many specific functions of proteins had already been identified. Variety and specificity of function were considered essential for hereditary material.

Alfred Hershey and Martha Chase chose to use the T2 bacteriophage to identify the genetic material. It has a coat composed entirely of protein, with DNA inside the coat. In the 1950s, it was known that a virus can transform a host cell so that it produces viral proteins; for this to happen, viral genes must have been injected into the host cell.

In their experiment, Hershey and Chase took advantage of the fact that DNA contains phosphorus but not sulfur while proteins contain sulfur but not phosphorus. They cultured some viruses that contained proteins with radioactive (^{35}S) sulfur and other viruses that contained DNA with radioactive (^{32}P) phosphorus. Then they infected separate groups of bacteria with the two viruses.

For each group of bacterial cells, they used a blender to separate the non-genetic component of the virus. Then they centrifuged the culture solution to concentrate the cells in a pellet. The cells were expected to contain the radioactive genetic component of the virus. Finally, Hershey and Chase measured the radioactivity in the pellet and the supernatant.

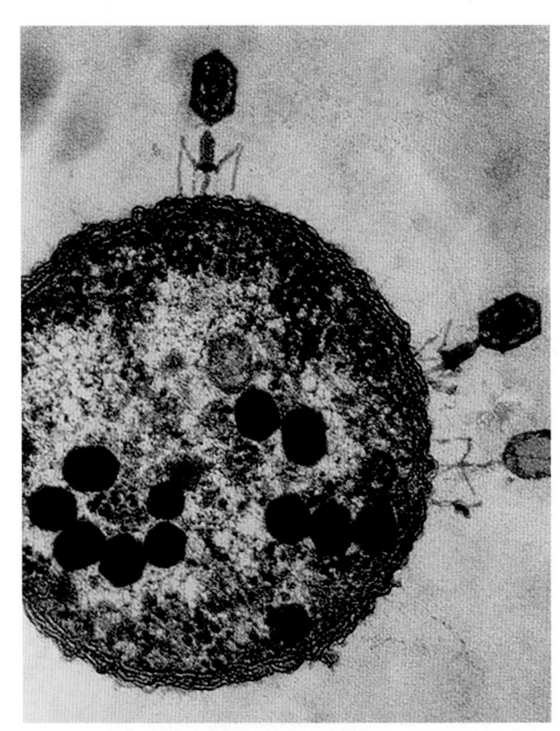

▲ Figure 23 Coloured transmission electron micrograph (TEM) of T2 viruses (blue) bound to an *Escherichia coli* bacterium. Each virus consists of a large DNA-containing head and a tail composed of a central sheath with several fibres. The fibres attach to the host cell surface, and the virus DNA is injected into the cell through the sheath. It instructs the host to build copies of the virus (blue, in cell)

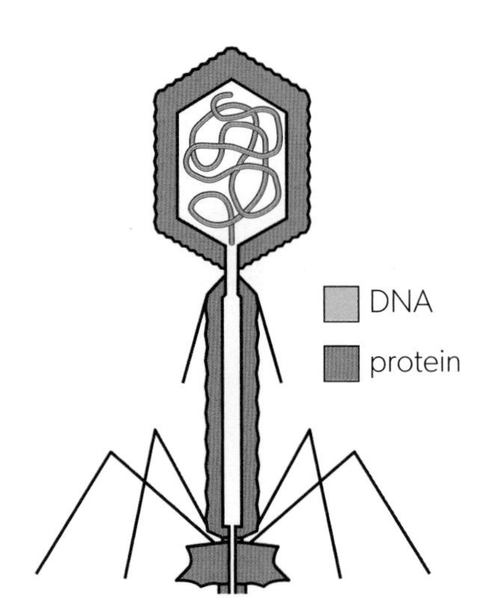

▲ Figure 24 Diagram illustrating the structure of the T2 virus

Data-based questions: The Hershey–Chase experiment

These diagrams show the process of the Hershey–Chase experiment.

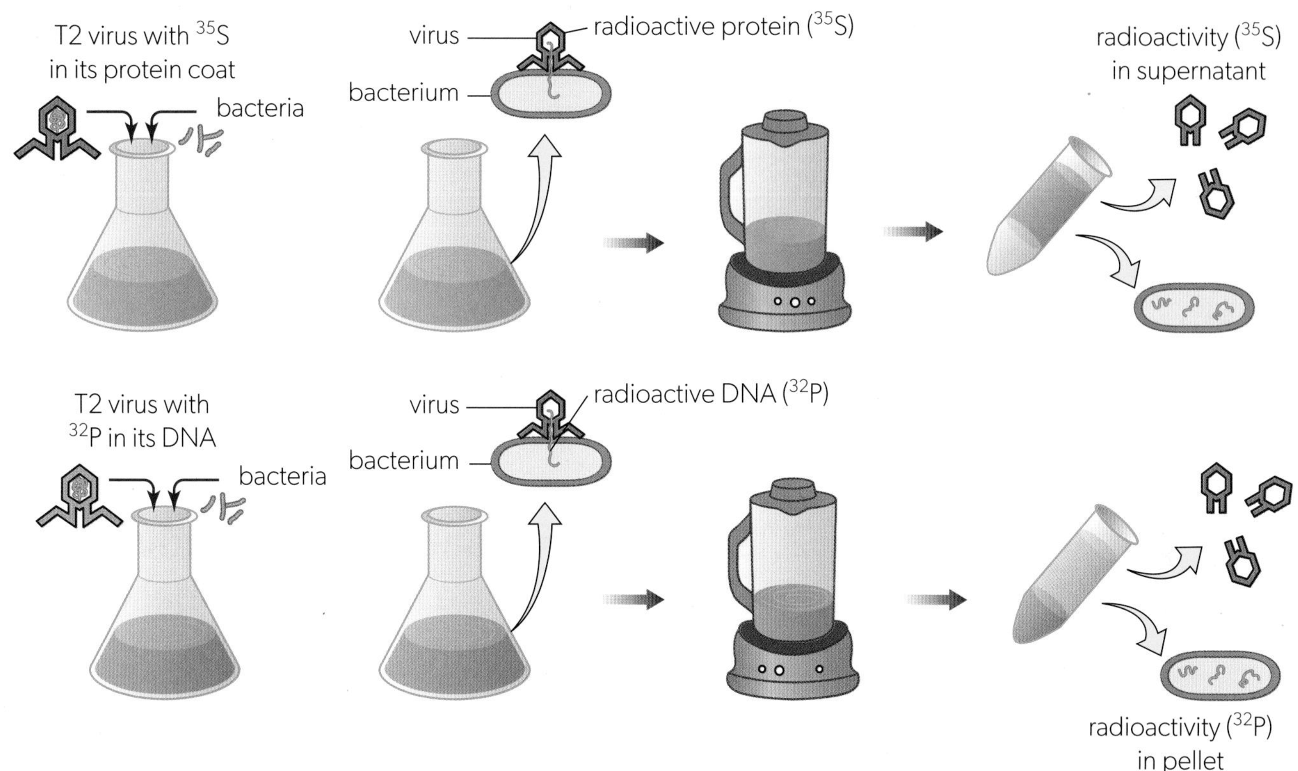

▲ **Figure 25**

Figure 26 shows the results of the experiment.

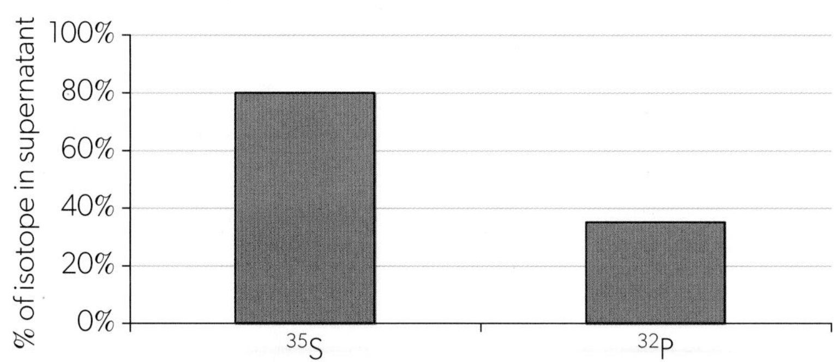

▲ **Figure 26**

1. Distinguish between a supernatant and a pellet. [2]
2. Explain why the genetic material should be found in the pellet and not the supernatant. [2]
3. State the percentage of ^{32}P that remains in the supernatant. [1]
4. Determine the percentage of ^{32}P that is spun down into the pellet. [2]
5. Discuss the evidence that DNA is the chemical which transforms the bacteria into infected cells. [3]

Experiments: Use of radioisotopes as research tools

Two atoms are different isotopes of the same element if they have the same number of protons but different numbers of neutrons in their atomic nucleus. The chemical properties of an atom are determined by the numbers of protons and electrons—not neutrons—so isotopes of an element have the same chemical properties. However, they may have different levels of nuclear stability. For example, ^{32}S has 16 protons and 16 neutrons and is stable but ^{35}S has 16 protons and 19 neutrons and is unstable.

Protons are positively charged and repel each other with an electrical force. However, at very close distances they attract each other with a nuclear force. Certain combinations of protons and neutrons are unstable because the attractive nuclear force cannot counterbalance the repulsive electric force. As a result, unstable atomic nuclei release energy in the form of radiation as they assume more stable forms. This radiation can be detected.

For scientists in the 1950s, radioisotopes were a valuable research tool. When introduced in minuscule quantities to biological systems, these unstable variants of common atomic elements can be traced as they move through biological systems. György Hevesy won the Nobel Prize in Chemistry in 1943 for pioneering the use of radioisotopes in biological research. In 1923, Hevesy published the first study using radioactive ^{212}Pb as a tracer to follow the absorption and translocation of minerals in plants.

A1.2.15 Chargaff's data on the relative amounts of pyrimidine and purine bases across diverse life forms

Before the structure of DNA was known, scientists hypothesized that it would contain a repeating sequence of the four bases. This would mean the four nucleotides occurred in equal numbers. The tetranucleotide hypothesis was formulated in 1910. However, if DNA had a tetranucleotide structure, it would not be able to vary enough to be the genetic material. This is why scientists thought it was more likely that the 20 amino acids making up proteins were the genetic material.

To test the tetranucleotide hypothesis, Erwin Chargaff and others analysed DNA samples from a range of species to find their nucleotide composition. A portion of their data is shown in Table 2.

Source of DNA	Group	Adenine	Guanine	Cytosine	Thymine
Human	Mammal	31.0	19.1	18.4	31.5
Cattle	Mammal	28.7	22.2	22.0	27.2
Salmon	Fish	29.7	20.8	20.4	29.1
Sea urchin	Invertebrate	32.8	17.7	17.4	32.1
Wheat	Plant	27.3	22.7	22.8	27.1
Yeast	Fungus	31.3	18.7	17.1	32.9
Mycobacterium tuberculosis	Bacterium	15.1	34.9	35.4	14.6
Bacteriophage T2	Virus	32.6	18.2	16.6	32.6
Polio virus	Virus	30.4	25.4	19.5	0.0

▲ Table 2

Data-based questions: Chargaff's data

Use the data in Table 2 to answer the following questions.

1. Compare the base composition of *Mycobacterium tuberculosis* (a prokaryote) with the base composition of the eukaryotes shown in Table 2. [2]
2. Calculate the base ratio A + G / T + C, for humans and for *Mycobacterium tuberculosis*. Show your working. [2]
3. Evaluate the claim that in the DNA of eukaryotes and prokaryotes, the amounts of adenine and thymine are equal and the amounts of guanine and cytosine are equal. [2]
4. Explain the ratios between the amounts of bases in eukaryotes and prokaryotes in terms of the structure of DNA. [2]
5. Explain how these results falsify the tetranucleotide hypothesis. [2]
6. Suggest reasons for the difference in the base composition of bacteriophage T2 and the polio viruses. [2]

Falsification: The nature of the genetic material

Knowledge claims are based on evidence gathered through the senses. Scientists make observations, detect patterns and then form generalizations. These generalizations are used to draw conclusions about things that have not yet been observed. This is known as inductive reasoning. The problem with induction is that there is no certainty that the unobserved things will conform to the generalization; thus, is there anything about the natural world of which we can be certain when we have not observed all cases?

Falsifiability is the idea that in science we can at least be certain of what is not the case, by finding a counter-example. Chargaff's analysis falsified the tetranucleotide hypothesis that the four DNA bases occur in equal amounts. The Hershey and Chase experiment falsified the hypothesis that protein is the genetic material. The work of these scientists provided certainty of what was **not** the case.

Linking questions

1. What makes RNA more likely to have been the first genetic material, rather than DNA?
   a. Explain the role of enzymes in the cellular processes associated with heredity. (D1.1.8)
   b. Outline the role of RNA as a catalyst. (A2.1.6)
   c. Compare and contrast the structure of DNA and RNA. (A1.2.7)
2. How can polymerization result in emergent properties?
   a. Distinguish between the properties of glucose and starch. (B1.1.5)
   b. Outline the role of condensation reactions in forming nucleic acids. (B1.1.2)
   c. Explain the relationship between the sequence of amino acids and the structure and function of proteins. (B1.2.10)

TOK

How can we know that current knowledge is an improvement on past knowledge?

Phoebus Levene made significant contributions to the development of our understanding of nucleic acids. He established the existence of the sugar–phosphate backbone in nucleic acids; he identified deoxyribose as the sugar in DNA; and he coined the name nucleotide.

He also incorrectly stated that DNA was made up of repeating units of the four DNA nucleotides stacked together. This was known as the tetranucleotide hypothesis. This hypothesis led Levene to state that DNA could not be the hereditary material, because the tetranucleotides were not sufficiently variable to be the basis of the code for the tremendous diversity of life that exists. This idea was widely accepted. Instead, proteins were thought to form the hereditary material, because they were known to have great variety of structure.

Figure 2 shows the results of an experiment carried out in 1928 by Griffiths involving viruses. Injecting mice with the rough strain of a virus did not cause the death of the mice. The smooth strain did cause the death of the mice. The heat-killed version of the smooth strain did not cause death. However, mixing the heat-killed smooth version with the living version of the non-deadly rough version did cause death. Somehow the genetic material of the heat-killed version was able to transform the living non-virulent version to the virulent version. In 1944, Avery, McCarty and McCleod took the experiment further. From Griffith's experiment, they were aware that dead virulent strains of bacteria could transform living strains to make them virulent. In different experiments, they attempted to establish what the transforming material was by adding enzymes that would break down different chemicals to determine if they

▲ **Figure 1** A tetranucleotide molecule

could interfere with the transformation process. They added RNAase, proteinase and DNAase. The DNAase was able to interfere with the transformation process. This established that the hereditary material was DNA.

In 1950, Erwin Chargaff analysed the nucleotide composition of cells from a number of different species. He found that the amount of adenine (A) was not equal to the amount of guanine (G), and the amount of thymine (T) was not equal to the amount of cytosine (C). This was sufficient evidence to establish that the tetranucleotide hypothesis was incorrect. The weight of evidence was beginning to support the theory that DNA was the genetic material.

The Hershey and Chase experiment in 1952 showed convincingly that DNA was the genetic material. The combination of the three experiments was enough to establish that the new knowledge was an improvement upon past knowledge.

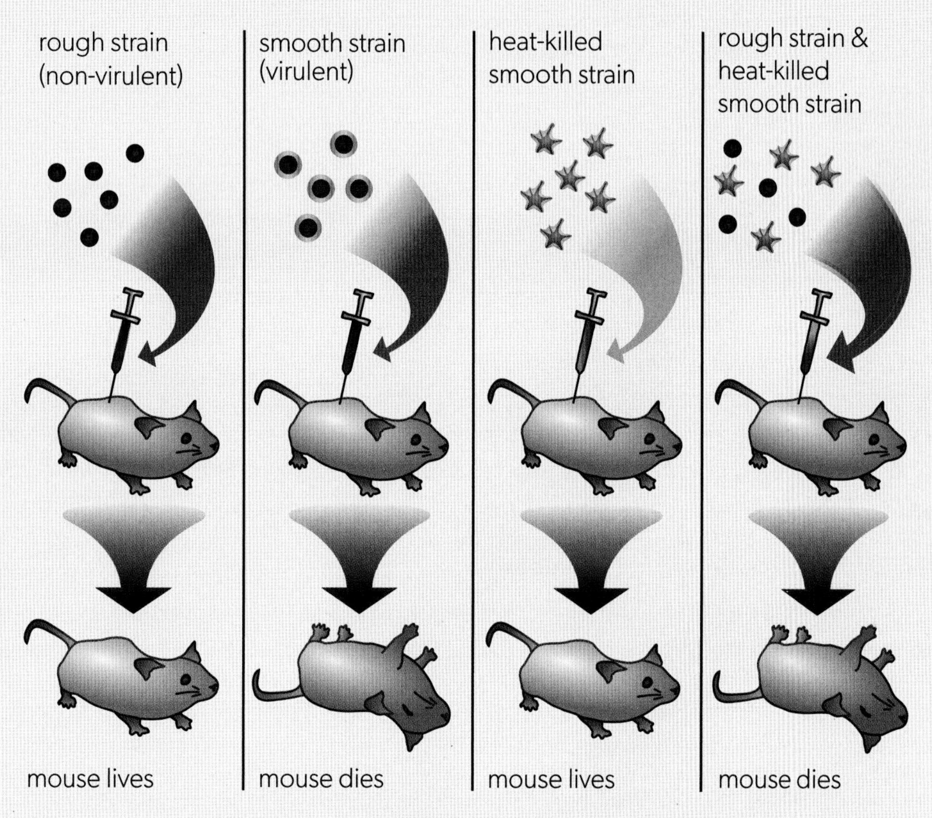

▲ Figure 2 Avery, McCleod and McCarty were aware that heat-killed virulent bacteria contained a chemical that could transform non-virulent bacteria. They worked to identify the transforming chemical was

End of chapter questions

1. Table 1 compares the physical properties of water and air.

Property	Water	Air	Condition
Specific heat	$4.18\,J°C^{-1}g^{-1}$	$1.01\,J°C^{-1}g^{-1}$	27°C
Thermal conductivity	$0.6\,W\,m^{-1}K^{-1}$	$0.028\,W\,m^{-1}K^{-1}$	16°C
Density	$1000\,kg\,m^{-3}$	$1.225\,kg\,m^{-3}$	15°C; sea level
Buoyancy	650 N	0.8 N	Assuming a body of volume $0.0664\,m^3$ at 15°C
Viscosity	$0.7978\times10^{-3}\,Pa\cdot s$	$18.6\times10^{-6}\,Pa\cdot s$	27°C

▲ Table 1

a. Specific heat refers to the amount of energy it takes to change the temperature of one gram of substance by 1°C.

   i. Identify which substance (air or water) is more resistant to changes in temperature. [1]

   ii. Identify the habitat that would have a more stable thermal environment and discuss the implications of this. [3]

b. Thermal conductivity is a measure of the degree to which the medium is a conductor of heat.

   i. Calculate the factor by which the thermal conductivity of water exceeds the thermal conductivity of air. [2]

   ii. Identify the medium in which it would be more difficult for organisms to sustain their internal temperatures above that of the surrounding medium. [1]

c. Buoyancy is a measure of the upward force provided by a medium to counteract gravity. It is a function of the density of the medium and the volume of the object.

   i. Which organism would have to expend more energy to counteract the effects of gravity: a swimming seal or a bird in flight? [1]

   ii. Suggest a possible relationship between fish having air bladders and buoyancy. [2]

d. Viscosity is the force per unit area resisting flow.

   i. Calculate the factor by which the viscosity of water exceeds the viscosity of air. [2]

   ii. Identify the medium that offers the greater resistance to movement. [1]

e. Construct a visual representation that compares the advantages and disadvantages of air and water as a habitat for a flying bird and a swimming seal. [4]

2. Most intertidal animals are adapted to a marine existence. However, during exposure to air at low tide, they are subjected to the heat stress that is characteristic of the terrestrial environment. This can affect the ability of intertidal organisms to survive and reproduce.

The limpet *Lottia gigantea* is an important food source for shore birds and for subsistence human harvesters. A study was carried out to model thermal stress. Some of the results are shown in Figure 2.

▲ Figure 1

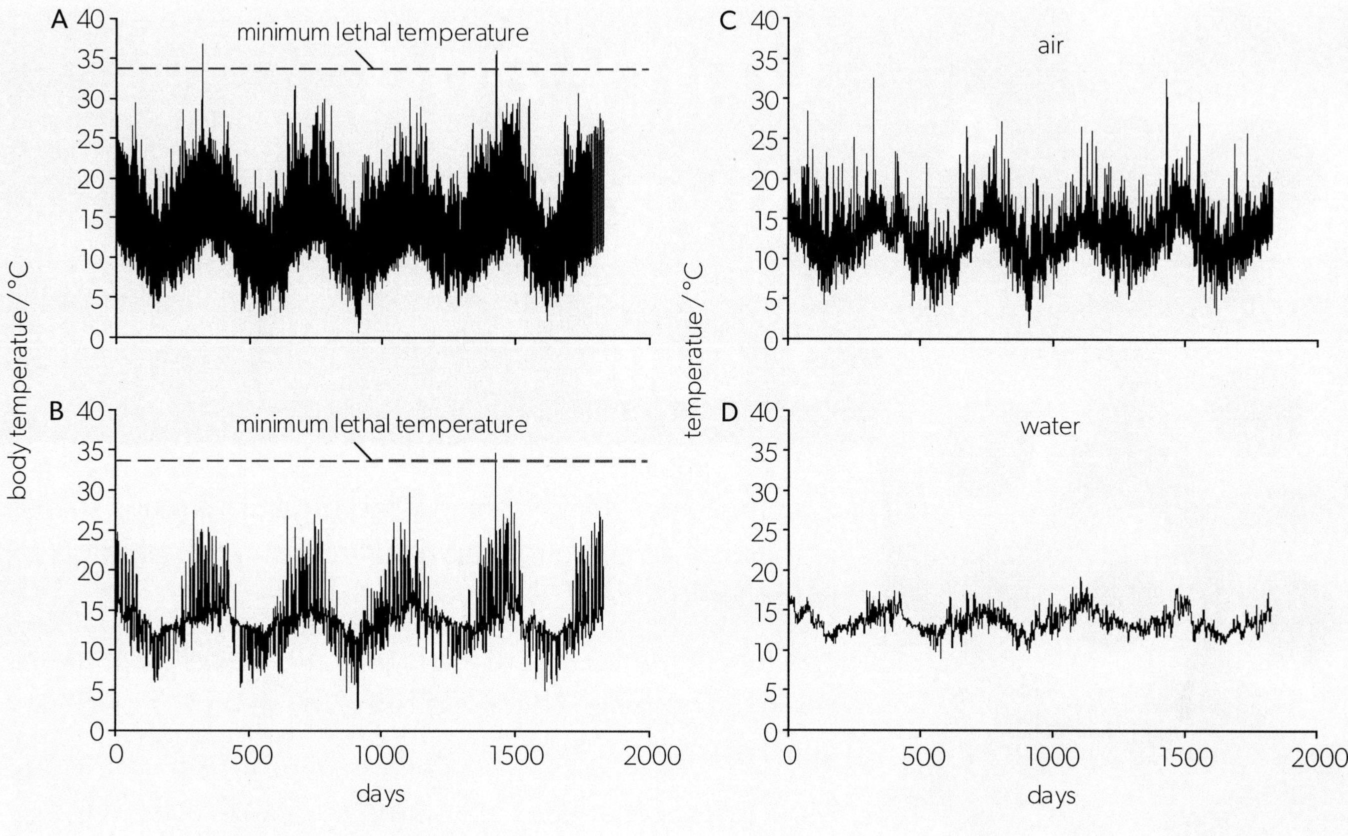

▲ **Figure 2** Source: J Exp Biol (2006) 209 (13): 2420–2431

a. Compare and contrast the temperature variations over time between air (graph C) and water (graph D). [2]

b. Identify the temperature that is viewed as being lethal to limpets (graph A and graph B). [1]

c. Graph A and Graph B represent the temperature over time in two areas: 0.5 m above the average low level water line and 1.5 m above the average low level water line.

   i. Deduce, with a reason, which graph represents which position. [2]

   ii. Deduce which location results in a higher mortality rate due to thermal stress. [1]

3. Analysis of the base composition of the polio virus shows that it is adenine 30.4%, guanine 25.4%, cytosine 19.5% and thymine 0.0%

   a. From this data, what is the evidence that the polio virus is

      i. an RNA virus [1]

      ii. single-stranded [1]

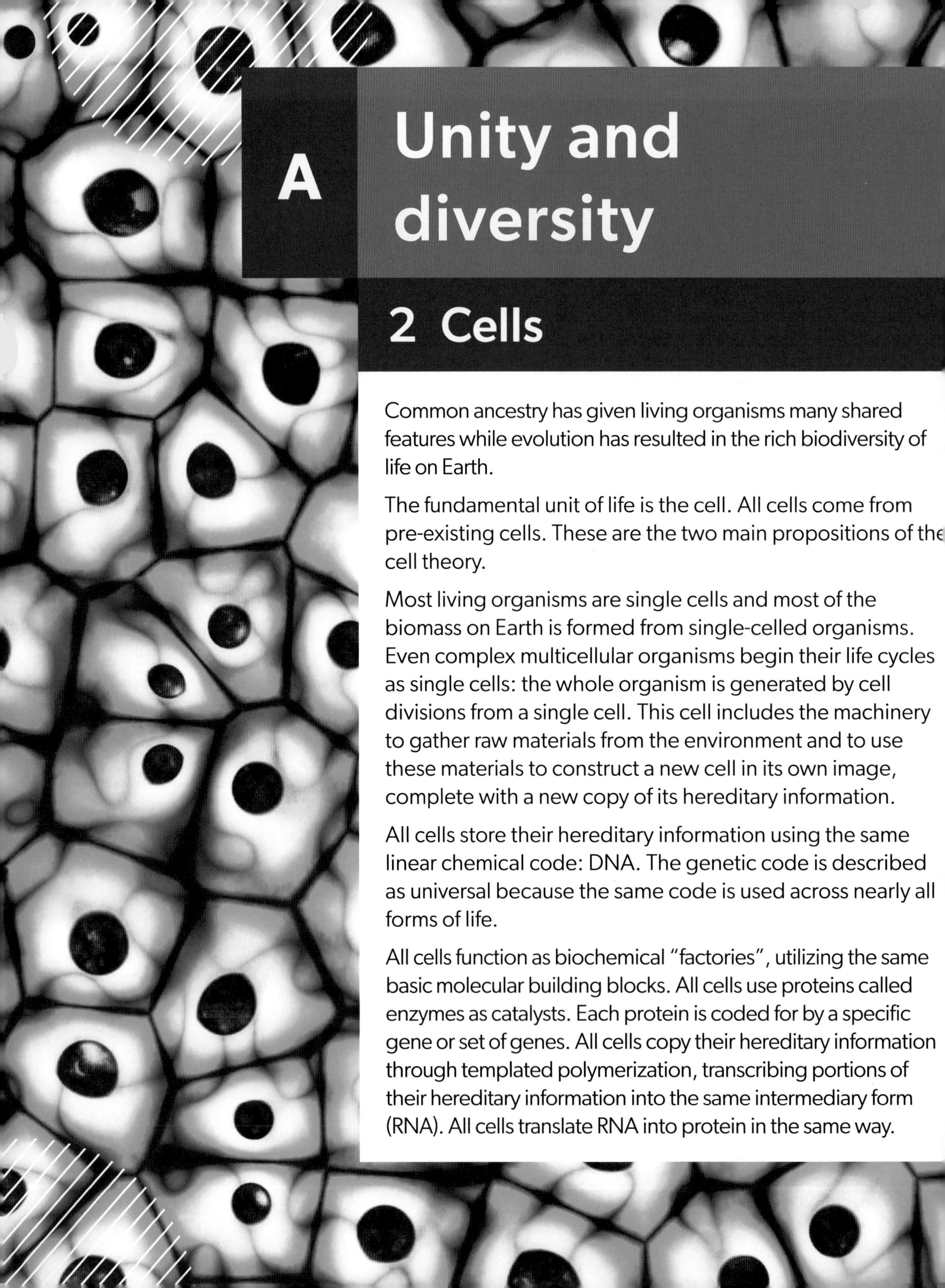

A Unity and diversity

2 Cells

Common ancestry has given living organisms many shared features while evolution has resulted in the rich biodiversity of life on Earth.

The fundamental unit of life is the cell. All cells come from pre-existing cells. These are the two main propositions of the cell theory.

Most living organisms are single cells and most of the biomass on Earth is formed from single-celled organisms. Even complex multicellular organisms begin their life cycles as single cells: the whole organism is generated by cell divisions from a single cell. This cell includes the machinery to gather raw materials from the environment and to use these materials to construct a new cell in its own image, complete with a new copy of its hereditary information.

All cells store their hereditary information using the same linear chemical code: DNA. The genetic code is described as universal because the same code is used across nearly all forms of life.

All cells function as biochemical "factories", utilizing the same basic molecular building blocks. All cells use proteins called enzymes as catalysts. Each protein is coded for by a specific gene or set of genes. All cells copy their hereditary information through templated polymerization, transcribing portions of their hereditary information into the same intermediary form (RNA). All cells translate RNA into protein in the same way.

A2.1 Origins of cells

What plausible hypothesis could account for the origin of life?

Is there a consensus view of the conditions that existed on the pre-biotic earth? If there is one, what were these conditions? How do they differ from the conditions that exist today? In what ways did living things cause some of the changes to conditions on the early earth?

Figure 1 shows a cross-section through a fossilized stromatolite, showing concentric layers of rock (white) and coal-like organic material (black). The layers of organic material were laid down by cyanobacteria (blue-green algae). How far back can such layers be found? Do they offer conclusive proof of life?

▲ Figure 1 Cross-section through a fossilized stromatolite

What intermediate stages could there have been between non-living matter and the first living cells?

Figure 2 shows a "white smoker" which is a hydrothermal vent. It is not known when or where life on Earth began. However, some of the earliest habitable environments may have been submarine hydrothermal vents. The oldest known fossils providing evidence of microbial life have been found in precipitates associated with seafloor hydrothermal vents. For the first living cells to have appeared, catalysis would have to have occurred. What were some of the some of the necessary developments that would need to have occurred for the first life to emerge? How might hydrothermal events provided the correct conditions for this emergence?

▲ Figure 2 A "white smoker" hydrothermal vent

AHL only

A2.1.1 Conditions on early Earth and the pre-biotic formation of carbon compounds
A2.1.2 Cells as the smallest units of self-sustaining life
A2.1.3 Challenge of explaining the spontaneous origin of cells
A2.1.4 Evidence for the origin of carbon compounds
A2.1.5 Spontaneous formation of vesicles by coalescence of fatty acids into spherical bilayers
A2.1.6 RNA as a presumed first genetic material
A2.1.7 Evidence for a last universal common ancestor
A2.1.8 Approaches used to estimate dates of the first living cells and the last universal common ancestor
A2.1.9 Evidence for the evolution of the last universal common ancestor in the vicinity of hydrothermal vents

A2.1.1 Conditions on early Earth and the pre-biotic formation of carbon compounds

The Sun formed about 4,500 million years ago, two-thirds of the way through the time that our Universe has existed. The Earth formed soon afterwards, as gravity caused gas and dust in the early solar system to come together. At first, there was no life—there was a pre-biotic period in the Earth's development. Gases accumulated but in very different concentrations to those in today's atmosphere. Evidence from ancient rocks has helped scientists to describe the pre-biotic atmosphere:

- There were only traces of oxygen because it reacted with other elements. For example, oxygen reacted with iron to produce iron oxide.
- Methane concentrations were higher than today due to intense volcanic activity and meteorite bombardment.
- Carbon dioxide concentrations were also probably higher due to emissions from volcanoes.

▲ Figure 3 An artist's impression of conditions on Proxima b. This planet orbits our nearest neighbouring star Proxima Centauri, in the habitable zone where water could exist as a liquid. What conditions, apart from liquid water, are needed for life to evolve on this planet?

Temperatures are likely to have been higher than they are now. Carbon dioxide and methane are heat-trapping greenhouse gases and although the Sun was emitting 20% less energy, comet and asteroid impacts will have raised temperatures. Estimates of temperatures on pre-biotic Earth vary widely. It is also uncertain what the pH of the oceans was; estimates range from pH 5 to pH 11.

The stratospheric ozone layer that currently protects us by absorbing ultraviolet radiation (UV) would not have existed, because of the lack of oxygen. Without this layer, more solar UV would have penetrated to the Earth's surface. UV is a high-energy form of radiation and provides the activation energy for chemical reactions. There may also have been more lightning on early Earth, triggering other chemical processes.

Because of these differences, reactions that are not possible today may have occurred spontaneously on pre-biotic Earth. As a result, a variety of carbon compounds may have formed in specialized environments such as hot springs on land or hydrothermal vents in the oceans. Carbon compounds may have formed in droplets of water in the atmosphere, creating what has been called an organic aerosol haze. These carbon compounds would then have been deposited by rainfall into pools, lakes or seas, creating a "soup" of carbon compounds.

There is much uncertainty about conditions on pre-biotic Earth so scientists are unsure which carbon compounds may have spontaneously formed. However, many of the building blocks of life—such as carboxylic acids, aldehydes, amino acids and the bases that are part of DNA and RNA—are all possibilities.

Once living organisms had evolved, they caused enormous changes to conditions on Earth. Over time, organisms increased the concentration of oxygen in the atmosphere from zero to 20%. They also reduced carbon dioxide to very low concentrations. As oxygen levels increased, an ozone layer formed, giving protection against UV. The greenhouse effect was reduced. It is ironic that once living organisms had evolved from non-living matter, the changes that they caused probably made it impossible for life to evolve again.

Data-based questions: Titan's atmosphere

Titan is Saturn's largest moon. The atmosphere around it formed by processes similar to those on Earth. Table 1 shows the gases that make up more than 1% of the atmosphere at the surface of the Earth and Titan. There is a thick orange smog in Titan's atmosphere due to ethane, propane, propene and other hydrocarbons.

Gas	Titan	Earth
nitrogen	95%	78%
methane	4.9%	<0.001%
oxygen	0.0001%	21%

▲ Table 1

1. What are the differences between the atmospheres of Earth and Titan? [5]
2. What are the reasons for the atmospheres being so different? [5]

▲ Figure 4 Image of Titan taken by NASA's Cassini spacecraft

A2.1.2 Cells as the smallest units of self-sustaining life

It is easy to recognize that a crying baby is alive while a lump of rock is not. It is not so easy to define what life is. It is common to use a checklist of "functions of life" (such as nutrition and respiration). However, these are just processes required to maintain life, not life itself.

A key difference between living and non-living things is that living things use energy to keep themselves in a highly ordered state. They are self-sustaining. This highly ordered state would be extremely difficult to achieve starting with non-living components. A complex device such as a smartphone can be assembled from individual components and work when it is switched on. In contrast, living organisms are just too complex. Currently, the only way to produce a living thing is by reproducing an existing one. Passing on the ability to maintain a highly ordered state to offspring is a second key difference between living and non-living things.

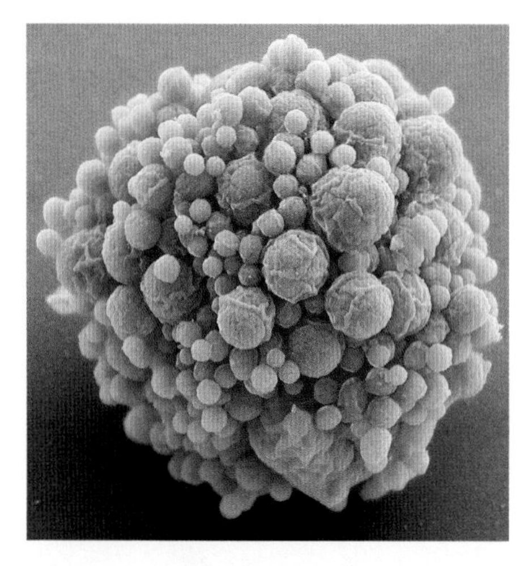

▲ Figure 5 The bacteria in this scanning electron micrograph have the smallest genome (473 genes) of any self-replicating organism. An artificial chromosome containing these genes, which have base sequences found in *Mycoplasma mycoides*, was transplanted into a cell of *Mycoplasma capricolum* whose own DNA had been removed. This pared-down bacterium, developed in 2016, enables scientists to study the roles of genes in cells. The function of 149 of the genes is not currently known. Is this an artificial cell or a modified natural cell?

An individual organism is certainly alive. Living organisms may be composed of one cell (unicellular) or many cells (multicellular). To decide whether each cell in a multicellular organism is alive, or just the whole organism, we can look at various types of evidence:

- Individual cells use energy to maintain a highly ordered state.
- Some cells in a multicellular organism may stop doing this. These cells are clearly dead, for example, hair cells or cells in the outer layers of skin.
- Cells can divide to produce more cells.
- Cells can be taken from the body and cultured. For example, HeLa cells have been kept in cultures since 1951.

The evidence is strong enough for us to regard cells as living. However, individual cell components are not self-sustaining. For that reason, cells are considered to be the smallest units of self-sustaining life.

A2.1.3 Challenge of explaining the spontaneous origin of cells

According to the theory of spontaneous generation, living organisms can be formed from non-living matter.

Experiments by Louis Pasteur and other biologists showed that this claim was false. Evidence was also gained from Robert Remak's observational studies of cells in chickens and frogs. He discovered that every cell is formed by division of a pre-existing cell. This principle became an essential part of the cell theory and from it some remarkable deductions can be made.

Consider the trillions of cells in your body. Each one was formed by division of a pre-existing body cell, starting with a zygote produced when one sperm from your father fused with an egg from your mother. The sperm and egg cells were produced by cell division in your parents. You can trace the origins of all the cells in your parents' bodies back to the zygote from which they developed—and so on through many generations of human ancestry. If you accept that humans evolved from pre-existing ancestral species, you can trace the origins of your cells back through vast numbers of generations over thousands of millions of years of life.

However, life has not always existed on Earth. If you keep going back through the generations, you must eventually reach the earliest cells to have existed. Unless they travelled to Earth from somewhere else in the universe, these first cells must have developed from non-living material. This deduction leads to a very difficult question: how could a structure as complex as a cell arise from non-living material by natural means?

It is argued that cells are too complex to have arisen by evolution. However, if there was a series of intermediate stages over a long period of time, evolution of the first cells becomes more feasible. These are developments required for the origin of cells:

- catalysis—to give control over which chemical reactions occur
- self-assembly—carbon compounds such as amino acids must assemble to form polymers

- compartmentalization—a membrane must develop to enclose cell contents
- self-replication of molecules—as a basis for inheritance and the persistence of successful variants.

There are hypotheses for how each development could have occurred.

Falsification: The origin of the first cells

Biologists understand that claims in science, including hypotheses and theories, must be testable. Unfortunately, it is very difficult to test hypotheses relating to the origin of cells, because this happened billions of years ago when conditions on Earth were very different. Well-preserved fossils of the first protocells are unlikely to be found. Some of the most ancient specimens described as fossils have been re-analysed and found to be crystals. Scientists have attempted to model conditions on pre-biotic Earth but it is not possible to replicate with certainty the conditions that would have existed.

These difficulties do not mean that hypotheses about the origin of cells are untestable and therefore unscientific. Other methods of testing can be used. For example, the genomes of living organisms contain vast amounts of data and one approach is to examine this data for information about the origins of the first cells.

▲ Figure 6 3.77 billion year old banded iron rock from Isua in western Greenland. Some scientists claim that there are signs of life in this rock. Others believe that the structures visible could be the result of metamorphosis of sedimentary rocks

A2.1.4 Evidence for the origin of carbon compounds

In 1929, the biologist J.B.S. Haldane wrote an article on the origin of life. In it, he described the pre-biotic ocean as a "hot dilute soup" in which a variety of carbon compounds could have formed. He based this hypothesis on an assumption that the atmosphere contained water vapour, carbon dioxide and ammonia. Haldane claimed that: "when ultraviolet light acts on a mixture of water, carbon dioxide and ammonia, a vast variety of organic substances are made, including sugars and apparently some of the materials from which proteins are built up".

This hypothesis for the origin of carbon compounds was tested experimentally in the early 1950s by Stanley Miller and Harold Urey. In a five-litre flask, they mixed methane, hydrogen and ammonia—the mixture of gases they thought was representative of the pre-biotic atmosphere. They added water vapour to the mixture by boiling water in another flask, then used electrical discharges to simulate lightning. A condenser cooled the substances produced, then the condensate was returned to the flask of boiling water.

After the experiment had been running for a day, the water turned pink. After a week, it was dark red. Analysis showed that a variety of carbon compounds had indeed been produced, including more than 20 different amino acids. This showed that it was possible for carbon compounds to form spontaneously on Earth before life had evolved, as long as the conditions in the Miller–Urey apparatus simulated pre-biotic conditions accurately enough.

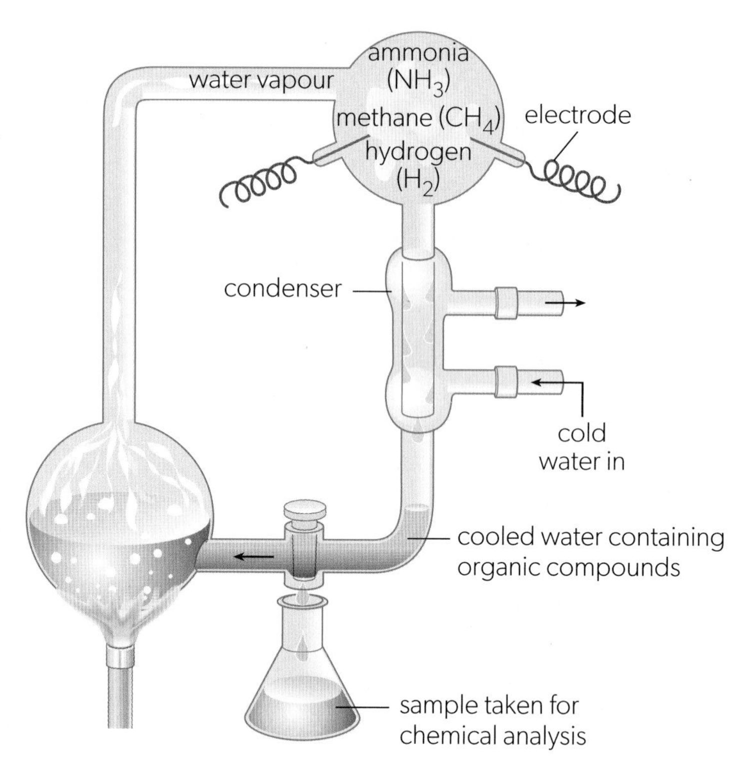

▲ Figure 7 Miller and Urey's apparatus

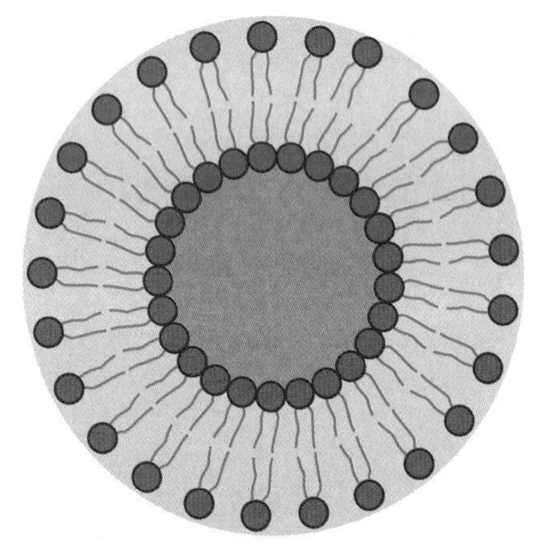

▲ Figure 8 Phospholipids naturally form spherical bilayers in water, with the hydrophilic heads (blue) facing out from the bilayer and the hydrophobic tails (green) forming the core

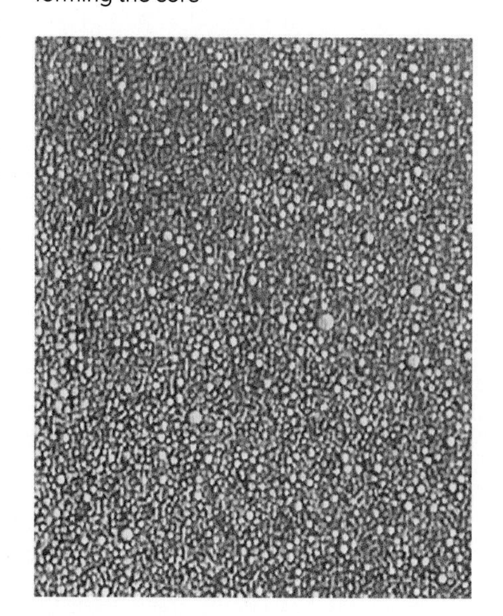

▲ Figure 9 Butterfat droplets

A2.1.5 Spontaneous formation of vesicles by coalescence of fatty acids into spherical bilayers

Vesicles are small droplets of fluid, enclosed in a membrane. They are very common structures inside cells. Some cells also produce extracellular vesicles, called exosomes. Vesicles probably played a part in the early evolution of cells.

The membrane of vesicles is mainly composed of phospholipids. One end of a phospholipid molecule is attracted to water (the hydrophilic head) because it is polar. The remainder is non-polar (two hydrophobic tails). These tails are more attracted to non-polar substances than to water. A molecule with both hydrophilic and hydrophobic parts is called amphipathic.

When mixed with water, phospholipids naturally assemble into bilayers. The hydrophilic heads face outwards so they are in contact with water. The hydrophobic tails face inwards, away from water. Experiments have shown that these bilayers spontaneously form stable spherical structures; such structures are the basis of vesicles.

If phospholipids or other amphipathic molecules were part of the "soup" of carbon compounds on pre-biotic Earth, they would have self-assembled into bilayers and vesicles would very likely have formed. Movement of polar molecules into and out of these spherical structures would have been limited by the hydrophobic membrane core. As a result, the vesicles could have developed their own internal chemistry, different from that of the surroundings. They would have been cell-like even though they were not yet proper cells.

A2.1.6 RNA as a presumed first genetic material

Living organisms today have genes made of DNA and use enzymes as catalysts. To replicate DNA and pass genes to offspring, living organisms need enzymes. However, to make enzymes, they need genes! At an earlier phase in evolution, RNA may have been the genetic material. RNA can store information in the same way as DNA but it is self-replicating and it can act as a catalyst.

Some viruses (usually considered to be non-living) use RNA as their genetic material. This supports the theory that RNA could have been used before genes made of DNA evolved. Viruses with RNA as their genetic material (for example, coronaviruses) tend to have a very high mutation rate, because the polymerase enzyme that copies the base sequence is much less accurate than the equivalent enzyme used to copy DNA. This does not matter much in a virus with only a few genes and a high reproduction rate. It may even be an advantage in helping the virus to evade the host's immune system. However, in a living organism with thousands of genes, genetic stability is much more important so a change to using DNA as the genetic material would have been beneficial.

Living cells produce hundreds or even thousands of enzymes, all of which are proteins. At one time it was thought that proteins were the only molecules whose three-dimensional structure was complex enough to act as a catalyst. However, a small number of processes in cells have been found to be catalysed by RNA. Consider the synthesis of polypeptides in ribosomes. The core of the large subunit of the ribosome is composed of two RNA molecules. Together, these molecules catalyse the formation of peptide bonds between amino acids. This process is repeated many times to produce a polypeptide. RNA can act as a catalyst because it can form complex three-dimensional structures that can undergo precise interactions with other molecules.

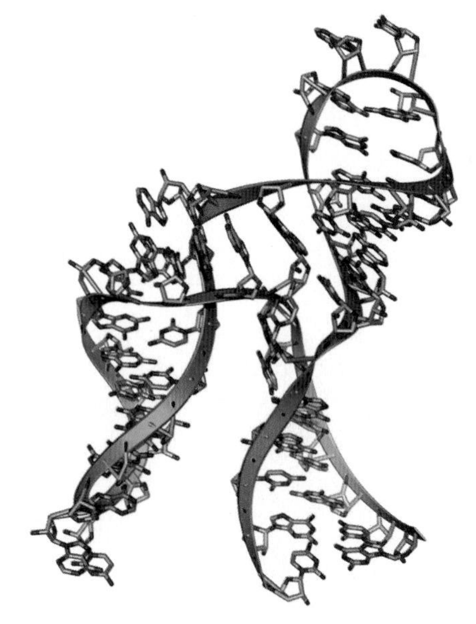

▲ Figure 10 RNA can form complex three-dimensional shapes, as in this hammerhead ribozyme which has two loops and three helices

Data-based questions: Protocells

A protocell is a compartment enclosed in a phospholipid membrane. Protocells are used to model how more complex biological cells or components of cellular organization may have originated. It is likely that protocells existed on pre-biotic Earth, with self-replicating molecules encapsulated by a membrane. It is also thought that such structures could grow and divide.

In an experiment, researchers used a simple protocell consisting of a self-replicating RNA molecule encapsulated by a membrane. For growth to occur, new membrane material is needed. However, there were no cellular mechanisms to manufacture new membrane material. It has been hypothesized that protocells containing RNA can grow by capturing new membrane material from the environment.

1. Protocells lacking RNA were radioactively labelled and mixed with unlabelled vesicles that contained RNA (lower curve in Figure 11) or did not contain RNA (upper curve in Figure 11).
   a. Compare and contrast changes in the size of the radioactively labelled protocells when they are mixed with vesicles containing RNA and not containing RNA.
   b. Suggest what might have happened to make the radioactively labelled protocells become smaller.

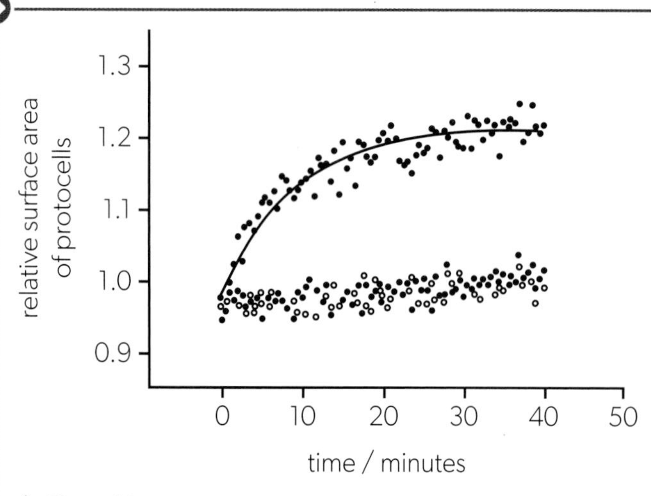

▲ Figure 11

2. The radioactively labelled protocells were artificially swollen, which makes them more likely to capture membrane material from other protocells. They were mixed with unswollen protocells containing RNA (lower curve in Figure 12) or not containing RNA (upper curve in Figure 12).

a. Analyse the results.

b. Evaluate the hypothesis that the presence of RNA inside vesicles increases the likelihood that they will capture membrane material from other protocells.

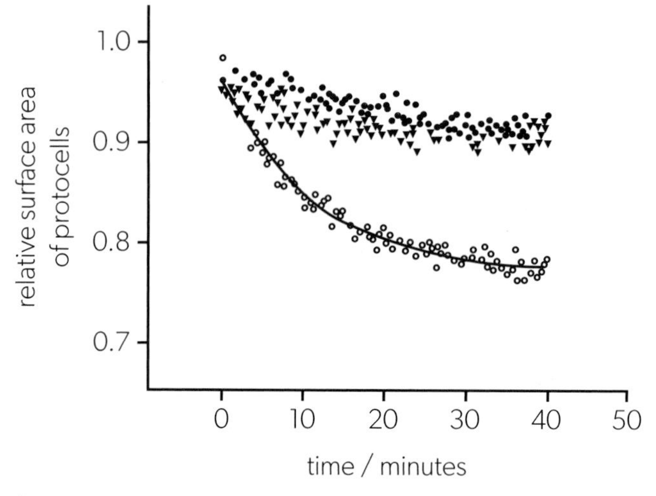

▲ Figure 12 Source of Figure 11 and Figure 12: Chen, I. A., Roberts, R. W., & Szostak, J. W. (2004). The emergence of competition between model protocells. Science (New York, N.Y.), 305(5689), 1474–1476. https://doi.org/10.1126/science.1100757

Applying technology: Computer modelling

Using computer modelling to view ribozymes

Visit the RCSB PDB website and search for "ribosome". Choose an image of the large subunit of the ribosome and view it. You should be able to rotate the image to investigate its protein and RNA components. There are two RNA molecules which together act as a ribozyme: 23S rRNA and 5S rRNA.

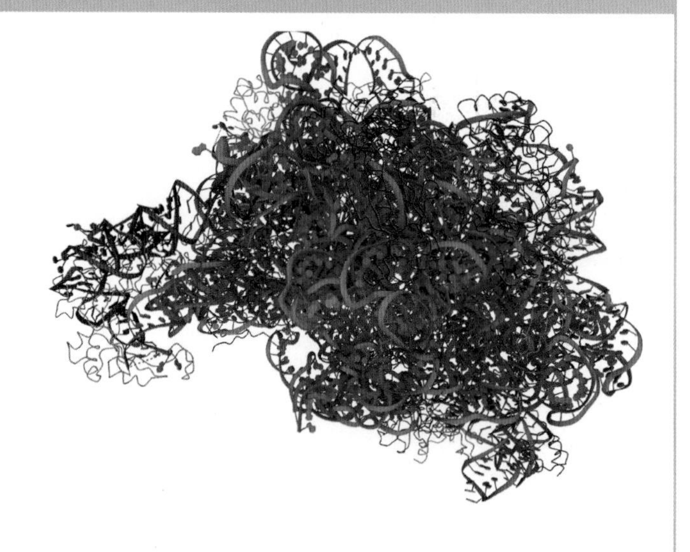

▶ Figure 13 In this image, 23S rRNA is shown in green and 5S rRNA is shown in orange

A2.1.7 Evidence for a last universal common ancestor

Living organisms store information using a genetic code. There are 64 "words" in the code, called codons, and each codon is a sequence of three bases. Every codon has a specific meaning; it is either an amino acid or a stop or start signal. It does not matter how meanings are assigned to codons, as long as there is consistency. The same is true of language—for example, the letters *f-i-s-h* could have referred to anything but now have a specific meaning.

It would be perfectly possible for different species to use different genetic codes, just as humans use different languages around the world. However, when the genetic code was investigated, it was found to be universal—it is the same in all species, with only a small number of minor variations. The meanings of the 64 different codons could be assigned in an almost limitless number of ways, having different meanings for different species. This makes it highly unlikely that two species would use the same genetic code by chance. Instead, the obvious explanation for species using the same code is that they inherited it from a common ancestor.

It is possible for strikingly similar structures to evolve in organisms that do *not* have a recent common ancestor. This is called convergent evolution. However, this is not thought to be the reason for the universal code because living organisms have so many other shared features. For example, key parts of structures within cells—such as the ribosome and the enzymes that synthesize DNA and RNA—are essentially the same in all organisms. More than 350 widely occurring protein families have been identified in prokaryotes, each with an evolutionary fingerprint that can be traced back to a common ancestor. The most recent common ancestor to have existed is called LUCA—the last universal common ancestor.

It is likely that other forms of life evolved. At some stage, however, these life forms became extinct, presumably due to competition from LUCA or species that evolved from LUCA. This is the process of natural selection, which has continued from the first evolution of life onwards. Figure 14 represents the evolution and extinction of life forms over time.

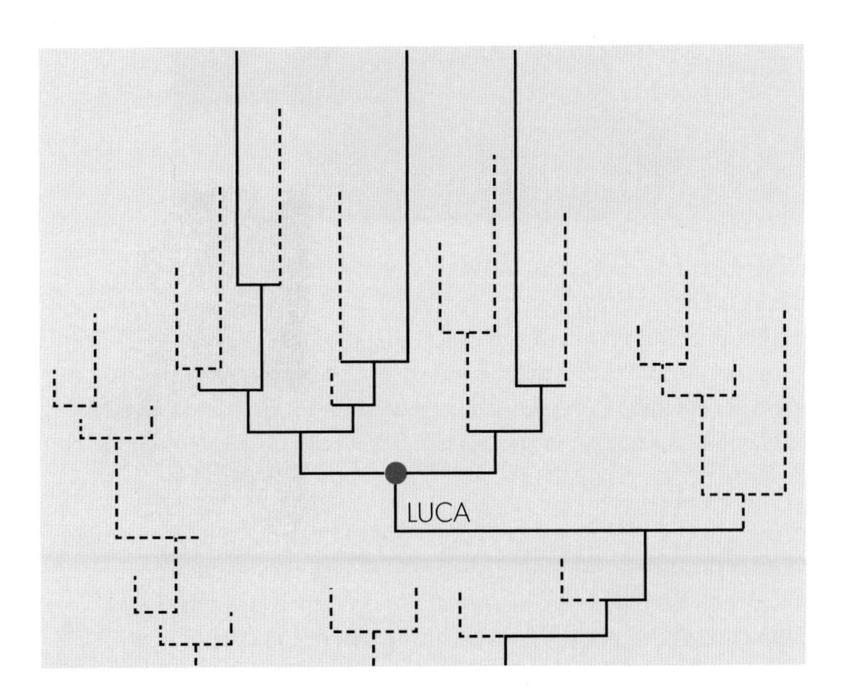

◀ Figure 14 On this tree diagram, extinct lineages are shown with dotted lines and extant lineages with continuous lines. Three origins of life from non-living matter are shown on this diagram but scientists are not yet sure how many origins there actually were

A2.1.8 Approaches used to estimate dates of the first living cells and the last universal common ancestor

Palaeontology has provided invaluable evidence about the pathways of evolution, so it is an obvious approach for dating the first living cells and the last common universal ancestor. There are relatively well-preserved rocks dating from 3.5 to 3.0 billion years ago (Gya). These rocks contain fossil-like structures with isotope ratios suggesting they are the remains of living organisms. However, alternative explanations have been suggested for many of these structures. The earliest uncontested evidence of life comes from rocks in Western Australia, known as the Strelley Pool Formation. In these rocks, large structures that resemble fossilized stromatolites have been discovered.

A stromatolite is formed when mats of cyanobacteria in shallow seawater trap sediments and secrete calcium carbonate, slowly building rocky mounds over thousands of years. No other plausible explanation has been proposed for the Strelley Pool structures, which date from 3.42 Gya. This is therefore a minimum date for the earliest cells and for LUCA, based on fossils. However, there are simpler forms of life than the bacteria that form stromatolites, so we can deduce that the earliest cells must have existed before 3.42 Gya.

▶ **Figure 15** Stromatolites in shallow water at Hamelin Pool Marine Nature Reserve, Shark Bay, Western Australia

The oldest rocks found on Earth have all been metamorphosed by heat and pressure, so they do not contain clearly recognizable fossils. The only evidence of life that we can hope to find comes from isotope ratios. Carbon originating from living organisms has a low $^{13}C/^{12}C$ ratio. Banded iron rock from Akila and Isua in west Greenland shows this ^{13}C depletion and dates from 3.70 to 3.85 Gya. These rocks may be the remains of stromatolites; if so, this would push the origin of the simplest cells further back in time.

The Earth formed about 4.5 Gya. No rocks older than about 4.0 Gya remain because tectonic processes continuously destroy rock by subduction and create new rock from magma. However, when rock is ground down by erosion, the hardest fragments can persist and become part of new sedimentary rocks. Zircon particles with depleted ^{13}C dating from 4.1 Gya have been found in younger rocks at Jack Hills, Western Australia. The $^{13}C/^{12}C$ ratio is consistent with a biogenic origin but this is far from proof of the existence of life.

Scientists trying to date the origins of life can also analyse genomic information. The number of differences between the genomes of two species is proportional to the time since they diverged from a common ancestor. A recent study using this approach suggested that LUCA and the first living cells existed nearly 4.5 Gya, soon after the Earth is thought to have been impacted by the planet Theia. This impact would have sterilized the Earth and led to the formation of the Moon.

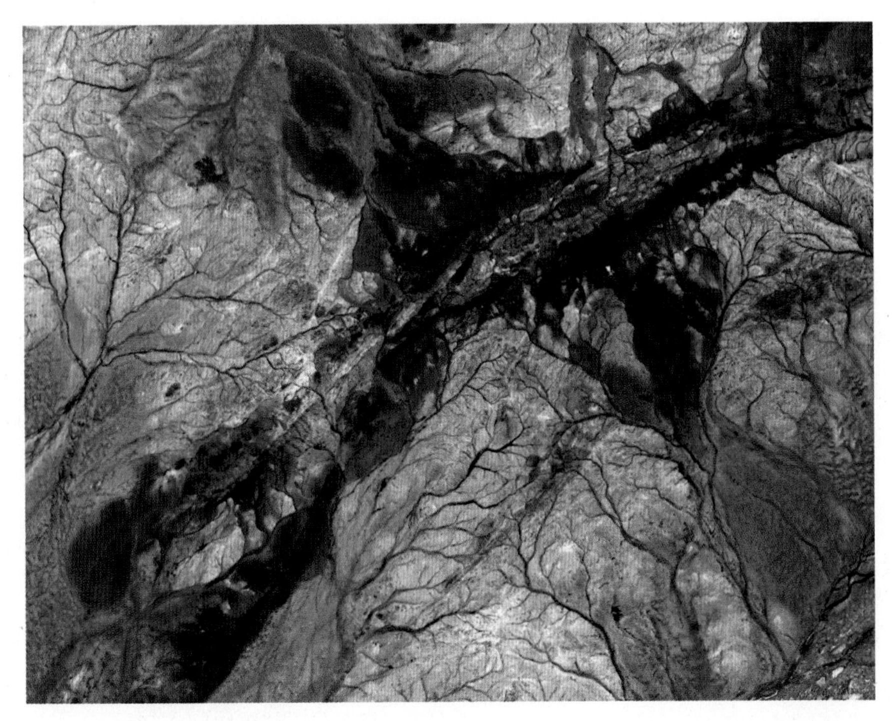

▲ Figure 16 Satellite view of Jack Hills, Western Australia, where there are zircon particles dating to 4.4 Gya. These are the oldest fragments of the Earth's crust so far discovered

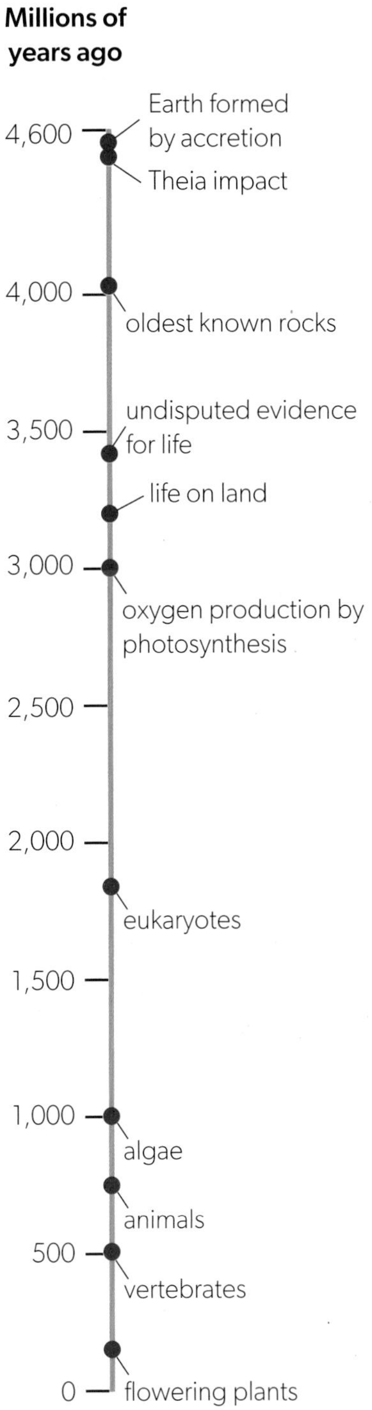

A2.1.9 Evidence for the evolution of the last common ancestor in the vicinity of hydrothermal vents

Even though LUCA probably lived more than 3,500 million years ago (>3.5 Gya), it is possible to investigate its genetic make-up using organisms alive today. Researchers have identified genes that occur widely among the groups of organisms that originated early in the history of life—bacteria and archaea. A wide distribution suggests that these genes were inherited from an early common ancestor. If an evolutionary tree is constructed for a shared gene and it matches the accepted evolutionary tree for bacteria and archaea, deductive reasoning tells us that a gene has been inherited from a common ancestor of bacteria and archaea. This suggests that LUCA had the gene.

Using this technique, researchers have identified 355 protein families that are likely to have been in LUCA's genome. They are genes needed for anaerobic metabolism and for fixing carbon dioxide and nitrogen. From this, we can deduce that LUCA lived in an environment with high concentrations of hydrogen, carbon dioxide and iron. These conditions are found in and around hydrothermal vents in the oceans.

Hydrothermal vents are cracks in the Earth's surface, characterized by gushing hot water carrying reduced (unoxidized) inorganic chemicals such as iron sulfide. There are various types of vent. However, alkaline hydrothermal vents (white smokers) have conditions most suited to the origin of life. The hydrothermal fluids emerge at temperatures of 60°C to 90°C and contain high concentrations of hydrogen, methane, ammonia and sulfides. These chemicals represent readily accessible supplies of energy, which early cells would have needed to assemble carbon compounds into polymers. Carbon dioxide would also have been

Millions of years ago

4,600 — Earth formed by accretion; Theia impact

4,000 — oldest known rocks

3,500 — undisputed evidence for life

life on land

3,000 — oxygen production by photosynthesis

2,500

2,000

eukaryotes

1,500

1,000 — algae

animals

500 — vertebrates

0 — flowering plants

▲ Figure 17 Timeline for life based on evidence from rocks. Most of these dates are still hotly debated

AHL

required; this was probably present in much higher quantities at the time when the first cells were evolving.

There are still many problems in understanding how the first cells evolved from non-living matter. However, it seems likely that hydrothermal vents were the site of this amazing event.

▲ Figure 18 Kawio Barat is a submarine volcano off the coast of Indonesia. It has vents releasing superheated, chemical-laden water into the ocean at a depth of over a mile. When this superheated water meets the cold surrounding water, minerals are precipitated producing pale "smoke". This is why this type of vent is known as a "white smoker". The minerals form porous deposits and it is in these pores that the first cells may have evolved from non-living matter, 4.5 billion years ago

Linking questions

1. For what reasons is heredity an essential feature of living things?
   a. Outline the processes that are dependent on cell division. (D2.1.8)
   b. Explain why meiosis is uniquely necessary for sexual reproduction. (D2.1.9)
   c. Discuss the relationship between heredity and natural selection. (D4.1)
2. What is needed for structure to be able to evolve by natural selection?
   a. Compare discrete and polygenic inheritance. (D3.2.14)
   b. Distinguish between intraspecific and interspecific competition. (C.4.1.10, C4.1.11)
   c. Discuss the role of diversity in the process of natural selection. (D2.1)

A2.2 Cell structure

What are the features common to all cells and the features that differ?

Figure 1 shows a hot spring extremophile community. This community thrives in 75°C water in the hills of New Mexico. The community in the picture is made up of sulfur bacteria (purple), algae and protozoa, all one celled organisms. How does the cell theory take into account the diversity of cell structure? What features of cells are universal? What are some examples of features that are unique to certain cells? What are the implications of the cell theory? What are the limits to what the cell theory predicts or explains?

▲ Figure 1 Scanning electron micrograph of a hot spring extremophile community

How is microscopy used to investigate cell structure?

The human eye has limited resolving power. What does resolution refer to with respect to optical devices? What is the actual limit to the resolving power of the human eye? How does this compare to a bird of prey like an eagle? How large are cells? Organelles? Membranes? What is the resolving power of the different type of microscopes like light and electron microscopes? A scanning electron microscope was used to prepare the image shown in Figure 2, which is an embryo on the head of a pin. What is the value of a SEM over a transmission electron microscope?

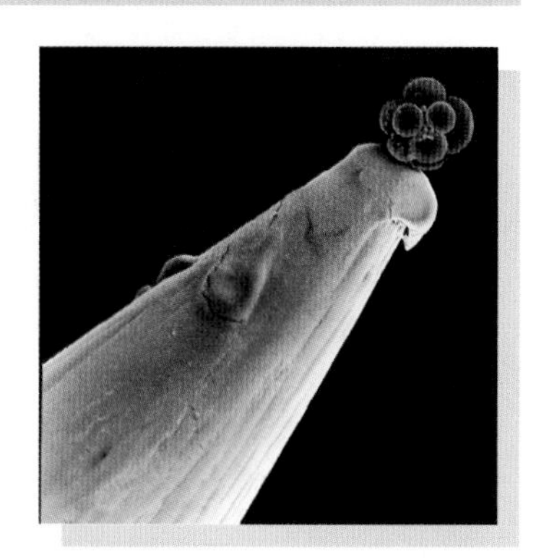

▲ Figure 2 Coloured scanning electron micrograph (SEM) of a human embryo on the tip of a pin

SL and HL	AHL only
A2.2.1 Cells as the basic structural unit of all living organisms A2.2.2 Microscopy skills A2.2.3 Developments in microscopy A2.2.4 Structures common to cells in all living organisms A2.2.5 Prokaryote cell structure A2.2.6 Eukaryote cell structure A2.2.7 Processes of life in unicellular organisms A2.2.8 Differences in eukaryotic cell structure between animals, fungi and plants A2.2.9 Atypical cell structure in eukaryotes A2.2.10 Cell types and cell structures viewed in light and electron micrographs A2.2.11 Drawing and annotation based on electron micrographs	A2.2.12 Origin of eukaryotic cells by endosymbiosis A2.2.13 Cell differentiation as the process for developing specialized tissues in multicellular organisms A2.2.14 Evolution of multicellularity

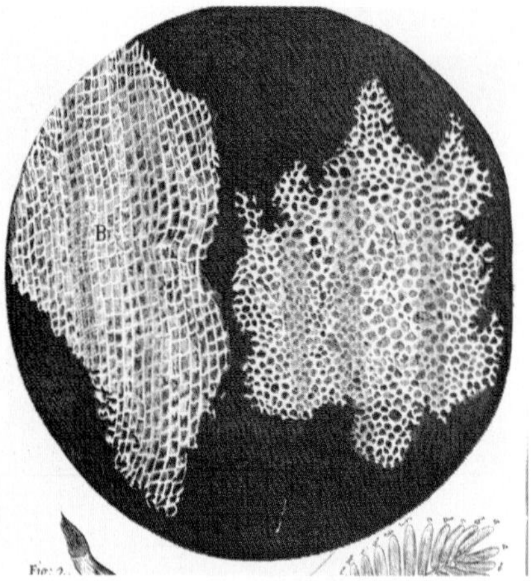

A2.2.1 Cells as the basic structural unit of all living organisms

Individual cells are fundamental units of life. Some small organisms consist of a single cell but larger organisms are multicellular. It has been estimated that a 70 kg human consists of 3.8×10^{13} cells—that is nearly 40 trillion cells. Large multicellular organisms have many different cell types, each specialized for a particular role.

The statement that living organisms consist of cells is an example of a theory. This theory was developed when Robert Hooke and other biologists from the 17th century onwards used microscopes to look at the structure of living organisms. Plant cells are relatively easy to view with a microscope. By the 19th century, animal tissues could also be examined. Both types of tissue were found to consist of cells. From this, scientists concluded that all organisms are made of cells. They had not looked at all parts of all organisms but they had found a trend that allowed them to make general predictions about the structure of organisms.

▲ Figure 3 Robert Hooke's drawing of cork cells

Since the development of the cell theory, researchers have discovered some structures in living organisms that do not consist of typical cells; some of these structures are described later. Despite these exceptions, however, the cell theory is still useful and it has not been rejected. If a new organism is discovered, we can be reasonably confident that some or all of it will consist of cells.

Observations, theories and inductive reasoning

Biologists are interested in the natural world and look carefully at it—they act as observers and make observations. Sometimes biologists notice a trend or pattern in their observations and from this they develop a general theory. Theories developed from specific observations are an example of inductive reasoning—going from the specific to the general. In the case of the cell theory, the specific discovery that parts of diverse organisms consisted of cells led to the generalization that all organisms consist of cells.

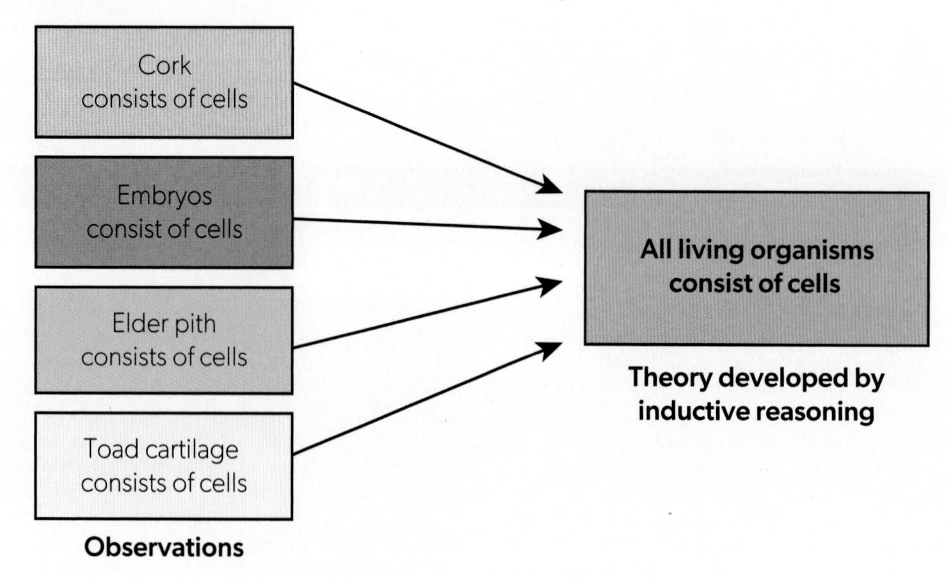

▲ Figure 4 The cell theory was developed by inductive reasoning

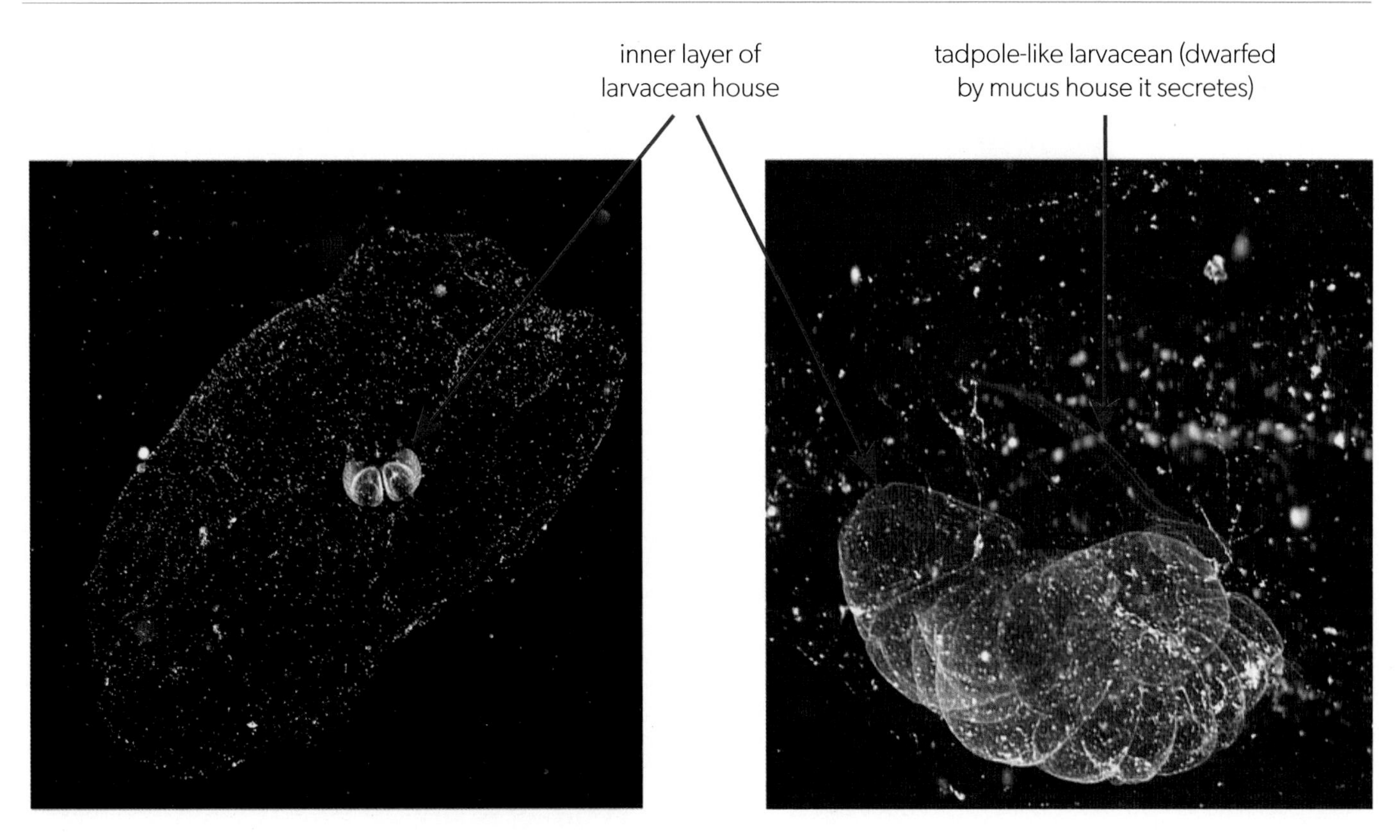

▲ Figure 5 A team of researchers at Monterey Bay Aquarium Research Institute (MBARI), led by Kakani Katija, has been researching marine organisms called larvaceans. The photograph on the left shows a larvacean's "house" which consists of two non-cellular mucus structures. The large coarse-mesh outer structure excludes coarse non-food particles. The inner fine-mesh structure captures smaller food particles. The larvacean itself is too small to be seen in this image. The photograph on the right shows a magnified view of a larvacean (the blue tadpole-like organism) adjacent to the inner part of its mucus house. By beating its tail, the larvacean pumps water through the inner and outer mucus filters to extract food particles from the surrounding seawater. The actual organism is about 3 to 5 centimetres long but the non-cellular house it makes and lives inside can be up to a metre in diameter. The MBARI research shows there are still exciting discoveries to be made about the natural world. It also shows that there are some exceptions to the theory that living organisms make everything out of cells

A2.2.2 Microscopy skills

Lenses allow us to look at structures that are too small to see with the naked eye—anything smaller than about 0.1 millimetres. A single convex lens is useful for magnifying up to 20 times (20×). However, this is not enough for studying the structure of cells. Microscopes with two or more lenses make much smaller structures visible because the magnification of the lenses is multiplied. For example, a 10× eyepiece lens combined with a 40× high-power objective lens gives a total magnification of 400×. This allows us to see structures as small as 0.0001 millimetres (0.1 micrometres). Using a microscope is an essential skill for biologists.

Using a light microscope

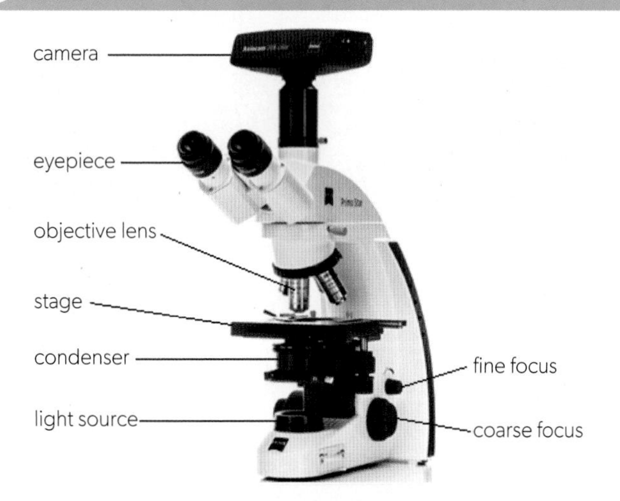

▲ Figure 6 Parts of a light microscope

Try to improve your skill at using microscopes as much as you can.

- Learn the names of parts of the microscope.
- Understand how to focus the microscope to get the best possible image.
- Look after your microscope so it stays in perfect working order.
- Know how to troubleshoot problems.

Look after your microscope by following these guidelines:

- Always focus by moving the lens and the specimen further apart. Never move them closer to each other.
- Make sure the upper and lower surfaces of the slide are clean and dry before putting it on the stage.
- Never touch the surfaces of the lenses with your fingers or anything else.
- Carry the microscope carefully with a hand under it to support its weight securely.

Course and fine focusing

- Put the slide on the stage, with the most promising region in the centre of the window in the stage that the light comes up through.
- Always focus at low power first, even if you need high-power magnification eventually.
- Focus with the larger coarse-focusing knobs first. When you have nearly got the image in focus, use the smaller fine-focusing knobs to make it really sharp.
- If you want to increase the magnification, move the slide so the most promising region is exactly in the middle of the field of view and then change to a higher magnification lens.

Use these hints to troubleshoot when you are focusing:

Problem	Solution
Nothing is visible when you try to focus	Make sure the specimen is actually under the lens, by carefully positioning the slide and using low power first.
You can see a circle with a thick black rim.	There is an air bubble on the slide. Ignore it and try to improve your technique for making slides so there are no air bubbles.
There are blurred parts of the image even when you focus it as well as you can.	Either the lenses or the slide have dirt on them. Ask your teacher to clean them.
The image is very dark.	Adjust the diaphragm to increase the amount of light passing through the specimen.
The image looks rather bleached.	Adjust the diaphragm to decrease the amount of light passing through the specimen.

▲ Table 1

Making temporary mounts of cells and tissues and using stains

The slides you examine with a microscope can be permanent or temporary. Making permanent slides is very skilled and takes a long time, so these slides are normally made by experts. Permanent slides of tissues are made using very thin slices of tissue.

Making temporary slides is quicker and easier, so you can do this for yourself.

- Place the cells on the slide, in a layer not more than one cell thick.
- Add a drop of water or stain. Stains help structures that are pale or transparent to show up more clearly.
- Carefully lower a cover slip onto the drop. Try to avoid trapping any air bubbles.

- Remove excess fluid or stain by putting the slide inside a folded piece of paper towel and pressing lightly on the cover slip.

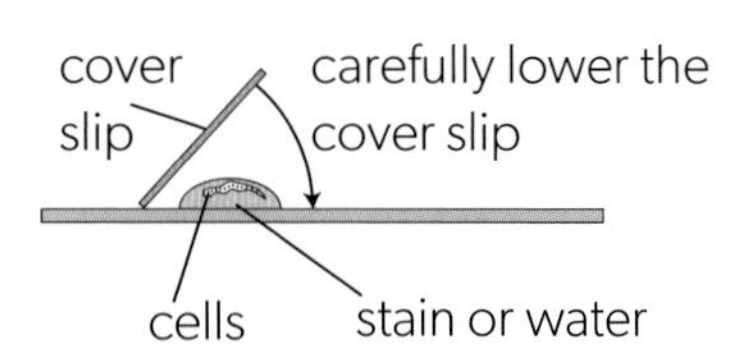

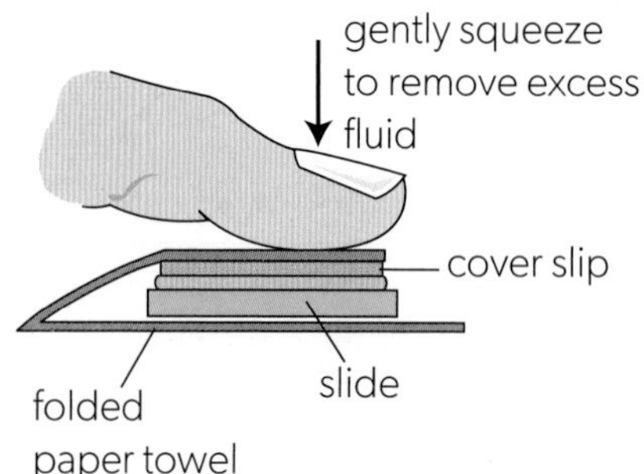

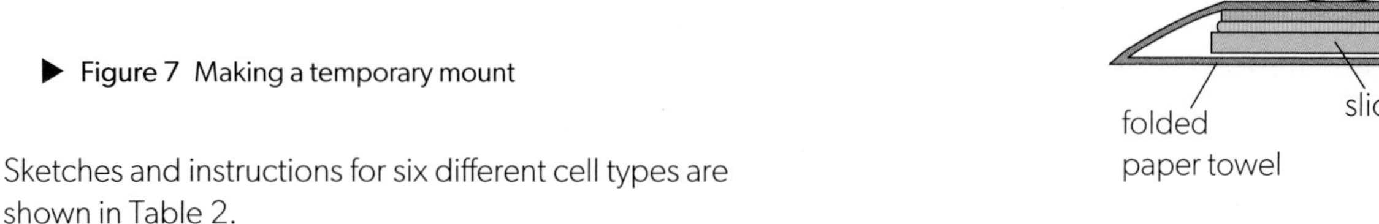

▶ **Figure 7** Making a temporary mount

Sketches and instructions for six different cell types are shown in Table 2.

1 Moss leaf Use a moss plant with very thin leaves. Mount a single leaf in a drop of water or methylene blue stain. 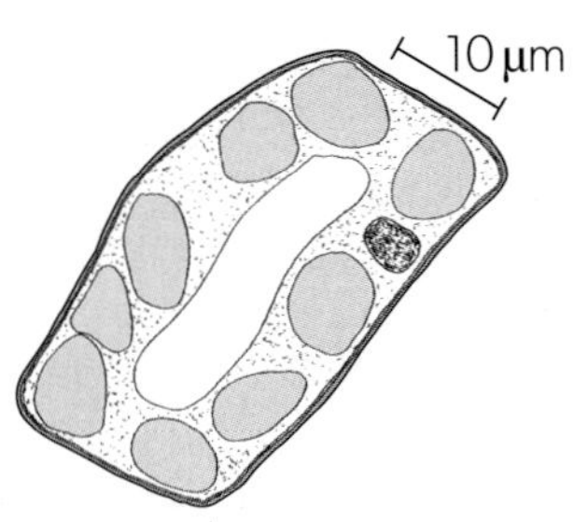 	**4 Leaf lower epidermis** Peel the lower epidermis off a leaf. The cell drawn here was from *Centranthus*. Mount in water or in methylene blue. 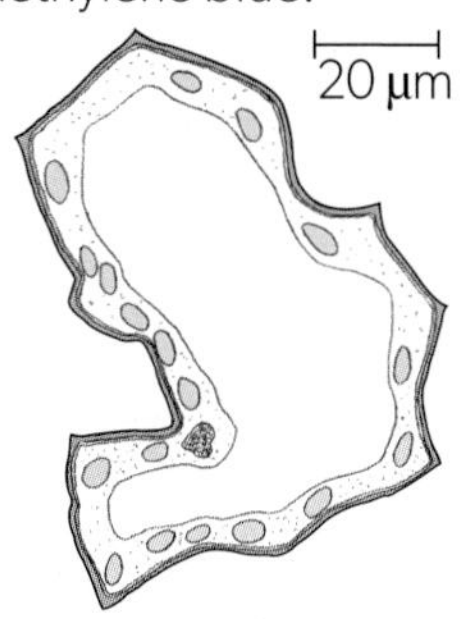
2 Banana fruit cell Scrape a small amount of the soft tissue from a banana and place on a slide. Mount in a drop of iodine solution. 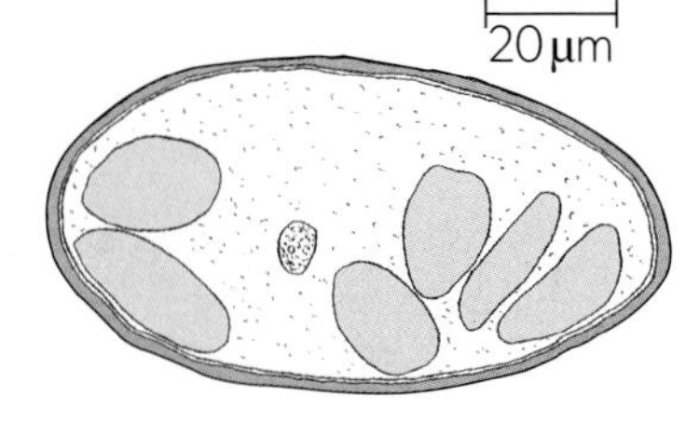 	**5 Human cheek cell** Use a cotton bud to scrape cells from the inside of your cheek. Smear them on a slide and add methylene blue to stain. 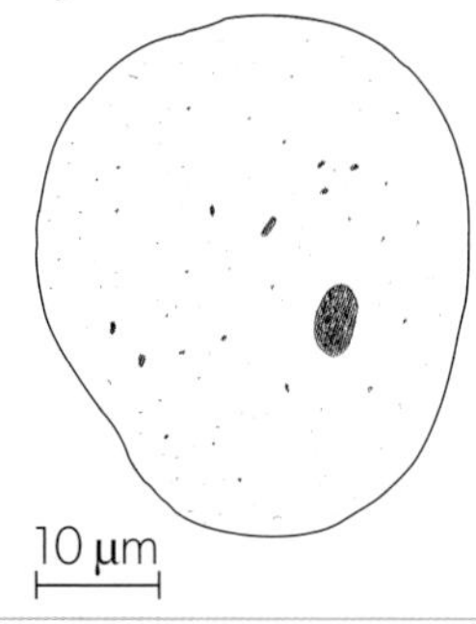
3 Mammalian liver cell Scrape cells from a freshly cut surface of liver (not previously frozen). Smear onto a slide and add methylene blue to stain. 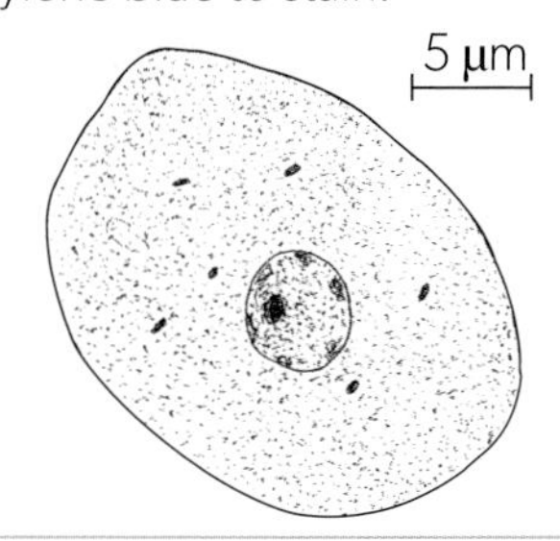 	**6 White blood cell** Smear a thin layer of mammalian blood over a slide and stain with Leishman's stain. 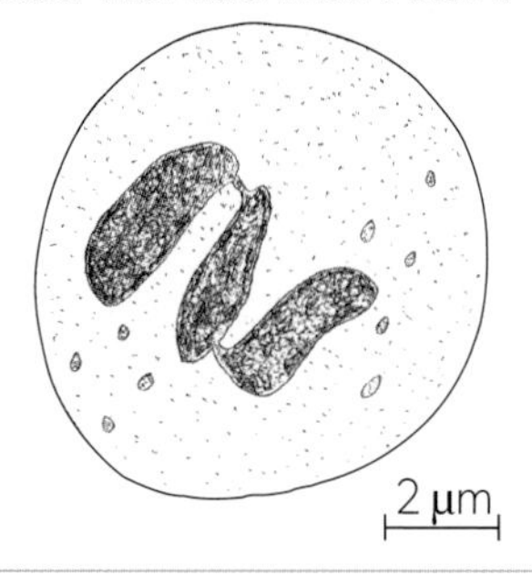

▲ Table 2

Observing, drawing and photographing cells

When you have focused at high power on plant or animal cells, it is useful to record your observations by drawing a typical cell. Alternatively, you could use a smartphone to take a photo through the microscope.

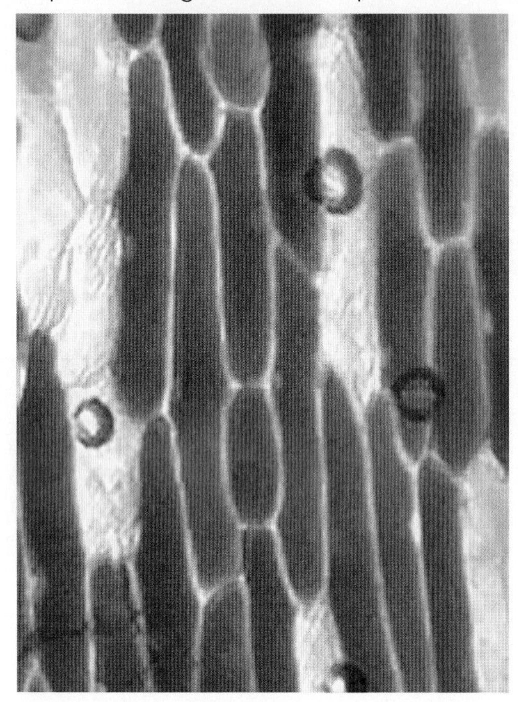

▲ Figure 8 Onion epidermis cells photographed with a smartphone. Some of the cells have a red pigment in their cytoplasm. The black-edged circles are air bubbles; this is a common fault in temporary microscope slides. They are a type of artefact—something not naturally present that was introduced during preparation of the slide

A biological drawing of a cell is a type of diagram because cell structure is shown with some features only, such as the nucleus. Usually the lines on a drawing represent the edges of structures. Do not show unnecessary detail and only use faint shading. The cell membrane is too thin to see, but you can deduce its position and you should include it in your drawing. For example, it forms the outer edge of a blood cell.

Drawings and diagrams are more informative if they are annotated. Use a ruler to draw a straight line from each structure of interest to a position off the diagram. Then add notes there. You might simply identify the structure with a name, or you can add other details such as the function.

Drawings of structures seen using a microscope are larger than the actual size—the drawing shows structures magnified. Everything on a drawing should be shown to the same magnification and the magnification should be calculated and indicated.

Photography is an alternative to drawings. The advantage of photographs is that they contain real data rather than one biologist's subjective interpretation of it. Digital microscopes are increasingly common and they make it very easy to take photos.

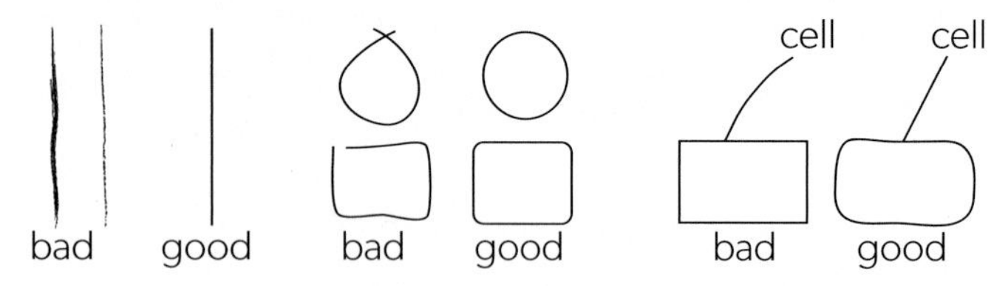

▲ Figure 9 Qualities in drawings

Measuring sizes using an eyepiece graticule

You can measure the actual sizes of structures visible through a microscope by using a scale inside the eyepiece, called a graticule. The graticule has to be calibrated so you know what size each unit on the scale indicates. This will be different for each objective lens. For example, if one unit on the scale represents 2.5 micrometres at 400× magnification (high power), it will represent 25 micrometres at 40× magnification (low power).

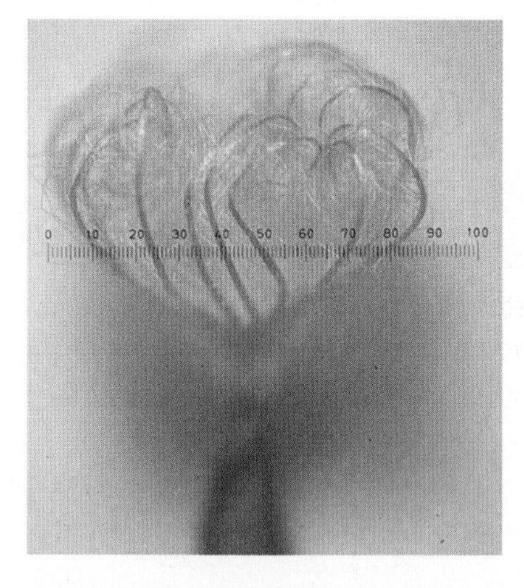

▲ Figure 10 Microscope image photographed using a smartphone, showing a fruit of *Centranthus ruber* and a graticule scale. 100 graticule units = 5 millimetres or 5,000 micrometres

Calculating actual size, magnification and scale

Structures seen using a microscope appear larger than they actually are. Most microscopes have more than one objective lens, so you can magnify specimens by different factors. A typical school microscope has three levels of magnification:

- 40× (low power)
- 100× (medium power)
- 400× (high power)

If you take a photo down a microscope, you can magnify the image even more. A photo of a microscope image is called a photomicrograph, often abbreviated to micrograph. Electron micrographs are taken using an electron microscope. When you draw a specimen, you can make the drawing larger or smaller, so the magnification of the drawing is not necessarily the same as the magnification of the microscope.

To find the magnification of a micrograph or a drawing, you need to know two things: the size of the image (in the drawing or the micrograph) and the actual size of the specimen. Then you use this formula:

$$\text{magnification} = \frac{\text{size of image}}{\text{actual size of specimen}}$$

If you know the size of the image and the magnification, you can calculate the actual size of a specimen.

When using this formula, you must make sure you use the same units for the size of the image and the actual size of the specimen. They could both be millimetres (mm) or micrometres (μm) but they must not be different, otherwise the calculation will be wrong. You can convert millimetres to micrometres by multiplying by 1,000. You can convert micrometres to millimetres by dividing by 1,000.

Scale bars may be put on micrographs or drawings, or shown alongside them. A scale bar is a straight line, labelled with the actual size that the bar represents. For example, a 10 mm long scale bar on a micrograph with a magnification of ×10,000 would be labelled 1 μm.

Worked example

The length of an image is 30 mm. It represents a structure that has an actual size of 3 μm. Determine the magnification of the image.

Either:

$30\,\text{mm} = 30 \times 10^{-3}\,\text{m}$

$3\,\mu\text{m} = 3 \times 10^{-6}\,\text{m}$

$$\text{magnification} = \frac{30 \times 10^{-3}}{3 \times 10^{-6}} = 10{,}000\times$$

Or:

$30\,\text{mm} = 30{,}000\,\mu\text{m}$

$$\text{magnification} = \frac{30{,}000}{3} = 10{,}000\times$$

Data-based questions: Size, magnification and scale

1. a. Determine the magnification of the *Thiomargarita* cells in Figure 11. [3]
   b. Determine the maximum diameter of the whole cell in the micrograph. [2]

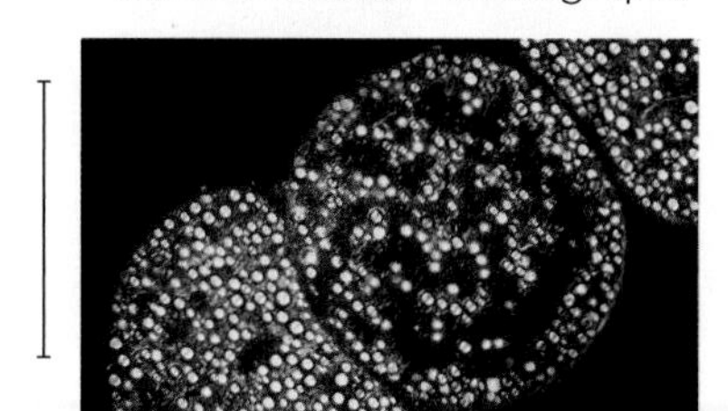

◀ Figure 11 *Thiomargarita* cells (one whole cell and two in part). The scale bar represents 0.2 μm

2. In Figure 12, the actual length of the mitochondrion is 8 μm.
   a. Determine the magnification of the electron micrograph. [2]
   b. Calculate the length of a 5 μm scale bar on this electron micrograph. [2]
   c. Determine the width of the mitochondrion. [1]

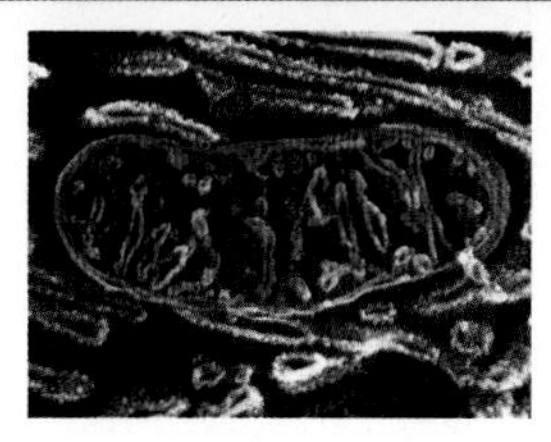

◀ Figure 12 Mitochondrion with false colour (red)

3. The magnification of the human cheek cell from a compound microscope (Figure 13) is 2,000×.
   a. Calculate the length of a 20 μm scale bar on the image. [2]
   b. Determine the maximum length of the cheek cell. [2]

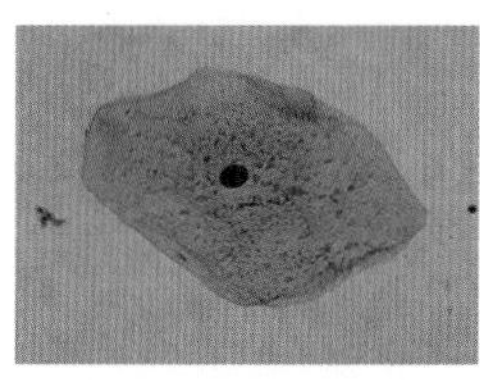

◀ Figure 13 Human cheek cell stained with methylene blue

4. a. Using the width of the hen's egg as a guide, estimate the actual length of the ostrich egg in

Figure 14. [2]

b. Estimate the magnification of the image. [2]

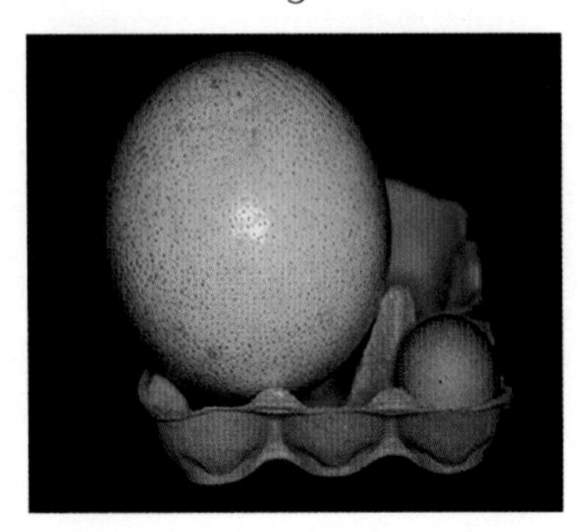

▲ **Figure 14** Ostrich egg

5. *Caenorhabditis elegans* has been widely used as a model organism in research. Most adults are hermaphrodite and have exactly 959 cells.
   a. Measure the maximum width and total length of the worm, in eyepiece units (EPU). [2]
   b. One unit on the scale indicates 9.5 micrometres. Calculate the actual dimensions of the worm in micrometres. [2]
   c. Convert the dimensions from micrometres to millimetres. Which length units are more convenient with an organism of this size? [3]

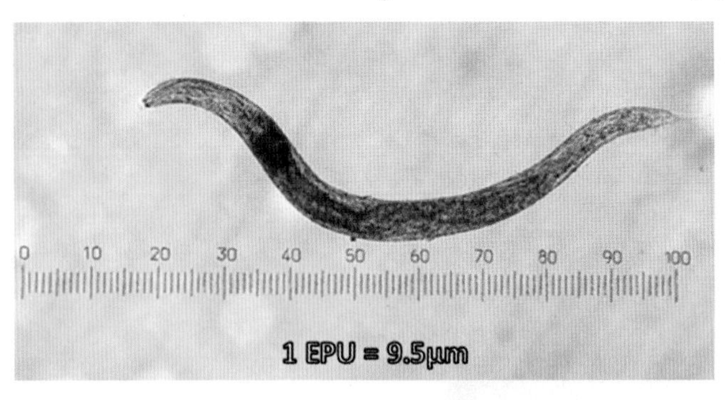

▲ **Figure 15** *Caenorhabditis elegans* together with an eyepiece scale

Quantitative versus qualitative observations

Quantitative data is numerical and is usually obtained with a measuring instrument. An eyepiece scale (graticule) in a microscope is an example of a measuring instrument. Sometimes quantitative data can be obtained simply by counting. For example, we might count how many legs a centipede has, to see if it is 100.

In contrast, qualitative observations involve descriptions that are likely to be more subjective. Consider the micrograph of pond water in Figure 16. An example of a qualitative observation is that the copepod larva (centre right) is transparent. Two types of algae are visible: *Synura* appears yellow and *Pandorina morum* appears green.

1. Create a data table with two columns headed "qualitative observations" and "quantitative observations".
2. Make observations about the organisms in the micrograph and record them in your table.
3. Compare your observations with those of a classmate.
4. Discuss the advantages of qualitative and quantitative observations.
5. Do all quantitative observations involve a measuring instrument? Or numbers? Or units?

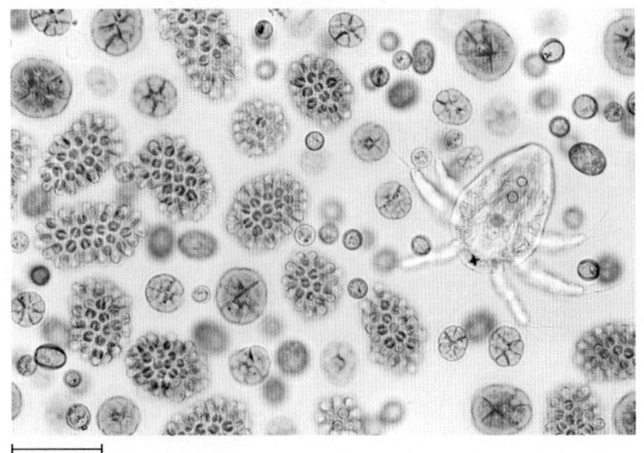

0.8 mm

▲ **Figure 16** A micrograph of pond water

A2.2.3 Developments in microscopy

Microscopes were first invented in the 17th century. Since then, there have been many technological developments in microscopy, which have made new and more detailed observations possible. Improved light microscopes in the second half of the 19th century allowed the discovery of bacteria and other unicellular organisms. Chromosomes were seen for the first time and the processes of mitosis, meiosis and gamete formation and fertilization were discovered. More advanced microscopes also revealed the complexity of organs such as the kidney, and the presence of mitochondria, chloroplasts and other structures in cells.

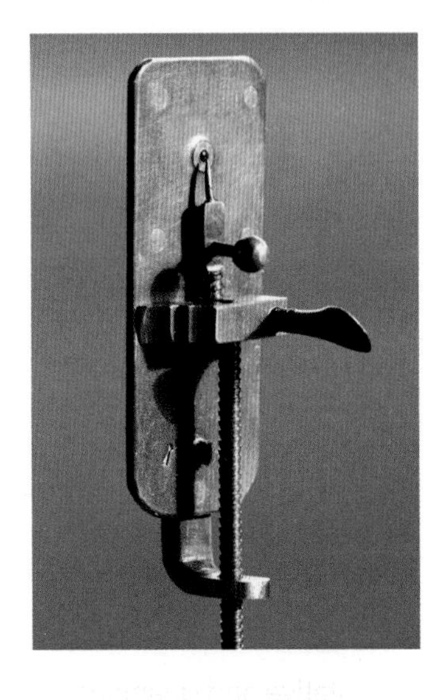

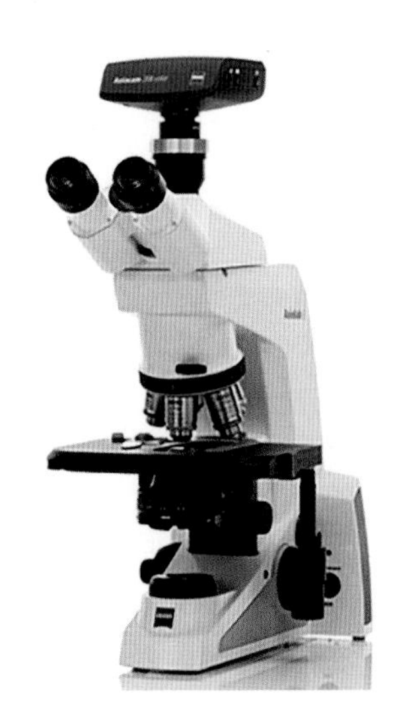

◀ Figure 17 Leeuwenhoek microscope of about 1670 (left) and Zeiss microscope of 2020 (right)

At magnifications of more than 400×, it becomes increasingly difficult to produce a focused image with a light microscope. Imagine a pair of dots that appear closer together as they become smaller. Eventually, it will be impossible to see them as separate dots. A hand lens or microscope allows smaller details to be distinguished but there is a limit because of distortions caused by the wavelength of light.

This problem was overcome by the development of microscopes that use beams of electrons instead of light. The wavelength of electrons is much shorter than the wavelength of visible light. The first electron microscopes were designed and constructed in Germany during the 1930s. They came into use in research laboratories in the 1940s and 50s. Some electron microscopes can give magnifications up to 1,000,000×.

Making the separate parts of an object distinguishable by eye is called resolution. Table 3 shows the maximum resolution of the unaided eye, the light microscope and the electron microscope, using three different SI size units.

	Resolution millimetres / mm	Resolution micrometres / µm	Resolution nanometres / nm
Unaided eyes	0.1	100	100,000
Light microscopes	0.0002	0.2	200
Electron microscopes	0.000001	0.001	1

▲ Table 3

30 point	....
20 point	....
10 point	....
5 point	....
4 point	....
3 point	....
2 point	....
1 point	....

▲ Figure 18 Size of printed periods (full stops) used for punctuation can be used to test the resolution of the naked eye, and of the eye aided by one or more lenses. Font size is the maximum height of letters and is measured in "points". In desktop publishing fonts, 1 point is 0.353 mm. The diameter of a period is just less than one-tenth of the overall font size, so a period at font size 30 has a diameter of approximately 1 mm. The table shows a row of 10 periods at font sizes from 30 to 1. Which sizes of period can you distinguish as individual dots?

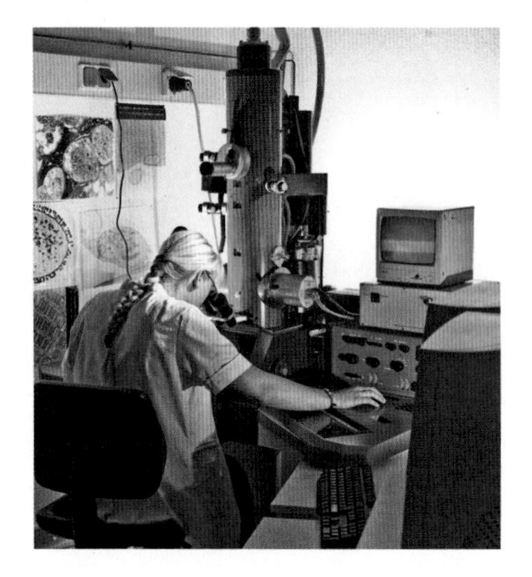

▲ Figure 19 An electron microscope in use at the Max-Planck Institut in Halle, Germany. Three of the many technological developments in microscopy are described here

Because electron microscopes have better resolution, they can give much higher magnification. This means much smaller structures can be seen than with a light microscope. Electron microscopes have allowed scientists to investigate the detailed structure (ultrastructure) of cells. Variations in ultrastructure between different cell types are described later.

Electron microscopes do have some disadvantages. They can only give black and white images, so any colour in electron micrographs has to be added artificially. The methods used to prepare material for the electron microscope always kill the cells. In any case, cells would die inside an electron microscope because there is a vacuum and the beams of electrons are very destructive. In contrast, light microscopes can be used to examine living material and produce images in colour. Therefore, both types of microscope are very useful in research and continue to be widely used.

1. Fluorescent stains and immunofluorescence

Most of the chemicals in cells are white or colourless so they are difficult to distinguish unless stained. Stains used in microscopy are coloured substances that bind to some chemicals but not others. For example, methylene blue binds to DNA and RNA so it stains the nucleus dark blue and cytoplasm a lighter blue.

Fluorescence is when a substance absorbs light and then re-emits it at a longer wavelength. Fluorescent stains have been used for over 100 years. Some absorb ultraviolet light and re-emit it as blue light, for example. Special fluorescence microscopes have been designed and built with intense light sources such as high power LEDs or lasers that emit a single wavelength. This light is absorbed and re-emitted by the sample, generating particularly bright images.

Immunofluorescence is a development of fluorescent staining. Antibodies that bind to particular chemicals (antigens) in the cell are produced. Then fluorescent markers of different colours are linked to these antibodies. A multicoloured fluorescent image can then be produced showing where the chemicals are located. There are many research applications of this technique; for example, it can be used to find out if one specific type of protein is being produced in a cell.

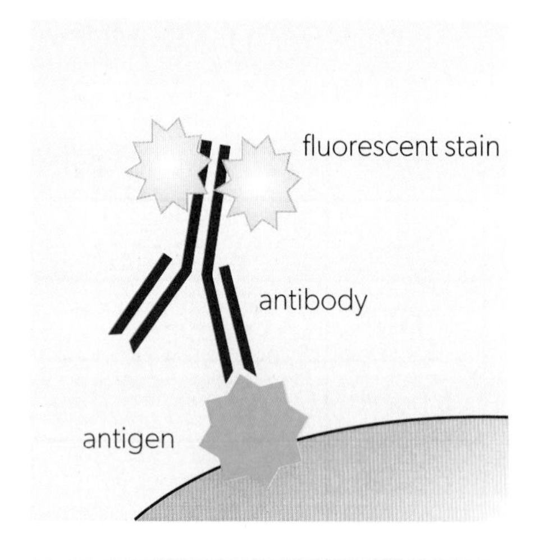

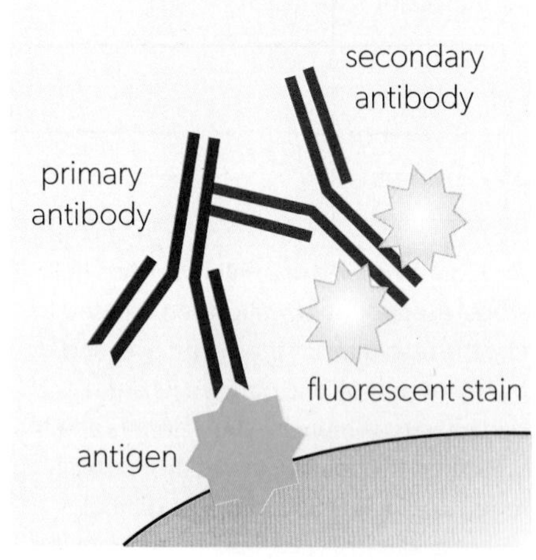

▲ Figure 20 The fluorescent stain (yellow) may be linked directly to the antibody that binds to the target antigen (green). Alternatively, it may be linked indirectly by an antibody that binds to the primary antibody

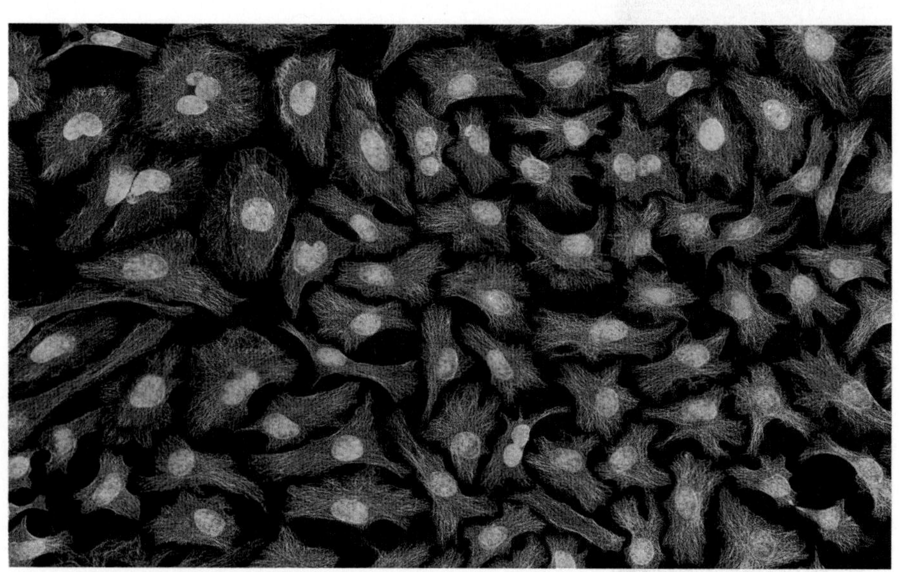

▲ Figure 21 In this image produced by immunofluorescence, DNA in the nuclei is stained cyan. Microtubule proteins of the cytoskeleton, which are normally invisible, are stained magenta. This image was produced using a Nikon RTS2000MP custom laser scanning microscope

2. Freeze-fracture electron microscopy

This technique is used to produce images of surfaces within cells. A sample is plunged into liquefied propane at −190°C so it rapidly freezes. A steel blade is then used to fracture the frozen sample. The fracture goes through the weakest points of the cells. Some of the ice at the fractured surface is removed by vaporization, to enhance the texture of the surface. This is called etching. Then a vapour of platinum or carbon is fired onto the fracture surface at an angle of about 35° to form a coating. This creates a replica of the fracture surface.

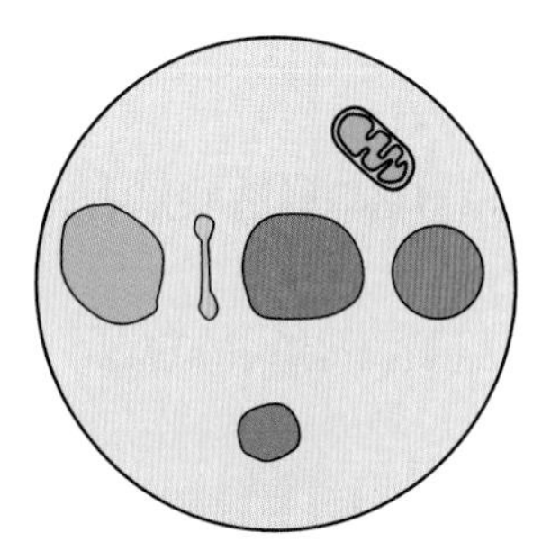

1. Cell is frozen

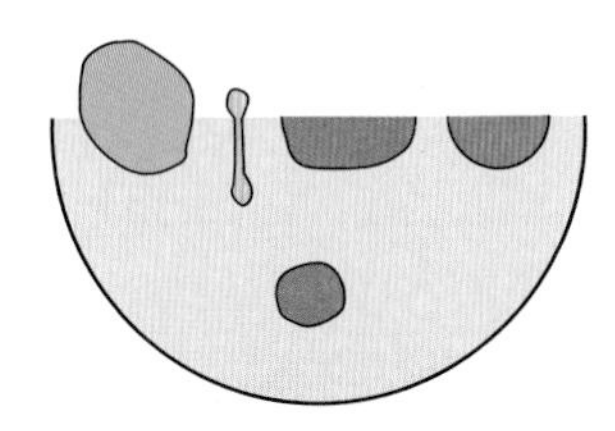

2. Cell is fractured

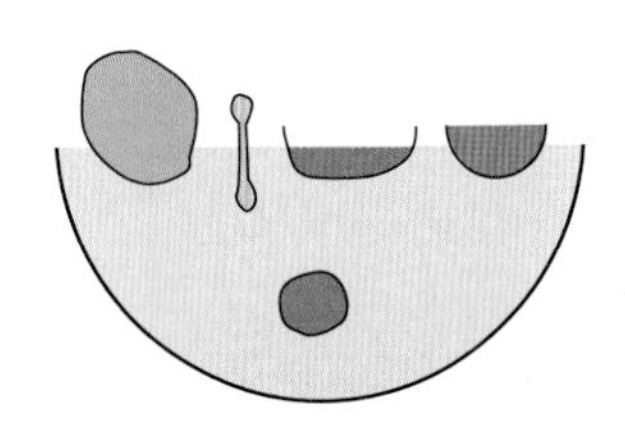

3. Cell is etched

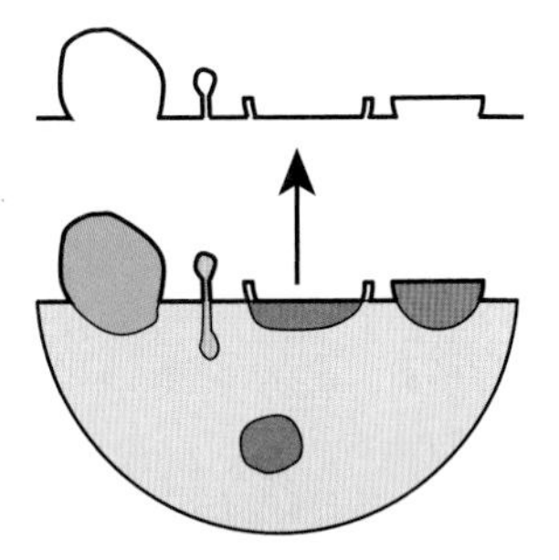

4. Fractured surface is replicated

▲ Figure 22 The freeze-fracture process

The replica is removed from the frozen sample and can be examined using an electron microscope. It is about 2 nanometres thick on average but the thickness varies because of the angle at which the coating is applied. This gives the impression of a 3D image through shadowing.

The weakest point in cells is usually the middle of membranes, between the two layers of phospholipid. The freeze-fracture process gives a unique image of this part of cells. When these images were first produced, they led to a fundamental change in theories about membrane structure. This is described in *Topic B2.1*.

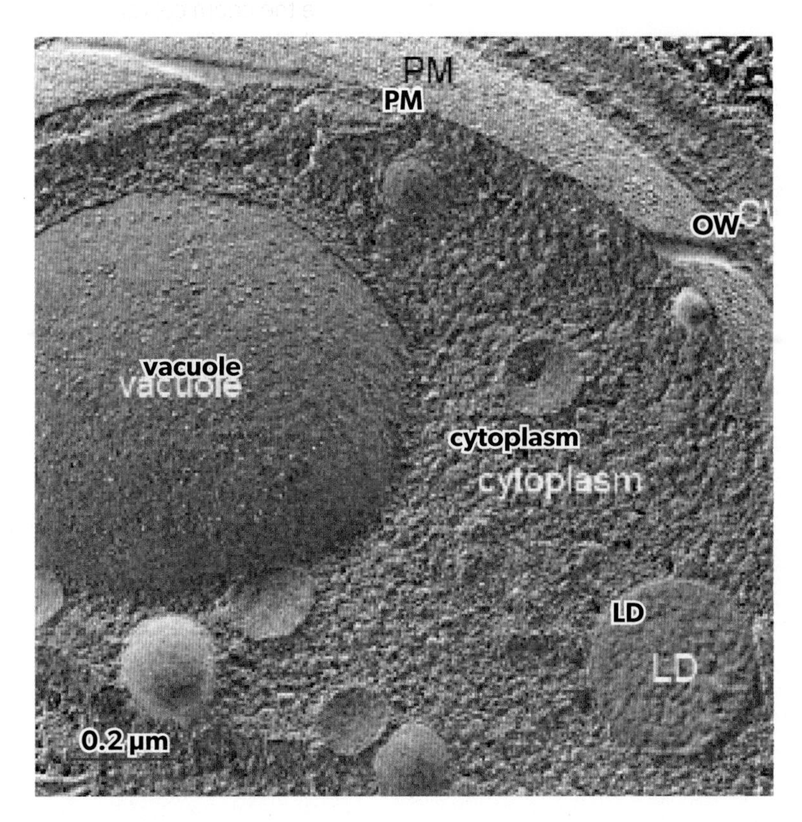

◀ Figure 23 Freeze-fracture image of a yeast cell, showing a large vacuole, smaller vesicles (unlabelled), plasma membrane (PM), cell wall (CW) and a lipid droplet (LD). The vacuole, vesicles and plasma membrane appear convex or concave because the fracture followed the centre of the membrane, which was curved. The lipid droplet is cross-fractured because it is not surrounded by a membrane

3. Cryogenic electron microscopy

This technique is often called cryo-EM. It is principally used for researching the structure of proteins. A thin layer of a pure protein solution is applied to a grid. The solution is flash-frozen, to create smooth vitreous ice and prevent the formation of water crystals. Liquid ethane just above its melting point of −182.6°C is usually used as the coolant.

The grid with the frozen protein solution is placed in an electron microscope and detectors record the pattern of electrons transmitted by individual protein molecules. Because the protein molecules are randomly orientated in the layer of frozen solution, many different patterns are produced. Using computational algorithms, these patterns are combined to produce a 3D image of the protein molecules.

Previous methods for analysing protein structure only produced images of a protein in its most stable form. However, cryo-EM analyses proteins at the instant in time when the water around them froze. This allows scientists to research proteins that change from one form to another as they carry out their function.

Since 2010, cryo-EM techniques have improved rapidly. They can now give resolutions of 0.12 nanometres. This allows the generation of images of individual atoms in a protein or other molecule. Over 10,000 protein structures have now been shared in the Electron Microscopy Data Bank (EMDB). The 2017 Nobel Prize in Chemistry was awarded to Jacques Dubochet, Joachim Frank and Richard Henderson for their work in developing cryo-EM. Figures 24 and 25 show an example of this.

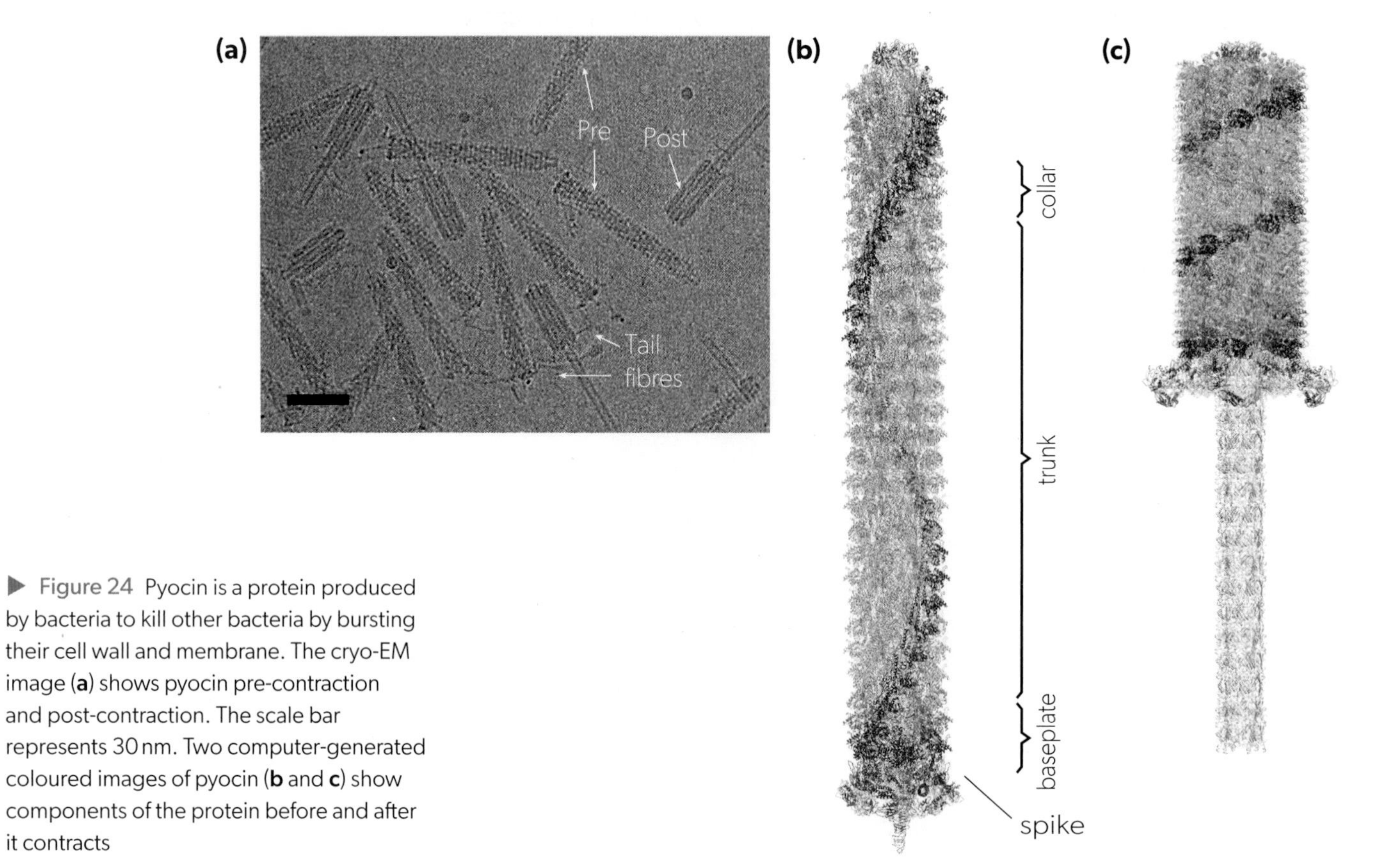

▶ Figure 24 Pyocin is a protein produced by bacteria to kill other bacteria by bursting their cell wall and membrane. The cryo-EM image (**a**) shows pyocin pre-contraction and post-contraction. The scale bar represents 30 nm. Two computer-generated coloured images of pyocin (**b** and **c**) show components of the protein before and after it contracts

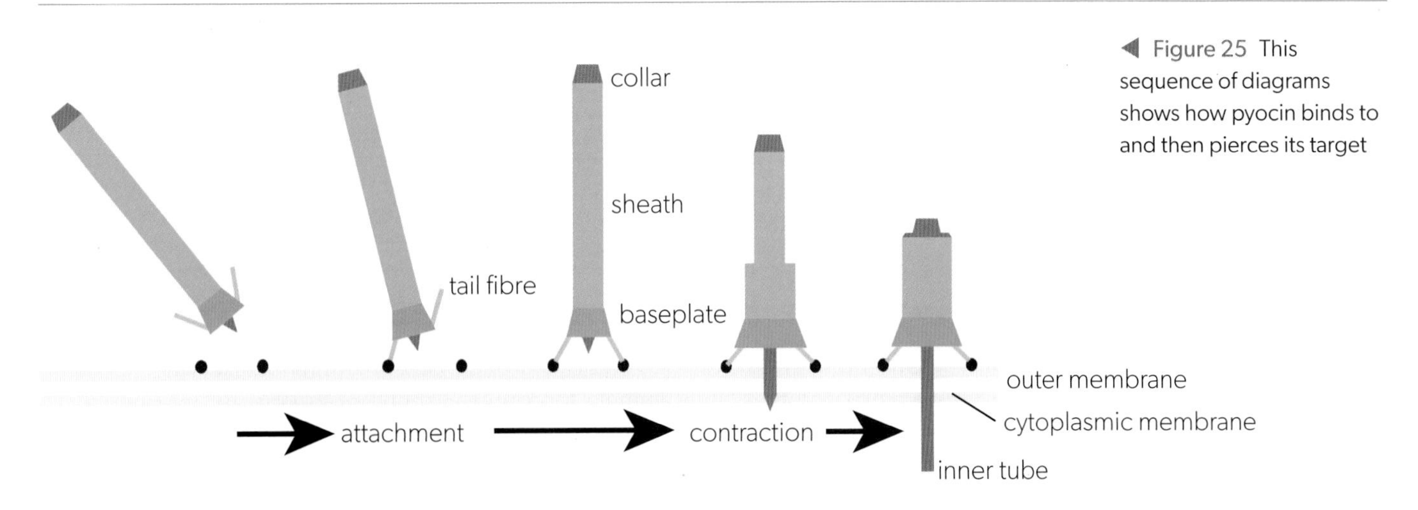

◀ Figure 25 This sequence of diagrams shows how pyocin binds to and then pierces its target

A2.2.4 Structures common to cells in all living organisms

Cells vary considerably in size, shape and structure, but they share some common features:

1. Plasma membrane

This is the outer boundary of the cell and encloses all of its contents. The plasma membrane controls the entry and exit of substances. It can pump substances in, even if the concentration outside the cell is very low. It is also very effective at preventing entry of unwanted or even toxic substances. It allows a cell to maintain concentrations of substances that are very different from those in the surrounding environment. The permeability of the plasma membrane relies on a structure based on lipids.

Occasionally the plasma membrane of a cell bursts. This is known as lysis and can be caused by excess pressure or by viruses. It can even be carried out by the cell itself (autolysis). Lysis always leads to the death of the cell; this shows that the plasma membrane is a vital structure.

2. Cytoplasm

Water is the main component of cytoplasm and there are many substances dissolved or suspended in this water. Enzymes in the cytoplasm catalyse hundreds or even thousands of different chemical reactions. These reactions are the metabolism of the cell.

Metabolism provides a cell with energy and produces all the proteins and other substances that make up the structure of a cell. Proteins are quite easily damaged, so even when a cell is not growing the cytoplasm must continuously break down and replace its proteins.

3. DNA

Genes, made of DNA, contain the information needed for a cell to carry out all its functions. Many genes hold the instructions for making a protein. Some proteins are structural so are needed for growth and repair. Others act as enzymes, without which a cell cannot control chemical reactions and does not have a functioning metabolism.

DNA can be copied and passed on to daughter cells, so the information it stores is heritable. Plant and animal cells have a nucleus that contains almost all their DNA. Bacteria do not have a nucleus and their DNA is in the cytoplasm instead. Use of DNA as a genetic material is therefore common to all cells, but the location of this DNA is not universal.

A2.2.5 Prokaryote cell structure

Organisms can be divided into two groups, prokaryotes and eukaryotes. Bacteria are prokaryotic. Plants, animals, fungi and a variety of other organisms (such as Amoeba) are eukaryotic. The key feature of eukaryotic cells is the nucleus, which contains chromosomes. This is bounded by a nuclear envelope consisting of a double layer of membrane. Prokaryotic cells do not have a nucleus.

Prokaryotes were the first organisms to evolve on Earth and they still have the simplest cell structure. They are mostly small in size and are found almost everywhere—in soil, in water, on our skin, in our intestines and even in pools of hot water in volcanic areas.

All cells have a plasma membrane. Some cells, including prokaryotic cells, also have a cell wall outside the cell membrane. This structure is thicker and stronger than the membrane. It protects the cell, maintains its shape and supports the plasma membrane to prevent it from bursting. In prokaryotes, the cell wall contains peptidoglycan.

There is no nucleus in a prokaryotic cell so the interior is entirely filled with cytoplasm. The cytoplasm is not divided into compartments by membranes; instead, it is one uninterrupted chamber. Prokaryotic cells are therefore structurally simpler than eukaryotic cells—although they still contain a very complex mixture of biochemicals including many enzymes.

The cytoplasm of eukaryotic cells contains organelles that are analogous to the organs of multicellular organisms. Both organs and organelles are distinct structures with specialized functions. Prokaryotes do not have cytoplasmic organelles apart from ribosomes. Prokaryote ribosomes are smaller than those of eukaryotes: they are 70S whereas eukaryote ribosomes are 80S. The S stands for Svedberg units, which are a measure of the rate at which a particle sinks during centrifugation.

In many electron micrographs of prokaryotes, part of the cytoplasm appears lighter than the rest. This region contains DNA. There is usually only a single molecule of DNA that forms a loop or circle. The DNA is "naked": unlike eukaryotic DNA, it is not associated with proteins. The lighter region of the cytoplasm is called the nucleoid. It is similar to a nucleus because it contains DNA, but it is not a true nucleus. Other parts of the cytoplasm appear darker in electron micrographs. They contain ribosomes, enzymes and other proteins.

Data-based questions: Ultrastructure of *Clostridium*

Figure 26 shows an electron micrograph of the Gram-positive bacterium *Clostridium botulinum*. This bacterium produces a neurotoxin that is the most poisonous protein so far discovered. This neurotoxin is used in cosmetic treatments under the brand name Botox®.

1. What causes the cytoplasm of *Clostridium* to appear so dark in the electron micrograph? [2]
2. This image is a longitudinal section: you can see a thin slice of the bacterium going from end to end. What shape would you see in a transverse section (going from side to side)? [1]
3. There are two nucleoids visible in the cytoplasm of this cell. What does this suggest it is getting ready to do? [2]
4. There is a scale bar on the micrograph. Use this to calculate the magnification of the micrograph. [3]
5. Use the magnification to calculate the actual length of the cell. [2]

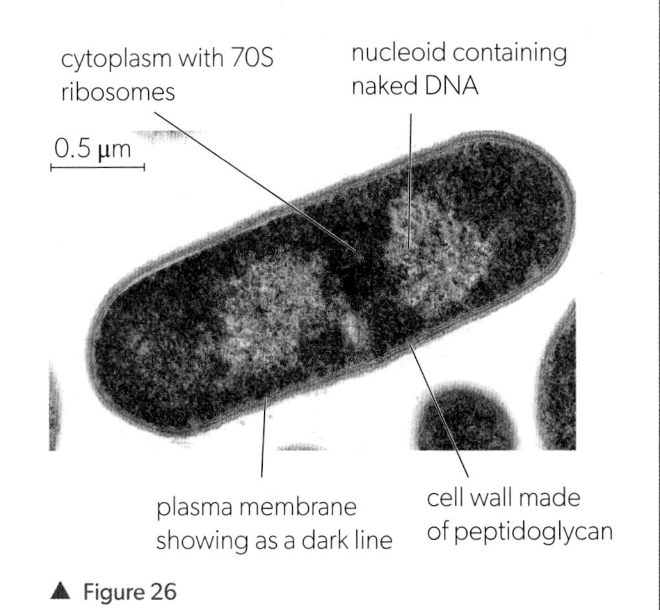

▲ Figure 26

Research skills: Using search engines effectively

Your task is to research the connection between *Clostridium botulinum* and cosmetic facial injections. What cellular processes are affected?

Your primary purpose is to use web-based sources to find information. You should use precise language in your search terms, including scientific language. For example, you might search "*Clostridium botulinum*" and "cosmetic facial injections". However, this is likely to return results for businesses offering cosmetic treatments. These will be sites with domain names ending in ".com". For this task, you want information from organizations whose primary purpose is education. Such sites have domains ending in ".edu". To filter your search results, include the search term "site: edu".

Enter the following terms in your search engine. Compare the results of the different searches.

a. Botox® treatment

b. *Clostridium botulinum* and cosmetic facial injections

c *Clostridium botulinum* and cosmetic facial injections site:edu

Which search terms enabled you to answer the questions: "What is the connection between *Clostridium botulinum* and cosmetic facial injections?" and "What cellular processes are affected by *Clostridium botulinum*?"

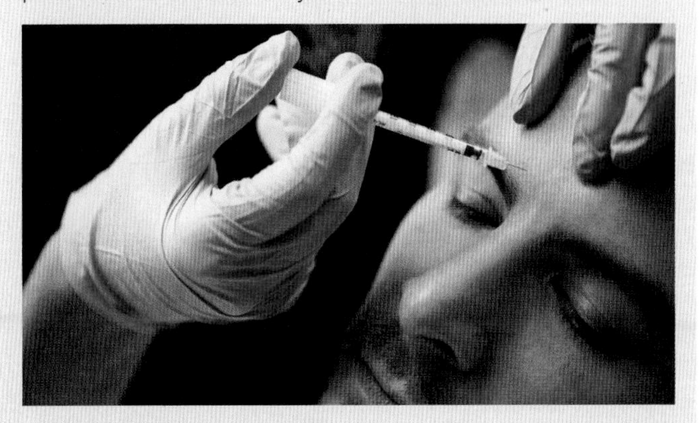

▲ Figure 27 Injecting Botox®

A2.2.6 Eukaryote cell structure

Eukaryotes, like all other living organisms, have a basic cell structure with cytoplasm inside a plasma membrane. In some eukaryotes, there is also a cell wall outside the membrane. Whereas the cytoplasm of a prokaryotic cell is one undivided space, eukaryotic cells are compartmentalized. Areas are separated from the rest of the cytoplasm by single or double membranes. The advantages of having compartments are described in *Topic B2.2*. Three other fundamental features distinguish eukaryotic cells from prokaryotic cells:

Nucleus

This compartment holds the cell's chromosomes. The nucleus has a double membrane with pores through it. Each chromosome consists of one long DNA molecule attached to proteins, except when a cell is preparing to divide and the DNA is replicated. The DNA molecules are linear rather than circular. The proteins are histones, arranged in globular groups like small beads, with the DNA wound around the outside.

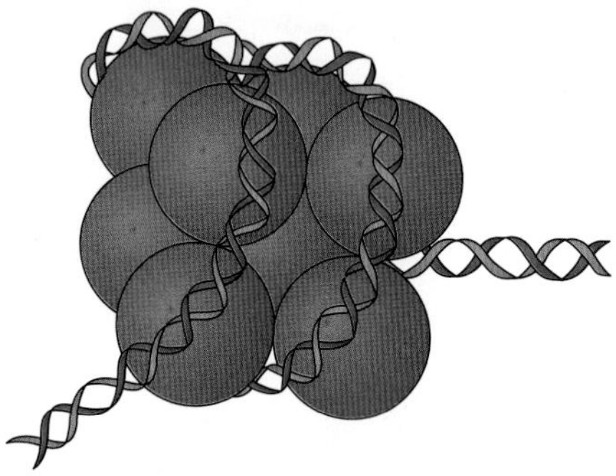

▲ Figure 28 Whereas the DNA of prokaryotes is naked, DNA in eukaryotes is attached to groups of proteins called histones. There are many of these histones along the chromosome, giving the overall appearance of a string of beads

Source: Caputi, Francesca & Candeletti, Sanzio & Romualdi, Patrizia. (2017). Epigenetic Approaches in Neuroblastoma Disease Pathogenesis. 10.5772/intechopen.69566

80S ribosomes

Ribosomes in eukaryotic cells synthesize proteins, as in prokaryotes, but there are structural differences and they are larger in size. This causes them to sink more quickly than prokaryotic ribosomes when centrifuged; this is quantified using Svedberg units (S). Eukaryotic ribosomes are 80S whereas those of a prokaryote are 70S.

Mitochondria

The cytoplasm of a eukaryotic cell contains mitochondria. A mitochondrion is surrounded by a double membrane. The inner membrane is usually folded inwards to increase the surface area. Mitochondria carry out aerobic cell respiration, so in eukaryotes they are only lacking in cells that never respire aerobically.

Nuclei, mitochondria and ribosomes are examples of organelles. All the important organelles of eukaryotic cells are described in *Section A2.2.10*.

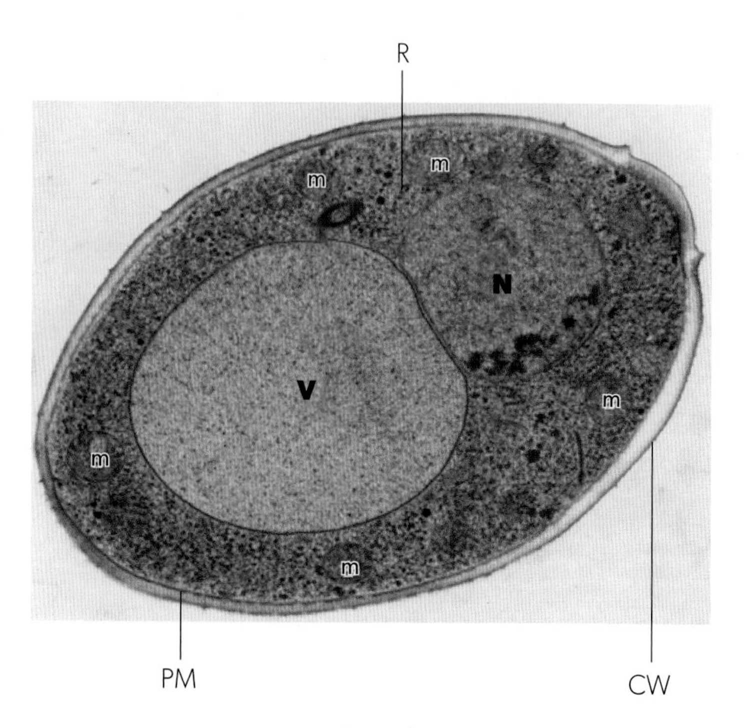

▶ Figure 29 An electron micrograph of the unicellular fungus, *Saccharomyces cerevisiae* (baker's yeast). The nucleus (N), mitochondria (m) and vacuole (V) are easily visible. The cell wall (CW) is the thicker pale outer layer. The plasma membrane (PM) is the thinner dark line inside the cell wall. 80S ribosomes (R) are smaller and more difficult to see, but many are present. The labelled ribosomes look like a string of beads. This cell is 8 μm long. What is the magnification of the micrograph? Remember that 1 μm = 1,000 nm

A2.2.7 Processes of life in unicellular organisms

Living organisms are very diverse in their activities. However, some vital processes are either universal or very widespread:

- **homeostasis**—maintenance of a constant internal environment in an organism
- **metabolism**—the sum of all the biochemical reactions that occur in a living organism
- **nutrition**—supplying the nutrients required for energy, growth and repair in an organism
- **excretion**—removal of waste products of metabolism from an organism
- **growth**—an increase in size or number of cells
- **response to stimuli**—perception of stimuli and carrying out appropriate actions in response
- **reproduction**—production of offspring, either sexually or asexually.

In a multicellular organism, different cell types are specialized to perform these functions, but the single cell of a unicellular organism must perform them all. The annotated diagrams of *Paramecium* and *Chlamydomonas* (Figure 30 and Figure 31) show how two unicellular organisms perform the functions of life.

▼ Figure 30 *Paramecium*

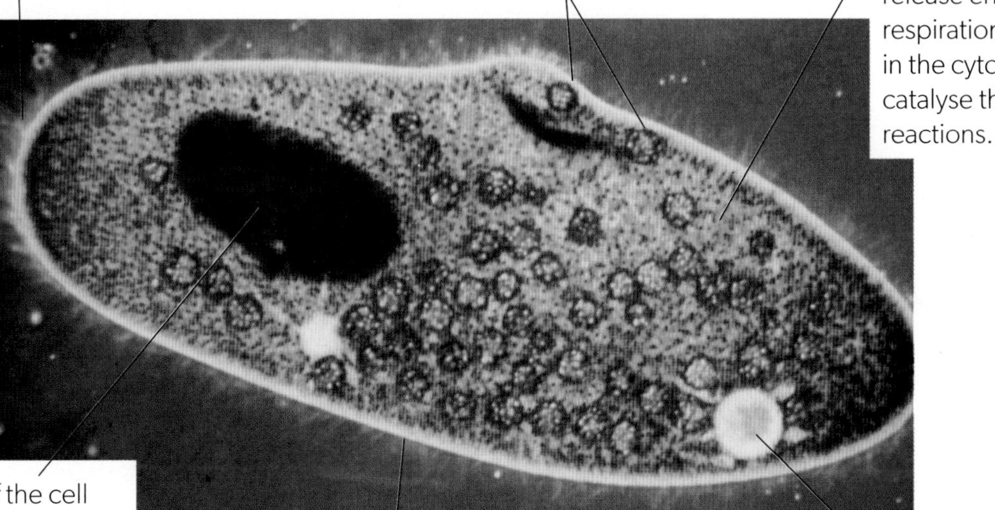

Beating of whip-like cilia moves Paramecium through the water. This can be controlled by the cell so that it moves in a particular direction in response to changes in the environment.
Food vacuoles contain smaller organisms that the Paramecium has consumed. These are gradually digested and the nutrients are absorbed into the cytoplasm, where they provide energy and materials needed for growth.
Metabolic reactions take place in the cytoplasm, including the reactions that release energy by respiration. Enzymes in the cytoplasm catalyse these reactions.
The nucleus of the cell can divide to produce the extra nuclei needed when the cell reproduces. Reproduction is often asexual, with the parent cell dividing to form two daughter cells.
The cell membrane controls what chemicals enter and leave. It allows the entry of oxygen for respiration. Excretion happens by waste products diffusing out through the membrane.
The contractile vacuoles at each end of the cell fill up with water and then expel it through the plasma membrane of the cell. This is a type of homeostasis, keeping the water content of the cell within tolerable limits.

▶ Figure 31 *Chlamydomonas*

The cell wall is freely permeable. The plasma membrane inside the wall controls which chemicals enter and leave the cell. For example, oxygen (a waste product of photosynthesis) is excreted by outward diffusion through the plasma membrane.

Photosynthesis occurs inside the large cup-shaped chloroplast. Carbon dioxide is converted into compounds needed for growth. Other metabolic reactions happen in the cytoplasm.

Contractile vacuoles at the base of the flagella fill up with water and then expel it through the plasma membrane. This is a type of homeostasis, which keeps the water content of the cell within tolerable limits.

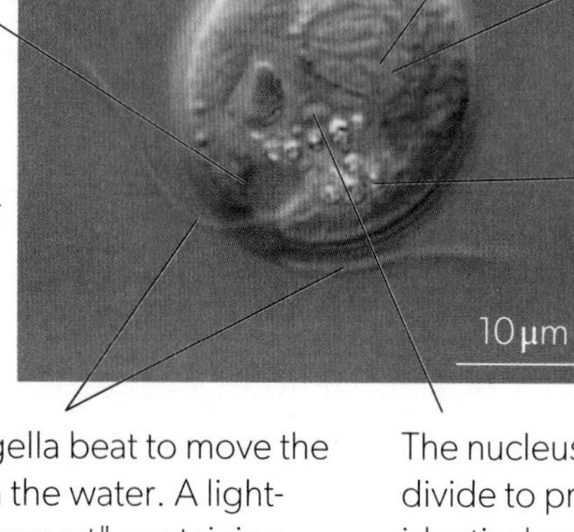

A store of starch is visible inside the chloroplast.

Food vacuoles are formed when other organisms are ingested by endocytosis.

The two flagella beat to move the cell through the water. A light-sensitive "eyespot" containing carotenoid pigments is visible nearby. It allows the cell to sense where the brightest light is and respond by swimming towards it.

The nucleus of the cell can divide to produce genetically identical nuclei by asexual reproduction. Nuclei can also fuse and then divide in a form of sexual reproduction. In this image, the nucleus is concealed by the cup-shaped chloroplast.

Making careful observations: Examining a community of life under a microscope

To see unicellular organisms, follow this procedure.

1. Collect some pond water.
2. If possible, centrifuge the sample to concentrate the organisms in it.
3. Place a drop of the concentrated pond water on a microscope slide.
4. Add a cover slip and view with a microscope.

You will almost certainly be able to see unicellular organisms. Alternatively, you could obtain a pure culture of unicellular organisms such as *Paramecium* or *Chlamydomonas*.

Data-based questions: Processes of life in a testate amoeba

Arcella gibbosa is a unicellular testate amoeba that lives in freshwater habitats. It has a hard outer coat made of chitin, with a light but strong structure resembling honeycomb. This is called the test. There is a circular aperture in the test and finger-like protrusions of cytoplasm can push out through this, then retract again. In Figure 32, five structures are labelled: nuclear membrane (nm), outer surface of test (to), plasma membrane (pm), contractile vacuole (cv) and aperture in the test (at). The unlabelled coloured structures are food vacuoles. A scale bar is shown lower right.

1. Calculate the diameter of the cell and the magnification of the micrograph. [3]
2. Deduce the maximum size of food particle that could be ingested. [2]
3. Explain the need for a contractile vacuole in this organism. [3]
4. Predict whether this cell was:
   a. growing [2]
   b. preparing to divide. [2]

5. Suggest a hypothesis for whether the cell has mitochondria in its cytoplasm. Give reasons and suggest how your hypothesis could be tested. [3]

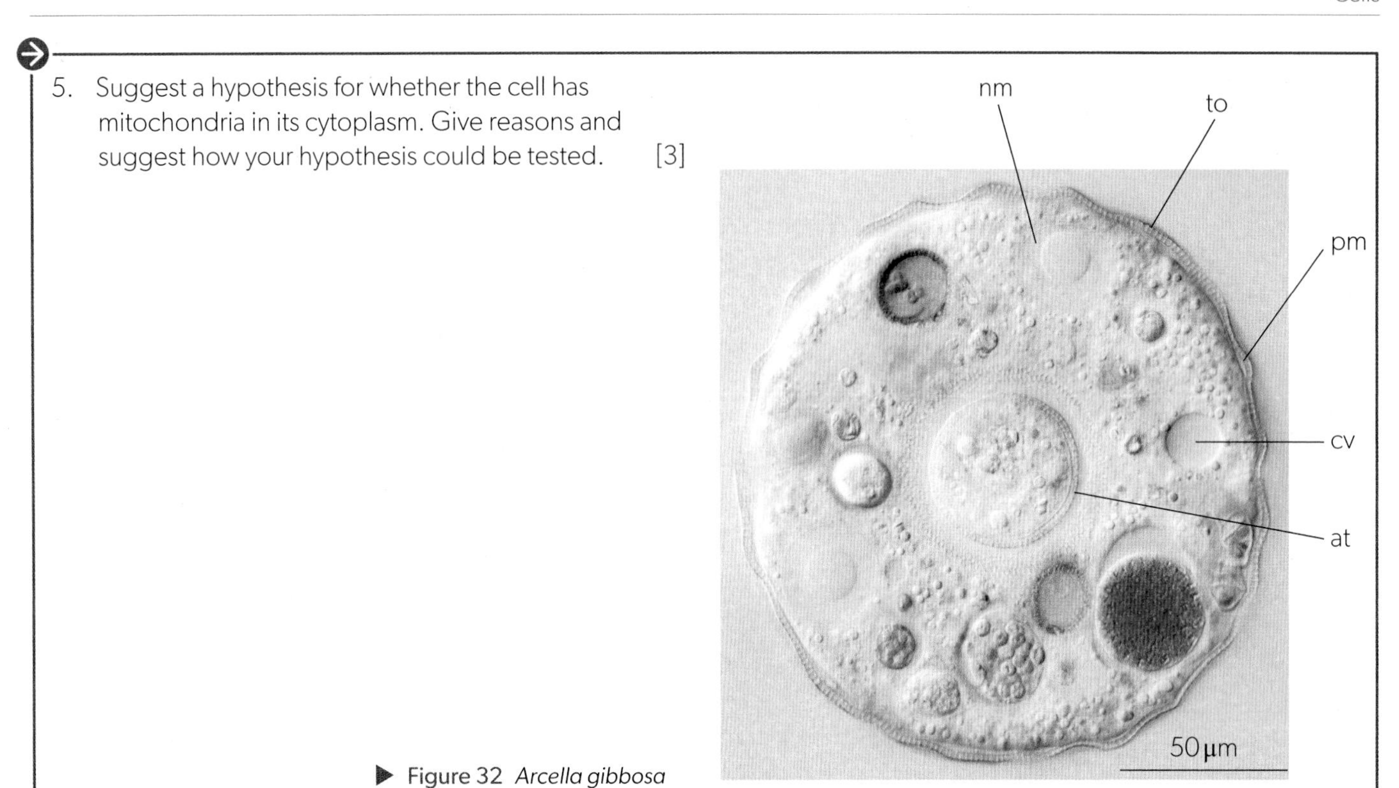

▶ Figure 32 *Arcella gibbosa*

A2.2.8 Differences in eukaryotic cell structure between animals, fungi and plants

Feature	Animals	Fungi	Plants
Plastids A family of organelles with two outer membranes and internal membrane sacs	None	None	Plastids of varied types such as chloroplasts (for photosynthesis) and amyloplasts (to store starch)
Cell wall A rigid layer outside the plasma membrane to strengthen and protect the cell	None	Cells of fungi and plants have walls, composed of chitin in fungi and cellulose in plants	
Vacuole A flexible fluid-filled compartment surrounded by a single membrane	Small temporary vacuoles expel excess water or digest food or pathogens taken in by endocytosis	There is often a large permanent vacuole in cells of fungi and plants, used for storage of substances and pressurizing the cell	
Centrioles Cylindrical organelles that organise the assembly of structures composed of microtubules	Used to construct the spindle that moves chromosomes in mitosis and the 9 + 2 microtubules in cilia and flagella	Absent, except in fungi and plants with swimming male gametes, which have a centriole at the base of the flagellum	
Undulipodia Cilia and flagella used to generate movement of a cell or movement of fluid adjacent to a cell	Cilia and flagella are present in many animal cells, including the tail of male gametes	Absent except in fungi and plants with male gametes that swim using flagella (tails)	

▲ Table 4

ATL Thinking skills: Reflecting on the reasonableness of results

It is claimed that, for every one of your human cells, there are 10 prokaryotic cells living in your body. Does this seem a reasonable claim? One way to test this claim is to model it using artist's clay.

1. Obtain some modelling clay.
2. Construct a model of a prokaryotic cell, with dimensions 10 mm × 5 mm × 5 mm.
3. Construct a model of a eukaryotic cell that is 10 times larger in every dimension: 100 mm × 50 mm × 50 mm.
4. Using these models, does the claim seem reasonable?

A2.2.9 Atypical cell structure in eukaryotes

According to the cell theory, all living organisms are made of cells. Each cell is expected to have one nucleus (unless it is preparing to divide, when there may be two). However, some structures in organisms do not follow the typical patterns. Some examples are red blood cells, phloem sieve tube elements, skeletal muscle and aseptate fungal hyphae.

Red blood cells

In mammals, these cells do not have a nucleus. At a late stage in their development in bone marrow, the nucleus is moved to the edge of the cytoplasm and the small part of the cell containing it is pinched off and destroyed by a phagocyte. Removal of the nucleus makes red blood cells smaller and more flexible, but they cannot repair themselves if they are damaged. For this reason, they have a lifespan of only 100 to 120 days.

Phloem sieve tube elements

Plants move sap through tubular vessels, made from columns of cylindrical cells. The flow of sap would be impeded if these cells had a typical structure. In xylem vessels, which conduct watery sap from the roots to the leaves, all the dividing walls between adjacent cells are removed and the plasma membrane and all of the cell contents break down. This creates a hollow tube that no longer consists of cells.

In phloem, which conducts sugary sap from the leaves to other parts, the conducting vessels are called sieve tubes. The dividing walls between adjacent cells are sieve-like, with large pores for the sap to pass through. During development of sieve tubes, the nucleus and most of the other cell contents break down, but the plasma membrane remains as it is essential for phloem transport. The subunits in a sieve tube are usually called elements rather than cells because of their atypical structure. Sieve tube elements are connected to adjacent companion cells, which have a nucleus and mitochondria. These companion cells help the sieve tube elements to survive and carry out their function.

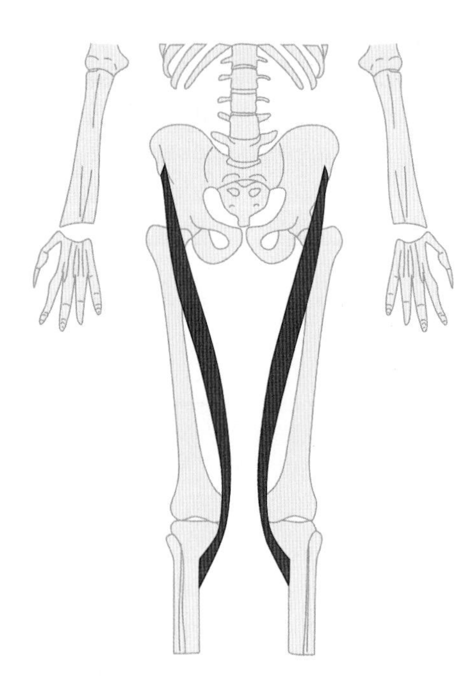

▲ Figure 33 The human sartorius muscle can be as much as 600 mm long. It contains muscle fibres that extend from one end to the other. Are these the longest cells in the human body?

Skeletal muscle

Some large multinucleate structures are formed when groups of cells fuse together. This type of structure is a syncytium. Muscle fibres develop in this way. Columns of cells, each with a nucleus, are formed by cell division. These cells then fuse together to form long muscle fibres.

Aseptate fungal hyphae

In some growing cells, the nucleus divides repeatedly without any subsequent cell division. This results in an unusually large multinucleate structure, known as a coenocyte. The thread-like hyphae of some fungi develop in this way. Walls that divide the hyphae of other types of fungi into uninucleate cells are called septa, so hyphae without these divisions are aseptate.

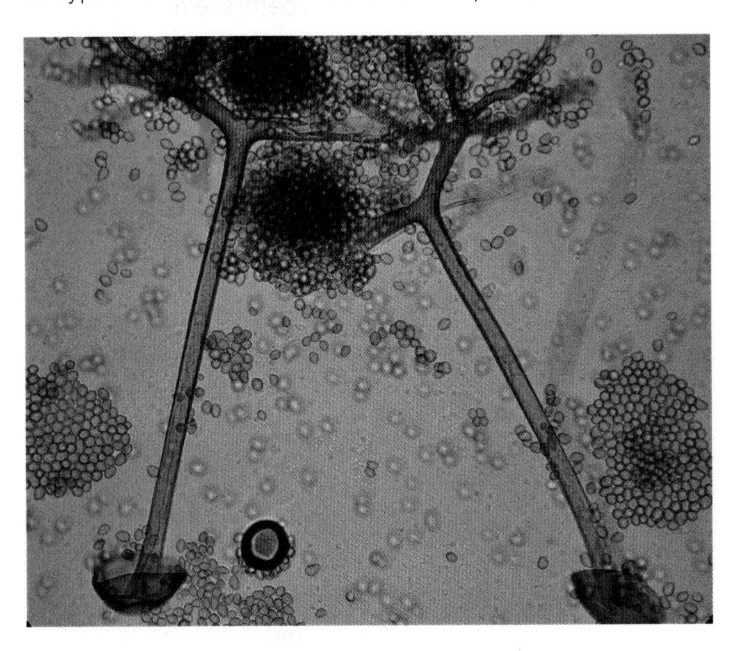

▲ Figure 34 A micrograph of aseptate hyphae of the fungus *Rhizopus arrhizus*, which is the most frequent cause of mucormycosis. Spores and the sporangia that produced them are also visible

A2.2.10 Cell types and cell structures viewed in light and electron micrographs

In light micrographs, these features help us to identify whether a cell is from a prokaryote, a plant or an animal.

Prokaryotic cells

- single cells, sometimes arranged in chains
- small size—cells usually less than 5 μm
- often rod-shaped (bacilli), spheroidal (cocci) or helical (spirilli)
- cell wall present
- no nucleus; instead there is a paler region of cytoplasm (nucleoid)
- simple internal structure with no membrane-bound organelles
- no vacuoles or other internal membranes

Plant cells

- always multicellular apart from gametes and zygotes
- larger size—cells usually more than 5 μm
- shape tends to be regular with flat sides and cell junctions easily visible in tissues
- cell wall present
- nucleus normally present but not always visible
- plastids present, such as chloroplasts, or amyloplasts storing starch
- large vacuole often present

Animal cells

- always multicellular apart from gametes, zygotes and blood cells
- larger size—cells usually more than 5 μm
- shape tends to be rounded with junctions between cells often hard to see in tissues
- no cell wall
- nucleus normally present but not always visible
- no chloroplasts or stored starch but cytoplasm contains many other organelles
- only small vacuoles are present

Table 5 describes the structure and functions of all the main organelles of eukaryotic cells.

▼ Table 5

Nucleus double nuclear membrane nuclear pores dense chromatin chromatin	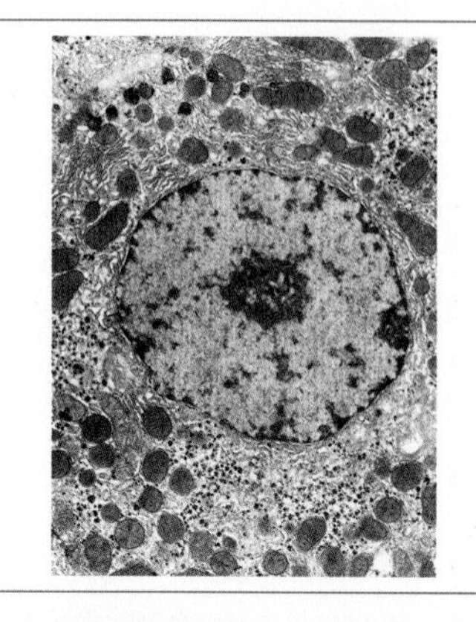	The nuclear membrane is double and has pores through it. The nucleus contains the chromosomes, consisting of DNA associated with histone proteins. Uncoiled chromosomes are spread through the nucleus in the areas that appear pale and grainy. The small areas that are more densely stained, mostly around the edge of the nucleus, contain parts of chromsomes that have remained coiled up (condensed). The nucleus is where DNA is replicated and transcribed to form mRNA, which is exported via the nuclear pores to the cytoplasm.
Rough endoplasmic reticulum 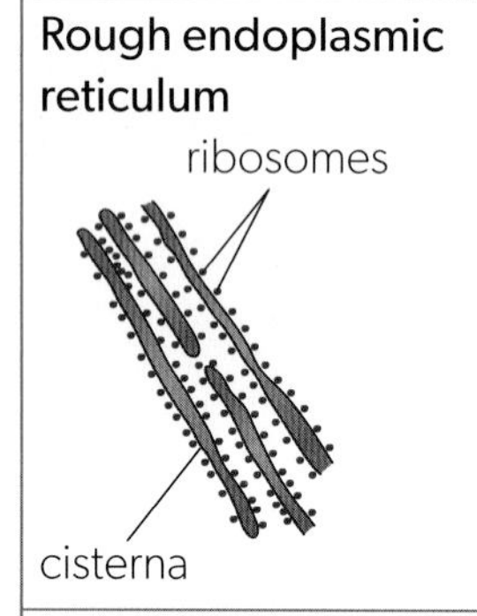	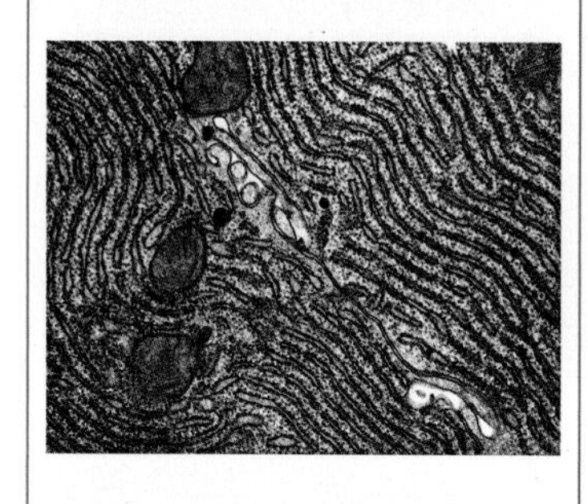	The rER consists of flattened membrane sacs, called cisternae. Ribosomes are attached to the outside of these cisternae. They are larger than in prokaryotes and are classified as 80S. The main function of the rER is to synthesize protein for secretion from the cell. Protein synthesized by the ribosomes of the rER passes into its cisternae. It is then carried by vesicles, which bud off and are moved to the Golgi apparatus.
Smooth endoplasmic reticulum 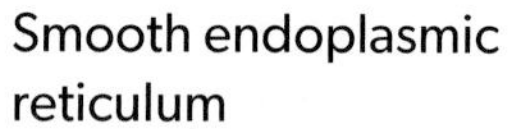 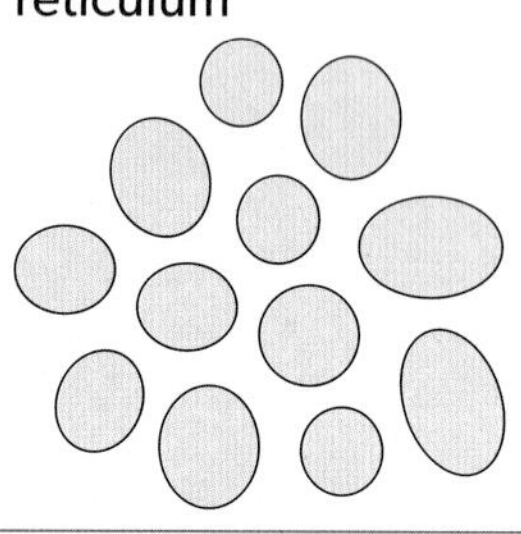	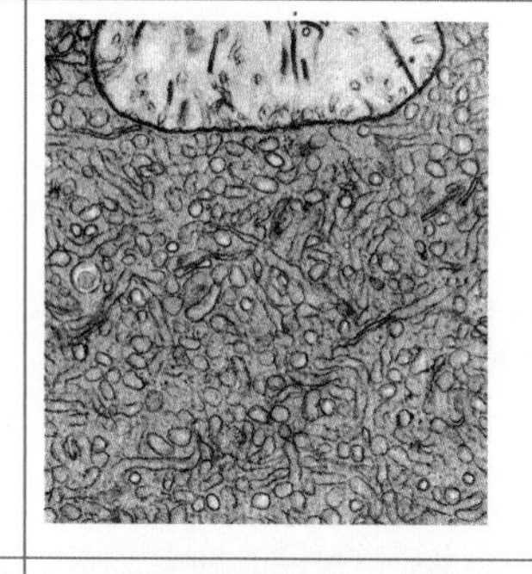	Smooth endoplasmic reticulum consists of a branched network of tubular membranes. In electron micrographs, it appears as circles or ovals of membrane. The membrane is smooth because there are no ribosomes attached. Smooth ER has a variety of functions. It is used to synthesize lipids, phospholipids and steroids. A special type of smooth ER stores calcium ions in muscle when it is relaxed.
Golgi apparatus 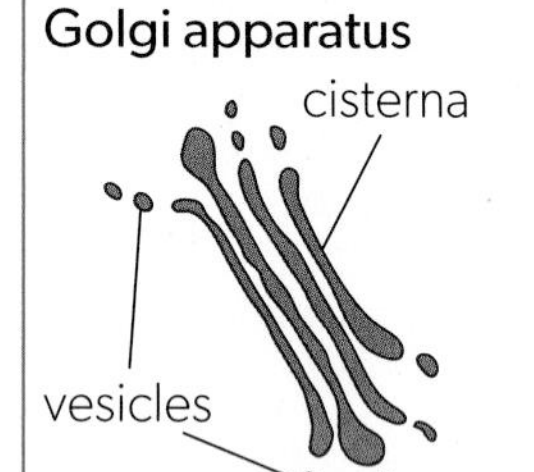	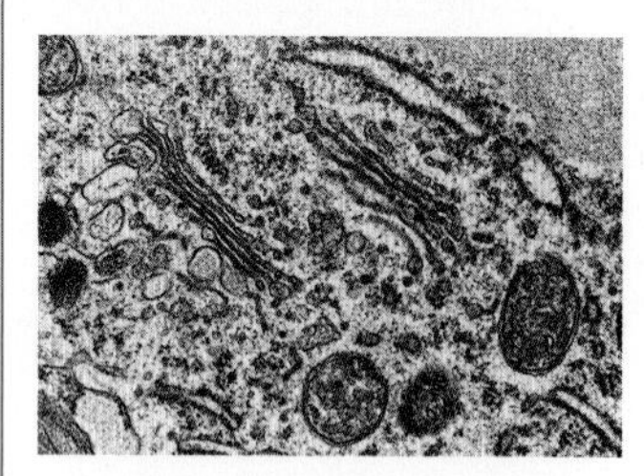	This organelle consists of flattened membrane sacs called cisternae (as in rER). However, these cisternae are not as long, are often curved, do not have attached ribosomes and have many vesicles nearby. The Golgi apparatus processes proteins brought in vesicles from the rER. Most of these proteins are then carried in vesicles to the plasma membrane for secretion.
Lysosome 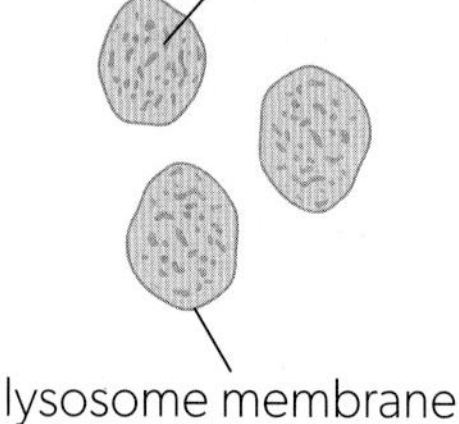	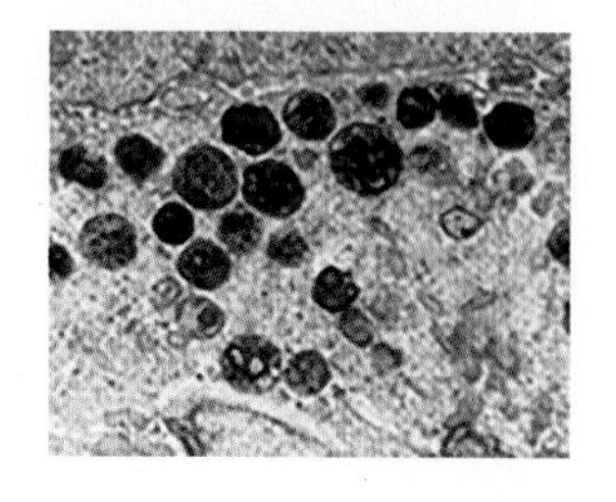	These are approximately spherical with a single membrane. They are formed from Golgi vesicles. They contain high concentrations of protein, which makes them densely staining in electron micrographs. They contain digestive enzymes, which can be used to break down ingested food in vesicles. These enzymes can also break down organelles or even whole cells.

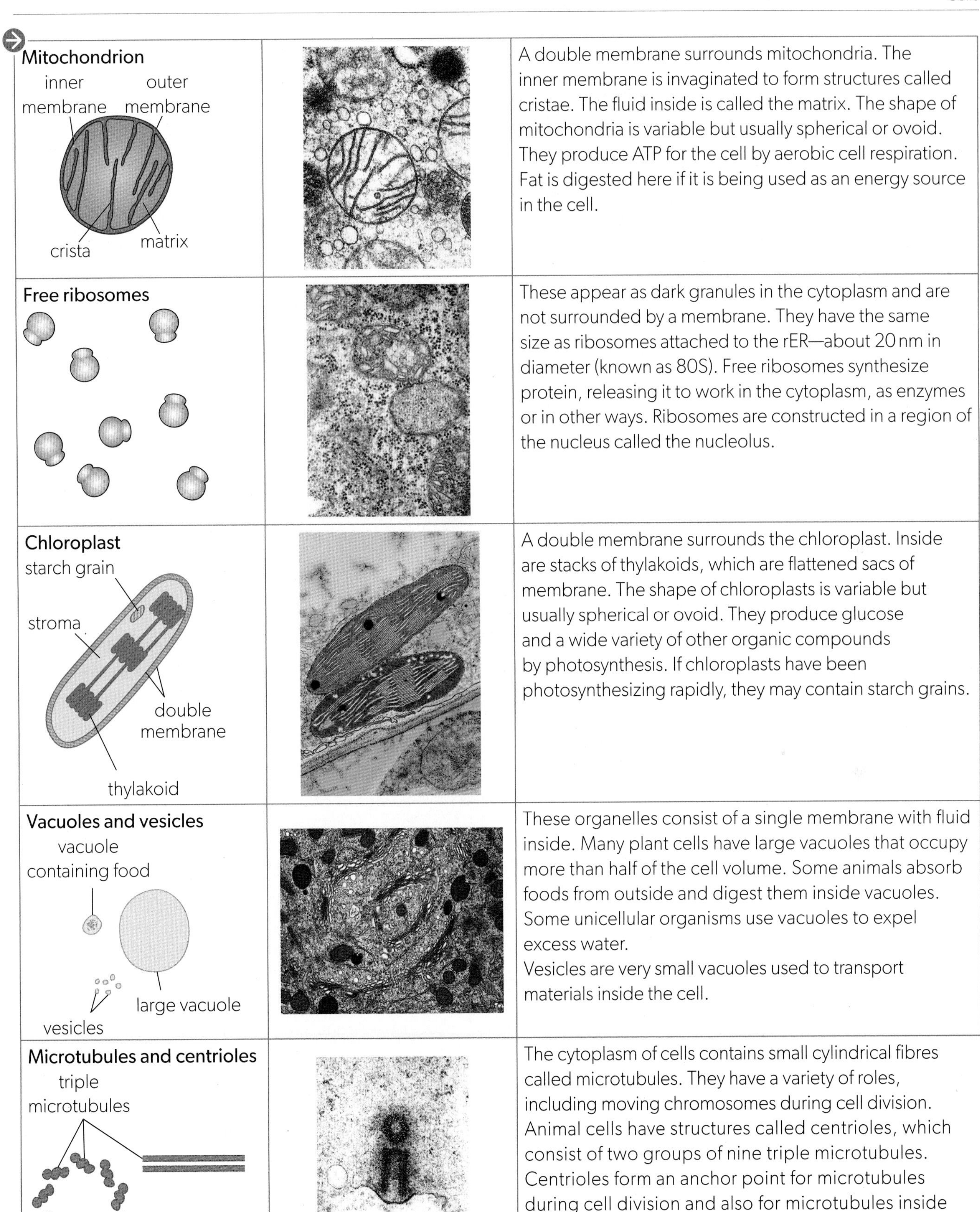

Mitochondrion		A double membrane surrounds mitochondria. The inner membrane is invaginated to form structures called cristae. The fluid inside is called the matrix. The shape of mitochondria is variable but usually spherical or ovoid. They produce ATP for the cell by aerobic cell respiration. Fat is digested here if it is being used as an energy source in the cell.
Free ribosomes		These appear as dark granules in the cytoplasm and are not surrounded by a membrane. They have the same size as ribosomes attached to the rER—about 20 nm in diameter (known as 80S). Free ribosomes synthesize protein, releasing it to work in the cytoplasm, as enzymes or in other ways. Ribosomes are constructed in a region of the nucleus called the nucleolus.
Chloroplast		A double membrane surrounds the chloroplast. Inside are stacks of thylakoids, which are flattened sacs of membrane. The shape of chloroplasts is variable but usually spherical or ovoid. They produce glucose and a wide variety of other organic compounds by photosynthesis. If chloroplasts have been photosynthesizing rapidly, they may contain starch grains.
Vacuoles and vesicles		These organelles consist of a single membrane with fluid inside. Many plant cells have large vacuoles that occupy more than half of the cell volume. Some animals absorb foods from outside and digest them inside vacuoles. Some unicellular organisms use vacuoles to expel excess water. Vesicles are very small vacuoles used to transport materials inside the cell.
Microtubules and centrioles		The cytoplasm of cells contains small cylindrical fibres called microtubules. They have a variety of roles, including moving chromosomes during cell division. Animal cells have structures called centrioles, which consist of two groups of nine triple microtubules. Centrioles form an anchor point for microtubules during cell division and also for microtubules inside cilia and flagella.

Cytoskeleton 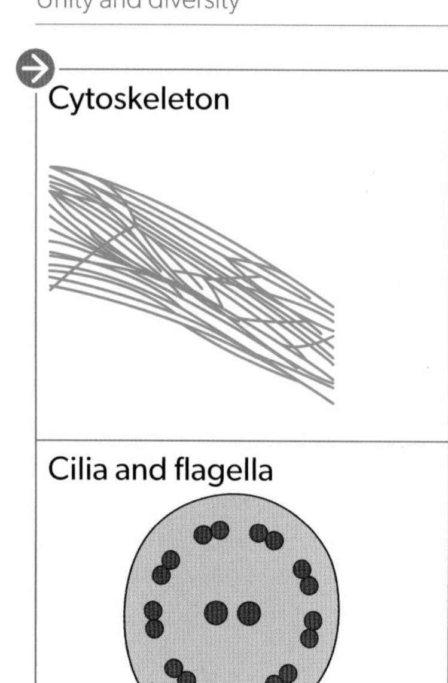	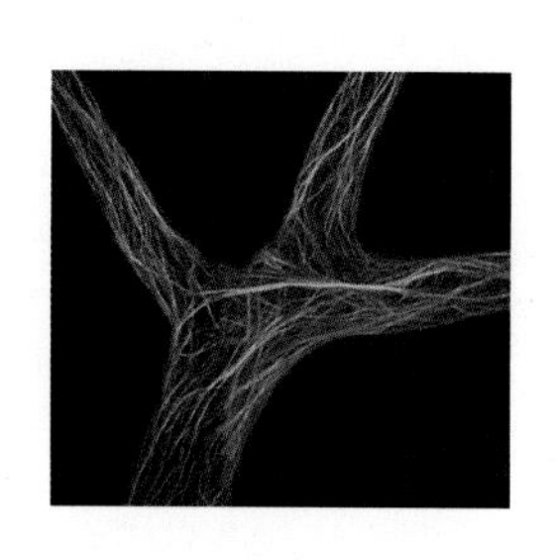	The cytoskeleton is constructed from several types of protein fibre. Tubulin is used to make microtubules and actin is used to make microfilaments. These structures can easily be constructed or deconstructed, so the cytoskeleton is dynamic. Microtubules guide the movement of components within the cell. They help plant cells to construct cell walls. A layer of microfilaments just inside the plasma membrane helps animal cells to maintain their shape.
Cilia and flagella	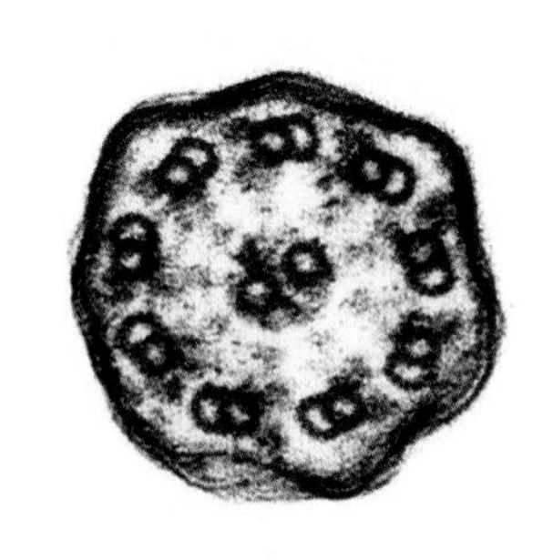	These are whip-like structures projecting from the cell surface. They contain a ring of nine double microtubules plus two central ones. Flagella are larger and usually only one is present, as in a sperm. Cilia are smaller and many are present. Cilia and flagella can be used for locomotion. Cilia can also be used to create a current in the fluid next to a cell.

A2.2.11 Drawing and annotation based on electron micrographs

Electron micrographs show cell structure in great detail. However, they sometimes include artefacts as well. (An artefact is something that is not naturally present but was introduced as the specimen was prepared by staining and sectioning.) Therefore, a drawing of an electron micrograph may show the structure more clearly. Basic drawing skills were described earlier and Table 5 shows how the structure of organelles can be shown in drawings.

Electron micrographs of a prokaryotic cell (Figure 35) and a eukaryotic cell (Figure 36) are shown. A drawing of the prokaryotic cell is also included, to show how its structure can be interpreted. Organelles in the electron micrograph of a eukaryotic cell are labelled. Using your knowledge of these organelles, you should be able to draw the whole cell to show its ultrastructure.

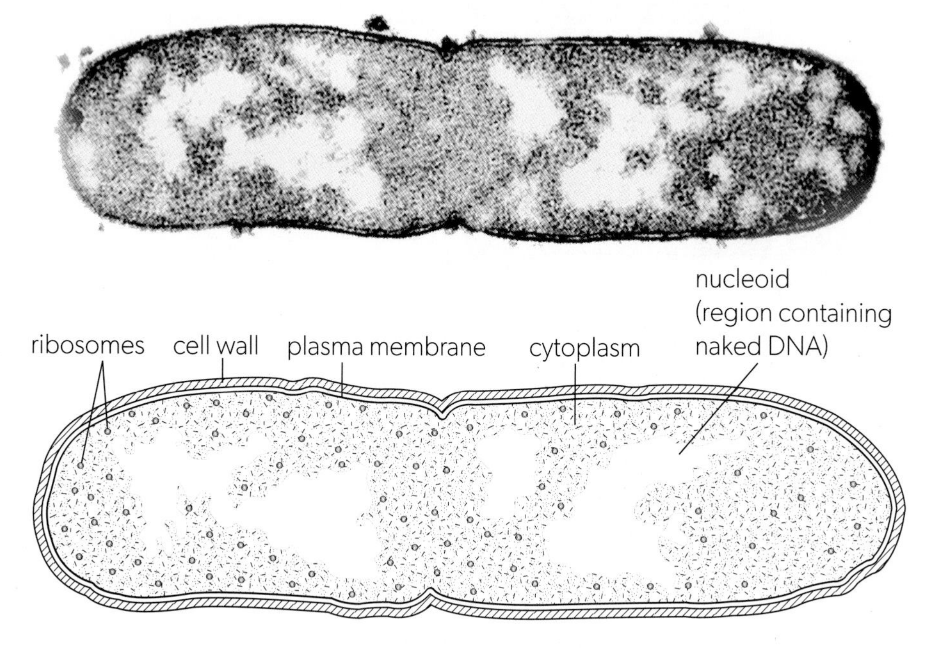

▲ **Figure 35** Electron micrograph of *Escherichia coli* (1–2 μm in length), with a drawing to help interpret the micrograph

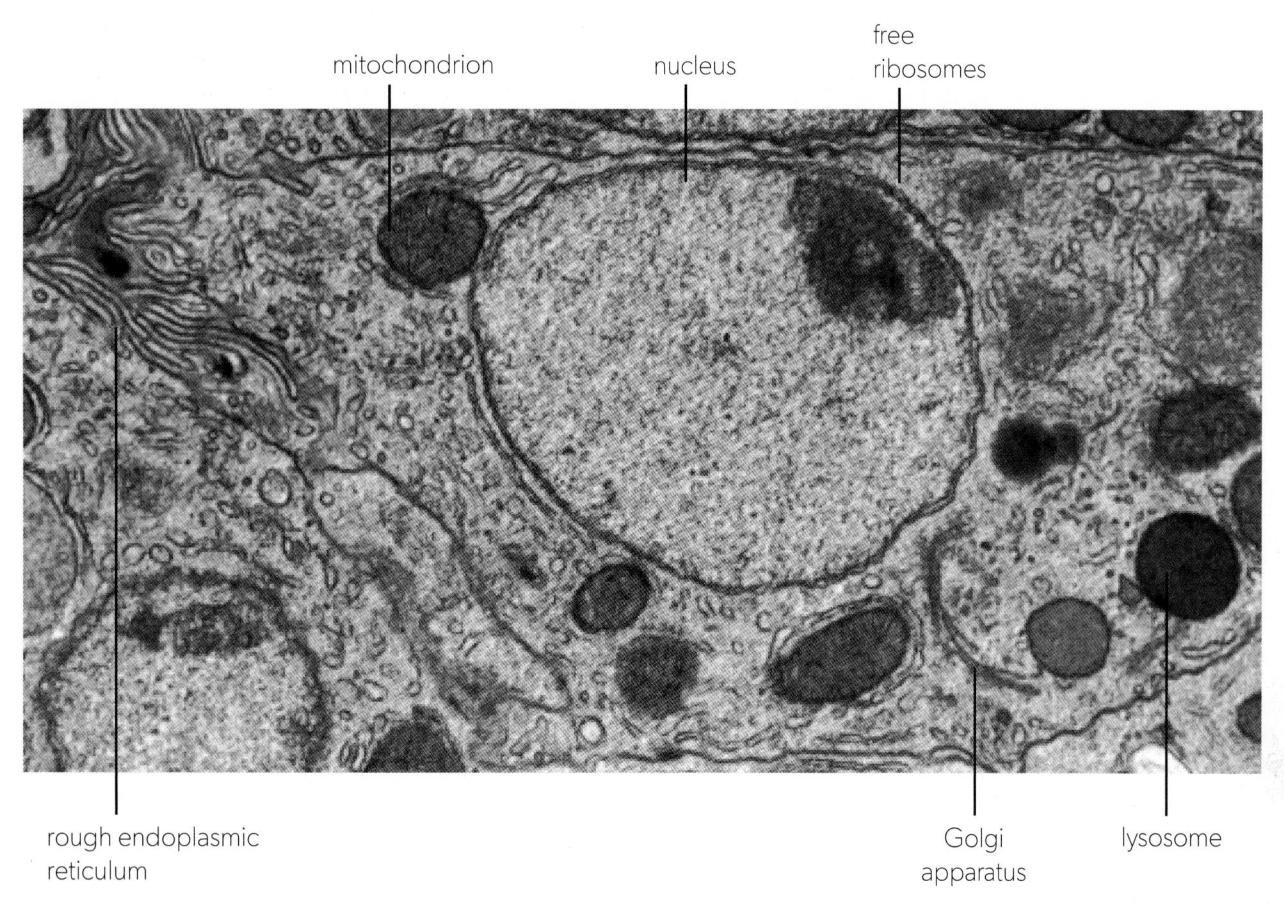

▲ **Figure 36** Electron micrograph of a liver cell. The plasma membrane is visible as a dark line. Part of the cell on the right is not visible

AHL

A2.2.12 Origin of eukaryotic cells by endosymbiosis

Symbiosis is living together in a close association. In endosymbiosis, one organism (the endosymbiont) lives inside another (the host). In the closest form of this, the endosymbiont lives inside a cell of the host. The endosymbiont enters the host cell by endocytosis. This is the process that cells use to make a vesicle or small vacuole by pinching off a piece of the plasma membrane. It is described fully in *Topic B2.1*.

Cells can use endocytosis to ingest other, smaller cells. For example, phagocytes in humans ingest viruses or bacteria, and unicellular organisms such as *Paramecium* ingest the organisms on which they feed. In those cases, digestive enzymes are added to the vacuole to break down the ingested organisms, which are therefore killed. In other cases, the host can gain more from the ingested organism if it is alive. The result is endosymbiosis. In a mutualistic relationship, both the host and the endosymbiont benefit. Examples of mutualisic endosymbiosis are studied in *Topic C4.1*.

Endosymbiosis almost certainly contributed to the evolution of eukaryotic cells. According to a well-established theory, mitochondria were once free-living prokaryotes that developed the process of aerobic respiration. Larger prokaryotes that could only respire anaerobically took in these smaller prokaryotes by endocytosis; instead of killing and digesting them, they allowed the engulfed cells to live in the cytoplasm as endosymbionts. Aerobic respiration in the endosymbiont supplied energy to the host, far more efficiently than the host's own anaerobic respiration. At the same time, the endosymbiont was supplied with food by the host. Natural selection therefore favoured cells that developed this mutualistic endosymbiotic relationship.

If the endosymbionts grew and divided as fast as the host cell, they could persist inside host cells for many generations. According to the endosymbiotic theory, we can deduce that they have persisted inside eukaryotic cells for hundreds of millions of years, evolving to become the mitochondria of eukaryotic cells alive today.

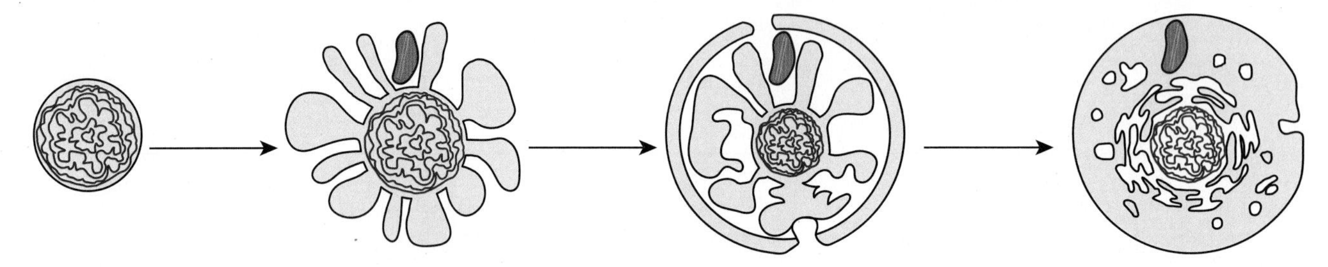

Shared features suggest that eukaryotic cells evolved from the cells of an archaean, usually known as Asgard.

Outgrowths of the plasma membrane expanded the cytoplasm. The archaean DNA remained in the centre and the membrane around it started to develop into the inner nuclear membrane.

Membrane invagination and vesicle formation generated organelles which became more complex and diverse. An association developed with an aerobically respiring eubacterium.

The aerobic eubacterium became totally enclosed by endosymbiosis and developed into the mitochondrion. In some cells, a cyanobacterium also became enclosed and developed into chloroplasts.

▲ **Figure 37** Origins of the nucleus, mitochondria and chloroplasts

The endosymbiotic theory also explains the origin of chloroplasts. If a prokaryote that had developed photosynthesis was taken in by a host cell and allowed to survive, grow and divide, it could have developed into the chloroplasts of photosynthetic eukaryotes—algae and plants. Again, both the endosymbiont and the host would benefit from the relationship.

This explanation for the evolution of mitochondria and chloroplasts remains a theory because it cannot be conclusively proved. However, the features of both mitochondria and chloroplasts provide strong evidence for it:

- They have a double membrane. This would be expected if a prokaryote with a single plasma membrane was ingested by endocytosis.
- They have their own genes, on a circular DNA molecule like that of prokaryotes.
- They transcribe their own DNA and use the mRNA to synthesize some of their own proteins.
- The ribosomes they use for protein synthesis have a size (70S) and structure more typical of prokaryotic cells than eukaryotic.
- They can only be produced by division of pre-existing mitochondria and chloroplasts.

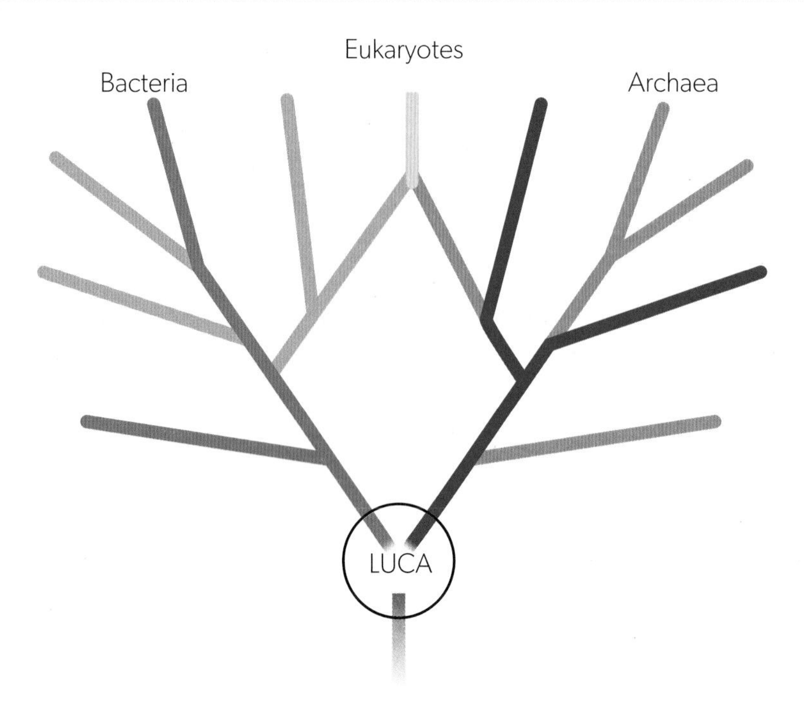

▲ Figure 38 Evidence suggests that the mitochondrion was originally a member of the domain Bacteria and the host cell that took it in was a member of the domain Archaea. In the tree of life, the domain Eukaryota was therefore formed by uniting two branches rather than by splitting off a branch

Theories: The theory of endosymbiosis

Use the theory of endosymbiosis to make predictions:

1. Mitochondria have a double membrane. Predict which membrane would have prokaryotic features and which would have eukaryotic features.
2. There are ribosomes within the matrix of mitochondria. Predict whether the ribosomes within mitochondria are 70S (like those of prokaryotes) or 80S (as in eukaryotes).

Use the theory of endosymbiosis to explain these features:

1. Mitochondria and chloroplasts have circular DNA, rather than linear DNA with two ends.
2. Human mitochondrial DNA has only 16,569 base pairs, compared with an average of 143,000,000 base pairs of human chromosomes located in the nucleus.
3. There are only 37 genes in human mitochondrial DNA, compared with more than 500 in free-living prokaryotic cells.

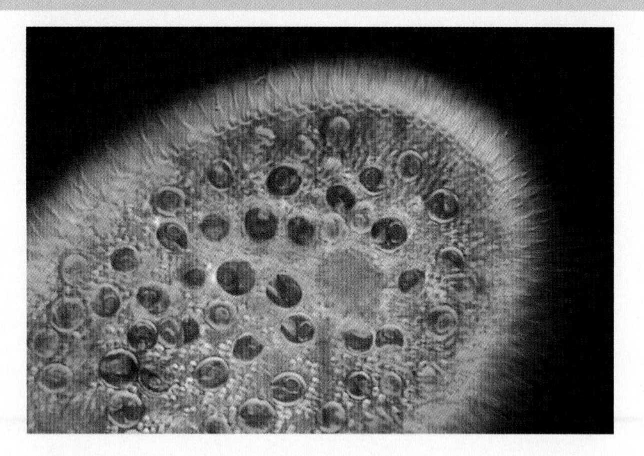

▲ Figure 39 Inside this protozoan (*Paramecium bursaria*), there are individual cells of a green alga. The two organisms have a mutualistic relationship. The algae photosynthesize inside the *Paramecium*, providing it with sugars and oxygen, while deriving protection and carbon dioxide from their host. In what way does this support the theory of endosymbiosis?

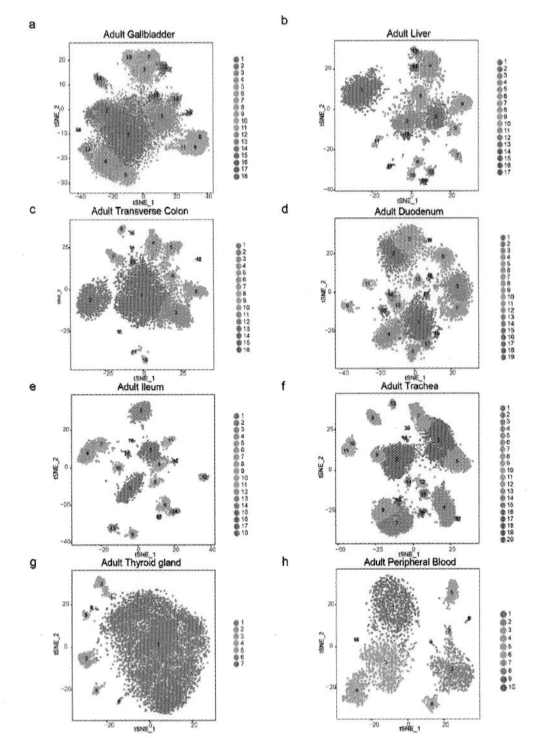

▲ Figure 40 Biologists recently analysed 600,000 different cells from all parts of the human body to find out which genes they were using. A total of 102 cell clusters were discovered, corresponding to different tissues or organs of the body. Each of these clusters contained different cell types. For example, 17 distinct cell types were found in the liver cluster. This image shows similarities in gene use between individual liver cells as a two-dimensional distribution

A2.2.13 Cell differentiation as the process for developing specialized tissues in multicellular organisms

Multicellular organisms have an advantage because cells can develop differently to perform different functions. Specialized cells develop only the features they need to carry out their functions, which makes them more efficient. For example, red blood cells transport oxygen using the protein haemoglobin. They produce large amounts of this protein but do not produce other proteins that they would not use.

Some activities are needed in all cells, such as release of energy by respiration. About 4,000 genes have been detected in human cells that are active in all cell types. They are called housekeepeing genes and they are not associated with specialized roles. Other genes vary in their expression and in some cases are only ever active in a single cell type.

The development of specialized cell types happens from a very early stage in the life of humans and other organisms. Even in a tiny embryo, different cells begin to take different pathways of development. This is the process of cell differentiation. In differentiated cells, different genes are "switched on" and expressed, so the cell makes particular proteins and other gene products. The control of which genes act in a cell is called gene expression.

A2.2.14 Evolution of multicellularity

All plants and animals are multicellular. Multicellularity has evolved independently more than once in the origins of plants and at least once in animals. Many fungi and eukaryotic algae are multicellular. Even some prokaryotes cooperate to form multicellular aggregates. Most cells within a multicellular organism have lost the ability to live independently or to divide.

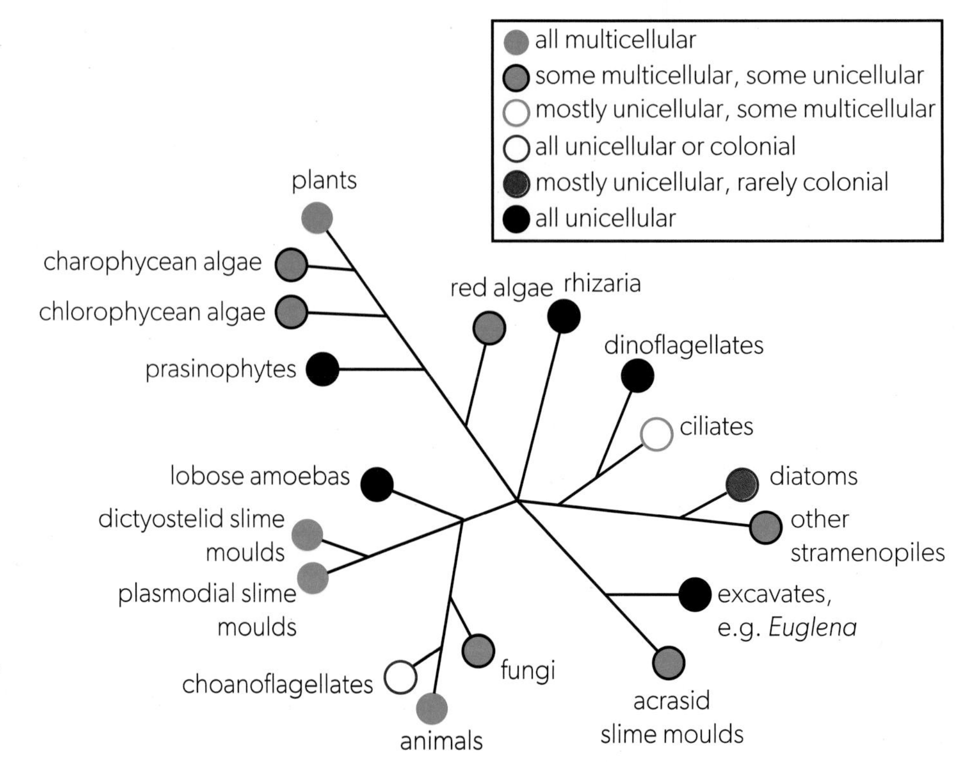

▶ Figure 41 This evolutionary tree diagram shows all the major groups of eukaryotes and shows that multicellularity evolved separately in different groups

There are several advantages to multicellularity. Multicellular organisms tend to have longer lifespans, because the death of one cell does not prevent the continued survival of the individual. Multicellular organisms are generally larger than unicellular organisms, so they can exploit niches that single-celled organisms cannot. Multicellularity also allows for complexity as there can be differentiation of cell types within an organism.

Nonetheless, most of the individual living organisms on Earth are single-celled and most of the biomass on Earth consists of single-celled organisms. This suggests that although some traits possessed by multicellular organisms (such as longer lifespans and differentiation) are advantages, unicellular organisms must have a relative advantage in some situations.

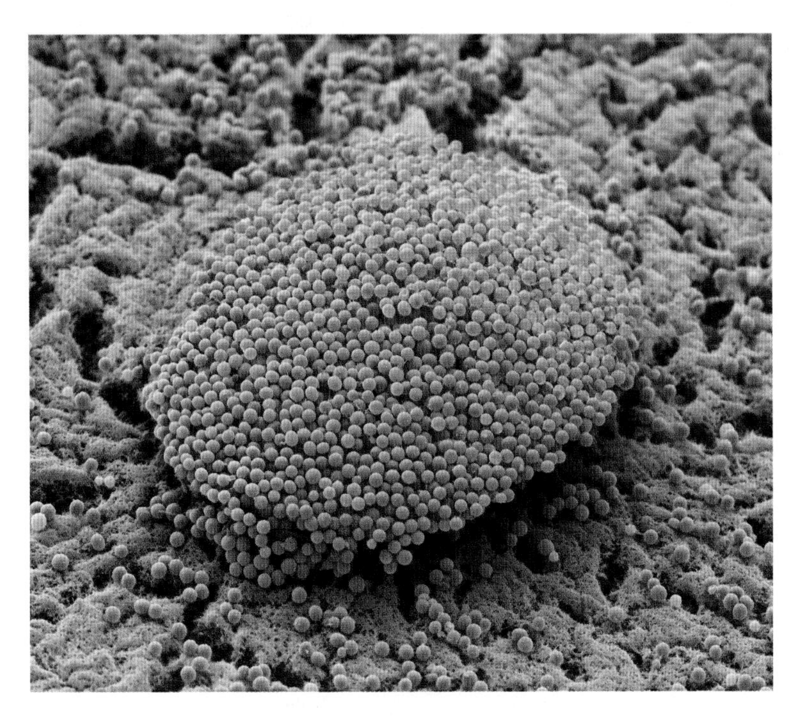

▲ Figure 42 *Myxococcus xanthus* is a rod-shaped Gram-negative bacterium that lives in the soil and feeds on other species of bacteria. It is found in clusters, called swarms, which act as a collective unit and show coordinated movement in response to environmental cues. The bacteria are also able to differentiate to form multicellular inactive (resting) spores that are resistant to drying out. They form when the availability of nutrients is limited. They have the adaptive advantage that when conditions become more favourable, the cells will reactivate as a swarm

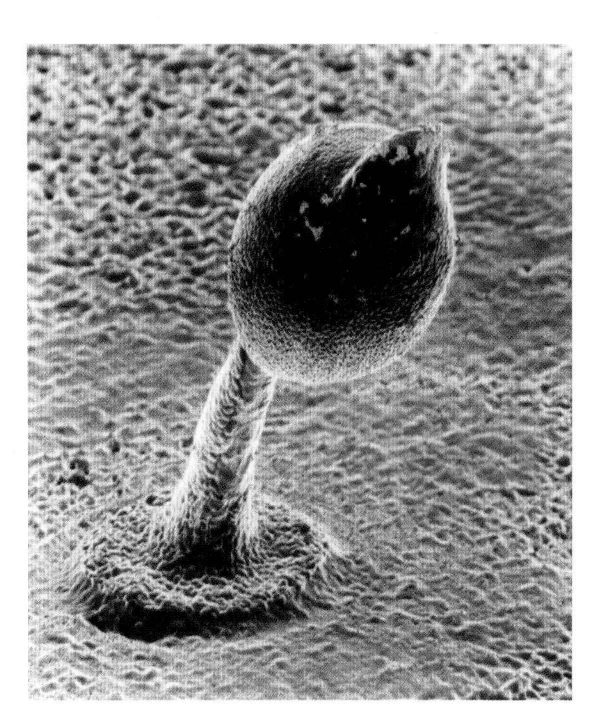

▲ Figure 43 A slime mould can exist as a number of slow-moving single-celled protists, each of which engulfs solid food particles. Under certain conditions, the single cells group together to form a multicellular body called the plasmodium. This can then form into a reproductive spore tower. Most slime moulds are saprophytes, feeding on dead or decaying organic matter

Data-based questions: Diversity in green algae

1. a. State the shape of the cells in the two species of algae. [2]

   b. Most of the cells of *K. klebsii* have only one chloroplast. Describe the features of these chloroplasts that can be seen in the micrograph. [3]

   c. Explain a reason for the hypothesis that some *K. klebsii* cells must contain two chloroplasts. [1]

   d. Spherical lipid droplets are visible in the cytoplasm of both species, but nuclei are not visible. Outline how the nuclei could be made visible. [1]

   e. Discuss whether *K. klebsii* and *C. fenestrata* are unicellular or multicellular. [3]

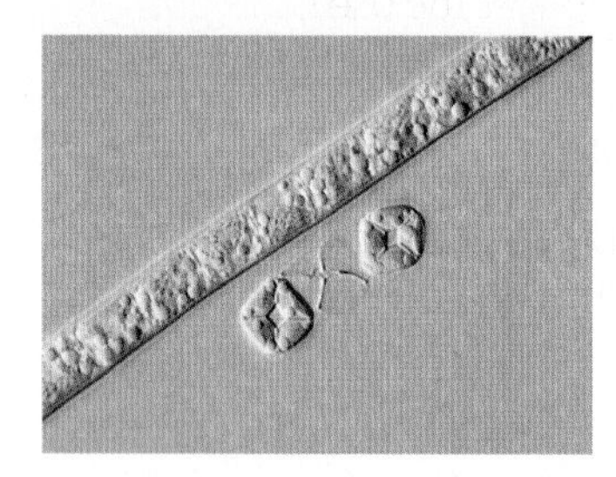

▲ Figure 44 Two species of green algae, with *Klebsormidium klebsii* above and *Crucigenia fenestrata* below

2. *Staurodesmus convergens* is a desmid. These algae have two symmetrical parts to their cells, linked by a bridge where the nucleus is located. Each of the two "semi-cells" contains one large chloroplast with a circular store of starch. There are two layers in the cellulose cell wall. In the outer layer the cellulose is impregnated with other substances and often forms spines or other protrusions.

   a. Calculate the maximum length of the cell, with and without the spines. [2]

   b. Suggest a function for the spines. [2]

   c. This alga secretes a mucus coat outside its cell wall. Cylindrical bacteria, which are always present, are visible in this mucus. Calculate the length of one of these bacteria. [1]

   d. Suggest benefits of the mucus to the bacterium, and also to the alga. [3]

   e. Discuss whether this alga is one cell or two. [2]

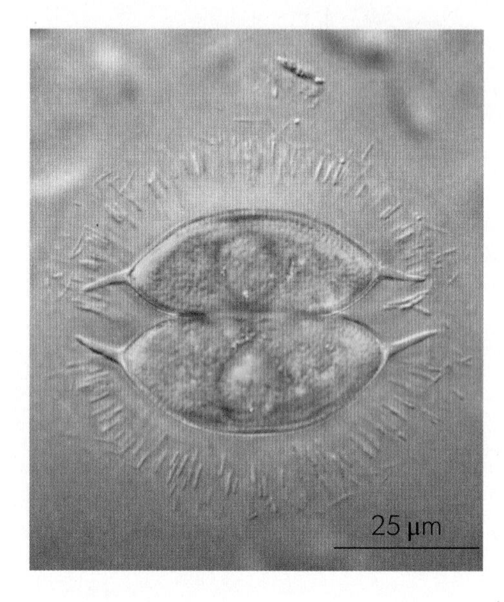

▲ Figure 45 *Staurodesmus convergens*

3. The alga on the left is the desmid *Bambusina brebissonii* and the alga on the right is the desmid *Staurastrum senarium*. Between them is a ciliated protozoan that has engulfed cells of the alga *Chlorella* by endocytosis.

   a. Identify three similarities and two differences between the cells of *B. brebissonii* and *S. senarium* that are visible in the micrograph. [5]

   b. The cilated protozoan has engulfed more than 15 *Chlorella* cells. Calculate the diameter of the largest of these *Chlorella* cells. [2]

   c. Discuss the relative advantages to the ciliated protozoan of digesting or not digesting the *Chlorella* cells. [4]

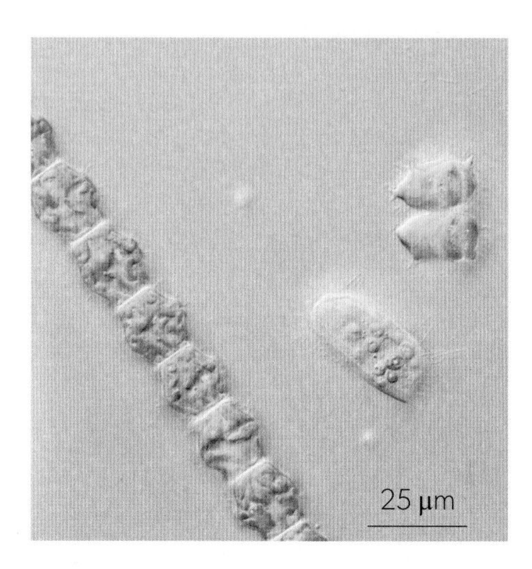

▲ Figure 46 *Bambusina brebissonii* (left), *Staurastrum senarium* (right) and a ciliated protozoan (centre)

Linking questions

1. What explains the use of certain molecular building blocks in all living cells?
   a. Outline the diverse roles of proteins in living cells. (B1.2.12)
   b. Explain how hydrophobicity contributes to compartmentalization in cells. (A2.1.5)
   c. Describe the diverse forms that the genetic material takes in cells. (A2.2.10)
2. What are the features of a compelling theory?
   a. A new multicellular plant is discovered. Predict the features that would be observed in the cells of the organism.
   b. Using the theory of endosymbiosis and the theory of evolution by natural selection, explain the evolution of:
      i. eukaryotic cells (A2.2.12)
      ii. multicellularity. (A2.2.14, A4.1.1)
   c. Using one theory, such as the theory of endosymbiosis or the theory of evolution by natural selection, discuss the extent to which the theory is useful for:
      i. explaining observations (A4.1.1, A2.2.12)
      ii. predicting observations. (A4.1.1, A2.2.12)

A2.3 Viruses

How can viruses exist with so few genes?

Figure 1 shows a human cell infected with influenza (flu) virus. Viruses vary in the total number of genes they have. For example, the influenza virus has just 8 genes while the human HHV-6 virus (Figure 2) has more than 100 genes. What is the minimum number of genes found in any cell? How does this compare to the virus with the largest number of genes? How can viruses endure with so few genes? In what ways are viruses dependent on their hosts? Are there any types of genes which are found in all viruses? Do RNA viruses have genes?

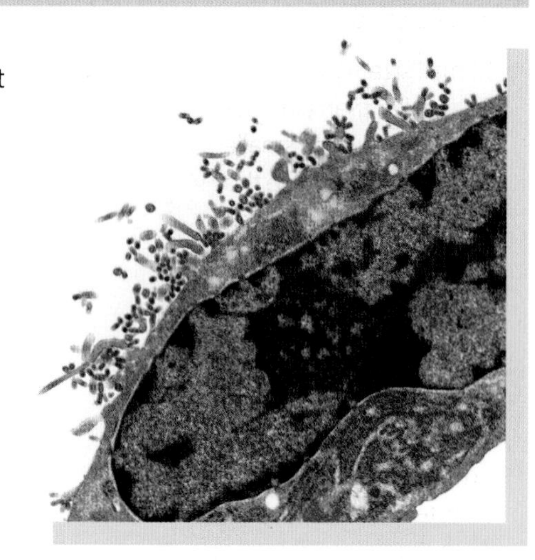

▲ Figure 1 The micrograph shows a human cell infected with influenza (flu) virus

In what ways do viruses vary?

The herpes viruses shown in Figure 2 are about to be taken up by a white blood cell which will become their host. This virus, HHV-6, infects nearly all humans in early childhood resulting in a variety of symptoms including a rash called roseola. The virus is also taken up by white blood cells called T-lymphocytes and B-lymphocytes. What generalizations can be made about all viruses? What are some of the ways that viruses show structural variability? What is the difference between the lytic cycle and the lysogenic cycle?

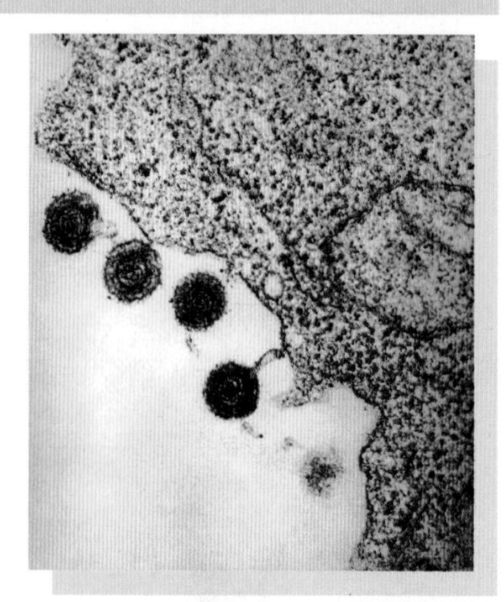

▲ Figure 2 Herpes viruses about to be taken up by a white blood cell

AHL only

A2.3.1 Structural features common to viruses
A2.3.2 Diversity of structure in viruses
A2.3.3 Lytic cycle of a virus
A2.3.4 Lysogenic cycle of a virus
A2.3.5 Evidence for several origins of viruses from other organisms
A2.3.6 Rapid evolution in viruses

A2.3.1 Structural features common to viruses

Viruses are non-cellular agents that infect cells and reproduce inside them. Unlike living organisms, which share common features because they are all descended from a single common ancestor (LUCA), viruses probably have multiple origins, as they share relatively few features. Features that they do have in common are examples of convergent evolution—they developed for functional reasons:

- Small size—Most viruses are between 20 and 300 nanometres in diameter. This is smaller than almost all bacteria and much smaller than plant or animal cells. Viruses must be smaller than their host cells so they can enter them. Viruses are also small because they lack cytoplasm and other structural features.
- Fixed size—Viruses do not grow so they do not increase in size. A virus is assembled inside a host cell, in a similar way to a car being assembled from components—both a virus and a car are their full size as soon as assembly is completed. Many viruses are composed of a fixed number of components, each with a fixed size, so this determines the overall size.
- Nucleic acid as genetic material—All viruses have genes made of DNA or RNA and they use the universal genetic code. This is essential as their proteins are synthesized by the nucleic acid-to-polypeptide translation mechanisms of their host cell.
- Capsid made of protein—Before viruses are released from their host cell, their genetic material is enclosed in a protein coat called the capsid. This is made of repeating protein subunits. A few viruses have only one type of protein in the capsid, but most have several. Self-assembly of the repeating subunits of the capsid gives viruses a symmetrical structure that is strikingly different from the shape of living cells.
- Viruses released from host cells have no cytoplasm and contain no (or very few) enzymes. Even when a virus has infected a host cell, relatively few viral enzymes are produced because viruses rely on the metabolism of the host. The viral enzymes that are produced are required for replication of the virus's genetic material, for infecting host cells or for bursting host cells to release the new viruses.

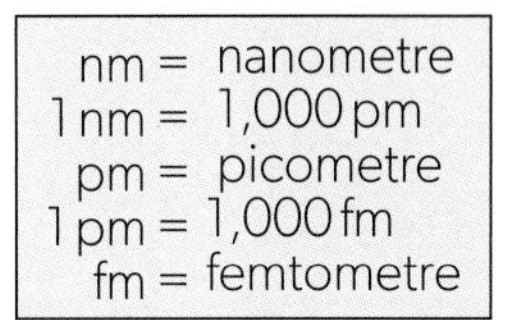

nm = nanometre
1 nm = 1,000 pm
pm = picometre
1 pm = 1,000 fm
fm = femtometre

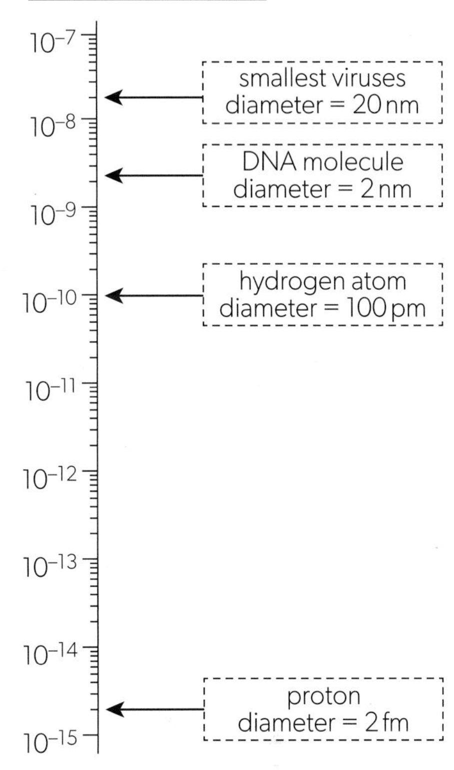

▲ Figure 3 This logarithmic scale shows the relative size of viruses

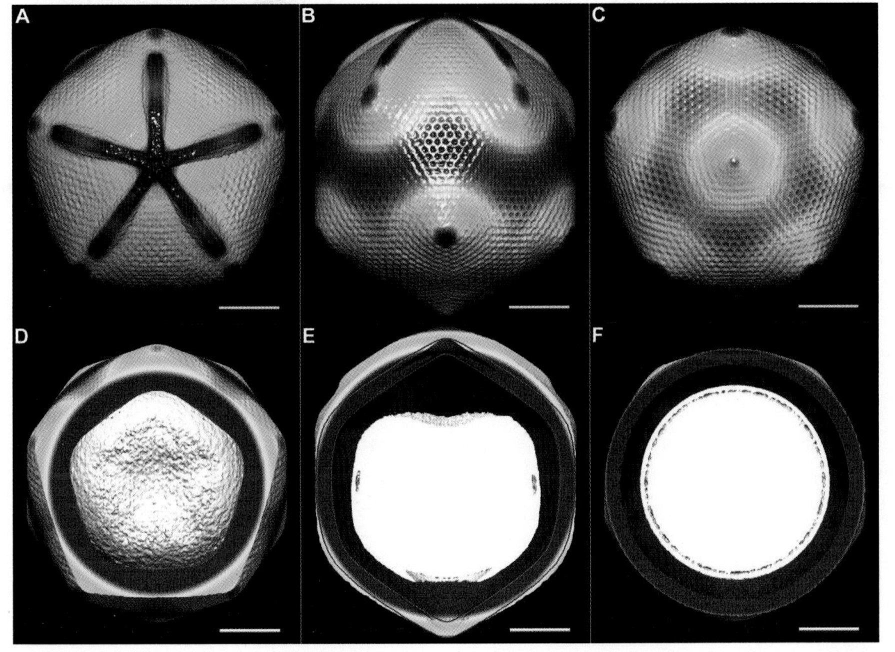

◀ Figure 4 Cryo-electron microscope images of mimivirus, an exceptionally large virus, that uses Amoeba as its host. Colouring indicates distances from the centre of the virus. The grey area (0 – 180 nm from the centre) holds double stranded DNA that is the genetic material of the virus. The rainbow colouring (red to blue =180 to 250 nm) shows the capsid. A distinctive feature of this virus is the starfish shaped vertex on the surface of the capsid

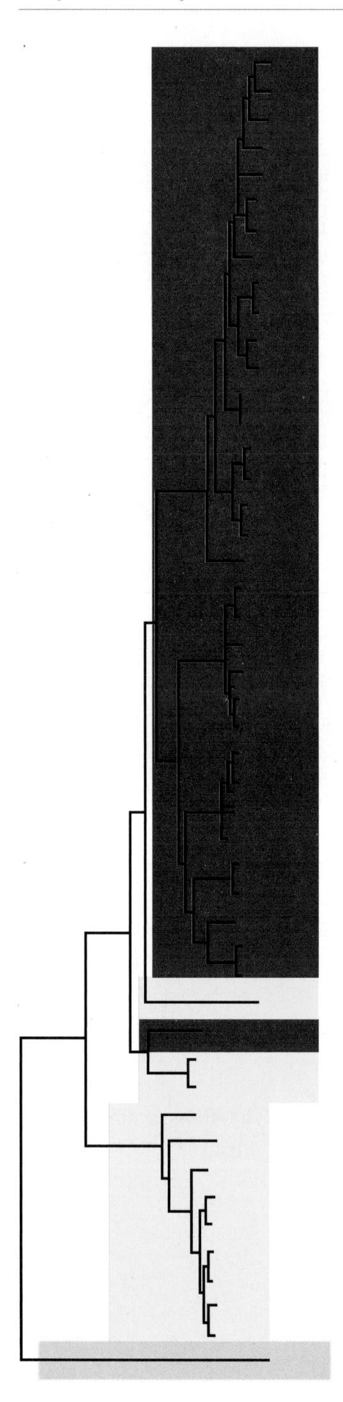

▲ Figure 5 A cladogram showing relationships between coronaviruses, based on base sequences of their RNA genomes. Different colours indicate the class of the host. How easy is it for a virus to change to a different class of host?

Source: Shi, M. et al. Nature 556, 197–202 (2018).

A2.3.2 Diversity of structure in viruses

Viruses are very diverse in shape and structure. No genes occur in all viruses. Based on this observation, scientists have deduced that viruses have multiple evolutionary origins.

1. **Diversity of genetic material**
   Viruses have genes made of either DNA or RNA. When a virus enters a host cell, the DNA or RNA could be single or double stranded. There is considerable variation in length of the nucleic acid molecule and it may be circular with no ends or linear with two ends. There is further variation in how viruses replicate their genetic material and use it during protein synthesis. For example, single-stranded RNA viruses use one of three different methods:
   - positive-sense RNA viruses use their genes directly as messenger RNA
   - negative-sense RNA viruses transcribe their genes to make messenger RNA
   - retroviruses make double-stranded DNA copies of their RNA genes and then transcribe the negative-sense strand of the DNA to produce mRNA.
2. **Enveloped and non-enveloped viruses**
   To be released from their host cell, viruses burst it in a process called lysis. Alternatively, some viruses are released by budding, where they become covered in a membrane. Budding is particularly common in viruses that infect animal cells. The phospholipids in the membrane around the virus are derived from the plasma membrane of the host cell. The proteins, mostly glycoproteins, come from the virus itself. The membrane helps the enveloped virus to make contact with a host cell and infect it.

Other viruses do not become enclosed in a membrane. They are called non-enveloped viruses. Most viruses that infect bacteria or plant cells are non-enveloped. Table 1 summarizes some key properties of three different viruses.

A2.3.3 Lytic cycle of a virus

Bacteriophage lambda has proteins at the tips of its tails which bind to the outer surface of its host, *Escherichia coli (E. coli)*. The DNA of this virus enters the host cell through its tubular tail. The viral DNA has single-stranded ends, which link by base pairing to convert the molecule from a linear to a circular form. Two alternative strategies can then be followed:

- Lysogenic cycle—The viral DNA becomes integrated into the bacterial DNA molecule, so new whole virus particles are not produced. This is described in *Section A2.3.4*.
- Lytic cycle—The virus reproduces and then bursts out of the host cell, killing it. This is illustrated in Figure 6.

Bacteriophage lambda is virulent when it follows the lytic cycle because it destroys its host. It can spread to more and more *E. coli* bacteria but as it kills them it must continue to find new host cells. If lambda or other bacteriophages kill an entire population of bacteria, they are at risk of dying out themselves.

Viruses that infect cells in plants or animals often follow a lytic cycle. As a result, they spread from cell to cell within the host organism. The viral infection becomes increasingly widespread within the body and the effects of the disease become more severe. Usually, an animal host will be able to fight off viruses multiplying by the lytic cycle. For example, humans produce antibodies that destroy all copies of a virus within the body. If a viral infection remains uncontrolled, however, it can become life-threatening for a multicellular host.

Virulence therefore has disadvantages for a virus. The virus may be detected and destroyed by the host, or it may lose its host by killing it. In either case, the virus can only persist if it spreads to another host.

	Bacteriophage lambda	COVID-19	HIV
Type of virus	Bacteriophage (a DNA virus that uses either a bacterium or an archaean as its host)	Coronavirus (an RNA virus with a crown-like shape that uses an animal cell as its host)	Retrovirus (a virus that converts its RNA genome to DNA after infecting a host)
Enveloped or non-enveloped	Non-enveloped	Enveloped	Enveloped
Genetic material	One double-stranded DNA molecule with positive and negative sense strands and 48,502 base pairs. There are 32 genes which code for 29 proteins including 4 enzymes.	One single-stranded positive-sense RNA molecule with 29,903 bases. The 16 genes code for 29 proteins, including 4 structural proteins and 6 enzymes.	Two copies of a single-stranded positive-sense RNA molecule of 9,749 bases. There are 9 genes, coding for 15 viral proteins, including 4 enzymes.
Distinctive features	The virus can follow either a lytic cycle (in which it reproduces and then kills the host cell as it bursts out) or a lysogenic cycle (in which it integrates its DNA and does not kill the host).	COVID-19 caused a pandemic, starting in 2020. It is an example of a zoonosis, because it was passed to humans from another species, probably a bat.	The virus contains the enzyme reverse transcriptase which makes a double-stranded DNA copy of the viral RNA genome. This is then integrated into a host cell chromosome.
Host	*Escherichia coli*— a Gram-negative bacterium	Human cells and possibly cells in other mammals	T-helper cells in the human immune system
	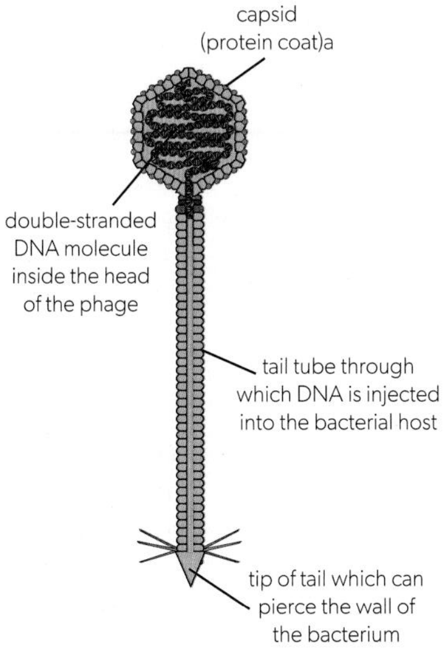 capsid (protein coat)a double-stranded DNA molecule inside the head of the phage tail tube through which DNA is injected into the bacterial host tip of tail which can pierce the wall of the bacterium	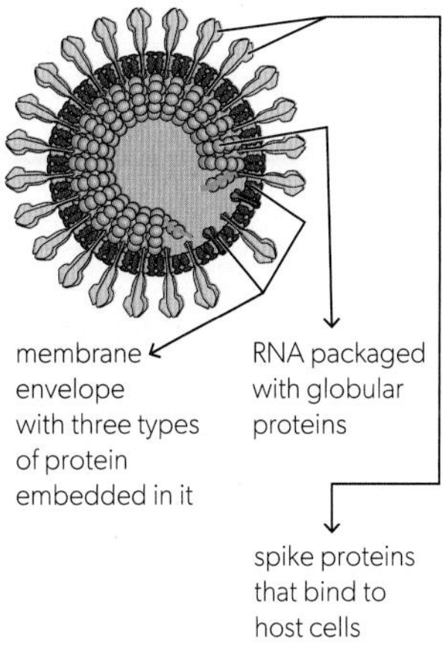 membrane envelope with three types of protein embedded in it RNA packaged with globular proteins spike proteins that bind to host cells	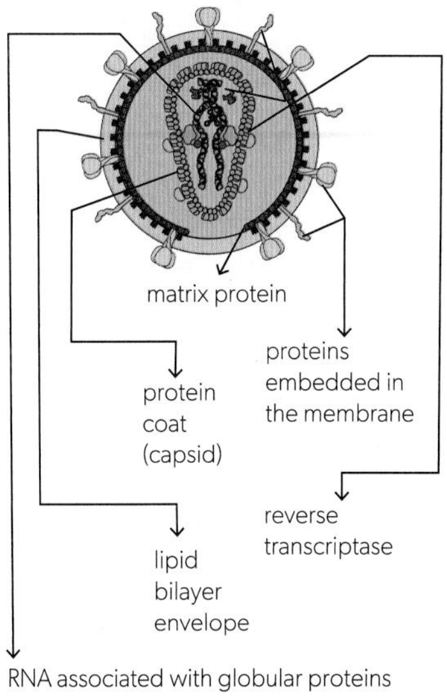 matrix protein protein coat (capsid) proteins embedded in the membrane reverse transcriptase lipid bilayer envelope RNA associated with globular proteins

▲ Table 1

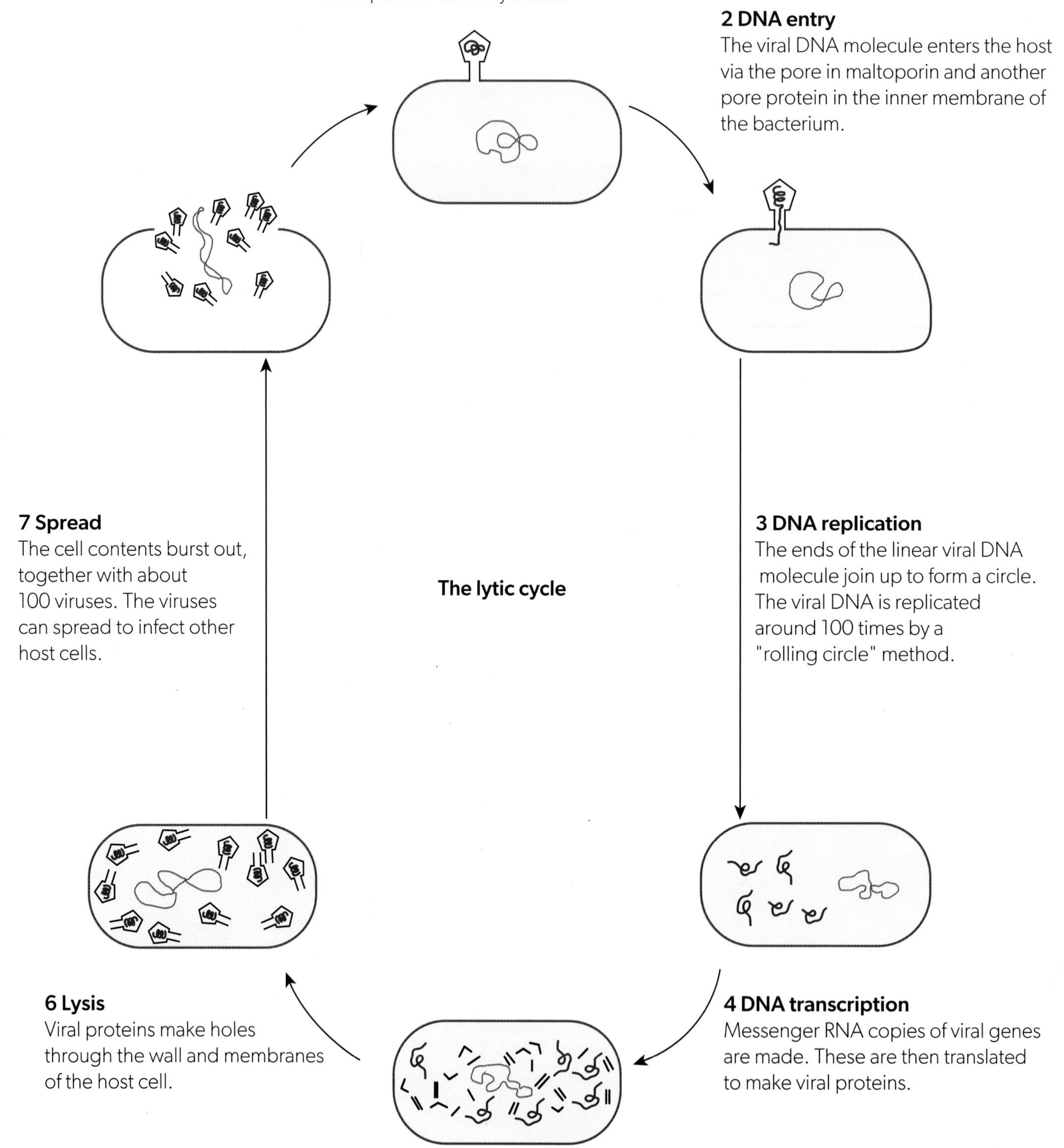

1 Attachment
Proteins in the tip of the tail bind to maltoporin, a protein in the outer membrane of *E. coli* used for absorption of carbohydrates.
2 DNA entry
The viral DNA molecule enters the host via the pore in maltoporin and another pore protein in the inner membrane of the bacterium.
3 DNA replication
The ends of the linear viral DNA molecule join up to form a circle. The viral DNA is replicated around 100 times by a "rolling circle" method.
4 DNA transcription
Messenger RNA copies of viral genes are made. These are then translated to make viral proteins.
6 Lysis
Viral proteins make holes through the wall and membranes of the host cell.
7 Spread
The cell contents burst out, together with about 100 viruses. The viruses can spread to infect other host cells.
The lytic cycle

5 Protein synthesis

Viral proteins are synthesized using host cell ribosomes. Initially, proteins are made for use during DNA replication and other functions while the virus is inside the host. Then large quantities of head and tail proteins are made. These self-assemble to form capsids, with one copy of the viral DNA molecule inside each capsid.

▲ Figure 6 The lytic cycle

A2.3.4 Lysogenic cycle of a virus

The lysogenic cycle, shown in Figure 7, is an alternative to the lytic cycle.

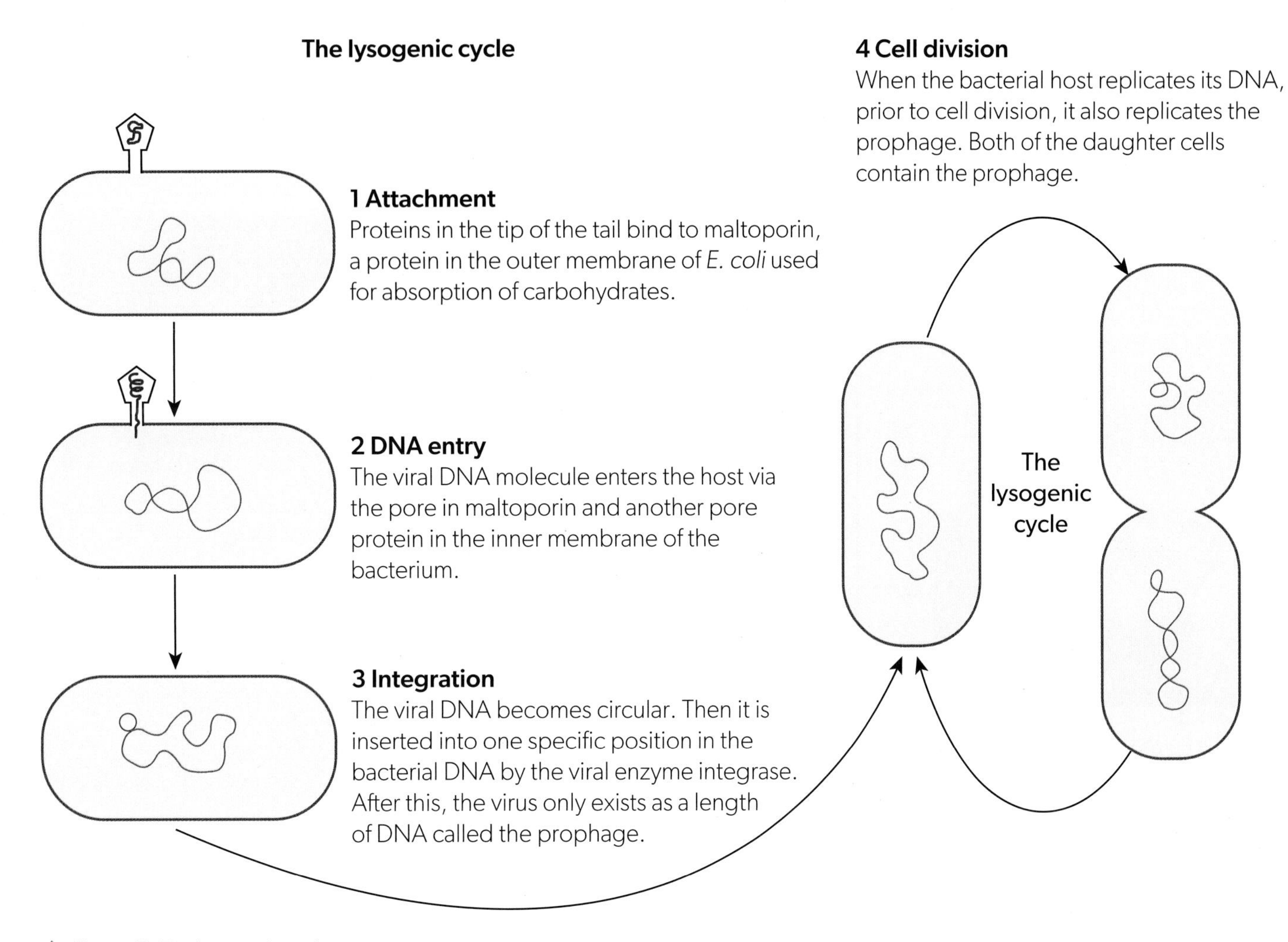

The lysogenic cycle
1 Attachment
Proteins in the tip of the tail bind to maltoporin, a protein in the outer membrane of E. coli used for absorption of carbohydrates.
2 DNA entry
The viral DNA molecule enters the host via the pore in maltoporin and another pore protein in the inner membrane of the bacterium.
3 Integration
The viral DNA becomes circular. Then it is inserted into one specific position in the bacterial DNA by the viral enzyme integrase. After this, the virus only exists as a length of DNA called the prophage.
4 Cell division
When the bacterial host replicates its DNA, prior to cell division, it also replicates the prophage. Both of the daughter cells contain the prophage.
The lysogenic cycle

▲ Figure 7 The lysogenic cycle

While a virus remains in the lysogenic cycle, it is "temperate": it does not kill its host and it causes minimal harm. The virus remains undetectable as a prophage in the bacterial DNA. It is inherited by daughter cells but cannot spread by infecting uninfected cells.

A temperate virus existing as a prophage is called "lysogenic" because it could change to the lytic state and then cause lysis. For this to happen, genes in the prophage must be activated in response to stimuli from inside or outside the bacterial cell.

Temperate viruses can benefit the host cell because their DNA may include genes transferred from a previous host. These genes become integrated into the bacterial DNA along with the viral genes. This increases the genetic diversity of the bacterial host, facilitating evolution.

AHL

ATL Justifying hypotheses: Causes of the switch to the lytic cycle in *Herpes simplex* viruses

In everyday language, a hypothesis is often referred to as an "educated guess". This means there is a reasonable theoretical justification for the hypothesis. Once a hypothesis is generated, it is worded as a testable statement that can be investigated through an experiment. A well-worded hypothesis will suggest the method that should be followed to test it.

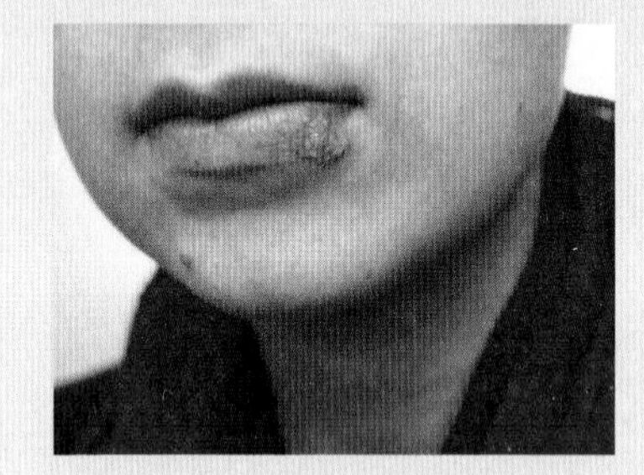

▲ Figure 8 A cold sore

Cold sores are caused by one variant of the *Herpes simplex* virus, known as HSV1. This virus alternates between a dormant lysogenic phase and an active lytic phase; during the lytic phase, it causes painful blisters. Affected individuals often spend a period of time symptom free. What causes the virus to convert to the lytic cycle?

Hypothesis 1: If the host is in poor health, then the virus will convert to the lytic cycle.

Hypothesis 2: If the host is in robust health, then the virus will convert to the lytic cycle.

While viruses are not living, they are subject to selection pressures and the most fit adaptations encoded in the viral DNA will determine successful reproduction. Use the theory of natural selection to justify both of these hypotheses.

Data-based questions: Marine viruses

Water samples were taken from the St Petersburg city pier in Tampa, Florida every two weeks for 13 months. The numbers of bacteria and viruses in the samples were counted. The concentrations of chlorophyll *a* were measured to give an estimate of the abundance of photosynthetic algae. Temperature and salinity of the water samples were also measured. Table 2 shows correlation coefficients between these variables. In the area where samples were taken, there is most rainfall in summer.

	Numbers of viruses	Numbers of bacteria	Chlorophyll *a* concentration	Temperature
Numbers of bacteria	0.561			
Chlorophyll *a* concentration	0.725	0.513		
Temperature	0.649	0.793	0.588	
Salinity	−0.803	−0.518	−0.750	−0.534

Source: Jiang and Paul. 1994. Marine Ecology Progress Series. Vol. 104. Pp 163–172.

▲ Table 2

1. Explain what is indicated by a correlation coefficient of 1.00. [1]
2. Some of the correlation coefficients are positive and some are negative. Explain what is indicated by:
   a. a positive correlation coefficient [1]
   b. a negative correlation coefficient. [1]
3. Numbers of viruses varied between 0.22×10^7 and 3.0×10^7 per cm^3. Suggest a reason for the correlations between the numbers of viruses and:
   a. numbers of bacteria [1]
   b. chlorophyll *a* concentration [1]
   c. salinity. [1]
4. Discuss the difficulties of analysing correlation coefficients. [2]
5. Bacteria from the samples were tested to find out whether they contained prophages. Four out of ten bacteria tested positive. Calculate the percentage occurrence of lysogeny in bacteria in seawater at St Petersburg city pier. [1]

A2.3.5 Evidence for several origins of viruses from other organisms

Viruses are simpler in structure than cells, suggesting the hypothesis that they evolved first. All viruses use the same genetic code, with a few insignificant differences. If they did evolve before cells, the universality of the genetic code implies a single ancestral virus with this code, from which all existing viruses are descended. However, the diversity in structure and genetic constitution of viruses (described in *Section A2.3.2*) suggests multiple origins rather than a single common ancestor.

Viruses are obligate parasites. They need a host cell in which to replicate. An obvious deduction is that cells must have evolved before viruses. All living organisms use essentially the same genetic code, inherited from LUCA. Viruses also use this code. It seems reasonable to deduce that viruses must have evolved from cells. There are two types of hypothesis for the mechanism of evolution.

1. **Progressive hypotheses**
   Viruses are built up in a series of steps by taking and modifying cell components. This fits with the observation that there are virus-like components in some cells, for example retrotransposons.

   Retrotransposons are sequences of nucleotides that occur widely in the genomes of eukaryotes. When a retrotransposon is transcribed to produce RNA and this RNA is translated, several enzymes are produced. These enzymes make more DNA copies of the transposon by reverse transcription of the RNA, then insert these copies into the cell's chromosomes in random positions.

   There are striking similarities between this method of propagating DNA in a eukaryotic cell and the method used by retroviruses such as HIV to integrate their genetic material into a host cell's chromosomes. For retroviruses to have evolved from retrotransposons, capsid proteins would also have had to evolve from host cell proteins.

2. **Regressive hypotheses**
   Viruses develop from cells in a series of steps by loss of cell components. This fits with the observation that both viruses and bacteria show variation in complexity and self-reliance.

   Some viruses are small and simple with few components, for example, the polio virus. Others are much larger and more complex, such as the smallpox virus. Mimivirus is an even larger example, with a diameter of 0.75 micrometres and a genome of 1.2 million base pairs. These large and complex viruses have some enzymes of their own and perform functions that most viruses leave to their host.

   The cells of bacteria are expected to be self-reliant but there are parasitic bacteria which replicate inside a host cell. Some types of bacteria have lost the ability to perform certain metabolic functions. For example, the bacterium *Chlamydia* has a diameter of only 0.6 micrometres and as few as 600 genes. At one time, *Chlamydia* bacteria were thought to be viruses but they are cells with a cell wall and membrane. They are likely to have evolved from an independent organism that became parasitic, entering host cells and reproducing inside them.

   These observations suggest that viruses might have originated from intraparasitic bacteria by loss of more and more life functions, including respiration and protein synthesis.

Viruses may have arisen by various progressive and regressive routes. This would help to explain their diversity. Shared features of viruses could be the result of convergent evolution—they are shared for functional reasons rather than because of ancestry.

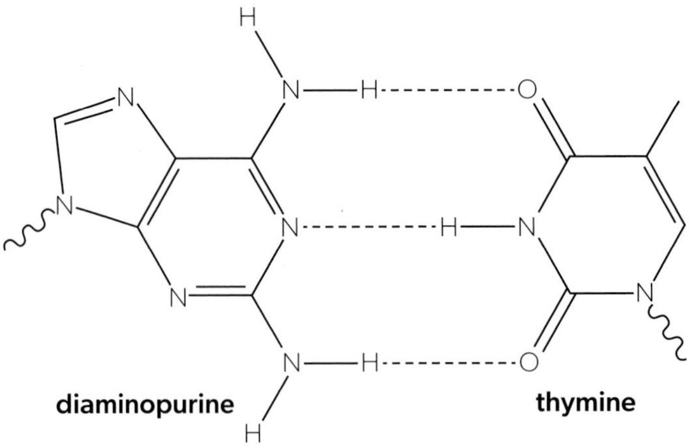

diaminopurine thymine

▲ Figure 9 Scientists have discovered a bacteriophage (S 2-L) that uses the universal genetic code with one difference: it has diaminopurine instead of adenine in its DNA. The letter Z is used for this modified base, so the DNA of the bacteriophage has Z–T base pairs where other organisms would have A–T. This modification makes the DNA more heat-stable and protects it from attack by the host. How does this affect our understanding of the origin of viruses and the genetic code?

A2.3.6 Rapid evolution in viruses

Viruses can show extremely rapid rates of evolution. Even during an infection of one person, a virus can undergo heritable changes—it can evolve. There are three main reasons for this rapidity.

1. Evolutionary change can only happen between one generation and the next, so it is limited by generation time. In humans, the average generation time is about 25 years but in viruses it can be less than an hour.
2. Evolution depends on genetic variation. The ultimate source of this variation is mutation and mutation rates tend to be high in viruses. This is particularly true in RNA viruses such as coronaviruses, which do not perform any checks or correct errors made during replication of their genetic material.
3. Evolution is the result of natural selection acting on variation in a population. The intensity of natural selection on viruses tends to be high. The host organism, whether a bacterium, plant or animal, has mechanisms for detecting and destroying invading viruses. For example, antibodies in humans target antigens such as proteins on the surface of the virus. If the antigen changes, the antibodies no longer recognize it. Viruses with a new variant of protein in the capsid or in the enveloping membrane can evade the immune system and multiply, whereas those with the previous form are destroyed. As a consequence, natural selection is powerful and this encourages rapid evolution.

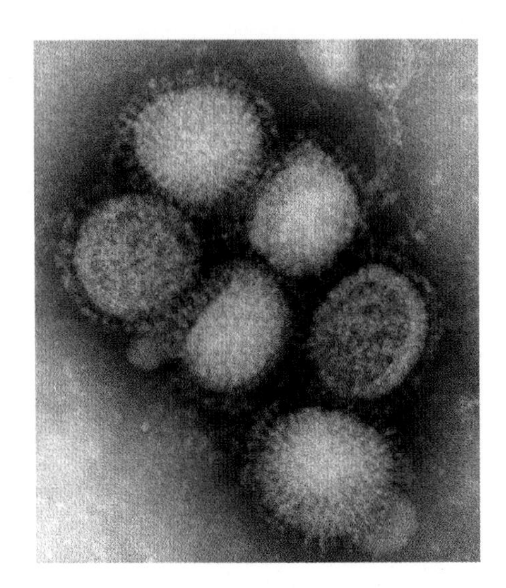

▲ Figure 10 H1N1 influenza viruses with haemagglutinin (H1) and neuraminidase (N1) proteins visible in the enveloping membrane

Two examples of rapid evolution are the influenza virus and HIV.

The influenza virus

Influenza is caused by an enveloped virus that uses negative-sense single-stranded RNA as its genetic material. It replicates its genetic material using RNA replicase which, unlike DNA polymerase, does not proofread or correct errors. This leads to a high mutation rate. Instead of a single RNA molecule, the genome consists of eight separate molecules. Because of this, a new strain of the virus can appear if a host cell is invaded by two different strains of the virus and some RNA molecules from each strain are combined. The influenza virus can also be transmitted between species, particularly between birds and humans. This is another cause of new strains appearing frequently.

Two proteins in the enveloping membrane of the influenza virus act as antigens: haemagglutinin is used to bind to a host cell, and neuraminidase helps with release from the host cell. These proteins can change and be put together in new combinations, creating novel strains of the virus that have the potential to cause a pandemic. Strains are referred to by the types of combination of these proteins. For example, Spanish flu in 1918 was caused by H1N1 and Hong Kong flu in 1968 was caused by H3N2. Rapid evolution of the influenza virus explains how a person can contract influenza repeatedly and also why protection depends on vaccination every year. Each vaccine contains several strains of influenza virus.

The HIV virus

HIV is a retrovirus that uses reverse transcriptase to convert its single-stranded RNA genome to DNA. This enzyme does not proofread or correct errors (unlike DNA polymerase), leading to many mutations. Mutations are also caused by cytidine deaminase, an enzyme made by the host that converts cytosine to uracil. These two factors together give HIV the highest known mutation rate of any virus. Even within a person infected by one strain of HIV, mutations will produce many genetically different strains. When a host cell is invaded by two of more different strains, the viral genes can combine leading to even more diversity.

Most of the mutations that occur in HIV are harmful to the virus, so the action of cytidine deaminase may be protective to the host. Even so, the rapid generation of new strains within a person helps the virus to evade the immune system. As a result, most infections are chronic rather than curable. HIV has a protein on its surface that it uses to bind to and enter a host cell. Mutations in the *env* gene that codes for this protein allow HIV to evolve to use different cell types in the human body as hosts. HIV can also evolve to become resistant to the antiretroviral drugs used to treat patients infected with HIV, so a combination of two or more drugs is necessary.

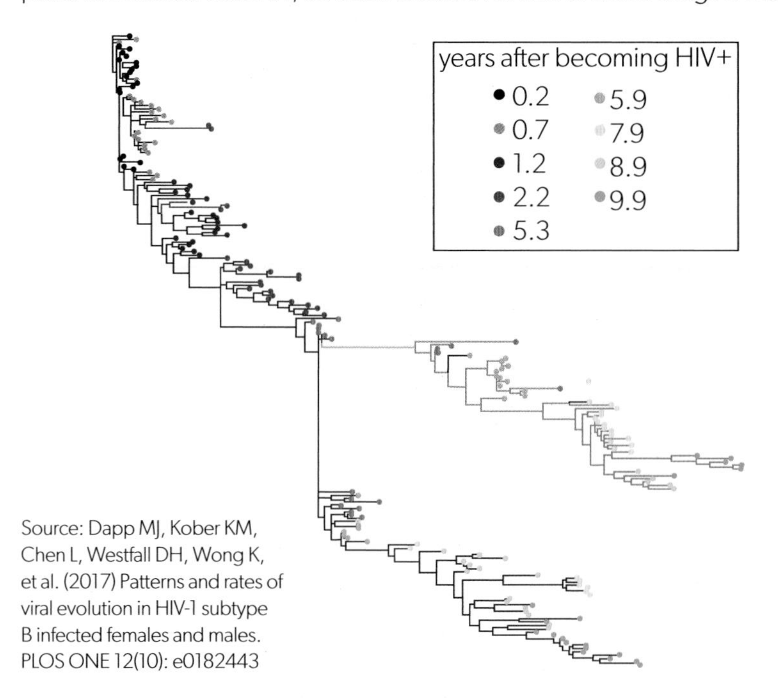

Source: Dapp MJ, Kober KM, Chen L, Westfall DH, Wong K, et al. (2017) Patterns and rates of viral evolution in HIV-1 subtype B infected females and males. PLOS ONE 12(10): e0182443

◀ Figure 11 This tree diagram shows how the *env* gene evolved in one patient over the 10-year period after they became HIV-positive. Each dot represents a new version of the gene, with the colour showing when it was first identified. A change in colour in the branches on the tree diagram shows that the *env* protein would bind to a different host cell protein

Data-based questions: Progression in HIV infection

HIV targets CD4 T-cells in humans. These cells are part of the immune system used to fight infectious disease. When the level of CD4 cells falls below 200 cells mm^{-3} of blood, the infected individual begins to display a number of relatively rare opportunistic infections. At this point, an HIV-infected individual is said to have AIDS (acquired immunodeficiency syndrome). Individuals vary in their response to HIV infection. The four graphs in Figure 12 show the CD4 concentration (thick curves with black squares) and level of HIV in the blood (thin curves) for patients typical of four different types of progression.

1. Describe the changes in CD4 T-cell numbers in a patient with typical progression of the infection. [3]
2. Compare and contrast the levels of virus found in the blood in typical progressors and long-term survivors. [3]
3. Determine the length of time it takes for AIDS to develop in:
   a. typical progressors [1]
   b. rapid progressors. [1]
4. Suggest two reasons for the differences in the progress of the disease in different individuals. [2]

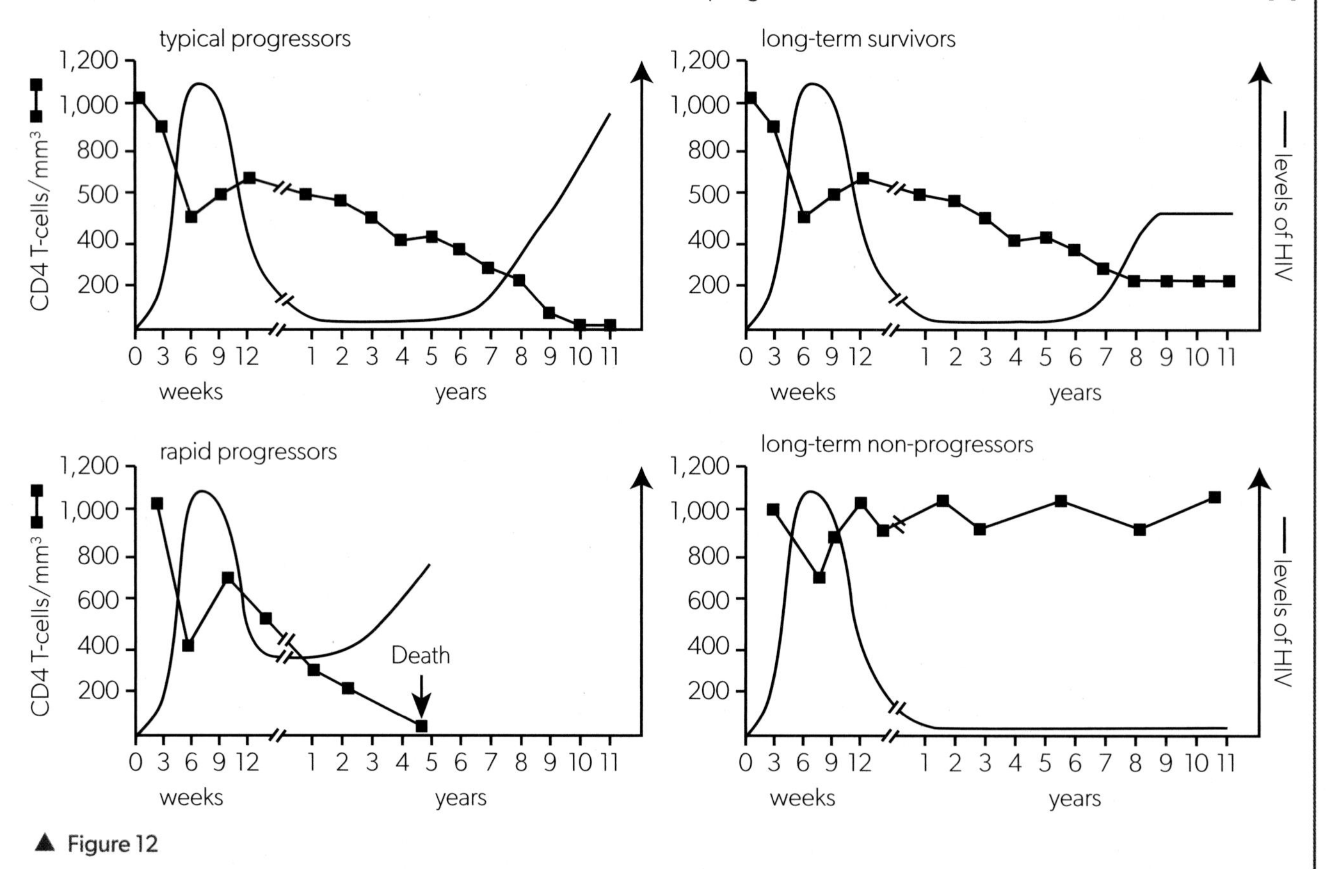

▲ Figure 12

Data-based questions: COVID-19

Figure 13 shows results of an investigation into the origin of COVID-19. This investigation was carried out during the early stages of the pandemic caused by this virus. The graph shows how similar COVID-19 is to five other coronaviruses. The chart above shows how the genome of COVID-19 is organized. Genes S, E, M and N code for the four structural proteins: spike, membrane, envelope and nucleocapsid. Other regions, called open reading frames (ORF), have numbers ranging from 1a to 8 and contain varying numbers of genes.

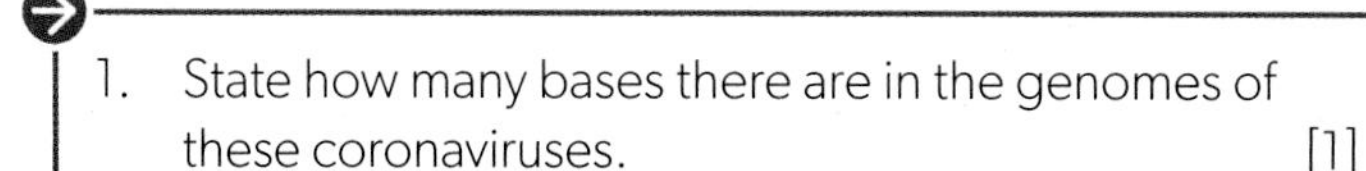

1. State how many bases there are in the genomes of these coronaviruses. [1]
2. Predict, with a reason, which part of the coronavirus genome contains the most genes. [2]
3. Deduce, with a reason, which of the other coronaviruses COVID-19 is most closely related to. [2]
4. Compare and contrast the genome of Bat SARSr-CoV HKU3-1 with the genomes of the other coronaviruses, including COVID-19. [3]
5. a. Deduce which part of the COVID-19 genome is least similar to that of the other viruses. [1]

   b. Suggest a reason for this part of the genomes varying the most. [1]

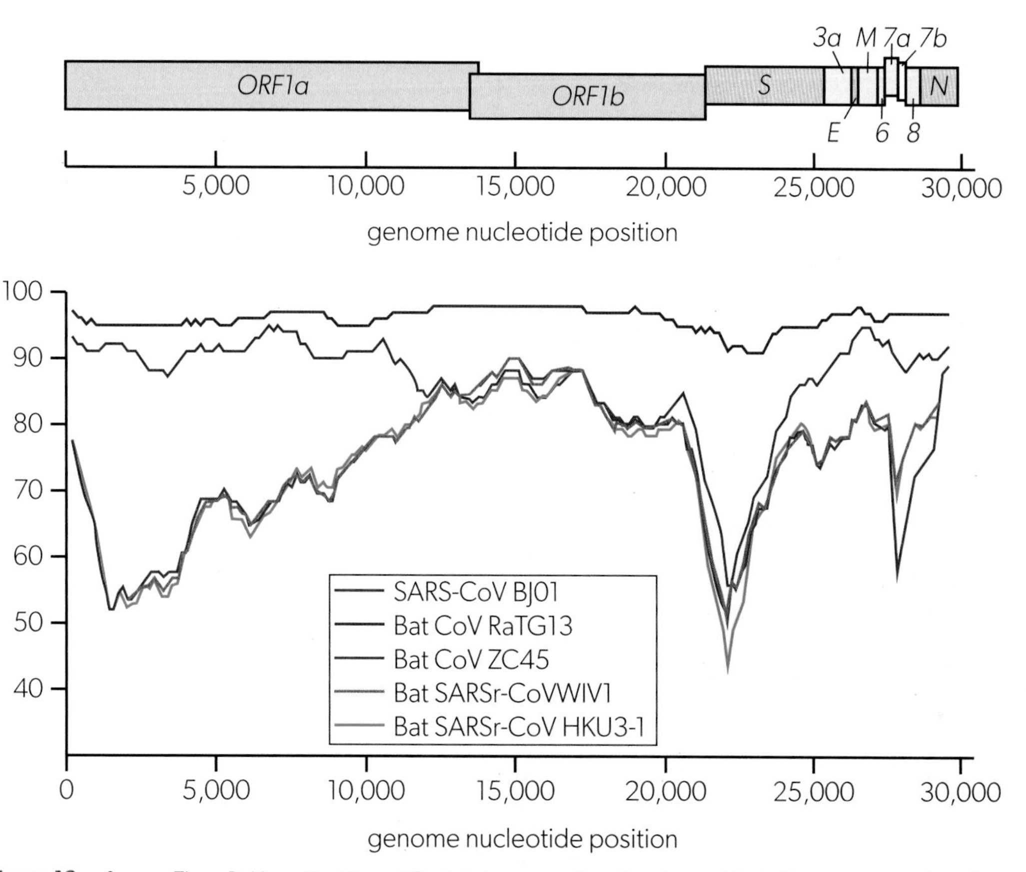

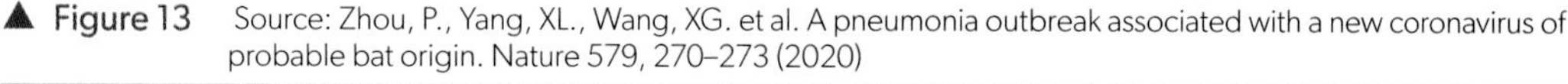

▲ **Figure 13** Source: Zhou, P., Yang, XL., Wang, XG. et al. A pneumonia outbreak associated with a new coronavirus of probable bat origin. Nature 579, 270–273 (2020)

Linking questions

1. What mechanisms contribute to convergent evolution?

   a. Using an example, explain the rapid evolution of viruses. (A2.3.5)

   b. Outline the concept of analogous structures. (A3.2.8)

   c. Explain what is meant by a selection pressure. (D4.1)

2. To what extent is the history of life characterized by increasing complexity or simplicity?

   a. Compare and contrast the structure of typical prokaryotic and eukaryotic cells. (A2.2.5 and A2.2.6)

   b. Outline the theory of endosymbiosis. (A2.2.12)

   c. Discuss the evidence for multiple origins of viruses. (A2.3.5)

TOK

Are some things unknowable?

In some cases, scientists have to struggle with hypotheses that are difficult to test. Abiogenesis is the process by which life arose from non-life. It is impossible for researchers to replicate the exact conditions on prebiotic Earth, because they are not fully known and the first protocells did not fossilize. In the early 1950s, Stanley Miller demonstrated that it was possible to form amino acids from simple inorganic precursors. However, in the 70 or so years since then, scientists have been unable to create a simple life form from non-living precursors—despite significant efforts.

To know something *a priori* is to know it through reason rather than by observing it. Through reason, scientists agree that abiogenesis occurred and that it involved the emergence of self-replicating molecules, compartmentalization, catalysis and polymerization. Researchers have been able to achieve all of these steps separately under laboratory conditions.

For example, researchers created solutions of nitrogenous bases, then dried the mixture on silicon wafers. They placed these samples in a simulation chamber (shown in Figure 1) and exposed them to wet-dry, day-night and seasonal cycles, as well as to moisture, high temperature, oxidizing environments, high levels of radiation and other conditions that are thought to have been present on the prebiotic Earth. The purpose of these experiments is to investigate the emergence of self-catalytic RNA molecules.

To know something *a posteriori* is to know it because it has

◀ Figure 1 The planetary simulator at McMaster University in Hamilton

◀ Figure 2 Sugars have been detected on two different meteorites: NWA 801 and the Murchison meteorite

been observed directly. It is not possible to know the biochemical features of life on other planets *a posteriori*. *A priori*, we know that life on other planets is likely to be based on carbon, associated with water and cellular. This reasoning is based on the following key factors:

- Molecular diversity is essential for life's functions and for the process of evolution, and no other element can form as many different compounds or types of structure as carbon.
- The subunits of the four major categories of biological molecule—including amino acids, nucleobases, the components of lipids, and sugars—have all been found in carbon-rich meteorites.
- No solvent dissolves a greater range of molecules than water. In addition, water is found throughout the universe. When planets form around stars, there is a tendency for them to contain large volumes of condensed water, and water exists as a liquid over a relatively large temperature range.
- The compartmentalization that cells achieve is essential to allow a living organism to maintain the conditions for life in chemically and physically diverse environments. The interface between the compartment and the surrounding environment would need to be semi-permeable to allow for exchange of waste and raw materials as well as communication.

The strength of a theory comes from the observations it can explain and the predictions it can support. The theory of endosymbiosis—used to account for the evolution of eukaryotic cells—accounts for a wide range of observations. The theory that mitochondria originated as intracellular mutualistic prokaryotes is supported by the observation of intracellular parasitic bacteria such as *Rickettsia*, the cause of Rocky Mountain spotted fever. Within mitochondria, prokaryotic type ribosomes, a prokaryotic type single circular chromosome and double membranes all provide empirical evidence for the theory of endosymbiosis. Since the original event occurred millions of years ago, the phenomenon is not directly observable. However, because the theory predicts and explains the observations, we hold it to be a pragmatic truth—one that "works".

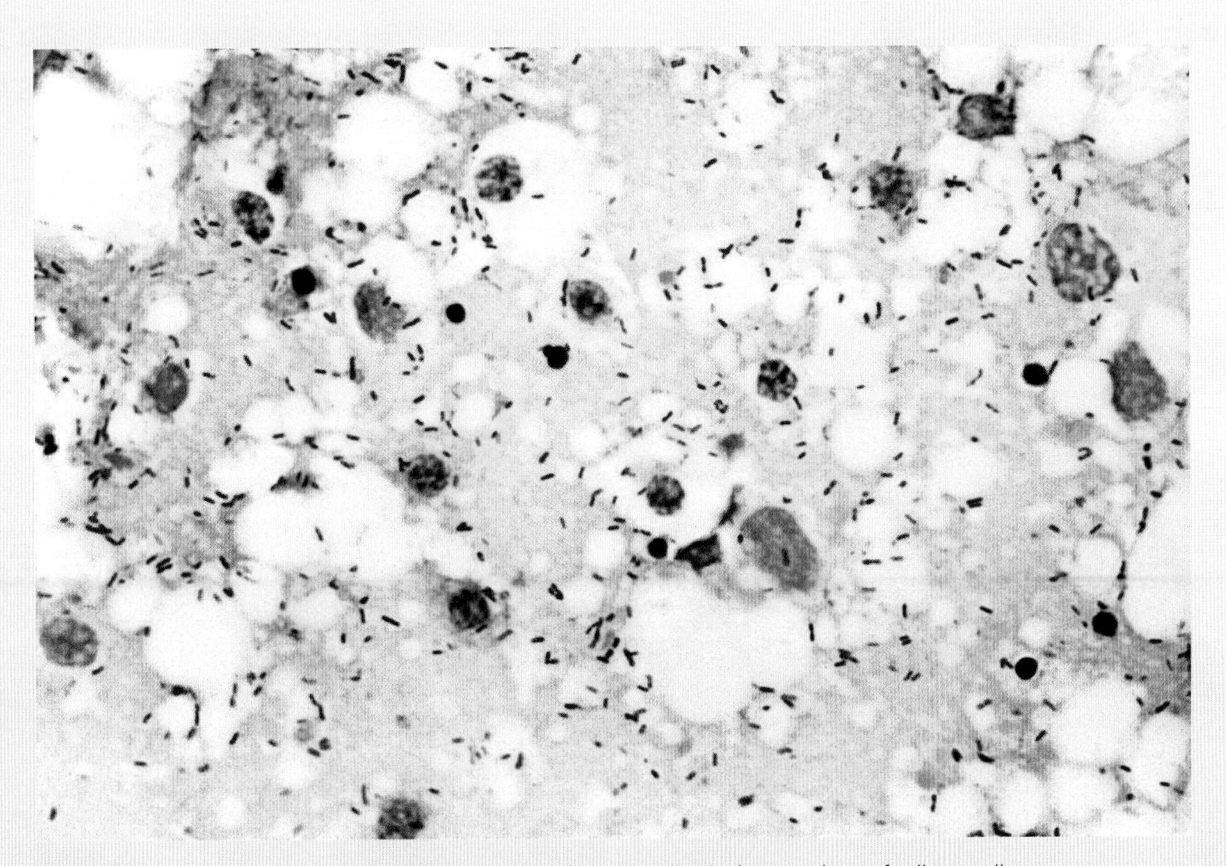

▲ Figure 3 The red cells in this micrograph are intracellular parasites in the cytoplasm of yolk sac cells

End of chapter questions

1. Figure 1 represents a cell from a multicellular organism.
   a. Identify, with a reason, whether the cell is:
      i. prokaryotic or eukaryotic [1]
      ii. part of a root tip or a finger tip. [1]
   b. The magnification of the drawing is 2,500×.
      i. Calculate the actual size of the cell. [2]
      ii. Calculate how long a 5 μm scale bar should be if it is added to the drawing. [1]

▲ Figure 1

2. The electron micrograph in Figure 2 shows a section through a single *Chlamydomonas reinhardtii* green alga. *C. reinhardtii* is a unicellular (single-celled) organism used as a model system in genetics and cellular motion studies.
   a. Eight organelles are shown. Deduce the identity of the organelles labelled C, D, G and H. [4]
   b. Organelle A is an eyespot. This is an organelle that aids in the detection of light. When exposed to light, *Chlamydomonas reinhardtii* moves towards the light. Suggest the adaptive advantage of this behaviour. [2]
   c. Organelle B is a contractile vacuole. Conduct research and state the function of a contractile vacuole. [1]
   d. Structure E is called a pyrenoid. Figure 3 shows the growth rates of *C. reinhardtii* normal cells (purple) and mutant cells that lack a pyrenoid (green), in air with 5% CO_2 and normal air with 0.04% CO_2.
      i. Compare and contrast the growth rates of the normal cells and the mutant cells at low CO_2 concentrations. [2]
      ii. State the reason photosynthetic organisms require CO_2. [1]

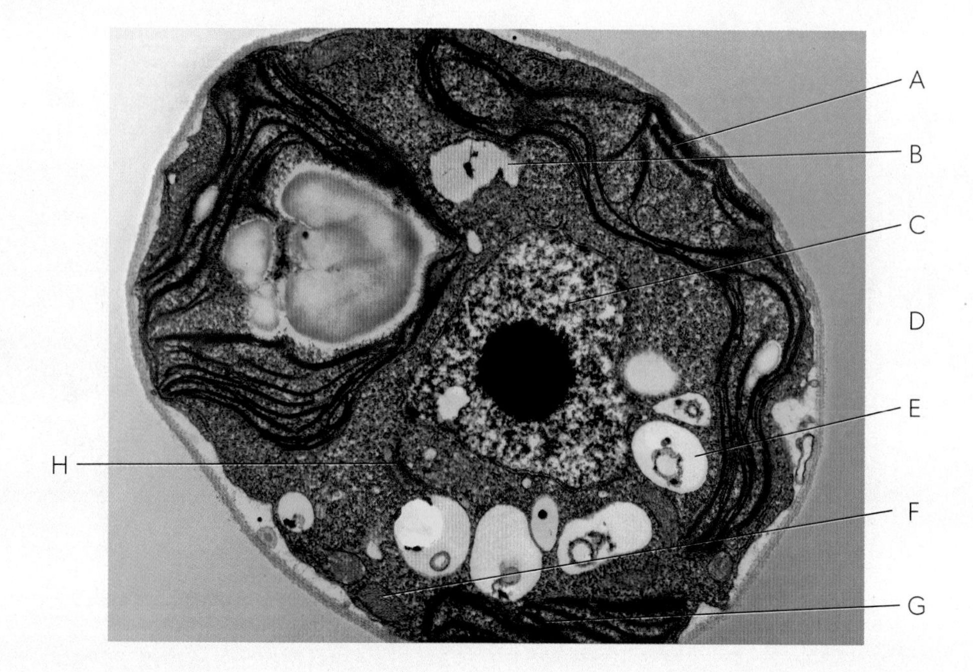

▲ Figure 2

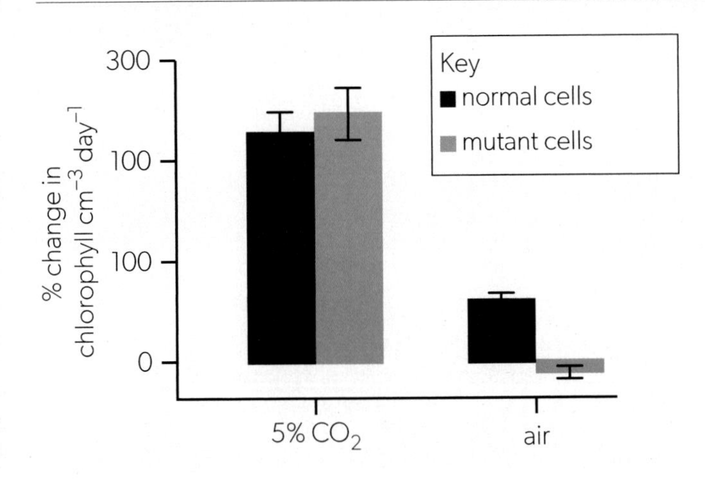

▲ **Figure 3** Source: Caspari, O. et al. Pyrenoid loss in Chlamydomonas reinhardtii causes limitations in CO2 supply, but not thylakoid operating efficiency, Journal of Experimental Botany, Volume 68, Issue 14, 8 September 2017, Pages 3903–3913

iii. Suggest a possible function for pyrenoids based on the data. [2]

e. Structure F in Figure 2 is a starch granule. Research and explain the reasons for storage of carbohydrate in photosynthetic cells as starch rather than as sugar. [2]

f. Discuss what is signified by the error bars in Figure 3. [3]

g. State the dependent and independent variables in this example. [3]

3. The microscope image in Figure 4 shows a rotifer (centre bottom) and a filament of *Spirogyra* (right). The longer numbered ticks on the scale are 122 μm apart.

a. The rotifer is multicellular, being composed of about 1,000 cells. Discuss the advantages and disadvantages of being multicellular. [4]

b. Outline two quantitative and two qualitative observations that can be made from the micrograph. [4]

c. Estimate the length of the main body of the rotifer [1]

d. Distinguish between the size of the rotifer cells and the size of the *Spirogyra* cells. [2]

e. Deduce a possible combination of ocular and objective lens that were used to obtain this field of view. [1]

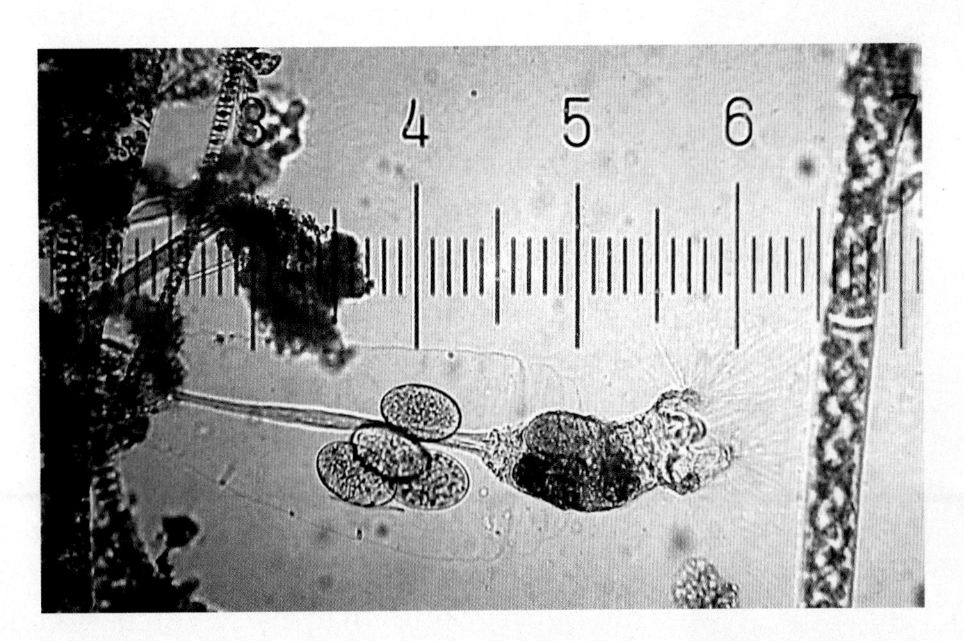

▲ **Figure 4**

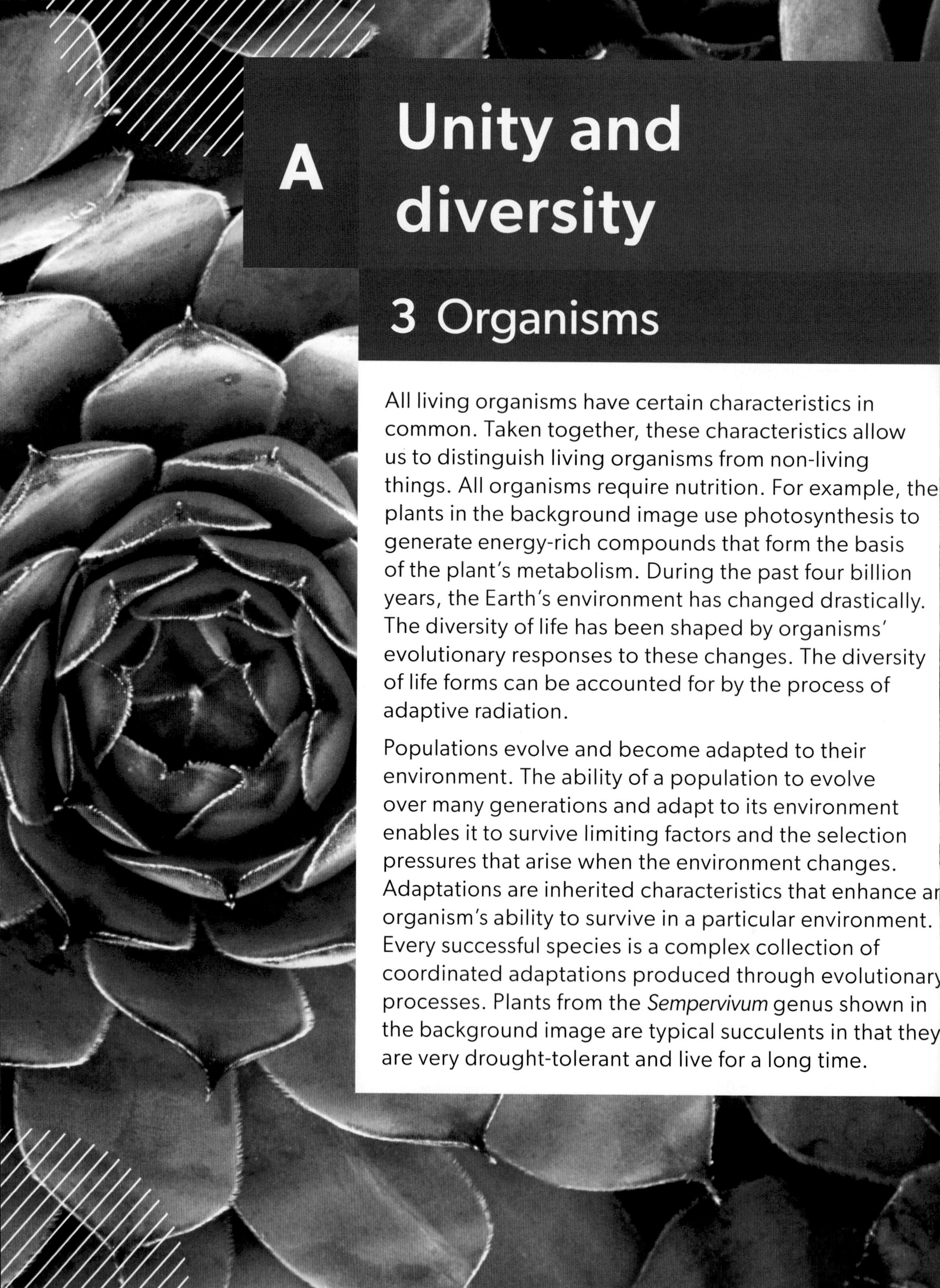

A Unity and diversity

3 Organisms

All living organisms have certain characteristics in common. Taken together, these characteristics allow us to distinguish living organisms from non-living things. All organisms require nutrition. For example, the plants in the background image use photosynthesis to generate energy-rich compounds that form the basis of the plant's metabolism. During the past four billion years, the Earth's environment has changed drastically. The diversity of life has been shaped by organisms' evolutionary responses to these changes. The diversity of life forms can be accounted for by the process of adaptive radiation.

Populations evolve and become adapted to their environment. The ability of a population to evolve over many generations and adapt to its environment enables it to survive limiting factors and the selection pressures that arise when the environment changes. Adaptations are inherited characteristics that enhance an organism's ability to survive in a particular environment. Every successful species is a complex collection of coordinated adaptations produced through evolutionary processes. Plants from the *Sempervivum* genus shown in the background image are typical succulents in that they are very drought-tolerant and live for a long time.

A3.1 Diversity of organisms

What is a species?

Biologists define a species as a group of organisms with shared traits that interbreed in the wild. How does this definition work for organisms that reproduce asexually? What other challenges exist with this definition? In 1859, Charles Darwin wrote, "No one definition has satisfied all naturalists; yet every naturalist knows vaguely what he means when he speaks of a species". What are the reasons that establishing a definition of a species is so difficult? What classification systems did early naturalists use? Consider the two jaguars in Figure 1. To what extent is it surprising that they are considered to be the same species?

▲ Figure 1 The light morph (left) and the "melanistic" or dark morph (right) of the jaguar (*Panthera onca*) interbreed in the wild

What patterns are seen in the diversity of genomes within and between species?

The genome is the whole of the genetic information of an organism; that is, the total amount of DNA. In what ways do genomes vary across the kingdoms of life in terms of structure, composition, association with proteins, location, size, number of chromosomes? In what ways does the genome within a species vary? The Red viscacha rat is one of the few identified polyploid animals. It has the highest chromosome number of any mammal, 102. Its closest living relative is *Octomys mimax*, the Andean viscacha rat (right), which has 56 chromosomes. Why is this condition more likely to be found in plants?

▲ Figure 2 The red viscacha (*Tympanoctomys barrerae*, left) and *Octomys mimax*, the Andean viscacha-rat of the same family (right)

SL and HL	AHL only
A3.1.1 Variation between organisms as a defining feature of life A3.1.2 Species as groups of organisms with shared traits A3.1.3 Binomial system for naming organisms A3.1.4 Biological species concept A3.1.5 Difficulties distinguishing between populations and species due to divergence of non-interbreeding populations during speciation A3.1.6 Diversity in chromosome numbers of plant and animal species A3.1.7 Karyotyping and karyograms A3.1.8 Unity and diversity of genomes within species A3.1.9 Diversity of eukaryote genomes A3.1.10 Comparison of genome sizes A3.1.11 Current and potential future uses of whole genome sequencing	A3.1.12 Difficulties in applying the biological species concept to asexually reproducing species and to bacteria that have horizontal gene transfer A3.1.13 Chromosome number as a shared trait within a species A3.1.14 Engagement with local plant or animal species to develop a dichotomous key A3.1.15 Identification of species from environmental DNA in a habitat using barcodes

▲ Figure 3 Even monozygotic twins show some differences at birth and accumulate more as they grow older

A3.1.1 Variation between organisms as a defining feature of life

An organism is an individual plant, animal, bacterium, or any other living thing. The variety of organisms alive today is immense. Consider the differences between humans, trees growing taller than 100 metres, fungi that consist of a network of narrow threads growing through the soil and brightly coloured bacteria inhabiting volcanic pools at temperatures above 80°C and pHs below 2. Even chimpanzees—animals to which we are closely related—are different from us in many ways.

There is less variation among the members of a single species, but there are still differences between all individuals. There is least variation when two individuals are genetically identical. In humans, monozygotic twins are formed when a zygote or early-stage embryo divides and develops into two individuals. Such twins start out with the same genes but even they acquire differences through mutations and because the environment in which they develop is never identical.

The diversity of organisms adds to the richness of the natural world and helps to make biology such a fascinating subject. Variation is also essential for the future of life because evolution by natural selection could not happen without it.

A3.1.2 Species as groups of organisms with shared traits

If organisms in an area are studied, it soon becomes obvious that each individual is a member of a group with recognizable traits or characteristics. These groups of organisms are often given a name in the local language, especially if they are used by people or have an impact in other ways. For example, when Māoris arrived in New Zealand about 800 years ago, they found tree ferns growing in the forests and used them to build the walls of their houses. They recognized seven different types of tree fern, which they named whekī, kuripaka, tuokura, mamuka, punui, ponga and kātote.

▲ Figure 4 The Māori name for these New Zealand tree ferns is ponga. The scientific name is *Alsophila dealbata*. In addition to the seven species of tree fern recognized by Māoris in New Zealand, biologists have described three more: *Alsophila colensoi*, *Alsophila milnei* and *Alsophila kermadecensis*

From the 17th century onwards, biologists used the term "species" for a group of organisms with shared traits. Biologists have been naming and classifying species ever since. Carl Linnaeus, who worked in the 18th century, was a pioneer of this research. Linnaeus and other biologists of his time described the outer form and inner structure of typical members of a species. This is known as morphology. The idea of a species as a group of organisms that share a particular outer form and inner structure is the morphological concept of a species.

If asked about the origins of species, Linnaeus and his contemporaries would probably have said that they were the work of a creator. They would have thought that each species was created from nothing and remained unchanged after its creation. When describing the morphology of species, early biologists believed they were looking at evidence of a creator's work.

A3.1.3 Binomial system for naming organisms

The international system that biologists use for naming species is called the binomial system. Each species name consists of two words, for example, *Linnaea borealis*. The first name is the genus name. A genus is a group of species that have similar traits. The second name is the species or specific name.

There are various rules about binomial nomenclature:

- The genus name begins with an uppercase (capital) letter.
- The species name begins with a lowercase (small) letter.
- In typed or printed text, a binomial is shown in italics.
- After a binomial or genus name has been used once in a piece of text, it can be abbreviated to the initial letter of the genus name with the full species name, for example, *L. borealis*.

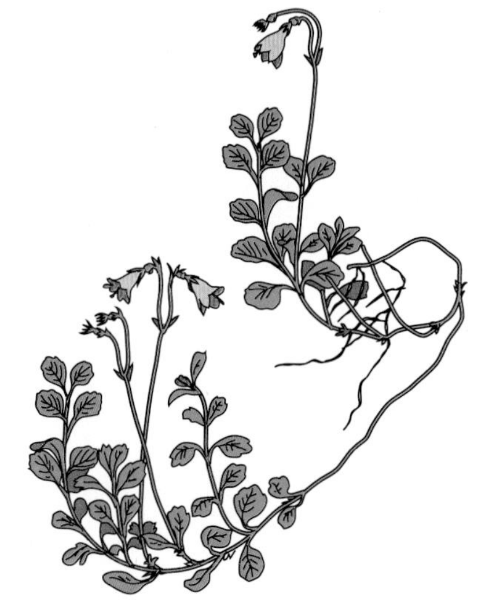

▲ Figure 5 Binomials are often chosen to honour a biologist, or to describe a feature of the organism. *Linnaea borealis* is a small woodland plant that was named in honour of Carl Linnaeus, the Swedish biologist who introduced the binomial system and named many plants and animals using it

A3.1.4 Biological species concept

According to the morphological species concept, a species is an unchanging group of organisms with clear differences in external form and internal structure between it and other species. However, this does not fit with the concept of evolution by natural selection proposed by Charles Darwin in 1857. Biologists have looked for a new concept to describe species, but it has proved extremely difficult to find a definition that fits all contexts. So far, at least 30 different definitions have been suggested!

The biological species concept defines a species as a group of organisms that can successfully interbreed and produce fertile offspring. This concept explains how a group of individuals can exist as a coherent unit—the members of a species interbreed and therefore share genes in a gene pool.

The biological species concept works well with some groups of organisms. For example, the genus *Allium* contains hundreds of species, including onion and garlic, but few interspecific hybrids have been reported in natural habitats and these hybrids are usually sterile. The garden variety "Globemaster" was deliberately bred by crossing *Allium christophii* with *Allium macleanii* and is sterile. Similarly, there are more than 600 species of conifer and interbreeding between these species is very unusual. Where interspecific conifer hybrids do occur, they are usually sterile. This is partly because many conifers have no close relatives, for example, *Ginkgo biloba*. In conifer genera where speciation is occurring rapidly, such as junipers and pines, there is some interspecific hybridization and it is less easy to identify species.

▲ Figure 6 *Allium christophii* (left), *Allium* Globemaster (centre) and *Allium macleanii* (right). Globemaster is a hybrid of *A. christophii* and *A. macleanii*

In other groups of plants and animals, the biological species concept is very difficult to apply, due to geographical separation and gradual divergence. This is described in *Section A3.1.5*. Further difficulties arise when migration brings closely related but apparently distinct species together.

According to the biological species concept, hybridization of two species suggests that those species are *not* distinct. For example, captive lions and tigers have sometimes hybridized, producing offspring known as ligers (male lion × female tiger) or tigons (male tiger × female lion). Male ligers and tigons are infertile but female hybrids are sometimes fertile. A rigorous interpretation of the biological species definition would therefore consider lions and tigers to be the same species, but this is not acceptable to biologists or the wider public.

▶ Figure 7 Polar bears (*Ursus maritimus*) and grizzly bears (*Ursus arctos horribilis*) are usually geographically separated but grizzly bears are spreading north. If polar bears and grizzly bears meet, they can mate and produce fertile offspring. The photo shows such a hybrid

A3.1.5 Difficulties distinguishing between populations and species due to divergence of non-interbreeding populations during speciation

A population is a group of organisms of the same species, living in the same area, at the same time. If two populations live in different areas, they are unlikely to interbreed with each other. This does not necessarily mean that they are different species. If they are physically and genetically similar, both populations are part of the same species.

However, if two populations of a species do not interbreed, they can diverge. Recognizable physical differences may develop as the populations become genetically more different. If differences continue to accumulate, the two populations may eventually become separate species. Because this process is usually very gradual, it can be difficult to decide whether two populations have become separate species and biologists sometimes disagree. (It would be inappropriate to carry out experiments with animal species to try to resolve these issues.) The natural process by which species diverge to form new species is called speciation. This is described in *Topic A4.1*.

▲ Figure 8 The sandwich tern *Thalasseus sandvicensis* (left) was first recognized as a species by John Latham in 1787. Cabot's tern (right) was classified as a subspecies of the sandwich tern but recent phylogenetic research suggests that it is a separate species, *Thalasseus acuflavidus*. Not all biologists agree. Populations of *T. sandvicensis* live in Europe whereas *T. acuflavidus* lives in North and South America

A3.1.6 Diversity in chromosome numbers of plant and animal species

A fundamental characteristic of any species is its chromosome number. During the evolution of a species, this number can change: it can decrease if chromosomes become fused together, or increase if splits occur. There are also mechanisms that can cause the chromosome number to double. However, changes to the chromosome number are rare and usually there is no change in a species over millions of years.

In most plants and animals, body cells have an even number of chromosomes. This is a consequence of sexual reproduction. A new life starts by fusion of a male gamete and a female gamete, with each gamete containing one set of chromosomes (9 in cabbages). This fusion of gametes produces a zygote with two sets of chromosomes (18 in cabbages). All cells produced from the zygote by mitosis inherit these two sets of chromosomes. Gametes with one set of chromosomes are haploid. Body cells with two sets are diploid.

▲ Figure 9 Who has more chromosomes—a dog or its owner?

There is immense diversity in chromosome number among plants and animals. It is useful to remember that humans have 46 chromosomes and chimpanzees, our nearest relatives, have 48. You can easily find other chromosome numbers by searching databases. They range from two to hundreds. Table 1 shows numbers for some species.

The Oxford English Dictionary consists of 20 large volumes, each containing information about the origins and meanings of words. This information could have been published in a smaller number of larger volumes or in a larger number of smaller volumes. There is a parallel with the numbers and sizes of chromosomes in plants and animals. Some have a few large chromosomes and others have many small ones. Researchers experimented by fusing the 16 chromosomes of yeast cells to reduce the chromosome number to 4 or even 2. Their findings suggested that the actual number of chromosomes in a species is not very significant, as long as all members of the species have the same number.

Data-based questions: Differences in chromosome number

Plants	Number	Animals
no plant species yet discovered	2	*Myrmecia pilosula* (jack jumper ant)
Haplopappus gracilis (in the aster family)	4	*Parascaris equorum* (horse threadworm)
Luzula purpurea (woodrush)	6	*Aedes aegypti* (yellow fever mosquito)
Crepis capillaris (in the aster family)	8	*Drosophila melanogaster* (fruitfly)
Vicia faba (field bean)	12	*Musca domestica* (house fly)
Brassica oleracea (cabbage)	18	*Chorthippus parallelus* (grasshopper)
Citrullus vulgaris (watermelon)	22	*Cricetulus griseus* (Chinese hamster)
Lilium regale (royal lily)	24	*Schistocerca gregaria* (desert locust)
Bromus texensis (Texas brome grass)	28	*Desmodus rotundus* (vampire bat)
Camellia sinensis (Chinese tea)	30	*Mustela vison* (mink)
Magnolia virginiana (sweet bay)	38	*Felis catus* (domestic cat)
Arachis hypogaea (peanut)	40	*Mus musculus* (mouse)
Coffea arabica (coffee)	44	*Mesocricetus auratus* (golden hamster)
Stipa spartea (porcupine grass)	46	*Homo sapiens* (modern human)
Chrysosplenium alternifolium (saxifrage)	48	*Pan troglodytes* (chimpanzee)
Aster laevis (Michaelmas daisy)	54	*Ovis aries* (domestic sheep)
Glyceria canadensis (manna grass)	60	*Capra hircus* (goat)
Carya tomentosa (hickory)	64	*Dasypus novemcinctus* (armadillo)
Magnolia cordata (a small deciduous tree)	76	*Ursus americanus* (American black bear)
Rhododendron keysii (evergreen shrub)	78	*Canis familiaris* (dog)

▲ Table 1

1. There are many different chromosome numbers in Table 1 but some numbers are not seen, for example, 5, 7, 11 and 13. Explain why none of the species has 13 chromosomes. [3]
2. Using the data in Table 1, discuss the hypothesis that there is a positive correlation between the number of chromosomes in a species and its complexity. [4]
3. Explain what makes it impossible to calculate the genome size of a species from its chromosome number. [1]
4. Using the data in Table 1, identify a change in chromosome structure that may have occurred during human evolution. [2]

A3.1.7 Karyotyping and karyograms

The chromosomes of an organism become visible when cells are dividing, with metaphase giving the clearest view. To study the chromosomes of an organism, cells are stained and placed on a microscope slide. They are burst to spread the chromosomes by pressing on the cover slip. The chromosomes often overlap each other but with careful searching, it is usually possible to find a cell with no overlaps. The stained chromosomes can then be photographed. Originally, analysis involved cutting out each chromosome from a print and arranging them manually. This process can now be done digitally.

Chromosomes are classified based on three types of difference:

- Some stains give chromosomes distinctive banding patterns, with different banding in each type of chromosome.
- Chromosomes vary in size. In humans, the largest (chromosome 1) is more than five times longer than the shortest (chromosome 21).
- Each chromosome visible in metaphase consists of two strands called chromatids, held together

by a centromere. The position of the centromere varies. In some chromosomes it is near the centre, so the arms of the chromosomes are equal length. In other chromosomes the centromere is nearer to one end, so the chromosome has a shorter and a longer arm.

The characteristic types of chromosome in a species are called the karyotype. An image showing the karyotype of an organism is called a karyogram. The chromosomes are arranged in pairs, starting with the longest pair and ending with the smallest.

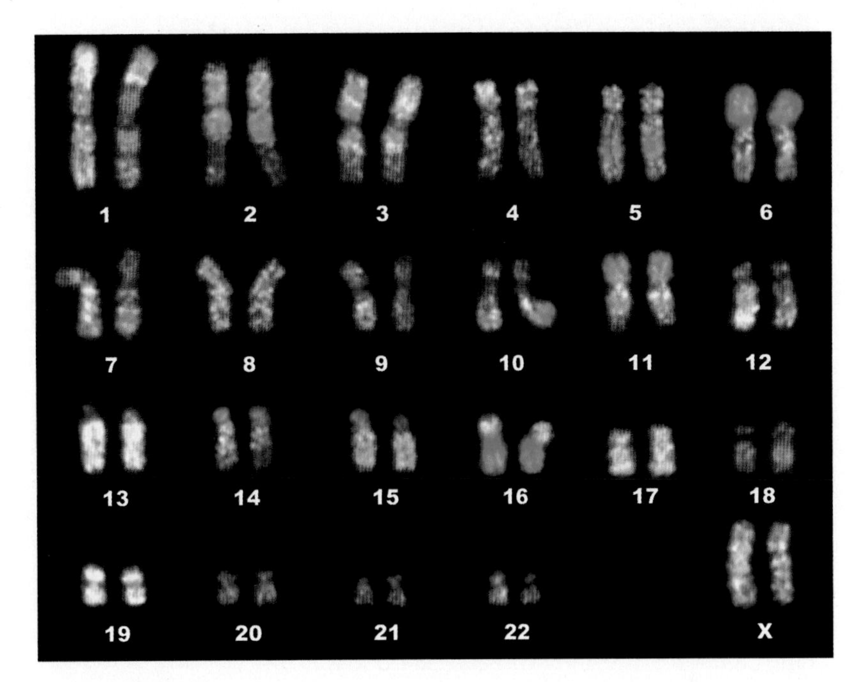

▲ **Figure 10** Karyogram of a human female, with fluorescent staining to generate banding patterns

Data-based questions: Primate chromosome numbers

Human somatic (body) cells have 46 chromosomes. Our closest primate relatives—chimpanzees, gorillas and orangutans—all have 48. Human chromosome types are numbered from 1 to 22. One hypothesis is that human chromosome 2 was formed from the fusion of two chromosomes in a primate ancestor. Figure 11 shows banding patterns of human chromosome 2 compared with chromosomes 12 and 13 from chimpanzees.

1. Compare human chromosome 2 with the two chimpanzee chromosomes. [3]
2. The ends of chromosomes, called telomeres, have many repeats of the same short DNA sequence. If the fusion hypothesis were true, predict what would be found in the region of the chromosome where the fusion is hypothesized to have occurred. [2]
3. Normally a chromosome has just one centromere, but in chromosome 2 there are remnants of a second centromere. Explain this observation. [2]
4. Discuss the strength of the evidence for a fusion of chimp chromosomes in the evolution of chromosome 2 in humans. [3]

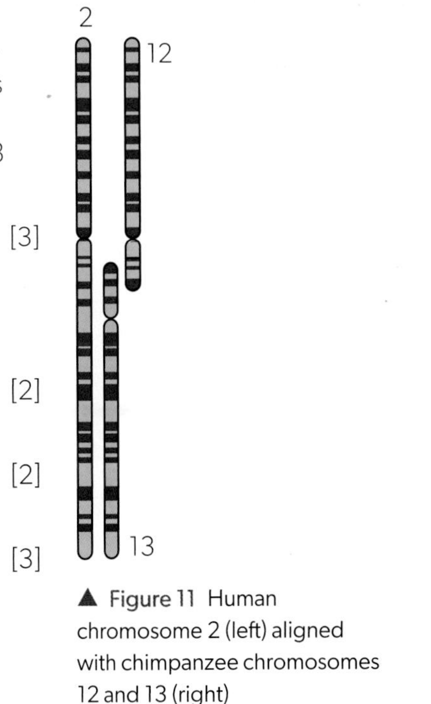

▲ **Figure 11** Human chromosome 2 (left) aligned with chimpanzee chromosomes 12 and 13 (right)

Falsification: Testable versus non-testable statements

The nature of scientific theories enables both explanations and predictions. When enough observations are predicted and explained by a theory, it becomes the consensus. If a new observation or experimental result is not well explained or predicted, the theory is either enhanced to address the observation, or considered falsified.

James Hutton (1726–1797) developed the theory that geological features were not fixed but underwent constant transformation over long periods of time. He argued that the Earth's history can be inferred from evidence in present-day rocks and that the Earth must be much older than predictions based on biblical chronology. No biblically reconstructed date for the creation of the Earth was early enough to fit with the huge timescale implied by geology, zoology and paleontology.

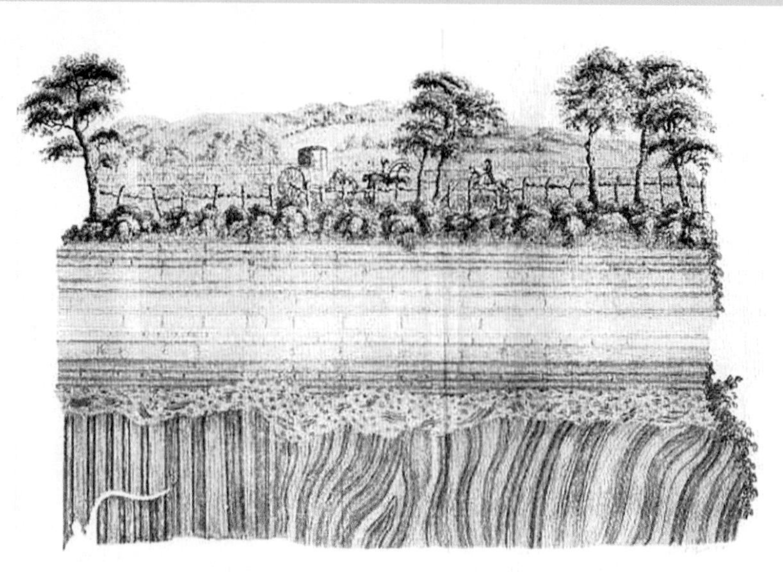

▲ Figure 12 The lower layers of rock in this drawing contain evidence of marine life. The striated appearance indicates many years of sediment being laid down, followed by a geological event that changed their orientation from vertical to horizontal. After being partly eroded away, these layers were subsequently covered by further layers of sediment. Hutton used observations of rock formations such as this to support the theory that geological features were not fixed but underwent transformation over long periods of time

Knowledge claims based on religious faith are often not falsifiable by observation or experimentation. This is not to say they are not valid; rather, they are not testable. For example, Hutton's observations were explained by the notion that the biblical "days" were metaphorical and corresponded to much longer periods of time (the "interval theory"). Another thinker proposed that God's omnipotence meant he could cause long geological ages to occur in short periods of time.

You have learned about the theory that the human chromosome 2 arose from fusion of chromosomes 12 and 13 in an ancestral primate. Is this theory testable?

A3.1.8 Unity and diversity of genomes within species

Among biologists today, the word "genome" means all of the genetic information of one individual organism or group of organisms. Genetic information is contained in DNA, so the genome is the entire base sequence of each of the DNA molecules (chromosomes).

A genome contains functional units called genes. A gene is a length of DNA carrying a sequence of hundreds or even thousands of bases. Typically, the members of a species have the same genes, in the same sequence, along each of their chromosomes. This allows parts of the chromosomes to be exchanged during meiosis, promoting genetic diversity in a species without any genes being omitted or duplicated. The genome of a species and the arrangement of genes on the chromosomes is thus an illustration of the unity in living organisms.

Diversity in the genomes of a species is largely due to variation in individual genes. Alternative forms of a gene, called alleles, often exist within a species. The alleles of a gene differ from each other in base sequence. Usually only one or a very small number of bases are different—for example, one allele might have adenine at a certain base position while another allele might have cytosine in that position. Sometimes larger sections of a gene become altered, but this usually results in loss of gene function.

Positions in a gene where more than one base may be present are called single-nucleotide polymorphisms, abbreviated to SNPs and pronounced "snips". Many thousands of individual human genomes have been sequenced, allowing researchers to assess the frequency of SNPs. More than 100 million different SNPs have been discovered so far in human genomes. This seems a huge number but remember there are over three billion base pairs in our genome. Most bases are therefore the same in all humans—another illustration of unity.

Within one individual, there are typically about 4,000–5,000 SNPs, so only about 1 base in 650,000 is different from that commonly occurring in humans. This may seem a low level of diversity but these SNPs are the main factor in making humans different from each other (unless we have an identical twin!).

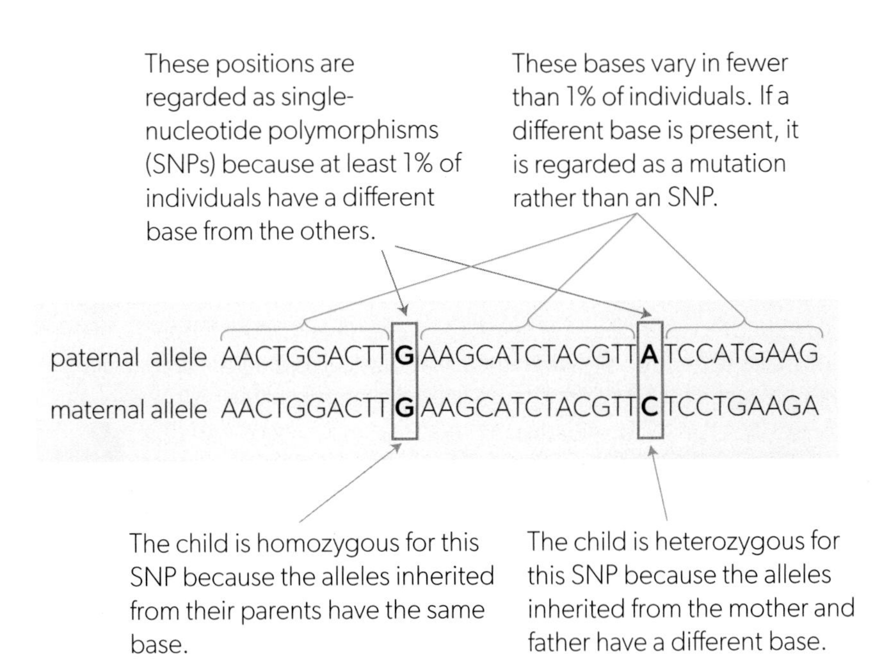
These positions are regarded as single-nucleotide polymorphisms (SNPs) because at least 1% of individuals have a different base from the others.
These bases vary in fewer than 1% of individuals. If a different base is present, it is regarded as a mutation rather than an SNP.
paternal allele AACTGGACTT G AAGCATCTACGTT A TCCATGAAG
maternal allele AACTGGACTT G AAGCATCTACGTT C TCCTGAAGA
The child is homozygous for this SNP because the alleles inherited from their parents have the same base.
The child is heterozygous for this SNP because the alleles inherited from the mother and father have a different base.

◀ Figure 13 SNPs are inherited from our parents

A3.1.9 Diversity of eukaryote genomes

The genomes of plants, animals and other eukaryotes vary by a huge amount, both in the overall size of the genome and in base sequences. Variation between species is far larger than genome variation within a species.

Variation in genome size

Overall genome size is measured in base pairs. There is a huge range in genome size and some species have a surprising amount of DNA. Large genomes can contain a lot of non-functional DNA, so they do not necessarily contain more functioning genes than smaller genomes. For example, about half of the human genome consists of transposons (transposable sequences), most of which have no known function. Transposons are sometimes referred to as "junk DNA".

Table 2 shows the range of genome sizes in different organisms.

Organism	Genome size / million base pairs	Description
Paramecium tetraurelia	27	Unicellular organism
Apis mellifera	217	Honey bee
Homo sapiens	3,080	Human
Pan troglodytes	3,175	Chimpanzee
Paris japonica	150,000	Woodland plant

▲ Table 2

Variation in base sequence

Two populations of a species will have some differences in base sequence. If these populations diverge to form separate species, more differences will accumulate over time. In some genes, changes in base sequence are infrequent. These tend to be genes with a vital function that does not change—for example, the gene for the protein cytochrome c, which has an essential role in respiration. As a result, there may be relatively few (or no) base sequence differences, even between distantly related species.

Different species also have different numbers and types of genes. Genes can be added to a genome or removed from it, so species that diverged from a common ancestor hundreds of millions of years ago have developed differences in their genetic make-up, especially when they are adapted to different ways of life.

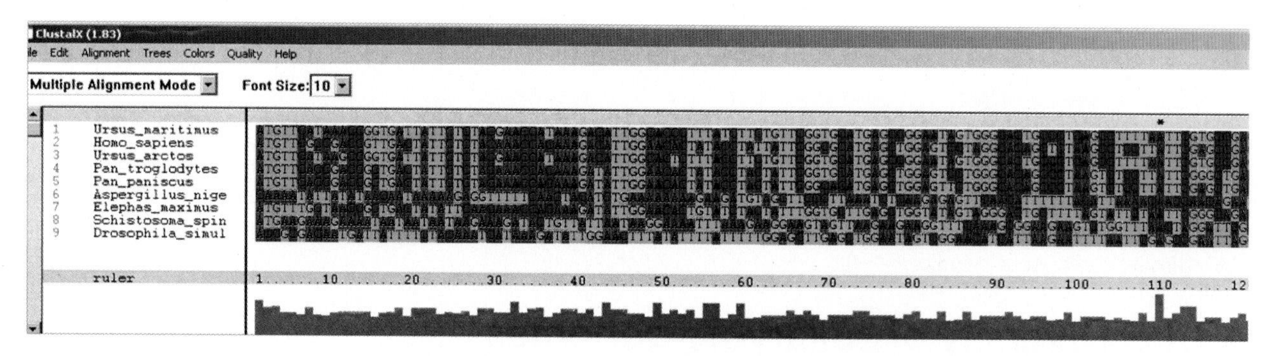

▲ Figure 14 You can use the GenBank website to compare base sequences of specific genes between species. This image shows the first 120 bases in the sequences of the gene that codes for cytochrome oxidase 1 in nine species. The sequences have been aligned using additional software to allow comparison

Data-based questions: Genome sizes

The graph in Figure 15 compares genome size with the number of genes that code for proteins in species of eukaryote.

1. What trend does the data in the graph show? [2]
2. The curve that has been fit to the graph shows that the number of protein-coding genes is not directly proportional to genome size.
   a. What trend line on the graph would indicate direct proportion between the variables? [1]
   b. Discuss the reasons for the number of protein-coding genes not being directly proportional to genome size. [2]
3. The statistic R^2 has been calculated for this data and is 0.919.
   a. What is the statistic R^2? [1]
   b. What does a value as high as 0.919 indicate? [2]
4. The scales on the axes are logarithmic.
   a. If the $\log_{10}$ protein-coding gene number is 4.0, what is the actual number of protein-coding genes? [1]
   b. If the $\log_{10}$ genome size (kbp) is 6.0, what is the actual number of base pairs? [2]

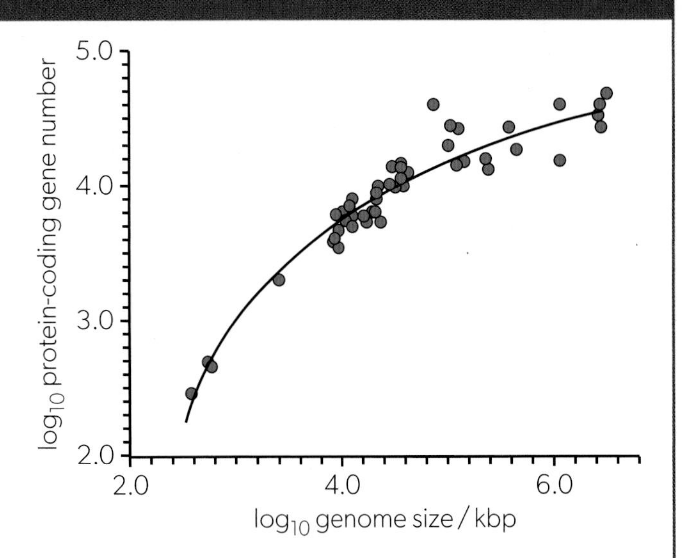

▲ Figure 15

Source: Hou Y, Lin S (2009) Distinct Gene Number-Genome Size Relationships for Eukaryotes and Non-Eukaryotes: Gene Content Estimation for Dinoflagellate Genomes. PLOS ONE 4(9): e6978.

A3.1.10 Comparison of genome sizes

Knowledge of the size of genomes can form the basis of research into genome evolution. It can also be used to estimate the cost and difficulty of future genome sequencing programmes. Genome sizes are typically given as nuclear DNA contents of a haploid cell such as a gamete (C-values), either in units of mass (usually picograms; 1 pg = 10^{-12} grams) or in number of base pairs or megabase pairs (1 Mbp = 10^6 base pairs).

Nuclear DNA content data for more than 10,000 species of plant, animal, fungus and microbe is available from these four independent databases:

a. Plant DNA C-values Database hosted by Kew Gardens (https://cvalues.science.kew.org.com)

b. Animal Genome Size Database (www.genomesize.com)

c. Fungal Genome Size Database (www.zbi.ee/fungal-genomesize.com)

d. Microbial Genomes (https://www.ncbi.nlm.nih.gov/genome/microbes.com)

ATL Thinking skills: Evaluating alternative perspectives

Knowledge claims are affected by criteria for judgment. For example, the answer to the question, "How does genome size correlate with complexity?" depends on our criteria for judgement. In particular, what do we mean by "complexity"?

Daniel W. McShea, a paleobiologist at Duke University quoted in *Scientific American*, discusses the problems with the term complexity: "It's not just that they don't know how to put a number on it. They don't know what they mean by the word".

For example:

- Plants are metabolically more complex than animals.
- Multicellular organisms have a greater diversity of cell types than prokaryotes, due to more complex regulation of gene expression. Are organisms with many tissue types more complex than organisms with fewer tissue types?
- Single-celled organisms carry out many different activities per cell, while a single cell within a multicellular organism carries out a smaller range of activities. Does that make singled-celled organisms more complex?
- More recently evolved organisms often have a greater number of novel adaptations than those which evolved longer ago. Does this make them more complex?

Until we agree on a definition of the term "complexity", cannot agree on an answer to the question.

ATL Thinking skills: Answering open-ended questions

A database is an organized collection of data stored electronically in a computer system. Data mining is a process used by researchers to turn raw data into useful information. Many successful internal assessment investigations and extended essays in biology have been based on inquiries carried out using databases.

In both types of project, it is essential to begin with an open-ended question. An open-ended question cannot be answered with a simple "yes" or "no" and the researcher does not know the answer before they start. The clearest questions are expressed so that the wording suggests the method to be followed. The dependent and independent variables should be easy to identify from the question. For example:

a. Do angiosperms, on average, have larger genomes than pteridophytes?

b. Do fungi and animals have similar genome sizes?

Generate a research question about genome size and test it using one or more databases.

A3.1.11 Current and potential future uses of whole genome sequencing

Whole genome sequencing is determining the entire base sequence of an organism's DNA. This was first done in the 1990s with bacteria and archaea, because their relatively small genomes made it easier. It is now feasible with most organisms. Some of the early landmarks in whole genome sequencing are shown in Table 3.

Year	Organism	Number of base pairs
1995	*Haemophilus influenzae* (a pathogenic bacterium)—first prokaryote	1.8 million
1996	*Saccharomyces cerevisiae* (yeast—a unicellular fungus)—first eukaryote	12 million
1998	*Caenorhabditis elegans* (a nematode worm)—first multicellular organism	100 million
2000	*Arabidopsis thaliana*—first plant	135 million
2003	*Homo sapiens*—complete sequence published	3,080 million

▲ Table 3

Look at the data in Table 3. In less than 10 years, the size of the genomes being sequenced increased by a factor of a thousand. This was made possible by technological developments which both increased the speed of sequencing and reduced the cost. These developments have continued. The cost of sequencing one human genome, for example, has dropped from $100 million in 2001 to less than $1,000 in 2020.

There has also been exponential growth in the number of species for which at least one sequence has been completed, so any figure quoted for the number of complete genome sequences will soon be exceeded. The Earth BioGenome Project aims to sequence the genomes of all known species.

A principal goal of sequencing the genomes of a wide range of species is investigation of evolutionary origins. Comparisons between genomes allow researchers to identify relationships between species and trace the diverging pathways from common ancestors. Knowledge gained from studying the genomes of different species will make it easier to conserve and protect biodiversity. Research into the genomes of pathogenic bacteria and viruses will help in the control and prevention of infectious diseases caused by these organisms.

There are ambitious aims for sequencing more genomes of individual humans. So far, over one million individual human genomes have been sequenced, and this number has been doubling about every eight months. This has increased understanding of human origins and migrations in all parts of the world. It is also providing more data than ever about genetic diseases and genes that affect human health. In future, it may be possible to sequence the genome of every person. This could lead to the development of personalized medicine. If it is known which SNPs and other genetic features are present in a person's genome, it will be easier to predict health problems and prescribe appropriate drugs and other treatments for that person.

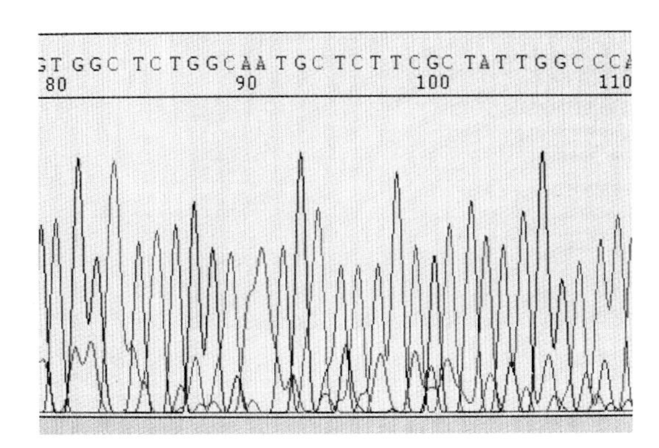

◀ Figure 16 Sequencing read from the DNA of the Pinot Noir variety of grape

A3.1.12 Difficulties in applying the biological species concept to asexually reproducing species and to bacteria that have horizontal gene transfer

The biological species concept works well with many groups of species. However, it works less well with species that reproduce asexually or have methods of horizontal gene transfer.

Asexually reproducing species

If members of a species interbreed by sexual reproduction, their traits are remixed every generation. This prevents the development of significant differences between individuals. Shared traits allow members of the species to be identified. Many species reproduce both sexually and asexually, but as long as they sometimes reproduce sexually, they will remain as a coherent and therefore unified group.

Some species—such as blackberries (*Rubus fruticosus*) and dandelions (*Taraxacum officinale*)—only reproduce asexually. Both these plant species produce flowers and look as though they are reproducing sexually but offspring are actually produced by mitosis and are genetically identical to the parents.

All offspring produced by asexual reproduction are clones of their parent. If a clone does not interbreed with other clones, it is a separate species according to the biological species concept. Many different clones may be recognized. Among blackberries, for example, hundreds of clones have been named as separate species. Only a few experts can distinguish between these "microspecies" and great efforts are made to conserve some of the rarer clones. A better policy is to recognize that blackberries, dandelions and other species that have abandoned sexual reproduction are no longer species according to the biological species concept.

▲ Figure 17 The yellow flower head of a dandelion develops into a spherical array of wind-dispersed fruits, each with a single seed. Because they have been produced asexually, all the seeds are genetically identical

▲ Figure 18 The dandelions in this field may all be members of the same clone. They are flowering and producing seed asexually so they are not a typical biological species

Species with horizontal gene transfer

The evolution of life is often thought to resemble a tree, starting with a single trunk from which branches emerge. Repeated branching eventually leads to individual species. Once formed, a branch remains separate and does not rejoin other branches. In the same way, species do not interbreed with other species so their genes remain separate.

However, genome sequencing has revealed that the separation between species is not always complete. Genes are sometimes transferred from one species to another, even between distantly related species. This process is called horizontal gene transfer, to distinguish it from vertical transfer from parent to offspring.

Horizontal gene transfer is frequent among bacteria. For example, it is how antibiotic resistance genes can move from one species to another. In fact, there is so much gene transfer between bacteria that it is debateable whether the biological species concept (or any other species concept) works with prokaryotes. Among eukaryotes, although horizontal gene transfer has occurred, it is less frequent and species are easier to define.

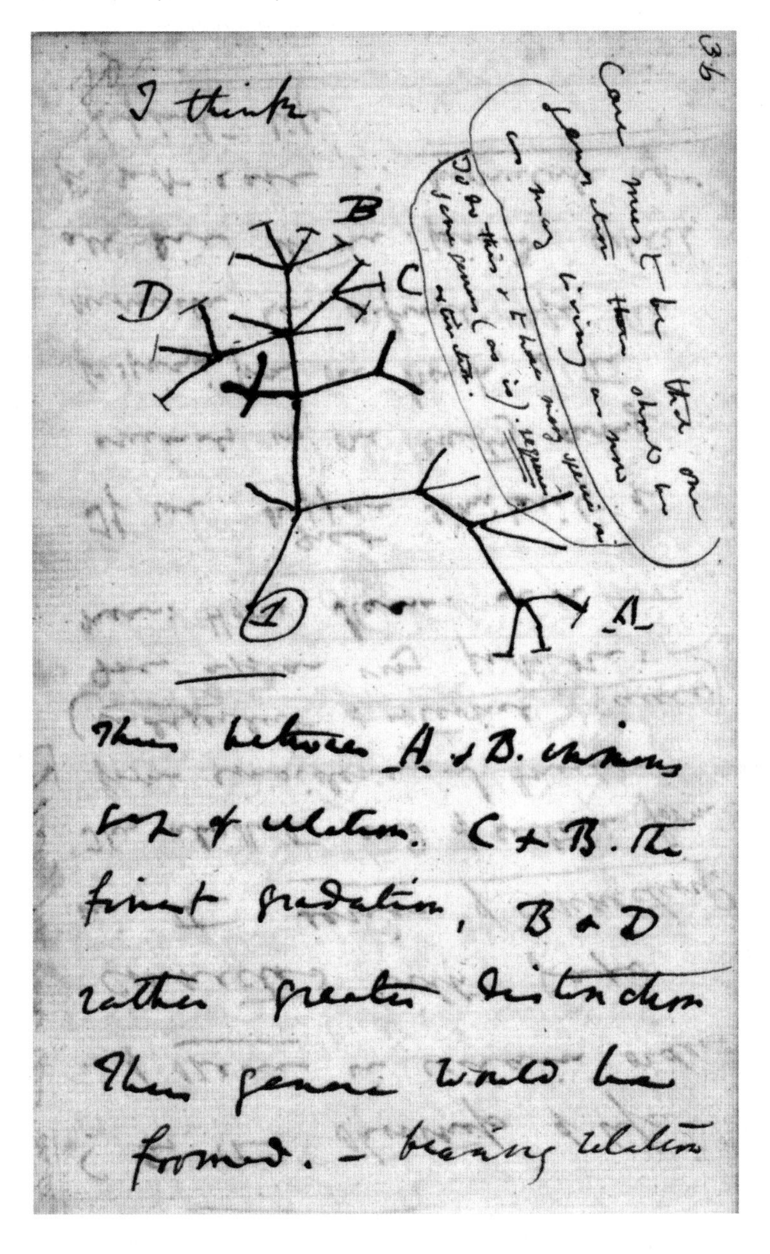

◀ Figure 19 Darwin's famous evolutionary tree diagram, drawn in about 1837 in one of his notebooks

A3.1.13 Chromosome number as a shared trait within a species

Earlier in this topic, you learned that the members of a species usually have the same number of chromosomes. This lack of diversity is a consequence of reproducing sexually rather than asexually.

For sexual reproduction to occur, males and females of the species produce gametes. These gametes have the haploid number of chromosomes (for example, 23 in human eggs and sperm). In eukaryotes, these gametes are formed my meiosis, which halves the chromosome number. Male and female gametes then fuse to produce a zygote with the diploid number of chromosomes (46 in humans).

In a diploid nucleus, there are two sets of chromosomes and each chromosome carries the same sequence of genes as one other chromosome. Two chromosomes carrying the same sequence of genes are said to be homologous. During meiosis, homologous chromosomes pair up with each other, so they can be reliably separated into different daughter cells. The separation of homologous chromosomes into separate daughter cells halves the chromosome number.

If two organisms with different chromosome numbers mated and produced offspring, the offspring would almost certainly have problems in carrying out meiosis. Some of the chromosomes would not be able to pair up because they would not be homologous to any other chromosome. As a result, there would not be an orderly segregation of chromosomes into two groups. The cells produced by meiosis would not be viable and gametes could not be produced. This is why offspring of parents with different chromosome numbers are usually infertile.

▲ Figure 20 These flowers are on a Bramley apple tree, which is triploid, with 51 chromosomes instead of the usual 34. Meiosis therefore fails and the anthers in the flowers produce no pollen, so a Bramley cannot pollinate any other apple tree

▲ Figure 21 All the cells in these Bramley apples are triploid, like the tree on which they grew. Bramley apple trees can produce fruit even though they cannot carry out meiosis, because cells in the fruit are produced by mitosis

Data-based questions: Chromosome numbers in *Sphagnum* mosses

Researchers can estimate the DNA content of cells by using a stain that binds specifically to DNA. A narrow beam of light is passed through a stained nucleus and the amount of light absorbed by the stain is measured. This gives an estimate of the quantity of DNA. Table 4 shows such estimates for leaf cells in eight species of bog moss (*Sphagnum*) on the Svalbard islands.

Sphagnum species	Mass of DNA / pg	Number of chromosomes
S. aongstroemii	0.47	19
S. arctium	0.95	
S. balticum	0.45	19
S. fimbriatum	0.48	19
S. olafii	0.92	
S. teres	0.42	19
S. tundrae	0.44	19
S. warnstorfii	0.48	19

▲ Table 4

1. Compare the DNA content of the bog mosses. [2]
2. Suggest a reason for six of the species of bog moss on the Svalbard islands having the same number of chromosomes. [2]
3. *S. arcticum* and *S. olafii* probably arose as new species when meiosis failed to occur in one of their ancestors.
   a. Deduce the number of chromosomes in a leaf cell nucleus of these species. Give two reasons for your answer. [3]
   b. Suggest a disadvantage to *S. arcticum* and *S. olafii* of having more DNA than other bog mosses. [1]
4. It is unusual for plants and animals to have an odd number of chromosomes in their nuclei. Explain how mosses can have odd numbers of chromosomes in their leaf cells. [2]

A3.1.14 Engagement with local plant or animal species to develop a dichotomous key

Dichotomous keys are constructed for identification of species within a group. A dichotomy is a division into two; a dichotomous key consists of a numbered series of pairs of descriptions. In each pair, one description should clearly match the species and the other should clearly be wrong. The features that the designer of the key chooses to describe must therefore be reliable and easily visible. Each pair of descriptions leads either to another numbered pair in the key, or to an identification. An example key is shown in Figure 22.

Keys are usually designed for use in a particular area. All the groups or species found in that area can be identified using the key. There may be a group of organisms in your area for which a key has never been designed. Choose from these suggestions or come up with your own idea:

- trees in the local forest or on your school campus, using descriptions of leaves or bark
- water plants in a local pond
- birds that visit bird-feeding stations in your area
- invertebrates that are associated with one particular plant species.

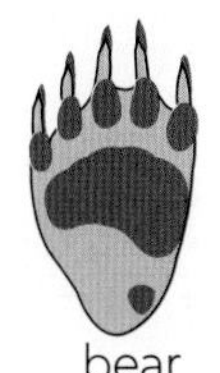

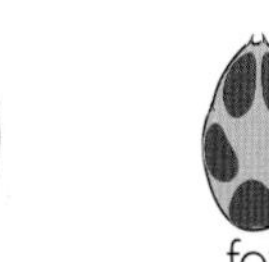

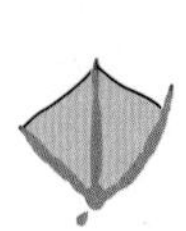

▲ **Figure 22** These images show the right front footprints of 10 types of mammal and bird (not to scale). They can be used to develop skills in constructing dichotomous keys

ATL Communication skills: Construction of dichotomous keys for use in identifying specimens

The distinguishing features described in a dichotomous key must be reliable and easily visible. An example key is shown in Figure 23. We can use it to identify the species in Figure 24. In step 1, you must decide if hind limbs are visible. They are not, so you are directed to step 6 of the key. You must now decide if the species has a blowhole. It does not, so it is a dugong or a manatee. A fuller key would have another step to separate dugongs and manatees.

1	Fore and hind limbs visible, can emerge on land.......... Only fore limbs visible, cannot live on land..................	2 6
2	Fore and hind limbs have paws................................. Fore and hind limbs have flippers..............................	3 4
3	Fur is dark.. Fur is white..	sea otters polar bears
4	External ear flap visible.. No external ear flap..	sea lions and fur seals 5
5	Two long tusks... No tusks..	walruses true seals
6	Mouth breathing, no blowhole.................................. Breathing through blowholes..................................	dugongs and manatees 7
7	Two blowholes, no teeth... One blowhole, teeth..	baleen whales dolphins, porpoises and whales

▲ Figure 23 A dichotomous key to groups of marine mammals

▲ Figure 24 A marine mammal, photographed in Florida

A3.1.15 Identification of species from environmental DNA in a habitat using barcodes

DNA barcodes are short sections of DNA from one gene, or at most several genes, which are distinctive enough to identify a species. For example, part of the gene for cytochrome oxidase subunit 1 is used as a barcode for animal species. DNA barcoding allows scientists to identify species from small pieces of tissue that might otherwise be difficult to recognize. For example, many plant species have leaves that are oval with a pointed end. Barcodes make it possible to distinguish these species.

Species identification is now possible using environmental DNA, collected from water, soil or any other part of the abiotic environment. Typically, this contains DNA from a wide diversity of organisms that have interacted with the sampled environment. DNA barcodes can be used to identify these organisms. This technological advance has many applications in ecology and conservation.

In a recent case, samples taken from waterholes in northern Australia were analysed using DNA barcodes. This analysis showed that Gouldian finches (*Erythrura gouldiae*), an increasingly rare bird species, had visited the waterholes. In another case, DNA left in snow tracks was used to confirm the presence of a small carnivorous mammal called a fisher (*Pekania pennanti*) in Idaho.

▲ Figure 25 Early warnings of the spread of diseases can be obtained by regular sampling of wastewater and testing for DNA of pathogens. This technique has been used to test for new strains of COVID-19 and for resurgence of polio

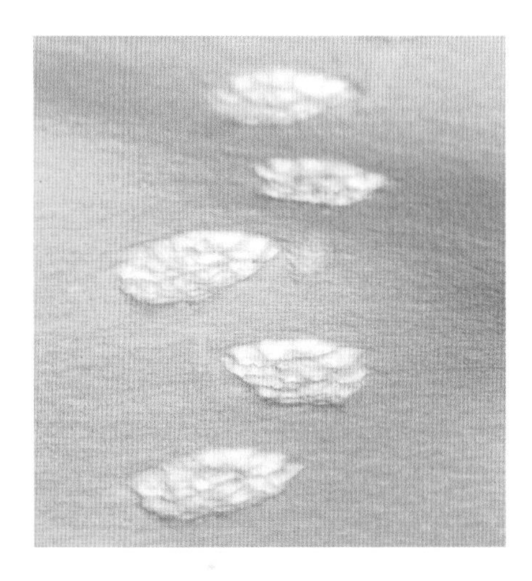

▲ Figure 26 Tracks of a fisher (*Pekania pennanti*) in fresh snow in winter

Linking questions

1. What might cause a species to persist or go extinct?
   a. Explain how adaptations demonstrate the correlation between form and function in plants. (B4.1.8)
   b. Distinguish, using examples, between top-down and bottom-up limiting factors. (C4.1.17)
   c. Outline two named examples of species extinction including causes and ecological consequences. (A4.2.3)
2. How do species exemplify both continuous and discontinuous patterns of variation?
   a. With reference to an example, outline what is meant by polygenic inheritance. (D3.2.14)
   b. Distinguish between codominance and incomplete dominance. (D3.2.9)
   c. With reference to a named example, explain the mechanism behind disruptive natural selection. (D4.1.12)

A3.2 Classification and cladistics

What tools are used to classify organisms into taxonomic groups?

Historically, scientists have used shared observable features to classify groups of organisms. The organism shown in Figure 1 is *Relicanthus daphneae*. If it looks like a sea anemone, does that make it one? It is unusually large for an anemone, with tentacles up to 7 feet long. Across several genes, its DNA sequence is distinct from all other anemones. It is categorized as a cnidarian. What features can we expect to see if that is the classification? Anemones are unique among cnidarians in having flaps over their stinging cells. How would the discovery of flaps affect the classification of this unusual animal?

▲ Figure 1

How do cladistic methods differ from traditional taxonomic methods?

In Figure 2, a marabou stork is waiting for an opportunity to capture any fish that might be dropped by the crocodile. What are the differences between birds and reptiles that have led them to be classified as separate classes of vertebrates? Molecular analysis has established that the bird is more closely related to the crocodile than the crocodile is to other reptiles such as snakes and turtles. Other than greater homology of DNA, what other morphological and physiological features do birds and crocodiles share? If birds descended from dinosaurs, what prevents birds from being reclassified as reptiles?

▲ Figure 2

AHL only

A3.2.1 Need for classification of organisms
A3.2.2 Difficulties classifying organisms into the traditional hierarchy of taxa
A3.2.3 Advantages of classification corresponding to evolutionary relationships
A3.2.4 Clades as groups of organisms with common ancestry and shared characteristics
A3.2.5 Gradual accumulation of sequence differences as the basis for estimates of when clades diverged from a common ancestor
A3.2.6 Base sequences of genes or amino acid sequences of proteins as the basis for constructing cladograms
A3.2.7 Analysing cladograms
A3.2.8 Using cladistics to investigate whether the classification of groups corresponds to evolutionary relationships
A3.2.9 Classification of all organisms into three domains using evidence from rRNA base sequences

A3.2.1 Need for classification of organisms

Millions of species have been named and described, and more are discovered every day. Biologists have accumulated huge amounts of knowledge about these species. This poses a considerable challenge in terms of information storage and retrieval. To make this easier, biologists have devised systems for classifying life. Classification involves placing organisms in groups according to their traits or evolutionary origins.

A hierarchical system of classification has been developed over the last 300 years. All organisms are divided into major groups; at present, domains are the broadest type of group. These large groups are subdivided again and again until we reach the basic level of classification—the species. Without this system, it would be very difficult to identify unknown species. Consider the organism pictured in Figure 3.

- It is obviously eukaryotic and an animal, so we immediately know the domain (eukaryotes) and the kingdom (animals). However, there are over a million possible animal species.
- We can see hair and we would be able to find mammary glands, so we can deduce that it is one of the 6,500 species of mammal.
- Other traits show that it is a member of the Carnivora. This limits the possibilities to 270 species.
- In a similar way, we can place the organism in the Mustelid family, which contains about 60 species.
- It then becomes relatively easy to identify the genus and species: *Pekania pennanti*—the fisher.

▲ Figure 3 What is this organism?

Once we know the name of the species, we can easily access large amounts of information about this organism and the groups to which it belongs. This is the power of classification.

ATL Thinking skills: Evaluating alternative perspectives

Are classification systems invented or discovered?

It is natural for humans to arrange things in groups, to make it easier to study them. The process of arranging things in groups is classification.

1. The clouds that we see in the sky appear in an infinite variety of forms. The World Meteorological Organization has developed a classification of clouds. Ten genera are recognized, such as cirrus, stratus and cumulus. These genera are subdivided into species and then varieties. This classification is worthwhile because it enables more accurate prediction of weather. For example, rainfall is unlikely if the clouds are cumulus. Is this classification invented or discovered?

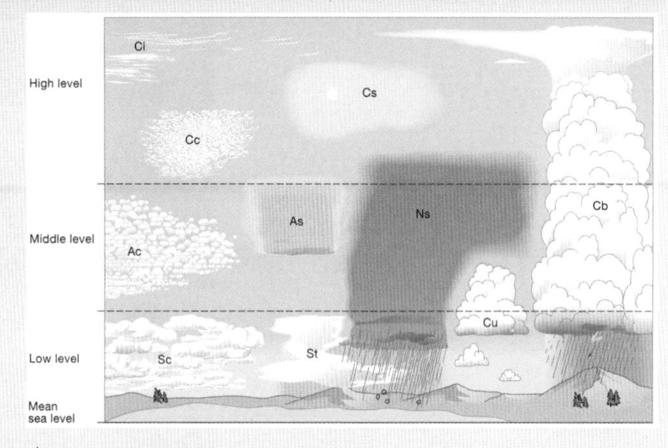

▲ Figure 4 The 10 genera of clouds

2. In how many ways can the oval, triangle and square be classified into two groups, based on their similarities and differences? Is one of these classifications better?

▲ Figure 5

3. The animals in Figure 6 both have a tail for aiding balance and a parachute-like membrane that stretches from wrist to ankle that allows them to glide between tree branches. They also have mammary glands and fur. Southern flying squirrel foetuses develop in the uterus with a placenta. Sugar glider foetuses develop in their mother's pouch. Which features are most important in assessing the relationship between the two gliding organisms?

▲ Figure 6 (left) Southern flying squirrel (*Glaucomys volans*) and (right) sugar glider (*Petaurus breviceps*)

A3.2.2 Difficulties classifying organisms into the traditional hierarchy of taxa

Any classificatory group is a taxon, for example, "phylum". The plural is taxa. Assigning organisms to groups is taxonomy. Biologists have developed a hierarchy of taxa with ranks from species up to kingdom. This traditional hierarchy is shown in Figure 7, with two examples. A genus contains one or more species, a family contains one or more genera and so on. Moving up through the hierarchy, the taxa contain larger and larger numbers of species that share fewer and fewer traits.

In practice, it can be difficult to classify organisms according to this hierarchy. Even when taxonomists agree over which species should be classified together, they often disagree over what taxonomic rank the grouping should have. One taxonomist might think the traits in a group of species are similar enough to form a genus; another might think they are different enough to be a family.

Taxon	Grey wolf	Date palm
Kingdom	Animals	Plants
Phylum	Chordates	Angiosperms
Class	Mammals	Monocotyledons
Order	Carnivores	Palmales
Family	Canidae	Arecaceae
Genus	*Canis*	*Phoenix*
Species	*lupus*	*dactylifera*

▲ Figure 7 Traditional classification in the hierarchy of taxa

These uncertainties are a result of the gradual divergence of species and larger groups over time. For example, as the species in a genus diverge from each other, there will eventually be sufficient diversity for the genus to be divided into two or more separate genera. As divergence continues over thousands or even millions of years, these genera will become different enough to be placed in different families. The instant in time when these separations should happen cannot be determined objectively. This is called the boundary paradox and, because of it, taxonomic rankings are inevitably rather arbitrary.

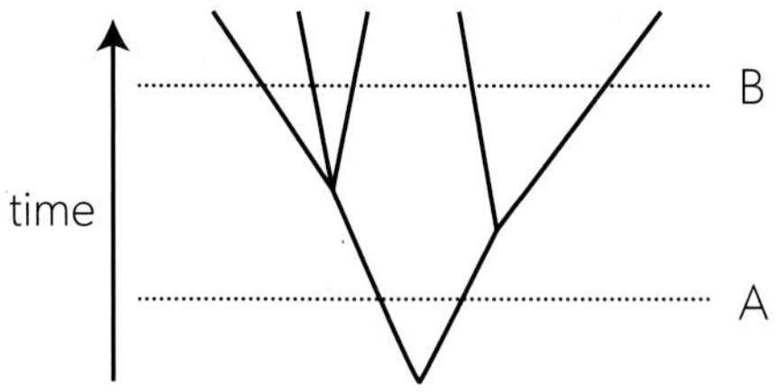

time B A

▲ Figure 8 Each line represents a species over time. How many genera are there at A and at B? How can you justify your answer?

A3.2.3 Advantages of classification corresponding to evolutionary relationships

Biologists agree that classification should mirror the evolutionary origins of species. Two criteria can be used to judge whether a classification achieves this:

- Every organism that has evolved from a common ancestor is included in the same taxonomic group.
- In each taxonomic group, all the species are evolved from the same common ancestor.

If these criteria are satisfied, all members of a taxonomic group will share traits that they have inherited from their common ancestor. Such shared traits are known as synapomorphies. This sharing of traits between members of a taxonomic group allows biologists to make predictions based on classification. Two examples are given here.

▲ Figure 9 *Murina beelzebub* was recently discovered in Vietnam. It is a tube-nosed bat, with a mass of only 5 to 6 grams. It is aggressive when captured, hence the species name

Species of bat

New species of bat are sometimes discovered. Because we know that bats are classified as mammals, we can immediately make predictions with reasonable certainty: a new species of bat will have a four-chambered heart, hair, mammary glands, a placenta and therefore a navel (belly button), plus many other mammalian features.

Species of daffodil

Some types of daffodil (*Narcissus* species) produce galanthamine. This substance has been used as a drug for treatment of Alzheimer's disease and is one of a group of compounds called alkaloids. There is strong evidence that all species in the genus *Narcissus* evolved from a common ancestor. It is therefore reasonable to predict that other alkaloids are synthesized by *Narcissus* species. Over 80 different alkaloids have now been found in species in the genus, some of which are likely to prove useful as drugs.

▲ Figure 10 *Narcissus poeticus*

A3.2.4 Clades as groups of organisms with common ancestry and shared characteristics

Species can evolve over time and split to form new species. With some highly successful species, this has happened repeatedly, so there are now large groups of species all derived from a common ancestor. These groups of species can be identified based on shared characteristics. A group of organisms evolved from a common (shared) ancestor is called a clade.

Clades include all the species alive today, together with the ancestral species. They also include any species that evolved from the common ancestor and then became extinct. Clades can be very large and include thousands of species, or they can be very small with just a few species. For example, birds form one large clade with about 10 thousand living species, because they have all evolved from a common ancestral species. In contrast, the tree *Ginkgo biloba* is the only living member of a clade that evolved about 270 million years ago. There have been other species in that clade but all are now extinct.

It is not always obvious which species have evolved from a common ancestor and should therefore be included in a clade. The most objective evidence comes from base sequences of genes or amino acid sequences of proteins. The genomes of organisms contain a huge amount of information, from which their evolutionary history can be deduced. Where sequence data is not available, morphological traits can be used to assign organisms to clades. This is particularly useful with species that have become extinct, where sequence data is not available and the only evidence comes from fossils.

Every species is in multiple clades, not just one. Smaller clades are "nested" within larger clades. For example, Figure 11 shows 10 species of gymnosperm (non-flowering seed plants). *Araucaria araucana* (the monkey puzzle tree) is in a clade with *Podocarpus totara* because they have a common ancestor. They are nested in a clade with *Taxus baccata* and the two species below it in the tree diagram. Those five species are nested in a clade with *Pinus radiata* and the three species below it—again because of common ancestry. Finally, those 9 species plus *Gingko biloba* are in a clade that includes all 10 of these species, plus all other gymnosperms.

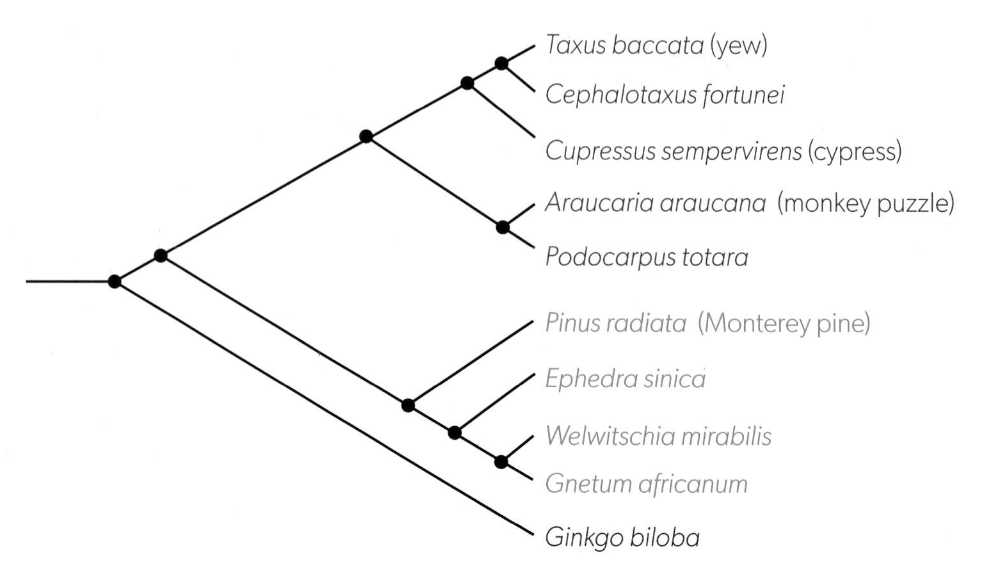

▲ Figure 11 Smaller clades are nested within larger ones

Paradigm shifts

A fixed ranking of taxa (kingdom, phylum and so on) is arbitrary as it does not reflect the gradation of variation. Cladistics offers an alternative approach to classification using unranked clades. This is an example of a paradigm shift in scientific thinking.

Increasingly, clades are being named and used in classification instead of the ranks in the traditional hierarchy of taxa. Figure 12 shows how date palms and grey wolves have been classified using clades. The taxa from species up to order, and the kingdom, are still assigned using traditional taxa. The intermediate levels in the hierarchy are all referred to as clades and the number of levels is not fixed.

1. What is the advantage of this approach to classification?
2. What makes this a paradigm shift rather than a modification?

Taxon	**Grey wolf**
Kingdom	Animals
Clade	ParaHoxozoa
Clade	Bilateria
Clade	Nephrozoa
Clade	Deuterostomia
Clade	Chordata
Clade	Olfactores
Clade	Vertebrata
Clade	Tetrapoda
Clade	Amniota
Class	Mammals
Order	Carnivores
Family	Canidae
Genus	*Canis*
Species	*lupus*

Taxon	**Date palm**
Kingdom	Plants
Clade	Tracheophytes
Clade	Angiosperms
Clade	Monocotyledons
Clade	Commelinids
Order	Arecales
Family	Arecaceae
Genus	*Phoenix*
Species	*dactylifera*

▲ **Figure 12** Classification using clades

A3.2.5 Gradual accumulation of sequence differences as the basis for estimates of when clades diverged from a common ancestor

Differences in the base sequence of DNA—and therefore in the amino acid sequence of proteins—are the result of mutations. These differences accumulate gradually over long periods of time. If we assume that this happens at a roughly constant rate, we can use the number of differences to estimate the time since two species diverged from a common ancestor. This method of estimating time is known as the "molecular clock". The larger the number of sequence differences between two species, the longer since they diverged from a common ancestor.

When considering timings based on the molecular clock, it is important to remember the assumption that has been made—that mutations accumulate at a constant rate. In fact, this rate can vary and is affected by the length of the generation time, the size of the population, the intensity of selective pressure and other factors. Thus, the molecular clock can only give estimates.

Base sequence differences have been used to estimate when humans split from the nearest living relatives. Based on a mutation rate of $10^{-9} yr^{-1}$, this happened 4.5 million years ago. Using the same assumptions, common chimpanzees and bonobos split more recently—around one million years ago. It is also possible to estimate when humans split from other hominid species and therefore when the most recent common ancestor of all humans existed. Using variations in the base sequence of mitochondrial DNA as a molecular clock, our most recent common ancestor is estimated to have lived 150,000 years ago.

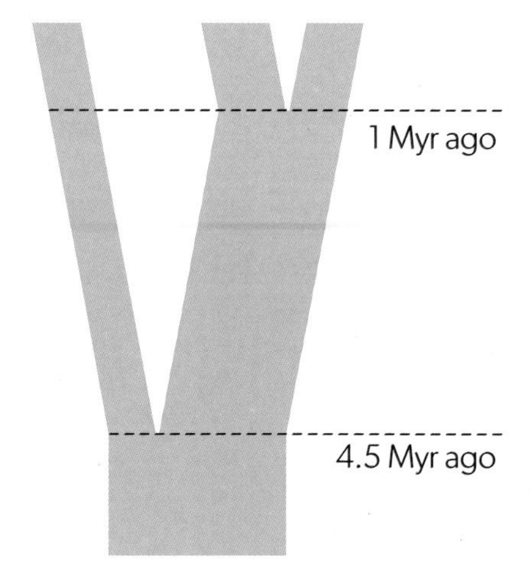

▲ Figure 13 **Estimated dates for the divergence of humans and chimpanzees, based on the base sequences of the entire genomes**

AHL

Applying technology to process data: Conducting a sequence alignment

Sequence similarities in the DNA or proteins from different organisms suggest evolutionary relationships. The greater the similarity, the closer the relationship. It is possible to compare two relatively short sequences visually. However, comparison of longer sequences or multiple sequences relies on the use of computer algorithms.

Various web-based applications can be used to carry out sequence alignment, such as the Multiple Sequence Alignment tools found at the European Bioinformatics Institute (EMBL-EBI). In addition, the BLAST search web page of the National Centre of Biotechnology Information (NCBI) will align two sequences. Figure 14 shows a DNA sequence alignment of nine different organisms, generated using the programme ClustalX.

```
Lion            ATGTTCATAAACCGCTGACTATTTTCAACCAATCACAAAGACATTGGAAC
                ATGTTCATAAACCGCTGACTATTTTCAACCAATCACAAAGATATTGGAAC
Snow leopard    ATGTTCATAAACCGCTGACTATTTTCAACCAATCACAAGGATATTGGAAC
                ATGTTCATAAATCGCTGACTGTTTTCAACTAATCATAAAGATATTGGCAC
Tiger           *********** ********.******** ***** **.** *****.**
```

▲ Figure 14 A DNA sequence alignment of the cytochrome oxidase gene for four different large cats, generated using the programme ClustalW

The NCBI website (http://www.ncbi.nlm.nih.gov/) has tools for carrying out a comparison of protein and DNA sequences. In this example, we will conduct a sequence alignment using the cytochrome oxidase (cox1) protein for two species of primate called tarsiers. Horsfield's tarsier, variously classified as *Cephalopachus bancanus* and *Tarsius bancanus*, is a threatened species that lives in Borneo and Sumatra. The cox1 sequence for this tarsier will be compared with the sequence of the same gene for the Philippine tarsier (*Carlito syrichta*). There is some uncertainty over the classification of Horsfield's tarsier and sequence comparison is often used to resolve this kind of controversy. Use the search term cytochrome c oxidase subunit I [Tarsius bancanus]. Click on the COX1 highlighted text to determine the accession numbers (They start with NC_ for DNA and NP_ for proteins). Repeat the procedure using the search term cytochrome c oxidase subunit I [Carlito syrichta]. Then go to the BLAST tool (https://blast.ncbi.nlm.nih.gov/) and decide whether you are going to align the DNA or protein sequences. Be sure to check the "align two or more sequences". Identify the differences in the protein and DNA sequences. Explain how such information can be used to determine evolutionary relationships.

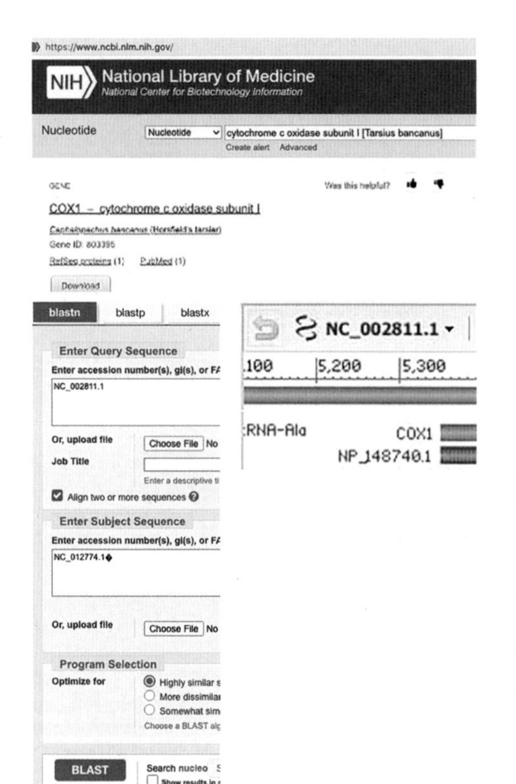

▲ Figure 17

▲ Figure 15 Horsfield's tarsier

▲ Figure 16 Philippine tarsier

A3.2.6 Base sequences of genes or amino acid sequences of proteins as the basis for constructing cladograms

Within a clade, some species will be more closely related than others, because they diverged relatively recently. They will have fewer differences in base or amino acid sequence. Conversely, species that diverged a longer time ago are likely to have more differences. By comparing base sequences, it is possible to estimate how long ago pairs of species diverged. These estimates can then be used to suggest the order in which the divergences occurred.

Much more sophisticated analysis can be done using computer software. Sequences for all species in a clade can be compared in combination. The software can then use calculations to determine how the species could have evolved with the smallest number of sequence changes. This is known as the parsimony criterion. It does not *prove* how a clade evolved but it indicates the most probable pattern of divergence.

Sequence analysis is used to construct a cladogram. A cladogram is a branching diagram that represents ancestor–descendant relationships. An example is shown in Figure 18.

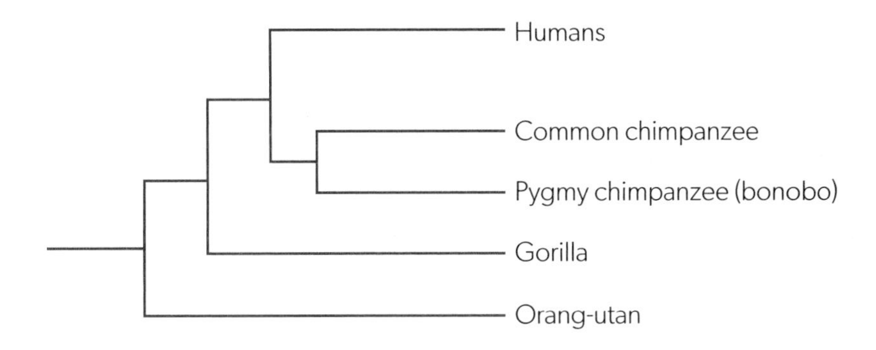

▲ Figure 18 A cladogram for humans and primates

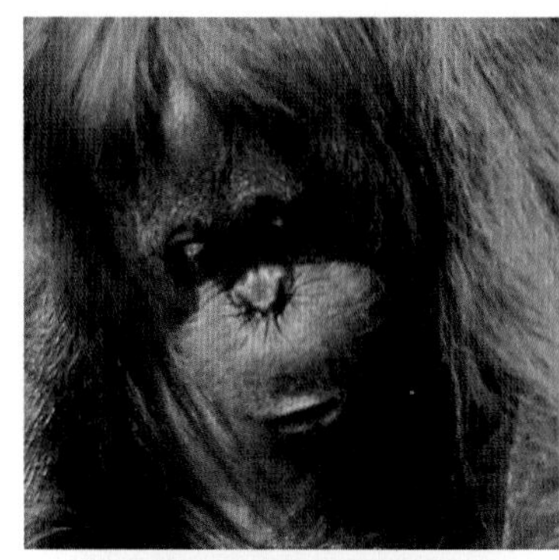

▲ Figure 19 Which of these three members of the family Hominidae are most closely related, according to the cladogram in Figure 18?

Sequence data for more than one gene can be used to produce multiple cladograms for a group of organisms. If the cladograms show the same pattern of divergence, this is strong evidence of how the group evolved.

Applying technology to process data: Using software to construct a phylogenetic tree

A phylogenetic tree created using the methods of cladistics is a cladogram. A hypothesis about evolutionary relationships can be tested using bioinformatics software. For example, consider large cats from around the world. Their evolutionary relationship is likely to be correlated with their current geographic distribution. More closely related cats are more likely to be closer to each other in geographic distribution. For example, African large cats are more likely to be related than African cats are to North American cats.

Follow these steps to construct a cladogram:

a. Search a database to find DNA sequences for a gene which is conserved across many species such as cytochrome c oxidase subunit 1. The National Center for Biotechnology Information site (NCBI) is freely searchable and is recommended. We are going

→

to construct a phylogenetic tree for the following species: puma (*Puma concolor*), tiger (*Panthera tigris*), snow leopard (*Panthera uncia*), lion (*Panthera leo*) and leopard (*Panthera pardus*).

b. Search the NCBI website for the DNA sequences using the search term: cytochrome c oxidase subunit I [Genus species] for example cytochrome c oxidase subunit I [Panthera leo].

c. Under the Genomic regions, transcripts and products, choose 'FASTA' which should display a nucleotide sequence. Highlight all of the DNA sequence including the title (for example '>NC_010638.1:6099-7643 Panthera uncia)

d. Open either Notepad from your PC or TextEdit on a Mac.

e. Paste your sequence into the text editing document.

f. Repeat with several other sequences from different organisms.

g. Edit the titles but remember to include the '>' symbol and to separate words in the title with an underscore. For example: >Panthera_uncia

h. Collate all of the DNA sequences representing the different organisms in the study. Depending on the software to be used in later steps, the sequences could be copied and pasted into a text file. In this example, you will use the Multiple Sequence Alignment tool (Clustal Omega found at the EMBL website).

i. Upload the file to Clustal Omega.

j. Perform the sequence alignment. Through observation can you determine the relationship between the cats?

k. Construct the tree using Clustal Omega.

l. Compare your tree to other trees available through an internet search. How does your tree compare?

Hypotheses: Parsimony as a criterion for judgement

A cladogram is a statement of hypothesized evolutionary relationships. Different criteria for judgement can lead to different hypotheses. Parsimony analysis is used to select the most probable cladogram, in which observed sequence variation between clades is accounted for with the smallest number of sequence changes.

Sequences could have reached their current order in many different ways. For example, the base cytosine could change to thymine, then back to cytosine and then to thymine again. In this case, parsimony analysis would presume that the change was simply cytosine to thymine. The tree that involves the smallest number of evolutionary changes is chosen: although there is no proof of how the clade actually evolved, the simplest explanation is most likely to be true.

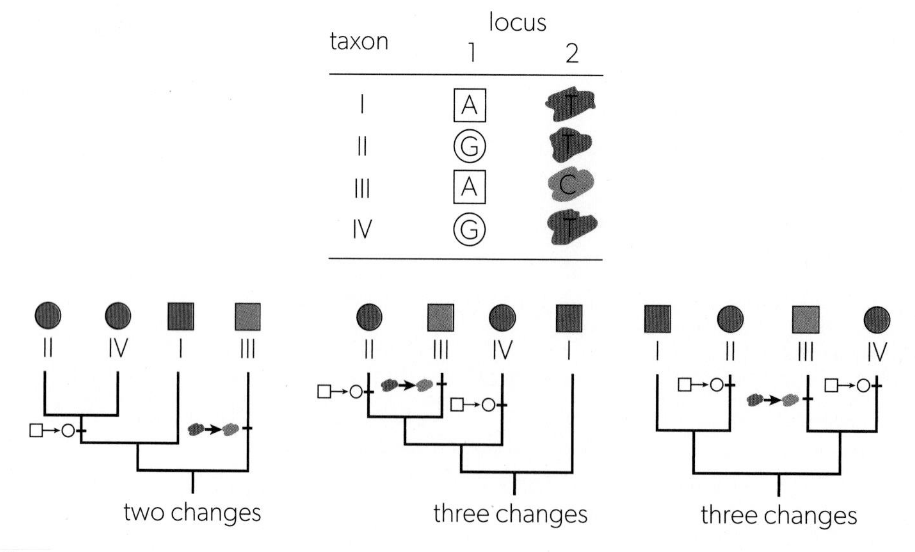

▲ **Figure 20** Cladograms showing DNA base changes: What base was present at locus 1 in the ancestor? At locus 2? Which cladogram is the best one if parsimony is the criterion for judgement?

A3.2.7 Analysing cladograms

When analysing cladograms, remember:

- A cladogram is a tree diagram with a number of branches.
- The **terminal branches** are ends that represent individual clades. These may be species or groups of species that are not subdivided on the cladogram.
- The branching points on a cladogram are called **nodes**. Usually, two clades branch off at a node but sometimes there are three or more. A node represents the point at which a hypothetical ancestral species split to form two or more clades.
- Two clades that are linked at a node are relatively closely related. Clades that are only connected via a series of nodes are less closely related.
- The **root** is the base of the cladogram. This is the hypothetical common ancestor of all the clades.
- Some cladograms include numbers to indicate numbers of sequence differences.
- Some cladograms are drawn to scale, based on estimates of the time since each split occurred.
- The pattern of branching in a cladogram is assumed to match phylogeny of the organisms—the evolutionary origins of each species.

Although cladograms can provide strong evidence for the evolutionary history of a group, they cannot be regarded as proof. Cladograms are constructed on the assumption that the smallest possible number of mutations occurred that can account for current base or amino acid sequence differences. Sometimes this assumption is incorrect and pathways of evolution were more convoluted. It is therefore important to be cautious when analysing cladograms. Where possible, you should compare several versions that have been produced independently using different genes.

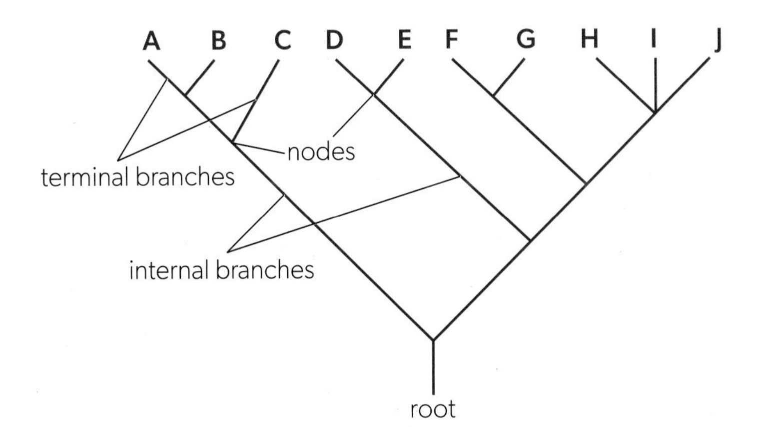

▲ Figure 21 Cladograms usually show only two branches forming at each node. Occasionally three or more branches are shown, as on the right of this diagram. With more research it would almost certainly be found that either H, I or J split off first, followed later by a split between the other two species

Data-based questions: Origins of turtles and lizards

Cladograms based on morphology suggest that turtles and lizards are not a clade. To test this hypothesis, microRNA genes have been compared for nine species of chordate. The results were used to construct the cladogram in Figure 22. The numbers on the cladogram show which microRNA genes are shared by members of a clade but not by members of other clades. For example, there are six microRNA genes that are found in humans and short-tailed opossums but not in any of the other chordates on the cladogram.

1. Deduce, using evidence from the cladogram, whether humans are more closely related to the short-tailed opossum or to the duck-billed platypus. [2]
2. Calculate how many microRNA genes are found in the mammal clade on the cladogram but not in the other clades. [2]
3. Discuss whether the evidence in the cladogram supports the hypothesis that turtles and lizards are not a clade. [3]
4. Evaluate the traditional classification of tetrapod chordates into amphibians, reptiles, birds and mammals using evidence from the cladogram. [3]

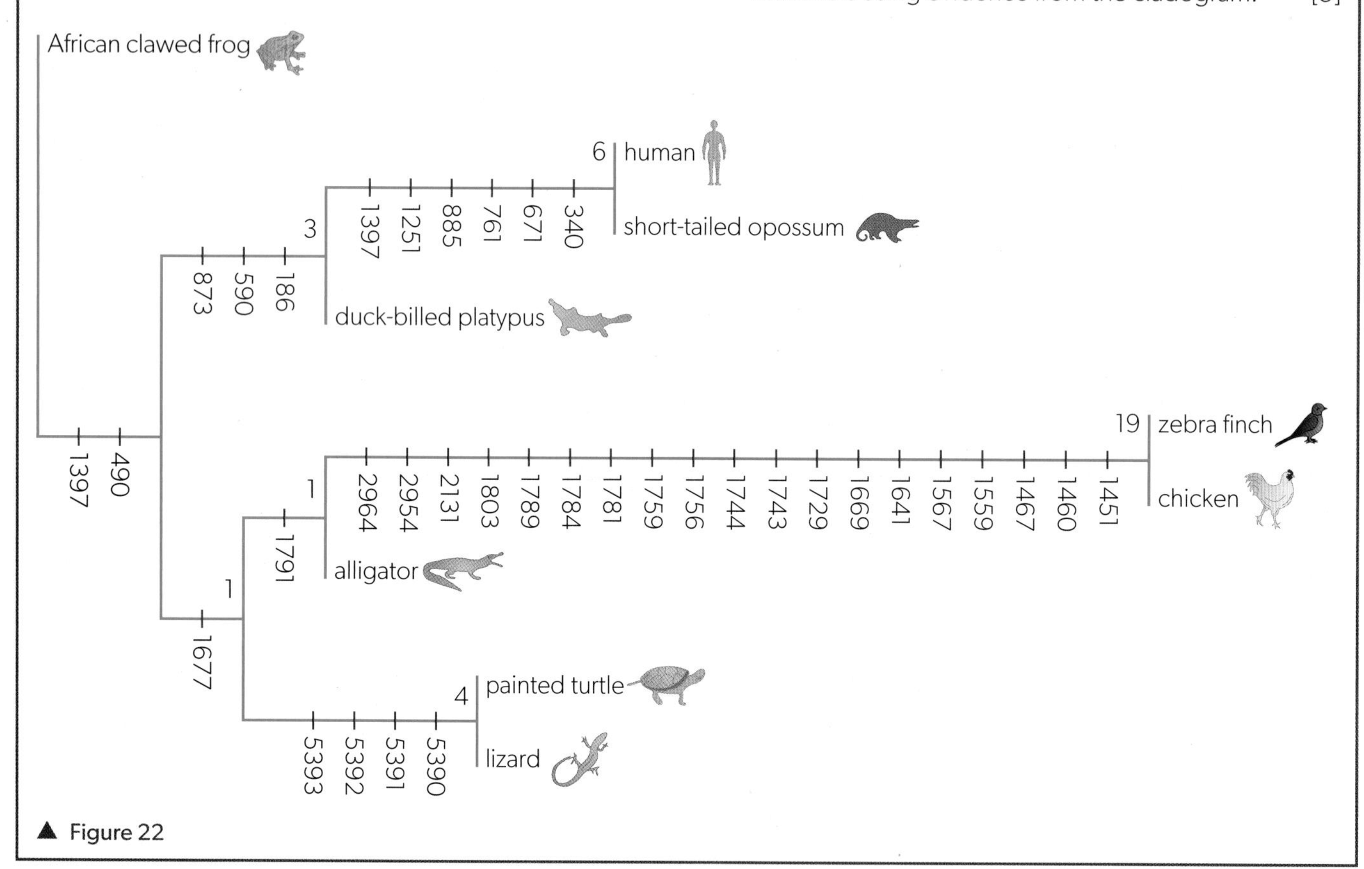

▲ Figure 22

A3.2.8 Using cladistics to investigate whether the classification of groups corresponds to evolutionary relationships

Since the 1990s, it has been relatively easy to find the base sequences of genes or even whole genomes. This sequence data has allowed researchers to check traditional classifications of plants and animals, using the objective method of cladistic analysis. In many cases, cladistics research has confirmed that the traditional classification matches the most probable pathways of evolution.

In some cases, however, the species placed in a taxonomic group do not all share a common ancestor. In other cases, species that have evolved from the same common ancestor have been placed in different groups. In such cases, reclassification is justified. The following case study illustrates this process. The details are not important but they will help you to understand the principles involved. You could study the reclassification of other groups, such as the Sterculiaceae, if you wish to find out more.

Reclassification based on cladistic analysis

There are more than 400 families of angiosperms (flowering plants). Until recently, the eighth largest was the Scrophulariaceae, commonly known as the figwort family. It included over 275 genera, with more than 5,000 species. Taxonomists used cladistics to investigate the evolutionary origins of the figwort family. They found that species in this family did not share a single common ancestor so were not a true clade.

A major reclassification was carried out. Five groups of species were moved to other families, leaving fewer than half of the original species in the figwort family. It is now only the 36th largest among the angiosperms. A summary of the changes is shown in Figure 23.

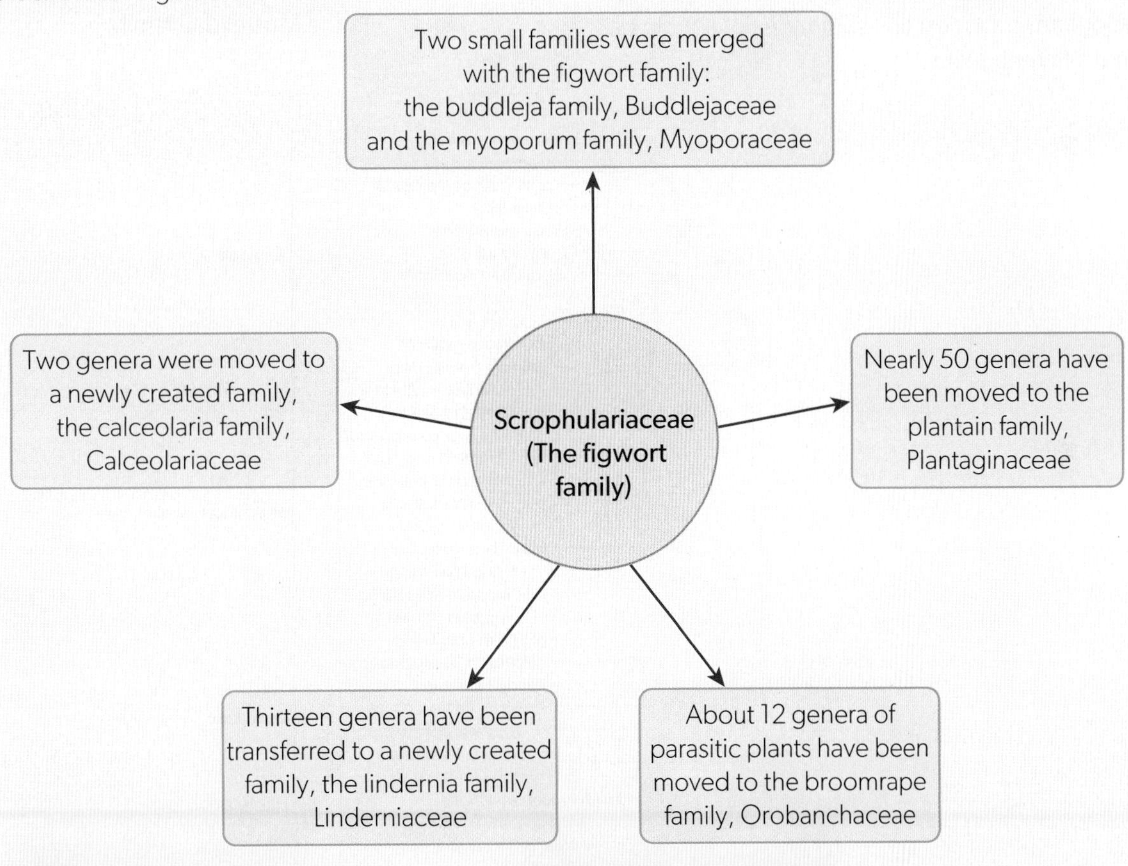

▲ Figure 23

▲ Figure 24 *Scrophularia chrysantha* (golden figwort) has remained in the figwort family

▲ Figure 25 *Veronica beccabunga* (European speedwell) has been transferred to the Plantaginaceae (plantain family)

Data-based questions: Mustelid classification

The Mustelidae is a family of 59 species of mammal, classified into 22 genera on the basis of morphology. Base sequences of 22 gene segments were analysed to produce thousands of different hypothetical cladograms. Figure 26 shows the consensus cladogram, based on maximum parsimony. *Bassariscus astutus* and *Procyon lotor*, at the bottom of the cladogram, are members of a different family and were used for reference purposes. Such species in a cladogram are called the outgroup. All the other species from *Aonyx capensis* to *Taxidea taxus* are members of the Mustelidae. According to the traditional classification, the Mustelidae family is subdivided into the Lutrinae and Mustelinae. A further proposed subdividsion is shown to the right of the cladogram. There is evidence in the cladogram to suggest that some species should be moved to a different genus.

1. The nodes on the cladogram have been numbered. What is indicated by a node? [2]
2. The scale bar shows how many base substitutions are indicated by each length of horizontal line. What can be estimated from the number of base substitutions since two species diverged? [2]
3. Using evidence from the cladogram, discuss whether:
   a. *Martes pennanti* should be moved to a different genus [2]
   b. the Mustelinae should be subdivided into seven smaller groups [2]
   c. *Bassariscus astutus* and *Procyon lotor* should be moved to the Mustelidae family. [2]

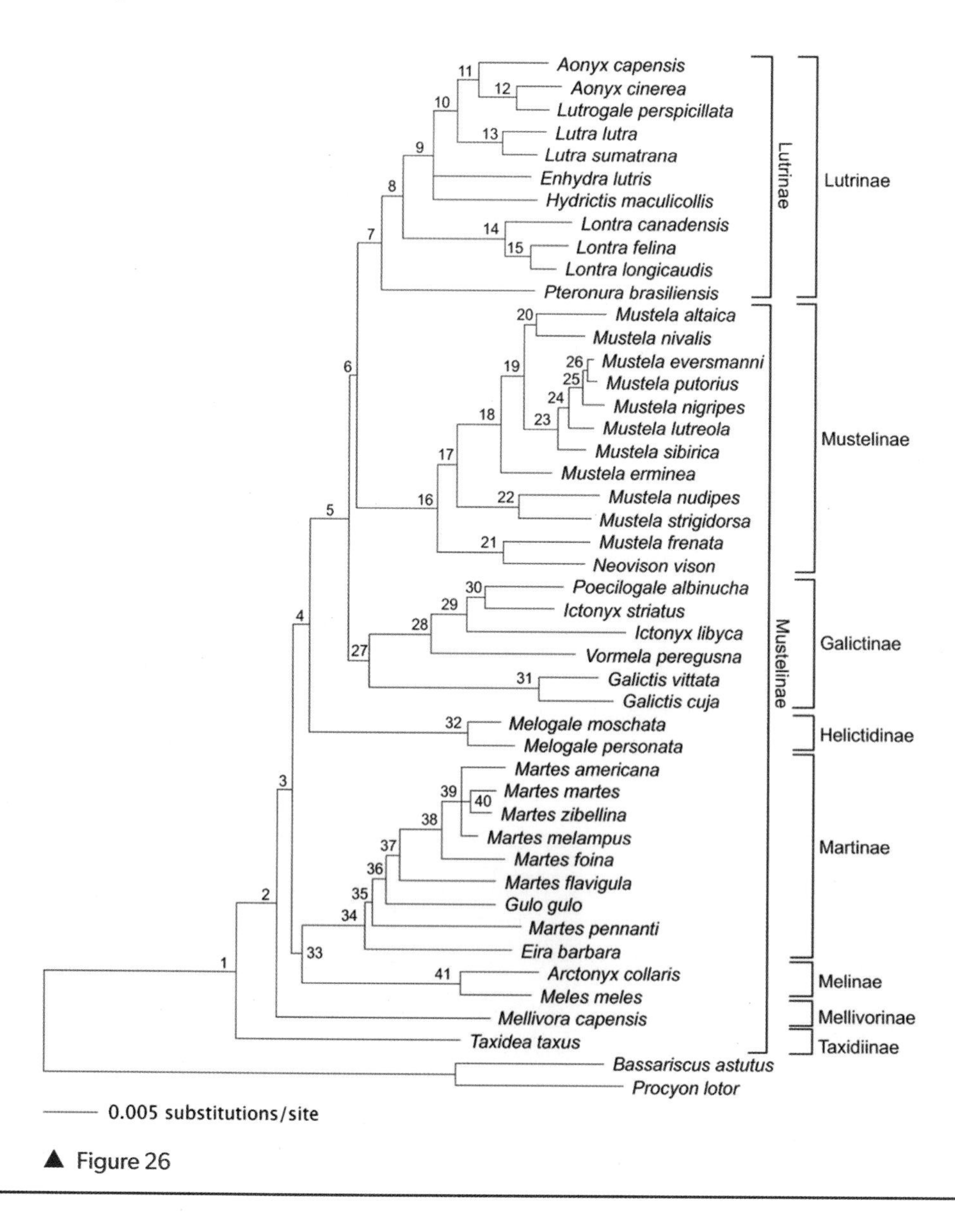

▲ Figure 26

Falsification: Reclassification based on phylogeny

A cladogram depicts a hypothesized evolutionary relationship. Because scientific knowledge claims are based on empirical evidence, it is not possible to claim with absolute certainty that they are true. Certainty becomes possible if we can find a counterexample and establish what is **not** the case. Karl Popper called this property of scientific knowledge claims "falsifiability". Hypotheses, theories and other scientific knowledge claims may eventually be falsified.

Elephant shrews are small insect-eating mammals native to Africa. Their common English name, elephant shrew, comes from a perceived likeness between their long noses and the trunk of an elephant, and their superficial similarity with shrews. Phylogenetic analysis has shown that elephant shrews should not be classified as true shrews; in fact, they are more closely related to elephants than to shrews!

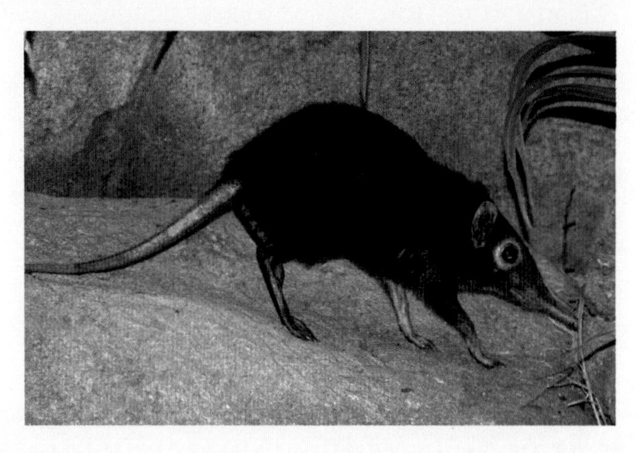

▲ **Figure 27** East African black and rufous elephant shrew or Sengi (*Rhynchocyon petersi*)

A3.2.9 Classification of all organisms into three domains using evidence from rRNA base sequences

Traditional classification systems have recognized two major categories of organisms based on cell types: eukaryotes and prokaryotes. This classification is now regarded as inappropriate because the prokaryotes are so diverse. In particular, when the base sequences of ribosomal RNA were determined, it became apparent that there are two distinct groups of prokaryotes. They were given the names Eubacteria and Archaea.

Most classification systems therefore now recognize three major categories of organism, Eubacteria, Archaea and Eukaryota. These categories are called domains, with all organisms classified into the three domains. Members of the domains are usually referred to as bacteria, archaeans and eukaryotes. Bacteria and eukaryotes are relatively familiar to most biologists but archaeans are often less well-known.

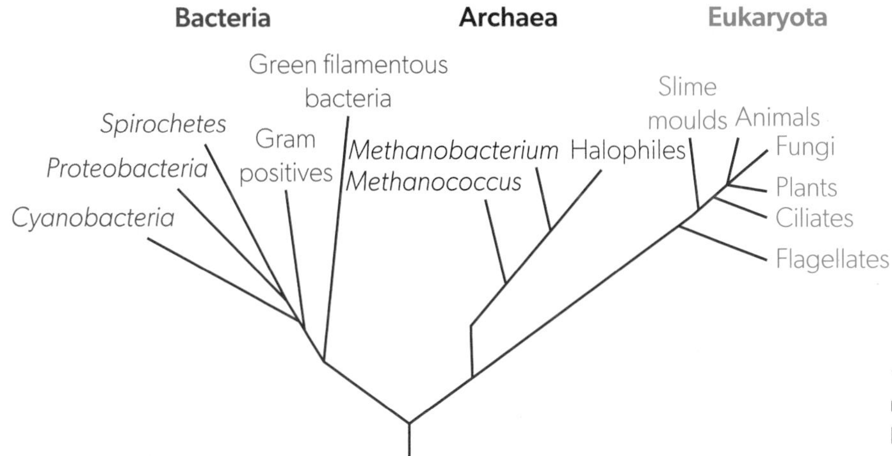

◀ **Figure 28** Tree diagram showing relationships between living organisms based on base sequences of ribosomal RNA

AHL

ATL Social skills: Actively considering the perspective of others

During debates, you may find yourself disagreeing with someone. If this happens, take time to be fair-minded and consider other people's perspective. Be sure that you are not using different criteria for judgement. For example, it is possible to argue that the elephant shrew and tree shrews are not related, or that they are related, depending on your criteria for judgement.

Approach conversations with the understanding that you might be mistaken and that other people might have valid ideas. The consensus is that classification should be based on evolutionary relationships, as this allows us to make predictions. For example, two closely related organisms should have similar metabolism. However, in some contexts it is also reasonable to classify an organism by its niche. In this case, the elephant shrew could be considered related to other shrews as they are all insectivores.

Data-based questions: Similarities and differences in microbial cell wall structure

Figure 29 shows the plasma membrane and cell wall structures in five groups of microorganism.

1. Compare the plasma membranes of the microorganisms. [2]
2. Compare the cell walls of the microorganisms. [2]
3. Distinguish between the cell wall of Group Z and the other groups. [2]
4. a. Compare and contrast the structures outside the cell wall in Groups V, W, X and Y. [4]
   b. Construct a cladogram based on these comparisons and contrasts. [4]
5. Deduce which of the five groups is the fungi. [1]

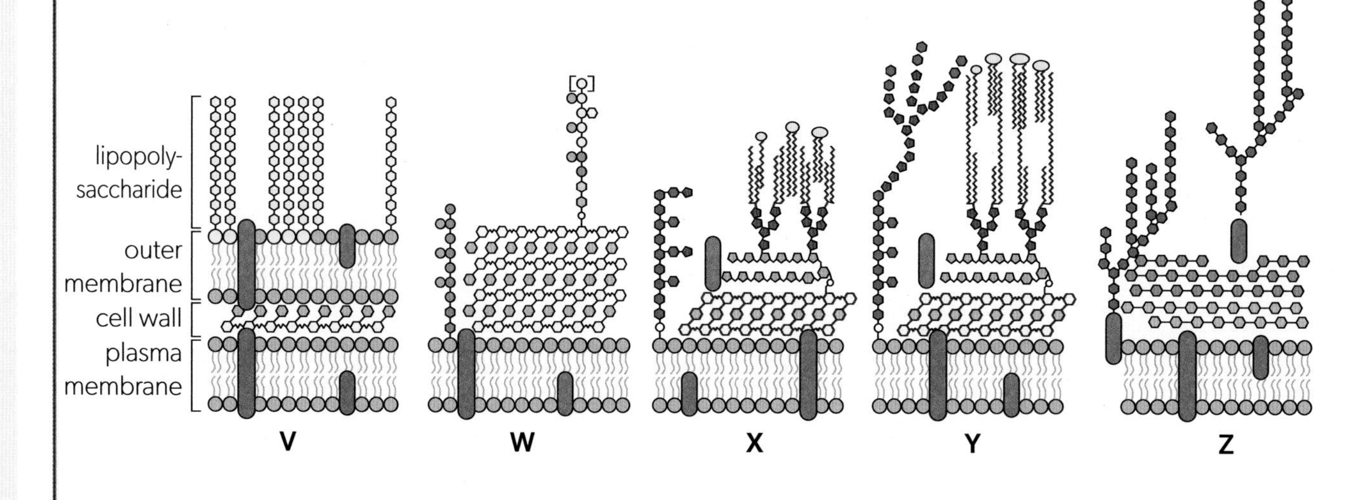

▲ Figure 29

Source: Chen, Y., Fischbach, M. & Belkaid, Y. Skin microbiota–host interactions. Nature 553, 427–436 (2018). https://doi.org/10.1038/nature25177

Linking questions

1. How can similarities between distantly related organisms be explained?
   a. Outline the concept of an ecological niche. (B4.2)
   b. Describe the mechanism of evolution by natural selection. (D4.1)
   c. Explain the mechanism of convergent evolution. (A4.1.5)
2. What are some examples of ideas over which biologists disagree?
   a. Describe the evidence that confirmed DNA and not protein was the genetic material. (A1.2.14)
   b. Explain how using the principle of parsimony could lead to an error in classification. (A3.2.6)
   c. Suggest why categorization of some populations as "evolutionarily distinct" and therefore a higher priority for conservation might be controversial. (A4.2.8)

TOK

How does the way in which we organize or classify knowledge affect what we know?

Perception is the act of interpreting sensory information. The environment presents you with endless sensory information, most of which you ignore as unimportant. What you do notice, you classify in a variety of ways—often without even realizing it. For example: When you hear a loud sound, you might classify it as representing a threat or not a threat; or you might identify the direction from which it came. When you see a fruit on a plant, you might classify it as ripe or not, or edible or not. In everyday language, you might classify organisms as domesticated or wild; dangerous or harmless; and so on. There are infinite ways to interpret and organize your observations.

Similarly, scientists classify organisms in a number of ways. For example, they might group them by morphology (physical similarity to other organisms), phylogeny (evolutionary history) or niche (ecological role).

Consider the four animals in Figure 1. They can be classified by trophic level; marsupial or placental; niche or habitat; patterned fur or monochrome; conservation status (threatened or not); and so on. Each categorization focuses on a different feature or features of the organism.

Natural classification looks at evolutionary relationships. Thus, the lynx and the musk deer are grouped together as placental mammals, while the spotted-tailed quoll (a marsupial that occupies the niche of a small cat) and the whiptail wallaby are grouped together as marsupials. In terms of trophic level, the deer and the wallaby are grouped together because they are both herbivores. In terms of habitat, the quoll and the lynx are both forest dwellers.

The names used can impact our perception, because different names predispose the listener to focus on a particular aspect of the organism. This is particularly apparent in folk taxonomy—the everyday names given to things. For example, insects in the family Pentatomidae have a shape which looks like a heraldic shield when viewed from above. This is why, in some folk taxonomies, they are referred to as "shield bugs". Because they release a strong-smelling spray when threatened, they are referred to in other folk taxonomies as "stink bugs". The Latin family name, Pentatomidae, refers to the fact that their antennae have five segments.

▶ **Figure 1** A Siberian musk deer (top left); a lynx (top right); a whiptail wallaby (bottom left) and a spotted-tailed quoll (bottom right)

The *Cerion* snail is endemic to the Caribbean. Folk taxonomies refer to it as the peanut snail or the honeycomb snail. Figure 3 shows some of the varieties of snail from the three islands of the Netherlands Antilles. Nineteenth century naturalists classified *Cerion* snails into a large number of different species based on physical differences (such as colour, lip thickness and number of grooves). In the 20th century, molecular biologists were able to show that the variety represented a much smaller number of species. All of the snails shown in Figure 3 belong to the species *Cerion uva*. The environment has an impact on gene expression and how the snails develop. On windy, wavy shores, the snails develop thicker and stronger shells. On low energy coastlines, colour differences are more pronounced.

Taxonomists are scientists who classify organisms. Taxonomists themselves can be classified! "Lumpers" are those who tend to see different individuals as varieties of the same species. "Splitters" are more prone to emphasize these differences as indicative of unique species.

▲ Figure 2 A shield bug or stink bug

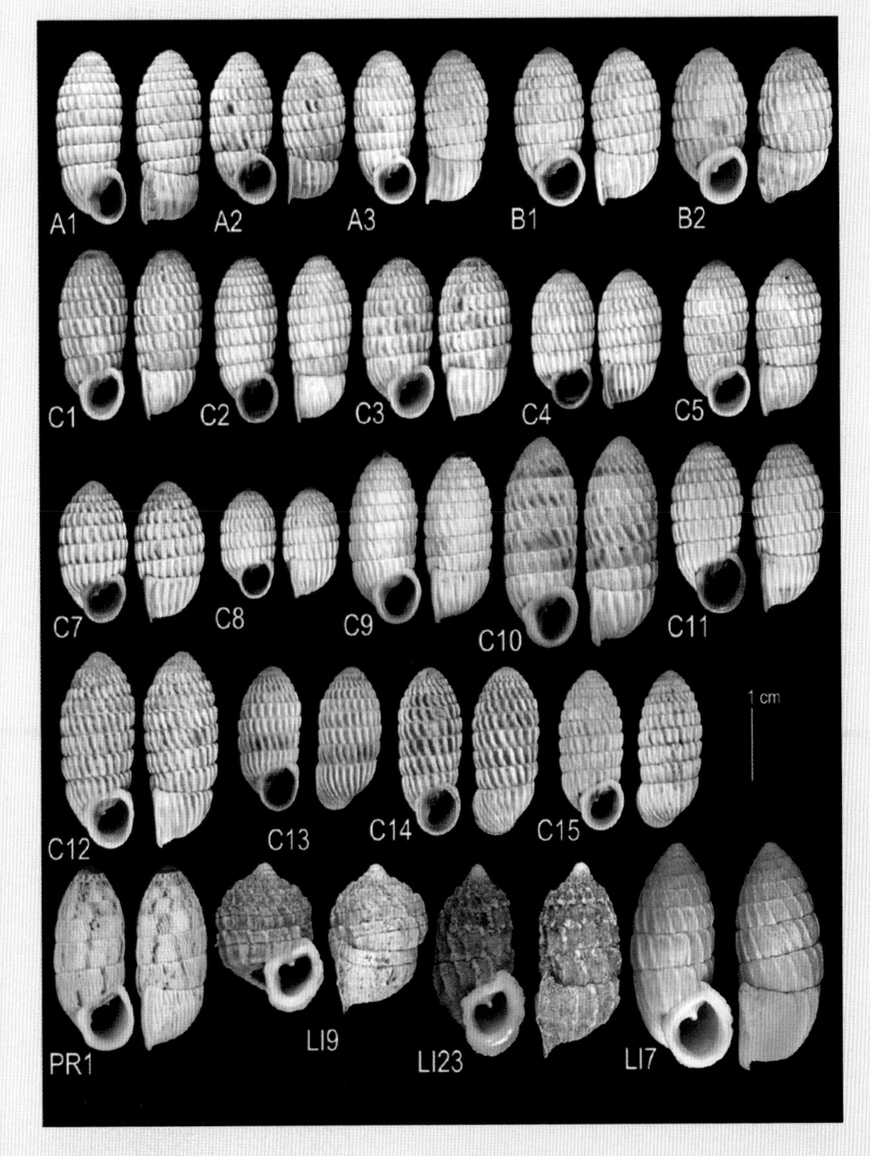

▲ Figure 3 *Cerion uva* snails

End of chapter questions

1. Evidence suggests that humans are descended from a species of African ape that has spread to colonize new areas. The species most closely related to humans are the chimpanzee and the gorilla. Studies of biochemistry and chromosome numbers provide conflicting evidence of the evolutionary relationship between these three primates. Three models showing possible relationships are shown in Figure 1. Some of the evidence on which the diagrams were constructed is shown in Table 1.

   a. Identify which model appears correct, based on the evidence from chromosome number. [1]

   b. Evaluate each of the models according to the biochemical evidence. [3]

   c. Canine teeth can be divided into two groups, large or small. The evidence from canine teeth supports Model C. Gorillas have large teeth. Deduce the type of teeth that chimpanzees and humans have.

   d. Humans, chimpanzees and gorillas all possess broad flat molar teeth for grinding plant matter. Lion molars are not adapted for grinding. Suggest why eating plant matter requires molar teeth while meat eating is less reliant on broad, flat molars. [2]

   e. Outline the evidence from chromosome 2 of the relatedness of humans, gorillas and chimpanzees. [3]

Model A

H G C

CA

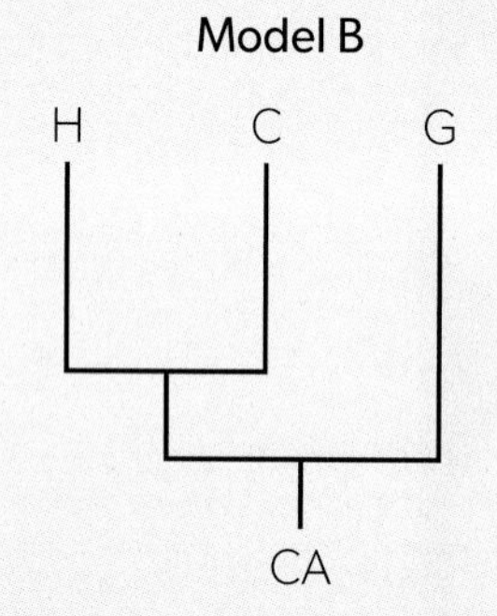

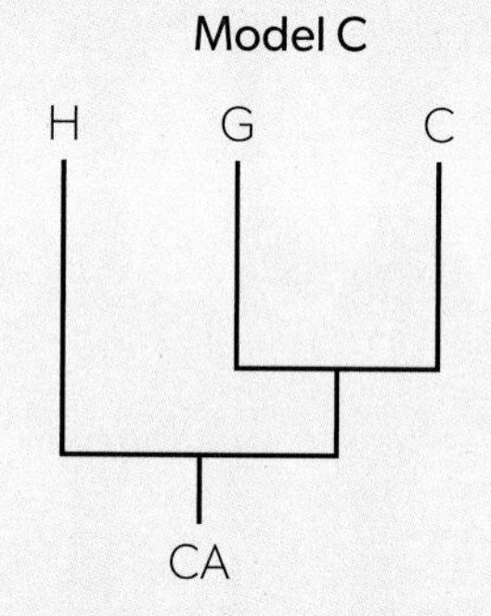

Key
CA common ancestor
H human
G gorilla
C chimpanzee

▲ Figure 1

Evidence	Human	Chimpanzee	Gorilla
Chromosome number	46	48	48
Plasma proteins	same as gorilla and chimpanzee	same as human and gorilla	same as human and chimpanzee
Myoglobin	differs from chimpanzee and gorilla by one amino acid	same as gorilla	same as chimpanzee
Haemoglobin	same as chimpanzee	same as human	differs from human and chimpanzee by one amino acid

▲ Table 1

2. One method used by microbiologists to distinguish between Archaea and Eubacteria is based on the conditions they need for survival. Both groups include thermophiles—species that are adapted to live at high temperatures. The graph in Figure 2 shows the optimum temperature and minimum pH required for growth by selected species of Archaea and thermophilic Eubacteria.

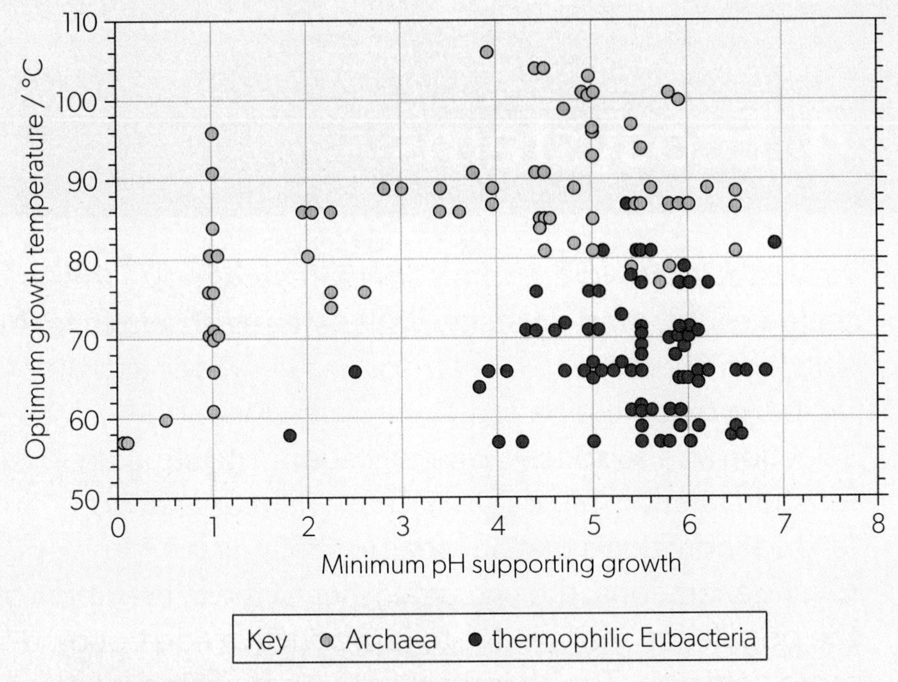

Source: Adapted from D. L. Valentine (2007), 'Adaptations to energy stress dictate the ecology and evolution of the Archaea', Nature Reviews Microbiology, 5, page 316.

▲ Figure 2

a. State the highest optimum growth temperature recorded for the thermophilic Eubacteria. [1]

b. State the relationship between minimum pH supporting growth and optimum growth temperature for Archaea. [1]

c. Compare the results for the Archaea with those for the thermophilic Eubacteria. [2]

d. With reference to the data, suggest why this method would not always be suitable for distinguishing between Archaea and thermophilic Eubacteria. [2]

e. State a possible habitat for methanogenic Archaea. [1]

A Unity and diversity

4 Ecosystems

Analogous and homologous structures exemplify the theme of unity and diversity. Both patterns arise due to the selection pressures within an ecosystem. Biotic and abiotic factors provide the pressures that contribute both to natural selection and to species diversification. The structure–function relationship exists because natural selection favours individuals that are adapted to their environment. Over generations, the characteristics of those members of the population that survive to reproductive age become more common. The change in species over time is known as evolution. When unrelated species encounter the same selection pressures, they can develop analogous structures that are similar in appearance and function but have different histories. For example, the long-eared jerboa (*Euchoreutes naso*) and kangaroos have distinct evolutionary histories. However, they have both developed long ears for heat exchange, movement by hopping and an extended tail for balance. These features are an example of convergent evolution, a result of both organisms living in arid habitats. Yet, they have significantly different reproductive strategies as the jerboa develops a placenta during pregnancy and the kangaroo is a marsupial which means its young complete development in a pouch outside of the uterus. These examples show that the habitat of an organism can drive both convergence and divergence. It follows that loss of habitat and unique environments is a significant factor increasing the threat of extinction of species. Biodiversity is the variety of life in all its forms, levels and combinations, including ecosystem diversity, species diversity and genetic diversity.

A4.1 Evolution and speciation

What is the evidence for evolution?

The theory that species change over time by the mechanism of natural selection has such strong predictive and explanatory power that it is unlikely to be falsified. Figure 1 shows both fossil and human pentadactyl limb evolution. The five-fingered pentadactyl hand is shared by humans with ancestors going back over 300 million years of evolution. Here, a human hand is shown with the foot bones of a small predatory North American Permian reptile called Captorhinus, which is approximately 280 million years old. Explain how the shared anatomy of limbs with diverse functions provides evidence for evolution.

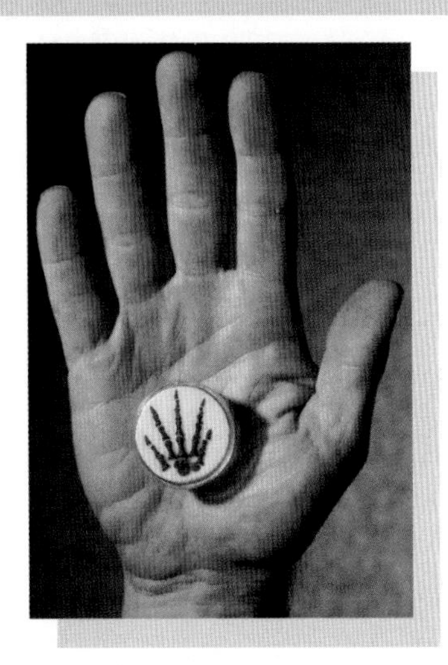

◀ Figure 1

How do analogous and homologous structures exemplify commonality and diversity?

The wings of a bat and a seagull are similar in form and function. If this is not a result of common ancestry, how did these analogous structures develop? What is the role of selection pressures in evolution? What is the distinction between analogous and homologous structures?

Why do organisms with a common ancestry that are subjected to different selection pressures become increasingly different? How do homologous structures provide evidence of evolution? Are the hands of a human and bonobo ape (*Pan paniscus*) homologous or analogous structures? What is the significance of the opposable thumb to evolution?

▲ Figure 2

SL and HL	AHL only
A4.1.1 Evolution as change in the heritable characteristics of a population A4.1.2 Evidence for evolution from base sequences in DNA or RNA and amino acid sequences in proteins A4.1.3 Evidence for evolution from selective breeding of domesticated animals and crop plants A4.1.4 Evidence for evolution from homologous structures A4.1.5 Convergent evolution as the origin of analogous structures A4.1.6 Speciation by splitting of pre-existing species A4.1.7 Roles of reproductive isolation and differential selection in speciation	A4.1.8 Differences and similarities between sympatric and allopatric speciation A4.1.9 Adaptive radiation as a source of biodiversity A4.1.10 Barriers to hybridization and sterility of interspecific hybrids as means of preventing the mixing of alleles between species A4.1.11 Abrupt speciation in plants by hybridization and polyploidy

A4.1.1 Evolution as change in the heritable characteristics of a population

There is strong evidence for the characteristics of populations changing over time. Biologists call this process evolution. It is how the diversity of life developed and lies at the heart of a scientific understanding of the natural world. Evolution only concerns heritable characteristics—traits that are inherited by offspring from parents. This is emphasized in the definition:

Evolution is change in the heritable characteristics of a population.

The mechanism of evolution is now well understood—it is natural selection (explained in *Topic D4.1*). Evolution by natural selection is also called Darwinism.

There is also strong evidence for the characteristics of individual organisms changing during their lifetimes. For example:

- trees can develop a very asymmetric form if they grow in a position exposed to wind
- birds are influenced by hearing their parents singing when they develop their song
- human tennis players develop stronger muscles and bones in the arm they use to hold the racket
- children learn the languages their parents speak.

These are known as acquired characteristics.

▲ Figure 3 Windswept hawthorn trees on a Welsh hilltop

Before Charles Darwin published *On the Origin of Species* in 1859, the leading theory for evolution was based on inheritance of acquired characteristics. The main proponent of this theory was Jean-Baptiste Lamarck, so it is known as Lamarckism. It is obvious that seeds from a tree growing asymmetrically will not grow into asymmetric offspring unless they are exposed to the same environment as the parent. Similarly, a tennis player's children will not develop stronger bones in one arm than the other. No mechanism has been discovered for the environment causing specific adaptive changes to the base sequence of genes, or for it causing the creation of new genes. Therefore, acquired characteristics are not inherited and do not lead to evolution. Lamarckism has been falsified again and again, despite repeated attempts to revive the theory.

Theories: Pragmatic truth

In everyday language, a "true" statement is one that everyone agrees corresponds to reality. However, the correspondence theory of truth is not the only possibility. For example, the pragmatic theory of truth holds an assertion to be true if it "works".

Knowledge claims in science are based on observations of a fraction of possible cases or instances. Scientists use their observations to form generalizations that are then tested. If the generalizations are supported, a theory emerges. If the theory can explain and predict future observations, it is said to be a pragmatic truth: a truth that works.

The theory of evolution by natural selection predicts and explains a broad range of observations, such as antibiotic and pesticide resistance and also the existence of homologous and analogous structures. Thus the theory is unlikely to be falsified. However, the nature of science makes it impossible to prove formally that the theory of evolution by natural selection is true. It is a pragmatic truth and is therefore regarded as a theory, despite all the supporting evidence.

A4.1.2 Evidence for evolution from base sequences in DNA or RNA and amino acid sequences in proteins

If evolution is a change in the heritable characteristics of a population, we can expect to see changes in genes whenever evolution occurs. These changes will happen in the base sequence of DNA or RNA and in the amino acid sequences of proteins made using those base sequences. Consider the evolution of the coronavirus that caused a pandemic starting in 2020 (COVID-19). Many base sequence changes occurred in the genes of this coronavirus, affecting the viral traits. Some new variants were more successful than earlier ones in spreading through the human population—the virus evolved.

Evidence for evolution also comes from comparing base sequences of the same gene in different species. A clear relationship is seen: the more closely related two species are, in their morphology and other traits, the fewer differences in base sequence there are. This trend is difficult to explain without evolution. It is convincingly explained by the theory that species develop over time, gradually diverging from a common ancestor as a result of differences in natural selection.

In addition, observed combinations of differences are only easily accounted for by repeated splitting of ancestral species by evolution. This is why cladograms based on sequence differences usually match closely with classifications based on morphology and the likely sequence of splits between lineages.

Evidence of evolution also comes from gene families that occur across diverse groups of organisms. For example, the Hox gene family occurs widely in animal genomes. Genes in this family help to determine the body plan during development. Similarities between Hox genes can only reasonably be explained by common ancestry, with duplication to give multiple copies of the gene and gradual modification for different functions in different lineages. Hox genes occur in cnidaria and in all animals with a clear head-to-tail axis, including annelids, arthropods and vertebrates; these species form a clade known as the bilateria.

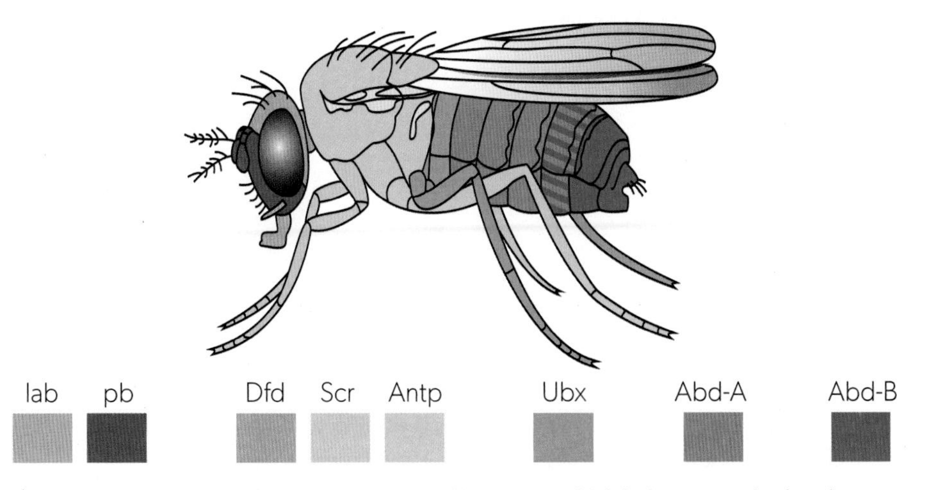

▲ Figure 4 The fruit fly *Drosophila* has eight Hox genes which help to organize head-to-tail development of different parts of the body. Humans have 39 Hox genes which help to organize our head-to-tail development

Data-based questions: Convergence and divergence of sequences

After a clade splits, there can be divergence of the base and amino acid sequences of the separated clades. The more time has passed since the split, the more differences we expect due to this evolution. It therefore follows that if we look back at the ancestry of two related clades, the closer we get to a common ancestor, the fewer sequence differences there will be. Figure 5 shows a theoretical cladogram, with a common ancestor (P) that split to produce two ancestral clades (Q and R), which then split repeatedly to form multiple clades.

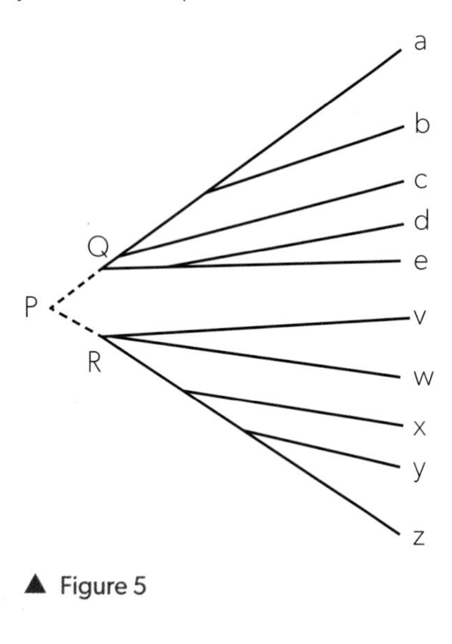

▲ Figure 5

The hypothesis that there is ancestral convergence in sequences was tested using two plant clades (monocots and eudicots). There is strong evidence for monocots and eudicots having a common ancestor. Amino acid sequences of 51 proteins in 24 species of monocot and 44 species of eudicot were compared. Sequence convergence in the ancestors of the two clades was found: the probability of the observed pattern of sequence differences being due to anything other than evolution was calculated as 1×10^{-132}. This is an infinitesimally small chance.

1. Explain how the most probable base sequences of ancestor P and ancestor Q can be determined. [2]
2. Discuss whether more base sequence differences are expected between clade e and clade v or between ancestor P and ancestor Q. [3]
3. Explain the reasons for using data from multiple species and multiple proteins. [2]
4. There are 10^{80} protons in the universe. What is the chance of picking the same proton at random twice in a row? [2]
5. This research is free to access online (https://journals.plos.org/plosone/article?id=10.1371/journal.pone.0069924). It is called "Beyond Reasonable Doubt: Evolution from DNA Sequences". Discuss whether this is a suitable name. [3]

A4.1.3 Evidence for evolution from selective breeding of domesticated animals and crop plants

Humans have bred animals selectively over thousands of years for a range of purposes, including:

- meat and milk production; for example, sheep
- transport; for example, horses
- pets; for example, cats.

If modern breeds of livestock are compared with the wild species that they most resemble, the differences are often huge. Consider modern egg-laying hens and the junglefowl of Southern Asia from which they have been developed; or consider Belgian Blue cattle and the aurochs (now extinct) of Western Asia. There is also much variation between different breeds of domesticated livestock, as shown by the diversity of dog breeds.

▲ Figure 6 Many breeds of dog have been developed by artificial selection, starting with grey wolves—perhaps as long ago as 30 to 40 thousand years

Similar patterns are observed among crop plants. Humans have selectively bred a range of plant species for various purposes, including:

- food for humans; for example, wheat
- fibres; for example, cotton
- cut flowers; for example, roses.

As with livestock, crop plants resemble wild species of plant but are markedly different. In addition, there are many different varieties of some crop plant species. It is obvious that domesticated animals and crop plants have not always existed in their current forms. The only credible explanation is that changes have been achieved simply by repeatedly selecting and breeding the individuals most suited to human uses. This process is called artificial selection.

The considerable changes that have occurred in domesticated animals and crop plants over relatively short periods of time show that artificial selection can cause rapid evolution. If artificial selection achieved this over the 12,000 or so years during which humans have grown crops and reared livestock, it seems reasonable to assume that natural selection could have caused major evolutionary changes over the billions of years of life on Earth.

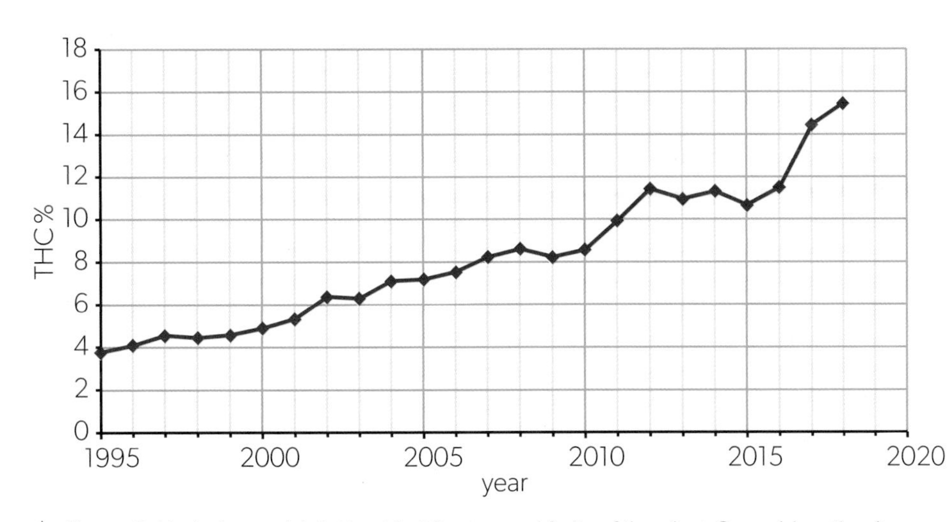

Source: University of Mississippi / National Institute on Drug Abuse

▲ Figure 7 Herbal cannabis is the dried flowers and fruits of the plant *Cannabis sativa*. It contains tetrahydrocannabinolic acid, which is converted to tetrahydrocannabinol (THC) by heat. Through artificial selection, the average THC content of cannabis sold to users has quadrupled in 23 years, increasing the risk of early-onset psychosis and schizophrenia

A4.1.4 Evidence for evolution from homologous structures

Darwin found it curious that the forelimbs of a human, mole, horse, porpoise and bat were apparently so different, yet inside them are the same bones in the same relative positions. Darwin called such similarities "unity of type". These limbs are pentadactyl, which means they have five digits (toes or fingers). Pentadactyl limbs are an excellent example of homologous structures—features with similar anatomical position and structure despite differences in function.

Pentadactyl limbs

The pentadactyl limb consists of these structures:

Bone structure	Forelimb	Hindlimb
single bone in the proximal part	humerus	femur
two bones in the distal part	radius and ulna	tibia and fibula
group of wrist or ankle bones	carpals	tarsals
series of bones in each of five digits	metacarpals and phalanges	metatarsals and phalanges

▲ Table 1

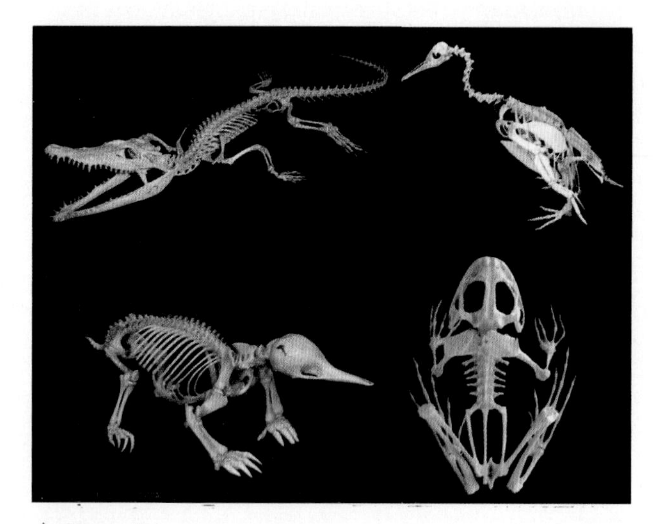

▲ Figure 8

All amphibians, reptiles, birds and mammals have the same pattern of bones (or a modification of it), whatever the function of their limbs. The photos in Figure 8 show the skeletons of one example from each of the vertebrate classes that have limbs: amphibians, reptiles, birds and mammals. All have pentadactyl limbs.

- Crocodiles walk or crawl on land and use their webbed hind limbs for swimming.
- Penguins use their hind limbs for walking and their forelimbs as flippers for swimming.
- Echidnas use all four limbs for walking and also use their forelimbs for digging.
- Frogs use all four limbs for walking and their hindlimbs for jumping.

You can see differences in the relative lengths and thicknesses of the bones. Some metacarpals and phalanges have been lost during the evolution of the penguin forelimb.

The explanation for homologous structures such as pentadactyl limbs is that they were inherited from a common ancestor but have evolved in diverse ways as they have become adapted for different functions. The common ancestor of all tetrapods (four-legged vertebrates) had pentadactyl limbs, which it probably used for walking on land. All of its descendants retain the same basic arrangement of limb bones—this is Darwin's "unity of type".

There are many examples of homologous structures. They do not prove that organisms have evolved into their present forms or that groups of organisms had common ancestry. Nor do they reveal anything about the mechanism of evolution. However, they are difficult to explain without evolution.

The structures that Darwin called "rudimentary organs" are particularly interesting. These are reduced structures that serve no function, now known as vestigial organs. Examples are the beginnings of teeth found in embryo baleen whales, despite adults being toothless; the small pelvis and thigh bones found in the body wall of whales and some snakes; and the appendix in humans. These structures are easily explained by evolution: they no longer have a function so are being gradually lost.

A4.1.5 Convergent evolution as the origin of analogous structures

There are similarities between the tails of fishes and the tail fins of whales. However, when we study these structures, we find that they are very different. The wings of birds and insects are also similar in some respects but close examination reveals that the similarities are superficial. Such features are known as analogous structures.

The evolutionary explanation of analogous structures is that they had different origins but became similar because they perform the same or a similar function. This is called convergent evolution.

It can be difficult to determine whether similar structures in different organisms are homologous or analogous.

Cladistics is increasingly used to deduce the evolutionary origins of organisms and their structures. Consider the central nervous systems (CNS) of annelids, arthropods and vertebrates as an example.

Activity

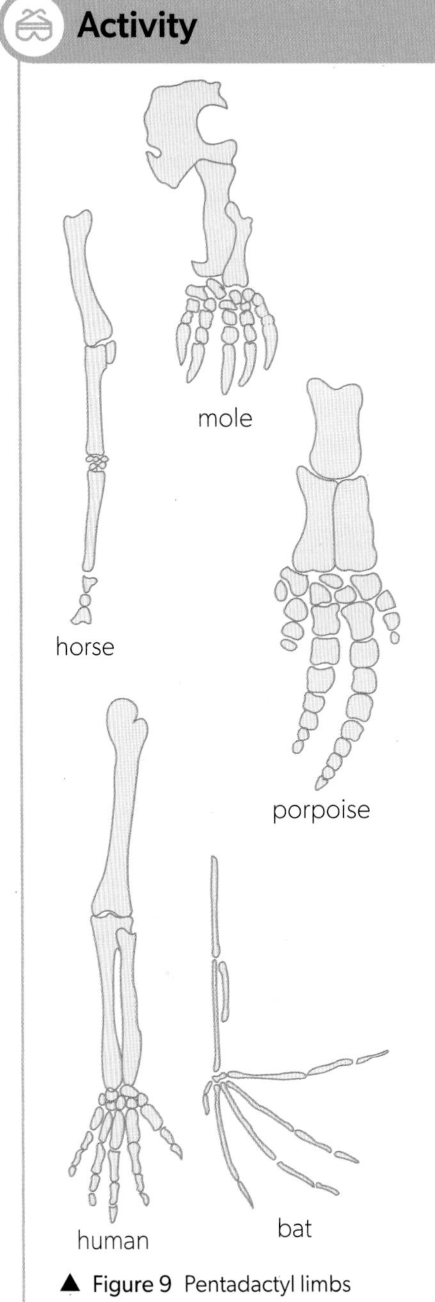

▲ **Figure 9** Pentadactyl limbs (not to scale)

Choose a different colour for each type of bone in a pentadactyl limb, then copy and colour the diagrams in Figure 9 to identify the bones. How is each limb used? What features of the bones in each limb make it well adapted to its use?

Are central nervous systems homologous or analogous?

There is a clade of animals with bilateral symmetry. Their bodies have left and right sides, anterior and posterior ends and a need for communication between anterior and posterior. Annelids, arthropods and vertebrates achieve this communication via a single nerve cord running along the midline of the organism, with an enlarged section at the anterior end. In vertebrates, there is a spinal cord and brain.

The development of nerve cords in annelids, arthropods and vertebrates is associated with a similar pattern of expression of a suite of genes called homeobox genes. This suggests that the nerve cords are homologous. However, nervous system development in other groups of bilaterians is markedly different. This suggests that the common ancestors of annelids, arthropods and vertebrates did not have the characteristic pattern of homeobox gene expression. The nerve cords must have evolved independently in these three phyla so they are an example of convergent evolution. They are analogous rather than homologous structures.

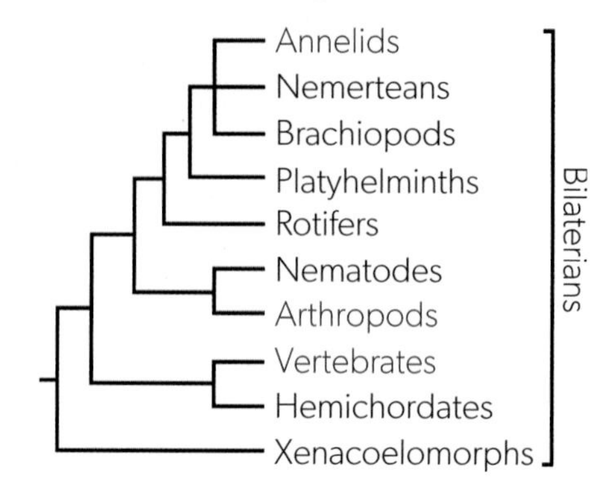

▲ Figure 10 In annelids, arthropods and vertebrates development of a central nervous system is associated with a similar pattern of homeobox gene expression but this has not been found in other bilaterian groups

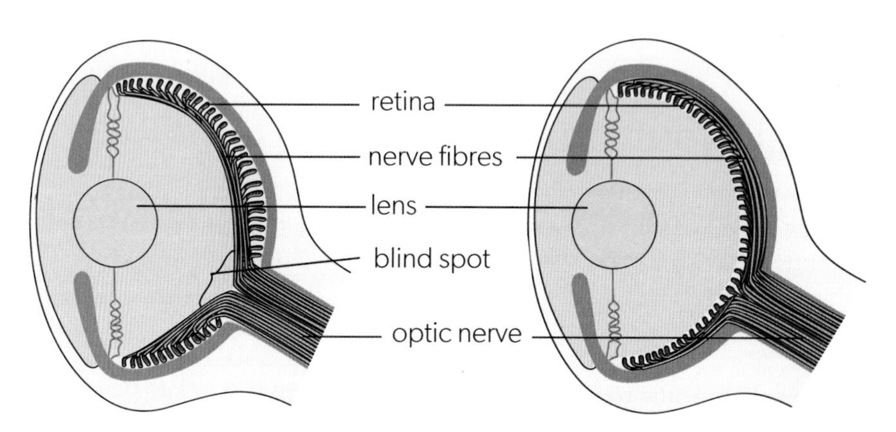

▲ Figure 11 The human eye (left) and the octopus eye (right) are strikingly similar in some respects. However, the human eye has nerve fibres in front of the retina and there is a blind spot whereas in the octopus the nerve fibres are behind the retina and there is no blind spot. These two types of eye are the product of convergent evolution so are analogous structures

A4.1.6 Speciation by splitting of pre-existing species

If two populations of a species become separated so they cannot interbreed and natural selection then acts differently on the two populations, they will evolve in different ways. The characteristics of the two populations will gradually diverge. After a time, they will be recognizably different. If the populations subsequently merged and had the chance of interbreeding, but did not actually interbreed, it would be clear that they had evolved into separate species. This process is called speciation.

▲ Figure 12 This fractal tree shows a sequence of splits. In what ways does it resemble speciation? How does the evolution of species differ from the pattern in the fractal?

▲ Figure 13 In some groups, speciation has happened many times, leading to large numbers of species spread over a wide area. This is known as explosive species diversification. This has occurred in *Zosterops*, a genus of birds called white-eyes. There are now over 100 species in this genus, from Africa though Asia to Australia and New Zealand. This is the Abyssinian white-eye, *Zosterops abyssinica*

A4.1.7 Roles of reproductive isolation and differential selection in speciation

Speciation is the formation of a new species by the splitting of an existing species. Two processes are required for this to happen: reproductive isolation of populations and differential selection.

Before two populations can split into separate species, they must stop interbreeding with each other. Interbreeding causes a mixing of genes and therefore a blending of traits, whereas speciation depends on separation and divergence. Biologists refer to the genes of a population as a gene pool. For speciation to occur, there must be barriers preventing gene flow between the gene pools of the two populations. This can be achieved by any method of reproductive isolation.

Geographical separation is the most obvious and probably the most common cause of reproductive isolation. There may be gaps in the range of a species, which divide it into separate populations. These gaps could be due to physical barriers that are difficult to cross—for example, a mountain range, a wide river or a stretch of ocean between two islands. Such barriers prevent interbreeding between populations, so the gene pools are separated. Geographical separation is usually associated with differences in selection pressures, which are also required for speciation.

Natural selection can cause the traits of a population to change. However, if it operates in the same way in two populations of a species, their traits will remain the same and they will not become separate species. Where there are significant differences in selection, this is called differential or divergent selection. Differential selection causes the traits of the populations to become more and more different; when this divergence is judged by taxonomists to be significant, the populations are classified as separate species.

▲ Figure 14 The bonobo (shown foraging for insects in the river) and the chimpanzee are both primates from the genus *Pan*. Bonobos are smaller and have markedly different behaviours from chimps. The range of the bonobo and the chimpanzee do not overlap as they are geographically separated by the Congo River which is renowned for being deep. Neither species is thought to be able to swim. It is thought that at one point in history the water level fell drastically for a time allowing chimpanzees to cross temporarily. It is thought that these migrants became geographically isolated from their ancestors when the water level of the Congo rose again. This founder population, being subject to different selection pressures, diverged from chimpanzees to become bonobos

To understand how there can be differential selection, consider a new population of a species that has been established by migration to an island. Any or all of these factors might be different from the other parts of the species range:

- climate—temperatures, rainfall and other aspects
- predation—there might be different predators or even no predators in some areas
- competition—there might be more or less competition for resources.

The lava lizards of the Galápagos archipelago are an example of geographical isolation and speciation.

Speciation in lava lizards

Speciation often occurs after a population of a species extends its range by migrating to an island. This explains the large numbers of endemic species on islands. An endemic species is one that is found only in a certain geographical area.

The lava lizards of the Galápagos Islands are an example. One species (*Microlophus albemarlensis*) is present on all the larger islands of the archipelago. On six smaller islands, there are six closely related but different species, formed by migration to an island, reproductive isolation and divergence due to differential selection.

Cladistics research suggests that there were two separate migrations of lava lizards from mainland South America to the Galápagos. One of these migrations populated San Cristóbal and Marchena; the other populated all the other islands apart from Genovesa.

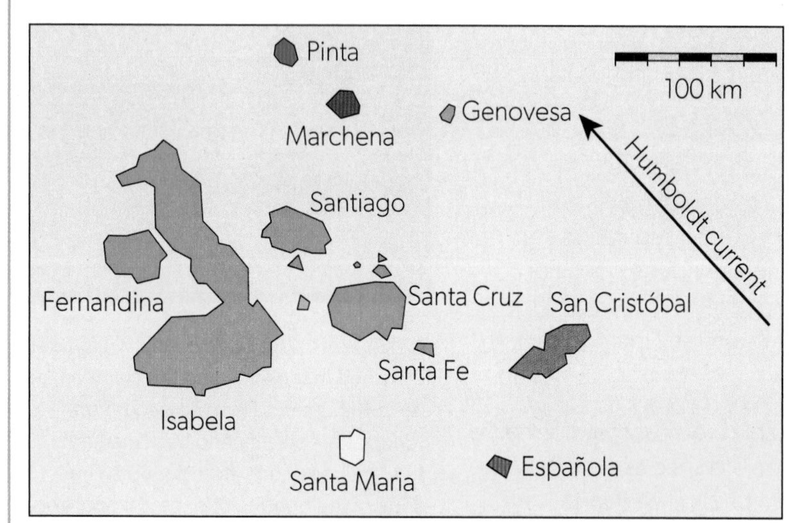

▲ **Figure 15** Distribution of lava lizards in the Galápagos Islands

▲ **Figure 16** Galápagos lava lizard *Microlophus albemarlensis* on Santa Cruz Island

Data-based questions: Flightless steamer ducks

Steamer ducks are members of the genus *Tachyeres* and inhabit southern Chile and Argentina. Recent research suggests that there are four species. Two of them are flightless and live on the coast of Chile and Argentina. A third species can fly and occurs both on the coast and inland, with its range overlapping those of the flightless species. The fourth species of steamer duck occurs on the Malvinas or Falklands islands to the east of Argentina. This species has flightless coastal populations and also populations that can fly and only breed on inland lakes. The map shows the ranges of the four species.

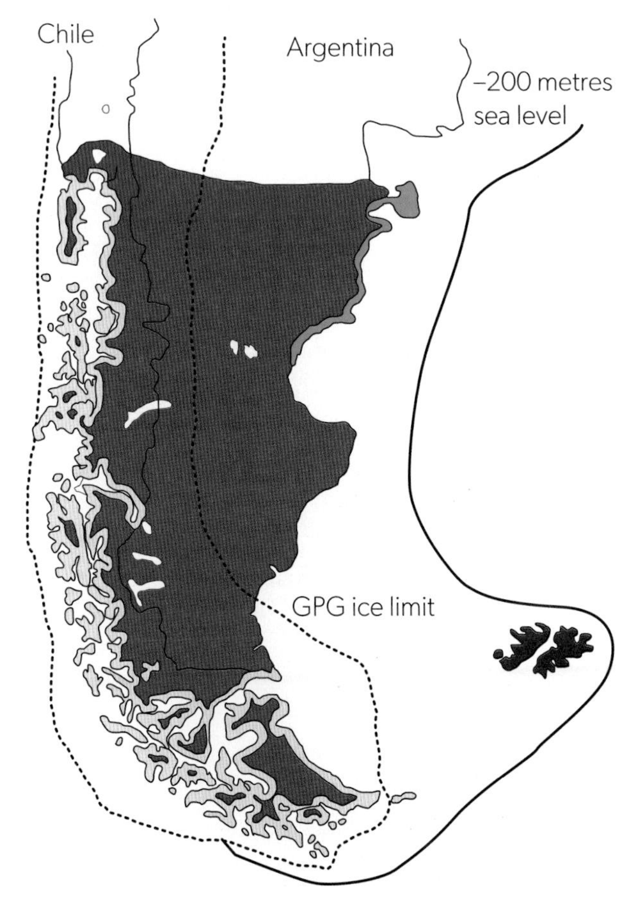

▲ Figure 17

Source: Fulton et al, Proc. R. Soc. B (2012) 279, 2339–2346 doi:10.1098/rspb.2011.2599

Analysis of mitochondrial DNA base sequences suggests that the species on the Malvinas diverged from the three continental species between 2.2 and 0.6 million years ago and that the continental species diverged from a common ancestor about 15,000 years ago.

There have been repeated glaciations with extensive ice cover and low sea levels in the areas inhabited by these ducks. During the Great Patagonian Glaciation (GPG) about a million years ago, much of southern Chile and Argentina was ice-covered and the sea level was 200 metres lower than it currently is. During the Last Glacial Maximum (LGM) 15,000 years ago, the ice cover was not as extensive and sea levels did not drop as much.

1. Suggest how populations of steamer duck could have become reproductively isolated, allowing *T. brachypterus* to diverge from the continental species. [2]
2. Discuss, with reasons, whether there is currently interbreeding between *T. pteneres* and *T. leucocephalus*. [2]
3. Suggest how *T. pteneres* and *T. leucocephalus* could have evolved from a common ancestor 15,000 years ago. [2]
4. Discuss whether *T. patachonicus* is likely to cross-breed with *T. pteneres* and *T. leucocephalus*. [2]
5. Predict whether *T. brachypterus* is likely to diverge into flightless and flying species. [2]

▲ **Figure 18** *T. brachypterus* showing the rapid method of swimming characteristic of steamer ducks, using legs and wings and resembling a paddle steamer. The genus name *Tachyeres* means "fast rower"

A4.1.8 Differences and similarities between sympatric and allopatric speciation

Speciation is the process by which one species splits into two or more separate species. It can only happen if populations of a species are reproductively isolated. Geographical separation, described earlier in this chapter, is an obvious means of reproductive isolation. When populations in different geographical areas become separate species, allopatric speciation has occurred.

It is also possible for a population of a species living together in one geographical area to split into two populations that do not interbreed. If they remain reproductively isolated, the populations could diverge to form separate species. This is called sympatric speciation. Sympatric means "same homeland" and allopatric means "different homelands".

Reproductive isolation in sympatric populations may be a consequence of behavioural differences in animals and temporal differences in animals or plants. Sympatric speciation is certainly much less common than allopatric speciation and it is difficult to be sure whether closely related species living in the same geographical area are the product of true sympatric speciation, or allopatric speciation followed by migration. Examples of reproductive isolation due to behavioural and temporal separation are given here.

Behavioural separation

Two forms of a species of cichlid fish (*Astatotilapia calliptera*) have been discovered in Lake Massoko, a 700 metre wide crater lake in Tanzania. One form prefers to feed near the shore (littoral) and the other in deeper water (benthic). The two forms have adaptations corresponding to these preferences—body size and shape, structure of the jaw and teeth, coloration of breeding males (bluer or yellower) and sensitivity of retinal pigments to different wavelengths of light. Genetic differences have been found between the two forms and experiments have shown that females tend to select a mate who is genetically similar to themselves. This is an example of behavioural separation, which reduces the mixing of genes between the two forms. Over time, this may result in speciation.

▲ **Figure 19** Lake Masoko with a male littoral (yellow) morph *Astatotilapia calliptera*, a male benthic (blue) morph and a female mouthbrooding eggs

Temporal separation

The winter pine processionary moth (*Thaumetopoea pityocampa*) lives in countries around the Mediterranean. Its life cycle takes one year to complete. Adults emerge in summer or early autumn and live for just two or three days. During that brief time they mate and females then lay 100–200 fertilized eggs. The eggs hatch into larvae that feed during the autumn and winter on leaves of pine and cedar trees. In February or March, the larvae migrate in head-to-tail processions down from the trees and to their pupation sites underground in soil. The pupae develop into the next generation of adults, emerging at the same time of year as did their parents.

In one area of Portugal, researchers have discovered a population of this species that has different timings for all stages in its life cycle. Adults emerge in May or June and larvae feed and grow through the summer, rather than the winter. In the warm summer conditions, the larvae grow quickly and are ready to pupate by the end of September.

The more common form of the moth, with winter larvae, also lives in this area of Portugal—they are sympatric. The timing of the life cycle is a heritable trait so must be determined genetically. The two forms never mate with each other as there is temporal separation: the two or three days of adult life happen at different times of year. It seems reasonable to assume that the two forms will diverge, because different adaptations are needed by larvae active in summer or winter. If the divergence becomes great enough, sympatric speciation will have occurred.

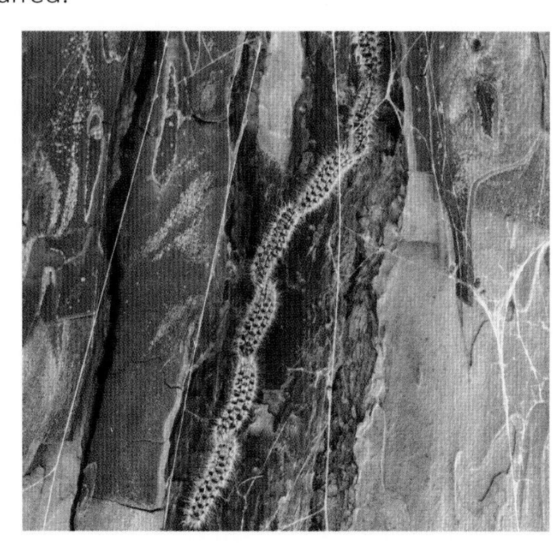

▲ **Figure 20** Winter processionary moth larvae in a procession

A4.1.9 Adaptive radiation as a source of biodiversity

Characteristics that make an individual suited to its environment or way of life are called adaptations. This term is used because the fit between structure and function is developed over time, by a process of modification. The process of modification is "adaptation" and a trait developed by this process is "an adaptation".

Species extend their range if a group of individuals migrates to a new area. These individuals are the founders of a new population. If they cannot interbreed with other populations, the traits of the new population will tend to diverge from the rest of the species. This is partly due to chance, often aided by the small initial number of founders. It is also partly a result of adaptation to differences in the environment. Another factor that can cause rapid adaptation in a new population is the availability of an ecological niche that is not being fully exploited by other species.

Speciation and adaptation to new niches have happened repeatedly in some groups. This is called adaptive radiation; the word "radiation" means spreading out. In this case, the radiation is ecological, rather than geographic. Adaptive radiation is defined as a pattern of diversification in which species that have evolved from a common ancestor occupy a range of ecological roles. It is a source of considerable biodiversity. Because of the diversity of ecological niches, adaptive radiation minimizes competition between species so they can coexist. Even if the process of speciation is allopatric, migration can occur and closely related species can then live sympatrically.

Darwin's finches

Galápagos finches are the best known example of adaptive radiation. Over the past 2.3 million years, 14 species of finch have evolved from a common ancestor on the islands of the Galápagos archipelago. These finches have become adapted to different food sources: leaves, fruits, pollen, nectar, small soft seeds, large hard seeds, insects on leaves and insects under bark. The beaks of the finches show particularly clear adaptations. Up to 10 species of Galápagos finch have been found living together in one locality. It is unlikely that this would be possible without adaptive radiation—there would be too much competition.

▶ **Figure 21** This statue depicts the young Charles Darwin stepping onto the Galápagos islands in 1835. He studied the finches on these islands and later wrote: "The most curious fact is the perfect gradation in the size of the beaks in the different species of *Geospiza*, from one as large as that of a hawfinch to that of a chaffinch, and ... even to that of a warbler... Seeing this gradation and diversity of structure in one small, intimately related group of birds, one might really fancy that from an original paucity of birds in this archipelago, one species had been taken and modified for different ends" (Darwin, 1839)

Brocchinias—adaptive radiation of bromeliads on the Guiana Shield

Brocchinia is a genus of bromeliads. The 20 species grow on the Guiana Shield in southern Venezuela and Guyana. The common ancestor of *Brocchinia* species diverged from all other bromeliads 20 million years ago and diversification within the genus has been happening for at least the last 13 million years.

The sandstone rock of the Guiana Shield yields nutrient-deficient soils that can limit plant growth. Plants which develop successful nutrient-capture strategies therefore have a competitive advantage. Brocchinias have developed a greater diversity of strategies than any other genus of plants. For example:

- *Brocchinia prismatica* relies solely on its roots growing through the soil for a supply of nutrients.
- *Brocchinia reducta* has curved vertical leaves that together form a tank in which water collects. Chemicals are secreted into the fluid to give it a sweet nectar-like smell, which attracts insects. The wax covering the leaves is particularly slippery, so insects fall into the fluid and cannot escape. The fluid is very acidic and contains digestive enzymes, so trapped insects are killed. Specialized hairs on the leaves absorb mineral nutrients released by digestion.

▲ **Figure 22** *Brocchinia reducta*, Mount Roraima, Venezuela

- *Brocchinia acuminata* has expanded leaf bases creating chambers in which ant colonies live. Dead ants and detritus from ant activities are digested in fluids at the base of these chambers. Roots grow into the fluid and absorb the nutrients released by digestion.
- *Brocchinia micrantha* grows very large and collects litres of rainwater at the base of each of its leaves. Dead leaves and other plant detritus falling into these water tanks are digested, providing a supply of nutrients to the plant.
- *Brocchinia tatei* also has leaf bases that collect water and falling plant debris. It can live on the ground or as an epiphyte growing on the trunks and branches of trees. Nitrogen-fixing cyanobacteria grow in its tanks and supply the plant with nitrogen compounds.

▲ **Figure 23** *Brocchinia micrantha*, Kaieteur National Park, Guyana

A4.1.10 Barriers to hybridization and sterility of interspecific hybrids as means of preventing the mixing of alleles between species

Interspecific hybrids are produced by cross-breeding members of different species. The hybrids combine traits of the species that were crossed. Hybridization is often done deliberately by plant or animal breeders. The mule was probably the first hybrid, produced by cross-breeding a horse with a donkey (*Equus caballus* × *Equus asinus*). Mules combine useful traits of those two species and also have what is known as hybrid vigour. They have therefore been deliberately bred for 5,000 years or more. Horses have 64 chromosomes and donkeys have 62 so a mule has 63. This causes problems in meiosis. For that reason and other genetic incompatibilities, mules are nearly always sterile.

Plant breeders often use interspecific hybridization to produce new varieties. The first person known to have done this was Thomas Fairchild who, in the early 18th century, crossed carnations with Sweet Williams (*Dianthus caryophyllus* × *Dianthus barbatus*). The hybrids showed traits of both parents and were nicknamed "Fairchild's Mule". Both parent species have 30 chromosomes, but even so Fairchild's Mule was sterile. This is very common in interspecific hybrids produced by breeders.

Interspecific hybridization sometimes happens naturally if the ranges of closely related species overlap in an ecosystem. Like artificial hybrids, natural interspecific hybrids are often totally or partially sterile so they cause little or no permanent mixing of alleles between the parent species.

In evolutionary terms, the resources that a parent expends on producing a sterile hybrid are wasted. It is not surprising, therefore, that many species have evolved barriers to prevent the development of hybrid offspring. A hybrid zygote may be produced but it is likely to die during development.

Even fewer resources are wasted if no mating takes place between different species. In animals, this is one of the functions of courtship behaviour: an individual can check whether a potential partner is a member of its own species by looking for distinctive behavioural features. There are often several stages in courtship, with rejection at any stage if the characteristic behaviour pattern of the species is not displayed. To prevent interspecific hybridization, courtship behaviour needs to be distinctive. This explains the immense diversity, particularly among birds—birds of paradise in Papua New Guinea, for example.

In some cases, closely related species do have ranges that overlap and these species produce fertile interspecific hybrids. This can happen where geographical separation has allowed speciation but migration brings the newly separated species back together again. If barriers to hybridization have not developed, there may be mixing of alleles and speciation may be reversed. This can also happen if humans bring species together that would naturally have remained geographically separated. For example, on some Hawaiian islands the native duck (*Anas wyvilliana*) is hybridizing with introduced non-native mallard (*Anas platyrhynchos*), forming hybrid swarms with a mixture of traits from both species. As a result, *Anas wyvilliana* faces extinction through hybridization, with a consequent loss of biodiversity.

Courtship in Clark's grebes

Clark's grebe (*Aechmophorus clarkii*) of western North America has an elaborate courtship display that can be viewed in video clips on the internet. However, the closely related Western grebe (*Aechmophorus occidentalis*) has the same display and these species do sometimes mate and produce hybrid offspring. Two possible hypotheses explain this apparent anomaly: either these two species have not had enough time since diverging to evolve differences in their courtship displays, or they should not be regarded as separate species.

▲ **Figure 24** Courtship in Clark's grebes involves a sequence of distinctive and coordinated actions, which establish and reinforce the bond between male and female

▲ Figure 25 In honey bees (*Apis mellifera*) fertile females (queens) only ever mate with fertile males (drones) on one mating flight. When queens are ready to mate, they fly in the early afternoon to special drone congregation areas, which are 10–20 m above the ground and about 100 m in diameter. Drones from different colonies have already assembled there. Both queens and drones typically fly 2–3 km from their colony to reach a drone congregation area. Queens release the pheromone (e)-9-oxo-2-decenoic acid (left) as a sex attractant for drones, which have receptors for this pheromone on their large antennae (right). Crowds of drones chasing queens resemble comets shooting across the sky. During the mating flight, queens copulate with up to 20 drones. This provides all the sperm needed to fertilize hundreds of thousands of eggs that a successful queen will lay over several years. How is interspecific hybridization prevented in honey bees?

A4.1.11 Abrupt speciation in plants by hybridization and polyploidy

A polyploid organism has more than two sets of homologous chromosomes. Polyploidy is a consequence of the duplication of chromosomes in a cell without a subsequent cell division, so it is whole-genome duplication. Genome sequencing studies show that it has happened many times in evolution.

If whole genome duplication happens in a diploid cell, the result is four sets of homologous chromosomes, so the cell is tetraploid. Because all the sets of chromosomes come from the same organism, it is called an autotetraploid. Autotetraploidy is usually associated with low fertility, because there are four homologous chromosomes of each type and mis-pairing is very likely during meiosis. Over time, there can be genetic changes that overcome this problem, allowing autotetraploid populations to become established.

Sand rock-cress (*Arabidopsis arenosa*) is an example of the establishment of autotetraploid populations. Diploid plants of this species (with 16 chromosomes) only grow in eastern and southeastern Europe. Part of this area, in the Balkan Peninsula and Western Carpathian mountains, has both diploids and autotetraploids. It is thought that the autotetraploids originated here before spreading to western Europe and Scandinavia, where only autotetraploid plants have been found.

Meiosis in an autotetraploid individual produces diploid cells and therefore diploid gametes. If these fuse with haploid gametes from an individual that is diploid, triploid offspring are produced (see Figure 26). These may grow vigorously but they are very unlikely to perform meiosis successfully, so are sterile.

According to the biological species definition, diploids and autotetraploids should be regarded as separate species. However, autotetraploids are usually very similar in morphology to diploids, so taxonomists may be reluctant to recognize them as new species. They are also relatively uncommon in nature, perhaps because the similarities with the original diploid population make competition likely.

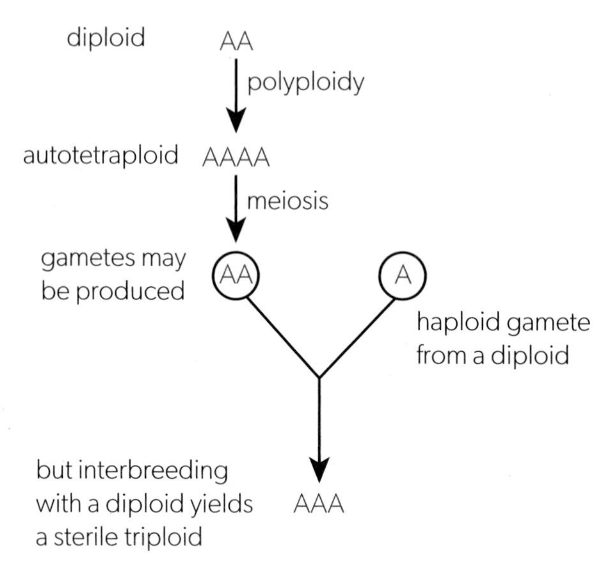

▲ Figure 26 Autotetraploidy: the symbol A represents one haploid set of chromosomes

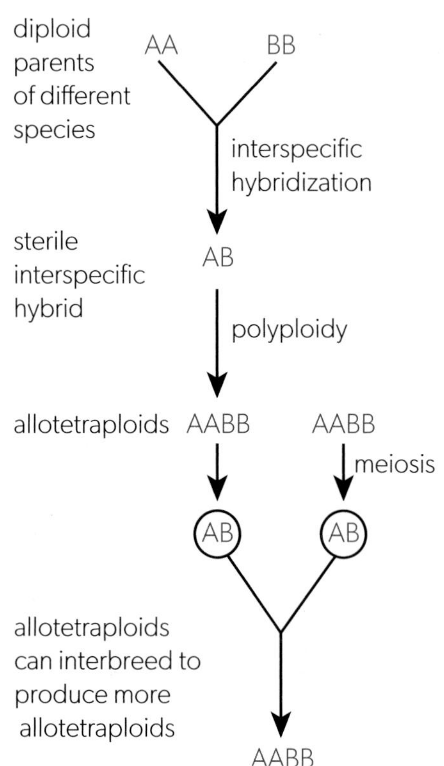

▲ Figure 27 Allotetraploidy: the symbols A and B represent different haploid sets of chromosomes

Another type of polyploidy is the result of a two-stage process.

1. Individuals from different species cross-breed. The resulting interspecific hybrid has two sets of chromosomes, with one set from each of the two different parent species. Unless these two species are very closely related, chromosomes will not form homologous pairs when meiosis is attempted, so the hybrids will be sterile.
2. If any cell in the sterile interspecific hybrid duplicates its chromosomes but does not then divide, the cell will have four sets of chromosomes. It is an allotetraploid because the four sets of chromosomes are from two different species. More of these allotetraploid cells can be produced by mitosis. It is likely that these allotetraploid cells will be able to divide by meiosis because there are two homologous chromosomes of each type, which can reliably form pairs. By becoming an allopolyploid, the interspecific hybrid will overcome its fertility problems.

Allotetraploids can interbreed with other allotetraploids, but not with either of the diploid parent species. They are therefore a new species and, as they have a mixture of traits from both parent species, they are usually recognized and named as a distinct species. Many species have been produced by this two stage process, especially in the plant kingdom—for example, in the genus *Persicaria*.

Horse chestnut trees

Horse chestnut trees (*Aesculus hippocastanum*) are native to northern Greece. Red buckeye trees (*Aesculus pavia*) are native to the southern United States. Both species were introduced to Germany, where they hybridized. The initial hybrid combined traits from both parents but was sterile. A shoot on the tree then developed that was fertile and produced seeds. These seeds germinated to produce more trees with the same traits as the original hybrid. What is the likely explanation for these observations? The hybrid trees are regarded as a new species, called *Aesculus* × *carnea*. Figure 28 shows a specimen in the author's garden.

◀ Figure 28

Hybridization and polyploidy in the genus *Persicaria*

There are over 100 species in the genus *Persicaria*, with species occurring in most parts of the world. There is evidence for at least 15 species in the genus having originated by allopolyploidy. One of these is *Persicaria maculosa* ($2n = 44$ chromosomes), which is native to Europe and Asia. (It is an introduced alien in other parts of the world.) Research indicates that this species arose by hybridization between *Persicaria foliosa* ($2n = 22$) and *Persicaria lapathifolia* ($2n = 22$), followed by a doubling of the chromosome number.

▲ Figure 29 *Persicaria maculosa*

Linking questions

1. How does the theory of evolution by natural selection predict and explain the unity and diversity of life on Earth?
   a. With reference to an example, outline the concept of adaptive radiation. (A4.1.9)
   b. Outline the features shared by all cells. (A2.2.4)
   c. Discuss the relationship between niche and convergent evolution. (B4.1.7)
2. What counts as strong evidence in biology?
   a. Explain how the emergence of antibiotic resistance is predicted and explained by the theory of evolution by natural selection. (A4.1.1)
   b. Discuss the experimental evidence that confirmed that DNA is the genetic material. (A1.2.14)
   c. Other than the emergence of antibiotic resistance, outline the evidence that supports the theory of evolution by natural selection. (A4.1)

A4.2 Conservation of biodiversity

What factors are causing the sixth mass extinction of species?

A number of factors threaten biodiversity, including the loss of habitat, exposure to pollution, overexploitation, threats from invasive species and climate change. The image shows a Red fox (*Vulpes vulpes*) consuming an Arctic fox (*Alopex lagopus*) it has killed. The Arctic fox faces a number of threats. What habitat changes is it facing due to climate change? How is climate change impacting the distribution of its prey, its predators and competitor species that occupy the same niche? Is the range of temperate species having an impact on the Arctic fox?

▲ Figure 1 Red fox (*Vulpes vulpes*) consuming an Arctic fox (*Alopex lagopus*) it has killed, Wapusk National Park, Cape Churchill, Manitoba, Canada

How can conservationists minimize the loss of biodiversity?

One approach to minimize the loss of biodiversity is to counteract the main factors that threaten extinction. How can conservationists respond to threats to biodiversity due to climate change, habitat loss and the spread of invasive species? What are examples of in situ and ex situ conservation measures? What challenges have to be surmounted by both? The image shows Young white storks (*Ciconia ciconia*) being fed fish. These birds are being raised for reintroduction to the wild. What type of conservation approach is this?

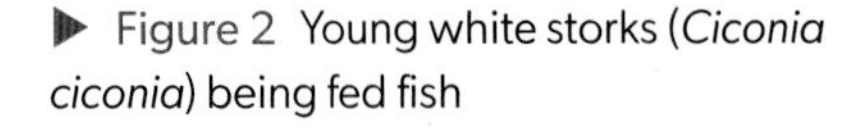

▶ Figure 2 Young white storks (*Ciconia ciconia*) being fed fish

SL and HL

A4.2.1 Biodiversity as the variety of life in all its forms, levels and combinations
A4.2.2 Comparisons between current number of species on Earth and past levels of biodiversity
A4.2.3 Causes of anthropogenic species extinction
A4.2.4 Causes of ecosystem loss
A4.2.5 Evidence for a biodiversity crisis
A4.2.6 Causes of the current biodiversity crisis
A4.2.7 Need for several approaches to conservation of biodiversity
A4.2.8 Selection of evolutionarily distinct and globally endangered species for conservation prioritization in the EDGE of Existence programme

A4.2.1 Biodiversity as the variety of life in all its forms, levels and combinations

The word biodiversity is an abbreviation for "biological diversity". Diversity has been defined as "variety or multiformity, a condition of being different in character and quality". It is the opposite of unity. Biology is the study of life, so biodiversity is the variety or multiformity of life. It exists at multiple levels, including:

- Ecosystem diversity—variety in the combinations of species living together in communities. This diversity is partly due to the very varied environments on Earth. It is also due to the geographical ranges of organisms.
- Species diversity—the many different species on the evolutionary tree of life. These species have varied body plans, internal structure, life cycles, modes of nutrition and more.
- Genetic diversity within species—variety in the gene pool of each species. There is variation both between geographically separated populations and within populations. Species with only a few surviving individuals inevitably have little genetic diversity and problems due to inbreeding.

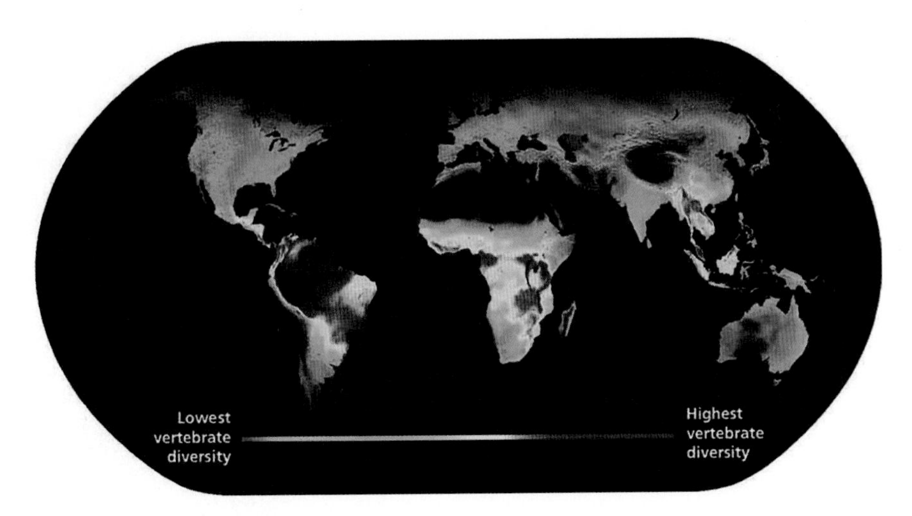

◀ Figure 3 Living land-based vertebrate species are not distributed evenly around the planet. The highest concentration of diversity is shown in red, in regions around the equator. Diversity decreases closer to the Earth's poles; this is shown by (in order) orange, yellow, green and blue shading

A4.2.2 Comparisons between current number of species on Earth and past levels of biodiversity

Fewer than two million species have been named and described. Many more remain undiscovered, so it is impossible to state with confidence the current number of eukaryotic species on Earth. Estimates vary widely but are mostly between 2 and 10 million. With prokaryotes, there are too many uncertainties for reliable estimates of numbers of species to be made.

It is even more difficult to estimate how many eukaryotic species lived on Earth in the past. Relative levels of biodiversity can be deduced from fossil evidence. This shows large variations. In particular, there have been five mass extinctions when many species disappeared. The most recent mass extinction was 66 million years ago, at the end of the Cretaceous period. This occurred when a huge asteroid

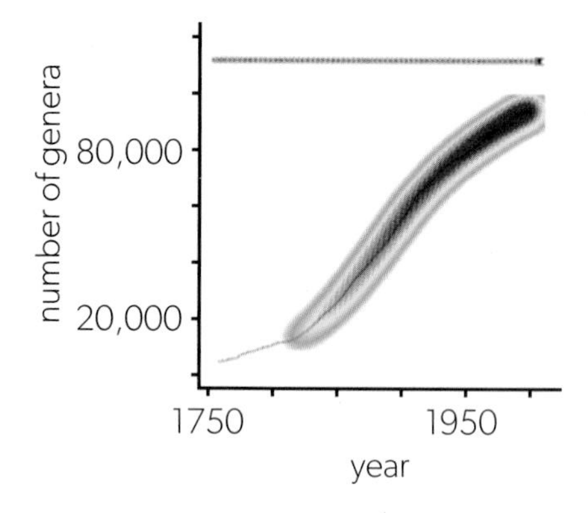

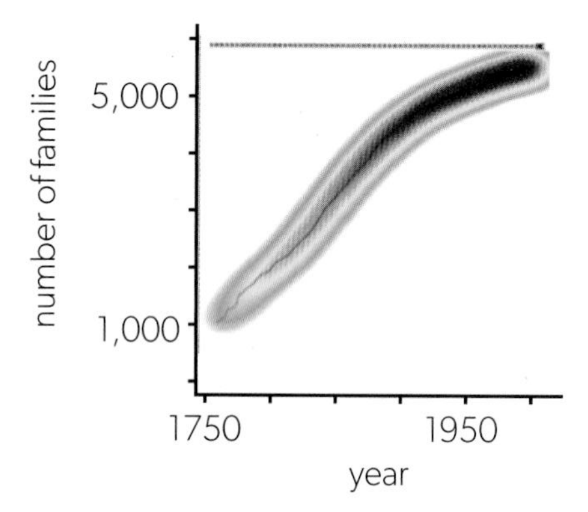

▲ Figure 4 One approach for predicting the number of species, genera, families and higher taxa is to extrapolate from past discovery rates. The graphs show numbers of genera and families in the animal kingdom. The lines showing an asymptote represent the expected total numbers

Source: Mora C, Tittensor DP, Adl S, Simpson AGB, Worm B (2011) How Many Species Are There on Earth and in the Ocean?. PLOS Biology 9(8): e1001127. https://doi.org/10.1371/journal.pbio.1001127

collided with the Earth. The consequent environmental disruption caused many species to die out, including all non-avian dinosaurs. The previous four mass extinctions have been attributed to volcanic activity and major changes to the atmosphere and global climate.

Between mass extinction events, biodiversity tends to rise gradually, with new forms of life evolving. For example, the extinction of the non-avian dinosaurs and other groups at the end of the Cretaceous was followed by evolution of many new species of birds and mammals.

There have been no mass extinction events for 66 million years now. As a result, biodiversity has been able to undergo a sustained increase: it is probably higher now than it has ever been. However, human activity is causing what is predicted to be the sixth mass extinction, so this peak of biodiversity is unlikely to be sustained.

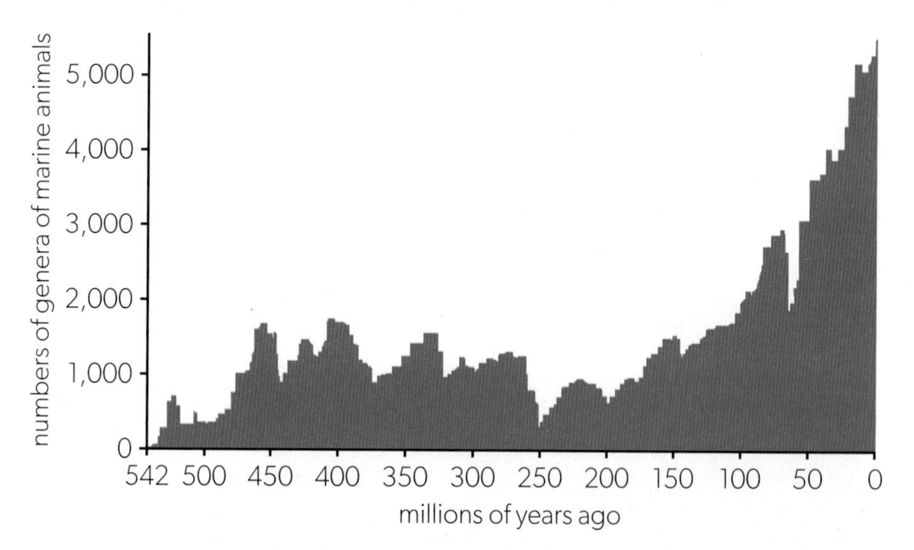

▲ Figure 5 This graph shows the numbers of marine animal genera known from fossil evidence over the 542 million years since the start of the Cambrian period. It shows that biodiversity of animals in marine habitats is probably higher now than it has ever been. Other groups of organisms are likely to have followed the same trend. During the Cambrian period, there was a major diversification of animals and other multicellular organisms, known as the Cambrian explosion

Patterns, trends and classification

Ornithologists are scientists who study birds. The graph in Figure 8 shows the incidence of "corrections" in the classification of North American bird species since the formation of the American Ornithologists Union in 1883. For example, Traill's flycatcher was split into two distinct species in 1973: alder and willow flycatchers. Over much of the 20th century, the trend was to reduce the number of species, by uniting those that showed strong similarities ("lumping"). Since the 1980s, that trend has changed towards more splitting due to recognition of the phylogenetic distinctiveness of populations that were previously considered a single species.

▲ Figure 6 Alder flycatcher

▲ Figure 7 Willow flycatcher

1. Suggest why the 1935 publication of a paper proposing the biological species concept might have led to the tendency towards "lumping".
2. Suggest why the improvement in DNA technology in the 1980s might have led to an increasing tendency towards splitting.
3. The graph is cumulative and shows some periods when there was relatively little "lumping" or "splitting". What were these periods and what could the reasons have been?

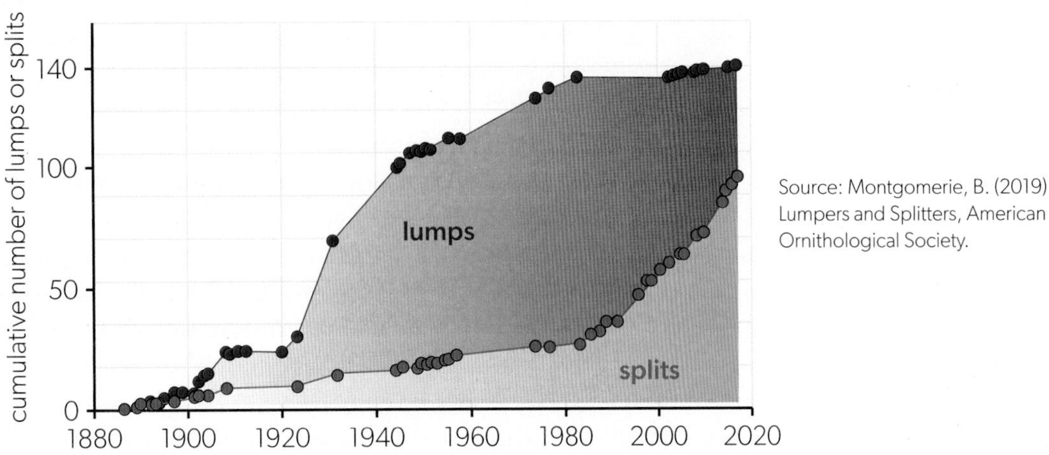

▲ **Figure 8** Cumulative graph showing the total number of "lumps" or "splits" in bird species in North America

A4.2.3 Causes of anthropogenic species extinction

Species extinction is a natural process. If it is balanced by evolution of new species, biodiversity does not decrease. Extinctions have been happening for billions of years, but current rates are very high. Five main types of cause can be identified and all of them are anthropogenic (due to human activities).

1. **Overharvesting**
   Humans take plants and animals from natural ecosystems by hunting animals, harvesting plants for food or medicines, logging forests to obtain timber, and fishing in freshwater or marine ecosystems. If this happens at a faster rate than a species can reproduce, extinction will occur. For example, current fishing rates are certainly unsustainable. Sometimes only part of an animal is used, for example, shark fins, elephant tusks and tiger bones.

2. **Habitat destruction**
   Agriculture began about 12,000 years ago in the Middle East. Today, over 13 billion hectares of land are cultivated or used for rearing livestock. Natural habitats such as forests or grasslands were destroyed so that land could be used for agriculture. This led to the loss of some species. About 6,000 years ago, humans began to establish towns and cities, causing more losses of natural habitat.

3. **Invasive species**
   When alien species are introduced to ecosystems, they can drive native species to extinction by predation, spreading of pests and diseases, or competition for resources. Endemic species can also become extinct if

they hybridize with aliens. Some introductions have been deliberate—for example, possums and domestic cats in New Zealand. In other cases, they are accidental and a consequence of transport on boats and airplanes.

4. **Pollution**
   Chemical industries produce a vast range of substances that are used and then discarded or released into the environment. No part of the world is unaffected by pollution—for example, lead from the Roman era can be detected deep in Arctic ice, and plastic waste washes up on beaches in the most remote parts of the world. Burning of fossil fuels, agriculture, mining, oil extraction and pharmaceuticals are all major sources of pollutants.

5. **Global climate change**
   Plants and animals adapt to the conditions that they experience. If conditions change gradually, they will evolve to survive. However, human activities are causing very rapid changes in temperature, rainfall, snow cover and other environmental variables on Earth. Some species will be able to adapt or migrate, but others face extinction. For example, coral species may not adapt quickly enough to survive in rising sea temperatures.

There are many well-understood cases of species extinction. Three examples are described here.

Giant moas (*Dinornis novaezealandiae*)

The megafauna of Africa has declined but most species still survive. In other parts of the world, many of the largest animals have become extinct during the last 20,000 years. In at least some cases, this happened as humans spread around the world and began hunting the local megafauna. In North America, there is evidence that humans hunted mammoths and wild horses to extinction about 13,000 years ago. Smilodons (sabre-toothed cats) also disappeared at this time, probably due to loss of their prey.

Moas were a group of flightless birds, native only to New Zealand. There were nine species, the largest of which were the giant moas—*Dinornis novaezealandiae* on the North Island and *Dinornis robustus* on the South Island. They grew up to 3.6 m tall and had a mass of approximately 230 kg. New Zealand was not settled by humans until the arrival of Polynesians in the 13th century, who became the Māori iwi. It then took less than 200 years for all species of moa to be hunted to extinction.

▲ **Figure 9** Painting of *Dinornis novaezealandiae* by William Frohawk

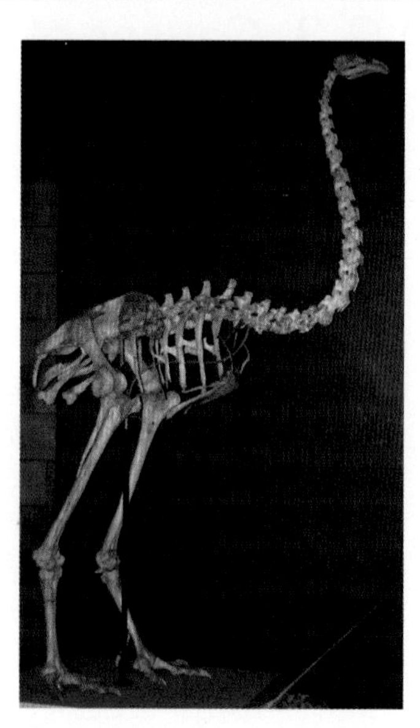

▲ **Figure 10** Skeleton of *Dinornis novaezealandiae* in the British Natural History Museum

Atitlán grebe (*Podilymbus gigas*)

Cats, rats and other alien species spread by humans are one of the major causes of extinction. Where it is not too late, eradication programmes can be very successful. On Henderson Island in the South Pacific, for example, three bird species have already been driven to extinction by Pacific rats, including the Henderson imperial pigeon (*Ducula harrisoni*). Four other endemic species of land bird are threatened. A rat-eradication programme, led by the RSPB, has reduced rat numbers. However, until all the rats have been removed, Henderson Island's endemic birds will remain at risk of extinction.

For the Atitlán grebe (*Podilymbus gigas*) it is too late. This large grebe was endemic to Lago de Atitlán in Guatemala, at an altitude of 1,700 m. It had small wings and was flightless. The decline of the Atitlán grebe began in 1958, when two species of black bass (*Micropterus dolomieu* and *Micropterus salmoides*) were introduced to the lake to try to promote tourism by anglers. These fish multiplied and competed for the crabs and fish that were foods for the grebes. The bass also predated grebe chicks. Numbers of Atitlán grebe declined from 200 in 1960 to 80 in 1965.

In 1966, a refuge was established for Atitlán grebes and numbers rose to 210 by 1973. Unfortunately an earthquake in 1976 fractured the lake bed and the water level fell. This caused a marked decline in bird numbers, with only 32 grebes surviving in 1983. Most of these were hybrids with the pied-billed grebe (*Podilymbus podiceps*). Two birds were seen in 1989 but they then disappeared and the Atitlán grebe was declared extinct.

▲ **Figure 11** Atitlán grebe, now extinct

Mount Glorious torrent frog (*Taudactylus diurnus*)

Amphibians breed in water but spend most of their adult life in terrestrial habitats, so they are particularly vulnerable to extinction due to habitat destruction. The Mount Glorious torrent frog lived in rainforests in three mountain ranges in northeastern Australia. When first described in 1966, it appeared to be relatively common in mountain streams. However, its populations declined rapidly and it was probably extinct by 1980.

The principal cause of extinction was deforestation: over 10,000 hectares of trees were cleared from the range of *T. diurnus* during the 20th century. This altered water flows and increased turbidity in the streams where the frogs bred. Alien feral pigs were another factor. They predated the frogs and contaminated the water of streams by churning up mud. Invasive alien plant species and an infectious chytrid fungal disease may also have contributed to this sudden and regrettable species extinction.

▲ **Figure 12** Mount Glorious torrent frog

A4.2.4 Causes of ecosystem loss

Over much of the Earth's surface, human activity has caused the loss of natural ecosystems. In many areas, the causes of loss were direct—the ecosystem was cleared to allow other forms of land use. In other areas, the causes of loss were indirect and unintentional. An ecosystem consists of interacting and

Activity

Species have become extinct in all parts of the world. Research an example in the area where you live.

interdependent components, so if key parts are removed, an entire ecosystem may collapse and be lost. This can happen if an environmental variable is changed to be outside the range of tolerance of keystone species in an ecosystem. Eight categories of direct or indirect cause of ecosystem loss are described here.

1. Land-use change for agricultural expansion is the main cause of ecosystem loss. In temperate zones, most areas suitable for farming were cleared of natural forests, grasslands and wetlands before the 1970s. For example, the prairies of North America were mostly plowed up in the 19th century. Since the 1970s, it is mostly old-growth tropical forest ecosystems that have been lost.

2. Urbanization is another major cause of land-use change and ecosystem loss. The urban area of the world has doubled since 1992, to accommodate the rapidly growing human population. Natural ecosystems have been cleared to allow building of homes, offices and factories, together with the associated infrastructure of roads and railways.

3. Overexploitation of natural resources has destroyed some ecosystems. Gathering of fuel wood, hunting of animals for bushmeat and fishing in freshwater and marine habitats are examples. Even harvesting of a single keystone species can threaten ecosystems. An example is the overfishing of cod on the Grand Banks of Newfoundland. This is explored in *Topic D4.2*.

4. Mining and smelting destroy areas of natural ecosystems directly through land-use change. In addition, pollution from these activities can cause much more widespread damage. For example, nickel mining and smelting in Ontario has caused damage to lakes and rivers over a wide area by acid rain and pollution of soils with copper, nickel and other metals. This has led to the loss of natural forests.

5. Building of dams and extraction of water for irrigation can lead to loss of natural river and lake ecosystems. For example, the Colorado River now rarely flows as far as the Pacific Ocean because of water extraction for agricultural, industrial and domestic uses. Similarly, the annual flooding of the Nile no longer occurs because a series of dams hold back water from monsoons in Ethiopia.

6. Drainage or diversion of water for human uses has caused the loss of swamps and other wetlands in many parts of the world. For example, the Mesopotamian Marshes in southern Iraq were drained by diversion of the Tigris and Euphrates rivers in the 1990s. About two-thirds of the two million hectares of these wetlands became desert.

7. Leaching of fertilizers into rivers and lakes causes eutrophication and algal blooms. Oligotrophic ecosystems, in which organisms are adapted to low nutrient concentrations, have been lost. Lake Erie for example has been severely affected, with excessive growths of algae every summer since the 1990s. Rivers carry the nutrient-enriched water out to sea, where algal blooms can also occur.

8. Perhaps the most widespread threat to natural ecosystems is climate change. Ecosystems are adapted to specific patterns of temperature, rainfall and other physical variables. When these variables change, entire ecosystems can be lost. Forest is replaced by scrubland or grassland if rainfall decreases. Tundra is replaced by forest if temperatures rise. The relationship between ecosystem types and climate is explored more fully in *Theme B* and the likely future effects of climate change are considered in *Theme D*.

Two specific examples of ecosystem loss are described here but wherever you live or attend school, there will be local examples that are worthy of study.

Mixed dipterocarp forest of southeast Asia

The Dipterocarpaceae is a family of about 700 tropical rainforest trees. They are tall-growing and produce valuable timber. Dipterocarps used to dominate large areas of rainforest in southeast Asia, including Brunei, Borneo and Papua New Guinea.

Mixed dipterocarp forest (MDF) has an extremely high diversity of tree species. On the island of Brunei, for example, there are 20 native species of dipterocarp and small areas of MDF often containing 10 or more of these species. Interspecific hybrids may be produced but they rarely grow to adult size. The highest diversity of tree species tends to occur in areas with nutrient-poor sandy soils.

MDF typically has particularly high quantities of merchantable timber per hectare. As a result, it has been widely targeted for logging, both legal and illegal. Since the 1970s, most areas of MDF have been lost; undisturbed areas are now largely found in upland sites where access is more difficult.

The areas of MDF that have suffered the greatest losses are on lowland sites, especially where nutrient-rich soils overlie deep peat. Large areas have been converted to oil palm plantations. This is particularly unfortunate as the peat in these areas can be up to 15 m deep; this peat, formed over the past 4,000–5,000 years, can store 250 tonnes of carbon per hectare. Drainage during land conversion causes the peat to decompose, releasing CO_2 into the atmosphere. This contributes to another threat—rising sea levels caused by global warming will flood deep-peat lowland areas with seawater, destroying what little MDF remains on such areas.

▲ **Figure 13** Lowland mixed dipterocarp forest, Lambir Hills National Park, Borneo, Malaysia

Loss of the Aral Sea—an ecological disaster

The Aral Sea, between Kazakhstan and Uzbekistan, was the fourth largest lake in the world. It was fed by rivers but it had no outflows; instead, it lost water by evaporation. As a result, it had higher salinity than a freshwater lake. In the 1960s, two major rivers that fed the Aral Sea were diverted to irrigate an area of desert. This led to falling water levels and much of the former lake is now desert.

Apart from the reduction in the area and depth of the lake, an increase in the water salinity was a major contributor to ecosystem collapse. In some of the remaining parts of the lake, the salinity has risen from 1% to more than 22%, compared with about 3.5% for normal seawater. Twenty-four species of fish were endemic to the Aral Sea, all of which are now extinct. Most invertebrate species have also disappeared.

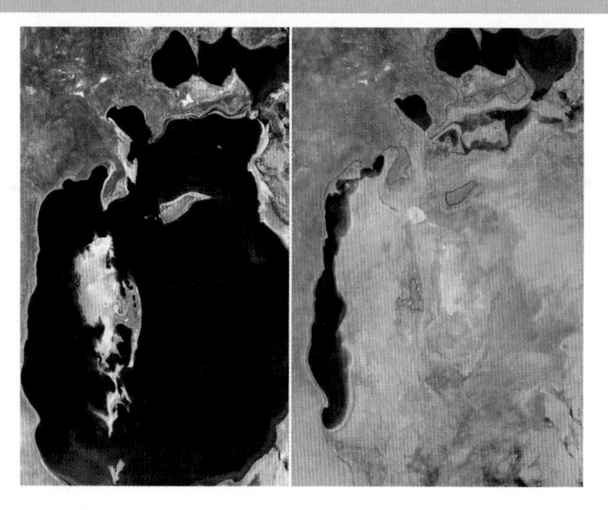

▲ **Figure 14** A comparison of the Aral Sea in 1989 (left) and 2014 (right)

A4.2.5 Evidence for a biodiversity crisis

Journalists use the term "biodiversity crisis" to describe the unprecedented losses of ecosystems and species occurring today. As scientists, we must always look for evidence before making a claim. In this case we need evidence of losses before declaring that there is indeed a biodiversity crisis.

One source of evidence is the Intergovernmental Science-Policy Platform on Biodiversity and Ecosystem Services (IPBES). This is an intergovernmental body which assesses the state of biodiversity and periodically produces reports.

A more active approach is to gather evidence directly by monitoring. Many types of variable can be monitored:

- population size of a species—for example, the number of pairs of gannets in a breeding colony each year
- range of a species—for example, the area over which rattlesnakes are found in North America
- diversity of species in an ecosystem—for example, the number of fish species on a coral reef
- richness and evenness of biodiversity in an ecosystem—these are two statistical measures of diversity
- area occupied by an ecosystem, such as Brazilian rainforest
- extent of degradation of an ecosystem—for example, fragmentation of forests on Brunei
- number of threatened species within a taxonomic group—for example, native bird species of Hawaii
- genetic diversity within a species.

Although expert scientists play a key role in monitoring biodiversity and identifying the most serious threats, there are opportunities for all citizens to contribute. This is an example of what is often called "citizen science". Some of the most useful data has been collected by individuals who have monitored a population or an ecosystem regularly over many years. This can allow the detection of harmful changes while there is still time for them to be reversed.

Applying techniques: Use of Simpson's diversity index

The Simpson's reciprocal index quantifies biodiversity by taking into account species richness and evenness. The greater the biodiversity in an area, the higher the value of D. The formula for Simpson's reciprocal index of diversity is:

$$D = \frac{N(N-1)}{(\Sigma\, n(n-1)}$$

where D = diversity index

N = total number of organisms of all species found

n = number of individuals of a particular species

The highest values occur where there are equal numbers of individuals in the species present and there are many species, so both evenness and richness are high.

You could compare the diversity of species found in an undisturbed forest with a clearing or glade in the same forest undergoing succession. You could use a phone app such as *Picture This* or *iNaturalist* to identify the species of individual plants.

Data-based questions: Using Simpson's diversity index

Groups of students studied the species diversity of the beetle fauna found on two upland sites in Europe. The same number of students searched for a similar length of time in each of the two sites. The two sites were of equal area.

The number of individuals of the four species found at each site is given in Table 1.

Species	Site A	Site B
Trichius fasciatus	10	20
Aphodius lapponum	5	10
Cicindela campestris	15	8
Stenus geniculatus	10	2

▲ Table 1

1. Calculate the reciprocal Simpson diversity index (D) for the beetle fauna of the two sites. [3]
2. Suggest a possible conclusion that can be formed. [2]

Activity

A population of silver-studded blue butterflies (*Plebejus argus*) at Prees Heath in Shropshire, England is monitored by local lepidopterists. The nearest other population of this species is about 100 km away. The graph in Figure 16 shows counts of silver-studded blue butterflies along a fixed-route 5 m wide transect at Prees Heath. The same methodology is used each year, to ensure the counts are comparable. What variables would need to be considered when recording numbers of butterflies? Is there a species in your area that could be monitored, to help in its conservation?

▲ Figure 15 Silver-studded blue butterfly

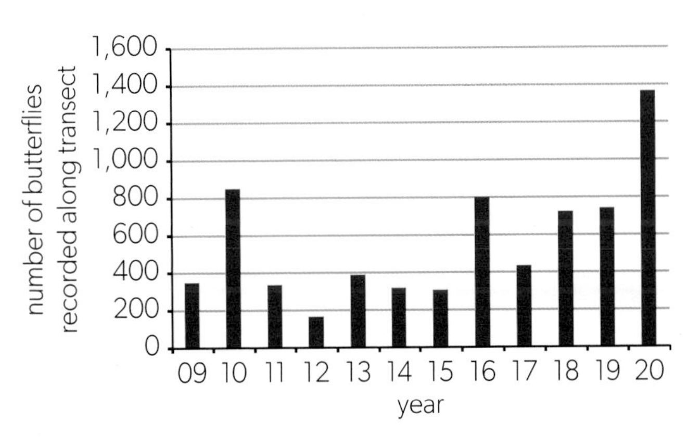

▲ Figure 16

Data-based questions: Satellite monitoring

Sentinel satellites are operated by the European Space Agency (ESA). They provide images of each part of Earth every five days. The data is made available worldwide at no charge and can be used for monitoring changes to land use.

The Sustainable Natural Resource Management Association (SUNARMA) in Ethiopia uses these images to monitor plowing and burning of vegetation.

The satellite images in Figures 17 and 18 show the same area in different seasons. In the wet season, grassland and arable crops in growth are light green and recently plowed land is dark brown. In the dry season, grassland and crops ready to harvest are light brown and recently burned land is darker brown. Forest with trees in leaf is darker green in both seasons.

The map in Figure 19 shows the same area as the satellite images. It was produced by analysing the image for 11 November 2019 to identify areas that had been burned during the dry season (brown) or were being burned when the satellite passed over (purple). Green outlines show designated forests managed by local community cooperatives.

1. Calculate the size of the area shown in the satellite images. [2]
2. Deduce what the state of the land is at these co-ordinates: 900, 110 in the dry season; 910, 116 in the wet season. [2]
3. Deduce what the designation is of land at 905, 116 and what the satellite images indicate about its use. [4]
4. Using the satellite images, suggest a hypothesis for the distribution of forest. [2]
5. (a) State two signs of burning that are visible in the dry season satellite image, with co-ordinates of an example of each. [4]

   (b) Estimate the percentage of the land area that was burned in the 2019 dry season. [1]
6. Suggest benefits of satellite monitoring of burning and plowing [5]

▲ **Figure 17** Dry season (11 November 2019)

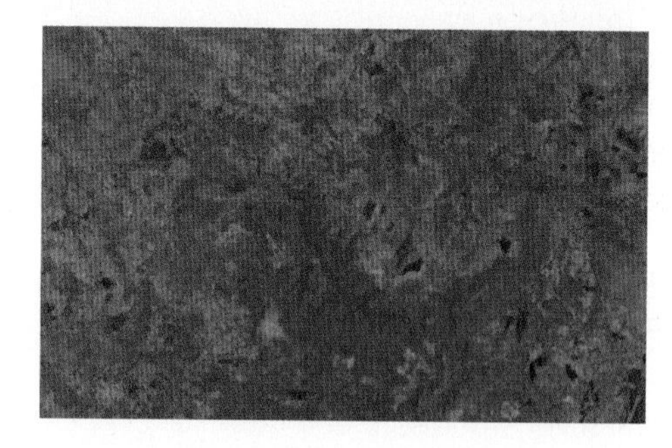

▲ **Figure 18** Wet season (2 August 2020)

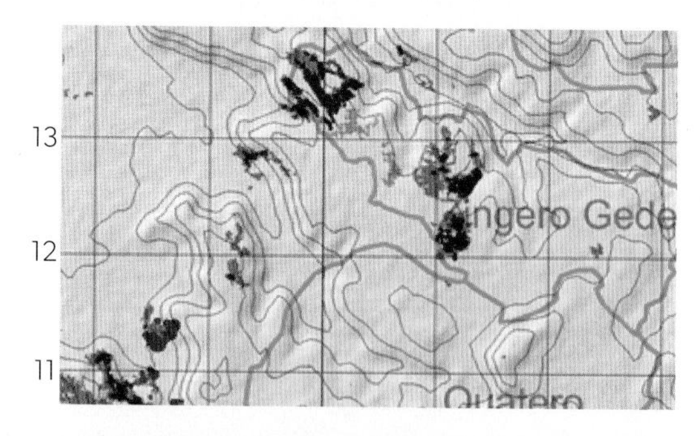

▲ **Figure 19** Map with 1 × 1 kilometre grid squares

Applying technology to collect data

The Audubon Society sponsors an annual bird count. Each year in December, volunteer bird-watchers from across North America report sightings of birds. All of these sightings all entered into a searchable database (https://netapp.audubon.org/cbcobservation/). The database contains records from more than 120 years of Audubon bird counts.

You can use the database to answer questions such as:

a. Has the average latitude where a particular species is observed shifted northward due to climate change over the past 40 years?

b. How has the presence in the Great Lakes Region of the invasive and destructive emerald ash borer beetle (*Agrilus planipennis*) impacted the population of birds that feed on ash seeds / birds that feed on beetles / birds that nest in dead trees?

Science as a shared endeavour

To be verifiable, data usually has to come from a published source, which has been peer-reviewed. This allows the methodology to be checked.

Consider the Audubon annual bird count as an example of data collected by volunteer "citizen scientists".

1. Discuss the strengths and limitations of data collected in this way.
2. Discuss the advantages and disadvantages of undertaking inquiries using data collected by other people.

A4.2.6 Causes of the current biodiversity crisis

Humans have been causing species extinctions and loss of ecosystems for thousands of years but a biodiversity crisis has developed since about 1970. According to a UN report, by 2019, 75% of the terrestrial environment and 66% of marine environments had been "severely altered" as a consequence of human actions. Current rates of species extinction are between 100 and 1,000 times higher than normal and they are rising. If this trend continues, the rate of species loss could become 10,000 times higher than normal within the next 100 years.

The Earth's four-billion-year history has been turbulent. During "Snowball Earth" phases, the surface was almost completely covered by ice. However, there were also phases when the entire surface was ice-free, even at the poles. At times, there was a single giant continent; at other times, land was made up of many isolated island continents. Five mass extinctions have occurred as a result of unstoppable forces such as asteroid strikes, volcanic activity and transformations in the atmosphere and climate patterns. Widespread ecosystem collapse has unfolded over hundreds of thousands of years, or even longer. The current and sixth mass extinction is happening much more rapidly. However, because it is largely caused by human actions, it is stoppable.

To avert the biodiversity crisis, we must appreciate how human activities cause species extinction and ecosystem loss. The principal causes, discussed in *Sections A4.2.3* and *A4.2.4*, are:

- hunting and other forms of over-exploitation
- urbanization, with towns and cities growing ever larger
- deforestation and clearance of land for agriculture, leading to loss of natural habitats
- pollution of land and sea throughout the world
- spread of alien invasive species due to global transport or deliberate introductions; such species may be pests, cause disease or compete with native species.

None of these causes is new but their intensity has increased significantly over the last 100 years. This is a consequence of the enormous rise in the number of people on Earth. Between 1920 and 2020 the human population more than quadrupled, from less than two billion to almost eight billion. Overpopulation is the overarching issue that makes human activities a threat to most other species and risks widespread ecosystem collapse.

Data-based questions: Human population increases

Figure 20 shows estimated worldwide human population growth between 1700 and 2100.

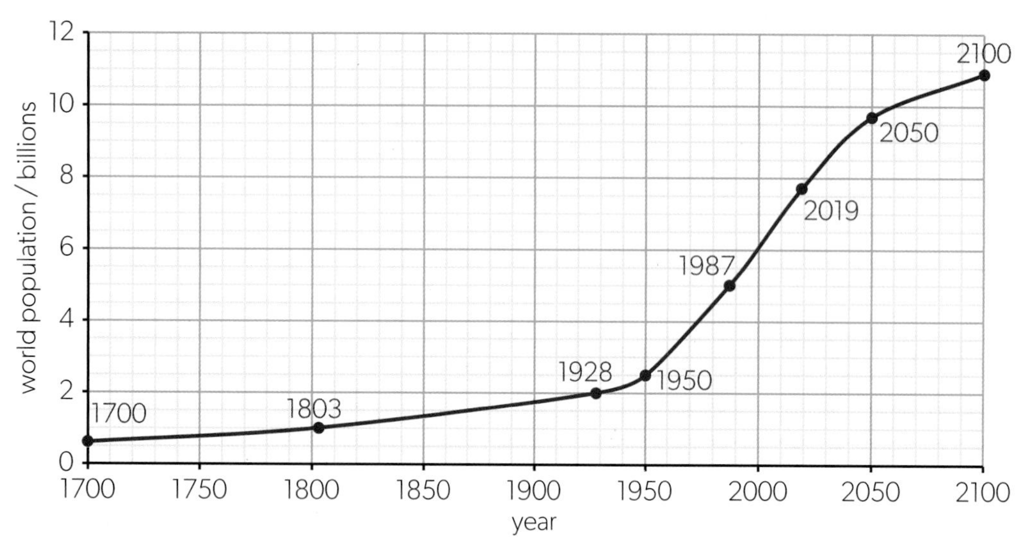

▲ Figure 20

1. According to the graph, in what decade was there:
   a. the greatest absolute rise in the human population [1]
   b. the greatest percentage rise in the human population? [1]
2. a. In what year did the population reach four billion? [1]
   b. When is the population predicted to reach eight billion? [1]
3. How many years did it take for the population to double from:
   a. one to two billion [1]
   b. two to four billion [1]
   c four billion to a predicted eight billion? [1]
4. Using the data in the graph, discuss whether the rise in the human population since 1700 has been:
   a. rapid [3]
   b. exponential. [3]
5. Discuss what the peak human population might be, assuming it does not continue to rise indefinitely. [2]

A4.2.7 Need for several approaches to conservation of biodiversity

The biodiversity crisis is acute and no single approach to tackling it will be enough. Any strategy that can help should be adopted, including all of those described here.

In situ methods conserve species in their natural habitats. The ideal approach is to leave areas of the Earth's surface in a state of pristine wilderness. Partially degraded pristine areas may still be extremely valuable for the purpose of nature conservation. Legislation or land purchase can be used to create national parks or nature reserves. The larger the protected area, the better. Terrestrial, freshwater and marine areas are now protected in many countries.

In situ conservation has some significant advantages. It ensures that a species lives in the abiotic environment to which it is adapted, so it does not start to adapt to different conditions. It allows the species to interact with other wild species, conserving more aspects of the organism's niche and the integrity of the ecosystem. Animal behaviour patterns can remain normal. In addition, costs are low if a wildlife reserve is in a good enough state for human intervention to be unnecessary.

Human influences are so pervasive around the world that most areas of wilderness are threatened with change. As a result, nature reserves often require active management. Depending on the type of ecosystem and the nature of the threats, management may involve removal of alien species, reintroduction of species that have become locally extinct, measures to increase or decrease population sizes of herbivores and predators, prevention of poaching, supplementary feeding of animals and control of access by humans.

In some cases, ecosystems have become so damaged that major interventions are needed. It is possible to reverse ecosystem collapse, and recovery is sometimes surprisingly rapid. During the 21st century there has been an increasing trend for rewilding, where degraded ecosystems are returned to as natural a state as possible and balance is maintained by natural processes rather than human intervention.

Ex situ conservation is the preservation of species outside their natural habitats. At the outset, organisms are removed from the wild. Traditionally, plant species were then grown in botanic gardens and animals were kept in zoos. Clearly, it is not acceptable to remove scarce plants and animals from wild populations repeatedly. Therefore, botanic gardens must propagate plants and zoos must breed animals. Increasingly, zoos carry out carefully planned captive breeding programmes, followed by release of the captive-bred individuals back into their natural habitats.

In some cases, removal and relocation of endangered species is justifiable because they cannot safely remain in their natural habitats. For example, populations of flightless native bird species in New Zealand have been moved to offshore islands to protect them from attacks by invasive alien predators. If the threat to a species in its natural habitat is eliminated, the species can be returned to its original site. In New Zealand, this may happen if programmes to eliminate rats, stoats and possums are successful.

A more radical approach to ex situ conservation is the long-term storage of living material that could be used for propagation in the future. This material is called germplasm. The usual approach with plants is to store seed in seed banks at low temperatures, so they can maintain viability for long periods. With animals, the stored material may be samples of tissue, eggs or sperm.

▲ Figure 21 Takahē (*Porphyrio hochstetteri*) are flightless birds that were presumed extinct for many years. A small population was discovered in a remote valley on the South Island of New Zealand in the 1940s. Since then, active conservation measures have increased numbers by as much as 10% per year and there are now more than 400 individuals. One of the conservation methods has been translocations to five small predator-free offshore islands

▲ Figure 22 A new approach to conservation is to create "mainland islands" by fencing off areas within which alien invasive species and other threats can be controlled. The fence in this photo is part of the boundary of Orokonui Ecosanctuary, a mainland island near Dunedin in New Zealand

A4.2.8 Selection of evolutionarily distinct and globally endangered species for conservation prioritization in the EDGE of Existence programme

The scale of the biodiversity crisis is so large that conservation efforts have to be targeted where the benefits are likely to be greatest. This raises the controversial question of which species are most worthy of our efforts to conserve them. The EDGE of Existence project uses two criteria to identify animal species that are most deserving of conservation.

- Does the species have few or no close relatives, so it is a member of a very small clade?
- Is the species in danger of extinction, because all of its remaining populations are threatened?

Lists are prepared of species that are both Evolutionarily Distinct and Globally Endangered, hence the name of the project.

Species on these lists can then be targeted for more intense conservation efforts than other species that are either not threatened or that have close relatives. Some species are the last members of a clade that has existed for tens or hundreds of millions of years and it would be tragic for them to become extinct as a result of human activities.

▲ Figure 23 Two species on the EDGE list: *Loris tardigradus tardigradus* (Horton Plains slender loris) from Sri Lanka and *Bradypus pygmaeus* (Pygmy three-toed sloth) from Isla Escudo de Veraguas, a small island off the coast of Panama. What species on EDGE lists are in your part of the world and what can you do to help conserve them?

Global impact of science

Red wolves are native to parts of the southeastern United States. They are intermediate in form between grey wolves (*Canis lupus*) and coyotes (*Canis latrans*). There has been disagreement whether to classify red wolves as a subspecies of wolf (*Canis lupus rufus*) or as a distinct species (*Canis rufus*). Because of this, it is sometimes excluded from endangered species lists despite its critically low numbers. It is listed under the US Endangered Species Act and therefore given legal protection. The International Union for the Conservation of Nature (IUCN) also lists it as a critically endangered species. However, it is not listed in the appendices of the Convention on Trade in Endangered Species (CITES). Some of the debate is based on the lack of a universally accepted species concept.

▲ **Figure 24** Red wolf (left) and coyote (right). If red wolves are as similar to coyotes as to grey wolves, should they be classified as a separate species or a subspecies of grey wolf?

ATL Thinking skills: Evaluating and defending ethical positions

When evaluating an ethical question, different criteria can yield different results. Consequentialism uses the standard that "the ends justify the means"; motivism uses the standard that understanding the intentions of an action is important in evaluating whether it is ethical.

Issues such as which species should be prioritized for conservation efforts have complex ethical, environmental, political, social, cultural and economic implications and therefore need to be debated. Of the 85,604 species identified in the IUCN Global Red List, 24,307 are classified as threatened with extinction. Resources to address this challenge are limited and priorities have to be set. However, conflicts of values can arise. Different groups will have different goals—such as defending against the most predictable species losses; maximizing phylogenetic diversity; conserving keystone species over others; or conserving species of cultural significance—and different goals will necessitate different approaches.

Smallpox viruses (*Variola major* and *Variola minor*)

The disease smallpox was caused by two types of virus, *Variola major* and *Variola minor*. It caused the deaths of hundreds of millions of people. Smallpox vaccines were introduced to give immunity to the disease. In 1967, the World Health Organization started a campaign to eliminate the disease completely by vaccination. The last case of smallpox was in 1978. Because there are no reservoirs of the viruses that cause the disease, it has been permanently eradicated.

The extinction of a virus that causes death and suffering in humans is uncontroversial. But should humans try to eradicate other groups of troublesome organisms—for example, bacterial pathogens, parasites of the human gut or skin, or pests or diseases of crop plants or livestock? These issues raise important ethical questions. Do we as humans have the right to eliminate species that are harmful—or simply not useful—to us?

Linking questions

1. In what ways is diversity a property of life at all levels of biological organization?
    a. Distinguish between prokaryotic cells and eukaryotic cells. (A2.2.5 and A2.2.6)
    b. Outline the mechanism of adaptive radiation. (A4.1.9; HL)
    c. Explain how DNA is able to code for an infinite variety of proteins. (A1.2.9)
2. How does variation contribute to the stability of ecological communities?
    a. Outline the concept of niche. (B4.2.1)
    b. With reference to named examples, construct an annotated food web. (C4.2.4)
    c. With reference to ecological succession, distinguish between the structure of a pioneer and climax community. (D4.2.12)

TOK

In what ways do values affect the production and acquisition of knowledge?

Individuals have different opinions about what is important, useful or worthy of attention: they have different values. Science is a human endeavour, so it is not surprising that scientists have different approaches to decision making.

Research funding is limited and the costs of scientific research are often met by grant agencies. But who decides how funds are allocated? Scientists submit research proposals to agencies and each application is reviewed by a funding panel. Some grant applications ask scientists to project outcomes or suggest applications of the research before it has even begun. Questions arise when the grant agency has a stake in the study's outcome. A sponsor may fund several different research groups, suppressing results that run counter to their interests and publishing those that support their industry. For example, in a 2008 report based on chemical industry studies, the US Food and Drug Administration (FDA) concluded that the bisphenol A (BPA) found in plastic containers was not a health risk. Independent research studies reported different conclusions. There have been claims of funding bias in many areas, including pharmaceutical research, nutrition research and climate change research.

According to the UN World Health Organization, there are 20 neglected tropical diseases (NTDs). These diverse conditions are common in tropical areas, mostly affecting poor communities and disproportionately affecting children and women. Many of the diseases have affected humanity since ancient times. It has been stated that these diseases are under-researched and efforts to eradicate them are under-funded, because the values of the research community lead to prioritization of other diseases.

In scientific investigations, scientists have to choose between hypotheses. Inevitably, these choices are influenced by human values such as a desire for simplicity, accuracy of data and explanatory power. In statistical testing, researchers test null and alternative hypotheses. The null hypothesis is a hypothesis that a given factor has no observable effect. Two types of error can occur in hypothesis testing:

1. An experimenter can mistakenly reject the null hypothesis when it is true. In medical testing, this would be known as a false positive. This is a type I error.
2. An experimenter can accept the null hypothesis when it is false. This is a type II error or a false negative.

It is not possible to minimize the likelihood of one type of error without increasing the likelihood of the other type of error. This choice involves a value judgement. For example, the null hypothesis might say that an introduced species does not have an effect on the host community. To minimize the risk of a type I error, the researcher might invest significantly in controlling the invader when it would not have become invasive. On the other hand, minimizing a type II error might lead them to ignore the threats represented by the invader. Their choice will be based on a value judgement.

Perhaps the most widely discussed false positives in medical screening come from the breast cancer screening procedure known as a mammogram (Figure 1). The US rate of false positive mammograms is the highest in the world; one study found it to be as high as 15%. Women are offered mammograms annually, starting in middle age. The consequence of the high false positive rate in the US is that, in any 10-year period, half of the American women screened receive at least one false positive mammogram. False positive mammograms often result in costly follow-up tests. They also cause women unnecessary anxiety. In contrast, the Netherlands has the lowest rate in the world, with just 1% false positives being reported. The lowest rates are generally in Northern Europe where mammography films are read twice and there are high thresholds for additional testing. When contrasting the two jurisdictions, what differences in values are evident?

One standard for making judgements is known as Occam's Razor. In everyday language, this is the idea that the simplest explanation or more probable cause is most likely to be true. In cladistics, a characteristic shared between two groups can indicate a shared ancestry. Alternatively, it could indicate that the characteristic has evolved twice. It is less likely that complex traits evolved twice or multiple times, so the criterion for judgement when constructing cladograms is parsimony—the history with the smallest number of rare events is assumed to be the most likely. This is not always the case; for example, the octopus eye and the vertebrate eye are remarkably similar but the consensus is that they evolved separately. Similarly, warm-bloodedness in birds and mammals is believed to have evolved separately. In general, however, the simplest theory is often true.

In medical diagnosis, an aphorism is: "If you hear the sound of hooves, it could be zebras, but it is most likely horses". The most rational approach is to start with the most likely scenario and progress toward rarer conditions. Sadly, extreme risk avoidance or profit motivation can lead practitioners to start with the least likely but worst case hypothesis and test towards the more likely explanation.

Onchocerciasis is known as river blindness. It is a disease caused by infection with the parasitic worm *Onchocerca volvulus*. It is one of the 20 neglected diseases as no vaccination is available despite the disease affecting humanity since ancient times. Figure 2 shows the parasite coming out of the antenna of an adult black fly. The black fly is a vector for the disease.

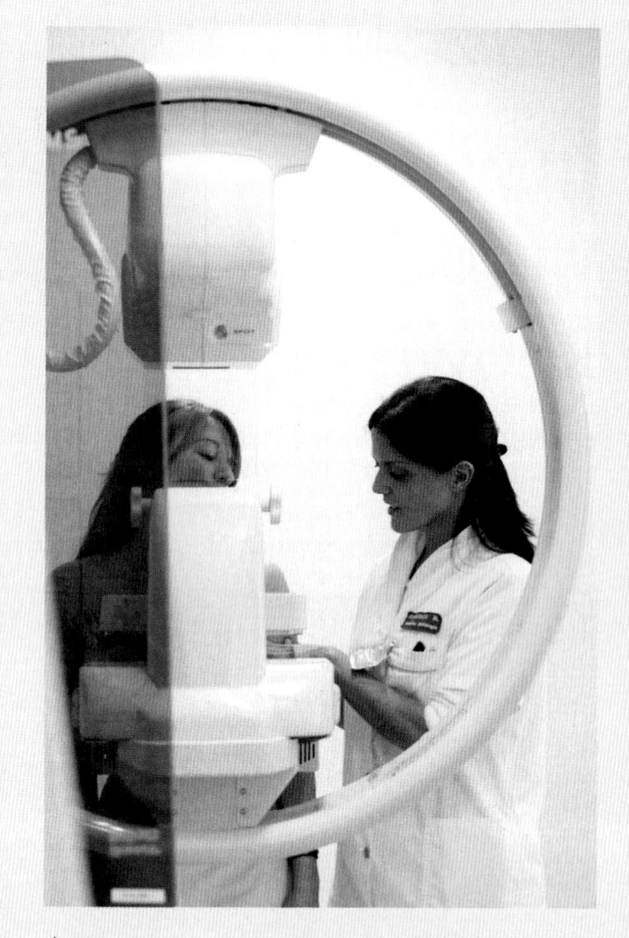

▲ Figure 1 A woman having a mammogram

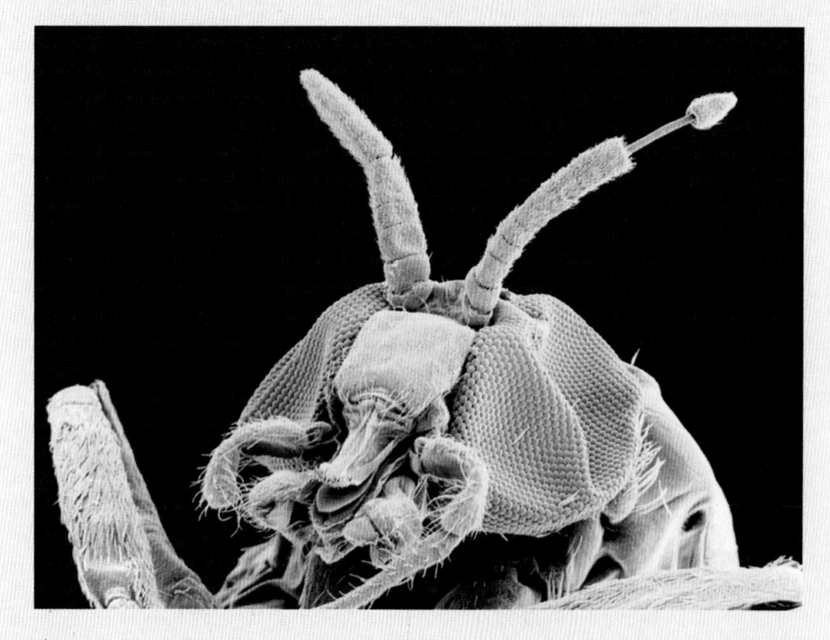

▲ Figure 2 *Onchocerca volvulus* parasite coming out of the antenna of a black fly

End of chapter questions

1. About 2.5 million years ago, falling sea levels resulted in the joining together of North America and South America through a narrow land bridge, the isthmus of Panama. This event allowed two-way traffic of land mammals between the formerly isolated continents. A redistribution of families and genera (plural form for genus) occurred. The graphs in Figure 1 show the total number of known native and immigrant families and genera in South America over a time span ranging from nine million years ago to the present.

   a. Compare the changes in the number of South American native families and the number of North American immigrant families in the one million years after the formation of the land bridge. [1]

   b. Suggest a reason for the decline in the number of South American native families and the number of North American immigrant families within the last 1.5 million years. [1]

   c. Using the data from nine million years ago and the present, calculate the percentage increase in the total number of genera found in South America. [1]

   d. Discuss why the percentage increase in genera is much greater than the apparent percentage increase in families. [2]

   e. State a form of evidence on which the data in the graphs is based. [1]

   f. Referring to the competitive exclusion principle, suggest why the number of native families and genera declined. [2]

   g. Many of the immigrant mammals were placental mammals and a number of the native mammals were marsupials. Suggest what is the adaptive advantage of placental rather than marsupial gestation. [2]

   h. With reference to this example, outline the concept of adaptive radiation. [3]

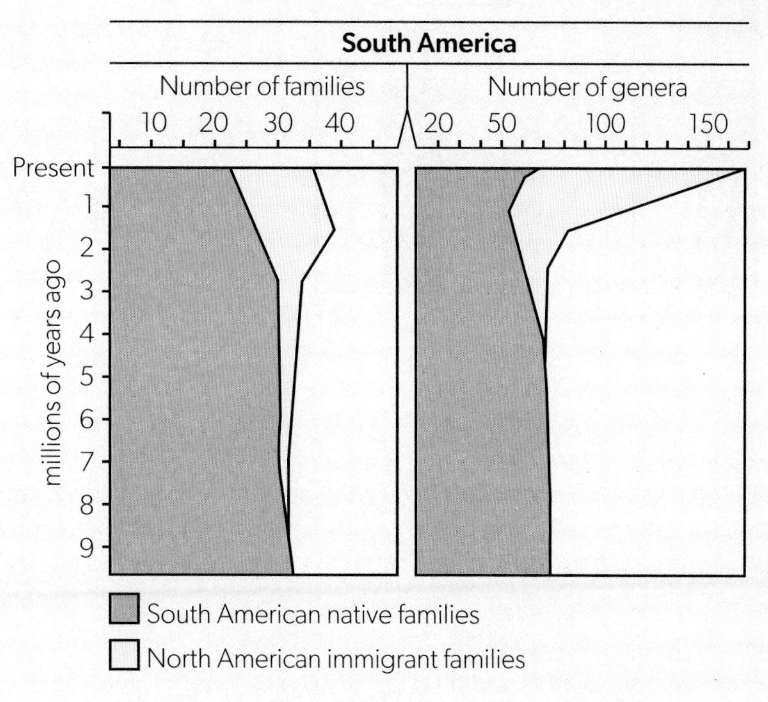

▲ Figure 1

2. The mountain yellow-legged frog (*Rana muscosa*) was once a common inhabitant of the Sierra Nevada (California, USA). It has declined during the past century due in part to the introduction of non-native fish, such as trout, into naturally fish-free habitats. The bar chart in Figure 2 shows the average number per lake of tadpoles.

    a. State the number of tadpoles per lake with and without trout. [1]

    b. Compare and contrast results for lakes with and without trout. [2]

    c. The trout might affect the number of frogs or tadpoles by competing for resources. Suggest one other way in which trout might affect the number of tadpoles or frogs in lakes. [1]

    d. The "without trout" group was achieved by the intentional removal of trout from the experimental waterways. Discuss the challenges of such a conservation measure. [2]

    e. Outline a method that could be used to determine the population of frogs in the lake. [3]

In order to restore the frog population, introduced trout were removed from the lakes. The map of the LeConte Basin study area shows the distribution of mountain yellow-legged frogs and trout populations just prior to the removal of the trout in 2001. The graphs show the population of tadpoles and frogs in the lakes before, during, and after the removal of the trout.

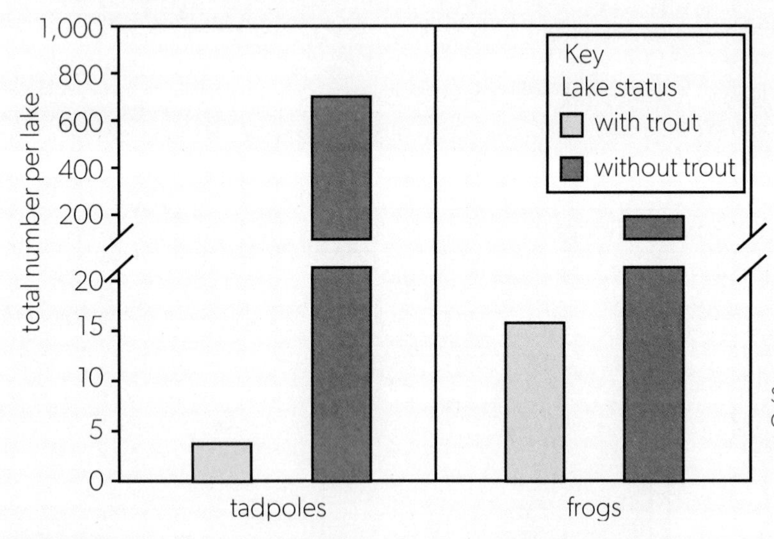

Source: V Vredenburg. 2004. *PNAS*. Vol. 101 Pp. 7646–7650. Copyright 2004 National Academy of Sciences. USA.

▲ Figure 2

3. One challenge associated with establishing nature reserves is concerns about "edge effects". The graph in Figure 3 shows that some edge effects in the Amazon rainforest are detected quite far in from the edge.

    a. Determine how far from the forest edge an increase in disturbance-adapted beetles would be detected. [1]

    b. With respect to the example of disturbance-adapted beetles, suggest what is meant by an indicator species. [2]

    c. Explain how this information about edge effects can influence the design of reserves. [3]

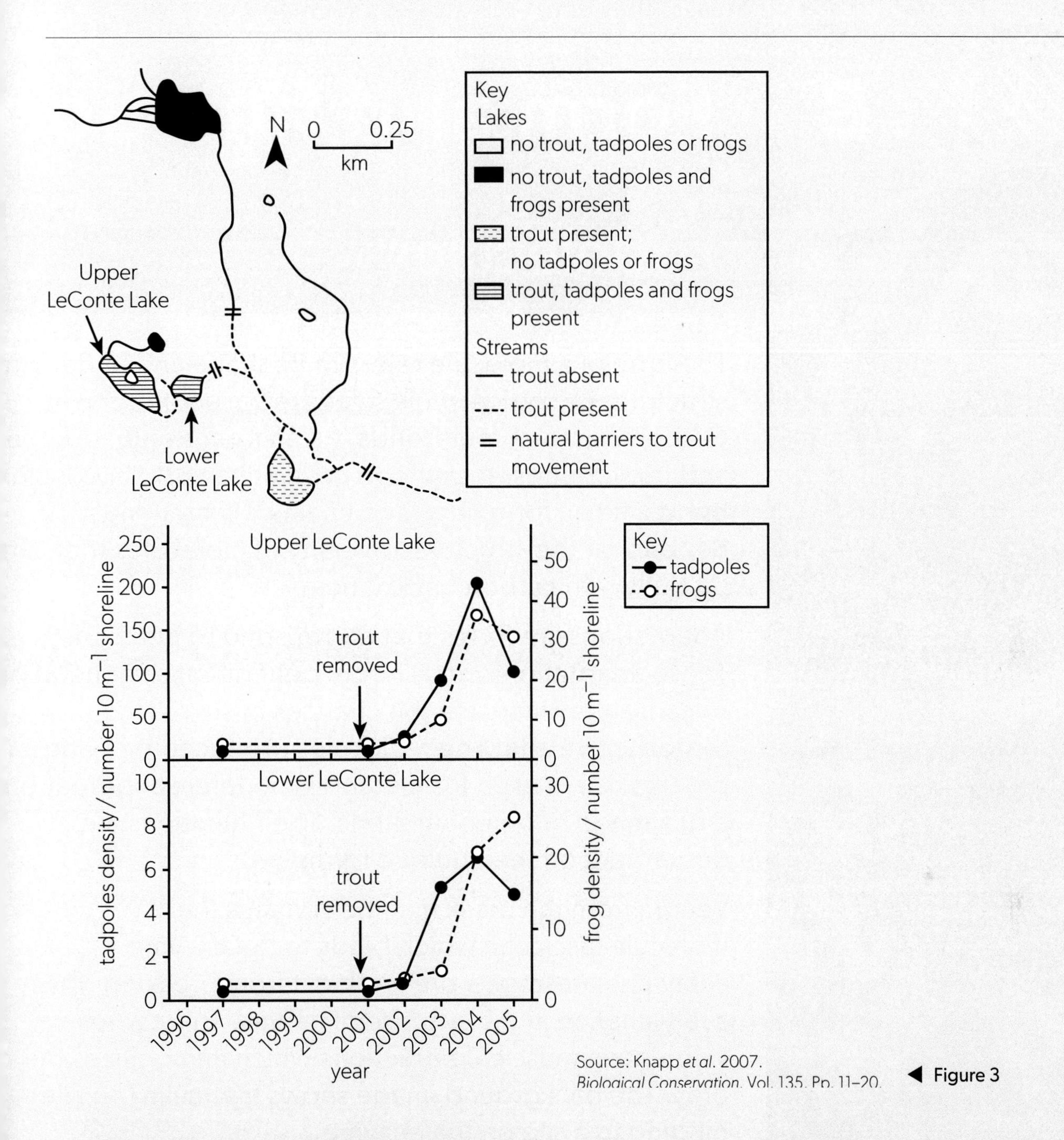

Source: Knapp *et al.* 2007. *Biological Conservation*. Vol. 135. Pp. 11–20.

◀ **Figure 3**

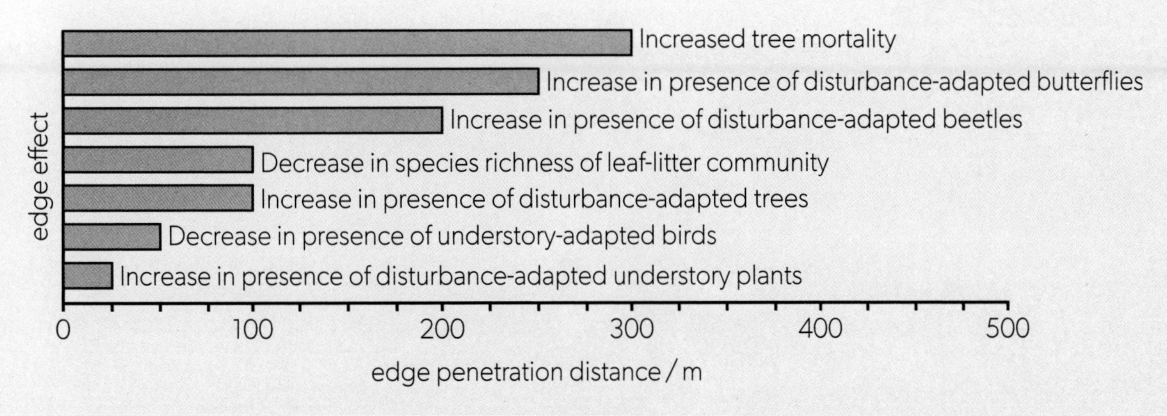

Source: WF Laurance. 2008. *Biological Conservation*. Vol. 141. Pp. 1731–1744.

▲ **Figure 4**

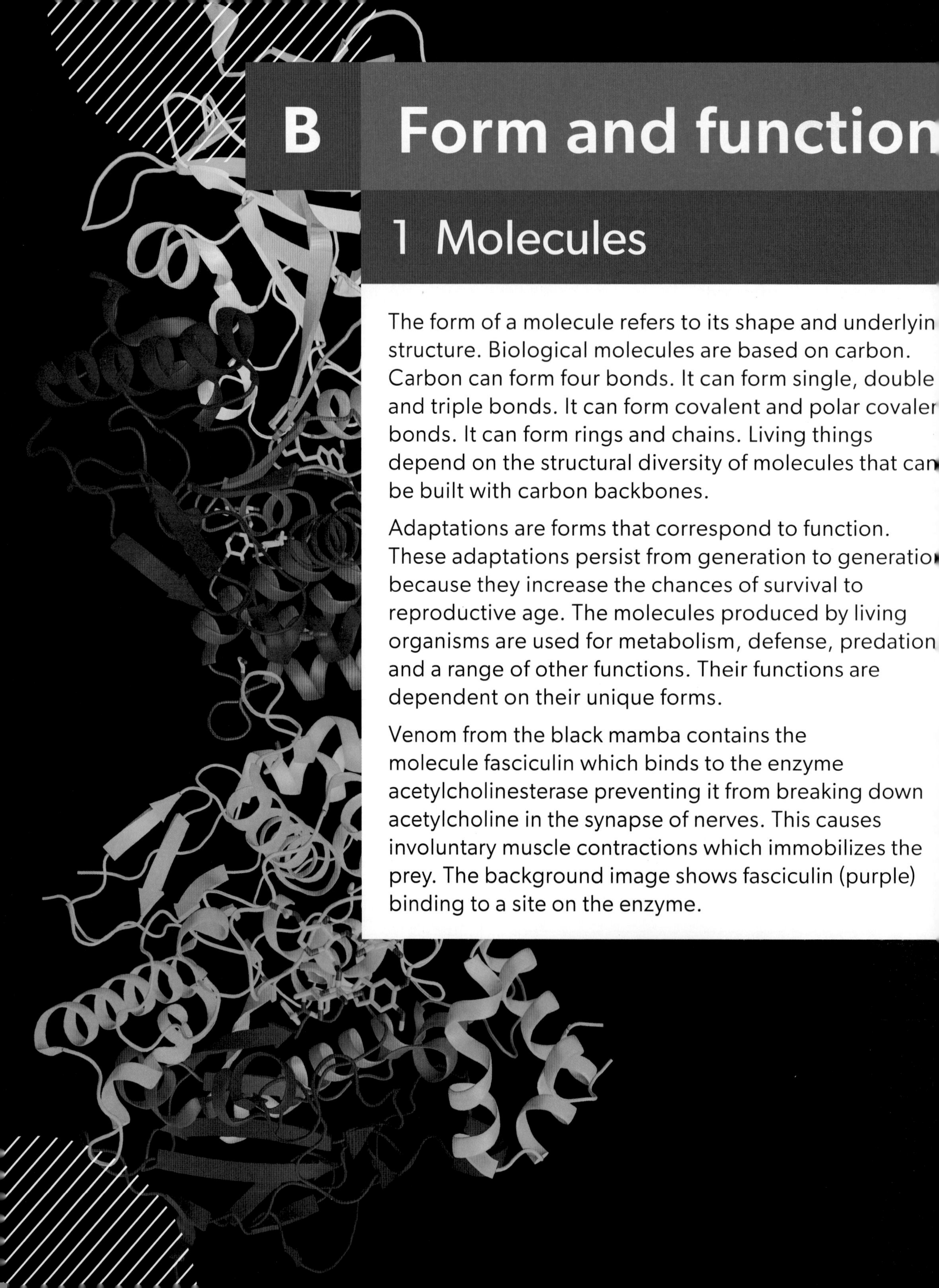

B Form and function

1 Molecules

The form of a molecule refers to its shape and underlyin structure. Biological molecules are based on carbon. Carbon can form four bonds. It can form single, double and triple bonds. It can form covalent and polar covaler bonds. It can form rings and chains. Living things depend on the structural diversity of molecules that can be built with carbon backbones.

Adaptations are forms that correspond to function. These adaptations persist from generation to generatio because they increase the chances of survival to reproductive age. The molecules produced by living organisms are used for metabolism, defense, predation and a range of other functions. Their functions are dependent on their unique forms.

Venom from the black mamba contains the molecule fasciculin which binds to the enzyme acetylcholinesterase preventing it from breaking down acetylcholine in the synapse of nerves. This causes involuntary muscle contractions which immobilizes the prey. The background image shows fasciculin (purple) binding to a site on the enzyme.

B1.1 Carbohydrates and lipids

In what ways do variations in form allow diversity of function in carbohydrates and lipids?

Carbohydrates and lipids are both composed of carbon, oxygen and hydrogen. However, they have very different properties because of differences in the form of their molecules. A total of 1,679 different molecules with a wide range of properties have been identified in watermelon plants. Figures 1 and 2 show some examples of both types of molecules. Compare and contrast the relative amounts of oxygen, carbon and hydrogen in carbohydrates and lipids. Why are lipids relatively insoluble in water compared to carbohydrates?

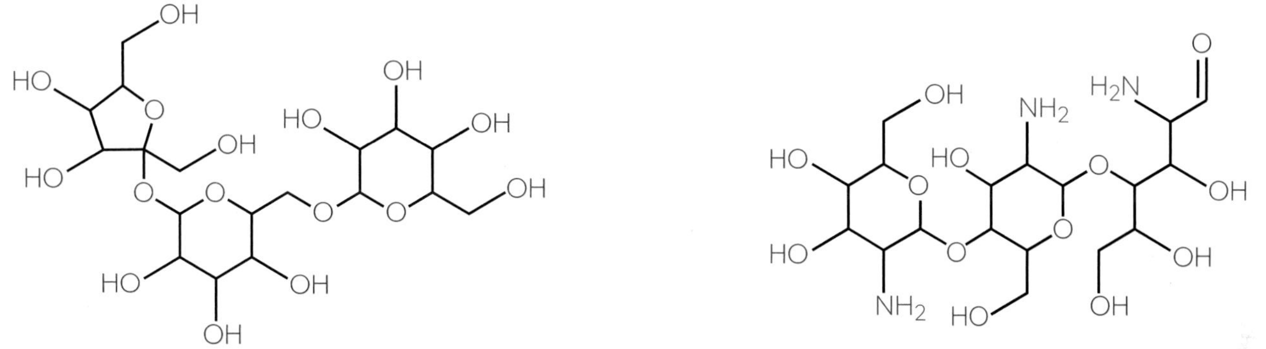

raffinose

chitotriose

▲ Figure 1 Some sugars

How do carbohydrates and lipids compare as energy storage compounds?

Carbohydrates in the form of starch or glycogen and lipids in the form of fats or oils can be used as energy stores. They are chemically stable and energy is released when they are oxidized by cell respiration. How does the relative amount of oxygen in the molecule affect how much energy per gram it releases? What advantages do fats and oils have as energy sources? What advantages do carbohydrates have? Can both carbohydrates and lipids be used in either aerobic or anaerobic respiration? Which energy form is more easily transported from one part of an organism to another?

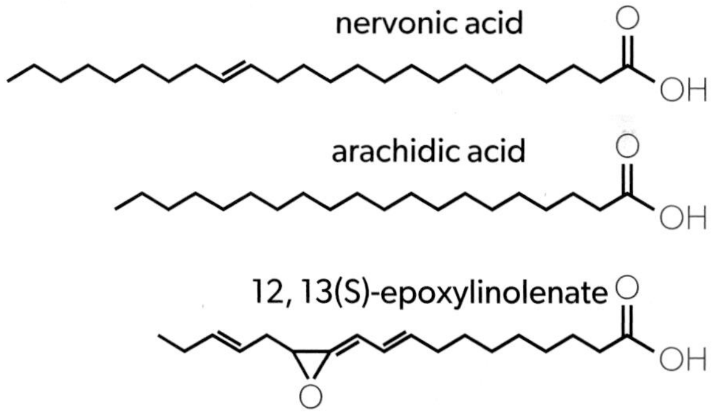

nervonic acid

arachidic acid

12, 13(S)-epoxylinolenate

▲ Figure 2 Some fatty acids

SL and HL

B1.1.1 Chemical properties of a carbon atom allowing for the formation of diverse compounds upon which life is based
B1.1.2 Production of macromolecules by condensation reactions that link monomers to form a polymer
B1.1.3 Digestion of polymers into monomers by hydrolysis reactions
B1.1.4 Form and function of monosaccharides
B1.1.5 Polysaccharides as energy storage compounds
B1.1.6 Structure of cellulose related to its function as a structural polysaccharide in plants
B1.1.7 Role of glycoproteins in cell–cell recognition
B1.1.8 Hydrophobic properties of lipids
B1.1.9 Formation of triglycerides and phospholipids by condensation reactions
B1.1.10 Difference between saturated, monounsaturated and polyunsaturated fatty acids
B1.1.11 Triglycerides in adipose tissues for energy storage and thermal insulation
B1.1.12 Formation of phospholipid bilayers as a consequence of the hydrophobic and hydrophilic regions
B1.1.13 Ability of non-polar steroids to pass through the phospholipid bilayer

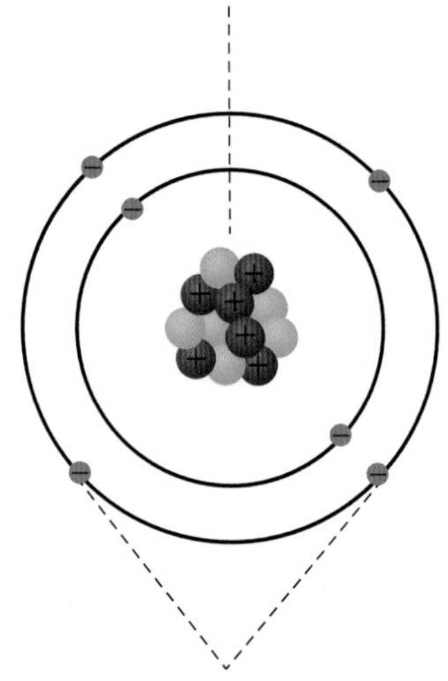

▲ Figure 3 Stylized drawing of a carbon atom

B1.1.1 Chemical properties of a carbon atom allowing for the formation of diverse compounds upon which life is based

Carbon is only the 15th most abundant element on Earth, but without it life would not exist. Its chemical properties allow many different forms of molecule to be produced, so the range of functions is almost limitless.

Carbon atoms can form covalent bonds with other atoms. A covalent bond is formed by sharing a pair of electrons between two adjacent atoms. The negatively charged shared electrons are attracted to the positively charged nuclei of both atoms. Covalent bonds are the strongest type of bond between atoms. This means stable molecules based on carbon can be produced.

Each carbon atom can form four covalent bonds, so molecules containing carbon can have complex structures. There can be four single covalent bonds or two single and one double covalent bond. Double covalent bonds are found, for example, in unsaturated fatty acids.

Carbon atoms can form covalent bonds with other carbon atoms or with atoms of other elements such as hydrogen, oxygen, nitrogen or phosphorus. Carbon atoms can bond with four atoms of one other element—for example, with four hydrogen atoms to form methane. They can also bond to more than one other element—for example, with oxygen and hydrogen to form ethanol.

Carbon atoms can be linked up by covalent bonds to form a chain of any length. Fatty acids contain unbranched chains of up to 20 carbon atoms. Chains can also be branched, with the branch often made using an oxygen atom.

Single covalent bonds allow both of the bonded atoms to rotate, but not to move further apart or nearer to each other. The covalent bonds formed by a carbon atom spread apart as much as possible so they form a tetrahedral shape. So, a chain of covalently bonded carbon atoms is not straight—the straightest it can be is a zig-zag. Because of the bond angles, chains of carbon atoms can form rings. The ring may be made entirely of carbon atoms—for example, in menthol which is synthesized by mint plants. Or contain an atom of another element, usually oxygen or nitrogen. A molecule may contain a single ring as in the base thymine,

▲ Figure 4 The plant *Chrysanthemum cinerariifolium* produces chrysanthemic acid, a very unusual molecule that has a ring of three carbon atoms. You could research the reasons for this molecule being rather unstable and the benefits to the plant of producing it

or two rings as in adenine, or more. Cholesterol molecules have four rings all composed entirely of carbon.

methane—a single carbon with four single covalent bonds all to hydrogen

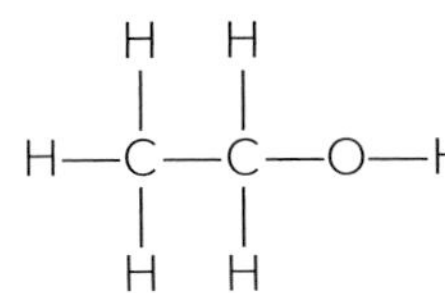

ethanol—two carbon atoms and bonds to two different other elements

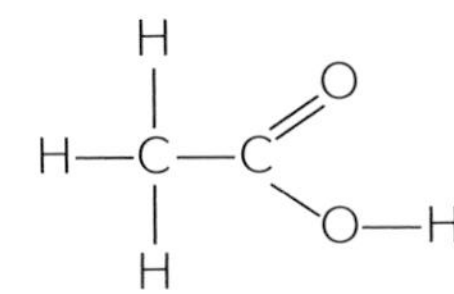

ethanoic acid—single covalent bonds and one double bond

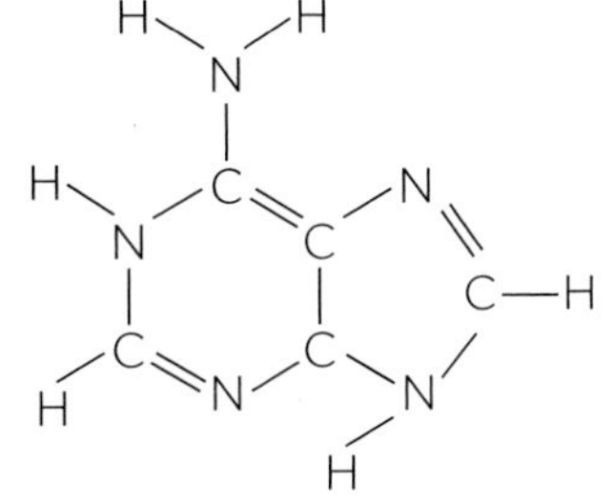

adenine—with two rings both with carbons and nitrogens and sharing of electrons in the ring

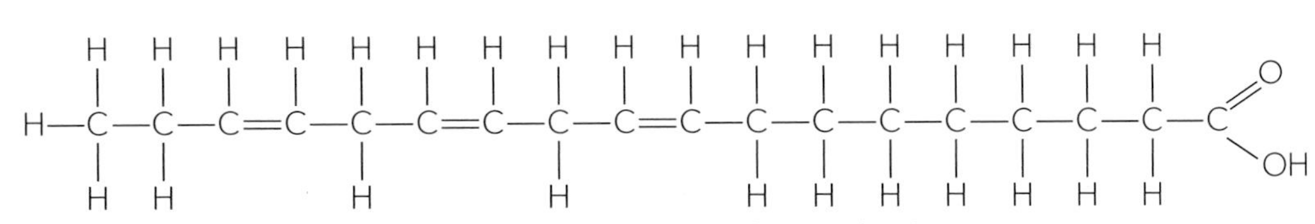

linolenic acid—an omega-3 fatty acid with a chain of 18 carbon atoms containing 3 double bonds

▲ Figure 5 Some common naturally occurring carbon compounds

Activity: Macromolecules

Titin is a giant protein that acts as a molecular spring in muscle. The backbone of the titin molecule is a chain of 100,000 atoms, linked by single covalent bonds. Can you find an example of a molecule in your body with a chain of over 1,000,000,000 atoms?

Science as a shared endeavour: SI units

The International System of Units (SI) is the scheme for metric units of measurement, agreed by scientists around the world in 1960. From time to time it is updated. There are 7 base units and 22 other units derived from the base units. Each unit can be given prefixes that make it larger or smaller. The preferred prefixes change the size by a factor of a thousand but the prefix "centi" is still sometimes used to indicate a hundredth.

Base units

- second (s) time
- metre (m) length
- kilogram (kg) mass
- ampere (A) electric current
- kelvin (K) temperature
- mole (mol) amount of substance
- candela (cd) luminous intensity

Examples of derived units

- newton (N) force ($kg\,m\,s^{-2}$)
- hertz (Hz) frequency (s^{-1})
- pascal (Pa) pressure ($N\,m^{-2}$)
- joule (J) energy (N m)
- watt (W) power ($J\,s^{-1}$)
- volt (V) voltage ($W\,A^{-1}$)
- lux (lx) illuminance ($cd\,m^{-2}$)

Metric prefixes

- giga (G) 10^9 (billion)
- mega (M) 10^6 (million)
- kilo (k) 10^3 (thousand)
- milli (m) 10^{-3} (thousandth)
- micro 10^{-6} (millionth)
- nano 10^{-9} (billionth)
- pico 10^{-12} (trillionth)

B1.1.2 Production of macromolecules by condensation reactions that link monomers to form a polymer

Macromolecules are molecules composed of a very large number of atoms, with a relative molecular mass above 10,000 atomic mass units. The main classes of macromolecule in living organisms are polysaccharides, polypeptides and nucleic acids. Each of these is made by linking together subunits into a chain. The subunits are monomers and the chain is a polymer. In each case, the chemical process that links another monomer onto the end of the polymer is a condensation reaction.

In a condensation reaction, two molecules are linked together and at the same time a smaller molecule is released. When polysaccharides, polypeptides and nucleic acids are constructed, the simpler molecule is always water. It is produced by removing a hydroxyl group (–OH) from one of the molecules being linked and a hydrogen from the other. This allows a bond to be made to bridge the two molecules.

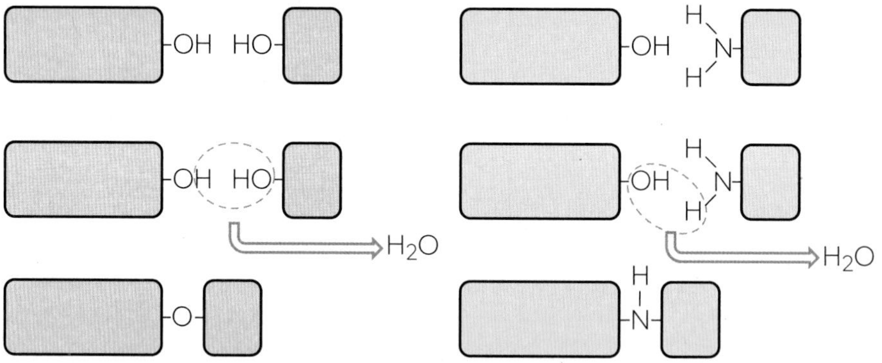

▶ Figure 6 Two methods of linking a monomer to a polymer by a condensation reaction

Energy is required to construct polysaccharides, polypeptides and nucleic acids by condensation. This energy is supplied by ATP. The synthesis of polysaccharides is described in detail here. Polypeptide synthesis is described in *Topic B1.2* and nucleic acid production in *Topic A1.2*.

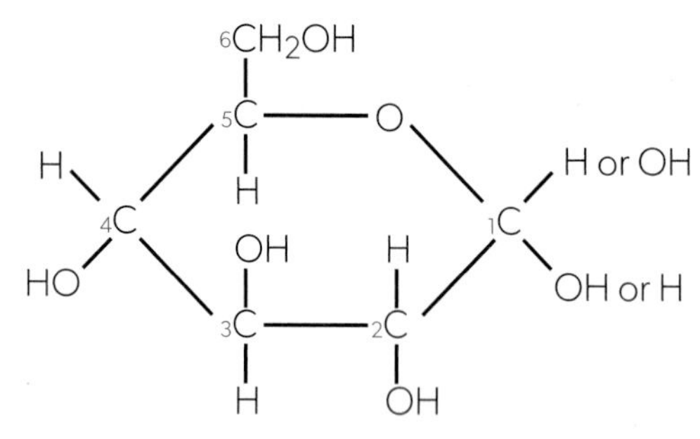

▲ Figure 7 Structure of glucose. There is always one –OH (hydroxyl) group and one –H group on C_1. The upper group is –H in α-glucose and –OH in β-glucose

A disaccharide is two monosaccharides linked together. A polysaccharide is a chain of monosaccharides. Glucose is the monosaccharide used to make the polysaccharides glycogen, starch and cellulose.

Glucose molecules are linked up with glycosidic bonds. These are C–O–C linkages formed by condensation, using hydroxyl groups. The hydroxyl on C_1 of a glucose is linked to the hydroxyl on C_4 at the end of the growing chain. In an unbranched chain, all the glycosidic bonds are 1→4. To form branches, the C_1 of a glucose is linked to the C_6 of a glucose already in the chain. This 1→6 linkage forms a side-branch, and more glucose molecules can be added to it with 1→4 bonds.

Cellulose molecules in plant cell walls are unbranched chains of β-glucose that can contain 15,000 or more glucose molecules. Glycogen molecules in liver or muscles cells are branched chains of α-glucose, with up to 60,000 glucose molecules.

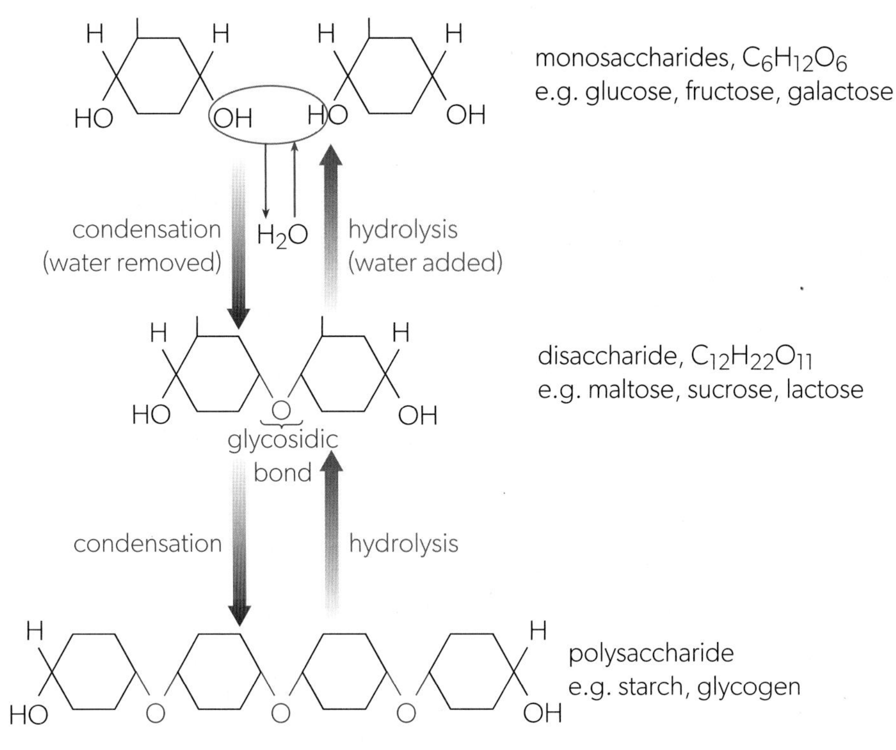

▲ Figure 8 Formation of 1–4 glycosidic bonds by condensation and their breakage by hydrolysis

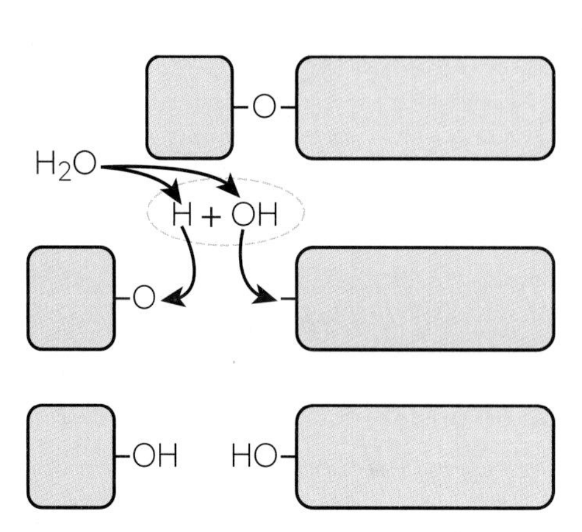

▲ Figure 9 In a hydrolysis reaction, water molecules are split to provide hydrogen and hydroxyl groups. These are used to make bonds to replace the bond that has been broken

B1.1.3 Digestion of polymers into monomers by hydrolysis reactions

Polymers are deconstructed so the monomers in them can be reused to build new polymers or used as a source of energy. Hydrolysis reactions are used to deconstruct polysaccharides, polypeptides and nucleic acids into monosaccharides, amino acids and nucleotides. These are the reactions that occur during digestion.

Digestion of polysaccharides, polypeptides and nucleic acids can be carried out by all cells. Digestion also happens outside the cell in the gut of animals. Decomposers release digestive enzymes into the environment around them in order to break down polymers by hydrolysis so they can absorb and use the monomers.

B1.1.4 Form and function of monosaccharides

Monosaccharides have between three and seven carbon atoms. Pentoses have five carbons and hexoses have six. Both pentoses and hexoses normally have molecules with a ring of atoms. There is one oxygen atom in the ring and four or five carbon atoms.

Monosaccharides have properties that allow them to be used in a variety of ways by living organisms. Glucose is a widely used monosaccharide.

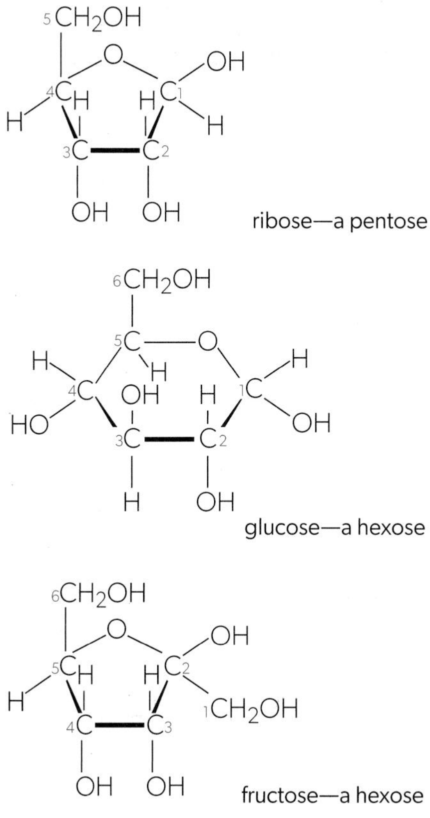

ribose—a pentose

glucose—a hexose

fructose—a hexose

▲ Figure 10

Use of molecular models: Modelling glucose

Pentoses and hexoses are unusual in that they can exist in straight-chain form as well as in ring form. They need to be in the ring form in order to form disaccharides and polysaccharides.

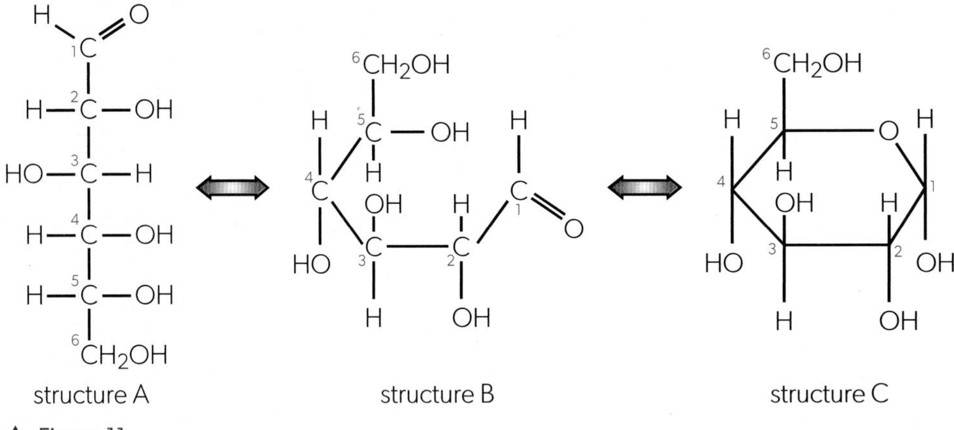

▲ Figure 11

1. Using a molecule model kit, construct a model of structure A. This is the straight-chain form.
2. Twist the model so that carbon 1 comes near the oxygen attached to carbon 5 as shown in structure B.
3. Break the double bond on carbon 1 and remove the hydrogen attached to the oxygen attached to carbon 5.
4. Attach carbon 1 to the oxygen on carbon 5 and reposition the detached hydrogen as shown in structure C.
5. Place the model on a table. Identify the plane of the ring. Which –OH groups are above the plane of the ring and which are below it?
6. Is your model α-glucose or β-glucose?

Data-based questions: Health consequences of the consumption of fructose

Obesity (excessive weight) is recognized as a global health problem and has been correlated with a large number of health issues, diseases and deaths. The increased consumption of fructose, now widely used as a sweetener, has been associated with the increase in obesity.

In a study, mice were divided into four groups. Each group was given the same amount of food and either a soft drink with a different sweetener or water.

1. Distinguish between the structure of sucrose and fructose. [1]
2. Use the graph in Figure 12 to compare and contrast the body fat accumulation in the four groups of mice. [3]

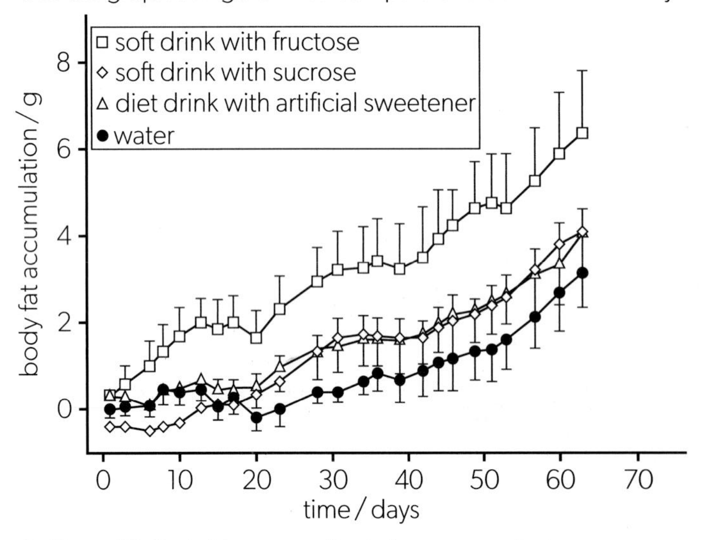

▲ Figure 12 Body fat accumulation in four groups of mice

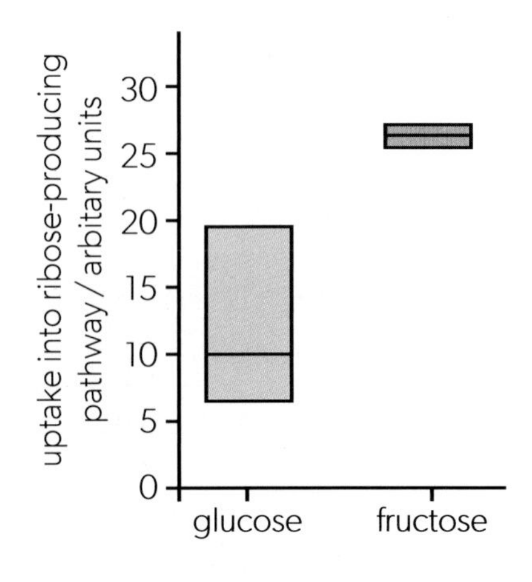

▲ Figure 13 Uptake of sugars in pancreatic cells

Studies investigated the role of glucose and fructose in the development of pancreatic cancer cells. Pancreatic cancer cells were grown in equal concentrations of each sugar and the uptake of each into ribose-producing pathways was measured. The graph in Figure 13 shows the range of uptake of sugars and the mean value.

3. Discuss if the results provide clear evidence of a difference in uptake of the two sugars. [2]
4. Determine which sugar is primarily used in the production of ribose by pancreatic cancer cells. [1]

Properties and uses of glucose

- Like all monosaccharides, glucose is soluble and is a relatively small molecule, so it is easily transported. It circulates in blood, dissolved in the plasma.
- Like most other carbohydrates, glucose is chemically very stable. This property is useful for food storage. However, glucose would cause osmotic problems if it was stored in cells in large quantities. Therefore, it is usually converted to glycogen or starch.
- Glucose yields energy when it is oxidized. It can therefore be used as a substrate for respiration.

▲ Figure 14 A black bear is feeding on wild berries. Foods containing glucose are attractive to animals because of their energy yield, so plants put glucose (and usually also fructose) in the flesh of animal-dispersed fruits and in the nectar of animal-pollinated flowers

B1.1.5 Polysaccharides as energy storage compounds

Starch and glycogen are used as energy stores. Starch is used in plants and glycogen in animals. Both of these substances are composed of large numbers of α-glucose molecules, which can be used a substrate in aerobic and anaerobic cell respiration.

There are two types of starch molecule.

- Amylose is an unbranched chain of α-glucose linked by 1→4 glycosidic bonds. Because of the bond angles, the chain is helical rather than straight.
- Amylopectin has the same structure as amylose but there are some 1→6 glycosidic bonds making the molecule branched.

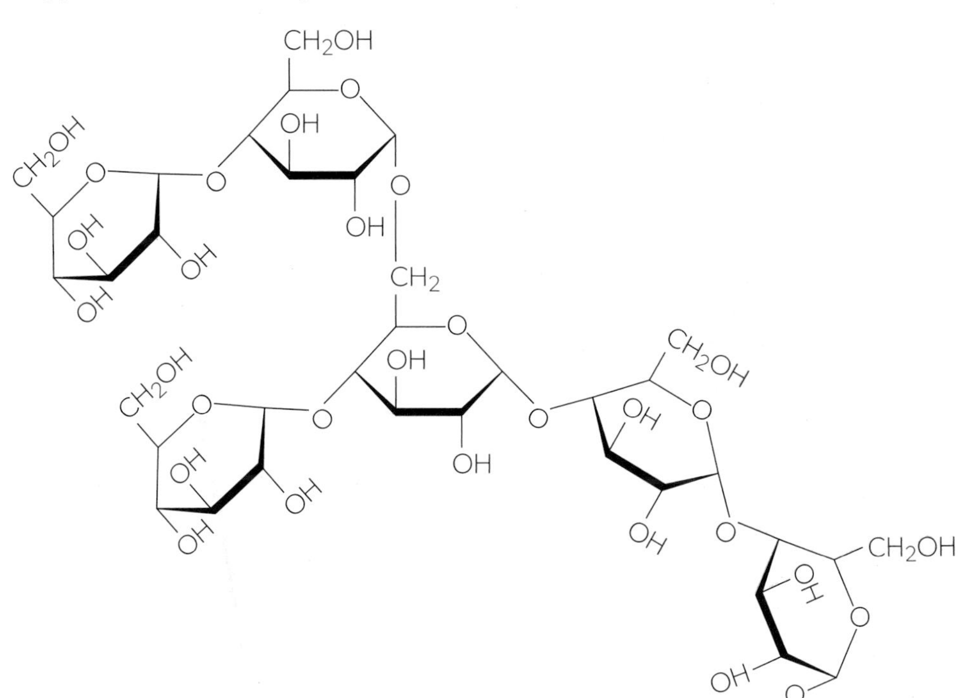

▲ Figure 15 Small portion of an amylopectin molecule showing six α-glucose molecules, all linked by 1→4 bonds apart from one 1→6 bond that creates a branch

Glucose can be removed from starch and glycogen molecules when it is needed. A hydrolysis reaction breaks a 1→4 glycosidic bond to separate one glucose molecule from the end of a chain. This allows it to be transported elsewhere or used in the cell. You can think of starch and glycogen as a sort of bank account because glucose can be deposited when there is a surplus and withdrawn when there is a shortage. Adding or removing glucose can happen more quickly with amylopectin than amylose because the branched structure provides more ends of chains.

Glycogen has a similar structure to amylopectin: α-glucose molecules linked by 1→4 glycosidic bonds and branched by 1→6 bonding. In glycogen, about 1 in 10 glucose molecules has a 1→6 bond, compared with about 1 in 20 in amylopectin, so glycogen molecules are more branched.

Glycogen can contain tens of thousands of glucose subunits and amylopectin can contain more than a hundred thousand. The very large size of these molecules gives them much lower solubility than glucose, so they contribute little to the osmotic concentration of cells. This means starch or glycogen can be used to store large amounts of glucose without the cell swelling up with water drawn in by osmosis. The branched structure of glycogen and amylopectin makes them relatively compact despite their huge molecular mass. This is another useful property in a storage compound.

A consequence of the limitless addition and removal of glucose is that starch and glucose do not have a fixed molecular mass, so molar solutions cannot be made. Concentrations have to be expressed in percentage terms (grams of substance per 100 cm^3 of solution).

B1.1.6 Structure of cellulose related to its function as a structural polysaccharide in plants

Cellulose, like starch and glycogen, is composed of glucose, but its properties are markedly different because it is a polymer of β-glucose rather than α-glucose. Condensation reactions link C_1 on a free β-glucose molecule to C_4 on the β-glucose at the end of the growing cellulose molecule. All the links in cellulose are 1→4 glycosidic bonds, so it is an unbranched chain. A cellulose molecule can contain more than 10,000 β-glucose molecules each with a size of about 1 nm, giving an overall length of more than 10 μm.

In β-glucose, the –OH group on C_1 is angled upwards and the –OH group on C_4 is angled downwards. To bring these –OH groups together and allow a condensation reaction to occur, each β-glucose added to the chain has to be inverted in relation to the previous one. The glucose subunits in the chain therefore face alternately upwards and downwards.

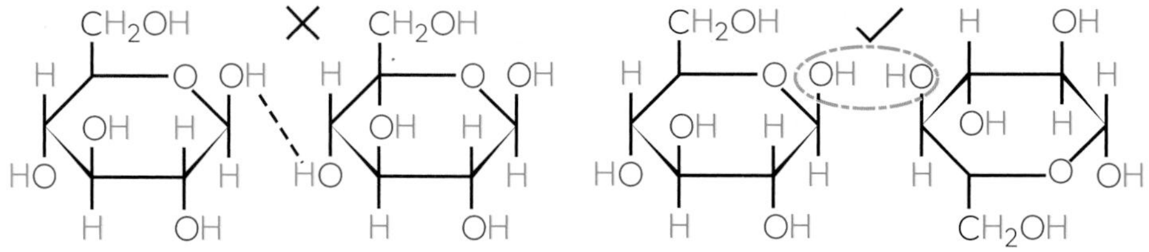

▶ Figure 16 Beta glucose molecules can only form a 1→4 glycosidic bond if one faces up and the other faces down

The chains of α-glucose in starch wind into a helix, but in cellulose the alternating orientation of β-glucose results in a straight chain. This allows formation of bundles of molecules arranged in parallel. Hydroxyl groups are regularly spaced along each cellulose molecule, allowing many hydrogen bonds to form between the molecules. These bundles of cellulose molecules are called microfibrils and are the basis of plant cell walls. Microfibrils have very high tensile strength because of the strong covalent bonds in the cellulose molecules, the number of molecules and the cross-links between them. The strength prevents plant cells from bursting, even when very high pressures have developed inside the cell due to entry of water by osmosis.

▲ Figure 17 Part of a cellulose molecule viewed from the side, showing eight glucose subunits. Carbon atoms are grey, oxygen red and hydrogen green

B1.1.7 Role of glycoproteins in cell–cell recognition

Glycoproteins are composed of polypeptides with carbohydrate attached. In most cases, the carbohydrate is an oligosaccharide—a short chain of monosaccharides linked by glycosidic bonds. Glycoproteins are a component of plasma membranes in animal cells and are positioned with the attached carbohydrate facing outwards. By displaying distinctive glycoproteins, cells allow other cells to recognize them. The glycoprotein on the surface of one cell is recognized by receptors on the surface of another cell.

Cell-to-cell recognition helps with the organization of tissues and can also allow foreign cells or infected body cells to be identified and destroyed. The ABO antigens in red blood cells are an example of glycoproteins providing the means of cell–cell recognition.

ABO glycoproteins

Red blood cells have glycoproteins in their membranes that do not have a known function, but that affect blood transfusion. Any of three possible types of oligosaccharide can be present on the glycoprotein. The oligosaccharides are called O, A and B. One or two of these types of glycoprotein are present in every person's blood, but not all three.

If blood containing glycoprotein A is transfused into a person who does not produce it themself, the blood will be rejected. Similarly, blood containing glycoprotein B is rejected if a person does not produce it themself. However, glycoprotein O does not cause rejection problems, because it has the same structure as A and B with one monosaccharide less, so is not recognized as foreign.

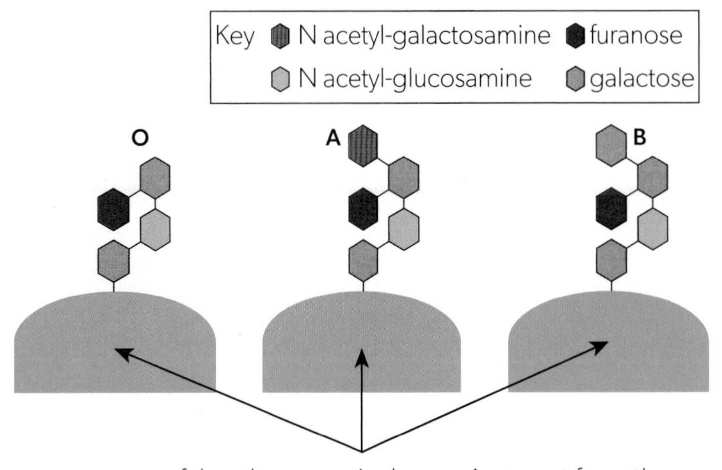

▲ Figure 18 O, A and B glycoproteins in red blood cells

B1.1.8 Hydrophobic properties of lipids

▲ Figure 19 Fatbergs are an increasing problem is sewers. Warm liquid fat from food waste cools and solidifies in sewers because it does not dissolve in water. This sewer under the Strand in London has large fat accumulations

Lipids are a diverse group of substances in living organisms that dissolve in non-polar solvents. Ethanol, toluene and propanone (acetone) are examples of non-polar solvents. Lipids are only sparingly soluble in aqueous (water-based) solvents. For this reason, they are said to be hydrophobic. This is a rather misleading term, because lipids are not repelled by water—they are just more attracted to non-polar substances.

Fats, oils, waxes and steroids are classes of commonly occurring lipids.

- Oils have a melting point below 20°C, so they solidify at low temperatures.
- Fats have a melting point between 20°C and 37°C so they are solid at room temperature and liquid at body temperature.
- Waxes have a melting point above 37°C, so they liquify at high temperatures.
- Steroids have molecules with a characteristic four-ring structure.

B1.1.9 Formation of triglycerides and phospholipids by condensation reactions

A triglyceride is made by combining three fatty acids with one glycerol. Each of the fatty acids is linked to the glycerol by a condensation reaction, so three water molecules are produced. The linkage formed between each fatty acid and the glycerol is an ester bond. This type of bond is formed when an acid reacts with the hydroxyl group (–OH) in an alcohol. In this case, the reaction is between the carboxyl (–COOH) group on a fatty acid and a hydroxyl on the glycerol. These groups are the only hydrophilic parts of fatty acid and glycerol molecules and are used up in the condensation reaction, so triglycerides are entirely hydrophobic. Depending on the type of fatty acids they contain, triglycerides may be oils or fats.

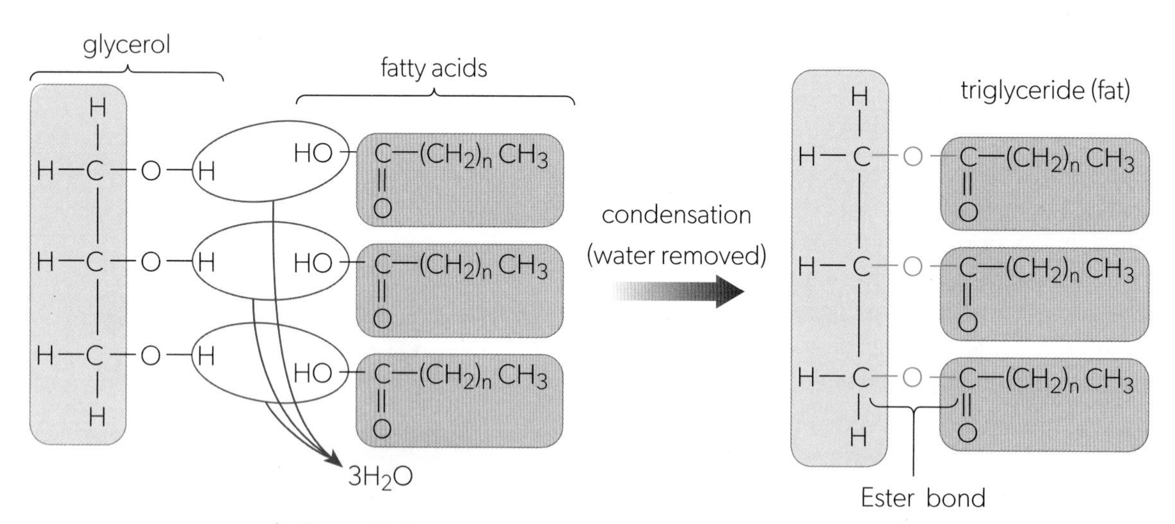

▲ Figure 20 Formation of a triglyceride from glycerol and three fatty acids

Phospholipids have a structure similar to triglycerides, but there are two fatty acids linked to glycerol, with a phosphate group instead of a third fatty acid. The phosphate is hydrophilic, so phospholipid molecules are partly hydrophilic and partly hydrophobic.

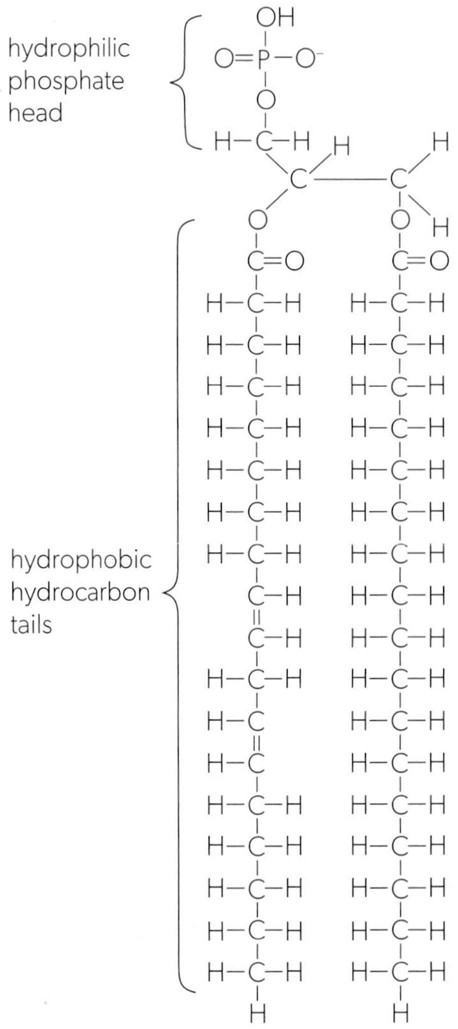

B1.1.10 Difference between saturated, monounsaturated and polyunsaturated fatty acids

Fatty acids have an unbranched chain of carbon atoms, with hydrogen atoms covalently bonded, so it is a hydrocarbon chain. The acid part of the molecule is a carboxyl group (–COOH) at one end of the chain. At the other end there is a methyl group (–CH_3). The length of the hydrocarbon chain is variable but most of the fatty acids used by living organisms have between 14 and 20 carbon atoms.

Another variable feature of fatty acids is the covalent bonding between the carbon atoms. Some have single bonds between all the carbon atoms, whereas others have double bonds between some pairs of carbon atoms in the chain. Carbon atoms linked by single bonds can also bond to two hydrogen atoms. Carbon atoms linked by a double bond to an adjacent carbon in the chain, can only bond to one hydrogen. A fatty acid with single bonds between all of its carbon atoms contains as much hydrogen as it possibly could and is called a saturated fatty acid. Fatty acids that have one or more double bonds are unsaturated because they contain less hydrogen than they could. If there is one double bond, the fatty acid is monounsaturated; if it has more than one double bond, it is polyunsaturated.

Nearly all unsaturated fatty acids in living organisms have the hydrogen atoms on the same side of the two double-bonded carbon atoms—these are called cis-fatty acids. The alternative arrangement is for the hydrogens to be on opposite sides—these are called trans-fatty acids. In cis-fatty acids, there is a bend in the hydrocarbon chain at the double bond. This makes triglycerides containing cis-unsaturated fatty acids less good at packing together in regular arrays than saturated fatty acids, so they have a low melting point. So, triglycerides with cis-unsaturated fatty acids are usually liquid at room temperature—they are oils.

Trans-fatty acids do not have a bend in the hydrocarbon chain at the double bond; they have straight chains and a higher melting point. They are solid at room temperature. They are produced artificially by partial hydrogenation of vegetable or fish oils. This is done to produce solid fats for use in margarine and some other processed foods, but serious health concerns have led the Food and Drug Administration (FDA) in the US to ban use of industrial trans-fats.

hydrophilic phosphate head

hydrophobic hydrocarbon tails

▲ Figure 21 The molecular structure of a phospholipid. The phosphate often has other hydrophilic groups attached to it, but these are not shown in this diagram

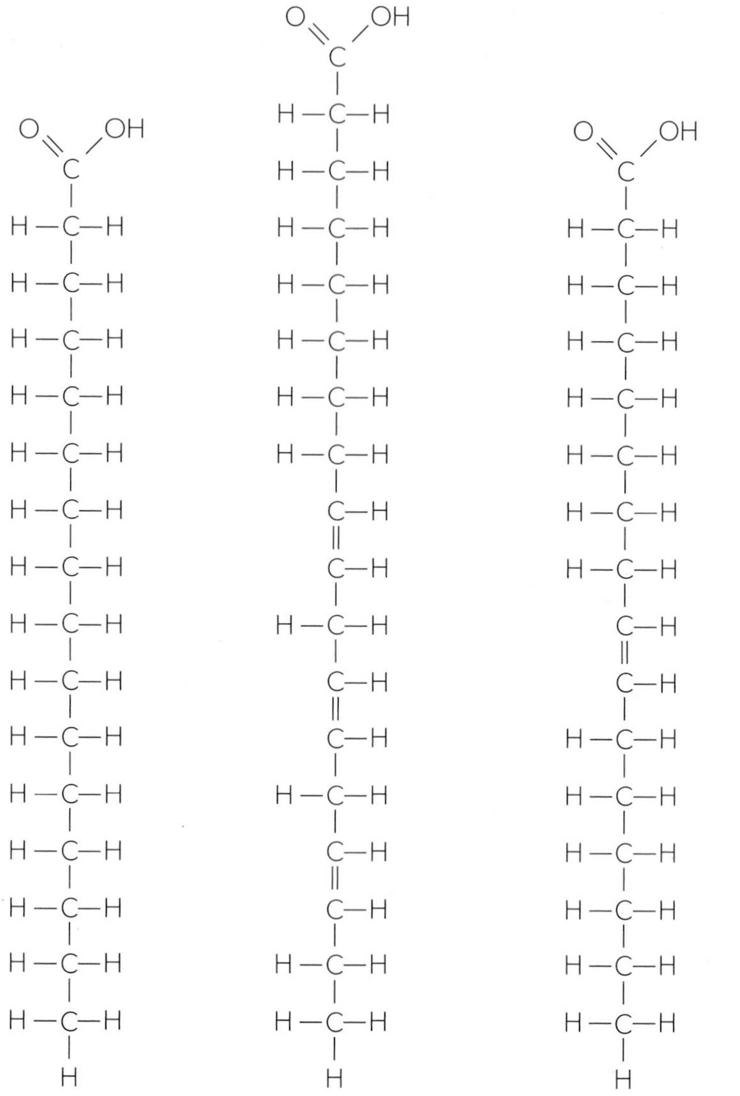

palmitic acid
- saturated
- non-essential

linolenic acid
- polyunsaturated
- all cis
- essential
- omega 3

palmitoleic acid
- monounsaturated
- cis
- non-essential
- omega 7

▲ Figure 22 Examples of fatty acids. The omega number indicates how far from the methyl group the first double bond is located

▲ Figure 23 Sunflower oil is pressed from the seeds. Two-thirds of the fatty acids in the oil are polyunsaturated and most of the rest are monounsaturated

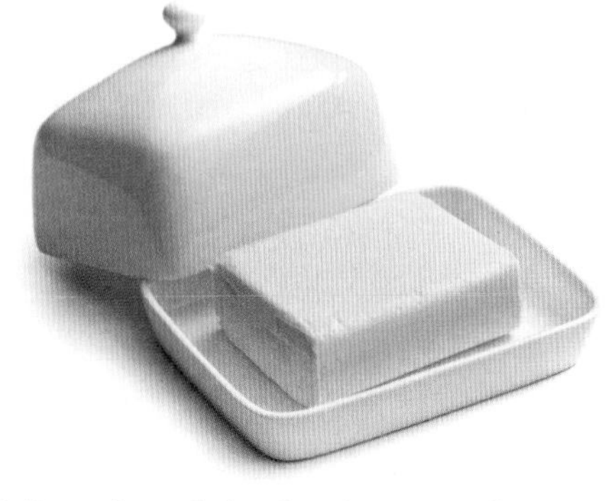

▲ Figure 24 Butter is made by churning cream from cow's milk. Two-thirds of the fatty acids in butter are saturated and most of the rest is monounsaturated

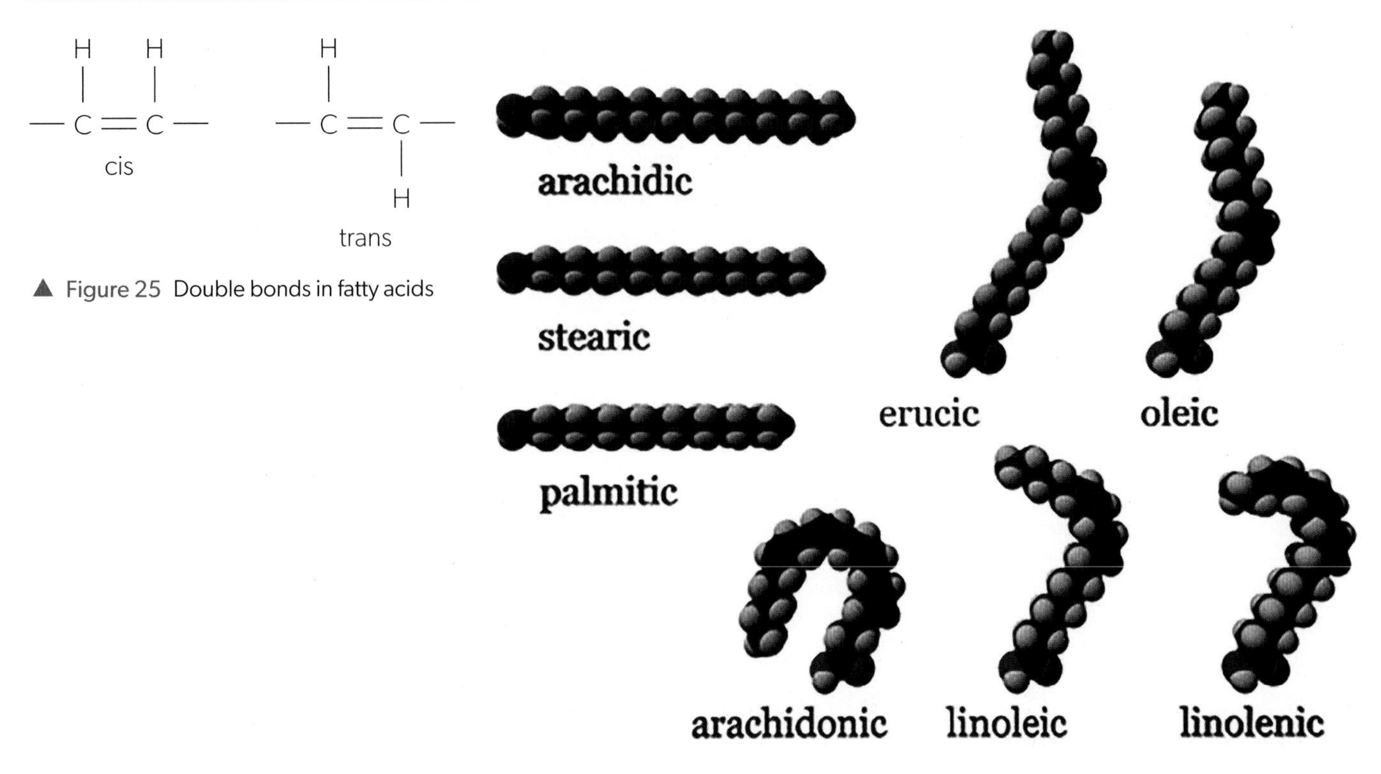

▲ Figure 25 Double bonds in fatty acids

▲ Figure 26 Saturated fatty acids have straight chains. Monounsaturated fatty acids have one kink in the chain and polyunsaturated fatty acids have more than one kink so they can become curved

ATL Research skills: Using Google Scholar for research

A 2017 journal article by the American Heart Association estimated that replacing saturated fat with polyunsaturated fat in the diet has the potential to reduce the risk of cardiovascular diseases by 30%. On the other hand, a small number of scientists have challenged the negative press received by saturated fats. Google Scholar is a search engine that allows a researcher to focus their search on scholarly literature. You might use it for your extended essay or for your internal assessment projects.

Figure 27 on the next page shows the Google Scholar entries for two different journal articles

1. Contrast the two papers in terms of citation frequency and currency.
2. Is there a consensus view on saturated fat in the diet? What challenges arise when the media reports both points of view without acknowledging which is the consensus of the scientific community?

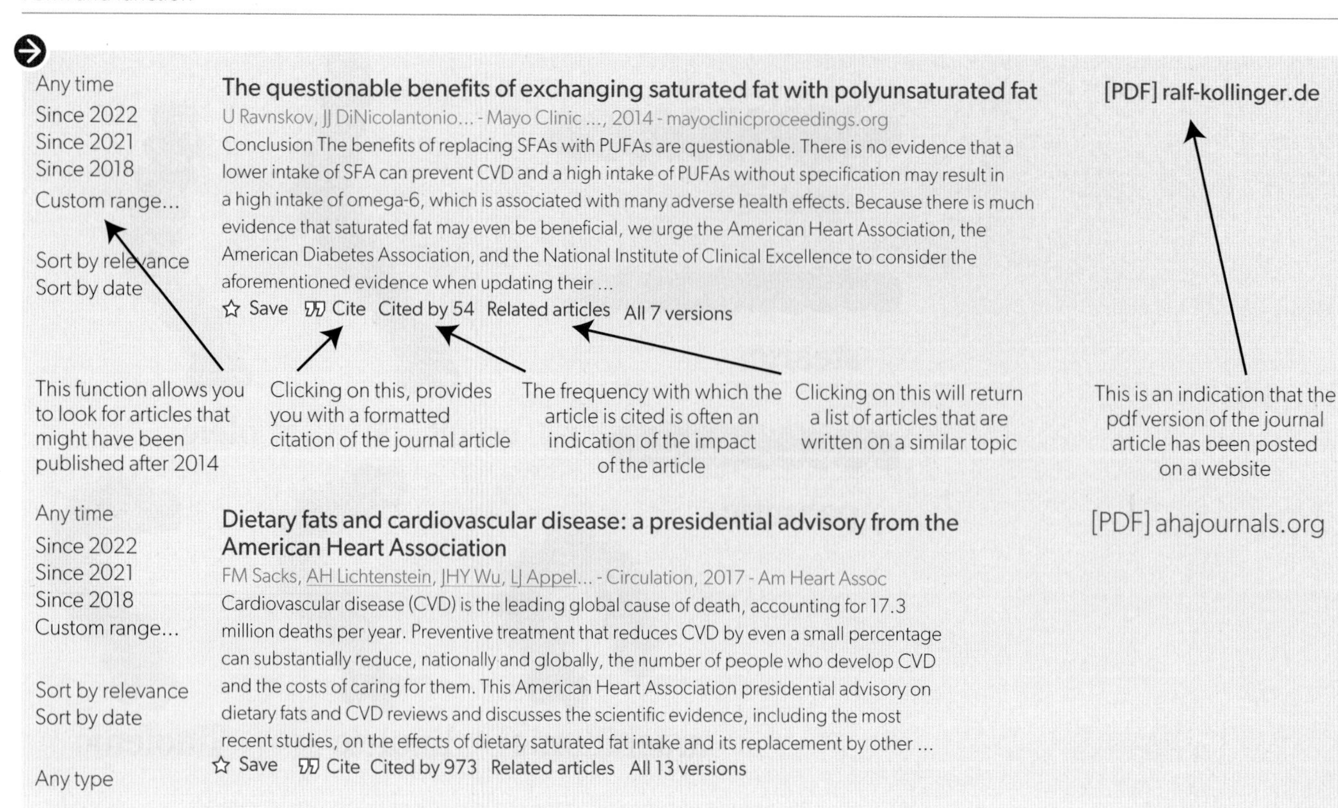

▲ Figure 27 This annotated page gives tips on using Google Scholar for research. The two different entries take a different view on the role of saturated fat in the diet

B1.1.11 Triglycerides in adipose tissues for energy storage and thermal insulation

Triglycerides are used for energy storage in plants and animals. In animals, the triglycerides are fats and are stored in specialized groups of cells called adipose tissue. In humans, adipose tissue is located immediately beneath the skin and also around some organs including the kidneys.

The properties of triglycerides make them particularly suitable for long-term energy storage.

- They are chemically very stable, so energy is not lost over time.
- They are immiscible with water, so they naturally form droplets in the cytoplasm which do not have osmotic or other effects on the cell.
- They release twice as much energy per gram in cell respiration as carbohydrate, so enough energy can be stored in half the body mass. This is important for animals that move and especially for birds and bats that fly.
- They are poor conductors of heat, so they can be used as a thermal insulator in animals that need to conserve body heat.
- They are liquid at body temperature, so they can act as a shock absorber—for example, around the kidneys.

Thermal insulation is needed most by animals that live in cold habitats and that maintain a body temperature much higher than the environment. Such animals have thick layers of subcutaneous adipose tissue. In marine mammals it is called blubber. In animals such as sea lions there are sometimes problems with overheating when adults emerge onto land to breed, because the thick layer of blubber impedes dissipation of heat produced by metabolism and the air is much warmer than the water in the ocean habitat.

▲ Figure 28 A typical 40 kg male emperor penguin has 12.7 kg of body fat at the start of winter, but only 2.4 kg at the end. Why do male emperor penguins need such a large amount of body fat at the start of the winter?

B1.1.12 Formation of phospholipid bilayers as a consequence of the hydrophobic and hydrophilic regions

Substances attracted to water are called hydrophilic. Other substances that are not attracted to water are called hydrophobic. Phospholipids are unusual because part of a phospholipid molecule is hydrophilic and part is hydrophobic. Substances with this property are described as amphipathic. The hydrophilic part of a phospholipid is the phosphate group. The hydrophobic part consists of the two hydrocarbon chains. The chemical structure of phospholipids is described in *Section B1.1.9*. The structure can be represented simply using a circle for the phosphate group and two lines for the hydrocarbon chains.

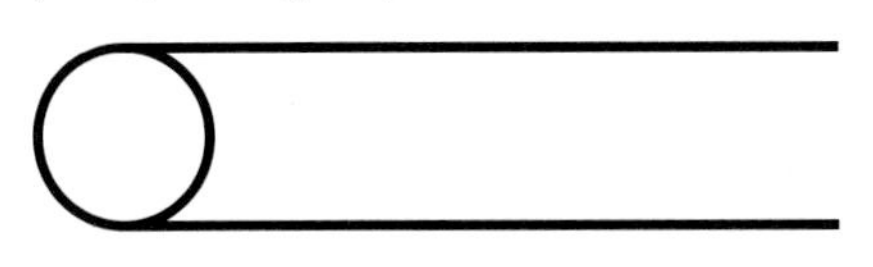

▲ Figure 29 Simplified diagram of a phospholipid molecule

The two parts of the molecule are often called the phosphate head and the hydrocarbon tails. When phospholipids are mixed with water the phosphate heads are attracted to the water but the hydrocarbon tails are attracted to each other more than to the water. Because of this the phospholipids become arranged into double layers, with the hydrophobic hydrocarbon tails facing inwards and the hydrophilic heads facing outwards to the water on either side. These double layers are called phospholipid bilayers. They are stable structures and they form the basis of all cell membranes.

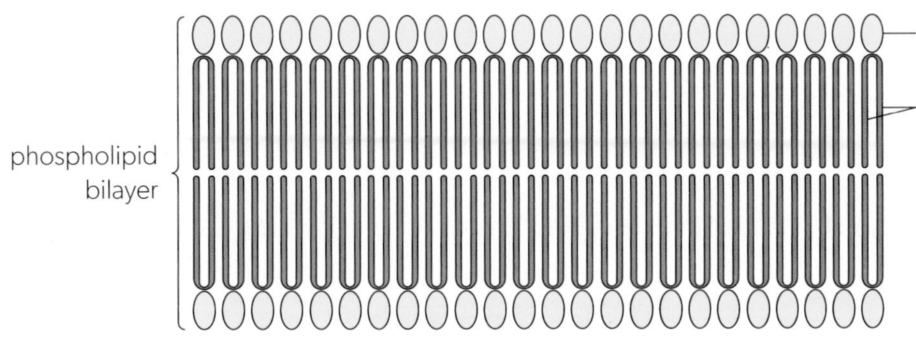

▲ Figure 30 Simplified diagram of a phospholipid bilayer

B1.1.13 Ability of non-polar steroids to pass through the phospholipid bilayer

Steroids are a group of lipids with molecules similar to sterol. They can be identified using these features:

- four fused rings of carbon atoms
- three cyclohexane rings (Figure 31; A, B and C) and one cyclopentane ring (Figure 31; D)
- 17 carbon atoms in total in the rings.

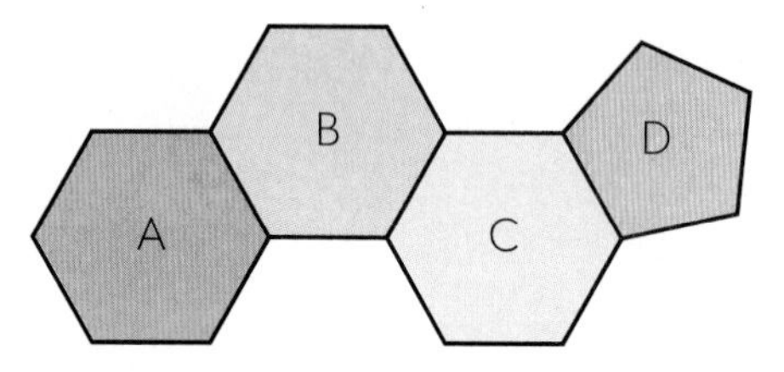

▲ Figure 31 The four-ring structure of steroids

There are hundreds of examples of steroids, which differ in the position of C=C double bonds and in the functional groups such as –OH that are added to the four-ring structure. Steroids are mostly hydrocarbon and therefore hydrophobic. This allows them to pass through phospholipid bilayers and enter or leave cells.

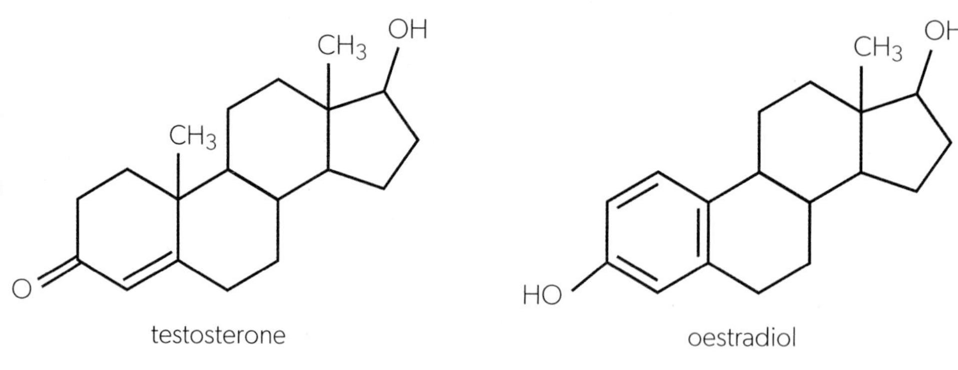

▶ Figure 32 Testosterone and oestradiol have very similar molecular structures despite their markedly different effects on the body. In these skeletal diagrams, the carbon atoms are not individually shown but the bonds between them are. Hydrogen atoms attached to the carbon are not shown but can be inferred because each carbon atom has a total of four bonds

Linking questions

1. How can compounds synthesized by living organisms accumulate and become carbon sinks?
   a. Describe the relationship between the process of ecological succession and changes in biomass. (D4.2.13)
   b. Outline the events of the Calvin cycle. (C1.3.17)
   c. Outline four sinks from which carbon is being released due to human activity. (C4.2.19)
2. What are the roles of oxidation and reduction in biological systems?
   a. Outline the role of oxidation in the release of energy from carbohydrates. (B1.1.4)
   b. Explain the role of NADP in photosystem 1 and the Calvin cycle. (C1.3.13)
   c. Outline one example of chemoautotrophy. (C4.2.7)

B1.2 Proteins

What is the relationship between amino acid sequence and the diversity in form and function of proteins?

Every protein contains one or more polypeptides. The 20 different amino acids that can be used to assemble polypeptides are chemically diverse. There are parallels with the linking of letters of the alphabet to form words: any length and sequence is possible, but only a small proportion of the possibilities are used. A difference is that polypeptides are much longer than words—most have hundreds or even thousands of amino acids. How does the sequence of amino acids in polypeptides determine their three-dimensional shape? How is the shape of a protein related to its function?

How are protein molecules affected by their chemical and physical environments?

Relatively weak interactions within protein molecules maintain the conformation needed for functions to be performed. These interactions are sensitive to the physical and chemical environment. How do changes of pH cause protein structure to be altered? How do heavy metals such as mercury cause misfolding of proteins? How does heat affect the structure of proteins? Why do some environmental changes lead to denaturation? What are some examples of proteins that function in extreme environments such as high or low temperatures? Low or high values of pH?

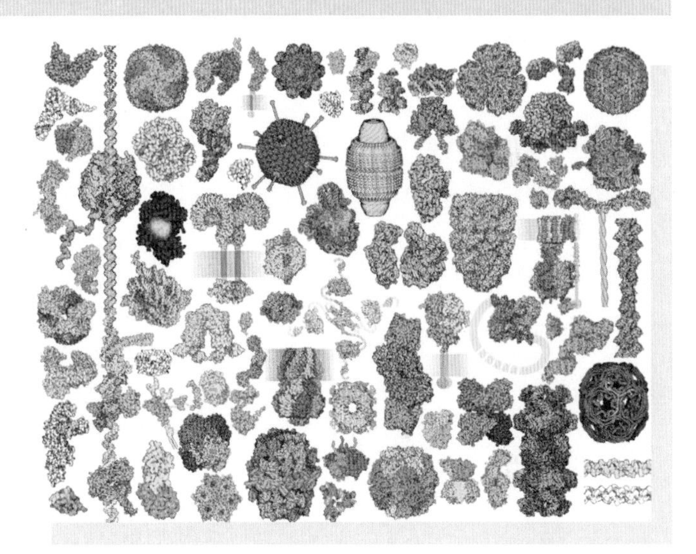

▲ Figure 1 Some protein structures

SL and HL	AHL only
B1.2.1 Generalized structure of an amino acid B1.2.2 Condensation reactions forming dipeptides and longer chains of amino acids B1.2.3 Dietary requirements for amino acids B1.2.4 Infinite variety of possible peptide chains B1.2.5 Effect of pH and temperature on protein structure	B1.2.6 Chemical diversity in the R-groups of amino acids as a basis for the immense diversity in protein form and function B1.2.7 Impact of primary structure on the conformation of proteins B1.2.8 Pleating and coiling of secondary structure of proteins B1.2.9 Dependence of tertiary structure on hydrogen bonds, ionic bonds, disulfide covalent bonds and hydrophobic interactions B1.2.10 Effect of polar and non-polar amino acids on tertiary structure of proteins B1.2.11 Quaternary structure of non-conjugated and conjugated proteins B1.2.12 Relationship of form and function in globular and fibrous proteins

B1.2.1 Generalized structure of an amino acid

Amino acids are the building blocks of proteins. Each amino acid molecule has a central carbon atom called the alpha carbon, with single covalent bonds to four other atoms. One of these is the nitrogen atom of an amine group and another is the carbon atom of a carboxyl group. The carboxyl group (–COOH) is acidic because it can donate a proton and the amine group is basic because it can accept one, so amino acids are amphiprotic.

The alpha carbon atom also has a single covalent bond to a hydrogen atom. The other covalent bond links the alpha carbon to a side chain, called the R-group. The R-group can be any one of a wide range of possibilities.

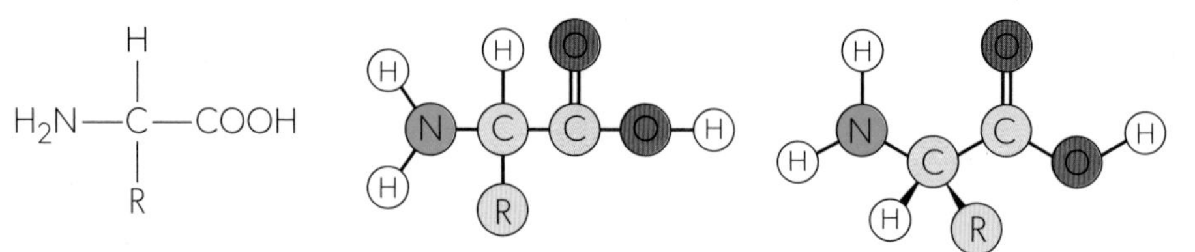

▶ Figure 2 The generalized structure of an amino acid can be represented in different ways. Which of these is most informative?

Activity: Researching residues

The R in "R-group" stands for residue.

1. What is a residue?
2. When does the R-group become a residue?

B1.2.2 Condensation reactions forming dipeptides and longer chains of amino acids

To form a dipeptide, two amino acids are linked by a condensation reaction. More amino acids can be linked by further condensation reactions to create a longer chain. Polypeptides can contain any number of amino acids, though chains of fewer than 20 amino acids are usually referred to as oligopeptides rather than polypeptides. Polypeptides are the main component of proteins.

Amino acids are linked with peptide bonds. These are C–N bonds formed by a condensation reaction between the amine group (–NH_2) of one amino acid and the carboxyl group (–COOH) of another. The reaction is catalysed in cells by ribosomes. It is a directional process: the amine group of a free amino acid is linked to the carboxyl group at the end of the growing chain. Because peptide bonds are made using groups that are part of all amino acids, the bond is the same, whatever the R-groups of the amino acids are.

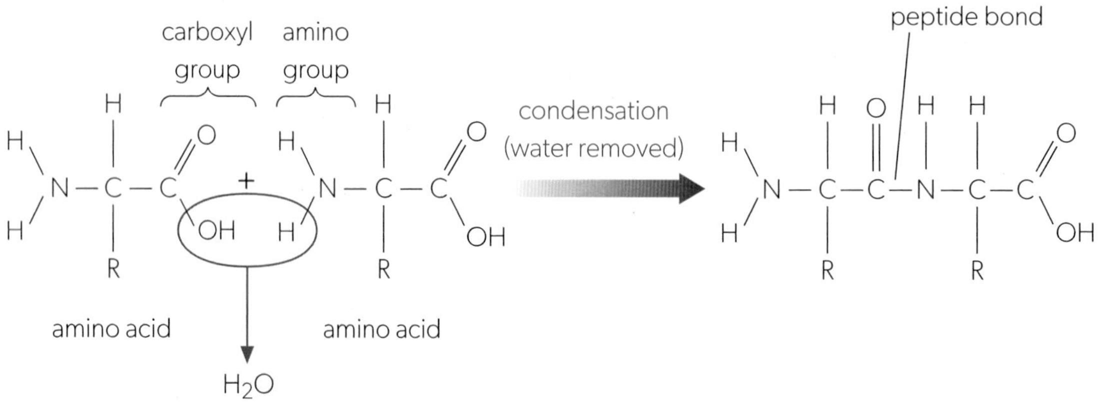

carboxyl group · amino group · amino acid · amino acid · H_2O · condensation (water removed) · peptide bond

▲ Figure 3 Condensation joins two amino acids with a peptide bond to produce a dipeptide

Activity: Drawing dipeptides and oligopeptides

To test your skill at showing how peptide bonds are formed, try showing the formation of a peptide bond between two of the amino acids in Figure 4. There are 16 possible dipeptides that can be produced from these four amino acids.

```
                        COOH
                         |
     OH                H-C-H                 H
     |                   |                   |
   H-C-H               H-C-H               H-C-H                  H
     |                   |                   |                    |
H2N-C-COOH          H2N-C-COOH          H2N-C-COOH           H2N-C-COOH
     |                   |                   |                    |
     H                   H                   H                    H

   serine          glutamic acid          alanine              glycine
```

▲ **Figure 4** Some common amino acids

You could also draw an oligopeptide of four amino acids, linked by three peptide bonds. If you do this correctly, you should see the following features.

- A chain of atoms linked by single covalent bonds forming the backbone of the oligopeptide, with a repeating sequence of –N–C–C–. If this is shown as a zigzag (N↗C↘C↗N↘C↗C) the bond angles are closer to being correct.
- A hydrogen atom is linked by a single bond to each nitrogen atom in the backbone and an oxygen atom is linked by a double bond to one of the two carbon atoms.
- The amine ($-NH_2$) and carboxyl (–COOH) groups are used up in forming the peptide bond and only remain at the ends of the chain. These are called the amino and carboxyl terminals.
- The R-groups of each amino acid remain and project outwards from the backbone.

Computer modelling

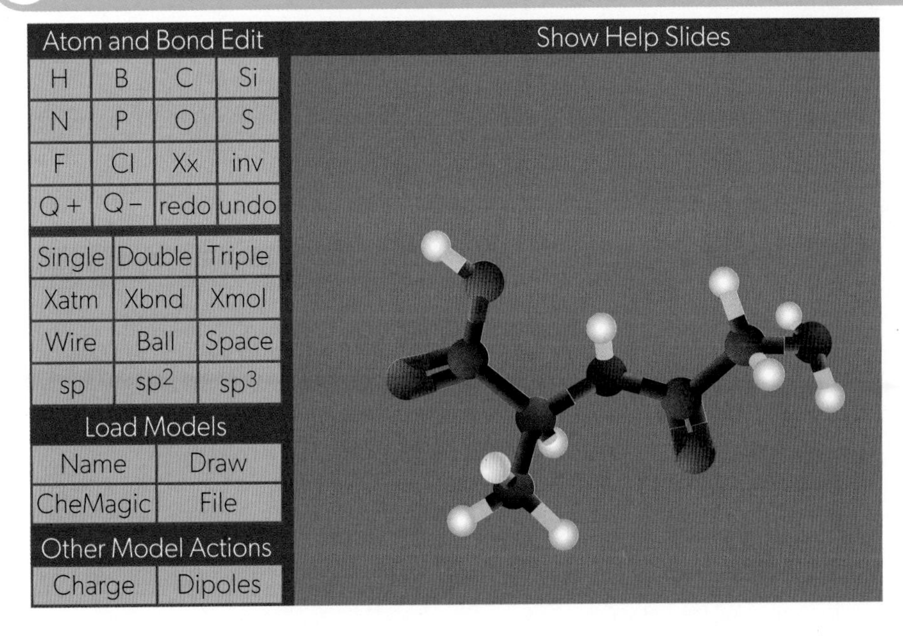

◀ **Figure 5** This image of the dipeptide glycine-alanine was constructed using the web-based computer application called ChemMagic from the University of Illinois. Similar molecule builder apps exist on other websites

The app begins with a molecule of methane. Students then substitute individual atoms with desired atoms.

1. Using Figure 5, deduce what colours represent oxygen, carbon, nitrogen and hydrogen.
2. Identify the R-groups of glycine and alanine.

B1.2.3 Dietary requirements for amino acids

Twenty different amino acids are used by ribosomes to make polypeptides. Plants can make all of these by photosynthesis. Animals obtain amino acids from their food. An essential amino acid is one that cannot be synthesized in sufficient quantities by the animal so must be obtained from the diet. A non-essential amino acid can be synthesized by an animal using metabolic pathways that transform one amino acid into another.

Nine of the 20 amino acids are essential in humans. The others are non-essential, though several become essential in special circumstances. For example, the amino acid phenylalanine is essential because it cannot be synthesized by the human body; tyrosine is non-essential because it can be made from phenylalanine.

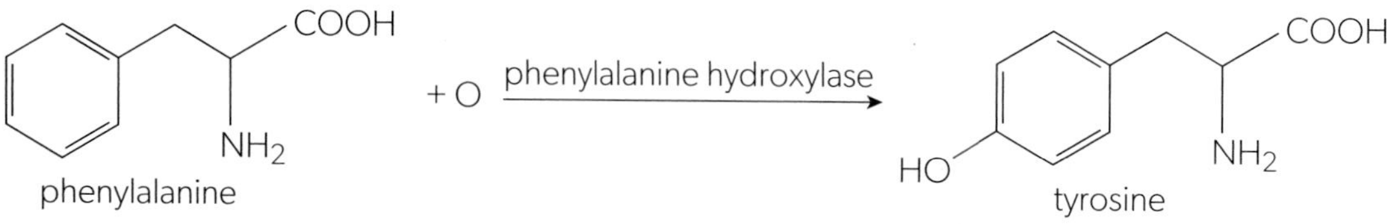

COOH
NH_2
phenylalanine
+ O
phenylalanine hydroxylase
COOH
HO
NH_2
tyrosine

▲ Figure 6 Conversion of an essential amino acid into a non-essential amino acid

Foods vary in their amino acid content. It is possible to eat a protein-rich diet and still be deficient in an essential amino acid. Animal-based foods (fish, meat, milk, eggs) have a balance of amino acids that is similar to what is needed in the human diet. Plant-based foods have a different balance and some are deficient in specific amino acids. For example, cereals such as wheat have a low lysine content, and peas and beans are low in methionine. Both lysine and methionine are essential amino acids for humans. So, people eating a vegan diet, must ensure that enough of each essential amino acid is consumed. Traditional plant-based diets in successful civilizations do provide such a balance.

Data-based questions: Essential amino acids

1. Table 1 summarizes the relative content of essential amino acids in different foods. Cysteine and tyrosine are classified as being "conditionally essential". The quantity of each amino acid in a hen egg is set as 1.0 and all other values are relative to the hen egg standard.

	hen's eggs	human milk	cow's milk
isoleucine	1.0	1.1	1.1
leucine	1.0	1.4	1.3
valine	1.0	1.0	1.0
threonine	1.0	1.0	0.9
methionine and cysteine	1.0	1.1	0.7
tryptophan	1.0	1.6	1.3
lysine	1.0	1.0	1.3
phenylalanine and tyrosine	1.0	1.0	0.9
histidine	1.0	0.9	1.1

◀ Table 1

[Source: Data obtained from Robert McGlivery, *Biochemistry: A Functional Approach*, 1970, W. B. Saunders.]

a. Outline what is meant by the term "essential amino acid". [2]

b. Evaluate human milk as an overall source of essential amino acids. [2]

c. Phenylalanine is converted to tyrosine by the enzyme phenylalanine hydroxylase.
Deduce the reason that tyrosine is considered a conditionally essential amino acid. [1]

d. When infants with the condition phenylketonuria (PKU) are left untreated, they have a build-up of phenylalanine in their blood and high levels of phenylalanine in their urine. Suggest the cause of this condition. [1]

B1.2.4 Infinite variety of possible peptide chains

Ribosomes link amino acids together one at a time, until a polypeptide is fully formed. The ribosome can make peptide bonds between any pair of amino acids, so all sequences are possible. Ribosomes do not make random sequences of amino acids. They receive instructions in the form of genetic code. Twenty different amino acids are included in the code.

The number of possible amino acid sequences can be calculated starting with dipeptides. Both amino acids in a dipeptide can be any of the 20, so there are 20 × 20 possible sequences (20^2). There are 20 × 20 × 20 possible tripeptide sequences (20^3). For a polypeptide of *n* amino acids, there are 20^n possible sequences. The number of amino acids in a polypeptide can be anything from 20 to tens of thousands.

For example, if a polypeptide has 400 amino acids, there are 20^{400} possible amino acid sequences. This is an incredibly large number, and some online calculators simply express it as infinity. Given that the number of amino acids in a polypeptide can be tens of thousands, the number of possible sequences is effectively infinite. But only an extremely small proportion are made by an organism. This is the organism's proteome.

Examples of polypeptides

- Beta-endorphin is natural pain killer secreted by the pituitary gland that is a polypeptide of 31 amino acids.
- Insulin is a small protein that contains two short polypeptides, one with 21 amino acids and the other with 30.
- Alpha amylase is the enzyme in saliva that starts the digestion of starch. It is a single polypeptide of 496 amino acids, with one chloride ion and one calcium ion associated.
- Titin is the largest polypeptide discovered so far. It is part of the structure of muscle. In humans, titin is a polypeptide of 34,350 amino acids, but in mice it is even longer with 35,213 amino acids.

Activity: Famous vegetarians and vegans

▲ Figure 7 Leonardo da Vinci is reported to have said that he did not want his body to be a tomb for other creatures and was probably vegetarian. Can you find other examples of great thinkers who became vegan or vegetarian?

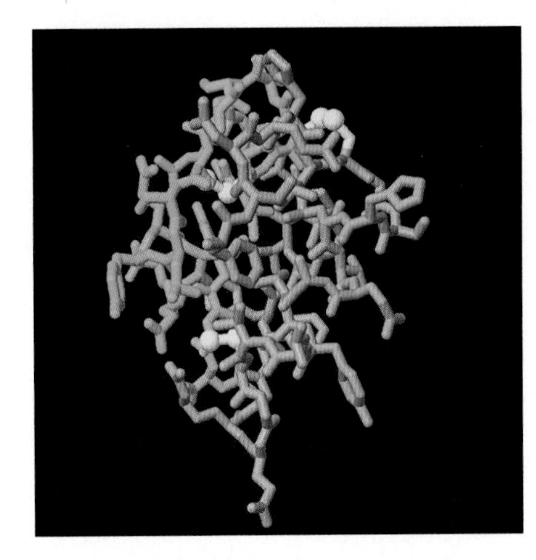

▲ Figure 8 This image of insulin can be viewed and rotated on the PDB Molecule of the Month website

B1.2.5 Effect of pH and temperature on protein structure

The three-dimensional conformation of proteins is stabilized by bonds or interactions between the R-groups of amino acids within the molecule. Most of these bonds and interactions are relatively weak and they can be disrupted or broken. This results in a change to the conformation of the protein and is called denaturation.

A denatured protein does not normally return to its former structure—the denaturation is permanent. Soluble proteins often become insoluble and form a precipitate. This is due to the hydrophobic R-groups in the centre of the molecule becoming exposed to water by the change in conformation.

▲ Figure 9 When eggs are heated, proteins that were dissolved in both the white and the yolk are denatured. They become insoluble so both yolk and white solidify

Heat can cause denaturation because it causes vibrations within the molecule that can break intermolecular bonds or interactions. Proteins vary in their heat tolerance. Some microorganisms that live in volcanic springs or in hot water near geothermal vents have proteins that are not denatured by temperatures of 80°C or higher. The best-known example is DNA polymerase from *Thermus aquaticus*, a prokaryote that was discovered in hot springs in Yellowstone National Park. It works best at 80°C and because of this it is widely used in biotechnology. Nevertheless, heat causes denaturation of most proteins at much lower temperatures.

Extremes of pH, both acidic and alkaline, can cause denaturation. This is because positive and negative charges on R-groups are changed, breaking ionic bonds within the protein or causing new ionic bonds to form. As with heat, the three-dimensional structure of the protein is altered and proteins that have been dissolved in water often become insoluble. There are exceptions: the contents of the stomach are normally acidic, with a pH as low as 1.5, but this is the optimum pH for the protein-digesting enzyme pepsin that works in the stomach.

Applying techniques: Using a colorimeter to measure turbidity: denaturation experiments

A colorimeter and a spectrophotometer are both instruments that measure the amount of light that passes through a sample. If a colorimeter is set at 500 nm or as close to the UV range as possible it means that it will emit blue light into the sample. If there is a function choice, choose transmittance to measure how much light has passed through the sample. If the function is set at absorbance, the machine will measure how much light has been absorbed.

Albumen is one of the main proteins in egg white. A solution of egg albumen in a test tube can be heated in a water bath to find the temperature at which it denatures. The effects of pH can be investigated by adding acids and alkalis to test tubes of egg albumen solution.

To quantify the extent of denaturation, a colorimeter can be used, as denatured albumen absorbs more light than dissolved albumen. The solution will become more turbid.

▲ Figure 10 These tubes contain increasing quantities of denatured albumin. The concentration of albumin in urine is an important diagnostic for determining kidney function. Albumin in urine is precipitated using sulphosalicylic acid, which results in turbidity. This can be measured using a colorimeter. Protein should not normally be present in urine

B1.2.6 Chemical diversity in the R-groups of amino acids as a basis for the immense diversity in protein form and function

The 20 amino acids that ribosomes use to make polypeptides are very varied in the chemical nature of their R-groups. The elements present in the R-groups are shown in Table 2.

Elements in R-group	Number of amino acids
H only	1
C and H only	5
C, H and S only	2
C, H and N only	5
C, H and O only	5
C, H, N and O	2

▲ Table 2 Variation in R-groups of amino acids

When amino acids are linked up into a polypeptide, their amine and carboxyl groups are used to make peptide bonds. This leaves an amine group ($-NH_2$) at one end of the chain and a carboxyl group (–COOH) at the other end. The hydrogen atom attached to the alpha carbon atom of each amino acid has little effect on the properties of the polypeptide; it is the R-groups that determine the chemical characteristics. Some of the R-groups are hydrophobic and some hydrophilic. Of the hydrophilic R-groups, some are polar and others become charged (+ or −) by acting as an acid or a base. This broad diversity of R-groups allows living organisms to make and use an amazingly wide range of proteins. Some of the differences between R-groups are shown in Table 3.

<table>
<tr><td colspan="2" rowspan="2">Nine R-groups are hydrophobic with between zero and nine carbon atoms</td><td colspan="3">Eleven R-groups are hydrophilic</td></tr>
<tr><td rowspan="2">Four hydrophilic R-groups are polar but never charged</td><td colspan="2">Seven R-groups can become charged</td></tr>
<tr><td>Three R-groups contain rings</td><td>Six R-groups do not contain rings</td><td>Four R-groups act as an acid by giving up a proton and becoming negatively charged</td><td>Three R-groups act as a base by accepting a proton and becoming positively charged</td></tr>
</table>

▲ Table 3 Classification of amino acids

Some proteins contain amino acids that are not in the basic repertoire of 20. In most cases this is due to one of the 20 being modified after a polypeptide has been synthesized. There is an example of modification of amino acids in collagen, a structural protein used to provide tensile strength in tendons, ligaments, skin and blood vessel walls. Collagen polypeptides made by ribosomes contain proline at many positions, but at some of these positions it is converted to hydroxyproline, which makes the collagen more stable.

B1.2.7 Impact of primary structure on the conformation of proteins

The structure of proteins has four levels of complexity: primary, secondary, tertiary and quaternary. Primary structure is the linear sequence of amino acids in a polypeptide.

The backbone of a polypeptide is a repeating sequence of atoms linked by covalent bonds (–C–C–N–C–C–N– and so on). The bond angles are all tetrahedral and there can be rotation about the bonds between the alpha carbon atoms and adjacent nitrogen and carbon atoms. This allows polypeptides to fold into almost any three-dimensional shape.

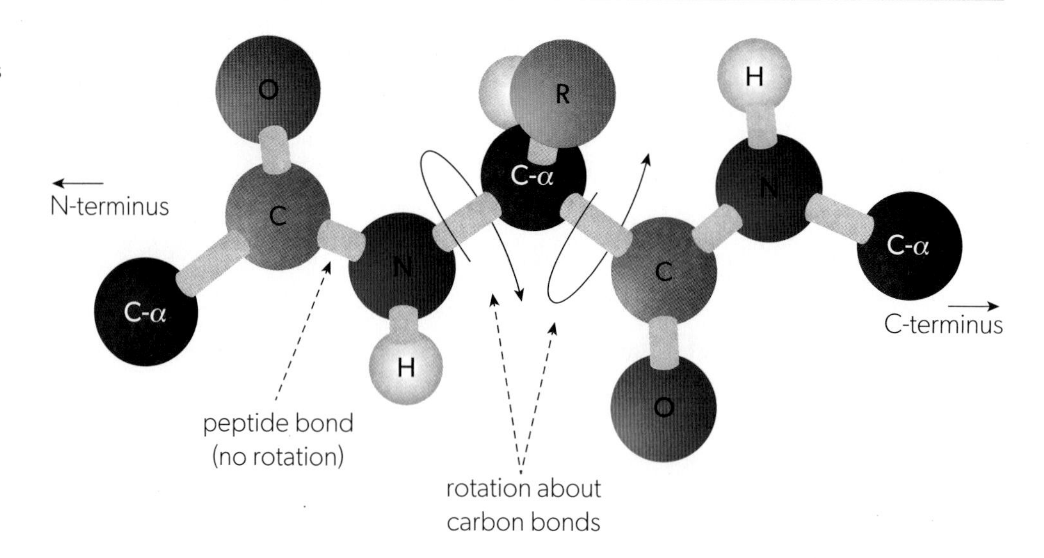

▶ Figure 11 Rotation about bonds in a polypeptide

The three-dimensional arrangement of atoms in a polypeptide or protein is its conformation. Most polypeptides self-assemble into a specific conformation determined by the sequence of amino acids and their R-groups. The conformation of proteins determines their functions and through this the behaviour of cells. This is why protein conformation is of great interest to biologists.

Since the 1970s, biologists have used experimental procedures to determine the structures of more than 180,000 proteins. These have been deposited in the Protein Data Bank, a freely available online resource. This is only a small fraction of the naturally occurring proteins. To speed up the working out of protein conformations, artificial intelligence and massive computing power is being used to make predictions based on primary structure. This will, for example, allow all protein conformations in the human proteome to be discovered relatively quickly.

B1.2.8 Pleating and coiling of secondary structure of proteins

At regular intervals along a polypeptide chain there are C=O and N–H groups. They are what remains of carboxyl and amine groups after they have been used to make peptide bonds. Both C=O and N–H are polar, with the oxygen having a slight negative charge and the hydrogen a slight positive charge. Due to this polarity, hydrogen bonds can form between these groups. Although hydrogen bonds are individually weak, the frequency of C=O and N–H groups along polypeptide chains allows many of them to form and collectively they are strong enough to stabilize distinctive conformational structures within protein molecules.

Two commonly occurring types of structure are stabilized by hydrogen bonding.

- The α-helix—the polypeptide is wound into a helical shape, with hydrogen bonds between adjacent turns of the helix.
- The β-pleated sheet—two or more sections of polypeptide are arranged in parallel with hydrogen bonds between them. The sections of polypeptide run in opposite directions, forming a sheet that is pleated because of the tetrahedral bond angles.

Regular structures stabilized by hydrogen bonding within polypeptides are the secondary structure of a protein.

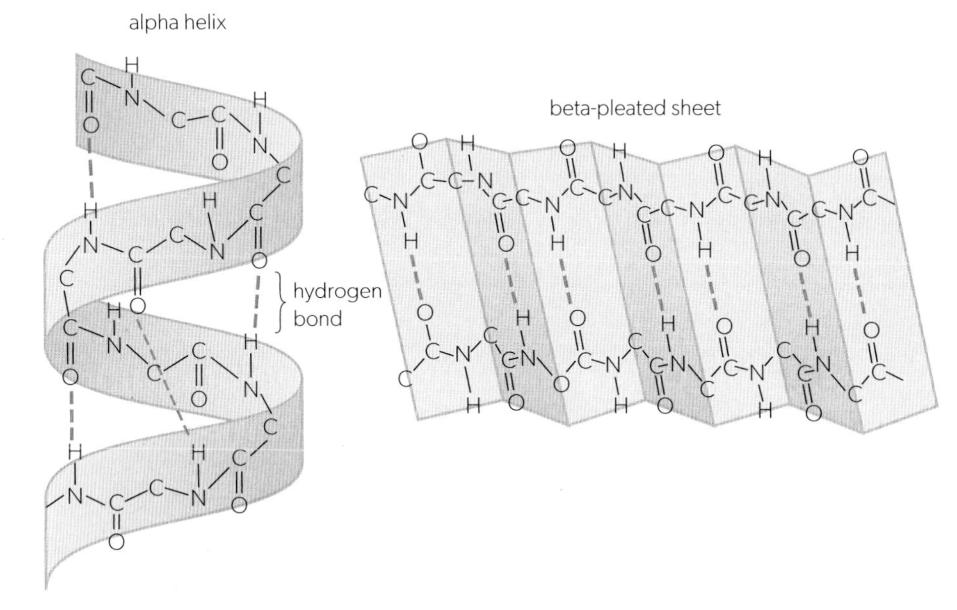

▲ Figure 13 The α-helix (left) and the β-pleated sheet (right) are examples of secondary structures

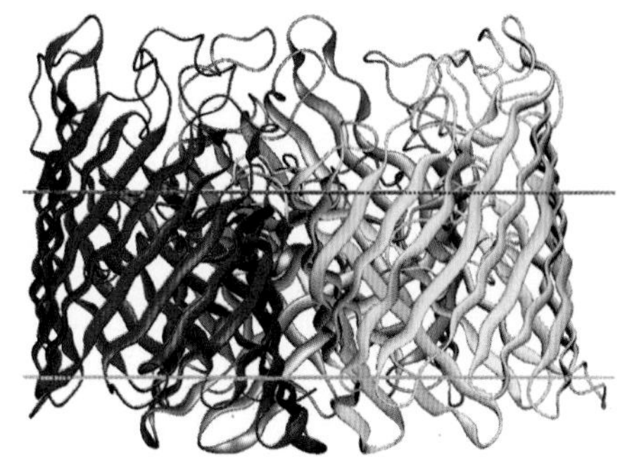

▲ Figure 12 Beta barrel proteins in membranes have large β-pleated sheets curved to form a cylinder. What functions can they perform?

B1.2.9 Dependence of tertiary structure on hydrogen bonds, ionic bonds, disulfide covalent bonds and hydrophobic interactions

Tertiary structure is the folding of a whole polypeptide chain into a three-dimensional structure. This structure is stabilized by interactions between R-groups. There are four main types of interaction.

- Ionic bonds between positively charged and negatively charged R-groups. Amine groups become positively charged by accepting a proton ($-NH^{2} + H^{+} \rightarrow -NH^{3+}$). Carboxyl groups become positively charged by donating a proton ($-COOH \rightarrow -COO^{-} + H^{+}$). Because of the involvement of protons (hydrogen ions), ionic bonds in proteins are sensitive to pH changes.
- Hydrogen bonds between polar R-groups. A hydrogen atom forms a link between two electronegative atoms such as O or N. It is covalently bonded to one of them, which results in the hydrogen having a slight positive charge, making it attractive to the other, which has a slight negative charge.
- **Disulfide bonds** between pairs of cysteines. This is a covalent bond and the strongest of all the interactions.
- **Hydrophobic interactions** between any of the non-polar R-groups.

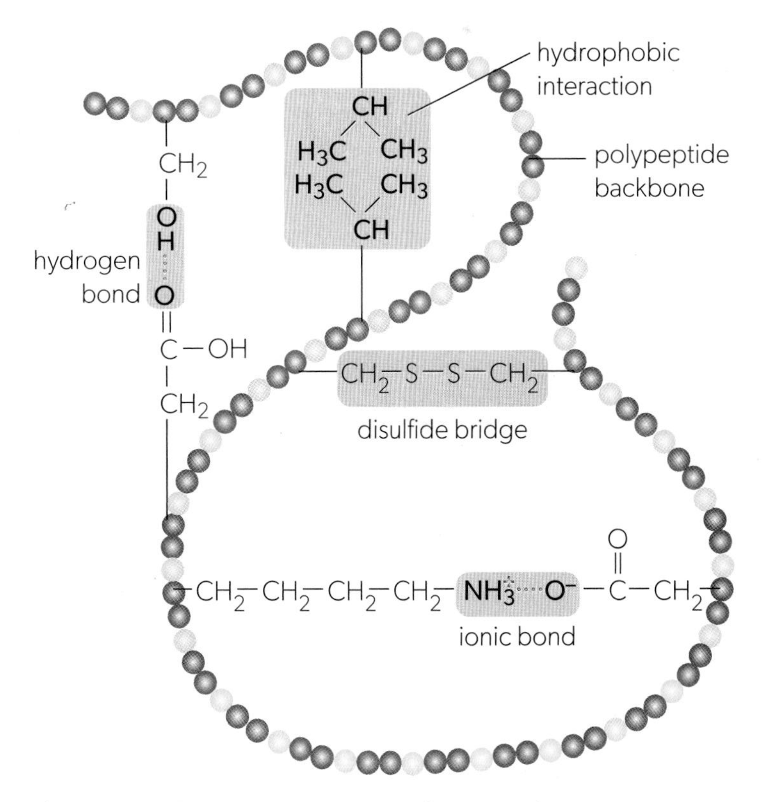

▲ Figure 14 R-group interactions contribute to tertiary structure

Tertiary structure develops as a polypeptide is synthesized by the ribosome. In some cases, a chaperone protein helps with this process to ensure that it results in a correctly folded and fully functional protein.

A wide range of three-dimensional shapes is produced, most of which are globular. Within these tertiary structures there are often parts with secondary structure—α-helices and/or β-pleated sheets.

Some polypeptides do not become folded and instead remain elongated—they do not have tertiary structure. These are fibrous proteins and have structural roles, which are described in *Section B1.2.12*.

B1.2.10 Effect of polar and non-polar amino acids on tertiary structure of proteins

Amino acids in proteins can be divided into two broad categories:

- non-polar and therefore hydrophobic
- polar or charged and therefore hydrophilic.

Many globular proteins need to be soluble in water because they carry out their function in the cytoplasm or in an aqueous solution outside the cell. These proteins have hydrophilic amino acids on their surface where they are in contact with water and hydrophobic amino acids clustered in the centre where water is excluded. This arrangement stabilizes the tertiary structure of the protein because it maximizes hydrophobic interactions between amino acids in the centre and hydrogen bonding between amino acids on the surface and the water around the protein.

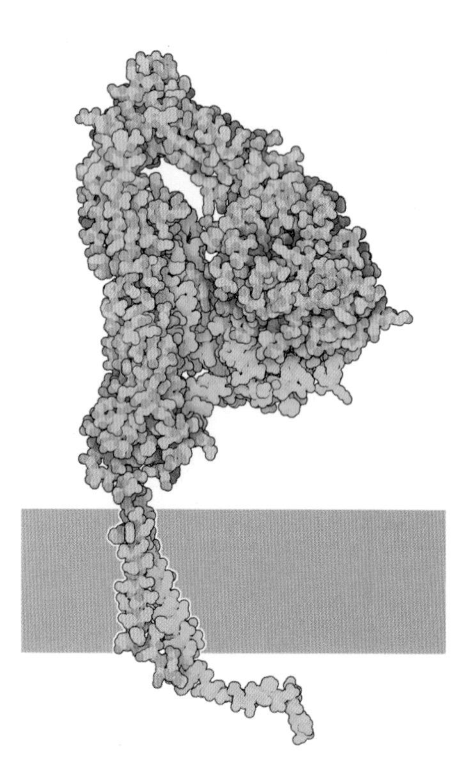

▲ Figure 15 Integrin is a transmembrane protein composed of two polypeptides (shown green and blue), each with an alpha helix of hydrophobic amino acids embedded in the core of the membrane (grey). Integrin connects the cytoskeleton inside the cell to components of the extracellular matrix, so helps bind the cells of a tissue

Some proteins are routinely in contact with non-polar substances over some or all their surface. Such proteins have hydrophobic amino acids on parts of their surface. Integral proteins embedded in membranes have hydrophobic amino acids where they contact the non-polar hydrocarbon core of the membrane. In transmembrane proteins this hydrophobic region is a belt, with hydrophilic regions inside and outside that are in contact with aqueous solutions inside and outside the cell. This arrangement both stabilizes the tertiary structure of the protein and ensures that it remains positioned correctly in the membrane where its function can be performed.

Channel proteins in membranes allow hydrophilic solutes or water to diffuse across the hydrophobic core of the membrane. They have hydrophilic regions with a hydrophobic region between, which holds them in a transmembrane position. In addition, they have a tunnel lined with hydrophilic amino acids through the centre of the protein. The width and charge distribution of this channel allows specific hydrophilic ions or molecules to pass through.

B1.2.11 Quaternary structure of non-conjugated and conjugated proteins

All proteins have at least one polypeptide, but many consist of two or more polypeptides linked together and some have one or more non-polypeptide components. In proteins that consist of more than a single polypeptide, the three-dimensional arrangement of subunits is the quaternary structure.

In a non-conjugated protein, there are only polypeptide subunits. To form the quaternary structure the polypeptides are linked by the same types of interaction as in tertiary structure. For example, insulin has two polypeptides, linked by disulfide bonds (shown in Figure 8 on page 199). Collagen is another non-conjugated protein. It consists of three polypeptides wound together to form a rope-like structure with high tensile strength. It is illustrated in Figure 19 on page 207.

Conjugated proteins have one or more non-polypeptide subunits in addition to their polypeptides. For example, the haemoglobin molecule consists of four polypeptide chains, each associated with a haem group. The inclusion of non-polypeptide components increases the chemical and functional diversity of proteins. The haem group of haemoglobin binds oxygen, allowing this protein to transport oxygen. Many enzymes have a non-polypeptide component that contributes to the catalytic activity of their active site.

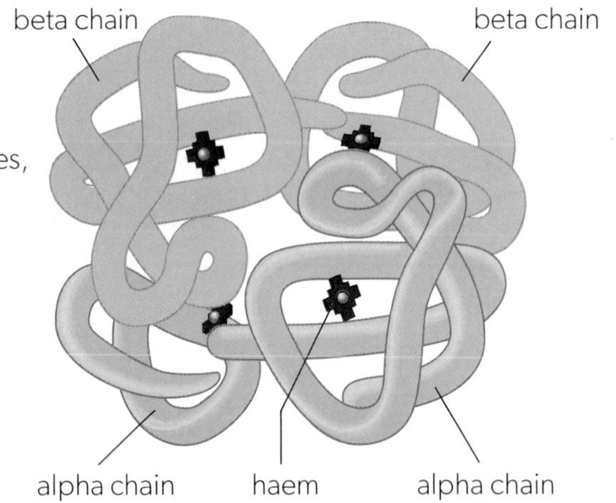

▲ Figure 16 The quaternary structure of haemoglobin in adults consists of four polypeptide chains (two α-chains and two β-chains) each of which is bound to an iron-containing haem group

Data-based questions: Haemoglobin subunits during development

During the process of development from conception through to six months after birth, human haemoglobin changes in composition. Adult haemoglobin is a protein composed of two subunits called globins. It has two alpha and two beta globin subunits. Four other polypeptides are found during development in different amounts: zeta, delta, epsilon and gamma globins. Figure 17 illustrates the changes in haemoglobin composition during gestation and after birth in a human.

1. State which two subunits are present in the highest amounts early in gestation. [1]
2. Distinguish between the changes in the amount of the gamma globin with the amount of beta globin. [2]
3. Determine the composition of the haemoglobin at 10 weeks of gestation and at 6 months of age. [2]
4. State the source of oxygen for the foetus. [1]
5. Suggest reasons for the differences in subunit composition during foetal development and after birth. [3]

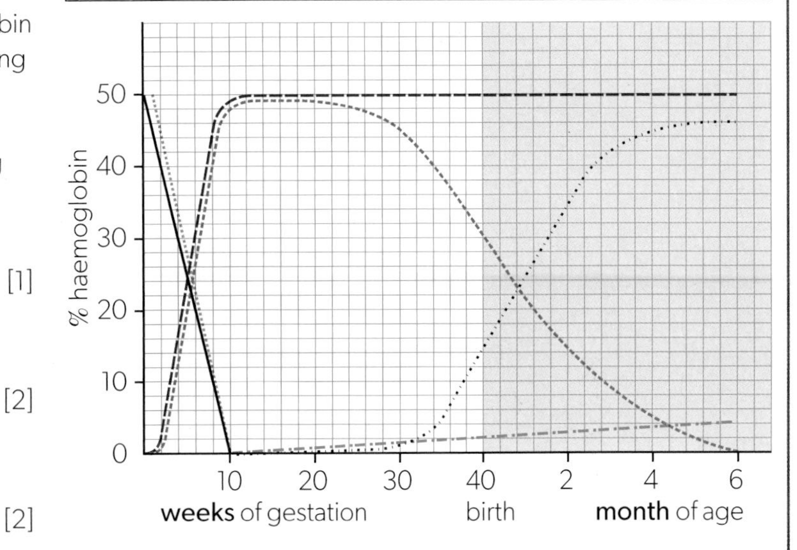

▲ Figure 17

Experiments: Cryo-electron microscopy

A haemoglobin molecule has a diameter of about 5 nm. This is far too small for a light microscope to produce an image. Even the images produced by electron microscopes of this size of molecule were until recently fuzzy blobs. Improvements in technology have revolutionized imaging of protein structures, so tertiary and quaternary structures can now be determined and interactions between proteins and other molecules. As so often with improvements in technology, this has led to a wave of discoveries from research labs around the world.

The new technique is cryo-electron microscopy (cryo-EM). It avoids the need to crystallize proteins, which is almost impossible in many cases, especially with integral membrane proteins. A protein sample is applied to a sample grid and is plunged into liquid ethane to flash-freeze it. The protein molecules are trapped in a thin layer of ice and images can then be obtained using a beam of electrons. Software has been developed for processing the images to increase the resolution. At the start of the 21st century, the highest resolution possible was about 1 nanometre. By 2020, cryo-EM had reduced this to 0.12 nm. This resolution allows the position of individual atoms in a protein to be discovered.

In addition to determining the form of proteins, cryo-EM enables function to be investigated. The freezing technique allows conformation changes to be revealed as a protein carries out its task.

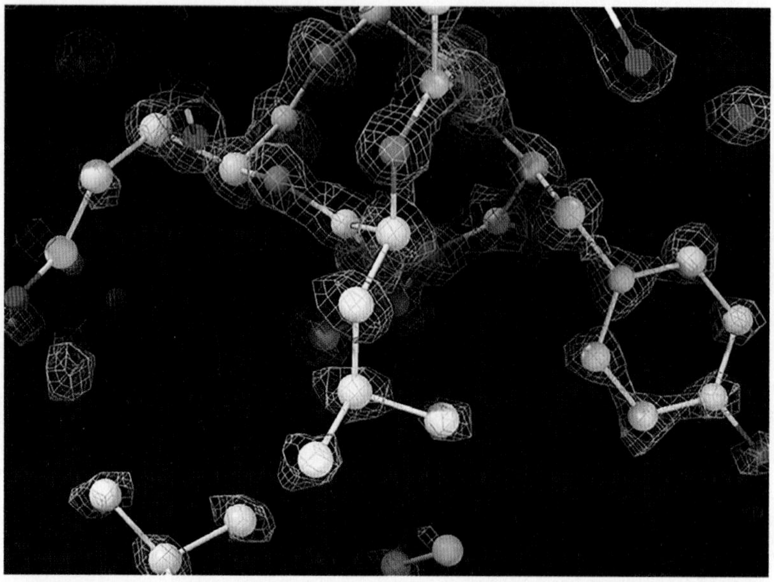

▲ **Figure 18** Part of an image of the protein apoferritin generated by cryo-electron microscopy

B1.2.12 Relationship of form and function in globular and fibrous proteins

The function of a protein depends on its form. This can be illustrated by considering the difference between fibrous and globular proteins. Fibrous proteins consist of elongated polypeptides that lack the folding of typical tertiary structure. Also, the polypeptides in fibrous proteins do not develop secondary structures such as alpha helices. Their quaternary structure is developed by linking together polypeptide chains into narrow fibres or filaments, with hydrogen bonds between the chains.

Collagen is an example of a fibrous protein. The quaternary structure is three polypeptides, wound together into a triple helix. The primary structure of the polypeptides is a repeating sequence of three amino acids: P–G–X. The P in this sequence is proline or hydroxyproline, which has the special property of preventing formation of an α-helix. The winding together of the three polypeptides would be impossible if they were α-helices. The R-group of every third amino acid faces inwards towards the centre of the triple helix and glycine is the only amino acid with an R-group small enough to fit: it is a single hydrogen atom.

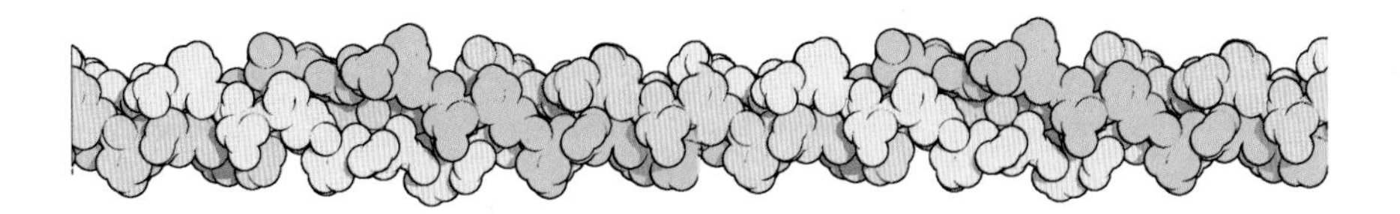

◀ Figure 19 Collagen—the quaternary structure consists of three polypeptides wound together to form a tough, rope-like protein

The rope-like structure of collagen gives it very high tensile strength. The R-group of amino acid X faces outwards and is variable, allowing many variations of collagen to be produced for use in skin, tendons, ligaments, cartilage, basement membranes of epithelia and the tough outer coat of the eye (visible at the front as the white of the eye).

Globular proteins have a rounded shape, formed by the folding up of polypeptides. The shape is very intricate and is stabilized by bonds between the R-groups of the amino acids that have been brought together by the folding. There are many examples of the precise position of each atom in a globular protein, known as the conformation, being critical to the protein's function. The active site of enzymes and the ligand-binding site of receptors show this relationship. Insulin is another example. Only an insulin molecule has the conformation needed to bind to a specific site on the insulin receptor. This allows a specific and unambiguous signal to be sent to body cells when blood sugar concentration is too high.

▲ Figure 20 The insulin receptor (blue) is an integral protein that is positioned in the plasma membrane of many body cells. It has a binding site for insulin. When insulin binds to the receptor (right) there is a conformational change in the receptor, which conveys a signal to the interior of the cell

Linking questions

1. How do abiotic factors influence the form of molecules?
   a. Explain the effect of temperature on enzyme activity. (C1.1.8)
   b. Explain why changes in pH affect the tertiary structure of proteins. (B1.2.9)
   c. Outline the relationship between light and phytohormone activity. (C3.1.19)
2. What is the relationship between the genome and the proteome of an organism?
   a. Explain the mechanisms behind the regulation of transcription. (D2.2.2)
   b. Outline the process of translation. (D1.2.5)
   c. Explain how a mutation in the genome can lead to a change in the structure of the resulting polypeptide. (D1.2.11)

TOK

What constraints are there on the pursuit of knowledge?

Prematurely born infants often have poorly developed digestive tracts and must be fed nutrients through their blood vessels. This is known as total parenteral nutrition (TPN). For the infant's skeleton to grow at the same rate that it would have grown in utero, very large quantities of calcium need to be dissolved in the TPN solution. Different calcium salts dissolve to different concentrations, some higher than others. The question is, are all these calcium salts equally bioavailable? Does all the calcium from a highly soluble salt end up in the bones of the infant? How can this be determined? Investigating this question is a challenge because there are constraints on the types of investigation that can be carried out.

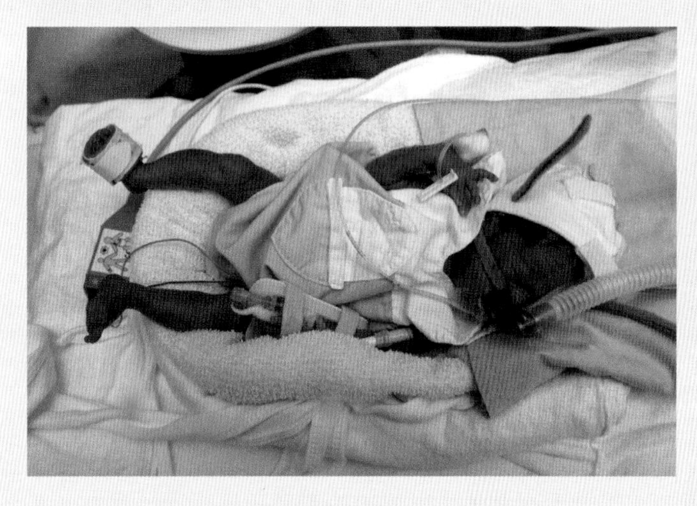

▲ **Figure 1** **The premature baby in an incubator is receiving nutrition through his blood vessels**

Parents of unwell infants may be under financial stress and participation in compensated research might seem attractive. They might feel beholden to medical experts or hopeful that the novel treatment might be superior to currently available treatments. Importantly, investigations involving human subjects are governed by research ethics committees that ensure patient's rights are protected. Informed consent procedures require transparent disclosure to subjects or their guardians of:

- the purposes of the experiment
- the limits on the ways in which their samples can be investigated.

The gold standard of investigation of the metabolic fates of elements is often through use of radioisotopes or samples enriched in a particular rare stable isotope. Legitimately, parents might find mention of such investigations worrisome.

For this reason, most human trials must be preceded by animal trials. For example, it has been argued that a piglet that is removed from its mother before it is weaned is a good model for prematurely born infants.

▲ **Figure 2** **A couple consults with a medical professional. Such a consultation would normally occur in human trials of medical interventions**

There are also restrictions placed on the use of animals in research. Ethics boards of research institutions often have guidelines that include:

- providing the justification for using animals
- reducing the total number of individuals used
- evidence that the experiment is not a duplication of previous research
- restrictions on the types of species that can be chosen
- providing details on the regimens used to eliminate or minimize pain.

End of chapter questions

1. Migrating birds must refuel along the way to continue flying. A field study was conducted among four different species of migrating birds known to stop at high-quality and low-quality food sites. Birds were captured and blood samples were taken at the two sites.

   Among birds, high triglyceride concentration in blood plasma indicates fat deposition whereas high butyrate concentration in blood plasma indicates fat utilization and fasting.

   The following data summarizes triglyceride levels and butyrate levels measured for the same groups of birds.

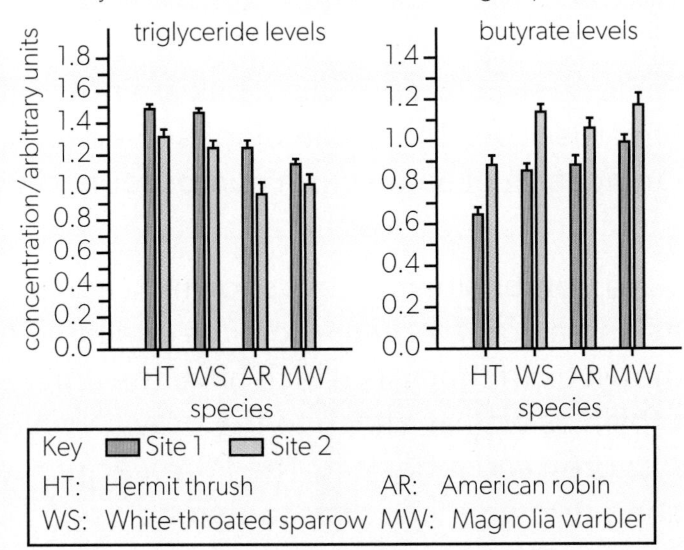

Source: Guglielmo, C.G., Cerasale, D.J. and Eldermire, C. (2005) Physiological and Biochemical Zoology, 78(1), pp. 116–125. https://doi.org/10.1086/425198.

   a. Butyrate is a fatty acid. Distinguish between fatty acids and triglycerides. [2]

   b. Describe, using the triglyceride levels graph, the results at Site 1 and Site 2 for all the birds. [2]

   c. Explain the differences in the triglyceride level and butyrate level for the hermit thrush at Site 1 and Site 2. [2]

   d. Scientists have hypothesized that the food quality is better at Site 1 than at Site 2.

      Evaluate this hypothesis using the data provided. [2]

2. The figure shows a tripeptide. Label **one** peptide bond in this molecule.

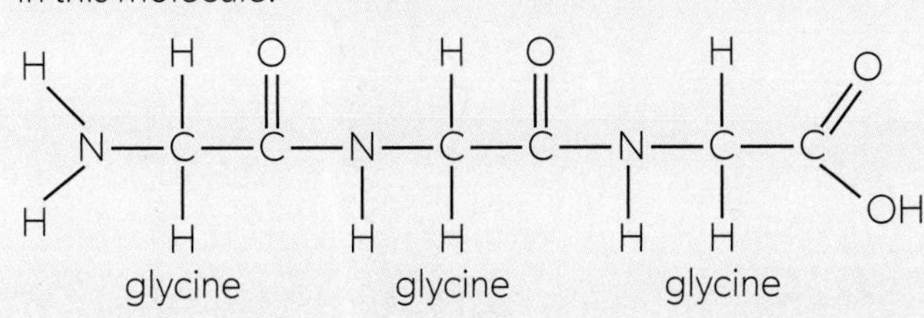

3. Outline the specific functions of **three named** proteins. [3]

4. Cellulose is the most abundant organic polymer on Earth. Describe the structure of cellulose. [3]

5. Compare and contrast cis-fatty acids and trans-fatty acids. [2]

6. Proteins such as keratin and myosin consist of two or more alpha helices winding around each other into a super-coil and are known as long coiled-coil (LCC) proteins. Such proteins are involved in a wide variety of structural and mechanical processes in cells.

   A study was carried out to compare the presence of LCC proteins in species from different kingdoms. The LCC proteins were grouped by similarities in primary structure. They were then analysed to show family relationships and homology.

   The diagram below shows the distribution of groups of LCC protein sequences by kingdom.

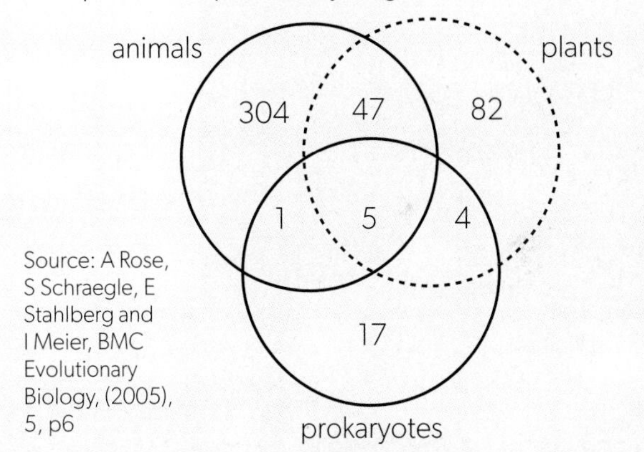

Source: A Rose, S Schraegle, E Stahlberg and I Meier, BMC Evolutionary Biology, (2005), 5, p6

   a. State what determines the primary structure of a protein. [1]

   b. Outline the concept of secondary structure of proteins. [2]

   c. State how many groups of LCC proteins are common to all the species studied. [1]

   d. Deduce the significance of these proteins being found in all of the species studied. [1]

   e. Calculate how many groups of LCC proteins are found in the prokaryote kingdom. [1]

   f. Calculate the percentage of groups analysed that are found in the animal kingdom only. [1]

   g. Deduce whether this data supports the hypothesis that plants are more closely related to animals than to prokaryotes. [2]

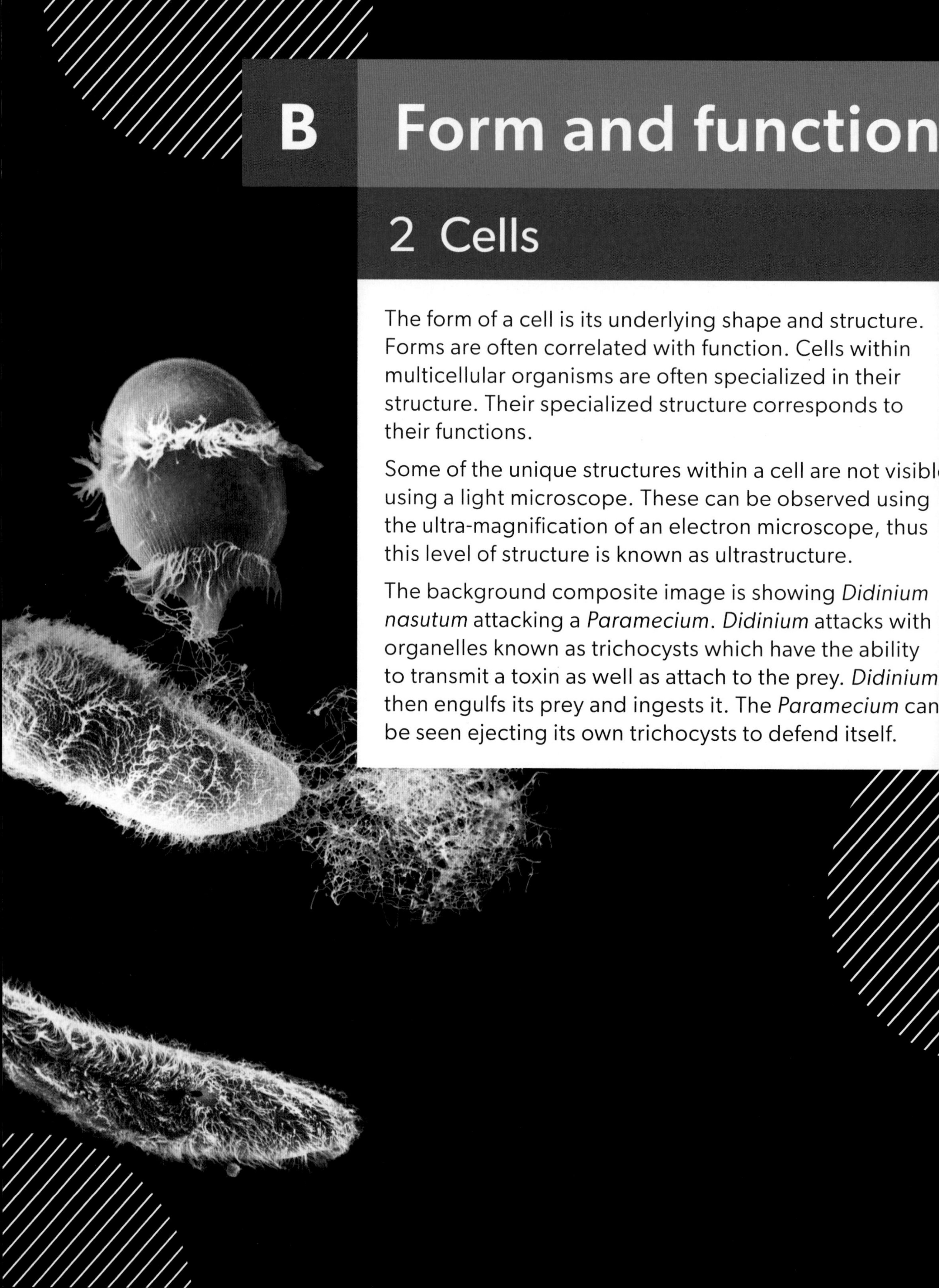

B Form and function

2 Cells

The form of a cell is its underlying shape and structure. Forms are often correlated with function. Cells within multicellular organisms are often specialized in their structure. Their specialized structure corresponds to their functions.

Some of the unique structures within a cell are not visible using a light microscope. These can be observed using the ultra-magnification of an electron microscope, thus this level of structure is known as ultrastructure.

The background composite image is showing *Didinium nasutum* attacking a *Paramecium*. *Didinium* attacks with organelles known as trichocysts which have the ability to transmit a toxin as well as attach to the prey. *Didinium* then engulfs its prey and ingests it. The *Paramecium* can be seen ejecting its own trichocysts to defend itself.

B2.1 Membranes and membrane transport

How do molecules of lipid and protein assemble into biological membranes?

The liquid blobs in a lava lamp are a mixture of oils and waxes, dissolved in a hydrophobic solvent. The other liquid, through which the blobs rise and fall is water with a dye to colour it. What prevents the two liquids from ever mixing? Biological membranes form a flexible frontier around every cell, separating the water-based cytoplasm inside from the water-based environment outside. Do you expect the interior of membranes to be hydrophobic or hydrophilic? Will the surfaces of the membrane be hydrophobic or hydrophilic? What hydrophobic/hydrophilic properties are needed for lipids and proteins in membranes?

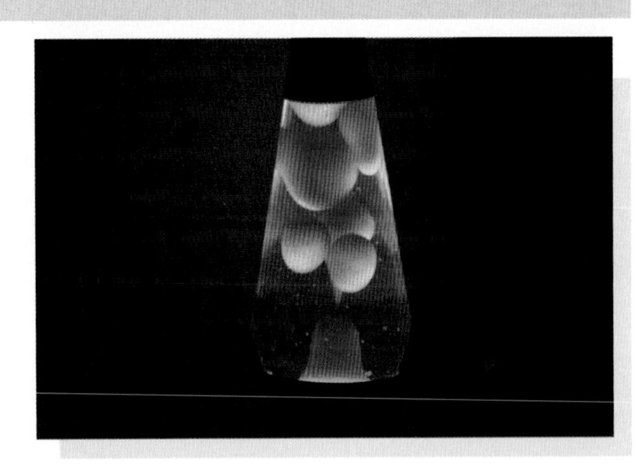

▲ Figure 1 A lava lamp

What determines whether a substance can pass through a biological membrane?

Face masks are porous and allow particles smaller than the pore size to pass through. What particles must be able to pass through a dentist's mask? What size should the largest pores in a dentist's mask be if it was intended to exclude virus particles? Biological membranes are very sophisticated and can distinguish between hydrophilic and hydrophobic particles. Which type will pass through a membrane more easily? Membranes can discriminate between ions such as Na^+, K^+ and Cl^- and only allow ions needed by the cell to enter. How could a membrane do this?

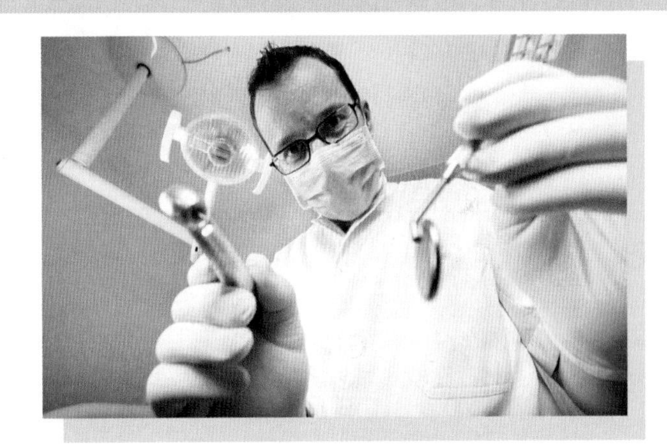

▲ Figure 2 Dentist wearing a surgical mask

SL and HL	AHL only
B2.1.1 Lipid bilayers as the basis of cell membranes B2.1.2 Lipid bilayers as barriers B2.1.3 Simple diffusion across membranes B2.1.4 Integral and peripheral proteins in membranes B2.1.5 Movement of water molecules across membranes by osmosis and the role of aquaporins B2.1.6 Channel proteins for facilitated diffusion B2.1.7 Pump proteins for active transport B2.1.8 Selectivity in membrane permeability B2.1.9 Structure and function of glycoproteins and glycolipids B2.1.10 Fluid mosaic model of membrane structure	B2.1.11 Relationships between fatty acid composition of lipid bilayers and their fluidity B2.1.12 Cholesterol and membrane fluidity in animal cells B2.1.13 Membrane fluidity and the fusion and formation of vesicles B2.1.14 Gated ion channels in neurons B2.1.15 Sodium–potassium pumps as an example of exchange transporters B2.1.16 Sodium-dependent glucose cotransporters as an example of indirect active transport B2.1.17 Adhesion of cells to form tissues

B2.1.1 Lipid bilayers as the basis of cell membranes

Membranes are an essential component of cells. The plasma membrane forms the border between a cell and its environment. Membranes inside eukaryotic cells divide the cytoplasm into compartments. The basic structure of all biological membranes is the same. A bilayer of phospholipids and other amphipathic molecules forms a continuous sheet that controls the passage of substances despite being 10 nanometres or less across. The structure of phospholipid molecules and their arrangement into bilayers is described in *Topic B1.1*.

▶ Figure 3 Freeze-fracture electron micrographs of cells show membranes very clearly. In this image, a nuclear membrane is visible at the top and below it many membrane-bound vesicles and cisternae that divide the cytoplasm into small compartments

B2.1.2 Lipid bilayers as barriers

Phospholipid molecules have a phosphate "head" and two hydrocarbon "tails". The tails of the phospholipids are hydrophobic and interact with each other to form the core of biological membranes. Due to this, the membrane core has low permeability to all hydrophilic particles, including ions with positive or negative charges and polar molecules such as glucose.

There are usually aqueous solutions on either side of cell membranes. These solutions are in a liquid state, so both water molecules and hydrophilic solutes are in continuous random motion. The solutes nearest to the membrane surface might penetrate between the hydrophilic phosphate heads of the phospholipids, but if they reach the hydrophobic core of the membrane they are drawn back to the aqueous solution outside the membrane. The hydrophobic hydrocarbon chains that form the core of the membrane do not repel hydrophilic solutes but they are more attracted to each other, and the solutes are much more attracted to water outside the membrane.

Molecular size also influences membrane permeability. The trend is that the larger the molecule, the lower the permeability. For example, water molecules which are only slightly larger than single oxygen atoms, pass through membranes more easily than large molecules such as glycogen or protein.

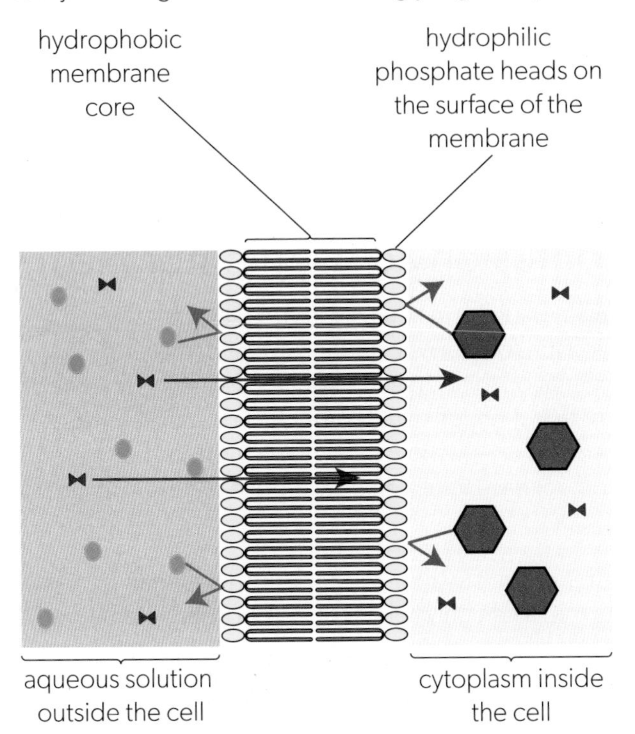

◀ Figure 4 The hydrophobic core of the membrane has low permeability to polar molecules such as glucose (blue) and charged particles such as chloride ions (green) so they can be kept either in or out of a cell, whereas small non-polar molecules such as oxygen (red) can pass through freely

Data-based questions: Membrane permeability

The graph in Figure 5 shows the energy level of six substances at different distances from the centre of a phospholipid bilayer. Progesterone is a hormone and the other substances are drugs. Free energy is reduced by bond formation.

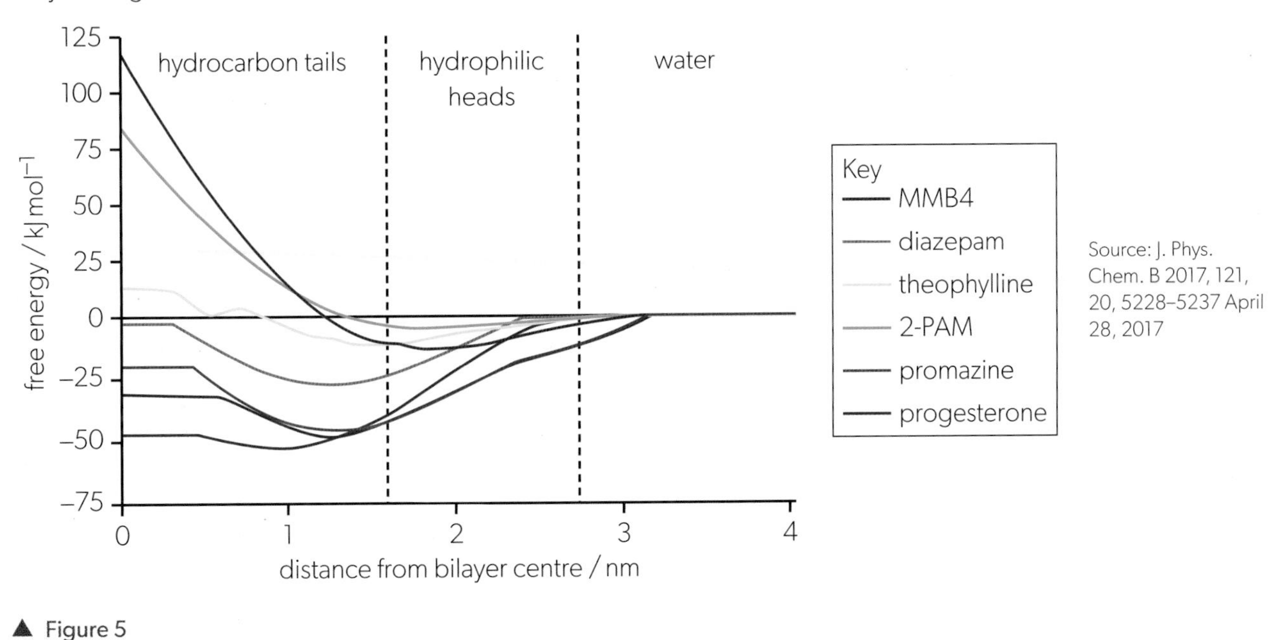

Source: J. Phys. Chem. B 2017, 121, 20, 5228–5237 April 28, 2017

▲ Figure 5

1. Compare and contrast the energy levels of:
   a. MMB4 and 2-PAM [2]
   b. promazine and progesterone [2]
2. Deduce from the curves in the graph, which of the six substances is:
   a. most hydrophobic [2]
   b. most hydrophilic [2]
3. Using the diagrams and the graph, explain whether hydroxyl (–OH) groups make molecules more or less hydrophilic. [2]
4. Four membrane permeability categories have been defined: impermeable, low, medium, and high. At least one of the five drugs is in each category. Suggest one drug for each category. [4]
5. Predict the permeability category for progesterone. [1]

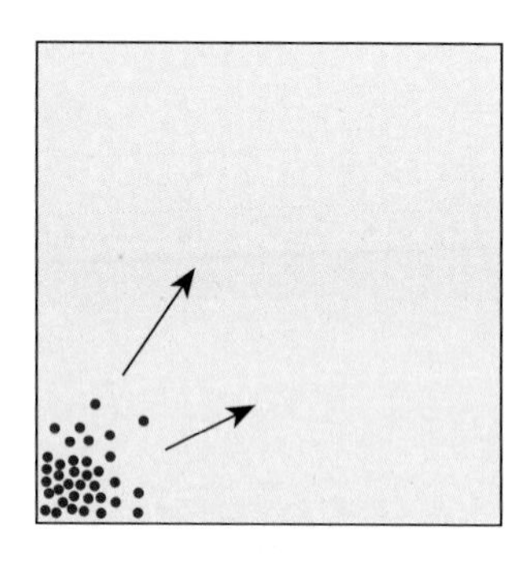

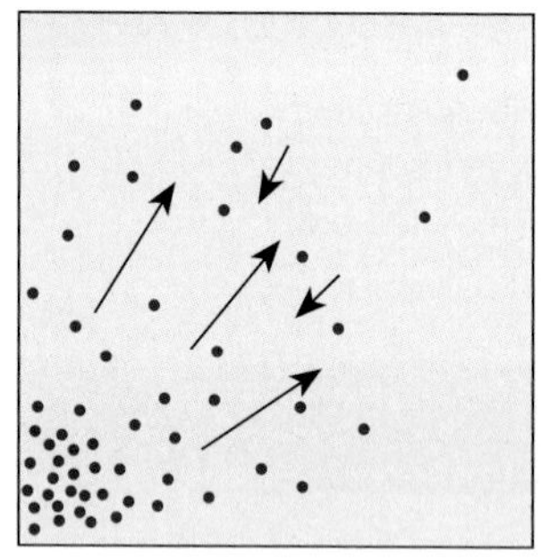

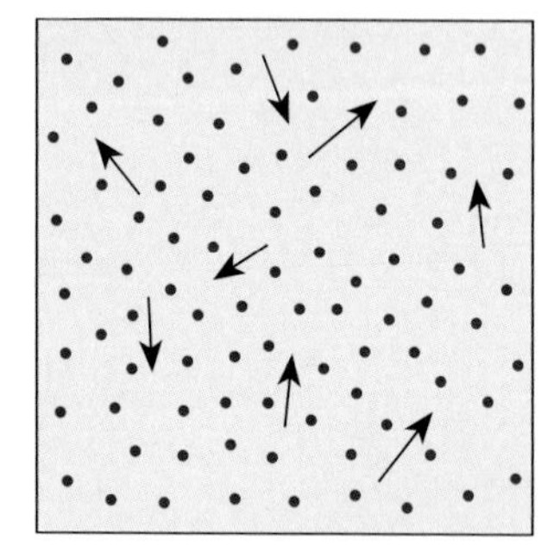

▲ Figure 6 Model of diffusion with dots representing particles

B2.1.3 Simple diffusion across membranes

Diffusion is the spreading out of particles in liquids and gases that happens because the particles are in continuous random motion. More particles move from an area of higher concentration to an area of lower concentration than move in the opposite direction. There is therefore a net movement of particles from the higher to the lower concentration—a movement down the concentration gradient. Living organisms do not have to use energy to make diffusion occur; it is a passive process.

Simple diffusion across membranes is due to particles passing between phospholipids in the membrane. It can only happen if the phospholipid bilayer is permeable to the particles. Non-polar particles such as oxygen can diffuse through easily. If the oxygen concentration inside a cell is reduced due to aerobic respiration and the concentration outside is higher, oxygen will pass into the cell through the plasma membrane by passive diffusion.

The centre of membranes is hydrophobic, so ions with positive or negative charges cannot easily diffuse through. Polar molecules, which have partial positive and negative charges over their surface, can diffuse at low rates between the phospholipids of the membrane. Small polar particles such as urea or ethanol pass through more easily than large particles.

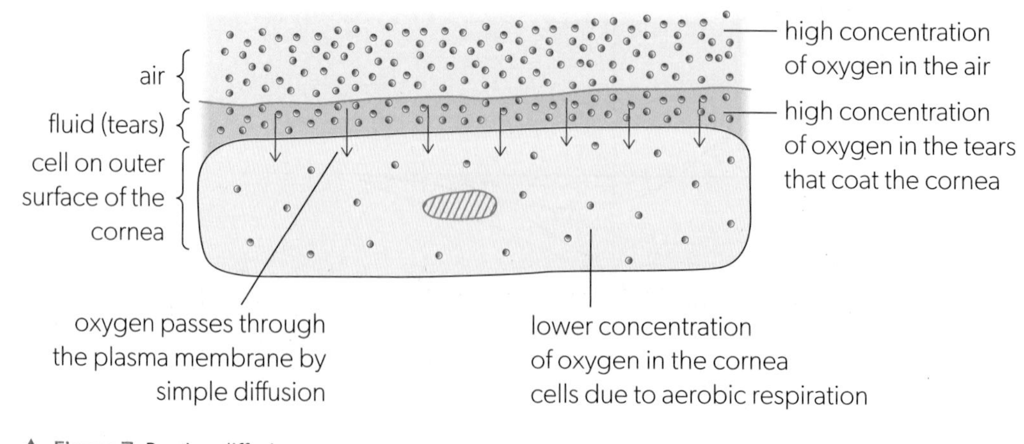

the cornea has no blood supply so its cells obtain oxygen by simple diffusion from the air
air
fluid (tears)
cell on outer surface of the cornea
high concentration of oxygen in the air
high concentration of oxygen in the tears that coat the cornea
oxygen passes through the plasma membrane by simple diffusion
lower concentration of oxygen in the cornea cells due to aerobic respiration

▲ Figure 7 Passive diffusion

Data-based questions: Diffusion of oxygen in the cornea

Oxygen concentrations were measured in the cornea of anesthetized rabbits at different distances from the outer surface. These measurements were continued into the aqueous humor behind the cornea. The rabbit's cornea is 400 micrometres (400 μm) thick. The graph in Figure 8 shows the measurements. You may need to look at a diagram of eye structure before answering the questions. The oxygen concentration in normal air is 20 kilopascals (20 kPa).

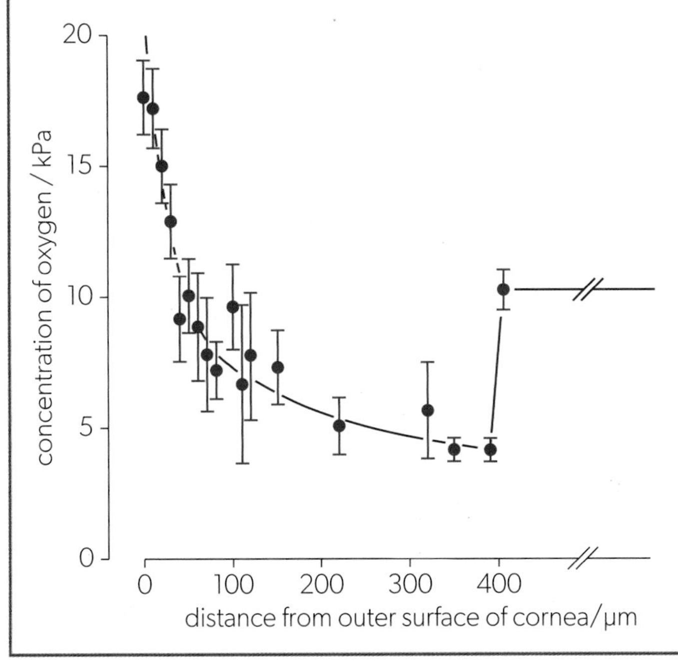

◀ Figure 8

1. Calculate the thickness of the rabbit cornea in millimetres. [1]
2. a. Describe the trend in oxygen concentrations in the cornea from the outer to the inner surface. [2]
   b. Suggest reasons for the trend in oxygen concentration in the cornea. [2]
3. a. Compare the oxygen concentrations in the aqueous humor with the concentrations in the cornea. [2]
   b. Using the data in the graph, deduce if oxygen diffuses from the cornea to the aqueous humor. [2]
4. Using the data in the graph, evaluate diffusion as a method of moving substances in large multicellular organisms. [2]
5. a. Predict the effect of wearing contact lenses on oxygen concentrations in the cornea. [1]
   b. Suggest how this effect could be minimized. [1]
6. The range bars for each data point indicate how much the measurements varied. Explain the reason for showing range bars on the graph. [2]

B2.1.4 Integral and peripheral proteins in membranes

Because of these varied functions, membrane proteins are very diverse in structure and in their position in the membrane. They can be divided into two groups.

- Integral proteins are hydrophobic on at least part of their surface and are therefore embedded in the hydrocarbon chains in the centre of the membrane. They may fit in one of the two phospholipid layers or extend across both. Many integral proteins are transmembrane proteins—they extend across the membrane, with hydrophilic parts projecting through the regions of phosphate heads on either side.
- Peripheral proteins are hydrophilic on their surface, so are not embedded in the membrane. Most of them are attached to the surface of integral proteins and this attachment is often reversible. Some have a single hydrocarbon chain attached to them which is inserted into the membrane, anchoring the protein to the membrane surface.

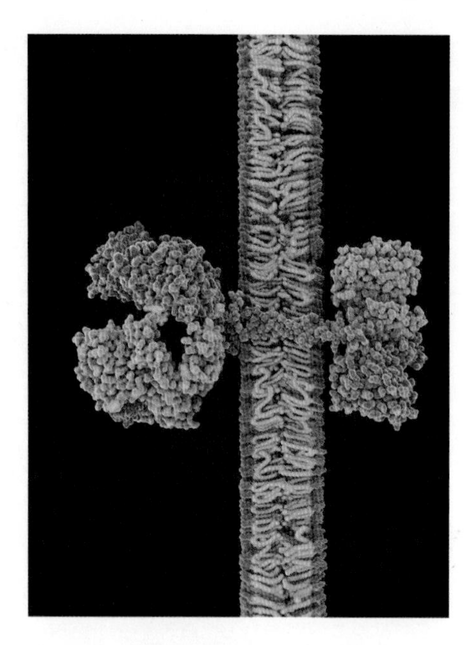

▲ Figure 9 The protein shown blue is a receptor for the hormone EGF (epidermal growth factor). It is an integral protein because it is embedded in the phospholipid bilayer (orange = hydrophobic region and purple = hydrophilic). EGF, shown red, is also a protein but as it binds to the exterior surface of the receptor, it is a peripheral rather than an integral protein.

Membranes all have an inner face and an outer face, and membrane proteins are oriented so that they can carry out their function correctly. For example, pump proteins in the plasma membranes of root cells in plants are oriented so that they pick up potassium ions from the soil and pump them into the root cell.

The protein content of membranes is very variable because the function of membranes varies. The more active a membrane, the higher is its protein content. Membranes in the myelin sheath around nerve fibres just act as insulators and have a protein content of about 18%. Most plasma membranes on the outside of the cell have a protein content of about 50%. The highest protein content—about 75%—is found in the membranes of chloroplasts and mitochondria, which are active in photosynthesis and respiration.

B2.1.5 Movement of water molecules across membranes by osmosis and the role of aquaporins

Water can move in and out of most cells freely. Sometimes, the number of water molecules moving in and out is the same and there is no net movement. At other times, more molecules move in one direction or the other. This net movement is osmosis.

Osmosis is due to differences in the concentration of substances dissolved in water (solutes). Substances dissolve by forming intermolecular bonds with water molecules. These bonds restrict the movement of the water molecules. This means that regions with a higher solute concentration have a lower concentration of water molecules that are free to move than regions with a lower solute concentration. Because of this, there is net movement of water from regions of lower solute concentration to regions with higher solute concentration. This movement is passive because no energy is directly expended to make it occur.

Osmosis can happen in all cells because water molecules, despite being hydrophilic, are small enough to pass through the phospholipid bilayer. Some cells have water channels called aquaporins, which greatly increase membrane permeability to water. Examples are kidney cells that reabsorb water, and root hair cells that absorb water from the soil.

At its narrowest point, the channel in an aquaporin is only slightly wider than water molecules, which therefore pass through in single file. Positive charges at this point in the channel prevent protons (H^+) from passing through.

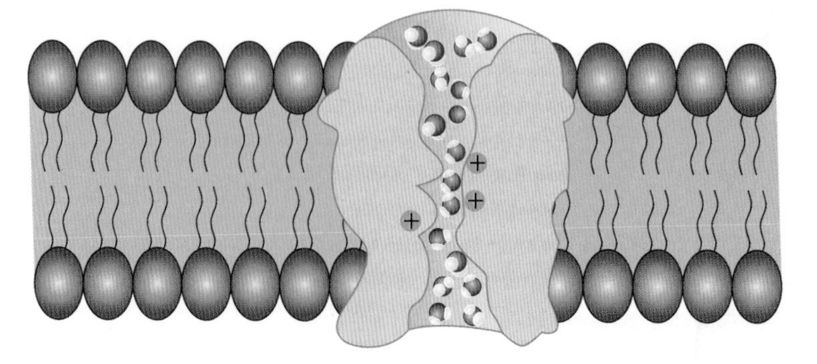

▲ Figure 10 Structure of an aquaporin

B2.1.6 Channel proteins for facilitated diffusion

Ions and polar molecules cannot easily pass between phospholipids, but diffusion of these substances across a membrane is still possible with the help of proteins acting as channels. A channel protein is an integral, transmembrane protein with a pore that connects the cytoplasm to the aqueous solution outside the cell. The diameter of a pore and the chemical properties of its sides ensure that only one type of particle passes through—for example, sodium ions or potassium ions, but not both.

Channel proteins allow particles to pass through in either direction, but more pass from the higher to the lower concentration than vice versa. There is therefore a net movement from the higher concentration to the lower. No energy is expended by the cell to cause this movement, so it is a type of diffusion. It is called facilitated diffusion because channel proteins are required for the movement to occur. Simple diffusion here would be movement between phospholipid molecules in the membrane.

Cells can select which hydrophilic substances diffuse in and out by the types of channel that are synthesized and placed in the plasma membrane. Some channels can be opened and closed, so permeability can be temporarily changed when necessary.

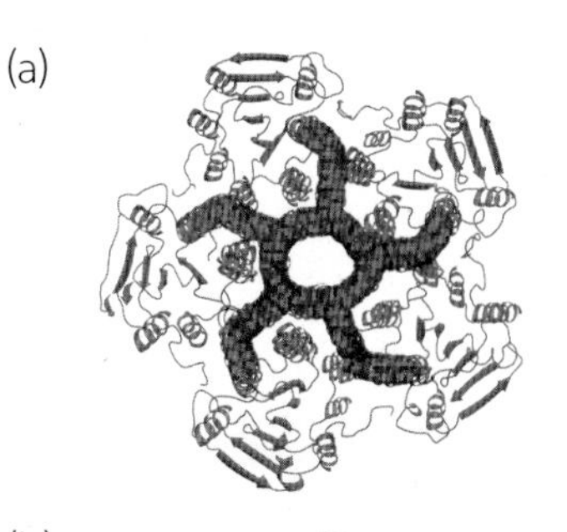

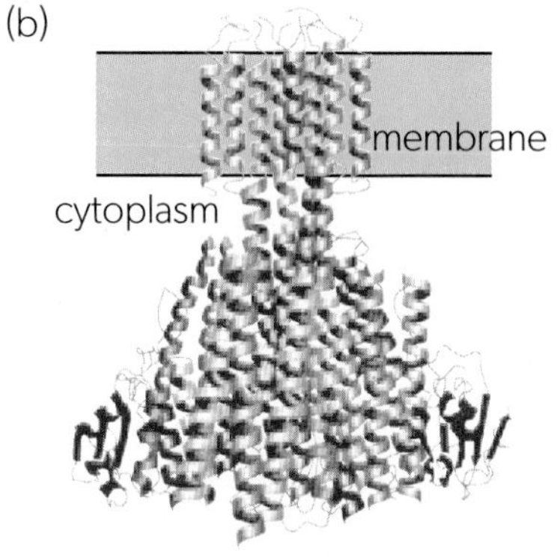

▲ Figure 11 Magnesium channel viewed from the side and from the outside of the membrane. The structure of the protein making up the channel ensures that only magnesium ions can pass through the pore in the centre

B2.1.7 Pump proteins for active transport

Cells absorb some substances, even though the concentration inside is already higher than outside. The substance is absorbed against the concentration gradient. Less commonly, cells sometimes pump substances out even though there is already a higher concentration outside. Pump proteins in the membrane carry out these transport tasks. Pump proteins differ in three ways from channel proteins in how they transport particles across membranes:

- pump proteins use energy so they carry out active transport, whereas diffusion through channel proteins is passive
- pump proteins only move particles across the membrane in one direction, whereas particles can move in either direction through a channel protein
- pump proteins usually move particles against the concentration gradient, whereas facilitated diffusion through channel proteins is always down the concentration gradient.

Pump proteins are interconvertible between two different conformations. In one conformation, the transported particle can enter the pump from one side of the membrane to reach a central chamber or a binding site. The pump protein then changes to the other conformation, which allows the ion or molecule to pass out on the opposite side of the membrane. The pump protein returns to its original conformation. Energy is used to change the protein from one of the conformations (the more stable) to the other (the less stable), but the reverse change does not require energy. Most pump proteins use ATP to supply the energy required for active transport. Every cell produces its own ATP by cell respiration.

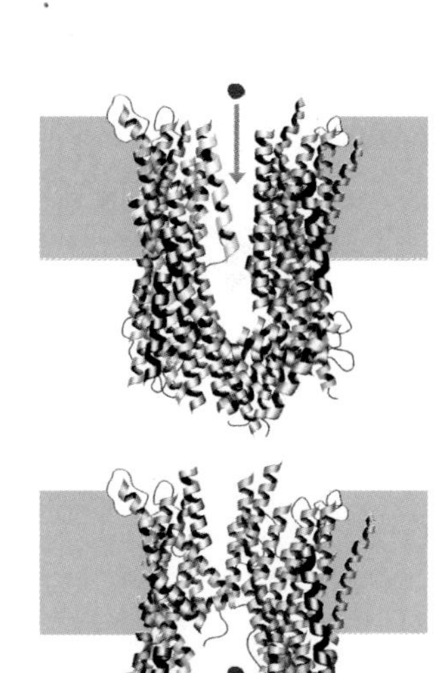

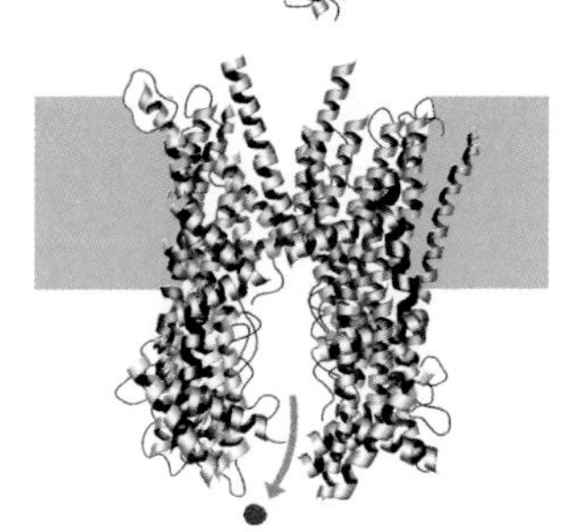

▲ Figure 12 Action of a pump protein that transports Vitamin B_{12} into *Escherichia coli*

The membranes of cells contain many different pump proteins, each of which transfers one specific type of particle across the membrane. This allows the cell to control the content of its cytoplasm precisely. It also allows specific solutes required by a cell to be absorbed even when they are in very low concentrations in the environment.

Data-based questions: Phosphate absorption in barley roots

Roots were cut off from barley plants and used to investigate phosphate absorption. Roots were placed in phosphate solutions and air was bubbled through. The phosphate concentration was the same in each case, but the percentages of oxygen and nitrogen were varied in the air bubbled through. The rate of phosphate absorption was measured. Table 1 shows the results.

Oxygen /%	Nitrogen /%	Phosphate absorption/ $\mu mol\ g^{-1}\ h^{-1}$
0.1	99.9	0.07
0.3	99.7	0.15
0.9	99.1	0.27
2.1	97.1	0.32
21.0	79.0	0.33

▲ Table 1

1. Describe the effect of reducing the oxygen concentration below 21.0% on the rate of phosphate absorption by roots. You should only use information from Table 1 in your answer. [3]
2. Explain the effect of reducing the oxygen concentration from 21.0% to 0.1% on phosphate absorption. In your answer, you should use as much biological understanding as possible of how cells absorb mineral ions. [3]

An experiment was done to test which method of membrane transport was used by the roots to absorb phosphate. Roots were placed in the phosphate solution as before, with 21.0% oxygen bubbling through. Varying concentrations of a substance called DNP were added. DNP blocks the production of ATP by aerobic cell respiration. Figure 13 shows the results of the experiment.

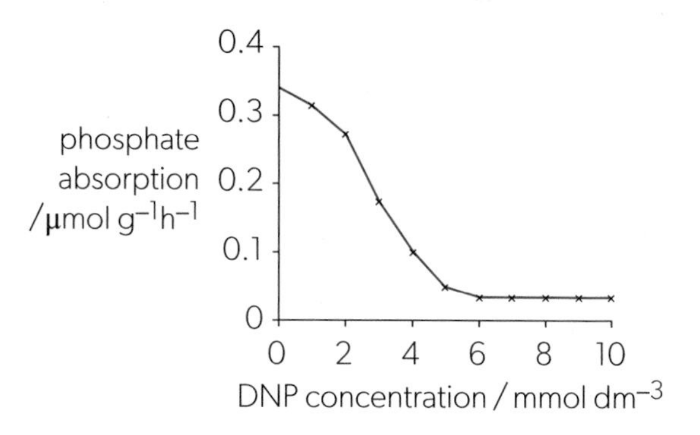

▲ Figure 13 Effect of DNP concentration on phosphate absorption

3. Deduce, with a reason, whether the roots absorbed the phosphate by diffusion or by active transport. [2]
4. Discuss the conclusions that can be drawn from the data in the graph about the method of membrane transport used by the roots to absorb phosphate. [2]

B2.1.8 Selectivity in membrane permeability

A semi-permeable membrane allows the passage of certain small solutes and is freely permeable to the solvent. This describes artificial membranes of the type that are used for kidney dialysis, but it does not match the permeability properties of cell membranes, which show more selectivity and have variable permeability to water.

A selectively permeable membrane allows the passage of particular particles, but not others. Facilitated diffusion and active transport allow selective permeability because channel proteins and pump proteins are specific to particular particles. A chloride channel, for example, allows only chloride ions to diffuse across the membrane. However, simple diffusion is not selective and depends only on the size and polarity of particles. Small hydrophobic particles cannot be prevented from passing across cell membranes.

Because cell membranes are partly semi-permeable and partly selectively permeable, they are sometimes described as partially permeable—all three of these terms are widely used.

B2.1.9 Structure and function of glycoproteins and glycolipids

Glycoproteins are conjugated proteins with carbohydrate as the non-polypeptide component. They are a component of the plasma membrane of cells, with the protein part embedded in the membrane and the carbohydrate part projecting out into the exterior environment of the cell.

Glycolipids are molecules consisting of carbohydrates linked to lipids. The carbohydrate part is usually a single monosaccharide or a short chain of between two and four sugar units. The lipid part usually contains one or two hydrocarbon chains, which naturally fit into the hydrophobic core of membranes. Glycolipids occur in the plasma membranes of all eukaryotic cells, with the attached carbohydrate projecting outwards into the extracellular environment of the cell.

The role of glycoproteins in cell-to-cell recognition is described in *Section B1.1.7*. Glycolipids also have a role in cell recognition. They help the immune system to distinguish between self and non-self cells, so pathogens and foreign tissue can be recognized and destroyed. Glycoproteins and glycolipids together form a carbohydrate-rich layer on the outer face of the plasma membrane of animal cells, with an aqueous solution in the gaps between the carbohydrates. This layer is called the glycocalyx. The glycocalyx of adjacent cells can fuse, binding the cells together and preventing the tissue from falling apart.

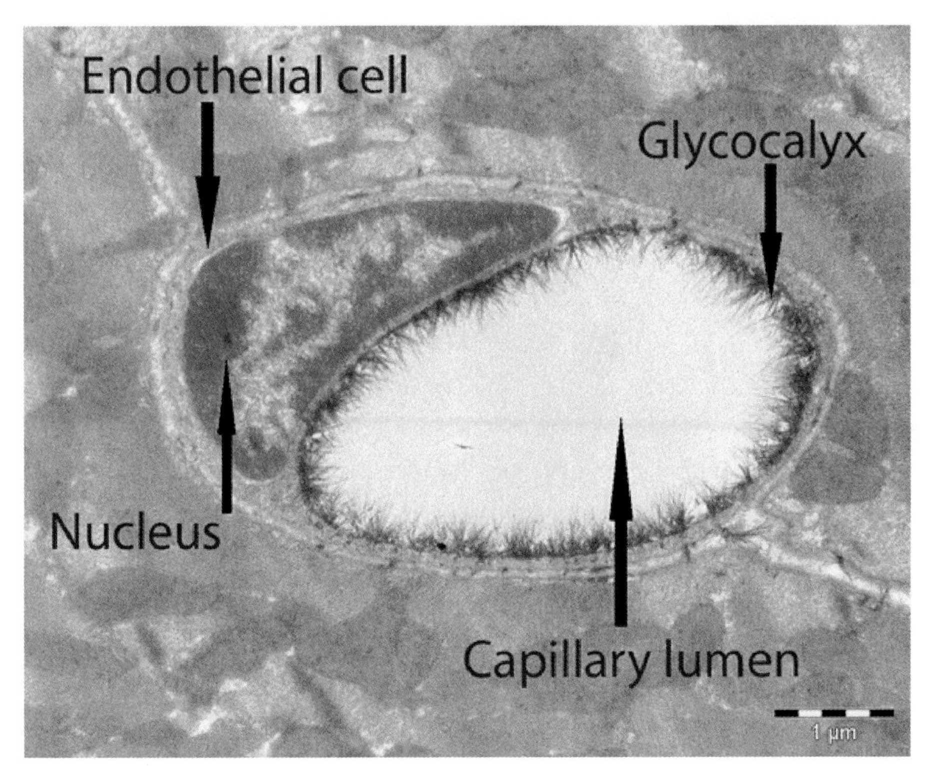

◀ Figure 14 The glycocalyx of endothelium cells in blood capillaries projects into the lumen. In this brain capillary the glycocalyx is particularly dense and forms part of the blood–brain barrier. It prevents plasma cells and circulating proteins from binding to the capillary wall, which reduces the chance of inflammation and blood clotting.

B2.1.10 Fluid mosaic model of membrane structure

Several models of membrane structure have been proposed but one particular model is now so strongly supported by evidence that it is unlikely to be replaced. In this model, there is a bilayer of phospholipids with proteins in a variety of positions. Peripheral proteins are attached to the inner or outer surface. Integral proteins are embedded in the phospholipid bilayer, in some cases with parts protruding on one or both sides. The proteins are likened to the tiles in a mosaic. Because the phospholipid molecules are free to move laterally in each of the two layers of the bilayer, the proteins can also move. This gives the model its name—the fluid mosaic model.

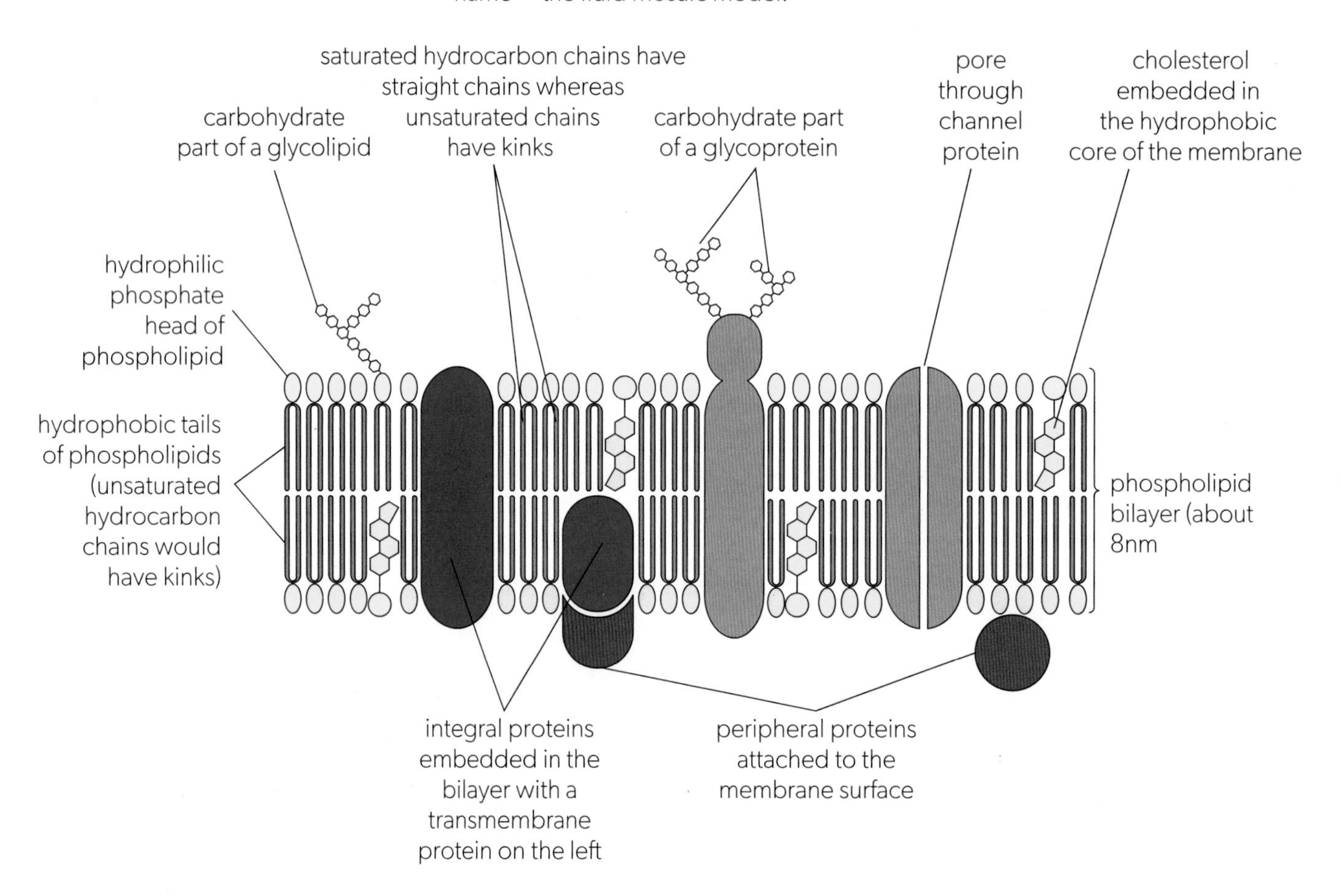

▲ Figure 15 Fluid mosaic model of membrane structure

AHL

B2.1.11 Relationships between fatty acid composition of lipid bilayers and their fluidity

Saturated fatty acids have straight chains and therefore pack together tightly in bilayers, giving a high density of phospholipids. This reduces the fluidity of the membrane and therefore its flexibility and permeability by simple diffusion. In contrast, unsaturated fatty acids have one or more kinks in their hydrocarbon chain, so they pack together more loosely. This makes the membranes more fluid, flexible and permeable.

Relative amounts of saturated and unsaturated fatty acids are regulated so that the membranes have the required properties. They must remain fluid but be strong enough to avoid becoming perforated. They must be permeable but not too porous. The ideal ratio of saturated to unsaturated fatty acids depends on the temperatures that a cell experiences. For example, fish from Antarctic waters have been found to have a higher percentage of unsaturated fatty acids in their membranes than fish from warmer waters.

▲ Figure 16 A membrane containing only saturated fatty acids (left) is thicker, more viscous, has a higher density of phospholipids and a higher melting point than a membrane containing both saturated and unsaturated fatty acids (right)

Data-based questions: Frost hardiness and double bonds in chickpeas

Freezing temperatures cause cytoplasm to leak out of leaf cells in chickpea plants (*Cicer arietinum*). This kills the cells. The effectiveness of two treatments preventing leakage was investigated. The treatments were:

- acclimatization of plants by keeping them at temperatures close to freezing point for two weeks
- spraying the outside of the leaves with ABA, a hormone produced by plants in response to stress.

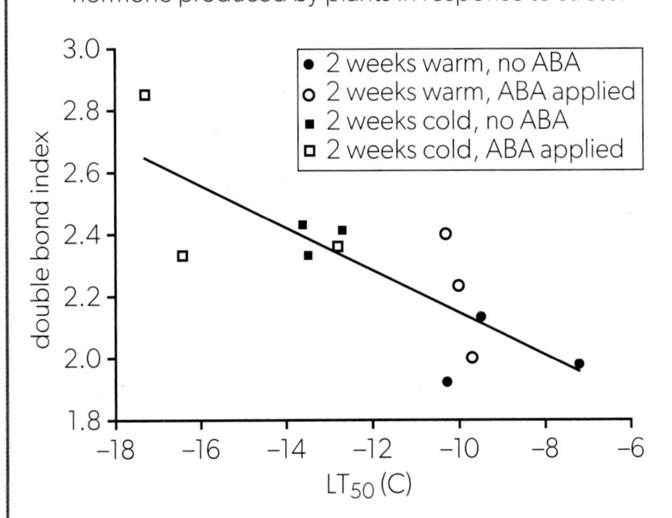

▲ Figure 17

The proportions of saturated and unsaturated membrane lipids were measured after the treatments (double bond index). Frost hardiness was assessed by finding the temperature that killed 50% of leaf cells. The graph in Figure 17 shows the results.

1. a. State the relationship between LT_{50} and double bond index. [1]

   b. Explain the relationship. [2]

2. Using only the data in the graph, outline the effects of ABA on the chickpea plants. [2]

3. Deduce the effects of cold treatment on the proportions of saturated and unsaturated membrane lipids in chickpea plants. [2]

4. Gardeners are advised to "harden off" plants that have been raised in a warm greenhouse before planting them outside in colder conditions. Discuss whether spraying with ABA or 2 weeks of cold acclimatization is likely to be more effective. [3]

AHL

B2.1.12 Cholesterol and membrane fluidity in animal cells

Figure 18 The structure of cholesterol

Cholesterol makes up between 20% and 40% of the lipids in the plasma membranes of eukaryotes. It is a steroid rather than a glyceride. Most of a cholesterol molecule is hydrophobic so it is attracted to the hydrophobic hydrocarbon tails in the centre of the membrane, but one end of the molecule has a hydroxyl (–OH) group which is hydrophilic. This is attracted to the phosphate heads on the periphery. Cholesterol molecules are therefore positioned between phospholipids in the membrane, with the hydroxyl group usually facing outwards. It preferentially intercalates between saturated rather than unsaturated hydrocarbon chains.

The fluidity of membranes needs to be carefully controlled. If membranes were too fluid, they would be less able to control what substances pass through. If they were too viscous and inflexible, cell movement would be restricted and the cell would be more likely to burst. Cell membranes do not correspond exactly to any of the three states of matter—they are in what is called a liquid-ordered phase. The lipid molecules are packed densely but are still free to move laterally. Cholesterol helps to maintain the necessary orderly arrangement of phospholipids. Cholesterol therefore stabilizes membranes at higher temperatures, maintaining impermeability to hydrophilic particles such as sodium ions and hydrogen ions. Cholesterol also helps to ensure that saturated fatty acid tails do not solidify at low temperatures, so preventing a stiffening of the membrane.

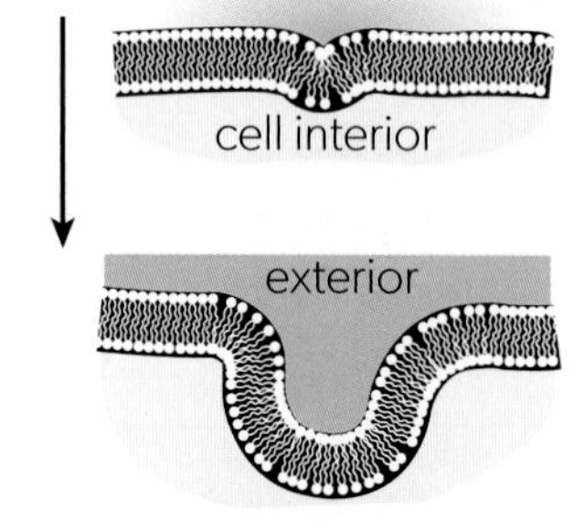

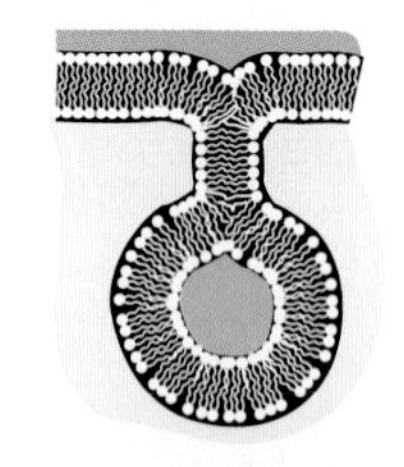

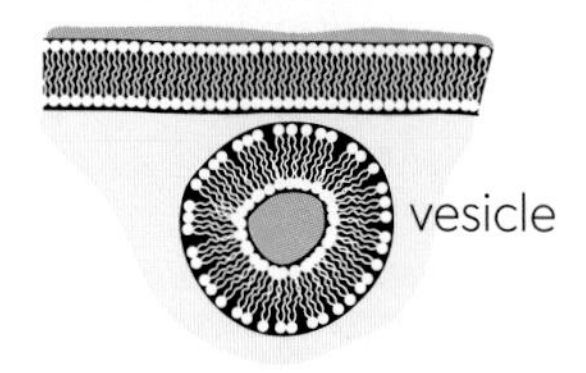

Figure 19 Formation of a vesicle by endocytosis

B2.1.13 Membrane fluidity and the fusion and formation of vesicles

A vesicle is a small sac of membrane with a droplet of fluid inside. Vesicles are spherical and most eukaryotic cells contain them. They are a very dynamic feature of cells as there is a continuous cycle of making vesicles, moving them within the cell to transport their contents and then unmaking them. This can happen because of the fluidity of membranes, which allows structures surrounded by a membrane to change shape and move.

To make a vesicle, a small region of a membrane is pulled from the rest of the membrane and is pinched off. Proteins in the membrane carry out this process, using energy from ATP. If a vesicle is made from the plasma membrane by pinching a small piece of it inwards, the vesicle will contain material that was outside the cell. This is method of taking materials into the cell and is called endocytosis.

Vesicles made by endocytosis contain water and solutes from outside the cell. Often, they contain larger molecules needed by the cell that cannot pass across the plasma membrane. For example, in the placenta, proteins from the mother's blood, including antibodies, are absorbed into the foetus by endocytosis. Some cells take in large undigested food particles by endocytosis. This happens in unicellular organisms including *Amoeba* and *Paramecium*. Some types of white blood cell take in pathogens including bacteria and viruses by endocytosis and then kill them. This is part of the body's response to infection.

Vesicles can be used to move materials around inside cells. In some cases, it is the contents of the vesicle that need to be moved. In other cases, it is proteins in the membrane of the vesicle that need to be moved. An example of moving the vesicle contents occurs in secretory cells. Protein is synthesized by ribosomes on the rough endoplasmic reticulum (rER) and accumulates inside the rER. Vesicles containing the proteins bud off the rER and carry them to the Golgi apparatus.

When they have reached their destination, vesicles fuse with a target membrane and disappear in the process. This has the effect of transferring all the contents of a vesicle across the membrane. If a vesicle fuses with the plasma membrane, the contents are expelled from the cell. This process is called exocytosis.

Exocytosis can also be used to expel waste products or unwanted materials. An example is the removal of excess water from the cells of unicellular organisms. The water is loaded into a vesicle, sometimes called a contractile vacuole, which is then moved to the plasma membrane for expulsion by exocytosis. Polypeptides that have been processed in the Golgi apparatus are carried to the plasma membrane in vesicles for exocytosis. In this case, the release is referred to as secretion, because a useful substance is being released, not a waste product. Digestive enzymes and protein hormones are secreted in this way.

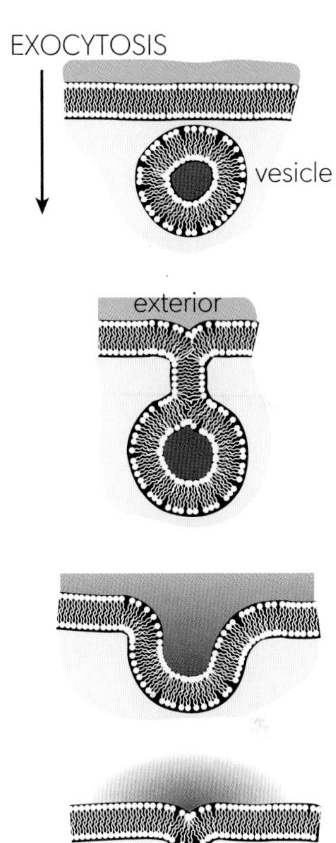

▲ Figure 20 Fusion of a vesicle with the membrane in exocytosis

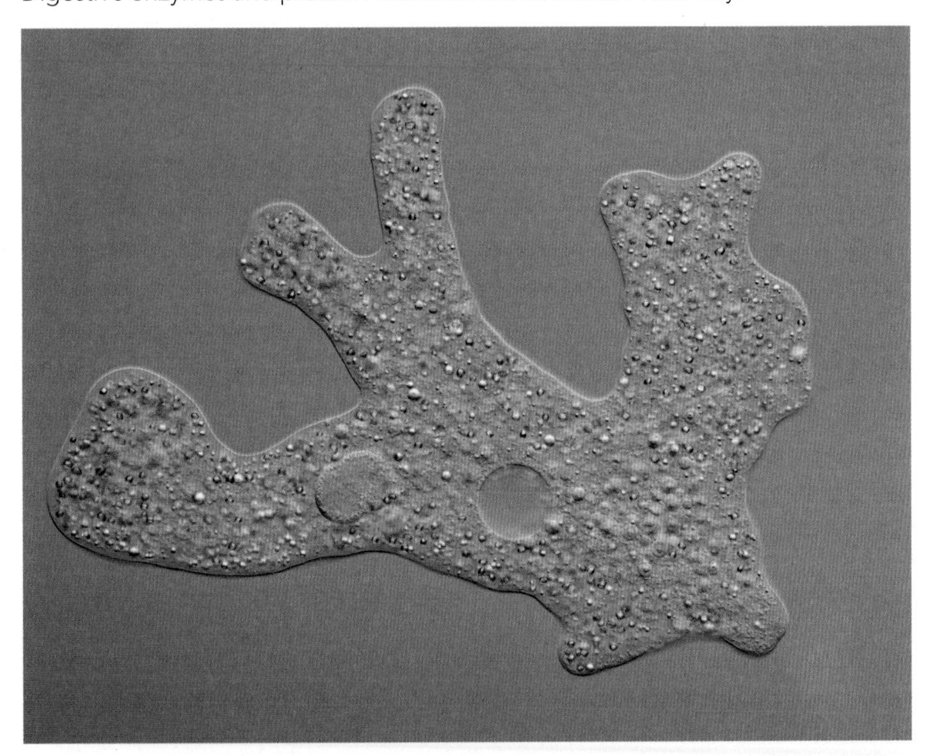

▲ Figure 21 The large vesicle in the centre of this Amoeba is a contractile vacuole. There are also many food vacuoles in the cytoplasm of the cell

In a growing cell, the area of the plasma membrane needs to increase. Phospholipids are synthesized and then inserted into the rER membrane. Ribosomes on the rER synthesize membrane proteins which are added to the membrane. Vesicles bud off the rER and move to the plasma membrane. They fuse with it, each increasing the area of the plasma membrane by a very small amount. The same method is used to increase the size of organelles such as mitochondria in the cytoplasm.

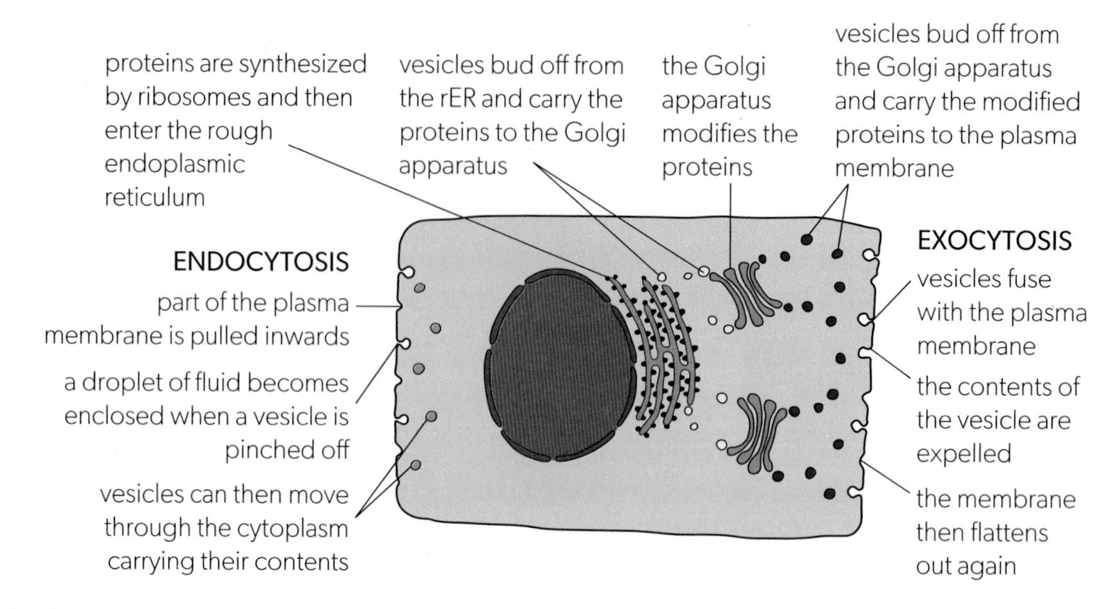
proteins are synthesized by ribosomes and then enter the rough endoplasmic reticulum
vesicles bud off from the rER and carry the proteins to the Golgi apparatus
the Golgi apparatus modifies the proteins
vesicles bud off from the Golgi apparatus and carry the modified proteins to the plasma membrane
ENDOCYTOSIS
part of the plasma membrane is pulled inwards
a droplet of fluid becomes enclosed when a vesicle is pinched off
vesicles can then move through the cytoplasm carrying their contents
EXOCYTOSIS
vesicles fuse with the plasma membrane
the contents of the vesicle are expelled
the membrane then flattens out again

▲ Figure 22 Vesicle movements in a cell

B2.1.14 Gated ion channels in neurons

Ion channels allow specific ions to pass across a membrane in either direction, resulting in a net movement from the higher to the lower concentration of the ion. This type of membrane transport is facilitated diffusion. Gated ion channels are able to open and close reversibly, allowing diffusion to be switched on and off. This is particularly useful in neurons (nerve cells) where there are voltage-gated sodium and potassium channels along nerve fibres and neurotransmitter-gated channels at synapses.

Voltage-gated sodium and potassium channels

A nerve impulse involves rapid movements of sodium and potassium ions across a neuron's membrane. These movements occur by facilitated diffusion through sodium and potassium channels, both of which are voltage gated. Voltages across membranes are due to an imbalance of positive and negative charges across the membrane. A negative voltage indicates that there are relatively more positive charges outside the neuron than inside. If the voltage is below −50 mV, sodium and potassium channels remain closed. If it rises above −50 mV sodium channels open, allowing sodium ions (Na^+) to diffuse in. This causes the voltage to rise more. When it reaches +40 mV, potassium channels open, allowing potassium ions (K^+) to diffuse out of the neuron.

The gating mechanism of both sodium and potassium channels involves reversible conformation changes with subunits. The subunits can be in either an open position with a narrow pore between them that allows ions to pass or in a closed position with no pore. The potassium channel has four subunits and an extra globular protein subunit that resembles a ball, attached by a flexible chain of amino acids. When the four subunits are in the open conformation, the ball can fit inside the open pore and does so within milliseconds of the pore opening. The ball remains in place until the potassium channel returns to its original closed state. There may be a similar mechanism in the sodium channel.

Sodium and potassium channels must be specific, despite Na^+ and K^+ ions both carrying a single positive charge. Sodium channels allow Na^+ ions to pass through but not the larger K^+ ions. Potassium channels do not allow Na^+ ions to pass through. The pore in a potassium channel is 0.3 nm wide at its narrowest. Potassium ions are slightly smaller than 0.3 nm but when they dissolve, they become bonded to a shell of water molecules. This makes them too large to pass through the pore. To pass through, the bonds between the potassium ion and the surrounding water molecules are broken and bonds form temporarily between the ion and a series of amino acids in the narrowest part of the pore. After the potassium ion has passed through this part of the pore, it can again become associated with a shell of water molecules. Sodium ions are too small to form bonds with the amino acids in the narrowest part of the pore, so they cannot shed their shell of water molecules.

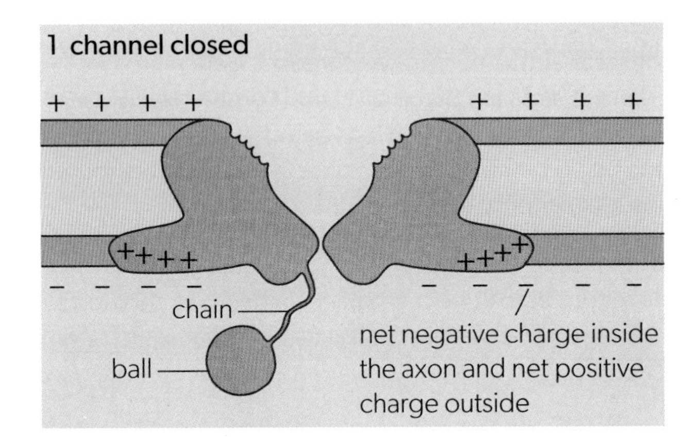

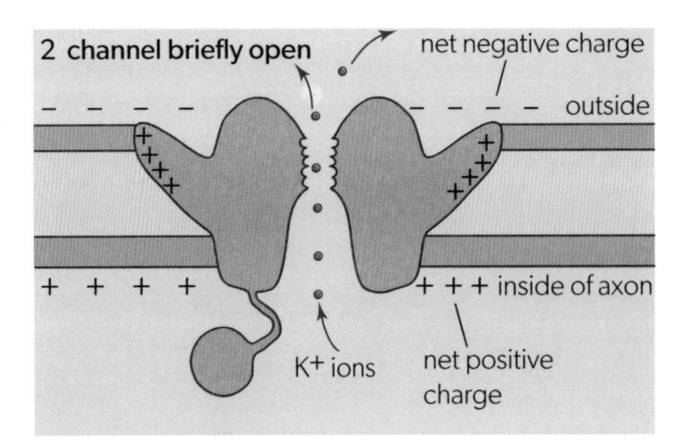

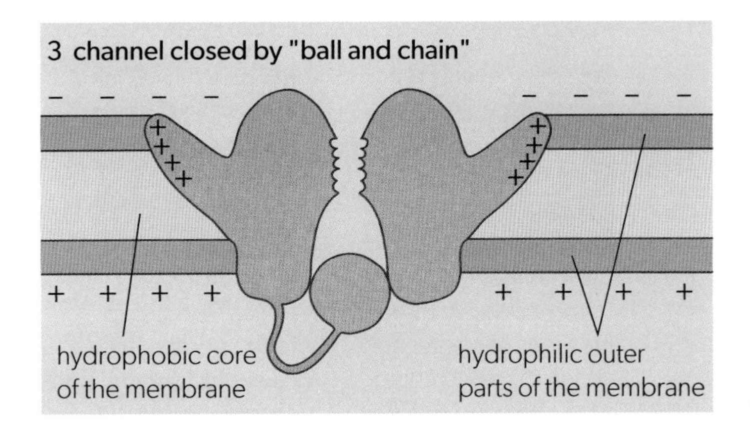

◀ Figure 23 Voltage-gating of potassium channels

Nicotinic acetylcholine receptors

Acetylcholine is the neurotransmitter in many synapses. At these synapses, there are receptors for acetylcholine. But both nicotine and acetylcholine bind to the receptors, hence they are called nicotinic acetylcholine receptors.

These receptors have five transmembrane subunits arranged symmetrically, with a binding site between two of the subunits for acetylcholine. Binding causes a conformational change, which opens a pore between the five subunits, through which cations (positively charged ions) including sodium can pass. Sodium diffuses into the postsynaptic neuron, changing its voltage and causing voltage-gated sodium channels to open. Binding of acetylcholine is reversible. When it dissociates from the receptor, the conformational change caused by binding is reversed and the pore in the receptor is closed.

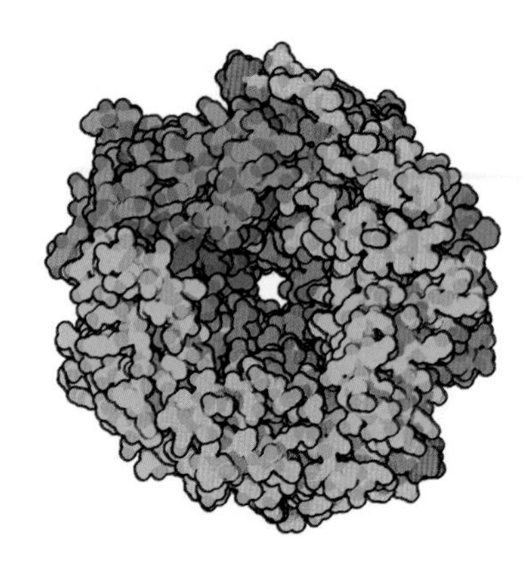

▲ Figure 24 Nicotinic acetylcholine receptor with acetylcholine bound (red) and the cation pore open

Activity: Sketching the nicotinic acetylcholine receptor

Figure 24 on the previous page shows the structure of a nicotinic acetylcholine receptor viewed from the outside of the plasma membrane. Sketch this protein in side-view within the membrane, to show the structure and position that you expect it to have, then check whether your sketch matches the actual structure by going online to the PDB Molecule of the Month website (molecule code number 2BG9).

B2.1.15 Sodium–potassium pumps as an example of exchange transporters

For a neuron to convey a nerve impulse there must be concentration gradients of sodium and potassium ions across the membrane. These are generated by active transport, using a sodium–potassium pump protein. This pump follows a repeating cycle of steps that result in three sodium ions being pumped out of the axon and two potassium ions being pumped in. Each time the pump goes round this cycle it uses one ATP to supply energy.

This pump is an example of an exchange transporter because it transports different ions in opposite directions across the membrane. In neurons, this helps to generate a charge imbalance and therefore a membrane potential, which is a voltage across the membrane.

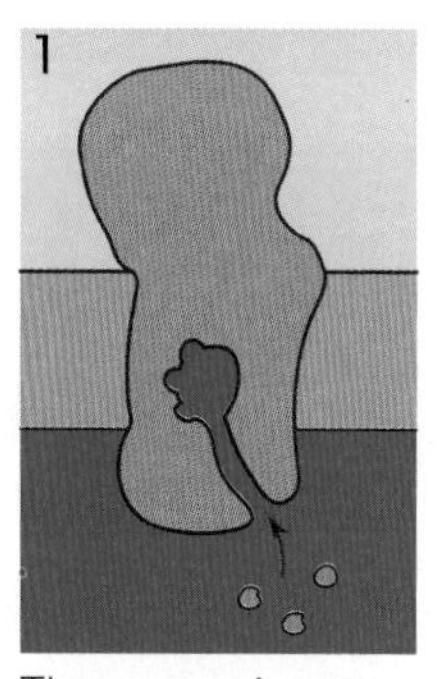

The pump is open to the inside, so three Na^+ ions can enter and attach to their binding sites, reducing the Na^+ concentration inside

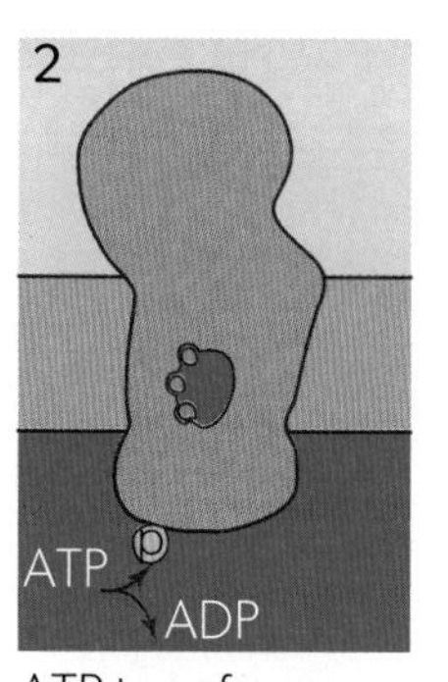

ATP transfers a phosphate group to the pump which causes a conformational change and closes the pump

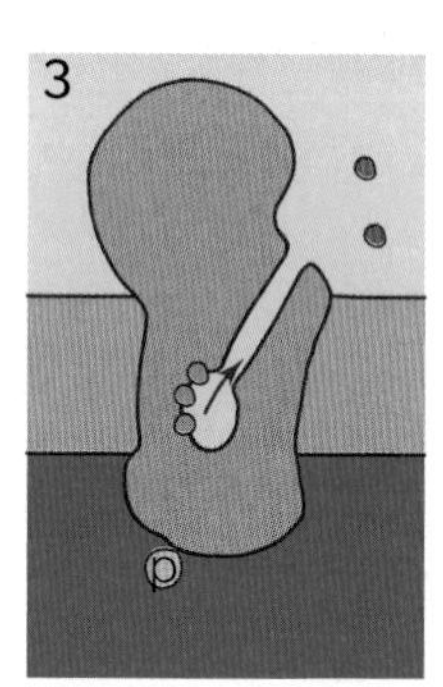

The pump opens to the outside and the Na^+ ions can exit, increasing the Na^+ concentration outside the neuron

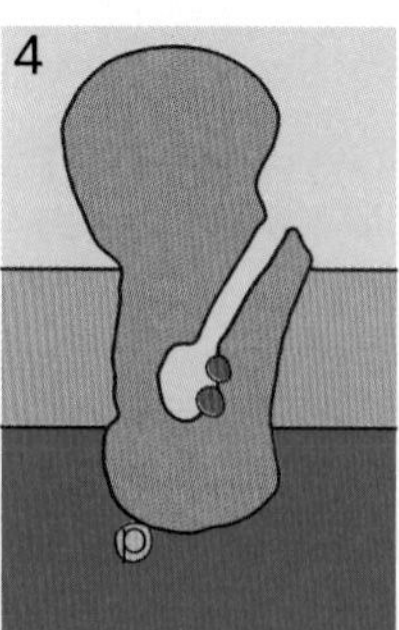

Two K^+ ions from outside enter and attach to their binding sites in the pump, reducing the K^+ concentration outside

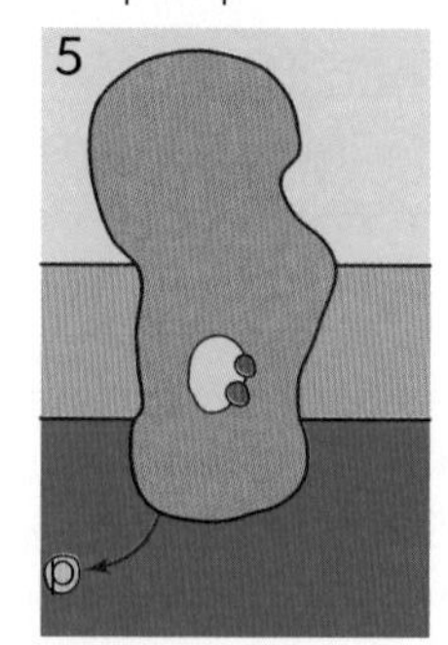

Binding of K^+ causes release of the phosphate group, which causes a conformational change and closes the pump

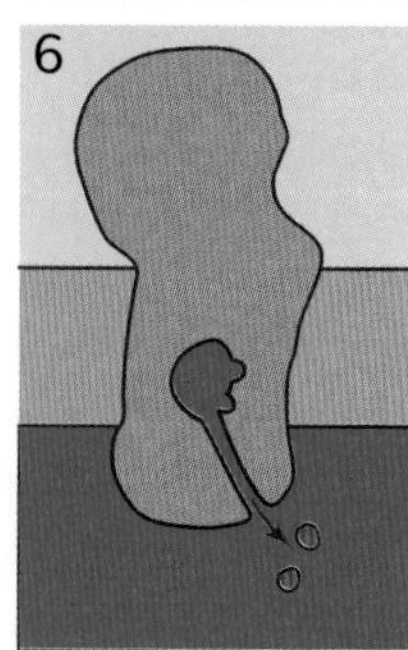

The pump opens to the inside and the K^+ ions can exit, increasing the K^+ concentration inside; more Na^+ ions can then enter

▶ Figure 25 The sodium–potassium pump

AHL

B2.1.16 Sodium-dependent glucose cotransporters as an example of indirect active transport

Sodium–glucose cotransporter proteins transfer a sodium ion and a glucose molecule together across a plasma membrane into a cell. The glucose molecule can move against its concentration gradient because the sodium ion is moving down its concentration gradient. The energy released by the movement of the sodium ion is greater than the energy needed to move the glucose.

Sodium-dependent cotransport is used by cells in the wall of the proximal tubule in the kidney. These cells reabsorb glucose that has been filtered out the blood to prevent it being lost in urine.

Glucose absorption into cells depends on the Na^+ ion concentration being greater outside than inside. The concentration gradient is maintained by active transport of Na^+ ions out of the cell. Sodium–potassium pumps in the plasma membrane on the inner (basal) side of the cell transfer Na^+ out of the cell towards nearby blood capillaries and transfer K^+ in. Sodium-dependent glucose cotransport depends on energy from ATP, so it is not passive. However, it is not typical active transport because the energy is not used directly by the cotransporter. This is called indirect or secondary active transport.

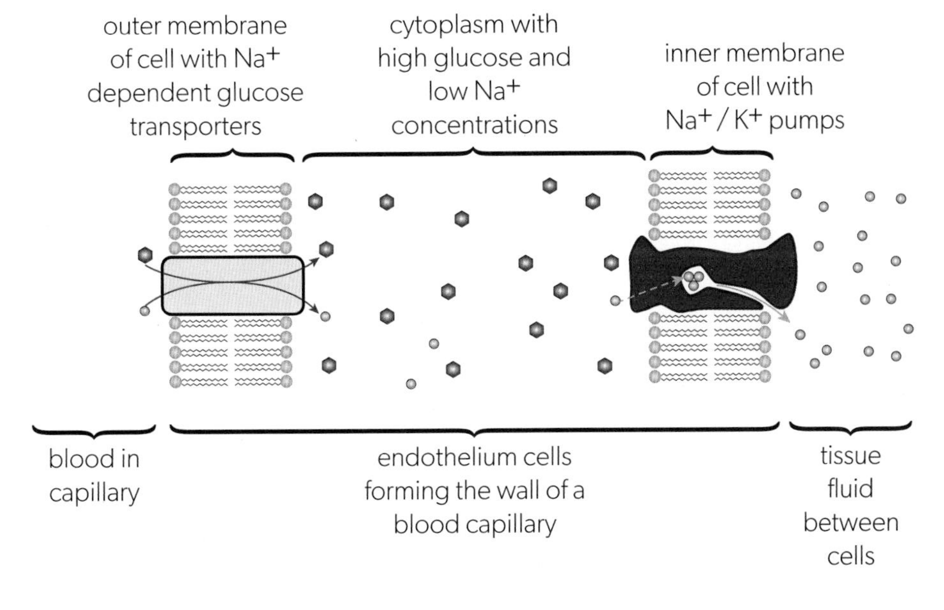

◀ Figure 26 Glucose uptake into a cell by cotransport

B2.1.17 Adhesion of cells to form tissues

Cells in a tissue are linked by cell-to-cell junctions. These junctions depend on cell-adhesion molecules (CAMs) in the plasma membranes of adjacent cells. A range of CAMs is found in different types of cell junction. CAMs are typically proteins with some domains embedded in the phospholipid bilayer and others protruding outwards into the extracellular environment. A junction is formed by the CAMs in adjacent cells binding together their extracellular domains. In some cases, the same type of CAM is present in both cells and these bind together to build a group of cells of the same type. In other cases, the CAMs are different and an asymmetrical junction is formed. This is useful in linking different cell types to form a more complex structure.

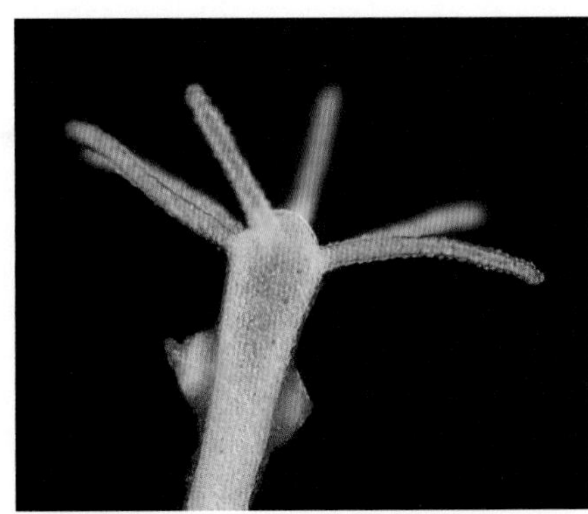

▲ Figure 27 If Hydra is broken up into single cells, the cells reaggregate into tissues by cell-to-cell adhesion, with the tissues arranging to form a new polyp

AHL

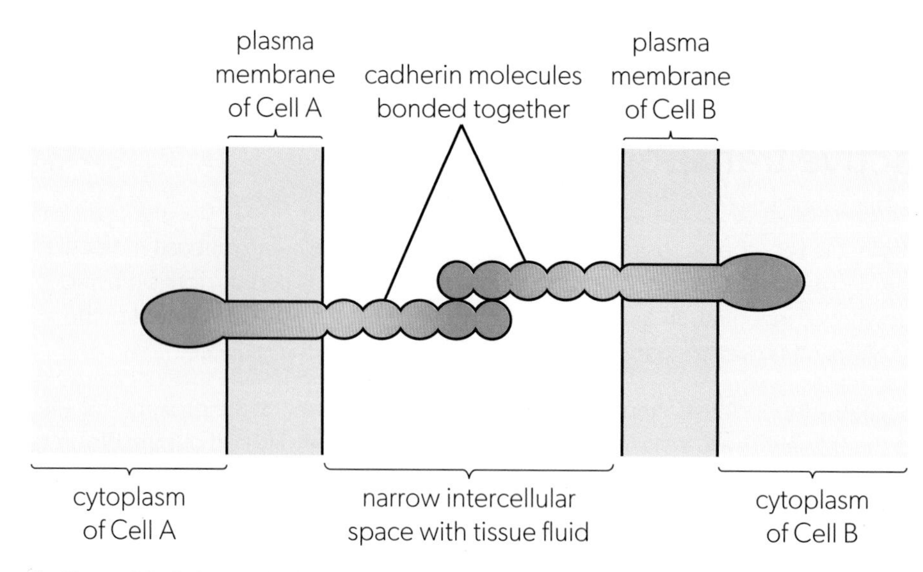

▲ Figure 28 Cell-to-cell adhesion

Cell adhesion maintains the architecture of tissues and organs. It is needed for functional relationships between adjacent cells. Some types of junction prevent extracellular movement of substances in a tissue and other types facilitate it. Cell adhesion has major roles in the immune system. In tumours, it prevents cells from becoming separated and migrating to form secondary tumours, so it prevents malignancy spreading (metastasis).

Linking questions

1. What processes depend on active transport in biological systems?
   a. Outline the role of active transport in the generation of root pressure in plants. (B3.2.16)
   b. Explain the role of auxin efflux carriers in maintaining concentration gradients of phytohormones. (C3.1.20)
   c. Describe how active transport plays a role in osmoregulation by the loop of Henle. (D3.3.9)
2. What are the roles of cell membranes in the interaction of a cell with its environment?
   a. Describe the role of the cell surface activation of B-lymphocytes by helper T-lymphocytes. (C3.2.8)
   b. Outline one example of the role of glycolipids in cell-to-cell recognition. (B1.1.7)
   c. Explain the process of tyrosine kinase activation. (C2.1.11)

B2.2 Organelles and compartmentalization

How are organelles in cells adapted to their functions?

A shoemaker who works by hand has many different tools. Each tool has a structure that makes it well suited to carry out a specific task such as cutting leather, making holes, or stitching the shoe together. What is the advantage of specialization in tools? Each organelle in a eukaryotic cell is specialized for a particular function and is adapted to the function by its structure. In what ways is the tool kit of a shoemaker similar and in what ways is it different from the organelles within a cell? How do tools evolve? Is the evolution of tools similar to the evolution of cells?

▲ Figure 1 Shoemaker's tools

What are the advantages of compartmentalization in cells?

Compartmentalization must have significant advantages to justify the large quantities of energy needed to construct the 100,000 compartments in a bee colony. What advantages does compartmentalization provide to the bee colony? What are the advantages of storing faeces, pollen, honey and larvae in different compartments? Compartmentalization also has great benefits in eukaryotic cells. What are some examples of compartmentalized functions in eukaryotic cells? What is the difference between the comb constructed by honeybees and the compartments of eukaryotic cells?

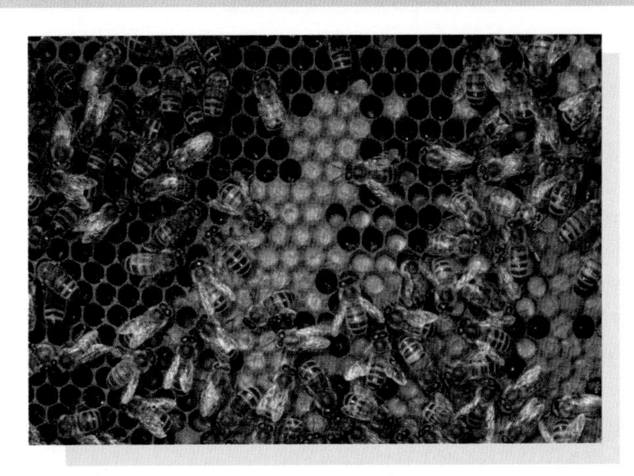

▲ Figure 2 Honeybees construct wax comb with hexagonal compartments that are used for storage of pollen or honey and protection for worker larvae and pupae

SL and HL	AHL only
B2.2.1 Organelles as discrete subunits of cells that are adapted to perform specific functions B2.2.2 Advantage of the separation of the nucleus and cytoplasm into separate compartments B2.2.3 Advantages of compartmentalization in the cytoplasm of cells	B2.2.4 Adaptations of the mitochondrion for production of ATP by aerobic cell respiration B2.2.5 Adaptations of the chloroplast for photosynthesis B2.2.6 Functional benefits of the double membrane of the nucleus B2.2.7 Structure and function of free ribosomes and of the rough endoplasmic reticulum B2.2.8 Structure and function of the Golgi apparatus B2.2.9 Structure and function of vesicles in cells

B2.2.1 Organelles as discrete subunits of cells that are adapted to perform specific functions

Organelles are discrete structures in cells that are adapted to perform one or more vital functions. Organelles are efficient because they are specialized for a limited range of functions. Eukaryotic cells have many organelles. Their structure and their appearance in electron micrographs are described in *Section A2.2.10*. In some cases a single membrane or double membrane encloses the fluid contents of an organelle. In other cases the organelle is a solid structure, largely composed of proteins or RNA, and it is not membrane-bound (Table 1).

No membrane	Single membrane	Double membrane
Ribosomes Centrioles Microtubules Proteasomes Nucleoli	Vesicles and vacuoles Rough endoplasmic reticulum Smooth endoplasmic reticulum Golgi apparatus Lysosomes	Nuclei Mitochondria Chloroplasts Amyloplasts Chromoplasts

▲ Table 1 Examples of organelles in eukaryotic cells

▲ Figure 3 Cells in ripe tomato fruits contain chromoplasts that give the fruit a bright red colour. Chromoplasts are double membraned and contain DNA. They develop from chloroplasts when the seeds in the tomato fruit are mature and ready to be dispersed. Chlorophyll is replaced by red pigments. Red coloration attracts animals that feed on the fruits and disperse the seeds

Nuclei, vesicles, ribosomes and the plasma membrane are all organelles. Some other structures are not considered to be organelles:

- cell walls are outside the plasma membrane so are extracellular structures rather than organelles
- cytoskeletons consist of narrow protein filaments spread through much of the cell so are not discrete enough to be an organelle
- cytoplasm is not a discrete structure as it includes many different structures and performs many functions.

Prokaryotic cells have fewer organelles than eukaryotes. This could be because their cells are smaller or because they concentrate on a more limited range of functions. It may also allow functions to be integrated and therefore carried out more rapidly—for example, transcription and translation.

ATL Thinking skills: Applying criteria for judgement

Do you think these organelles are part of the cytoplasm or not?

a. vacuoles in plant cells

b. mitochondria and chloroplasts

During cell division, mitosis refers to the division of the nucleus and cytokinesis refers to the division of the cytoplasm. Does the duplication of mitochondria and chloroplasts constitute a third process? What happens to plant vacuoles during cytokinesis?

One definition of the cytoplasm is everything enclosed by the plasma membrane excluding the nucleus. In what ways is this definition useful? How does it affect your answer to (a) and (b)?

Another definition of the cytoplasm is the thick solution that fills each cell and is enclosed by the cell membrane. The organelles are embedded in the cytoplasm. In what ways is this definition useful? How does this definition affect your answer to (a) and (b)?

Discuss the statement: "The criteria for judgement we use affect the truth of the knowledge claims we make."

Experimental techniques: Differential centrifugation

Separating cells into types of organelle is called cell fractionation. The first stage is to mix the cells with ice-cold extraction buffer. The cold temperature slows down degeneration and the buffer prevents problems caused by pH differences and osmosis. The mixture is then gently blitzed in a scientific version of a food blender to burst open the cells and release the organelles. The resulting homogenate is filtered to remove whole cells and other structures larger than organelles. It is then centrifuged and because organelles are denser than the extraction buffer, they sediment to the bottom of the centrifuge tube to form a "pellet". The remaining liquid, called the supernatant, is discarded. The pellet is mixed with another solution to resuspend the organelles.

This new mixture is then centrifuged again, with the speed and duration carefully chosen so that the required organelles separate from everything else. This is called differential centrifugation. Larger organelles sink to the bottom of the tube at a faster rate and at lower centrifugation speeds than smaller organelles.

The density of the liquid can be varied to separate organelles of different density. For example, to separate chromoplasts from other organelles in tomato cells, three layers of 0.5 mol dm^{-3}, 0.9 mol dm^{-3} and 1.45 mol dm^{-3} sucrose solution are placed in the centrifuge tube. Centrifugation at high speed (62,000 g) causes the chromoplasts to become concentrated between the 0.9 mol dm^{-3} and 1.45 mol dm^{-3} sucrose layers. Only when such protocols are developed can there be rapid progress in determining and investigating the functions of individual organelles. This is an example of progress in science following development of new techniques.

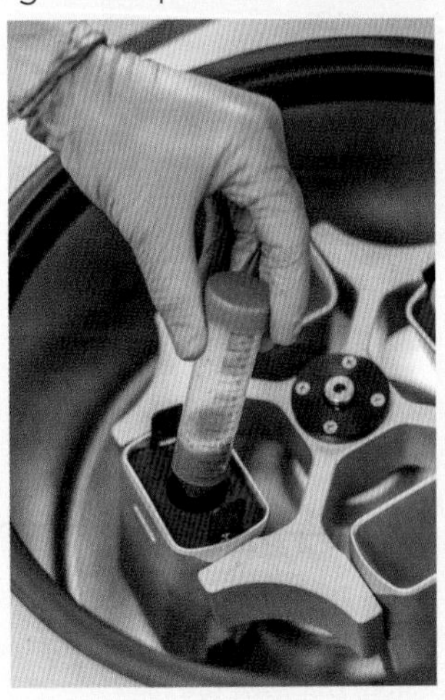

▲ Figure 4 Ultracentrifuge tubes are spun at high revolutions per minute by a rotor

B2.2.2 Advantage of the separation of the nucleus and cytoplasm into separate compartments

In eukaryotes, keeping chromosomes inside the nucleus safeguards the DNA. Eukaryotes gain another advantage in having the nucleus and cytoplasm as separate compartments. In prokaryotic cells, there is no nucleus, so DNA and ribosomes are together in the cytoplasm and translation can happen immediately after transcription. In eukaryotic cells, translation cannot begin until messenger RNA (mRNA) has passed out of the nucleus via the pores in the nuclear membrane. This allows mRNA to be modified after it has been produced by transcription in the nucleus, but before it is translated. The process is called post-transcriptional modification and is described in more detail (for HL only) in *Section D1.2.15*, and both transcription and translation are described (for SL and HL) in *Topic D1.2*.

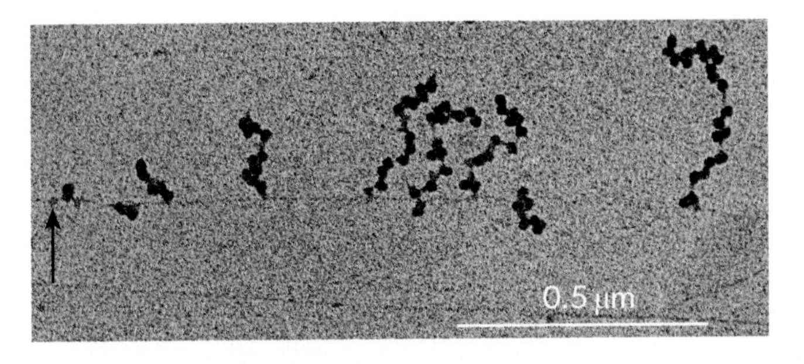

◀ Figure 5 In this electron micrograph of part of a prokaryotic cell, the arrow points to an RNA polymerase attached to a strand of DNA at the point where transcription of a gene is initiated. The black spherical structures are ribosomes. They are translating the mRNA produced by a series of RNA polymerases that are moving to the right along the DNA. This image shows that translation can begin before transcription of a gene has been completed in prokaryotes

B2.2.3 Advantages of compartmentalization in the cytoplasm of cells

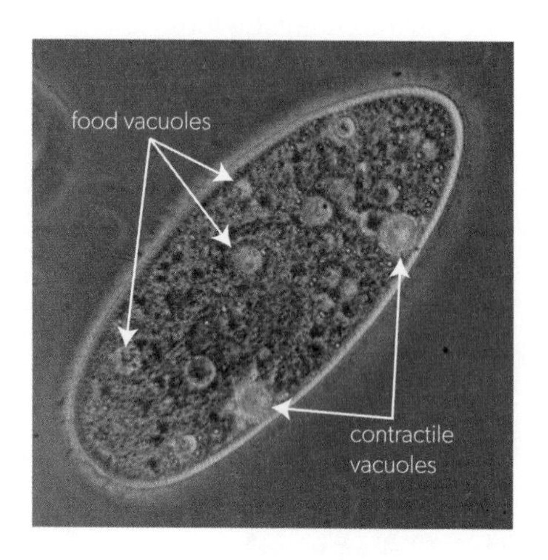

▲ Figure 6 *Paramecium* is a unicellular eukaryote that feeds by endocytosis, forming vacuoles in which ingested food is digested. After formation of a vacuole, the pH inside drops below pH2 and later rises above pH7, to provide optimum conditions for the series of enzymes that digest the food. This is only possible because the vacuole is a separate compartment. Can you suggest two other advantages of food vacuoles as separate compartments?

The cytoplasm of eukaryotic cells is divided into compartments by membrane-bound organelles. There are several advantages of being compartmentalized.

- Enzymes and substrates for a particular process can be much more concentrated than if they were spread throughout the cytoplasm.
- Substances that could cause damage to the cell can be kept inside the membrane of an organelle. For example, the digestive enzymes of a lysosome could digest and kill a cell, if they were not safely stored inside the lysosome membrane.
- Conditions such as pH can be maintained at an ideal level for a particular process, which may be different from the levels needed for other processes in a cell.
- Organelles with their contents can be moved around within the cell.
- There is a larger area of membrane available for processes that happen within or across membranes.

Activity: Garlic cells and compartmentalization

Garlic cells store a harmless sulfur-containing compound called alliin in their vacuoles. They store an enzyme called alliinase in other parts of the cell. Alliinase converts alliin into a compound called allicin, which has a very strong smell and flavour and is toxic to some herbivores. This reaction occurs when herbivores bite into garlic and damage cells, mixing the enzyme and its substrate. Many humans like the flavour but to get it, garlic must be crushed or cut, not used whole. You can test this by smelling a whole garlic bulb, then cutting or crushing it and smelling it again.

B2.2.4 Adaptations of the mitochondrion for production of ATP by aerobic cell respiration

Mitochondria produce ATP by aerobic cell respiration. They are adapted to this function by their structure.

- The outer membrane separates the contents of the mitochondrion from the rest of the cell, creating a compartment specialized for the biochemical reactions of aerobic respiration.
- The inner mitochondrial membrane is the site of oxidative phosphorylation. It contains electron transport chains and ATP synthase which together generate a proton gradient and use it to produce ATP. Cristae are projections of the inner membrane that increase the surface area available for oxidative phosphorylation.
- The intermembrane space between the inner and outer membranes is where a high concentration of protons is generated by the electron transport chains. The volume of this space is very small, so a concentration gradient across the inner membrane builds up rapidly.
- The matrix is the fluid filling the compartment inside the inner mitochondrial membrane. It contains all the enzymes and substrates for the Krebs cycle and the link reaction. By concentrating enzymes and substrates in the small volume of the matrix, the reactions of these two parts of aerobic respiration can be performed more rapidly than if they were dispersed in the cytoplasm.

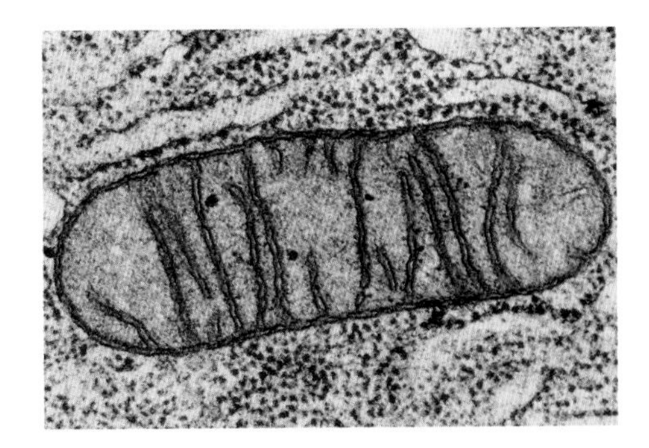

outer mitochondrial membrane separates the contents of the mitochondrion from the rest of the cell, creating a cellular compartment with ideal conditions for aerobic respiration

matrix contains enzymes of the Krebs cycle and link reaction

inner mitochondrial membrane contains electron transport chains and ATP synthase

intermembrane space into which protons are pumped by the electron transport chain, with a rapid concentration buildup due to the small volume

cristae are projections of the inner membrane which increase the surface area available for oxidative phosphorylation

ribosomes and DNA for expression of mitochondrial genes

▲ Figure 7 Electron micrograph of mitochondrion with annotated diagram of its structures and their functions

Data-based questions: Structure and function in mitochondria

Study the electron micrographs in Figure 8 and then answer the questions.

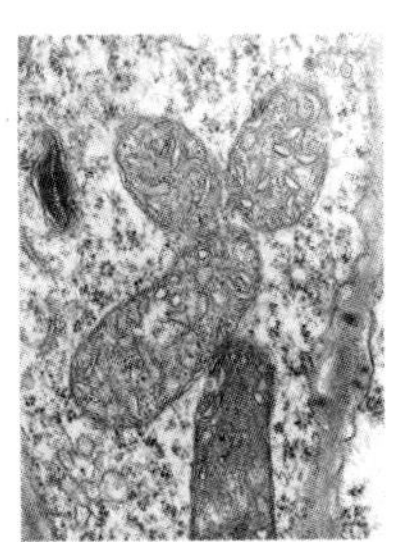

a)

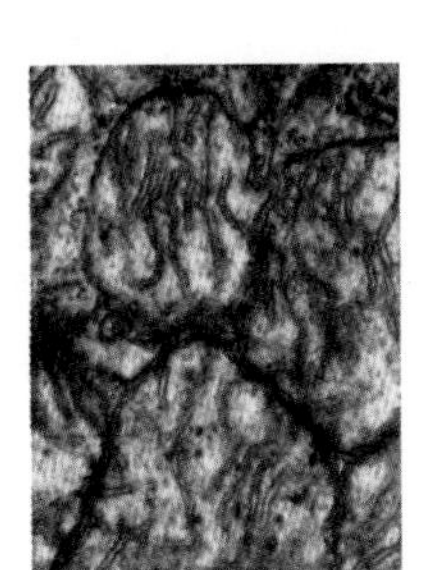

b)

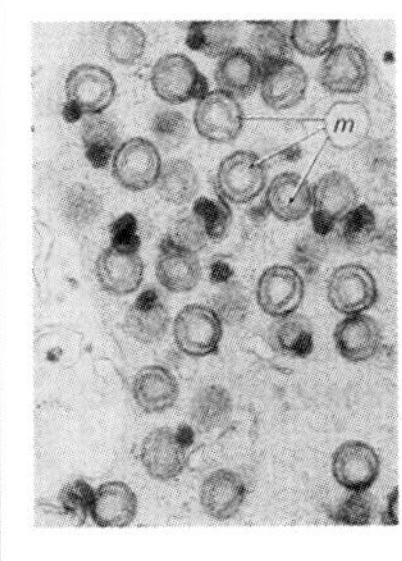

c)

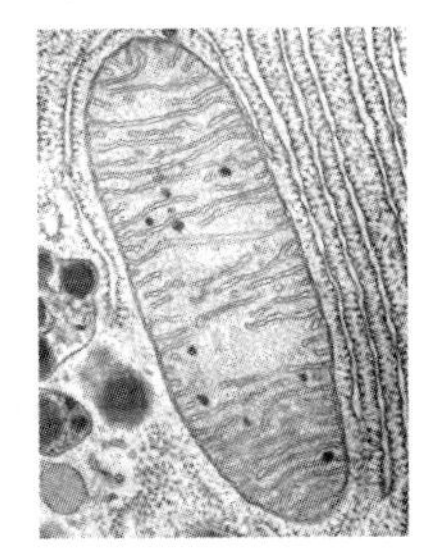

d)

▲ **Figure 8** Electron micrographs of mitochondria: (a) from a bean plant, (b) from cardiac muscle, (c) from axolotl sperm, (d) from bat pancreas

1. The fluid-filled centre of the mitochondrion is called the matrix. What separates the matrix from the cytoplasm around the mitochondrion? [1]

   A One wall
   B One membrane
   C Two membranes
   D One wall and one membrane

2. The mitochondrial matrix contains 70S ribosomes, whereas the cytoplasm of eukaryotic cells contains 80S ribosomes. Which of these hypotheses is consistent with this observation? [1]

   i. Protein is synthesized in the mitochondrion.
   ii. Ribosomes in mitochondria have evolved from ribosomes in bacteria.
   iii. Ribosomes are produced by aerobic cell respiration.

   A (i) only
   B (ii) only
   C (i) and (ii)
   D (i), (ii) and (iii)

3. Discuss the claim that the mitochondria in Figure 8b and Figure 8c are spherical. [2]

4. Predict, with reasons, which of the four types of mitochondria produces most ATP per unit time. [3]

5. Identify these other structures in the micrographs:
   a. to the right of the mitochondria in Figure 8a [1]
   b. to the right of the mitochondrion in Figure 8d. [1]

AHL

B2.2.5 Adaptations of the chloroplast for photosynthesis

Chloroplasts are quite variable in structure but share certain features:

- a double membrane forming the outer chloroplast envelope
- an extensive system of internal membranes called thylakoids, which are an intense green colour due to chlorophyll
- small fluid-filled spaces inside the thylakoids
- a colourless fluid around the thylakoids called stroma that contains many different enzymes.

In most chloroplasts there are stacks of thylakoids, called grana. If a chloroplast has been photosynthesizing rapidly then there may be starch grains or lipid droplets in the stroma.

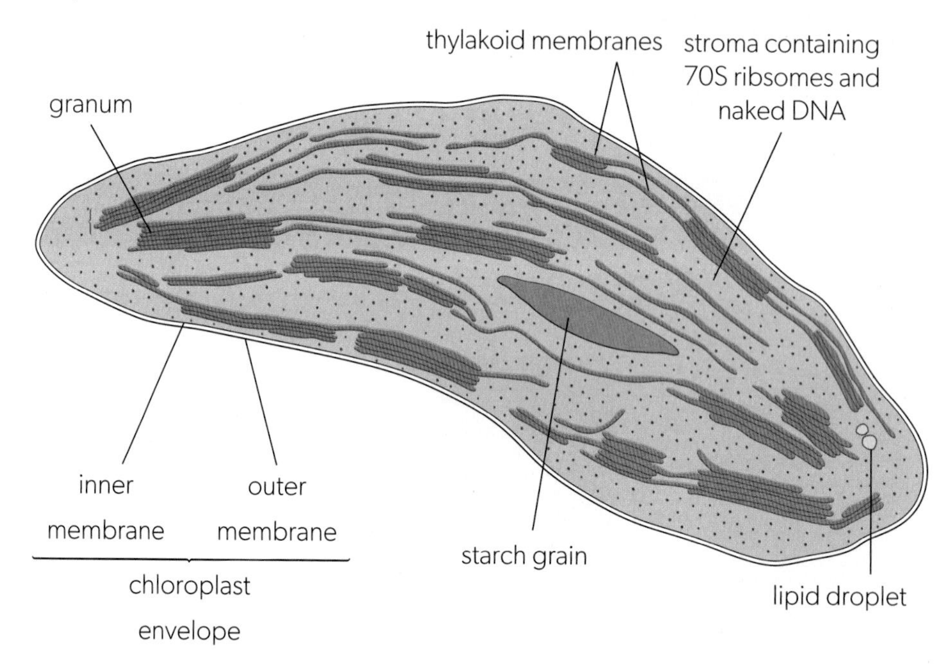

▲ Figure 9 Chloroplast structure

There is a clear relationship between the function of chloroplasts (described in *Topic C1.3*) and their structure.

- **Chloroplasts absorb light.** Pigment molecules, arranged in photosystems in the thylakoid membranes, carry out light absorption. The large area of thylakoid membranes ensures that the chloroplast has a large light-absorbing capacity. The thylakoids are often arranged in stacks called grana. Leaves that are brightly illuminated typically have chloroplasts with deep grana composed of many thylakoids, which allow more light to be absorbed.
- **Chloroplasts produce ATP by photophosphorylation.** A proton gradient is needed. This develops between the inside and outside of the thylakoids. The volume of fluid inside the thylakoids is very small, so when protons are pumped in, a proton gradient develops after relatively few photons of light have been absorbed. This allows ATP synthesis to begin.

- **Chloroplasts carry out the many chemical reactions of the Calvin cycle.** The stroma is a compartment of the plant cell in which the enzymes needed for the Calvin cycle are kept together with their substrates and products. This concentration of enzymes and substrates speeds up the whole Calvin cycle. ATP and reduced NADP are needed for the Calvin cycle and are easily available because the thylakoids, where they are produced, are distributed throughout the stroma.

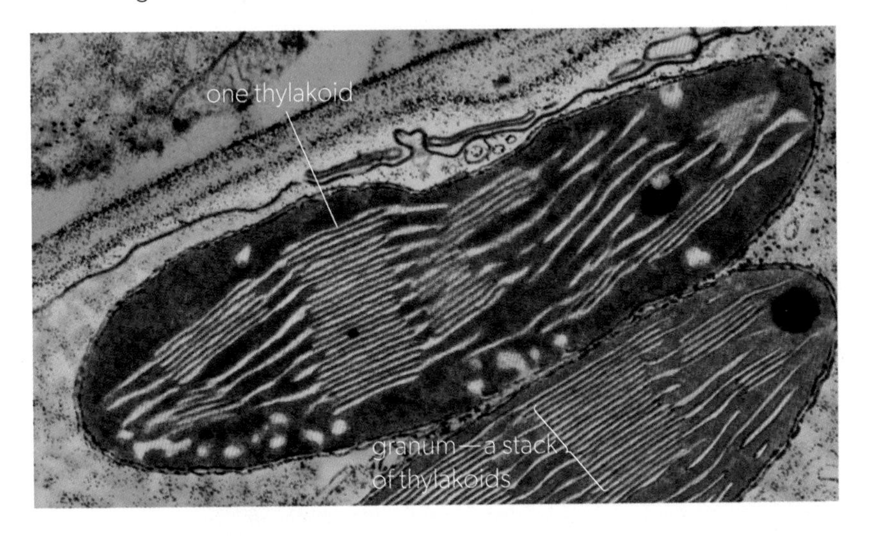

▲ Figure 10 Electron micrograph of pea chloroplast

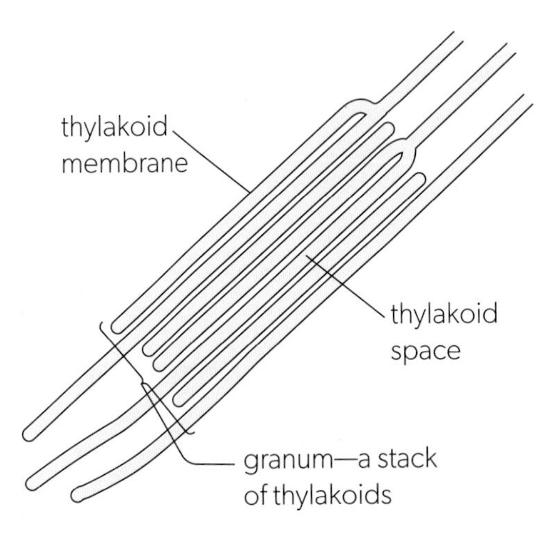

▲ Figure 11 Drawing of part of the pea chloroplast to show the arrangement of the thylakoid membranes

B2.2.6 Functional benefits of the double membrane of the nucleus

A general principle with phospholipid bilayers is that the hydrophobic core is never exposed to water. If a membrane became perforated by phospholipid molecules moving apart, they would be drawn back together rapidly, closing the perforation. Areas of membrane in water naturally adopt shapes such as spheres or flattered cisternae that avoid edges where the hydrophobic core is exposed. Pores are formed using integral proteins, allowing specific molecules to pass through, but larger holes through single membranes only occur when there has been catastrophic damage to a cell, for example when red blood cells burst after being bathed in pure water.

Proteins synthesized by ribosomes in the cytoplasm are needed in the nucleus to form part of the structure of chromosomes. They also regulate gene expression by promoting or repressing gene transcription. These proteins must be able to enter from the cytoplasm. Messenger RNA, transfer RNA (tRNA) and ribosomes produced in the nucleus are exported to the cytoplasm. The RNA molecules are large, and ribosomes are even larger because they are assemblages of ribosomal RNAs (rRNAs) and proteins. This means there is a need for unusually large pores through the nuclear membrane—larger than the pores through channel proteins in membranes. A double membrane is used to make a larger pore, with the inner and outer membrane connected to form a circular hole. The rims of these nuclear pores are lined with proteins that can control whether or not a protein passes through.

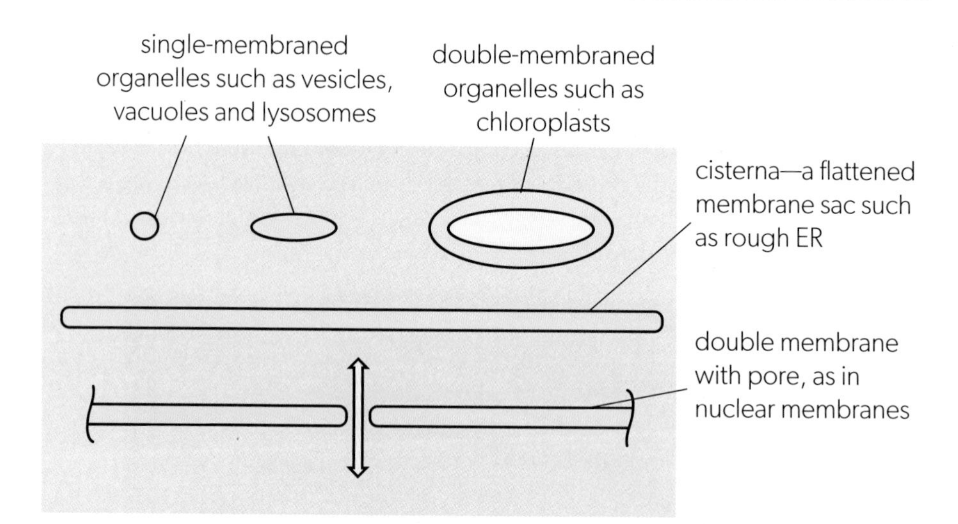

▲ **Figure 12** Membranes can form a variety of shapes, but never begin or end

The double nuclear membrane has another functional benefit. During both mitosis and meiosis, the nuclear membrane breaks down to allow the chromosomes to be moved to the poles of the cell. Nuclear membranes then reform around the new groupings of chromosomes. This can easily be achieved with a double membrane. Vesicles bud off, progressively breaking the whole nuclear membrane up into vesicles, which are moved to the sides of the cell. Later, these vesicles can be used to make new nuclear membranes by fusing together.

▲ **Figure 13** Freeze-etched electron micrograph of the double nuclear membranes, with nuclear pores visible and vesicles in the surrounding cytoplasm

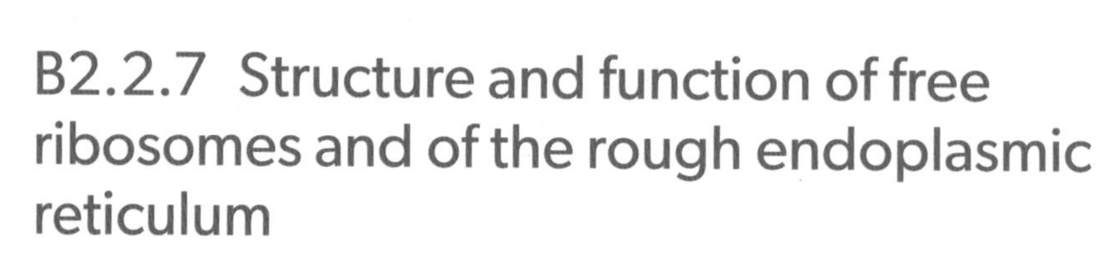

B2.2.7 Structure and function of free ribosomes and of the rough endoplasmic reticulum

Ribosomes are large assemblages of rRNA and proteins. Eukaryote ribosomes have a diameter of nearly 30 nanometres. There are two subunits, one large and one small. The small subunit has a binding site for mRNA. The large subunit has three binding sites for tRNA molecules, an area that catalyses the formation of peptide bonds and an exit tunnel for the synthesized polypeptide. Protein synthesis by ribosomes is described in *Topic D1.2*.

Ribosomes that are not attached to membranes in the cytoplasm are known as free ribosomes. Polypeptides synthesized by them are released into the cytoplasm and either remain there or enter the nucleus. The cytoplasm of a typical cell contains a wide range of proteins, some carrying out housekeeping roles, for example enzymes that catalyse glycolysis, and others performing the specialized functions of the cell.

If a ribosome synthesizes a polypeptide that must be transported to a specific location, the ribosome becomes attached to the rough endoplasmic reticulum. This organelle consists of cisternae, which are flattened sacs bounded by a single membrane. The polypeptide passes into the lumen of the rER and is then transported elsewhere in the cell by a vesicle that buds off from the rER. The usual initial destination for these polypeptides is the Golgi apparatus, with many of them ultimately secreted from the cell.

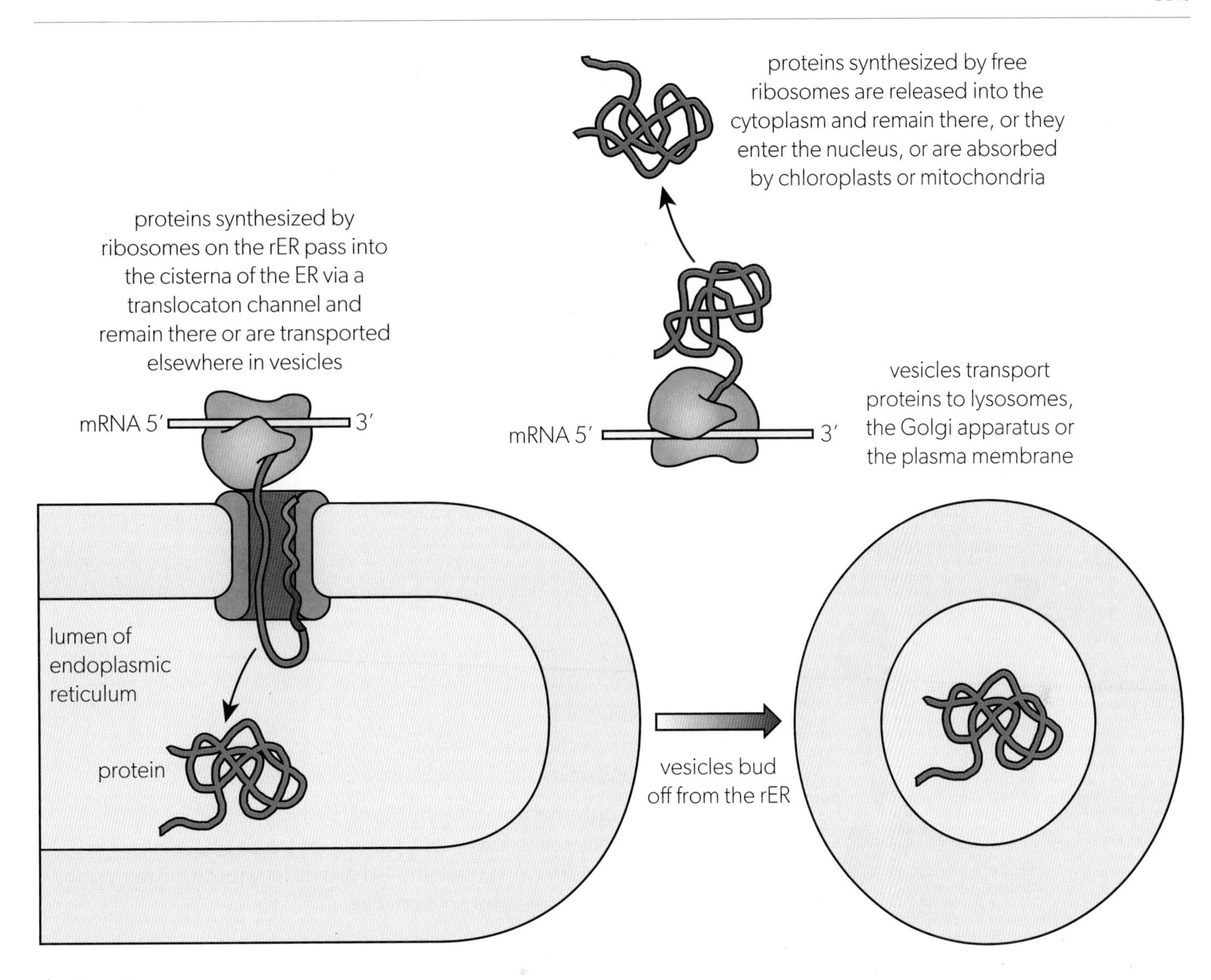
proteins synthesized by free ribosomes are released into the cytoplasm and remain there, or they enter the nucleus, or are absorbed by chloroplasts or mitochondria
proteins synthesized by ribosomes on the rER pass into the cisterna of the ER via a translocaton channel and remain there or are transported elsewhere in vesicles
mRNA 5′
3′
mRNA 5′
3′
vesicles transport proteins to lysosomes, the Golgi apparatus or the plasma membrane
lumen of endoplasmic reticulum
protein
vesicles bud off from the rER

▲ Figure 14

Applying technology

The protein data bank (PDB) is a public database containing data regarding the three-dimensional structure for many biological molecules. In 2000, structural biologists Venkatraman Ramakrishnan, Thomas A. Steitz and Ada E. Yonath made the first data about ribosome subunits available through the PDB. In 2009, they received a Nobel Prize for their work on the structure of ribosomes. Visit the RCSB protein data bank to obtain images of the *Thermus thermophilus* ribosome. Use the search function to locate image 1jgo. Using the structure view, rotate the image to visualize the small subunit and the large subunit, the associated tRNA molecules and mRNA molecules.

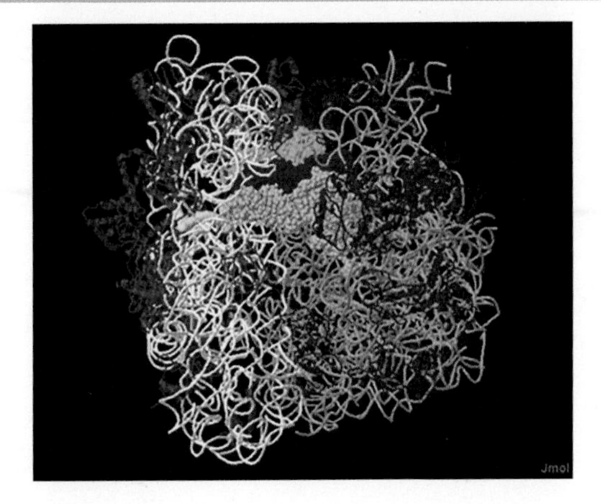
Jmol

▲ **Figure 15** Molecular visualization of a functioning ribosome. Key: mRNA yellow; tRNAs pink, purple and blue; rRNA white in the small subunit and grey in the large subunit; proteins violet in the small subunit and red in the large subunit

AHL

B2.2.8 Structure and function of the Golgi apparatus

The Golgi apparatus is a stack of flattened sacs (cisternae) in which polypeptides made by the rough endoplasmic reticulum are processed. The polypeptides are transported from the rER in vesicles. Enzymes inside the cisternae can change the polypeptide in numerous ways—for example, by adding carbohydrate to make glycoprotein or by adding phosphate or sulfate groups. The quaternary structure of proteins can be established by assembling polypeptides and other subunits.

When the processing of a protein is completed, it is transported from the Golgi apparatus to its destination in a vesicle. This may be a lysosome or a food vacuole formed by endocytosis. For proteins that are being secreted, the destination is the plasma membrane of the cell. For example, digestive enzymes are secreted by pancreas cells when vesicles move from to the plasma membrane and fuse with it to release the enzymes from the cell.

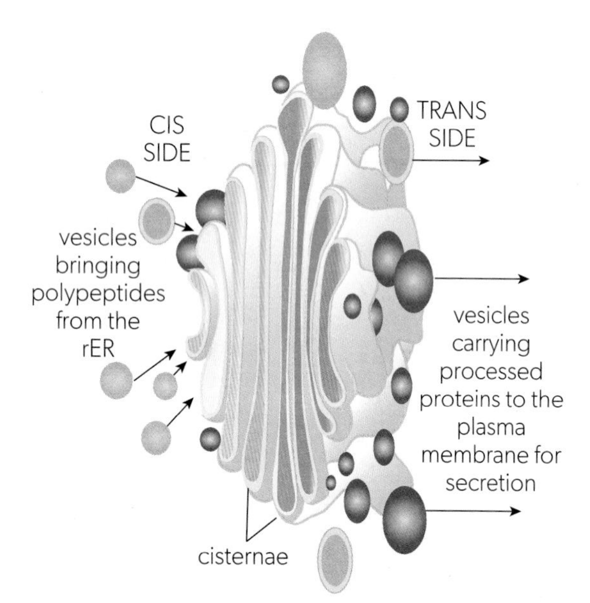

▲ Figure 16 According to the cisternal maturation model, cisternae are formed on the cis side from coalescing vesicles and the move through the stack. When they reach the trans side they fragment into vesicles

Processing proteins in the Golgi apparatus is sequential, with each protein gradually moving through the cisternae from the side nearest the rER (the cis side) to the opposite side (the trans side). Two models have been proposed to explain how proteins could move through the Golgi apparatus.

- **The vesicle transport model** in which the cisternae do not move and vesicles transfer proteins between them.
- **The cisternal maturation model** in which vesicles from the rER coalesce to form new cisternae on the cis side, which then gradually move through the Golgi until they reach the trans side, where they break up into vesicles.

Currently, evidence for the cisternal maturation model is stronger, but many questions remain about the functioning of the Golgi apparatus, including the reason for the cisternae needing to be kept together in a stack.

B2.2.9 Structure and function of vesicles in cells

Vesicles are rounded sacs made of a single layer of membrane, and the material inside it. They are typically small and dynamic structures that are continuously made, moved and merged within cells. They are made by pinching off a small area of membrane from a larger area. This happens in endocytosis in order to take in a small droplet of fluid from outside the cell. The protein clathrin helps with this process.

Clathrin is a three-legged protein (Figure 17) that becomes positioned on the inner face of the plasma membrane when a vesicle is being made. Adjacent clathrin molecules bind to each other to form a lattice of pentagons and/or hexagons. This process helps the plasma membrane to become indented and eventually to detach to form a sphere of membrane with a clathrin cage around it.

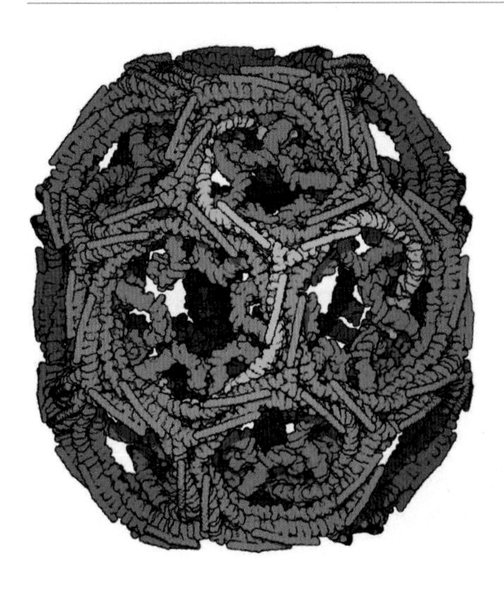

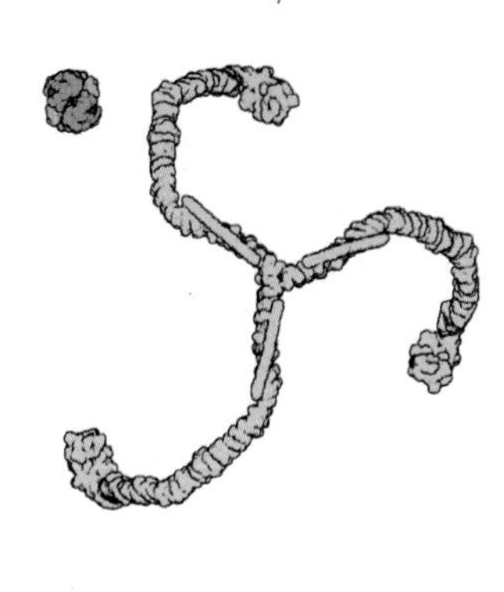

▲ Figure 17 Three-legged clathrin molecules bind to each other forming a cage to support a vesicle

Vesicles can be used to move materials around inside cells. There are two general reasons for this.

- In some cases, it is the contents of the vesicle that need to be moved. The transport of neurotransmitters to the presynaptic membrane of a neuron is an example of moving materials in vesicles.
- In other cases, it is the membrane of the vesicle or the proteins in the membrane that are the reason for vesicle movement. In a growing cell, the area of the plasma membrane needs to increase. Phospholipids are synthesized next to the rER and become inserted into the rER membrane. Ribosomes on the rER synthesize membrane proteins which also become inserted into the membrane. Vesicles bud off the rER and move to the plasma membrane. They fuse with it, each increasing the area of the plasma membrane by a very small amount. This method can also be used to increase the size of organelles in the cytoplasm such as lysosomes and mitochondria.

Linking questions

1. What are examples of structure–function correlations at each level of biological organization?
   a. Outline the relationship between the structure and function of glycoproteins and glycolipids. (B2.1.9)
   b. Explain the role of the Golgi apparatus in processing and secretion of protein. (B2.2.8)
   c. Draw the anatomy of the female reproductive system and annotate the structures with their functions. (D3.1.4)
2. What separation techniques are used by biologists?
   a. Outline the technique of gel electrophoresis. (D1.1.5)
   b. Discuss the relationship between the definition of organelles and the techniques of cellular fractionation and ultracentrifugation. (B2.2.1)
   c. Explain the separation and identification of photosynthetic pigments by chromatography. (C1.3.4)

B2.3 Cell specialization

What are the roles of stem cells in multicellular organisms?

Some lizards escape from a predator by shedding their tail. In the weeks that follow, a tail can regrow. When the gecko's tail regrows, it has skin, muscle and nerve fibres, but the lost vertebrae are not replaced. Why does this happen? If a person loses a finger in an accident, why does the finger not re-grow? An early embryo divides repeatedly. Up to the eight-cell stage, the cells can be separated and still give rise to an entire organism. Gradually they become more restricted in their potential to become all cell types. What role do stem cells play in these examples?

▲ Figure 1 Gecko that has lost its tail

How are differentiated cells adapted to their specialized functions?

Figure 2 shows the ultrastructure of a hepatocyte while Figure 3 shows the ultrastructure of several neurons. What is the identity of the red dots in the nerve cell? Can you deduce which is the pre-synaptic cell and which is the post-synaptic cell? What are the blue organelles? How does the structure of the neuron differ from the hepatocyte? How are the structures of the organelles related to their functions? What do neurons and hepatocytes have in common?

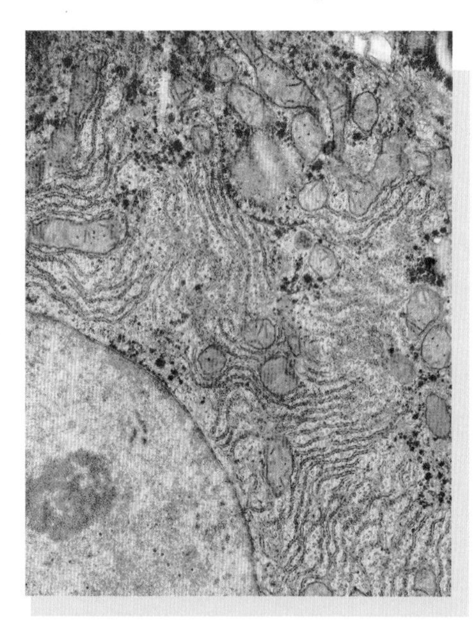

▲ Figure 2 Liver cells

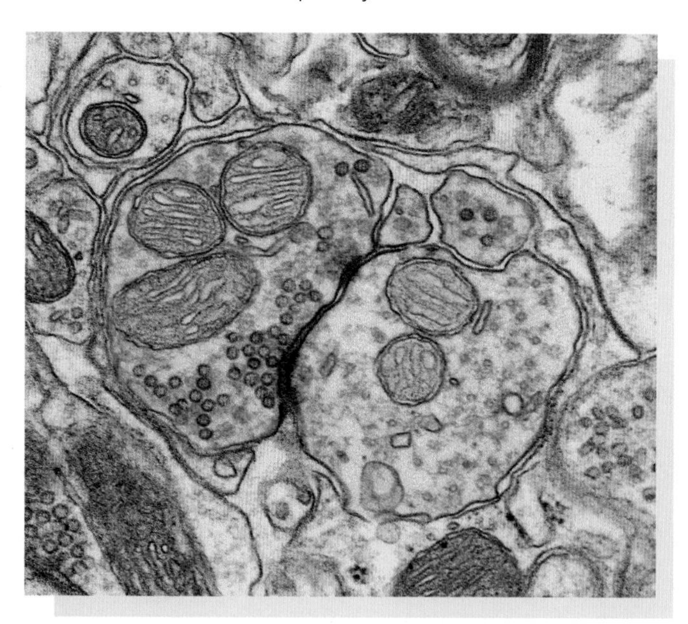

▲ Figure 3 Cells in the cerebellum of the brain

SL and HL	AHL only
B2.3.1 Production of unspecialized cells following fertilization and their development into specialized cells by differentiation B2.3.2 Properties of stem cells B2.3.3 Location and function of stem cell niches in adult humans B2.3.4 Differences between totipotent, pluripotent and multipotent stem cells B2.3.5 Cell size as an aspect of specialization B2.3.6 Surface area-to-volume ratios and constraints on cell size	B2.3.7 Adaptations to increase surface area-to-volume ratios of cells B2.3.8 Adaptations of type I and type II pneumocytes in alveoli B2.3.9 Adaptations of cardiac muscle cells and striated muscle fibres B2.3.10 Adaptations of sperm and egg cells

B2.3.1 Production of unspecialized cells following fertilization and their development into specialized cells by differentiation

Fertilization is the fusion of a male and female gamete to produce a single cell. In multicellular organisms, this cell divides repeatedly to generate an embryo of many cells. Mitosis ensures that the cells in an embryo are all genetically identical. They have all the genes in the organism's genome and could develop in any way.

In an early-stage embryo, the cells are unspecialized. As an embryo grows, its cells develop along different pathways and become specialized for specific functions. This allows each cell to carry out its function more efficiently than if it had multiple roles. The cell can develop the ideal structure, with the enzymes needed to carry out all the chemical reactions associated with its function. The development of cells in different ways to carry out specific functions is called differentiation. In humans, there are 220 distinctively different highly specialized cell types, all of which develop by differentiation.

When a gene is being used in a cell, we say that the gene is being expressed. In simple terms, the gene is switched on and the information in it is used to make a protein or other gene product. The development of a cell involves switching on and expressing particular genes but not others. Cell differentiation happens because a different sequence of genes is expressed in different cell types.

In a multicellular organism, there must be enough cells of each type and they must all be in the positions within the body where they are needed. The position of a cell in the embryo must therefore determine how it differentiates. Gradients of signalling chemicals indicate a cell's position in the embryo and determine which pathway of differentiation it follows. These chemicals are regulators of gene expression. For example, gradients of retinoic acid guide differentiation of cells in the development of forelimbs, pancreas, lungs, kidneys and other organs.

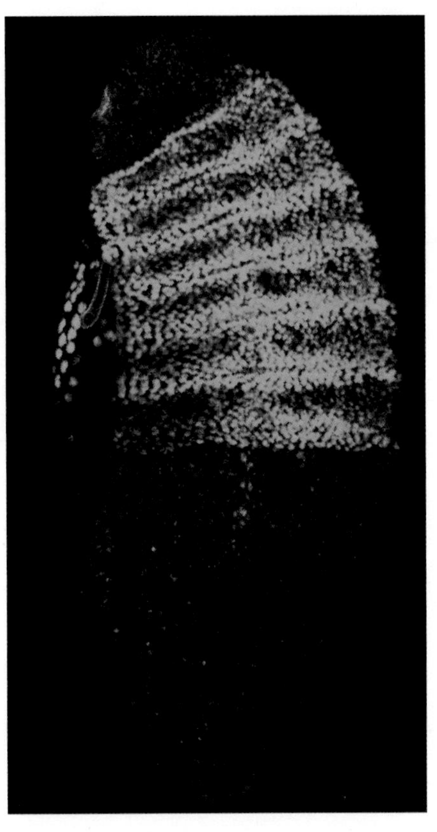

▲ Figure 4 In this *Drosophila* embryo, a fluorescent marker shows cells that are expressing the genes for the two proteins AbdA and Exd. These proteins form a complex that binds to DNA to regulate expression of specific genes. The marker shows that this is happening in some segments of the embryo but not others

B2.3.2 Properties of stem cells

In the 19th century, the term "stem cell" was given to the zygote and the cells of the early embryo because all the tissues of the adult stem from them. Stem cells have been intensively researched because of their role in development and because they have many potential therapeutic or regenerative uses.

A stem cell can divide repeatedly. In theory, there is no limit to the number of times it can split into two cells. Stem cells in the skin divide repeatedly throughout our lives, allowing us to replace lost skin cells. Stem cells in the testes also divide endlessly. This is the first stage of gamete production and allows males to produce vast numbers of sperm throughout adulthood.

Cells produced by division of a stem cell might remain as stem cells or differentiate into a specific cell type. If they differentiate, they are no longer stem cells. A stem cell is either undifferentiated or partially differentiated; they are always capable of differentiating along different pathways.

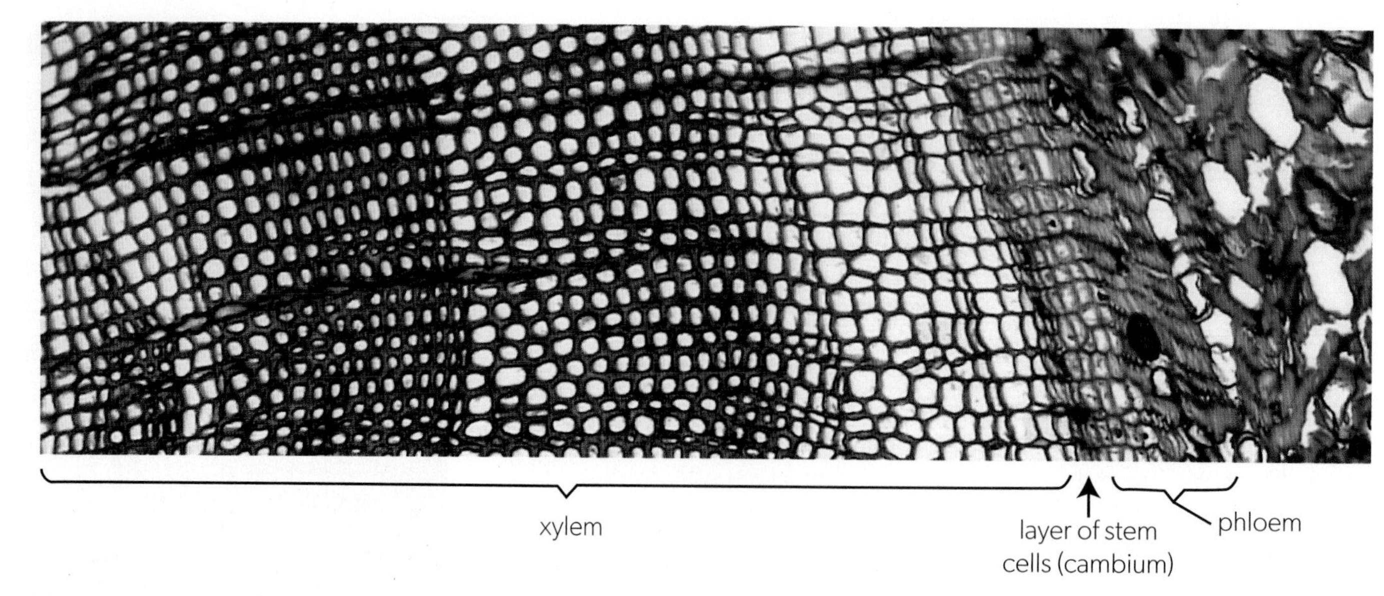

▲ Figure 5 Tree trunks have a cylinder of stem cells on the inner side of their bark. Unlike the layer of stem cells in human skin which produces cells on one side, the layer of stem cells in the trunk generates cells on both sides. On the inner side, the new cells differentiate into xylem; on the outer side, they become phloem. Xylem and phloem are both used for transport but they transport different substances by different methods. For this reason, they are dissimilar in structure. Xylem tissue is the wood that supports the tree. Because the stem cells in the bark divide endlessly, more xylem is added to the trunk every year and it continues to widen throughout the tree's life. That may be thousands of years in long-lived trees

B2.3.3 Location and function of stem cell niches in adult humans

Some stem cells remain in the adult body. They are present in many human tissues, including bone marrow, skin and liver. Stem cells give these tissues considerable powers of regeneration and repair. The precise location of stem cells within a tissue is called the stem cell niche. It must provide a microenvironment with conditions needed for the stem cells to remain inactive and undifferentiated over long periods of time, and also for them to proliferate rapidly and differentiate when required.

In striated (skeletal) muscle, there are stem cells that remain inactive unless there is muscle injury. Changes in the stem cell niche then cause these cells to proliferate and differentiate, to replace damaged muscle tissue. Striated muscle is highly regenerative after damage. Bone marrow and hair follicles are two examples of stem cell niches where the microenvironment promotes continuous stem cell proliferation and differentiation. This results in production of replacement blood cells and in hair growth.

Stem cell niches are of research interest because if they can be simulated outside the body for a particular stem cell type, it should be possible to generate human tissue in vitro (literally, in glass; that is, in the laboratory) and use it in restorative surgery. There are also non-therapeutic uses for stem cells if appropriate microenvironments can be created. One possibility is to use them to produce large quantities of striated muscle fibres (that is, meat) for human consumption. The beef burgers of the future may therefore be produced from stem cells, without the need to rear and slaughter cattle.

Data-based questions: Adaptations of the Western spadefoot toad

The Western spadefoot toad (*Scaphiopus hammondii*) lives in desert areas in California and lays its eggs in pools formed by rain. When the egg first hatches, its body form is referred to as the tadpole stage. At some point, it undergoes metamorphosis (a change in body form) to develop into the adult toad. If the pool where the eggs have been laid shrinks due to a lack of rain, the tadpoles quickly develop into small adult toads. If there is sufficient rain and the pool persists, the tadpoles develop more slowly and grow large before developing into adult toads.

1. Suggest how undergoing metamorphosis at different times in response to high and low water levels helps the survival of the toad. [3]

An experiment was carried out to determine what hormones might be involved in triggering development in response to pond drying. Tadpoles were raised in a constant high-water environment. They were then divided into two groups. One group was transferred to a tank containing 10 dm^3 of water—a high-water environment. The other group was transferred to a tank of the same size containing only 1 dm^3 of water—a low-water environment. The concentrations of thyroxine and corticosterone were measured in each group. The results are shown in Figure 6.

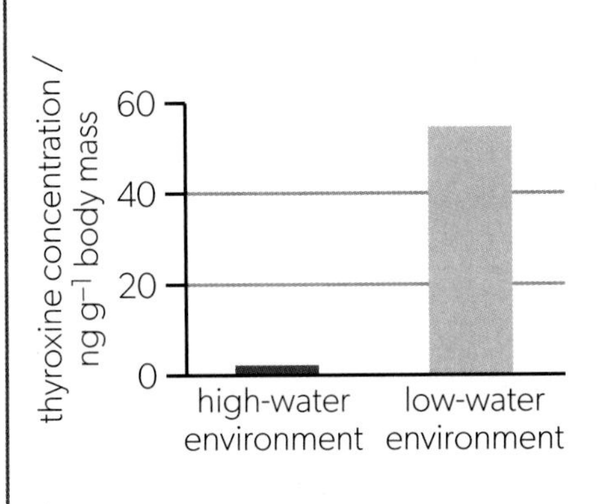

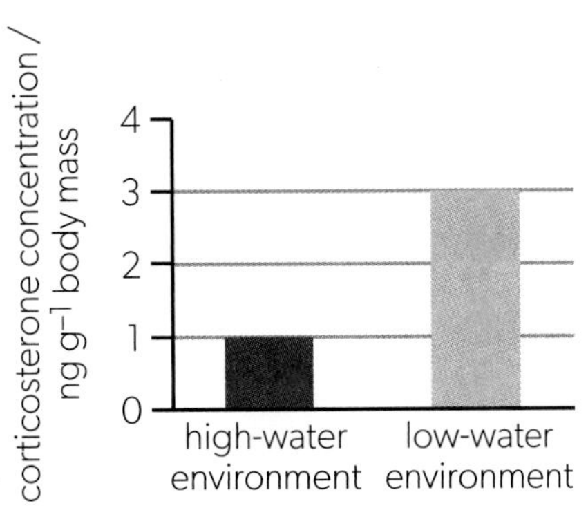

▲ Figure 6

2. Compare and contrast the concentrations of thyroxine and corticosterone in the two groups. [2]
3. Suggest how cellular differentiation plays a role in metamorphosis in the spadefoot toad. [2]

B2.3.4 Differences between totipotent, pluripotent and multipotent stem cells

Early-stage embryos are entirely composed of stem cells. These cells are totipotent. This means that they can differentiate into any cell type. Totipotent embryonic stem cells are potentially very useful, for example, in the growth of whole replacement hearts, kidneys or other organs.

Cells gradually commit to particular pathways of differentiation during embryo development. This involves a series of points at which a cell commits to develop along one pathway or another. Embryonic stem cells change from being totipotent to pluripotent. A pluripotent cell is still capable of differentiating into a range of cell types, but not every cell type.

The stem cells that remain in the adult body are more restricted in potential, but if they can differentiate into several types of mature cell, they are considered to be multipotent. Haematopoietic stem cells in bone marrow are multipotent because they can generate different types of blood cell, but not other cell types.

ATL Social skills: Actively seeking and considering the perspective of others

When considering debatable questions, individuals bring different perspectives to the conversation. To be open-minded means that you recognize that your own view does not exhaust the possibilities. It also requires a genuine curiosity regarding other possibilities and a willingness to suspend judgement until all the facts have been considered.

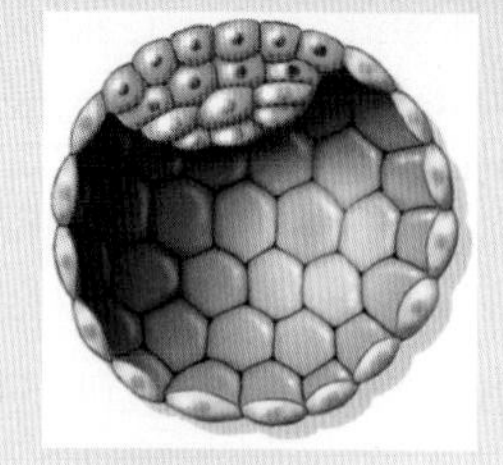

▲ Figure 7

Embryonic stem cells are totipotent and have unique properties that make them valuable tools for therapy and medical treatments. Figure 7 shows a diagram of a blastocyst, the fluid-filled ball of cells that forms after several divisions of a fertilized egg. The orange cells represent the inner cell mass, the part of the embryo that will become the foetus. It is these cells that are used in stem cell therapies.

1. Both skin cells and embryos can be cultured in vitro. Do they have the same ethical status or should there be a difference in the rules that govern their use? Would an accident causing the destruction of embryos be of equal concern to an accident causing the destruction of a culture of epidermal cells?
2. Hyperovulation can be medically induced to increase the production of eggs in women to be used in the creation of embryos. What concerns, if any, arise from compensating women for egg donation?

B2.3.5 Cell size as an aspect of specialization

The size of a mature differentiated cell is one way in which it is adapted to perform its function. Evidence for this in humans is provided by the examples in Table 1.

Cell type	Adaptation to function by cell size
sperm	50 μm long, which is longer than most cells but sperm are extremely narrow so they have one of the smallest volumes of any human cell. Narrowness and small volume reduce resistance and allow sperm to swim to the egg more easily.
egg	110 μm in diameter and spherical in shape, so egg cells have the largest volume of any human cell. This allows large quantities of food reserves to be stored in the cytoplasm. In birds, egg cells are even larger, with huge amounts of stored food (yolk).
red blood cells	6 μm to 8 μm in diameter but indented on both sides and only about 1 μm thick in the middle. The small size and shape allow passage along narrow capillaries and gives a large surface area-to-volume ratio, so loading and unloading of oxygen is faster.
white blood cells	B-lymphocytes are only about 10 μm in diameter when inactive but enlarge to as much as 30 μm if they are activated and become antibody-secreting plasma cells. The extra volume is cytoplasm with rER and Golgi apparatuses for protein synthesis.
cerebellar granule cells	The cell body is only 4.0 μm in diameter, but twin axons extend for about 3 mm (3,000 μm) in the cerebellar cortex. The very small volume of these neurons allows the cerebellum to accommodate 50 billion of them—75% of the brain's neurons.

motor neurons	The cell body is about 20 μm in diameter. This large size allows enough proteins to be synthesized to maintain the immensely long axon. It can extend for a metre or more (a million μm), so can carry signals from the central nervous system to a distant muscle.
striated muscle fibres	Striated muscle fibres are larger than normal cells, with a diameter of 20 μm to 100 μm and lengths that can exceed 100 mm (100,000 μm). These dimensions allow the fibre to exert greater force and contract by a greater length than smaller muscle cells.

◀ Table 1 Cell sizes

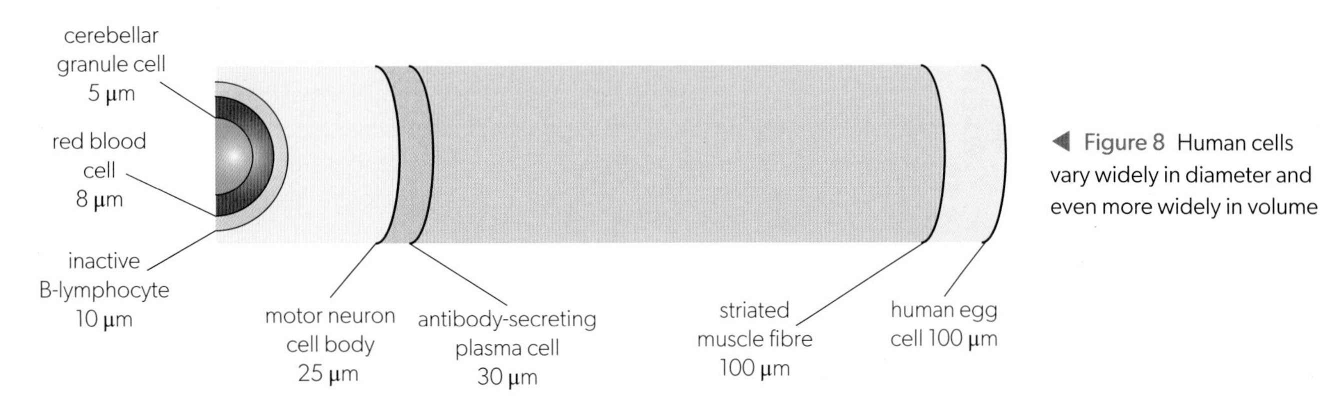

◀ Figure 8 Human cells vary widely in diameter and even more widely in volume

B2.3.6 Surface area-to-volume ratios and constraints on cell size

Many chemical reactions take place in the cytoplasm of cells. These reactions are known collectively as the metabolism of the cell. The rate of these reactions (the metabolic rate of the cell) is proportional to the volume of the cell. For metabolism to continue, substances used in the reactions must be absorbed by the cell and waste products must be removed. Substances move into and out of cells through the plasma membrane at the surface of the cell. The rate at which substances cross this membrane depends on its surface area.

The relative amounts of surface area and volume can be expressed mathematically as a ratio:

$$\text{surface area-to-volume ratio} = \frac{\text{surface area (mm}^2\text{)}}{\text{volume (mm}^3\text{)}}$$

Surface area-to-volume ratio is very important to a cell. If it is too small, substances will not enter the cell as quickly as they are required. Also waste products will accumulate because they are produced more rapidly than they can be excreted. Surface area-to-volume ratio is also important in relation to heat production and loss. If the ratio is too small, cells may overheat because the metabolism produces heat faster than it is lost over the cell's surface.

Activity:

Effect of size on surface area-to-volume ratio

Calculate the volume and surface area and then the surface area-to-volume ratio of cubes with sides 1 mm, 10 mm and 100 mm. What is the trend?

same cube unfolded

▲ Figure 9

Mathematical models: Surface area-to-volume ratio

Models are simplified versions of complex systems. The effect of size on the surface area-to-volume ratio of cells can be modelled using cubes with sides of different lengths. Although the cubes have a simpler shape than real organisms, scale factors operate in the same way, so the trend for the cubes will also operate in cells of more rounded or irregular shape, as long as the shape stays the same as the dimensions increase.

B2.3.7 Adaptations to increase surface area-to-volume ratios of cells

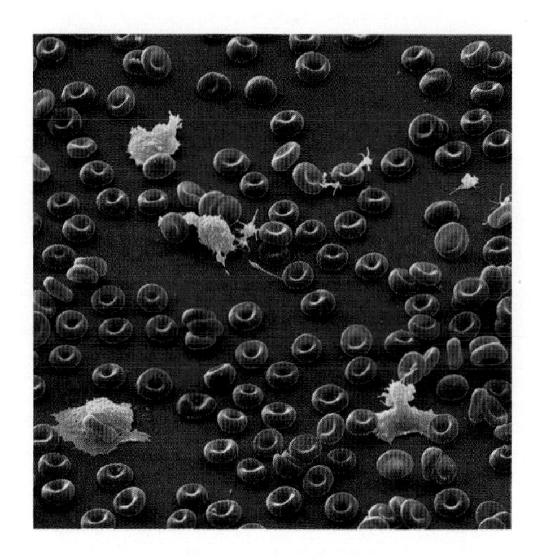

▲ Figure 10 Red and white blood cells

Some cells are specialized for exchange processes. Examples are proximal convoluted tubule cells in the kidney that reabsorb useful substances from glomerular filtrate, and red blood cells (erythrocytes) which transport oxygen from the lungs to respiring tissues. These cells must be able to transport substances rapidly in and out across their plasma membrane. They all show adaptations that increase their surface area-to-volume ratio.

Red blood cells

The shape and small size of red blood cells gives them a large surface area-to-volume ratio, which helps them load and unload millions of oxygen molecules rapidly. They are only about 8 μm in diameter and their biconcave disc shape gives them a lower volume than a sphere of the same diameter and a smaller maximum distance from any part of the cytoplasm to the plasma membrane.

Proximal convoluted tubule cells

Near the outer surface of the kidney there are large numbers of narrow, coiled tubes, called proximal convoluted tubules. These tubules receive the large volumes of fluid that are filtered out of the blood in the kidney. They reabsorb most of this filtrate, including all molecules of useful substances such as glucose. The wall of the proximal convoluted tubules is only one cell thick, with the inner apical membrane in contact with the filtrate and the outer basal membrane close to blood capillaries. To be reabsorbed from the filtrate to the blood, a molecule or ion must pass through both the apical membrane and the basal membrane. Channel proteins and pump proteins in these membranes ensure that only substances required by the body are reabsorbed, with waste products such as urea remaining in the filtrate. The apical membrane has large numbers of microvilli, which provide a large surface area. The basal membrane has infoldings (invaginations), which also increase the surface area. Both the apical and basal membranes therefore have ample space for the channel and pump proteins that carry out selective reabsorption.

B2.3.8 Adaptations of type I and type II pneumocytes in alveoli

The lungs contain huge numbers of alveoli. These air sacs provide a very large total surface area for diffusion. The wall of the alveolus is one cell thick and is an epithelium. There are two types of cell in the alveolar epithelium.

- Type I pneumocytes (AT1 cells) are adapted for diffusion of oxygen and carbon dioxide. This is a passive process, so there is little need for mitochondria or other organelles and the volume of cytoplasm is small. These are very wide but extremely thin cells. The thickness is only about 0.15 μm, widening slightly where the nucleus is located. The wall of the adjacent capillaries also consists of a single layer of very thin cells. The air in the alveolus and the blood in the alveolar capillaries are less than 0.5 μm apart. The distance over which oxygen and carbon dioxide must diffuse is thus very small, increasing the rate of gas exchange.

- Type II pneumocytes (AT2 cells) are more numerous than type I cells (they represent 90% of alveolar cells) but they occupy only about 5% of the alveolar surface area. They are about 10 µm across with a dense cytoplasm containing mitochondria, rough endoplasmic reticulum and lysosomes. Large amounts of phospholipid are synthesized in the cytoplasm and stored in lamellar bodies, which are vesicles containing many layers of phospholipid and some proteins. The contents of the lamellar bodies are secreted by exocytosis.

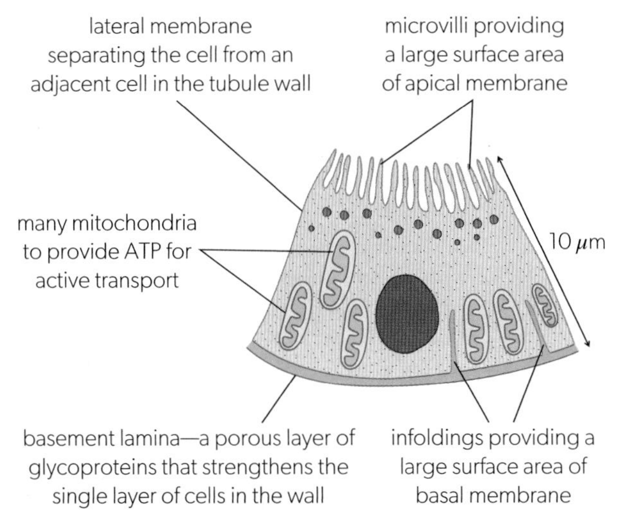

▲ **Figure 11** Proximal convoluted cell with microvilli and invaginations

The alveolus is lined by a film of moisture, which allows oxygen in the alveolus to dissolve and then diffuse to the blood in the alveolar capillaries. It also provides an area from which carbon dioxide can pass into the air and be exhaled. Phospholipids secreted by the lamellar bodies spread to form a single layer of molecules on the outer surface of the film of moisture, with the hydrophilic heads of the phospholipids inwards and the hydrophobic tails facing outwards to the air in the alveolus. Proteins secreted from the lamellar bodies are dispersed between the phospholipid molecules. The layer of phospholipids and proteins acts as a surfactant, reducing surface tension. Without surfactant, the alveolus might collapse on itself with the sides adhering due to hydrogen bonding between water molecules.

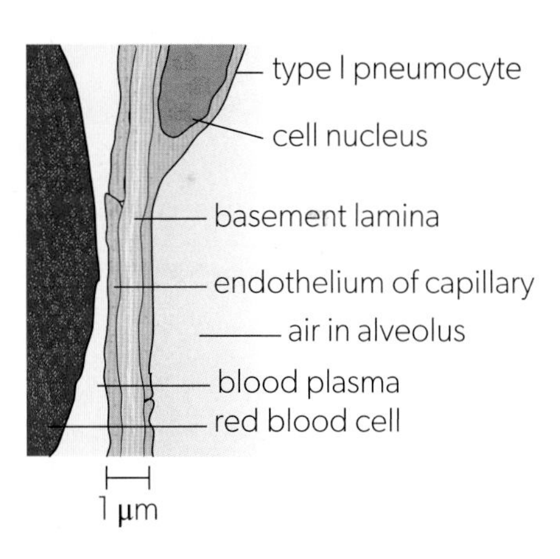

▲ **Figure 12** Flattened cells in the walls of the alveolus and blood capillary

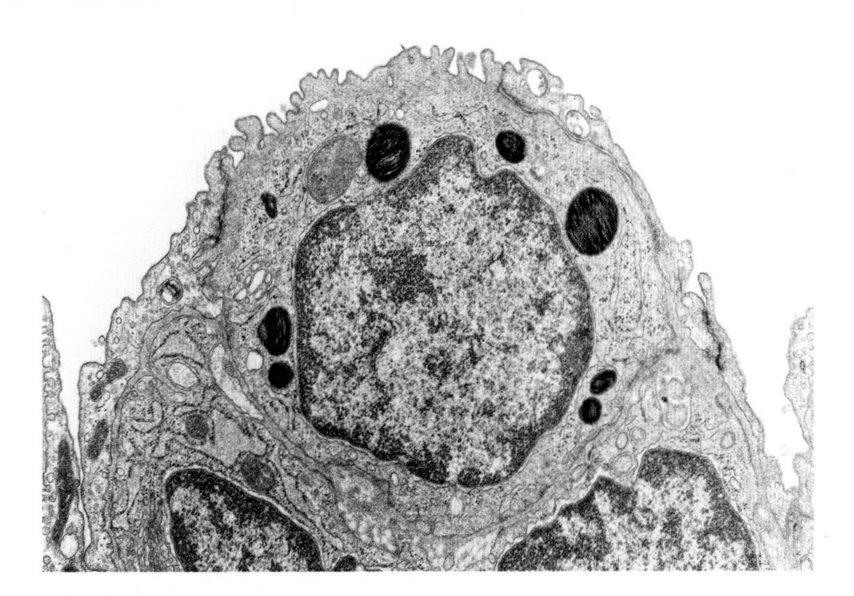

◀ **Figure 13** Type II pneumocyte with a prominent nucleus (blue). The cytoplasm contains many organelles including lamellar bodies (brown). Layers of phospholipid are visible in the lamellar bodies. The irregular surface of the cell is due to recent exocytosis

▲ Figure 14 Light micrograph of striated muscle fibres with stripes and multiple nuclei visible

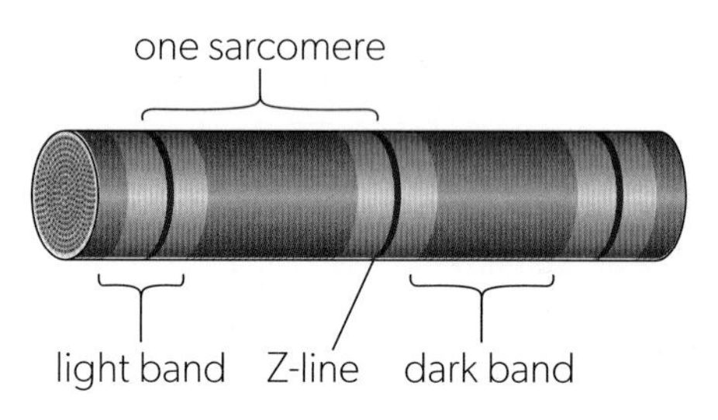

▲ Figure 15 Structure of a myofibril

▶ Figure 16 Part of one striated muscle fibre is visible in this electron micrograph, with mitochondria between the myofibrils coloured blue. Light bands (green) and dark bands (red) are aligned in all the myofibrils within the muscle fibre. It is this that gives the striated appearance

B2.3.9 Adaptations of cardiac muscle cells and striated muscle fibres

Muscle tissue is contractile—it can shorten in length. When it does this it exerts a pulling force that can be used to cause movement. To return to its original length, a pulling force must be exerted on the muscle. This is usually provided by another muscle, so many muscles work in antagonistic pairs—the contraction of one causes the lengthening of the other.

The muscles that are used to move the body are attached to bones, so they are called skeletal muscles. When their structure is viewed using a light microscope, stripes are visible. Because of this, they also called striated muscle. Striated muscle is composed of many long, unbranched cylindrical structures known as muscle fibres, which are arranged in parallel. Although a single plasma membrane surrounds each muscle fibre, there are many nuclei present and muscle fibres are much longer than typical cells. These features are because embryonic muscle cells fuse together to form muscle fibres.

Electron microscopes reveal that within each muscle fibre there are many parallel, cylindrical structures called myofibrils. These have alternating light and dark bands. In the centre of each light band is a disc-shaped structure, referred to as the Z-line. Muscle contraction is explained in *Topic B3.3*.

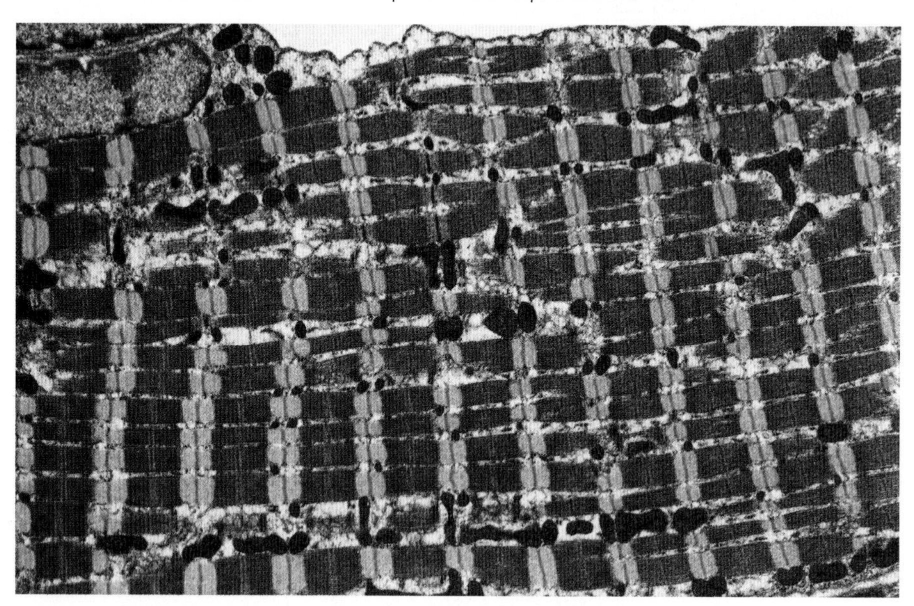

Cardiac muscle forms the wall of the heart. As in skeletal muscle, there are myofibrils, with the light and dark bands aligned, so cardiac muscle has a striated appearance. However, unlike the elongated fibres of skeletal muscle, cardiac muscle is composed of much shorter cells, most of which only have one nucleus. Where the end of one cell contacts the end of another cell, there is a specialized junction called an intercalated disc. Cardiac muscle cells are branched, so intercalated discs connect each cell at both ends with several other cells. In an intercalated disc, there are connections between the plasma membranes and cytoplasm of adjacent cardiac muscle cells, allowing electrical signals to be propagated rapidly from cell to cell. If one cardiac muscle cell in the wall of the heart is stimulated to contract, the stimulus is passed on to all the other cells, so there is a synchronization of muscle contraction and blood is pumped quickly out of the heart.

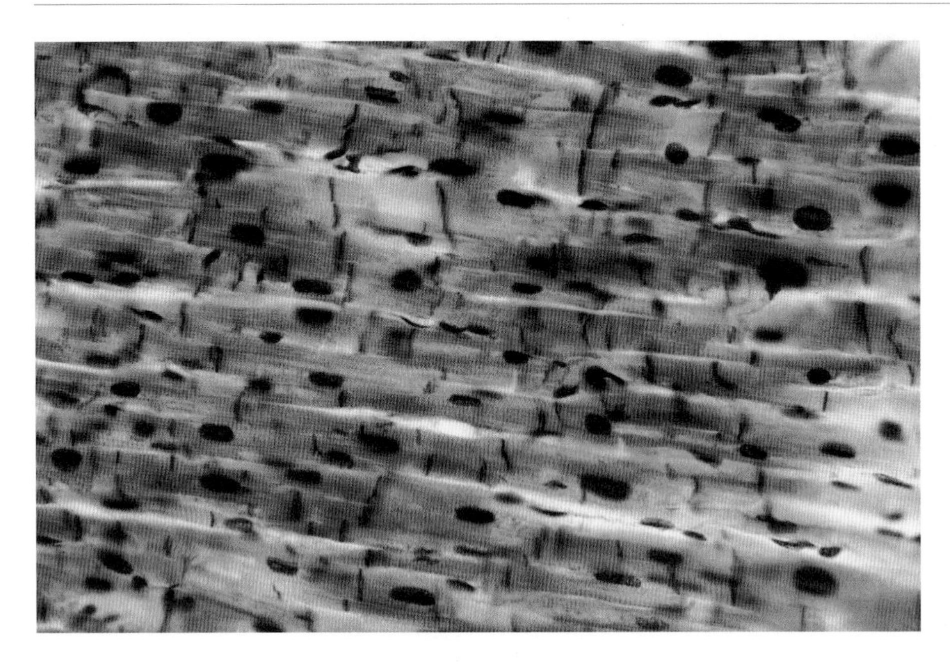

Figure 17 Light micrograph of cardiac muscle tissue with short, branched cells and striations (pink). Intercalated discs (purple) are clearly visible where cells meet. Each cell contains a single nucleus (purple)

Whether or not a muscle fibre should be classed as a cell is debatable. They are enclosed in a plasma membrane, but have many nuclei, rather than just one. They are much larger than most animal cells, with an average length in humans of about 30 mm, whereas other human cells are mostly less than 0.03 mm in length.

B2.3.10 Adaptations of sperm and egg cells

Egg cells and sperm cells are gametes each containing a haploid nucleus that passes on genes from parent to offspring. Beyond these similarities, male and female gametes in humans are radically different. Sperm move actively and rapidly whereas egg cells are moved passively and relatively slowly. Egg cells have food reserves needed for embryo development whereas sperm have little or no stored food.

Egg cells have structures that enable them to receive one sperm (and no more) in the process known as fertilization.

- Zona pellucida—a layer of glycoproteins containing ZP3 to which sperm bind and which a sperm can penetrate, but which later can be chemically altered to prevent any more sperm from penetrating.
- Binding proteins in the plasma membrane which help it to fuse with the membrane of the sperm, allowing the sperm nucleus to enter the egg.
- Cortical granules—vesicles of enzymes near the plasma membrane of the egg cell which are released into the zona pellucida and make it impenetrable after the nucleus of a sperm has entered the egg.

Egg cells also have structures which provide the resources needed for the zygote and then the embryo to develop.

- Yolk—this is the large volume of cytoplasm inside the egg cell that contains stores of lipids and other foods.
- Mitochondria—these produce ATP and divide repeatedly to generate all the mitochondria in an adult's body.
- Centrioles—these are needed for mitosis.

AHL

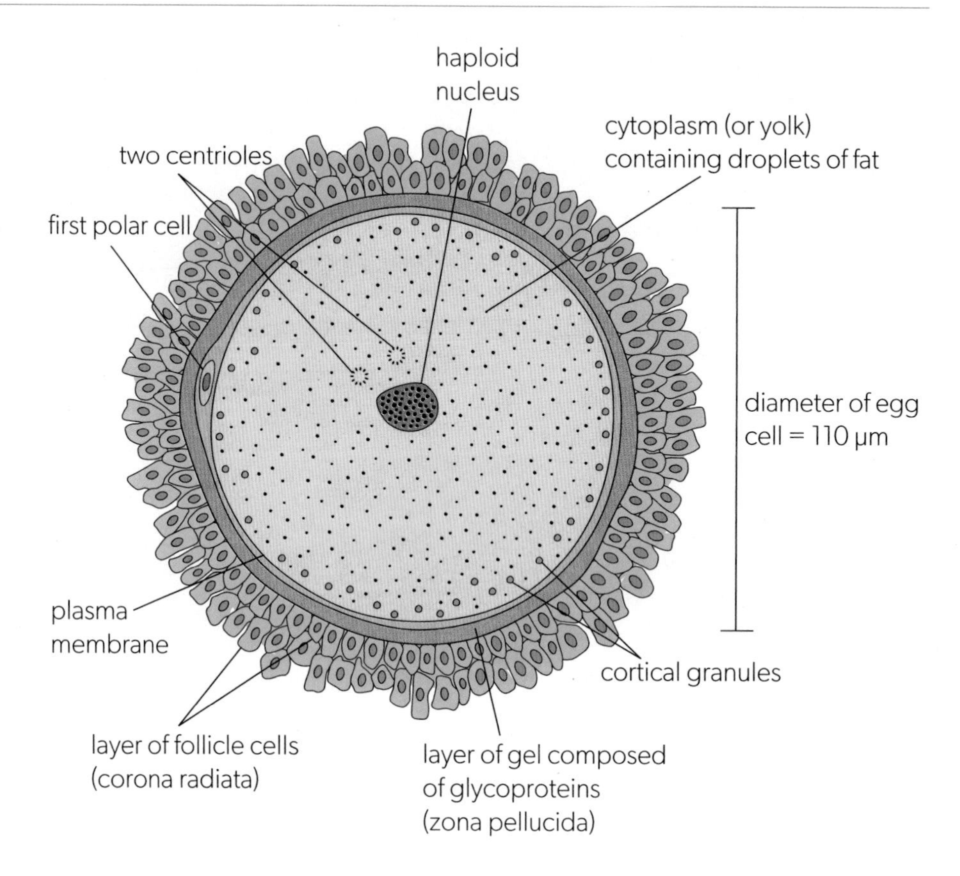

▶ Figure 18 Structure of a human egg cell

Sperm cells are adapted to transfer a haploid nucleus from the testis of a male to the cytoplasm of an egg cell in the oviduct of a female. This is a competitive process—the first sperm to reach and penetrate an egg cell is the only one to achieve its role. Sperm cells have structures which allow them to swim rapidly.

- Tail—a very long flagellum with the characteristic arrangement of 9 + 2 microtubules that generates the force needed for forward motion with its beating action.
- Midpiece with mitochondria—multiple mitochondria are wound round the microtubules at the base of the tail, where they can supply the large quantities of ATP needed for the motion of the tail.
- Head—streamlined in shape and very narrow due to the nucleus having tightly packed chromosomes and the volume of cytoplasm being very small, so resistance to movement is minimized.

Sperm also have structures that they use to insert their nucleus into the egg cell.

- Receptors in the plasma membrane for ZP3 glycoproteins in the zona pellucida to which the sperm binds.
- Acrosome—a sac of enzymes that digest proteins and polysaccharides in the protective zona pellucida so the sperm can reach the plasma membrane of the egg cell.
- Binding proteins in the inner acrosomal membrane, revealed after exocytosis of the acrosome, which bind to proteins in the plasma membrane of the egg cell, leading to fusion of the membranes and entry of the sperm nucleus to the egg.

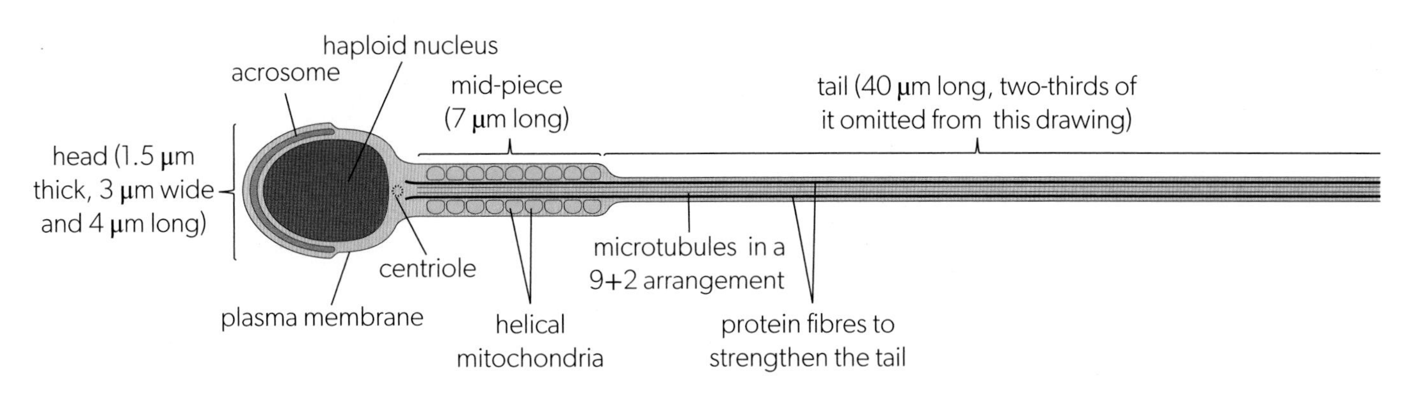

▲ Figure 19 Structure of a human sperm cell

Linking questions

1. What are the advantages of small size and large size in biological systems?
   a. Explain the differences in the size of cells in three different kingdoms. (A2.2.8)
   b. Explain the challenges associated with gas exchange as cells become larger. (B3.1.1)
   c. Outline the factors that lead to variation in the speed of nerve impulses. (C2.2.4)
2. How do cells become differentiated?
   a. Outline the role of methylation and acetylation in gene expression. (A2.2.13)
   b. Distinguish between totipotent, pluripotent and multipotent cells. (B2.3.4)
   c. Discuss the process of cellular differentiation with respect to the process of spermatogenesis. (D3.1.4)

TOK

Should some knowledge not be sought on ethical grounds?

Ethics is the academic study that looks into questions of right and wrong. Scientists are sometimes confronted with situations where their investigations raise ethical questions. For example, they might want to explore a topic using methods that some people consider wrong or the investigation itself might result in an outcome that some people consider wrong.

There are many scientific activities from the 20th century involving human subjects in experiments that should not have been carried out on ethical grounds. These experiments include:

- exposure of humans to debilitating diseases (the Tuskegee Syphilis experiment)
- deliberate exposure of humans to radiation
- surgical experiments
- tests with mind-altering substances.

Some of the most famous examples of previous unethical research practices would be unlikely to happen today because of the existence of institutional review boards (IRBs). An IRB is a committee within a university or research organization that reviews research proposals to ensure they follow the ethical principles necessary to protect human subjects. An IRB has the authority to approve, reject or require modifications to research proposals within the institution in which it operates.

Henrietta Lacks was a patient at Johns Hopkins University in the early 1950s. She had a cancerous cervical tumour that ultimately led to her death. A sample of her cancer cells taken during the course of her medical treatment was sent to a tissue lab. It was discovered in the laboratory that the cells survived and were able to multiply in vitro. The cells were nicknamed HeLa cells. HeLa cells have been highly useful for studying a range of phenomena, though they were used for this purpose without her permission. While the collection and use of Henrietta Lacks' cells in research was a legal practice in the 1950s, it would not happen today. Today, she would be required to provide informed consent to authorize their use for any purpose.

HeLa cells were the first human cell line to be cultured for research into cancer. The source of the cells was a carcinoma biopsy from her cervix. At the time, scientists had no success growing cancer cells in the laboratory. HeLa cells, however, thrived. In the 70 years since 1952, HeLa cells have been used in research around the world.

◀ Figure 1 Henrietta Lacks standing outside her home several years before her death from cervical cancer at the age of 31

The Tuskegee Syphilis Study was carried out from 1932 to 1972 by the United States Public Health Service. The objective of the study was to discern the natural course of untreated syphilis in black men. Of the 600 male participants, about 400 had syphilis. There was none of the transparency needed to ensure informed consent and the men were never given the option of leaving the study. Participants were not told their diagnosis and were only treated with placebos, although during the study it became well known that penicillin was an effective treatment for syphilis.

Based on the lack of opportunity to get treatment when available, the Tuskegee study was "ethically unjustified". Of the original participants, 28 died as a result of syphilis and a further 100 died of complications of the disease. Forty wives were infected and 19 children were born with congenital syphilis.

In modern experiments, if a treatment is found to be more effective than the placebo, a study is typically stopped and access to the successful treatment is provided.

◀ Figure 2 Tuskegee Syphilis Study participants

End of chapter questions

1. The table shows the area of membranes in a rat liver cell.

Membrane component	Area/μm²
plasma membrane	1,780
rough endoplasmic reticulum	30,400
mitochondrial outer membrane	7,470
mitochondrial inner membrane	39,600
nucleus	280
lysosomes	100
other components	18,500

a. Calculate the total area of membranes in the liver cell. [2]

b. Calculate the area of plasma membrane as a percentage of the total area of membranes in the cell. Show your working. [3]

c. Explain the difference in area of the inner and outer mitochondrial membranes. [3]

d. Using the data in the table, suggest two of the main activities of liver cells. [2]

2. In human secretory cells, for example in the lung and the pancreas, positively charged ions are pumped out, and chloride ions follow passively through chloride channels. Water also moves from the cells into the liquid that has been secreted. In the genetic disease cystic fibrosis, the chloride channels malfunction and too few ions move out of the cells. The liquid secreted by the cells becomes thick and viscous, with associated health problems.

a. State the names of the processes that:

i move positively charged ions out of the secretory cells [1]

ii move chloride ions out of the secretory cells [1]

iii move water out of the secretory cells. [1]

b. Explain why the fluid secreted by people with cystic fibrosis is thick and viscous. [4]

3. A study was carried out to determine the relationship between the diameter of a molecule and its movement through a membrane. The graph shows the results of the study.

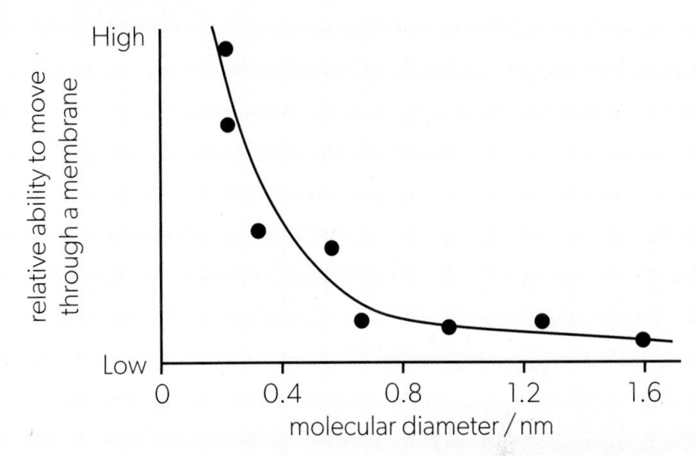

a. From the information in the graph alone, describe the relationship between the diameter of a molecule and its movement through a membrane. [2]

A second study was carried out to investigate the effect of passive protein channels on the movement of glucose into cells. The graph shows the rate of uptake of glucose into erythrocytes by simple diffusion and facilitated diffusion.

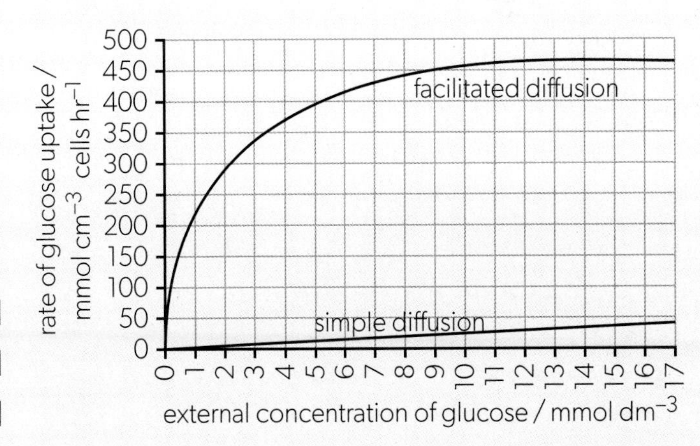

b. Identify the rate of glucose uptake at an external glucose concentration of 4 mmol dm^{-3}

i simple diffusion. [1]

ii facilitated diffusion. [1]

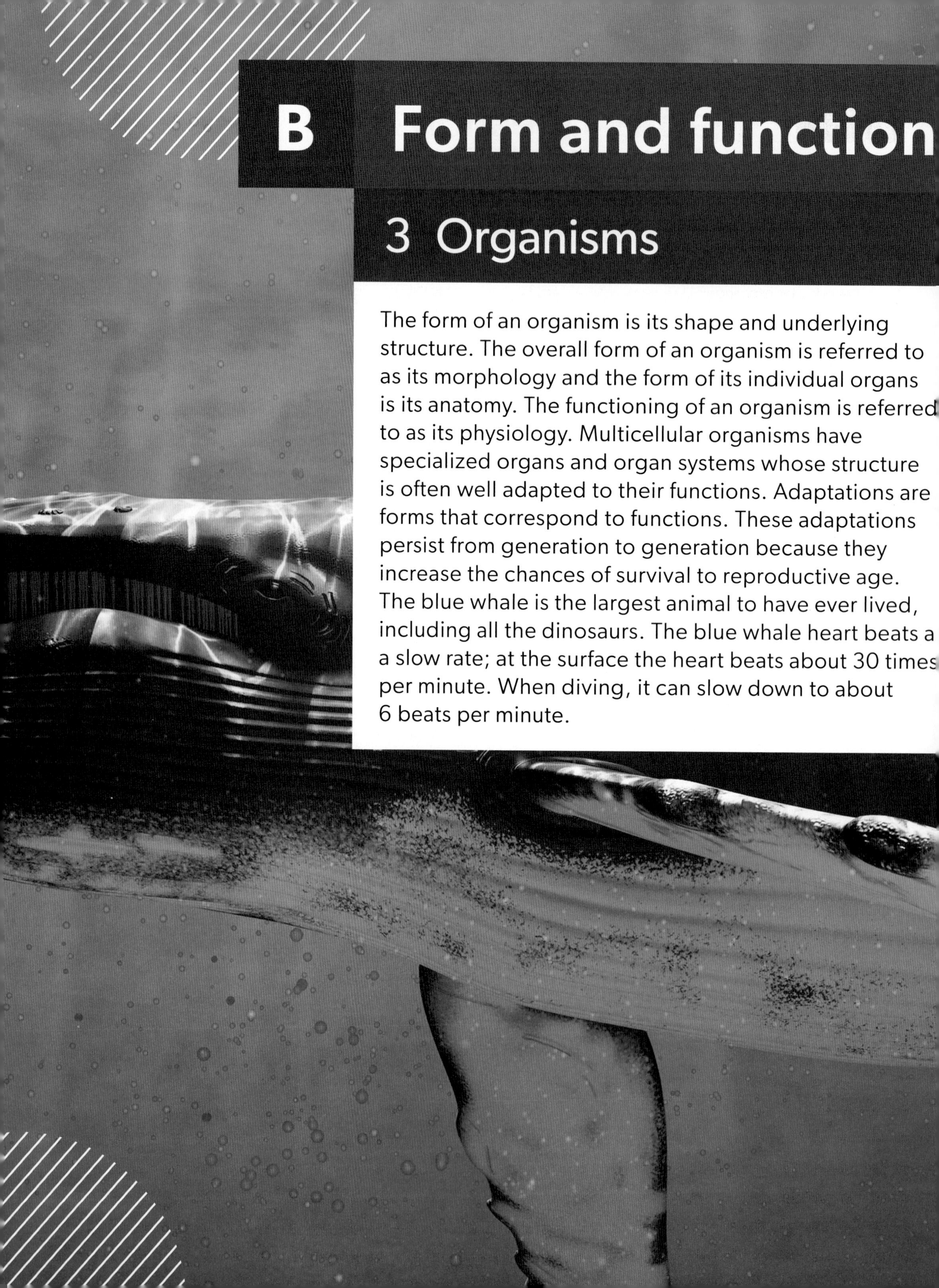

B Form and function

3 Organisms

The form of an organism is its shape and underlying structure. The overall form of an organism is referred to as its morphology and the form of its individual organs is its anatomy. The functioning of an organism is referred to as its physiology. Multicellular organisms have specialized organs and organ systems whose structure is often well adapted to their functions. Adaptations are forms that correspond to functions. These adaptations persist from generation to generation because they increase the chances of survival to reproductive age. The blue whale is the largest animal to have ever lived, including all the dinosaurs. The blue whale heart beats a a slow rate; at the surface the heart beats about 30 times per minute. When diving, it can slow down to about 6 beats per minute.

B3.1 Gas exchange

How are multicellular organisms adapted to carry out gas exchange?

All organisms absorb the gasses they need from their environment and release other gases into the environment as waste products. What are the gases that need to be exchanged? Which processes require these gases and which processes generate them? What is the importance of surface area-to-volume ratio for cells? What challenges do multicellular organisms have when it comes to gas exchange? How is it that the fish in the water can freely exchange gasses with the water but the snorkeler needs a connection to the atmosphere?

▲ Figure 1 Snorkelers can swim with fish but still exchange gases with the air above

What are the similarities and differences in gas exchange between a flowering plant and a mammal?

The glass beads in water are spheres. Spheres have the smallest surface area-to-volume ratio of any shape. Shapes approximating to spheres provide the gas-exchange surface in both mammals and flowering plants. By making the spheres very small, a huge surface area relative to volume can be generated. Although there is similarity in shape, the gas-exchange surfaces of mammals and flowering plants are markedly different. In what ways do these structures differ?

▶ Figure 2 Glass beads in water

SL and HL	AHL only
B3.1.1 Gas exchange as a vital function in all organisms B3.1.2 Properties of gas-exchange surfaces B3.1.3 Maintenance of concentration gradients at exchange surfaces in animals B3.1.4 Adaptations of mammalian lungs for gas exchange B3.1.5 Ventilation of the lungs B3.1.6 Measurement of lung volumes B3.1.7 Adaptations for gas exchange in leaves B3.1.8 Distribution of tissues in a leaf B3.1.9 Transpiration as a consequence of gas exchange in a leaf B3.1.10 Stomatal density	B3.1.11 Adaptations of foetal and adult haemoglobin for the transport of oxygen B3.1.12 Bohr shift B3.1.13 Oxygen dissociation curves as a means of representing the affinity of haemoglobin for oxygen at different oxygen concentrations

B3.1.1 Gas exchange as a vital function in all organisms

All organisms absorb one gas from the environment and release another one. This is gas exchange. Redwood trees absorb carbon dioxide for use in photosynthesis and release oxygen produced in the process. Humans absorb oxygen for cell respiration and release the carbon dioxide produced. Terrestrial organisms exchange gases with the air. Aquatic animals such as fish exchange gases with water.

Diffusion is the basis of gas exchange. Because the molecules move randomly, diffusion is a relatively slow process. Gas exchange is only rapid enough if it occurs over a large surface area and the distance across which the gases must diffuse is short. Unicellular and other small organisms have a large surface area-to-volume ratio and the distance between the centre of the organism and the exterior environment is small. They can therefore use their outer surface for gas exchange. In larger organisms, the surface area-to-volume ratio is smaller and the distance between the centre of an organism and the exterior is greater. A specialized gas-exchange surface is required that is much larger than the outer surface, for example alveoli in lungs or the spongy mesophyll in a leaf.

▲ Figure 3 Axolotls develop lungs but also retain their feathery external gills into adulthood. The gill filaments have a large total surface area with a distance of 10 μm between blood in capillaries and the water outside. Axolotls are critically endangered in their habitat in Mexico

B3.1.2 Properties of gas-exchange surfaces

Gas-exchange surfaces share four properties. They are:

- permeable—oxygen and carbon dioxide can diffuse across freely
- large—the total surface area is large in relation to the volume of the organism
- moist—the surface is covered by a film of moisture in terrestrial organisms so gases can dissolve
- thin—the gases must diffuse only a short distance, in most cases through a single layers of cells.

ATL Thinking skills: The reasonableness of knowledge claims: The surface area of axolotl gills

The total surface area of alveoli in the lungs is often said to be the size of a tennis court, without any evidence for this. To estimate the surface area, we need to know how many alveoli there are in the lungs and what their average surface area is.

- Can you find evidence-based values for the number and mean surface area of human alveoli?
- What numbers are needed to calculate an estimate for the total surface area of an axolotl's gills?

B3.1.3 Maintenance of concentration gradients at exchange surfaces in animals

Diffusion of gases only happens if there are concentration gradients. For example, oxygen diffuses from air in the alveoli to the adjacent capillaries because the oxygen concentration of blood in the capillaries is lower than in air. Carbon dioxide diffuses from the blood to air in the alveoli because there is a lower concentration of carbon dioxide in the air. Diffusion evens out concentration gradients, which could slow and then stop gas exchange. For gases to continue to diffuse across exchange surfaces, concentration gradients must be maintained.

In small, aerobically respiring organisms that use their outer surface for gas exchange, it is cell respiration that maintains concentration gradients. This process continuously uses oxygen and produces carbon dioxide, so the oxygen concentration within the organism remains lower than outside and the carbon dioxide concentration remains higher. In larger organisms such as fish or mammals, blood flows continuously through dense capillary networks in the organs specialized for gas exchange. Due to aerobic respiration, this blood has a low concentration of oxygen and a high concentration of carbon dioxide.

Ventilation also helps to maintain concentration gradients. This term was originally used for the movement of air in and out of the lungs, but it is now also used to refer to movement of water across gills.

- Mammals periodically expel air from the alveoli by exhaling and then replace it by inhaling fresh air. This prevents the oxygen concentration from dropping too low for diffusion from the air to the blood and also prevents the carbon dioxide concentration from rising too high. The rate of ventilation is adjusted according to the carbon dioxide concentration of the blood.
- Fish take in fresh water through their mouth and pump it over their gills and then out through the gill slits. This one-way flow of water, combined with blood flow in the opposite direction, ensures that the oxygen concentration in the water adjacent to the gills remains high and the carbon dioxide concentration remains low.

Data-based questions: Concentration gradients

Figure 4 shows the typical composition of atmospheric air, air in the alveoli and gases dissolved in air returning to the lungs in the pulmonary arteries.

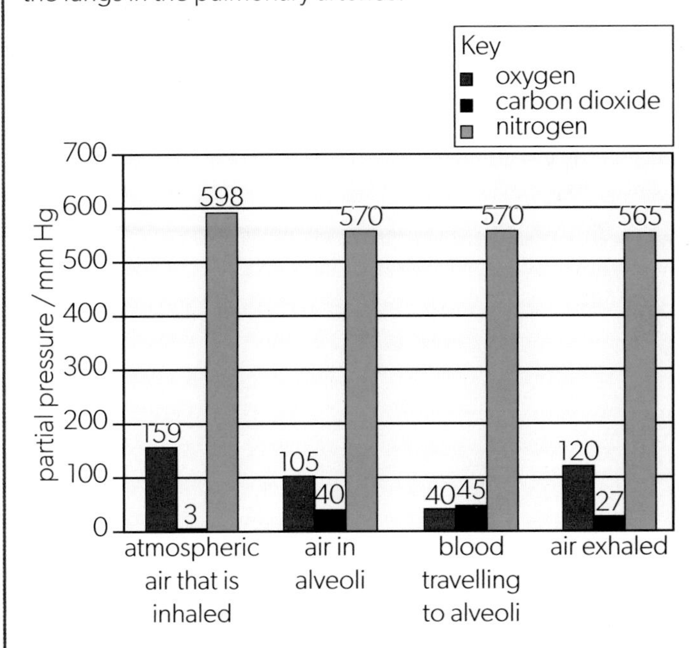

▲ Figure 4

1. Explain why the oxygen concentration in the alveoli is not as high as in fresh air that is inhaled. [2]
2. a. Calculate the difference in oxygen concentration between air in the alveolus and blood arriving at the alveolus. [1]
   b. Deduce the process caused by this concentration difference. [1]
   c. i. Calculate the difference in carbon dioxide concentration between air inhaled and air exhaled. [1]
      ii. Explain this difference. [2]
   d. Despite the high concentration of nitrogen in air in alveoli, little or none diffuses from the air to the blood. Suggest reasons for this. [2]

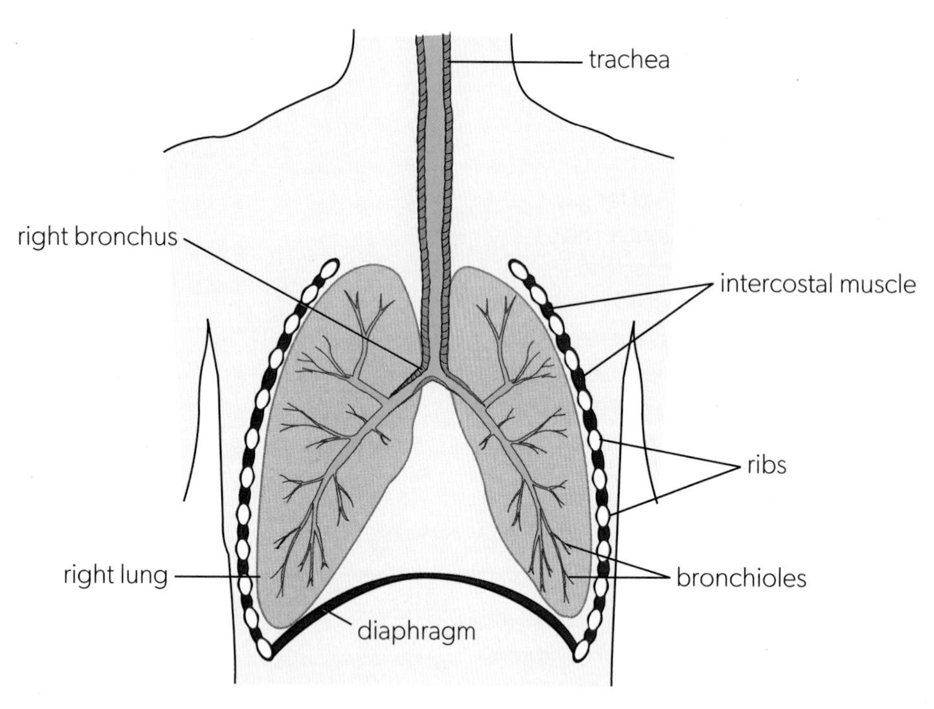

▲ Figure 5 Airways, lungs and associated muscles in the human thorax

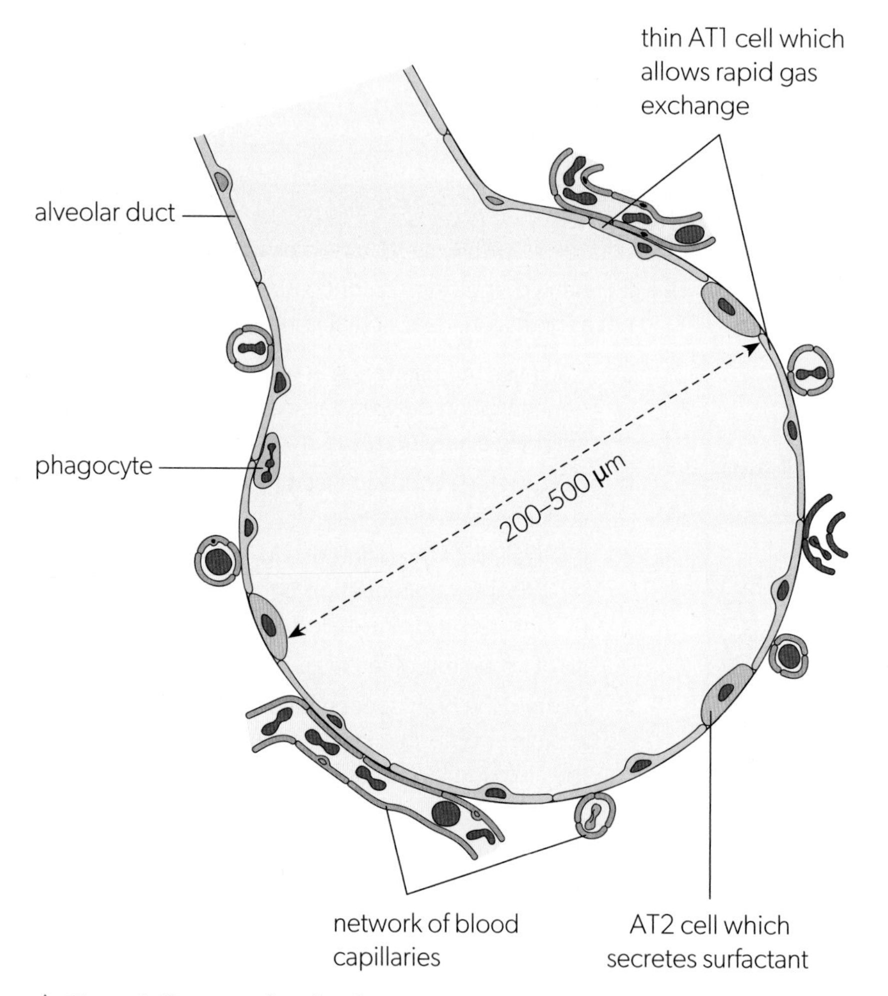

▲ Figure 6 Structure of an alveolus

B3.1.4 Adaptations of mammalian lungs for gas exchange

All mammals use lungs for gas exchange, even marine species such as whales and dolphins. Air is drawn into the lungs through the trachea (windpipe) and then the left and right bronchi (singular, bronchus). In each lung, the bronchus branches repeatedly to form bronchioles. Alveolar ducts branch off from the bronchioles, each leading to a group of five or six alveoli (air-sacs).

A pulmonary alveolus has a diameter of 0.2 mm to 0.5 mm, but its wall is a single layer of cells, much of which is only about 0.2 μm thick. Alveoli are surrounded by a dense capillary network. The capillary wall is also extremely thin and consists of a single layer of cells. Air and blood are therefore a very short distance apart. The capillaries cover much of the surface of the alveoli but there are also some other cells. Some of these have collagen fibres to strengthen the lung tissue and elastic fibres to help limit inhalation and cause passive exhalation.

An individual alveolus provides a small surface area for gas exchange, but because there are so many of them—about 300 million in a pair of adult lungs—the total area is very large: about 40 times greater than the outer surface of the body. The surface area of the basket-like networks of blood capillaries around the alveoli is almost as large as that of the alveoli.

Cells in the wall secrete a pulmonary surfactant. Its molecules have a structure similar to that of phospholipids in cell membranes. They form a monolayer on the surface of the moisture lining the alveoli, with the hydrophilic heads facing the water and the hydrophobic tails facing the air. This reduces the surface tension and prevents the water from causing the sides of the alveoli to adhere when air is exhaled from the lungs. This helps to prevent collapse of the lung.

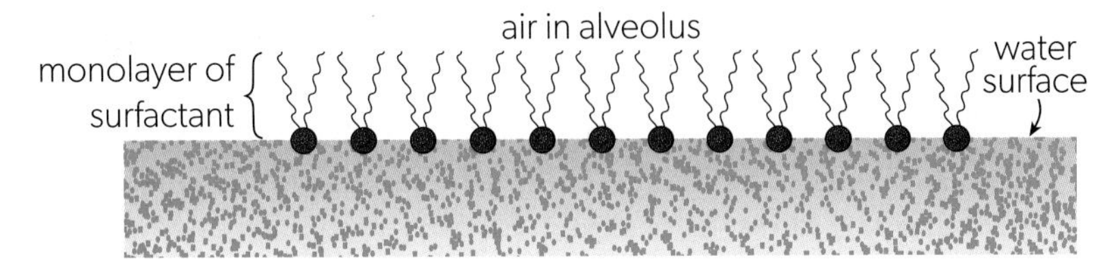

▲ **Figure 7** Pulmonary surfactant molecules on the surface of the film of moisture lining the alveoli

Data-based questions: COPD and gas exchange

In healthy lung tissue, there are open airways and groups of small thin-walled alveoli. Chronic obstructive pulmonary disease (COPD) is a group of irreversible lung conditions that are associated with smoking and air pollution. Patients with COPD have airways that have become narrow and a smaller number of larger air sacs.

1. a. Place a ruler across each micrograph and count how many times the edge of the ruler crosses a gas-exchange surface. Repeat this several times for each micrograph, in such a way that the results are comparable. Tabulate your results and calculate means. [3]

   b. Explain the conclusions that you draw from the results. [3]

2. People who have COPD feel tired all the time. Explain the reasons for this. [3]

3. People with COPD often have an enlarged and strained right side of the heart. Suggest a reason for this. [1]

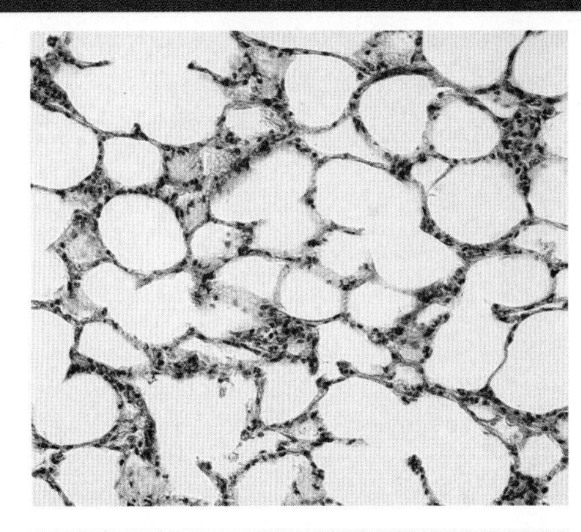

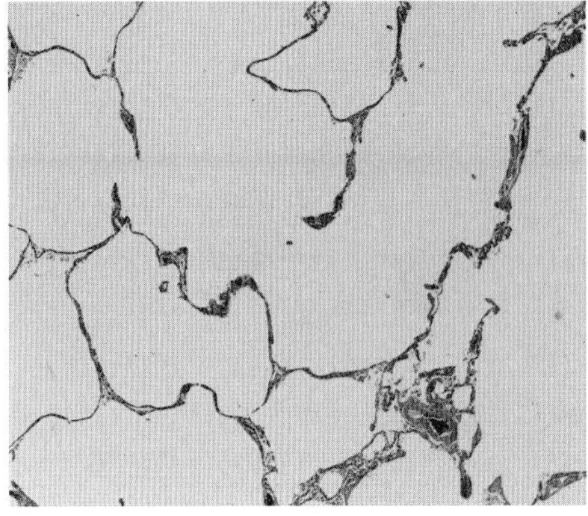

▶ **Figure 8** Healthy lung tissue (top) and lung tissue from a person with COPD (bottom)

B3.1.5 Ventilation of the lungs

The airways that connect the lungs to the air outside the body consist of the nose, mouth, trachea, bronchi and bronchioles. The trachea and bronchi have cartilage in their walls to ensure they remain open. The bronchioles have smooth muscle fibres in their walls, allowing the width of these airways to vary.

Ventilation of the lungs involves some basic physics. If particles of gas spread out to occupy a larger volume, the pressure of the gas becomes lower. Conversely, if a gas is compressed to occupy a smaller volume, the pressure rises. If gas is free to move, it will always flow from regions of higher pressure to regions of lower pressure.

During ventilation, muscle contractions cause the pressure inside the thorax to drop below atmospheric pressure. As a consequence, air is drawn into the lungs from the atmosphere (inspiration) until the lung pressure has risen to atmospheric pressure. Other muscle contractions then cause pressure inside the thorax to rise above atmospheric, so air is forced out from the lungs to the atmosphere (expiration), helped by the recoil of elastic fibres in lung tissue that become stretched during inspiration.

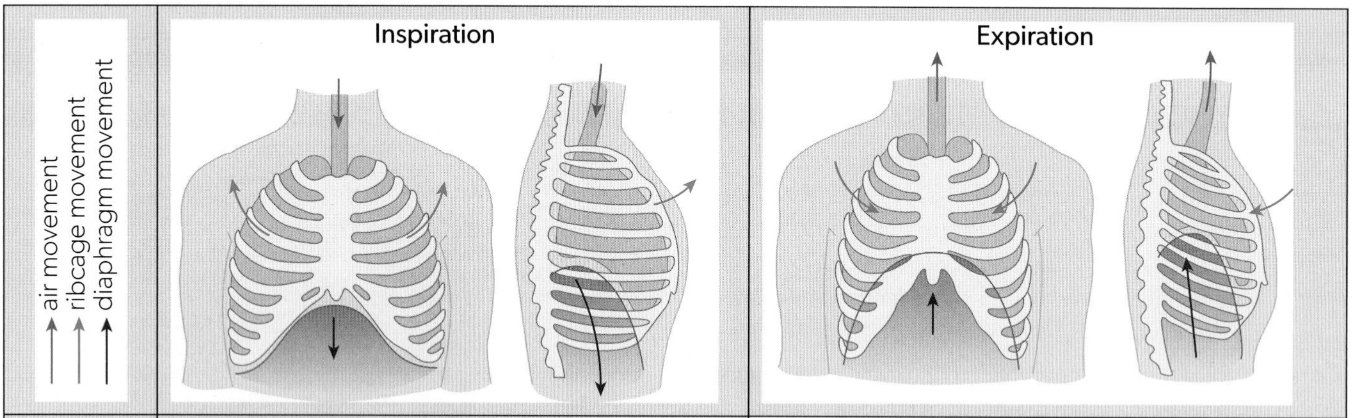

→ air movement → ribcage movement → diaphragm movement	Inspiration	Expiration
Diaphragm	The diaphragm contracts and so it moves downwards and pushes the abdomen wall out	The diaphragm relaxes so it can be pushed upwards into a more domed shape
Abdomen wall muscles	Muscles in the abdomen wall relax allowing pressure from the diaphragm to push it outwards	Muscles in the abdomen wall contract pushing the abdominal organs and diaphragm upwards (but only during forced expiration)
External intercostal muscles	The external intercostal muscles contract, pulling the ribcage upwards and outwards	The external intercostal muscles relax and are pulled into their elongated state
Internal intercostal muscles	The internal intercostal muscles relax and are pulled into their elongated state	The internal intercostal muscles contract, pulling the ribcage inward and downwards (but only during forced expiration)
Volume or pressures	The volume inside the thorax increases and consequently the pressure decreases, sucking air in	The volume inside the thorax decreases and consequently the pressure increases, forcing air out

▲ **Figure 9** Muscle actions that cause inspiration and expiration

B3.1.6 Measurement of lung volumes

- Tidal volume is the volume of fresh air that is inhaled and also the amount of stale air that is exhaled with each ventilation. Ventilation rate is the number of times that air is drawn in or expelled per minute.
- Vital capacity (or forced vital capacity) is the total volume of air that can be exhaled after a maximum inhalation or the total volume of air that can be inhaled after a maximum exhalation.
- Inspiratory reserve volume is the amount of air a person can inhale forcefully after normal tidal volume inspiration.
- Expiratory reserve volume is the amount of air a person can exhale forcefully after a normal exhalation.

These lung volumes can be measured using either simple apparatus or specialized meters. Simple apparatus is shown in Figure 10. One normal breath is exhaled through the delivery tube into a vessel and the volume is measured. It is not safe to use this apparatus for repeatedly inhaling and exhaling air as the carbon dioxide concentration will rise too high.

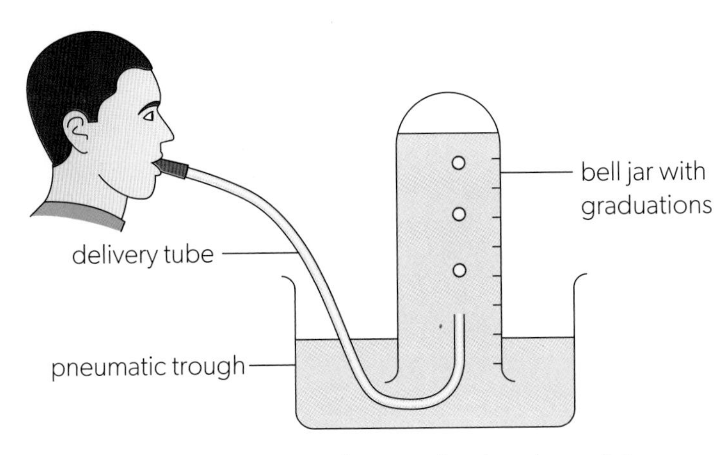

▲ Figure 10 Simple apparatus for measuring the volume of air exhaled. How could it be modified to measure volumes inhaled?

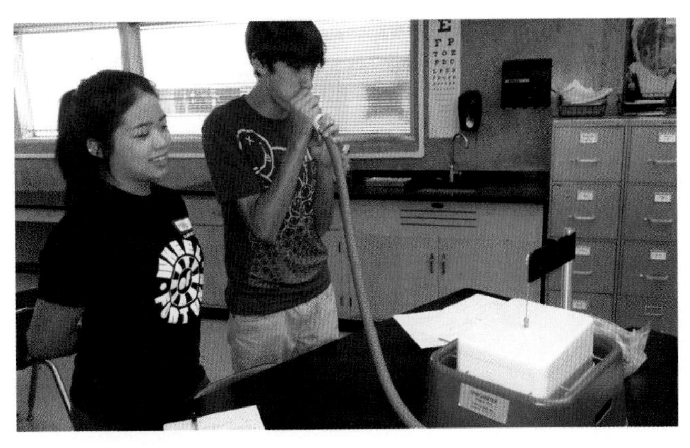

▲ Figure 11 A float spirometer has an air reservoir floating on water. Air in the reservoir can be breathed in and out through a tube with a mouthpiece on the end. Exhaled air passes back into the reservoir via alkali that absorbs carbon dioxide. This prevents the concentration of carbon dioxide from rising with repeated breaths

Specially designed spirometers measure flow rate into and out of the lungs and from these measurements lung volumes can be deduced. There are many different designs, some of which can be used with data logging software.

ATL Communication skills: Deciphering meanings by knowing etymology

Etymology is the study of the origin of words. What biological terms are derived from these Latin words?

- *pulmonarius* means "lung"
- *alveus* means "hollow vessel"
- *ventus* means "wind"
- *fundere* means "to pour"
- *diffundere* means "to pour apart"

Why do so many terms in biology and medicine have Latin origins?

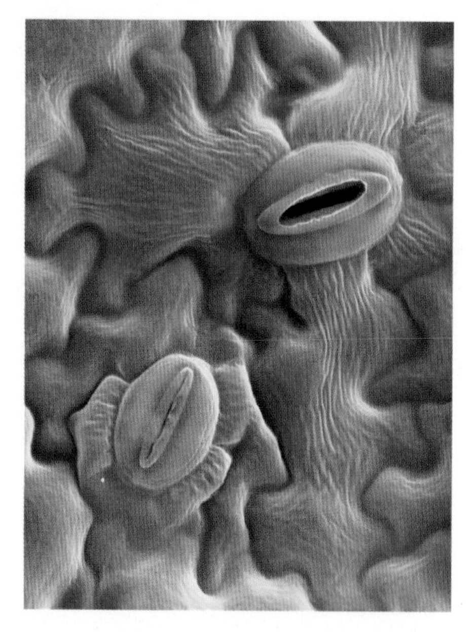

▲ Figure 12 One of the two stomata is open and the other is closed in this scanning electron micrograph of the outer surface of a lavender plant leaf. The epidermis cells have sinuous edges, with ridges of thick wax visible on their surfaces

B3.1.7 Adaptations for gas exchange in leaves

Chloroplasts need a supply of carbon dioxide for photosynthesis. The oxygen produced during the process of photosynthesis must be removed. A large area of moist surface is required over which carbon dioxide can be absorbed and oxygen excreted—this is provided by the leaves. A challenge for plants is to avoid excessive loss of water from the surface, so leaves are adapted for both gas exchange and water conservation.

The outer surface of the leaf is covered in a layer of wax, secreted by the epidermis cells. This waterproof layer is called the waxy cuticle. It varies in thickness between plants and is particularly thick on the upper surface of leaves and in plants adapted to dry habitats.

The waxy cuticle has low permeability to gases, but within the epidermis there are pairs of guard cells, which can change their shape either to open up a pore or to close it. The pore is called a stoma (plural, stomata) and it allows carbon dioxide and oxygen to pass through. The guard cells usually close the stomata at night when photosynthesis is not occurring and gas exchange is not required. They also close the stomata if a plant is suffering water stress and is in danger of dying from dehydration.

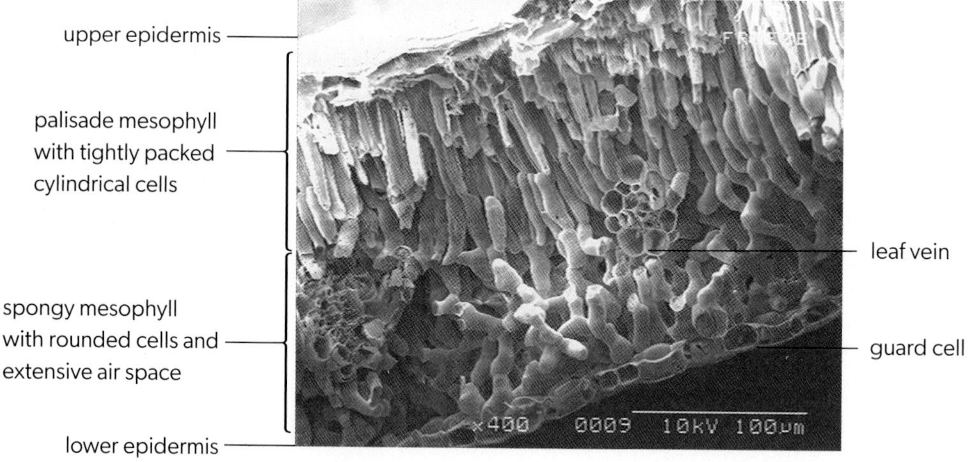

▲ Figure 13 Scanning electron micrograph of a leaf of *Prunus*. The leaf was frozen and then fractured so the interior structure is visible

The stomata connect the air outside to a network of air spaces in the spongy mesophyll of the leaf. Carbon dioxide and oxygen can diffuse through these air spaces. The walls of the spongy mesophyll cells provide a very large total surface area for gas exchange. Because these walls are permanently moist, carbon dioxide in the air spaces can dissolve and then diffuse through the mesophyll cells. Photosynthesis uses carbon dioxide, generating a concentration gradient from the air outside the leaf to the chloroplasts in mesophyll cells. Photosynthesis raises the concentration of oxygen in chloroplasts, so it diffuses to the surfaces of spongy mesophyll cells and then into the air spaces and out of the leaf.

▲ Figure 14 Veins in a leaf of *Gunnera manicata*

Inevitably, there is some loss of water by evaporation from the moist spongy mesophyll cell walls and diffusion out through the stomata. There is also some use of water in photosynthesis. Water is supplied to the leaf in xylem vessels, located in the leaf veins (Figure 14).

B3.1.8 Distribution of tissues in a leaf

Plan diagrams show the areas of tissues, but not individual cells. The lines indicate the junctions between tissues.

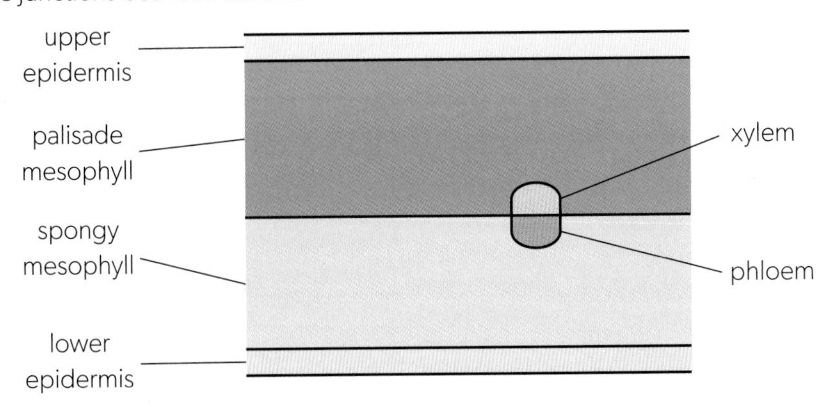

◀ Figure 15 An example of a tissue plan of a typical dicotyledonous leaf, with the leaf in transverse section

Data-based questions: Sun and shade leaves

On many trees there are differences in structure between leaves on upper branches that are in full sunlight and leaves on the same tree that are lower down and shaded.

1. Draw plan diagrams of the tissues in a representative part of each of the *Prunus* leaves. [7]
2. Compare and contrast the structure of the two leaves, including overall thickness, thickness of waxy cuticle and the structure of the palisade mesophyll and spongy mesophyll. [4]
3. a. Deduce which leaf grew in the sun and which in the shade. [1]

   b. Discuss reasons for the differences between sun and shade leaves. [3]

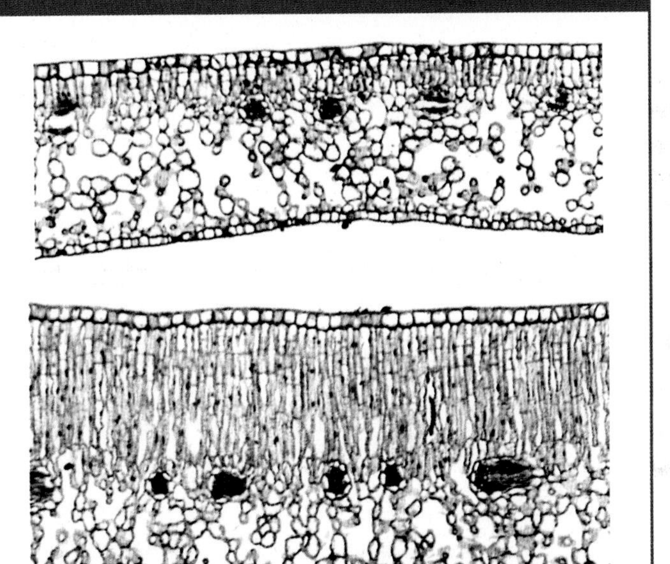

▶ Figure 16 Micrographs of two leaves of *Prunus caroliniana*, one that grew in the sun and one that grew in the shade. The micrographs have the same magnification

B3.1.9 Transpiration as a consequence of gas exchange in a leaf

Exchange of oxygen and carbon dioxide only works effectively if the gas-exchange surface is moist. Water evaporates from a moist surface unless the air is already saturated. Evaporation is the separation of individual water molecules from liquid water, so the molecules become part of a gas. Molecules in this separated state are called water vapour. The opposite process is condensation—water vapour molecules join others to become liquid.

If the air is very humid and the number of water molecules evaporating is equal to the number condensing, we say the air is saturated with water vapour. There will be no net loss from a gas-exchange surface. The amount of water vapour that air can hold when it is saturated varies with temperature. This is because there is more energy available to break hydrogen bonds between water molecules at higher temperatures.

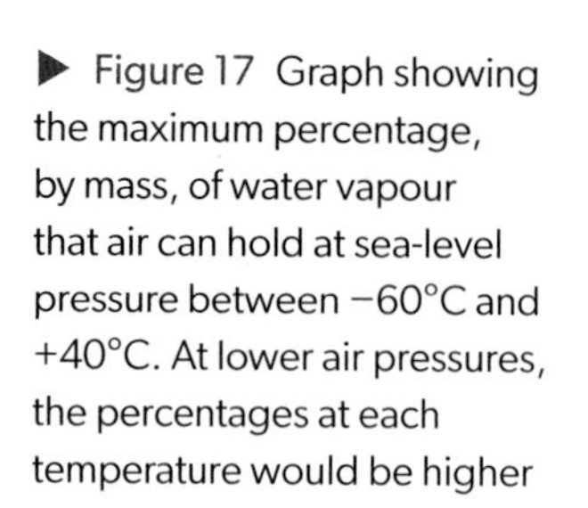

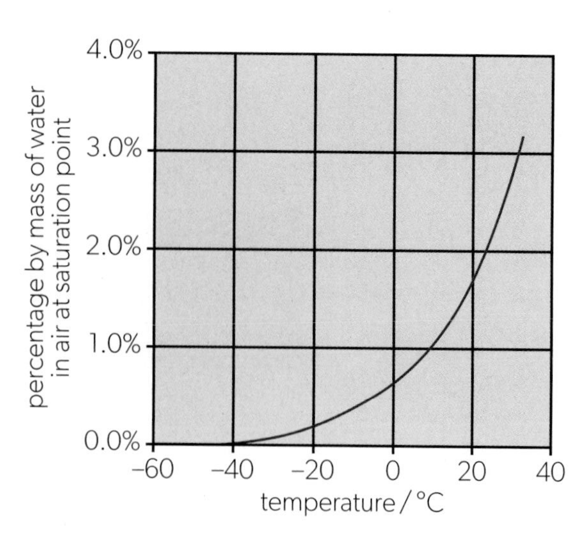

▶ Figure 17 Graph showing the maximum percentage, by mass, of water vapour that air can hold at sea-level pressure between −60°C and +40°C. At lower air pressures, the percentages at each temperature would be higher

The walls of the spongy mesophyll inside a leaf are kept moist for gas exchange. Some of this water will evaporate unless the air spaces are already saturated with water vapour. Unless the concentration of water vapour in the air outside the leaf is as high as the concentration in the air spaces, water vapour will diffuse out through the stomata. This causes the humidity of the air spaces to drop below the saturation point, so more water evaporates from spongy mesophyll cell walls. The loss of water vapour from the leaves and stems of plants is called transpiration.

Transpiration rates are affected by environmental factors.

- Temperature (positive correlation): at higher temperatures there is more energy available for evaporation. Also warmer air can hold more water vapour before becoming saturated.
- Humidity (negative correlation): the higher the relative humidity of the air, the smaller the concentration gradient of water vapour between the air spaces inside the leaf and the air outside, so the lower the rate of diffusion. There is no transpiration if the air outside the leaf is saturated with water vapour.

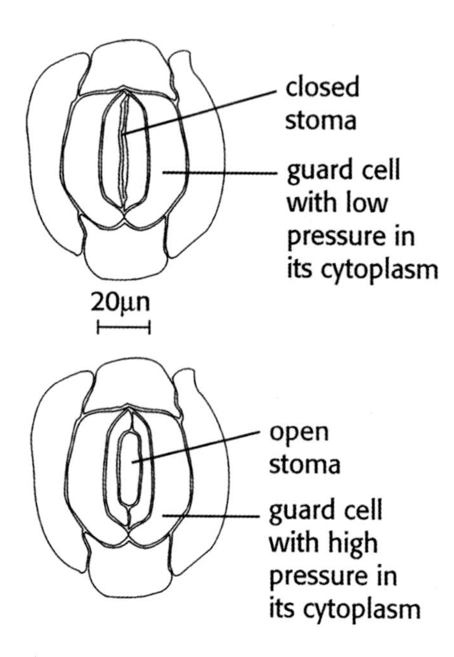

▲ Figure 18 Open and closed stomata

Plants minimize water losses using guard cells. These cells are found in pairs, one on either side of a stoma. Guard cells control the aperture of the stoma and can adjust from wide open to fully closed. Plants can prevent nearly all transpiration by closing their stomata. Most plants do this routinely at night when there is no photosynthesis. The disadvantage of closing the stomata in daylight is that little or no carbon dioxide can be absorbed, so the rate of photosynthesis is limited. Control mechanisms in the guard cells allow the aperture of the stomata to be varied according to the carbon dioxide concentration inside the leaf. Rising atmospheric carbon dioxide concentrations due to human activity are allowing plants to open stomata less widely, easing the problem of water loss a little.

B3.1.10 Stomatal density

Stomatal density is the number of stomata per unit area of leaf surface. To find the density, the number of stomata in a known area must be counted. Guard cells and stomata are too small to be seen with the naked eye but are easily visible with a microscope.

Two techniques can be used.

1. A sample of epidermis is peeled off the leaf. This is easy with *Commelina* and *Tradescantia*. Other species are worth trying. The leaf can be folded across to break all the tissues apart from the lower epidermis and then the epidermis can be peeled off or the leaf can be torn in half obliquely which often separates areas of epidermis. Small areas of epidermis are then mounted in water on a microscope slide and are examined.
2. Another technique can be used if the leaf is non-hairy and smooth. Colourless nail varnish is painted on to a small area of upper epidermis and lower epidermis. When it is dry, the nail varnish is peeled off, mounted on a microscope slide and examined. The nail varnish forms a cast of the leaf surface, with the margins of the cells and the stomata clearly visible.

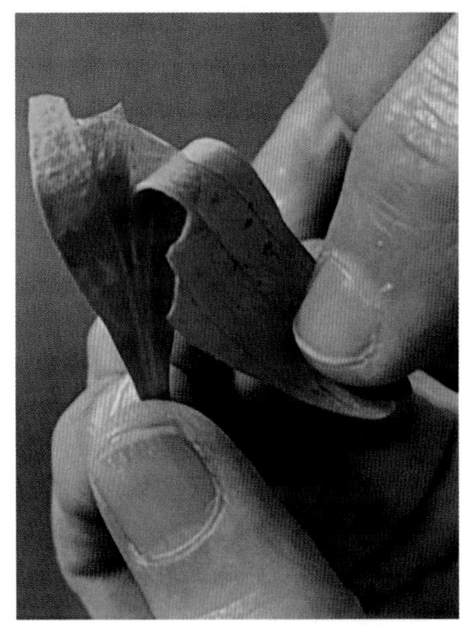

▲ Figure 19 The lower epidermis can usually be peeled more easily than the upper epidermis

The microscope slide should be moved until the field of view is filled by the peeled epidermis or leaf cast. The number of stomata can then be counted. Repeat counts should be carried out and a mean number of stomata calculated. If the area of the field of view is determined, the stomatal density can be calculated.

$$\text{stomatal density (mm}^{-2}) = \frac{\text{mean number of stomata}}{\text{area of field of view (mm}^2)}$$

Applying techniques: Using a potometer to measure rates of transpiration

Mechanisms involved in water transport in the xylem can be investigated using apparatus and materials that show similarities in structure to plant tissues. Figure 20 shows a potometer. This is a device used to measure water uptake in plants. The apparatus consists of a leafy shoot in a tube (right), a reservoir (left of shoot), and a graduated capillary tube (horizontal). A bubble in the capillary tube marks the zero point. As the plant takes up water through its roots, the bubble will move along the capillary tube. The distance the bubble travels and the time taken are measured. The tap below the reservoir allows the bubble to be reset to carry out new measurements.

▲ Figure 20 A potometer

Measurement: Repeat measurements to improve reliability

It is standard practice in scientific research to repeat measurements and replicate trials. In this case, samples should be taken from as many leaves on the plant and as many plants of the species as possible. For each leaf, as many areas as possible should be examined and a count of the number of stomata taken in each area. The counts will not be the same, but if done carefully, each count will be correct. The variability is natural in biological material.

Repeating the counts has several advantages. It helps avoid the danger of an outlier having a disproportionate effect on the conclusions. It increases reliability because it allows a mean to be calculated which will be closer to the true stomatal density than a single count is likely to be. It also allows the reliability of the mean to be assessed statistically. The less variation between the repeats, the more reliable the mean.

AHL

B3.1.11 Adaptations of foetal and adult haemoglobin for the transport of oxygen

Haemoglobin is the oxygen transport protein carried by red blood cells. Oxygen binds reversibly to haemoglobin. Each of the four subunits in a haemoglobin molecule has a haem group which acts as a binding site, so up to four molecules of oxygen can be transported per haemoglobin molecule.

Binding is cooperative in a haemoglobin molecule, because when oxygen binds to one haem group, conformational changes are caused that increase the oxygen affinity of the other haem groups. Conversely, when an oxygen dissociates, it causes conformational changes that reduce affinity in other haem groups. The two most probable states for a haemoglobin molecule are therefore fully saturated with four oxygens bound (the R state), or unsaturated with no oxygen bound (the T state).

The oxygen saturation level of haemoglobin is positively correlated with oxygen concentration. The oxygen concentrations are given as partial pressures, with kilopascals as the pressure units. As partial pressure of oxygen rises, percentage saturation rises until the partial pressure reaches 10 kPa, above which haemoglobin becomes 100% saturated. This happens as blood flows through the capillaries surrounding the alveoli, which have an oxygen concentration of between 10 kPa and 13 kPa (in normal healthy lungs). Fully oxygenated blood leaving the lungs is carried to all other organs of the body, where the partial pressure is usually below 10 kPa. At least some of the oxygen carried by the haemoglobin therefore dissociates (separates) and diffuses into the tissues of the organ.

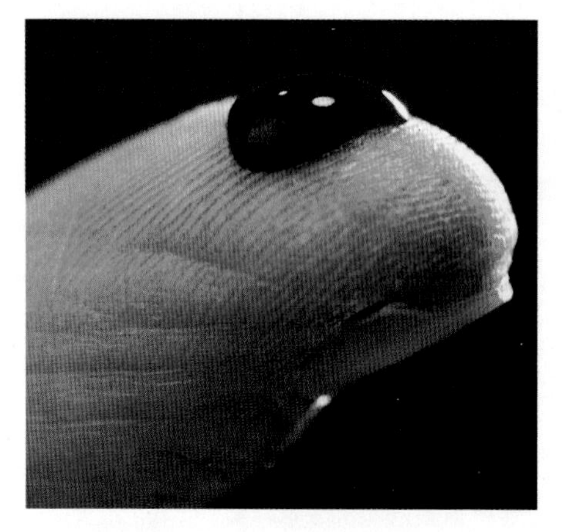

▲ Figure 21 Less than 50 cm^3 of oxygen can dissolve in 1 dm^3 of water at 37°C, but blood can hold over 200 cm^3 because of the oxygen binding capacity of haemoglobin

Because of cooperative binding, oxygen saturation of haemoglobin is not directly proportional to oxygen concentration. Instead, it changes from fully saturated to unsaturated over a relatively narrow range of oxygen concentrations. This ensures that haemoglobin unloads oxygen very readily in a tissue where aerobic respiration has reduced the oxygen concentration. Without this adaptation, concentrations of oxygen could not be kept as high as they are in respiring tissues, potentially reducing the activity of muscle and other tissues. Haemoglobin is also adapted for oxygen transport by interacting with carbon dioxide. The mechanisms for this are described in *Section B3.1.12*.

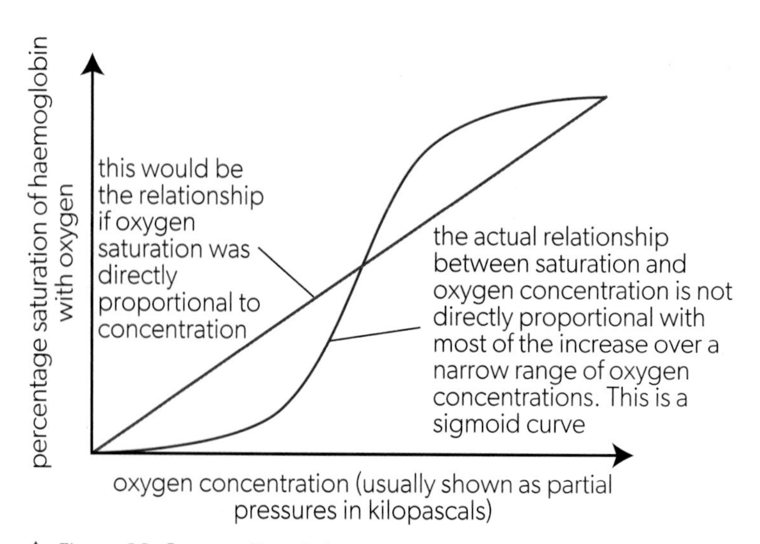

▲ Figure 22 Oxygen dissociation curve

Humans produce different types of haemoglobin before and after birth. At birth, a baby still has red blood cells with foetal haemoglobin. It takes several months for all the red blood cells carrying foetal haemoglobin to be replaced with cells carrying adult haemoglobin. Foetal haemoglobin has a stronger affinity for oxygen than adult haemoglobin. At any partial pressure of oxygen, foetal haemoglobin is therefore more saturated with oxygen than adult haemoglobin. During pregnancy a foetus obtains oxygen via the placenta. Oxygen dissociates from haemoglobin in maternal blood in the placenta and binds to haemoglobin in foetal blood. This can only happen because foetal haemoglobin has a stronger affinity for oxygen than adult haemoglobin.

B3.1.12 Bohr shift

Increased aerobic respiration in active tissues results in greater release of carbon dioxide into the blood. Increases in carbon dioxide concentration decrease the affinity of haemoglobin for oxygen and therefore increase dissociation of oxygen from haemoglobin. Two mechanisms cause the decrease in affinity.

1. Carbon dioxide and water are converted in red blood cells into hydrogen ions and hydrogen carbonate ions.

   $$CO_2 + H_2O \rightarrow H^+ + HCO_3^-$$

   This reduces the pH of the blood, which reduces the affinity of haemoglobin for oxygen. In the lungs where the concentration of carbon dioxide is low, the pH is 7.4. In active muscle, there is a higher carbon dioxide concentration and the pH typically is about 7.2. This small pH difference is enough to promote oxygen binding to haemoglobin in the lungs and dissociation in active respiring tissues.

2. Each of the four subunits of haemoglobin can react reversibly with carbon dioxide at the amino terminal of the polypeptide. The amine group is converted to carbamate and the haemoglobin becomes carbaminohaemoglobin.

   $$\text{haemoglobin} + 4CO_2 \rightleftarrows \text{carbaminohaemoglobin}$$

   This reaction reduces the affinity of haemoglobin for oxygen. Due to the high carbon dioxide concentration of actively respiring tissues, haemoglobin is converted to carbaminohaemoglobin, promoting release of oxygen. Carbaminohaemoglobin changes back to haemoglobin in the lungs, due to the low carbon dioxide concentration. The haemoglobin then becomes 100% saturated as it is carried in red blood cells through the alveolar capillaries. Another consequence of this mechanism is that haemoglobin molecules can each remove four carbon dioxide molecules from respiring tissues and transport them to the lungs.

   The reduction in the affinity of haemoglobin for oxygen in high carbon dioxide concentrations is known as the Bohr shift or Bohr effect. It helps to ensure that respiring tissues have enough oxygen when their need for oxygen is greatest.

B3.1.13 Oxygen dissociation curves as a means of representing the affinity of haemoglobin for oxygen at different oxygen concentrations

Oxygen dissociation curves show the percentage oxygen saturation of haemoglobin at different oxygen concentrations. Normal atmospheric pressure is 101.3 kPa and as 21% of air is oxygen, the partial pressure of oxygen is 21.2 kPa. The oxygen concentration inside the alveoli is lower, so oxygen dissociation curves usually only cover a range from 0 kPa to 15 kPa. Figure 23 shows the oxygen dissociation curve for adult haemoglobin. The sigmoid form of the curve is due to cooperative binding.

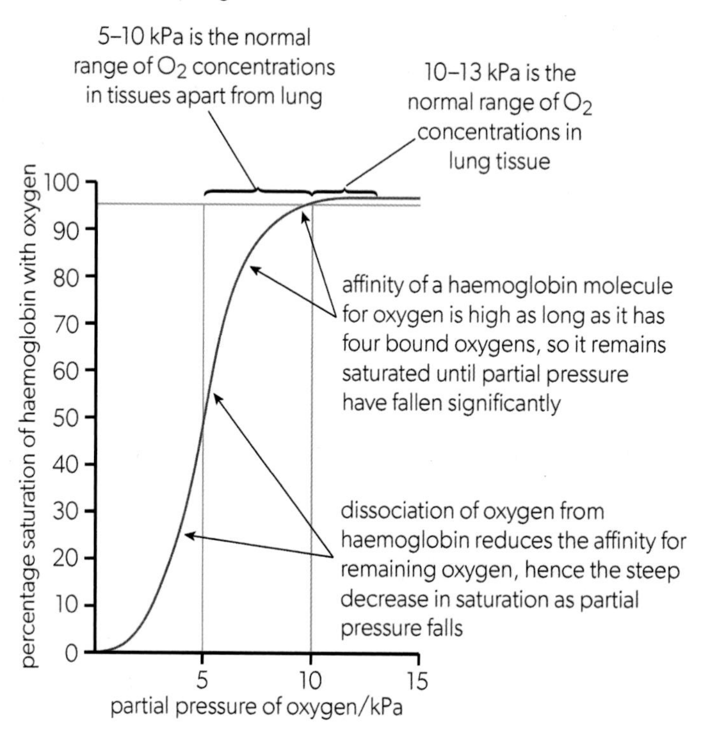

▶ Figure 23 The oxygen dissociation curve for adult haemoglobin is sigmoid because of the effects of cooperative binding

Figure 24 shows that the dissociation curve for foetal haemoglobin is sigmoid, like adult haemoglobin, but the curve is further to the left. An oxygen saturation curve that is displaced to the left indicates increased affinity for oxygen. Foetal haemoglobin has a stronger affinity than adult haemoglobin, so percentage saturation is higher at every partial pressure of oxygen.

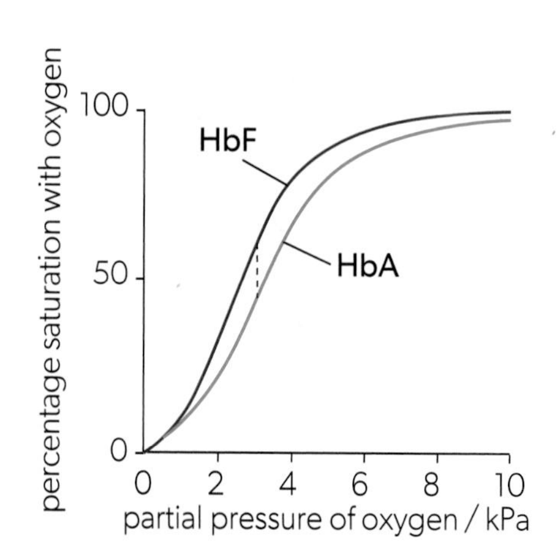

▶ Figure 24 A comparison of the oxygen dissociation curves of foetal (HbF) and adult haemoglobin (HbA)

Figure 25 shows the oxygen dissociation curve for adult haemoglobin at two different concentrations of carbon dioxide. The curve for the higher carbon dioxide concentration is displaced to the right, showing a decreased affinity of haemoglobin for oxygen. This is called the Bohr shift (see *Section B3.1.12*) and results in a greater release of oxygen from haemoglobin at the same partial pressure of oxygen.

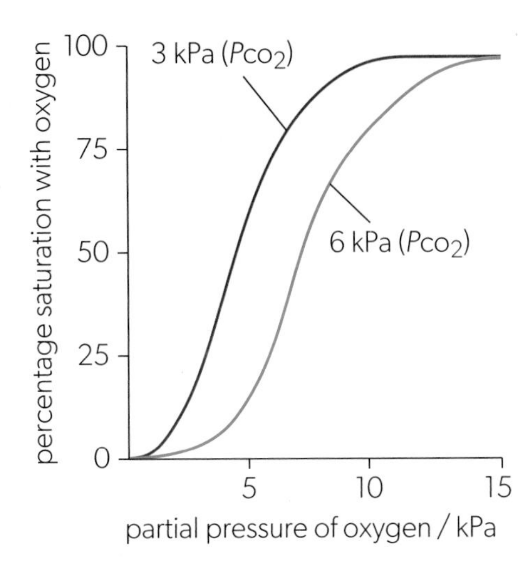

▲ Figure 25 The Bohr shift

Linking questions

1. How do multicellular organisms solve the problem of access to materials for all their cells?
   a. Outline the relationship between surface area and volume and the exchange of materials between cells and their environment. (B2.3.6)
   b. Outline the role of the circulatory system in the distribution of materials in the body. (B3.2.1)
   c. Describe the role of diffusion in gas exchange. (B2.1.3)

2. What is the relationship between gas exchange and metabolic processes in cells?
   a. Outline the movement of gases between capillaries and alveoli. (B3.1.1)
   b. List metabolic processes that produce carbon dioxide in the human body. (C1.2.11)
   c. Explain the role of oxygen in the mitochondrial electron transport chain. (C1.2.16)

B3.2 Transport

What are the differences and similarities between transport in animals and plants?

Figure 1 shows a "nodding donkey"—a tool for drawing underground oil to the surface. Initially oil flows up a well because of pressure in the oil-bearing rock, but eventually a pump is needed at the bottom of the well to pull the oil out of the well. Are fluids only pushed in mammalian circulatory systems or are they sometimes drawn? Why is fluid, like blood or sap, necessary for transport materials? To what extent is the movement of fluid in xylem and phloem like the closed circulatory system found in organisms like mammals? Can phloem and xylem be thought of as one circulatory system or two? What is the role of pressure differences in plant and animal circulatory systems?

▲ Figure 1 A "nodding donkey" for bringing oil to the surface

What adaptations facilitate transport of fluids in animals and plants?

Figure 2 is a satellite image showing the delta of the Lena River in Siberia. The reticulate pattern of channels in the Lena delta is reminiscent of blood capillaries or venation in a leaf. What role do pressure gradients play in the movement of fluids in living things and in rivers? What is the reason that vessel walls are typically structurally strengthened? Are there common features that plant and animal vessels have to reduce resistance to flow? What mechanisms exist to ensure that all parts of a plant or animal receive sufficient quantities of the materials that they require?

◀ Figure 2 The delta for the Lena River in Siberia

SL and HL	AHL only
B3.2.1 Adaptations of capillaries for exchange of materials between blood and the internal or external environment	B3.2.11 Release and reuptake of tissue fluid in capillaries
B3.2.2 Structure of arteries and veins	B3.2.12 Exchange of substances between tissue fluid and cells in tissues
B3.2.3 Adaptations of arteries for the transport of blood away from the heart	B3.2.13 Drainage of excess tissue fluid into lymph ducts
B3.2.4 Measurement of pulse rates	B3.2.14 Differences between the single circulation of bony fish and the double circulation of mammals
B3.2.5 Adaptations of veins for the return of blood to the heart	B3.2.15 Adaptations of the mammalian heart for delivering pressurized blood to the arteries
B3.2.6 Causes and consequences of occlusion of the coronary arteries	B3.2.16 Stages in the cardiac cycle
B3.2.7 Transport of water from roots to leaves during transpiration	B3.2.17 Generation of root pressure in xylem vessels by active transport of mineral ions
B3.2.8 Adaptations of xylem vessels for transport of water	B3.2.18 Adaptations of phloem sieve tubes and companion cells for translocation of sap
B3.2.9 Distribution of tissues in a transverse section of the stem of a dicotyledonous plant	
B3.2.10 Distribution of tissues in a transverse section of the root of a dicotyledonous plant	

B3.2.1 Adaptations of capillaries for exchange of materials between blood and the internal or external environment

Capillaries are the narrowest blood vessels with a diameter of about 10 μm. They branch and rejoin repeatedly to form a capillary network with a huge total length. Capillaries transport blood through almost all tissues in the body. Two exceptions are the lens and the cornea of the eye—these tissues must be transparent so there are no blood vessels.

Many narrow capillaries have a total surface area that is greater than fewer wider blood vessels. This means that the capillary network in any tissue increases the scope for diffusion between the blood and the tissue cells. The density of capillary networks in different tissues depends on the needs of the cells, but all active cells in the body are close to a capillary.

The capillary wall consists of one layer of endothelium cells (see Figure 3 on the next page). This layer of cells has a coating of extracellular fibrous proteins which are crosslinked to form a gel. The gel is called the basement membrane and it acts as a filter that allows small or medium-sized particles to pass through, but not macromolecules. There are pores between the epithelium cells, so the capillary wall is very permeable. The pores allow part of the blood plasma, but not the red blood cells, to leak out through the basement membrane.

The fluid that leaks out is very similar but not identical in composition to blood plasma. It is called tissue fluid. Tissue fluid contains oxygen, glucose and all other substances in blood plasma except large protein molecules, which are too large to pass through the basement membrane. The fluid flows between the cells in a tissue, allowing them to absorb useful substances and excrete waste products. The tissue fluid then re-enters the capillary network.

In some tissues, there are greater numbers of very large pores in the capillary walls. These are called fenestrated capillaries. Fenestrated capillaries allow larger volumes of tissue fluid to be produced, which speeds up exchange between the tissue cells and the blood. The glomerulus (filter unit) of the kidney has fenestrated capillaries so it can produce large volumes of filtrate in the first stage of urine production.

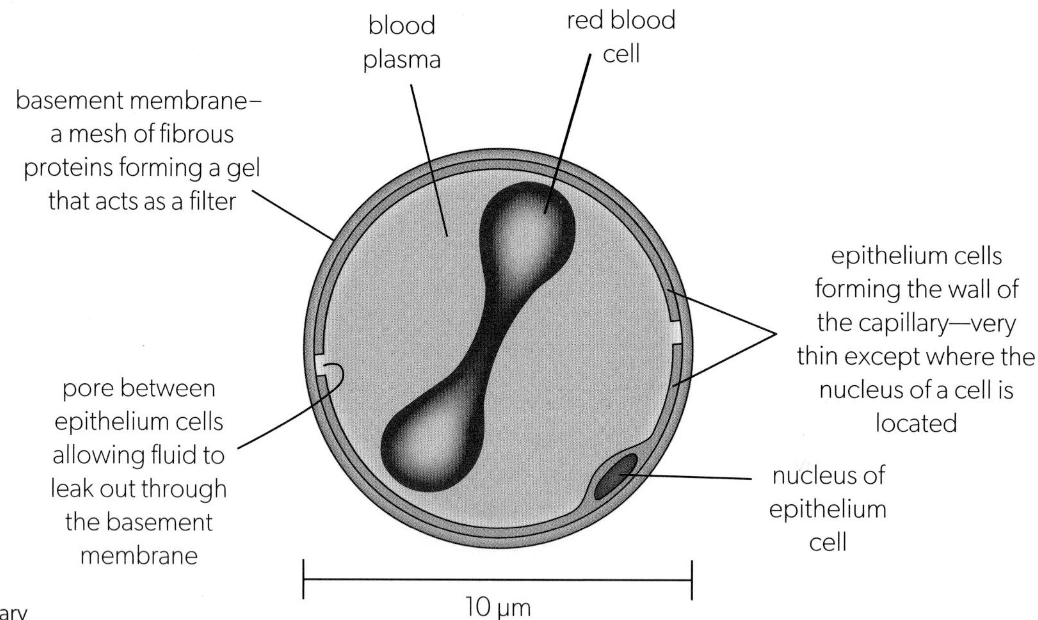

▶ Figure 3 Structure of a capillary

B3.2.2 Structure of arteries and veins

Arteries carry pulses of high-pressure blood away from the heart to the organs of the body. Veins carry a stream of low-pressure blood from the organs to the heart. Because of the difference in function, these two types of blood vessel have a different structure to their walls (Figure 4).

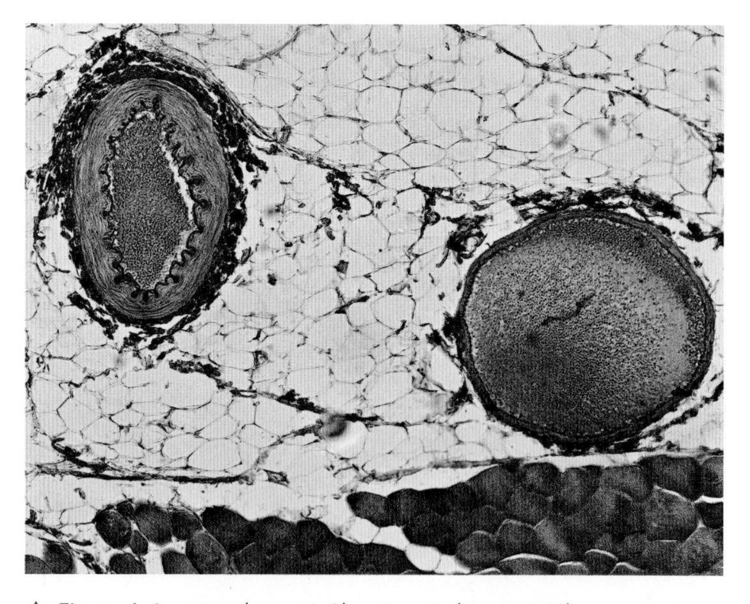

▲ Figure 4 An artery (upper left) and a vein (lower right) are surrounded by adipose (fat storage) tissue. The tissue below is muscle

▼ Table 1 Differences in structure of the walls of arteries and veins

Arteries	Veins
Thicker wall	Thinner wall
Narrower lumen	Wider lumen
Circular in section	Circular or flattened in section
Inner surface corrugated	No inner surface corrugation
Fibres visible in the wall	Few or no fibres visible in the wall

Table 1 lists the distinguishing features of arteries and veins that are easily distinguishable in micrographs.

B3.2.3 Adaptations of arteries for the transport of blood away from the heart

The wall of the artery is composed of several layers:

- tunica externa—a tough outer layer of connective tissue with collagen fibres
- tunica media—a thick layer containing smooth muscle and elastic fibres made of the protein elastin
- tunica intima—a smooth endothelium forming the lining of the artery; in some arteries the tunica intima also includes a layer of elastic fibres.

Each time the ventricles of the heart pump, a burst of blood under high pressure enters the arteries and flows along them. The pressure then declines until the next heartbeat. Arteries have relatively narrow lumens, which helps them to maintain high blood pressures and high velocities of blood flow.

Artery walls are relatively thick and contain elastic fibres and tough collagen fibres. The elastic fibres are proteins that can stretch and then recoil. Collagen fibres are tough rope-like proteins with high tensile strength. These features make arteries strong enough to withstand high and variable blood pressures without bulging outwards (known as an aneurysm) or bursting.

Elastic fibres make up as much as 50% of the dry mass of artery walls. Peak pressure in an artery (systolic pressure) causes the wall of an artery to be pushed outwards, widening the lumen and stretching the wall. When stretched, elastic fibres store potential energy. At the end of each heartbeat the pressure in arteries falls and the stretched elastic fibres return the energy by recoiling and squeezing the blood in the lumen. In this way, the elastic fibres help to reduce the amount of energy expended in transporting blood to the organs of the body.

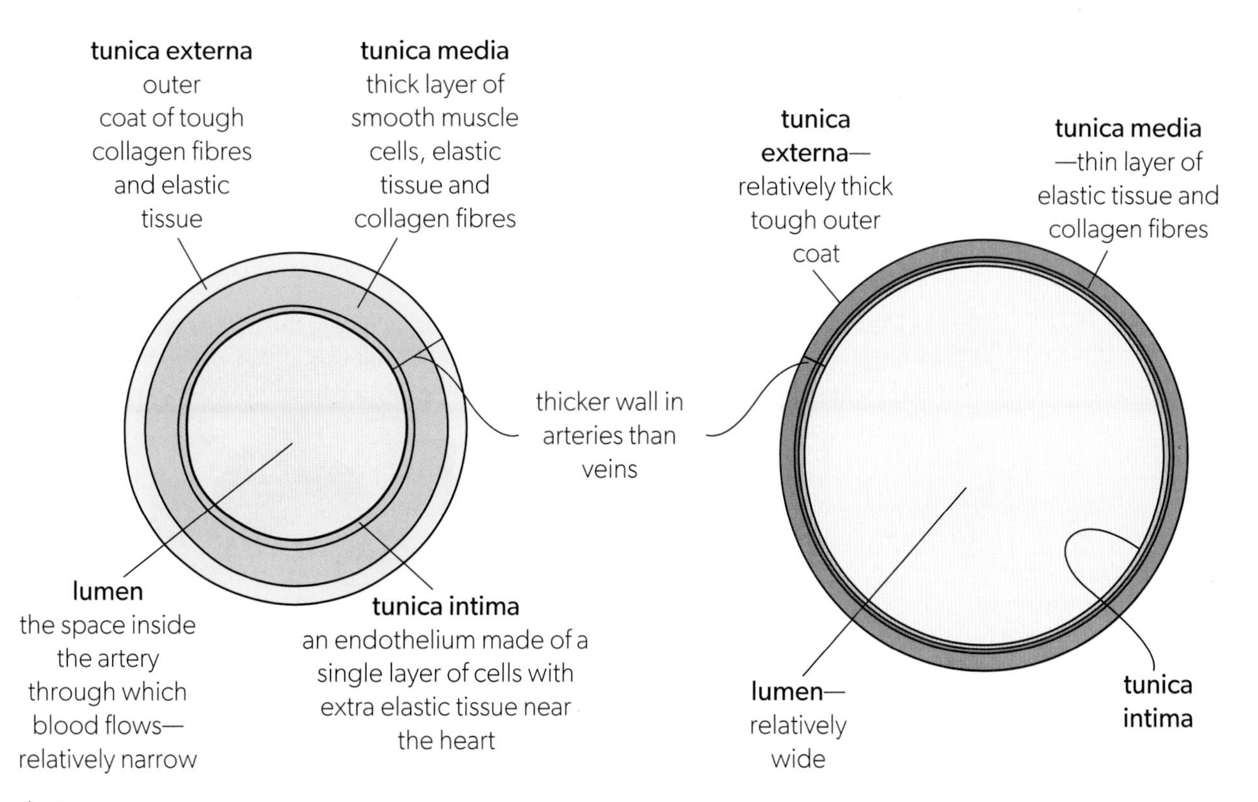

▲ **Figure 5** Tissue plan diagrams of artery and vein in transverse section

When the elastic fibres are recoiling and pushing on blood in an artery, the semilunar valves at the exit of the ventricles are closed. This means that blood cannot flow back towards the heart, it is forced onwards towards the organs. Elastic fibres therefore help to pump blood along the arteries and prevent the minimum pressure inside the artery (diastolic pressure) from becoming too low. They help to even out blood flow in the arteries.

Artery walls also contain smooth muscle cells with a particularly high density in the branches of arteries (called arterioles). The smooth muscle cells are circular (rather than radial or longitudinal) so when they contract, the diameter of the lumen is narrowed. This is called vasoconstriction and it reduces flow of blood along an artery or arteriole. When the smooth muscle cells relax, the lumen widens and blood flow is increased. This is vasodilation. The smooth muscle cells respond to hormone and neural signals and enable the body to adjust the flow rate of blood to tissues in each organ depending on availability and need.

ATL Communication skills: Drawing plan diagrams

A plan diagram is a drawing that shows the distribution of tissues in an organ. It does not show individual cells. Each line on the drawing represents the interface between two tissues. The low power objective of a microscope is usually used to observe the distribution of tissues, so that a plan diagram can be drawn. Unnecessary detail is avoided and if any of the areas of tissue are shaded, this is done very faintly. Everything on the drawing should be shown to the same magnification.

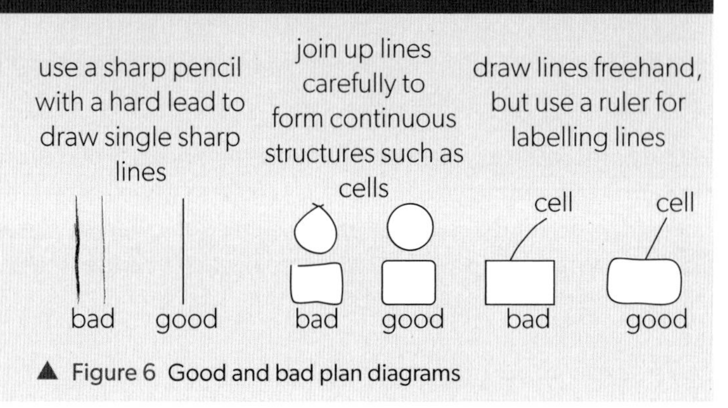

▲ Figure 6 Good and bad plan diagrams

B3.2.4 Measurement of pulse rates

Every time the heart beats, a wave of blood under high pressure passes along the arteries. Where an artery is close to the body surface, this pressure wave can be felt as a pulse. This is because the artery wall becomes stretched and then recoils. There is one pulse per beat of the heart, so measurement of pulse rate allows heart rate to be deduced. Pulse and heart rate are counted in beats per minute.

The wrist and the neck are two parts of the body where the pulse can often be felt. Two or three fingertips are pressed lightly against the skin where the artery is located. The thumb should not be used because it has a pulse which could cause confusion.

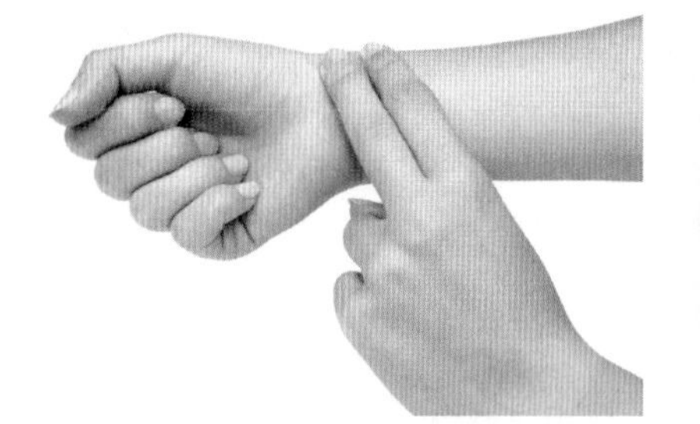

▶ Figure 7 The radial pulse is on the thumb side of the wrist. There is a carotid pulse on either side of your neck in the groove next to the windpipe. Counts can be done for a whole minute or for 30 seconds and then doubled

It is also possible to use digital meters to calculate pulse rate. Pulse oximeters are usually clipped to a fingertip. They have LEDs that shine red and infrared light through the finger and detectors to measure how much of the light passes through the tissues of the finger. This enables detection of variation in the amount of blood in the tissues each time the heart beats, and from this the heart rate is calculated. The percentage saturation of the blood with oxygen can also be deduced because deoxygenated blood absorbs red light whereas oxygenated blood absorbs infrared light.

Assessing reliability and accuracy of tools: Traditional versus digital estimation of heart rate

Try the traditional method of estimating heart rate by measuring pulse rate and also the more modern approach of using a pulse oximeter.

- Do you get the same estimates for your heart rate?

Devise a procedure for assessing the reliability and accuracy of the traditional and modern methods.

- Which method is more reliable?

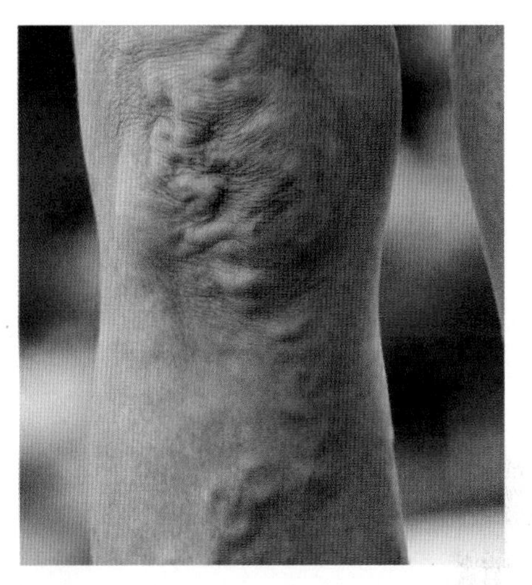

▲ Figure 8 Varicose veins develop in the legs when pocket valves become weakened or damaged. Blood can then flow backwards in the vein and accumulate, causing swelling and enlargement. Varicose veins usually develop in the legs because venous return to the heart is usually against gravity

B3.2.5 Adaptations of veins for the return of blood to the heart

Veins collect blood from all organs of the body and convey it back to the heart. Blood drains out of capillaries into veins continuously. This means there is no pulse. The wall of a vein contains far fewer elastic fibres than the wall of an artery. There are also far fewer smooth muscle cells because veins are not used to adjust blood flows to different parts of the body.

Blood in veins is at much lower pressure than in arteries, so the wall does not need to be thick to prevent bursting. The potential problem from the low blood pressure in a vein is backflow towards the capillaries and insufficient return of blood to the heart. To maintain circulation, veins contain pocket valves. These consist of three cup-shaped flaps of tissue projecting into the vein in the direction of blood flow.

- If blood starts to flow backwards, it gets caught in the flaps of the pocket valve, which fill with blood and close the valve. This blocks the lumen of the vein.
- When blood flows towards the heart, it pushes the flaps to the sides of the vein. The pocket valve therefore opens and blood can flow freely.

Blood flow in veins is assisted by gravity and by pressures exerted by adjacent tissues, especially skeletal muscles. Contraction makes a muscle shorter and wider so it squeezes on adjacent veins like a pump. The relatively thin walls of veins help because they allow a vein to be squeezed into a flatter shape. Walking, sitting or even just fidgeting greatly improves venous blood flow. Around 80% of the blood in a person at rest is in the veins, but this is reduced during vigorous exercise.

▲ Figure 9 Discuss the pattern of venous return that is occurring in the gymnast during this manoeuvre

Standing on your head

Pocket valves and vein walls become less efficient with age, causing poor venous return to the heart. Have you ever performed gymnastic moves such as headstands or handstands, or experienced very high g-forces on a ride at an amusement park? Young people can mostly do any of these activities easily but older people may not be able to. What is the explanation?

▼ Figure 10 A normal artery (top) has a much wider lumen than an artery that is occluded by atheroma (bottom)

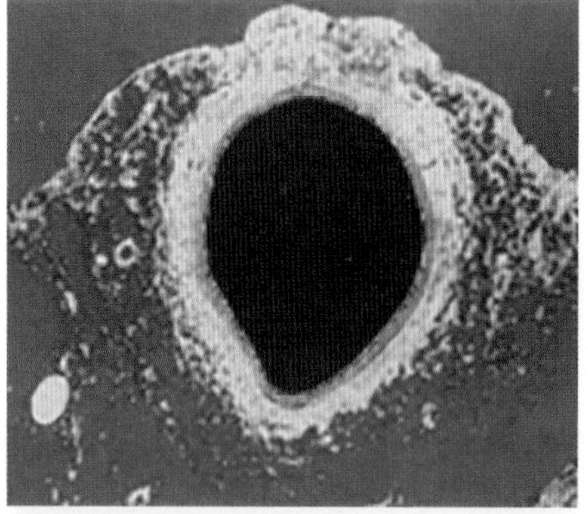

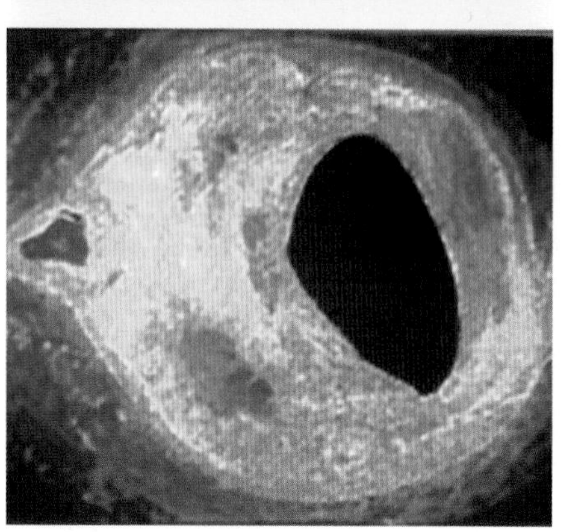

B3.2.6 Causes and consequences of occlusion of the coronary arteries

The aorta carries blood pumped by the left side of the heart to all organs of the body apart from the lungs. Two arteries branch off from the aorta close to its origin at the semilunar valve. They are the right coronary artery that supplies the right side of the heart and the left coronary artery, which branches into two arteries that supply the left anterior and left posterior regions of the heart wall. There are thus three main coronary arteries, each of which branches repeatedly to provide oxygenated blood to all parts of the muscular wall of the heart.

The coronary arteries can become narrowed or totally blocked by fatty deposits. The fatty deposits are called atheroma (plaque) and the blockage is an occlusion. The deposits build up in the wall of the artery and contain a variety of lipids including cholesterol. They restrict blood flow to the downstream region of the heart wall, often causing pain in the chest (angina) or shortness of breath, especially during exercise.

Fatty deposits in the artery wall can become impregnated with calcium salts, which harden the artery and make the inner surface rough. This tends to trigger the formation of a blood clot (thrombosis). Hypertension (high blood pressure) increases the risk of thrombosis. Blood clots can block the flow of blood to part of the muscular wall of the heart, depriving it of oxygen and preventing normal contractions. This is known as a heart attack.

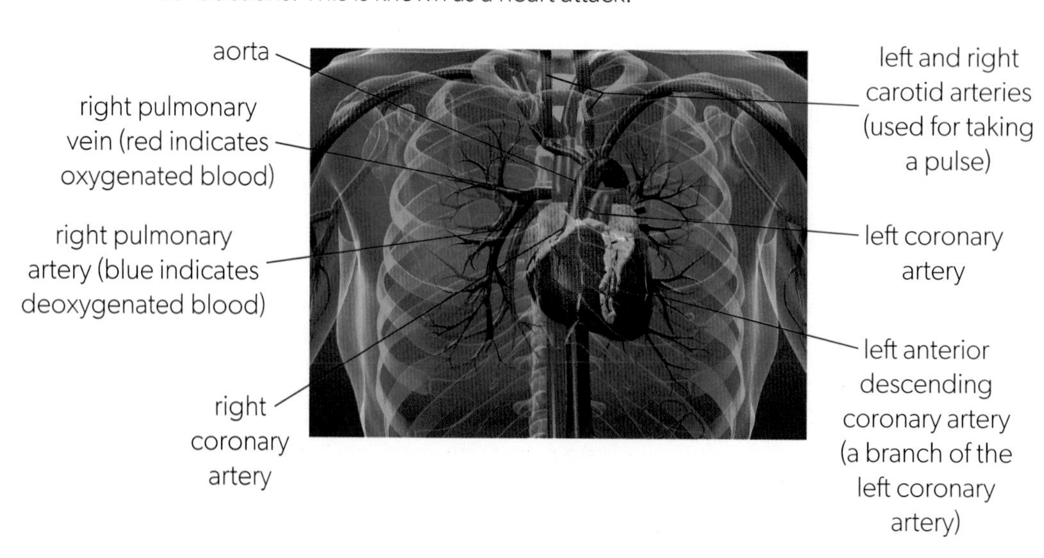

▲ Figure 11 The coronary arteries are the first branches off the aorta

If a blockage persists there will be tissue death and therefore permanent damage to the heart. Tissue death in heart muscle due to inadequate blood supply is called a myocardial infarction. The conditions associated with narrowed or blocked coronary arteries are collectively known as coronary heart disease (CHD).

Coronary heart disease is very common and there have been many epidemiological studies to try to identify risk factors and causes. Epidemiology is the study of nature and spread of diseases in the human population. Multiple risk factors have been identified:

- hypertension—raised blood pressure increases the chance of blood clot formation
- smoking—raises blood pressure because nicotine causes vasoconstriction
- eating too much saturated fat and cholesterol—promotes plaque formation
- obesity—associated with raised blood pressure and high blood cholesterol concentrations
- high salt intake—a large quantity of sodium chloride in the diet raises blood pressure
- drinking excessive amounts of alcohol—associated with raised blood pressure and obesity
- sedentary lifestyles—a lack of exercise is correlated with obesity and prevents the return of venous blood from the extremities leading to a greater risk of clot formation
- genetic predisposition—some genes increase the risk of hypertension and thrombosis
- old age—blood vessels become less flexible.

Data-based questions: Hypertension

Hypertension is a major risk factor for coronary heart diseases. In a major study, more than 316,000 males were followed for 12 years to investigate the effects of high blood pressure. Figure 12 shows the relationship between systolic and diastolic blood pressure and the effect on the death rate per 10,000 people per year. Systolic pressure is the maximum pressure reached in arteries after the ventricles have contracted. Diastolic pressure is the minimum pressure in the arteries just before the ventricles contract again.

1. Determine the death rate for a systolic blood pressure between 140 mmHg and 159 mmHg and a diastolic blood pressure between 75 mmHg and 79 mmHg. [1]
2. Describe the effect of systolic blood pressure and diastolic blood pressure on the death rate. [2]
3. Calculate the minimum difference between systolic and diastolic blood pressure where the death rate is highest. [1]
4. Evaluate the impact of differences between systolic and diastolic pressure on death rate. [3]

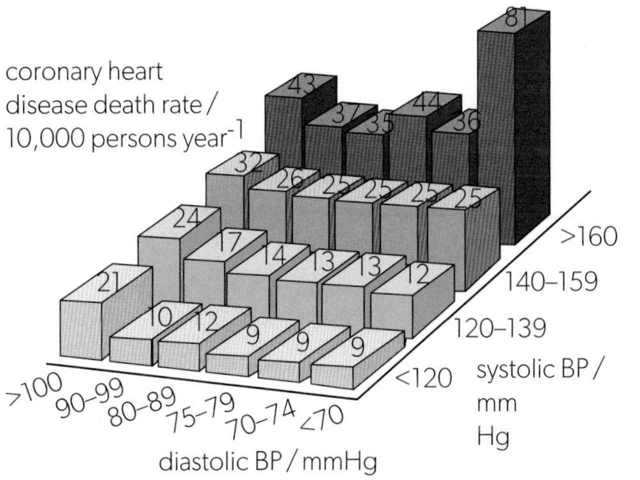

▲ **Figure 12** The effect of blood pressure on coronary heart disease

Patterns and trends: Correlation coefficients

A correlation is an association between two numerical variables. For example, we might expect a correlation between wingspan and body mass in blue jays (*Cyanocitta cristata*) or between average saturated fat intake per person and CHD rate of different countries. Two variables are **positively correlated** if higher values of one variable tend to correspond to higher values of the other variable. The variables are **negatively correlated** if higher values of one variable tend to correspond to lower values of the other variable. If there is no relationship between the variables, they are **uncorrelated** or **independent**.

Data can be displayed in a scatter diagram to show the extent to which two variables are correlated. If the points are fairly widely scattered, there is weak correlation. If the variables lie very close to a straight line with positive gradient there is strong positive correlation. If the variables lie very close to a straight line with negative gradient, there is strong negative correlation. A scatter diagram only shows a sample taken from the whole population, and with a small sample it is usually hard to reach conclusions about the whole population.

A correlation coefficient is a numerical measure of the level of association between two variables. It provides an objective assessment of the strength of the correlation. If the points on a scatter diagram lie exactly on a straight line with positive gradient, the correlation coefficient is 1. If the points lie exactly on a straight line with negative gradient, the correlation coefficient is −1. If there is no correlation, the correlation coefficient is 0. For most relationships, the correlation coefficient is somewhere between 1 and −1.

These numerical measures of correlation are used in testing hypotheses. High correlation coefficients can provide significant evidence of association between two factors, for example mean intake of saturated fatty acids per person and CHD rate of different countries. However, even a strong correlation such as this one does not prove a causal link. Scientific and medical research is needed to establish a biological mechanism by which saturated fat intake *causes* increased risk of CHD. However, the time and expense involved in finding such a mechanism would not be invested in unless a high correlation had been established first.

B3.2.7 Transport of water from roots to leaves during transpiration

Xylem is the tissue in plants that is used to transport water. Water is absorbed by roots and lost from leaves in transpiration, so the main flow of water is from the roots to the leaves. If suction is applied to the top of an air-filled tube with its base in water, the water is effectively pushed up the tube by atmospheric pressure and reaches a maximum height of 10.4 m. Trees can grow to more than 10 times this height and water flows to the top of them, so the mechanism used in plants must be different from just atmospheric pressure.

Xylem vessels are normally filled by xylem sap, which consists of water with relatively low concentrations of potassium, chloride and other ions. In a transpiring leaf, water is lost by evaporation from the cell walls of spongy mesophyll cells and then diffusion of water vapour out through the stomata. Cell walls contain a mesh of cellulose molecules which are hydrophilic and form hydrogen bonds with water. There is adhesion between the water and the cellulose of the cell walls. Loss of water therefore causes water to be drawn through the interconnected leaf cell walls in the pores between cellulose molecules. This process is a type of capillary action and is similar to the way that water is drawn through filter paper (also mostly composed of cellulose). The source of the water that is drawn through leaf cell walls is a xylem vessel in the nearest vein.

As the cell walls of leaf cells draw out water from xylem, they generate tensions (pulling forces). As long as there is a continuous column of water in a xylem vessel, these tensions are transmitted from the leaves down to the roots. This is called transpiration pull and is strong enough to move water upwards, against the force of gravity, to the top of the tallest tree. For the plant, it is a passive process; all the energy needed comes from the thermal energy (heat) that causes transpiration.

The pulling of water upwards in xylem vessels depends on the cohesion that exists between water molecules. Many liquids would be unable to resist the very low pressures in xylem vessels and the column of liquid would break. This is called cavitation and it does occasionally happen even with water, but it is unusual. Even though water is a liquid, it can transmit pulling forces in the same way that a solid length of rope does.

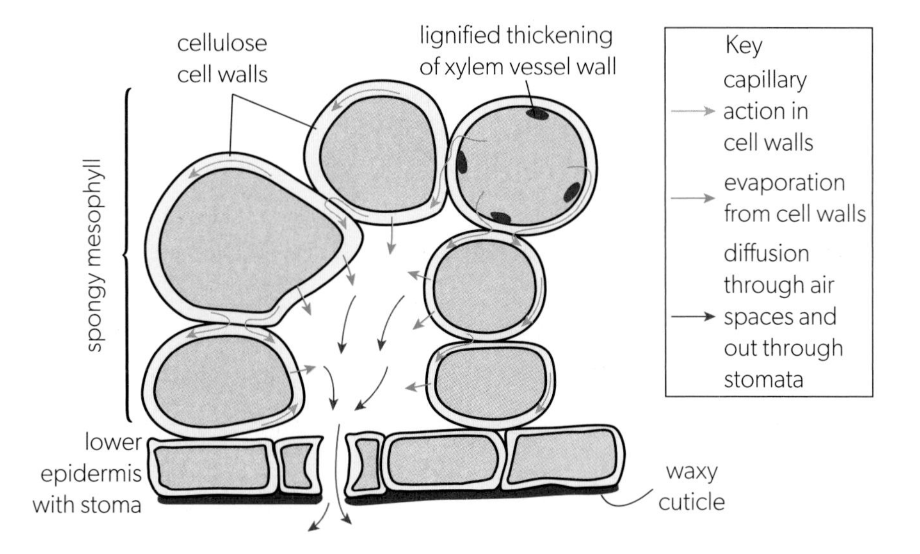

▲ Figure 13 Water movements in the leaf

Modelling water movement through cell walls

- How high do you expect water to rise in the paper towel?
- Do you expect any difference between the stoppered and unstoppered tubes?
- What other factors could be varied?

Test your hypotheses by setting up experiments.

- How does this model help us to understand water movement in plants?

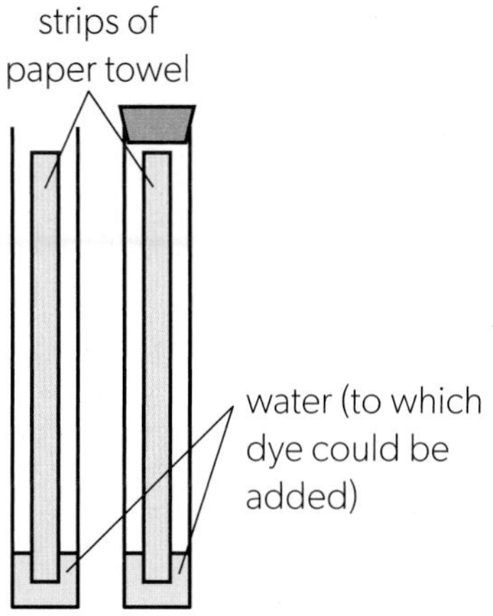

strips of paper towel

water (to which dye could be added)

◀ **Figure 14** Tall tubes with paper towel strips dipping into water, one sealed and one open to the air

▲ Figure 15 Light micrograph of a vertical section of the primary wood or xylem of a tree showing wood vessels with lignified supporting thickenings

B3.2.8 Adaptations of xylem vessels for transport of water

The structure of xylem vessels allows them to transport water inside plants very efficiently. They are formed from columns of cells, arranged end-to-end. The cell wall material between adjacent cells in the column is largely removed and the plasma membranes and contents of the cells break down. This creates long continuous tubes, with minimal resistance to the flow of xylem sap. When mature, xylem vessels are non-living, so the flow of water along them must be a passive process.

The vessel walls are thickened, and the thickenings are impregnated with a polymer called lignin. The pressure inside xylem vessels is usually much lower than atmospheric pressure and there is commonly tension (negative pressure potential) but the strength of the walls prevents the xylem vessel from collapsing.

The lignified wall thickenings are impermeable to water but there are always gaps in the thickening through which water can enter and exit. In the xylem vessels formed by young plants, the wall thickenings are in rings or helices with large gaps for water passage. In older plants, the wall thickenings are more extensive, with holes called pits through which water can pass.

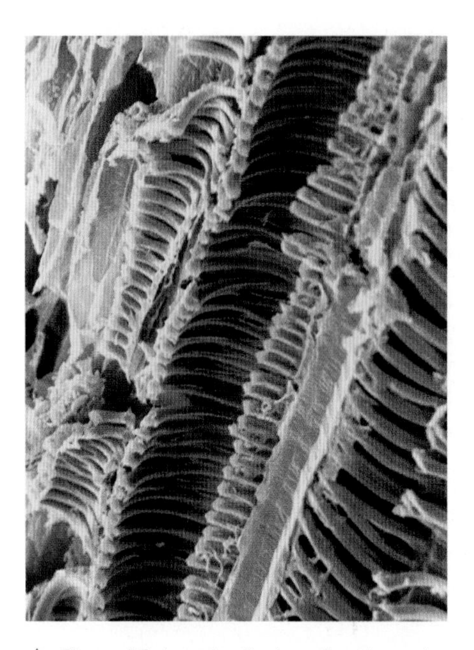

▲ Figure 16 Longitudinal section through a rhubarb stem, *Rheum rhaponticum*. Cut xylem vessels are coloured brown. Xylem vessels are reinforced and strengthened with spiral bands of lignin. Spiral bands allow xylem vessels to elongate and grow lengthwise and also retain their flexibility

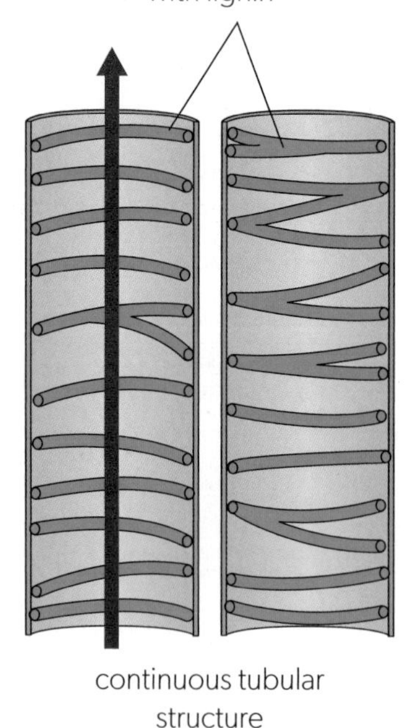

▲ Figure 17 Structure of xylem vessels in growing parts of the plant such as roots and stems

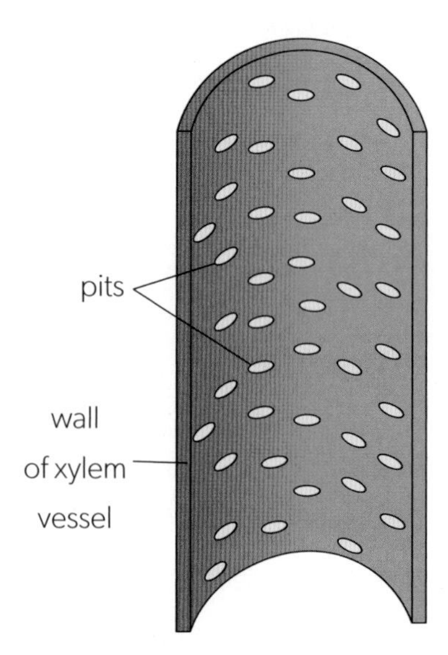

▲ Figure 18 Structure of xylem vessels in parts of the plant that are thickening such as the trunk or roots of a tree

B3.2.9 Distribution of tissues in a transverse section of the stem of a dicotyledonous plant

The outer layer of cells in all parts of a young plant is the epidermis. The stems of dicotyledonous plants (dicots) typically have transport tissue in vascular bundles near the epidermis. Dicots have two seed leaves and include sunflowers, peas and oaks and most other flowering plants. The xylem is usually on the inner side of a vascular bundle and the phloem on the outer side, with cambium consisting of stem cells between. The other tissues that usually occur in a dicot stem are pith in the centre and cortex near the epidermis.

▼ Table 2 Plant tissues and their functions

Tissue	Main functions in stems
Xylem	Transport of water from roots to leaves
Phloem	Transport of sugars from leaves to roots
Cambium	Production of more xylem and phloem
Epidermis	Waterproofing and protection
Cortex	Support and photosynthesis
Pith	Bulking out the stem

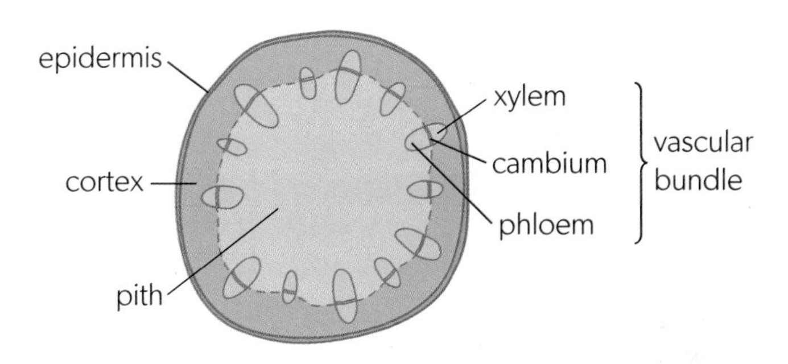

▲ Figure 19 Tissue plan of a stem in a typical dicotyledonous plant

Application of skills: Drawing stem tissue plans

Full instructions for drawing tissue plans are given in *Section B3.1.8*. The arrangement of tissues is not the same in all plant stems, so it is essential either to view a stem section with a microscope while drawing a plan, or view a micrograph, such as that in Figure 20.

There are many variations in stem structure. Some dicots such as elder (*Sambucus nigra*) have a hollow centre to the stem instead of pith. Some dicots such as cucumber (*Cucumis sativa*) have two areas of phloem in each vascular bundle, instead of just one. Some dicots such as sunflowers (*Helianthus annuus*) have sclerenchyma tissue (fibres) adjacent to each vascular bundle. These variations in stem structure can be shown by drawing tissue plans. Figure 20 is a micrograph showing one vascular bundle in part of a young sunflower stem. Draw a tissue plan based on this micrograph.

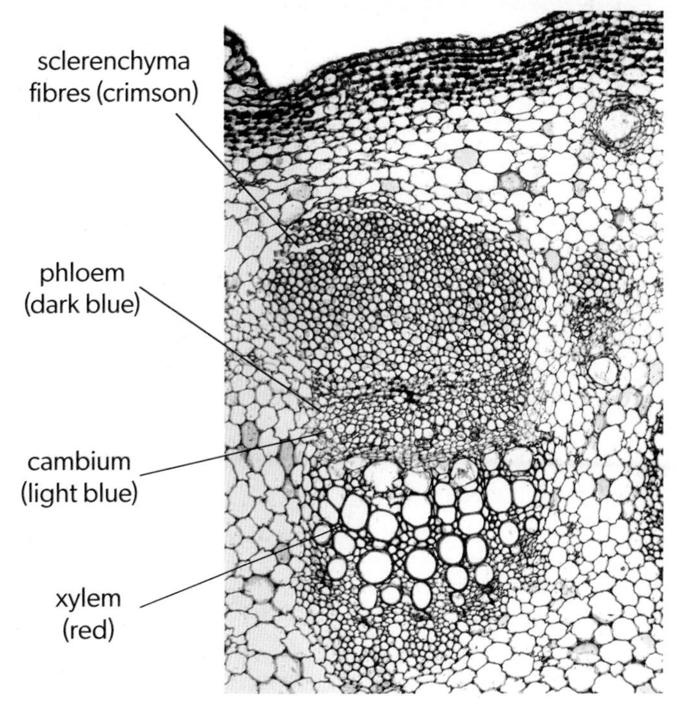

▲ Figure 20 Light micrograph of a section through a young stem from a sunflower (*Helianthus annuus*), showing one of the many vascular bundles

B3.2.10 Distribution of tissues in a transverse section of the root of a dicotyledonous plant

The distribution of tissues in stems and roots of dicotyledonous plants is different. All the vascular tissue is grouped in the centre of a root, with xylem in a star-shaped area and phloem between the points of the star. The xylem vessels can be identified by their large size, thick walls and rounded shape in transverse section. Xylem walls may be stained red in microscope images because they are lignified. Other cells in the root are unlignified and are usually stained blue. Phloem cells are smaller than xylem with thinner walls. The outer layer of cells in the root is epidermis, with small cells that may have root hairs protruding. Between the vascular tissue and the epidermis there is cortex, with relatively large and thin-walled cells.

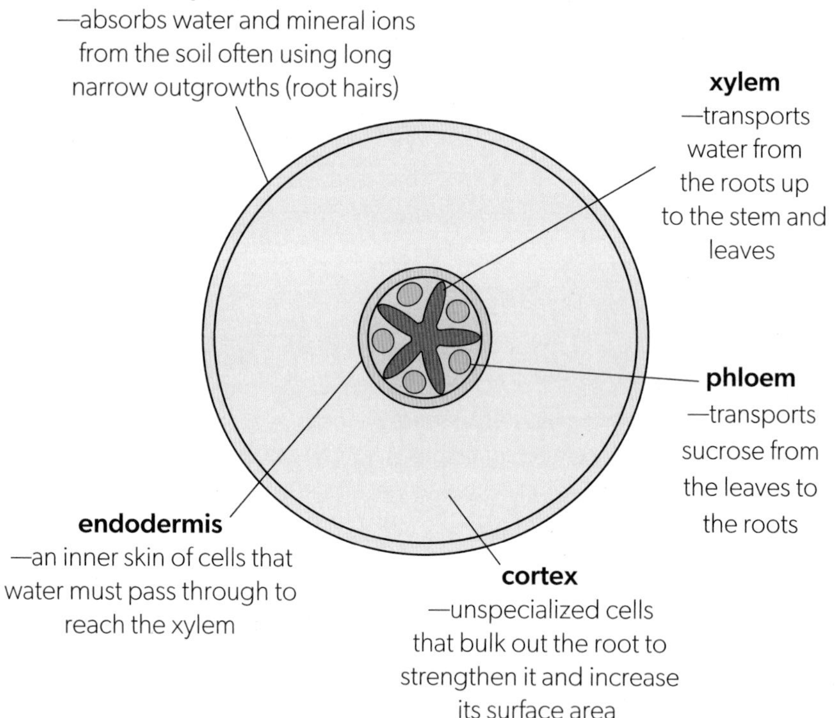
epidermis
—absorbs water and mineral ions from the soil often using long narrow outgrowths (root hairs)
xylem
—transports water from the roots up to the stem and leaves
phloem
—transports sucrose from the leaves to the roots
endodermis
—an inner skin of cells that water must pass through to reach the xylem
cortex
—unspecialized cells that bulk out the root to strengthen it and increase its surface area

▲ Figure 21 Tissue plan of a dicot root

Application of skills: Drawing root tissue plans

Figure 22 shows part of a root of *Allium tuberosum* in transverse section. Draw a tissue plan to identify the tissues.

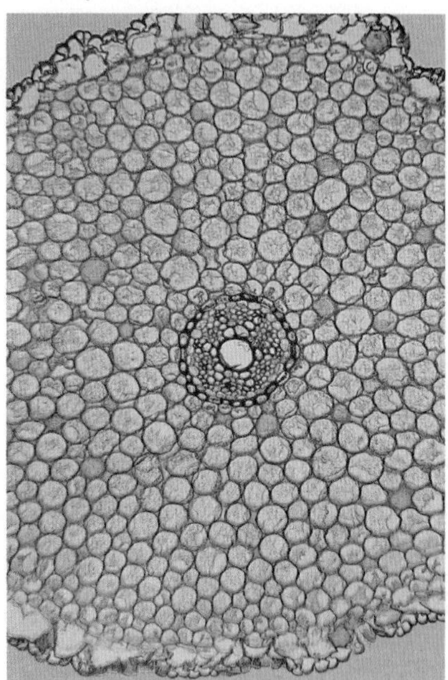

▲ Figure 22

B3.2.11 Release and reuptake of tissue fluid in capillaries

Plasma is the fluid in which the blood cells are suspended. It consists of water with many different dissolved substances: glucose, amino acids, mineral ions such as chloride and sodium, vitamins, hormones and plasma proteins. Lipids are carried in lipoprotein droplets. The structure of the capillary wall (described in *Section B3.2.1*) is adapted to let part of the blood plasma leak out into spaces between the cells in a tissue. Most protein molecules are too large to pass through the basement membrane so are retained in the plasma, but other substances can pass out through the capillary wall.

The fluid that leaks out of capillaries is a type of extracellular fluid, called tissue fluid. At any time, there are about 14 dm^3 of this tissue fluid in the tissues of a 70 kg human, so it constitutes about 20% of body mass. There is a continual process of release and reuptake of tissue fluid. Capillaries that are close to an arteriole tend to release tissue fluid because the blood supplied by the arteriole is at high pressure. Reuptake tends to happen in capillaries that are close to a venule, where the blood pressure is much lower.

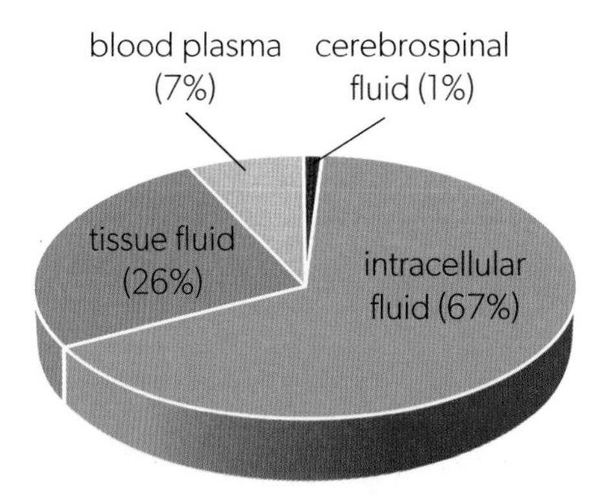

▲ Figure 23 Distribution of water in the human body

B3.2.12 Exchange of substances between tissue fluid and cells in tissues

Tissue fluid contains oxygen, glucose and all other substances in blood plasma apart from large protein molecules, which cannot pass through the capillary wall. The fluid flows between the cells in a tissue, allowing the cells to absorb useful substances. Oxygen is absorbed from the tissue fluid by diffusion because the oxygen concentration in cells is lower due to aerobic respiration. Glucose, also used in aerobic respiration, is absorbed by sodium–glucose cotransporters. Growing cells absorb amino acids by active transport.

Carbon dioxide, produced by cell respiration, diffuses out of cells into the tissue fluid, along with other waste products of metabolism. As tissue fluid flows between the cells of a tissue it accumulates dissolved waste products. The tissue fluid then re-enters the capillary network, becoming part of the blood plasma. The capillaries merge to form venules, which carry the waste products out of the tissue. Carbon dioxide is excreted by the lungs and other waste products are detoxified by the liver or excreted by the kidneys.

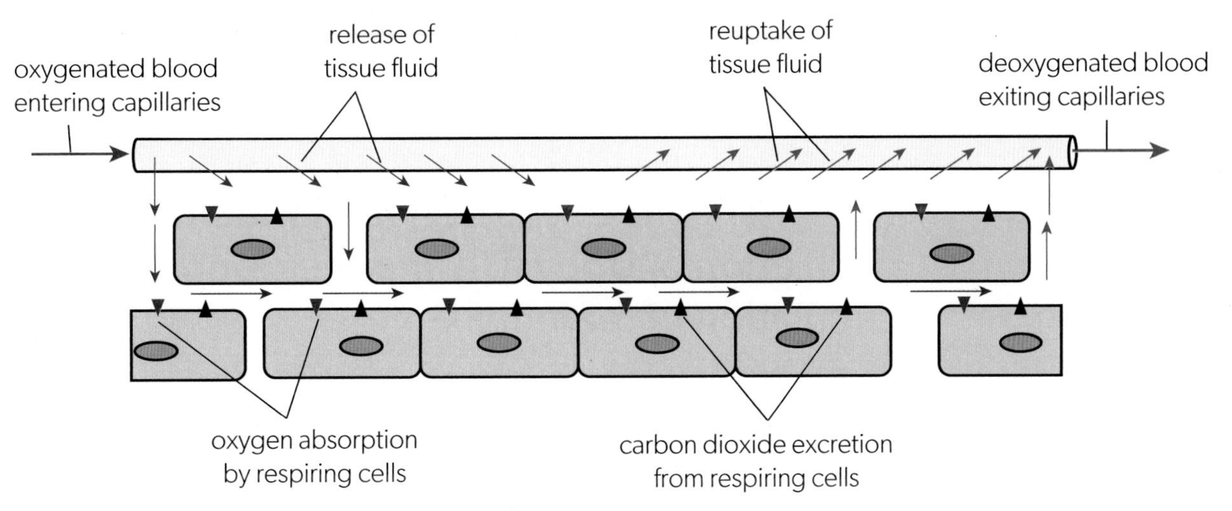

▲ Figure 24 Exchange processes in respiring tissues

AHL

B3.2.13 Drainage of excess tissue fluid into lymph ducts

Most of the tissue fluid released by capillaries returns to them, but some does not. Of the 20 dm^3 of tissue fluid produced per day in an average adult's body, 17 dm^3 return to the capillaries. If the other 3 dm^3 of fluid stayed in tissues it would cause swelling, called oedema. This is prevented by the drainage of tissue fluid into vessels of the lymphatic system.

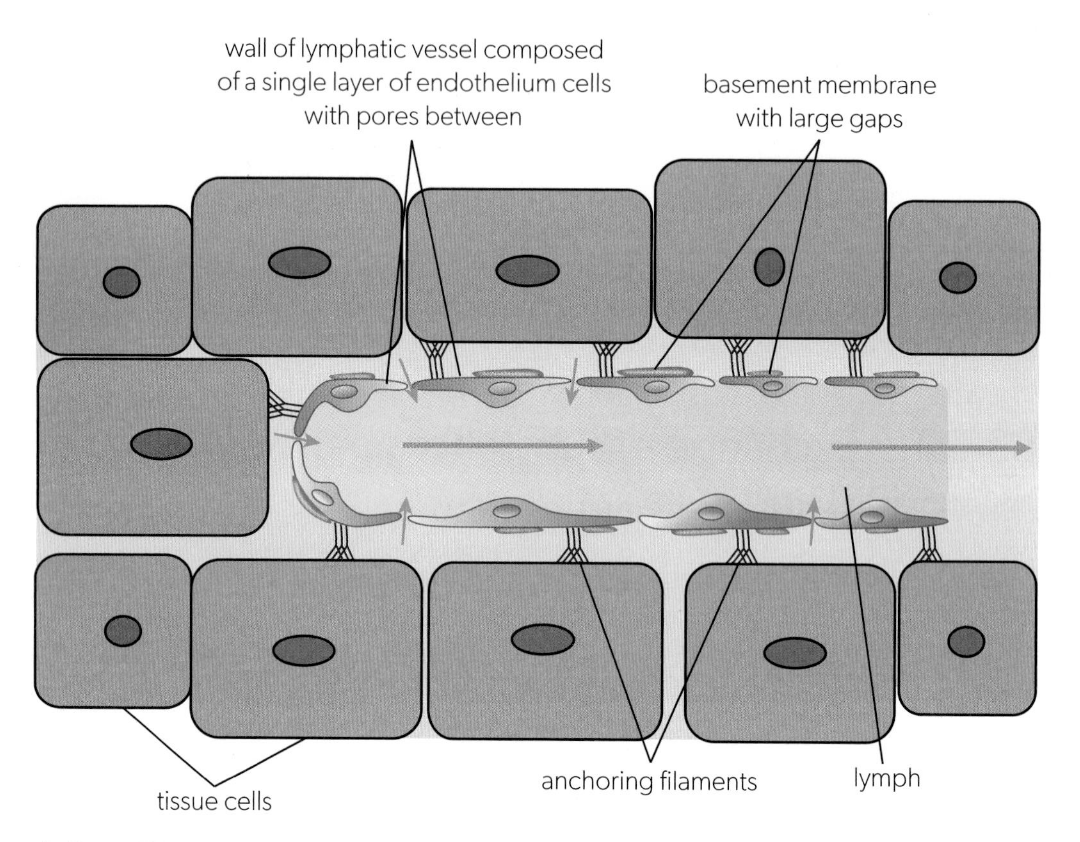

▲ Figure 25 Narrow ending of a lymph vessel into which excess tissue fluid drains

In all tissues, there are narrow blind-ended lymphatic vessels with permeable walls through which tissue fluid can pass. After entering the lymphatic vessels, the fluid is known as lymph rather than tissue fluid. The narrow vessels join up repeatedly to form wider lymphatic vessels. At the end of this system of vessels, there are just two—the left and right lymphatic ducts. These merge with the subclavian veins. Lymph is therefore drained off from all tissues of the body and is returned to the blood system. Blood in the subclavian veins flows into the vena cava and on to the right side of the heart.

B3.2.14 Differences between the single circulation of bony fish and the double circulation of mammals

There are valves in mammalian veins and heart that ensure a one-way flow, so blood circulates through arteries, capillaries and veins. Mammals pump blood to the lungs to be oxygenated. The blood must be at relatively low pressure to prevent capillaries in the alveoli from bursting. After flowing through the alveolar capillaries, the residual pressure is too low for the blood to flow on to other organs of the body, so it returns to the heart to be re-pumped. Oxygenated blood returning from the lungs must not mix with deoxygenated blood being pumped to the lungs, so the heart has separate left and right sides.

The left side of the heart receives oxygenated blood and pumps it to all organs of the body apart from the lungs. This requires relatively high blood pressure. The kidneys in particular carry out pressure filtration of blood, so need much higher blood pressure than the lungs. With a few notable exceptions, oxygenated blood pumped by the left side of the heart flows through capillaries in only one organ of the body and then returns to the heart deoxygenated and at much lower pressure. It returns to the right side of the heart, which pumps blood to the lungs.

Mammals have a double circulation, with the blood passing twice through the heart to make a full circuit. The heart is a double pump, delivering blood under different pressures to different organs of the body. The two circulations are known as the pulmonary and systemic circulations. The pulmonary circulation receives deoxygenated blood that has returned from the systemic circulation, and the systemic circulation receives blood that has been oxygenated by the pulmonary circulation.

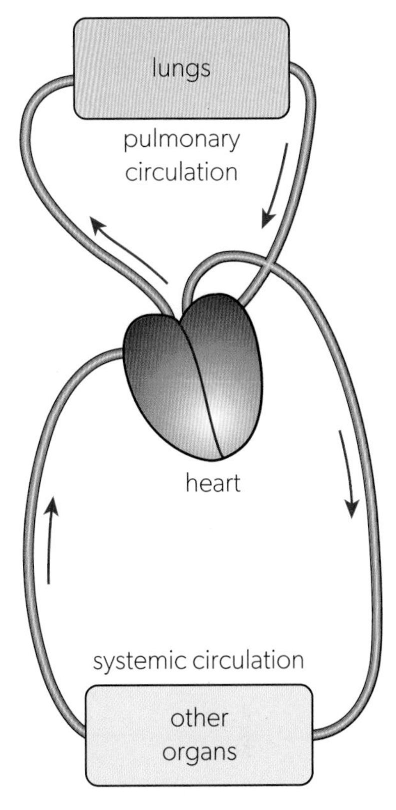

▲ **Figure 26** The double circulation of mammals

Fish pump blood to their gills to be oxygenated. The blood flows through capillaries in narrow gill filaments. Water is pumped over the gill filaments and oxygen diffuses from water to the blood; carbon dioxide moves in the opposite direction. The blood can be pumped at high pressure to the gills because the surrounding water provides support and reduces the risk of capillaries bursting. After flowing through the gills, the blood is oxygenated and still has enough pressure to flow directly to another organ of the body. While passing through capillaries in one organ, the blood becomes deoxygenated and its pressure falls, so it must return to the heart for re-pumping to the gills. Fish thus have a single circulation.

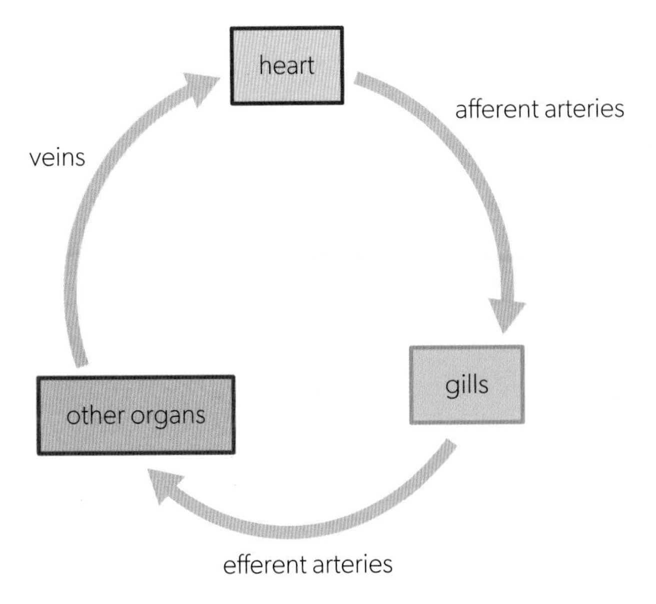

▲ **Figure 27** The single circulation of fish

B3.2.15 Adaptations of the mammalian heart for delivering pressurized blood to the arteries

The mammalian heart has evolved to pump pressurized blood to the organs of the body continuously throughout our lives. It is well adapted through its form to carry out this function.

- Ventricles—chambers with a strong muscular wall that can generate high blood pressure when it contracts, pumping blood out into the arteries.
- Atria—chambers with a thinner muscular wall that collect blood from the veins and pump it on to the ventricle, so the ventricle is as full as possible when it contracts and the atrium is as empty as possible so it can collect more blood from the veins.
- Atrioventricular valves between the atria and the ventricles. These valves close to prevent backflow of blood to the atria when the ventricles contract and open to allow blood to flow from the atria to the ventricles when the ventricles relax.
- Semilunar valves between the ventricle and the artery. These valves close to prevent backflow of blood to the ventricles when the ventricles relax and open to allow blood to flow from the ventricles to the arteries when the ventricles contract.
- Cardiac muscle—specialized muscle tissue that forms the wall of the ventricles and atria. Cardiac muscle has branched cells and connections between plasma membranes of adjacent cells that allow electrical signals to be propagated throughout the wall of the heart. This enables coordinated contractions. Cardiac muscle is unique in the body because it can contract without stimulation from motor neurons. The contraction is called myogenic; this means that it is generated in the muscle itself. The membrane of a heart muscle cell depolarizes when the cell contracts and this activates adjacent cells, so they also contract. A group of cells therefore contracts almost simultaneously at the rate of the fastest.

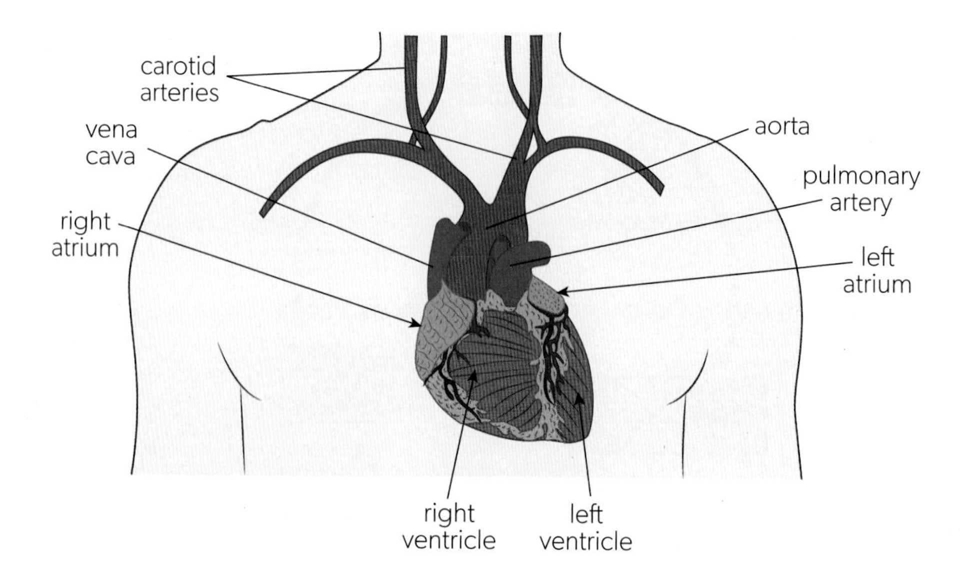

▲ **Figure 28** Exterior view of the heart in its position in the thorax

- Pacemaker—the sinoatrial node in the wall of the right atrium initiates each heartbeat by sending an electrical signal into the atria. The interval between signals determines the rate (pace) of the heartbeat. The sinoatrial node is the region of the heart with the fastest rate of spontaneous beating. The cells of the sinoatrial node have few of the proteins that cause contraction in other muscle cells, but they have extensive membranes. The sinoatrial node can initiate each heartbeat because the membranes of its cells are the first to depolarize in each cardiac cycle.
- Septum—the wall of the heart between the left and right ventricles and between the left and right atria. It prevents oxygenated blood in the left side of the heart from mixing with deoxygenated blood in the right side.
- Coronary vessels—coronary arteries and veins in the wall of the heart. Coronary arteries carry oxygenated blood from the aorta to all parts of the heart wall, supplying oxygen and glucose. Coronary veins collect deoxygenated blood from the heart wall and return it to the right atrium.

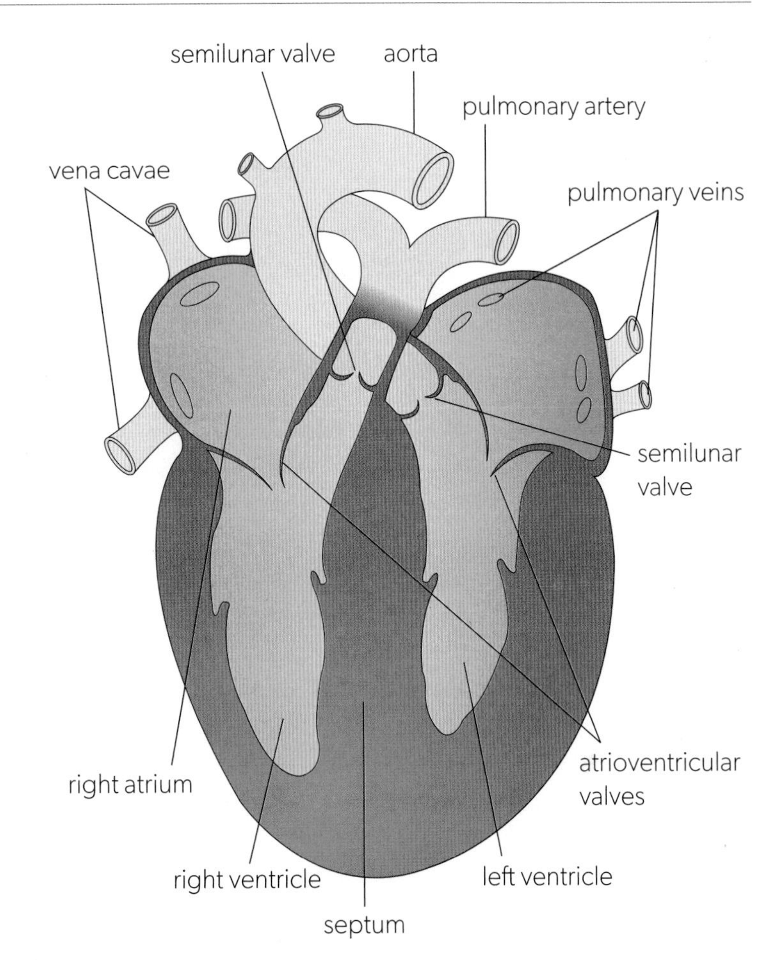

▲ Figure 29 Structure of the heart

Structure and function of the heart

Discuss the answers to these questions.

1. How do you explain the walls of the atria being thinner than the walls of the ventricles?
2. What prevents the atrioventricular valve from being pushed into the atrium when the ventricle contracts?
3. How do you explain the left ventricle wall being thicker than the right ventricle wall?
4. Does the left side of the heart pump oxygenated or deoxygenated blood?
5. Why does the wall of the heart need its own supply of blood, brought by the coronary arteries?
6. Does the right side of the heart pump a greater volume of blood per minute, a smaller volume per minute, or the same volume per minute as the left?

B3.2.16 Stages in the cardiac cycle

The heart follows a repeating sequence of actions, known as the cardiac cycle. The sinoatrial node initiates each turn of the cycle by sending out an electrical signal that spreads throughout the walls of the atria. It takes less than a 10th of a second for all cells in the atria to receive the signal. This propagation of the electrical signal causes the whole of both left and right atria to contract. After a time-delay of about 0.1 seconds, the electrical signal is conveyed to the ventricles. The delay allows the atria to pump the blood that they are holding into the ventricles. The electrical signal is then propagated throughout the walls of the ventricles, stimulating them to contract and pump blood out into the arteries.

Stages in the cardiac cycle can be deduced from a graph showing pressure changes during a heartbeat in the atrium, ventricle and artery on one side of the heart. Figure 30 shows the state of heart chambers and valves during the stages of a heartbeat. Vertical arrows show flows of blood.

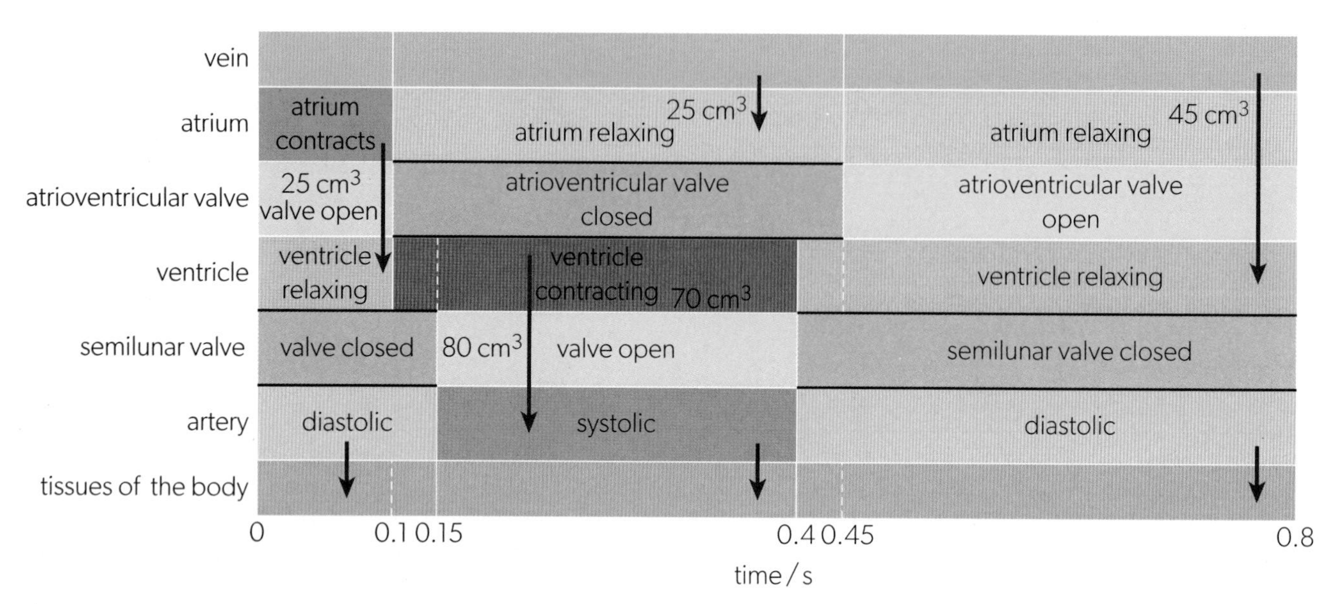

▲ Figure 30 Summary of stages of the cardiac cycle with arrows to indicate blood flow between the vein, atrium, ventricle and artery, including typical volumes

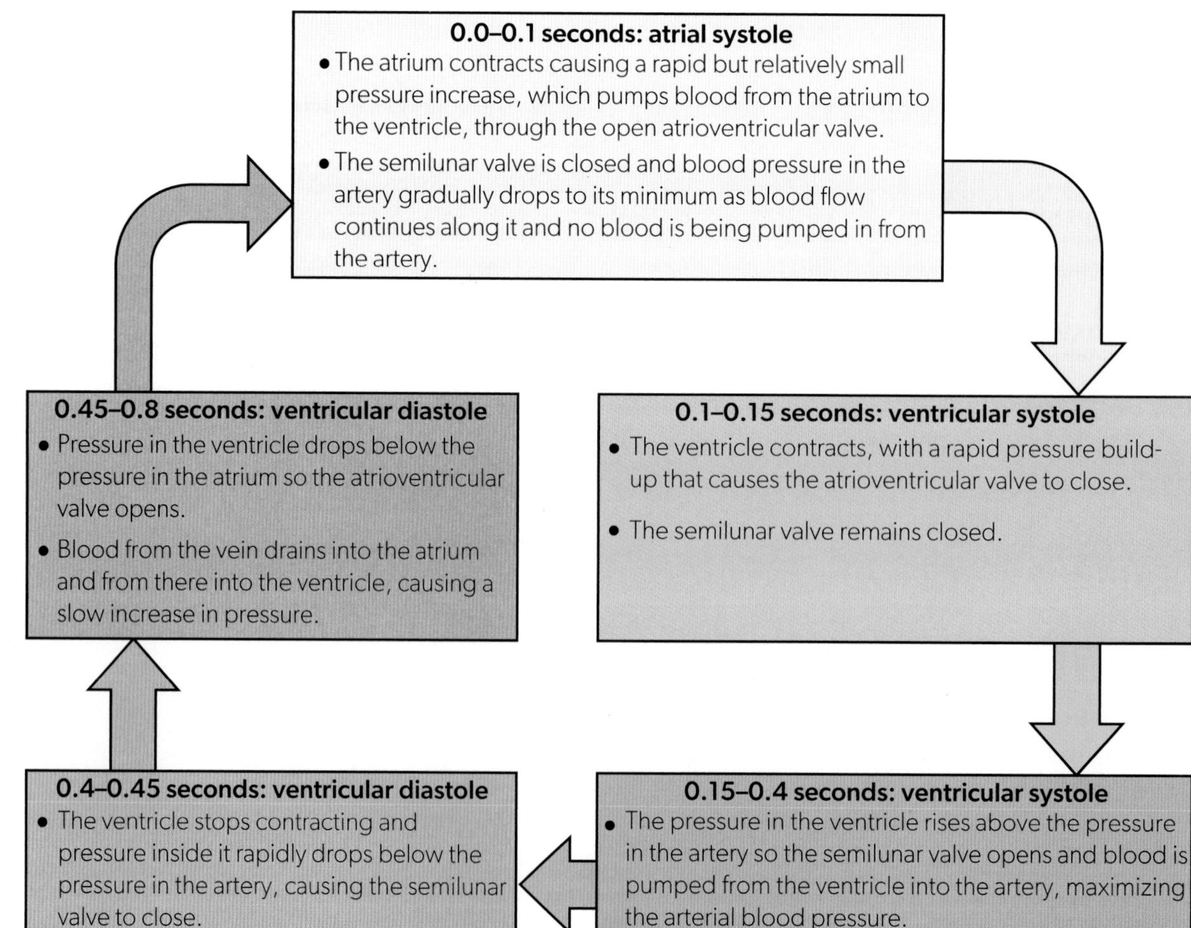

0.0–0.1 seconds: atrial systole
• The atrium contracts causing a rapid but relatively small pressure increase, which pumps blood from the atrium to the ventricle, through the open atrioventricular valve.
• The semilunar valve is closed and blood pressure in the artery gradually drops to its minimum as blood flow continues along it and no blood is being pumped in from the artery.
0.1–0.15 seconds: ventricular systole
• The ventricle contracts, with a rapid pressure build-up that causes the atrioventricular valve to close.
• The semilunar valve remains closed.
0.15–0.4 seconds: ventricular systole
• The pressure in the ventricle rises above the pressure in the artery so the semilunar valve opens and blood is pumped from the ventricle into the artery, maximizing the arterial blood pressure.
• Pressure slowly rises in the atrium as blood drains in from the vein and the atrium fills.
0.4–0.45 seconds: ventricular diastole
• The ventricle stops contracting and pressure inside it rapidly drops below the pressure in the artery, causing the semilunar valve to close.
• The atrioventricular valve remains closed.
0.45–0.8 seconds: ventricular diastole
• Pressure in the ventricle drops below the pressure in the atrium so the atrioventricular valve opens.
• Blood from the vein drains into the atrium and from there into the ventricle, causing a slow increase in pressure.

▶ Figure 31 Timing of actions in the cardiac cycle

Data-based questions: Heart action and blood pressures

Figure 32 shows the pressures in the atrium, ventricle and artery on one side of the heart, during one second in the life of the heart.

1. Deduce when blood is being pumped from the atrium to the ventricle. Give both the start and the end times. [2]
2. Deduce when the ventricle starts to contract. [1]
3. The atrioventricular valve is the valve between the atrium and the ventricle. State when the atrioventricular valve closes. [1]
4. The semilunar valve is the valve between the ventricle and the artery. State when the semilunar valve opens. [1]
5. Deduce when the semilunar valve closes. [1]
6. Deduce when blood is being pumped from the ventricle to the artery. Give both the start and the end times. [2]
7. Deduce when the volume of blood in the ventricle is:
   a. at a maximum [1]
   b. at a minimum. [1]

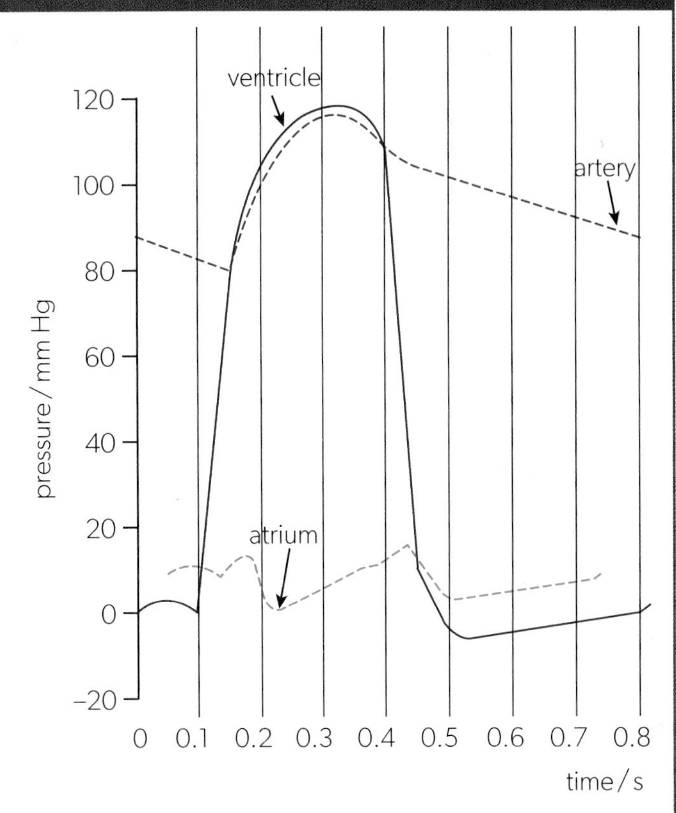

▲ Figure 32 Pressure changes in the atrium and ventricle of the heart and the aorta during a cardiac cycle. The timings assume a heart rate of 75 beats per minute

B3.2.17 Generation of root pressure in xylem vessels by active transport of mineral ions

Water has to pass through at least one root cell before it enters a xylem vessel. If the plant is transpiring, xylem vessels will be filled with sap that is under tension. The tension is strong enough to draw water out of root cells and into xylem vessels, even though the xylem sap is hypotonic to root cells and water might be expected to move in the opposite direction by osmosis. If the plant is not transpiring, there is likely to be positive pressure rather than tension in the sap and another mechanism is needed to cause water to move into xylem vessels from root cells.

There are various reasons for transpiration not occurring.

- High atmospheric humidity may prevent diffusion of water vapour through stomata from air spaces inside the leaf to the atmosphere outside.
- At night, most plants close their stomata, preventing transpiration and in some cases allowing xylem vessels to become air-filled as the sap sinks back down to the roots.
- In deciduous trees that have been leafless in winter, xylem vessels are air-filled but must become refilled with sap before new leaves have grown and started transpiring in spring.

▲ Figure 33 If grape vines are pruned too late in winter when sap has started to rise due to root pressure, there can be "bleeding" of sap from the cut ends of the stems

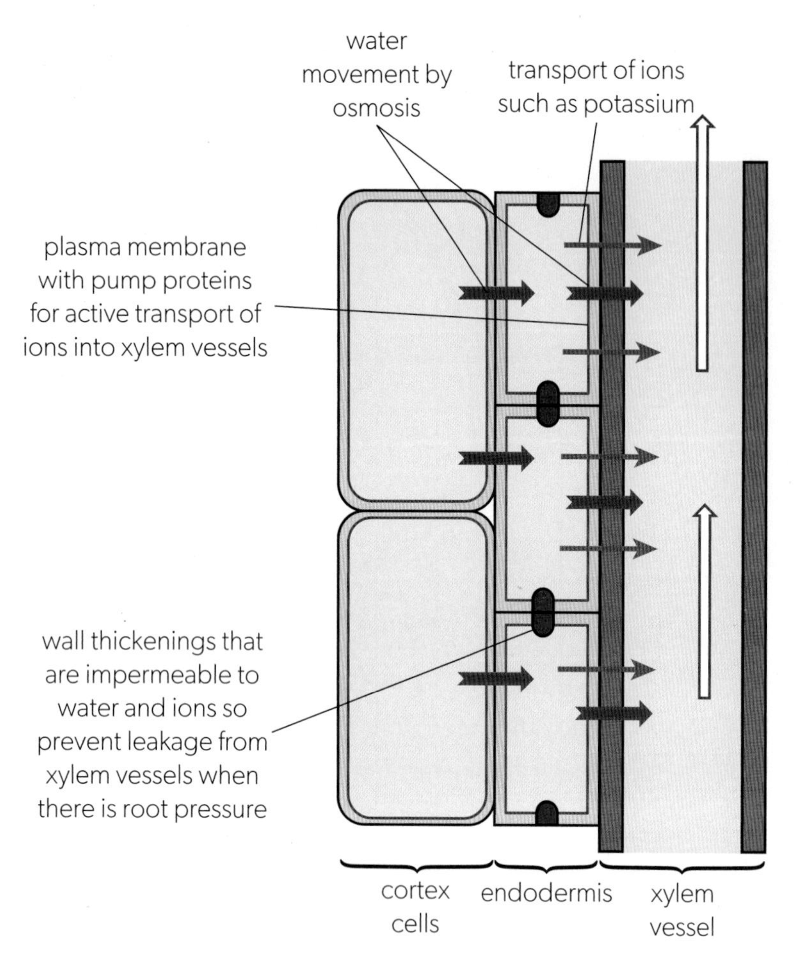

▲ Figure 34 Generation of root pressure by active transport of ions and water movement by osmosis

Root pressure is the mechanism that is used to refill xylem vessels with sap when they have been air-filled. It also causes sap to rise when the xylem sap is not under tension. Root cells adjacent to xylem vessels load mineral ions into the vessels by active transport. The pump proteins that carry out active transport are in the plasma membranes of these living cells. The xylem vessels are dead and do not have plasma membranes. Active transport makes the xylem sap hypertonic compared with the cytoplasm of the adjacent cells, so water moves from them to the xylem vessels by osmosis. This raises the pressure inside the vessels and pushes the sap upwards, against the force of gravity. Unlike pumps that cause fluids to rise by creating a vacuum above them, root pressure is not dependent on atmospheric pressure so there is no limit to the height to which xylem sap can rise.

Data-based questions: Modelling root pressure

Water moves from an area of higher water potential to an area of lower water potential. The contributions of solute potential (Ψ_s) and pressure potential (Ψ_p) to water potential (Ψ_w) are described in *Section D2.3.10*.

Table 3 shows a model scenario for generating root pressure by active transport of ions and water movement by osmosis.

compartment	Ψ_p/kPa	Ψ_s/kPa	Ψ_w/kPa
soil	0	−250	
cytoplasm of root cells	300	−650	−350
xylem sap	100	−750	

▲ Table 3

1. Calculate the water potential of the soil and the xylem sap. [2]
2. Deduce the direction of water movement between the soil, the cytoplasm of root cells and the xylem sap. Give reasons for your answer. [3]
3. The solute potentials are due almost entirely to dissolved ions. Deduce, giving a reason for your answer, how ions are moved from the cytoplasm of root cells to the xylem sap. [2]
4. State what is indicated by a pressure potential of zero in the soil. [1]
5. Discuss the reasons for the difference in pressure potential between the cytoplasm of the root cells and the xylem sap. [2]

B3.2.18 Adaptations of phloem sieve tubes and companion cells for translocation of sap

Sucrose and other carbon compounds can be transported from one part of a plant to another. The transport is from sources to sinks. Sources are tissues of the plant where the compounds are produced by photosynthesis or are being unloaded from a store. Sinks are tissues that need to be supplied with substrates for cell respiration or anabolic reactions as a part of growth or tissues where starch or lipids are being stored. Leaves are sources because they produce carbon compounds by photosynthesis. Roots are sinks because they need substrates for cell respiration.

Phloem is the tissue that transports carbon compounds. Of the several cell types within phloem, it is the sieve tubes that provide channels through which transport can occur. Sieve tubes develop from columns of cells. Adjacent cells in the column become connected by large perforations (holes) in the end walls, which are then called sieve plates. Nearly all the contents of the cells break down during differentiation, including the nucleus. The cell contents are replaced by phloem sap, which is a solution of sucrose and other compounds being transported. The loss of cell contents and the perforations of the end walls make it easier for the sap to flow through the sieve tube.

▲ Figure 35 The holes in a capsule used in a coffee-making machine are reminiscent of the pores in a sieve plate

It is debatable whether the remaining subunits of sieve tubes are still cells, but they do have a membrane and use ATP for processes requiring energy, so they are certainly alive. They are sometimes called sieve tube elements rather than cells. The main energy-requiring process is loading and unloading sucrose, which is carried out by active transport. Sieve tube elements have few or no mitochondria of their own and rely on adjacent companion cells for a supply. Companion cells have many mitochondria and fine cytoplasmic connections called plasmodesmata, through which the ATP can pass. These plasmodesmata have a larger diameter than those elsewhere in the plant. In addition to ATP, sucrose loaded into the companion cell by active transport can pass through to the sieve tube.

High solute concentrations develop in sieve tubes of the leaf and other sources. This draws water in by osmosis, increasing the hydrostatic pressure. This could not happen without the plasma membrane that is maintained in sieve tube elements.

Although not as thick as in xylem, the cellulose cell walls of the sieve tube elements can withstand the high pressures that develop during transport. In particular, the sieve plates brace the tube, preventing bulges or bursts.

Roots generally act as sinks because they need substrates for cell respiration. In roots and other sinks, compounds required by the tissue are unloaded by active transport. This lowers the solute concentration in the sieve tubes, so water exits by osmosis and the hydrostatic pressure drops. The difference in pressure between phloem sap in sources and sinks drives the flow of sap from source to sink. Sieve tubes can conduct sap in either direction (but not both directions at once). As an example, sieve tubes transport carbon compounds into a growing leaf and then in the opposite direction out of the leaf when it is producing more carbon compounds by photosynthesis than it is using in growth and respiration.

AHL

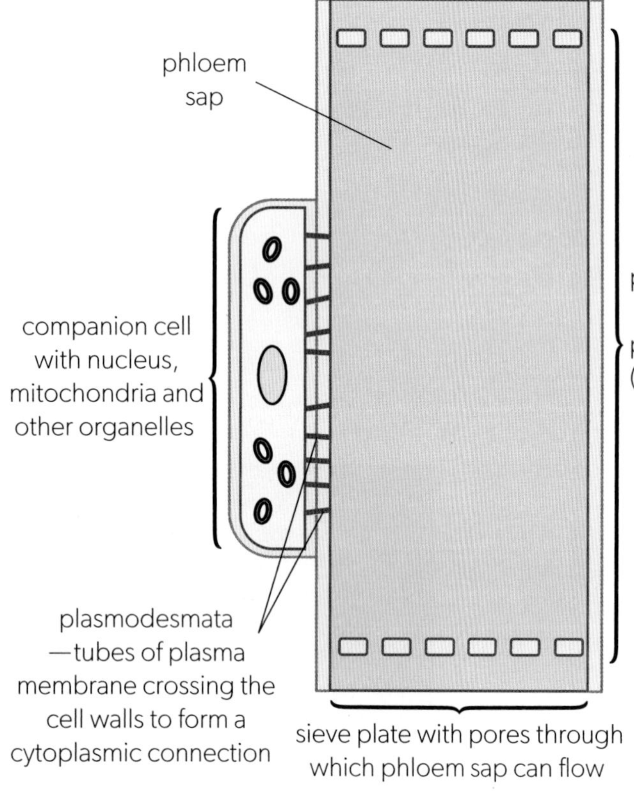

▲ Figure 36 Structure of a sieve tube element and adjacent companion cell

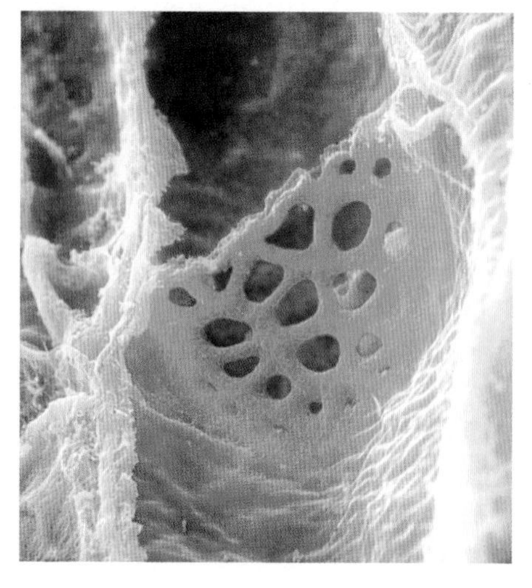

▲ Figure 37 Freeze–fracture electron micrograph showing cell walls in a sieve tube

Linking questions

1. How do pressure differences contribute to the movement of materials in an organism?
   a. Describe the adaptations of arteries for high pressure transport of blood away from the heart. (B3.2.3)
   b. Explain the role of pressure filtration in the formation of tissue fluid in capillaries. (B3.2.10)
   c. Explain the role of the generation of root pressure in xylem vessels by active transport of mineral ions. (B3.2.16)
2. What processes happen in cycles at each level of biological organization?
   a. Compare and contrast cellular reproduction with the lysogenic cycle of a virus. (A2.3.4)
   b. Outline the stages of the cardiac cycle. (B3.2.15)
   c. Using examples, explain the role of oxidation and reduction in biological systems. (B1.1.4)

B3.3 Muscle and motility

How do muscles contract and cause movement?

The rowers in Figure 1 develop powerful muscles for use in the power stroke, when the blade of the oar is pulled through the water. The oar rests on a rowlock, which is positioned 1 m from the end of a 3.7 m racing oar. In which direction will this boat move? In what ways is rowing a good analogy for the mechanism within muscles that causes contraction and generation of force? Muscle tissue is contractile—it can shorten itself and exert a pulling force as it does so. What is the relative role of actin and myosin in muscle contraction?

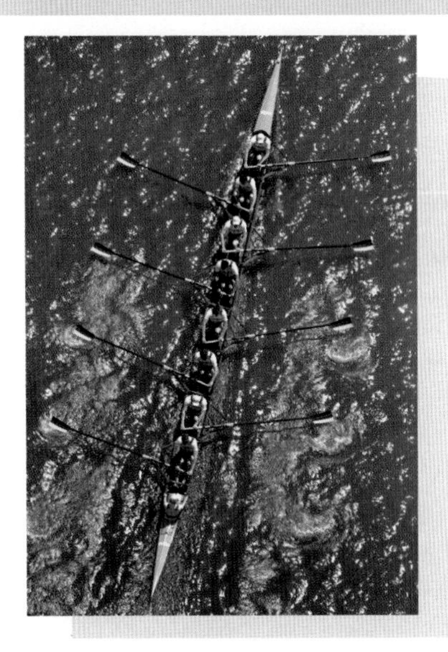

▶ Figure 1

What are the benefits to animals of having muscle tissue?

The golden silk orb-weaving spider (*Nephila clavipes*) applies a potent neurotoxin to its web. Insects that become caught tremble and then move increasingly slowly before becoming paralysed. The spider injects enzymes and later sucks the digested contents out of the insect's exoskeleton. Without working muscle, the prey's survival chances are nil.

Which taxonomic group of animals lack muscle? What role does muscle play in the organs of mammals such as digestive systems or circulatory systems? What is the role of skeletal muscle? How metabolically expensive is muscle contraction?

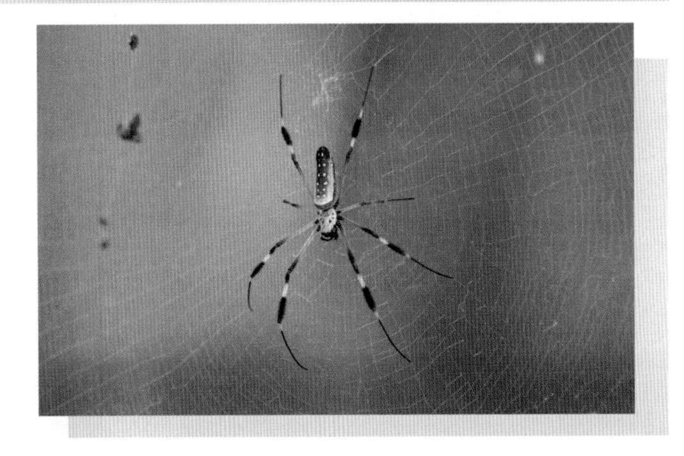

▲ Figure 2 The golden silk orb-weaving spider (*Nephila clavipes*)

AHL only

B3.3.1 Adaptations for movement as a universal feature of living organisms
B3.3.2 Sliding filament model of muscle contraction
B3.3.3 Role of the protein titin and antagonistic muscles in muscle relaxation
B3.3.4 Structure and function of motor units in skeletal muscle
B3.3.5 Roles of skeletons as anchorage for muscles and as levers
B3.3.6 Movement at a synovial joint
B3.3.7 Range of motion of a joint
B3.3.8 Internal and external intercostal muscles as example of antagonistic muscle action to facilitate internal body movements
B3.3.9 Reasons for locomotion
B3.3.10 Adaptations for swimming in marine mammals

▲ Figure 3 Bar-tailed godwits (*Limosa lapponica*) have flight muscles that they use to beat their wings, generating lift and velocities of up to 90 km^{-1} h

B3.3.1 Adaptations for movement as a universal feature of living organisms

Movement is one of the functions of life. Two types of movement can be distinguished:

- movements within the body of an organism such as peristalsis in the gut or ventilation of the lungs
- locomotion, which is the movement of an organism from one place to another.

The former happens in all living organisms. Even in a unicellular organism there are movements within the cytoplasm. The latter happens in some organisms but not all.

An organism that moves from place to place is motile. Many animals move around while feeding within their territory. Some animals move much greater distances when they migrate. For example, bar-tailed godwits (*Limosa lapponica*) migrate 10,400 km from eastern Siberia to New Zealand in 7–8 days. They double their body weight with fat reserves before the journey.

An organism that remains in a fixed position is sessile. Most plants are sessile, with roots growing into the soil. Most animals are motile. There are some sessile animals, particularly in aquatic or marine habitats. For example, a coral consists of a colony of sessile polyps. In hard corals, the polyps construct a rigid skeleton around themselves. This allows them to extend their tentacles into the water when they are filter-feeding, but they cannot to move to a new location.

▲ Figure 4 Adult barnacles remain attached to a solid surface, so they are sessile. There are larval stages in the barnacle life cycle which swim, so they are motile. This photo shows adult goose barnacles (*Lepas anatifera*). They are filter-feeding using modified legs that are not required for motility

B3.3.2 Sliding filament model of muscle contraction

Muscle fibres contain many parallel myofibrils. Each myofibril consists of a series of sarcomeres linked end-to-end at Z-discs. There are light bands at either end of a sarcomere and a dark band in the centre. In relaxed muscle, the Z-discs are further apart, the light bands are wider and overall the sarcomere is longer. Z lines, dark bands and light bands are visible in electron micrographs.

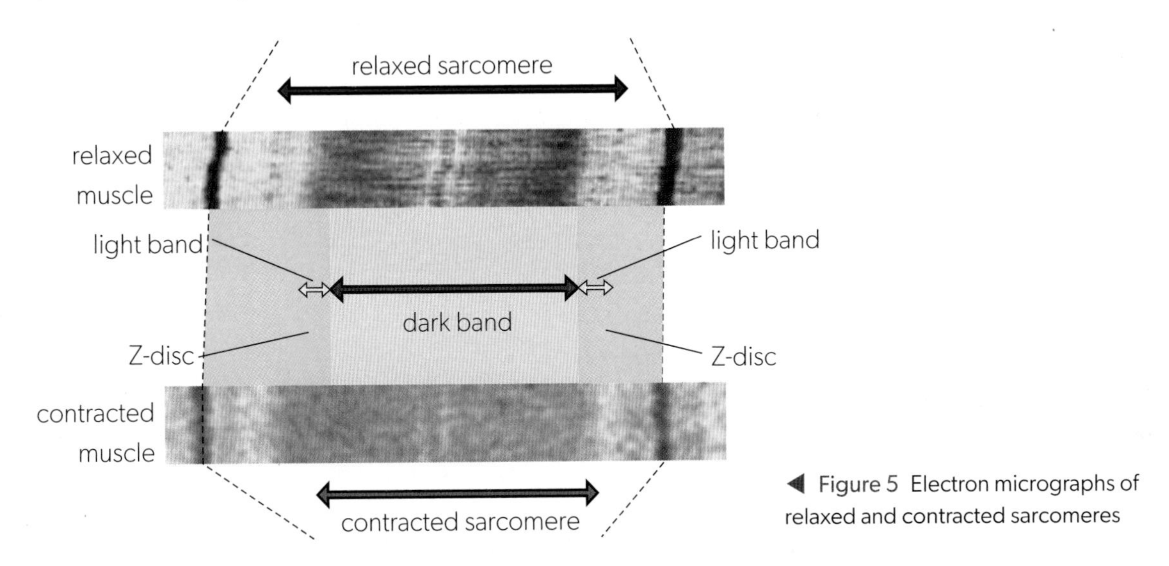

◀ **Figure 5** Electron micrographs of relaxed and contracted sarcomeres

The pattern of light and dark bands in a sarcomere is due to a precise and regular arrangement of two types of protein filament—thin actin filaments and thick myosin filaments. Actin filaments are attached to a Z-disc at one end. The myosin filaments occupy the centre of the sarcomere and interlock like fingers with the actin filaments at both ends. Each myosin filament is surrounded by six actin filaments and can form cross-bridges with them.

The contraction of sarcomeres, and therefore of muscle, is due to the sliding of actin and myosin filaments. Myosin filaments cause this. At regular intervals they have "heads" that form cross-bridges by binding to sites along actin filaments, spaced at the same regular intervals. The cross-bridges can exert a force, using energy from ATP. This pushes the actin filament towards the centre of the sarcomere, making the actin and myosin filaments overlap more.

Each time the myosin heads bind, they swivel, exerting force which pushes the actin filaments a short distance (8–10 nm) towards the centre of the sarcomere. The heads then detach, then swivel back and reattach to the next binding site on the actin. This is sometimes referred to as a ratchet mechanism. Because of the regular spacing of the heads and the binding sites on the actin, many heads along a myosin filament bind at the same time. And there are many myosin filaments, in many sarcomeres, in many myofibrils, in many muscle fibres. There are therefore very large numbers of myosin heads in muscle. So although each myosin head exerts only a small force, collectively they can exert very powerful forces.

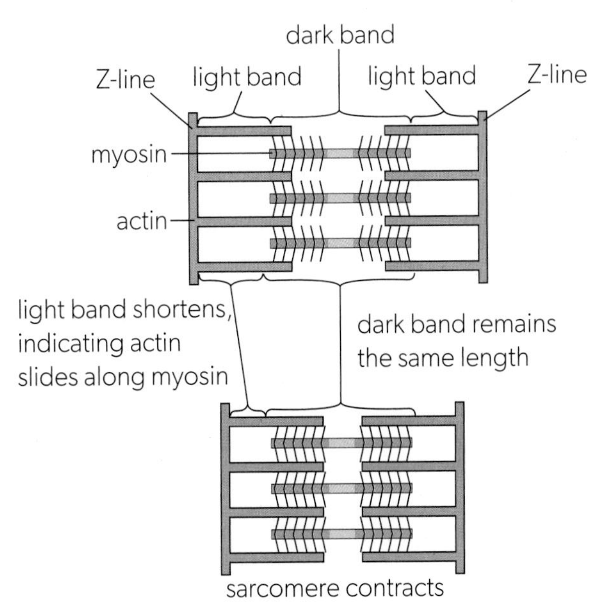

▲ **Figure 6** Diagram of a sarcomere in relaxed muscle (above) and contracted muscle (below)

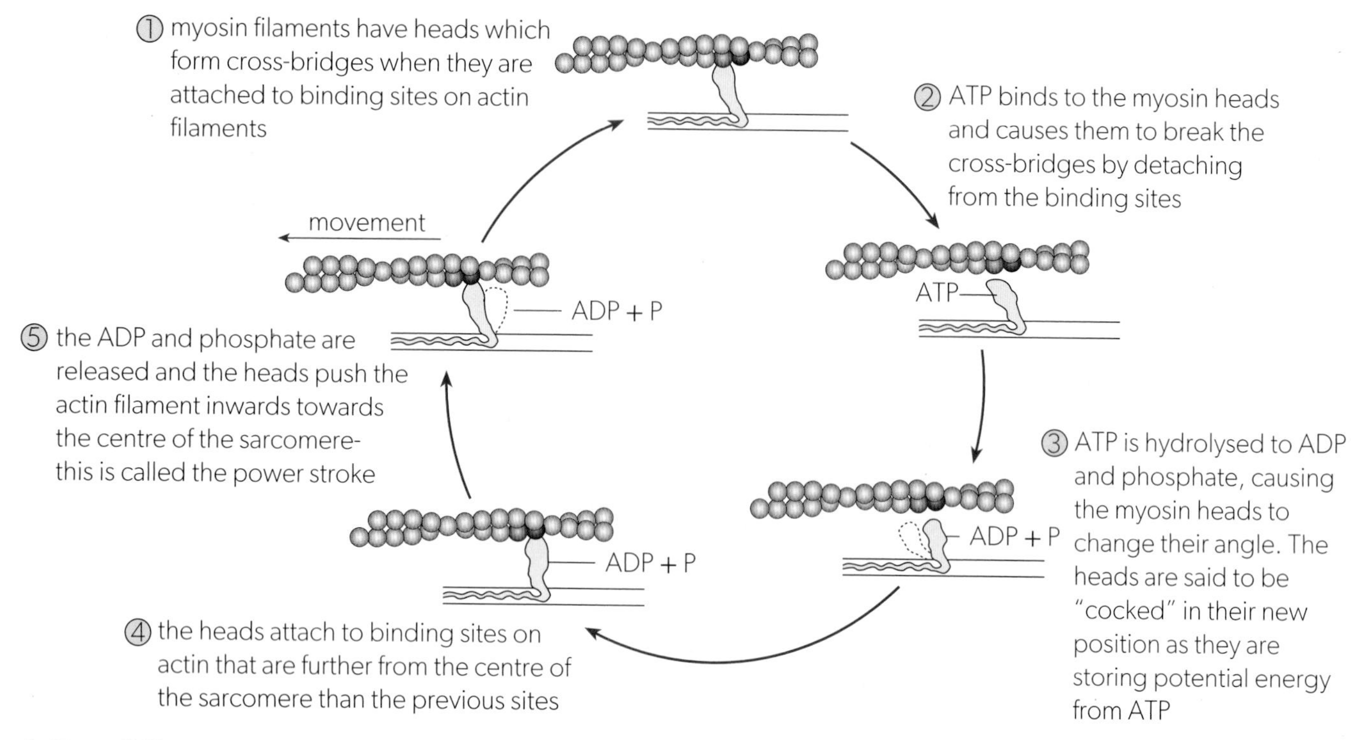

① myosin filaments have heads which form cross-bridges when they are attached to binding sites on actin filaments
② ATP binds to the myosin heads and causes them to break the cross-bridges by detaching from the binding sites
ATP
③ ATP is hydrolysed to ADP and phosphate, causing the myosin heads to change their angle. The heads are said to be "cocked" in their new position as they are storing potential energy from ATP
ADP + P
④ the heads attach to binding sites on actin that are further from the centre of the sarcomere than the previous sites
ADP + P
⑤ the ADP and phosphate are released and the heads push the actin filament inwards towards the centre of the sarcomere- this is called the power stroke
ADP + P
movement

▲ Figure 7 The cycle of stages in the ratchet mechanism that causes an actin filament to slide over myosin filament

Data-based questions: Transverse sections of striated muscle

The drawings in Figure 8 show small parts of a myofibril, as seen in an electron micrograph of a transverse section of muscle tissue.

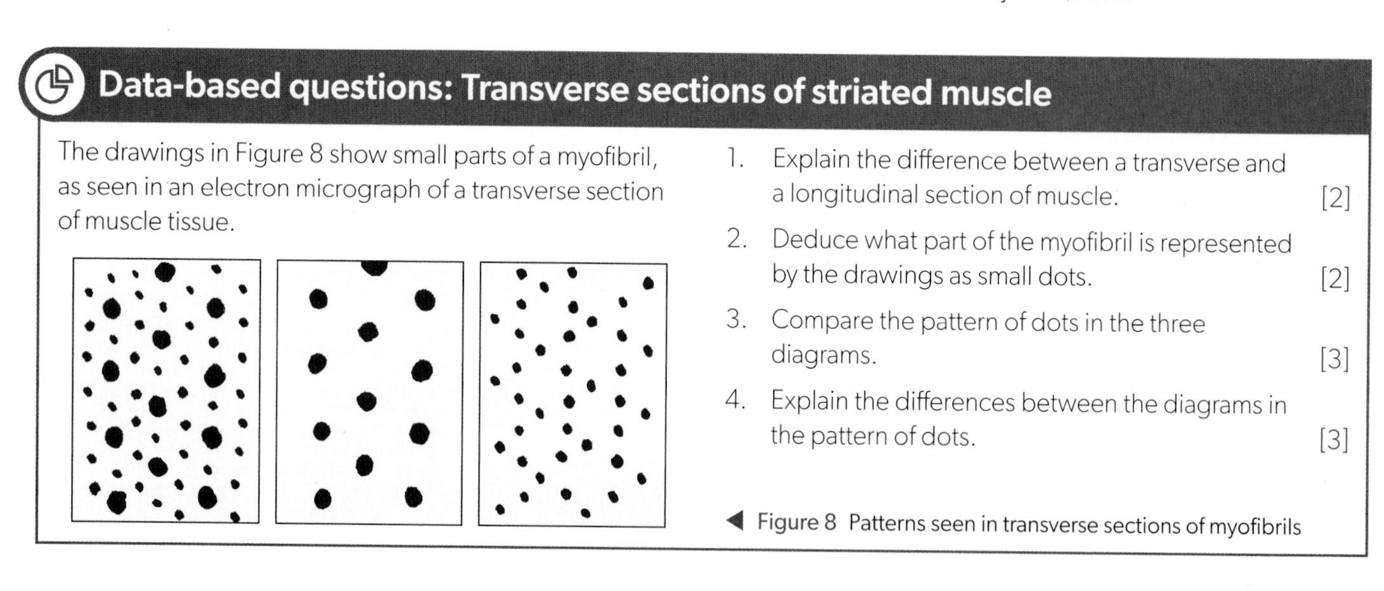

1. Explain the difference between a transverse and a longitudinal section of muscle. [2]
2. Deduce what part of the myofibril is represented by the drawings as small dots. [2]
3. Compare the pattern of dots in the three diagrams. [3]
4. Explain the differences between the diagrams in the pattern of dots. [3]

◀ Figure 8 Patterns seen in transverse sections of myofibrils

B3.3.3 Role of the protein titin and antagonistic muscles in muscle relaxation

Titin is the largest polypeptide so far discovered. In humans, it is a chain of 34,350 amino acids, but in mice it is even longer with 35,213 amino acids. Titin is elastic and acts like a molecular spring, storing potential energy when it is stretched and releasing this energy when it recoils. Titin connects the end of myosin filaments to the Z-disc and has several functions.

- It holds each myosin filament in the correct position in the centre of six parallel actin filaments.
- It prevents overstretching of the sarcomere.
- It adds to the force of contraction by releasing energy as it recoils.

Energy is needed to stretch titin and therefore to lengthen a muscle. Lengthening of muscles happens when they relax. Muscles can only exert force when they contract, so a muscle cannot supply the energy it needs to lengthen. The energy has to be provided by another muscle that is known as the antagonist. Despite the name, an antagonistic pair of muscles work together, with the contraction of each member of the pair providing the energy needed for lengthening the titin molecules in the other as it relaxes.

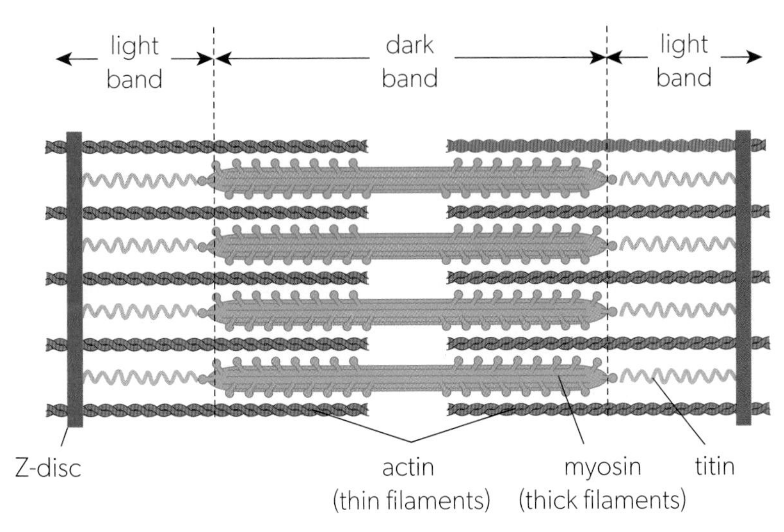

▲ Figure 9 Structure of a sarcomere showing titin filaments

B3.3.4 Structure and function of motor units in skeletal muscle

Skeletal muscles are composed of striated muscle fibres. These contract when stimulated by a motor neuron. The stimulus is passed from the neuron to the muscle fibre at a synapse, using the neurotransmitter acetylcholine. These synapses are called neuromuscular junctions.

There are many more muscle fibres than motor neurons in a typical skeletal muscle, because each motor neuron has branches that stimulate different muscle fibres. One motor unit consists of a single motor neuron together with all the muscle fibres that it stimulates via neuromuscular junctions. There are usually hundreds of muscle fibres in a motor unit. These muscle fibres are not clumped together in a single group but are mingled with muscle fibres of other motor units.

When a nerve impulse passes along the main axon of a motor neuron and then along all the branches to the multiple muscle fibres in the motor unit, all of the muscle fibres respond at the same time by contracting. This helps to achieve a coordinated contraction of the muscle with as few motor neurons as possible.

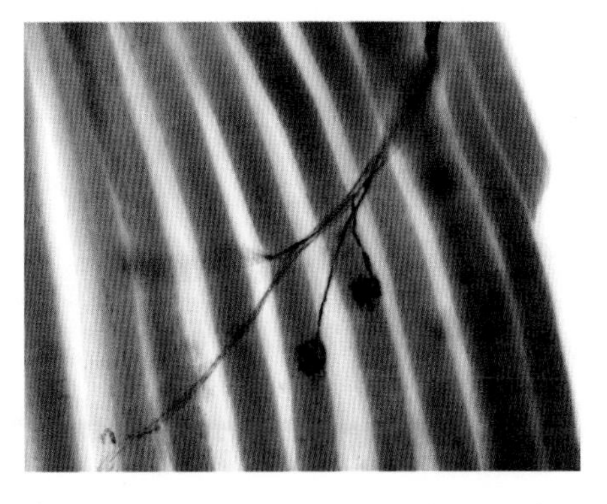

▲ Figure 10 Motor end plates (boutons) at the ends of branches of a motor neuron forming synapses with muscle fibres

AHL

B3.3.5 Roles of skeletons as an anchorage for muscles and as levers

A skeleton is a hard framework that supports and protects an animal's body. Exoskeletons are on the outside of the body whereas endoskeletons are internal. Arthropods such as spiders, crustaceans and insects have exoskeletons consisting of tough plates of chitin that cover most of the body surface. Vertebrates have endoskeletons consisting of bones.

Skeletons facilitate movement by providing an anchorage for muscles and acting as levers. Typically, a muscle is attached to two parts of the skeleton. One attachment is the insertion, where muscle contraction causes movement. The other is the origin and is fixed, so contraction does not cause movement. For example, the insertion of the masseter muscle is on the jawbone (mandible) and the origin is on the cheek bone, which is part of the skull. Contraction of this muscle moves the jawbone, not the fixed cheek bone enabling biting and speech.

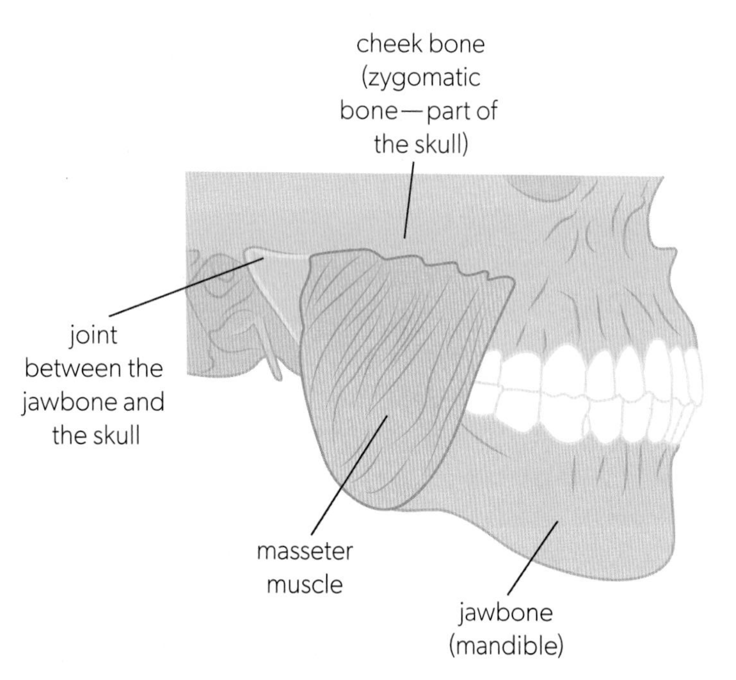

▲ Figure 11 Contraction of the masseter muscle causes the jawbone to move upwards, closing the mouth

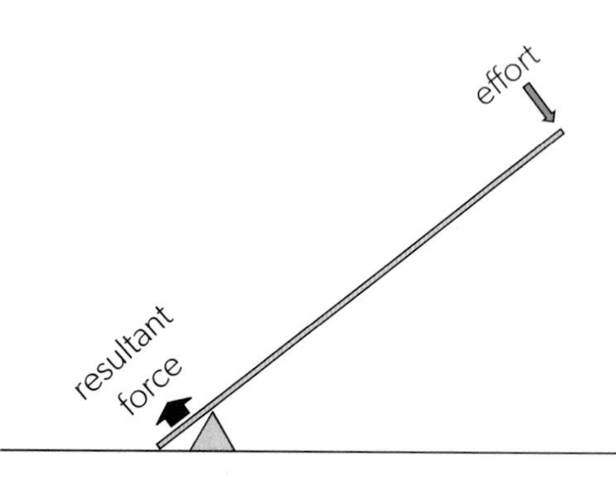

▲ Figure 12 The effort is further from the fulcrum than the resultant force so this lever increases the size of the force but decreases the distance moved. Because the forces are on opposite sides of the fulcrum, the direction of the force is reversed

By acting as levers, bones can change the size and direction of a force (Figure 12). A lever has a fixed point called the fulcrum, which is the pivot point. The force applied to the lever is the effort. When the effort is applied, a resultant force is exerted at a position on the other side of the fulcrum. If the effort is applied further from the fulcrum than the resultant force, the lever increases the size of the force, but decreases the distance moved. Conversely, if the effort is applied nearer to the fulcrum than the resultant force, a lever decreases the size of the force, but increases the distance moved.

For a bone acting as a lever, the fulcrum is the joint where the bone meets another bone. The effort is applied to the bone by one or more muscles, via tendons.

▲ Figure 13 Skeleton of European mole (*Talpa europaea*), with short, wide forelimb bones that are used for digging

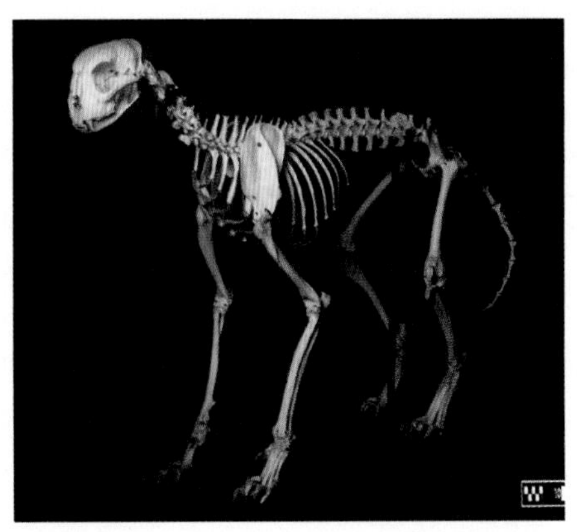

▲ Figure 14 Skeleton of cheetah (*Acinonyx jubatus*) with long narrow limb bones for fast running

B3.3.6 Movement at a synovial joint

Bones meet at joints. Fixed joints, such as the sutures between bones of the skull, do not allow any movement. Most joints allow bones to move in relation to each other—this is called articulation. Most articulated joints have a similar structure and are called synovial joints.

Synovial joints have the following components.

- Bones provide an anchorage for muscles and ligaments. By their shape, bones guide the types of movement that can occur at a joint.
- Cartilage is tough, smooth tissue that covers bone at the joint. It helps to prevent friction by preventing contact between regions of bone that might otherwise rub together. It also absorbs shocks that might cause bones to fracture.
- Synovial fluid fills a cavity in the joint between the cartilages on the ends of the bones. It lubricates the joint, helping to prevent the friction that would occur if the cartilages were dry and touching.
- Ligaments are tough cords of tissue containing large quantities of the protein collagen. They prevent aberrant movements that would dislocate or damage the joint. The joint capsule is a tough ligamentous covering to the joint. It seals the joint and holds in the synovial fluid and it helps to prevent dislocation.
- Muscles provide the forces that cause movement at the joint.
- Tendons attach muscle to bone. Like ligaments they are composed of living tissue, with large quantities of extracellular collagen, which has high tensile strength. Some tendons are long and cord-like so forces can be transmitted over the distance between the muscle and the bone to which the tendon attaches it.

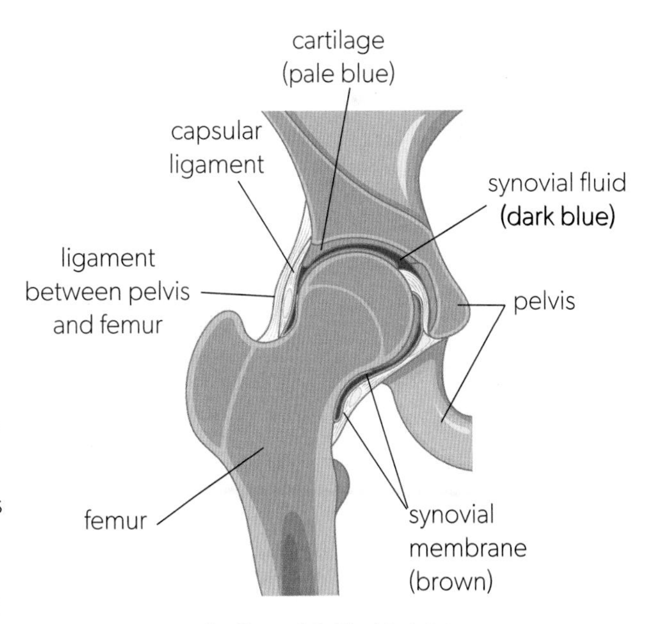

▲ Figure 15 The hip joint

Poultry wing dissection

The anatomy of a poultry wing such as a chicken or turkey wing is homologous to the human arm. In this dissection, focus on the elbow joint of the poultry wing.

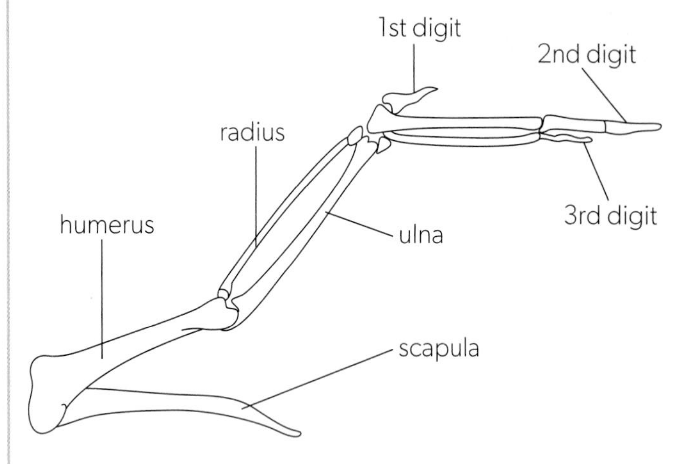

▲ Figure 16 Bird wing bones

1. Rinse the wing under running water and thoroughly dry it with a paper towel because the surface may be contaminated with *Salmonella* bacteria.
2. Cut the skin along the entire length of the wing, pointing the scissors up so as not to cut the tissues underneath.
3. Remove the skin from the wing by placing your finger under the skin and lightly tearing at the connective tissue below it.
4. Use a blunt probe to separate the individual muscles from each other without tearing them.
5. Pull on each of the muscles and note the movement that results. Determine pairs of muscles that are antagonistic.
6. Follow a muscle to where it connects to the bones. Note the appearance of the tendons.
7. Carefully remove the muscles and tendons to expose the ligaments which are white in appearance.
8. Identify the humerus, the radius and the ulna.
9. Separate the bones at the joint and note the appearance of the cartilage.
10. Note the oily texture of the surface of the cartilage. This is due to synovial fluid.

B3.3.7 Range of motion of a joint

The structure of a joint, including the ligamentous joint capsule and the ligaments, determines the range movements that are possible. The elbow joint and the knee joint are hinge joints allowing movements in one plane: flexion (bending) and extension (straightening). The hip joint, between the pelvis and the femur, is a ball-and-socket joint. It has a greater range of movement than the elbow joint: it can protract and retract, abduct and adduct, and rotate.

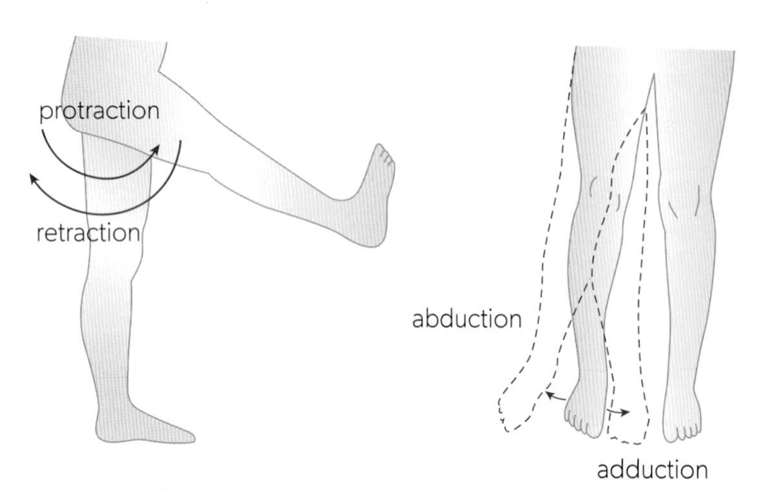

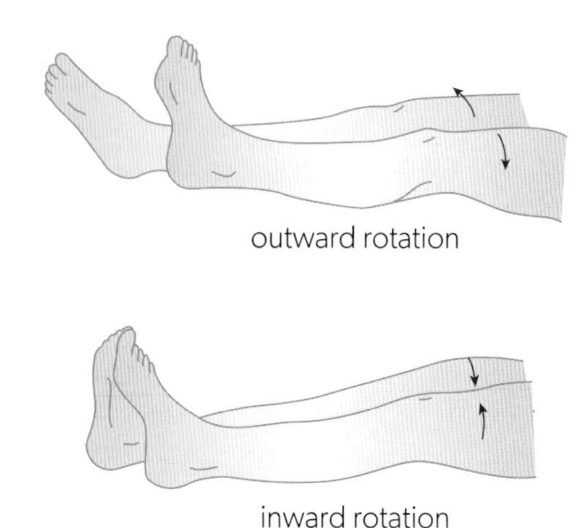

▲ Figure 17 The hip is a ball-and-socket joint that allows movements in all three planes

The range of movements at a joint can be investigated by measuring joint angles. The simplest method is to use a goniometer consisting of two rulers connected to a protractor (Figure 18). Digital goniometers are available as phone apps. There are also computer programs that analyse images to obtain measurements of joint angles.

Making careful measurements: Using a goniometer

Muscle stretching can increase the range of motion at a joint. There are a number of different types of stretching regime including dynamic stretching and isometric stretching. Here are some possible research questions.

- Does the effect of stretching on the range of motion of a joint persist one day later?
- Do the different types of stretching regime differ in terms of the increases they allow in the range of motion of a joint?
- Do different joints and types of motion see an increased range of motion after stretching?
- How does the range of motion at hip joints compare between boys and girls?

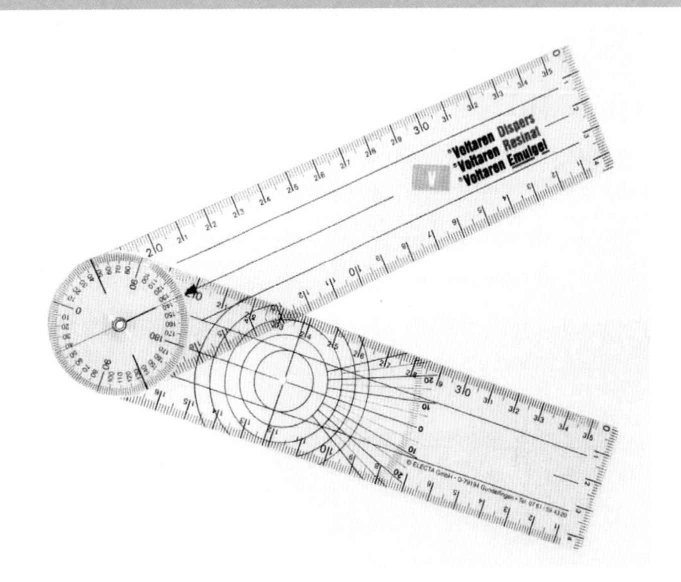

▲ **Figure 18** A goniometer

B3.3.8 Internal and external intercostal muscles as an example of antagonistic muscle action to facilitate internal body movements

The intercostal muscles are the muscles between the ribs. They are made up of external and internal layers and the muscle fibres in these layers are orientated differently. This means that alternating contraction of the different layers moves the ribcage in opposite directions. Contraction of the external intercostal muscles expands the ribcage allowing inhalation and stretching of the internal intercostal muscles. This stretching stores potential energy in the titin in the internal intercostal muscles. During exhalation, the internal intercostal muscles contract and stretch the external intercostal muscles.

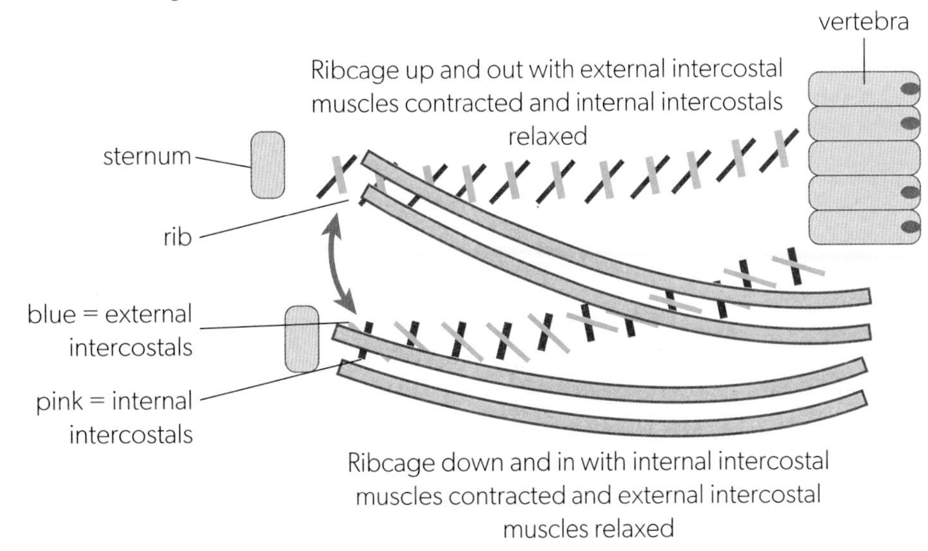

▲ Figure 19 The internal and external intercostal muscles are antagonistic. They form continuous layers of striated muscle between the ribs. In this diagram, pairs of ribs are viewed from the side, with muscle fibres shown at intervals

B3.3.9 Reasons for locomotion

Locomotion requires expenditure of energy, so will only occur if there are benefits for the animal. There are multiple reasons for moving from place to place.

Foraging for food

Herbivores move to find the plant foods that they need. Bees fly from flower to flower searching for nectar and pollen. Grazing animals move across grassland to find the best pastures. Frugivores move to find abundant sources of ripe fruit. Predators move to catch and kill their prey.

Escaping from danger

Prey move to escape from potential predators or from hostile members of their own species. There is therefore strong selective pressure for rapid movement and/or stamina. Many animals have a roosting site that they return to during times when they are inactive. Jackdaws (*Corvus monedula*), for example, gather at dusk in roosts with many individuals, often in treetops, and disperse at dawn.

Searching for a mate

Animals in dispersed populations must travel to find a mate. By leaving its home territory, an individual animal can find an unrelated individual to mate with thus avoiding inbreeding. Young male lions (*Panthera leo*), for example, leave the pride of their birth and travel to find another pride. When they find a new pride, they will attempt to displace the dominant male so they can mate with all the adult females.

Migration

Many species of bird migrate in both spring and autumn. Some migrate between the northern and southern hemispheres to avoid the food scarcities of winter. Others migrate closer to the poles for the summer and back towards the equator for the winter. For example, snow geese (*Anser caerulescens*) migrate between Arctic North America and the Atlantic coast of the US, breeding in the north and overwintering in the south. Some species such as salmon, carry out a once-in-a-lifetime migration to their breeding grounds, with the young migrating back to the region inhabited by the adults.

▲ Figure 20 A diademed sifaka (*Propithecus diadema*) eating wild guava fruit in Mantadia National Park, eastern Madagascar

▲ Figure 21 Springbok antelope (*Antidorcas marsupialis*) in southern Africa can escape from cheetahs and other predators by running at speeds of over 80 km h^{-1}. They sometimes jump repeatedly to a height of up to 2 m—a behaviour known as "pronking". This may alert other springbok to dangers or confuse predators chasing them

▲ Figure 22 Male American moon moths (*Actias luna*) have huge antennae that can detect pheromones given off by unmated females hundreds of metres away. This allows them to fly in the direction of a female in order to mate with her

▲ Figure 23 Salmon migrate from the oceans to their spawning grounds in the headwaters of rivers. Young salmon remain in the river for up to three years and then migrate out to sea

B3.3.10 Adaptations for swimming in marine mammals

Early mammals were all terrestrial, but about 50 million years ago some evolved adaptations for life in water. Water is about a thousand times denser than air and much more viscous. Swimming therefore requires different adaptations from those needed for locomotion on land or aloft in the air.

- Streamlining—marine mammals are shaped to minimize resistance to motion by these features:
  - shaped to be widest near the front and tapering towards the rear, which causes less drag than other shapes
  - flippers, flukes and dorsal fin with an elongated teardrop profile in transverse section which reduces drag
  - body surface smooth due to even distribution of blubber and absence of hind limbs and ear flaps
  - skin without hair, reducing friction.
- Adaptations for locomotion:
  - flippers, which are used for steering, in place of front legs
  - flukes on the tail, which are lobes to left and right that increase thrust when the tail is moved up and down
  - dorsal fin to provide stability by preventing rolling
  - blubber, which provides buoyancy, allowing a dolphin to cease moving and float just below the water surface, for example when it is sleeping.
- Airways to allow ventilation of the lungs:
  - blowhole leading from the larynx to the upper surface of the head, through which marine mammals breathe
  - no connection between the mouth and lungs to avoid water entering the lungs.

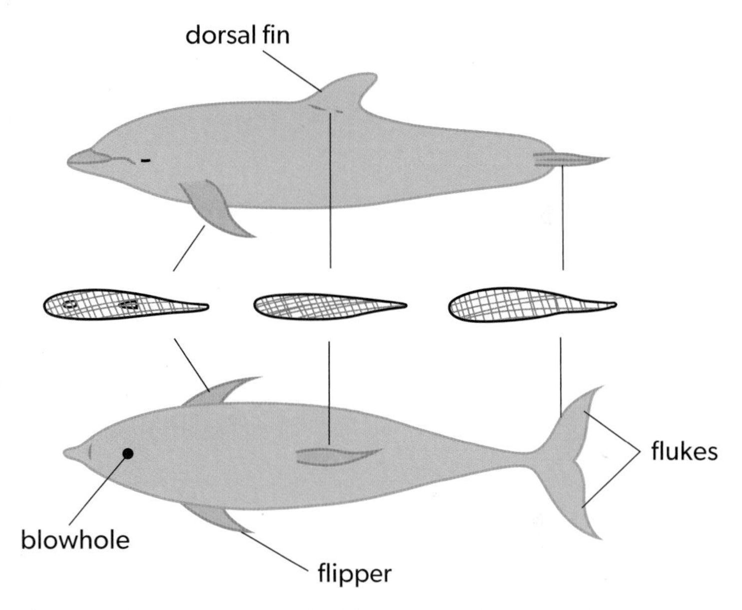

▲ Figure 24 Streamlining of a dolphin viewed from the side (upper drawing) and from above (lower drawing) with transverse sections through the flipper, dorsal fin and fluke

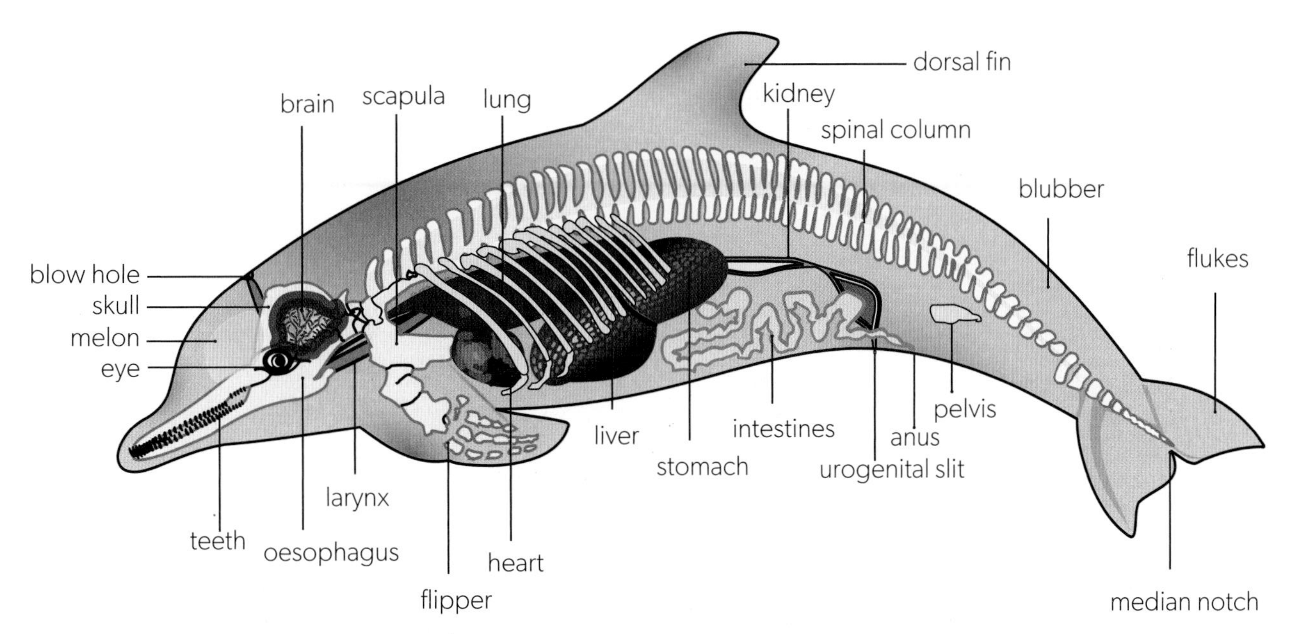

▲ Figure 25 Internal anatomy of dolphins

AHL

Data-based questions: Relationship between movement and food sources

Basking sharks (*Cetorhinus maximus*) filter feed on zooplankton (small floating marine animals) in temperate coastal seas. Marine biologists recorded the swimming paths taken by two basking sharks about 8 km off the coast of Plymouth (UK). At the same time, the densities of zooplankton (in g m^{-3}) were recorded within 3 m of the swimming path of the sharks.

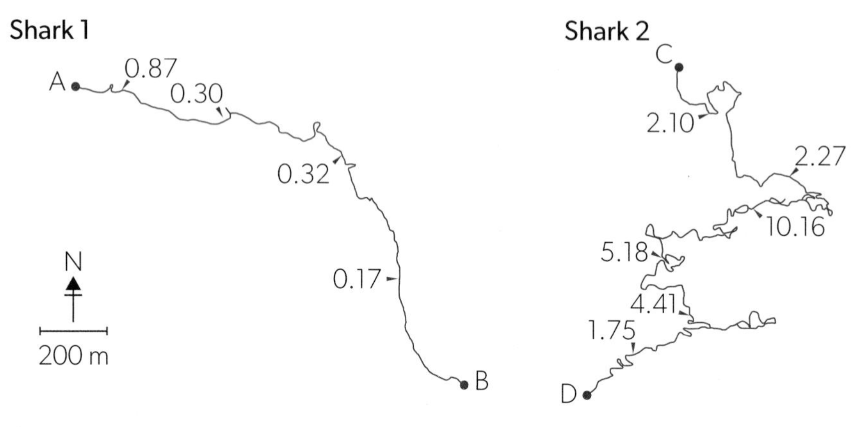

▲ Figure 26 Paths taken by two basking sharks with densities of zooplankton along the paths

1. Using the scale given, calculate the straightline distance:
   a. from point A to point B [1]
   b. from point C to point D [1]
2. Distinguish between the swimming behaviour of shark 1 and shark 2. [1]
3. Using the data given, suggest reasons for the difference in the swimming behaviour of the two sharks. [3]
4. State two factors other than food which may affect the distribution of the basking sharks. [2]

Linking questions

1. What are the advantages and disadvantages of dispersal of offspring from their parents?
   a. List examples of intraspecific competition. (C4.1.10)
   b. Distinguish between seed dispersal and pollination in plants. (D3.1.12)
   c. Outline reasons for locomotion in living things. (B3.3.9)
2. What are the relative advantages of versatility and specialization in biological mechanisms?
   a. Distinguish between stem cells and differentiated cells. (B2.3.2)
   b. Explain how specific immunity develops. (C3.2.8)
   c. Compare and contrast fundamental and realized niche. (D4.2.12)

TOK

What counts as a good justification for a claim?

Knowledge claims require justification for others to accept them. In biology, knowledge claims are justified by reference to empirical evidence. In other words, they are justified by observations. Repeated observations lead to hypotheses that can be tested by experimentation. A hypothesis is a form of generalization. What makes a good generalization? One requirement of good generalization is the number of times the phenomenon is observed under similar conditions. The reliability of quantitative data is increased by repeating measurements. For example, Table 1 shows that increasing the number of respondents in a survey reduces the margin of error.

Sample size/N	Margin of error/ %
10	31.6
20	22.4
50	14.1
100	10.0
500	4.5
1,000	3.2

▲ Table 1 Margin of error is reduced by increased sample size

Gas exchange occurs through the stomata of leaves. Does the number of stomata vary within members of a population whose microhabitats vary?

Repeated counts of the number of stomata visible in the field of view at high power illustrate the variability of biological material and the need to replicate trials from the same plant. Good generalizations apply in a variety of circumstances. In the case of this experiment, it would be important to observe many different individual plants under the same conditions as well as making replicate observations from each individual.

▲ Figure 1 Light micrograph of the epidermis (the upper cell layer) of a tulip leaf

Experimenters need to have an open mind because even the act of observation has associated problems. For example, an observer might consider that a particular field of view is showing too many or too few stomata. They might then reject it as an outlier without including it in the sample. In other words, our expectations can lead us to be biased of an observation so that we deem it to be unworthy of inclusion in the data set. This can be remedied by having a quantitative standard for rejecting outliers.

Sometimes a generalization is very surprising. It is possible, but unlikely, that no stomata are observed on either side of a leaf in a number of samples. In this case, it is not reasonable to conclude that this is a new species without stomata. It is more likely that there is a flaw in the technique used to detect stomata. This is known as the "coherence test".

Generalizations that are found to have predictive power can lead to the establishment of a theory. Sometimes, reason and theory are reasonable supplements to observation when it comes to justifying knowledge claims. Plants that come from extreme habitats, such as desert plants and water plants, are expected to have unique distributions of stomata. In what ways do they differ from mesophytes (plants that live in medium water environments)? Knowledge of the theory of evolution by natural selection leads to explanations of the unique adaptations found in different plants for gas exchange.

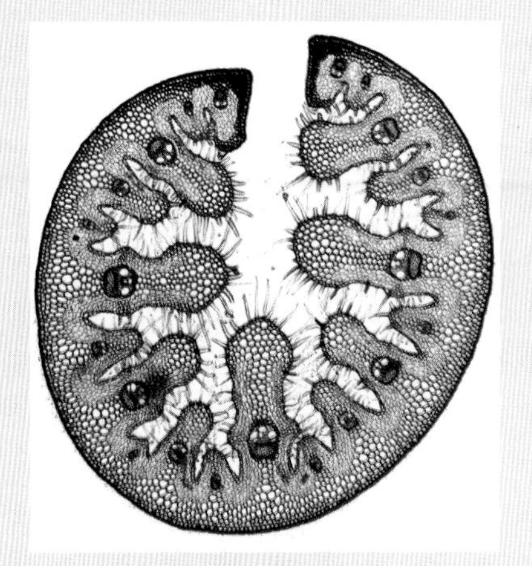

▲ Figure 2 Marram grass (*Ammophila arenaria*) is a xerophyte (a plant adapted to dry conditions). It has unique adaptations such as rolled leaves with stomata isolated within folds of the rolled leaves

End of chapter questions

1. European robins (*Erithacus rubecula*) migrate south in the autumn (fall) and north in the spring. They orient their direction of flight using the local magnetic field, which they detect through magnetoreceptors in the upper beak. The orientation of the birds in a captive environment was studied in spring and autumn, which are the times of year when the birds normally migrate. The response of the birds to green light, red light and total darkness was investigated. In the figure, triangles on the edge of circles indicate the mean direction flown by individual birds while the arrows indicate the overall mean direction of flight.

   a. Identify the season and light conditions which result in the strongest northerly direction flown by the robins. [1]

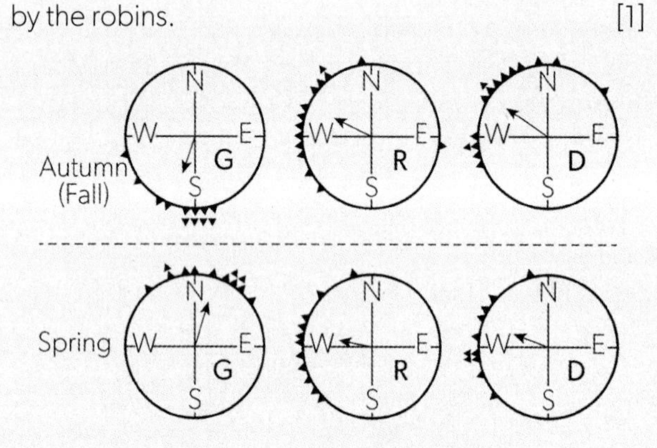

Source: R Wiltschko et al, The Journal of Experimental Biology, 211 (20), 3344-3350 2008

   b. Distinguish between the effect of red light and green light on the behaviour of the robins in spring and autumn (fall). [2]

   c. Based on the results of these experiments, suggest one possible conclusion that could be drawn regarding the effect of red light on the behaviour of robins. [1]

   d. Using the data in the diagram, deduce, with a reason, whether European robins migrate during the daytime or at night. [2]

   e. Scientists anesthetized the beaks of some robins in order to deactivate the magnetoreceptors. Predict how this would affect their orientation in red light. [1]

2. The graph shows the ventilation rate and tidal volume of a well-trained runner during exercise on a treadmill. The tidal volume is the volume of air being moved in and out of the lungs in each breath.

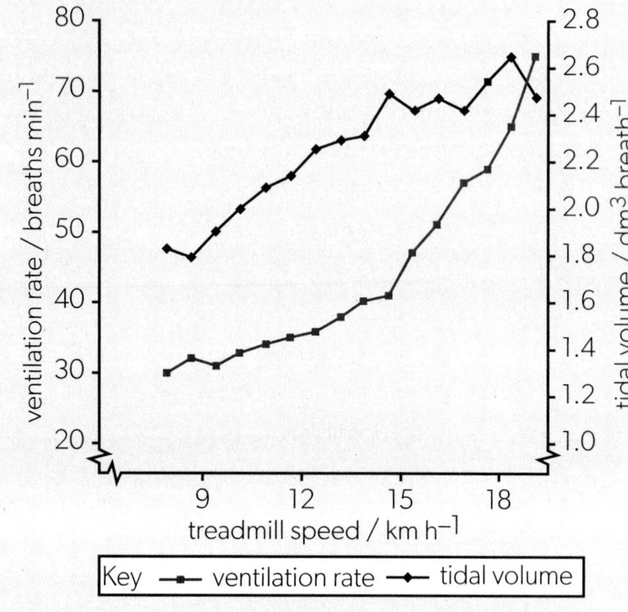

Source: Amonette, W. and Dupler, T. The effects of respiratory muscle training on VO 2 max, theventilatory threshold and pulmonary function. JEPonline 2002 5(2): 29-35

   a. Outline the process of ventilation. [3]

   b. Distinguish between ventilation and respiration as processes. [2]

   c. Distinguish between the role of external and internal intercostal muscles in ventilation. [2]

   d. State the apparatus used to measure the tidal volume. [1]

   e. Calculate the total volume of air inhaled in 1 minute during the highest velocity of the treadmill in this test, giving the units. [2]

   f. Compare and contrast the effect of increasing treadmill speed on the ventilation rate and tidal volume in this runner. [2]

3. The body mass index (BMI) is defined as the body mass divided by the square of the body height, and is expressed in units of $kg\,m^{-2}$ (resulting from mass in kilograms and height in metres). A long-term study followed nearly 40,000 apparently healthy young men for coronary heart disease (CHD) from adolescence through adulthood. The results show how the BMI at adolescence and adulthood affect the risk of CHD. The BMIs are divided into five groups (quintiles), Q1 being

the lowest BMI and Q5 the highest. A risk factor of 2 or less is desirable.

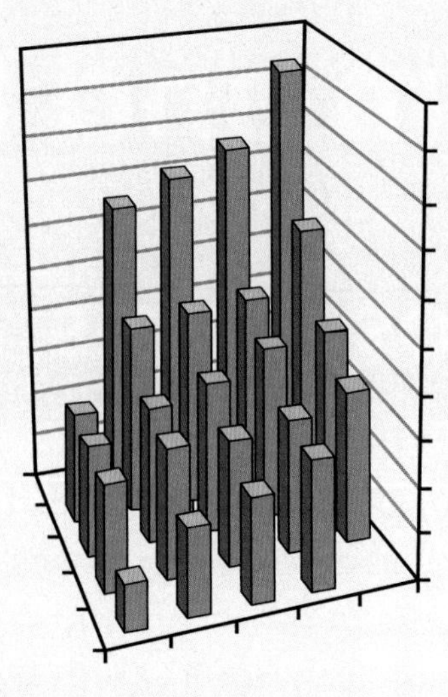

a. Determine the BMI of a person with height of 1.85 m and mass of 100 kg. [1]

b. Using the graph, discuss the hypothesis that a high BMI in adolescence is more dangerous than a high BMI in adulthood. [2]

c. State one factor, other than BMI, that increases the risk of CHD. [1]

Source: Tirosh et al NEJM 2011 364, 1315

4. Sometimes the ventilation of the lungs stops. This is called apnea. One possible cause is the blockage of the airways by the soft palate during sleep. This is called obstructive sleep apnea. It has some potentially harmful consequences, including an increased risk of accidents during the daytime due to disrupted sleep and tiredness. The figure shows the percentage oxygen saturation of arterial blood during a night of sleep in a patient with severe obstructive sleep apnea.

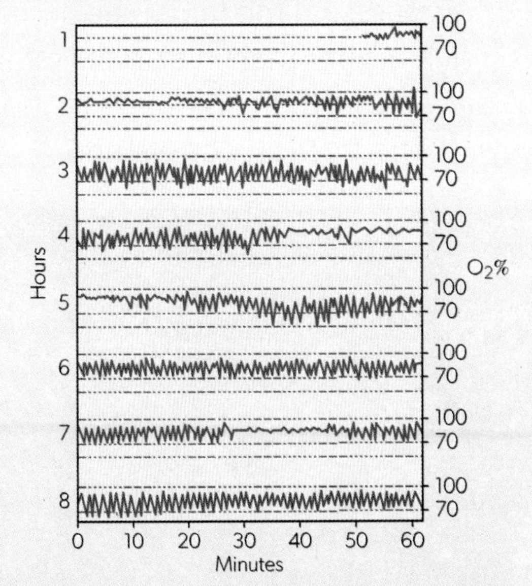

a. Hour 8 shows a typical pattern due to obstructive sleep apnea.

i Explain the causes of falls in oxygen saturation (%). [2]

ii Explain the causes of rises in saturation. [2]

iii Calculate how long each cycle of falling and rising saturation takes. [2]

b. Estimate the minimum oxygen saturation that the patient experienced during the night, and when it occurred. [2]

c. Deduce the sleep patterns of the patient during the night when the trace was taken. [2]

5. In one research project, pigeons (*Columba livia*) were trained to take off, fly 35 metres and land on a perch. During the flight, the activity of two muscles, the sternobrachialis (SB) and the thoracobrachialis (TB), was monitored using electromyography. The traces are shown below. The spikes show electrical activity in contracting muscles. Contraction of the SB muscle causes a downward movement of the wing.

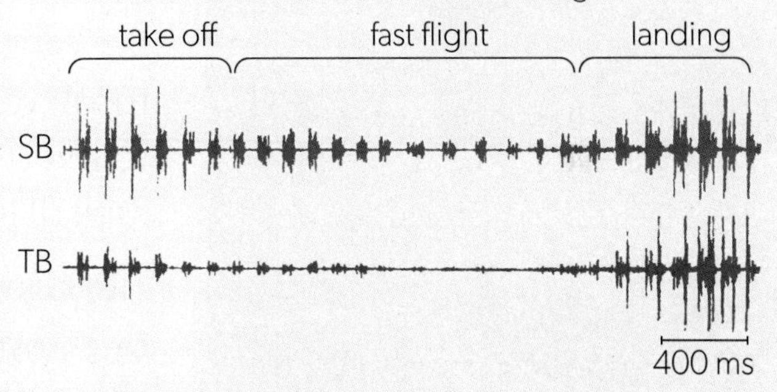

a. Outline, using an example, what is meant by antagonistic muscles. [3]

b. Using the data, deduce the number of downstrokes of the wing during the whole flight. [1]

c. Compare the activity of the SB muscle during the three phases of the flight. [3]

d. Deduce from the data in the electromyograph how the TB muscle is used. [1]

e. Another muscle, the supracoracoideus, is antagonistic to the SB muscle. State the movement produced by a contraction of the supracoracoideus. [1]

f. Predict the pattern of the electromyograph trace for the supracoracoideus muscle during the 35-m flight. [2]

g. Explain the role of ATP in vertebrate skeletal muscle contraction. [4]

B Form and function

4 Ecosystems

The structure of an ecosystem is its form. Ecosystems consis of biotic and abiotic components. The biotic community structure refers to the organisms that are present and the web of interactions between them. Organisms interact in feeding relationships, mutualistic relationships and competitive relationships. Abiotic factors also contribute to the overall form of an ecosystem. High levels of rainfall ofte result in the development of a forest, moderate levels lead to the development of a grassland ecosystem and sparse rainfall leads to the development of a desert.

The community structure of the taiga is influenced by temperature and rainfall. Taiga is a forest biome characterized by high levels of precipitation and cold average annual temperatures. In the taiga, the storage of nutrients in litter is much higher than in tropical rainforests. The rate of flow of nutrients from the biomass to the litter is relatively higher than in the tropical ecosystem.

B4.1 Adaptation to environment

How are the adaptations and habitats of species related?

The thick coat of a musk ox is correlated with the low temperatures of its northerly habitats. The water storage tissue in the stem of a cactus is related to infrequent rainfall in desert habitats. In biology, characteristics such as these that make an individual suited to its habitat are called adaptations. How do adaptations come to exist? What are other examples of adaptations of organisms to extreme environments? What is the reason that we avoid implying a purpose to an adaptation?

▶ Figure 1 Musk ox (*Ovibos moschatus*) during the autumn, Dovrefjell National Park, Norway

What causes the similarities between ecosystems within a terrestrial biome?

The Wallace line marks a division between species present in similar environments despite their geographical proximity. Islands to the east and west of the line have similar environments but very different species of plant and animal. Islands to the west of the line have Asian species and islands to the east have Australasian species. What is a possible explanation for this phenomenon? Wherever any particular type of environment occurs in the world, despite geographic separation, the same forms of plant and animal tend to evolve independently. What is the mechanism that leads to this convergence?

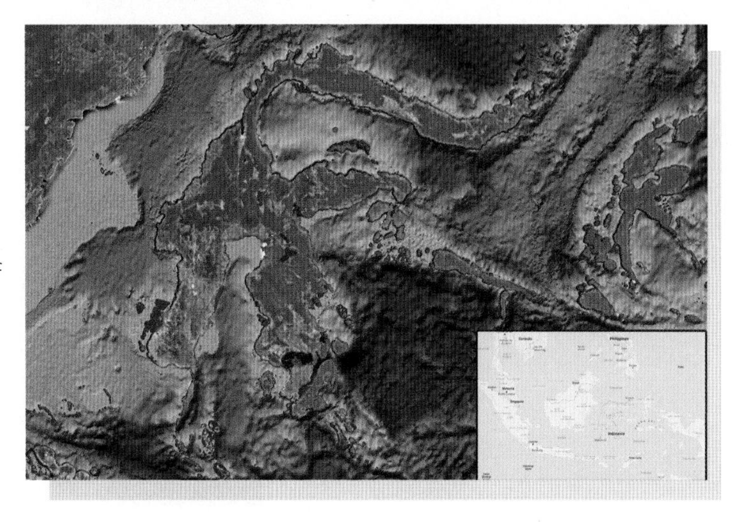

▲ Figure 2 The Wallace line runs between Borneo and the K-shaped island of Sulawesi

Every terrestrial environment poses challenges and adaptations are needed for plants and animals to survive and thrive. Wherever any particular type of environment occurs in the world, despite geographic separation, the same forms of plant and animal tend to evolve and therefore similar ecosystems. What is the mechanism that leads to this convergence?

SL and HL

B4.1.1 Habitat as the place in which a community, species, population or organism lives
B4.1.2 Adaptations of organisms to the abiotic environment of their habitat
B4.1.3 Abiotic variables affecting species distribution
B4.1.4 Range of tolerance of a limiting factor
B4.1.5 Conditions required for coral reef formation
B4.1.6 Abiotic factors as the determinants of terrestrial biome distribution
B4.1.7 Biomes as groups of ecosystems with similar communities due to similar abiotic conditions and convergent evolution
B4.1.8 Adaptations to life in hot deserts and tropical rainforest

B4.1.1 Habitat as the place in which a community, species, population or organism lives

Habitat means "he lives" or "she lives" in Latin. In biology, it means the place where an organism lives. This could be the geographical location—where in the world. More usually, it means the type of place inhabited: the physical conditions, the type of ecosystem and where within the ecosystem. It can apply to one organism or a whole population, species or community.

As an example, the habitat of *Ranunculus glacialis* is at very high altitude in the Alps and other mountains in Europe, on sites that are snow-covered through the winter and where there is little competition from other plants. These sites have a short growing season with intense sunlight. *R. glacialis* grows on acidic soils that are moist but also well drained.

▲ Figure 3 Glacier crowfoot (*R. glacialis*) growing at over 2,400 m altitude on a northeast facing slope of limestone rock on the Massif des Diablerets in the Alps

B4.1.2 Adaptations of organisms to the abiotic environment of their habitat

The environment of an organism is everything that is around it. This includes other living organisms and non-living materials such as air, water and rock. Living things are referred to as biotic factors and non-living things are called abiotic factors. Biotic factors dominate in ecosystems where there are dense communities of organisms—for example, in tropical rainforests. Abiotic factors have more influence in extreme habitats where population densities are low—for example, desert or taiga. All organisms are adapted to their abiotic environment. This is clearly seen in plants that live in extreme habitats such as sand dunes and mangrove swamps.

Adaptations of grasses to sand dunes

Sand dunes are mounds of sand that form from wind-blown sand in deserts and at the top of beaches. The challenges for plants on beach dunes are water conservation and tolerance of high salt concentrations and sand accumulation. Sand retains little water after rainfall and dunes initially contain little organic matter (which helps to store water in soils). Also, sand on beach dunes can contain high salt concentrations which hinders water uptake by osmosis. For these reasons, most types of plant would die of dehydration on sand dunes, so special adaptations are required for growth. Grasses are the dominant plant on beach dunes in many parts of the world. Lyme grass (*Leymus mollis*) occurs where sand is accumulating at the seaward edge of dunes in North America.

▲ Figure 4 Lyme grass on the Ma-le'l Dunes, Humboldt Bay, California

Lyme grass has these adaptations:

- thick waxy cuticle on leaves to reduce transpiration
- stomata in indentations (furrows) where humid air can remain even in windy conditions
- leaves that can roll up during droughts, creating a humid chamber and reducing the surface area exposed to wind
- tough sclerenchyma to prevent wilting during droughts
- rhizomes (underground stems) that grow upwards as sand accumulates and extend deep into the dune to obtain water
- accumulation of carbohydrates known as fructans in root and leaf cells to increase osmotic potential and thus water uptake.

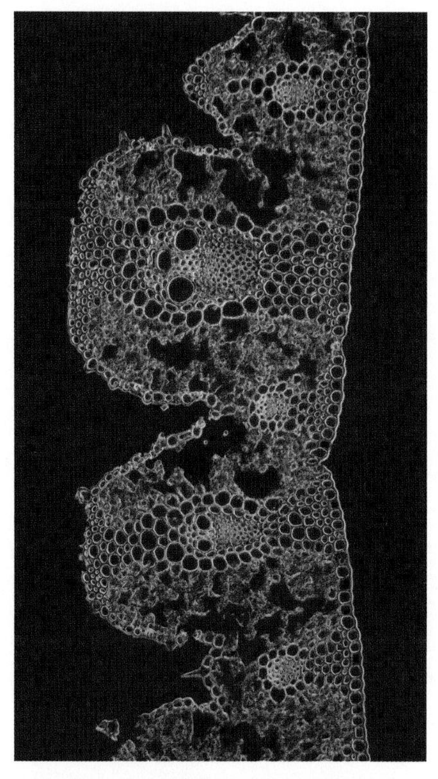

▲ Figure 5 Part of a lyme grass leaf showing tough sclerenchyma (pink) and furrows (visible on the left side in this micrograph) with stomata at their base

Adaptations of trees to mangrove swamps

Mangrove swamps develop on the coast in the tropics and subtropics where there are sheltered conditions and mud accumulates. These swamps are flooded with seawater at high tide. The dominant species are trees. The environmental challenges are waterlogged anaerobic soils and high salt concentrations. The salt concentration of the mud can be twice as high as that of seawater. This is due to the daily flooding with seawater and evaporation concentrating the salt in the mud.

Mangrove trees have the following adaptations that allow them to thrive in a habitat that would be intolerable for most species:

- secretion of excess salt from salt glands in the leaf
- root epidermis coated in suberin (cork) which reduces permeability to salt and prevents excessive absorption
- cable roots growing close to the soil surface where there is most oxygen
- pneumatophores, which are vertical root branches that grow up into the air and can absorb oxygen for use in roots
- stilt roots that grow out in a downward arch from the central trunk to buttress the tree in the soft mud
- large buoyant seeds that can be carried by the ocean to distant muddy shores
- accumulation of mineral ions and carbon compounds such as mannitol, which increases the osmotic potential of root and leaf cells, allowing water absorption from the very saline environment.

▲ Figure 6 The mud in mangrove swamps is deficient in oxygen. Mangrove trees have evolved vertical roots called pneumatophores which they use to obtain oxygen from the air

B4.1.3 Abiotic variables affecting species distribution

The distribution of a species is where it lives in the world, so it can be shown on a map. Distribution maps reflect the factors that affect species, especially abiotic factors. The adaptations of plants and animals suit them for living in some physical environments but not others.

Plant distributions are affected by temperature, water availability, light intensity, soil pH, soil salinity, and the availability of mineral nutrients. Every plant species has a range of tolerance for each of these factors. This means that a plant cannot grow in areas that are outside its range for one or more of the factors. For example, plant species from the tropics are not adapted to survive frosts so they would not survive in northern regions. Plants from these northern regions have chemicals in their cells that act like antifreeze and prevent frost damage caused by the formation of ice crystals. However, the northern plant species do not have adaptations for growth in the tropics. They would transpire excessively, and their method of photosynthesis would be very inefficient at high temperatures.

▲ Figure 7 Distribution of *Ranunculus glacialis* in Scandinavia. The species range is shown in green and the centre of gravity of the range in red

Animal distributions are affected by abiotic factors such as water availability and temperature. Extremes of temperature require special adaptations. The large ears of elephants with their dense networks of surface blood help to dissipate heat in hot climates, whereas polar bears have relatively small ears, minimizing heat loss in Arctic habitats. Some animals have adaptations for life in arid conditions. For example, desert rats have longer loops of Henle in their kidneys to minimize water loss. The adaptations required by aquatic animals are very different.

In some cases, animal distribution is limited by requirements for one stage in the life cycle. Salmon require fast flowing freshwater streams no more than 3 m deep for spawning. They must have gravel substrates with particle size between 10 mm and 100 mm and a water pH of between 5.5 and 8.0. As with plants, animals have a range of tolerance for each abiotic factor, based on their adaptations.

▲ Figure 8 Ecologists surveying on a transect extending from unburned to burned woodland at Backhouse Tarn, Tasmania

B4.1.4 Range of tolerance of a limiting factor

Plant and animal species have ranges of tolerance for abiotic variables. For example, many plant species will only grow in soils within a specific pH range; some require full sunlight and others only grow in shade. Animal species also have ranges of tolerance for variables such as salt concentration in aquatic habitats and temperature.

Ranges of tolerance can be investigated experimentally, or by finding correlations between the distribution of a species and abiotic variables. For example, a study in Taiwan of the mosquito, *Aedes aegypti* found it requires a minimum night-time temperature of 13.8°C.

Correlations between the distribution of a species and an abiotic variable can be investigated by mapping the entire species range, by random sampling for example using quadrats, or by sampling along transect lines.

Transects

Transects can be used to investigate the tolerance ranges of species to abiotic variables. A transect used for this purpose should span different levels of the variables of interest. For example, a line taken down a slope from woodland to peat bog might reveal correlations between the distribution of plant species and temperature, light intensity and soil pH. These and other abiotic variables can be measured using electronic sensors and portable data loggers. There are several different methods of sampling using a transect.

- Line intercept sampling—a tape is laid along the ground between two poles and all organisms that touch the line are recorded.
- Belt transects—the abundance of species is estimated in the area between two lines separated by a fixed distance, often 0.5 m or 1.0 m. Abundance can be assessed using quadrats placed at regular intervals along the belt.
- Observational transects—the observer walks along a defined route at a defined pace and records sightings of target species. This method can be used to investigate ranges of tolerance and is also used for monitoring changes in population size over time.

Observations: Making observations with sensors

A sensor is a device that records the level of a parameter. Electronic sensors are now available for many parameters that are of interest in ecological research. A log is a permanent record of measurements taken at regular intervals. Data logging is digital storage of measurements from electronic sensors. Compact, portable data loggers have been designed with a sensor to monitor an environmental condition such as temperature, light intensity or pH and an internal memory to record and store the digital data. These data loggers have many advantages:

- less expensive than many older designs and easy to operate
- designed to be compact and portable with battery power
- available for measuring hundreds of different parameters
- can take repeated measurements very rapidly
- can be left to take measurements automatically over long periods
- stored data can be transferred easily to a computer for analysis or long-term storage.

Because of their advantages, data loggers are widely used, both for ecological research and other purposes such as:

- medical diagnostics—in all settings from Intensive Care Units to remote areas far from hospitals
- industries such as food and drink production—for example, monitoring of fermentation in wineries
- flight recorders on aircraft.

Data-based questions: Intertidal zonation

The kite diagram in Figure 9 illustrates the distribution of common intertidal species 300 m south of Bembridge Lifeboat Station on the Isle of Wight, UK. The thickness of the shaded region indicates whether the organism was abundant, common, frequent, occasional or rare (ACFOR is a scale of abundance).

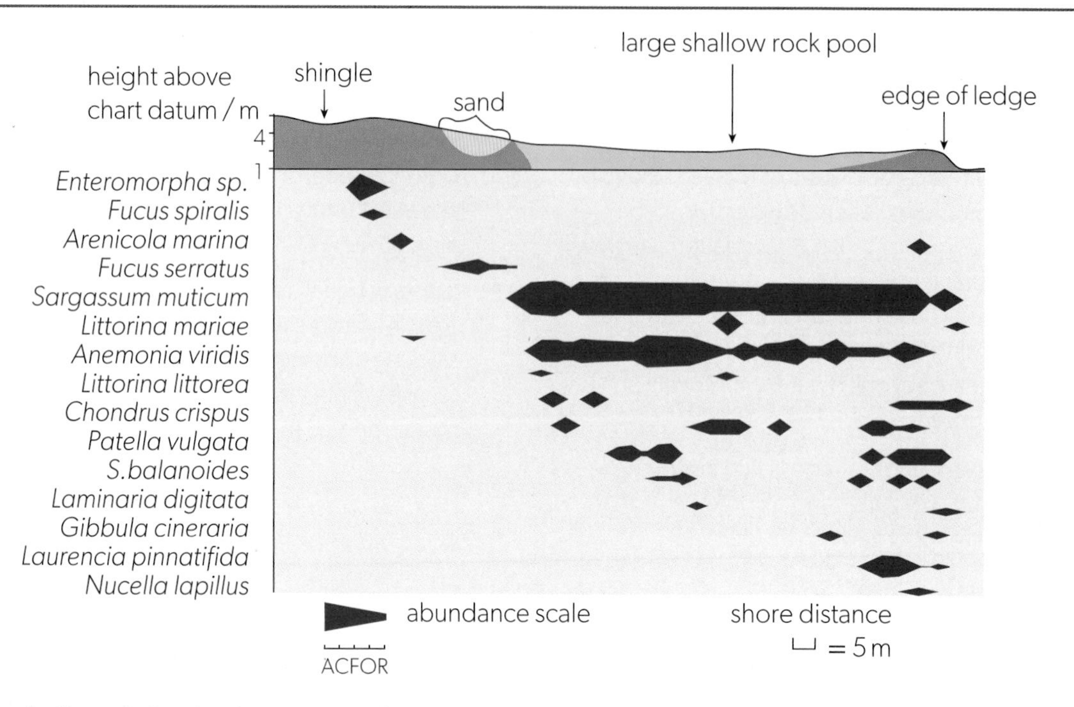

▲ Figure 9 Species abundance as a function of distance from spring tide high water mark (the highest point normally reached by salt water)

1. Examine the kite diagram and explain the methods used to collect the data. [3]
2. State the species that is most abundant in the survey area. [1]
3. Using the scale bar, determine the length of the large shallow rock pool. [2]
4. Deduce one species adapted to:
   a. shingle b. sand c. rock pools. [3]
5. Several species are only found near the lower edge of the intertidal zone. Suggest reasons for them being absent from the upper parts of the intertidal zone. [3]
6. Using the data in the kite diagram, predict two species that are adapted to the same abiotic environment. [2]
7. Suggest one way in which the objectivity of the research could have been improved. [1]

pH changes in rock pools

The pH in natural pools or in artificial aquatic mesocosms can be monitored using data loggers. Ecologists have monitored pH in rock pools on seashores that contain animals and photosynthesizing algae. The pH of the water rises and falls in a 24-hour cycle, due to changes in carbon dioxide concentration in the water. The lowest values of about pH 7 are found during the night and the highest values of about pH 10 when there is bright sunlight during the day. What are the reasons for these maxima and minima?

▲ Figure 10 Rock pool at Limerick Point, Ireland

Data-based questions: Data-logging pH in an aquarium

Figure 11 shows the pH and light intensity in an aquarium containing a varied community of organisms including pondweeds, newts and other animals. The data was obtained by data logging using a pH electrode and a light meter. The aquarium was illuminated artificially to give a 24-hour cycle of light and dark using a lamp controlled by a timer.

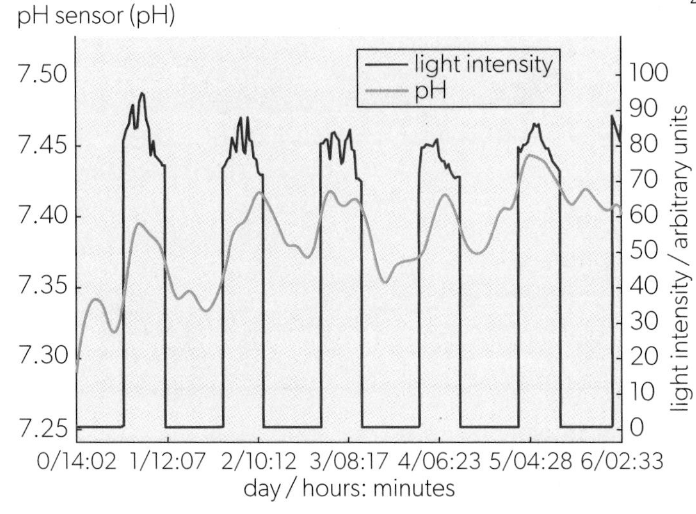

▲ Figure 11 Changes in pH and light intensity in an aquarium

1. Explain the changes in light intensity during the experiment. [2]
2. Determine how many days the data logging covers. [2]
3. a. Deduce the trend in pH in the light. [1]
   b. Explain this trend. [2]
4. a. Deduce the trend in pH in darkness. [1]
   b. Explain this trend. [2]

B4.1.5 Conditions required for coral reef formation

Coral reefs are biodiverse marine ecosystems. They can only develop where conditions are suitable for hard corals, whose skeletons form the rocky structure of the reef. Hard corals contain mutualistic zooxanthellae, which need light for photosynthesis. These are the conditions required.

- Depth—less than 50 m depth of water, so enough light penetrates.
- pH—above 7.8 to allow deposition of calcium carbonate in the skeleton.
- Salinity—between 32 and 42 parts per thousand of dissolved ions to avoid osmotic problems.
- Clarity—turbidity would prevent penetration of light so the water must be clear.
- Temperature—23–29°C so both the coral and its zooxanthellae remain healthy.

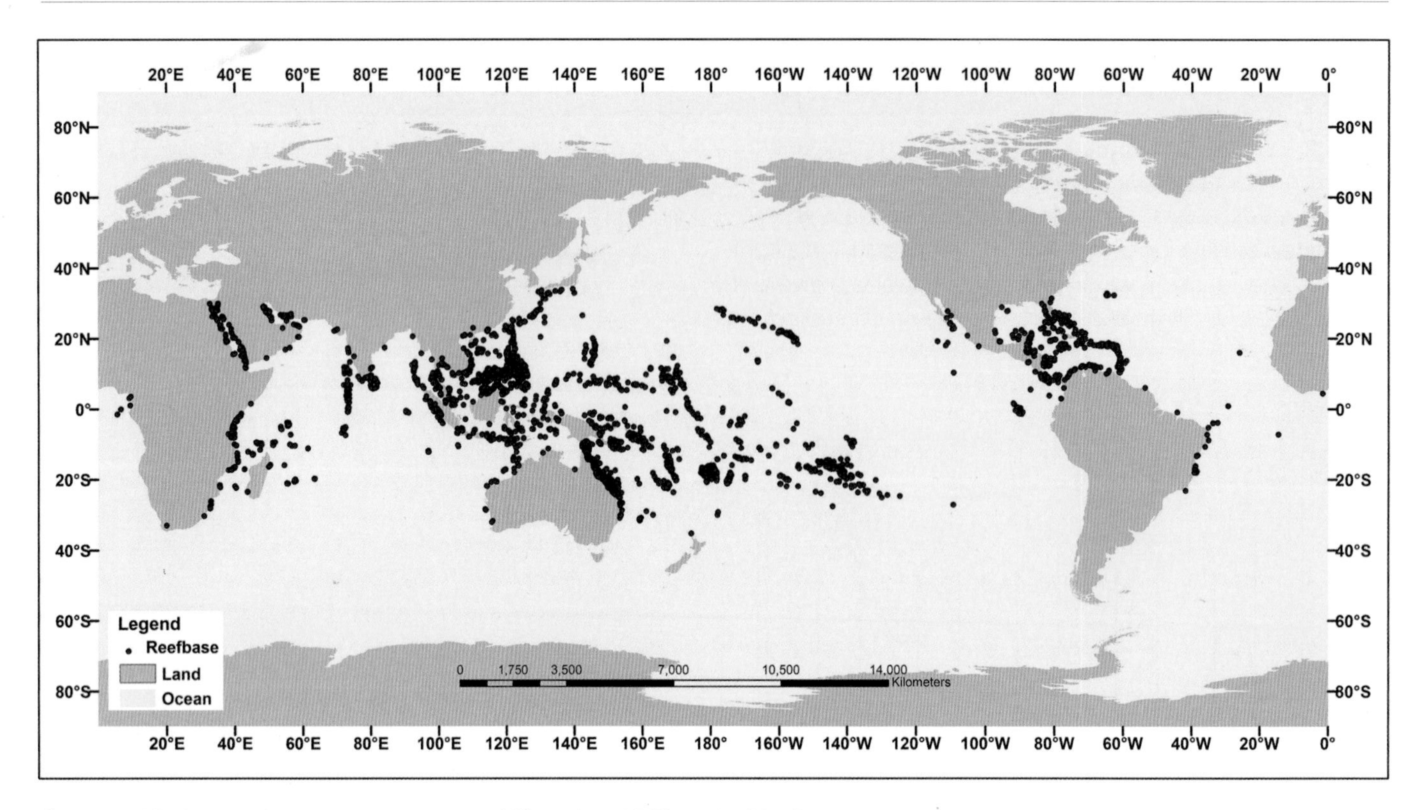

▲ Figure 12 Coral reefs can develop between 35° north and 35° south of the Equator

▲ Figure 13 Hard corals build the reef and provide a habitat for many other species. The blue-green fish on this Pacific reef are *Chromis viridis*

B4.1.6 Abiotic factors as the determinants of terrestrial biome distribution

With any combination of abiotic factors, one particular type of ecosystem is likely to develop. The species composition of the ecosystem will vary depending on the geographical location, but the adaptations of the species are likely to be similar. All ecosystems of a specific type are a biome.

Two abiotic factors are the principal determinants of biome distribution on Earth: temperature and rainfall. The most likely ecosystem given any particular combination of these factors can be shown using a graph, with mean annual precipitation on one axis and mean annual temperature on the other.

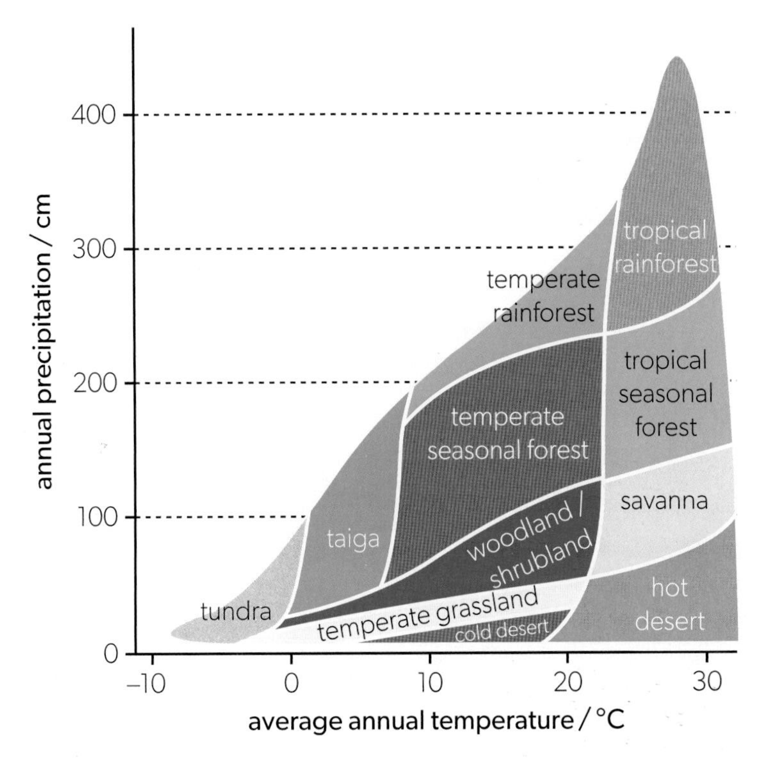

▲ Figure 14 Relationship between temperature, rainfall and biomes

B4.1.7 Biomes as groups of ecosystems with similar communities due to similar abiotic conditions and convergent evolution

Biomes are groups of ecosystems that resemble each other, even though they may be widely separated in the world. The resemblance is due to the similar abiotic conditions, with plants and animals evolving similar adaptations in response to the conditions. This is an example of convergent evolution. By natural selection, distantly related species that face the same problems find the same solutions. For example, plants in deserts develop adaptations for water conservation and storage. Cacti in America and euphorbias in Africa have very similar adaptations, despite not being closely related. In some cases, it is only when they produce flowers that these desert plants can be distinguished.

Gymnocalycium baldianum

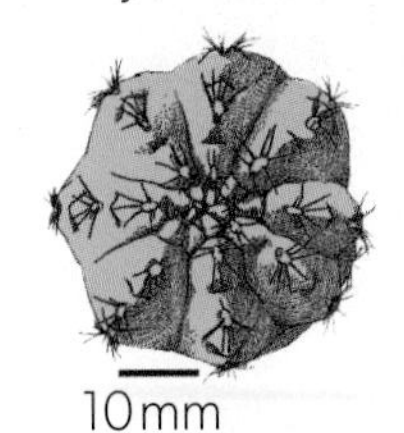

Euphorbia obesa

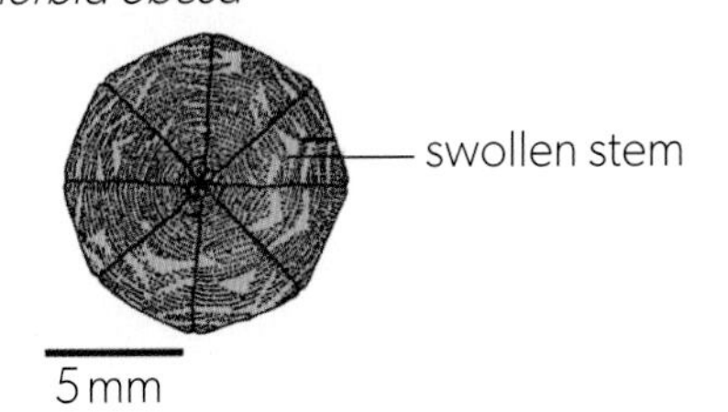

▲ Figure 15 *Gymnocalycium baldianum* (a cactus) and *Euphorbia obesa* (a euphorbia), both viewed from above

	Tropical forest	Temperate forest	Taiga (boreal forest)	Hot desert	Grassland	Tundra
Temperature	high	medium	low	high	high/medium	very low
Precipitation	high	high/medium	high/medium	very low	medium	medium/low
Light intensity	high	medium	medium/low	high	high/medium	low
Seasonal variation	minimal in rainforests	warm summers colder winters	short summers; long, cold winters	minimal variation	variation with a dry season or cold season	very short summer; very cold winter

▲ Table 1 Each of the major biomes is characterized by particular climatic conditions

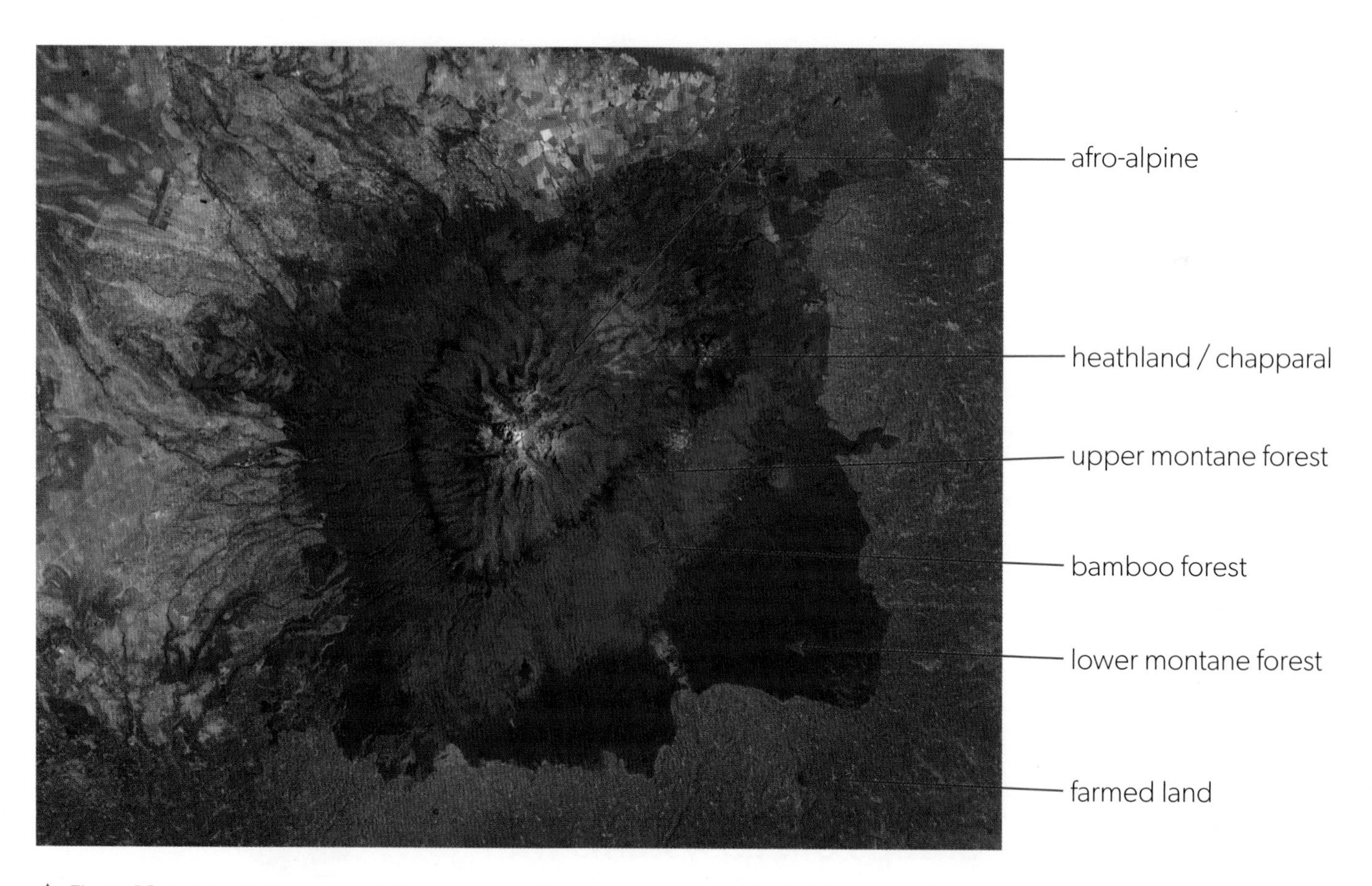

▲ Figure 16 In this satellite image of Mount Kenya, different ecosystems are visible. The summit is snow covered. Descending eastwards, the ecosystems are afro-alpine (light brown), heathland/chaparral (grey-green), upper montane forest (dark green), bamboo forest (light green), lower montane forest (dark green) with farmed land (mixed colours) outside the boundary of the protected area

B4.1.8 Adaptations to life in hot deserts and tropical rainforest

Hot deserts are characterized by very high daytime temperatures and much colder nights. Rainfall totals per year are very low and there can be long periods without any precipitation. Soil development is very limited, with little organic matter of soil organisms. The saguaro and fennec fox are examples of organisms adapted to these conditions.

The saguaro is a species of cactus that is adapted to life in hot deserts. It has the following adaptations:

- a wide-spreading root system to collect water up to 30 m from the stem
- deep tap roots that collect water from up to 1 m down in the subsoil
- fat stems with storage tissue to conserve water after infrequent desert rains
- pleated stems that allow shrinkage in droughts and swelling after rains
- vertical orientation of stems to reduce interception of sunlight at midday and maximize it at cooler times of day
- a thick waxy cuticle on the stem epidermis to reduce transpiration
- leaves reduced to spines, to reduce the surface area for transpiration and prevent herbivores from eating the slow-growing cactus
- CAM metabolism allowing stomata to open at night and close during the heat of the day, reducing transpiration.

▲ Figure 17 Saguaro (*Carnegiea gigantea*) in hot desert in Arizona

The fennec fox is a species of mammal that it adapted to life in hot deserts. It has the following adaptations:

- nocturnal so it avoids the highest temperatures during daylight hours
- it builds an underground den where it can stay cool during the day
- long thick hair to provide heat insulation both for the cold nights and hot days
- hairs covering the pads of the feet to provide insulation when walking on very hot sand
- a pale-coloured coat that reflects sunlight (a darker coat would absorb it)
- large ears that radiate heat and help keep body temperature down
- a variable ventilation rate that can be increased to more than 600 breaths per minute (panting) to cause heat loss by evaporation.

▲ Figure 18 Fennec fox (*Vulpes zerda*) at Farafra oasis, Egypt

Tropical rainforests are characterized by high temperatures, high precipitation and high light intensity. The yellow meranti and the spider monkey are examples of organisms adapted to these conditions.

The meranti (*Shorea faguetiana*) is a species of tree that is adapted to life in tropical rainforests. It has the following adaptations:

- it can grow to over 100 m high, overtopping other trees and avoiding competition for light
- trunk of hard dense wood to provide support especially against the wind stress

▲ Figure 19 Yellow meranti (*Shorea faguetiana*) near the Kinabatangan River in Borneo

- trunk is buttressed at the base to provide increased support because rainforest soils are shallow
- smooth trunk to shed rainwater rapidly
- broad oval leaves with pointed tips that shed rainwater rapidly
- evergreen leaves which take advantage of ideal conditions for photosynthesis throughout the year
- enzymes of photosynthesis adapted to tolerate temperatures as high as 35°C
- flowers and seed produced in large quantities about one year in five, with none in other years to deter species that eat the seeds.

▲ Figure 20 Spider monkey (*Ateles geoffroyi*) in Belize

The spider monkey (*Ateles geoffroyi*) is a species of mammal that is adapted to life in rainforests. It has the following adaptations:

- long arms and legs for climbing and reaching for fruit
- flexible shoulders allowing swinging from tree to tree
- large hook-like hands without thumbs that can grasp branches and lianas and pick fruit
- feet that can act like extra hands, grasping branches and allowing the arms to be used for feeding or other purposes
- long tail that can grip onto branches and act like a fifth hand
- highly developed larynx allowing a wide range of sounds to be made to communicate in the dense rainforest canopy
- sleeping at night and active in the daytime when vision is most acute and distances can be judged between branches.
- breeding at any time of year as there is a constant supply of fruit, nuts, seeds, buds, flowers, insects and eggs.

Linking questions

1. What are the properties of the components of biological systems?
   a. Explain the interactions between auxin and cytokinin as a means of regulating root and shoot growth. (C3.1.22)
   b. Discuss the statement: "integration results in emergent properties". (C3.1.2)
   c. Outline the role of feedback control in the regulation of the human heart rate. (C3.1.14)
2. Is light essential for life?
   a. Explain why the energy content of each trophic level decreases through a food chain. (C4.2.14)
   b. Explain how photosynthesis results in an increase in biomass. (C4.2.15)
   c. Outline how chemosynthesis represents an exception to the rule that ecosystems are dependent on light as a source of energy. (B4.2.6)

B4.2 Ecological niches

What are the advantages of specialized modes of nutrition to living organisms?

When Charles Darwin was sent a Madagascar star orchid (*Angraecum sesquipedale*), which has a 300 mm long nectar tube, he predicted that a moth with equally long tubular mouthparts must exist in the same ecosystem to act as the orchid's pollinator. The moth was finally discovered 21 years after Darwin's death and named *Xanthopan praedicta*. Its mouthparts are indeed 300 mm long and have to be coiled up when not in use. They unroll like a party blower when the moth is about to insert them into the nectar tube. On what do you think Darwin based his prediction?

▲ Figure 1 *Xanthopan praedicta*

How are the adaptations of a species related to its niche in an ecosystem?

Kettlehole ponds fill landscape features created by retreating glaciers. Circular zones of different plants can be seen at different depths in the pond. The deeper the water, the less light penetrates and the lower the oxygen concentration in mud at the base of the pond. Drying in summer is the main challenge in shallow water near the margins of the pond, so different structural adaptations are beneficial, thus different species dominate at the edges. Which types of plants would thrive in the centre of the pond? Which would thrive at the edges?

▲ Figure 2 Kettlehole pond

SL and HL

B4.2.1 Ecological niche as the role of a species in an ecosystem
B4.2.2 Differences between organisms that are obligate anaerobes, facultative anaerobes and obligate aerobes
B4.2.3 Photosynthesis as the mode of nutrition in plants, algae and several groups of photosynthetic prokaryotes
B4.2.4 Holozoic nutrition in animals
B4.2.5 Mixotrophic nutrition in some protists
B4.2.6 Saprotrophic nutrition in some fungi and bacteria
B4.2.7 Diversity of nutrition in archaea
B4.2.8 Relationship between dentition and the diet of omnivorous and herbivorous representative members of the family Hominidae
B4.2.9 Adaptations of herbivores for feeding on plants and of plants for resisting herbivory
B4.2.10 Adaptations of predators for finding, catching and killing prey and of prey animals for resisting predation
B4.2.11 Adaptations of plant form for harvesting light
B4.2.12 Fundamental and realized niches
B4.2.13 Competitive exclusion and the uniqueness of ecological niches

B4.2.1 Ecological niche as the role of a species in an ecosystem

One of the central hypotheses of ecology is that every species in an ecosystem fulfils a unique role, called its ecological niche. Ecological niches have both biotic and abiotic elements.

- Zones of tolerance for abiotic variables determine the habitat of a species—where it lives in the ecosystem.
- Food is obtained either by synthesis using light, water and carbon dioxide or by taking it in from other organisms. To minimize competition, species must specialize. To compete effectively, they must develop adaptations for the mode of nutrition that is their specialism.
- Other species are utilized to provide a diverse range of services—for example, the supply of mineral elements by recycling, pollination of flowers or dispersal of seeds; the support provided by the trunks and branches of trees.

The ecological niche of a species is made up of very many factors—it is multidimensional. Unless all the dimensions of the niche are satisfied in an ecosystem, a species will not be able to survive, grow or reproduce.

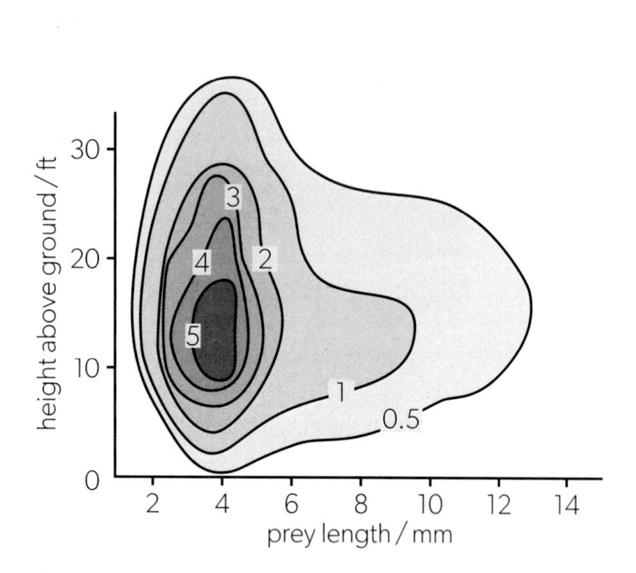

▲ Figure 3 Two aspects of the niche of the blue-gray gnatcatcher (*Polioptila caerulea*) are prey size and foraging height. The graph shows the percentage of the diet according to these two variables in oak woodland, in California. There are many other aspects of this bird's ecological niche

▶ Figure 4 *P. caerulea* eating a gnat

B4.2.2 Differences between organisms that are obligate anaerobes, facultative anaerobes and obligate aerobes

Animals and plants require oxygen for aerobic cell respiration, but some other organisms do not have this requirement. Some microorganisms can only live in the total absence of molecular oxygen (O_2) including some species of bacteria, archaea and protozoa. Anoxic (lack of oxygen) conditions occur in swamps, water-logged soil or muds, intestinal tracts (guts) of animals and deep in lakes or seas.

Living organisms can be placed in three categories according to their oxygen requirements (Table 1).

Category	Requirements	Examples
obligate aerobes	require a continuous oxygen supply so only live in oxic environments	All animals and plants; *Micrococcus luteus* (a skin bacterium)
obligate anaerobes	inhibited or killed by oxygen so only live in anoxic environments	*Clostridium tetani* (tetanus bacterium), methanogenic archaea
facultative anaerobes	use oxygen if available so live in oxic or anoxic environments	*Escherichia coli* (a gut bacterium), *Saccharomyces* (yeast)

▲ Table 1 Oxygen requirements of organisms

Activity: Winogradsky columns

To make a Winogradsky column, mud and water from a pond is placed in a large bottle or measuring cylinder, with a range of other materials. The column is sealed and placed in the light. Concentration gradients for oxygen and other substances develop in the column, with coloured bands due to groups of bacteria and archaea growing where the concentrations suit them.

▲ Figure 5 Winogradsky column in a glass bottle

B4.2.3 Photosynthesis as the mode of nutrition in plants, algae and several groups of photosynthetic prokaryotes

In photosynthesis, energy from sunlight is used for fixing carbon dioxide and using carbon from it to produce sugars, amino acids and the many other carbon compounds on which life is based. There are three groups of photosynthesizers:

- plants, including mosses, ferns, conifers and flowering plants
- eukaryotic algae including seaweeds that grow on rocky shores and unicellular algae such as *Chlorella*
- several groups of bacteria including cyanobacteria (blue–green bacteria) and purple bacteria.

Photosynthesis therefore occurs in two of the three domains of life: in eukaryotes and bacteria, but not in archaea.

B4.2.4 Holozoic nutrition in animals

Animals obtain supplies of carbohydrates, amino acids and other carbon compounds by consuming food. They are heterotrophic, because the carbon compounds come from other organisms. Molecules such as polysaccharides and proteins must be digested before they can be absorbed. Digestion in most animals happens internally, after the food has been ingested. This is holozoic nutrition, meaning that whole pieces of food are swallowed before being fully digested.

This is the sequence of stages in holozoic nutrition:

1. **ingestion**—taking the food into the gut
2. **digestion**—breaking large food molecules into smaller molecules
3. **absorption**—transport of digested food across the plasma membrane of epidermis cells and thus into the blood and tissues of the body
4. **assimilation**—using digested foods to synthesize proteins and other macromolecules and thus making them part of the body's tissues
5. **egestion**—voiding undigested material from the end of the gut.

Some animals digest their food externally so they are not holozoic. Spiders, for example, inject digestive enzymes into their prey and suck out the liquids produced. They absorb the products of digestion in their gut and then assimilate them.

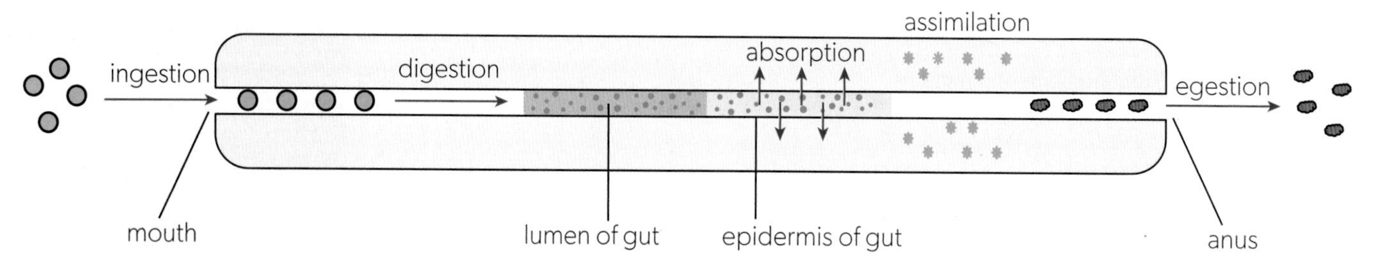

▲ Figure 6 Holozoic nutrition

B4.2.5 Mixotrophic nutrition in some protists

Autotrophs make their own carbon compounds from simple substances including carbon dioxide. Heterotrophs obtain their carbon compounds from other organisms. Some unicellular eukaryotes (protists) use both methods of nutrition. Organisms that are not exclusively autotrophic or heterotrophic are mixotrophic. Facultative mixotrophs can be entirely autotrophic, entirely heterotrophic, or use both modes. *Euglena gracilis*, for example, has chloroplasts and carries out photosynthesis when there is sufficient light, but it can also feed on detritus or smaller organisms by endocytosis, so it is a facultative mixotroph.

Obligate mixotrophs cannot grow unless they utilize both autotrophic and heterotrophic modes of nutrition. This may be because the food that they consume supplies them with a carbon compound that they cannot themselves synthesize. In other cases, a protist that does not have its own chloroplasts obtains them by consuming algae. It uses the "klepto-chloroplasts" obtained in this way for photosynthesis until they degrade and have to be replaced.

▲ Figure 7 *Arabidopsis thaliana*—the autotroph that molecular biologists use as a model plant

▲ Figure 8 Humming birds are heterotrophic; the plants from which they obtain nectar are autotrophic

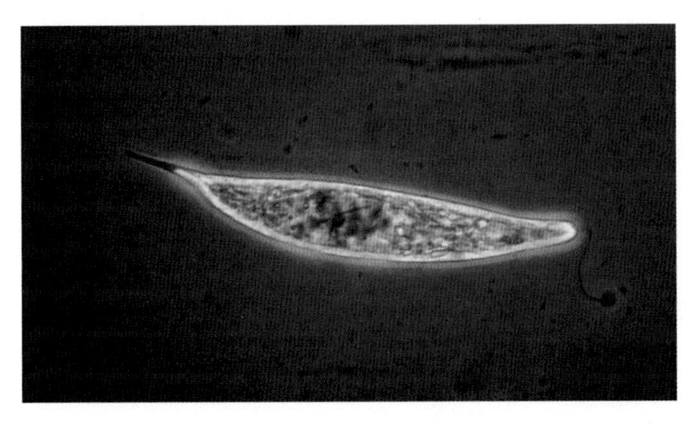

▲ Figure 9 *Euglena*—a facultative mixotroph. Organisms such as *Euglena* do not fit into the plant or animal kingdoms so are placed in another kingdom, called either Protista

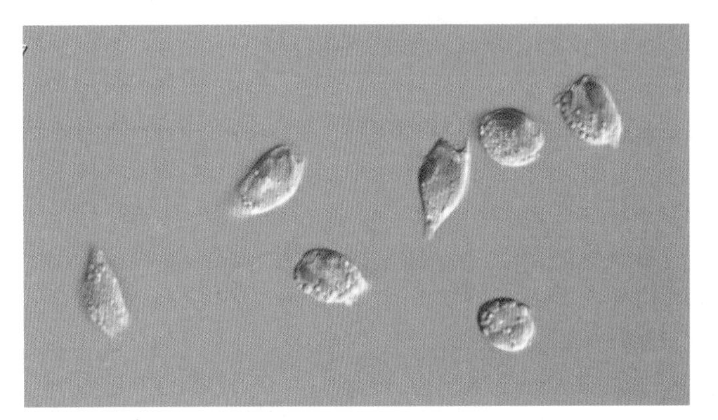

▲ Figure 10 *Ochromonas sp.* can make their own food through photosynthesis, but can also uptake both dissolved organic nutrients and particulate organic matter, including intact cells

Data-based questions: Mixotrophy in golden algae

Two strains of the golden alga *Ochromonas* were isolated, from the Atlantic Ocean east of New Jersey (isolate 1393) and from the Pacific Ocean east of Taiwan (isolate 2951). Their growth rates were measured in three combinations of conditions: light but no prey (autotrophic), prey but no light (heterotrophic) and both prey and light (mixotrophic). The prey supplied to the algae were *Vibrio* bacteria.

1. Compare and contrast the growth rates for the two isolates. [4]
2. Deduce, with reasons, whether isolate 1393 is an obligate mixotroph, a facultative mixotroph, or not a mixotroph. [3]
3. Discuss the nutrition of isolate 2951. [3]

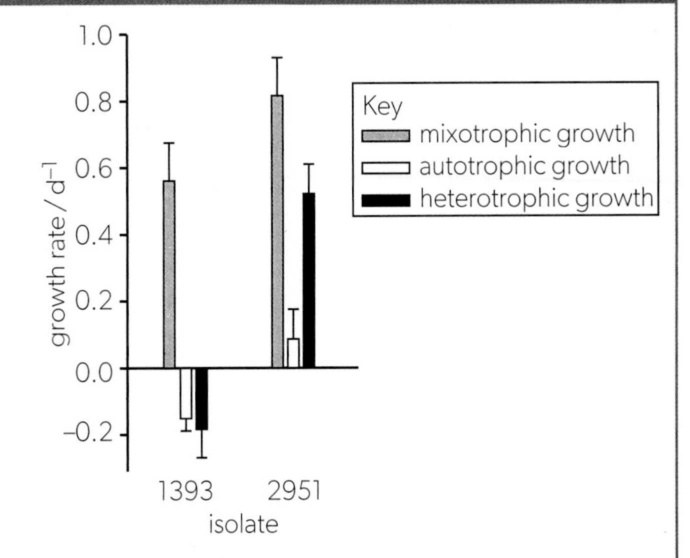

▲ Figure 11 Growth rates of two isolates of the alga *Ochromonas*

▲ Figure 12 Saprotrophic fungi growing over the surfaces of dead leaves and decomposing them by secreting digestive enzymes

B4.2.6 Saprotrophic nutrition in some fungi and bacteria

Saprotrophs secrete digestive enzymes into the dead organic matter and digest it externally. They then absorb the products of digestion. Many types of bacteria and fungi are saprotrophic. They are also known as decomposers because they break down carbon compounds in dead organic matter and release elements such as nitrogen into the ecosystem, allowing them to be used again by other organisms.

Activity: Determining trophic level

By answering a series of simple questions about an organism's mode of nutrition it is usually possible to deduce what trophic group it is in. The questions are presented in Figure 13 as a dichotomous key, which consists of a series of pairs of choices. The key works for unicellular and multicellular organisms, but not for parasites such as tapeworms or fungi that cause diseases in plants.

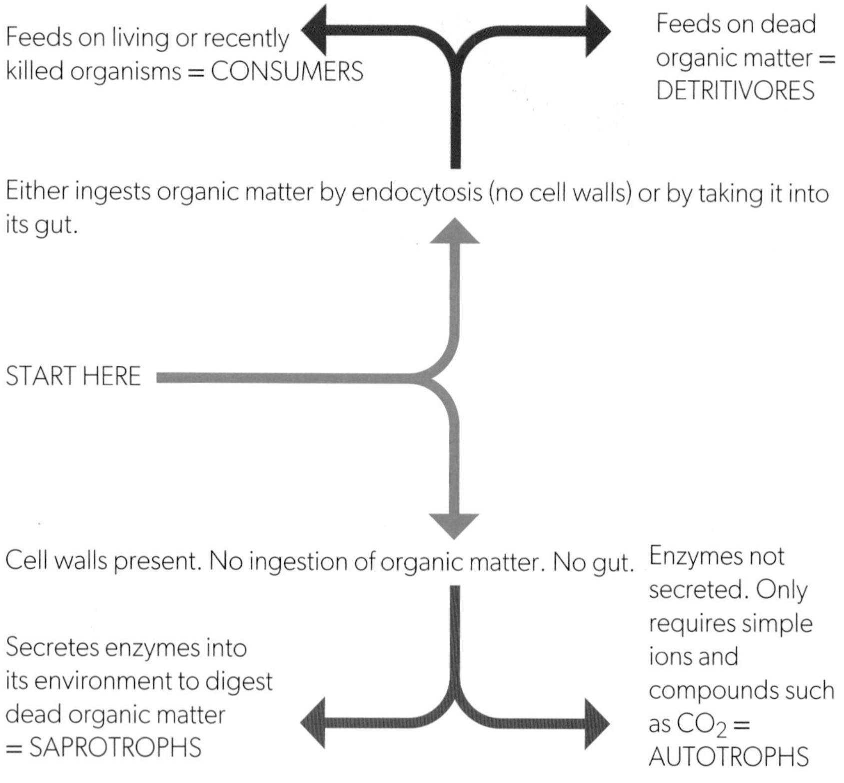

Feeds on living or recently killed organisms = CONSUMERS

Feeds on dead organic matter = DETRITIVORES

Either ingests organic matter by endocytosis (no cell walls) or by taking it into its gut.

START HERE

Cell walls present. No ingestion of organic matter. No gut.

Secretes enzymes into its environment to digest dead organic matter = SAPROTROPHS

Enzymes not secreted. Only requires simple ions and compounds such as CO_2 = AUTOTROPHS

▲ Figure 13 A dichotomous key

Data-based questions: Fishing down marine food webs

Trophic levels can be represented by a number indicating the position of a species within an ecosystem. By definition, the producers occupy the first trophic level (TL) and so have a TL of 1. For primary consumers, TL = 2, and so on. The higher the number, the more energy-transfer steps between the organism and the initial fixing of the Sun's energy. Trophic levels are not always stated as whole numbers. Fish and other animals that feed at more than one level often have estimated mean trophic levels.

One effect of commercial over-fishing is the reduction in the number of fish that feed at higher trophic levels (i.e. long-lived fish). The phrase "fishing down marine food webs" refers to the increased tendency for marine landings to consist of animals that feed at lower trophic levels (Figure 14).

1. Suggest a method that might be used to deduce the trophic level of a fish once it is captured. [2]
2. a. Compare the changes in mean trophic level of landed fish from marine and freshwater fisheries since 1970. [3]
   b. Suggest why there is a difference in the two trends. [2]
3. Explain why the mean trophic level might increase with age in an individual fish. [2]
4. Deduce the change in age of captured fish over the period shown. [2]
5. Explain two advantages of humans catching and consuming fish at a lower mean trophic level. [4]

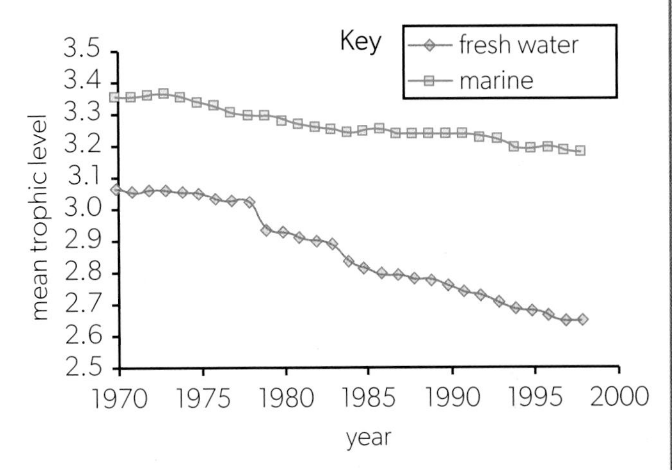

▲ **Figure 14** How the mean trophic level of landed fish has changed over a 30-year period

B4.2.7 Diversity of nutrition in archaea

There are three domains of life: archaea, bacteria and eukaryotes. The archaea are unicellular and have no nucleus, which is a similarity with bacteria. In other respects archaea are closer to eukaryotes.

Some types of archaea are adapted to extreme environments such as hot springs, salt lakes and soda lakes. Many are difficult to culture in the laboratory, so they are less well researched than the other domains of life.

Archaea are extremely diverse in the energy sources used for ATP production. There are three main categories:

- phototrophic—absorption of light energy by pigments—but pigments other than chlorophyll are used
- chemotrophic—oxidation of inorganic chemicals, for example Fe^{2+} ions to Fe^{3+}
- heterotrophic—oxidation of carbon compounds obtained from other organisms.

B4.2.8 Relationship between dentition and the diet of omnivorous and herbivorous representative members of the family Hominidae

The family Hominidae includes the genera that contain humans (*Homo*), orang-utans (*Pongo*), gorillas (*Gorilla*), and chimpanzees (*Pan*). Some members of the Hominidae have an exclusively herbivorous diet and others are omnivorous—some animal prey is included in the diet. Living members of the Hominidae show a relationship between diet and dentition. This can be studied using physical collections of skulls in natural history museums or digital collections available online such as those found at eSkeletons.org, a database sponsored by the University of Texas at Austin.

▲ Figure 15 Chimpanzees have much larger canines than humans

The teeth of herbivores tend to be large and flat to grind down fibrous plant tissues. Omnivores tend to have a mix of different types of teeth to break down both meat and plants in their diet. Humans have flat molars in the back of their mouth to crush and grind food, and sharper canines and incisors than herbivores to tear tougher food, like meat.

Once the structure–function relationships have been established, the diet of extinct species in the Hominidae can be inferred from their dentition—for example, in *Homo floresiensis* and *Paranthropus robustus*.

Activity: Deducing diet

Figure 16 shows the fossilized jaw and teeth of an individual of *Australopithecus anamensis*, who lived about 4.1 million years ago. The jaw in Figure 17 is from a female *Homo neanderthalensis* who lived more than 110,000 years ago (before the last glaciation).

What, if anything, can be deduced about their diet?

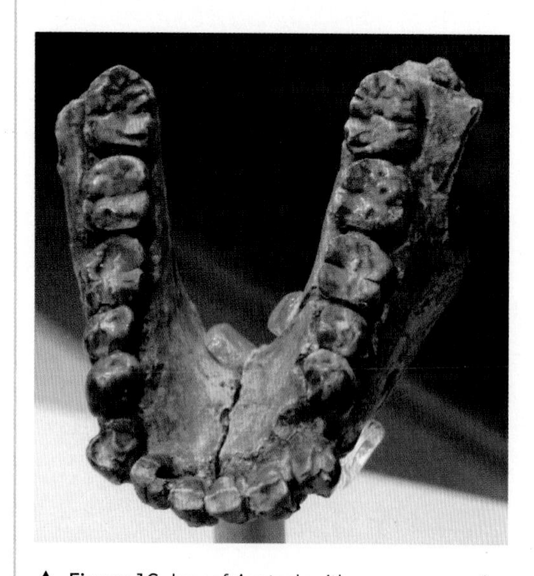

▲ Figure 16 Jaw of *Australopithecus anamensis*

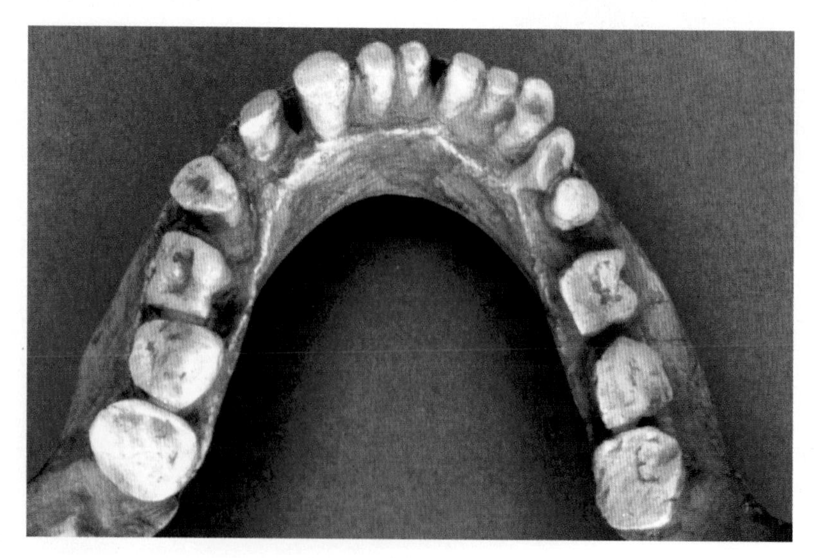

▲ Figure 17 Jaw of *Homo neanderthalensis*

Theories: Making deductions about diet from fragmentary evidence

A theory in science is a general explanation that is widely applicable. Theories can be based on observed patterns. Predictions can be generated from these theories by deductive reasoning. If observations are made of dentition in animals with known diets, including herbivores, carnivores and omnivores, theories can be developed about the structure–function relationships of teeth. These theories can be tested by predicting the diet of living animals from the characteristics of their teeth and then checking whether the actual diet matches the prediction. This may corroborate the theories or show that they are false and should be rejected.

Theories about dentition can also be used to infer the diet of extinct species of hominid; however, it is not possible to verify these predictions—we cannot be sure what the diet of an extinct hominid was. Are such predictions therefore non-scientific? What if skeletons of prey species are found in the vicinity of the extinct fossil hominid? Can this increase the certainty?

B4.2.9 Adaptations of herbivores for feeding on plants and of plants for resisting herbivory

Animals that feed exclusively on plants are herbivores. They have structural features that adapt them to their diet. Insect mouthparts show great diversity, but are all homologous—they have been derived by evolution from the same ancestral mouthparts. Most insects are herbivores. Insects that feed on leaves can be divided into two broad groups:

- beetles and other insects with jaw-like mouthparts for biting off, chewing and ingesting pieces of leaf
- aphids and other insects with tubular mouthparts for piercing leaves or stems to reach phloem sieve tubes and feed on the sap.

▲ Figure 18 Frog beetle (*Sagra buqueti*) has chewing mouthparts for feeding on leaves

▲ Figure 19 Rose aphids (*Macrosiphum rosae*) have piercing mouthparts

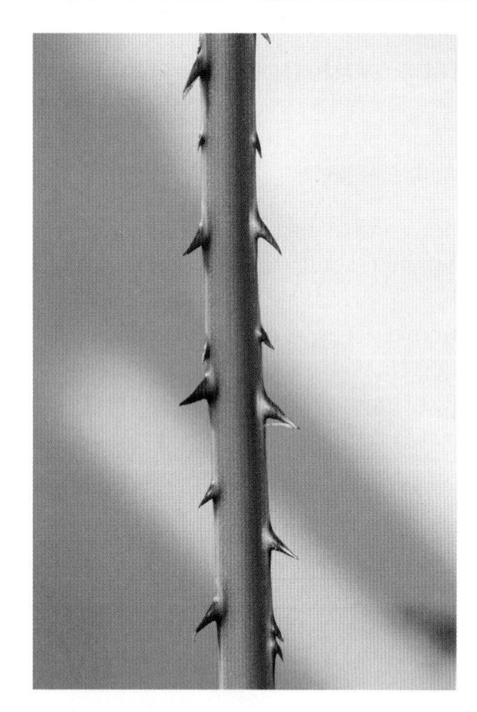

▲ Figure 20 Spines on a leaf stalk of the fan palm (*Saribus rotundifolius*)

Plants show a variety of adaptations for deterring herbivore attacks. Some have tough sharp-pointed spines, so herbivores risk injury when eating it. Others have stings to cause pain. Many plants synthesize substances that are toxic to herbivores. These substances are secondary metabolites. (Primary metabolites are substances that are part of the basic metabolic pathways of a cell.) They may be stored in any part of a plant, particularly seeds, which are attractive to herbivores because of their high concentrations of protein and starch or oil.

▲ Figure 21 Stings on the tree nettle (*Urtica ferox*) which is endemic to New Zealand

In some cases, herbivores have responded to toxic compounds in plants by developing metabolic adaptations for detoxifying them. This has resulted in plant–herbivore specificity, with only a few species of herbivore adapted to feed on a particular plant.

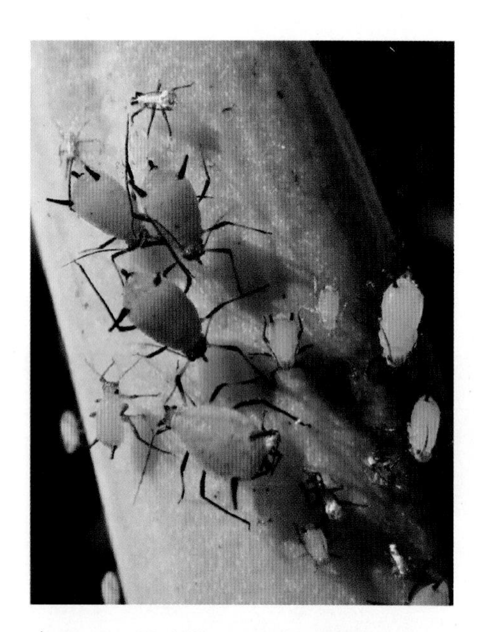

▲ Figure 22 Milkweed plants (*Asclepias tuberosa*) synthesize toxic glycosides to deter herbivores, but the milkweed aphid (*Aphis nerii*) not only tolerates these glycosides, it also makes itself toxic to predators by accumulating them

B4.2.10 Adaptations of predators for finding, catching and killing prey and of prey animals for resisting predation

Predators are adapted to find suitable prey and then catch and kill it. The prey may be killed before it is ingested, or it may die inside the predator's digestive system. Prey species are adapted to resist predation. Selected examples of adaptations are shown in Table 2, but there are many others. These adaptations may be structural, chemical or behavioural.

▼ Table 2 Some adaptations of prey and predators

Type	Predators	Prey
Physical	 ▲ Figure 23 Vampire bats (*Desmodus rotundus*) have unique dentition, with small premolars and no molars, but relatively large incisors and canines on their upper jaw that are pointed and razor-sharp. These are used to pierce prey, so the vampire can feed on the blood	 ▲ Figure 24 Buff-tip moths (*Phalera bucephala*) resemble broken birch twigs, giving them camouflage when roosting during daylight hours on twigs or on the ground. This is the time when the night-flying moths are most vulnerable to predation
Chemical	 ▲ Figure 25 Black mambas (*Dendroaspis polylepis*) produce venom containing a mixture of neurotoxins, including an inhibitor of the enzyme acetylcholinesterase. The venom paralyzes prey when injected via poison fangs. The snake can then swallow the prey without it resisting	 ▲ Figure 26 Caterpillars of the cinnabar moth (*Tyria jacobaeae*) feed on ragwort and accumulate toxic alkaloids from it. Their black and yellow stripes are warning coloration which deters predators. Adults are day-flying, with red and black warning coloration, indicating that they retain toxins obtained when the larvae fed on ragwort
Behavioural	 ▲ Figure 27 Grizzly bears (*Ursus arctos*) learn ambush strategies for catching migrating salmon either by trial and error or copying others. Some bears wait at the top of waterfalls for a fish to jump out of the water. Others put their heads underwater and watch for a fish swimming past	 ▲ Figure 28 Blue-striped snappers (*Lutjanus kasmira*) swim in a tight group, often with sudden changes of direction. This "schooling" behaviour reduces the chance of predation, because threats are more likely to be detected and it is difficult for a predator to catch any one individual in the bewildering shoal

▲ Figure 29 Blue tit (*Cyanistes caeruleus*) feeding on cream after pecking through the foil cap of a milk bottle

Behavioural adaptations can change relatively quickly. For example, in the 1920s blue tits started feeding on cream from milk bottles delivered to doorsteps. This behaviour spread rapidly across Europe, but disappeared as rapidly when deliveries of bottled milk with cream diminished in the 1990s. Structural adaptations take longer to develop because there must be genetic change, but research on seed-eating finches on the Galápagos Islands shows that their beaks soon start to change in size and shape when the size of seeds available on an island changes. Chemical adaptations are usually the slowest to change, because new enzymes may be needed or new ways of regulating enzymes and this may take millions of years.

B4.2.11 Adaptations of plant form for harvesting light

In environments where there is enough water for abundant plant growth and temperatures are suitable for photosynthesis, plants compete for light. Forest ecosystems develop in such environments. Plants use a variety of strategies in forests for obtaining light, so show great diversity of form.

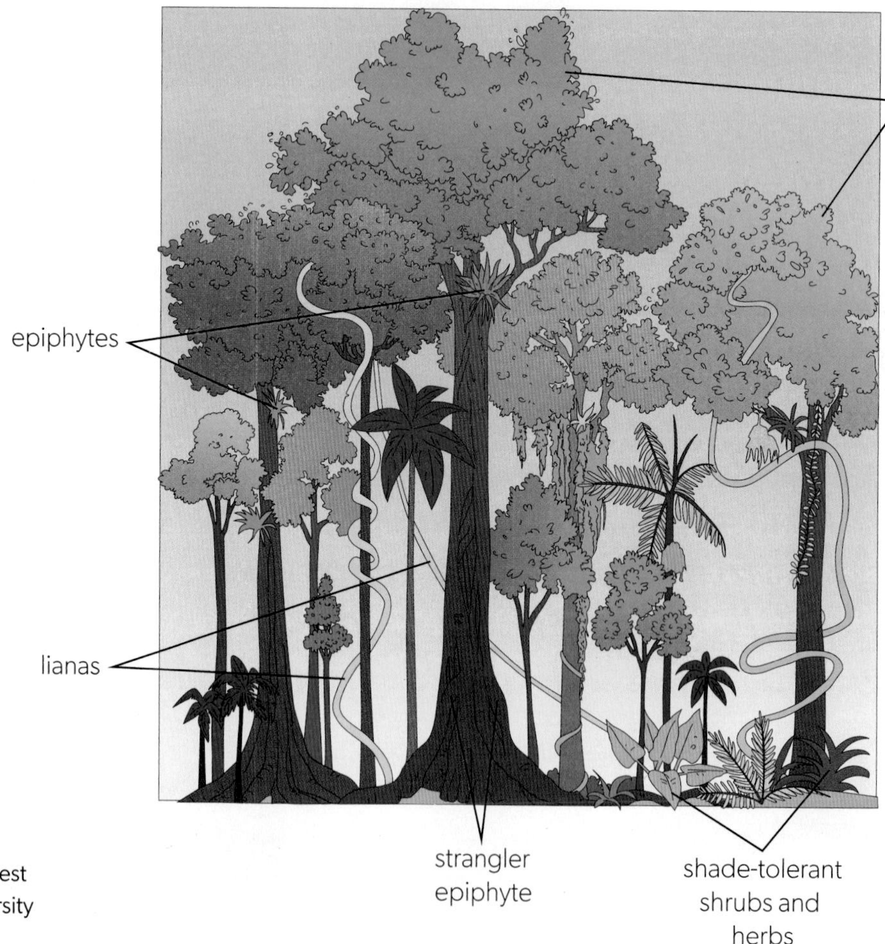

▶ Figure 30 Tropical rainforest is characterized by great diversity of plant form

- Trees have a dominant leading shoot that grows rapidly to great height to reach the forest canopy where they are unshaded by other trees.
- Lianas climb through other trees, using them for support. This means lianas do not need to produce as much xylem tissue (wood) as free-standing trees.

- Epiphytes grow on the trunks and branches of trees, so they receive higher light intensity than if they grew on the forest floor, but there is minimal soil for their roots.
- Strangler epiphytes climb up the trunks of trees encircling them and outgrowing their branches, to shade out the leaves of the tree. Eventually the tree dies leaving only the epiphyte.
- Shade-tolerant shrubs and herbs absorb light reaching the forest floor.

B4.2.12 Fundamental and realized niches

Living organisms tolerate a range of biotic and abiotic conditions, but their adaptations do not allow them to survive outside this range. The range of tolerance is the fundamental niche of the species. If the species were living without any competitors, it would occupy the entire fundamental niche. In natural ecosystems, there is competition and typically a species is excluded from parts of its fundamental niche by competitors. The actual extent of the potential range that a species occupies is its realized niche.

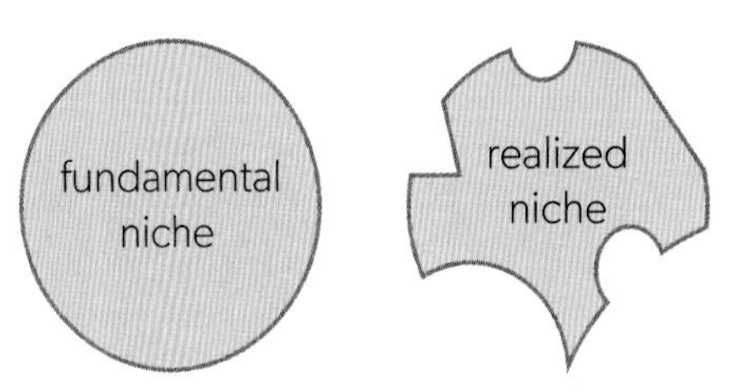

▲ Figure 31 A realized niche is a subset of a fundamental niche. The reductions are due to overlaps with the niches of other competitor species

B4.2.13 Competitive exclusion and the uniqueness of ecological niches

Where the fundamental niches of two species overlap, one species is expected to exclude the other from that part of its range by competition. This was demonstrated experimentally with the flour beetles *Tribolium castaneum* and *Tribolium confusum*. When reared together at different combinations of temperature and humidity, *T. castaneum* usually excluded *T. confusum* in some combinations but *T. confusum* was more successful in other combinations. In the pie charts in Table 3, blue segments indicate the percentage of trials where *T. confusum* excluded *T. castaneum* and orange segments indicate the converse.

▼ Table 3 Competition between *T. castaneum* and *T. confusum* at various temperatures and humidity levels

Temperature/°C	Humidity	
	30%	70%
24		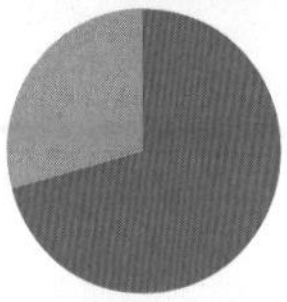
29	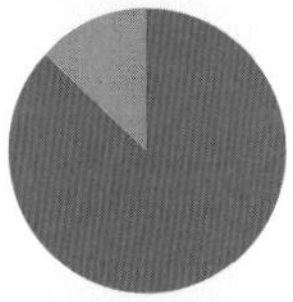	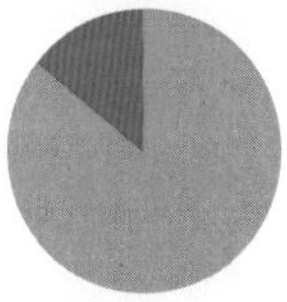
34	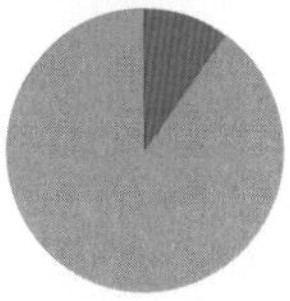	

If two species in an ecosystem have overlapping fundamental niches and one species outcompetes the other in all parts of the fundamental niche, the outcompeted species does not have a realized niche and will be competitively excluded from the whole ecosystem. According to ecological theory, every species must have a realized niche that differs from the realized niches of all other species if it is to survive in an ecosystem.

Data-based questions: Competitive exclusion in cat-tails

Typha latifolia and *Typha angustifolia* are two species of plant that grow on the margins of lakes. The upper graph shows primary production of each species when growing together in a natural ecosystem. The lower graph shows the biomass of transplants of the two species when grown without any competition. Negative depth means growing out of the water.

1. Compare and contrast the growth of *T. angustifolia* and *T. latifolia* in the absence of competition. [4]
2. Distinguish between the growth of *T. angustifolia* with and without competition from *T. latifolia*. [2]
3. Analyse the data in the graphs using the concepts of fundamental and realized niches. [4]

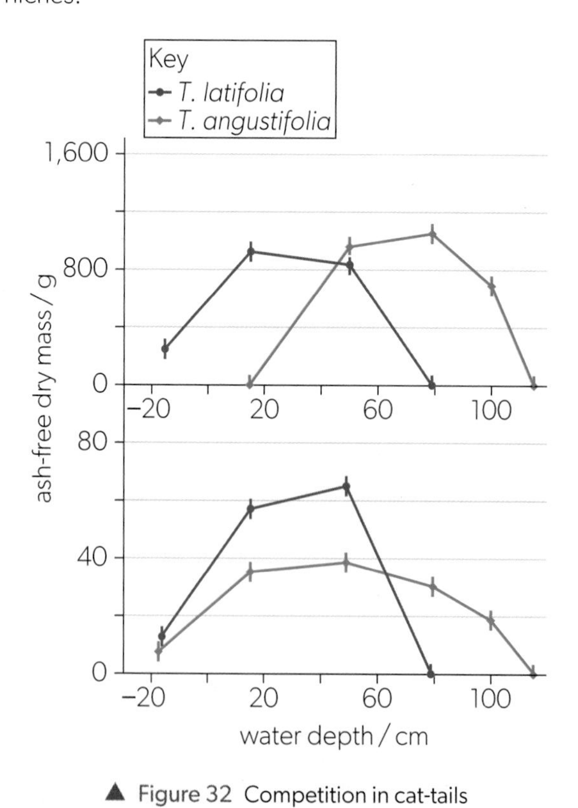

▲ Figure 32 Competition in cat-tails

Linking questions

1. What are the relative advantages of specificity and versatility?
   a. Outline the role of specificity in enzyme function. (C1.1.7)
   b. Explain the evidence for evolution provided by the pentadactyl limb. (A4.1.4)
   c. Explain what is meant by the universality of the genetic code. (A2.1.7)
2. For each form of nutrition, what are the unique inputs, processes and outputs?
   a. Explain what is meant by holozoic nutrition. (B4.2.4)
   b. Distinguish between the mechanisms of digestion of detritus feeders and saprotrophs. (C4.2.12)
   c. Outline one example of mixotrophy. (B4.2.5)

TOK

Are some types of knowledge less open to interpretation than others?

In everyday language, the word "interpretation" implies more than one possibility. In biology, there can be competing explanations for a phenomenon, particularly if there is fragmentary evidence because the subject material is uncontrollable. In such situations, there is greater tolerance for more than one interpretation. However, knowledge such as the structure of skeletal muscle fibres is not open to interpretation. It is a straightforward matter to verify the consensus view about their structure. This is because it is possible to obtain further samples to examine, and the methods of exploring cellular ultrastructure are reliable.

▲ Figure 1 The structure of skeletal muscle fibres is clearly visible through microscopy

When a palaeoanthropologist uncovers a skull, it is an artefact of the natural world. Scientists would participate in the analysis of such a bone. However, it is not possible to expand the study by looking at more skulls from the population if only one has been found. This is the case of the short-statured *Homo floresiensis* discovered on the island of Flores in Indonesia. Only one skull from this species has ever been found, though bones and teeth from 14 other individuals have been uncovered. Tools and bones of prey species have been found along with the *H. floresiensis* skeletal fragments. Some of the skeletal features suggest a relationship to more primitive species while others suggest it is a more modern species. One interpretation is that *H. floresiensis* is a descendant of *H. erectus* that underwent a process known as island dwarfism.

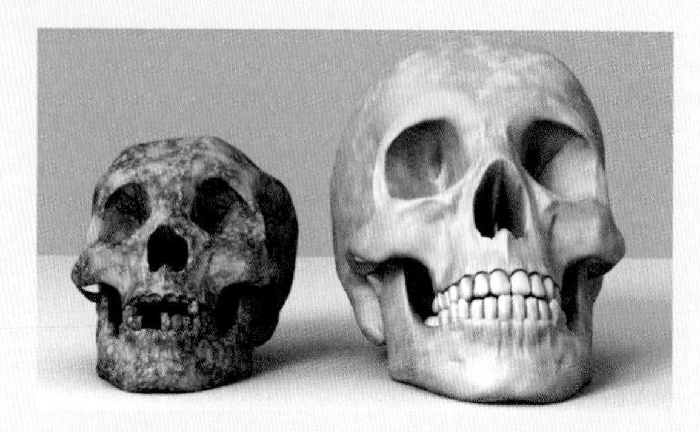

▲ Figure 2 *H. floresiensis* skull (left) next to computer artwork of a human (*H. sapiens*) skull (right)

An example of an area where disagreements due to different interpretations persist is in the phylogeny of the red wolf (*Canis rufus* or *C. lupus rufus*). The modern population is very small and has hybridized with coyotes. Efforts to bring the population back from extinction have depended on the introduction of animals from other populations. Viewpoints differ as to whether or not the present very small population of red wolves is worth preserving. Conservation efforts are expensive and financial resources for conservation are scarce. The US Fish and Wildlife Service currently recognizes the red wolf as an endangered species and grants protected status. The International Union for the Conservation of Nature has listed the red wolf as a critically endangered species. However, the red wolf is not recognized in the CITES appendices of endangered species. One interpretation is that the historical red wolf is a distinct species whereas the modern red wolves trace some of their ancestry to historic red wolves but hybridization with other wolves and coyotes has continued, threatening the distinctiveness of the population.

▲ Figure 3 A modern red wolf

End of chapter questions

1. One method used by microbiologists to distinguish between the Archaea and Eubacteria is the conditions they need for survival. Both groups include thermophiles, which are species that are adapted to live at high temperatures. The graph shows the optimum temperature and minimum pH required for growth by selected species of Archaea and thermophilic Eubacteria.

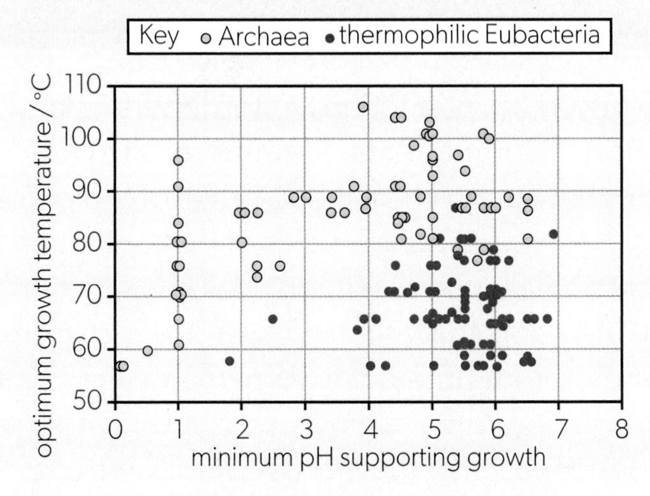

Source: DL Valentine (2007), Nature Reviews Microbiology, 5, p316

a. State the highest optimum growth temperature recorded for the thermophilic Eubacteria. [1]

b. State the relationship between minimum pH supporting growth and optimum growth temperature for Archaea. [1]

c. Compare the results for the Archaea with those for the thermophilic Eubacteria. [2]

d. With reference to the data, suggest why this method would not always be suitable for distinguishing between Archaea and thermophilic Eubacteria. [2]

e. State a possible habitat for methanogenic Archaea. [1]

2. New technologies such as dental topographic analysis are being used to help understand how early Hominids lived. This technique allows the pattern of wear of teeth over a lifetime to be analysed, revealing what types of food were eaten. Teeth from early humans and *Australopithecus afarensis* were compared. The upper surfaces of the teeth were analysed for slope. The teeth examined were in groups of similar stages of wear to ensure consistency of results. The lower the slope, the flatter the teeth. Flat teeth are best suited to crushing hard, brittle foods. More shaped teeth are better suited to eating elastic foods such as meat.

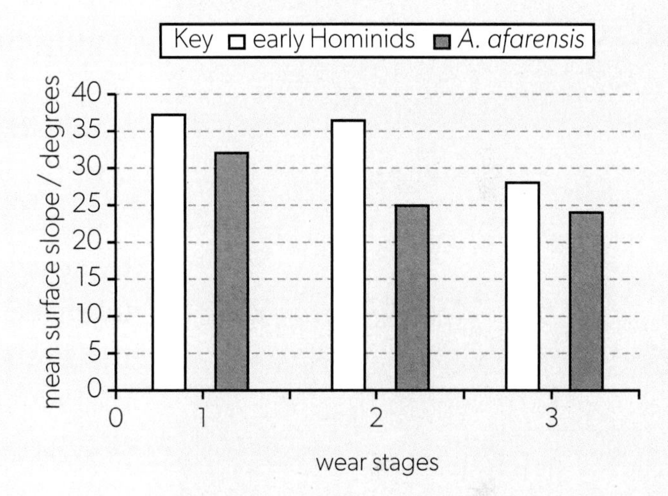

a. i State what changes occurred to all teeth with wear. [1]

ii Compare the teeth of early Hominids with those of *A. afarensis* [2]

b. Using the data, suggest how the diets of early Hominids and *A. afarensis* differed. [2]

c. Suggest what other evidence would help scientists to determine what food was eaten by early Hominids. [2]

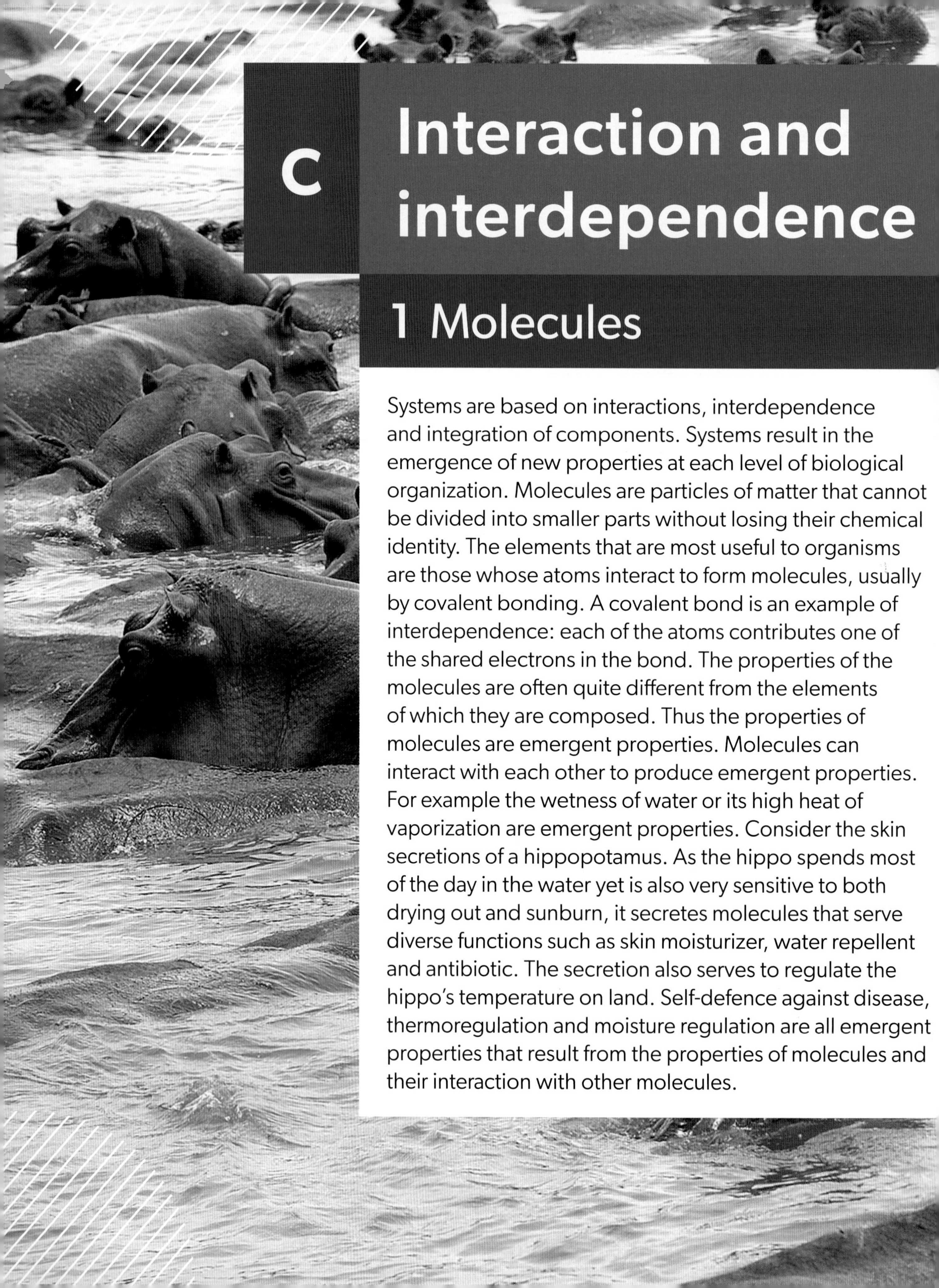

C Interaction and interdependence

1 Molecules

Systems are based on interactions, interdependence and integration of components. Systems result in the emergence of new properties at each level of biological organization. Molecules are particles of matter that cannot be divided into smaller parts without losing their chemical identity. The elements that are most useful to organisms are those whose atoms interact to form molecules, usually by covalent bonding. A covalent bond is an example of interdependence: each of the atoms contributes one of the shared electrons in the bond. The properties of the molecules are often quite different from the elements of which they are composed. Thus the properties of molecules are emergent properties. Molecules can interact with each other to produce emergent properties. For example the wetness of water or its high heat of vaporization are emergent properties. Consider the skin secretions of a hippopotamus. As the hippo spends most of the day in the water yet is also very sensitive to both drying out and sunburn, it secretes molecules that serve diverse functions such as skin moisturizer, water repellent and antibiotic. The secretion also serves to regulate the hippo's temperature on land. Self-defence against disease, thermoregulation and moisture regulation are all emergent properties that result from the properties of molecules and their interaction with other molecules.

C1.1 Enzymes and metabolism

In what ways do enzymes interact with other molecules?

Enzymes interact with a range of molecules including substrates, competitive and non-competitive inhibitors, and molecules within cellular structures such as membranes. Cells control metabolism by regulating enzyme activity. The end products of enzyme-catalysed reactions often act as inhibitors. The build-up of substrate can increase the expression of genes responsible for generating new enzymes, to reduce the concentration of the substrate again. Fireflies (family Lampyridae) are a group of species of beetles that are able to produce light as a result of an enzyme-catalysed reaction in their abdomen. How does the firefly regulate the emission of light?

▲ Figure 1 A firefly

What are the interdependent components of metabolism?

Metabolism is the sum of all of the interdependent chemical reactions within an organism. Some of these chemical reactions are involved in breaking down molecules to yield usable energy in the form of ATP. It includes the breakdown of macromolecules into their subunits and their re-assembly to make new molecules. The breakdown of toxins and waste products is also part of metabolism. Processes such as growth, maintenance and repair depend on a complex network of interdependent chemical reactions.

The coloured 3D combined positron emission tomography (PET) and computed tomography (CT) scan of a healthy human body (Figure 2) shows variations in metabolic activity in the body's internal organs such as the brain. Blue shows low activity, green shows intermediate activity and red shows high activity. Suggest why part of the leg is green.

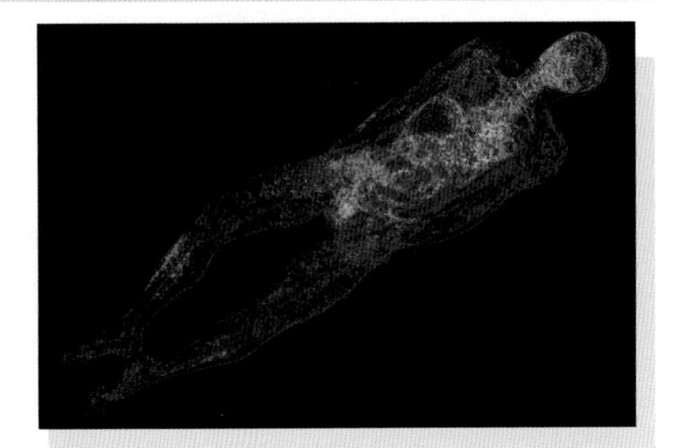

▲ Figure 2

SL and HL	AHL only
C1.1.1 Enzymes as catalysts C1.1.2 Role of enzymes in metabolism C1.1.3 Anabolic and catabolic reactions C1.1.4 Enzymes as globular proteins with an active site for catalysis C1.1.5 Interactions between substrate and active site to allow induced-fit binding C1.1.6 Role of molecular motion and substrate–active site collisions in enzyme catalysis C1.1.7 Relationships between the structure of the active site, enzyme–substrate specificity and denaturation C1.1.8 Effects of temperature, pH and substrate concentration on the rate of enzyme activity C1.1.9 Measurements in enzyme-catalysed reactions C1.1.10 Effect of enzymes on activation energy	C1.1.11 Intracellular and extracellular enzyme-catalysed reactions C1.1.12 Generation of heat energy by the reactions of metabolism C1.1.13 Cyclical and linear pathways in metabolism C1.1.14 Allosteric sites and non-competitive inhibition C1.1.15 Competitive inhibition as a consequence of an inhibitor binding reversibly to an active site C1.1.16 Regulation of metabolic pathways by feedback inhibition C1.1.17 Mechanism-based inhibition as a consequence of chemical changes to the active site caused by the irreversible binding of an inhibitor

C1.1.1 Enzymes as catalysts

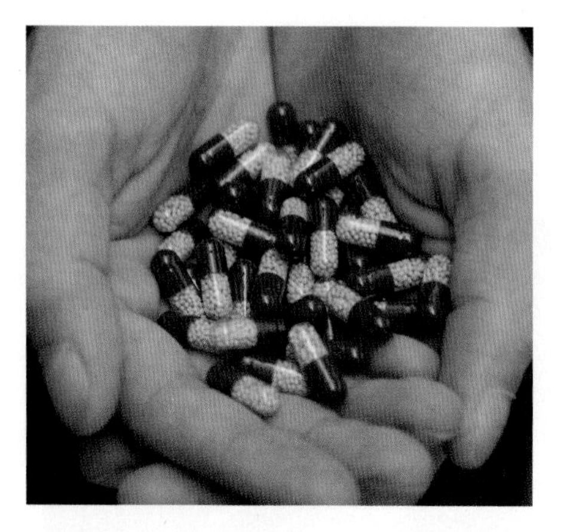

▲ Figure 3 Cystic fibrosis causes the pancreatic duct to become blocked with mucus. This prevents digestive enzymes produced by the pancreas from reaching the small intestine. Digestion is therefore much slower than normal. Pills containing a mixture of enzymes can help a person with cystic fibrosis to digest their food. The photograph shows one day's supply for a person with cystic fibrosis

A catalyst is a substance that increases the rate of a chemical reaction but is not changed by the reaction. Because catalysts are not used up, they can catalyse reactions many times. This means only small amounts are needed in relation to the quantity of reactants.

Platinum is an example of an inorganic catalyst. It is used in the catalytic converters fitted to vehicles with combustion engines, to help convert unburned hydrocarbons in exhaust gases to carbon dioxide and water.

Enzymes are biological catalysts. They are made by living cells to speed up biochemical reactions. In these reactions, enzymes convert substrates into products. A general equation for an enzyme-catalysed reaction is:

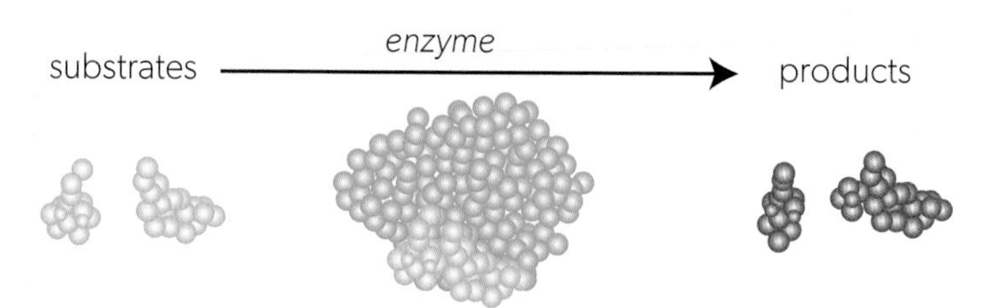

If cells did not make enzymes, the chemical reactions on which life is based would happen very slowly at ambient temperatures. Life processes such as respiration, digestion, growth and movement would all be very slow.

Data-based questions: The effectiveness of enzymes

Different enzymes increase rates of reactions by different amounts. Ratios comparing the rate of reaction with and without an enzyme allow comparison of the effectiveness of different enzymes. Table 1 shows the rates of four reactions with and without an enzyme. The ratio between these rates has been calculated for the first reaction.

Enzyme	Reaction rate / s^{-1} Without enzyme	Reaction rate / s^{-1} With enzyme	Ratio
Carbonic anhydrase	1.3×10^{-1}	1.0×10^{6}	7.7×10^{6}
Ketosteroid isomerase	1.7×10^{-7}	6.4×10^{4}	
Nuclease	1.7×10^{-13}	9.5×10^{6}	
OMP decarboxylase	2.8×10^{-16}	3.9×10^{8}	

▲ Table 1

1. Define the term "rate of reaction". [2]
2. State which reaction has the slowest rate without an enzyme. [1]
3. State which reaction has the fastest rate with an enzyme. [1]
4. Calculate the ratios between the rates of reaction with and without an enzyme for the second, third and fourth reactions. [2]
5. Discuss which of the enzymes is the most effective catalyst. [2]
6. Explain how the enzymes increase the rate of the reactions they catalyse. [2]

C1.1.2 Role of enzymes in metabolism

Metabolism is the complex network of interdependent and interacting chemical reactions that occurs in living organisms. Most of these reactions happen inside cells but there are also some extracellular reactions, for example, digestion of foods in the intestine.

There are thousands of metabolic reactions. They form pathways in which one type of molecule is transformed into another by a series of small steps. Most of these pathways are chains of reactions but there are also some cycles. An example of a cycle is shown in Figure 4. Maps showing all the pathways of metabolism are very complex. They are available on the internet, for example in the Kyoto Encyclopedia of Genes and Genomes.

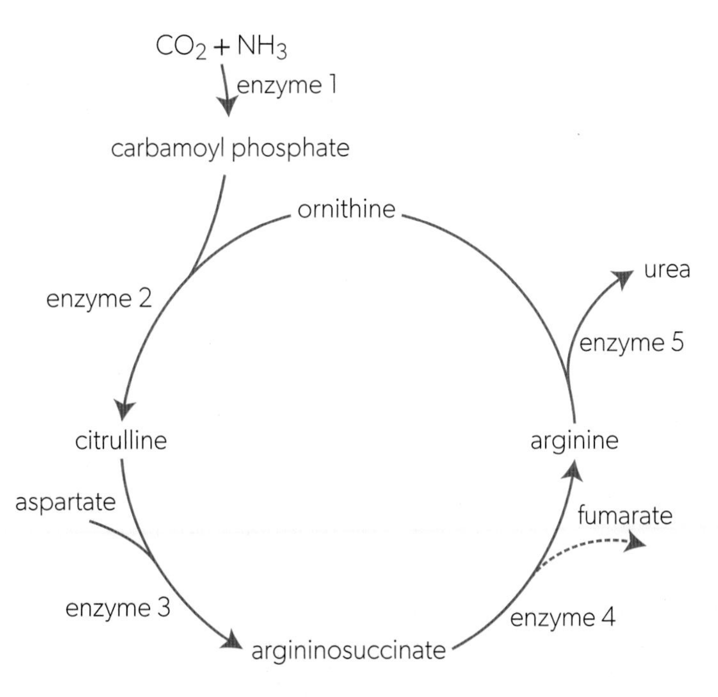

Figure 4 This cycle of metabolic reactions is used to synthesize urea in liver cells. There are five reactions, so five different enzymes are required. Can you find out what these enzymes are?

Almost all metabolic reactions are catalysed by an enzyme. One of the properties of enzymes is their specificity. Each enzyme catalyses one specific reaction, or a specific group of reactions. This is a significant difference between enzymes and non-biological catalysts such as platinum, which can catalyse many different reactions. Because of enzyme specificity, living organisms have to make large numbers of different enzymes. Even a relatively simple prokaryotic cell makes hundreds of different enzymes. Cells with more complicated metabolism, such as liver cells, make thousands of different enzymes.

Enzyme specificity has many benefits. It allows organisms to control metabolism. If a cell produces an enzyme, it can drive a particular reaction that would otherwise happen extremely slowly or not at all. By making more or less of an enzyme, cells can control the rate of a reaction. There are also mechanisms for temporarily stopping particular enzymes from working if a reaction is not required for a while. In summary, enzymes give living organisms considerable control over their metabolism and therefore over their activities and chemical composition.

C1.1.3 Anabolic and catabolic reactions

Metabolism has two parts: anabolism and catabolism. Anabolic reactions build up smaller molecules into larger ones. These reactions require energy. Photosynthesis is an example of anabolism, because carbon dioxide, water and other small molecules are combined to produce larger molecules, using energy from light. In anabolic reactions, macromolecules are produced from monomers, using energy from ATP. They are condensation reactions because water is a by-product. Examples of anabolic reactions include:

- protein synthesis (translation) by ribosomes
- DNA synthesis (replication)
- synthesis of complex carbohydrates including starch, cellulose and glycogen.

▲ Figure 5 An easy way to remember that anabolic reactions build smaller molecules into larger ones is to think of the anabolic steroids that are sometimes misused to promote "body building"

Catabolic reactions break down larger molecules into smaller ones, releasing energy. In some cases, this energy is captured by coupling the catabolic reaction to the synthesis of ATP, which can then be used in the cell. Examples of catabolic reactions include:

- digestion of food—in humans this happens in the mouth, stomach and small intestine
- cell respiration—in aerobic respiration, glucose or lipids are oxidized to carbon dioxide and water
- digestion of complex carbon compounds—decomposers do this with dead organic matter.

▲ Figure 6 Anabolic reactions in trees can lead to a huge accumulation of biomass, as in the Moor Park Oak in Shropshire (left). The tree on the right was recently blown over in a storm and catabolic reactions, mainly carried out by fungi, are now breaking down its macromolecules

C1.1.4 Enzymes as globular proteins with an active site for catalysis

Enzymes are globular proteins, with precise three-dimensional structure and chemical properties that allow them to function as catalysts. For a reaction to be catalysed, the substrate or substrates must bind to a special region on the surface of the enzyme called the active site (see Figure 7). The shape and chemical properties of the active site and the substrate match each other. This allows the substrate to bind with the enzyme while most other substances cannot. While the substrate is bound to the active site, it is converted into products. The products are then released, leaving the active site free to catalyse another reaction.

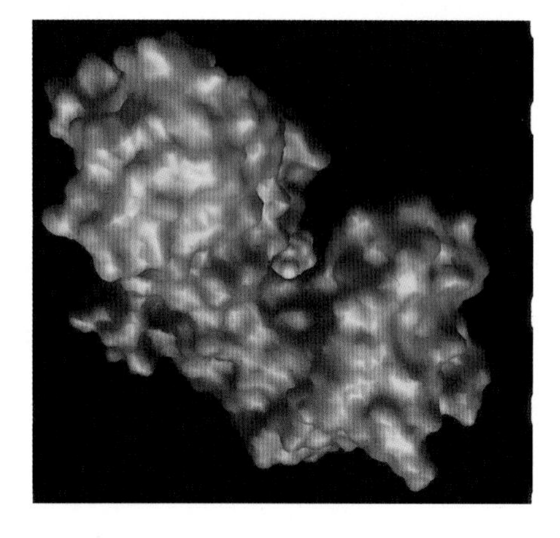

▲ Figure 7 Computer-generated image of the enzyme hexokinase (blue), with a molecule of its substrate glucose (yellow) bound to the active site. A second substrate, phosphate, binds to the active site and the two substrates are linked to make glucose phosphate

Active sites vary in size, depending on the size of the substrates. Typically, just a few amino acids at the active site are essential to create the chemical conditions that change the substrates enough to convert them into products. Often the amino acids that form the active site are not next to each other in the polypeptides that make up the enzyme. They are brought together by the folding of the polypeptides. For that reason the overall three-dimensional structure of the enzyme is crucial. If any part of the enzyme is altered, the structure of the active site may change and catalysis is unlikely to happen.

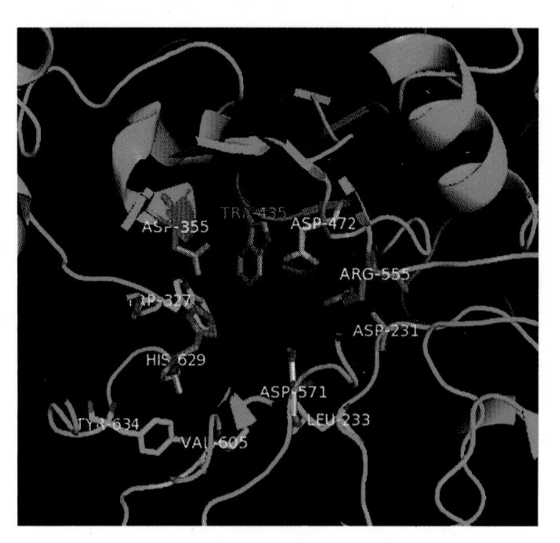

▲ Figure 8 Twelve amino acids form the active site of sucrase isomaltase. The numbers show where each amino acid comes in the sequence of the polypeptide. For example, LEU-233 shows that the amino acid leucine is 233rd in the sequence. Are any of the 12 amino acids that form the active site next to each other in the polypeptide?

C1.1.5 Interactions between substrate and active site to allow induced-fit binding

Interactions between the substrate and the active site of an enzyme are the basis of catalysis.

- A substrate approaches the active site. Until it is near to the enzyme, the substrate's direction of movement is random. When it is close enough to interact, the chemical properties of the enzyme surface attract the substrate molecule towards the active site.
- The substrate binds to the active site. This used to be compared with a key fitting into a lock. However, that model is inappropriate because interactions between the substrate and the active site cause both to change: bond angles and bond lengths are altered, changing the three-dimensional molecular shapes of the substrate and the active site. This is called induced-fit binding.
- If there is a second substrate, it approaches and binds to another part of the active site. Again, the substrate and the active site cause changes in each other to allow binding (Figure 9).

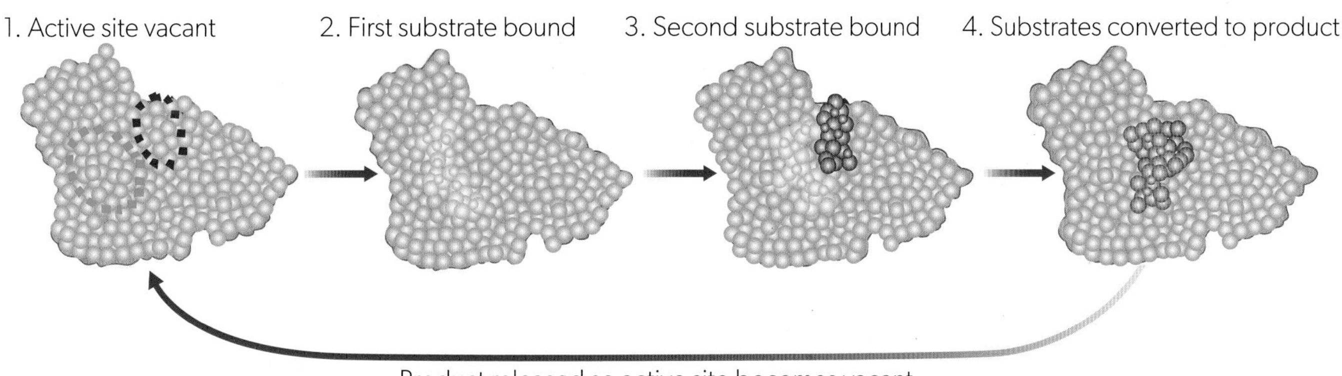

▲ Figure 9

- Changes to substrate molecules make it easier for bonds within them to break and new bonds to form, converting substrates into products.
- The products detach from the active site. Without substrates or products interacting with it, the enzyme's active site returns to its original state. It is now empty and available for more substrates to bind, so the catalytic cycle can be repeated.

C1.1.6 Role of molecular motion and substrate–active site collisions in enzyme catalysis

A substrate molecule can only bind with the active site of an enzyme if it moves very close to it. This happens as a result of molecular motion. When a substrate and an active site come together, this is known as a substrate–active site collision. However, it is not like the high velocity impacts that can happen between vehicles on a road. To understand how substrate–active site collisions occur, we need to think about molecular motion in liquids.

In a liquid, the molecules are packed closely together but they are free to move. The direction of each molecule changes repeatedly and at random. If the liquid contains both substrate and enzyme molecules, they will occasionally come together. The rate at which this happens will increase if there are more (a higher concentration of) substrate or enzyme molecules or if the temperature increases, leading to faster molecular motion.

When a collision occurs, the substrate may be at any angle to the active site. Successful collisions are ones in which the substrate and active site are aligned, so binding can take place. Some enzymes have chemical properties that draw substrates towards the active site or adjust their orientation. However, the forces involved only work over short distances so they only promote binding when a substrate molecule is already very close to the active site.

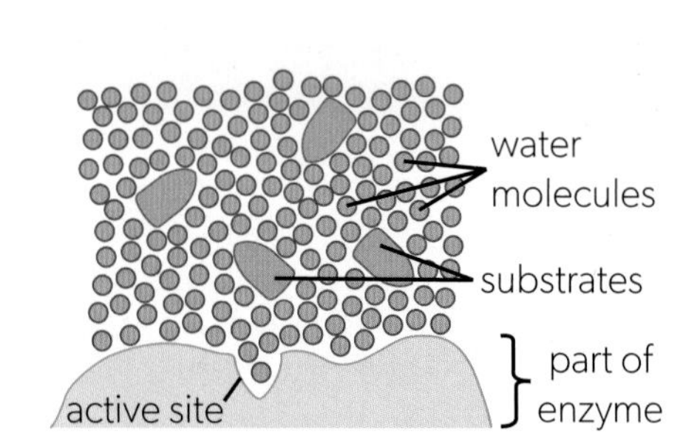

▲ Figure 10 Enzyme–substrate collisions. If random movements bring any of the substrate molecules close to the active site with the correct orientation, the substrate can bind to the active site

There is some variation in the molecular motion of substrates and enzymes:

- Many enzyme-catalysed reactions happen in the cytoplasm. Substrate and enzyme are both dissolved in water, so are free to move. In most cases, however, the substrate is a smaller molecule than the enzyme so it moves more.
- Some substrates are very large and do not move much. In these cases, the enzyme has to move in relation to the substrate. Enzymes that replicate or transcribe DNA do this.
- Some enzymes are embedded in membranes and cannot move—they are immobilized. In these cases, the substrate has to do all the movement.

C1.1.7 Relationships between the structure of the active site, enzyme–substrate specificity and denaturation

The shape and chemical properties of an enzyme's active site allow substrate molecules to bind, but not other substances. This is called enzyme–substrate specificity.

Some enzymes are absolutely specific and always bind the same substrate. For example, glucose is the only substrate that binds to the active site of the enzyme glucokinase. Other enzymes are less specific. For example, hexokinase can bind with any one of a group of hexose sugars. Proteases also have broad substrate specificity, so a few types of protease can between them digest polypeptides with any amino acid sequence.

Enzymes are proteins with a precise three-dimensional shape and an intricate chemical structure. This structure depends on relatively weak interactions between amino acids within the protein, including hydrophobic and hydrogen bonds. These interactions are affected by factors such as heat and acidity, so enzymes are easily altered. Even if changes happen at a distance from the active site, interactions within the enzyme are likely to affect the active site. Even small changes to the active site can prevent binding of substrates, or prevent catalysis after binding. As a result, the enzyme will no longer work as a catalyst. If the changes are too great to be reversed, the enzyme is denatured.

Data-based questions: Biosynthesis of glycogen

In 1947, the Nobel Prize in Physiology or Medicine was won by Gerty Cori and her husband Carl. They isolated two enzymes that convert glucose phosphate into glycogen. Glycogen is a polysaccharide. It is composed of glucose molecules bonded together in two ways, called 1,4 and 1,6 bonds (see Figure 11).

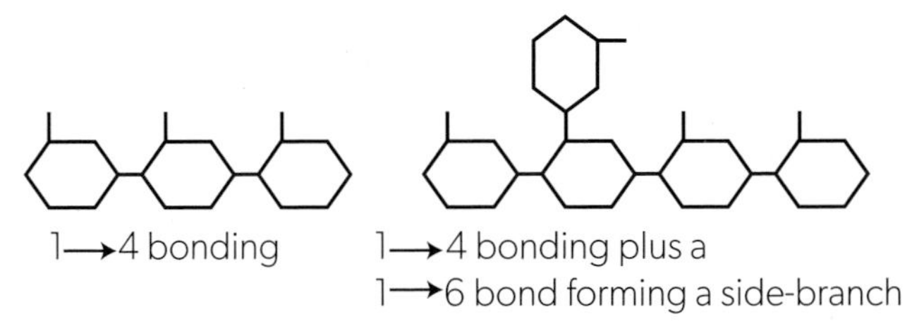

▲ Figure 11 Bonding in glycogen

1. Deduce whether the production of glycogen is catabolic or anabolic. Give a reason. [1]
2. Explain why two different enzymes are needed for the synthesis of glycogen from glucose phosphate. [2]
3. The formation of side-branches increases the rate at which glucose phosphate molecules can be linked to a growing glycogen molecule. Explain the reason for this. [2]
4. Curve A in Figure 12 was obtained using heat-treated enzymes. Explain the shape of curve A. [2]
5. Curve B in Figure 12 was obtained using enzymes that had not been heat-treated.
   a. Describe the shape of curve B. [1]
   b. Explain the shape of curve B. [2]

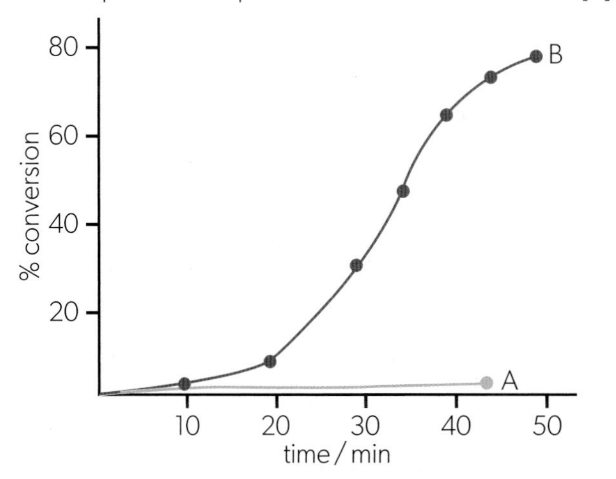

▲ Figure 12 Graph showing the percentage conversion of glucose phosphate to glycogen by the two groups of enzymes, over a 50-minute period

C1.1.8 Effects of temperature, pH and substrate concentration on the rate of enzyme activity

1. Effects of temperature

Enzyme activity is affected by temperature in two ways.

In liquids, the particles are in continual random motion. When a liquid is heated, the particles gain kinetic energy. As a result, enzyme and substrate molecules

rate at which reaction decreases owing to denaturation of enzyme molecules

optimum temperature, which is not always 40 °C and can be much higher in the enzymes of organisms adapted to high ambient temperatures

rate at which reaction increases owing to increased kinetic energy of substrate and enzyme molecules

actual rate of reaction

rate of reaction

0 10 20 30 40 50 60

temperature / °C

▲ Figure 13 Temperature and rate of enzyme activity

move around more quickly and the chance of a substrate molecule colliding with the active site of the enzyme is increased. Enzyme activity therefore increases.

When enzymes are heated, bonds in the enzyme vibrate more and the chance of these bonds breaking is increased. When bonds in the enzyme break, the enzyme structure changes. Changes to the active site will mean it can no longer bind with substrate molecules: the enzyme is denatured. Different enzyme molecules denature at slightly different temperatures. However, as temperature rises, more and more enzyme molecules in a solution will be denatured and enzyme activity will fall. Eventually, all enzyme molecules will be denatured and catalysis will stop completely.

As temperature rises, there are reasons for both increases and decreases in enzyme activity. Figure 13 shows the overall effect of temperature on a typical enzyme.

2. Effects of pH

Enzymes are sensitive to their chemical environment. In particular they are affected by how acidic or alkaline it is. Acidity is due to the presence of hydrogen ions (protons). The higher the hydrogen ion concentration, the more acidic a solution is. The pH scale is a measure of hydrogen ion concentration and therefore acidity. Lower pH values indicate higher hydrogen ion concentrations and therefore greater acidity.

The pH scale is logarithmic. This means that reducing the pH by one unit makes a solution 10 times more acidic. A solution at pH 7 is neutral. A solution at pH 6 is slightly acidic; pH 5 is 10 times more acidic than pH 6, pH 4 is 100 times more acidic than pH 6, and so on.

Most enzymes have an optimum pH at which their activity is highest. If pH increases or decreases from the optimum, ionic bonds between the amino acids in the enzyme are altered. This changes the structure of the enzyme, including its active site. As a result, the active site will no longer bind substrates or convert them to products. Beyond a certain pH, the enzyme will be irreversibly denatured.

Key
stomach
acidic hot springs
decaying plant matter
large intestine
small intestine
alkaline lakes

1 2 3 4 5 6 7 8 9 10

▲ Figure 14 pH variation in enzyme environments

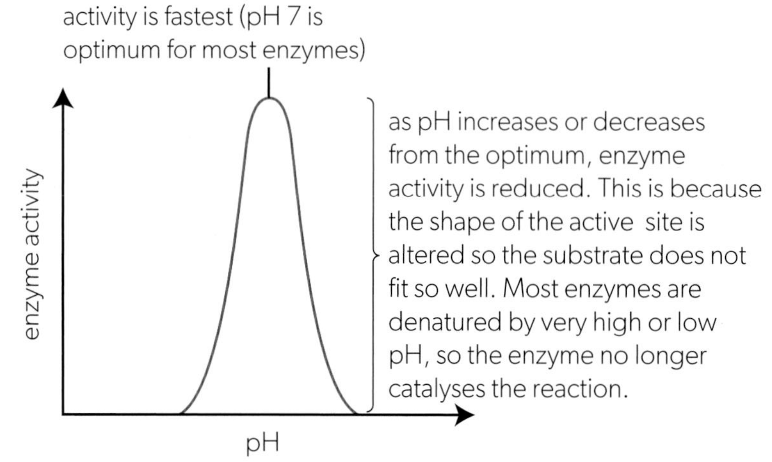
optimum pH at which enzyme activity is fastest (pH 7 is optimum for most enzymes)
as pH increases or decreases from the optimum, enzyme activity is reduced. This is because the shape of the active site is altered so the substrate does not fit so well. Most enzymes are denatured by very high or low pH, so the enzyme no longer catalyses the reaction.
enzyme activity
pH

▲ Figure 15 pH and enzyme activity

Not all enzymes have the same optimum pH—in fact, there is a wide range. This reflects the varied environments in which enzymes work. For example, the protease secreted by *Bacillus licheniformis* has a pH optimum between 9 and 10. This bacterium is cultured to produce its alkaline-tolerant protease for use in biological laundry detergents, which are alkaline.

3. Effects of substrate concentration

Enzymes cannot catalyse a reaction until the substrate binds with the active site. Collisions between substrates and active sites occur due to random movements of molecules in liquids. If the concentration of substrate molecules is increased, substrate–active site collisions will occur more frequently and the rate at which the enzyme catalyses its reaction will increase.

However, there is another trend that affects the rate of reaction. Once a substrate has bound to an active site, the active site is occupied and unavailable to other substrate molecules until products have been formed and released. As the substrate concentration rises, more and more of the active sites are occupied at any moment. Therefore, a greater and greater proportion of substrate–active site collisions are blocked. For this reason, the increases in the rate at which enzymes catalyse reactions get smaller and smaller as substrate concentration rises.

If the relationship between substrate concentration and enzyme activity is plotted on a graph, a distinctive curve is seen (Figure 16): the graph rises less and less steeply as substrate concentration increases, but never quite reaches a maximum.

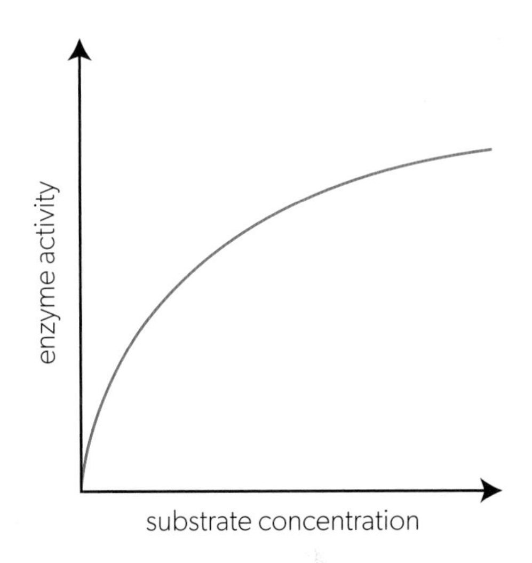

▲ **Figure 16** The effect of substrate concentration on enzyme activity

Data-based questions: Adenylate kinase

The enzyme adenylate kinase consists of a single polypeptide of 214 amino acids. It catalyses reversible reactions in which a phosphate is transferred between nucleotides:

ATP + AMP → 2ADP

The graph in Figure 17 shows the effect of temperature on the activity of five forms of adenylate kinase. WT is the wild type form. The other forms are mutants in which one amino acid (either valine or alanine) has been changed to glycine.

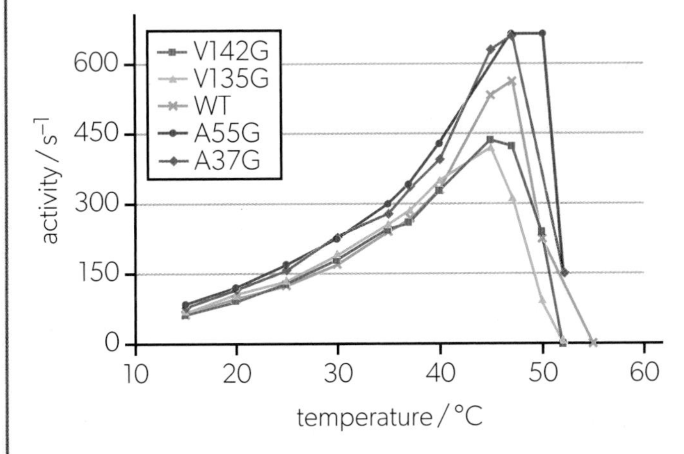

▲ **Figure 17** Source: Saavedra, H.G., Wrabl, J.O., Anderson, J.A. et al. Dynamic allostery can drive cold adaptation in enzymes. Nature 558, 324–328 (2018). https://doi.org/10.1038/s41586-018-0183-2

1. For the wild type form of the enzyme, identify:
   a. the optimum temperature [1]
   b. the temperature above which the enzyme is fully denatured. [1]
2. a. State the activity of the wild type form of the enzyme at each of these temperatures: 15°C, 25°C, 35°C and 45°C. [2]
   b. Using this data, evaluate the hypothesis that enzyme activity doubles with every 10°C rise in temperature. [2]
3. Explain the reasons for the change in activity of the wild type form of the enzyme between:
   a. 15 and 45°C [2]
   b. 50 and 55°C. [2]
4. Compare and contrast the curves for WT and V135G. [3]
5. Using the data in the graph, describe the effect of changing the 55th amino acid in adenylate kinase from alanine to glycine. [2]

Models: Use of graphs to show relationships between variables

A graph is used to show the relationship between two variables. Two axes are needed to do this. The x-axis goes across the graph from left to right and the y-axis goes up the graph. An easy way to remember this is that the x-axis goes across because the letter x is a cross! The independent variable in an experiment is plotted on the x-axis and the dependent variable is plotted on the y-axis.

The first step in any scientific investigation is to formulate a hypothesis. If the hypothesis is an expected relationship between two variables, it can be shown using a sketch graph. This type of graph is a model—a simple representation of something more complex. To test the hypothesis, an experiment is performed and the results are compared with predictions based on the hypothesis.

When the results of an enzyme experiment are plotted on a graph, each data point shows the level of the dependent variable at one particular level of the independent variable. It may be an individual result, or a mean result if repeat measurements were made. Often the data points are joined with straight lines, as in Figure 17 on the previous page. This indicates that it is uncertain what the values would have been at other levels of the independent variable.

A graph of experimental results can be used to evaluate the hypothesis—do the actual results match the model shown in the sketch graph? If they do not, then a new hypothesis may be needed. If the data points on the graph suggest an overall relationship between the dependent and independent variables, a line may be added to show this. This line is called a curve, whether it is curved or straight. It is usually a "line of best fit" that goes as close as possible to the data points but does not necessarily pass through them all. You should be able to look at the shape of a graph and deduce the relationship between the variables; this will be a useful skill in science. Some examples are shown below in Figure 18.

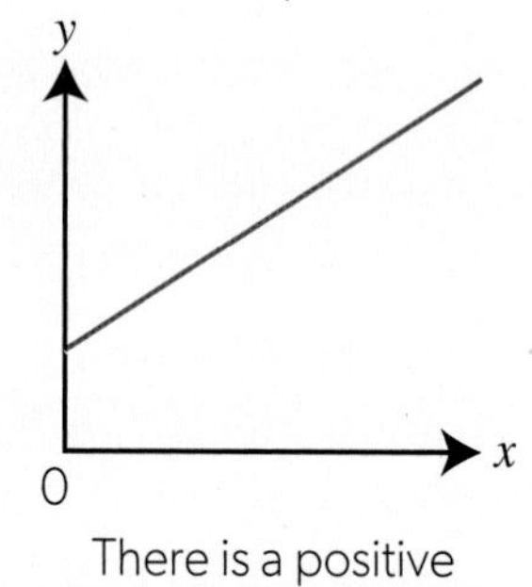

There is a positive correlation between x and y

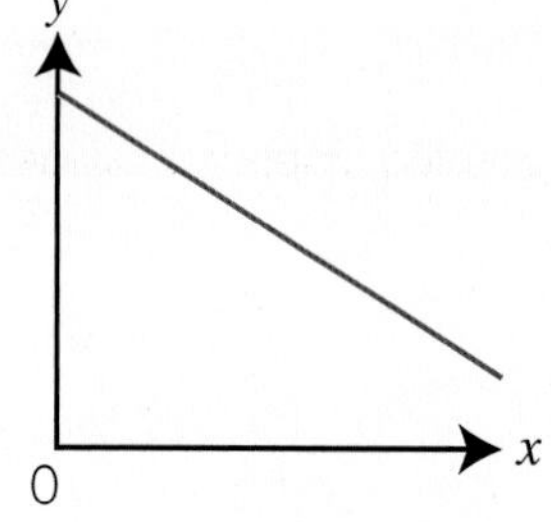

There is a negative correlation between x and y

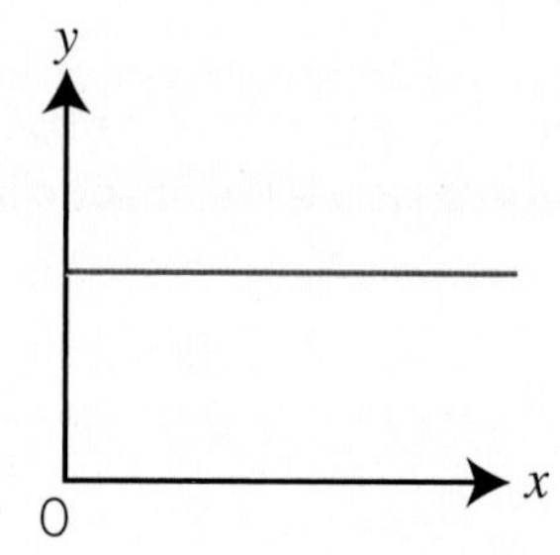

There is no relationship between between x and y

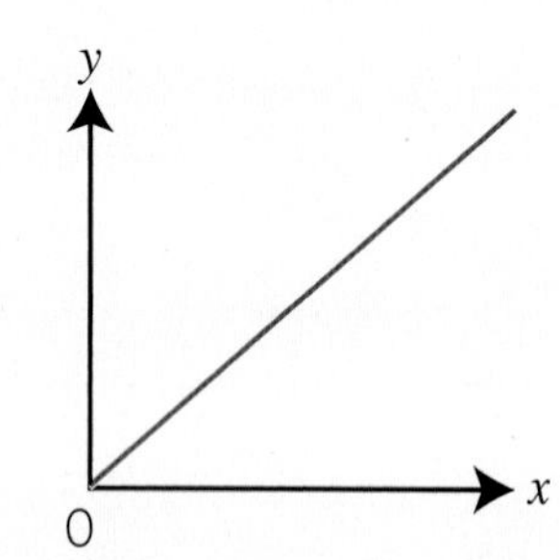

There is a positive correlation between x and y and y is directly proportional to x

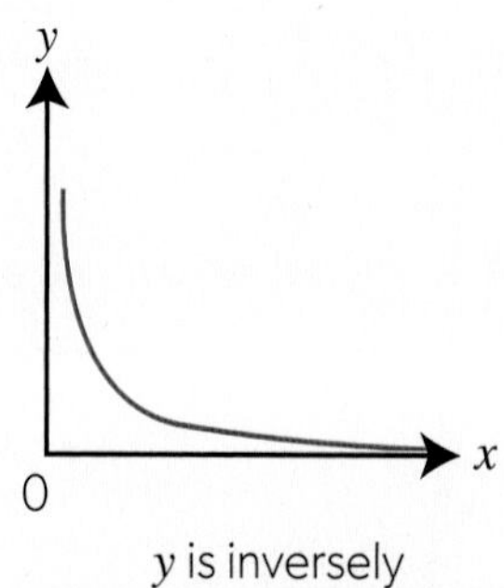

y is inversely proportional to x

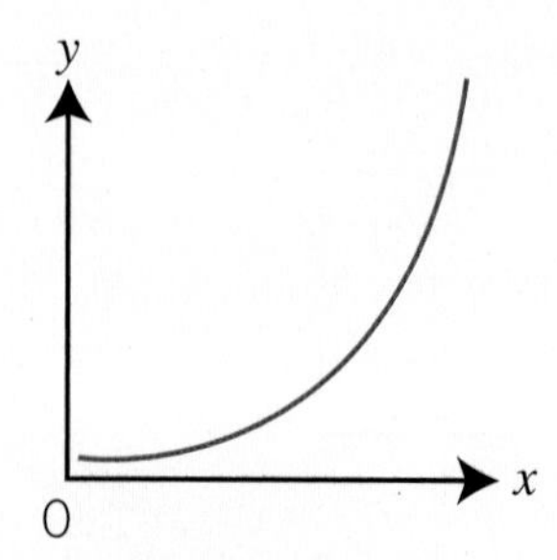

As x increases there is an exponential increase in y

▲ Figure 18

C1.1.9 Measurements in enzyme-catalysed reactions

Enzyme experiments require accurate measurements, to ensure the results are reliable. Reliability is demonstrated by repeating an experiment and showing that the results are consistent. There are different types of variable in an experiment:

- Independent variables—these are factors that are being investigated, so they are deliberately varied to see what the effect is. Often there is just one independent variable, making data from the experiment easy to analyse. Variables are independent if the researcher has a free choice of what levels to use. In enzyme experiments, the independent variable is commonly temperature, substrate concentration, enzyme concentration or pH.
- Control variables—these are factors that must be kept constant to ensure the experiment is a "fair test". Control variables should be monitored regularly to ensure they do not change. In a properly designed enzyme experiment, all factors that could affect enzyme activity—apart from the independent variable—are control variables.
- Dependent variables—these are the results of the experiment. In an enzyme experiment, the dependent variable is the quantity that is measured to calculate the reaction rate. Only changes to the independent variable should affect the level of the dependent variable. This rate is often called enzyme activity.

Calculation of reaction rates is an important skill, whether you do this using data from your own experiment or secondary data from an experiment carried out by someone else. Reaction rate is the speed at which substrates are converted to products, so the units are the change in the amount of chemical divided by time, for example, millimoles per second (mmol s^{-1}). There are two approaches to finding the reaction rate:

1. Allow the reaction to happen for a fixed time and measure the amount of substrate used up or product formed. The time should be relatively short, so the substrate concentration remains high.
2. Start with a known amount of substrate and allow the reaction to continue until all the substrate has been converted to products. Measure the time taken for the reaction to go to completion.

With both approaches, the quantity of product or substrate is divided by the time.

With enzymes, starch or other macromolecules, concentration is usually measured in grams per cubic decimetre or grams per 100 cm^3 of solution. Grams per 100 cm^3 can be expressed as a percentage. For example, 100 cm^3 of 1% starch solution contains 1 gram of starch.

Quantity measured	Units	Method of measurement
mass of enzyme or reagents	grams (g) or milligrams (mg)	electronic balance
volume of solutions	cubic decimetre (dm^3) or cubic centimetre (cm^3)	pipettes or syringes; measuring cylinders
molar concentration	moles per cubic decimetre (mol dm^{-3})	indirect measurement of mass and volume
mass concentration	grams per cubic decimetre (g dm^{-3})	indirect measurement of mass and volume
temperature	Celsius (°C)	thermometer; digital temperature probe
acidity	pH	pH meter; universal indicator
light absorbance	percentage absorbance (%)	colorimeter

▲ Table 2 Possible ways of measuring or determining the level of variables in enzyme experiments

The following example describes one experiment to investigate the rate of an enzyme-catalysed reaction. However, there are many other enzymes and ways of measuring reaction rate.

Collecting and processing data: Measuring catalase activity

The apparatus shown in Figure 19 can be used to investigate the activity of catalase. Yeast cells contain catalase. Yeast mixed with water is injected into the test tube to start the reaction. Catalase catalyses the conversion of hydrogen peroxide, a toxic by-product of metabolism, into water and oxygen. Catalase is one of the most widespread enzymes and other sources of the enzyme could be used (for example, liver tissue, kidney tissue or germinating seeds). These sources would have to be macerated and then mixed with water before being injected.

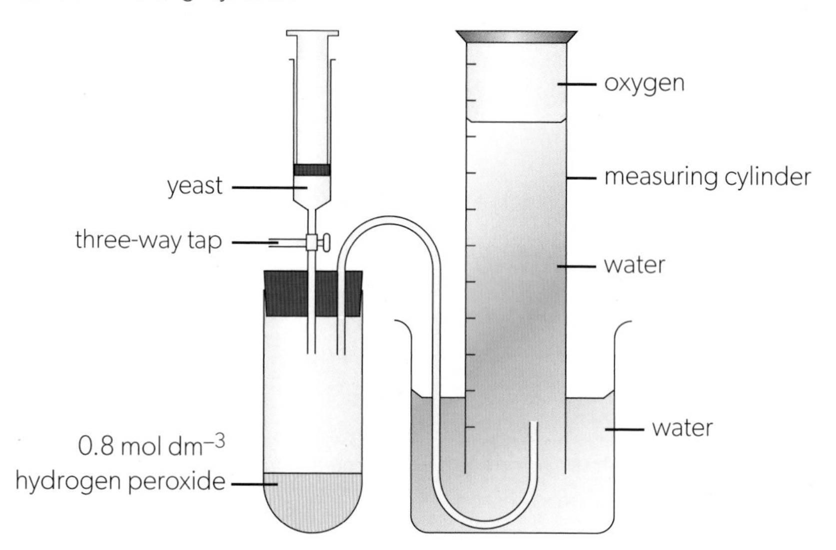

▲ **Figure 19** Apparatus for measuring catalase activity

To investigate the effect of substrate concentration, you could measure the reaction rate repeatedly using the same concentration of yeast but different hydrogen peroxide concentrations. Alternatively, you could investigate the effect of varying catalase concentration.

1. How can the activity of catalase be measured using the apparatus shown in Figure 19? Include suitable SI units for the reaction rate.
2. What factors should be kept constant if investigating the effect of substrate concentration?
3. How can the 0.8 mol dm^{-3} hydrogen peroxide solution be diluted to make concentrations of 0.2, 0.4 and 0.6 mol dm^{-3}?
4. Why is it necessary to macerate other catalase sources such as liver tissue before measuring catalase activity in them?

Safety goggles must be worn if this experiment is performed. Care should be taken not to get hydrogen peroxide on the skin.

Data-based questions: Calculating rates of reaction

1. $10\,cm^3$ of 1% starch solution was mixed with $1\,cm^3$ of 0.1% amylase solution. The reaction mixture was kept at 40°C. A test for starch was done every 30 seconds, using iodine solution. The first test that showed no starch was present after 8 minutes.
   a. Calculate the mass of starch in the reaction mixture in grams. [2]
   b. Convert this mass to milligrams. [1]
   c. Calculate the mass of starch digested per minute by the amylase. [1]
   d. Convert this rate of reaction from "per minute" to "per second". [1]

2. Ten drops of a commercial catalase solution were added to four reaction vessels containing a 1.5% hydrogen peroxide solution. Each of the solutions had been kept at a different temperature. The % oxygen in the reaction vessel was determined using a data logger in a set-up similar to Figure 20.
   a. Explain the variation in oxygen percentage at time zero. [1]
   b. Use the graph to determine the rate of reaction at each temperature. [4]
   c. Plot a graph of reaction rate against temperature. [3]
   d. Discuss whether a logarithmic scale for the *y*-axis should be used instead of a linear scale. [2]

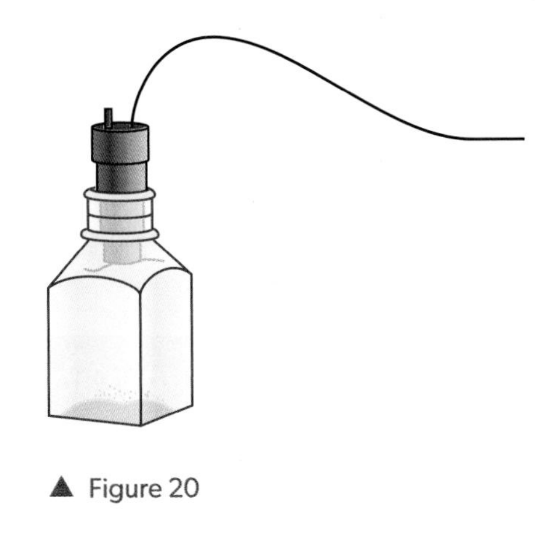

▲ Figure 20

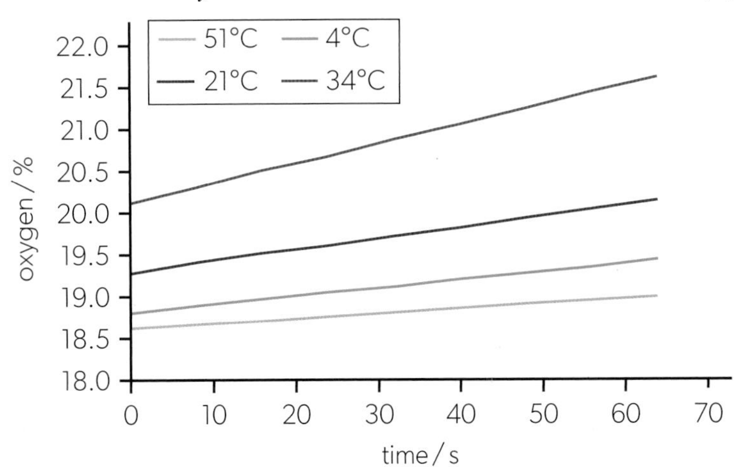

▲ Figure 21 Percentage of oxygen concentration over time at various temperatures after adding catalase to a 1.5% hydrogen peroxide solution

ATL Thinking skills: Choosing a method to determine reaction rate

Data processing involves converting raw data into a form that is easier to interpret. Reaction rate is measured as change in the amount of reactant or product per second.

For each of these enzyme experiments, describe how the rate of reaction can be determined:

a. Paper discs soaked in the enzyme catalase are added to different concentrations of hydrogen peroxide. The reaction produces oxygen bubbles.

b. Lipase catalyses the breakdown of triglycerides to fatty acids and water. Fatty acids affect the pH of the reaction as the reaction proceeds.

c. Gelatin cubes are digested by papain, which is a protease that can be extracted from papaya fruits.

d. The enzyme catechol oxidase can be extracted from bananas. It converts catechol to a yellow pigment in cut fruit. The yellow pigment reacts with oxygen in the air to turn brown.

Mathematics: Using logarithmic scales

Most graphs have linear scales on the axes. This means that the intervals between values on the axes are equal—for example, the values 0, 1, 2, 3 are spaced equally. Sometimes a logarithmic scale is more useful. Each interval is 10 times larger than the previous one—for example, a scale with 1, 10, 100 and 1,000 equally spaced is logarithmic.

Logarithmic scales are useful when a variable can take a very wide range of values and it is difficult to plot the small values or see the differences between them on a graph. They are also useful for testing whether the increase or decrease in a variable is truly exponential. With a linear scale, the curve for an exponential relationship gets steeper and steeper; with a logarithmic scale it should be a straight line.

On a graph, either one or both scales can be logarithmic. In a log-linear plot, one scale is logarithmic and the other is linear. In a log-log plot, both scales are logarithmic. Special graph paper is available for both types of plot. Alternatively, you can plot graphs with logarithmic scales by converting the data to logarithms (logs).

Logs are exponents. Usually 10 is used as the base number for the exponent.

$100 = 10^2$, so $\log 100 = 2$

$1{,}000 = 10^3$, so $\log 1{,}000 = 3$.

To find the logarithm for intermediate numbers, use a calculator or an online tool.

For example, $352 = 10^{2.5465}$, so log 352 is 2.5465.

Example

The enzyme experiment described in the data-based question on page 351 gave these results for enzyme activity at different temperatures: 68 at 15°C, 123 at 25°C, 243 at 35°C and 536 at 45°C. Is there an exponential increase in enzyme activity with rising temperatures?

1. Convert the values for enzyme activity to base 10 logs.
2. Plot the log values on a graph with temperature on the x-axis and the log values for enzyme activity on the y-axis. You can use an axis break at the origin.
3. Is it possible to draw a straight line through (or close to) all the points? If so, what conclusions can you draw?

Data-based questions: Interactions between temperature and enzyme activity

Changes to the amino acid sequence of enzymes can affect how heat-stable they are and how quickly they convert substrates to products. The graph in Figure 22 shows data for the same enzyme (isopropylmalate dehydrogenase) from six different species of microbe, three of which are adapted to live in hot conditions. The graph shows the relationship between the temperature at which half of the enzyme molecules have denatured due to unfolding, and enzyme activity at 25°C. The graph is a log-linear plot.

1. What is the advantage of using a logarithmic scale for the y-axis on this graph? [2]
2. a. Which microbe has the most heat-stable form of the enzyme? [1]
   b. Which microbe has a form of the enzyme that catalyses the reaction most quickly at 25°C? [1]
3. What trend does the graph show? [2]
4. Using the data in the graph, predict the problems that would be experienced by:
   a. *Bacillus subtilis* in a hot spring at 80°C or higher [2]
   b. *Sulfolobus tokodaii* in the human intestine at 37°C. [2]

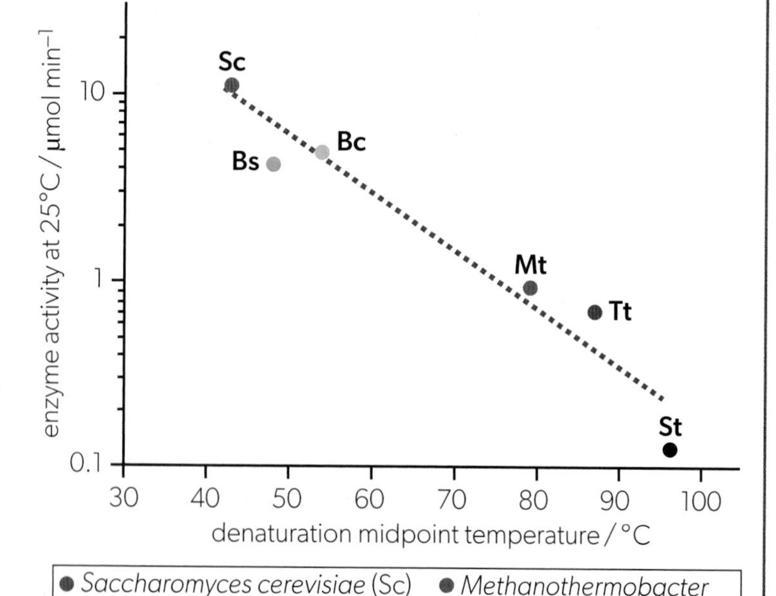

▲ Figure 22

Source: Akanuma, S., Bessho, M., Kimura, H. et al. Establishment of mesophilic-like catalytic properties in a thermophilic enzyme without affecting its thermal stability. Sci Rep 9, 9346 (2019). https://doi.org/10.1038/s41598-019-45560-x

ATL Communication skills: Generating citations

There are many electronic tools to support students in writing references in a standardized format such as APA or MLA. Google Scholar is one example. Search through Google Scholar to find the publication you are looking for. For example, you might search for Daniel Koshland's article in the journal *PNAS*, "Application of a theory of enzyme specificity to protein synthesis". Entering this title into Google Scholar returns a single entry. At the base of the entry, you will see a menu similar to the following:

☆ 99 Cited by 3078 Related articles All 9 versions

Clicking on the quotation marks will give you a range of choices. For example, you can copy the APA citation and paste it into a report:

Koshland Jr, D. E. (1958). Application of a theory of enzyme specificity to protein synthesis. *Proceedings of the National Academy of Sciences of the United States of America, 44*(2), 98–104.

1. To practise referencing, find the MLA formatted citation for the article used in the data-based question on page 352: *Establishment of mesophilic-like catalytic properties in a thermophilic enzyme without affecting its thermal stability.*
2. What are some of the ways in which APA and MLA citations differ? Which format is more commonly used in the bibliographies of scientific papers?
3. What rules do you need to follow when providing citations in your IB work? Check with your school librarian or your IB coordinator.

C1.1.10 Effect of enzymes on activation energy

Chemical reactions are not single-step processes. Substrates have to pass through a transition state before they are converted into products. Energy is required to reach this transition state. This is called the activation energy and is used to break bonds in substrate molecules. Energy is also released as new bonds are made and the product is formed.

The left-hand graph in Figure 23 shows these energy changes for a reaction carried out without an enzyme. The reaction is exothermic because there is a net release of energy. The energy released as new bonds are made is greater than the activation energy needed to break bonds and reach the transition state.

The right-hand graph in Figure 23 shows energy changes for the same reaction when it is catalysed by an enzyme. The net amount of energy released is unchanged but the activation energy is smaller. This is because the bonds in the substrate are weakened as it binds to the active site, so less energy is needed to break them. As a result, the rate of the reaction increases—typically by a factor of a million or more.

Questions

1. State the effect on the activation energy when an enzyme is present.
2. Compare the net change in energy content between substrates and products for the two systems shown in Figure 23.
3. Suggest why there are two peaks rather than one in the graph for a reaction catalysed by an enzyme.
4. Explain how lowering the activation energy increases the rate of the enzyme-controlled reaction.

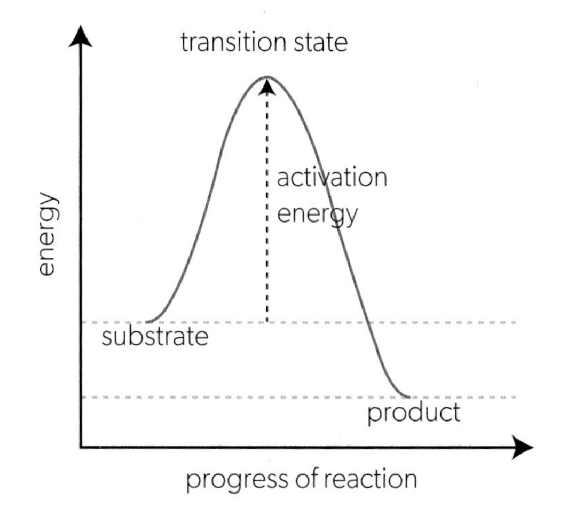

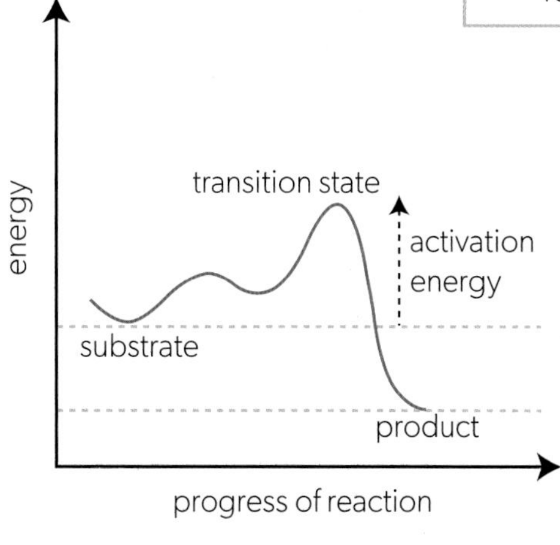

◀ Figure 23 Graphs showing activation energy without an enzyme (left) and with an enzyme (right)

AHL

C1.1.11 Intracellular and extracellular enzyme-catalysed reactions

Enzymes are synthesized by ribosomes. Extracellular enzymes (also known as exoenzymes) are released from the cell and work outside it. They are synthesized by ribosomes attached to the endoplasmic reticulum. Intracellular enzymes, for use inside the cell, are synthesized by free ribosomes in the cytoplasm.

Intracellular enzymes catalyse metabolic pathways such as glycolysis. The first reaction in glycolysis is the addition of a phosphate group to the glucose molecule. This reaction is catalysed by the enzyme hexokinase in the cytoplasm. Some intracellular enzymes work inside organelles—or example, enzymes of the Krebs cycle work in the mitochondrial matrix. One of the Krebs cycle enzymes is fumarase, which catalyses the conversion of fumarate to malate.

Many exoenzymes catalyse the breakdown of larger macromolecules. The monomers produced can then pass through the plasma membrane and enter cells. For humans and other multicellular organisms, this process occurs in the digestive system, where solid food is digested by extracellular enzymes.

Unicellular heterotrophic archaea, bacteria and fungi cannot take in macromolecules because their cell walls form a barrier so they cannot perform endocytosis. To feed on macromolecules, these microorganisms secrete exoenzymes. These enzymes work outside the cell to convert the macromolecules into monomers, which can then be absorbed across the plasma membrane.

Questions

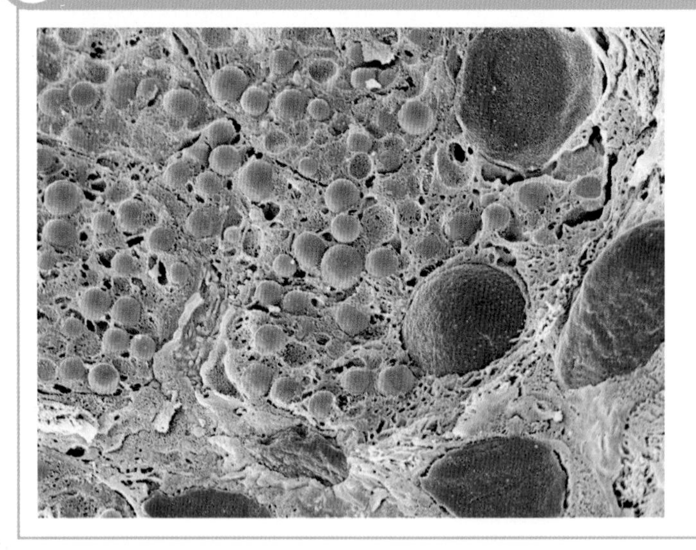

1. Suggest what the dark blue organelles are in the image.
2. Suggest a reason for storing digestive enzymes in vesicles, instead of secreting them immediately after they are produced.
3. Suggest reasons for the variation in size of the vesicles in these cells.

◀ **Figure 24** False-colour scanning electron micrograph of enzyme-secreting cells in the pancreas, with the cytoplasm coloured pale blue. Before secretion, the enzymes are stored inside membrane-bound sacs called vesicles (coloured yellow). The enzymes are secreted into the small intestine via the pancreatic duct and after activation they help the digestion of carbohydrates, fats and proteins

C1.1.12 Generation of heat energy by the reactions of metabolism

The conversion of energy from one form to another is never 100% efficient. For example, in metabolic reactions, the products contain less energy than the reactants. The additional energy is ultimately converted to heat.

Birds and mammals use the heat generated by metabolism to maintain a body temperature greater than that of their environment. Sometimes metabolism releases more heat than is needed for this purpose. For example, during exercise, the human body produces sweat and uses evaporative cooling to dissipate excess metabolic heat.

In very cold conditions, emperor penguins huddle together in groups to take advantage of the metabolic heat released by their neighbours. Birds and mammals raise their overall metabolic rate when basal metabolism does not release enough heat. They can do this in several ways. A human whose core temperature is falling will experience involuntary muscle contractions known as shivering. This is an adaptation to produce heat by contracting the muscles to raise the core temperature. Many mammals have brown fat tissue. Cells in this tissue have large numbers of mitochondria that carry out uncoupled respiration. This allows them to generate heat by oxidizing substrates without producing ATP.

▲ Figure 25 Decomposing manure or compost can become very hot, as this steaming heap shows. What organisms are responsible for this metabolic heat generation?

Questions

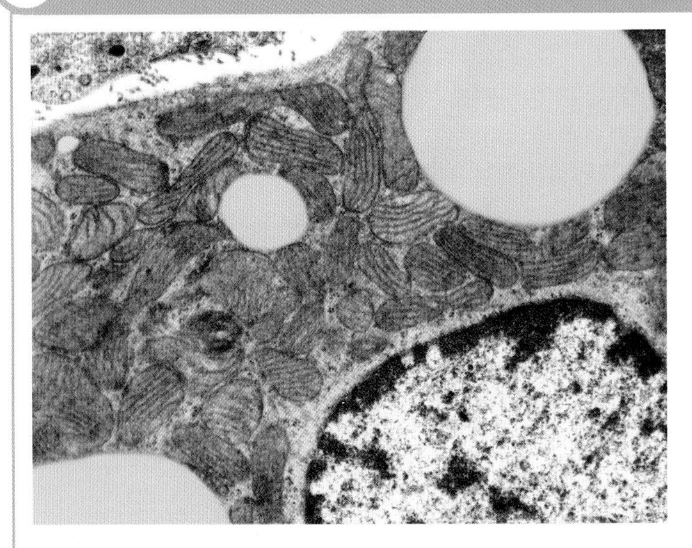

1. The electron micrograph in Figure 26 has been given false colour, to make the different structures easier to see. What would the natural colour of this cell have been?
2. Explain why stored fat forms large droplets in the cytoplasm.
3. Suggest why infants having relatively more brown adipose tissue than adults.
4. Suggest when brown adipocytes will be active in animals that hibernate.

◀ Figure 26 Electron micrograph of part of a heat-generating brown adipose cell with many mitochondria (dark green) in the cytoplasm (pale green) and droplets of fat (pale yellow). The cell's nucleus (purple) is partly visible (lower right)

C1.1.13 Cyclical and linear pathways in metabolism

Cells can perform a huge range of chemical reactions and have thousands of different types of enzyme to do this. Most enzyme-catalysed reactions cause a small change in a substrate. Large chemical transformations happen not in one large jump but in a sequence of small steps. Together, these steps form a metabolic pathway. The word game in Figure 27 is an analogy.

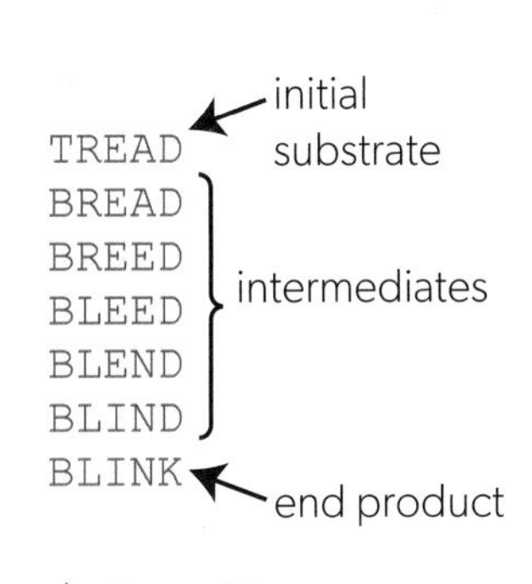

▲ Figure 27

Most metabolic pathways involve a chain of reactions. For example, glycolysis is used by cells to convert glucose into pyruvate (see Figure 28). The metabolic pathway for glycolysis involves nine different chemical reactions, catalysed by nine different enzymes. Glycolysis means "glucose-splitting" and is part of cell respiration. In all but one of the reactions, one molecule of substrate is converted to one molecule of product. In the other reaction, a six-carbon sugar is split

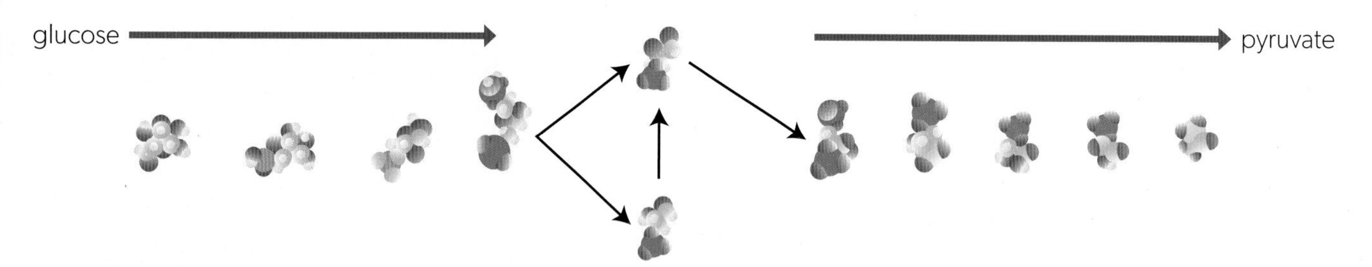

▲ Figure 28 Glycolysis

into two different three-carbon sugars. One of these is converted into a second molecule of the other sugar (glyceraldehyde-3-phosphate), which is the substrate for the next reaction in the chain. This means the latter stages of the chain happen twice per glucose molecule.

Branching of metabolic pathways is very common and metabolism as a whole is a network of interlinked and interdependent pathways. More unusual are cycles of reactions. In a cycle, every intermediate is a product of one reaction and a substrate of another reaction. Two examples of cycles, which will be described more fully in the next two chapters, are the Calvin cycle and the Krebs cycle.

▼ Figure 29 Simplified versions of the Calvin cycle and the Krebs cycle. These cycles are described in detail in *Sections C1.3.17* and *C1.2.12*

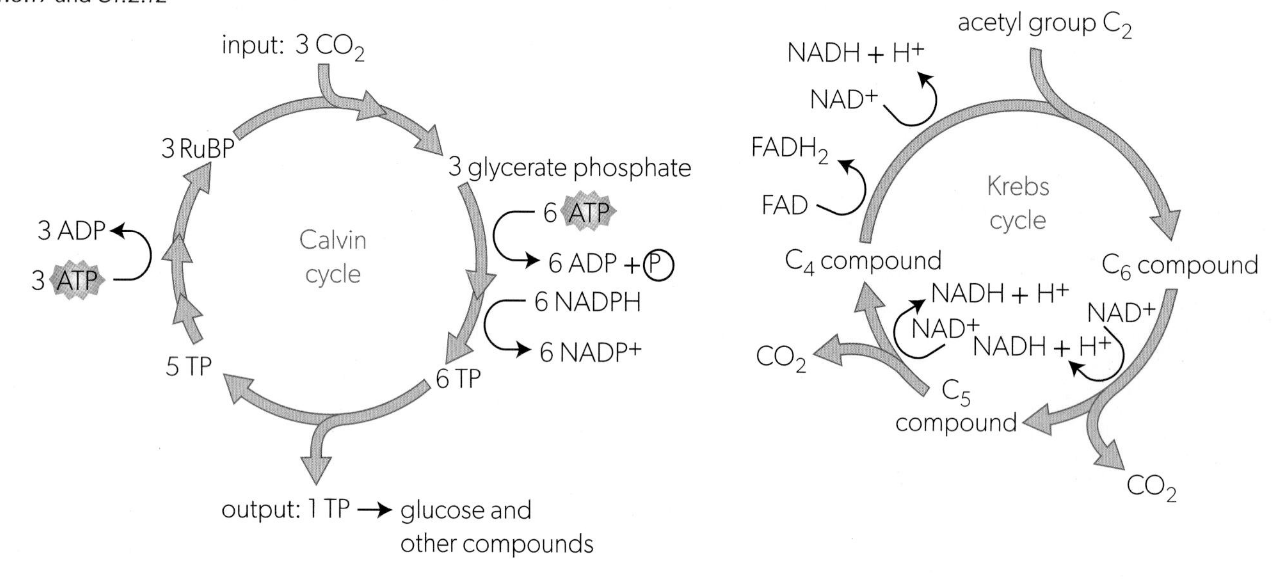

C1.1.14 Allosteric sites and non-competitive inhibition

Every enzyme has an active site, to which the substrate binds. Many enzymes have a second active site where a different specific substance can bind and unbind. Binding and unbinding cause the enzyme to change shape, so the second binding site is called an allosteric site.

Allosteric sites on enzymes have evolved because they allow the activity of an enzyme to be regulated. Switching between the two alternative states alters the structure and properties of the enzyme's active site. In some cases, binding to the allosteric site activates an enzyme so it will catalyse a reaction. In other cases, binding prevents catalysis and the enzyme is reversibly inhibited. Substances that inhibit an enzyme by binding to the allosteric site rather than the active site do not compete with substrates, so they are known as non-competitive inhibitors.

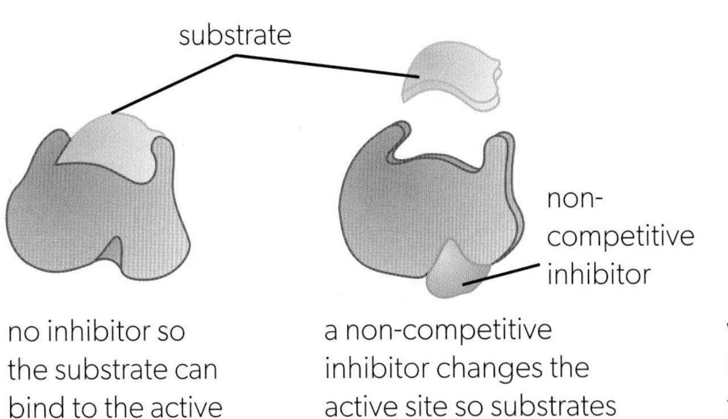

▲ Figure 30

C1.1.15 Competitive inhibition as a consequence of an inhibitor binding reversibly to an active site

Competitive enzyme inhibitors bind to the active site of an enzyme. As long as the inhibitor remains bound, the substrate cannot bind and the enzyme cannot catalyse its reaction. Competitive inhibitors are structurally similar to the substrate so they can bind to the same active site. However, unlike the substrate, they are not converted into products and so remain bound for longer than the substrate.

When the active site of an enzyme is vacant, either a substrate or an inhibitor molecule could bind. Whichever molecule arrives first and binds successfully with the active site will be the "winner" in this competition. The extent of inhibition becomes greater if the concentration of the competitive inhibitor increases.

If the inhibitor concentration is relatively low and the substrate concentration is increased, the extent of inhibition will reduce until the enzyme is effectively uninhibited. With many more substrate molecules than inhibitor molecules, substrates almost always arrive at the active site first. This is not the case with non-competitive inhibitors, because increases in substrate concentration cannot prevent a non-competitive inhibitor from binding to the allosteric site. This can be shown on a graph.

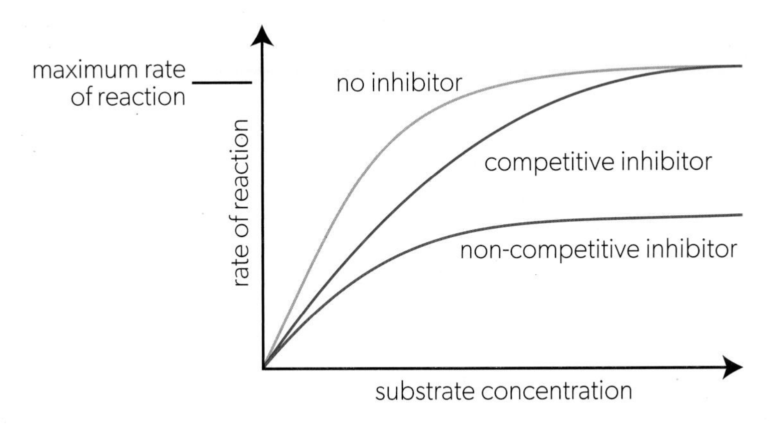

▲ Figure 31 Effect of substrate concentration on the rate of an enzyme-catalysed reaction with no inhibitor (orange) and with a fixed low concentration of a competitive inhibitor (red) or non-competitive inhibitor (blue)

Statins—an example of a competitive inhibitor

Statins are medicines that work by competitive enzyme inhibition. They are used to treat high blood cholesterol, which can contribute to heart disease. Statins bind to the active site of the enzyme HMG-CoA reductase. The full name of this enzyme is 3-hydroxy-3-methyl-glutaryl-coenzyme A reductase. It catalyses one of the reactions in the metabolic pathway used to synthesize cholesterol in liver cells. This reaction is the rate-limiting step in the pathway, so if statins lower the rate, less cholesterol is produced by the body. A person with high-blood cholesterol must receive the correct dose of statins, so cholesterol levels are reduced enough but not too much.

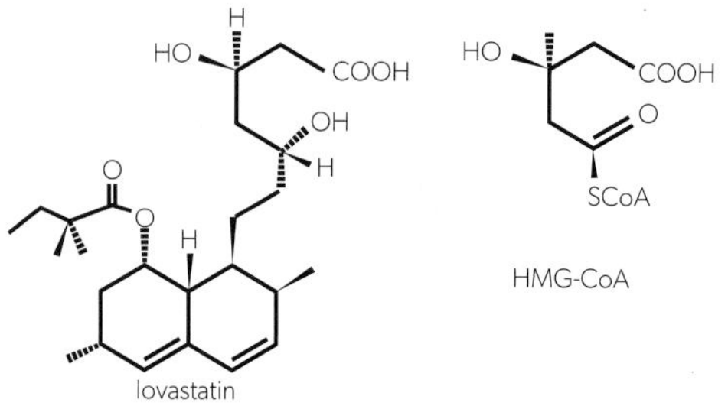

▲ **Figure 32** Structural similarities between HMG-CoA (the substrate of HMG-CoA reductase) and statins (such as lovastatin shown here) allow statins to bind to the active site and act as a competitive inhibitor

ATL Thinking skills: Evaluating and defending ethical positions

Chemical weapons would not exist without the activities of scientists. For example, sarin is a competitive inhibitor of the enzyme acetylcholinesterase and has a critical role in the functioning of the nervous system. It was discovered by German chemist Gerhard Schrader and developed as an insecticide. Later however, it was used as a chemical weapon. Despite being banned by several treaties, sarin has been used in the 21st century. Does Schrader deserve to be described as "the father of nerve agents"?

Fritz Haber received the 1918 Nobel Prize in Chemistry for his work in developing the chemistry behind the industrial production of ammonia fertilizer. Such fertilizers have been essential in increasing agricultural yields. However, some scientists boycotted the award ceremony because Haber had been instrumental in encouraging and developing the use of chlorine gas as a chemical weapon during the First World War. Haber is quoted as saying: "During peacetime a scientist belongs to the World, but during war time he belongs to his country". Should Haber have been recognized for important contributions to science?

C1.1.16 Regulation of metabolic pathways by feedback inhibition

The complex network of metabolic pathways in a cell must be regulated to ensure they produce enough of each substance, but not too much. Many pathways are controlled by a feedback inhibition system. In these systems, the product of the last reaction in the pathway—called the end product—inhibits the first reaction.

The enzyme that is inhibited has an allosteric site to which the end product binds. This binding changes the shape of the active site, preventing catalysis for as long as the end product remains bound. This is an example of non-competitive inhibition and also negative feedback. If too much of the end product is made, it will increasingly inhibit the first enzyme in the pathway. This effectively switches off the whole pathway and prevents synthesis of more end product. If there is too little of the end product, there will be minimal inhibition of the first enzyme. The metabolic pathway will be open to produce more of the end product.

This is a very efficient method of regulating a metabolic pathway. Because the first enzyme is inhibited, the products of the intermediate steps in the pathway—

which are only used as steps in making the end product—do not accumulate in a cell if the end product is not being used. The metabolic pathway in cells that converts the amino acid threonine into isoleucine is an example of feedback inhibition by an end product.

Through a series of five reactions, the amino acid threonine is converted to isoleucine. As the concentration of isoleucine builds up, it binds to the allosteric site of the first enzyme in the chain, threonine deaminase. As a result, it acts as a non-competitive inhibitor.

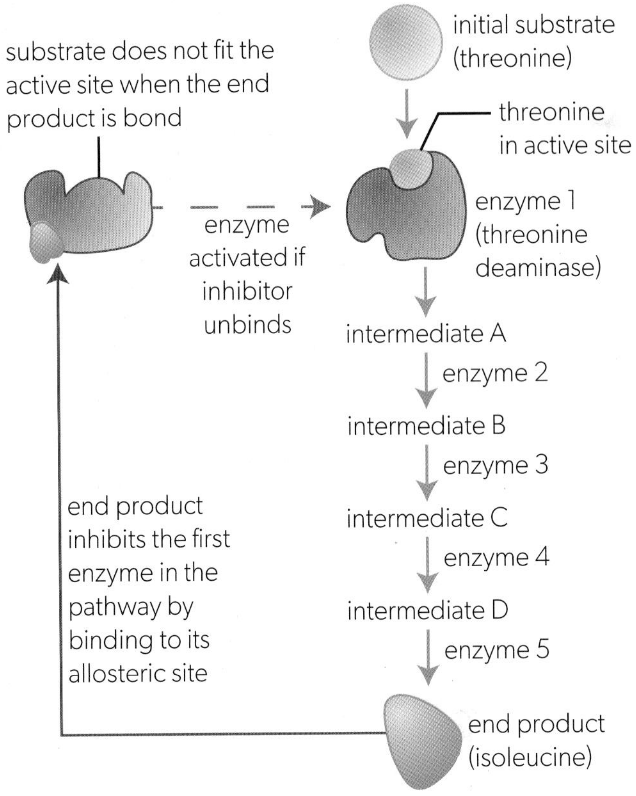

▲ Figure 33

C1.1.17 Mechanism-based inhibition as a consequence of chemical changes to the active site caused by the irreversible binding of an inhibitor

The examples of enzyme inhibition described so far have all been reversible. However, inhibition can also be irreversible. Heavy metals such as mercury and lead are non-specific inhibitors of a wide range of enzymes because they bind irreversibly to –SH groups in the amino acid cysteine wherever it occurs in the structure of an enzyme. For this reason, these heavy metals are very toxic if they enter the body and they are dangerous pollutants of the environment.

Some irreversible inhibitors target one specific enzyme. Such inhibitors are structurally similar to the substrate, so they bind to the enzyme's active site. The substrate would be converted to a product and released, leaving the active site open. The inhibitor, however, becomes permanently bound to the active site by the formation of a covalent bond. This produces a stable inhibitor–enzyme complex that can never work as a catalyst again. This is called mechanism-based inhibition.

Mechanism-based inhibitors cause harm to an organism, because every molecule of inhibitor can permanently inactivate one enzyme molecule. The inhibitor may kill an organism if the function of the inhibited enzyme is vital and there is a lethal concentration of the inhibitor. Some living organisms synthesize a mechanism-based enzyme inhibitor in order to kill another organism—for example, penicillin uses this method to kill gram-positive bacteria. In addition, some chemical weapons are mechanism-based enzyme inhibitors and they are some of the most toxic substances ever discovered.

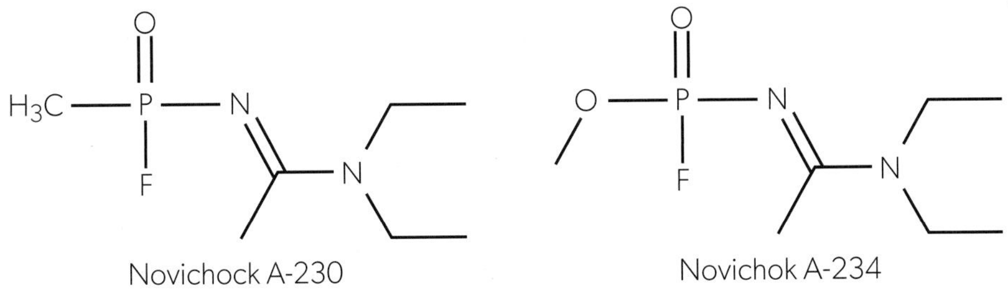

◀ Figure 34 Novichok agents are mechanism-based inhibitors of the enzyme acetylcholinesterase. They are lethal in minute quantities and have been implicated in some high-profile attacks on individuals

AHL

Penicillin—an example of mechanism-based inhibition

The cell walls of bacteria prevent them from bursting when low external solute concentrations cause water to enter by osmosis and hydrostatic pressures inside the cell become very high. The enzyme transpeptidase is very important in the process of cell wall formation, because it cross-links strands of carbohydrate into one huge peptidoglycan molecule that forms the entire cell wall. When bacteria grow, one enzyme breaks these links, allowing the wall to expand. Transpeptidase then remakes the links.

Saprotrophic bacteria and fungi compete for food because they both secrete enzymes for extracellular digestion of carbon compounds in dead matter and then absorb the products of digestion. The fungus *Penicillium notatum* produces a chemical known as penicillin, which is a mechanism-based inhibitor. It binds to the active site of transpeptidase in the cell walls of bacteria and prevents the substrate of the enzyme from binding. Penicillin forms a permanent covalent bond with a particular amino acid in the active site, binding irreversibly with the enzyme. The enzyme that breaks cross-links in the bacterial cell wall continues working but the transpeptidase enzyme cannot work to reform these links. As a result, the cell wall is weakened and the bacteria are killed by bursting (lysis). *Penicillium* can then monopolize the food source. The fungus only does this when food supplies are limited, because resources are needed for the synthesis and secretion of penicillin.

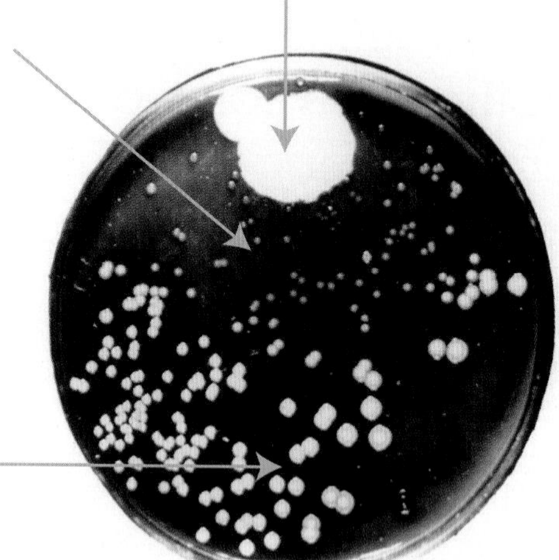

▲ **Figure 35** Alexander Fleming's petri dish which first showed the inhibition of bacterial growth by penicillin from a mycelium of *Penicillium*

Linking questions

1. What are examples of structure–function relationships in biological macromolecules?
   a. Explain the relationship between allosteric enzyme shape and enzyme activity. (C1.1.14)
   b. Explain the role played by complementary base pairing in semi-conservative replication. (D1.1.2)
   c. Outline the structure of phospholipids and their role in membrane structure. (B2.1.1)
2. What biological processes depend on differences or changes in concentration?
   a. Outline the concentration changes that must occur when the axon of a nerve is returned to resting potential. (C2.2.2)
   b. Explain how H^+ concentration can be used to power the production of ATP. (C1.2.15)
   c. Explain how the homeostatic regulation of blood glucose concentration can be achieved. (D3.3.3)

C1.2 Cell respiration

What are the roles of hydrogen and oxygen in the release of energy in cells?

In living systems, oxidation reactions are a type of energy-releasing reaction while reduction reactions are a type of energy-absorbing reaction. Bacteria called methanophiles oxidize methane as an energy source. At certain locations on the seabed, methane seeps out from the ocean floor. As a result, a food chain can emerge in the absence of light. In Figure 1, the white substrate is frozen methane. The ice worms (*Sirsoe methanicola*) consume the bacteria that grow on the methane.

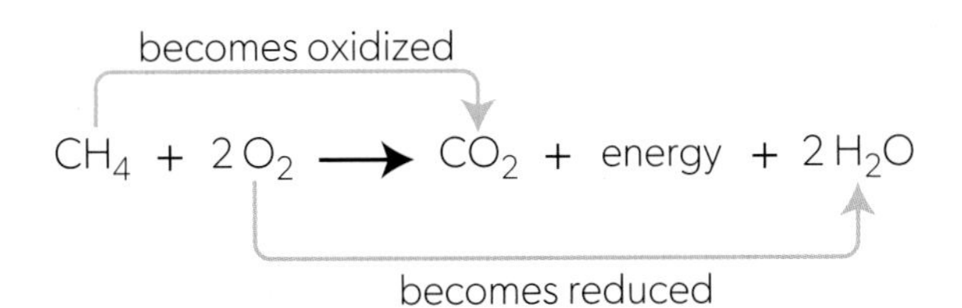

▲ Figure 1 These ice worms (Sirsoe methanicola) are at a depth of 800 m in the Gulf of Mexico

How is energy distributed and used inside cells?

Energy-rich molecules need to be converted into a usable form of energy such as ATP. Alternatively, they can be stored for later use in storage molecules such as glycogen or starch. The breakdown of energy-rich molecules such as glucose is done in stages, leading to the production of ATP and reduced NAD and FAD. Reduced molecules transfer their electrons to the electron transport chain, powering the production of more ATP. In Figure 2, the nucleus of the parenchyma cell is surrounded by an organelle called an amyloplast, which contains starch grains. Many mitochondria are also visible.

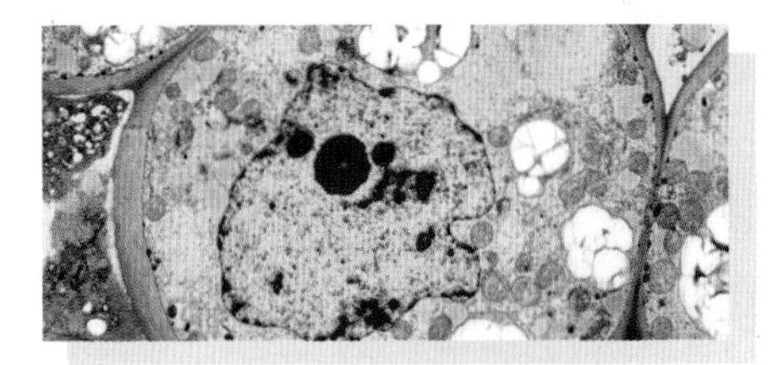

▲ Figure 2 A parenchyma cell from a voodoo lily

SL and HL	AHL only
C1.2.1 ATP as the molecule that distributes energy within cells C1.2.2 Life processes within cells that ATP supplies with energy C1.2.3 Energy transfers during interconversions between ATP and ADP C1.2.4 Cell respiration as a system for producing ATP within the cell using energy released from carbon compounds C1.2.5 Differences between anaerobic and aerobic cell respiration in humans C1.2.6 Variables affecting the rate of cell respiration	C1.2.7 Role of NAD as a carrier of hydrogen and oxidation by removal of hydrogen during cell respiration C1.2.8 Conversion of glucose to pyruvate by stepwise reactions in glycolysis with a net yield of ATP and reduced NAD C1.2.9 Conversion of pyruvate to lactate as a means of regenerating NAD in anaerobic cell respiration C1.2.10 Anaerobic cell respiration in yeast and its use in brewing and baking C1.2.11 Oxidation and decarboxylation of pyruvate as a link reaction in aerobic cell respiration C1.2.12 Oxidation and decarboxylation of acetyl groups in the Krebs cycle with a yield of ATP and reduced NAD C1.2.13 Transfer of energy by reduced NAD to the electron transport chain in the mitochondrion C1.2.14 Generation of a proton gradient by flow of electrons along the electron transport chain C1.2.15 Chemiosmosis and the synthesis of ATP in the mitochondrion C1.2.16 Role of oxygen as terminal electron acceptor in aerobic cell respiration C1.2.17 Differences between lipids and carbohydrates as respiratory substrates

C1.2.1 ATP as the molecule that distributes energy within cells

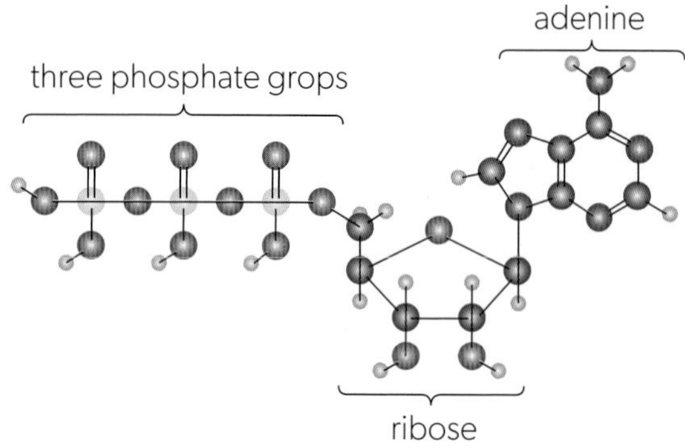

▲ Figure 3 Adenosine triphosphate (ATP). Black atoms are carbon, oxygen red, hydrogen white, nitrogen blue and phosphorus yellow

Nucleotides are the subunits RNA and DNA. They consist of three parts:

- a nitrogen-containing base
- a five-carbon sugar
- one or more phosphate groups.

ATP is a nucleotide because it consists of the base adenine, the five-carbon sugar ribose and three phosphate groups. The phosphate groups are in a chain and each of them is negatively charged.

ATP is often described as the energy currency of the cell, because it is used for temporary storage of energy and for energy transfer between processes and between different parts of the cell. The properties of ATP make it suitable for this role:

- ATP is soluble in water so it can move freely through the cytoplasm and other aqueous solutions in the cell.
- ATP is stable at pH levels close to neutral (as in cytoplasm).
- ATP cannot pass freely through the phospholipid bilayer of membranes. This means it cannot diffuse out of a cell and its movement between membrane-bound organelles within cells can be controlled.
- The third phosphate group of ATP can easily be removed and reattached by hydrolysis and condensation reactions:

$$ATP + H_2O \rightleftharpoons ADP + \text{phosphate} + \text{energy}$$

- Hydrolysing ATP to ADP and phosphate releases a relatively small amount of energy. This is enough for many processes within the cell. If more energy was released, there would often be an excess. This would be wasted by conversion to heat.

▲ Figure 4 What are the differences between using cash to buy something and using ATP to supply energy for a process?

C1.2.2 Life processes within cells that ATP supplies with energy

Cells need energy for three main types of activity.

1. Synthesizing macromolecules

Anabolic reactions that link monomers together into large polymers would be endothermic and therefore unlikely to happen without coupling them to conversion of ATP to ADP. One or more ATP molecules is used every time a monomer is linked to the growing polymer. Synthesis of DNA during replication, RNA in transcription and proteins in translation all require energy from ATP.

2. Active transport

Pumping of ions or other particles across a membrane against the concentration gradient requires energy from ATP. The energy is used to cause reversible changes in the conformation (shape) of the pump protein. When the pump is in one conformation, the particle can enter it from one side of the membrane. When the pump is in the other conformation, the particle can exit on the other side of the membrane. One of the two shapes is more stable than the other. ATP is used to cause the change from the more stable to the less stable conformation. The change back to the more stable conformation happens without the need for energy.

3. Movements

Cells require energy from ATP for movement. Components of cells are moved—for example, chromosomes are moved to the poles during mitosis and vesicles move to transport materials within cells. Larger amounts of energy—and therefore more ATP molecules—are needed to change the shape of a cell, for example, when a dividing cell pinches apart during cytokinesis. Some cells use changes of shape for movement from place to place (locomotion)—for example, phagocytes in the human blood system move to sites of infection. Muscle cells can contract powerfully using large arrays of actin and myosin filaments, which exert force by sliding across each other. The energy for these movements is provided by ATP.

C1.2.3 Energy transfers during interconversions between ATP and ADP

ATP contains more chemical potential energy than ADP. Therefore, energy is released when ATP is converted to ADP and phosphate. The amount of energy released is relatively small, but sufficient for many processes within the cell. In some cases, the phosphate group is linked to another molecule, such as a protein pump in the membrane or a substrate in a metabolic reaction. When the phosphate detaches from this molecule, energy is released. This energy causes a change in the molecule—for example, a conformational change in a membrane pump or a chemical change that converts a substrate into a product.

Energy is required to convert ADP and phosphate back to ATP. This energy can come from:

- cell respiration, in which energy is released by oxidizing carbohydrates, fats or proteins
- photosynthesis, in which light energy is converted to chemical energy
- chemosynthesis, in which energy is released by oxidizing inorganic substances such as sulfides.

The quantity of ATP within a cell at any time is very small; if it is all used up, processes that require energy stop. For example, neurons in the nervous system are unable to convey impulses and muscle cells stop contracting, causing cramp. Without ATP, cells start to degrade within minutes. This damage is soon irreparable, leading to cell death. However, this is normally prevented by continual regeneration of ATP from ADP and phosphate.

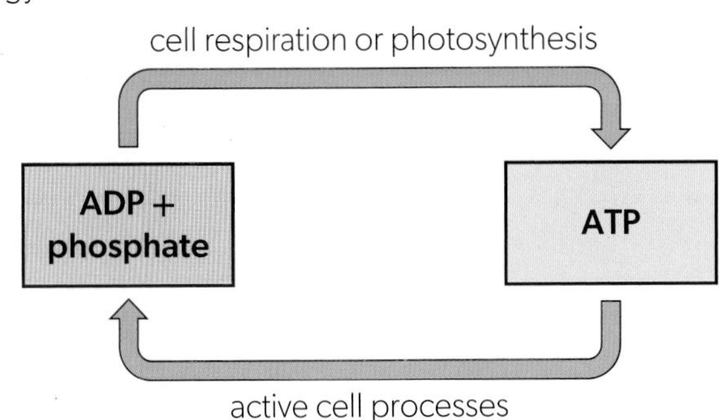

▲ Figure 5 ATP is converted to ADP + phosphate and back again by different cell processes

Energy transfers during interconversions between ATP and ADP are not 100% efficient, so some of the energy is transformed into heat.

Activity: Calculating the rate of conversion between ATP and ADP

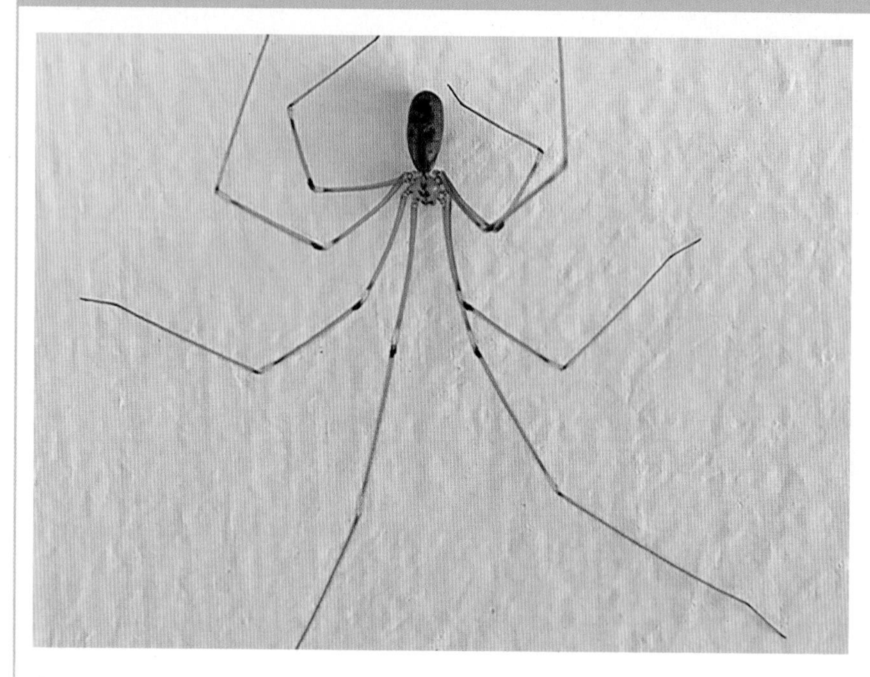

▲ **Figure 6** This six-eyed spider (*Pholcus phalangioides*) can remain motionless on a wall for days, conserving ATP while waiting for prey. Humans are typically much more active, using about 120 moles of ATP per day. At any moment, there is only about 0.2 moles of ATP in the body. How many times per day is an average ATP molecule converted to ADP and back again?

C1.2.4 Cell respiration as a system for producing ATP within the cell using energy released from carbon compounds

Cell respiration is a function of life that is performed by all living cells. In respiration, carbon compounds are oxidized to release energy and this energy is used to produce ATP. A wide range of carbon compounds can be used as respiratory substrates, but glucose and fatty acids are the main ones in many cells. In humans, the food we eat is the source of respiratory substrates. Plants use carbohydrates or lipids previously made by photosynthesis.

In many cells, respiration uses oxygen and produces carbon dioxide. It is therefore necessary for oxygen to enter cells through the plasma membrane while, at the same time, carbon dioxide exits the cells. Together these movements are known as gas exchange, although they do not involve direct one-for-one swapping of molecules. Instead both carbon dioxide and oxygen move across the plasma membrane independently, by simple diffusion.

Gas exchange and cell respiration are different processes but they are interdependent. Without gas exchange, cell respiration could not continue because there would soon be a lack of oxygen and a harmful excess of carbon dioxide inside the cell. Without cell respiration, gas exchange could not continue because the use of oxygen and production of carbon dioxide in respiration create the concentration gradients which cause the gases to diffuse.

Data-based questions: Energy in respiratory substrates

TYPICAL VALUES	Per 100ml	Per tablespoon (15ml)	%*RI
Energy	3397kJ/826kcal	509kJ/124kcal	6%
Fat	92g	14g	20%
of which saturates	5.8g	0.9g	5%
of which monounsaturates	59g	8.9g	
of which polyunsaturtaes	22g	3.4g	
Carbohydrate	<0.5g	<0.5g	
of which sugars	<0.5g	<0.5g	<1%
Fibre	<0.5g	<0.5g	
Protein	<0.5g	<0.5g	
Salt	<0.01g	<0.01g	<1%

NUTRITION INFORMATION	
TYPICAL VALUES	PER 100g
Energy	1380kJ 325kcal
Fat	0g
of which saturates	0g
Carbohydrate	80.5g
of which sugars	80.5g
Protein	0.5g
Salt	0.83g

▲ Figure 7 Nutritional content of canola (rapeseed) oil (left) and syrup (right) as shown on food labels

1. Compare and contrast the nutritional content of the two foods. [3]
2. The SI units for energy are joules. Using the data on the syrup label, calculate the number of kilojoules of energy in one kilocalorie, giving your answer to one decimal place. [1]
3. The density of the canola oil is 0.93 g per cubic centimetre. Calculate the energy and fat content of the canola oil per 100 g. [2]
4. In the syrup, only the sugar contains significant amounts of energy. Calculate the energy content of 100 g of sugar. [2]
5. Deduce whether fats or carbohydrates are a richer source of energy when used as a respiratory substrate. [2]

C1.2.5 Differences between anaerobic and aerobic cell respiration in humans

Cell respiration can be performed using a variety of alternative metabolic pathways. Some pathways are aerobic (they use oxygen). Others are anaerobic (no oxygen is needed). Simple word equations summarize different types of cell respiration using glucose as a substrate:

Aerobic respiration in humans and many other animals and plants

glucose + oxygen

ADP ATP

carbon dioxide + water

Anaerobic respiration in humans, other animals and some bacteria

glucose

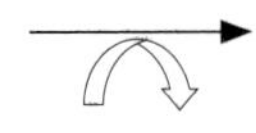

ADP ATP

lactate

Anaerobic respiration in yeast and other fungi

glucose

ADP ATP

ethanol + carbon dioxide

The features of aerobic and anaerobic respiration are compared in Table 1.

Aerobic cell respiration	Anaerobic cell respiration
Oxygen is used as an electron acceptor in oxidation reactions	Oxygen is not used—other substances act as oxygen acceptors in oxidation reactions
Carbohydrates such as glucose, lipids including fats and oils and amino acids after deamination can be used	Only carbohydrates can be used
Carbon dioxide and water are waste products	Carbon dioxide plus either lactate or ethanol are the waste products; water is not produced
The yield of ATP is much higher—more than 30 ATP molecules per glucose	The yield of ATP is lower—only 2 ATP per glucose
Initial reactions are in the cytoplasm, but more occur in mitochondria including use of oxygen	All reactions happen in the cytoplasm; mitochondria are not required

▲ Table 1

In humans, the lungs and blood system supply oxygen to most organs of the body rapidly enough for aerobic respiration. Sometimes, however, anaerobic cell respiration is used in muscles. The advantage of anaerobic respiration is that it can supply ATP very rapidly over a short time period. It is used when we need to maximize the power of muscle contractions.

▲ Figure 8 Short bursts of intense exercise are fuelled by ATP from anaerobic cell respiration

In our ancestors, maximally powerful muscle contractions will have been needed for survival by allowing escape from a predator or catching prey during times of food shortage. These events rarely occur in our lives today. Instead anaerobic respiration is more likely to be used during training or sport—for example, by:

- weight lifters during a lift
- short-distance runners in races up to 400 metres
- long-distance runners, cyclists and rowers during a sprint finish.

Lactate (lactic acid) is a waste product of anaerobic respiration in muscles. There is a limit to the concentration that the human body can tolerate and this restricts how much anaerobic respiration can be done. This is the reason for the short timescale over which the power of muscle contractions can be maximized. We can only sprint for a short distance—not more than 400 metres.

After vigorous muscle contractions, the lactate must be broken down. This requires oxygen. It can take several minutes for enough oxygen to be absorbed to break down all the lactate. The demand for oxygen that builds up during a period of anaerobic respiration is called the oxygen debt.

Data-based questions: Oxygen consumption in tobacco hornworms

Tobacco hornworms are the larvae of the moth *Manduca sexta*. Larvae emerge from the eggs laid by the adult female moths. There is a series of larval stages called instars. Each instar grows and then changes into the next one by shedding its exoskeleton and developing a new larger one. The exoskeleton includes the tracheal tubes that supply oxygen to the tissues.

The graphs in Figure 9 show measurements of the respiration rate of 3rd, 4th and 5th instar larvae, made using a simple respirometer.

Each data point on the graphs shows the body mass and respiration rate of one larva. For each instar, the results have been divided into younger larvae with low to intermediate body mass and older larvae with intermediate to high body mass. The results are plotted on separate graphs. The intermediate body mass is referred to as the critical weight.

Details of the methods are given in the paper published by the biologists who carried out the research. The reference to the research is Callier V and Nijhout H F 2011. "Control of Body Size by Oxygen Supply Reveals Size-Dependent and Size-Independent Mechanisms of Molting and Metamorphosis." *PNAS*. Vol. 108. Pp 14664–14669. This paper is freely available on the internet at http://www.pnas.org/content/108/35/14664.full.pdf+html.

1. a. Predict, using the data in the graphs, how the respiration rate of a larva will change as it grows from moulting until it reaches the critical weight. [1]

   b. Explain the change in respiration rate that you have described. [2]

2. a. Discuss the trends in respiration rate in larvae above the critical weight. [2]

   b. Suggest reasons for the difference in the trends between the periods below and above the critical weight. [2]

The researchers reared some tobacco hornworms in air with reduced oxygen content. They found that the instar larvae moulted at a lower body mass than larvae reared in normal air with 20% oxygen.

3. Suggest a reason for earlier moulting in larvae reared in air with reduced oxygen content. [2]

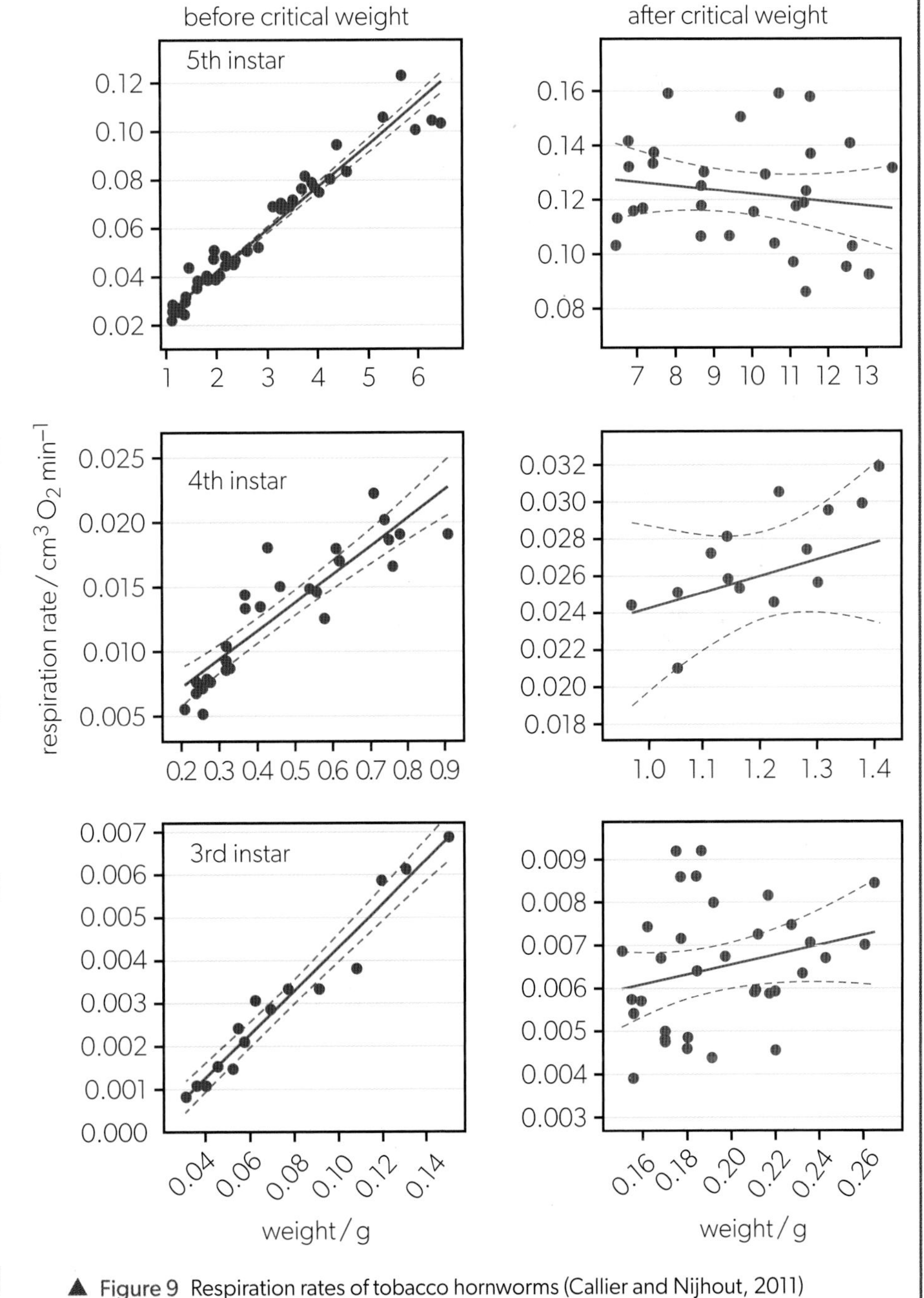

▲ **Figure 9** Respiration rates of tobacco hornworms (Callier and Nijhout, 2011)

C1.2.6 Exploring and designing: Variables affecting the rate of cell respiration

The rate of cell respiration can be determined from several types of measurement:

- oxygen uptake
- carbon dioxide production
- consumption of glucose or other respiratory substrates.

Oxygen uptake is usually used to determine the rate of aerobic respiration. It cannot be measured by finding how much air is breathed into the lungs because air contains gases other than oxygen and only a small proportion of air in each breath is absorbed by the body.

Measuring oxygen uptake

There are many possible designs of apparatus, all known as respirometers. They have these parts:

- a sealed glass or plastic container in which the organism or tissue is placed with enough air for it to remain healthy
- a base (alkali) such as potassium hydroxide to absorb carbon dioxide produced during respiration
- a capillary tube containing fluid connected to the container to measure changes in the volume of air inside the respirometer.

Carbon dioxide production normally adds to the volume of air in the atmosphere as oxygen intake reduces it. In a respirometer, all carbon dioxide is absorbed by the base, so any volume changes should be due to oxygen intake alone. Volume changes recorded by movements of the fluid in the capillary tube are therefore a measure of oxygen consumption.

Converting measurements to volumes

If the movement of the fluid in the capillary tube is measured in millimetres or other units of distance, it should ideally be converted to units of volume. This is simple if the diameter of the space inside the tube is known.

Calculating rates

A rate is an amount per unit time. Therefore, to find the respiration rate, the volume of oxygen used must be divided by the time. For example, if 50 mm^3 of oxygen was used in ten minutes, the rate would be 5 mm^3 oxygen per minute. This calculated rate is the dependent variable in an experiment using a respirometer.

Control of variables

Respirometers can give accurate results but it is essential to control variables carefully. In particular, temperature and pressure inside the respirometer must be controlled. This is because pressure, volume and temperature interact. For example, if heat generated by respiring organisms increases the air temperature inside a respirometer, the volume of the air will increase. If possible, temperature should be kept constant using a thermostatically controlled water bath. If air pressure outside the respirometer changes, the fluid in the capillary tube of some types of respirometer will move without any oxygen uptake.

Experimental investigation of variables

Respirometers can be used to perform experiments in which one variable is deliberately changed, while all others are kept constant. The variable that is changed is called the independent variable. There are many possible experiments but they must not cause harm to any organisms used. For example:

- the respiration rate of different organisms could be compared
- the effect of temperature on respiration rate could be investigated
- respiration rates could be compared in active and inactive organisms.

Data-based questions: Respirometers

Part 1: Experimental design

One possible design of respirometer is shown in Figure 10.

1. Explain the need for a base inside the respirometer. [2]
2. Deduce, giving a reason, the direction in which the fluid will move in the right-hand side of the capillary tube. [2]
3. Predict, with a reason, the change in the amount of oxygen inside the tube during the experiment. [2]
4. Explain how the following changes would improve the reliability of results from the experiment:

   a. putting the test tube in a thermostatically controlled water bath [2]

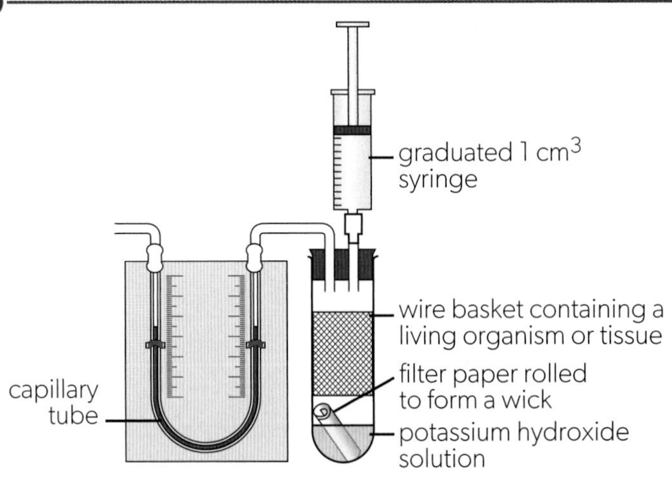

▲ Figure 10 Respirometer with a manometer for measuring oxygen uptake

b. attaching another test tube to the left-hand side of the capillary tube, that is identical to the right-hand tube but does not contain respiring tissue. [2]

Part 2: Using secondary data

Table 2 shows the results of an experiment in which the effect of temperature on respiration in germinating pea seeds was investigated.

Temperature / °C	Movement of fluid in respirometer / mm min^{-1}		
	1st reading	2nd reading	3rd reading
5	2.0	1.5	2.0
10	2.5	2.5	3.0
15	3.5	4.0	4.0
20	5.5	5.0	6.0
25	6.5	8.0	7.5
30	11.5	11.0	9.5

▲ Table 2

5. Discuss whether the repeats at each temperature are close enough to indicate that the results are reliable. [1]

6. a. The diameter inside the capillary tube of the respirometer was 2 mm. Convert the distances of movement at each temperature to volumes of oxygen used. [2]

   b. Calculate the mean volume for each temperature. [2]

7. Plot a graph of the mean volumes. [3]

8. Using your graph, deduce the relationship between temperature and the respiration rate of the germinating pea seeds. [2]

Part 3: Anaerobic respiration in yeast

The apparatus in Figure 11 was used to monitor mass changes during the brewing of wine. The flask was placed on an electronic balance which was connected to a computer for data logging. Results are shown in Figure 12.

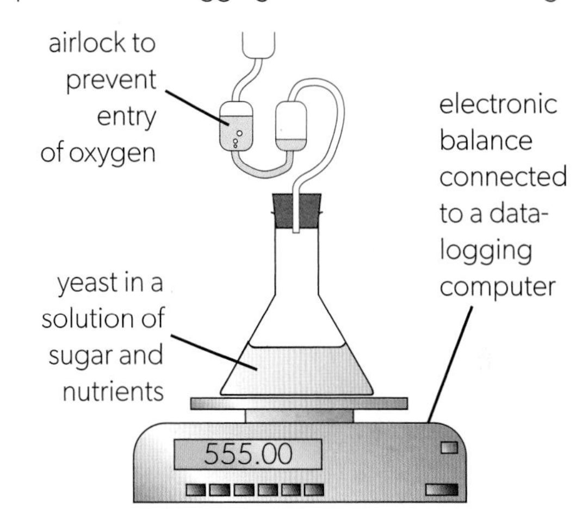

▲ Figure 11 Yeast data-logging apparatus

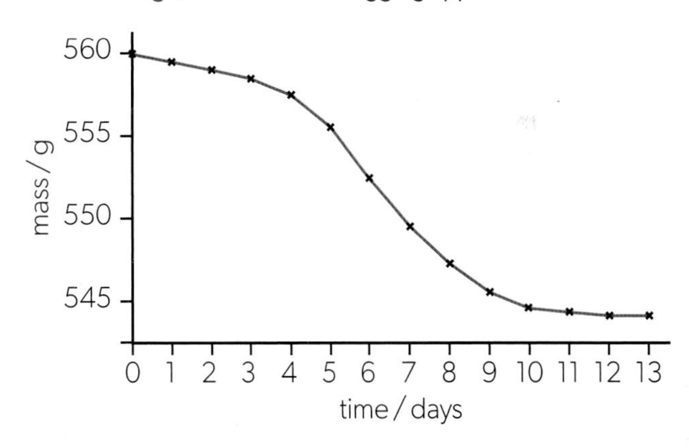

▲ Figure 12 Monitoring anaerobic cell respiration in yeast

9. a. Calculate the total loss of mass during the experiment and the mean daily loss. [2]

   b. Explain the loss of mass. [3]

10. Suggest two reasons for the increasing rate of mass loss from the start of the experiment until day 6. [2]

11. Suggest two reasons for the mass remaining constant from day 11 onwards. [2]

12. Suggest how the rate of respiration can be calculated from the data. [1]

ATL Thinking skills: Designing questions for experimental investigation

To plan an investigation in biology you need a focused research question. Here are some examples of research questions involving cell respiration in yeast.

1. Which foods can yeast use in anaerobic cell respiration?
2. Which monosaccharides is yeast able to use in anaerobic cell respiration?
3. Does yeast carry out anaerobic cellular respiration faster at 0°C or 100°C?
4. If oxygen is available, does yeast use aerobic or anaerobic cellular respiration?

What are some of the strengths and limitations of each question? What criteria do you think should be met for an inquiry question to be a good one?

Consider these criteria for a good IA inquiry question:

1. The answer should not be known before starting. Answering the question must depend on you carrying out the investigation.
2. The phrasing of the question should suggest the method to be followed.
3. The variables that are directly measured should be stated in the question.
4. The question should not include reference to significance. Students often muddle the intent of the experiment with the statement of the question.
5. Consider the significance of your investigation to an area that interests you. Are you interested in alternative fuel sources, enzymes or baking bread?

AHL

C1.2.7 Role of NAD as a carrier of hydrogen and oxidation by removal of hydrogen during cell respiration

Oxidation and reduction are chemical processes that always occur together. This happens because they involve transfer of electrons from one substance to another. Oxidation is the loss of electrons from a substance and reduction is the gain of electrons.

A useful example to help visualize this in the laboratory is in the Benedict's test—a test for certain types of sugar. The test involves the use of copper sulfate solution, containing copper ions with a charge of two positive (Cu^{2+}). Cu^{2+} ions often impart a blue or green colour to solutions. Sugar molecules cause a reduction reaction that changes these copper ions to atoms of copper by giving them electrons. Copper atoms are insoluble and form a red or orange precipitate. The sugar molecules are oxidized, because they have lost electrons.

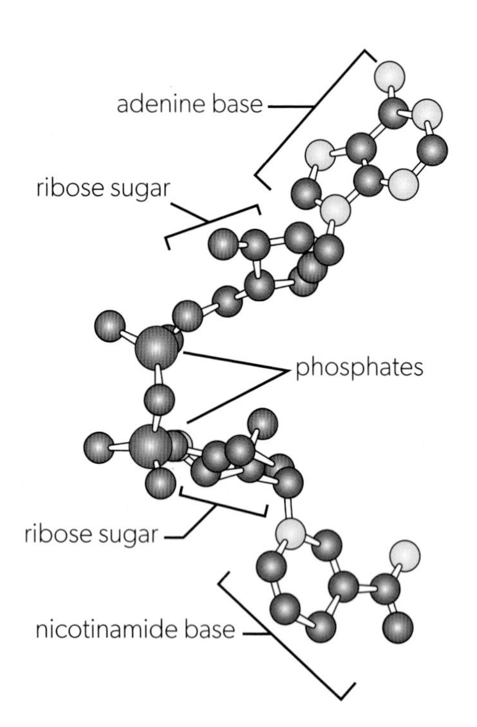

▲ Figure 13 Structure of NAD

Electron carriers are substances that can accept and lose electrons reversibly. They often link oxidations and reductions in cells. The main electron carrier in respiration is NAD (nicotinamide adenine dinucleotide). The structure of the NAD molecule is shown in Figure 13.

The equation below shows the basic reaction.

$$\text{NAD} + 2\text{ electrons} \rightarrow \text{reduced NAD}$$

The chemical details are a little more complicated. NAD initially has one positive charge and exists as NAD^+. Substances are oxidized in respiration by removing two hydrogen atoms. Each hydrogen consists of an electron and a proton. NAD^+ accepts two electrons and one proton from the hydrogen atoms, becoming NADH. The other proton (H^+) is released:

$$NAD^+ + 2H^+ + 2\text{ electrons }(2e^-) \rightarrow NADH + H^+$$

For simplicity, NAD in its reduced form is often shown as reduced NAD, rather than as $NADH + H^+$.

Reactions involving NAD show that reduction can be achieved by accepting atoms of hydrogen, because they hold an electron. Oxidation can therefore be achieved by losing hydrogen atoms.

Oxidation and reduction can also occur through loss or gain of atoms of oxygen. There are fewer examples of this in biochemical processes, perhaps because oxygen was absent from the atmosphere during the early evolution of life. A few types of bacteria can oxidize hydrocarbons using oxygen:

$$\underset{\text{n-octane}}{C_7H_{15}CH_3} + \frac{1}{2}O_2 \rightarrow \underset{\text{n-octanol}}{C_7H_{15}CH_2OH}$$

Nitrifying bacteria oxidize nitrite ions to nitrate:

$$NO_2^- + \frac{1}{2}O_2 \rightarrow NO_3^-$$

Adding oxygen atoms to a molecule or ion is oxidation, because the oxygen atoms have a high affinity for electrons and so tend to draw them away from other parts of the molecule or ion. In a similar way, losing oxygen atoms is reduction.

C1.2.8 Conversion of glucose to pyruvate by stepwise reactions in glycolysis with a net yield of ATP and reduced NAD

Glycolysis is the first part of aerobic respiration when glucose or another monosaccharide is the substrate. It happens in the cytoplasm of cells. In glycolysis, glucose is converted to pyruvate by a chain of reactions, each of which is catalysed by a different enzyme. A useful outcome of glycolysis is the production of a small yield of ATP without any oxygen being consumed. For this reason, the first stage of glycolysis may seem rather perverse as ATP is used rather than being produced.

Stage 1: Phosphorylation of glucose

Phosphorylation is the addition of phosphate (PO_4^{3-}) to a molecule. This requires energy but makes a molecule more unstable and therefore more likely to participate in subsequent reactions. In cells many phosphorylations are carried out by transfer of a phosphate from ATP.

In the first stage of glycolysis, glucose is phosphorylated. The reaction is usually shown in this way, with phosphorylation of glucose coupled to the conversion of ATP to ADP:

glucose ⟶ glucose-6-phosphate

ATP ADP

The -6- shows that the phosphate is linked to the sixth carbon atom of the glucose molecule. In the next reaction glucose is converted to fructose, creating a symmetrical molecule that can be split in half:

glucose-6-phosphate ⟶ fructose-6-phosphate

Then there is a second phosphorylation:

fructose-6-phosphate → fructose-1,6-bisphosphate

ATP ADP

Stage 2: Lysis

Fructose bisphosphate is now split to form two molecules of triose phosphate:

fructose-1,6-bisphosphate → 2 triose phosphate

Stage 3: Oxidation

Each of these triose phosphates is oxidized by removing hydrogen. Note that hydrogen atoms are removed, not hydrogen ions. If hydrogen ions were removed (H^+), no electrons would have been lost by the triose phosphate so it would not have been oxidized. The hydrogen is accepted by NAD, which becomes reduced NAD. Oxidation of a sugar produces an organic acid. In this case the sugar is triose with one phosphate group and the organic acid is glycerate carrying two phosphates. Energy released by the oxidation of triose allows a second phosphate group to become attached, so the product is bisphosphoglycerate rather than phosphoglycerate.

NAD reduced NAD

triose phosphate + phosphate → bisphosphoglycerate

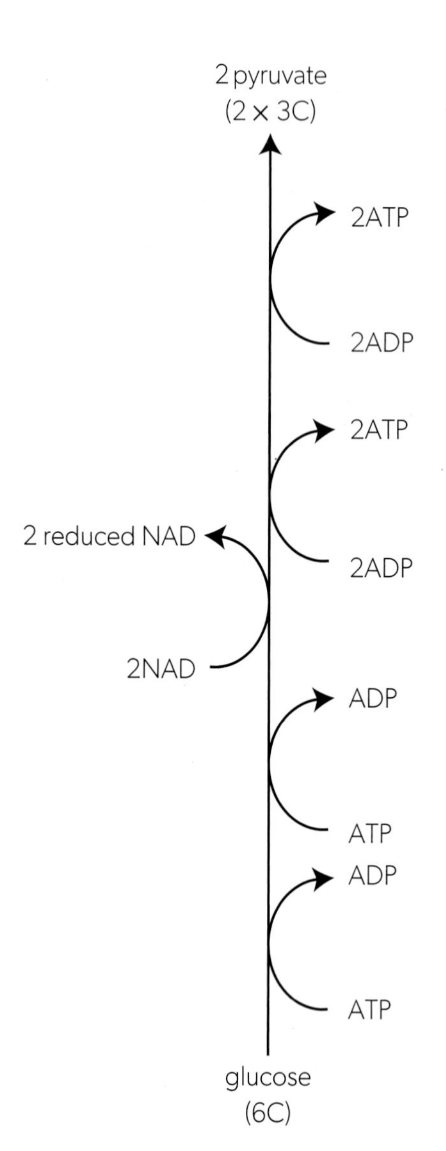

▲ Figure 14 An overview of glycolysis

Stage 4: ATP formation

ATP is produced in the final reactions of glycolysis, by transfer of phosphate groups to ADP. This can happen twice because bisphosphoglycerate has two phosphates. In these reactions, the glycerate is converted to another organic acid, pyruvate. This is the end product of glycolysis.

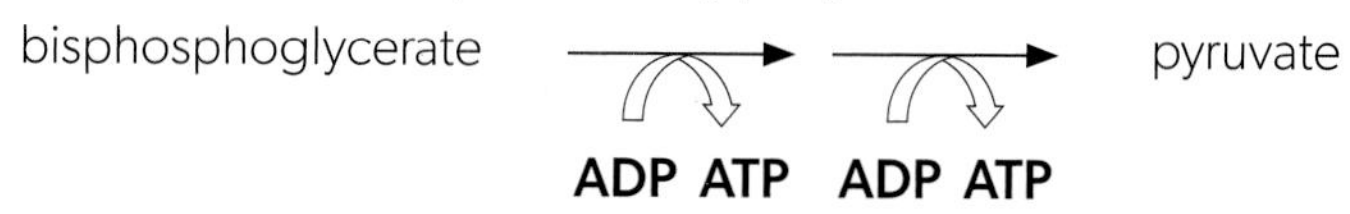

Two bisphosphoglycerate molecules are produced per glucose and each of them yields two ATPs. Four ATPs are therefore produced per glucose in these final reactions of glycolysis.

The overall outcomes of the four stages of glycolysis per glucose molecule are as follows:

- One glucose containing six carbon atoms is converted into two pyruvates each containing three carbon atoms.
- Two NAD molecules are converted to reduced NAD.
- There is a net yield of two ATPs. This is because two are used in the first stage of glycolysis and four are produced in the final stage, so the net yield is $4 - 2 = 2$ per glucose. Although this is a relatively small yield, it does not require the use of oxygen so is useful when the supply of oxygen is limited.

C1.2.9 Conversion of pyruvate to lactate as a means of regenerating NAD in anaerobic cell respiration

The summary equation for glycolysis (Figure 14) shows that supplies of glucose, ADP and NAD must be replenished for the process to continue in a cell.

- Glucose should not run out as long as there are stores in a cell or it is transported from elsewhere.
- ADP will only run out if all of it has been converted to ATP, in which case there is no need to carry out glycolysis.
- NAD will run out unless it is regenerated by oxidation of reduced NAD.

There are several methods of regenerating NAD. In each case, two hydrogen atoms (two protons and two electrons) are transferred to another molecule, oxidizing reduced NAD. In some human cells and also some other animal and bacterial cells, hydrogen is transferred from reduced NAD to pyruvate, converting it into lactate. This happens in the cytoplasm of cells.

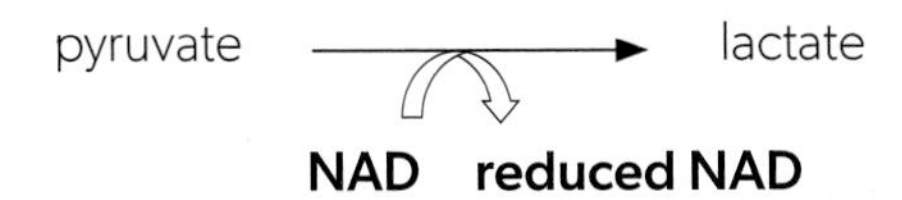

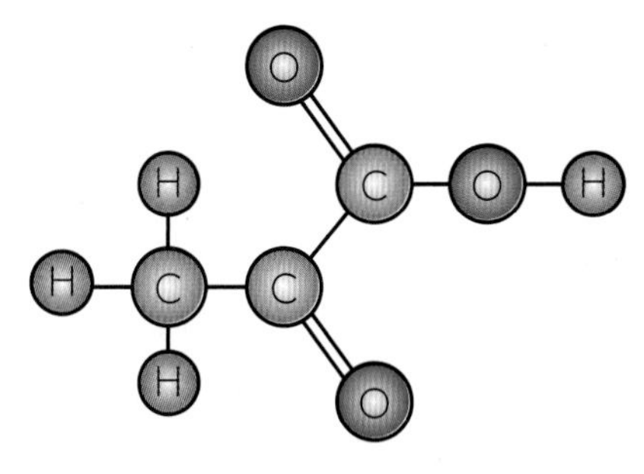

▲ Figure 15 This is pyruvic acid. It readily gives up a proton (H^+) from the hydrogen atom shown on the right-hand side, keeping the electron from the hydrogen. It is then pyruvate, with one negative charge. Pyruvate can accept a proton to become pyruvic acid. Organic acids such as pyruvic acid and lactic acid normally exist in their dissociated state, so are referred to with the '–ate' name rather than the '–ic acid' name

Two NAD molecules are used as each glucose is converted by glycolysis to pyruvate. Two pyruvates are produced and each of them can be used to convert a reduced NAD back to NAD, so all the NAD that was used in glycolysis is regenerated. For this reason, cells should not run out of NAD: as long as glucose is available and lactate concentrations do not rise too high, anaerobic cell respiration carried out in this way should be able to continue indefinitely.

Anaerobic cell respiration by glycolysis with conversion of pyruvate to lactate is called lactic fermentation. It is used in some methods of food preservation, because an accumulation of lactic acid (which dissociates to form lactate) lowers the pH and prevents decomposition by bacteria or fungi. Yoghurt, kimchi, sauerkraut and silage (for cattle) are all foods made by lactic fermentation.

▲ Figure 16 Kimchi was traditionally made from cabbage and other vegetables in South Korea, using large underground pots. It provided food for the winter when fresh vegetables were not available. The cool anaerobic conditions in the pots encourage *Lactobacillus* and other bacteria to carry out lactic fermentation, eventually killing themselves with low pH and preserving the vegetables

C1.2.10 Anaerobic cell respiration in yeast and its use in brewing and baking

NAD used in glycolysis can be regenerated by conversion of pyruvate to ethanol and carbon dioxide instead of lactate. This is a two-stage process. In the first stage, carbon dioxide is removed from the pyruvate in a decarboxylation reaction. The product is ethanal. In the second stage, two hydrogens are transferred from reduced NAD to ethanal, converting it to ethanol. The number of NADs regenerated is the same as the number used per glucose in glycolysis, so this process allows production of ATP by anaerobic respiration to continue indefinitely, as long as glucose is available and ethanol concentrations do not rise too high.

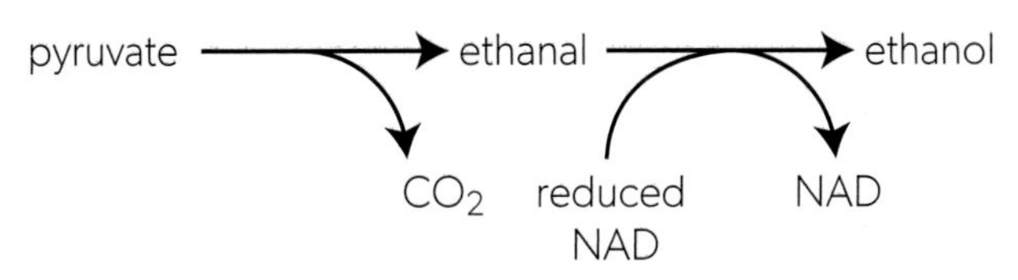

▲ Figure 17 Conversion of pyruvate to ethanol

▲ Figure 18 Kneading mixes the ingredients and stretches out gluten fibres, making bread dough more elastic

▲ Figure 19 Brandy that is burned on fruit puddings in some countries is distilled from wine. What does the release of heat by the flame tell us about ethanol as a waste product of anaerobic cell respiration?

▲ Figure 20 Bioethanol is currently an alternative to fossil fuels. What is the best energy source for cars in a zero-carbon future?

Anaerobic cell respiration by glycolysis which converts pyruvate to ethanol and carbon dioxide is known as ethanol fermentation or alcoholic fermentation. It is used in baking and brewing. In both cases, the organism that carries out the fermentation is yeast. Yeast is a unicellular fungus that occurs naturally in habitats where glucose or other sugars are available, such as the surface of fruits. Yeast is a faculative anaerobe—this means it can respire either aerobically or anaerobically.

Bread is made by adding water to flour, kneading the mixture to make dough and then baking it. To give the bread a lighter texture, something must be added to the dough to create bubbles of gas. Yeast is often used for this purpose. If the dough is kept warm, the yeast will grow and respire. Initially it respires aerobically but once all the oxygen in the dough has been used up, the yeast starts to respire anaerobically. Because the dough is very viscous (sticky), the carbon dioxide produced by anaerobic cell respiration cannot escape. Instead, it forms bubbles within the dough. These bubbles cause the dough to swell, or "rise". Ethanol is also produced by anaerobic cell respiration but it evaporates during baking.

Yeast is also used when brewing drinks such as beer and wine. Here, however, the aim is to produce ethanol rather than carbon dioxide. Wine is made from grape juice, which naturally has a high sugar concentration. Beer is made from barley grains, mixed with water. The grains contain large amounts of starch but little sugar. This starch must first be converted to sugar using amylase, because yeasts cannot metabolize starch. Brewing of wine or beer is carried out in large tanks, so diffusion of oxygen into the liquid in the tank is limited. The yeasts rapidly use up any oxygen present and then respire anaerobically. The ethanol produced remains dissolved but most of the carbon dioxide bubbles to the surface and escapes. Depending on the amount of sugar present at the start, ethanol fermentation ends either when all the sugar has been used up or when the ethanol concentration becomes toxic to the yeast (about 15% by volume).

As well as brewing drinks, ethanol fermentation can also be used to produce bioethanol, a renewable energy source. Any plant matter can be utilized as a feedstock, but most bioethanol is produced from sugar cane and corn (maize), using yeast. Sugars are converted into ethanol in large fermenters. The ethanol is purified by distillation. Bioethanol is used as a fuel in vehicles, sometimes in a pure state and sometimes mixed with gasoline (petrol).

C1.2.11 Oxidation and decarboxylation of pyruvate as a link reaction in aerobic cell respiration

If oxygen is available, pyruvate can be oxidized to carbon dioxide and water. This gives a much higher yield of ATP than anaerobic cell respiration. Most of the reactions are part of the Krebs cycle, but an initial reaction is conversion of pyruvate from glycolysis into a two-carbon acetyl group. This conversion forms a link between glycolysis and the Krebs cycle, so is referred to as the link reaction.

In the link reaction a complex of three enzymes carries out these processes:

- decarboxylation by removal of carbon dioxide, to change three-carbon pyruvate into a two-carbon molecule
- oxidation by removal of two electrons; these electrons are accepted by NAD, converting it to reduced NAD

- binding of the acetyl group (produced by the previous two process) to a complex carrier molecule called coenzyme A. The product is acetyl coenzyme A.

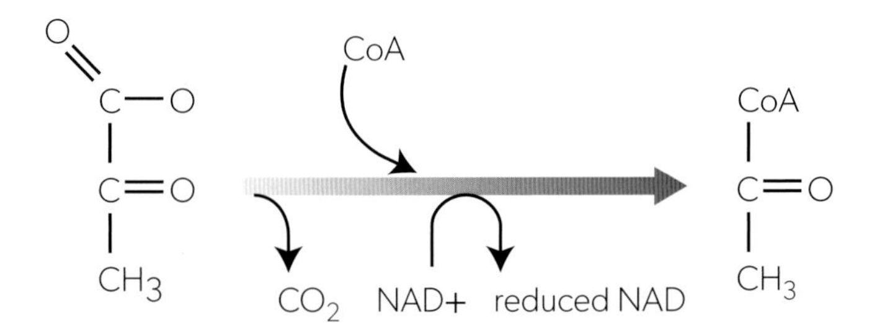

▲ Figure 21 The link reaction

Pyruvate is produced by glycolysis in the cytoplasm, but both the link reaction and the Krebs cycle take place in the matrix of the mitochondrion. A transporter protein in the outer membrane of the mitochondrion moves pyruvate from the cytoplasm into the mitochondrial matrix.

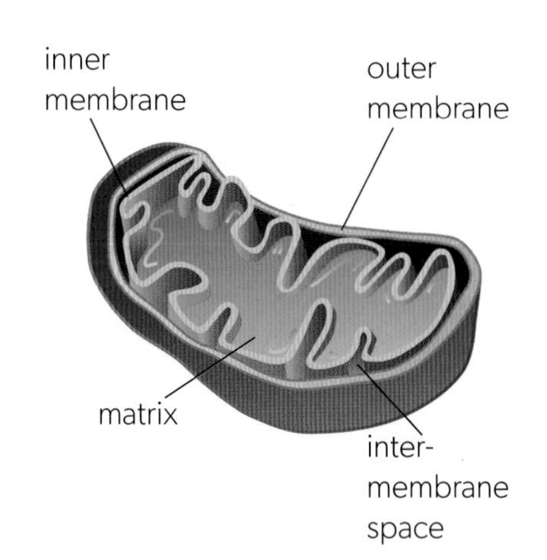

▲ Figure 22 Membranes and compartments of a mitochondrion

C1.2.12 Oxidation and decarboxylation of acetyl groups in the Krebs cycle with a yield of ATP and reduced NAD

Acetyl groups produced by the link reaction are oxidized in a cycle of reactions that happens in the matrix of the mitochondrion. This cycle has several names but is commonly called the "Krebs cycle" in honour of the biochemist who was awarded a Nobel Prize for its discovery. Acetyl groups are fed into the cycle by transfer from coenzyme A to oxaloacetate, producing the organic acid citrate.

Oxaloacetate has four carbon atoms and citrate has six. Citrate is converted back into oxaloacetate by a series of enzyme-catalysed reactions. The number of carbon atoms is decreased by two decarboxylation reactions, in which carbon and oxygen are removed, producing carbon dioxide. In aerobic cell respiration, all the carbon in substrates such as sugar or fat is removed by decarboxylation in the Krebs cycle or the link reaction. Carbon dioxide is a waste product in most cells and is excreted.

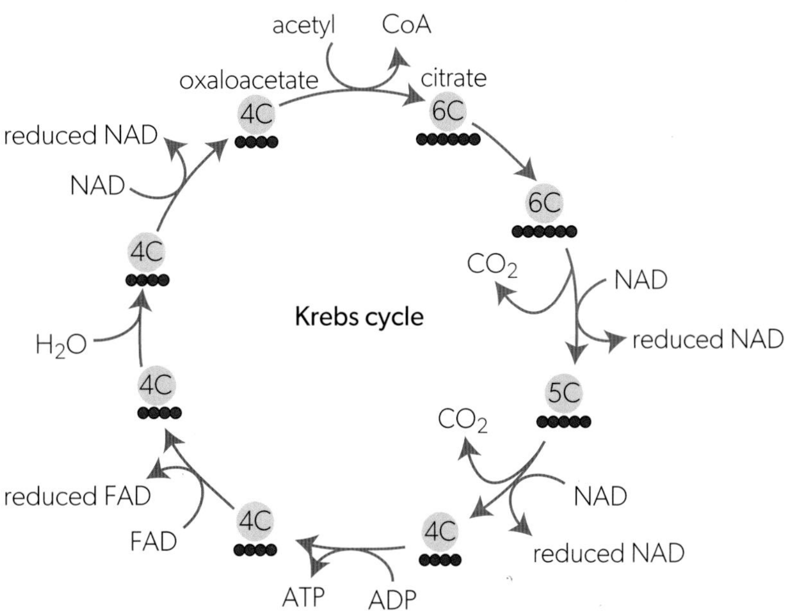

▲ Figure 23 The Krebs cycle

Four reactions in the Krebs cycle are oxidations and release energy. Much of the released energy is held by the electrons that are removed in the oxidations. These electrons are transferred either to NAD or to FAD. Both these molecules act as carriers of electrons; they also accept protons, so they are hydrogen carriers as well. NAD was described earlier in this topic. FAD functions in a similar way. When FAD or NAD accept a pair of electrons, they become reduced. Reduced NAD and reduced FAD transfer electrons and the energy they are holding to the electron transport chain in the inner mitochondrial membrane.

The net effects of one turn of the Krebs cycle are:

- one acetyl group is consumed
- three NADs are converted to reduced NAD and one FAD to reduced FAD
- two molecules of carbon dioxide are released
- one ADP is converted to ATP.

Activity: Turns of the cycle

Discuss the answers to these questions with your classmates:

1. Two carbon dioxide molecules are released for each acetyl group fed into the Krebs cycle. How many carbon dioxide molecules are produced by the cycle for each glucose?
2. Linoleic acid is the commonest fatty acid in sunflower oil. It has 18 carbon atoms that are catabolized into 9 two-carbon acetyl groups during aerobic respiration. How many molecules of carbon dioxide are released by the Krebs cycle when a molecule of linoleic acid is used in aerobic respiration?
3. The maze diagram in Figure 24 represents the route of carbon atoms through the Krebs cycle. Can you explain it?

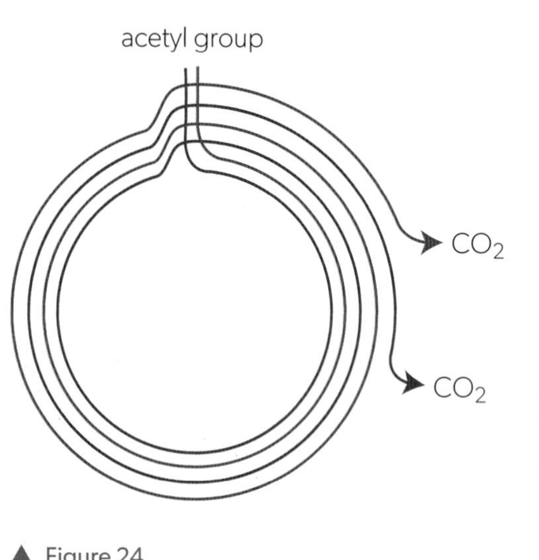

▲ Figure 24

C1.2.13 Transfer of energy by reduced NAD to the electron transport chain in the mitochondrion

In the inner mitochondrial membrane, there are groups of proteins that act as electron carriers by accepting and then passing on pairs of electrons (see Figure 25). Together, this sequence of carriers forms the electron transport chain (ETC). The first carrier in the chain accepts a pair of electrons from reduced NAD. This changes the carrier from an oxidized state to a reduced state and converts the reduced NAD back to NAD. The carrier gains chemical energy by this transfer of electrons.

Reduced NAD is produced in glycolysis, the link reaction and the Krebs cycle. Oxidation reactions in these processes are the source of the energy that is transferred by reduced NAD to the ETC. The Krebs cycle also produces reduced FAD, which transfers a pair of electrons to the ETC. The electrons transferred by reduced FAD carry less energy than those from reduced NAD, so they are accepted by a carrier part way along the chain which has a higher affinity for electrons than the first carrier.

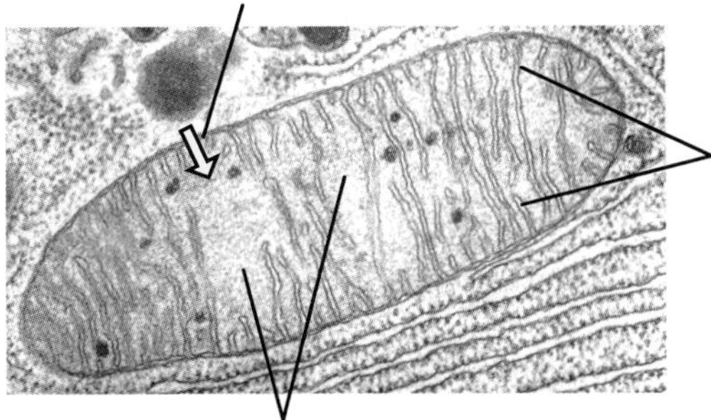
Reduced NAD from glycolysis enters the mitochondrion.
Reduced NAD transfers electrons carrying energy to electron transport chains in the inner mitochondrial membrane. Infoldings of this membrane called cristae increase the area of membrane and the number of ETCs it can accomodate.
Reduced NAD is produced in the matrix of the mitochondrion by the link reaction and Krebs cycle. It only has a short distance to travel to reach the inner mitochondrial membrane.

▲ Figure 25

C1.2.14 Generation of a proton gradient by flow of electrons along the electron transport chain

Electrons brought to the inner mitochondrial membrane by reduced NAD are accepted by the first carrier in the electron transport chain. They then pass along the chain from carrier to carrier. Energy is released at every stage in this flow of electrons.

The three main carriers in the electron transport chain each act as proton pumps. They use energy released by the flow of electrons to pump protons across the inner mitochondrial membrane, from the matrix to the intermembrane space between the inner and outer mitochondrial membranes. The first and second main electron carriers each pump four protons per pair of electrons; the third carrier pumps two. This gives a total of 10 protons pumped from the matrix to the intermembrane space per pair of electrons from reduced NAD.

Electrons brought by reduced FAD also fuel proton pumping. However, these electrons are fed into the chain after the first carrier so only 6 protons are pumped per pair of electrons, rather than 10.

Energy is needed to pump protons across the inner mitochondrial membrane because they are being moved against the concentration gradient. This energy is not lost—it is stored temporarily in the form of the proton gradient. This stored energy can then be used to generate ATP.

In summary, the role of the electron transport chain is to generate and maintain a proton gradient across the inner mitochondrial membrane. It does this by pumping protons across the membrane using energy released by the flow of electrons. The electrons from which this energy is obtained are brought to the ETC by reduced NAD and reduced FAD.

▲ Figure 26 Energy is required to pump water up to the tank at the top of a water tower. What are the similarities and differences between this and proton pumping in the mitochondrion?

C1.2.15 Chemiosmosis and the synthesis of ATP in the mitochondrion

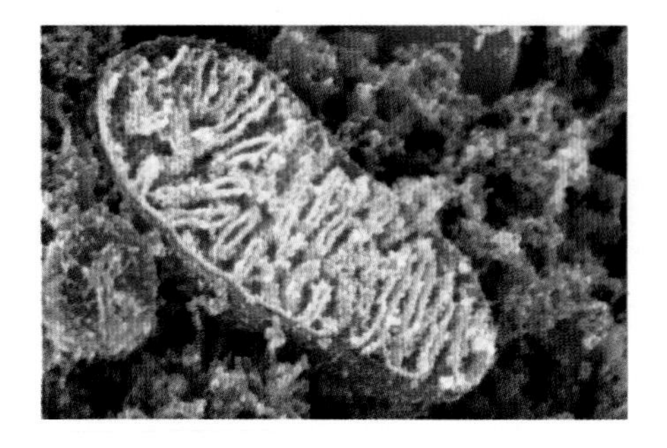

▶ Figure 27 Many cristae are visible in this electron micrograph of a mitochondrion. Cristae are infoldings of the inner mitochondrial membrane. The small blobs visible on some of the cristae are the globular part of ATP synthase that projects from the membrane (Magnification ×73,000)

ATP synthase is a large and complex protein that phosphorylates ADP to produce ATP. This is an endergonic (energy absorbing) reaction so a source of energy is needed. This energy is provided by the proton gradient created by the electron transport chain. The process used to couple the proton gradient to synthesis of ATP is called chemiosmosis.

In osmosis, a concentration gradient causes water to move across a membrane, but the energy released by this process is not utilized. In chemiosmosis, protons move down their concentration gradient from the high concentration in the intermembrane space to the lower concentration in the matrix. The energy released is used to link a phosphate group to ADP, producing ATP.

ATP synthase has two main regions. One of these is made of transmembrane subunits that are embedded in the inner mitochondrial membrane. This region allows protons to pass across the membrane, releasing energy. The other main region is globular and projects into the matrix. It has active sites that use energy released by the protons to catalyse production of ATP.

ATL Communication skills: Creating biological diagrams

Examine the electron micrograph of a mitochondrion in Figure 27. Create your own drawing of the mitochondrion. Consider the following guidelines regarding biological drawings.

- Leave no gaps in the lines when drawing closed shapes.
- Include clear and unambiguous labels.
- Ensure the size of the diagram is proportional to its complexity. If your drawing is too small, it will be difficult to distinguish detail or to identify what a label is pointing to.
- Draw structures in proportion to one another—for example, do not draw the mitochondrial cristae too thickly, the ribosomes too large, or the intermembrane space too wide.
- Ensure every line has meaning. For example, do not draw a line across the base of the cristae—this would suggest that they are partitioned off from other parts of the intermembrane space.

Activity: Evaluating and annotating a diagram of a mitochondrion

Consider this student drawing of a mitochondrion.

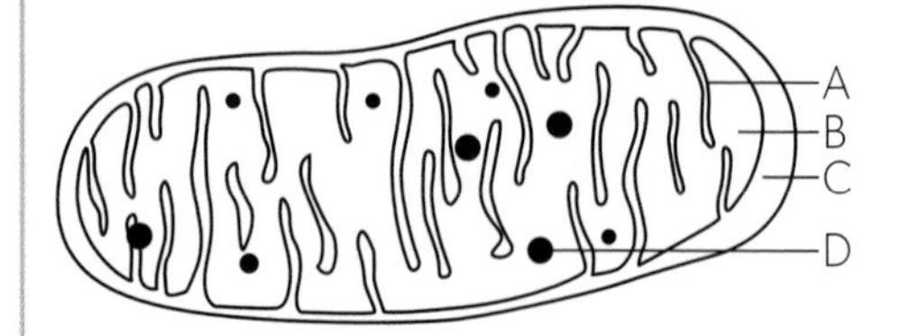

▲ Figure 28

1. What feedback would you offer to the student regarding their drawing?
2. Look at the structures labelled A, B, C and D. State:
   a. the function of structure A
   b. the location where protons build up as a result of the electron transport chain
   c. the location of the enzyme ATP synthase
   d. which letter indicates the matrix
   e. which letter indicates the cristae
   f. which letter indicates the intermembrane space.

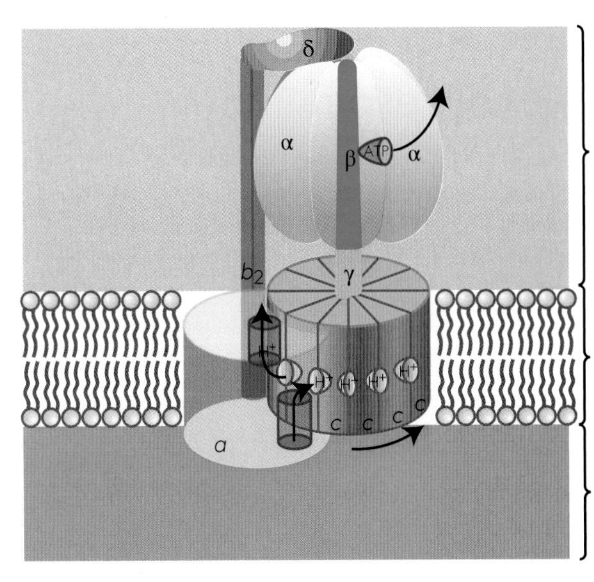

Figure 29 The structure of ATP synthase

Chemiosmosis—the mechanism used by ATP synthase to make ATP—has only been understood in recent years. Some details are given here, using letters to denote the parts of ATP synthase.

- The drum-shaped part of ATP synthase located in the membrane consists of identical subunits (c), each of which has a binding site for a proton. Next to this is another structure in the membrane (a) with two half-channels for protons. One of these half-channels allows protons from the intermembrane space to enter and bind to a subunit of the drum. The other half-channel allows protons that were bound to a subunit to exit to the matrix. These two half-channels are not aligned so, for protons to pass through, the drum has to rotate. The energy released by movement of protons down their concentration gradient is transformed into kinetic energy. Each proton is carried by the drum for almost a full rotation before it is released.
- The drum is tightly connected to a stalk (γ) that projects into the matrix. Because of the tight connection, rotation of the drum causes the stalk to rotate at the same rate. The drum and stalk together are known as the rotor of ATP synthase.
- The stalk is surrounded by the globular part of ATP synthase, which consists of a ring of alternating subunits (α and β). Each of the β subunits has an active site for catalysing the phosphorylation of ADP to ATP. When the stalk inside the ring of α and β subunits rotates, it causes conformational changes to the β subunits. These changes first cause ADP and a phosphate group to bind to the active site; then they cause the phosphate to link to the ADP to produce ATP; and finally they cause the ATP to be released into the matrix. This is how rotation of the stalk provides energy for ATP production.
- Rotation of the ring of α and β subunits is prevented by a structure adjacent to it called the rotor arm. This consists of two parts (b_2 and δ) linked to the proton channel (a).

With an understanding of how ATP synthase works, we can predict rates of ATP production per reduced NAD and per reduced FAD:

- Three β subunits in ATP synthase each make one ATP per turn of the rotor.
- One proton must pass through each of the 10 c subunits in the rotor of ATP synthase to cause one rotation.
- Each pair of electrons from reduced NAD that passes along the ETC results in ten protons pumped into the intermembrane space, and each pair of electrons from reduced FAD results in pumping of six protons.
- Production of ATP is therefore 2.5 per reduced NAD and 1.5 per reduced FAD.

AHL

▲ Figure 30 Cryo-electron microscopy allows generation of detailed images of ATP synthase

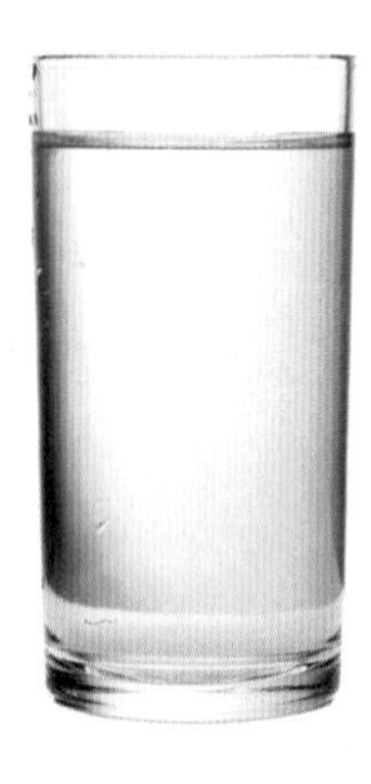

▲ Figure 31 Mitochondria in humans produce about half a litre of water per day as oxygen combines with electrons from the ETC and protons from the matrix

C1.2.16 Role of oxygen as terminal electron acceptor in aerobic cell respiration

ATP production by mitochondria can only continue when there is electron flow and proton pumping. This depends on reduced NAD supplying pairs of electrons to the start of the electron transport chain and the electrons being removed at the end of the chain. Each electron carrier in the ETC has a stronger affinity for electrons than the previous one, so removal of electrons from the last electron carrier can only be done by a substance that has a very strong affinity for electrons. Most organisms use molecular oxygen for this purpose. It is known as the terminal electron acceptor. Molecules of oxygen accept electrons from the final electron carrier and hydrogen ions from the matrix, producing water.

Use of oxygen is the last stage in aerobic cell respiration. However, all the previous stages apart from glycolysis depend on oxygen. If oxygen runs out, electrons are not removed from the end of the ETC, so all the carriers in the ETC become reduced. This means electrons cannot be accepted from reduced NAD at the start of the ETC. Reduced NAD therefore accumulates. When all the NAD in the mitochondrion has been converted to reduced NAD, oxidations in the link reaction and Krebs cycle are impossible so these processes stop. Anaerobic cell respiration can continue in the cytoplasm by regenerating NAD without using oxygen. However, the yield of ATP is far smaller—only 2 ATP per glucose, compared with 32 from aerobic cell respiration.

Activity: Interpreting a diagram

Examine the diagram of interactions in the inner mitochondrial membrane (Figure 32). Deduce the effect of the production of water on the concentration gradient across the membrane.

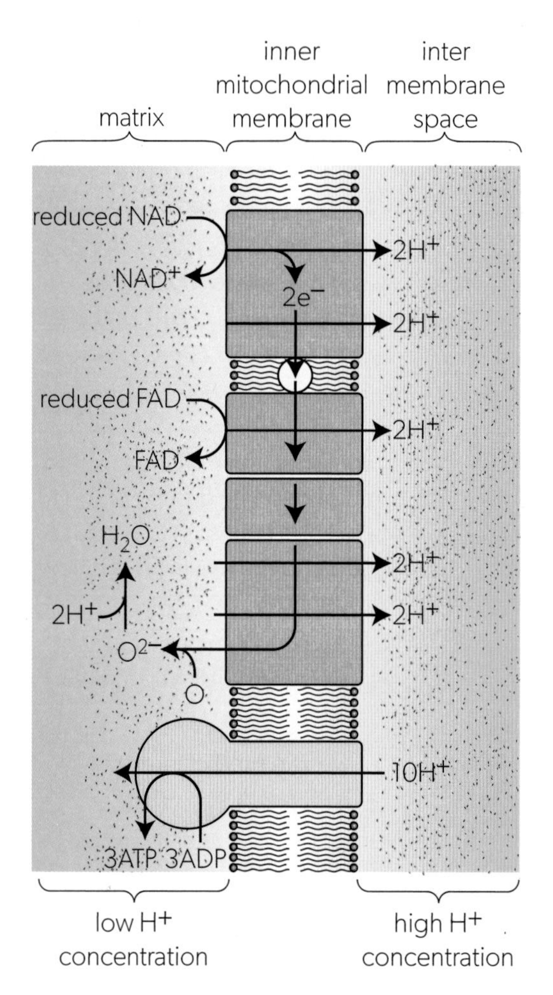

◀ Figure 32 Interactions in the inner mitochondrial membrane

Data-based questions: Water production in kangaroo rats

Kangaroo rats are small mammals that live in arid habitats where no water is available to drink and the foods eaten are mainly dry seeds. Table 3 gives experimental data for the kangaroo rat *Dipodomys spectabilis*, which lives in Mexico and the south west of the US. All the values are shown per gram of food consumed (g^{-1}).

Food type eaten	Metabolic water production / g water g^{-1} food	Oxygen used in respiration / $cm^3 g^{-1}$	Dry air (20% relative humidity)		Humid air (66% relative humidity)	
			Evaporative water loss / $g g^{-1}$	Net water yield / $g g^{-1}$	Evaporative water loss / $g g^{-1}$	Net water yield / $g g^{-1}$
starch	0.556	828	0.472	0.084	0.270	0.286
sugar	0.600	746	0.425	0.175	0.243	0.357
lipid	1.071	2019	1.151	−0.080	0.658	0.413
protein	0.396	967	0.551	−1.613	0.315	−1.377

Source of data: Frank, C. 1988. "Diet Selection by a Heteromyid Rodent: Role of Net Metabolic Water Production". *Ecology*. Vol. 69. Pp 1943–1951

▲ Table 3

1. a. Distinguish between metabolic water production when the substrate for cell respiration is carbohydrate or lipid. [2]

   b. Explain the difference in metabolic water production. [2]

2. a. Distinguish between the amount of oxygen used when the substrate for cell respiration is carbohydrate or lipid. [2]

   b. Explain the difference in the amount of oxygen used. [2]

3. The kangaroo rats lost water by evaporation at different rates in dry and humid air. Explain how air humidity affects rates of evaporative water loss. [2]

4. Evaluate the food types, using the data in Table 3. [3]

5. The kangaroo rats are nocturnal. Suggest advantages of this, using the data in Table 3. [2]

▲ Figure 33 Kangaroo rat

C1.2.17 Differences between lipids and carbohydrates as respiratory substrates

Both lipids and carbohydrates can be used as substrates in respiration. This is why they are both suitable as energy-yielding foods in the diet and also as energy stores in the body. Both lipids and carbohydrates are oxidized in respiration to release energy. However, there are differences in the metabolic pathways and in the energy yield per gram. These are shown in Table 4.

◀ **Figure 34** A traditional doughnut has a mass of 46 grams and contains 11.0 grams of fat and 18.6 grams of carbohydrate. Does the fat or the carbohydrate provide more energy?

	Carbohydrates	Lipids
Is anaerobic respiration possible?	The first stage of respiration with sugars such as glucose and fructose as the substrate is glycolysis, which generates some ATP and does not require oxygen. Anaerobic respiration is therefore possible. Pyruvate can be converted to acetyl groups by the link reaction and the acetyl groups can then be fed into the Krebs cycle. These stages can only happen if oxygen is available.	The first stage of respiration with lipids such as fats and oils is the breakdown of fatty acids to acetyl groups in the matrix of the mitochondrion. The acetyl groups are then fed into the Krebs cycle. These stages only happen when oxygen is available, so anaerobic respiration is not possible with lipids.
What is the energy yield?	The energy yield per gram of carbohydrate is only 17 kilojoules per gram. This is about half that from lipids. Energy is released from a substrate by oxidizing carbon and hydrogen and in carbohydrates more than 50% of the mass is oxygen, which does not yield energy.	The energy yield per gram of lipids is 37 kilojoules per gram. This is nearly twice as much as that from carbohydrates. Nearly 90% of the mass of lipids is carbon and hydrogen, from which there is a yield of energy in respiration.

▲ Table 4

Linking questions

1. In what forms is energy stored in living organisms?
    a. Explain the concepts of primary and secondary production. (C4.2.15)
    b. Outline the use of polysaccharides as energy storage molecules in animals and plants. (B1.1.5)
    c. Describe the production of ATP by chemiosmosis in the mitochondrion. (C1.2.15)
2. What are the consequences of respiration for ecosystems?
    a. Outline how carbon dioxide enrichment experiments have been used to predict future rates of photosynthesis and plant growth. (C1.3.8)
    b. Explain how the process of cellular respiration sets limits to the length of food chains. (C4.2.12)
    c. Describe an experiment to measure a variable affecting the rate of production of CO_2 as a result of cellular respiration. (C1.2.6)

Data-based questions: Cellular respiration in shaker muscles

The end of the tail of a rattlesnake consists of a series of hollow, interlocked segments made of keratin, which are created by modifying the scales that cover the tip of the tail. The rattle serves as a warning to predators. The contraction of special "shaker" muscles in the tail causes these segments to vibrate against one another, resulting in the rattling noise. Table 5 shows the results of continuous stimulation of the tailshaker muscle of eight western diamond rattlesnakes *(Crotalus atrox)*.

▲ Figure 35 Western diamond rattlesnake (*Crotalus atrox*)

	O_2 content in arteries / mmol dm^{-3}	Lactate content in arteries / mmol dm^{-3}
At rest	2.4 ± 0.5	2.8 ± 1.2
Rattling	2.8 ± 0.1	4.8 ± 0.8

▲ Table 5

The graph in Figure 36 shows ATP demand and sources of ATP supply in the tailshaker muscle.

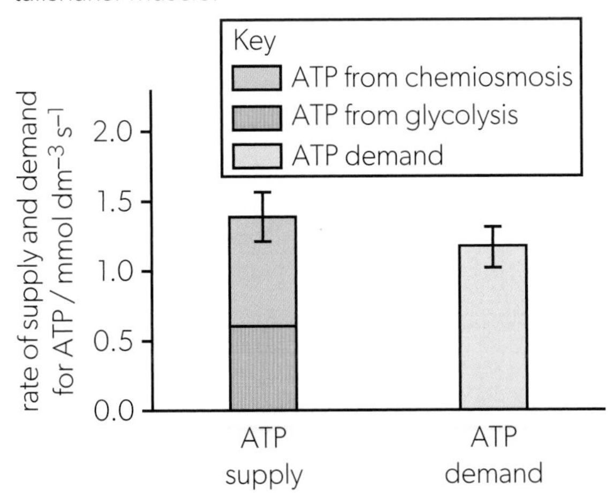

Source: Kemper, W.F. et al. (2000) "Shaking up glycolysis: Sustained, high lactate flux during aerobic rattling," Proceedings of the National Academy of Sciences, 98(2), pp. 723–728. Available at: https://doi.org/10.1073/pnas.98.2.723

▲ Figure 36

1. Using the bar chart, determine the amount of ATP produced by chemiosmosis, giving the units. [1]
2. Compare the changes in oxygen and lactate content in the blood when a resting rattlesnake starts rattling. [2]
3. Using the data, deduce, with reasons, whether anaerobic respiration provides some or all of the ATP used in rattling. [3]
4. Comment on the variability of the data. [2]

C1.3 Photosynthesis

How is energy from sunlight absorbed and used in photosynthesis?

Light energy is absorbed by pigments that transform and transfer the energy to a complex system of molecules known as photosystems. The molecular model shown represents photosystem II. These are collections of proteins found on the thylakoid membranes of cyanobacteria, algae and plants. They use the energy from light to reduce other molecules. What is the mechanism by which they achieve this? The final products of the activities of the photosystems is ATP and NADP. How are these used to produce carbohydrates?

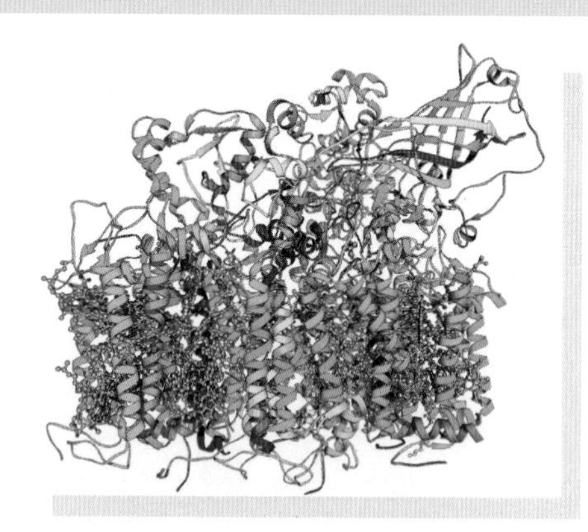

◀ Figure 1 A molecular model of photosystem, which absorbs light and uses the energy to drive the oxidation of water, creating oxygen as a by-product

How do abiotic factors interact with photosynthesis?

Abiotic factors that influence the rate of photosynthesis include: the availability of raw materials such as carbon dioxide; the amount of light; the availability of mineral nutrients; and climate factors such as temperature and precipitation. Antarctica's freezing temperatures, poor soil quality, lack of sunlight and frozen water restrict rates of photosynthesis. Large plants are therefore absent. In areas where the snow melts during the short summer, only algae, mosses, lichens and fungi grow.

▲ Figure 2

SL and HL	AHL only
C1.3.1 Transformation of light energy to chemical energy when carbon compounds are produced in photosynthesis	C1.3.9 Photosystems as arrays of pigment molecules that can generate and emit excited electrons
C1.3.2 Conversion of carbon dioxide to glucose in photosynthesis using hydrogen obtained by splitting water	C1.3.10 Advantages of the structured array of different types of pigment molecules in a photosystem
C1.3.3 Oxygen as a by-product of photosynthesis in plants, algae and cyanobacteria	C1.3.11 Generation of oxygen by the photolysis of water in photosystem II
C1.3.4 Separation and identification of photosynthetic pigments by chromatography	C1.3.12 ATP production by chemiosmosis in thylakoids
C1.3.5 Absorption of specific wavelengths of light by photosynthetic pigments	C1.3.13 Reduction of NADP by photosystem I
C1.3.6 Similarities and differences of absorption and action spectra	C1.3.14 Thylakoids as systems for performing the light-dependent reactions of photosynthesis
C1.3.7 Techniques for varying concentrations of carbon dioxide, light intensity or temperature experimentally to investigate the effects of limiting factors on the rate of photosynthesis	C1.3.15 Carbon fixation by Rubisco
C1.3.8 Carbon dioxide enrichment experiments as a means of predicting future rates of photosynthesis and plant growth	C1.3.16 Synthesis of triose phosphate using reduced NADP and ATP
	C1.3.17 Regeneration of RuBP in the Calvin cycle using ATP
	C1.3.18 Synthesis of carbohydrates, amino acids and other carbon compounds using the products of the Calvin cycle and mineral nutrients
	C1.3.19 Interdependence of the light-dependent and light-independent reactions

C1.3.1 Transformation of light energy to chemical energy when carbon compounds are produced in photosynthesis

Living organisms require complex carbon compounds to build the structure of their cells and to carry out life processes. Some organisms are able to make all the carbon compounds they need using only light energy and simple inorganic substances such as carbon dioxide and water. They do this using the process of photosynthesis.

Photosynthesis is an energy conversion, as light energy is converted into chemical energy in carbon compounds. The main groups of carbon compounds produced are carbohydrates, proteins, lipids and nucleic acids. This transformation supplies most of the chemical energy needed for life processes in ecosystems.

▲ Figure 3 The trees in one hectare of redwood forest in California can have a biomass of more than 4,000 tonnes. Most of this mass is carbon compounds produced by photosynthesis

C1.3.2 Conversion of carbon dioxide to glucose in photosynthesis using hydrogen obtained by splitting water

Plants convert carbon dioxide and water into carbohydrates by photosynthesis. The simple equation below summarizes the process, with glucose as the carbohydrate produced:

carbon dioxide + water → glucose + oxygen

Hydrogen is needed for the reduction reaction that converts carbon dioxide into glucose. This hydrogen comes from photolysis (a reaction that splits molecules of water). This reaction only happens when light is available to provide energy: "photo" means light and "lysis" means to make loose. Hydrogen is released from water as separated protons (hydrogen ions) and electrons.

$$2H_2O \rightarrow 4e^- + 4H^+ + O_2$$

Oxygen is a waste product of this reaction. It diffuses out of photosynthesizing cells.

▲ Figure 4 Leaves absorb carbon dioxide and light and use them in photosynthesis

C1.3.3 Oxygen as a by-product of photosynthesis in plants, algae and cyanobacteria

Oxygen is a by-product of photosynthesis, usually a waste product. It comes from the splitting of water (photolysis). Prokaryotes were the first organisms to perform photosynthesis, starting about 3,500 million years ago. Millions of years later, algae and plants also began carrying out photosynthesis using chloroplasts.

Photolysis increases the concentration of oxygen inside chloroplasts. This causes oxygen to diffuse out of chloroplasts and then out of leaf cells to air spaces inside the leaf. The oxygen then diffuses through stomata to the air outside the leaf.

▲ Figure 5 Oxygen produced by photosynthesis diffuses out through the stomata of leaves. This cannot be seen with leaves in the air but bubbles are sometimes visible on leaves of underwater plants when the water is saturated with oxygen

Water, water everywhere

All the oxygen that diffuses into the air from plant leaves, or emerges as bubbles from aquatic plants, comes from photolysis. For each glucose molecule that is produced, 12 water molecules are split giving 24 hydrogen ions. Figure 6 shows why so much hydrogen is needed to produce one molecule of glucose.

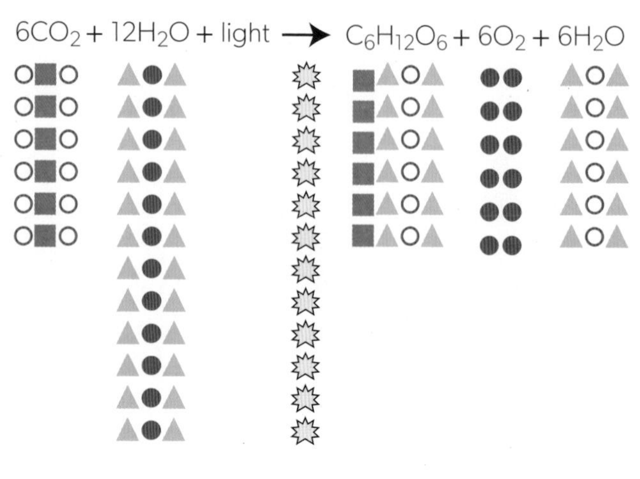

▲ Figure 6

C1.3.4 Experimental techniques: Separation and identification of photosynthetic pigments by chromatography

▲ Figure 7 Thin layer chromatography (TLC)

Chloroplasts contain several types of chlorophyll, along with other pigments called accessory pigments. These pigments absorb different ranges of wavelength of light, so they look different colours to us. Pigments can be separated by chromatography. You may be familiar with paper chromatography but thin layer chromatography gives better results. This is done with a plastic strip that has been coated with a thin layer of a porous material. A spot containing pigments extracted from leaf tissue is placed near one end of the strip. A solvent is allowed to run up the strip, to separate the different types of pigment. (At the end of the procedure, methods used for disposal of solvents must comply with regulations.)

1. Tear up a leaf into small pieces and put them in a mortar.
2. Add a small amount of sand for grinding.
3. Add a small volume of propanone (acetone).
4. Use the pestle to grind the leaf tissue and dissolve out the pigments.
5. If all the propanone evaporates, add a little more.
6. When the propanone has turned dark green, allow the sand and other solids to settle. Then pour off the propanone into a watch glass.
7. Use a hairdryer to evaporate off the propanone and the water from the cells' cytoplasm.
8. When you have just a smear of dry pigments in the watch glass, add 3–4 drops of propanone and use a paint brush to dissolve the pigments.
9. Use the paint brush to transfer a very small amount of the pigment solution to the TLC strip. Your aim is to make a very small spot of pigment in the middle of the strip, 10 millimetres from one end. This spot should be very dark. To achieve this, put a small drop onto the strip and allow it to dry before adding another drop. Repeat as many times as necessary. You can speed up the drying process by blowing on the spot or using the hairdryer.
10. When the spot is dark enough, slide the other end of the strip into the slot in a cork or bung that fits into a tube that is wider than the TLC strip. The slot should hold the strip firmly.
11. Insert the cork and strip into a specimen tube. The TLC strip should extend nearly to the bottom of the tube, without touching it.
12. Mark the outside of the tube just below the level of the spot on the TLC strip.
13. Take the strip and cork out of the tube.
14. In a well-ventilated room (or in a fume cupboard), pour running solvent into the specimen tube up to the level that you marked.
15. Place the specimen tube on a lab bench where it will not be disturbed. Carefully lower the TLC strip and cork into the tube, so the tube is sealed and the TLC strip is just dipping into the running solvent. The solvent must NOT touch the pigment spot.
16. Leave the tube alone for about five minutes, to allow the solvent to run up through the TLC strip. You can watch the pigments separate but DO NOT TOUCH THE TUBE.
17. When the solvent has nearly reached the top of the strip, remove the strip from the tube and separate it from the cork.
18. Rule two pencil lines across the strip, one at the level reached by the solvent and one at the level of the initial pigment spot.
19. Draw a circle around each of the separated pigment spots. Then mark a cross in the centre of each circle.
20. Use a ruler with millimetre markings to measure the distance moved by the running solvent (the distance between the two lines) and the distance moved by each pigment (the distance between the lower line and the cross in the centre of the circle).

Pigment	Colour of pigment	R_f
Carotene	orange	0.98
Chlorophyll *a*	blue green	0.59
Chlorophyll *b*	yellow green	0.42
Phaeophytin	olive green	0.81
Xanthophyll 1	yellow	0.28
Xanthophyll 2	yellow	0.15

▲ Table 1 Table of standard R_f values

21. Calculate the R_f value for each pigment, using the equation:

$$R_f = \frac{\text{distance run moved by pigment}}{\text{distance run moved by solvent}}$$

22. Show all your results in a copy of Table 2, starting with the pigment that moved least far.

Spot number	Colour	Distance moved / mm	R_f	Name of pigment
1				
2				
3				
4				
5				
6				
7				
8				

▲ Table 2

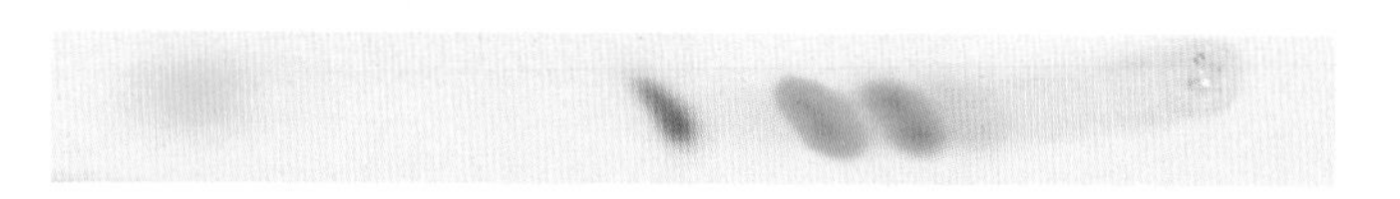

▲ Figure 8 A chromatogram of leaf pigments

Data-based questions: Determining R_f values for photosynthetic pigments

1. Use the data in Table 1 to sketch a coloured chromatogram. Include the location of the solvent front and the origin in your diagram. [5]

2. Thin layer chromatography was used to separate photosynthetic pigments from three eukaryotic organisms: spinach (*Spinacia oleracea*), a red alga (*Porphyra*) and a brown alga (*Fucus*). Figure 9 shows diagrams of the resulting chromatograms.

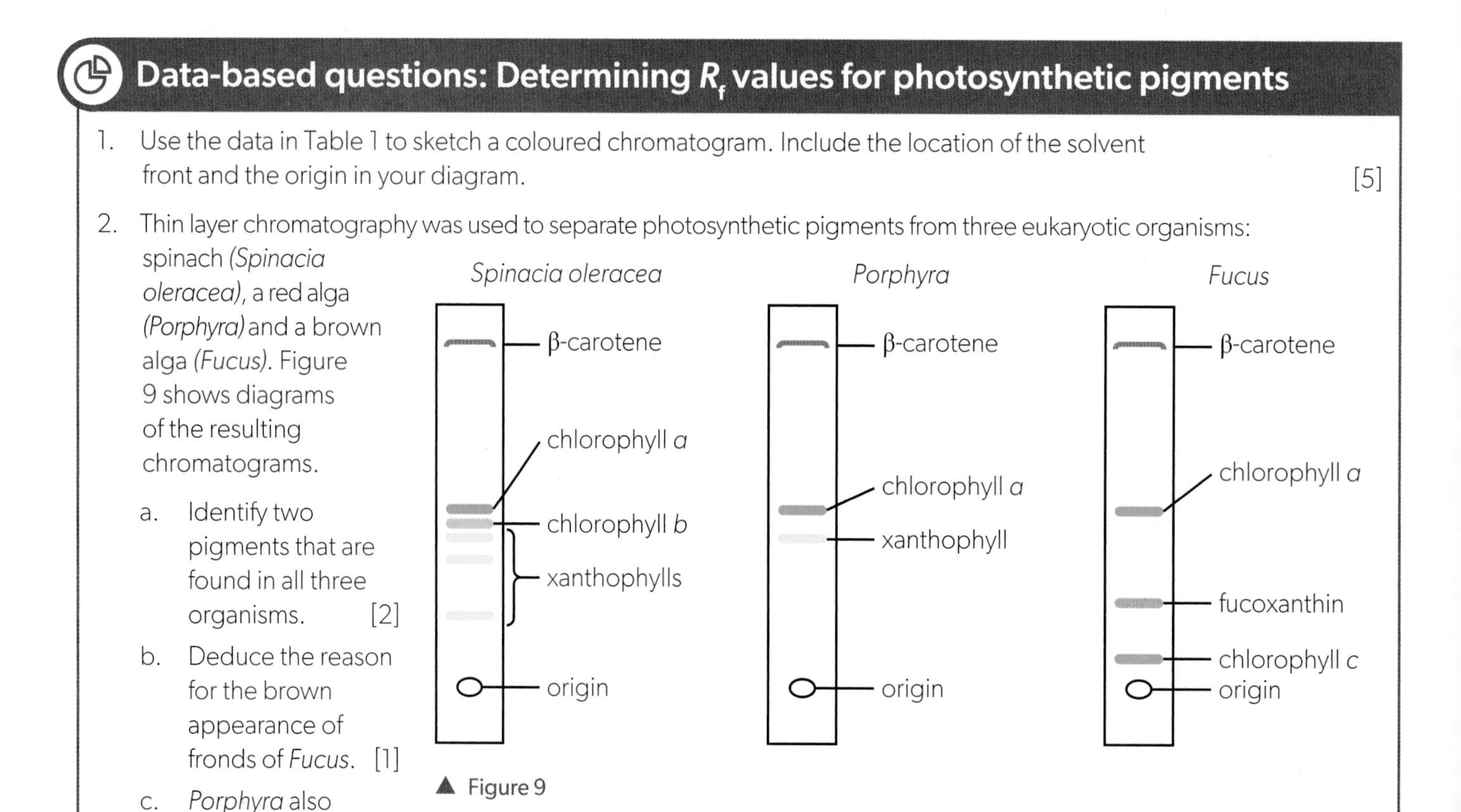

▲ Figure 9

a. Identify two pigments that are found in all three organisms. [2]

b. Deduce the reason for the brown appearance of fronds of *Fucus*. [1]

c. *Porphyra* also contains phycoerythrin, which is a red pigment.

i. Suggest why phycoerythrin is absent from the *Porphyra* chromatogram. [1]

ii. Predict one colour of light that will be absorbed efficiently by phycoerythrin. [1]

C1.3.5 Absorption of specific wavelengths of light by photosynthetic pigments

The first stage in photosynthesis is the absorption of sunlight. This involves chemical substances called pigments. Pigments absorb light and so appear different colours to us: the colours we see depend on the wavelengths (colours) of light the pigment absorbs and transmits.

- White and transparent substances are not pigments. White substances reflect all wavelengths of visible light, while transparent substances allow all wavelengths to pass through.
- Pigments that absorb all wavelengths of light appear black. (These pigments transform the light energy into other forms of energy, mostly heat.)
- Other pigments absorb some wavelengths of visible light but not others. For example, the pigment in a gentian flower absorbs all colours except blue. The flower appears blue to us, because this part of the sunlight is reflected and can pass into our eye, to be detected by cells in the retina.

▲ Figure 10 Gentian flowers contain the pigment delphinidin, which reflects blue light and absorbs all other wavelengths. The surrounding leaves absorb red and blue light, so they appear green

A photon is a particle or unit of light. Photons are discrete quantities of energy. The energy of a photon is related to its wavelength: the longer the wavelength, the less energy a photon holds. Photons are absorbed by pigment molecules if the energy they hold causes an electron in an atom of the pigment molecule to jump to a higher energy level (excitation). A specific amount of energy is required for this to happen and this energy is only supplied by certain wavelengths of light.

Photosynthesis involves a range of pigments but the main photosynthetic pigments are chlorophylls. All forms of chlorophyll appear a shade of green to us. This is because photons in the red and blue parts of the spectrum can excite an electron in chlorophyll, but wavelengths in the green parts of the spectrum (between red and blue) cannot. Therefore, most green light is reflected. This is why green is the dominant colour in ecosystems dominated by plants.

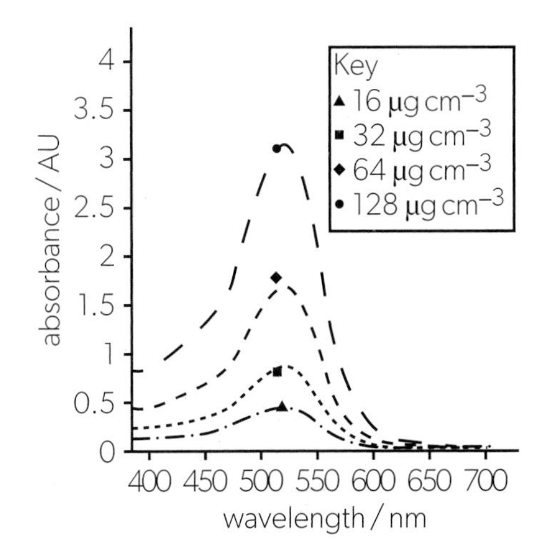

▲ Figure 11 Absorption spectra for pigments extracted from mulberry fruits at concentrations from 16 to 128 μg of pigment per cubic centimetre of solvent. The spectra use arbitrary units (AU) for absorbance, which show relative amounts of light absorption at the different wavelengths rather than percentage or absorbance measured with SI units

Source: Qin C., Li Y., Niu W., Ding Y., Zhang R., Shang X. (2010) Czech J. Food Sci., 28: 117-126.

The wavelengths of light absorbed by a pigment are shown on a graph called an absorption spectrum.

- The horizontal x-axis shows wavelength of light, in nanometres. The scale extends from 400 to 700 nanometres, reflecting the range of wavelengths in visible light. It is helpful to show the colours as well.
- The y-axis shows absorption, often as a percentage.

C1.3.6 Similarities and differences of absorption and action spectra

An absorption spectrum (Figure 13) is a graph showing the percentage of light absorbed at each wavelength by a pigment or a group of pigments.

An action spectrum (Figure 14) is a graph showing the rate of photosynthesis at each wavelength of light.

▲ Figure 12 Ripe mulberry fruits *(Morus nigra)*

- When plotting both action and absorption spectra, the horizontal x-axis should show wavelength of light, in nanometres. The scale should extend from 400 nm (violet) to 700 nm (red).
- On an action spectrum, the y-axis should be used for a measure of the relative amount of photosynthesis. This is often given as a percentage of the maximum rate, with a scale from 0 to 100%.
- On an absorption spectrum, the y-axis is used for absorption of light, either with a percentage scale or with arbitrary units. The spectra for more than one pigment can be shown on the same graph.

It is not difficult to explain the similarities between the curves on action and absorption spectra: photosynthesis can only occur in wavelengths of light that are absorbed by chlorophyll or other photosynthetic pigments.

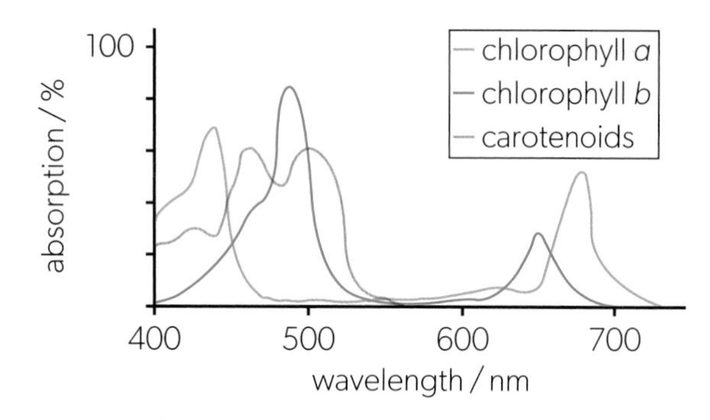

▲ Figure 13 Absorption spectra of plant pigments

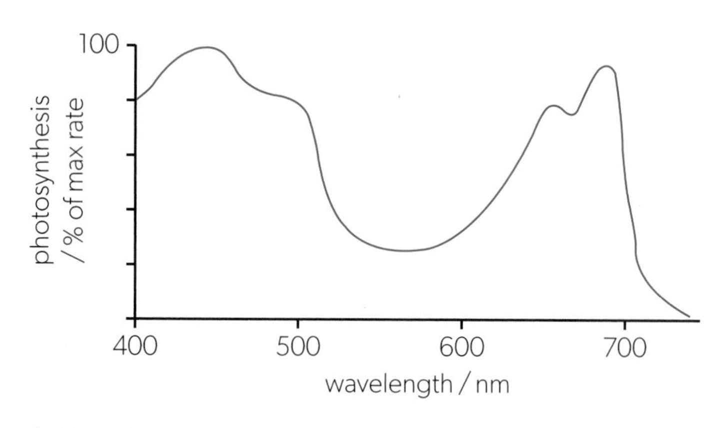

▲ Figure 14 Action spectrum for a plant pigment

Data for plotting an action spectrum for photosynthesis can be obtained experimentally, using either oxygen production or CO_2 consumption as a measure of the rate of photosynthesis. As many wavelengths as possible should be tested, so the data points can be linked by interpolation to give a curve.

C1.3.7 Experimental techniques: Techniques for varying concentrations of carbon dioxide, light intensity or temperature experimentally to investigate the effects of limiting factors on the rate of photosynthesis

Part 1: Varying carbon dioxide concentration

If a stem of pondweed such as *Elodea*, *Cabomba* or *Myriophyllum* is placed upside down in water and the end of the stem is cut, bubbles of gas may be seen to escape. If these bubbles are collected and tested, they are found to be mostly oxygen, produced by photosynthesis. The rate of oxygen production can be measured by counting the bubbles. Factors that might affect the rate of photosynthesis can be varied to find out what effect this has. In the method below, carbon dioxide concentration is the independent variable.

a. Boil enough water to fill a large beaker, then allow it to cool. This removes carbon dioxide and other dissolved gases.

b. Repeatedly pour the water from one beaker to another, to oxygenate it. Very little carbon dioxide will dissolve.

→

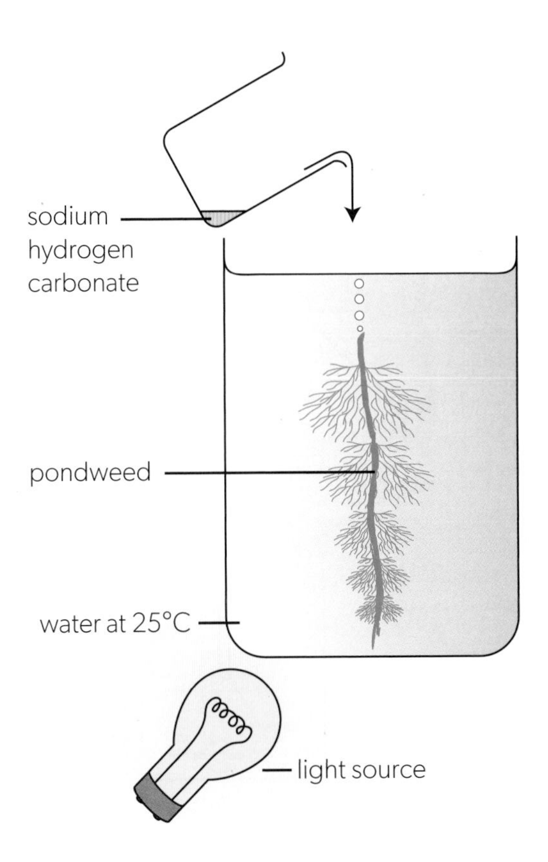

▲ Figure 15 Apparatus for measuring photosynthesis rates in different concentrations of carbon dioxide

c. Cut the end of a stem of pondweed and place it upside down in the water. The water contains almost no carbon dioxide, so no bubbles are expected to emerge. The temperature of the water should be about 25°C and the water should be very brightly illuminated. Figure 15 shows suitable apparatus.

d. Add enough sodium hydrogen carbonate to the beaker to raise the carbon dioxide concentration by 0.01 mol dm^{-3}. If bubbles emerge, count them for 30 seconds; repeating the counts until two or three consistent results are obtained.

e. Add enough sodium hydrogen carbonate to raise the concentration by another 0.01 mol dm^{-3}. Count the bubbles in the same way.

f. Repeat the procedure above until further increases in carbon dioxide do not affect the rate of bubble production. (Note that, non-native pondweeds should not be disposed of in ponds or other aquatic habitats as they may become invasive.)

Questions

1. Why are the following procedures necessary?
   - Boiling and then cooling the water before the experiment.
   - Keeping the water at 25°C and brightly illuminating it.
   - Repeating bubble counts until several consistent counts have been obtained.
2. What other factors could be investigated using bubble counts with pondweed? How would you design each experiment?
3. How could you measure the rate of oxygen production more accurately?

Part 2: Varying light intensity

Photosynthesis rates can be measured in leaf discs, cut out of young healthy leaves using a cork borer or the end of a plastic drinking straw. If the discs are placed in a suitable environment and prevented from drying out, they will continue to photosynthesize for at least a few hours.

a. Remove the barrel from a 10 cm^3 plastic syringe and cover the end of the nozzle with a finger. Then pour 10 cm^3 of 0.2 mol mm^{-3} sodium hydrogen carbonate solution into the barrel of the syringe. This will provide a supply of carbon dioxide.

b. Put 10 leaf discs into the sodium hydrogen carbonate solution. Then replace the plunger of the syringe.

c. Hold the syringe vertically with the nozzle pointing upwards. Squeeze all the air out of the syringe, then put a finger over the nozzle and pull on the barrel of the syringe. This will create suction, which will draw air out of the air spaces inside the leaf discs. Stop applying suction and take your finger off the end of the nozzle. Tap the syringe so the bubbles of gas rise into the nozzle.

d. Repeat the previous stage several times, until the leaf discs become denser than the sodium hydrogen carbonate solution and sink to the bottom.

e. Place the syringe in a vertical position at a measured distance from a light source. Make sure the discs are fully illuminated. As the discs photosynthesize, they will produce oxygen. This will cause them to become less dense than the solution, so they rise to the top of the syringe. Time how long it takes for each disc to rise to the surface.

f. Measure the light intensity using a lux meter.

g. Repeat with the syringe at different distances from the light source. The relative light intensity is $\frac{1}{\text{distance}^2}$ so a suitable range of distances might be 22 mm, 35 mm, 41 mm, 50 mm and 71 mm. Plot a graph of the results for all the light intensities tested.

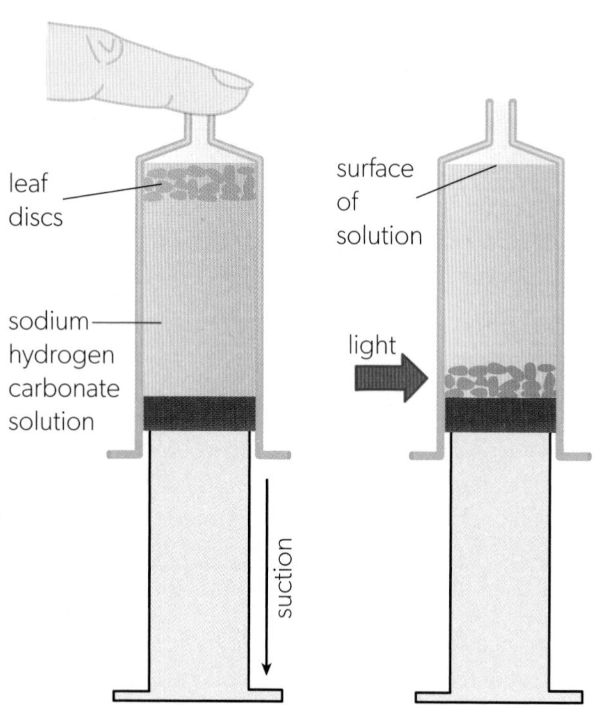

▲ Figure 16 Using leaf discs to measure the rate of photosynthesis

Part 3: Varying temperature

Temperature can be varied using a water bath, heat blocks or a fermenter. Photosynthesis rates in algae inside a fermenter can be measured using a data logger and an oxygen electrode to measure oxygen concentration. Alternatively, pH can be monitored: as carbon dioxide is absorbed from the water and used in photosynthesis, the pH will increase.

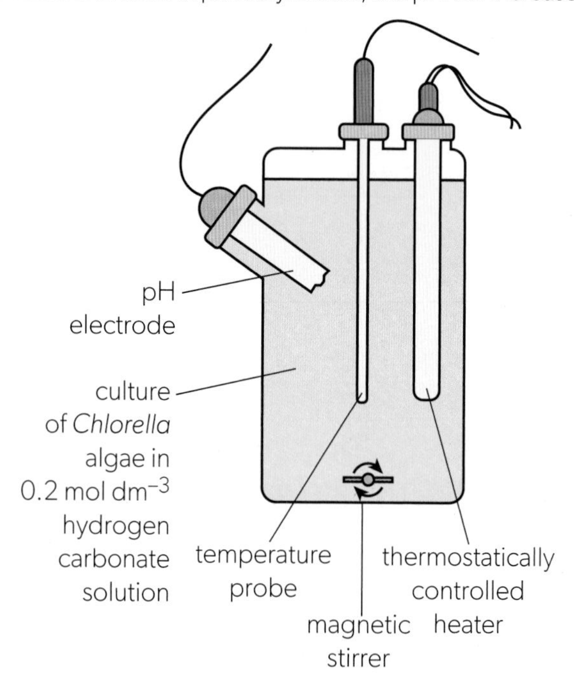

▲ Figure 17 Fermenter used for measuring the rate of photosynthesis at different temperatures by data logging

Hypothesis: Choosing relevant variables

Hypotheses are provisional explanations, which require repeated testing. During scientific research, hypotheses can either be based on theories and then tested in an experiment, or be based on evidence from an experiment that has been carried out already.

You can set hypotheses for the effects of limiting factors on photosynthesis before beginning an experiment (based on your understanding of the theory) or using data from previous experiments.

Through experimentation, a researcher may begin to wonder about an observed phenomenon. They might then develop testable explanations for their observations; these are new hypotheses.

When devising a testable hypothesis, you need to define the dependent and independent variables. The dependent variable is the outcome variable. The independent variable is the cause variable. You usually set the level of the independent variable and then measure the resulting dependent variable. The variables defined in the research question should be the ones that are directly measured.

Questions

In each of the following examples, two different variables are typed in **bold**. Identify which is the dependent variable and which is the independent variable.

1. the **cause** and the **effect**
2. what **I change** and what **I observe**
3. the ***y*-variable** on a graph and the ***x*-variable**
4. the effect of **humidity** on **carbon dioxide absorption** by leaves
5. the **amount of oxygen produced** per hour and the **number of chloroplasts** in a unicellular alga
6. the **length of time to solve an order of operations maths problem** versus **gender**.

Data-based questions: Photosynthesis rates in red light

Figure 18 shows the results of an experiment in which *Chlorella* cells were exposed to light of wavelengths from 660 nm (red) up to 700 nm (far red). The rate of oxygen production by photosynthesis was measured and the yield of oxygen per photon of light was calculated. This gives a measure of the efficiency of photosynthesis at each wavelength.

The experiment was then repeated. This time, at each of the wavelengths from 660 to 700 nm, the cells were also exposed to light with a wavelength of 650 nm. The overall intensity of light was kept the same as in the first experiment.

1. Describe the relationship between wavelength of light and oxygen yield, when there was no supplementary light. [2]
2. Describe the effect of the supplementary light. [2]
3. The bars for each data point on the graph show standard error. Explain how error bars such as these help in drawing conclusions from an experiment. [2]
4. The probable maximum yield of oxygen was 0.125 molecules per photon of light. Calculate how many photons are needed to produce one oxygen molecule in photosynthesis. [2]
5. Oxygen production by photolysis involves this reaction:

   $4H_2O \rightarrow O_2 + 2H_2O + 4H^+ + 4e^-$

   Each photon of light is used to excite an electron (raise it to a higher energy level). Calculate how many times each electron produced by photolysis must be excited during the reactions of photosynthesis. [2]

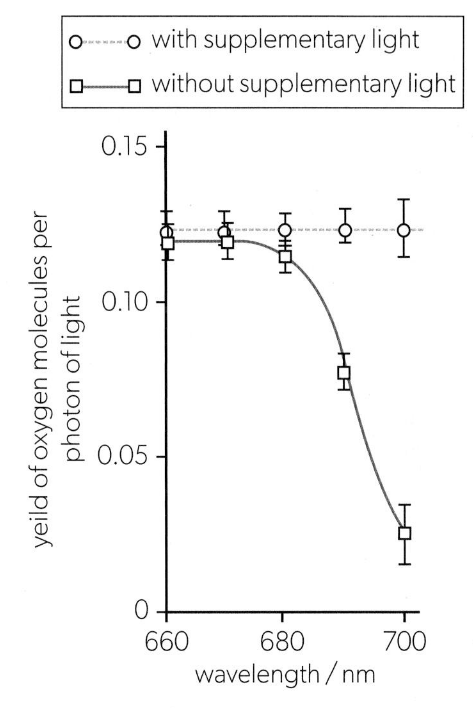

▲ Figure 18 Photon yield of photosynthesis in different light intensities

C1.3.8 Carbon dioxide enrichment experiments as a means of predicting future rates of photosynthesis and plant growth

In high light intensity and warm temperatures, rates of photosynthesis are frequently limited by carbon dioxide concentration. This has been demonstrated in greenhouse experiments: both temperature and light intensity are kept constant at optimal levels while CO_2 concentration is varied (the independent variable). Increasing carbon dioxide concentration above current atmospheric levels of about 400 ppm has been found to increase rates of photosynthesis and plant growth. When growing crops such as tomatoes, it is now common practice to raise CO_2 levels in greenhouses above normal atmospheric concentrations on sunny days.

The extra carbon dioxide can come from boilers that burn natural gas to produce heat or electricity for the greenhouse. More sustainably, it can come from compost-making where plant wastes or animal manures are decomposed by bacteria and fungi.

Before the start of the Industrial Revolution (in about 1780), the carbon dioxide concentration of the atmosphere was about 270 ppm. It is forecast to rise beyond

▲ Figure 19 This greenhouse at Thanet Earth in southeast England is dosed with carbon dioxide as shown in Table 3. The carbon dioxide is produced by combined heat and power engines (CHP) which burn natural gas. The heat is used to warm the greenhouse. Some of the electricity generated by the engines provides power for lighting and other purposes in the greenhouse, but much is sold for use elsewhere. Yields of tomatoes are increased significantly by dosing with CO_2

Light intensity / $J m^{-2}$	CO_2 concentration / ppm
0	400
100	400
200	500
300	600
400	700
500	800

▲ Table 3

550 ppm during the 21st century. This will have wide-ranging consequences. One obvious hypothesis is that higher atmospheric CO_2 concentrations will increase rates of photosynthesis and plant growth, as they do in greenhouses. This could happen with field crops, tree plantations and natural ecosystems. If this increases plant biomass, it will help to moderate increases in atmospheric carbon dioxide. This hypothesis is being tested experimentally.

The aim of the experiments is to increase CO_2 concentration, while keeping other factors unchanged. These experiments cannot be done in laboratories or greenhouses, where many factors differ from those in open conditions. Instead, they must be conducted "in the free air" so they are called free air carbon dioxide enrichment experiments (FACE).

The first series of FACE experiments investigated plant growth in agricultural crops and young tree plantations. A second series of large-scale FACE experiments is being set up in natural or semi-natural forests; the first of these was in a Eucalyptus forest in Australia and the second was in an oak forest in England. Circles of towers are built and carbon dioxide is then released from these towers. Concentrations of carbon dioxide in the air inside these circles are monitored; whenever they drop below 550 ppm, more CO_2 is released on the upwind side. This is done so that any wind blows the carbon dioxide into the circle rather than away from it. Each experiment also includes control plots, where air is released from the towers rather than carbon dioxide. The questions being asked in this series of FACE experiments include:

1. Does increased atmospheric CO_2 increase the carbon storage within a mature woodland ecosystem?
2. Do mineral nutrients, especially nitrogen or phosphorus, become limiting factors on uptake of carbon?
3. What aspects of biodiversity and ecosystem structure and function alter when the ecosystem is exposed to elevated CO_2?
4. How can lessons from the global network of second-generation FACE experiments be generalized to other woodlands and forests?

You can follow the progress of these experiments online by searching for the PhenoCam Network and finding the cameras for "millhaft".

▶ Figure 20 This oak wood in England is being used for one of the second-generation FACE experiments. Six circles of towers are visible. In three of them, the concentration of CO_2 is enriched to 550 ppm while the trees are in leaf. The other three circles are control plots, with air released from the towers instead of carbon dioxide

Experiments: Controlled experiments versus controlled variables

When designing an experiment, you must find methods to control variables. Two locations for biological research are "in the field" or "in the lab". It is easier to control variables in the laboratory but some experiments can only be done in the field. Field research relating to natural ecosystems is done in the habitats of organisms, including human communities. It is particularly useful for areas of biology such as behavioural ecology or epidemiology.

Laboratory research is used for areas such as biochemistry, molecular biology, cell biology and physiology. It can involve whole organisms that have been taken out of their natural habitats. More commonly, it involves material such as biological molecules, organelles, cells, tissues or organs.

There are two broad approaches to research in science: observations and experiments.

- Structured observations are particularly suited to ecology, where the aim is to investigate what happens in natural ecosystems without any human intervention. They are also commonly used with aspects of human biology where experimentation is problematic or unethical. A potential weakness of research based on observation is that multiple factors may affect a variable, so correlations between variables do not show that there is a causal link.
- Experiments form the basis of much biological research. The main benefit of experiments with effective control of variables is that it is possible to be sure about causation. For example, in an enzyme experiment where temperature is deliberately varied while all other factors are kept constant, any change in enzyme activity must be due to the changes in temperature.

You need to understand the difference between control experiments and controlled variables:

- A control experiment—or just control—is any object or system used as a standard of comparison in an experiment. A control is prepared or carried out exactly as the other parts of an experiment, except for one variable which is different. This allows the investigator to assess the significance of that variable. For example, in an enzyme experiment, a control experiment might use an enzyme that has been denatured by boiling, rather than an active (unboiled) enzyme, to show that enzyme catalysis is involved.
- A controlled variable is any factor that could vary but is kept at the same level in all parts of the experiment, so it does not influence the results of the experiment. For example, in an experiment to investigate the effect of temperature, substrate concentration and pH would be controlled variables. Any experiment in which all variables except the independent variable are carefully controlled is called a fair test.

Generally, research in labs is experimental and research in the field is often based on observations. However, this is not an absolute rule. Biologists sometimes plan research that combines the benefits of experiments and field conditions, such as the FACE experiments described in *Section C1.3.8*.

AHL

C1.3.9 Photosystems as arrays of pigment molecules that can generate and emit excited electrons

Photosystems are pigment–protein complexes located in the thylakoid membranes of chloroplasts. In a typical photosystem, there are about 100 chlorophyll molecules and 30 accessory pigment molecules arranged in a precise molecular array; however, there is a lot of variation between the photosystems that have evolved in different organisms. Carotene and xanthophyll are examples of accessory pigments.

Each photosystem has a core complex, connected to light-harvesting antenna complexes. Pigment molecules within antenna complexes absorb light because it causes an electron in one atom of the pigment to become excited and jump to a higher energy level. A specific amount of energy is required for this to happen; this precise amount of energy is only supplied by certain wavelengths. The amount of energy decreases as the wavelength increases so, for example, photons of blue light have more energy than photons of red light.

The light energy that is absorbed by a pigment can be re-emitted as light when the electron drops back down to its original energy level. This is called fluorescence.

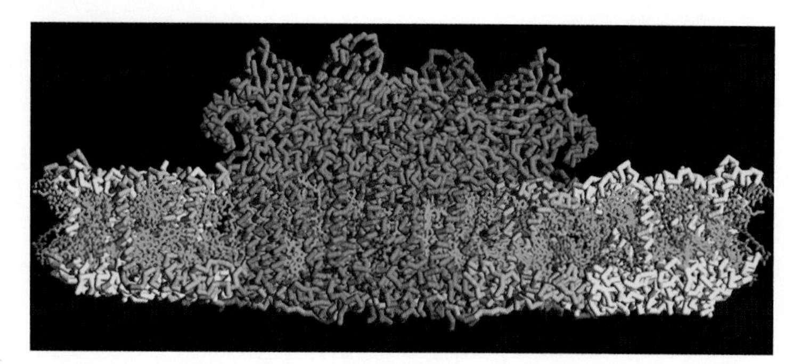

▲ Figure 21 This image of photosystem II is from the PDB-101 website; on this site, you can rotate the image to study the details of the structure. Chlorophyll molecules are shown bright green and accessory pigments such as carotene are pink. Protein chains in the different antenna complexes within the light-harvesting complex are shown in yellow and blue. In the core complex they are dark green

However, something different happens in a light-harvesting complex. When the excited electron in a pigment molecule drops back down to its original level, the energy emitted is absorbed by an electron in the adjacent pigment molecule, causing it to become excited. This process is called excitation energy transfer and it is repeated across the light-harvesting complex. In this way, energy is transferred from pigment to pigment until it reaches the reaction centre in the core complex. This process happens very rapidly, taking only a few femtoseconds. For this energy transfer to happen, the pigment molecules must be held in a precise array, in terms of both the distances between them and their relative orientations. This is achieved by the protein subunits in the light-harvesting complex.

Light energy absorbed by any of the pigments in the light-harvesting complex is funnelled into the core complex. Eventually, it reaches a special pair of chlorophyll molecules in the reaction centre. These molecules are able to donate pairs of excited electrons to electron acceptors. This completes the task of the photosystem. Light energy has been absorbed, generating excited electrons. These electrons are then emitted from the photosystem, carrying the energy needed for later stages of photosynthesis.

In low light intensities, this process is very efficient—more than 99% of photons that are absorbed result in excited electrons being emitted from the photosystem. In high light intensities, other factors make harvesting less efficient and some of the light energy absorbed is re-emitted by fluorescence.

There are two types of photosystem in the chloroplast of a plant, with different functions. The differences between them are shown in Table 4.

	Photosystem I	Photosystem II
Location in the chloroplast	Mostly located in thylakoid membranes between grana, called stroma lamellae	Mostly located in thylakoid membranes in grana, which are cylindrical stacks of thylakoids
Primary electron donor in the reaction centre	P700, containing a pair of chlorophyll molecules with peak light absorbance at 700 nm	P680, containing a pair of chlorophyll molecules with peak light absorbance at 680 nm
Transfer of excited electrons from the primary electron donor	To the enzyme NADP reductase, which uses the electrons to reduce NADP	To plastoquinone which transfers the electrons on to a chain of electron carriers
Source of replacement electrons	Two electrons from plastocyanin	Photolysis of water

▲ Table 4

C1.3.10 Advantages of the structured array of different types of pigment molecules in a photosystem

There are significant advantages in having pigment molecules arranged in the structured array of a photosystem:

- Photons of light are scattered. Even in high-intensity light, one pigment molecule would only intercept a few photons per second. A photosystem combines over 100 pigment molecules, increasing the number of photons absorbed per second by two orders of magnitude.

AHL

- Individual pigment molecules only absorb light in a narrow range of wavelengths. The range varies between pigments. For example, chlorophylls do not absorb green wavelengths but carotene does. A photosystem combines different types of pigment in one array, so a greater proportion of the energy in sunlight can be used.
- Energy is only transferred from one pigment molecule to another when the molecules are in a close and precise orientation. Otherwise, light energy is lost by fluorescence. The structured array also ensures that absorbed energy is funnelled to the reaction centre of the photosystem.

The pigment molecules in the structured array of a photosystem are interdependent. Individually they could not perform any part of photosynthesis; together, they can harvest light energy very efficiently, allowing photosynthesis.

The functioning of photosystems is increasingly understood but there are still many unanswered questions. The mechanisms used do not involve molecular motion or enzyme–substrate collisions. The interactions that occur within photosystems are complex and can only be explained using the principles of quantum mechanics. This is therefore a biological topic that has become the domain of biophysicists rather than biochemists.

Data-based questions: Photosynthesis in artificial light

The light-source spectra in Figure 22 show relative amounts of light of different wavelengths in sunlight and in artificial sources of light such as LED lamps.

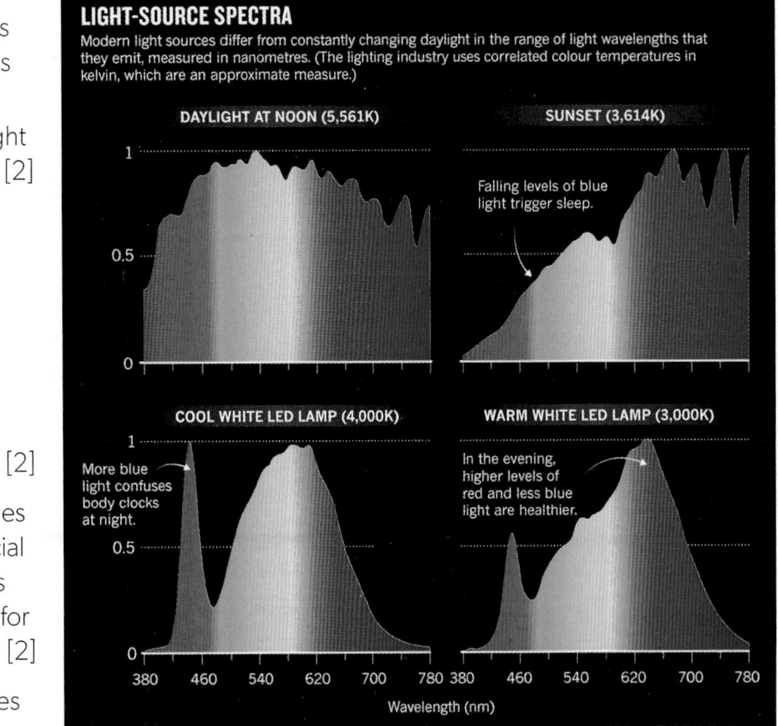

▲ Figure 22

1. Compare and contrast the spectra for daylight at noon and for white LED lamps. [2]
2. The action spectrum for photosynthesis, Figure 14 on page 390, shows that photosynthesis happens most rapidly in blue or red light. A combination of red and blue light usually gives higher rates of photosynthesis than red or blue light only. Suggest reasons for low levels of photosynthesis at sunset. [2]
3. White LED lamps are used to illuminate homes and other places where humans need artificial light. Discuss whether cool white LED lamps or warm white LED lamps are more suitable for homes in the evening. [2]
4. LEDs are widely used as artificial light sources for crops grown in artificial conditions. Suggest reasons for using a combination of red and blue LEDs, rather than white LEDs, for photosynthesizing plants. [2]
5. Blue light promotes the opening of stomata. Explain how rates of photosynthesis might be restricted in plants grown in light only from red LEDs. [4]

AHL

Activity: Leaf colours

Suggest a hypothesis for most naturally occurring plants not having black leaves (that is, explain why plant leaves do not absorb all the colours of light).

▲ **Figure 23** Growing leaves on *Viburnum tinus* are orange-red and then turn green. Many trees produce leaves that are red while they are young and growing, but green when they are mature. Can you suggest a hypothesis for this? Can you find any local plants with leaves that are flush red at first? Why do leaves not flush blue?

▲ **Figure 24** There are a few garden plants with very dark leaves, but they are selected varieties that would not thrive in natural ecosystems. The plant shown is the "Black Scallop" variety of the decorative plant *Ajuga reptans*

C1.3.11 Generation of oxygen by the photolysis of water in photosystem II

Absorption of photons of light by photosystem II causes a special chlorophyll (P680) in the reaction centre to become oxidized by emitting excited electrons. P680 is a powerful reducing agent, which is able to regain electrons from water. This happens in the oxygen-evolving complex (OEC) of photosystem II. The OEC contains a group of manganese, calcium and oxygen atoms and is in the core complex of the photosystem, next to the thylakoid space.

The OEC binds two water molecules and splits them to release four electrons and four protons. The remaining two oxygens bond together to produce a molecule of oxygen (O_2).

$$2H_2O \rightarrow O_2 + 4H^+ + 4e^-$$

This splitting of water is called photolysis because it only happens in the light, when the P680 chlorophyll is oxidized.

Photolysis happens in the OEC on the inner surface of thylakoid membranes. The electrons are transferred to the reaction centre, to replace those emitted by the P680 chlorophyll. The protons are released into the thylakoid space, contributing to a proton gradient across the thylakoid membrane. Oxygen molecules produced by photolysis are a waste product. They diffuse out from the thylakoids to the stroma of the chloroplasts. From there, they diffuse through the cytoplasm of the cell and eventually out of the organism. In plants with leaves, the oxygen diffuses out through the stomata.

For hundreds of millions of years, the Earth's atmosphere contained little or no oxygen, so elements such as iron existed mostly in a reduced state. The production of oxygen by photolysis in cyanobacteria led to the oxidation of iron and other elements. For example, 2,500 to 2,800 million years ago, iron dissolved in the oceans was oxidized to iron oxide and precipitated. This led to the formation of sedimentary rocks known as banded iron formations.

Once iron and other elements had been oxidized, oxygen started to accumulate in the atmosphere. This allowed aerobic respiration to evolve in bacteria. Later in the history of the Earth, chloroplasts evolved by endosymbiosis from cyanobacteria. Eukaryotic algae and then plants also began to contribute oxygen to the atmosphere, so concentrations continued to rise.

▲ Figure 25 The increase in oxygen concentrations in the oceans between 2,800 and 2,500 million years ago caused dissolved iron in the water to oxidize to insoluble iron oxide, which precipitated on the sea bed. This led to the formation of distinctive banded iron formations, with layers of bright red iron oxide alternating with other minerals

Data-based questions: Changes in atmospheric oxygen

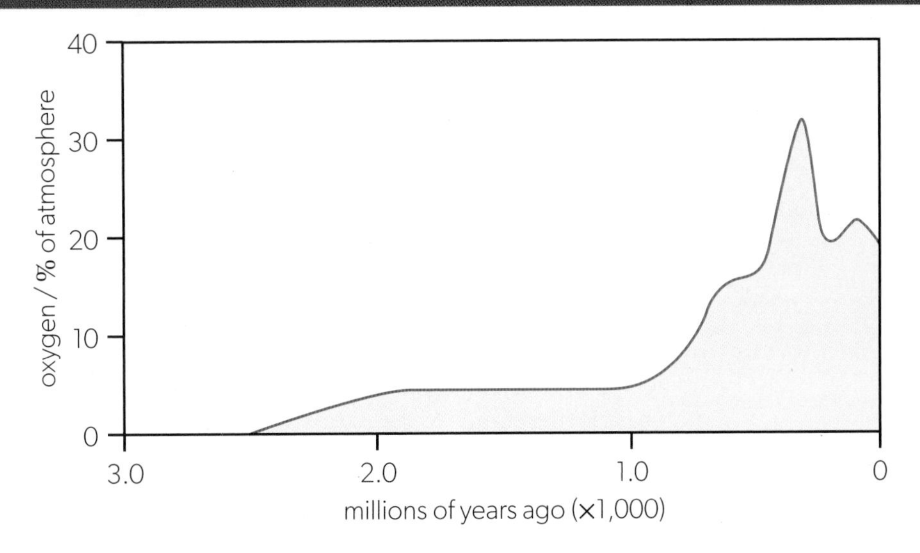

▲ Figure 26 Changes in atmospheric oxygen concentration over time

1. Use the graph in Figure 26 to describe the changes in atmospheric oxygen concentration over the history of the Earth. [3]
2. Determine the highest level of atmospheric oxygen and the time when it occurred. [2]
3. Research and suggest an explanation for these changes. [4]
4. Look at the data in Table 5. Compare the composition of the Earth's atmosphere with that of Venus and Mars. Include only the largest differences in your answer. [3]
5. Suggest the causes for these differences between the planets. [3]

Planet	Composition of atmosphere / %				
	CO_2	N_2	Ar	O_2	H_2O
Venus	98	1	1	0	0
Earth	0.04	78	1	21	0.1
Mars	96	2.5	1.5	2.5	0.1

▲ Table 5

C1.3.12 ATP production by chemiosmosis in thylakoids

Excited electrons generated by photosystem II are passed to plastoquinone, an electron carrier in the thylakoid membrane. Plastoquinone accepts two electrons and also two protons from the stroma, becoming plastoquinol. This happens at binding sites in the reaction centre of photosystem II.

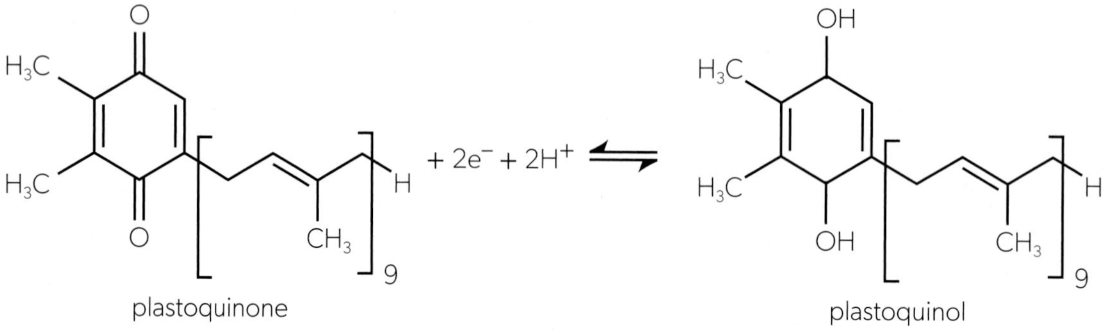

$$\text{plastoquinone} + 2e^- + 2H^+ \rightleftharpoons \text{plastoquinol}$$

O, H_3C, H_3C, O, CH_3, H, 9 — plastoquinone

OH, H_3C, H_3C, OH, CH_3, H, 9 — plastoquinol

Plastoquinol then moves through the thylakoid membrane to the cytochrome b_6f complex. It passes two electrons to the complex and releases two protons into the thylakoid space, contributing to the proton gradient across the thylakoid membrane. This converts plastoquinol back to plastoquinone, which can return to photosystem II to collect more excited electrons and protons.

The cytochrome b_6f complex contains electron transport chains, which transfer electrons from plastoquinol to another electron carrier, called plastocyanin. While plastoquinone is hydrophobic and remains in the thylakoid membrane, plastocyanin is water-soluble and is dissolved in the fluid space inside the thylakoid, where it is free to move. Plastocyanin picks up an electron from the b_6f complex and transfers it to the reaction centre of photosystem I.

The electrons reaching photosystem I carry less energy than they did when emitted by photosystem II. Energy from the electrons has been used to pump protons from the stroma to the thylakoid space, generating a proton gradient. Photolysis also contributes to this gradient by releasing protons inside the thylakoid space. The concentration gradient of protons across the thylakoid membrane is a store of potential energy.

ATP synthase in thylakoid membranes can generate ATP using the proton gradient. Protons travel across the membrane, down the concentration gradient, by passing through the enzyme ATP synthase. The energy released by the passage of protons is used to make ATP from ADP and inorganic phosphate. This method of producing ATP is very similar to the process that occurs inside the mitochondrion and is given the same name: chemiosmosis.

The ATP produced by ATP synthase is released into the stroma. Here, it provides energy for the synthesis of sugars and other carbon compounds from carbon dioxide, in later stages of photosynthesis.

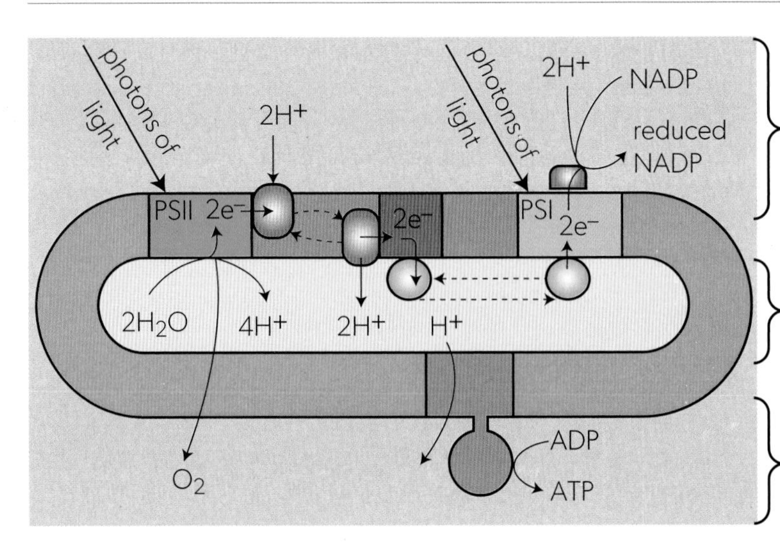

▲ Figure 27 Production of ATP is based on interaction between components of thylakoids

Data-based questions: Evidence for chemiosmosis

One of the first experiments to give evidence for ATP production by chemiosmosis was performed in the summer of 1966 by André Jagendorf. Thylakoids were incubated for several hours in darkness, in acids with a pH ranging from 3.8 to 5.2. The lower the pH of an acid, the higher its concentration of protons. During the incubation, protons diffused into the space inside the thylakoids, until the concentrations inside and outside were equal. The thylakoids were then transferred, still in darkness, into a solution of ADP and phosphate that was more alkaline. There was a brief burst of ATP production by the thylakoids. The graph shows the yield of ATP at three acid incubation pHs and a range of pHs of the ADP solution.

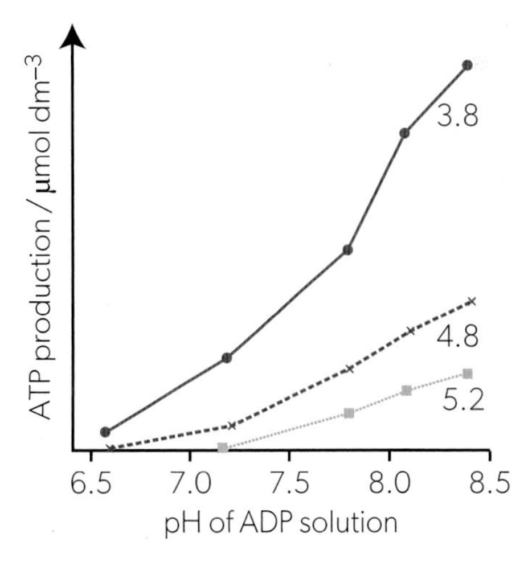

▲ Figure 28 Results of the Jagendorf experiment

1. a. Describe the relationship between pH of ADP solution and ATP yield, when acid incubation was at pH 3.8. [2]

   b. Explain why the pH of the ADP solution affects the ATP yield. [2]

2. Explain the effect of changing the pH of acid incubation on the yield of ATP. [2]

3. Explain why there was only a short burst of ATP production. [2]

4. Explain the reason for performing the experiment in darkness. [2]

C1.3.13 Reduction of NADP by photosystem I

Production of carbon compounds (such as glucose) by photosynthesis requires a supply of electrons. This is provided by NADP (nicotinamide adenine dinucleotide phosphate). NADP is identical to NAD, which is used in cell respiration, except that it has one extra phosphate group. Like NAD, NADP can exist in either a reduced or an oxidized state. It is converted to the reduced state by accepting two electrons:

$$\text{NADP} + 2e^- \rightarrow \text{reduced NADP}$$

AHL

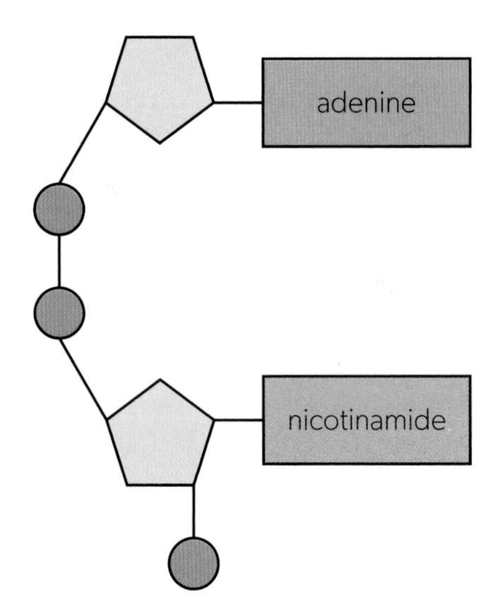

▲ Figure 29 In this diagram of NADP, ribose is shown as a blue pentagon and phosphate as a violet circle. The structure of NAD is the same, without the extra phosphate on the nicotinamide nucleotide

Reduced NADP is produced by photosystem I. Energy from photons of light is absorbed by pigment molecules in the photosystem and passed to the reaction centre. Here, it reaches a special pair of chlorophyll molecules (P700) that act as the primary electron donor. An electron in one of these chlorophyll molecules is excited and then emitted from the reaction centre. It is passed via a short chain of electron carriers to the enzyme NADP reductase. This enzyme is positioned on the stroma side of the thylakoid membrane where it can receive electrons from photosystem I. When NADP reductase has received two excited electrons, it can convert a molecule of NADP in the stroma to reduced NADP.

Electrons from photosystem I that are used to reduce NADP are replaced by electrons carried by plastocyanin. Photosystems I and II are therefore linked: electrons excited in photosystem II are passed via plastoquinone and cytochrome b_6f to plastocyanin, which transfers them to photosystem I.

The supply of NADP in a chloroplast sometimes runs out, because it has all been converted to reduced NADP. When this happens, excited electrons from photosystem I are diverted to plastoquinone instead of being passed to NADP. As the electrons flow back to photosystem I via cytochrome b_6f and plastocyanin, they cause proton pumping which allows ATP production by chemiosmosis. This process is cyclic photophosphorylation. It allows ATP to be produced when production of reduced NADP is impossible or unnecessary.

C1.3.14 Thylakoids as systems for performing the light-dependent reactions of photosynthesis

A thylakoid is a sac-like vesicle that performs the light-dependent reactions of photosynthesis. In these reactions, light energy is absorbed and used to split water by photolysis, reduce NADP and produce ATP by chemiosmosis.

Cyanobacteria have thylakoids that are variable in shape and are attached to the plasma membrane. Eukaryotic algae and plants have two types of thylakoid inside their chloroplasts:

- disc-shaped thylakoids are arranged in stacks called grana
- unstacked thylakoids, known as stroma lamellae, form connections between thylakoids in grana.

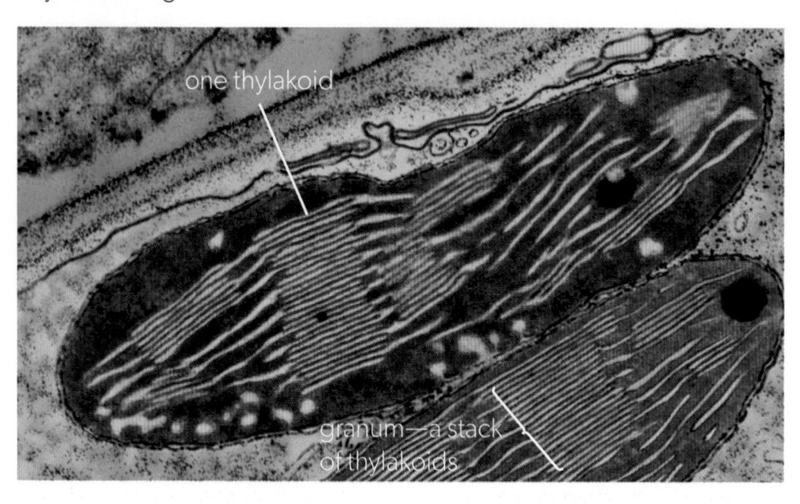

▶ Figure 30 An electron micrograph of a pea chloroplast

A thylakoid is a system, because it contains interacting components that individually would not be able to carry out their functions.

- Thylakoid membranes separate the fluid inside the lumen of the thylakoid from fluid in the surrounding stroma, so a proton gradient can be maintained.
- ATP synthase located in the thylakoid membrane uses the proton gradient to synthesize ATP on the stroma side of thylakoid membranes.
- The oxygen-evolving complex of photosystem II splits water in the lumen of the thylakoid by photolysis, providing a supply of electrons.
- Photosystem II absorbs light and uses the energy from it to excite electrons, which are passed to plastoquinone.
- Plastoquinone and the cytochrome b_6f complex use energy carried by excited electrons to pump protons from the stroma to the lumen of the thylakoid.
- Plastocyanin transfers electrons from the cytochrome b_6f complex to photosystem I.
- Photosystem I absorbs light energy and uses energy from it to excite electrons. These electrons are used to reduce NADP on the stroma side of the thylakoid membranes.

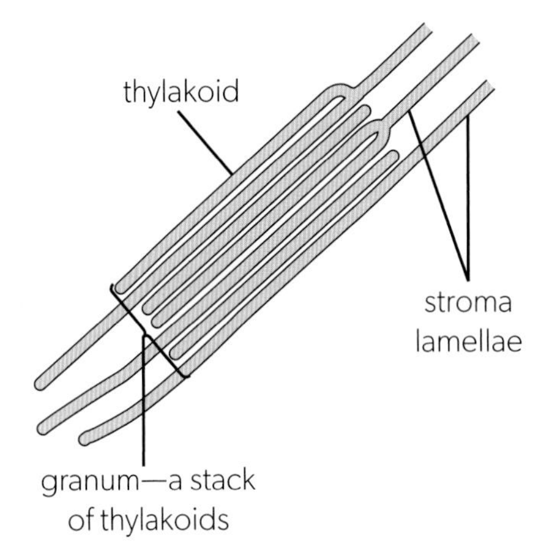

▲ Figure 31 Stroma lamellae are surrounded by stroma whereas most thylakoids in the grana have minimal contact with the stroma. Photosystem 1 arrays and ATP synthases are mostly in the stroma lamellae and photosystem II arrays and cytochrome B_6f complexes are mostly in the grana. Can you suggest reasons for this distribution of components?

There is evidence that components are not evenly distributed between grana and stroma lamellae. Photosystem II and cytochrome b_6f complexes are mostly in the grana. Photosystem I and ATP synthases are mostly in stroma lamellae, making synthesis of ATP and reduced NADP easier due to greater exposure to the stroma.

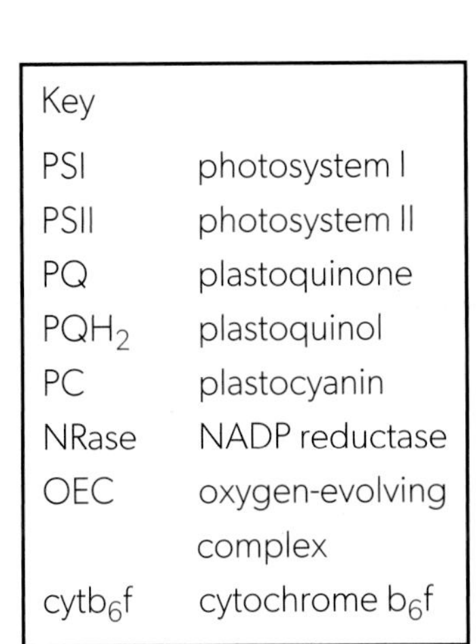

Key
PSI photosystem I
PSII photosystem II
PQ plastoquinone
PQH_2 plastoquinol
PC plastocyanin
NRase NADP reductase
OEC oxygen-evolving complex
$cytb_6f$ cytochrome b_6f

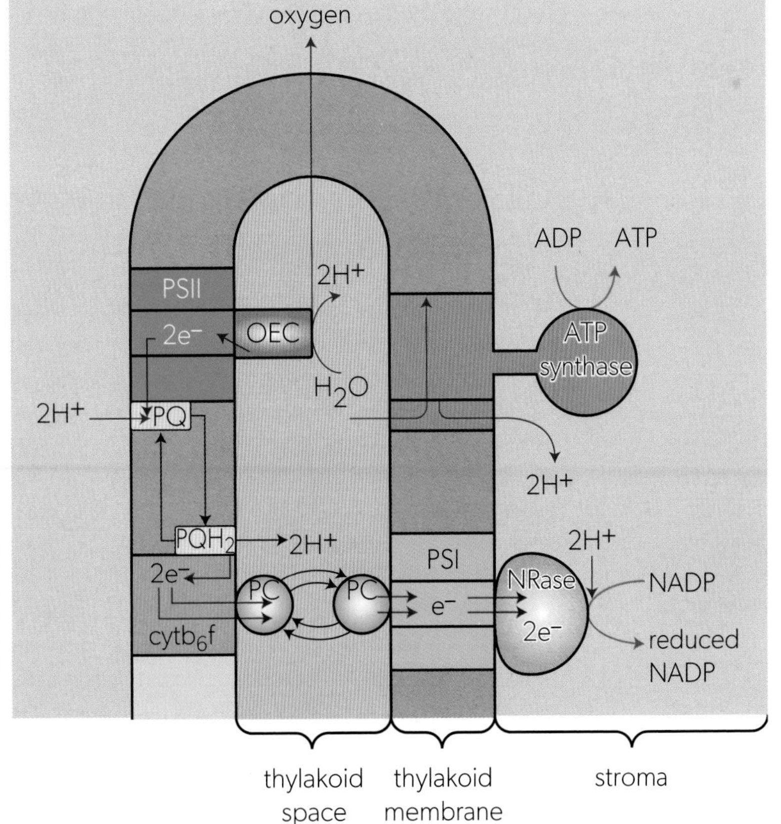

▲ Figure 32 This diagram shows part of one thylakoid and how the four main complexes within it interact through transfers of protons and electrons to produce ATP and reduced NADP

AHL

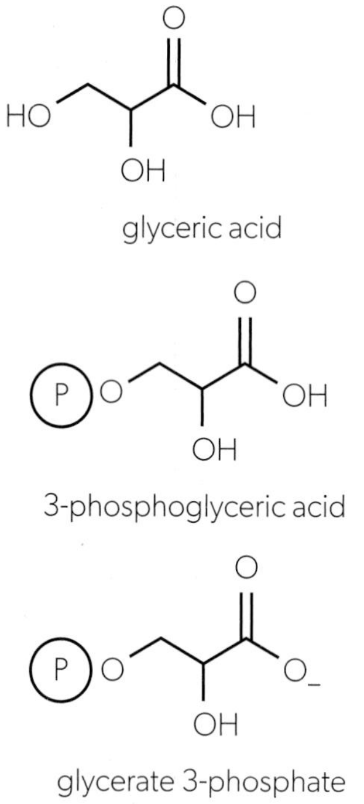

HO OH OH O

glyceric acid

P O OH OH O

3-phosphoglyceric acid

P O O_ OH O

glycerate 3-phosphate

▲ Figure 33 What are the differences between the three substances shown here? What are the similarities between glyceric acid and glycerol (see Figure 20 on page 188; *Section B1.1.9*)?

C1.3.15 Carbon fixation by Rubisco

Carbon dioxide is the carbon source for all organisms that carry out photosynthesis. It readily dissolves in water and passes across cell walls and membranes, so it diffuses from the atmosphere into photosynthesizing cells. It is also able to diffuse out of cells and evaporate. Escape of carbon dioxide from photosynthesizing cells is prevented by carbon fixation.

In the carbon fixation reaction, CO_2 is converted into a more complex carbon compound. This is arguably the most important chemical reaction in all living organisms. In plants and algae, it occurs in the stroma—the fluid that surrounds the thylakoids in the chloroplast. The product of this carbon fixation reaction is a three-carbon compound: glycerate 3-phosphate.

When the details of carbon fixation were discovered, they were a surprise. CO_2 does not react with a two-carbon compound to produce glycerate 3-phosphate. Instead, it reacts with a five-carbon compound called ribulose bisphosphate (RuBP), producing two molecules of glycerate 3-phosphate. The enzyme that catalyses this reaction is called ribulose-1,5-bisphosphate carboxylase-oxygenase—usually abbreviated to Rubisco.

ribulose bisphosphate + carbon dioxide —Rubisco→ 2 glycerate 3-phosphate

●●●●● ● ●●● ●●●

Rubisco is surprisingly inefficient. Most enzymes convert thousands of molecules of substrate to product per second. However, Rubisco only fixes about three CO_2 molecules per second. To compensate for this, there are very high concentrations of Rubisco in the stroma. It is thought to be the most abundant enzyme: the total mass on Earth was recently estimated to be 0.7 Gt (700 billion kilograms).

C1.3.16 Synthesis of triose phosphate using reduced NADP and ATP

In sugars and other carbohydrates, there are twice as many hydrogen atoms as oxygen atoms. RuBP is a five-carbon sugar derivative. It is converted to glycerate 3-phosphate by the addition of carbon and oxygen, but not hydrogen; as a result, the amount of hydrogen relative to oxygen becomes less than two to one. Hydrogen has to be added to glycerate 3-phosphate by a reduction reaction to produce carbohydrate. The carboxyl group in glycerate 3-phosphate is replaced by an aldehyde group.

This conversion involves both ATP and reduced NADP, produced by the light-dependent reactions of photosynthesis. ATP provides the energy needed to perform the reduction and reduced NADP provides the electrons (contained in hydrogen atoms). The product is a three-carbon sugar derivative, triose phosphate. Oxygen removed from the carboxyl group combines with hydrogen from reduced NADP to produce water.

Conversion of glycerate 3-phosphate to triose phosphate happens in the stroma of the chloroplast. It is part of the light-independent reactions of photosynthesis because light is not directly used. However, it can only continue for a short time in darkness as ATP and reduced NADP are required and they quickly run out.

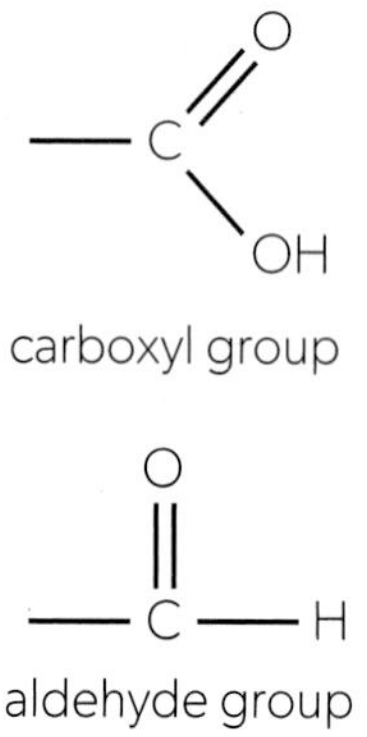

—C O OH

carboxyl group

O —C—H

aldehyde group

▲ Figure 34 Carboxyl and aldehyde groups

C1.3.17 Regeneration of RuBP in the Calvin cycle using ATP

The first carbohydrate produced by the light-independent reactions of photosynthesis is triose phosphate. Two triose phosphate molecules can be combined to form hexose phosphate. Hexose phosphate molecules can be combined by condensation reactions to form starch. When conditions in a leaf are suitable for photosynthesis, starch rapidly accumulates in chloroplasts.

If all of the triose phosphate produced by photosynthesis was converted to hexose or starch, the supplies of RuBP in the chloroplast would soon be used up. This would cause carbon fixation to stop. Therefore, some triose phosphate has to be used to regenerate RuBP. This process is a conversion of a three-carbon sugar into a five-carbon sugar and it cannot be done in a single step. Instead a series of reactions take place.

The light-independent reactions of photosynthesis form a cycle, in which RuBP is both consumed and produced. This cycle was named the Calvin cycle to honour Melvin Calvin, who was given the Nobel Prize in Chemistry in 1961 for his work in elucidating this process.

For the Calvin cycle to continue indefinitely, as much RuBP must be produced as consumed. When RuBP and CO_2 are combined by Rubisco, only one of the six carbon atoms is newly fixed. For this reason, only one-sixth of the triose phosphate molecules that are produced can be taken out of the Calvin cycle. Five-sixths of the triose phosphate must be used to regenerate RuBP. For a net gain of one molecule of hexose, the Calvin cycle must happen six times to fix six carbon atoms.

Regeneration of RuBP requires the use of ATP. This is because triose phosphate is converted into ribulose phosphate and this must be converted to RuBP.

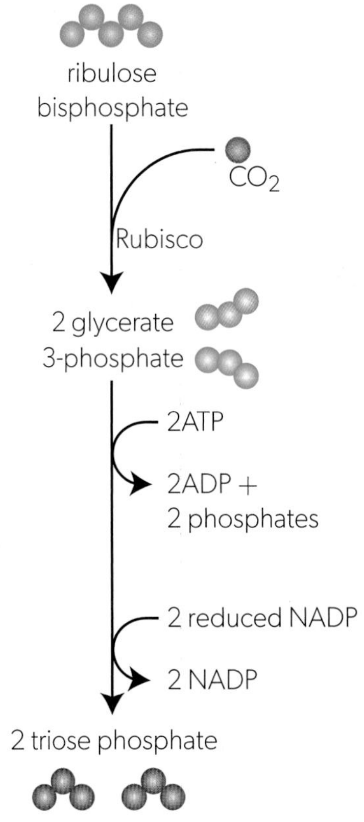

▲ **Figure 35** Summary of carbon fixation reactions

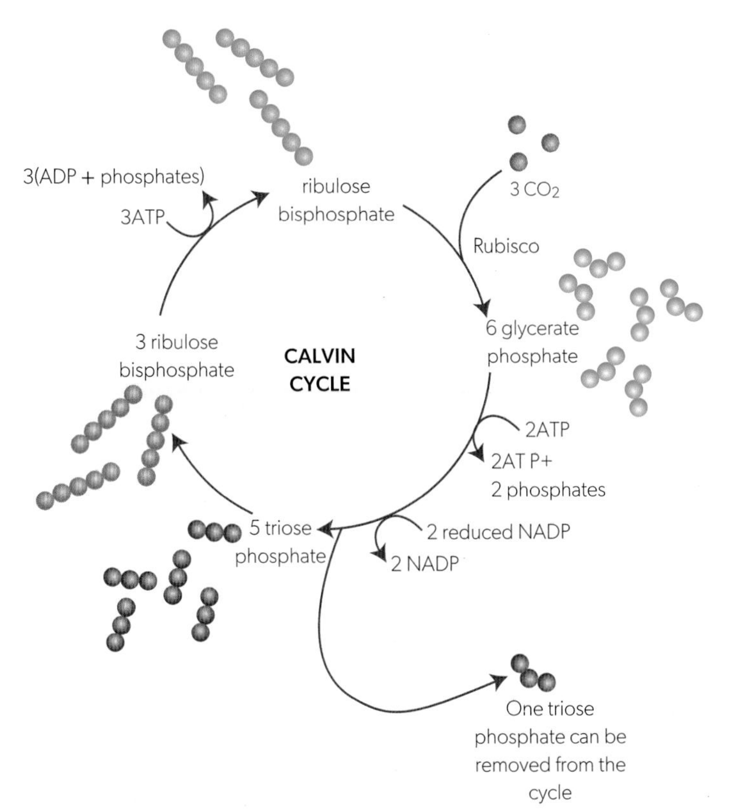

▲ **Figure 36** The Calvin cycle

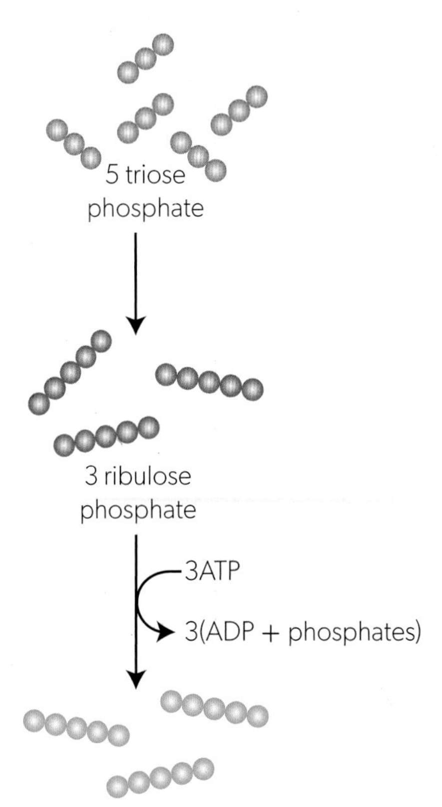

▲ **Figure 37** Summary of RuBP regeneration

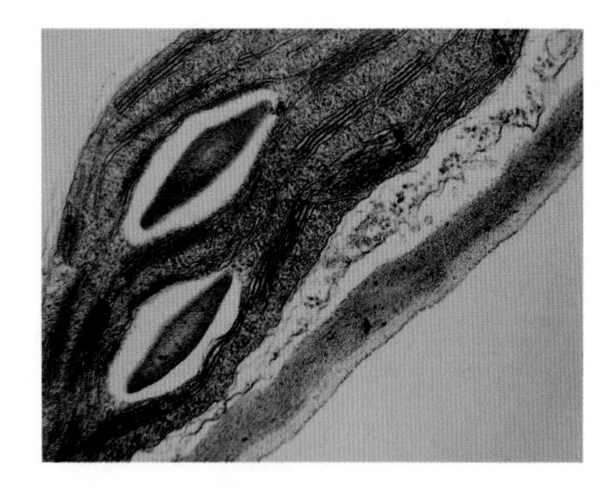

▲ Figure 38 Two large starch grains can be seen inside this chloroplast, the larger of which is over a micrometre long. There are also some dark spheres which are oils. Outside the chloroplast, parts of the plasma membrane, cell wall and vacuole of the cell are visible.

C1.3.18 Synthesis of carbohydrates, amino acids and other carbon compounds using the products of the Calvin cycle and mineral nutrients

Simple equations for photosynthesis usually show glucose as the end product. Plants require large quantities of glucose for cell respiration and for making cellulose. Six turns of the Calvin cycle are needed to produce one molecule of glucose; each turn of the cycle contributes one of the fixed carbon atoms in glucose.

Glucose is usually converted to sucrose for transport from leaves to other parts of the plant. At times, glucose is produced more quickly than it can be transported. At these times, it is converted to starch and stored temporarily inside chloroplasts. At night, when photosynthesis has stopped, this starch is broken down and the carbohydrate is exported from the leaf.

Chloroplasts can also convert triose phosphate from the Calvin cycle into fatty acids, using enzymes of the glycolysis pathway and link reaction to produce acetyl coenzyme A and then linking together two-carbon acetyl groups. Glycerol can also be made from triose phosphate and linked to fatty acids to produce triglycerides. Droplets of stored oil are often visible in chloroplasts.

Many other carbon compounds can be produced in photosynthesizing cells, starting either with glycerate 3-phosphate or triose phosphate from the Calvin cycle or with intermediates from pathways used for aerobic respiration. Mineral nutrients such as phosphate or sulfate are also needed to make compounds containing elements other than carbon, hydrogen and oxygen. All 20 amino acids are synthesized in photosynthesizing organisms, using branching metabolic pathways; nitrogen is supplied by nitrate or ammonium ions.

C1.3.19 Interdependence of the light-dependent and light-independent reactions

Light-dependent reactions in the thylakoid membranes or on the surface of them	Light-independent reactions in the stroma
o photolysis o light absorption by generation of excited electrons o transport of electrons by carriers o ATP synthesis by chemiosmosis o reduction of NADP	o carbon fixation o synthesis of triose phosphate and other carbon compounds o regeneration of RuBP

▲ Table 6 Comparison of the two parts of photosynthesis

Despite the name, light-independent reactions can only continue for a few seconds in darkness. This is because they are dependent on substances produced by the light-dependent reactions, which rapidly run out if they are not produced continuously. Similarly, light-dependent reactions cannot continue indefinitely without substances produced by the light-independent reactions. The two parts of photosynthesis are interdependent.

Light intensity affects which part of photosynthesis limits the overall rate at which carbon compounds are produced.

- In low light intensity, the production of ATP and reduced NADP are restricted. Therefore, the conversion of glycerate 3-phosphate in the Calvin cycle is the rate-limiting step.
- In high light intensity, carbon fixation is usually the rate-limiting step. Use of reduced NADP is restricted, so supplies of NADP limit the light-dependent reaction. Some photons of light absorbed by the photosystems are re-emitted as fluorescence.

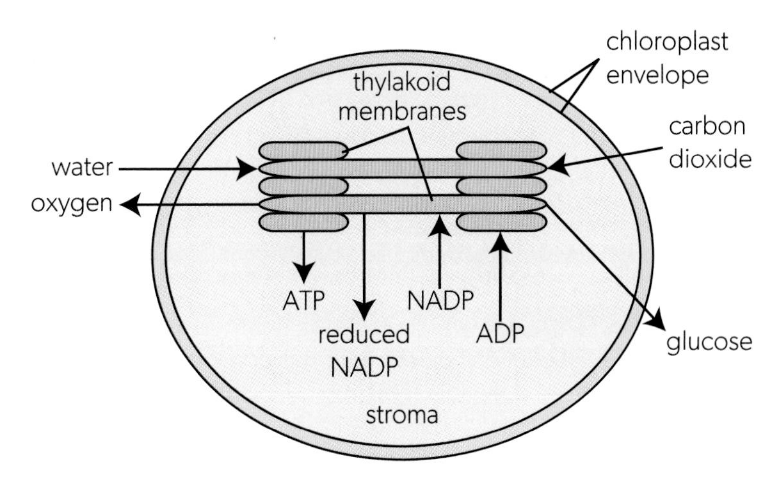

▲ Figure 39 NADP/reduced NADP and ATP/ADP are exchanged between the two parts of photosynthesis

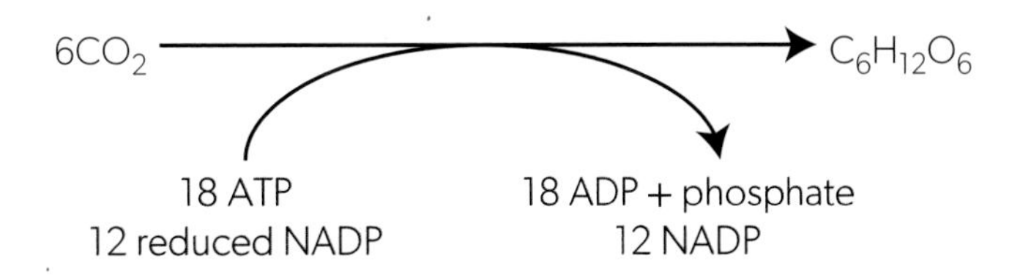

▲ Figure 40 Three ATP and two reduced NADP molecules allow one turn of the Calvin cycle in the stroma. Six times as much reduced ATP and reduced NADP are needed to synthesize one molecule of glucose

Data-based questions: The effect of light and dark on carbon dioxide fixation

One of the pioneers of photosynthesis research was James Bassham. The results of one of his experiments are shown in Figure 41. Concentrations of RuBP and glycerate 3-phosphate were monitored in a culture of cells of the alga, *Scenedesmus*. The algae were initially kept in bright light and then in darkness.

1. Compare the effects of the dark period on the concentrations of RuBP and glycerate 3-phosphate. [2]
2. Explain the change that took place in the 25 seconds after the start of darkness, to the concentration of:
   a. glycerate 3-phosphate [3]
   b. RuBP. [1]
3. Predict the effect of turning the light back on after the period of darkness. [2]
4. Predict the effect of reducing the carbon dioxide concentration from 1.0% to 0.003%, instead of changing from light to darkness:
   a. on glycerate 3-phosphate concentration [2]
   b. on RuBP concentration. [2]

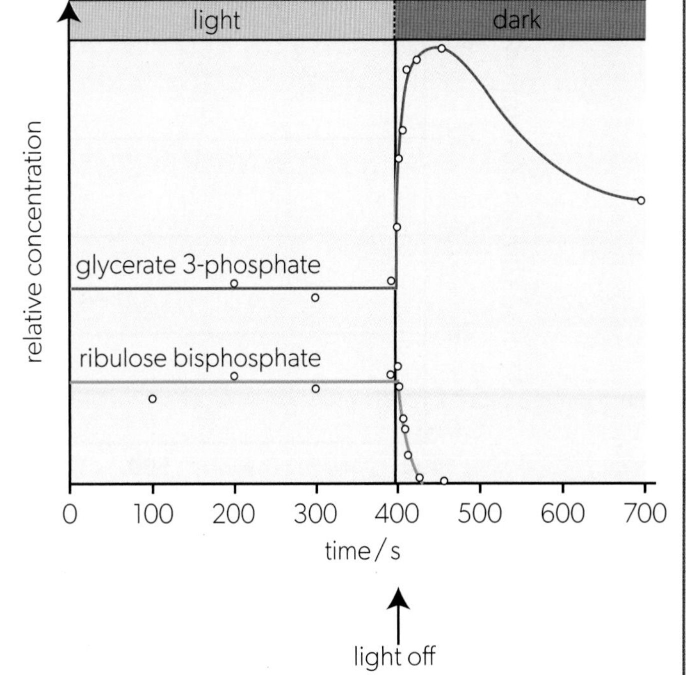

▲ Figure 41 Results of the Bassham experiment

Experimental techniques: Creating and interpreting absorption spectra

To test the light-absorbing properties of plant pigment:

a. isolate the pigment in solution

b. place the solution in a glass cuvette

c. place the cuvette in a spectrophotometer or colorimeter

d. pass light of different wavelengths through the solution, from violet (400 nm) to red (700 nm) and back again

e. record the values from the detector on the other side of the cuvette, which will determine the proportion of light of each wavelength absorbed by the solution

f. plot these values to show the absorption spectrum for the pigment.

To isolate the pigment, grind spinach leaves in an organic solvent such as propanone with some sand for abrasion. (This must be done in a well-ventilated space.) You can use this solution to produce an absorption spectrum for the combined pigments of the leaves.

Alternatively, carry out paper chromatography with the spinach leaf extract. After separation has occurred, cut out the different coloured bands on the chromatogram and immerse them in isopropyl alcohol in separate tubes. This will allow you to determine the absorption spectrum of a single pigment. (You will obtain better results if you combine your coloured bands with those of several other students in the test tube.)

Figure 42 shows the absorption spectra for pigments from two types of leaf.

1. Compare the absorption of the pigment samples shown in Figure 42, including similarities and differences. [3]

2. Deduce, with reasons, which curve shows the absorption of the pigments from the *Fagus* leaf and which shows the absorption of the pigments from the *Acer* leaf. [2]

3. Suggest why plants use pigments that absorb light in the range 400–700 nm and not higher or lower wavelengths. [3]

4. Some algae growing on rocky beaches have a brown colour. Predict, with reasons, the absorption spectrum for the pigments in these algae. [2]

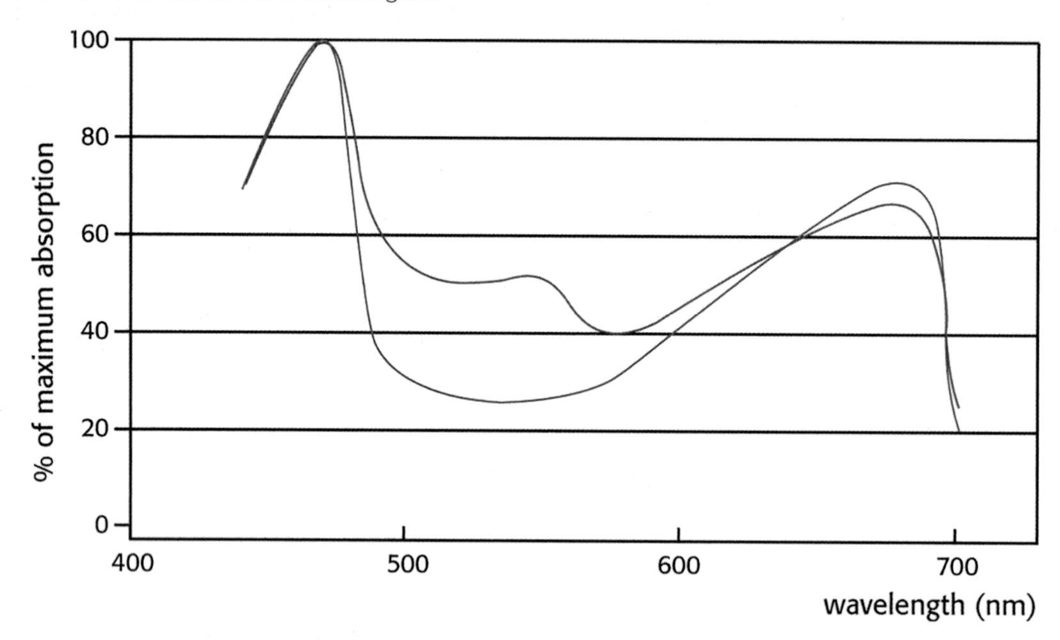

▲ **Figure 42** Absorption spectra for pigments from two types of leaf: a green leaf of *Fagus sylvatica* (beech) and a leaf of *Acer palmatum* (Japanese maple), which contained a red pigment in addition to chlorophyll

Experimental techniques: Investigating photosynthesis with immobilized algae

Alginate beads containing immobilized unicellular algae can easily be prepared.

1. Algae are cultured in a nutrient-rich liquid medium. *Chlorella*, *Scenedesmus* and some species of cyanobacteria are suitable.
2. The algae are concentrated by centrifugation.
3. A mixture of 2 cm^3 of concentrated algae with 8 cm^3 of 2% sodium alginate solution is made in a 10 cm^3 syringe.
4. The algae–alginate mixture is dripped into 2% calcium chloride solution, to create spherical beads, which harden in 5 minutes.
5. The beads are then separated and rinsed to remove the calcium chloride solution.

Many experiments could be performed using the algal beads—you may choose.

- You should vary one factor affecting photosynthesis. This is the independent variable.
- You will need a method of measuring the rate of photosynthesis. This is the dependent variable.

Possible independent variables include:

- light intensity
- light wavelength
- number of beads.

Possible dependent varaibles include:

- change in pH of fluid around the beads
- oxygen concentration of fluid around the beads
- colour change of a redox or pH indicator.

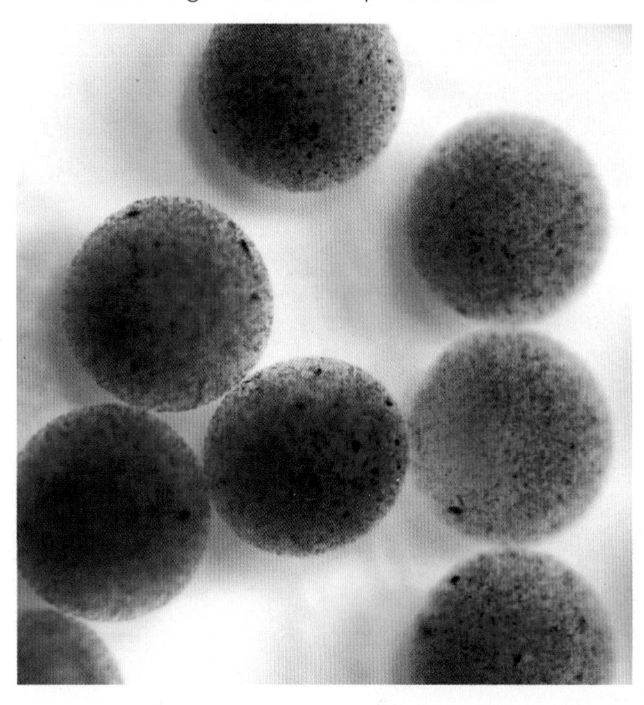

▲ **Figure 43** Alginate–algae beads

Linking questions

1. What are the consequences of photosynthesis for ecosystems?
   a. Outline the carbon cycle. (C4.2.17)
   b. Explain the mechanism of oxygen generation by plants. (C1.3.11)
   c. Distinguish between primary production and secondary production in ecosystems. (C4.2.15)
2. What are the functions of pigments in living organisms?
   a. The plumage of a bird of paradise is an adaptation that contributes to mate selection. Use the example of mate selection to describe the concept of a selection pressure. (D4.1.7)
   b. Outline the role of colour in pollination. (D3.1.9)
   c. Explain how an action spectrum is determined experimentally. (C1.3.6)

TOK

Can new knowledge change established values or beliefs?

Planck (1949, pp. 33–34) states:

> A new scientific truth does not triumph by convincing its opponents and making them see the light, but rather because its opponents eventually die, and a new generation grows up that is familiar with it.

In 1962, the physicist Thomas Kuhn published a book called *The Structure of Scientific Revolutions*. In it, he argued for a revision in our understanding of how science makes progress. At that time, Karl Popper had been one of the most influential writers on the philosophy of science. According to Popper, science is based on objective efforts to refute rather than confirm theories. Science makes progress by falsification—by establishing with certainty what is **not** true.

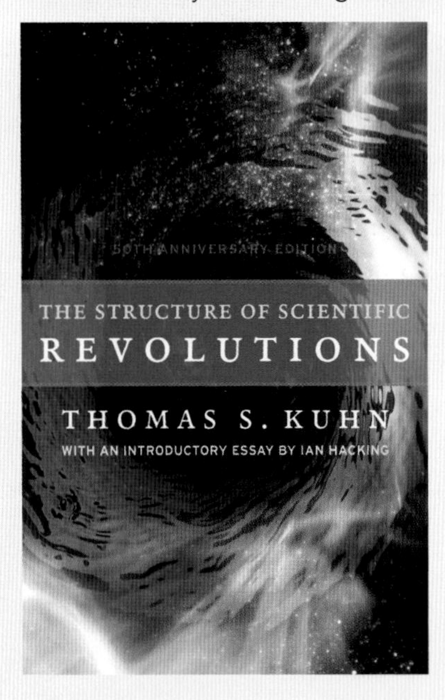

▲ Figure 1 The front cover of the first edition of *The Structure of Scientific Revolutions*

A framework for interpreting observations such as a theory is known as a paradigm. Kuhn suggested that scientists are often deeply embedded in an existing paradigm. As a result, they may be reluctant to refute the theories around which they have built their careers. Instead, many scientists commit themselves to a paradigm even when mounting evidence cannot be explained by it.

In 1961, Peter Mitchell proposed the chemiosmotic hypothesis to explain the coupling of electron transport in the inner mitochondrial membrane to ATP synthesis. This hypothesis was a radical departure from the dominant paradigm and was not generally accepted for many years. Mitchell was awarded the Nobel Prize in Chemistry in 1978. In his Nobel Prize speech, Mitchell stated that art and science share something in common in that they often involve an "imaginative jumping forward" followed by a critical view on what has gone before. His own experience of this phenomenon led him to caution that the imaginative leap forward is a hazardous activity:

> "In the experimental sciences, the scientific fraternity must test a new theory to destruction, if possible. Meanwhile, the originator of a theory may have a very lonely time, especially if his colleagues find his views of nature unfamiliar, and difficult to appreciate."

▲ Figure 2 Peter Mitchell and Jennifer Moyle. Moyle designed many of the experiments that were fundamental in testing the theory of chemiosmosis

End of chapter questions

1. The rate of carbon dioxide uptake by the green succulent shrub *Aeonium goochiae* can indicate the amount of photosynthesis taking place in the plant. This rate was measured at 15°C and 30°C over a 24-hour period. The units of carbon dioxide absorption are mg CO_2 h^{-1}. The results are shown below. The centre of the graph corresponds to −2 mg CO_2 h^{-1} and the outer ring is +2.5 mg CO_2 h^{-1}.

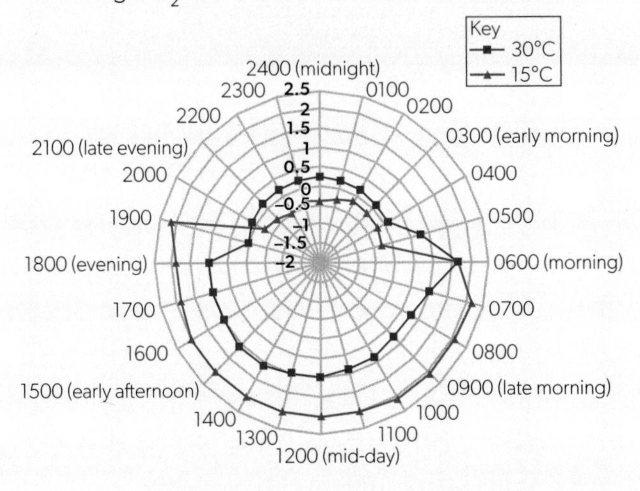

a. Identify a time when carbon dioxide uptake was the same at both temperatures. [1]

b. State the maximum rate of carbon dioxide uptake at 15°C. [1]

c. Compare the rate of carbon dioxide uptake at each temperature in daylight and in darkness. [3]

d. Suggest why carbon dioxide uptake is negative at times. [1]

2. Alcohol dehydrogenase is an enzyme that catalyses the reversible reaction of ethanol and ethanal according to the equation below.

$$NAD^+ + CH_3CH_2OH \rightleftharpoons CH_3CHO + NADH + H^+$$

ethanol ethanal

The initial rate of reaction can be measured according to the time taken for NADH to be produced.

In an experiment, the initial rate at different concentrations of ethanol was recorded (no inhibition). The experiment was then repeated with the addition of 1 mmol dm^{-3} 2,2,2-trifluoroethanol, a competitive inhibitor of the enzyme. A third experiment using a greater concentration of the same inhibitor (3 mmol dm^{-3}) was also performed.

The results for each experiment are shown in the graph.

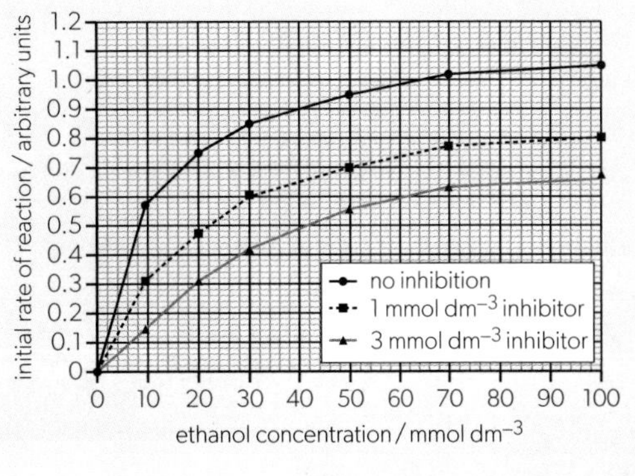

Source: Taber, R.L. (1998), The competitive inhibition of yeast alcohol dehydrogenase by 2,2,2-trifluoroethanol. Biochem. Educ., 26: 239-242. https://doi.org/10.1016/S0307-4412(98)00073-9

a. Outline the effect of increasing the substrate concentration on the control reaction (no inhibition). [2]

b. State the initial rate of reaction at an ethanol concentration of 50 mmol dm^{-3} in the presence of the inhibitor at the following concentrations:

i. 1 mmol dm^{-3} [1]

ii. 3 mmol dm^{-3}. [1]

c. Explain the mechanism of competitive inhibition. [3]

3. A key reaction in photosynthesis occurs when ribulose bisphosphate carboxylase (Rubisco) catalyses the fixation of carbon dioxide to ribulose bisphosphate (RuBP). To be effective, Rubisco must be activated by another enzyme called activase. The activities of Rubisco and activase (each isolated from tobacco leaves) were independently investigated in a laboratory, under conditions of increasing temperature.

a. State the relationship between Rubisco activity and temperature. [1]

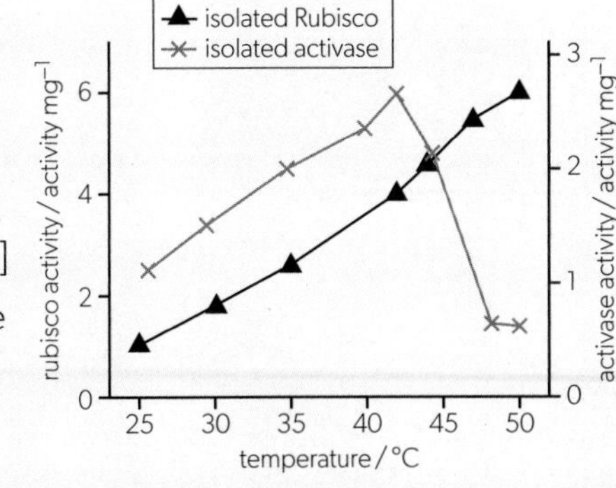

b. Calculate the percentage decrease of activase activity from the optimum temperature to 50°C. [1]

c. Determine which enzyme has the greater overall activity over the temperature range of 25 to 42°C. [1]

d. Explain the observation of activase activity at temperatures higher than 42°C. [2]

e. In a leaf, both enzymes are present together. Predict, with a reason, how the rate of photosynthesis would change from 35°C to 50°C. [2]

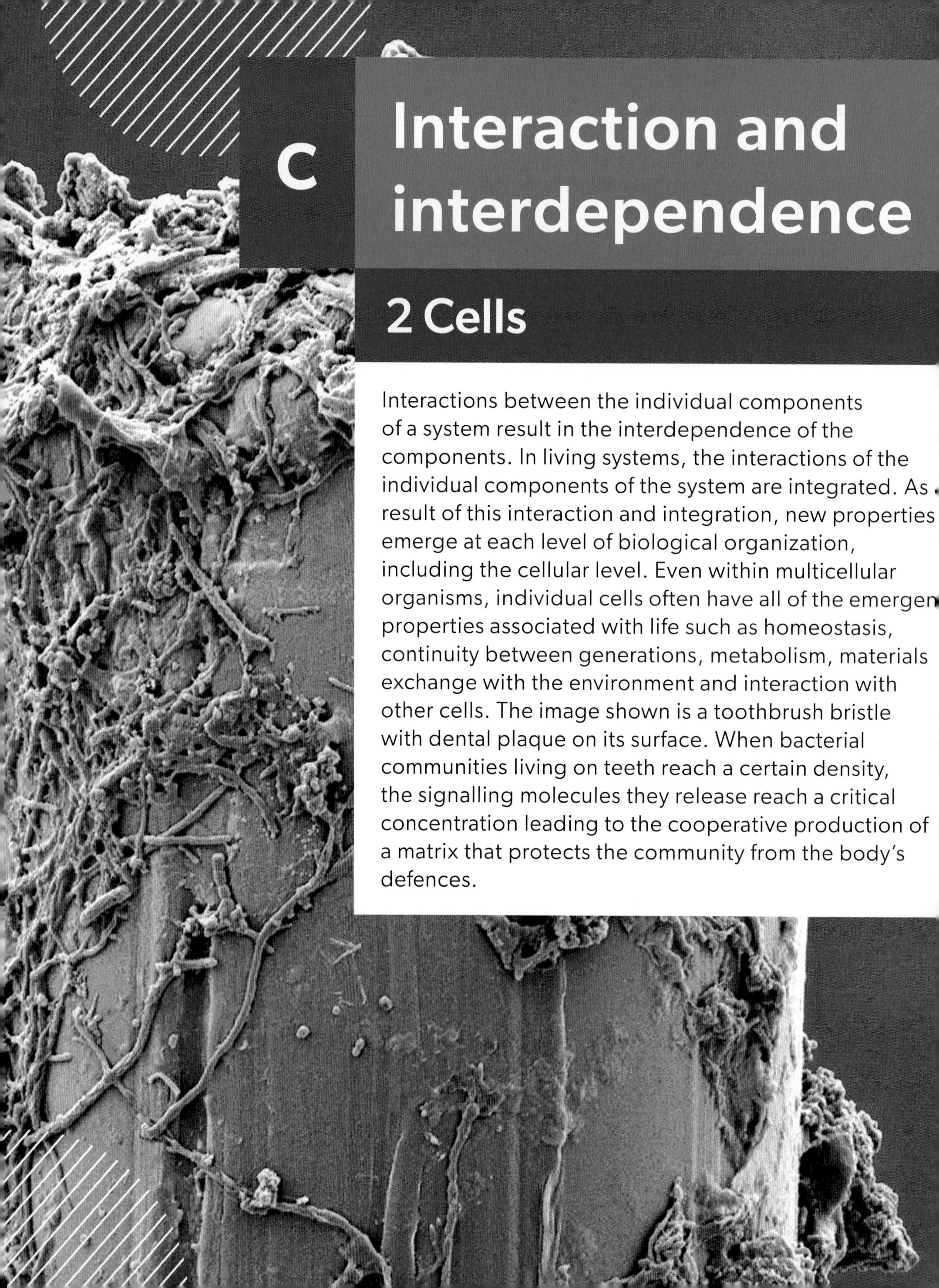

C

Interaction and interdependence

2 Cells

Interactions between the individual components of a system result in the interdependence of the components. In living systems, the interactions of the individual components of the system are integrated. As a result of this interaction and integration, new properties emerge at each level of biological organization, including the cellular level. Even within multicellular organisms, individual cells often have all of the emergent properties associated with life such as homeostasis, continuity between generations, metabolism, materials exchange with the environment and interaction with other cells. The image shown is a toothbrush bristle with dental plaque on its surface. When bacterial communities living on teeth reach a certain density, the signalling molecules they release reach a critical concentration leading to the cooperative production of a matrix that protects the community from the body's defences.

C2.1 Chemical signalling

How do cells distinguish between the many different signals they receive?

Signalling molecules bind to cellular receptors that are either membrane bound or, in the case of steroid hormones, in the cytoplasm of the cell. The receptors have shapes that match certain signal molecules but not others.

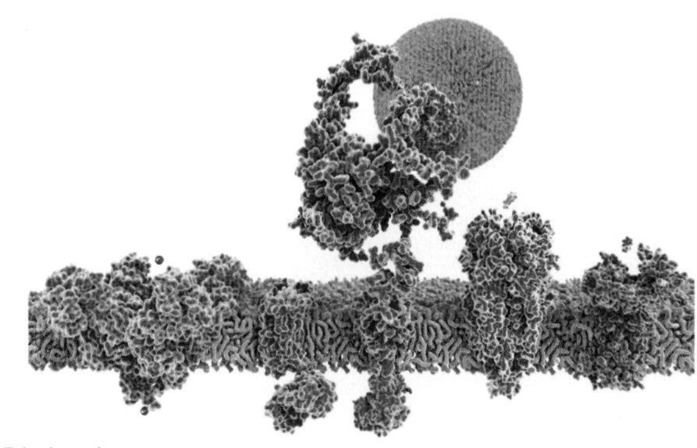

▶ Figure 1 This Illustration shows five transmembrane proteins with their ligands (left to right): a potassium channel, a delta-opioid receptor, an LDL (low-density lipoprotein) receptor, an acetylcholine receptor and a histamine receptor. Each channel or receptor is shown with its associated ligand: a potassium ion (purple sphere), an endorphin molecule (shown in yellow), an LDL droplet (spherical lipid particle), an acetylcholine molecule (small pink molecule) and a histamine molecule (orange)

What interactions occur inside animal cells in response to chemical signals?

Cellular activity is regulated by chemical signals, through various mechanisms. Some of these mechanisms involve signalling molecules binding to receptors in the membrane, triggering a cascade of intracellular responses.

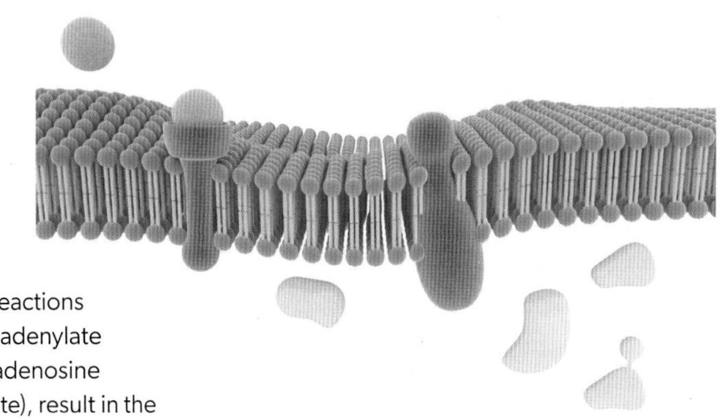

▶ Figure 2 The orange sphere represents a chemical signal. It is shown attaching to a receptor (red) in the cell membrane (blue). This causes the receptor to bind to a G protein (yellow). A series of reactions is triggered and the G protein activates a membrane enzyme called adenylate cyclase (red, centre right). Further reactions involving cAMP (cyclic adenosine monophosphate), various enzymes, and ATP (adenosine triphosphate), result in the target protein (bottom right, yellow) being phosphorylated, controlling its activity

AHL only

C2.1.1 Receptors as proteins with binding sites for specific signalling chemicals
C2.1.2 Cell signalling by bacteria in quorum sensing
C2.1.3 Hormones, neurotransmitters, cytokines and calcium ions as examples of functional categories of signalling chemical in animals
C2.1.4 Chemical diversity of hormones and neurotransmitters
C2.1.5 Localized and distant effects of signalling molecules
C2.1.6 Differences between transmembrane receptors in a plasma membrane and intracellular receptors in the cytoplasm or nucleus
C2.1.7 Initiation of signal transduction pathways by receptors
C2.1.8 Transmembrane receptors for neurotransmitters and changes to membrane potential
C2.1.9 Transmembrane receptors that activate G protein
C2.1.10 Mechanism of action of epinephrine (adrenaline) receptors
C2.1.11 Transmembrane receptors with tyrosine kinase activity
C2.1.12 Intracellular receptors that affect gene expression
C2.1.13 Effects of the hormones oestradiol and progesterone on target cells
C2.1.14 Regulation of cell signalling pathways by positive and negative feedback

C2.1.1 Receptors as proteins with binding sites for specific signalling chemicals

Cells interact with each other by sending and receiving signals. Signals can be sent using a chemical substance. Molecules of the chemical signal are produced by one cell and bind to receptors in another cell. Receptors are proteins, with a site to which the signalling chemical can bind. Binding causes changes in the receptor, which stimulate a response to the signal by the target cell.

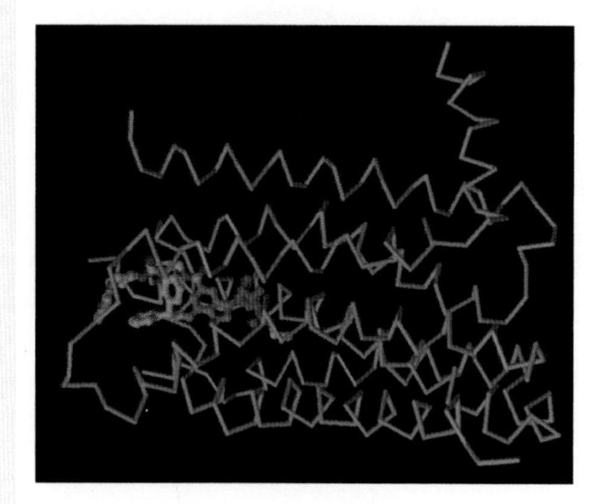

▲ Figure 3 In this model, the adenosine receptor is shown in "wireform", with the adenosine ligand pink (ribose) and green (adenine). The receptor is located in the plasma membrane, with an exposed pocket that acts as the ligand-binding site. The receptor is a single polypeptide with seven alpha helices that straddle the membrane

A molecule that binds selectively to a specific site on another molecule is known as a ligand. The site on a receptor to which the signalling chemical binds is therefore its ligand-binding site. The selectivity or specificity of binding is similar to enzyme–substrate specificity in enzymes:

- In both enzymes and receptors, binding of the ligand occurs at a specific site.
- The shape and chemical properties of the ligand-binding site match those of the ligand, preventing other substances from binding.
- Both enzymes and receptors are unchanged by the binding of a ligand, even if there are temporary changes to induce fit.

There are also key differences between enzymes and receptors.

- When a substrate binds to the active site of an enzyme, the substrate changes. It is converted chemically into the product and released. Another substrate can then bind to the active site, and this cycle can repeat many times per second. Binding is very brief.
- In contrast, a signalling chemical may remain bound to a receptor for a long time because the ligand-binding site does not act as a catalyst and does not convert the signal chemical into a product. The signalling chemical is eventually released unchanged.

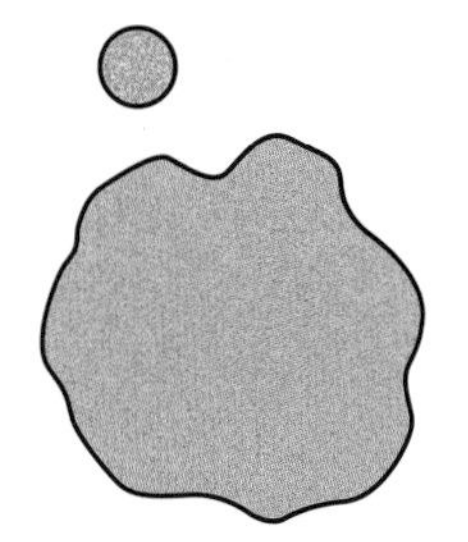

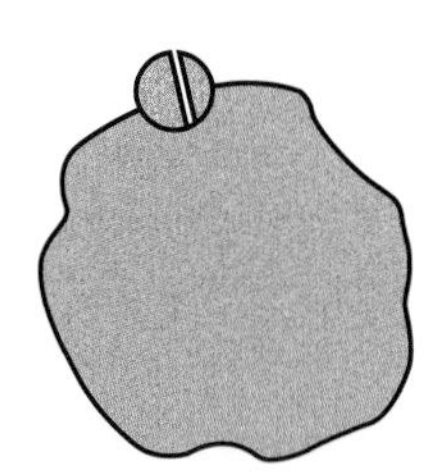

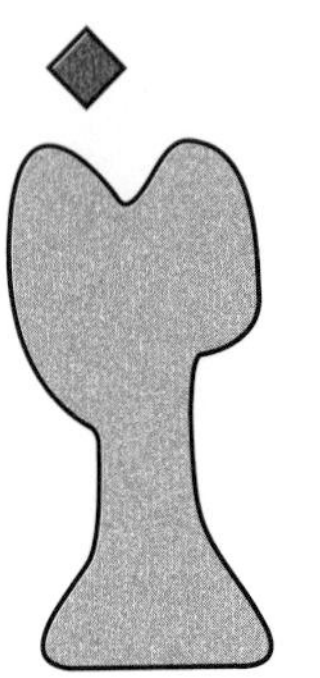

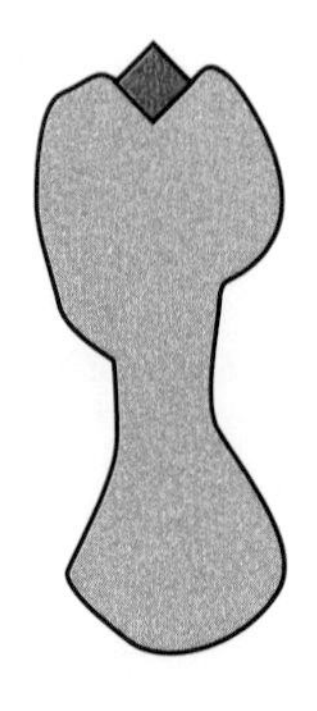

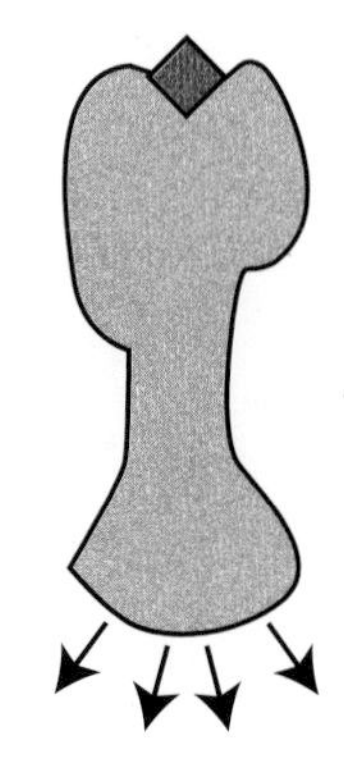

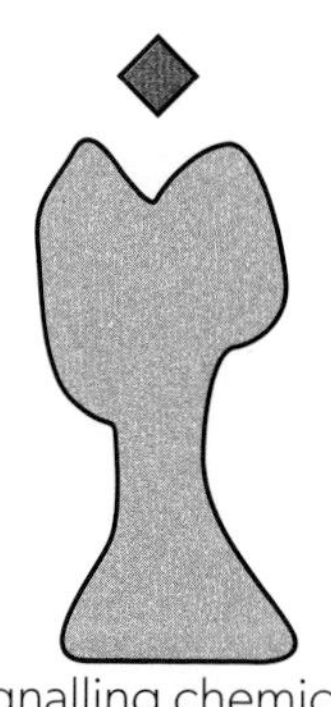

▶ Figure 4 Enzymes (top) and receptors (bottom) both bind a ligand, but the outcome is different

C2.1.2 Cell signalling by bacteria in quorum sensing

A quorum is a fixed number of individuals needed for a meeting to go ahead. For example, in the UN Security Council, two-thirds of the members must be present at any given meeting. A numerical count decides if there is a quorum. Other methods have evolved to assess whether a population is large enough for a group activity, for example, quorum sensing. This is based on intercellular communication and has been observed in a wide range of bacteria. A switch in activity or behaviour is triggered when the population density rises above a certain threshold.

In quorum sensing, signalling molecules are secreted at a low rate by all cells in the population. These molecules diffuse freely between cells and bind to receptors in each cell. When there has been sufficient binding of the signalling molecules to receptors in a cell, gene expression is changed. This causes a switch in activities.

As population density rises, all cells receive more of the signalling chemical from other cells. Above a certain density, every cell in the population receives enough to cause the change in gene expression and the resulting switch in activity—they have sensed that there is a quorum.

Quorum sensing is an example of interaction, because signalling molecules pass from cell to cell. The activities promoted by quorum sensing are examples of interdependence, because they are only effective if more than one cell participates. For example, high densities of bacteria on teeth secrete glue-like chemicals onto the tooth surface. Bacteria adhere (stick) to these chemicals in a thin layer called a biofilm. In other bacteria, bioluminescence is only switched on when there is a high population density capable of producing bright light.

Bioluminescence: An example of quorum sensing

Bioluminescence in the marine bacterium *Vibrio fischeri* was the first case of quorum sensing to be discovered. *V. fischeri* cells produce and secrete a type of signalling molecule known as an autoinducer. The autoinducer can diffuse between cells. It binds to a receptor protein in the cytoplasm, known as LuxR. When this binding has occurred, the LuxR–autoinducer complex binds to the DNA of *V. fischeri* at a position where it induces the transcription of genes. Expression of these genes results in the production of the enzyme luciferase.

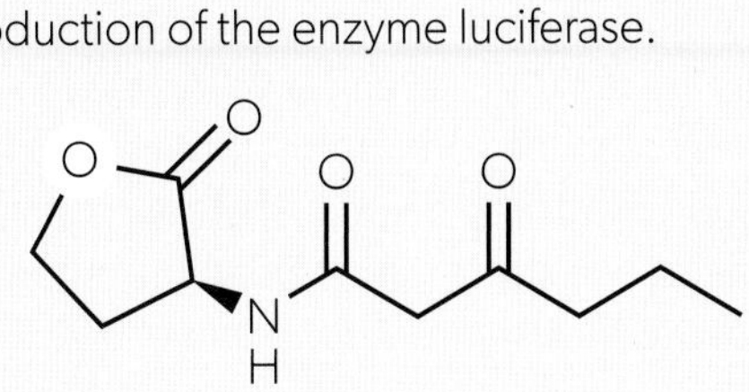

▲ **Figure 5** The signalling molecule that acts as an autoinducer of bioluminescence in *Vibrio fischeri*

Luciferase catalyses an oxidation reaction that releases energy. Over 80% of this energy is transformed to greenish-blue light. Free-living *V. fischeri* are at a low population density so they do not receive enough of the autoinducer for luciferase to be produced. Light is not emitted, as it would have no function and would be a waste of energy.

V. fischeri forms a mutualistic relationship with various animal hosts such as the bobtail squid. In the squid, the bacteria colonize a structure called the light organ. A high population density of the bacteria inside the light organ leads to a high concentration of autoinducer, so bioluminescence is induced. The light emitted from the squid helps to camouflage it in moonlight, reducing the risk of predation. Bacteria in the light organ are supplied with sugar and amino acids by the squid.

◀ **Figure 6** Bioluminescence in the bobtail squid *Euprymna berryi*

C2.1.3 Hormones, neurotransmitters, cytokines and calcium ions as examples of functional categories of signalling chemical in animals

Signalling chemicals in animals are very varied chemically, so they are usually classified according to their function rather than their structure.

Hormones

Hormones are signalling chemicals produced in small amounts by a group of specialized cells in the body and transported by the bloodstream. Organs that are specialized for secretion are called glands. Most hormones are secreted into blood capillaries in the gland tissue. Because of this internal secretion, glands that secrete hormones are called endocrine glands. Exocrine glands have a duct leading out of the organ to transport the secretion.

The bloodstream transports hormones to all parts of the body. However, they only have effects on target cells which have receptors for the hormone. The hormone regulates the activities of the target cells by promoting or inhibiting specific processes. Hormones can persist in the body for hours after being secreted, so the activities of target cells can be affected for much longer than with nerve impulses. Transport in the bloodstream means the secreting and target cells can be far apart and one hormone can have very widespread effects.

Insulin, thyroxin and testosterone are example of hormones.

Neurotransmitters

Neurotransmitters are chemicals that transmit signals across synapses. A synapse is a junction between two neurons in the nervous system: the presynaptic neuron secretes the neurotransmitter and the postsynaptic neuron receives it. The neurotransmitter is secreted when a nerve impulse reaches the end of the presynaptic neuron. It diffuses across the gap between the two neurons and then binds to receptors in the plasma membrane of the postsynaptic neuron. This binding influences whether a nerve impulse is initiated in the postsynaptic neuron. Excitatory neurotransmitters stimulate nerve impulses; inhibitory neurotransmitters have the opposite effect.

The gap between the two neurons at a synapse is between 20 and 40 nanometres and most neurotransmitters only travel this very short distance. This happens in a fraction of a second, so neurotransmitters convey their signal far more quickly than hormones. Neurotransmitters are rapidly broken down in the synaptic gap or reabsorbed into the presynaptic neuron, so they only persist for a fraction of a second. In consequence, their effects are short-lived. Rapid removal of neurotransmitter from the synaptic gap ensures it only affects one specific postsynaptic neuron; it does not usually diffuse out of the synapse to have more widespread effects.

Acetylcholine, norepinephrine and dopamine are examples of neurotransmitters.

Cytokines

Cytokines are small proteins that act as signalling chemicals. They are secreted by a wide range of cells. The same cytokine may be secreted by different cell types and one cell type may secrete several different cytokines. Certain cytokines can be secreted by almost all cells in the body.

Cytokines are not usually transported as far as hormones. Instead, they act either on the cell that produced them or on a nearby cell. Cytokines cannot enter cells so they bind to receptors in the plasma membrane of a target cell. This binding causes cascades of signalling inside the target cell, leading to changes in gene expression and thus in cell activity. One type of cytokine can bind to several types of receptor and so have multiple effects.

Cytokines have cell signalling roles in inflammation and in other responses of the immune system. They also have roles in the control of cell growth and proliferation and in the development of embryos.

Erythropoietin (EPO), interferon and interleukin are examples of cytokines.

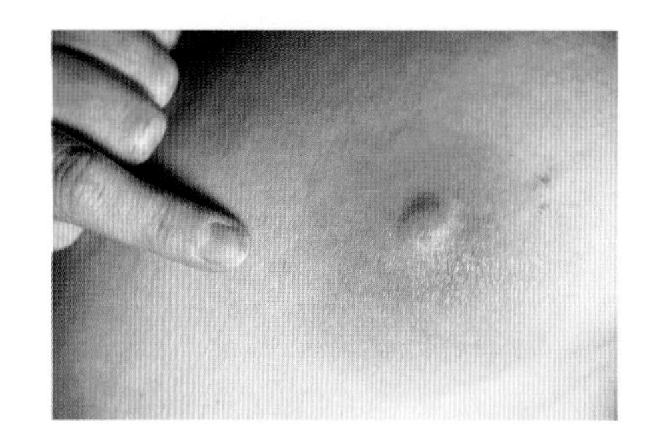

▲ Figure 7 Cytokines have caused inflammation in the region around a skin infection as part of the body's normal immune response. Sepsis is the poorly understood and life-threatening condition in which interactions in the immune system cause a cytokine storm, leading to widespread organ failure

Calcium ions

Calcium ions are used for cell signalling in both muscle fibres and neurons.

In muscle fibres, calcium ions are pumped into a specialized form of endoplasmic reticulum called the sarcoplasmic reticulum, generating a high concentration. When the muscle fibre receives a nervous impulse, calcium channels open in the membrane of the sarcoplasmic reticulum and the ions can diffuse out. They bind to proteins that block muscle contraction, causing the proteins to change position; this allows muscle contraction to occur. If the muscle fibre does not receive more nerve impulses, these changes are reversed and the calcium is pumped back into the sarcoplasmic reticulum.

In neurons, the arrival of a nerve impulse at a presynaptic membrane causes calcium channels to open, allowing inward diffusion. Inside the presynaptic neuron, Ca^{2+} ions cause secretion of neurotransmitter into the synaptic gap by exocytosis. The calcium ions are rapidly pumped back out into the synaptic cleft.

C2.1.4 Chemical diversity of hormones and neurotransmitters

Signalling systems using hormones and neurotransmitters have evolved repeatedly and a wide range of chemical substances have become signalling chemicals. A signalling chemical must:

- have a distinctive shape and chemical properties so the receptor can distinguish between it and other chemicals
- be small and soluble enough to be transported.

Table 1 shows chemical categories of hormone and neurotransmitter.

Hormones		Neurotransmitters	
Amines	• melatonin • thyroxin • epinephrine	**Amines**	• dopamine • norepinephrine
Peptides	• insulin • glucagon • ADH	**Gases**	• nitrous oxide
Steroids	• oestradiol • progesterone • testosterone	**Amino acids**	• glutamate • glycine
		Esters	• acetylcholine

◀ Table 1

AHL

Data-based questions: Nitrous oxide and mating in newts

The courtship of the crested newt (*Triturus carnifex*) involves the following stages:

1. male approaches female and sniffs her head
2. male waves his tail towards the female's head
3. male hits female on her head with his tail
4. male deposits sperm next to the female
5. female picks up sperm.

A receptive female responds to the courtship by remaining motionless until picking up the sperm. The male or female can stop the courtship at any stage by moving away.

1. Using only the information given above, suggest ways in which:
   a. the male can find out if the female is of the correct species [2]
   b. the female can decide whether or not to select the male for mating. [2]

Nitric oxide (NO) regulates sexual behaviour in some animals. Newt brains contain nitric oxide synthase, which catalyses the formation of nitric oxide. The amount of this enzyme was measured in the brains of male newts at various stages of courtship. The results are shown in Figure 8. (Source of data: Zerani and Gobetti. 1996. *Nature*. Vol. 382. P 31.)

2. Outline the changes in the amount of nitric oxide synthase in the male during the stages of normal courtship and after completing it. [3]

3. Using the data in the question, deduce the effects of the female newt on nitric oxide levels in the brains of male newts. [3]

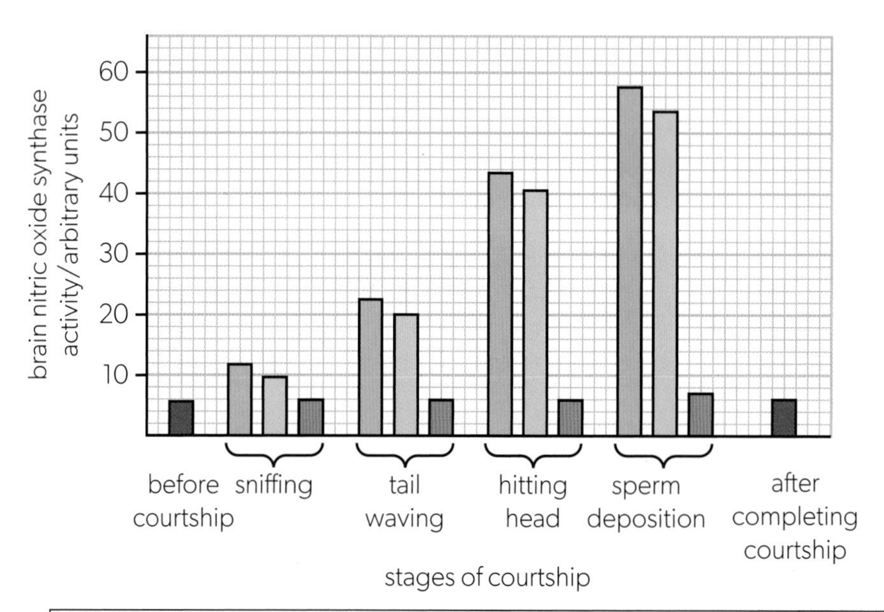

▲ **Figure 8** Source: Zerani, M., Gobbetti, A. Nature 382, 31 (1996). https://doi.org/10.1038/382031a0.

ATL Communication skills: Responding to command terms

In experiments, when students propose a hypothesis, it is not certain that the hypothesis is correct. However, it is good scientific practice that there be some basis for believing the hypothesis is correct. Sometimes a hypothesis is referred to as an "educated guess". Similarly, the command term "suggest" requires you to propose a solution, hypothesis or other possible answer. There can be more than one correct answer but not every answer would be correct. Mark schemes for "suggest" questions often give the most likely hypothesis but also include the statement "or other reasonable suggestion". The emphasis is on the word "reasonable"—there should be a basis for your "educated guess".

C2.1.5 Localized and distant effects of signalling molecules

Some signalling molecules are only transported a very short distance and therefore have very localized effects. For example, neurotransmitters are released by presynaptic neurons and may only have to diffuse 20 nanometres to reach the one postsynaptic neuron that they affect.

Other signalling molecules are transported long distances in the body, from the cells that secrete them to the target cells. Hormones are transported in the blood from the gland that produces them to all parts of the body; the target cells could be in any part of the body. For example, luteinizing hormone (LH) is secreted by the pituitary gland adjacent to the brain. In males, the target cells of LH are in the testes and in females they are in the ovaries, so the effects of this hormone are very distant from its source.

Activity: Kaiten sushi

In Kaiten sushi, customers help themselves to sushi as it circulates on a conveyor belt. Are there any similarities with chemical signalling in the body?

▲ Figure 9

C2.1.6 Differences between transmembrane receptors in a plasma membrane and intracellular receptors in the cytoplasm or nucleus

Signalling chemicals can be divided into two groups, according to whether they enter the target cell or not. Receptors for signalling chemicals that do pass through the plasma membrane are located in the cytoplasm or nucleus of the cell; they are intracellular. Receptors for chemicals that do not penetrate are located in the plasma membrane of the target cell, with the binding site exposed to the exterior. These receptors extend across the membrane with a region extending into the cytoplasm; they are transmembrane receptors.

The location of receptors is determined by the distribution of hydrophilic and hydrophobic amino acids on the surface of the receptor protein.

- Intracellular receptors have hydrophilic amino acids so they remain dissolved in the aqueous fluids of the cytoplasm or nucleus.
- Transmembrane receptors have a band of hydrophobic amino acids on their surface that is attracted to the apolar tails of phospholipids in the core of the membrane. On either side of this band, there are hydrophilic amino acids which are in contact with aqueous solutions inside and outside the cell.

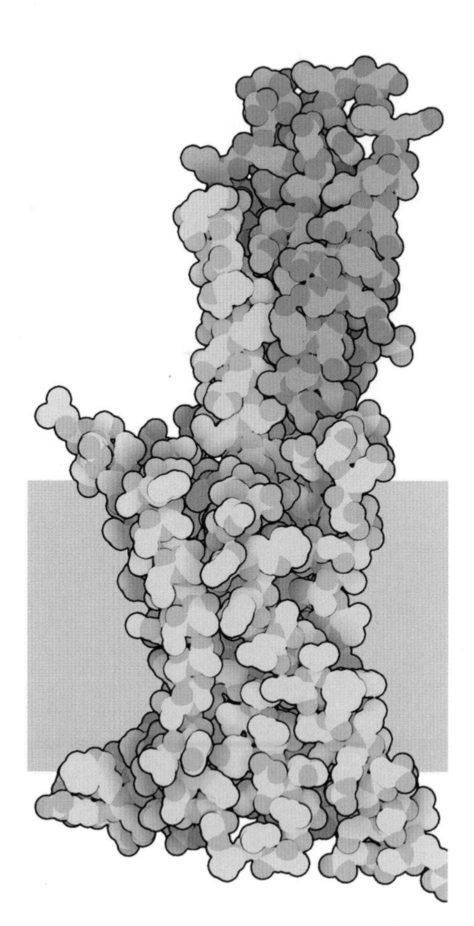

▲ Figure 10 A glucagon receptor (blue) with its position in the membrane shown by the grey slab. The hormone glucagon binds to an extracellular protein (darker blue) which delivers it to the binding site on the receptor

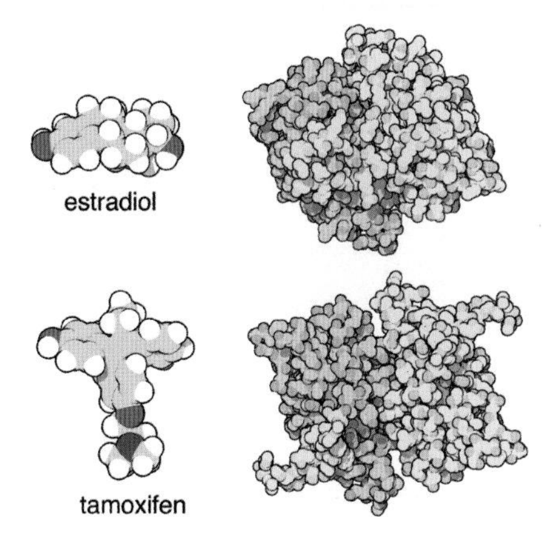

◀ Figure 11 Oestradiol (pink) binds to the oestradiol receptor (blue). Pairs of oestradiol–receptor complexes then jointly bind to DNA (orange and red), activating transcription of specific genes. Chains of blue dots indicate parts of the receptor that are not shown in this image

AHL

C2.1.7 Initiation of signal transduction pathways by receptors

Binding of a signalling chemical to a receptor causes a sequence of interactions in the cell, called a signal transduction pathway. These pathways are very varied as they have evolved repeatedly, rather than having a common origin. Some signalling chemicals such as proteins cannot pass through the plasma membrane; instead, they bind to receptors in the plasma membrane. Other signalling chemicals, for example steroids, pass through the membrane and bind to intracellular receptors. Transmembrane and intracellular receptors use different transduction pathways.

When a signalling chemical binds to the outer side of a transmembrane receptor, it changes the structure of the receptor. The inner side of the receptor becomes catalytically active and causes production of a secondary messenger within the cell. This conveys the signal to effectors within the cell that carry out the responses.

Binding of signalling chemicals to intracellular receptors results in the formation of an active ligand–receptor complex. In most cases, this complex regulates gene expression by binding to DNA at specific sites, promoting or inhibiting the transcription of particular genes.

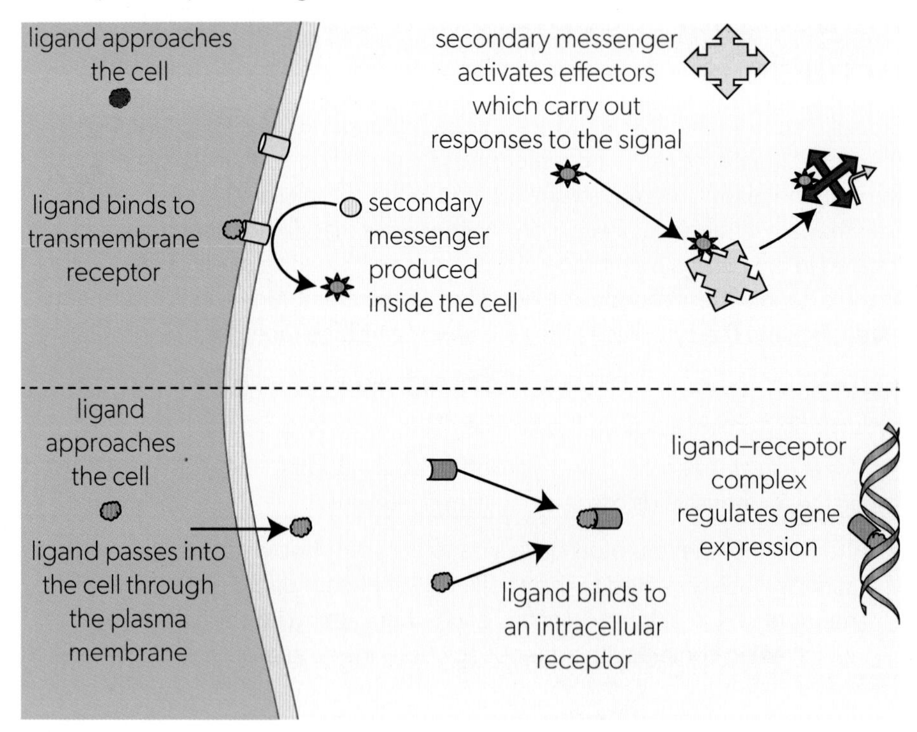

▲ Figure 12 An overview of signalling via transmembrane and intercellular receptors

Data-based questions: Treatments for hypoglycemia

Hypoglycemia is low blood sugar concentration. In people with type 1 diabetes, severe hypoglycemia sometimes requires urgent treatment. The hormones glucagon and epinephrine can both cause blood glucose concentration to increase in certain circumstances. Glucagon is a protein and epinephrine is an amine.

Ten children with type 1 diabetes that was being treated with insulin were allowed to develop a blood glucose concentration of 2.8 mmol dm^{-3}. Then they were given injections of either epinephrine from an epipen or glucagon. The graphs in Figure 13 show the concentrations of these two hormones in the children's blood and the blood glucose concentration.

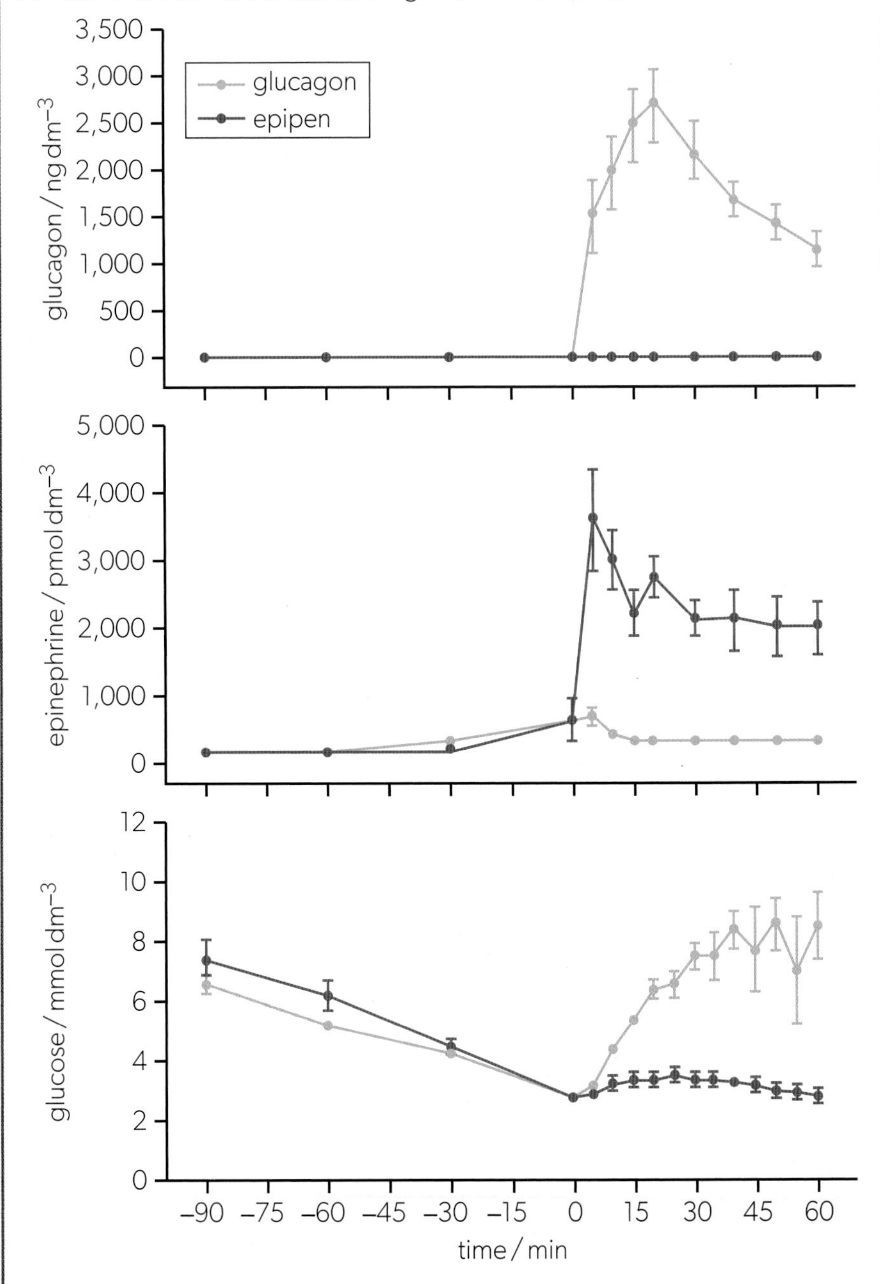

▲ **Figure 13** Source: T.P.C. Monsod et al;. Diabetes Care 1 April 2001; 24 (4): 701–704.

1. a. Compare and contrast the changes in glucagon and epinephrine following the injections. [3]

   b. Deduce whether these changes are statistically significant. [1]

2. a. Discuss how rapidly a glucagon injection causes blood glucose concentration to rise. [2]

   b. Explain the processes that occur after injection of glucagon. [2]

3. Evaluate the use of epipens for treatment of hypoglycemia in children with type 1 diabetes, using the data in the graphs. [2]

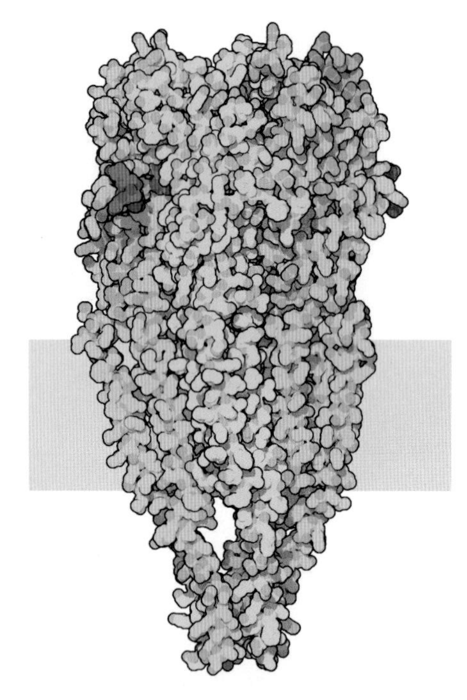

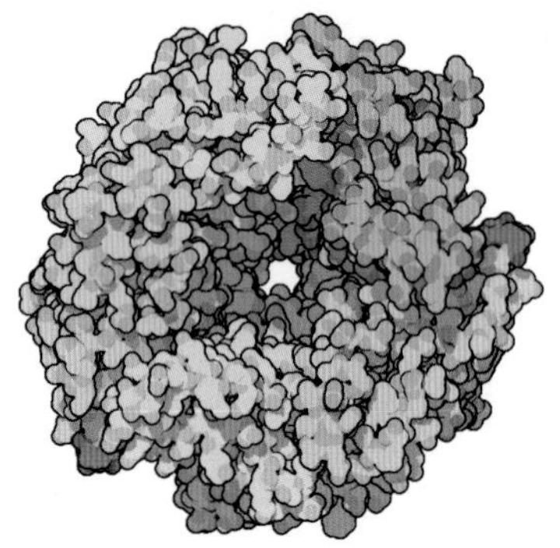

▲ Figure 14 Acetylcholine receptor viewed from the side and from the end. The binding site is shown in red and the membrane is shown in grey. The pore through which sodium ions pass is only open when acetylcholine is bound to the receptor

C2.1.8 Transmembrane receptors for neurotransmitters and changes to membrane potential

Neurotransmitters convey signals between neurons, and between neurons and muscle. Neurotransmitters are released into the synaptic gap and diffuse to the membrane of the postsynaptic neuron or the muscle fibre. There, they bind to receptors in these membranes. The receptors are transmembrane proteins. Binding causes membrane channels to open and ions move through these channels by facilitated diffusion, changing the membrane potential. This change in potential is a signal that stimulates or inhibits either a nerve impulse in a postsynaptic neuron, or a contraction in a muscle fibre.

Acetylcholine is used as a neurotransmitter in many synapses, including those between neurons and muscle fibres. When acetylcholine binds to the binding site on an acetylcholine receptor, the conformation (shape) of the receptor changes. A channel opens, allowing sodium ions to pass into the cell. This leads to a local depolarization that triggers an action potential. Synaptic transmission and action potentials are described in *Topic C2.2*.

C2.1.9 Transmembrane receptors that activate G protein

G-protein-coupled receptors (GPCRs) are a large and diverse group of transmembrane receptors. They convey signals into cells using a second protein located in the plasma membrane, called G protein. The three subunits of G protein (α, β and γ) assemble on the receptor. A molecule of guanosine diphosphate (GDP) bound to the α subunit keeps the G protein in an inactive state.

When a ligand binds to the binding site on the receptor, the receptor changes shape. This causes changes in the coupled G protein, so the GDP detaches from the α subunit. This allows guanosine triphosphate (GTP) to bind in its place. Binding of GTP activates the G protein, which separates into its subunits and dissociates from the receptor. The activated G protein subunits cause further changes within the cell, triggering the cell's response to the signal brought by the ligand.

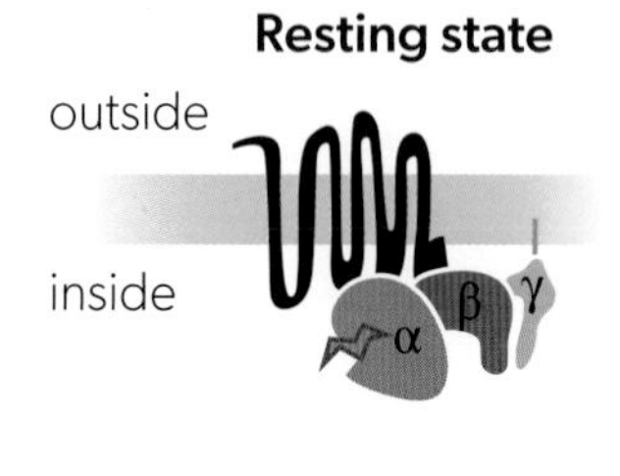

The alpha subunit of the G protein is inactive because GDP is bound to it.

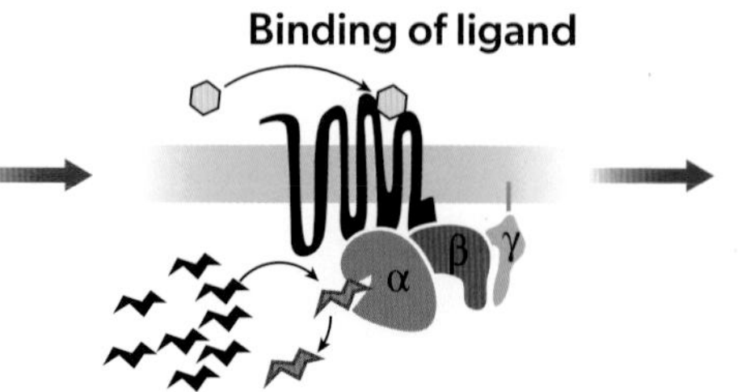

A ligand binds to the receptor, causing conformational changes which displace GDP from the alpha subunit.
This allows GTP to bind, activating the G protein.

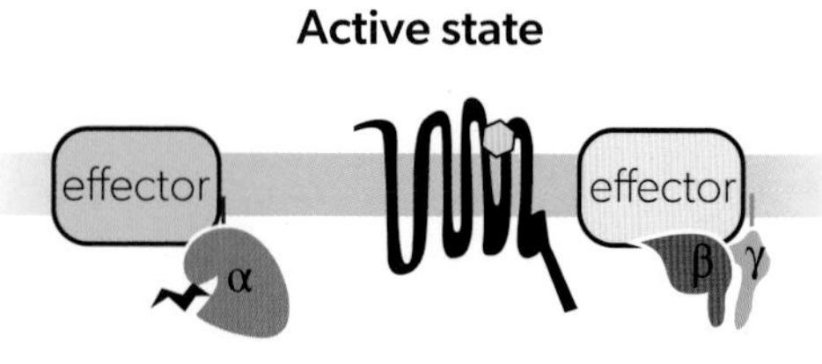

The activated G protein splits into alpha, beta and gamma subunits which convey signals to effectors within the cell.

▲ Figure 15 Signal transmission into cells by G-protein-coupled receptors

A broad range of receptor functions are mediated by G-protein-coupled receptors and their associated G proteins. The ligands that bind to these receptors are diverse and include light-sensitive compounds, odours, pheromones, hormones and neurotransmitters.

Activity: Structure comparisons

1. a. Compare and contrast the structures of GTP and ATP.

   b. Compare and contrast the structures of ATP and AMP.

   c. Compare and contrast the structures of AMP and cAMP.

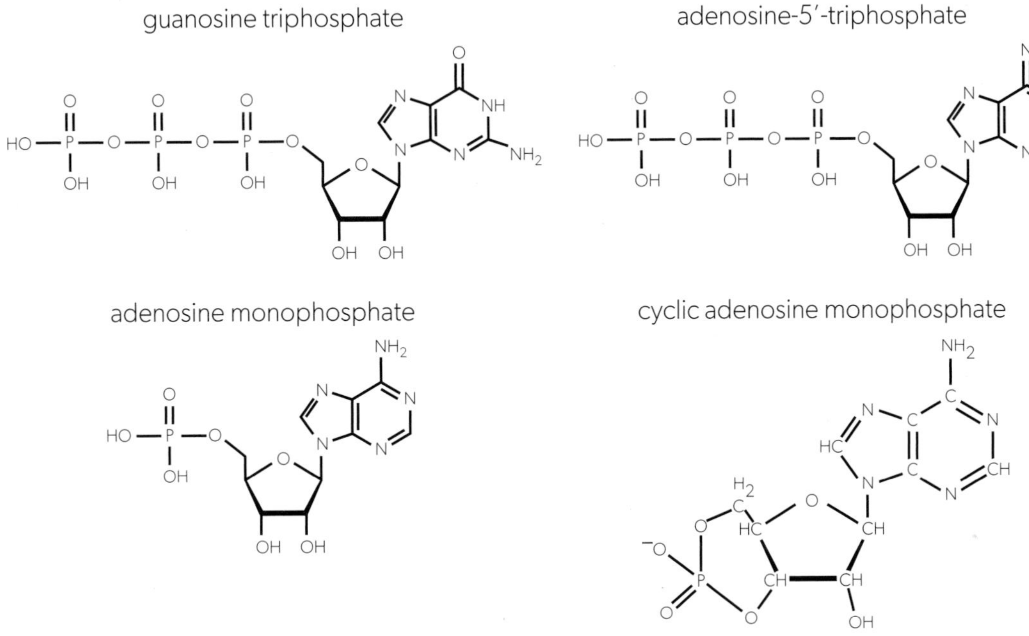

guanosine triphosphate

adenosine-5′-triphosphate

adenosine monophosphate

cyclic adenosine monophosphate

▲ **Figure 16** Structures of GTP, ATP, AMP, cAMP

C2.1.10 Mechanism of action of epinephrine (adrenaline) receptors

Epinephrine binds to a transmembrane receptor in the plasma membrane of target cells. This changes the shape of the receptor, activating G protein within the membrane. Activated G protein activates the enzyme adenylyl cyclase in the membrane and this converts ATP in the cytoplasm into cyclic AMP (cAMP). cAMP is the second or secondary messenger. Secondary messengers start a sequence of responses within the cell, amplifying the signal until a large-scale process is triggered. This happens very rapidly—for example, liver cells break down glucose and release glucose into the blood within seconds of receiving an epinephrine signal.

AHL

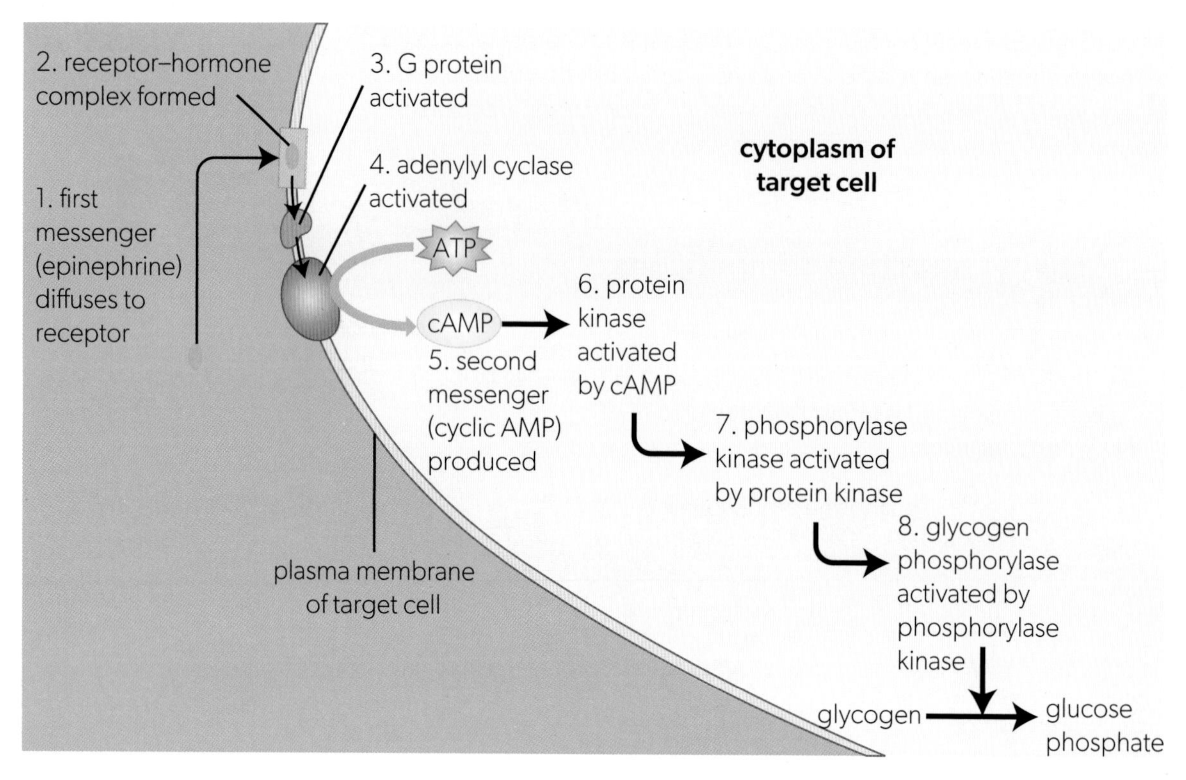
2. receptor–hormone complex formed
3. G protein activated
cytoplasm of target cell
4. adenylyl cyclase activated
1. first messenger (epinephrine) diffuses to receptor
ATP
6. protein kinase activated by cAMP
cAMP
5. second messenger (cyclic AMP) produced
7. phosphorylase kinase activated by protein kinase
8. glycogen phosphorylase activated by phosphorylase kinase
plasma membrane of target cell
glycogen
glucose phosphate

▲ Figure 17 Signal transduction pathway used in liver cells in response to epinephrine; this results in rapid release of glucose into the bloodstream

Science as a shared endeavour: Naming conventions

Naming conventions are an example of international cooperation in science for mutual benefit. Adrenaline and epinephrine are the same hormone. The term "adrenaline" relates to the adrenal gland, which produces the hormone. The term "epinephrine" refers to the position of the adrenal gland—above (epi-) the kidney (nephros). The chemical name of this hormone, based on the rules of the International Union of Pure and Applied Chemistry, is (R)-1-(3,4-dihydroxyphenyl)-2-methylaminoethanol.

In most of the world, the term "adrenaline" is more common. However, "epinephrine" is more commonly used in North America. Some people prefer to use a non-proprietary (generic) name—in this case, "epinephrine". However, there is a risk of confusion with the stimulant drug ephedrine, so other people prefer the term "adrenaline".

References

James, RH. 1998. "Ephedrine/Epinephrine Drug Label Confusion". *Anaesthesia*. Vol. 53. P 511.

"'Looks' like a Problem: Ephedrine – Epinephrine." *Institute For Safe Medication Practices*, 17 Apr 2003, https://www.ismp.org/resources/looks-problem-ephedrine-epinephrine.

C2.1.11 Transmembrane receptors with tyrosine kinase activity

A kinase is an enzyme that adds a phosphate group from ATP to a specific molecule. This process is called phosphorylation. For example, the enzyme tyrosine kinase transfers phosphate from ATP to the amino acid tyrosine in a protein.

Look at Figure 18. The insulin receptor (blue) is a transmembrane protein that is activated by the binding of insulin (orange). The two tails of the protein that extend into the cytoplasm are tyrosine kinase enzymes The binding of insulin causes structural changes in the receptor, so the two tails connect to form a dimer. Then each tail phosphorylates the other tail. These changes trigger a biochemical chain of events inside the cell (signal transduction). Vesicles containing glucose transporters move to the plasma membrane and fuse with it, inserting transporters (shown in red) into the membrane. These transporters are channel proteins that allow glucose (yellow) to enter the cell by facilitated diffusion. The glucose can then be used as a substrate in cell respiration.

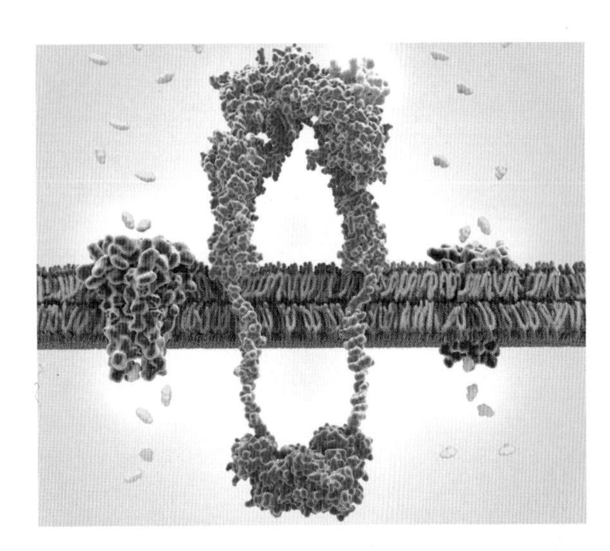

▲ Figure 18 Transmembrane insulin receptors

C2.1.12 Intracellular receptors that affect gene expression

Steroid hormones are hydrophobic. This means they are soluble in lipids and able to pass through the cell membrane. Once inside the cell, they bind to receptors in the cytoplasm. The hormone–receptor complex enters the nucleus and attaches to the DNA. This activates the production of a particular polypeptide. For example, the androgen receptor binds testosterone and the resulting complex increases production of the FADS1 gene. This in turn increases production of important fats in prostate cells.

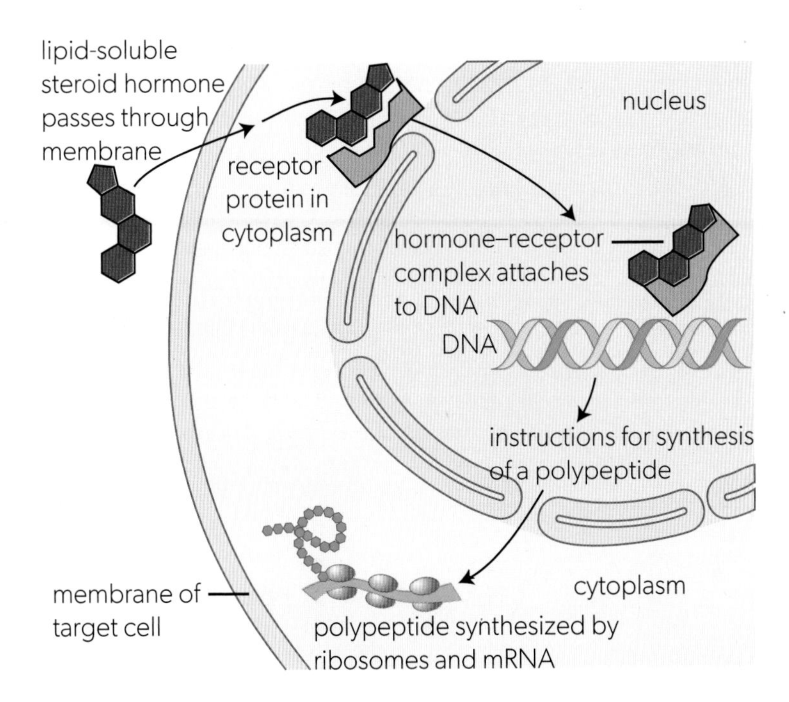

◀ Figure 19 Signal transduction pathway in cells that respond to steroid hormones

C2.1.13 Effects of the hormones oestradiol and progesterone on target cells

The hormones oestradiol and progesterone are involved in reproduction.

Oestradiol has a broad range of effects in the ovary and the uterus. It also acts on the brain, helping to regulate the release of reproductive hormones. Within the hypothalamus of the brain, the hormone gonadotropin-releasing hormone (GnRH) is produced and released. This hormone triggers the release of the sex hormones luteinizing hormone (LH) and follicle-stimulating hormone (FSH) from the anterior pituitary. At different stages of the human menstrual cycle, oestradiol can either inhibit or promote the release of GnRH by the hypothalamus. Just before and during ovulation, oestradiol has a stimulating effect by binding to a receptor within the cytoplasm of the hypothalamus cell. Once bound, the hormone–receptor complex moves to the nucleus where it acts as a transcription factor, enhancing the transcription of GnRH mRNA.

The hormone progesterone is produced by the ovary and maintains the uterine lining so that it can support a developing foetus. Progesterone is a steroid hormone, able to diffuse directly through the plasma membrane of uterine cells and bind to a receptor in the cytoplasm. The hormone–receptor complex then enters the nucleus where it interacts with DNA as a transcription factor. This affects gene expression. For example, one of the genes activated is insulin-like growth factor which contributes to the cellular proliferation necessary for maintaining the lining of the endometrium.

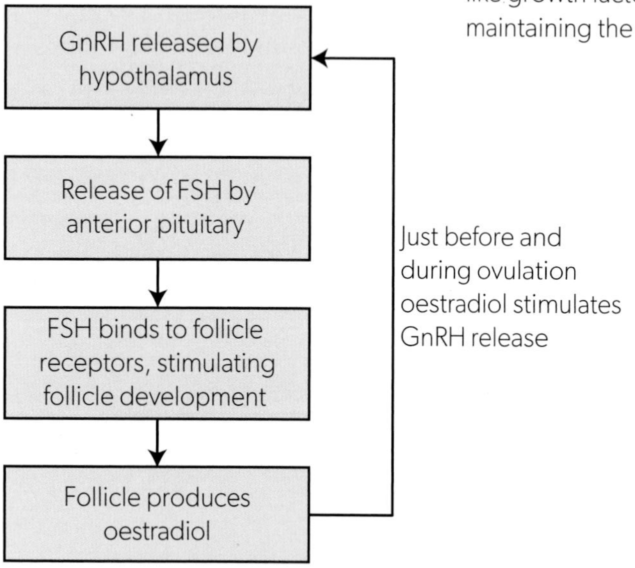

GnRH released by hypothalamus
↓
Release of FSH by anterior pituitary
↓
FSH binds to follicle receptors, stimulating follicle development
↓
Follicle produces oestradiol

Just before and during ovulation oestradiol stimulates GnRH release

▲ Figure 20 An example of positive feedback

C2.1.14 Regulation of cell signalling pathways by positive and negative feedback

In a positive feedback process, the end-product of a pathway amplifies the starting point so that more product is created. For example, in muscle, the endoplasmic reticulum (ER) stores calcium. The binding of an inositol trisphosphate (IP_3) molecule to an IP_3 receptor causes the partial release of calcium from the ER. This increase in Ca^{2+} activates the IP_3 receptor on a neighbouring calcium channel, causing further increases in Ca^{2+}. This process is known as calcium-induced calcium release.

In a negative feedback process, an increase in the end-product of a pathway shuts off the start of the pathway. In other words, the

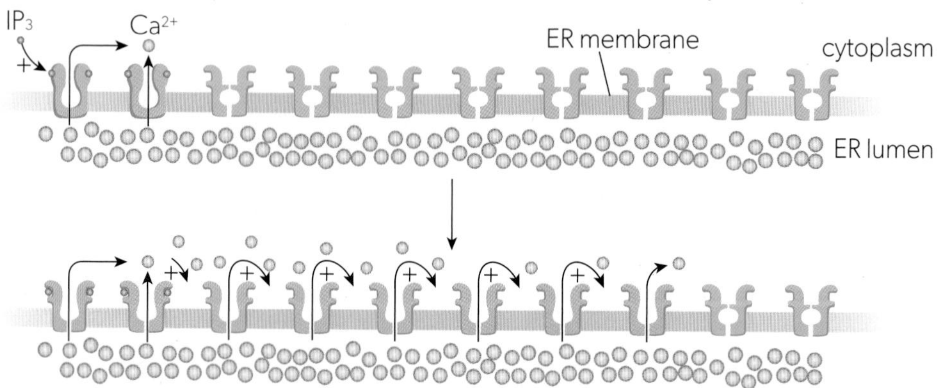

▶ Figure 21 Calcium-induced calcium release

end-product inhibits its own production. For example, testosterone production is regulated by negative feedback. GnRH released by the hypothalamus acts on the hypothalamus, stimulating the release of luteinizing hormone (LH). LH stimulates the release of testosterone from Leydig cells in the testes. Increasing testosterone levels have two effects:

- signals to the anterior pituitary decrease the release of LH
- signals to the hypothalamus stop the release of GnRH.

Exploring and designing: Testing a hypothesis

Bacteria in the mouth are in danger of being swallowed and then killed by stomach acid. They can prevent this by becoming attached to surfaces of the teeth and gums. Glycoproteins in saliva adhere to the surface of teeth, allowing bacteria to attach and then multiply. The bacteria use quorum sensing to detect when a high population density has developed. This triggers changes in gene expression. In particular, the bacteria secrete an extracellular polysaccharide that forms a layer of slime over the teeth; bacterial cells are immobilized in this slime. This type of biofilm is known as dental plaque. It can grow thicker as more bacteria adhere and multiply and more polysaccharide is secreted. Plaque formation has several advantages for bacteria:

- less risk of being washed off the tooth
- harder for white blood cells, antibodies and other antibacterials to penetrate and kill the bacteria
- different species of bacteria in the plaque can interact and benefit from each other's activities.

Bacteria in dental plaque secrete acids, which react with minerals in tooth enamel, causing dental decay. Removal of plaque is therefore advisable.

Disclosing tablets can be bought over the counter at drug stores/pharmacies or online. They contain a dye that stains dental plaque red or blue. This means they can be used to investigate the development or removal of plaque.

1. Suggest a hypothesis relating to factors affecting development of plaque or methods of removal.
2. Design an experiment to test your hypothesis.
3. Perform your experiment.
4. Use the results of the experiment to evaluate your hypothesis.

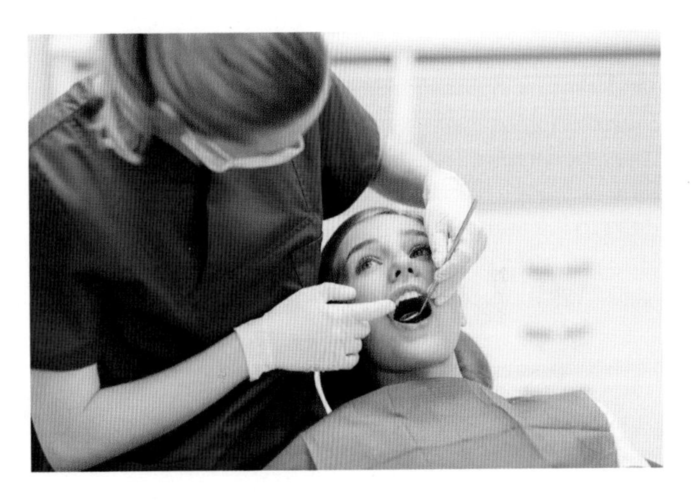

▲ **Figure 22** Dental inspection to check for decay

Linking questions

1. What patterns exist in communication in biological systems?
   a. Outline the process whereby heightened levels of glucose in the blood impact intracellular signalling. (C2.1.11)
   b. Explain the role of chemoreceptors in regulating ventilation rate. (C3.1.15)
   c. Outline, using an example, how an individual can signal reproductive fitness during mate selection. (D4.1.7)
2. In what ways is negative feedback evident at all levels of biological organization?
   a. Explain the mechanisms involved in the maintenance of body temperature. (D3.3.6)
   b. Outline the negative feedback control of population size by density-dependent factors. (C4.1.6)
   c. Outline the role of negative feedback in one example of cell signalling. (C2.1.14)

C2.2 Neural signalling

How are electrical signals generated and moved within neurons?

Neurons are cells within the nervous system that carry electrical impulses. As shown in Figure 1, each neuron has a large cell body with structures extending from it. The extensions usually consist of one thicker axon and several thinner dendrites. In a living system, the dendrites collect information. This information is interpreted by the cell body and then passed on to the axon. How is a resting electrical potential established? How is the nervous signal transmitted down the length of the axon? How is the neuron reset, ready for the next signal?

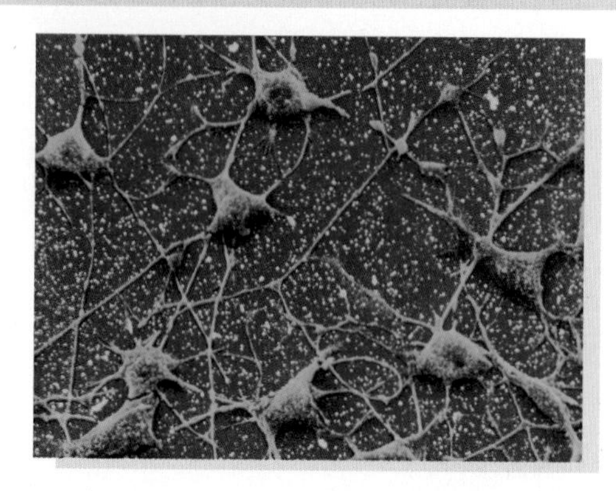

◀ Figure 1

How can neurons interact with other cells?

Figure 2 shows the junction between a nerve cell (green) and a muscle fibre (red). Such junctions are known as synapses. Neurons both receive and transmit information from other cells. How do neurons interact with sensory receptors? How do they then transmit it to the dendrites of neurons? The exchange of ions across axon membranes can lead to the transmission of electrical signals along an axon. How is the signal transmitted between cells? What processes occur at the synapse?

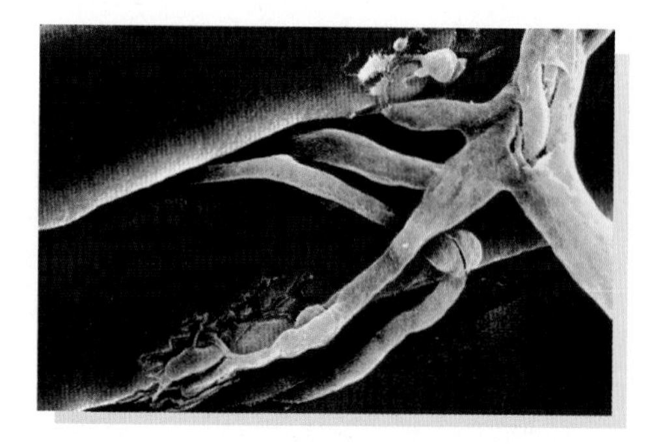

◀ Figure 2

SL and HL	AHL only
C2.2.1 Neurons as cells within the nervous system that carry electrical impulses C2.2.2 Generation of the resting potential by pumping to establish and maintain concentration gradients of sodium and potassium ions C2.2.3 Nerve impulses as action potentials that are propagated along nerve fibres C2.2.4 Variation in the speed of nerve impulses C2.2.5 Synapses as junctions between neurons and between neurons and effector cells C2.2.6 Release of neurotransmitter from a presynaptic membrane C2.2.7 Generation of an excitatory postsynaptic potential	C2.2.8 Depolarization and repolarization during action potentials C2.2.9 Propagation of an action potential along a nerve fibre/axon as a result of local currents. C2.2.10 Oscilloscope traces showing resting potentials and action potentials C2.2.11 Saltatory conduction in myelinated fibres to achieve faster impulses C2.2.12 Effects of exogenous chemicals on synaptic transmission C2.2.13 Inhibitory neurotransmitters and generation of inhibitory postsynaptic potentials C2.2.14 Summation of the effects of excitatory and inhibitory neurotransmitters in a postsynaptic neuron C2.2.15 Perception of pain by neurons with free nerve endings in the skin C2.2.16 Consciousness as a property that emerges from the interaction of individual neurons in the brain

C2.2.1 Neurons as cells within the nervous system that carry electrical impulses

Two body systems are used for internal communication: the endocrine system and the nervous system. The endocrine system consists of glands that release hormones. The nervous system consists of nerve cells called neurons. There are about 85 billion neurons in the human nervous system. Neurons help with internal communication by transmitting nerve impulses. A nerve impulse is an electrical signal.

Neurons have a cell body with cytoplasm and a nucleus. They also have narrow outgrowths called nerve fibres along which nerve impulses travel.

- Dendrites are short branched nerve fibres—for example, those used to transmit impulses between neurons in one part of the brain or spinal cord.
- Axons are very elongated nerve fibres—for example, those that transmit impulses from the tips of the toes or the fingers to the spinal cord.

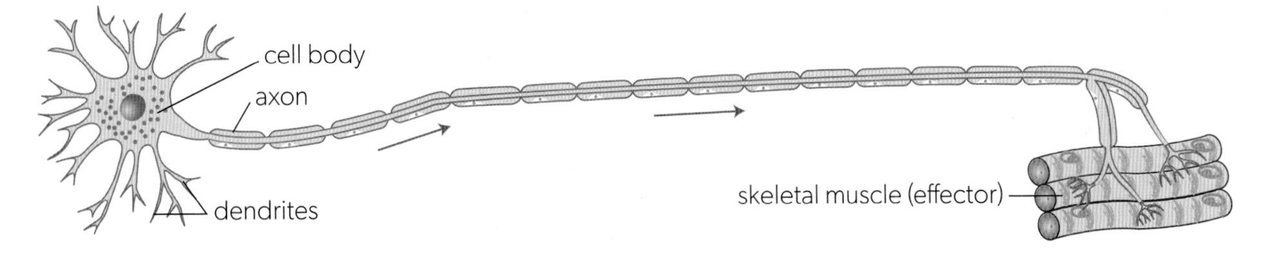

▲ Figure 3 A motor neuron with dendrites that transmit impulses to the cell body and an axon that transmits impulses a considerable distance to muscle fibres (axon length not to scale)

C2.2.2 Generation of the resting potential by pumping to establish and maintain concentration gradients of sodium and potassium ions

If microelectrodes are placed inside and outside any living cell, a voltage across the membrane will be detected. This voltage is usually between 10 and 100 millivolts; it is known as the membrane potential. This potential is due to an imbalance between the net charge (negative or positive) of cytoplasm and the fluid outside. The cytoplasm of cells is generally electrically negative compared with the fluid outside. For this reason, the membrane potential is expressed as a negative value—for example, liver cells have a potential of −40 mV.

When a neuron transmits an impulse, its membrane potential changes suddenly. When it is not transmitting an impulse, the potential across the membrane usually remains close to −70 mV. This is called the resting potential. Three factors contribute to it.

- Sodium–potassium pumps in the membrane transfer sodium ions (Na^+) out of the neuron and at the same time transfer potassium ions (K^+) in. This is active transport and uses energy from ATP. The numbers of ions

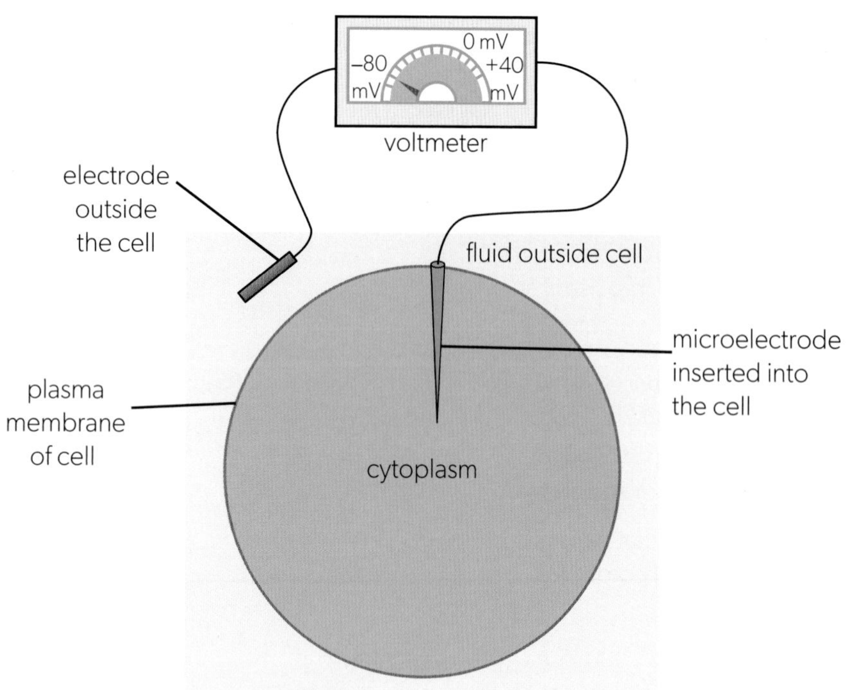

▲ Figure 4 Measuring membrane potential

pumped are unequal: when three Na^+ ions are pumped out, only two K^+ ions are pumped in. This creates a charge imbalance and concentration gradients for both ions.

- The pumped ions leak back across the membrane by diffusion. The diffusion rates are slow but the membrane is about 50 times more permeable to K^+ than to Na^+. Therefore, leakage of K^+ ions is faster. This increases the difference between the Na^+ and K^+ concentration gradients, increasing the overall charge imbalance across the membrane.
- There are negatively charged proteins inside the nerve fibre (organic anions), which also contribute to the charge imbalance.

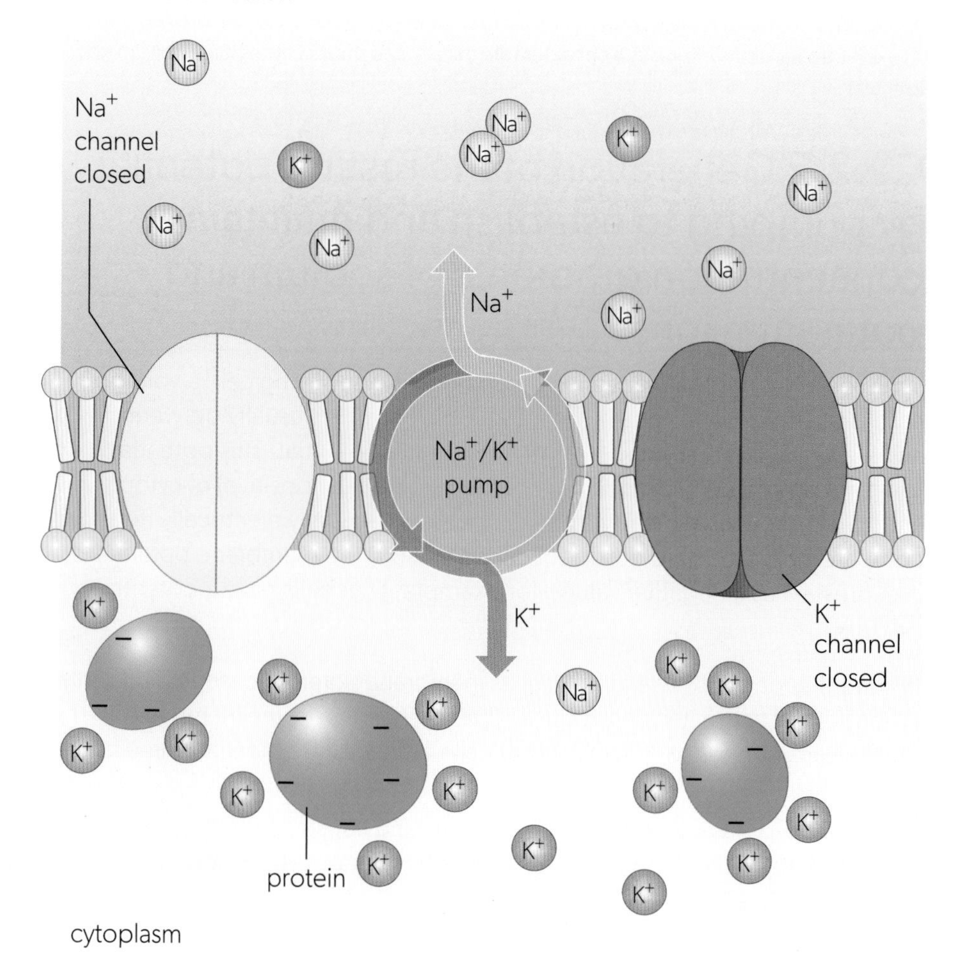

▶ Figure 5 The resting potential of a neuron is maintained by continual operation of the sodium–potassium pump

Data-based questions: Energy for the sodium–potassium pumps in the brain

Sodium–potassium pumps establish and maintain the resting potential in neurons of the brain. Samples of rabbit brain tissue were treated with ouabain, which inhibits the sodium–potassium ion pump. The oxygen consumption of the brain tissue was measured at different ouabain concentrations. The graph in Figure 6 shows the results. Four of the data points are mean values with error bars. The number of replicates is indicated in brackets next to the data point. The other data points on the graph are for concentrations where only a single measurement was taken.

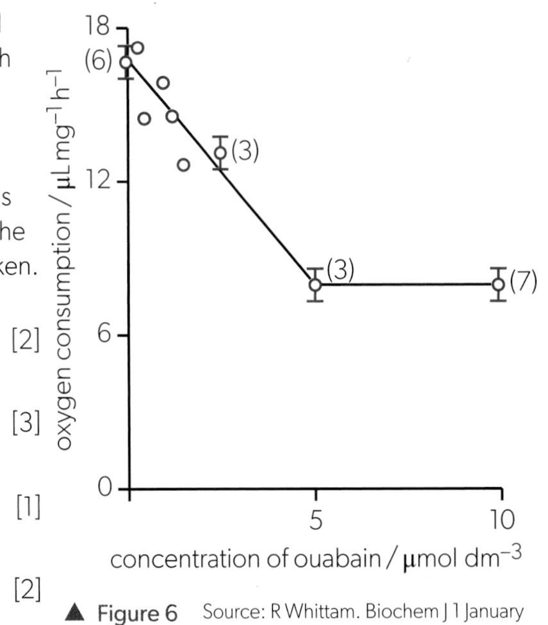

▲ **Figure 6** Source: R Whittam. Biochem J 1 January 1962; 82 (1): 205–212. doi: https://doi.org/10.1042/bj0820205.

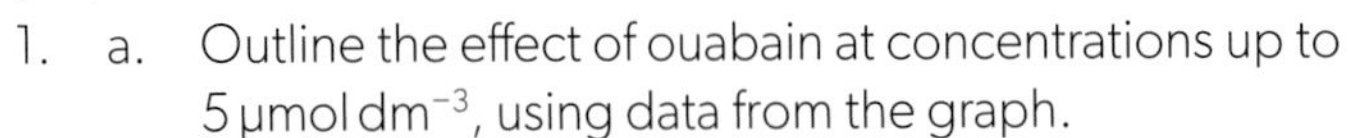

1. a. Outline the effect of ouabain at concentrations up to 5 μmol dm^{-3}, using data from the graph. [2]

   b. Explain this effect of ouabain, using your knowledge of cell respiration and active transport. [3]

2. a. Outline the results of the experiment at ouabain concentrations between 5 and 10 μmol dm^{-3}. [1]

   b. Deduce conclusions from the results between 5 and 10 μmol dm^{-3}. [2]

3. Use the data in the graph to estimate the proportion of energy expenditure in rabbit brain tissue that is used by sodium–potassium pumps. [1]

4. At some ouabain concentrations, there were no replicates; at other concentrations, there were only three. Suggest a reason for this. [1]

C2.2.3 Nerve impulses as action potentials that are propagated along nerve fibres

Electrodes can be used to monitor the membrane potential at one position along a nerve fibre. The potential can be displayed on a screen, with time on the x-axis and voltage on the y-axis. A horizontal line at about −70 mV represents the resting potential. A sudden spike represents an action potential. This is an all-or-nothing sequence of changes in membrane potential, with two main phases:

- depolarization—a change in membrane potential from negative to positive
- repolarization—a change back from positive to negative.

Both depolarization and repolarization are due to movement of positively charged ions across the membrane—not to movement of electrons.

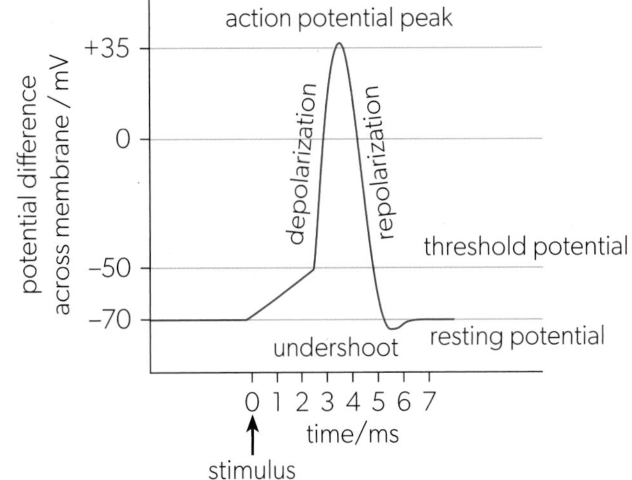

▲ Figure 7 Changes in membrane potential during an action potential

Depolarization is due to the opening of sodium channels in the membrane, allowing Na^+ ions to diffuse into the neuron down the concentration gradient. The concentration of sodium ions outside is about 10 times as high as that inside. The entry of Na^+ ions reverses the charge imbalance across the membrane, so the inside is positive relative to the outside. The raises the membrane potential, typically from about −70 mV to a positive value of about +30 mV.

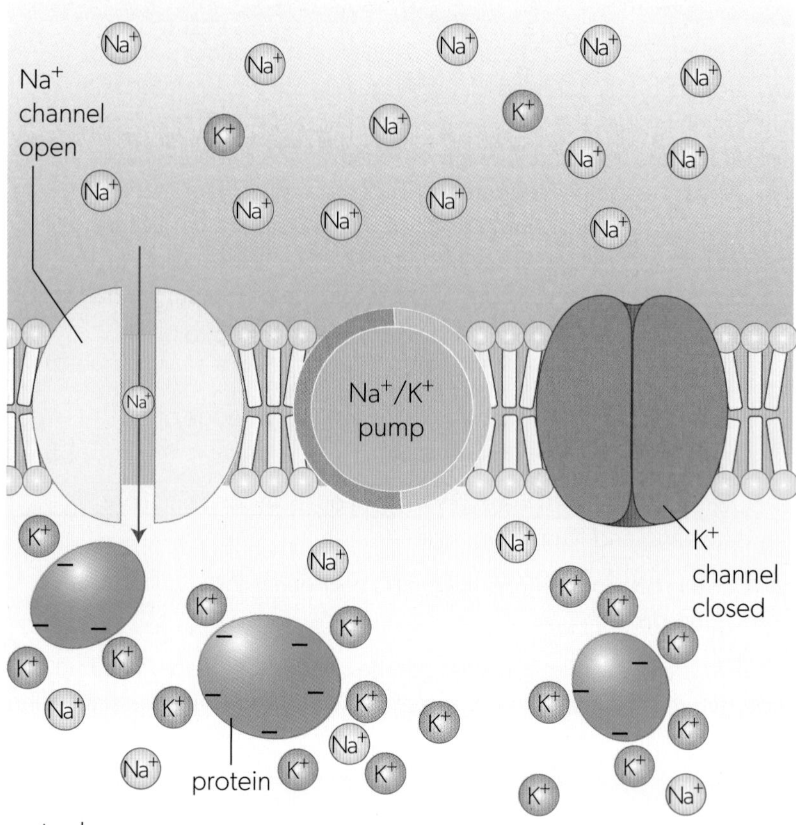

▶ Figure 8 Neuron depolarizing

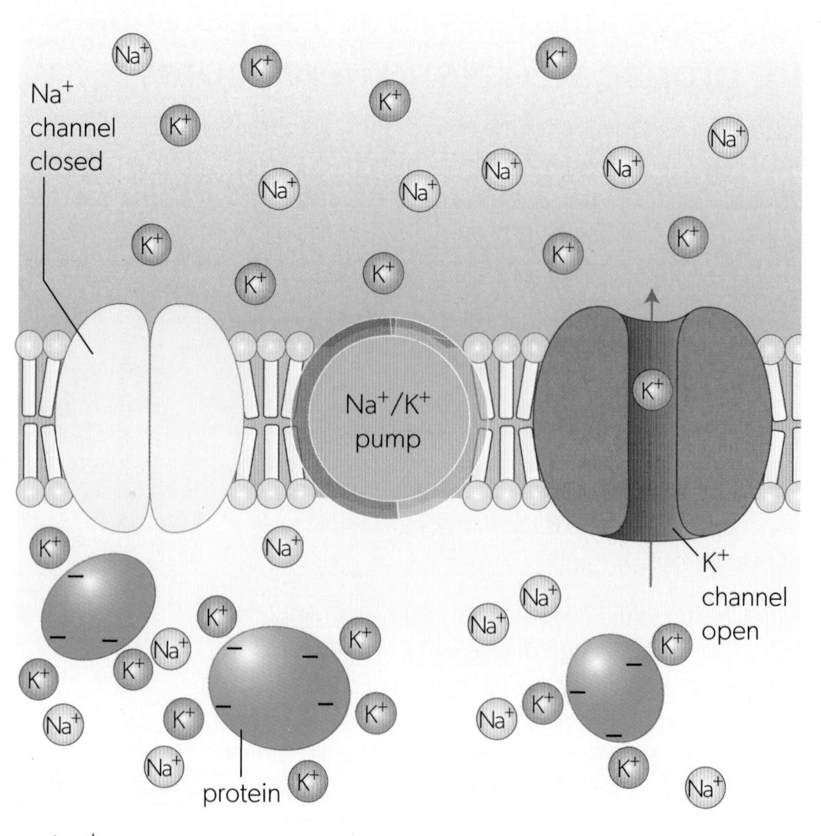

▶ Figure 9 Neuron repolarizing

Repolarization happens rapidly after depolarization and is due to the closing of the sodium channels and opening of potassium channels in the membrane. Potassium ions diffuse out of the neuron down their concentration gradient and no more sodium ions diffuse in. As a result, the inside of the neuron becomes negative again relative to the outside. The potassium channels remain open until the membrane potential has fallen to close to −70 mV.

The diffusion of potassium repolarizes the neuron, but it does not fully restore the resting potential as the concentration gradients of sodium and potassium ions have not yet been re-established. This takes a few milliseconds of actively pumping Na^+ out and K^+ in, before there can then be another action potential.

Action potentials are propagated along nerve fibres, because the ion movements that depolarize one part of the fibre trigger depolarization in the neighbouring part of the fibre. This is how neural signals pass along nerve fibres. A nerve impulse is an action potential that starts at one end of a neuron and is propagated along the axon to the other end of the neuron.

In humans and other vertebrates, nerve impulses always move in one direction along neurons. This is because an impulse can only be initiated at one terminal of a neuron and can only be passed on to other neurons or different cell types at the other terminal. Also, there is a refractory period after a depolarization that prevents propagation of an action potential backwards along an axon.

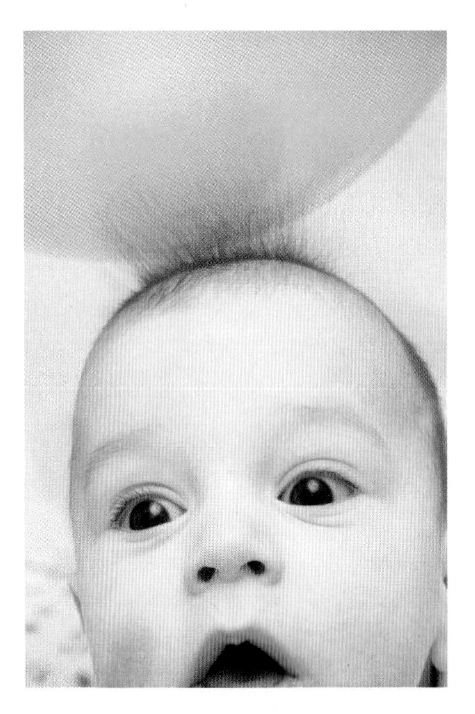

▲ Figure 10 Nerve impulses are electrical but they do not involve movements of electrons. Remember, electricity is energy associated with any negatively or positively charged particles, not just electrons

C2.2.4 Variation in the speed of nerve impulses

Nerve fibres are circular in cross-section, with a plasma membrane enclosing cytoplasm. In humans, the diameter is typically about 1 µm, although some nerve fibres are wider. Nerve impulses are conducted along nerve fibres at a speed of about 1 metre per second.

Some animals have nerve fibres with larger diameters. An increase in diameter reduces resistance, so impulses are transmitted along wider fibres more quickly. For example, giant axons in squid have a diameter up to 500 µm and conduct impulses at 25 metres per second. These axons are used to coordinate a rapid jet-propulsion escape response when a squid is in danger. Animals do not have the space or resources for many giant axons, so they can only use them to coordinate actions where speed is vital. For example, earthworms have just three giant axons that they use for an escape response to predator attacks.

Myelination is another modification of nerve fibres that increases the speed of nerve impulses. This is a coating of nerve fibres that consists of a series of Schwann cells, with gaps between called nodes of Ranvier. In myelinated nerve fibres the nerve impulse can jump from one node of Ranvier to the next, speeding up transmission along the nerve fibre to as much as 100 metres per second.

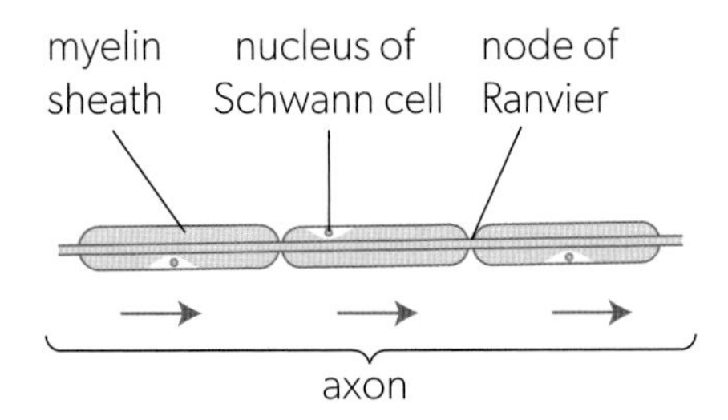

◀ Figure 11 Detail of a myelinated nerve fibre showing the gaps between adjacent Schwann cells (nodes of Ranvier)

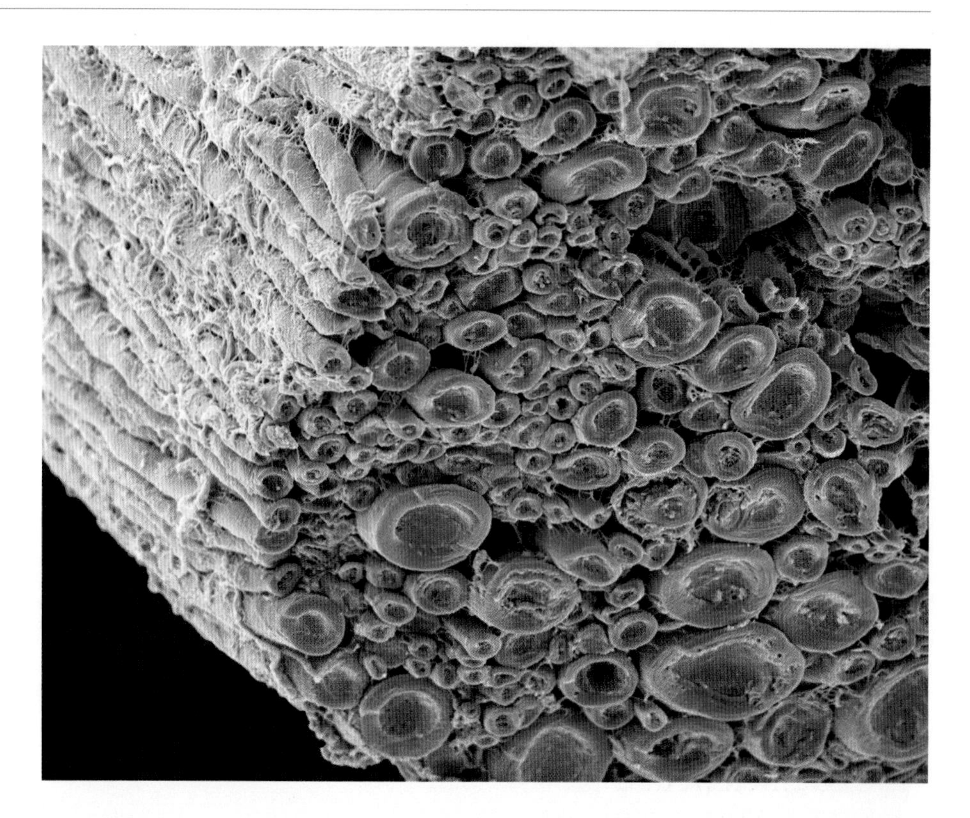

▶ Figure 12 An electron micrograph of the cut end of a nerve, which is a bundle of axons. The outer coat of the nerve has been removed. Myelin sheaths (purple) can be seen surrounding the axons (brown). The magnification of the micrograph is ×2,000. Calculate the range of diameters of the axons, not including the myelin sheaths around them

Data-based questions: Conduction velocities of nerve fibres and muscle

Table 1 gives data about the diameter of nerve fibres and muscle and the velocity of propagation of impulses.

Type	Diameter /μm	Velocity / m s^{-1}
Non-myelinated axons	1.0	1.3
	1.5	2.0
Squid giant axon (non-myelinated)	500	25
Myelinated axons	5	30
	12	70
	20	120
Smooth muscle cells	5	0.05
Cardiac muscle cells	15	0.5
Skeletal (striated) muscle fibres	50	6.0

▲ Table 1

1. Use an appropriate graph or chart to display the data in Table 1, so that the relationship between diameter, myelination and velocity of propagation can be evaluated. [5]
2. Evaluate the evidence provided by the data for the hypothesis that:
   a. there is a positive correlation between diameter and velocity of propagation, and calculate an r^2 value to support your answer [3]
   b. myelination increases the velocity of propagation of impulses. [2]

Mathematics: The correlation coefficient and the coefficient of determination

The correlation coefficient

In statistical terms, a correlation is any statistical association. It most often refers to the degree of linear association between a pair of variables. Correlation can be negative or positive as well as strong or weak.

The correlation coefficient (r) is a mathematical tool used to determine the strength of a correlation. The closer the absolute value of r is to 1, the stronger the correlation.

1. Figures 13 and 14 show the correlation between height and the speed of conduction of two nerves found in the arm, the median sensory nerve and the ulnar nerve.
   a. State the type of correlation (positive or negative) in each graph.
   b. Analyse the r values provided with respect to conduction velocity versus height for the two nerves.

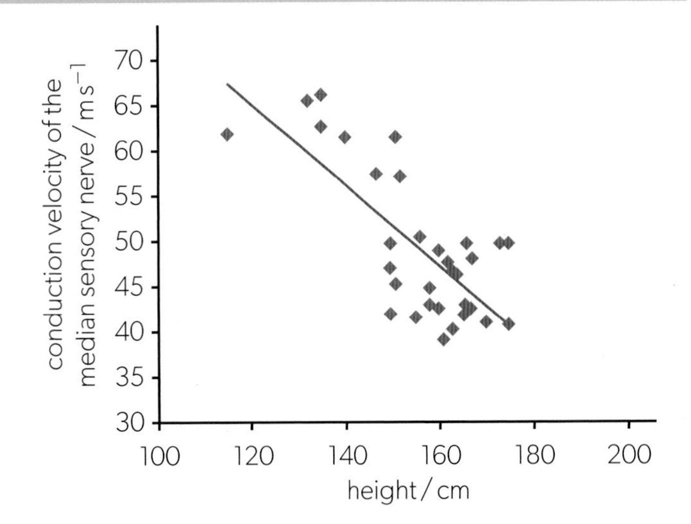

▲ **Figure 13** Body height and conduction velocity of the median sensory nerve in the arms of 37 people. $r = -0.740$, mean = 49.15 ± 8.11

The coefficient of determination

The coefficient of determination (R^2) is a statistic designed to evaluate the degree to which variation in the independent variable explains the variation in the dependent variable. It can be expressed as the percentage of the variation in the dependent variable that is explained by the variation in the independent variable. Another way to explain it is that it represents the percentage of the data that is closest to the line of best fit. For example, for the data in Figure 13, $R^2 = 0.1523$. This means that only 15% of the variation in conduction velocity is explained by the variation in height of the subject; many of the data points are far from the line of best fit. If the line of best fit passed through every point of the scatter plot, R^2 would be 1 and the variation in height would explain all of the variation in conduction velocity.

2. The data points in Table 2 represent the relationship between conduction velocity and axon diameter in mammalian nerves that are surrounded by a myelin sheath.

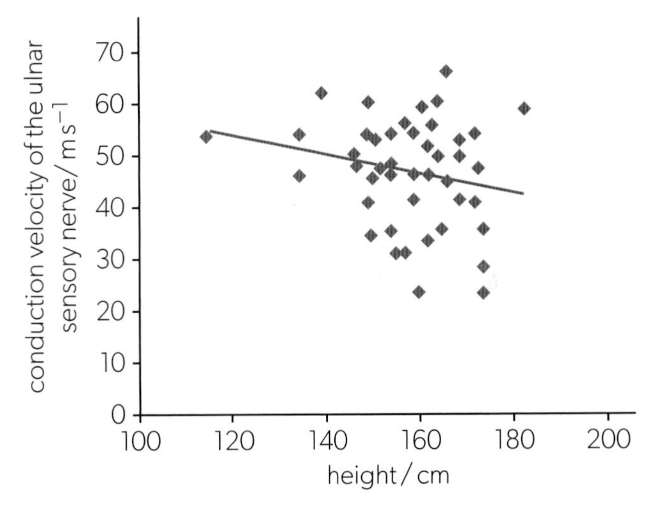

▲ **Figure 14** Body height and conduction velocity of the ulnar sensory nerve in the arms of 37 people. $r = -0.220$, mean = 46.67 ± 10.62

Source of Figures: Sulaxane, Y. and Bhavasar, R. (2017). National Journal of Physiology, Pharmacy and Pharmacology, p. 1. Available at: https://doi.org/10.5455/njppp.2017.7.0410317042017.

Axon diameter / μm	Conduction velocity / m s^{-1}	Axon diameter / μm	Conduction velocity / m s^{-1}
2.1	9	6.5	38
2.2	15	11.0	53
3.0	10	12.8	81
2.8	17	12.9	82
16.1	98	13.0	86
20.0	120	8.0	41
14.0	70	13.9	76
10.0	58	10.3	51

▲ Table 2

Follow these steps to find the correlation coefficient:

- Enter the data into a spreadsheet program.
- Click in an empty cell where you would like the *r* value to appear.
- From the functions menu, choose "CORREL" and then highlight the first column. The associated range should appear in the *r* value cell. Add a comma, then highlight the second column.
- Press return and the *r* value will appear in the cell.

Use a suitable software program to graph the data points. Choose a scatter plot to represent the data. Add a trendline and click in the box indicating that R^2 should be shown.

a. Determine the *r* and R^2 values.

b. Analyse the *r* value.

c. Analyse the R^2 value.

Collecting and processing: Calculating reaction time

The speed of transmission of nerve impulses allows rapid responses to stimuli. A simple method of measuring reaction time involves dropping a ruler.

Two students working together can assess reaction time. The subject rests their elbow on a table with their hand extended over the edge. The other student holds a metre stick with the 0 cm mark between the subject's thumb and index finger. They let go of the ruler and the subject attempts to catch it as quickly as possible.

The distance the ruler falls gives a measure of reaction time. This allows investigation of factors that might affect reaction time, such as the effect of auditory distraction or whether the subject has one or both eyes open. Variables must be carefully controlled.

Reaction time, *t*, can be calculated using the formula:

$$t = \sqrt{\frac{2d}{g}}$$

where *g* is the acceleration due to gravity (980 cm s^{-2}) and *d* is the distance measurement from the ruler.

Computer-based reaction timers are also available. You can find these by searching for "online reaction timer". Compare and contrast the results from the two methods.

C2.2.5 Synapses as junctions between neurons and between neurons and effector cells

A synapse is a junction between two cells in the nervous system. There are three main types of junction.

- synapses between sensory receptor cells and neurons, in sense organs
- synapses between neurons, in both the brain and spinal cord
- synapses between neurons and muscle fibres or gland cells. Muscles and glands are called effectors, because they effect (carry out) a response to a stimulus.

Signals can only pass in one direction across a synapse. The presynaptic neuron brings the signal to the synapse in the form of a nerve impulse or action potential. The postsynaptic neuron carries the signal away from the synapse, again in the form of a nerve impulse. Chemicals called neurotransmitters carry signals across a narrow fluid-filled gap between the presynaptic and postsynaptic neurons. This gap is only about 20 nm wide.

C2.2.6 Release of neurotransmitter from a presynaptic membrane

Synaptic transmission occurs very rapidly as a result of these events:

- A nerve impulse is propagated along the presynaptic neuron until it reaches the end of the neuron and the presynaptic membrane.
- Depolarization of the presynaptic membrane causes calcium ions (Ca^{2+}) to diffuse through channels in the membrane into the neuron.
- Influx of Ca^{2+} causes vesicles containing neurotransmitter to move to the presynaptic membrane and fuse with it.
- Neurotransmitter is released into the synaptic gap by exocytosis.

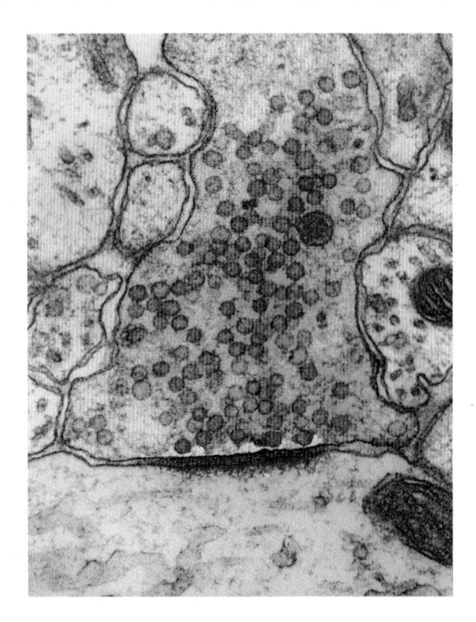

▲ Figure 15 Electron micrograph of a synapse. False colour has been used to indicate the presynaptic neuron (red) with vesicles of neurotransmitter (purple) and the postsynaptic neuron (green). The synaptic gap (orange) is very narrow. The slight swelling at the end of the presynaptic neuron is called the synaptic knob

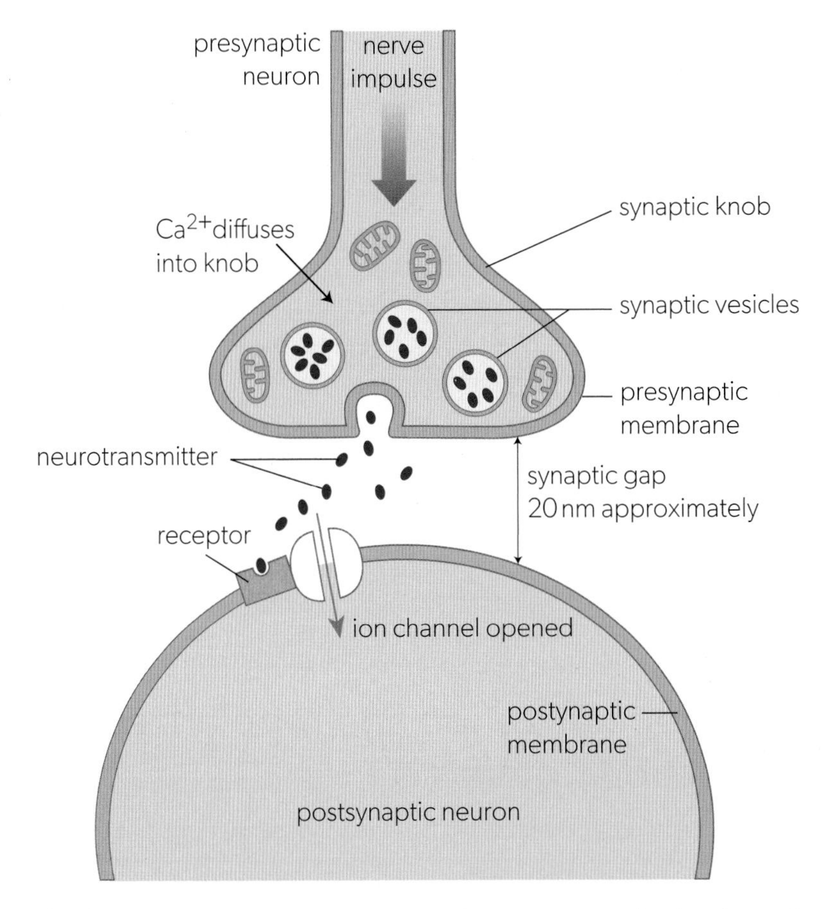

▲ Figure 16 A nerve impulse is propagated across a synapse by the release of neurotransmitter, followed by diffusion and binding to receptors

C2.2.7 Generation of an excitatory postsynaptic potential

Release of neurotransmitter from a presynaptic neuron leads to a series of events that trigger an action potential in the postsynaptic neuron.

- Neurotransmitter molecules diffuse across the synaptic gap. This happens extremely rapidly because the distance is so short (20–40 nm). The gap between the membranes is only two to four times the thickness of a typical phospholipid bilayer.

- The neurotransmitter binds to receptors in the postsynaptic membrane, causing ion channels to open. Some receptors have an ion channel as part of their structure. Other receptors cause an ion channel to open in a separate membrane protein.
- Ions diffuse down their concentration gradient into the postsynaptic neuron, causing the membrane potential to change. In most cases, the potential rises (becomes less negative)—this is called an excitatory postsynaptic potential.
- If the excitatory postsynaptic potential is strong enough, it triggers an action potential which propagates away from the synapse.
- The neurotransmitter is rapidly broken down and removed from the synaptic gap.

Many different neurotransmitters are used at synapses, with different effects. For example, acetylcholine is used as the neurotransmitter in many synapses, including neuromuscular junctions (synapses between neurons and muscle fibres). In the presynaptic neuron, choline (absorbed from the diet) is combined with an acetyl group produced by aerobic respiration. This produces acetylcholine, which is loaded into vesicles and then released into the synaptic gap during synaptic transmission.

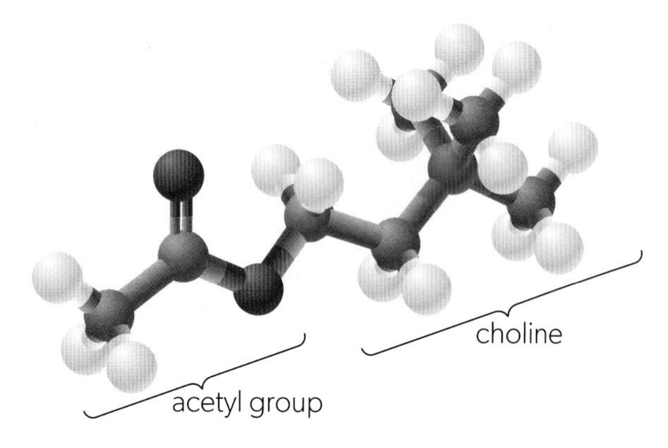

▲ Figure 17 Acetylcholine

When acetylcholine binds to its receptor in the postsynaptic membrane, a channel opens in the receptor. Sodium ions diffuse through this channel and into the postsynaptic membrane, causing an excitatory postsynaptic potential.

The acetylcholine only remains bound to the receptor for a short time and only one action potential is initiated in the postsynaptic neuron. This is because the enzyme acetylcholinesterase is present in the synaptic gap and rapidly breaks acetylcholine down into choline and acetate. The choline is reabsorbed into the presynaptic neuron, where it is converted back into acetylcholine by recombining with an acetyl group.

AHL

C2.2.8 Depolarization and repolarization during action potentials

Opening of the sodium and potassium channels that cause depolarization and repolarization is triggered by changes in the transmembrane voltage. This is called voltage-gating. If the resting potential of −70 mV increases to −50 mV, sodium channels in the membrane start to open. This allows sodium ions to diffuse into the axon, further reducing the membrane potential and causing more sodium channels to open. This is an example of positive feedback and causes the very rapid change in membrane potential from −50 to +30 mV that characterizes depolarization during an action potential.

The voltage that causes sodium channels to open is called the threshold potential. Depolarization will not occur unless the threshold potential is reached; instead, the sodium–potassium pump will re-establish the resting potential of −70 mV. A nerve impulse is "all-or-nothing" because the threshold potential must be reached.

Sodium channels remain open for a very short time—one to two milliseconds—before they close again. Their opening allows a pulse of sodium ions to diffuse out. The resulting depolarization causes voltage-gated potassium channels to open. These channels also remain open for one to two milliseconds, before closing. Even in this short time, enough potassium ions diffuse out of the axon to repolarize the axon. The membrane potential returns to −70 mV; it may briefly become more negative than this, before the sodium–potassium pump re-establishes concentration gradients.

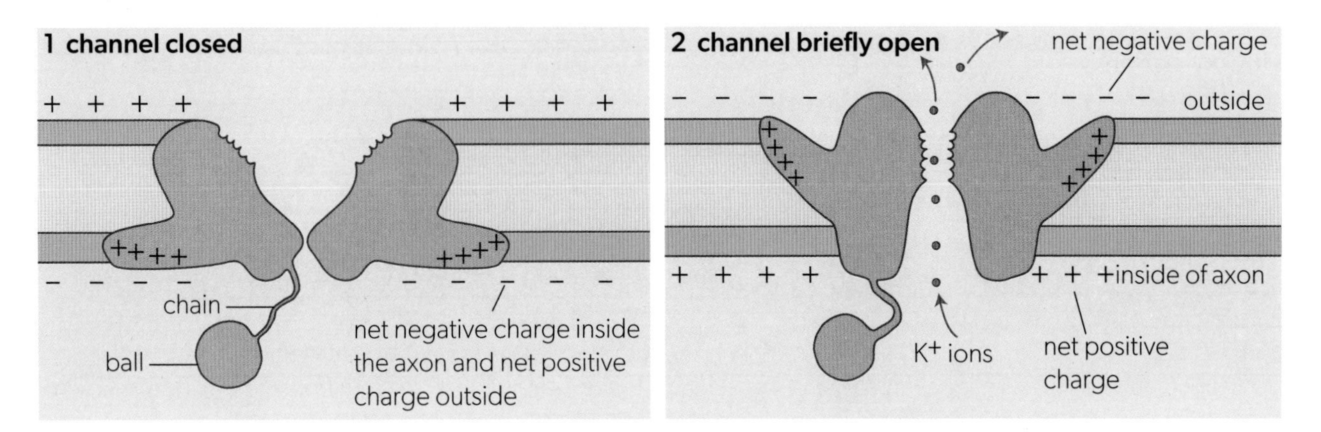

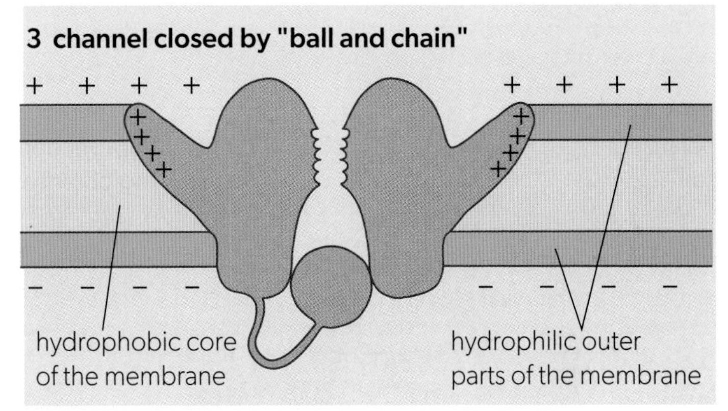

◀ Figure 18 Depolarization of the axon caused by inward diffusion of Na^+ causes a conformation change, opening the potassium channel and allowing outward diffusion, which repolarizes the axon (stage 2). After a few milliseconds, the protein ball attached to the potassium channel blocks it (stage 3). The channel then returns to its former closed conformation (stage 1)

C2.2.9 Propagation of an action potential along a nerve fibre/axon as a result of local currents

The propagation of an action potential along an axon is due to movements of sodium ions. Depolarization of part of the axon is due to diffusion of sodium ions into the axon through sodium channels. This reduces the concentration of sodium ions outside the axon and increases it inside. The depolarized part of the axon therefore has different sodium ion concentrations to the neighbouring part of the axon that has not yet depolarized. As a result, sodium ions diffuse between these regions both inside and outside the axon.

Inside the axon, there is a higher sodium ion concentration in the depolarized part of the axon. As a result, sodium ions diffuse along inside the axon to the neighbouring part that is still polarized. Outside the axon, the concentration gradient is in the opposite direction so sodium ions diffuse from the polarized part back to the part that has just depolarized. These movements are called local currents (see Figure 19).

Local currents reduce the concentration gradient in the part of the neuron that has not yet depolarized. This makes the membrane potential rise from the resting potential of −70 mV to about −50 mV. Sodium channels in the axon membrane are voltage-gated and open when a membrane potential of −50 mV is reached. This is the threshold potential. Opening of the sodium channels causes depolarization. Thus local currents cause a wave of depolarization and then repolarization to be propagated along the axon at a rate between 1 and 100 (or more) metres per second.

▼ Figure 19 Local currents

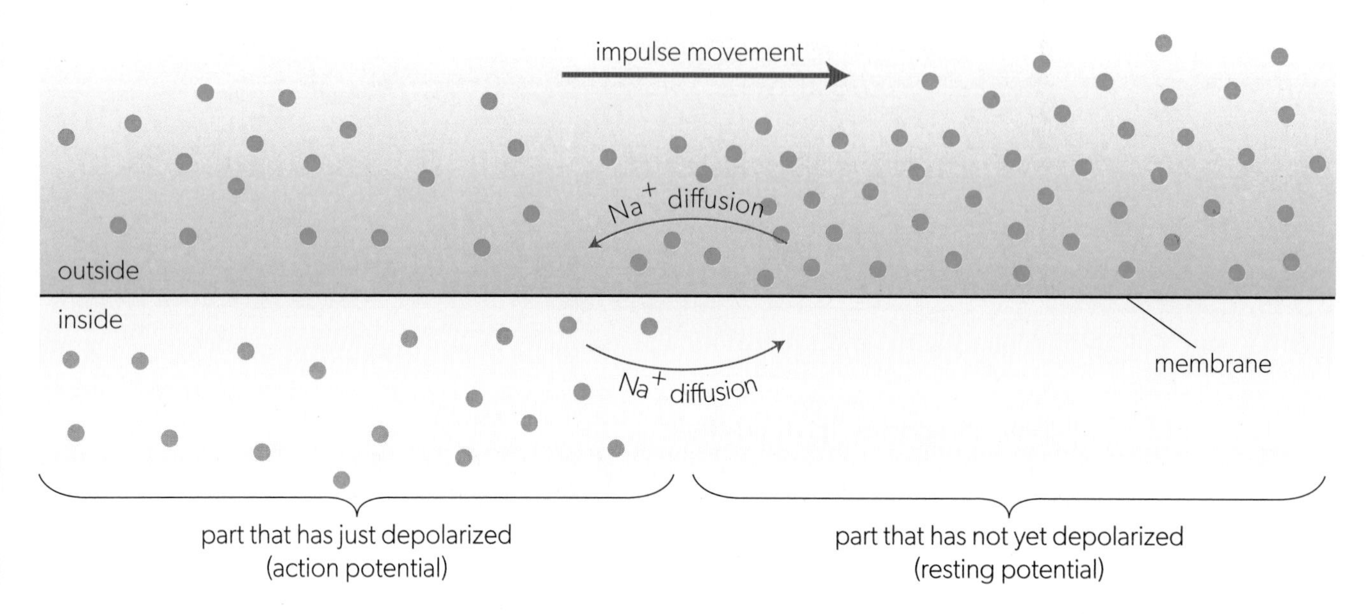

C2.2.10 Oscilloscope traces showing resting potentials and action potentials

Membrane potentials in neurons can be measured by placing electrodes on each side of the membrane. The potentials can be displayed using an oscilloscope. The display is similar to a graph with time on the x-axis and the membrane potential on the y-axis. If there is a resting potential, a horizontal line appears on the oscilloscope screen at a level of −70 mV (assuming this is the resting potential of the neuron). If an action potential occurs, a narrow spike is seen. The rising and falling phases of this spike show the depolarization and repolarization.

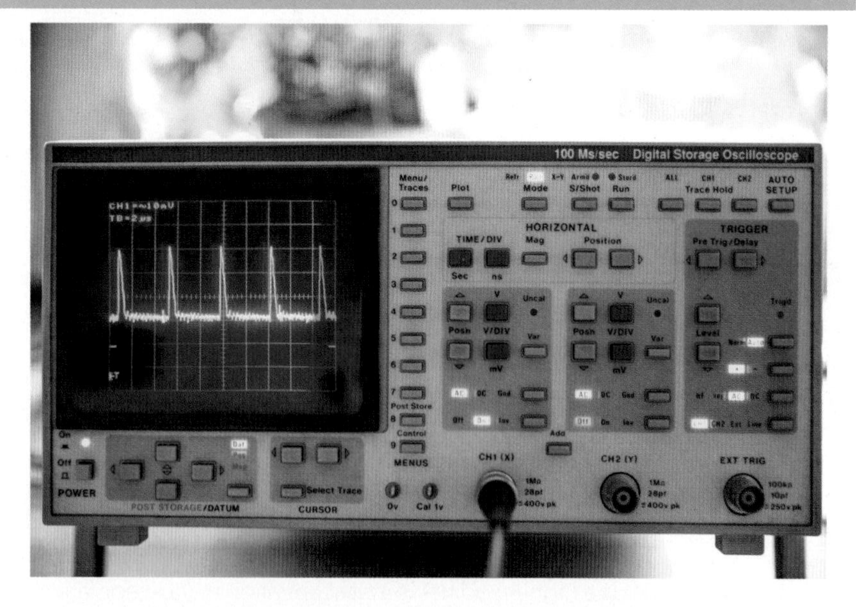

▲ Figure 20 Oscilloscope showing a trace with action potentials

Data-based questions: Analysing an oscilloscope trace

The oscilloscope trace in Figure 21 was taken from a digital oscilloscope. It shows an action potential in a mouse hippocampal pyramidal neuron that happened after the neuron was stimulated with a pulse of current.

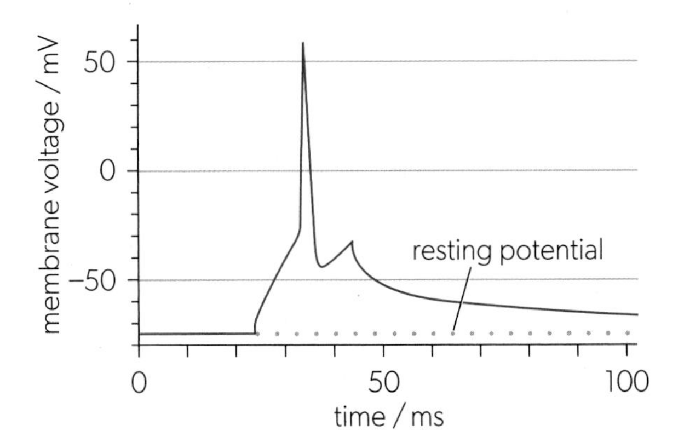

▲ **Figure 21** Oscilloscope trace taken from a digital oscilloscope, showing an action potential in a mouse hippocampal pyramidal neuron after it was stimulated with a pulse of current

1. State the resting potential of the mouse hippocampal pyramidal neuron. [1]
2. Deduce the threshold potential needed to open voltage-gated sodium channels in this neuron. Give a reason. [2]
3. Estimate the time taken for the depolarization, and the repolarization. [2]
4. Predict the time taken from the end of the depolarization for the resting potential to be regained. [2]
5. Discuss how many action potentials could be stimulated per second in this neuron. [2]
6. Suggest a reason for the membrane potential rising briefly at the end of the repolarization. [1]

C2.2.11 Saltatory conduction in myelinated fibres to achieve faster impulses

The basic structure of a nerve fibre along which a nerve impulse is transmitted is very simple: the fibre is cylindrical in shape, with a plasma membrane enclosing a narrow region of cytoplasm. The diameter in most cases is about 1 μm, though some nerve fibres are wider than this. A nerve fibre with this simple structure conducts nerve impulses at a speed of about 1 metre per second.

Myelination of nerve fibres is described in *Section C2.2.4*. Myelin is multiple layers of phospholipid membrane that are deposited around the nerve fibre, as Schwann cells grow round and round it. Each time they grow around the nerve fibre a double layer of phospholipid bilayer is deposited. There may be 20 or more layers when the Schwann cell stops growing.

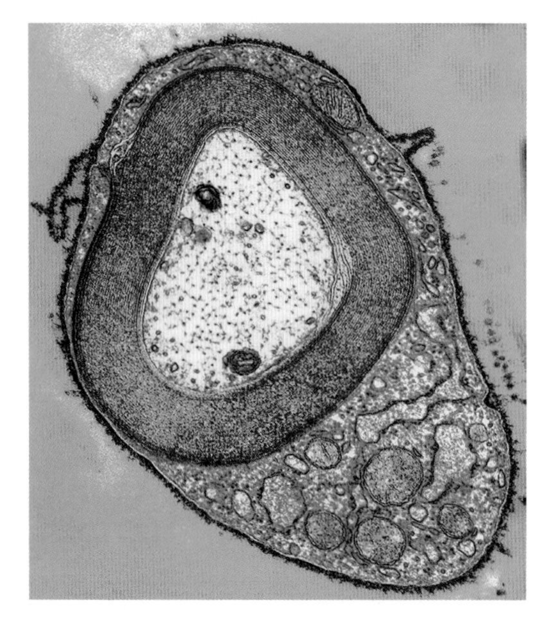

▲ Figure 22 Transverse section of axon showing the myelin sheath (red) formed by the Schwann cell's membrane wrapped round the axon many times. The cytoplasm of the Schwann cell (green) contains mitochondria (orange)

The myelin sheath prevents ion movements, so action potentials only occur at nodes of Ranvier. Sodium–potassium pumps and both sodium and potassium channels are clustered at nodes, with very few where the axon is coated in myelin. Local currents allow the nerve impulse to jump from one node of Ranvier to the next. This is called saltatory conduction and gives speeds of transmission of the nerve impulse as high as 100 metres per second.

C2.2.12 Effects of exogenous chemicals on synaptic transmission

An exogenous chemical is one that enters the body of an organism from an outside source. These chemicals can enter through the skin, the lungs or the gut, or they can be injected. Some exogenous chemicals affect synaptic transmission, either by blocking it or by promoting it. Two examples are described here.

Neonicotinoids

Neonicotinoids are synthetic compounds similar to nicotine. They bind to the acetylcholine receptor in cholinergic synapses in the central nervous system of insects. Acetylcholinesterase does not break down neonicotinoids, so the binding is irreversible. Because the receptors are blocked, acetylcholine is unable to bind and synaptic transmission is prevented. This leads to paralysis and death of the insects. Neonicotinoids are therefore very effective insecticides.

One of the advantages of neonicotinoids as pesticides is that they are not highly toxic to humans and other mammals. This is because insects have a much greater proportion of cholinergic synapses in their central nervous system, compared with mammals. Neonicotinoids also bind much less strongly to acetylcholine receptors in mammals than insects.

Neonicotinoid pesticides are now used on huge areas of crops. One neonicotinoid in particular, imidacloprid, is the most widely used insecticide in the world. However, concerns have been raised about the effects of these insecticides on other non-pest species. This has led to considerable controversy and the evidence of harm is disputed by the manufacturers and some government agencies.

Cocaine

Cocaine acts at synapses that use dopamine as a neurotransmitter. It binds to dopamine reuptake transporters, which are membrane proteins that pump dopamine back into the presynaptic neuron. Because cocaine blocks these transporters, dopamine builds up in the synaptic gap and the postsynaptic neuron is continuously excited. Cocaine is therefore an excitatory or stimulant psychoactive drug that gives feelings of euphoria that are not related to any particular reward activity (such as eating).

Activity: Updates on neonicotinoids

There are currently intense research efforts to try to discover whether neonicotinoids are to blame for collapses in honeybee colonies. What are the most recent research findings and do they suggest that these insecticides should be banned?

▲ Figure 23 Bumblebees on an *Allium* flower

Activity: Cholinergic receptor ligands

Can you see any molecular similarities that explain why all these structures can bind to acetylcholine receptors?

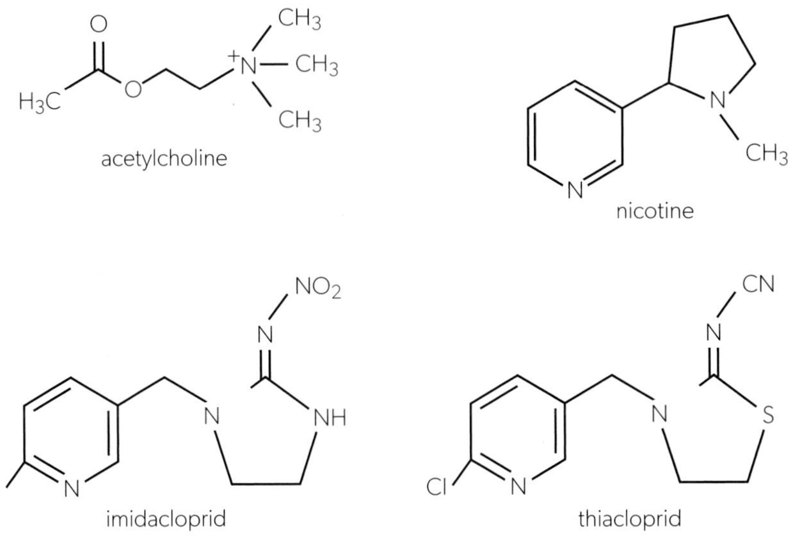

O, H_3C, O, ^{+}N, CH_3, CH_3, CH_3

acetylcholine

N, N, CH_3

nicotine

NO_2, N, N, NH, Cl, N

imidacloprid

CN, N, N, S, Cl, N

thiacloprid

▲ **Figure 24** The molecular structures of acetylcholine, nicotine and two neonicotinoids: imidacloprid and thiacloprid

C2.2.13 Inhibitory neurotransmitters and generation of inhibitory postsynaptic potentials

Not all neurotransmitters stimulate action potentials in the postsynaptic neuron. When some neurotransmitters bind to the postsynaptic membrane, the membrane potential becomes more negative. This hyperpolarization makes it more difficult for the postsynaptic neuron to reach the threshold potential. Nerve impulses are inhibited, rather than stimulated.

GABA (γ-aminobutyric acid) is an inhibitory neurotransmitter. When it binds to its receptor a chloride channel opens, causing hyperpolarization of the postsynaptic neuron by entry of chloride ions. In contrast, acetylcholine is an excitatory neurotransmitter because it causes entry of positively charged ions to the postsynaptic neuron, reducing polarization.

Transient changes to membrane potential caused by neurotransmitters such as GABA and acetylcholine are known as inhibitory and excitatory postsynaptic potentials. Figure 25 shows single inhibitory and excitatory potentials and possible effects of combined potentials.

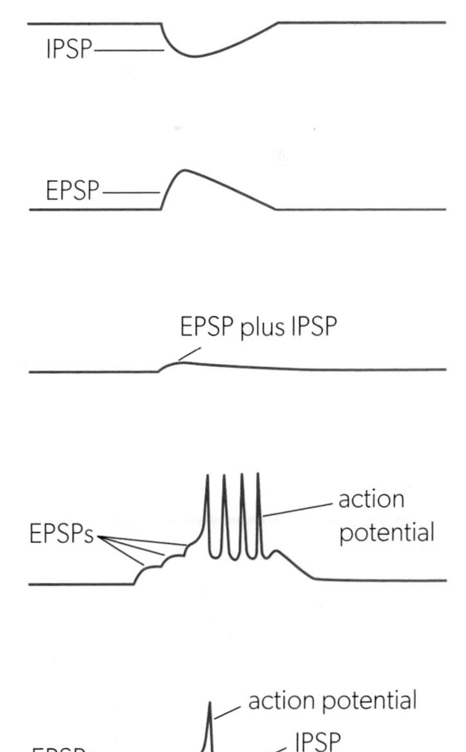

▶ **Figure 25** Sketches of typical oscilloscope traces for single inhibitory (IPSP) or excitatory postsynaptic potentials (EPSPs) and combinations of potential

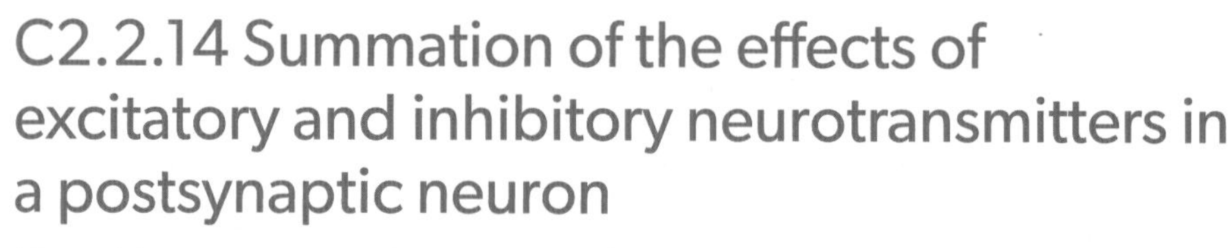

C2.2.14 Summation of the effects of excitatory and inhibitory neurotransmitters in a postsynaptic neuron

More than one presynaptic neuron can form a synapse with the same postsynaptic neuron—especially in the brain, where there may be hundreds or even thousands of presynaptic neurons. Usually a single release of excitatory neurotransmitter from one presynaptic neuron is insufficient to trigger an action potential, because one excitatory postsynaptic potential does not reach the threshold potential. Either one presynaptic neuron must repeatedly release neurotransmitter, or several adjacent presynaptic neurons must release neurotransmitter more or less simultaneously. When multiple releases of excitatory neurotransmitter combine to cause an action potential, this is called summation.

▲ Figure 26 Climbers heading for a summit must decide whether to continue or descend. They will consider factors such as weather conditions, tiredness, hours of daylight and proximity to the summit, before making the binary decision

Summation can also combine the effects of inhibitory and excitatory neurotransmitters. Whether or not an action potential is initiated in the postsynaptic neuron depends on the balance between the effects of these two types of neurotransmitter. Inhibitory neurotransmitters counter the effects of excitatory neurotransmitters so the threshold potential is not reached. The threshold potential will only be reached if there are many more excitatory neurotransmitters than inhibitory neurotransmitters.

The synapses integrate signals from many different sources. This is the basis of decision-making processes in the central nervous system.

C2.2.15 Perception of pain by neurons with free nerve endings in the skin

Pain receptors in the skin and other parts of the body detect stimuli such as the chemical substances in a bee's sting, excessive heat or the puncturing of skin by a hypodermic needle. These receptors are the endings of sensory neurons that convey impulses to the central nervous system. The nerve endings associated with pain receptors have channels for positively charged ions. These channels open in response to a stimulus such as high temperature, acid or certain chemicals (such as capsaicin in chili peppers). Entry of positively charged ions causes the threshold potential to be reached and nerve impulses then pass through the sensory neuron to the spinal column. Interneurons in the spinal cord relay the impulse to the cerebral cortex.

When impulses reach sensory areas of the cerebral cortex, we experience the sensation of pain. Signals are transmitted to the prefrontal cortex, allowing us to become fully aware of the pain and evaluate the situation. This will often result in a signal from the brain to the effectors of behaviour, reducing exposure to the stimulus. For example, you might move your hand away from a hot surface.

◀ Figure 27 Carolina Reaper chillies (*Capsicum chinense*) are among the "hottest". They have very high concentrations of capsaicin, which gives a sensation of heat or burning when eaten

Communication skills: Taking care with word choice

Equivocation involves treating words as though they have the same meaning, when they do not. This is a common mistake made by ToK students who use the terms perspective and perception interchangeably. Another language error is to use metaphors which misrepresent mechanisms. For example, a student might say that the nucleus is the "control centre" of the cell, but this implies that DNA has an active role in cellular metabolism. In fact, DNA is a relatively inert molecule and depends on other molecules interacting with it.

Taxonomists sometimes make mistakes in their naming that can lead to misunderstanding. The jalapeno pepper (*Capsicum annuum*) is not an annual plant as it can grow for more than one year. The habanero pepper (*Capsicum chinense*) is native to the Americas not China, so this species name is misleading.

C. annuum and *C. chinense* can be used to illustrate another situation where words must be chosen carefully. The seeds of these plants contain the chemical capsaicin, which is released by chewing and causes intense pain in the mouth. An animal that chews the seeds when eating jalapeno or habanero peppers kills the seeds and senses pain. An animal that swallows without chewing the seeds does not experience pain and the seeds pass through the gut undamaged, to be egested in faeces. The seeds are dispersed and provided with fertilizer from the faeces. Mammals including humans chew, but birds do not.

The teleological view is that nature has definite intentions or purposes. These are teleological statements:

- Habanero peppers are "designed" so the seeds are not killed by chewing.
- The pepper "wants to be eaten" by a bird.
- Humans "are not meant to eat the seed".
- Peppers make capsaicin "to stop mammals chewing their seeds".

Critics of teleological viewpoints argue that evolution by natural selection is not a directed process. Instead, mutations arise by chance and mutations that give an advantage are more likely to persist in the population.

C2.2.16 Consciousness as a property that emerges from the interaction of individual neurons in the brain

If we are conscious of something, we are aware of it. We do not have to be actively thinking about something to be aware of it, so we can be simultaneously aware of many things. This state of complex awareness is known as consciousness. There is agreement that it exists, but philosophers and scientists have not yet agreed on how to define it.

Sleep is a state of reduced or partial consciousness. General anaesthetics, used during surgery, make us unconscious. However, scientists do not fully understand how these drugs work, so they do not reveal much about the physiological basis of consciousness. Perhaps the most we can say with certainty is that consciousness emerges from the interaction of individual neurons in the brain: it is an example of an emergent property.

An emergent property is caused by interactions between the elements of a system. It is not a property of any one component; rather, it is a property of the system as a whole. When we recognize that a system is more than the sum of its parts, we are acknowledging the existence of emergent properties. Two biological examples are the catalytic activity of enzymes and flight in birds.

▲ Figure 28 Auguste Rodin's sculpture "Le Penseur"—a study of consciousness?

Linking questions

1. In what ways are biological systems regulated?
   a. Describe an example where the concentration of a chemical in the blood triggers a homeostatic mechanism. (D3.3.1)
   b. Explain the role of negative feedback in enzyme function. (C1.1.16)
   c. Explain the role of nerves in regulating a body system. (C3.1)
2. How is the structure of specialized cells related to function?
   a. Outline the role of proteins in nerve function. (C2.2.2)
   b. Distinguish between the structure and function of type I and type II pneumocytes. (B2.3.8)
   c. Compare and contrast the functions of phagocytes and lymphocytes. (C3.2)

TOK

What challenges are raised by the dissemination and/or communication of scientific knowledge?

Language presents challenges to scientists as they try to communicate ideas objectively and in ways that can be understood by others.

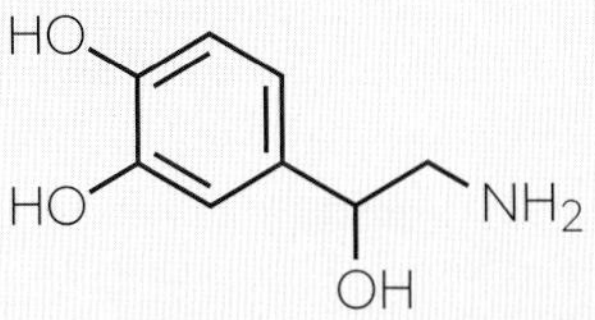

▲ **Figure 1** Can you recognize the pattern that leads to this molecule (epinephrine) being called 4-(1-hydroxy-2-(methylamino) ethyl)benzene-1,2-diol?

International agreements aim to ensure systematic naming procedures. For example, the International Union of Biochemistry directs that enzymes should generally be named for the substrate on which they act and the type of the reaction they catalyse, with the suffix –ase. For example, the systematic name for α-amylase is 1,4-α-D-glucan glucanohydrolase. Sometimes non-standard or trivial names persist, particularly where a chemical was identified some time before the development of naming conventions. For example, 4-(1-hydroxy-2-(methylamino)ethyl)benzene-1, 2-diol is far more commonly known as epinephrine.

Some names have arbitrary origins. For example, the enzyme that activates light emission by fireflies is known as luciferase. This comes from the Latin *lucifer*, meaning light-bearing.

Biochemicals such as the hormones insulin, epinephrine and cortisol were named according to where they are released from. Other chemicals are named for what they act on, such as amylase or follicle-stimulating hormone.

Whenever we name something, we predispose the listener to focus on one aspect of the complexity of the named thing. Language affects perception by leading us to focus on a narrow range of perceivable features. This is common in "folk taxonomy" or the everyday names for things. For example, the insect shown in Figure 3 is known as a "Daddy Long Legs" in North America. The same name applies to very different organisms in different English-speaking countries of the world. In all cases, it refers to animals whose "legs" are remarkably long. The animal shown in Figure 3 is also commonly referred to as the harvestman spider even though it is an insect rather than an arachnid. It is in fact an insect called a mite.

▲ **Figure 2** The "fly" is actually a small beetle. The light from its abdomen is not from a fire. The production of light in this case is by bioluminescence. The firefly does this by using an enzyme (luciferase) to oxidize a chemical in a chamber in its abdomen

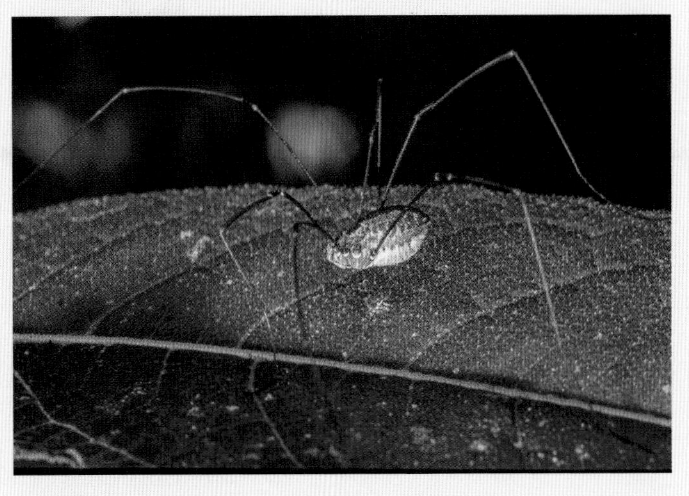

▲ **Figure 3** A "Daddy Long Legs" or harvestman mite

End of chapter questions

1. Odorants are substances which can be detected by chemoreceptors in the nose. Many different odorants can be detected but each chemoreceptor cell is sensitive to only one type. The diagrams below show the mechanism used in the chemoreceptor.
   a. Deduce which part of the mechanism is different in chemoreceptor cells that are sensitive to different odorants. [2]
   b. When the odorant binds to the receptor protein, the receptor protein starts activating G protein. Using the information shown in the diagrams, outline the effects of activated G protein. [3]
   c. Predict the effect of entry of calcium ions and exit of chloride ions on the chemoreceptor cell. [1]

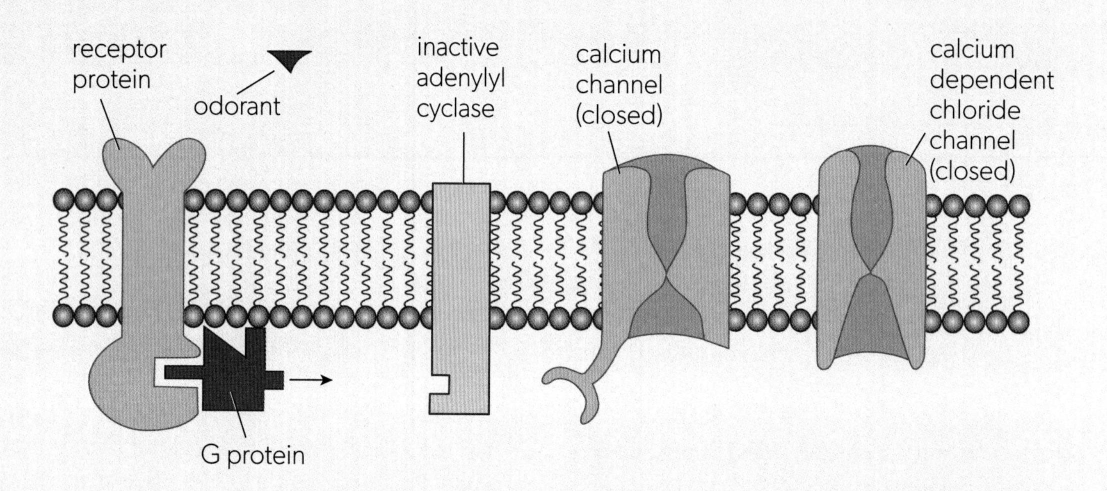

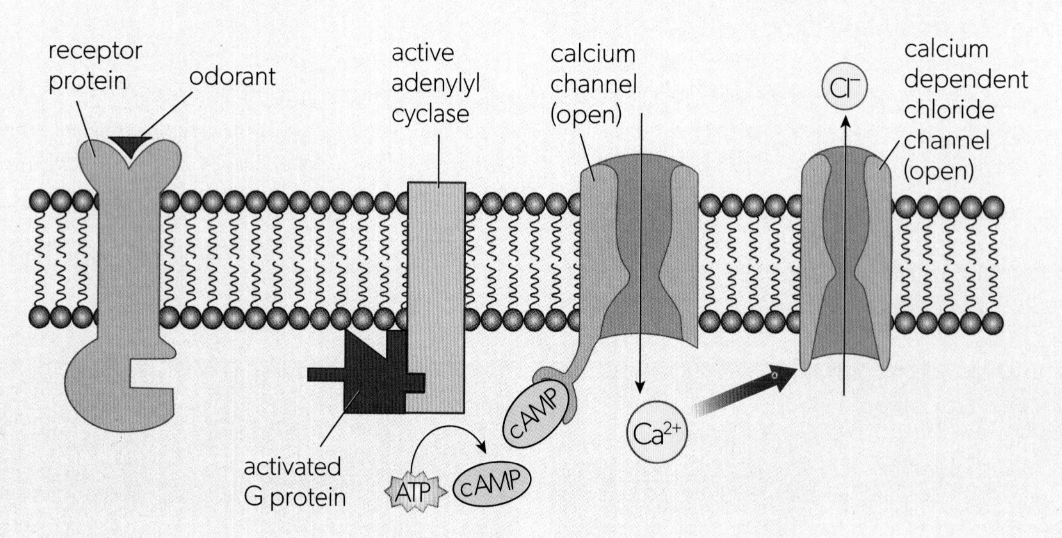

▲ **Figure 1**

Source: Gold et al, Nature (1997), 385 p 677

2. The scatter diagram below shows the relationship between brain size and total body mass in species of mammal. Primate species are shown as red circles and other species of mammal as blue circles.
   a. Using the data in the scatter diagram:
      i. state the relationship between body mass and brain size in mammals [1]
      ii. compare the brain size in relation to body mass of primates with that of other mammals [2]
      iii. explain briefly how the scatter diagram can be interpreted to show that human brains are larger than those of other primates. [2]
   b. Increases in brain size in relation to body mass could be due either to increases in brain size or to decreases in body mass. Suggest one advantage to primates of reduced body mass. [1]

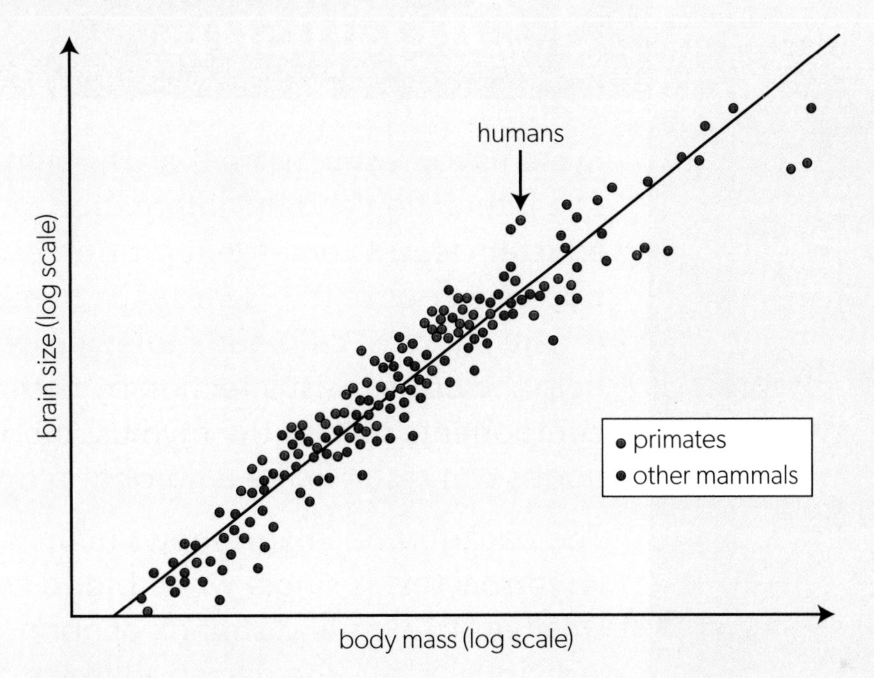

▲ **Figure 2** Source: Encylocpedia of Human Evolution, Cambridge University Press

3. The effects of neurotransmitters can be impacted by other molecules. Agonists are substances that bind to synaptic receptors and increase the effect of a neurotransmitter. Antagonists are substances that bind to synaptic receptors and decrease the effect of a neurotransmitter.

   Two molecules that impact the effects of acetylcholine (ACh) in the heart have become modern day pharmaceuticals, though they were used originally in indigenous medicine. Atropine is an extract from the Belladonna plant that was used historically for the cosmetic effect of dilating pupils. Muscarine is extracted from the mushroom *Amanita muscaria* that was used in Siberian religious rituals. The diagram includes the + symbol to represent agonists and the – to represent an antagonist.

   Predict the effects of each of these chemicals on heart rate.

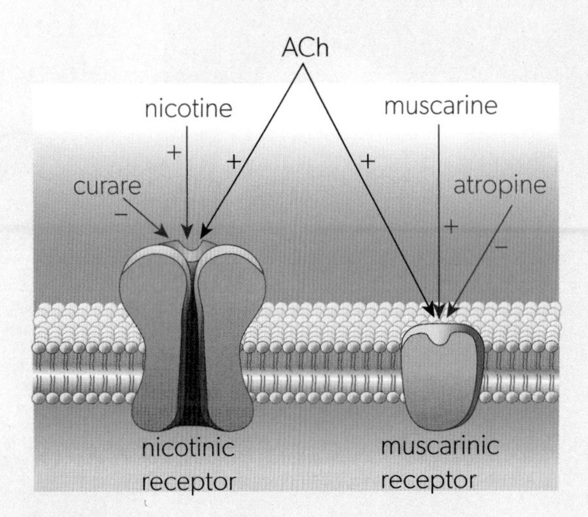

▲ **Figure 3** Nicotinic and muscarinic receptors

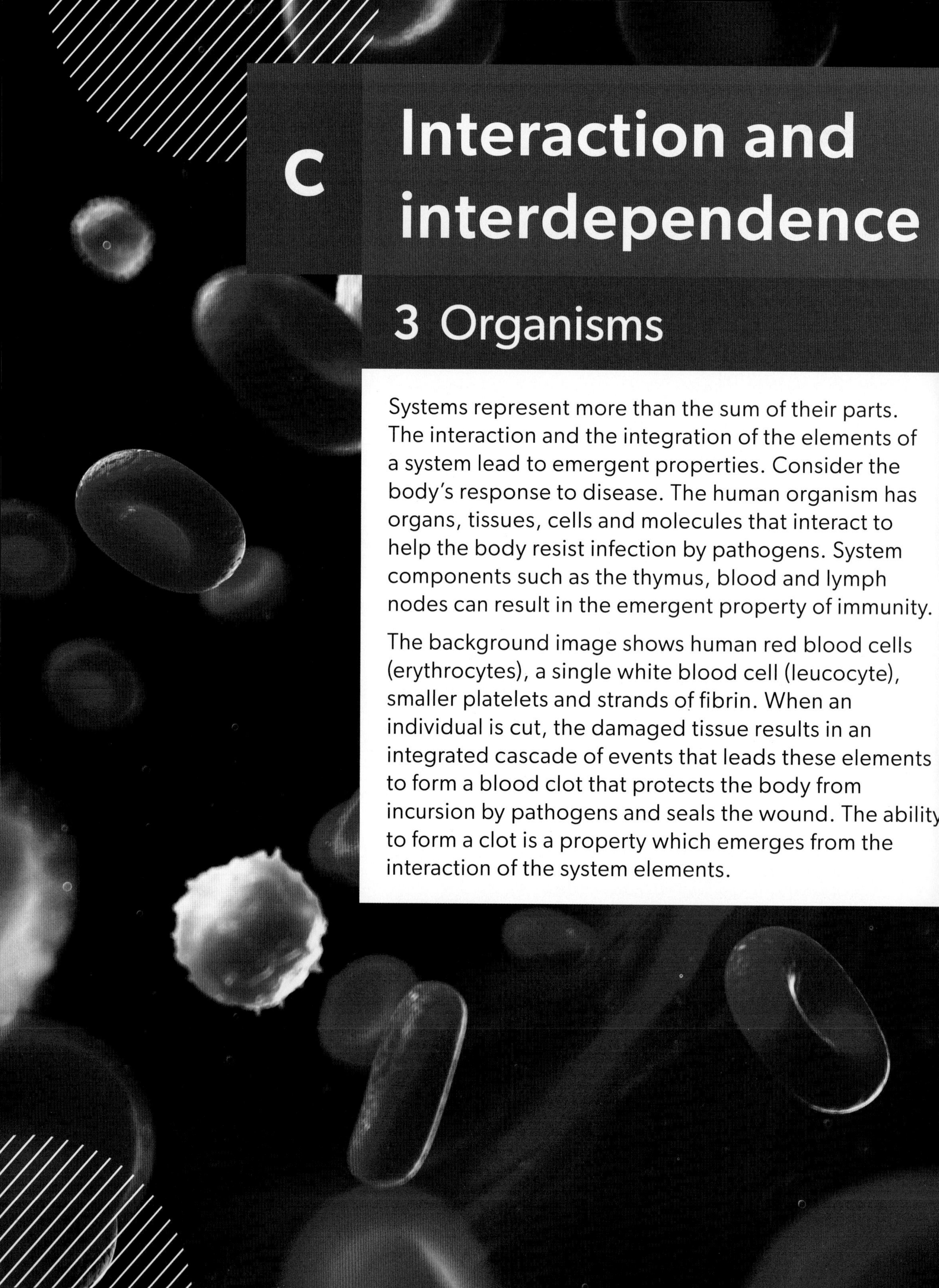

C Interaction and interdependence

3 Organisms

Systems represent more than the sum of their parts. The interaction and the integration of the elements of a system lead to emergent properties. Consider the body's response to disease. The human organism has organs, tissues, cells and molecules that interact to help the body resist infection by pathogens. System components such as the thymus, blood and lymph nodes can result in the emergent property of immunity.

The background image shows human red blood cells (erythrocytes), a single white blood cell (leucocyte), smaller platelets and strands of fibrin. When an individual is cut, the damaged tissue results in an integrated cascade of events that leads these elements to form a blood clot that protects the body from incursion by pathogens and seals the wound. The ability to form a clot is a property which emerges from the interaction of the system elements.

C3.1 Integration of body systems

What are the roles of nerves and hormones in integration of body systems?

Cheetah (*Acinonyx jubatus*) cubs aged 6–7 weeks old are shown involved in play. Play requires processing of sensory information and the transmission of effector responses to muscles. System integration is essential in living systems. Coordination allows the component parts of a system to work together to perform an overall function. This integration is responsible for emergent properties. Both the nervous and endocrine systems play a role in messaging. Blood plays a role in transporting endocrine messages between components of a system. The central nervous system integrates messages and sends responses.

◀ Figure 1

What roles do feedback mechanisms play in regulation of body systems?

In body systems, end points of processes are reached which trigger regulatory processes. Towards the end of pregnancy, a decline in the release of progesterone triggers an increase in the hormone oxytocin. Oxytocin triggers the uterus to contract, pressing the baby's head against the cervix. The stretching of the cervix triggers the release of more oxytocin, triggering uterine contractions which further stretch the cervix. This process of amplification continues until the baby is delivered. What type of feedback mechanism does this represent?

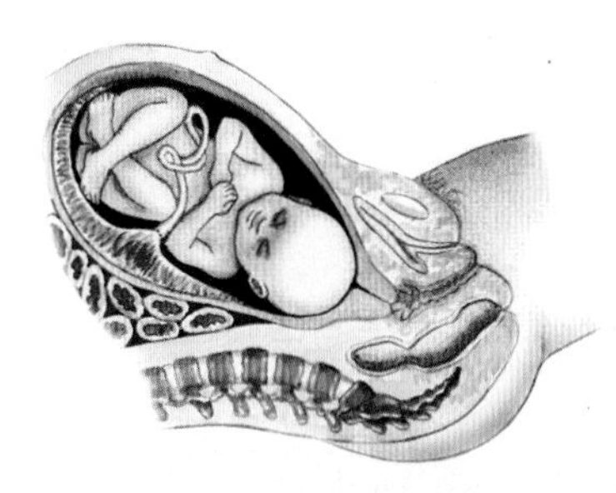

◀ Figure 2

SL and HL	AHL only
C3.1.1 System integration C3.1.2 Cells, tissues, organs and body systems as a hierarchy of subsystems that are integrated in a multicellular living organism C3.1.3 Integration of organs in animal bodies by hormonal and nervous signalling and by transport of materials and energy C3.1.4 The brain as a central information integration organ C3.1.5 The spinal cord as an integrating centre for unconscious processes C3.1.6 Input to the spinal cord and cerebral hemispheres through sensory neurons C3.1.7 Output from the cerebral hemispheres to muscles through motor neurons C3.1.8 Nerves as bundles of nerve fibres of both sensory and motor neurons C3.1.9 Pain reflex arcs as an example of involuntary responses with skeletal muscle as the effector C3.1.10 Role of the cerebellum in coordinating skeletal muscle contraction and balance C3.1.11 Modulation of sleep patterns by melatonin secretion as a part of circadian rhythms C3.1.12 Epinephrine (adrenaline) secretion by the adrenal glands to prepare the body for vigorous activity C3.1.13 Control of the endocrine system by the hypothalamus and pituitary gland C3.1.14 Feedback control of heart rate following sensory input from baroreceptors and chemoreceptors C3.1.15 Feedback control of ventilation rate following sensory input from chemoreceptors C3.1.16 Control of peristalsis in the digestive system by the central nervous system and enteric nervous system	C3.1.17 Observations of tropic responses in seedlings C3.1.18 Positive phototropism as a directional growth response to lateral light in plant shoots C3.1.19 Phytohormones as signalling chemicals controlling growth, development, and response to stimuli in plants C3.1.20 Auxin efflux carriers as an example of maintaining concentration gradients of phytohormones C3.1.21 Promotion of cell growth by auxin C3.1.22 Interactions between auxin and cytokinin as a means of regulating root and shoot growth C3.1.23 Positive feedback in fruit ripening and ethylene production

C3.1.1 System integration

All organisms use multiple systems to perform the various functions of life. Within these systems, there are interdependent subsystems that work together to perform an overall function. At every level in the functioning of an organism, there must be coordination between and within systems. This is achieved by system integration.

System integration depends on effective communication between components so they can interact. The interactions may be as simple as negative or positive feedback between two components. More commonly however, they are complex and multifactorial, with many loops and branches. Comparisons can be made with systems integration in engineering projects or software design, such as the control of wind turbines.

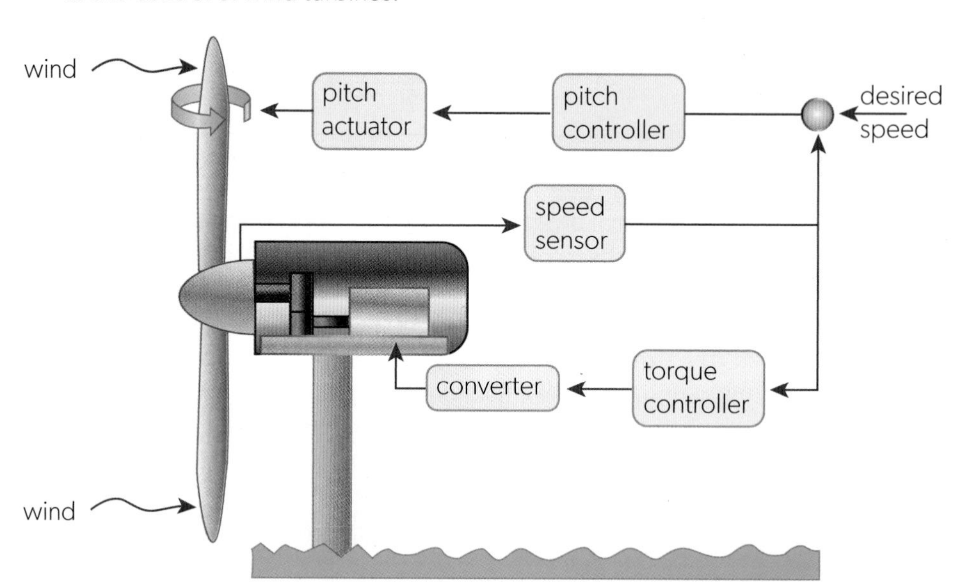

▶ Figure 3 Wind turbines can be kept rotating at constant rates, despite varying wind speeds, by altering the pitch of the blades

C3.1.2 Cells, tissues, organs and body systems as a hierarchy of subsystems that are integrated in a multicellular living organism

Cells within a multicellular organism interact with each other at multiple levels in a hierarchy of organization:

System	Component subsystems
A cell	Organelles
A tissue	Cells
An organ	Tissues
An organ system	Organs
An organism	Organ systems

▲ Table 1

1. Tissues

Cells within multicellular organisms are specialized to perform specific functions. Their structure is adapted to their function. One cell by itself cannot usually carry out its function on a large enough scale to meet the needs of the organism. Instead organisms use groups of cells of the same type to carry out a function. These groups of cells are called tissues. Large organisms tend to have more cells in each tissue, rather than larger cells (because of surface area-to-volume issues).

Tissues may contain two or more cell types, which specialize for different aspects of the function of the tissue. For example, the epithelium that forms the wall of alveoli in lungs has two cell types:

- AT1 (alveolar Type 1) cells make up 95% of the respiratory surface. They are extensive but very thin, allowing diffusion of gases.
- AT2 cells are cuboidal with dense cytoplasm. They secrete a surfactant that prevents the collapse of alveoli.

The cells in a tissue adhere (stick) to each other. Plant cells do this with a middle lamella between the cell walls that is rich in gluey pectin. Animal cells use transmembrane proteins that form strong links between neighbouring cells. If blood is regarded as a tissue, it is unusual because blood cells do not stick together.

Cells within a tissue communicate with each other. For example, plant tissues may use efflux pumps to transfer auxin, to coordinate growth; and heart muscle tissue transmits electrical impulses which trigger contraction. Cells within tissues also communicate with cells elsewhere in an organism.

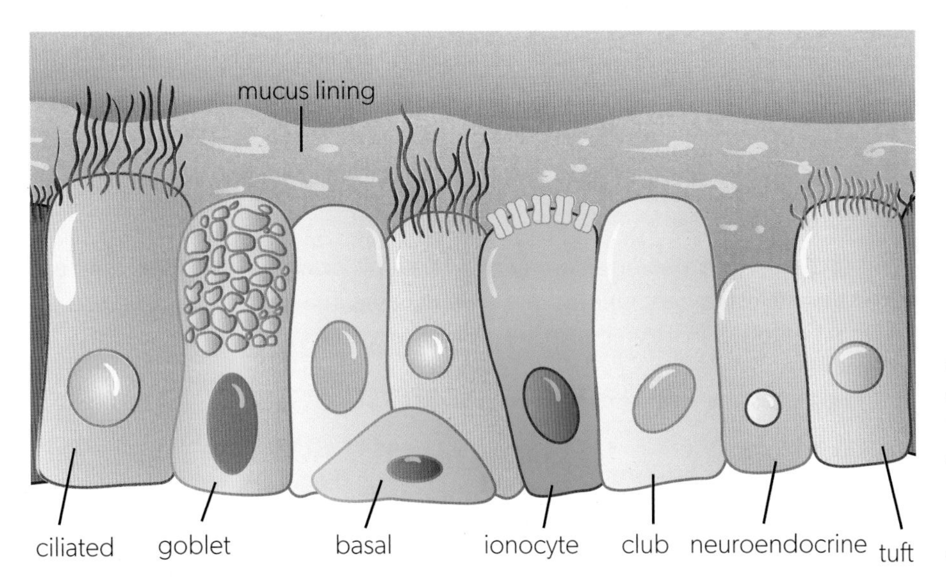

◀ Figure 4 The trachea is lined with epithelium tissue containing multiple cell types, including recently discovered ionocytes. Together these cells act as a cleaning system by secreting mucus and moving it up and out of the airways. Dirt in inhaled air is trapped in the mucus, so it does not reach the lungs

2. Organs

An organ is a group of tissues in an animal or plant that work together to carry out a specific function of life. For example, the kidney is an organ of excretion and the leaf is an organ of photosynthesis. The tissues within an organ are interdependent. For example, within a leaf:

- the spongy mesophyll is adapted for gas exchange; it depends on concentration gradients of carbon dioxide and oxygen, created by photosynthesis in the palisade mesophyll
- the palisade mesophyll is adapted for photosynthesis; it depends on the spongy mesophyll for a supply of carbon dioxide and for removal of oxygen.

▶ Figure 5 The trachea (windpipe) is an organ composed of five main tissues. The submucosa is one of these tissues. It contains blood vessels, branches of nerves, smooth muscle cells and elastic fibres

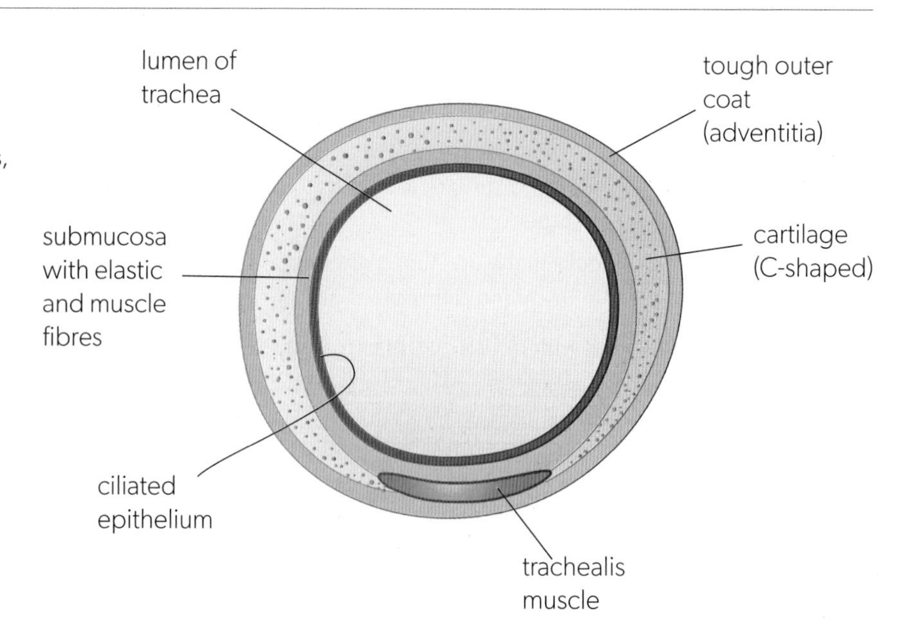

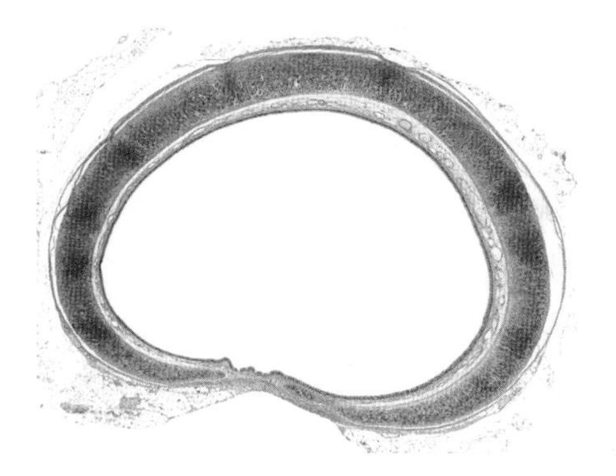

▲ Figure 6 How many tissues can you recognize in this transverse section of trachea?

3. Organ systems

Groups of organs interact with each other to perform an overall function of life. These groups are known as organ systems. In humans, 11 organ systems are recognized:

circulatory system	digestive system	endocrine system	gas exchange system
integumentary system	lymphatic system	muscular system	
nervous system	reproductive system	skeletal system	urinary system

In most cases, the organs in an organ system are physically linked—for example, in the digestive and nervous systems. In other cases, the organs are dispersed around the body—for example, in the endocrine system.

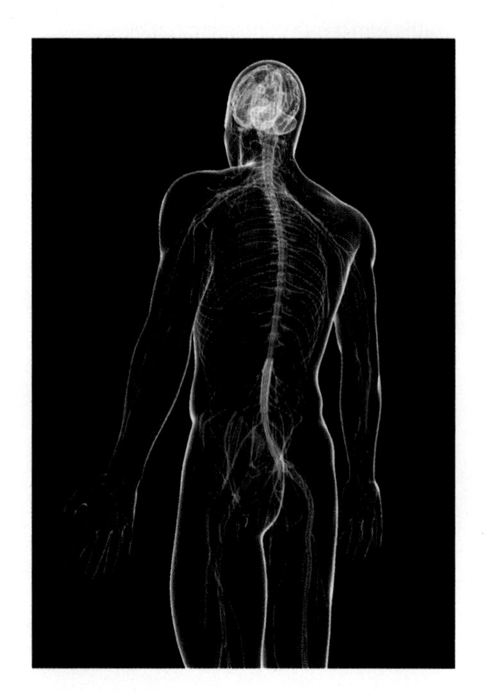

▲ Figure 7 Nervous system

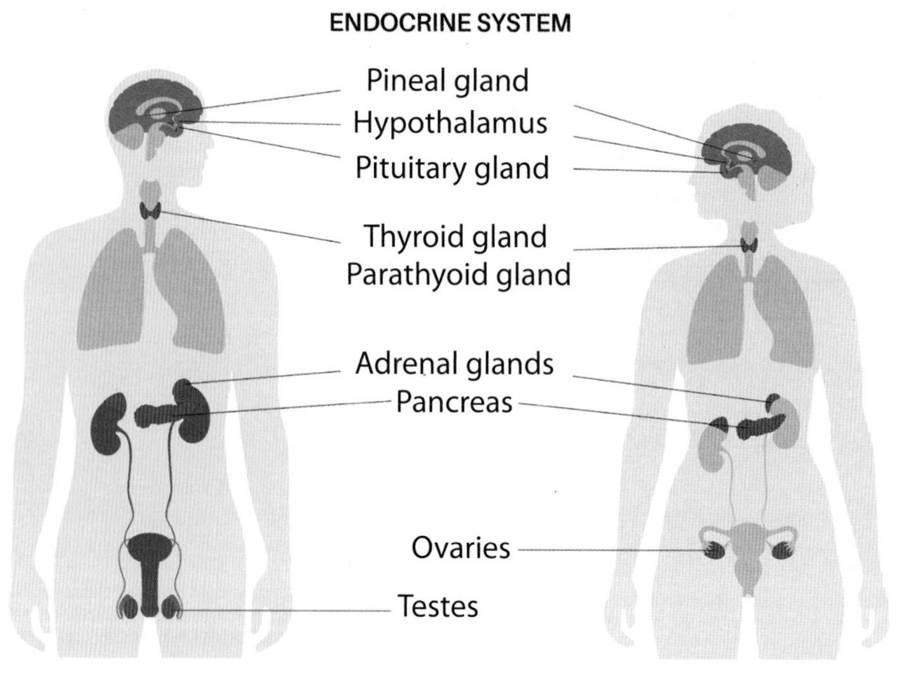

▲ Figure 8 Male and female endocrine systems

4. Organisms

An organism is a living individual made up of interconnected parts—organ systems, composed of organs, made up of tissues with constituent cells. These parts are interdependent so failure of a single group of cells in a tissue can cause an organism to die.

The parts of an organism interact and integration of body systems results in emergent properties. For example, a cheetah is an effective predator. However, this might be difficult to predict if each component is studied separately. To understand the emergent properties of organisms, you must consider systems as a whole.

▲ Figure 9 Cheetahs must decide whether or not it is worth investing time and energy in a chase. What body systems do they use before starting to hunt and during hunting?

C3.1.3 Integration of organs in animal bodies by hormonal and nervous signalling and by transport of materials and energy

The organs of animal bodies are integrated to form a functioning whole. This integration involves effective communication and the transport of materials and energy.

Two systems of the body are used for internal communication: the endocrine system and the nervous system. The endocrine system is composed of glands that release hormones. The nervous system contains neurons, which transmit nerve impulses. Communication between cells using nerves and hormones is described in *Topic C2.2*. The roles of these two methods of communication are explored in Table 2.

	Hormonal signalling	Nervous signalling
Type of signal	Chemical	Electrical, by passage of cations across membranes
Transmission of signal	In the bloodstream	In neurons
Destination of signal	Widespread—to all parts of the body that are supplied with blood, but only certain cells respond	Highly focused—to one specific neuron or group of effector cells
Effectors	Target cells in any type of tissue	Muscles or glands
Type of response	• Growth • Development, including puberty • Reproduction, including gamete production and pregnancy • Metabolic rate and heat generation • Solute concentrations in blood, including glucose and salts • Mood, including stress, thirst, sleep/wakefulness and sex drive	Responses due to contraction of muscle: • striated muscle, e.g. locomotion • smooth muscle, e.g. peristalsis and sphincter opening/closing • cardiac muscle, e.g. heart rate Response or secretion from a gland: • exocrine glands, e.g. sweat or saliva secretion • endocrine glands, e.g. epinephrine secretion
Speed of response	Slower	Very rapid
Duration of response	Long—until the hormone is broken down	Short—unless nerve impulses are sent repeatedly

▲ Table 2

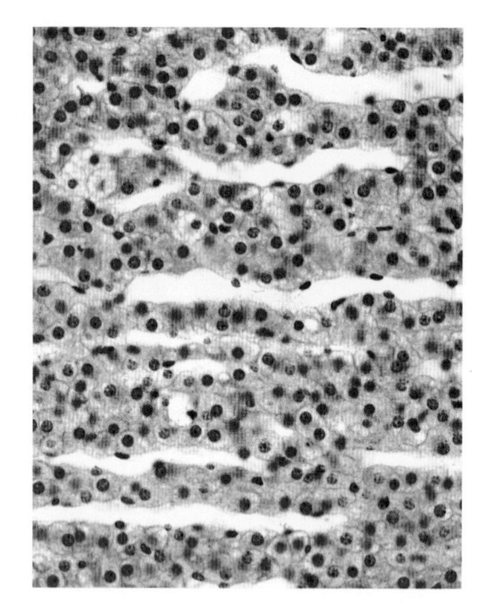

▲ Figure 10 There is interaction between nervous and hormonal communication in the adrenal gland. In this micrograph, adrenal gland cells are stained pink. The clear spaces are blood capillaries (emptied of blood during the making of the microscope slide) that carry out transport functions for the tissue. Neurons are connected to the adrenal gland cells via synapses. These neurons stimulate the gland cells to secrete epinephrine into the capillaries

Materials and energy are transported between organs by the circulatory (blood) system. All organs of the body are served by this system, with blood circulating through almost all tissues. Living cells need a constant supply of energy, provided by cell respiration. Cells therefore require a respiratory substrate—commonly glucose—and oxygen (assuming respiration is aerobic). The bloodstream supplies both these substances, along with water and carbon compounds needed for growth or repair. It also removes waste products, including carbon dioxide produced in respiration.

Some organ systems resemble a production line, in which material is transferred from organ to organ so that functions are performed in the correct sequence. This happens in the digestive, gas exchange, urinary and reproductive systems.

C3.1.4 The brain as a central information integration organ

The brain is the central integrating organ of our body. It receives information, processes it, stores some of it and sends instructions to all parts of the body to coordinate life processes. The information received by the brain comes from sensory receptors, both in specialized sense organs such as the eye and also from receptor cells in other organs (for example, pressure receptors in blood vessels).

The brain can store information, for the short term or longer term—and sometimes for the rest of life. The capacity to store information is called memory. It is essential for learning. Processing of information leads to decision making by the brain. This may result in signals being sent to muscles or glands, which cause these organs to carry out a response.

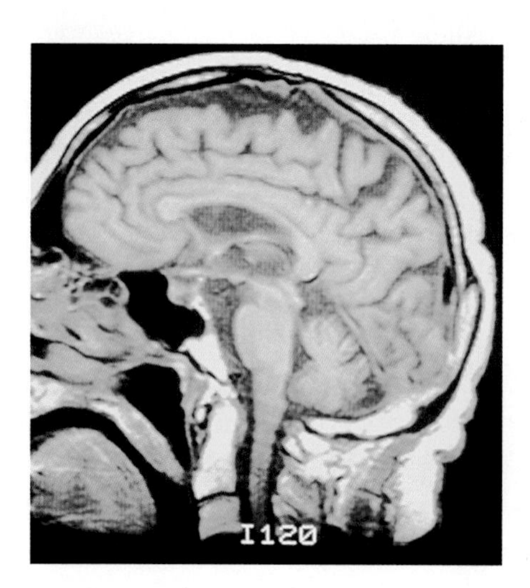

▲ Figure 11 Memory and learning are functions of the cerebrum—the folded upper part of the brain

C3.1.5 The spinal cord as an integrating centre for unconscious processes

The nervous system is made up of the central nervous system (CNS) and nerves that connect the CNS to all other organs of the body. There are two organs in the CNS: the brain and spinal cord. The spinal cord is located inside the vertebral column (backbone). It is widest at its junction with the brain and tapers going downwards towards the pelvis. Pairs of spinal nerves branch off to the left and right between the vertebrae. In humans, there are 31 pairs of spinal nerves, each serving a different region of the body.

The spinal cord has two main tissues:

- white matter containing myelinated axons and other nerve fibres, which convey signals from sensory receptors to the brain and from the brain to the organs of the body
- grey matter containing the cell bodies of motor neurons and interneurons, with many synapses between these neurons.

Synapses in the grey matter are used for processing information and for decision-making so the spinal cord is also an integrating centre. The spinal cord only coordinates unconscious processes, especially reflexes; in some cases, it can do this more quickly than if signals were conveyed to and from the brain. Table 3 shows differences between unconscious and conscious processes.

Unconscious	Conscious
Performed when awake or asleep	Only performed when awake
Performed involuntarily—we do not have to think about the actions and cannot normally prevent them through thought	Performed voluntarily—we can think about the action and decide whether or not to carry it out
Secretion by glands and contractions of smooth muscle (not attached to bones) are unconscious and therefore involuntary	Contraction of striated muscle (attached to bones) can be done consciously and is therefore voluntary
Coordinated by the brain and spinal cord	Coordinated only by the cerebral hemispheres of the brain
Example: swallowing of food once it has entered the oesophagus (gullet) and vomiting when stomach contents are regurgitated	Example: initiation of swallowing when food is pushed from the mouth cavity into the pharynx

▲ Table 3

There are exceptions to the statements in Table 3. Many of our actions are non-binary—we may consciously choose to carry them out but the processing then used is unconscious. Striated muscle can be controlled consciously or unconsciously. For example, we consciously choose to stand up and use striated muscles for this action, but the unconscious postural reflexes that keep us standing use the same muscles. When asleep in bed, we might turn over unconsciously; we use striated muscle to do this. The anus is a sphincter composed of smooth muscle cells, but early in our lives we achieve conscious control of it so we no longer have to wear diapers (nappies). Can you find other exceptions?

▲ Figure 12 A transverse section of spinal cord showing two main tissue layers: a butterfly-shaped area of grey matter and paler surrounding white matter. Protective membranes called meninges surround the spinal cord

C3.1.6 Input to the spinal cord and cerebral hemispheres through sensory neurons

Changes in the external environment can act as stimuli to the nervous system, if perceived by sensory receptors. These receptors are located in the skin and sense organs. Nerve endings of some sensory neurons act as receptors for touch and heat. Other stimuli are perceived by specialized receptor cells that pass impulses to sensory neurons—for example, light-sensitive rod and cone cells in the retina of the eye.

There are also receptors inside the body that monitor internal conditions. Stretch receptors in striated muscle sense the state of contraction, allowing the brain to deduce the posture of the body. Stretch receptors in the walls of arteries give a measure of blood pressure. Chemoreceptors in the walls of blood vessels detect whether concentrations of oxygen, carbon dioxide and glucose are low or high.

Signals from all receptor cells and from nerve endings that perceive stimuli directly, are conveyed to the central nervous system by sensory neurons. The signals are in the form of nerve impulses, carried along the axons of sensory neurons. These axons vary in length depending on the distance between the receptor cell and the brain or spinal cord. They might be a metre or more in length if, for example, the receptor cell is at the end of a toe.

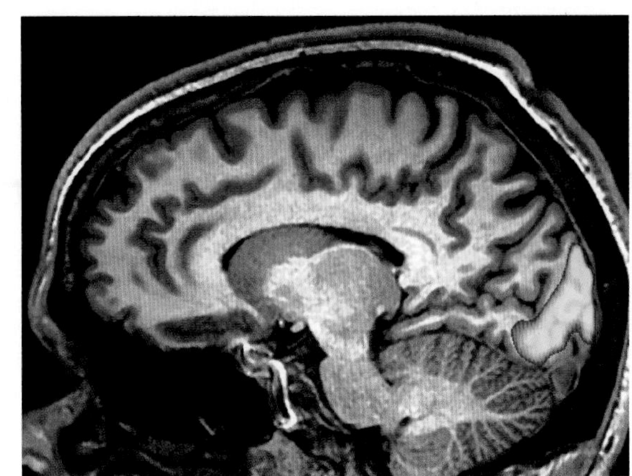

▲ Figure 13 Functional magnetic resonance image (fMRI) during visual stimulation of the brain. The active region is the primary visual cortex

The brain receives all the signals from the main sense organs located in the head: the eyes, ears, nose and tongue. The spinal cord receives signals from other organs of the body including skin and muscles. Sensory inputs to the brain are received by specialized areas in the cerebral hemispheres. For example, the visual cortex that receives signals from rod and cone cells is in the posterior part of the cerebrum.

The axons of sensory neurons enter either the spinal cord through one of 31 pairs of spinal nerves, or the brain by one of the 12 pairs of cranial nerves. For example, signals from rods and cones in the eye enter the brain via the left or right optic nerve (cranial nerve II) and signals from the ear enter via the vestibulocochlear nerve (cranial nerve VIII).

Exploring and designing: Two-point discrimination

Bend a paper clip into a U-shape so that the two tips are separated by 2 cm. If the tips of the paper clip are touched to the tip of a person's finger, the person will sense this as two points. If the tips of the paper clip are touched to the back of the person's shoulder, the person is likely to sense this as a single point.

Each sensory neuron connects to the central nervous system (CNS). A given CNS neuron responds to all information from its input area—for example, an area of the skin—as if it is coming from one point. This input area is called the receptive field of the CNS neuron. On the shoulder, each sensory receptor gathers information from a much larger area than a receptor on the finger tip; that is, the receptive field is larger on the shoulder. For a person to feel two points, the two tips of the paper clip must stimulate the receptive fields of two separate CNS neurons.

1. Design an experiment to investigate an aspect of two-point discrimination. Is it fixed or can it be influenced by external variables?
2. Can you hypothesize why there is better discrimination in some areas than in others? Is it possible to test this hypothesis? How many times will you repeat the test in each area of skin? How large must the difference be in the results for two areas to give reliable evidence of a difference in two-point discrimination?

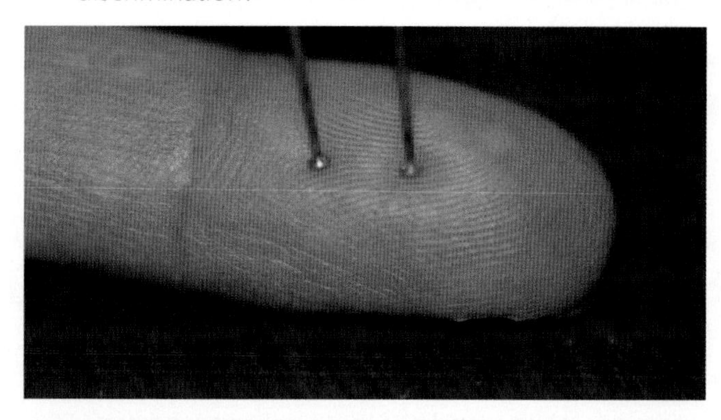

▲ **Figure 14** Fingertip (magnified) showing two-point discrimination test

C3.1.7 Output from the cerebral hemispheres to muscles through motor neurons

The cerebral hemispheres of the brain have a major role in the control of striated muscles and certain glands. In particular, the primary motor cortex sends signals via motor neurons to each striated muscle in the body. Striated muscle is attached to bone. It is used for locomotion and controlling posture and can be controlled consciously. For example, to stand up from a sitting position, signals are sent from parts of the motor cortex via motor neurons to muscles in the legs.

The signals in motor neurons are nerve impulses. The cell body and dendrites of many motor neurons are located in the grey matter of cerebral hemispheres. Typically there are many dendrites, receiving signals from different relay neurons and transmitting them to the cell body. One axon leads from the cell body out of the brain and down the spinal cord. There it forms a synapse with a second motor neuron, whose axon leads to one specific striated muscle. The axons of these two motor neurons may in total extend to a metre or more, depending on the location of the muscle. The axons of motor neurons are bundled up in nerves, often together with the axons of sensory neurons. When a nerve impulse reaches the end of the axon, it stimulates the muscle fibres to contract and gland cells to secrete.

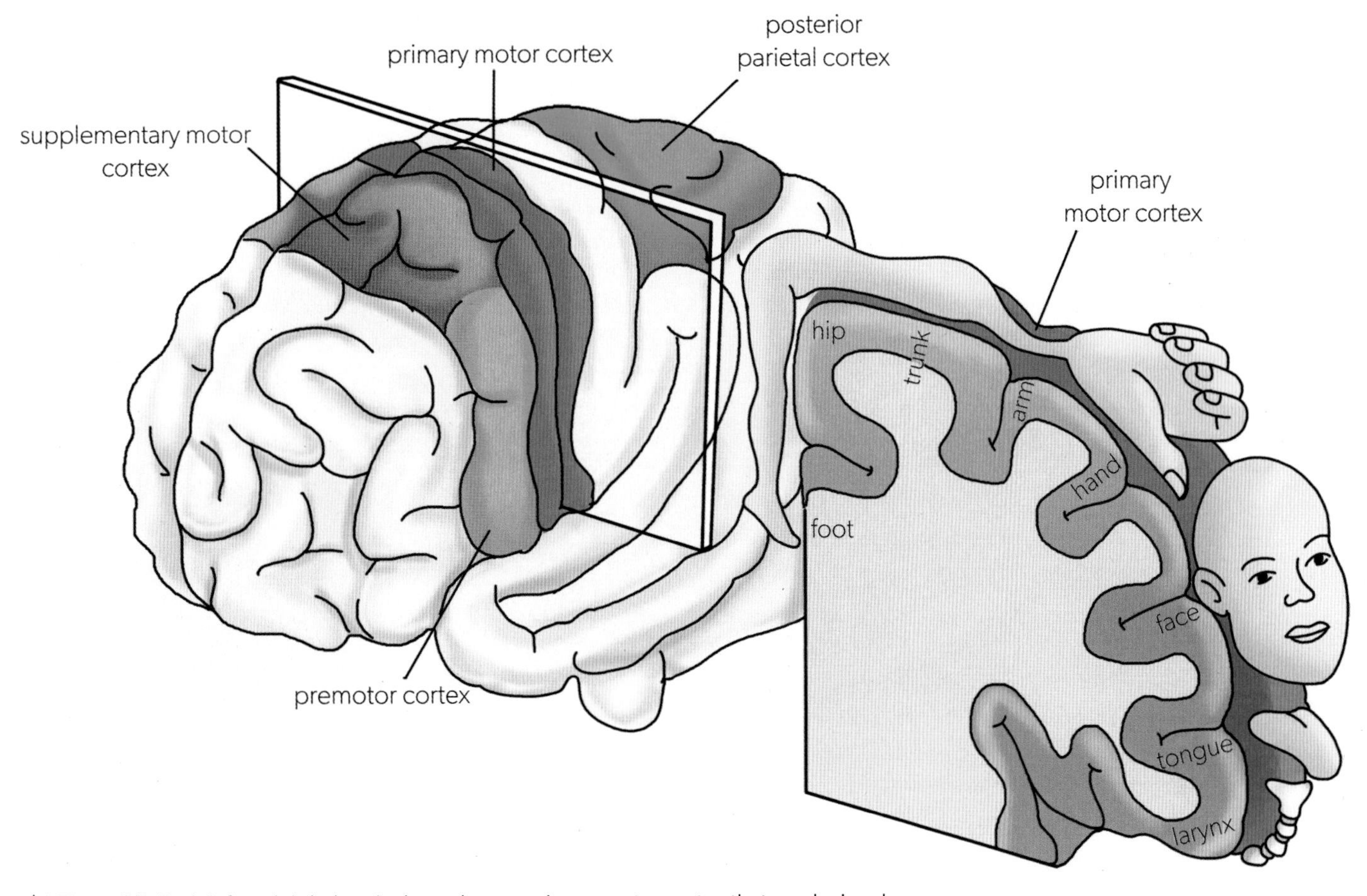

▲ Figure 15 Both left and right hemispheres have a primary motor cortex that sends signals to striated muscles in different parts of the body

C3.1.8 Nerves as bundles of nerve fibres of both sensory and motor neurons

A nerve is a bundle of nerve fibres enclosed in a protective sheath. Nerves vary in size depending on the number of nerve fibres and the proportion of them that are myelinated. The widest is the sciatic nerve, which is approximately 20 mm across. The optic nerve is estimated to contain between 770,000 and 1.7 million nerve fibres. Small nerves may contain fewer than a hundred fibres.

Most nerves contain nerve fibres of both sensory and motor neurons. However, some contain only sensory neurons (for example, the optic nerve) and some contain only motor neurons (for example, the oculomotor nerve). All organs of the body are served by one or more nerves.

Data-based questions: Nerves and nerve fibres

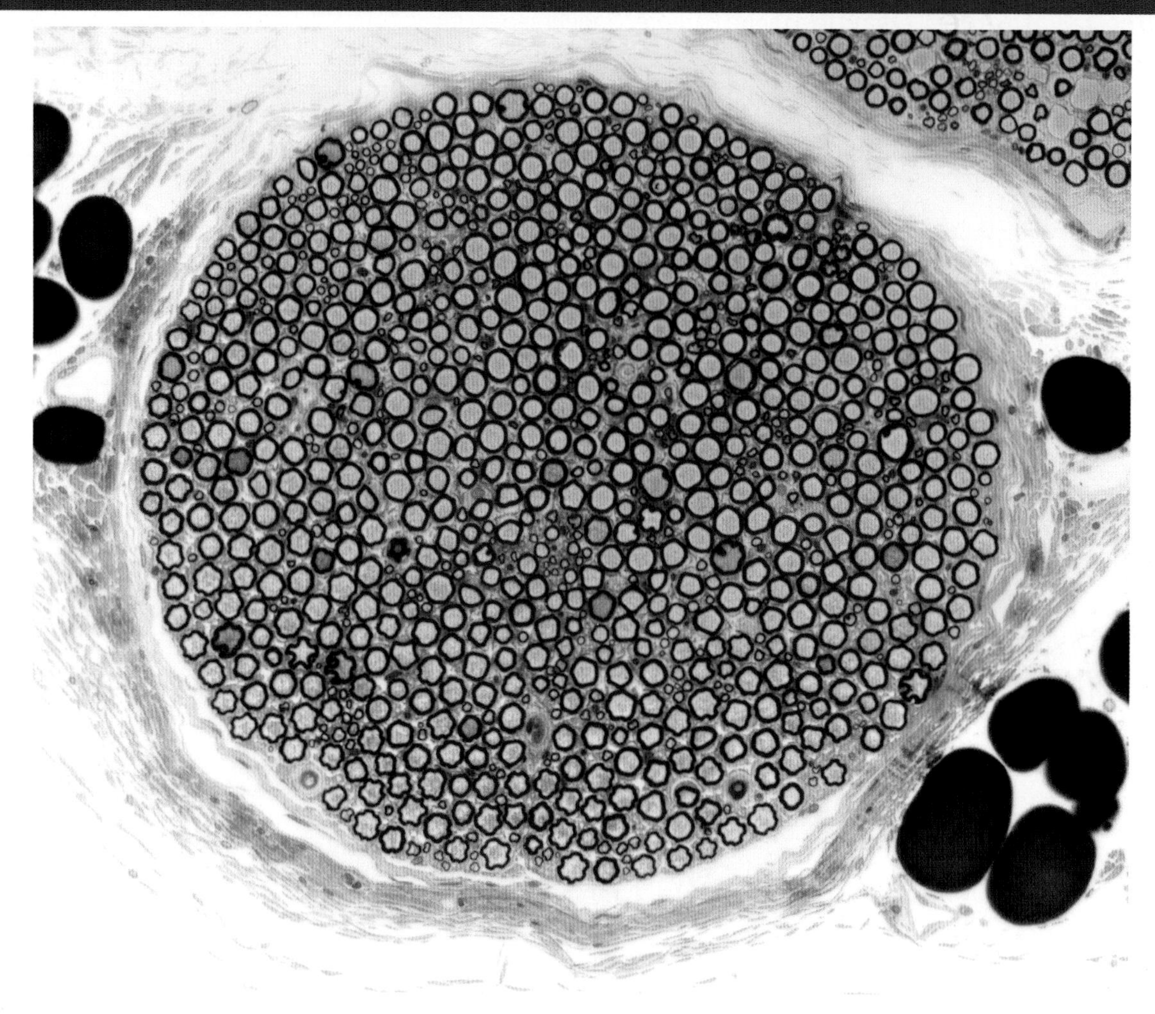

▲ **Figure 16** A micrograph showing a branch of the sciatic nerve, with the protective sheath around it. Myelin sheaths (dark blue circles) can be seen surrounding the axons (pale blue). There is some adipose (fat) tissue around the nerve, with fat droplets stained dark brown

1. The widest diameter of the nerve is 400 micrometres. What is the magnification of the micrograph? [1]

   A. ×0.3 B. ×30

   C. ×300 D. ×3,000

2. The wider nerve fibres inside the nerve have a diameter in the micrograph of about 3.6 mm. What is their actual diameter? [1]

   A. 0.009 μm B. 9 μm

   C. 0.36 cm D. 1,440 mm

3. Some nerve fibres in the nerve have a diameter of 2–3 μm. How wide will they appear in the micrograph? [1]

   A. 0.8–1.2 mm B. 0.8–1.2 cm

   C. 0.005–0.0075 mm D. 0.5–0.75 cm

4. Determine the volume of fat in the upper droplet on the right hand side of the nerve. Assume the droplet is spherical. [1]

   A. 905 μm^3 B. 7,238 μm^3

   C. 14,137 μm^3 D. 113,143 μm

5. Explain the variation in size of the nerve fibres inside the nerve. [3]

6. Suggest methods for estimating the number of nerve fibres in this nerve, without counting each fibre. [3]

C3.1.9 Pain reflex arcs as an example of involuntary responses with skeletal muscle as the effector

A reflex action is a rapid, involuntary response to a specific stimulus. Reflexes are the simplest type of coordination by the nervous system, as the signals pass through the smallest number of neurons. This helps to speed up reflexes, which is an advantage if the response prevents harm to the body.

Some reflex actions are coordinated by the spinal cord, such as the pain reflex when we lift our foot after treading on a sharp object. Other reflexes are coordinated by the brain—for example, constriction of the pupil in the eye in response to bright light, which protects the retina from damage. Reflexes depend on parts of the nervous, skeletal and muscular systems working together:

1. **Receptors**
   Receptor cells sense a change in conditions, known as a stimulus. Each receptor cell detects only one type of stimulus. The nerve endings of some sensory neurons can perceive a stimulus directly, so there is no need for a separate receptor cell. Pain and heat are detected in this way by nerve endings in the skin.

2. **Sensory neurons**
   Sensory neurons receive signals, either from receptor cells or from their own sensory nerve endings, and pass them to neurons in the central nervous system. To do this they have long axons that carry nerve impulses from the receptor to the spinal cord or brain. These axons end at synapses with interneurons in the grey matter of the spinal cord or brain. Grey matter is the tissue containing many cell bodies of interneurons and motor neurons.

3. **Interneurons**
   These cells are located inside the central nervous system. They typically have many branched fibres called dendrites, along which nerve impulses travel. Interneurons process signals brought by sensory neurons and make decisions about appropriate responses. They do this by combining impulses from multiple inputs and then passing impulses to specific other neurons. The decision-making process that results in a reflex action is very simple because there may be only one interneuron connecting a specific sensory neuron to the motor neuron that can cause an appropriate response.

4. **Motor neurons**
   The structure of motor neurons was described in *Topic C2.2*. Motor neurons receive signals via synapses with interneurons. If a threshold potential is achieved in a motor neuron, an impulse is passed along the axon which leads out of the central nervous system to an effector. The axon does not change its position or connections, so the impulse always travels to the same effector cell or small group of effector cells.

5. **Effectors**

   These carry out the response to a stimulus when they receive the signal from a motor neuron. There are two types of effector:

   - Muscles respond by contracting. For example, muscles in the leg contract to lift the foot off a sharp object.
   - Glands respond by secreting. For example, the smell of food may cause glands in the head to secrete saliva.

A reflex arc is a sequence of cells that participate in coordination of a reflex. Figure 17 shows a spinal reflex arc. If the stimulus is pain or heat, nerve endings of the sensory neuron act as the receptor, rather than specialized receptor cells.

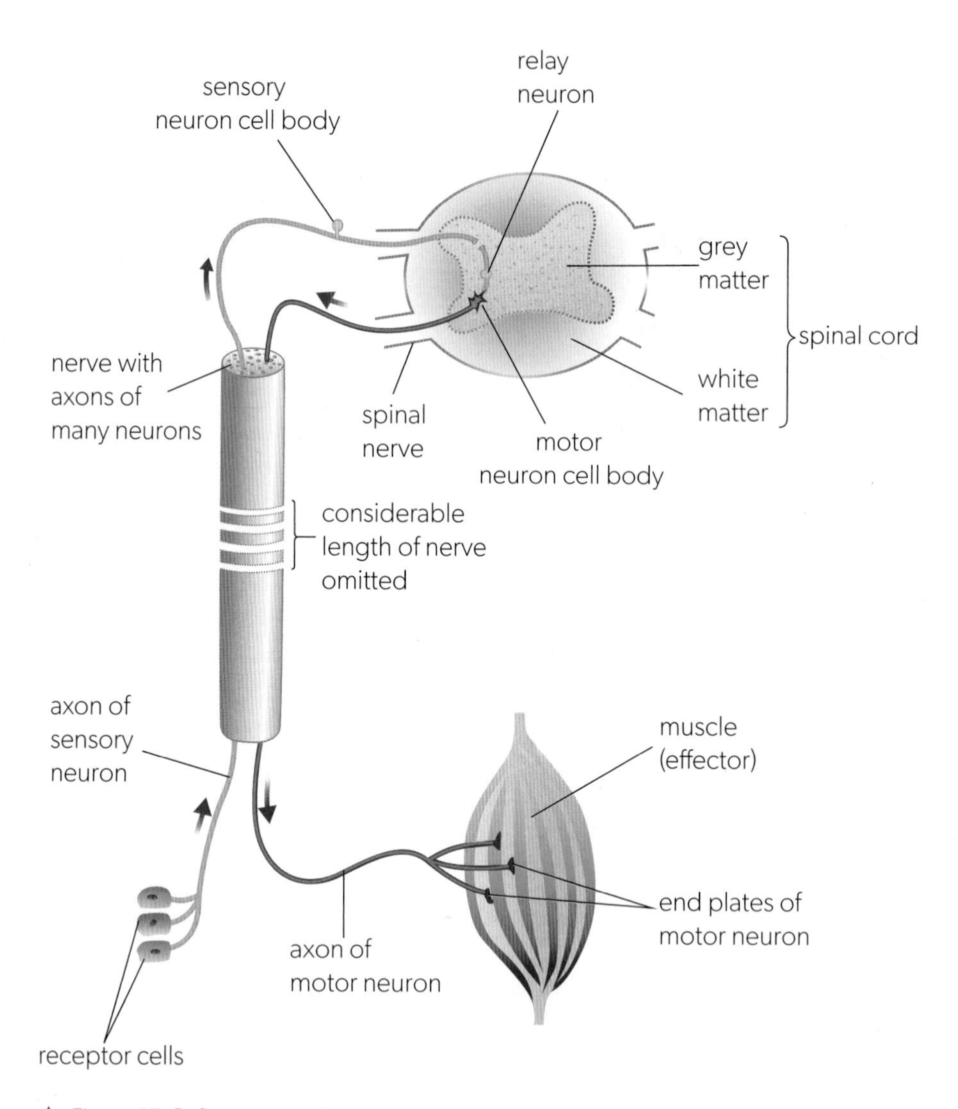

▲ Figure 17 Reflex arc used for spinal reflexes where the effector is a muscle

Activity: Investigating the patellar reflex

1. With a classmate, test the patellar (knee jerk) reflex. One person sits on the edge of a table or lab bench so their legs do not touch the ground and crosses their left leg over their right leg. The second person uses the edge of a narrow book or a similar shaped object to strike the first person's left leg, just below the knee cap. The left leg should straighten slightly at the knee, so the foot jerks upwards a little. What does it feel like when this reflex happens in you?
2. Below the knee, there is a tendon linked to the quadriceps muscle in the thigh. Striking this tendon stretches the quadriceps and this is detected by stretch receptors in the muscle. The response is contraction of the same muscle—the quadriceps. Stand up and try straightening one of your legs as much as you can at the knee. Feel your quadriceps muscle. Is it contracted or relaxed? What can you deduce about the movement the quadriceps causes when it contracts?
3. Doctors use a special reflex hammer to test the patellar reflex it in a patient. By testing this reflex, doctors can find out whether a region of the spinal cord in the lower back is healthy. How does this test work?
4. Many reflex actions increase our chances of survival, by helping us to avoid harm (for example, from touching a hot object). A small upward kick in your leg when you are sitting on a table does not increase your chances of survival! There must be some other reason for the quadriceps to contract when it starts to be stretched. Can you discover the real reason for the knee jerk reflex?

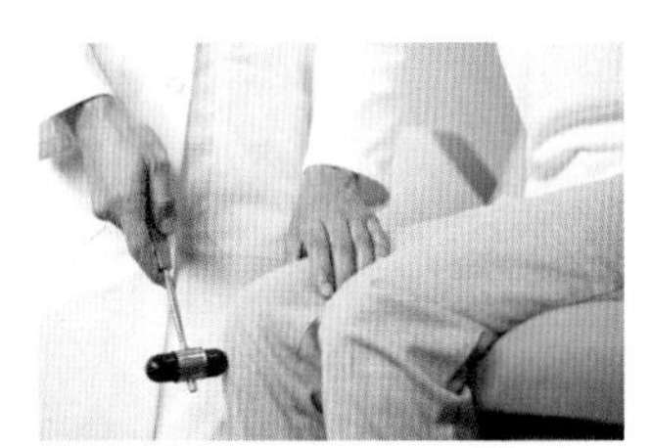

◀ **Figure 18** Doctors test the patellar reflex using a reflex hammer and feeling for a twitch in the quadriceps muscle

C3.1.10 Role of the cerebellum in coordinating skeletal muscle contraction and balance

The cerebellum has important roles in the control of skeletal muscle contraction and balance. It does not make decisions about which muscles will contract but it fine-tunes the timing of contractions. It allows very precise coordination of movements and helps us to maintain posture, for example when we are standing. It also helps us with activities requiring motor memory, such as riding a bike or typing on a keyboard.

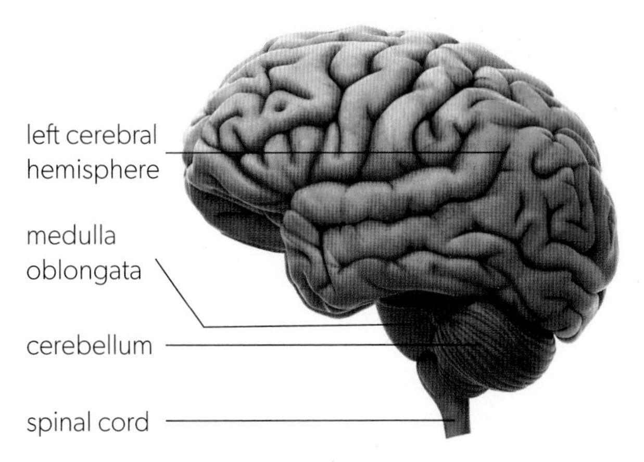

▲ **Figure 19** The cerebellum is located below the cerebral hemispheres in the rear part of the skull

C3.1.11 Modulation of sleep patterns by melatonin secretion as a part of circadian rhythms

Humans are adapted to live in a 24-hour cycle and have rhythms in behaviour that fit this cycle. These are known as circadian rhythms. These rhythms can continue even if a person is placed experimentally in continuous light or darkness, because they are controlled by an internal system.

Circadian rhythms in humans depend on two groups of cells in the hypothalamus called the suprachiasmatic nuclei (SCN). These cells set and follow a daily rhythm even when grown in an in vitro culture, with no external cues about the time of

▲ Figure 20 Until a baby is about three months old it does not develop a regular day–night rhythm of melatonin secretion so sleep patterns do not fit those of the baby's parents

day. In the brain they control the secretion of the hormone melatonin by the pineal gland. Melatonin secretion increases in the evening and drops to a low level at dawn. As the hormone is rapidly removed from the blood by the liver, blood concentrations rise and fall rapidly in response to these changes in secretion.

The most obvious effect of melatonin is the sleep–wake cycle. High melatonin levels cause feelings of drowsiness and promote sleep through the night. Falling melatonin levels encourage waking at the end of the night. Experiments have shown that melatonin contributes to the night-time drop in core body temperature: blocking the rise in melatonin levels at night reduces how much temperature drops, and giving melatonin artificially during the day causes a drop in core temperature. Melatonin receptors have been discovered in the kidney, suggesting that decreased urine production at night may be another effect of this hormone.

When humans are placed experimentally in an artificial environment with no light cues to indicate the time of day, the SCN and pineal gland usually maintain a rhythm of slightly longer than 24 hours. This shows that timing of the rhythm is normally adjusted by a few minutes or so each day so that it is synchronized with the diurnal cycle. A special type of ganglion cell in the retina of the eye detects light of wavelength 460–480 nm and passes impulses to cells in the SCN. This signals to the SCN the timing of dusk and dawn and allows it to adjust melatonin secretion so that it corresponds to the day–night cycle.

Mathematics: Regression analysis

Students often go without sleep to gain time to prepare for school assessments. This disrupts their circadian rhythms. There is evidence that, over the long-term, this can interfere with academic performance.

A study was carried out with 100 university students to determine the correlation between average sleep length and sleep length variability on semester grades for a particular course. Wearable technology was used to track sleep patterns. The first graph shows the correlation between average sleep duration and course grade. The second graph shows the correlation between standard deviation of mean sleep duration and course grade.

1. Outline the relationship between sleep duration and overall score in the course. [1]
2. The *p*-value indicates the probability that there is no correlation between the variables, that is, the slope of the regression line is zero. Explain the concept of *p*-value in relation to these two graphs. [2]
3. a. Outline what is indicated by a high standard deviation. [1]

   b. Suggest what would lead to a high value for the standard deviation of sleep duration. [2]
4. Discuss whether the data supports the conclusion that good sleep habits improve academic performance. [3]

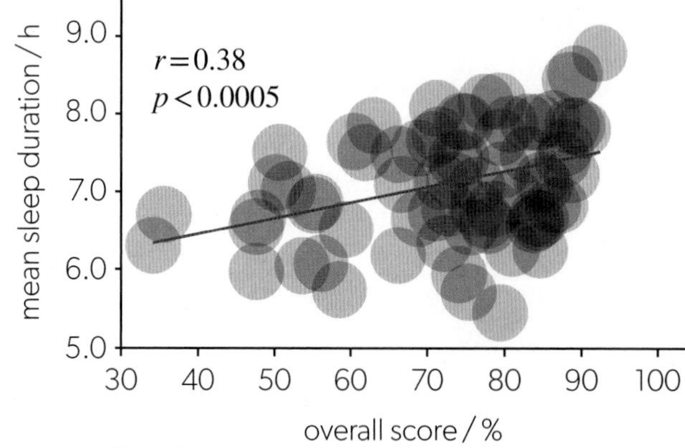

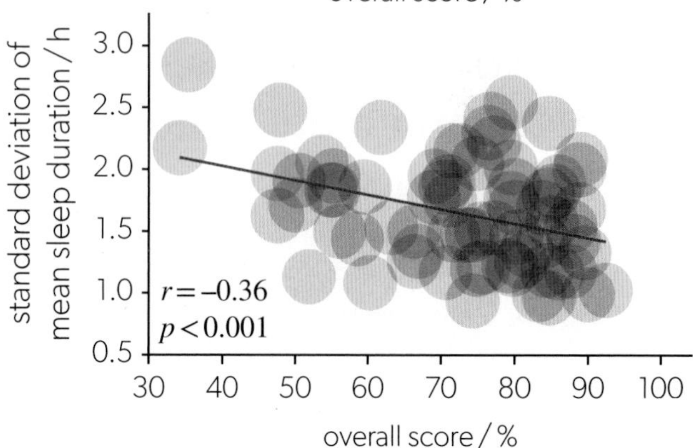

▲ Figure 21 Correlation between sleep duration and average course grades

ATL Communication skills: Taking effective notes

Health lies in healthy circadian habits

- Before you watch a video or a lecture, ask, "What do I need to learn by listening?" For example, watch the video: *Health Lies in Healthy Circadian Habits* by Dr Satchin Panda with these questions in mind:
    - What modern habits impact circadian rhythms negatively?
    - What recommendations does the speaker make to promote health related to circadian rhythms?
- Choose your method of note-taking carefully. Verbatim notes record everything, word-for-word, so they must be written as quickly as a speaker talks. Hand-writing is slower than speech, so you will have to type if you want verbatim notes. However, studies have shown that people who type verbatim notes are less likely to remember what the speaker said, compared with those who take selective handwritten notes. This is because choosing what to record requires us to think about what we are writing down.
- We can think more quickly than a person can talk. This is why our mind wanders during lectures. To maintain your focus, try these supplementary activities while recording the speaker's ideas:
    - Think about questions that the presentation does not answer.
    - Form opinions about what has been said.
    - Evaluate the evidence provided for any claims.
- If you know you need to answer specific questions in relation to a lecture, write down these questions ahead of time. Leave a space below each question so you can add the answers as you listen.
- Leave spaces between entries in your notes so you can add things later, when you are reviewing your notes. Take notes even if you do not understand what the speaker is saying: you can ask your teacher or do some research later to find out more.
- Focus more on the "big ideas" than the specific facts given.

When you have finished watching the video, compare your notes with those of a classmate.

C3.1.12 Epinephrine (adrenaline) secretion by the adrenal glands to prepare the body for vigorous activity

Epinephrine, also called adrenaline, is a hormone that prepares the body for vigorous physical activity. It is secreted by the adrenal glands. When epinephrine reaches tissues where it has an effect, it binds to adrenergic receptors in the plasma membrane of target cells. This triggers responses inside these cells.

Epinephrine has effects on most tissues. A common theme among the responses is preparation for vigorous physical activity. In particular, epinephrine increases the supply of oxygen and glucose to skeletal muscles, maximizing their production of ATP by respiration:

- Muscle cells break down glycogen into glucose, which can be used in aerobic or anaerobic respiration.
- Liver cells also break down glycogen into glucose, which is released into the bloodstream.
- Bronchi and bronchioles dilate due to relaxation of smooth muscle cells, so the airways become wider and ventilation is easier.
- The ventilation rate increases so a larger total volume of air is breathed in and out per minute.
- The sinoatrial node speeds up the heart rate, so cardiac output increases.

▲ Figure 22 Usain Bolt of Jamaica competing in the men's 200 m heats at the 2016 Olympic Games

- Arterioles that carry blood to muscles and to the liver widen due to relaxation of smooth muscle cells (vasodilation), so more blood flows to them.
- Arterioles that carry blood to the gut, kidneys, skin and extremities become narrower due to contraction of smooth muscle cells (vasoconstriction), so less blood flows to them.

As a result of these responses, the skeletal muscles used during vigorous activity receive a greater volume of blood per minute and this blood carries more glucose and oxygen.

The secretion of epinephrine in the adrenal glands is controlled by the brain. It increases when vigorous physical activity may be necessary because of a threat or an opportunity. For this reason, epinephrine is known as the "fight or flight" hormone. In the past, when humans were hunter-gatherers rather than farmers, epinephrine would have been secreted when they were hunting for prey or were threatened by a predator. In the modern world, many athletes use pre-race routines to stimulate epinephrine secretion so their heart rate is already increased when vigorous physical activity begins.

Data-based questions: Exercise and training

The effects of exercise were investigated using two groups of men. One group trained intensely for racing events by running or cycling long distances. The other group completed minimal training, by gentle running or racket sports. Table 4 shows the results of this investigation.

	Training group	At rest	After 4 mins of warm-up exercise	At maximal intensity of exercise
Heart rate / beats minute^{-1}	Minimal	79.5±4.3	149.6±5.4	186.9±2.1
	Intense	73.7±3.0	129.6±5.1	182.2±2.4
Mean arterial blood pressure / mm Hg	Minimal	89.6±1.3	101.7±1.6	109.2±1.9
	Intense	99.3±2.8	105.0±1.8	114.1±1.9
Oxygen uptake / ml min^{-1} kg^{-1}	Minimal	5.8±0.3	26.7±1.4	42.5±3.1
	Intense	4.3±1.1	26.7±1.6	55.2±2.2
Epinephrine concentration / ng ml^{-1} blood	Minimal	0.05±0.01	0.10±0.02	0.63±0.14
	Intense	0.09±0.02	0.17±0.05	1.03±0.23

Source: Rogers et al. 1991. "Catecholamine Metabolic Pathways and Exercise Training". *Circulation*. Vol 84. Pp 2346–2356

▲ Table 4

1. Display the data in Table 4 using the format that you think is most suitable. [5]
2. Using the data in Table 4, analyse the effect of training on exercise physiology. [5]
3. Discuss the conclusions that can be drawn from the data in Table 4 about the role of epinephrine in exercise physiology. [5]

C3.1.13 Control of the endocrine system by the hypothalamus and pituitary gland

The hypothalamus is a small region of the brain that has major roles in the integration of body systems. It consists of a thin wall of tissue located on the left and right sides of the third ventricle and below it. Ventricles are spaces inside the brain that are filled with cerebrospinal fluid. The hypothalamus links the nervous system to the endocrine system via the pituitary gland.

Within the hypothalamus there are specialized areas called nuclei. Each nucleus operates one or more specific control systems, using information from a variety of sources. Some nuclei have sensors for blood temperature, blood glucose concentration, osmolarity and the concentrations of various hormones. Many nuclei receive signals from sense organs, either directly or indirectly via the cerebral hemispheres. There are also inputs from other parts of the brain, such as the medulla oblongata, the hippocampus and the amygdala.

There are close relationships between the hypothalamus and the pituitary gland, which is located directly below it and connected to it by a narrow stalk. The pituitary gland has two distinct parts: the anterior lobe and the posterior lobe. These two lobes operate in different ways, but both of them secrete hormones into blood capillaries under the direction of nuclei in the hypothalamus. Table 5 shows hormones secreted into capillaries in each lobe.

Anterior pituitary
• HGH (human growth hormone) • TSH (thyroid-stimulating hormone) • LH (luteinizing hormone) • FSH (follicle-stimulating hormone) • prolactin
Posterior pituitary
• ADH (antidiuretic hormone) • oxytocin

▲ Table 5

Osmoregulation and puberty are two processes based on system integration by the hypothalamus and pituitary gland.

- Osmoreceptors in the hypothalamus constantly monitor the solute concentration of the blood. This and other inputs influence how much antidiuretic hormone (ADH) is produced by neurosecretory cells in the hypothalamus. The axons of the neurosecretory cells transport the ADH to the pituitary gland, where it is secreted into blood capillaries.
- The hypothalamus initiates puberty by secreting GnRH, a neurohormone that stimulates secretion of LH and FSH by the pituitary gland. These hormones in turn stimulate the secretion of testosterone in males and oestradiol and progesterone in females, leading to the changes associated with puberty.

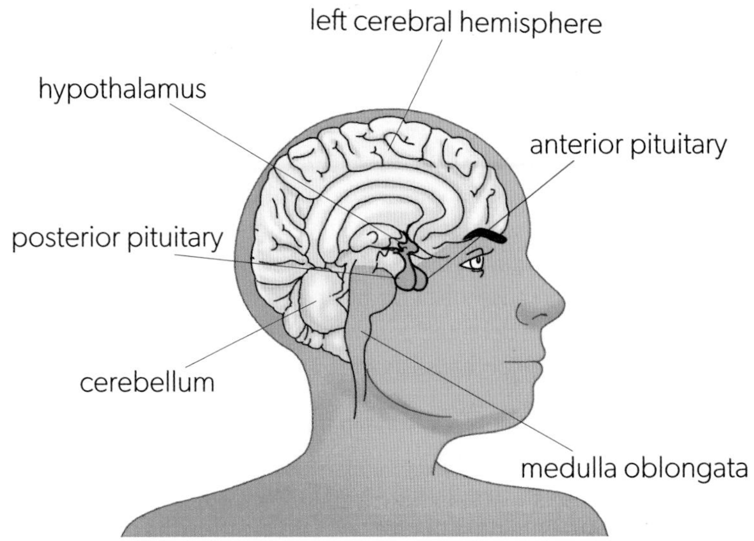

◀ Figure 23 The position of the hypothalamus and pituitary

C3.1.14 Feedback control of heart rate following sensory input from baroreceptors and chemoreceptors

The sinoatrial node (SAN) is a special group of cardiac muscle cells in the wall of the right atrium. It acts as a pacemaker for the heartbeat. The pacemaker receives signals from the cardiovascular centre, in the medulla oblongata of the brain. These signals reach the pacemaker via branches of two nerves:

- Signals from the sympathetic nerve cause the pacemaker to increase the frequency of heartbeats. In healthy young people, the heart rate can increase to three times the resting rate.
- Signals from the vagus nerve cause the pacemaker to decreases the heart rate.

These two nerves act rather like the throttle and brake of a car.

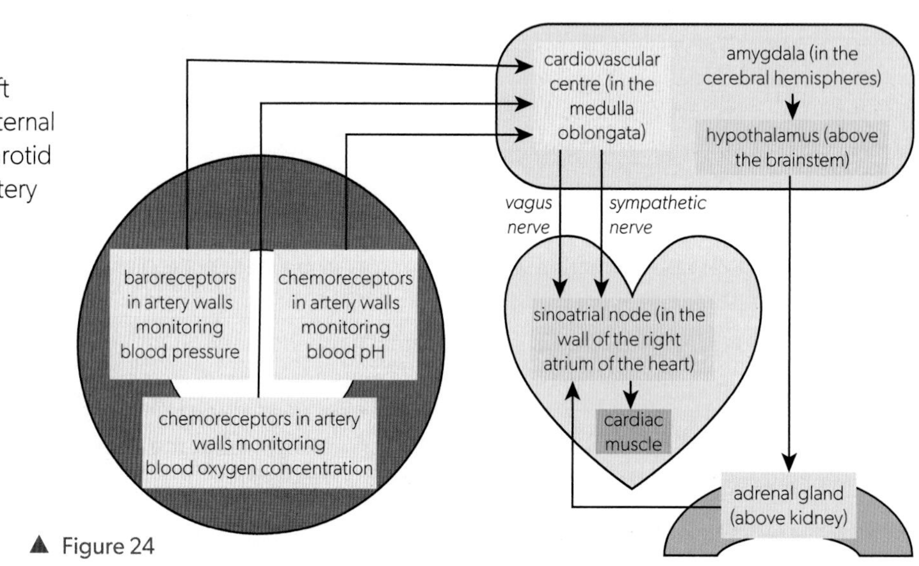

▲ Figure 24

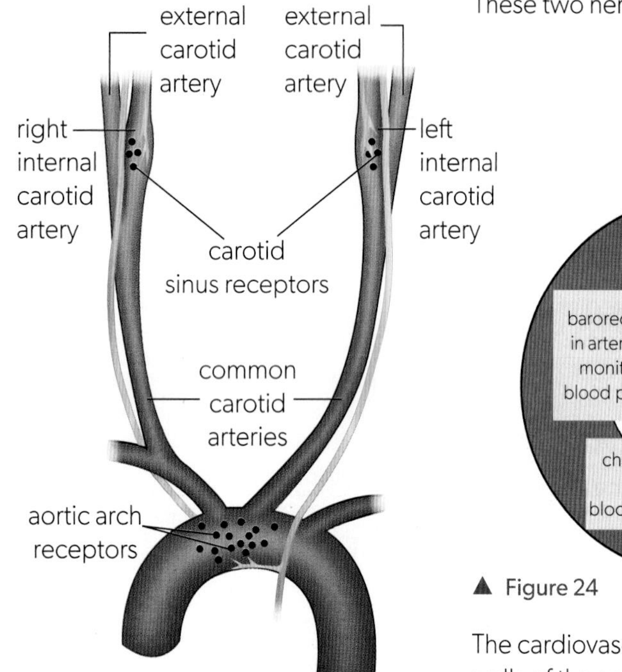

▲ Figure 25 Chemoreceptors and baroreceptors are located in the aorta, where it arches over the heart, and also in the carotid arteries, where inner and outer branches separate. Nerves that carry signals from these receptors to the brain are shown in yellow

The cardiovascular centre receives sensory inputs from baroreceptors in the walls of the aorta and the carotid arteries, which monitor blood pressure. This allows control of blood pressure by negative feedback. The response to low blood pressure is an increase in heart rate, which increases blood pressure. The response to high blood pressure is a decrease in heart rate, which reduces blood pressure.

The cardiovascular centre also receives sensory inputs from chemoreceptors in the aorta and carotid arteries. Some chemoreceptors monitor blood oxygen concentration and others monitor pH, which varies with carbon dioxide concentration. Again, negative feedback is used as the method of control. Low oxygen concentration and low pH cause the heart rate to speed up, increasing blood flow to the tissues so more oxygen is delivered and more carbon dioxide is removed. High oxygen concentration and high pH cause the heart rate to slow down.

The sinoatrial node also responds to epinephrine in the blood, by increasing the heart rate. This happens when the amygdala sends distress signals to the hypothalamus, which sends signals directly via nerve fibres to the cells in the adrenal gland that secrete epinephrine. Epinephrine can override the normal feedback control mechanisms while the body responds to a threat (often via vigorous physical activity).

C3.1.15 Feedback control of ventilation rate following sensory input from chemoreceptors

The bloodstream supplies oxygen to respiring cells and removes carbon dioxide. The overall rate of cell respiration in the body depends on energy use and reaches a maximum during vigorous physical exercise. As a result, the requirements for oxygen supply and carbon dioxide removal vary over time. Carbon dioxide concentration in blood must not be allowed to rise too high, because this causes a decrease in pH known as acidosis, with harmful consequences. The normal range for blood pH is 7.35 to 7.45. Levels below 6.8 can be life-threatening. Changes in the ventilation rate are the main method used to regulate blood pH.

Ventilation rate is the number of times air is inhaled and exhaled per minute. It is regulated by respiratory centres in the brainstem. Nerves carry signals from the respiratory centres to the muscles used to inhale: the diaphragm at the base of the thorax, and the external intercostal muscles between the ribs. Contraction of these muscles causes the lungs to expand. This is detected by stretch receptors, which send signals to the respiratory centres; this causes them to stop sending signals to inhale. Exhalation follows. This is usually passive but contractions of internal intercostal muscles and abdominal wall muscles may be used to increase the volume of air exhaled. After a short time, the respiratory centres trigger another inhalation.

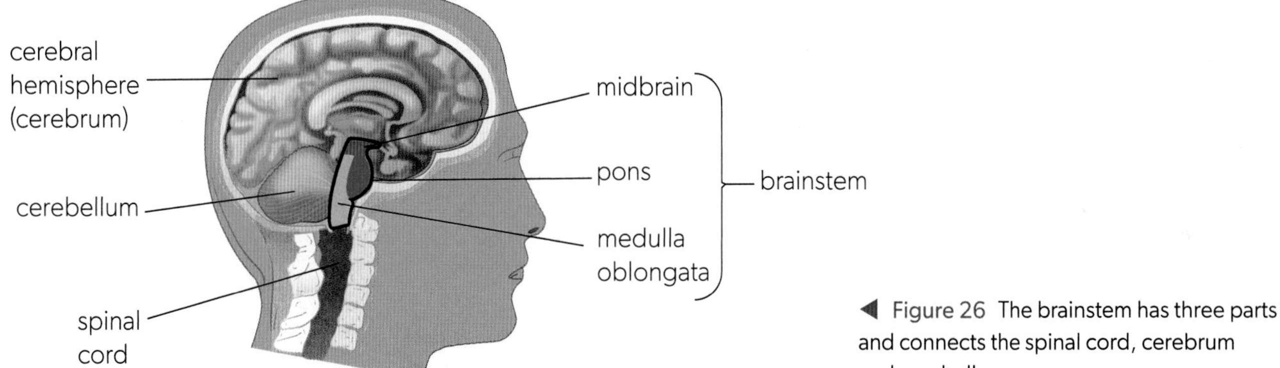

◀ Figure 26 The brainstem has three parts and connects the spinal cord, cerebrum and cerebellum

A negative feedback mechanism regulates ventilation rate. Chemoreceptors in the aorta and carotid arteries monitor blood pH, using acid-sensing ion channels. An increase in blood carbon dioxide concentration causes a decrease in blood pH. When chemoreceptors detect this decrease, they send signals to the respiratory centres, which decrease the interval between inhalations. This increases ventilation rate, which reduces the carbon dioxide concentration in the alveoli and increases the diffusion rate from blood in alveolar capillaries to air in the alveoli. Blood pH increases. When pH returns to the target range of 7.35 to 7.45, or rises above it, signals are no longer sent by the chemoreceptors and the respiratory centres allow the ventilation rate to decrease.

If carbon dioxide is removed from the body by the lungs rapidly enough, sufficient oxygen will usually be transported to respiring tissues as well. However, there is a back-up mechanism for oxygen supply. Chemoreceptors in the carotid arteries monitor the oxygen concentration of the blood flowing to the head. If these receptors detect hypoxia (a lack of oxygen), they send signals to respiratory centres in the brainstem, leading to an increase in ventilation rate. These signals override the signals from chemoreceptors monitoring blood pH and carbon dioxide concentration, helping to prevent the brain from becoming starved of oxygen.

Exploring and designing: Changes in ventilation rate in response to exercise

Figure 27 shows three phases in the response of the ventilation rate to vigorous exercise. The vertical axis is VO_2 which signifies the volume of O_2 consumed. Breaths per minute can be used to approximate VO_2.

1. Suggest a hypothesis relating to one specific response of ventilation rate to vigorous exercise.
2. Design an experiment to test your hypothesis.
3. Perform the experiment and use the results to evaluate your hypothesis.

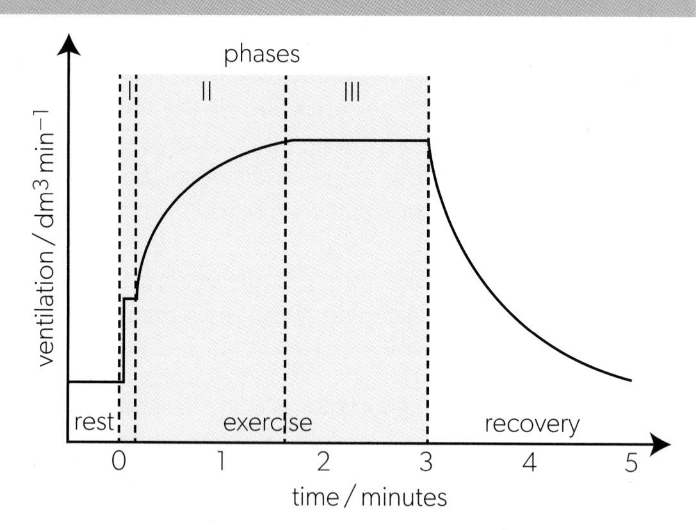

▶ Figure 27

Data-based questions: Exercise and ventilation

In an experiment, five healthy subjects carried out an incremental cycle ergometer test, twice each. During each test, the intensity of exercise increased until the subject was fatigued and decided to end the trial. Blood pH, minute ventilation (V_E), oxygen uptake (VO_2) and carbon dioxide output (VCO_2) were all monitored. The first test was used to determine the stage at which blood pH starts to decrease due to the build-up of lactate. This stage is shown on the graph with an arrow. During the second test, an alkali (HCO_3^-) was injected into a vein from this stage onwards, to prevent acidosis.

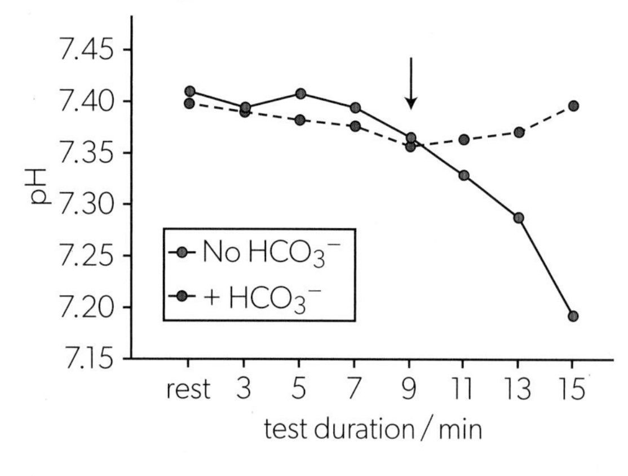

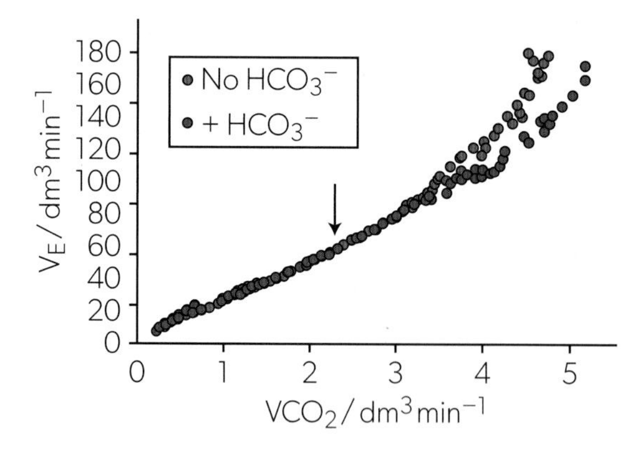

▲ Figure 28 Graphs showing results for one of the people tested. Results for the four other people tested were similar

Source: Meyer T, Faude O, Scharhag J, *et al.* 2004. "Is Lactic Acidosis a Cause of Exercise Induced Hyperventilation at the Respiratory Compensation Point"? *British Journal of Sports Medicine*. Vol 38. Pp 622–625.

1. a. State the changes during the first 9 minutes of the "No HCO_3^-" trial, in:
   i. blood pH [1]
   ii. carbon dioxide output [1]
   iii. minute ventilation. [1]

   b. Explain the interrelationships between these changes. [4]

2. a. Suggest reasons for increasing rates of blood acidification in the "No HCO_3^-" trial after 9 minutes. [4]

   b. Explain the conclusions that can be drawn from the results of the two trials after 9 minutes. [4]

Collecting and processing data: Interpreting spirometer data

A spirometer is used to measure the volume of air inspired and expired by the lungs. Data logging spirometers often gather time data as well. Using a spirometer and suitable software, it is possible to determine a number of different quantities experimentally, including:

- forced vital capacity (how much air a person can exhale during a forced breath)
- tidal volume (the volume of air that moves in and out between normal inhalation and exhalation)
- ventilation rate (the number of breaths taken per minute).

The graphs in Figure 29 show spirometer data collected from an individual at rest and the same individual running at 22 km h^{-1} on a treadmill. Use these graphs to answer the questions.

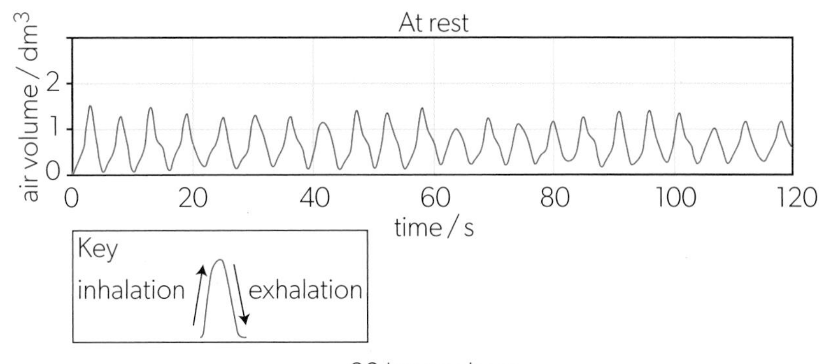

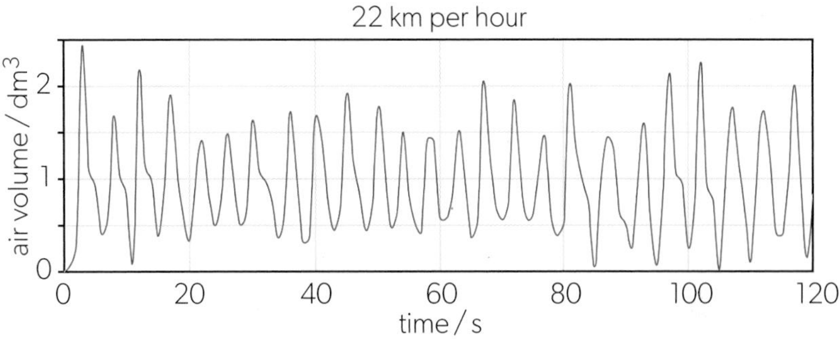

▲ Figure 29

1. Determine the resting ventilation rate.
2. Determine the running ventilation rate.
3. Determine the peak air flow rate while running. (This is the average of the three steepest slopes of individual breath cycles.)
4. Outline a procedure for determining average resting tidal volume.
5. State the effect of exercise on tidal volume.
6. Asthma is a condition in which airways swell and the bronchioles in the lung become narrower. Predict the effect of asthma on forced vital capacity.

Technology: Using a smartphone as a data logger

The regulation of ventilation is normally under involuntary control. However, we can control our ventilation rate by conscious effort. For example, we can choose to lower our breathing rate or even hold our breath for a while. How would this affect heart rate?

Many modern smartphones contain internal sensors, such as motion and magnetic field sensors. If you lie on your back, with your phone resting on your abdomen, the accelerometer will detect motion associated with breathing. In this way, it could be used to measure ventilation rate.

The graph in Figure 30 shows the data from a smartphone magnetic field sensor. A volunteer taped a magnet to their abdomen and then placed their phone near the magnet for about 70 seconds. The oscillation of the magnetic field shown in the graph gives a rough indication of the ventilation rate.

1. How many breaths were taken over the 70 seconds?
2. At which two times did the volunteer hold their breath?

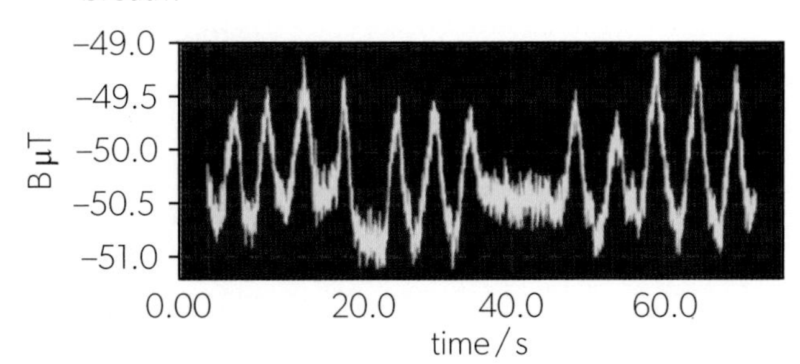

▲ Figure 30

C3.1.16 Control of peristalsis in the digestive system by the central nervous system and enteric nervous system

The gut is a tube that extends from the mouth to the anus. As food is passed along the gut, secretions are added to it from gland cells in the gut wall and from accessory glands such as the pancreas. Enzymes in these secretions hydrolyse molecules that are too large or insoluble to pass across the membrane of the epithelium cells that line the gut. Sugars, amino acids and other relatively small food molecules are then absorbed into the bloodstream.

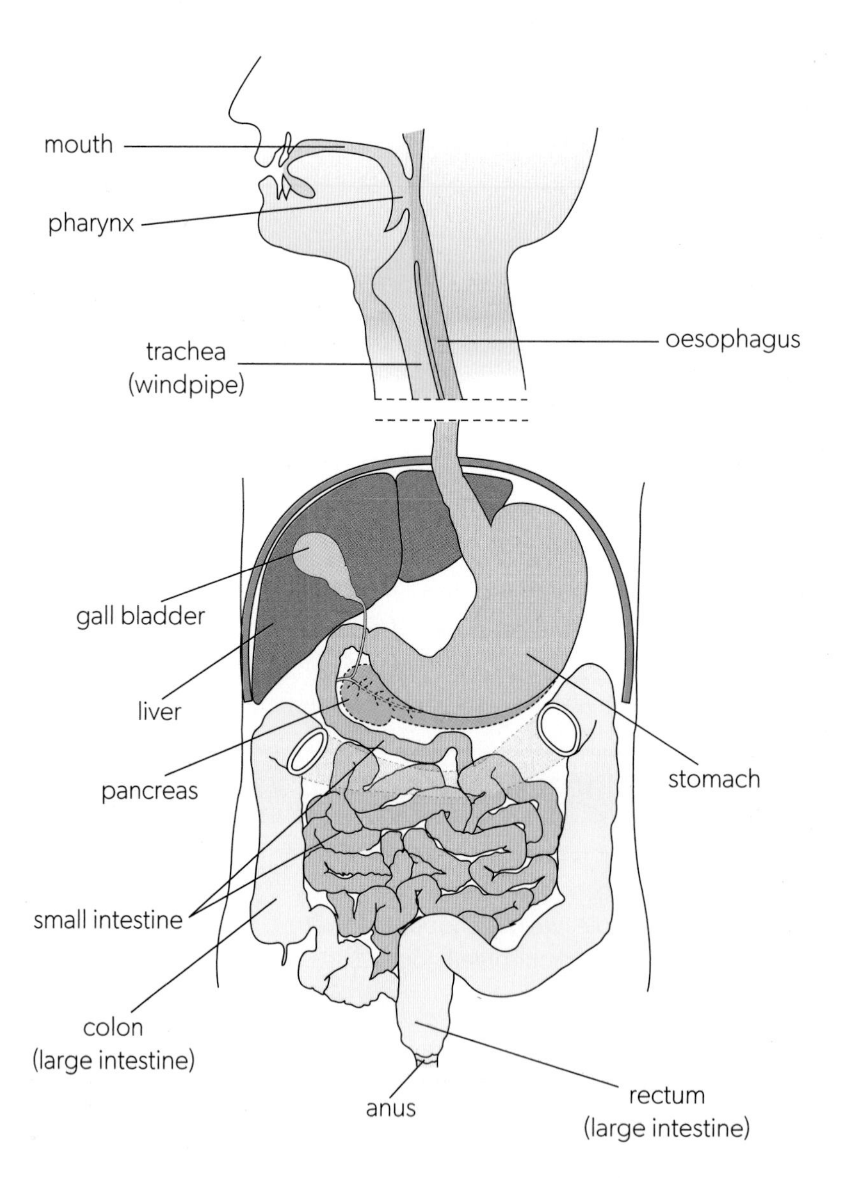

▶ Figure 31 The digestive system

The wall of the gut contains two layers of muscle tissue. Both are smooth muscle rather than skeletal (striated) muscle, so consist of relatively short cells rather than elongated fibres. The outer layer is longitudinal muscle, with cells orientated along the axis of the gut. The inner layer is circular muscle, with cells arrange in rings.

The muscle in the wall of the gut is typical of smooth muscle in that it can exert continuous moderate force, with short periods of more vigorous contraction. (In contrast, striated muscle remains relaxed unless it is stimulated to contract.)

Waves of vigorous contraction, called peristalsis, pass along the intestine. Contraction of circular muscles behind the food constricts the gut, so food is not pushed back towards the mouth. Contraction of longitudinal muscle where the food is located moves it along the gut.

Swallowed food moves quickly down the oesophagus to the stomach in one continuous peristaltic wave. Peristalsis occurs in one direction only, away from the mouth. When food is returned to the mouth from the stomach during vomiting, abdominal muscles are used rather than the circular and longitudinal muscle in the gut wall.

In the intestines, food is moved only a few centimetres at a time. This slower progression allows time for digestion. The main function of peristalsis in the intestine is churning of the semi-digested food to mix it with enzymes and thus speed up the process of digestion.

Peristaltic muscle contractions are controlled unconsciously by the enteric nervous system (ENS), which is extensive and complex. The ENS has intrinsic microcircuits that allow control of the stomach and intestines without inputs from the central nervous system (CNS).

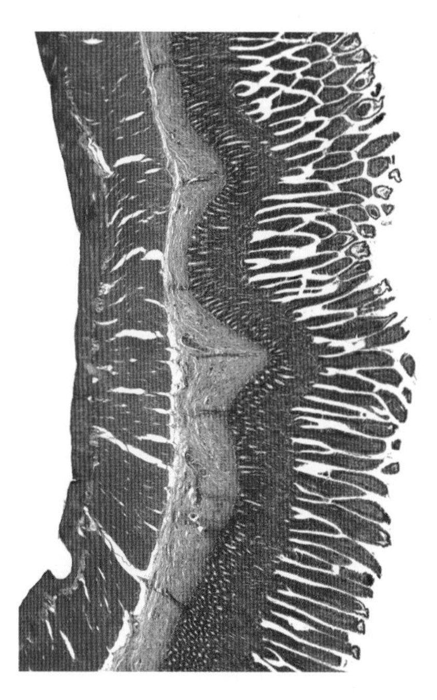

▲ Figure 32 In this longitudinal section of the gut, the outer part of the gut wall is on the left. Villi are visible on the inside of the gut wall, showing that it is part of the small intestine

Two gut movements are **not** involuntary and under the control of the ENS:

- The process of swallowing begins when the tongue pushes food to the back of the mouth cavity. This is a voluntary action because the tongue is composed of striated muscle. When food reaches the back of the mouth cavity, it stimulates touch receptors in the pharynx. Signals from these receptors pass to the brainstem, which stimulates contractions that push the food down the pharynx and into the oesophagus. Once food reaches the oesophagus, peristalsis is controlled by the ENS and is entirely involuntary.
- Defecation is the removal of faeces from the rectum via the anus. The anus contains a ring of smooth muscle (a sphincter). The wall of the rectum contains layers of circular and longitudinal smooth muscle. During defecation, the anus relaxes and widens and the wall of the rectum contracts. In babies this is controlled involuntarily, but during the early years of life humans achieve voluntary control of the process.

C3.1.17 Observations of tropic responses in seedlings

Plants control the direction of growth of their roots and shoots. If one side of a root or shoot grows more quickly than the other side, the root or shoot will become curved. The root or shoot tip will then be pointing in a new direction. This type of growth happens in response to external stimuli perceived by the plant, such as the direction of gravity or sunlight.

▲ Figure 33 A seed of the rock oak (*Quercus montana*) has fallen to the forest floor and germinated. Whichever way the seed is facing, the shoot will grow upwards towards the light and the root will grow downwards into the soil due to tropisms

Differential growth responses to directional stimuli are called tropic responses or tropisms.

- Positive tropism is growth towards the stimulus.
- Negative tropism is growth away from the stimulus.

Most roots are positively gravitropic (geotropic). This means they grow downwards, in the same direction as gravity. Most shoots are positively phototropic and negatively gravitropic. This means they grow towards the source of light and, in darkness, they grow upwards in the opposite direction to gravity.

Investigating tropisms

The roots and shoots of plants do not all show the same tropic responses. You can investigate these responses qualitatively—by recording observations in drawings or photos—or quantitatively. The angle of curvature of a root or shoot that has carried out a tropic response is an example of a quantitative measure.

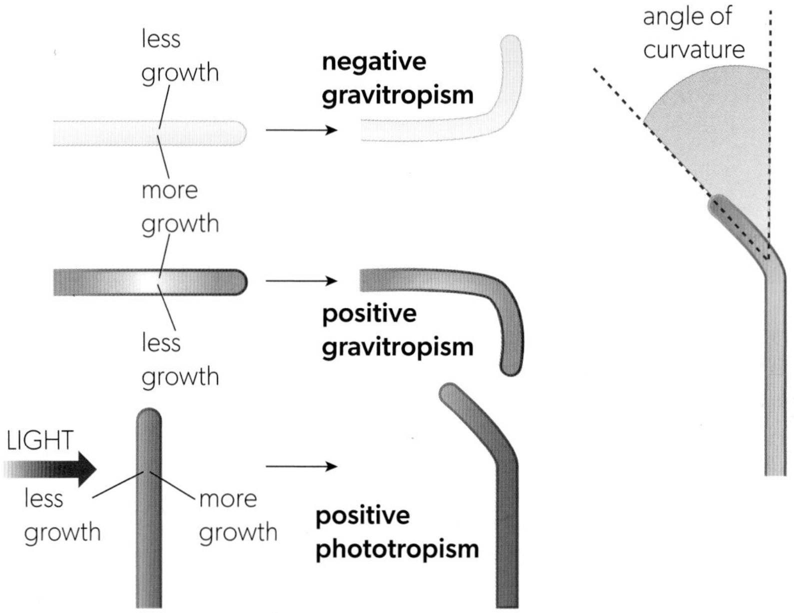

▲ **Figure 34** Types of tropism

Observations: Qualitative and quantitative data

Qualitative observations are made directly with the senses and do not involve measured variables. For example, stating that the colour of this text box is green is a qualitative observation.

Quantitative observations are made by counting or using measuring devices.

Accuracy is how close a measured value is to its true value. A metaphor for this is how close a dart is to the centre of a dart board.

Precision is how explicit or definitive a measured value is. It is the smallest difference in size that can be distinguished with certainty. For example, an electronic balance may be able to measure mass to the nearest gram, or tenth, hundredth or thousandth of a gram. Before anything is placed on the balance, the reading will be 0 g, 0.0 g, 0.00 g or 0.000 g depending on the precision of measurement.

Measurements which are both accurate and precise are reliable. To increase the accuracy of measurements, you must use measuring devices correctly and carefully. Measuring devices should be calibrated to ensure they read zero when nothing is being measured. When measuring lengths using a ruler, you must acknowledge the uncertainty associated with both the starting point and the ending point of a measurement.

Exploring and designing: Colour of light and phototropism

Green plants appear green because they reflect green light, so green light does not promote photosynthesis. This leads to the question: does the colour of light influence phototropism?

1. How can the degree of bending in tropic responses be quantified?
2. What are the differences between precision, accuracy and reliability in taking measurements?
3. The colour of light can be varied by using coloured LED lights to illuminate seedlings. Alternatively, a hole can be cut into the side of a canister and the opening can be covered by translucent filters (gels) of different colours. How can the colour of light be quantified?

C3.1.18 Positive phototropism as a directional growth response to lateral light in plant shoots

Shoots usually grow towards the highest light intensity in their environment. This is positive phototropism. If a shoot tip detects that it is not growing towards the brightest light, it responds by differential growth. The side of the shoot facing the brighter light is stimulated to grow at a slower rate than the shadier side. When the shoot has curved towards the direction of maximum light intensity, growth becomes equal on all sides so the shoot carries on growing in that direction.

The benefit of positive phototropism is obvious: it increases the amount of light absorbed by a shoot's leaves for use in photosynthesis. This is particularly important for plants growing in forests or other communities where there is competition with other plants for light and the brightest light may be to one side rather than directly above.

In the 1920s, it was discovered that a plant hormone was responsible for the differential growth that causes curvature towards the light. Since then, scientists have identified the pigment that detects light in the tip of the shoot and learned much more about the other mechanisms of phototropism.

◀ Figure 35 Seedlings growing on a window sill rapidly demonstrate positive phototropism

AHL

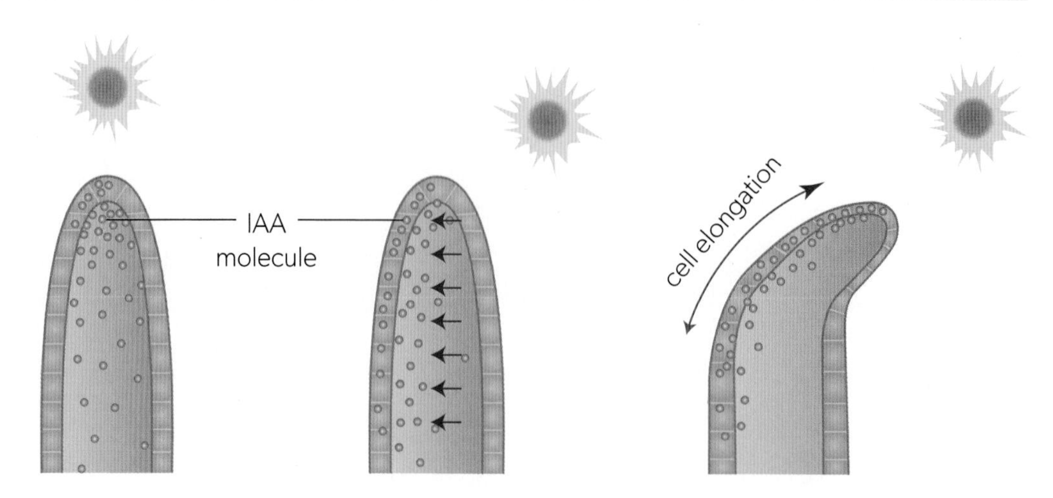

▲ Figure 36 IAA (auxin) is a plant hormone that promotes stem growth. It is moved from the sunnier to the shadier side of shoot tips to cause the differential growth that happens in phototropism

C3.1.19 Phytohormones as signalling chemicals controlling growth, development and response to stimuli in plants

A hormone is a chemical message that is produced and released in one part of an organism to have an effect in another part of the organism. Plants and animals produce, transport and use hormones in different ways, so plant hormones are called phytohormones. Phytohormones help to control growth, development and responses to stimuli in plants.

▼ Figure 37 Molecular structures of the main types of phytohormone

$CH_2{=}CH_2$

ethylene

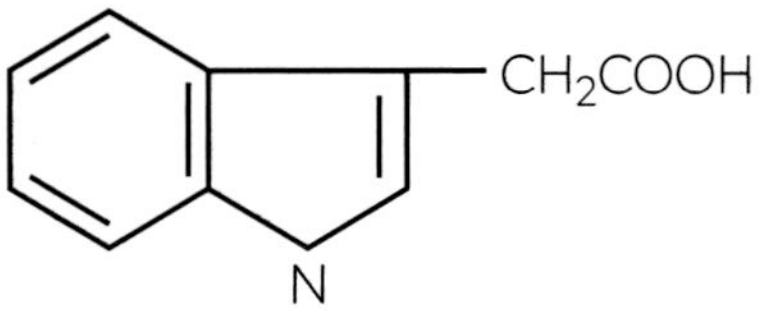

CH_2COOH

auxin

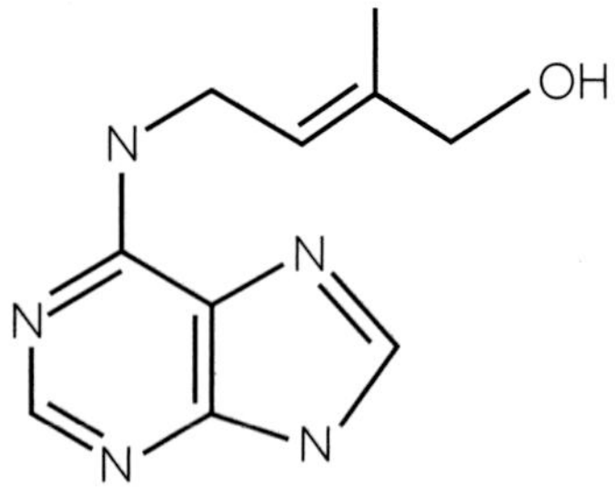

cytokinins

COOH

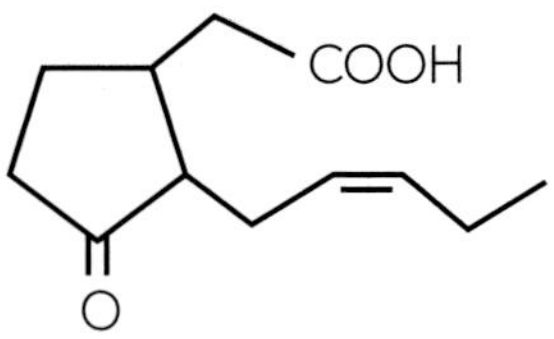

jasmonic acid

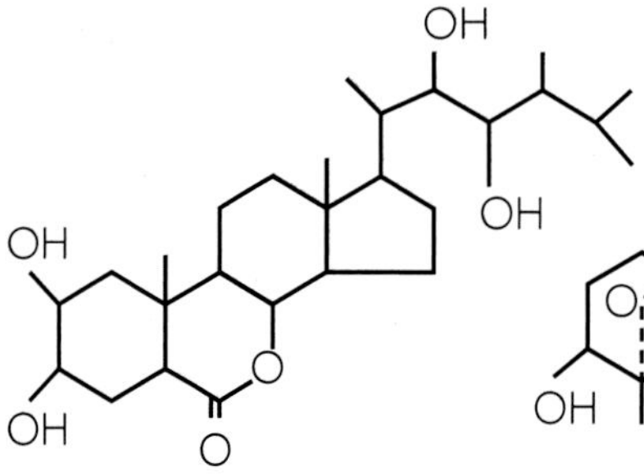

brassinosteroids

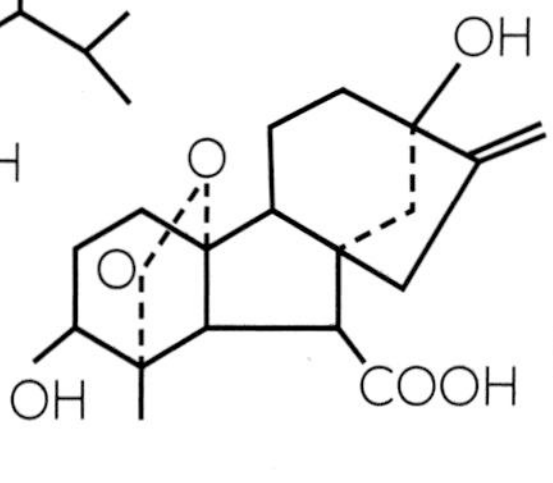

gibberellins

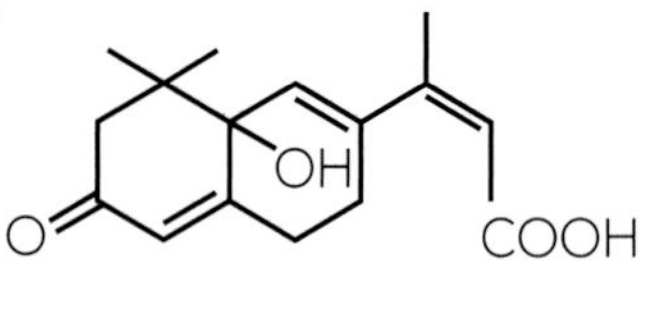

abscisic acid

1. Growth

Phytohormones can either promote or inhibit growth, by affecting rates of cell division and cell enlargement. For example, gibberellin promotes stem growth. Wheat and other crop plants have been bred with short stems by introducing alleles that make the plant less responsive to gibberellin.

2. Development

Phytohormones can promote or inhibit aspects of development—for example, whether a bud starts to grow to produce a side shoot, or whether the apex of a stem produces more leaves or changes to produce flowers. Ripening in the fruits of many plants is promoted by the phytohormone ethylene.

3. Responses to stimuli

Tropic responses are controlled by phytohormones. Tendrils of climbing plants respond to touch stimuli by coiling around a potential support. Communication using electrical signals is used for rapid responses, such as the capture of an insect by a Venus flytrap plant. However, a phytohormone called jasmonic acid triggers the subsequent secretion of enzymes to digest the fly.

C3.1.20 Auxin efflux carriers as an example of maintaining concentration gradients of phytohormones

Auxin is a phytohormone that promotes stem growth and causes the differential growth response of phototropism. Auxin can enter cells by passive diffusion as long as its carboxyl group (COOH) remains undissociated. The cytoplasm of plant cells is slightly alkaline, so once auxin has entered a plant cell the carboxyl group dissociates by losing a proton. This leaves it with a negative charge (COO^-), so the auxin is trapped inside the plant cell. However, plant cells produce membrane proteins known as auxin efflux carriers. These proteins can pump auxin in its charged state across the plasma membrane into the surrounding cell wall. The cell wall is slightly acidic, so the auxin reverts to its uncharged state. It can then diffuse into an adjacent cell.

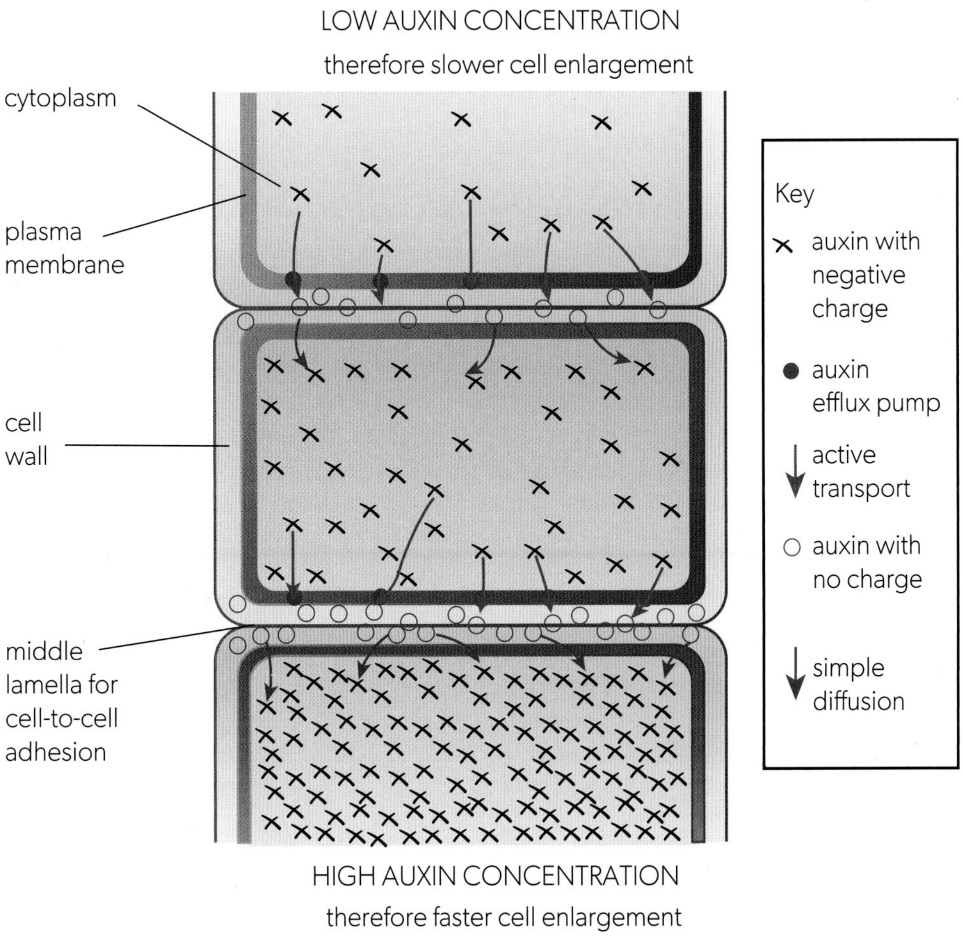

▲ Figure 38 Auxin can be concentrated in plant tissues by auxin efflux carriers

AHL

Plant cells can control the distribution of auxin efflux carriers. To transport auxin across a tissue and generate a concentration gradient, the carriers are moved to the same side of each cell. Auxin is therefore pumped out on that side and tends to enter the cell by diffusion on the other side.

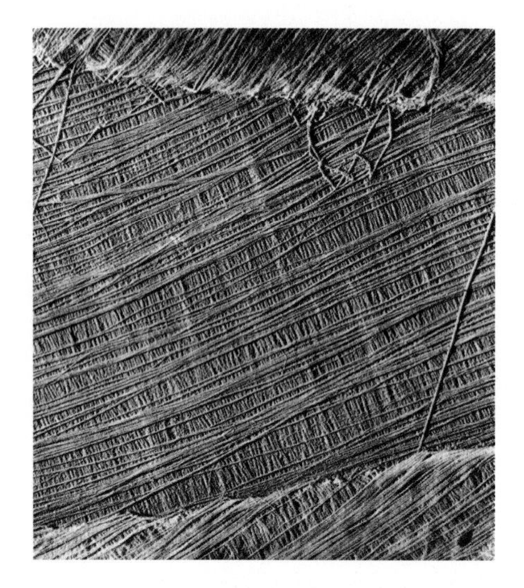

▲ Figure 39 Electron micrograph of cellulose microfibrils in a cell wall. The cell can control the orientation of microfibrils added to the wall. This affects the direction in which the wall can most easily expand in future

C3.1.21 Promotion of cell growth by auxin

The cell wall allows high turgor pressures to develop inside plant cells without them bursting. Turgidity helps shoots to resist the forces of gravity and wind. It gives roots strength to push through the soil. When plant tissue has finished growing, cell walls can be thickened to provide extra strength and resilience. Until then, they must remain thin enough to allow cells to increase in size.

Cell walls are constructed using bundles of cellulose molecules, called microfibrils. When the wall needs to be thickened, extra microfibrils are made and passed out through the plasma membrane. Cellulose molecules are inelastic, so a microfibril cannot stretch or extend in length. Therefore, extension of cell walls involves microfibrils moving further apart or sliding past each other.

The cellulose microfibrils are crosslinked by a variety of other carbohydrates, including pectin. The strength of these crosslinks is influenced by pH. Decreases in pH weaken the links, allowing the wall to extend. Auxin promotes the synthesis of proton pumps, which are inserted into the plasma membrane. These pumps transport H^+ ions from inside the cell (the protoplast) to the cell wall outside (the apoplast), acidifying the apoplast. This allows the wall to expand so the cell can elongate. Concentration gradients of auxin cause gradients of apoplastic pH, leading to differential cell wall extension and growth—as in phototropism.

C3.1.22 Interactions between auxin and cytokinin as a means of regulating root and shoot growth

Auxin is produced in shoot tips and transported down shoots and into roots in phloem. Cytokinin, another type of phytohormone, is produced in root tips and transported up roots and into shoots in xylem. The amounts of auxin and cytokinin produced in a plant provide a means of balancing root and shoot growth. In some cases, auxin and cytokinin work together to stimulate a process (synergism). In other cases, they have opposing effects (antagonism). Table 6 shows some of the effects of these two phytohormones.

▼ Table 6

	Cell division in the apices (tips) of stems and roots	Cell enlargement in the apices (tips) of stems and roots	Development of branches of roots or new roots	Development of lateral buds into branches of the stem
auxin	Stimulated if cytokinin is present	Stimulated	Stimulated	Inhibited
cytokinin	Stimulated	Stimulated if auxin is present	Inhibited	Stimulated
interaction	Synergistic	Synergistic	Antagonistic	Antagonistic

If, for example, the main shoot of a plant is eaten by an animal, less auxin will be produced. This will slow root growth and allow growth of one or more lateral buds to replace the main shoot. If the main shoot of the plant is not eaten and it continues to grow, more auxin will be produced. This will inhibit development of lateral buds and promote the root growth that is needed to supply extra water and nutrients.

Axillary buds are shoots that form at a node—a junction between the stem and the base of a leaf. As the shoot apical meristem grows and forms leaves, regions of meristem are left behind in buds at each node. Growth of these buds is inhibited by auxin produced by the shoot apical meristem. This is called apical dominance. The greater the distance between a node and the shoot apical meristem, the lower the concentration of auxin; as a result, growth in the axillary bud will be less inhibited by auxin. In addition, cytokinins promote axillary bud growth. The relative ratio of cytokinins and auxins determines whether an axillary bud will develop. Gibberellins are another category of hormones that contribute to stem elongation.

▲ Figure 40 Hedges are created by repeatedly cutting off shoot tips, removing the sources of auxin so side shoots can grow. This photo shows very old hedges at Powis Castle in Wales, which have grown enormous with bulges

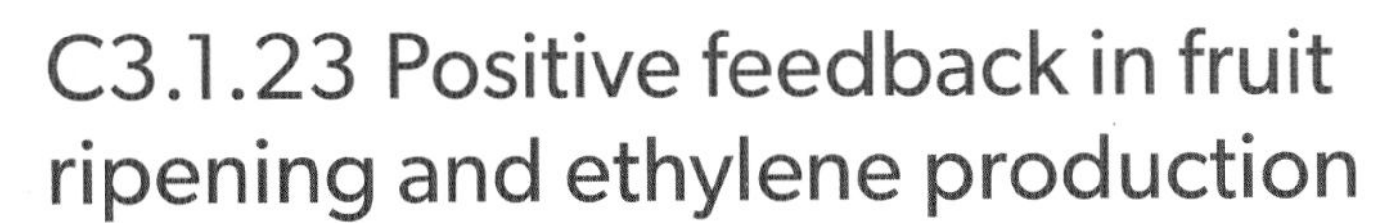

C3.1.23 Positive feedback in fruit ripening and ethylene production

Succulent fruits such as peaches rely on animals for seed dispersal. The flesh of the immature fruit is hard and acidic, so it is unpalatable. The fruits remain green and lacking in scent until the seeds they contain are fully developed, so animals are not attracted to them.

The process of fruit ripening happens over a relatively short time. These are typical changes:

- The colour of the fruit changes from green.
- Cell walls are partly digested, softening the flesh of the fruit.
- Acids and starch are converted to sugar, making the fruit palatable.
- Volatile substances are synthesized to give the fruit a distinctive scent.

In many plant species, these changes are stimulated by ethylene, which acts as a phytohormone. Once the seeds are ready for dispersal, a positive feedback mechanism causes rapid ripening. Ethylene stimulates ripening, and ripening fruits produce ethylene.

Ethylene is volatile so it is released as a vapour by ripening fruits and can diffuse to other fruits, initiating their ripening. This helps to synchronize the ripening of fruits on a plant such as a peach tree. It is useful for fruit farmers but in wild plants, it encourages animals that disperse the seeds to visit a plant by ensuring that plenty of ripe fruits are available at the same time.

▲ Figure 41 Tomatoes picked green can be encouraged to ripen by placing them in a plastic bag with a ripe banana. How does this trick work?

Technology: Using a smartphone as a data logger (2)

RGB colour sensors used to quantify fruit ripening

As tomatoes ripen, the relative levels of red and green pigments change. These changes can be detected by photography apps for smartphones that use RGB sensors.

Choose two samples of unripe tomatoes with similar red and green RGB values. Place the first treatment group in a bag with a ripe banana. Place the second treatment group in a bag without the banana. What impact does the presence of the banana have on RGB values over time?

Linking questions

1. What are examples of branching (dendritic) and net-like (reticulate) patterns of organization?
   a. Food webs are an example of a net-like pattern of energy flow. Construct a diagram of a representative food web from a local biological community. (C4.2.4)
   b. Outline the branching that occurs within the lung system. (B3.1.4)
   c. Construct an annotated diagram showing the relationship between arteries, arterioles, capillaries, venules and veins. (B3.2.10)
2. What are the consequences of positive feedback in biological systems?
   a. Outline an example of positive feedback cycles playing a role in climate change. (D4.3.2)
   b. Outline an example of the regulation of cell signalling pathways by positive feedback (C2.1.14).
   c. Explain the role of positive feedback in fruit ripening. (C3.1.23)

C3.2 Defence against disease

How do body systems recognize pathogens and fight infections?

Immune systems depend on the ability to distinguish between self and foreign molecules. A molecule which triggers an immune response is known as an antigen. The genetic changes that cause a cell to become cancerous lead to the presentation of tumour antigens on the cell's surface. In Figure 1, a T-lymphocyte cell (orange) is attacking a cancer cell (grey). A pathogen is normally recognized as a collection of antigens. What other mechansims are involved in mounting an immune response to infection by a pathogen?

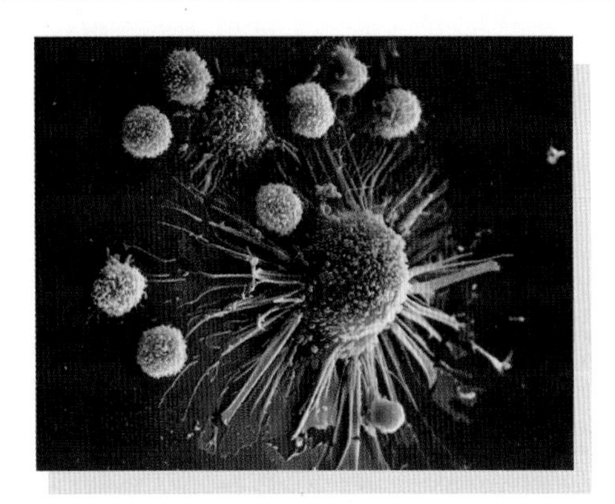

◀ Figure 1

What factors influence the incidence of disease in populations?

The photograph in Figure 2, taken in 1918, shows men in a makeshift influenza (flu) ward at the US Army's Eberts Field facilities in Lonoke, Arkansas, during the Spanish flu pandemic. This pandemic occurred in several waves between 1918 and 1920. It infected one-fifth of the world population and killed between 20 and 50 million people—more than had been killed in the First World War. The second wave of the pandemic, from August 1918, was much deadlier than the first. Mortality rates were high among young healthy adults including soldiers. What were the factors that made this such a deadly pandemic?

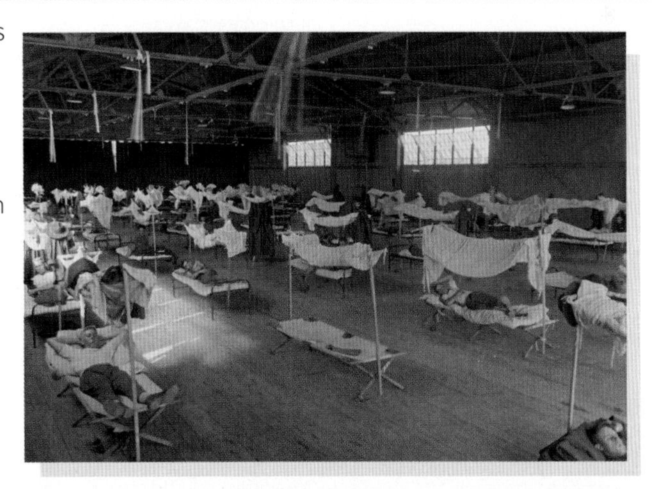

◀ Figure 2

SL and HL

C3.2.1 Pathogens as the cause of infectious diseases
C3.2.2 Skin and mucous membranes as a primary defence
C3.2.3 Sealing of cuts in skin by blood clotting
C3.2.4 Differences between the innate immune system and the adaptive immune system
C3.2.5 Infection control by phagocytes
C3.2.6 Lymphocytes as cells in the adaptive immune system that cooperate to produce antibodies
C3.2.7 Antigens as recognition molecules which trigger antibody production
C3.2.8 Activation of B-lymphocytes by helper T-lymphocytes
C3.2.9 Multiplication of activated B-lymphocytes to form clones of antibody-secreting plasma cells
C3.2.10 Immunity as a consequence of retaining memory cells
C3.2.11 Transmission of HIV in body fluids
C3.2.12 Infection of lymphocytes by HIV with AIDS as a consequence
C3.2.13 Antibiotics as chemicals that block processes occurring in bacteria but not in eukaryotic cells
C3.2.14 Evolution of resistance to several antibiotics in strains of pathogenic bacteria
C3.2.15 Zoonoses as infectious diseases that can transfer from other species to humans
C3.2.16 Vaccines and immunization
C3.2.17 Herd immunity and the prevention of epidemics
C3.2.18 Evaluation of data related to the COVID-19 pandemic

C3.2.1 Pathogens as the cause of infectious diseases

A disease is a particular kind of illness, with characteristic symptoms. For a long time, the causes of disease were not understood, because they are not visible to the naked eye. Many theories were suggested and later proved wrong, for example the idea that malaria was caused by breathing bad air. Scientific research into disease has identified three main types of cause:

- genetic—the alleles a person has
- environmental—for example, toxic chemicals or radiation
- infection with a pathogen.

Pathogens are organisms that cause disease. They are passed from one infected organism to another, either directly or indirectly. They enter the organism, multiply there and cause harm. Organisms that invade the body and can be seen with the naked eye, such as tapeworms, are usually considered to be parasites rather than pathogens. Pathogens are therefore microorganisms.

Pathogens must be able to multiply inside the body. Living organisms do this by reproduction. There are examples of pathogens in most of the principal groups of microorganisms—Table 1 gives human examples. No diseases are known to be caused by bacteria from the domain Archaea; all known examples of bacterial pathogens are in the domain Eubacteria.

Viruses are not usually considered to be alive because they cannot reproduce themselves. However, they are replicated by host cells and they certainly cause diseases so they are included among pathogens.

Group	Examples
bacteria	tuberculosis, gonorrhea, leprosy
fungi	athlete's foot, thrush, aspergillosis
Protista	malaria, toxoplasmosis, sleeping sickness
viruses	influenza, measles, Ebola

▲ Table 1

▼ Figure 3 A patient with symptoms of tuberculosis is examined in a hospital in the Himalayas, Nepal

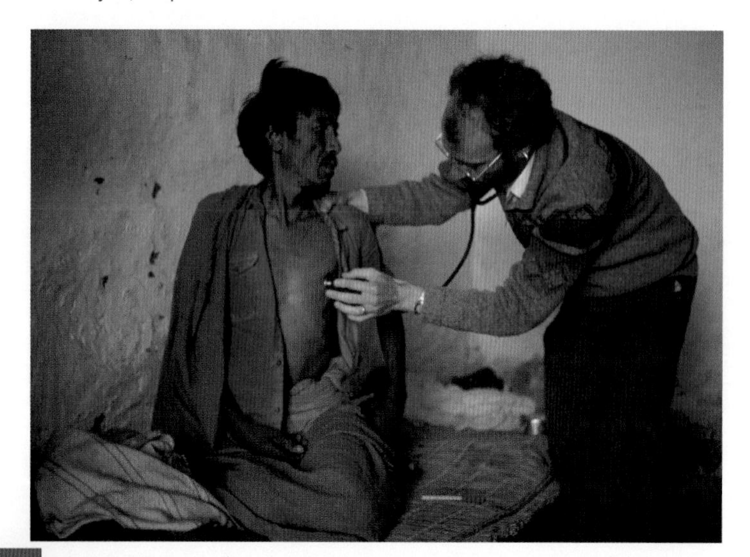

One group of infectious diseases has a surprising cause. These diseases affect the brain and are called spongiform encephalopathies. CJD (Creutzfeld–Jakob disease) is the human version. The causative agent has been found to be a prion, rather than a microorganism or virus. Prions are protein molecules. If they enter the body and reach brain tissue, they cause more proteins already there to turn into prion proteins. This causes brain cells to malfunction, leading to the development of the disease.

Observations: Infection control

1. **Ignaz Semmelweis and puerperal fever**. Puerperal fever is a disease that occurs in women during the days following childbirth. Research the causes of puerperal fever and the contribution of Hungarian physician Ignaz Semmelweis to its prevention.
2. **John Snow and cholera**. Cholera induces severe diarrhoea, leading to dehydration and sometimes death. Research the cause of cholera and the contribution of English physician John Snow to its identification.

C3.2.2 Skin and mucous membranes as a primary defence

Many different microbes can grow inside the human body and cause a disease. Some microorganisms are opportunistic and although they can invade the body they also commonly live outside it. Others are specialized and can only survive inside a human body.

The primary defence of the body against entry of pathogens is the skin. Its outermost layer provides a physical barrier. Much of the body is covered by a tough layer of dead cells containing large amounts of the protein keratin. This makes it very hard for pathogens to pass through.

Sebaceous glands are associated with hair follicles. They secrete a chemical called sebum, which maintains skin moisture and slightly lowers skin pH. Low pH (acidity) inhibits the growth of bacteria and fungi.

Mucous membranes are a thinner and softer type of skin. They are found in areas such as the vagina, the foreskin and head of the penis and the airways leading to the lungs. The skin in these areas secretes mucus, a sticky solution of glycoproteins. Mucus is a type of physical barrier: pathogens and harmful particles become trapped in it and are then swallowed or expelled. It also has antiseptic properties because of the presence of the anti-bacterial enzyme lysozyme.

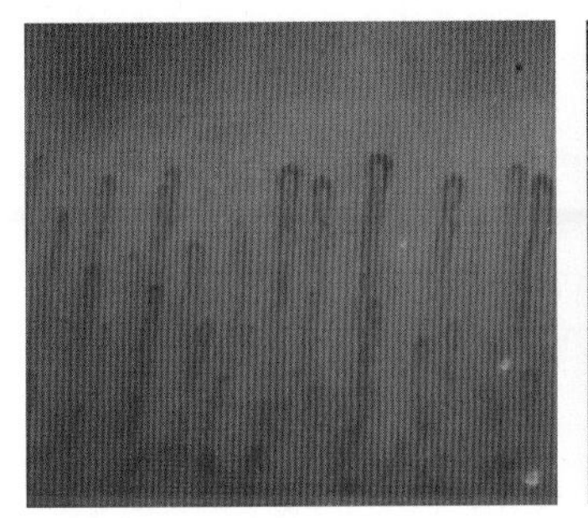

▲ Figure 4 A digital microscope can be used to produce images of the different types of skin covering the human body. Here are four images produced in this way

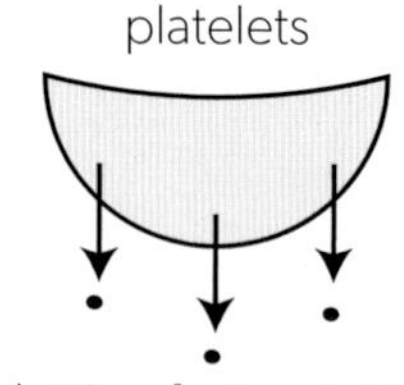
platelets
clotting factors initiate
a cascade of reactions

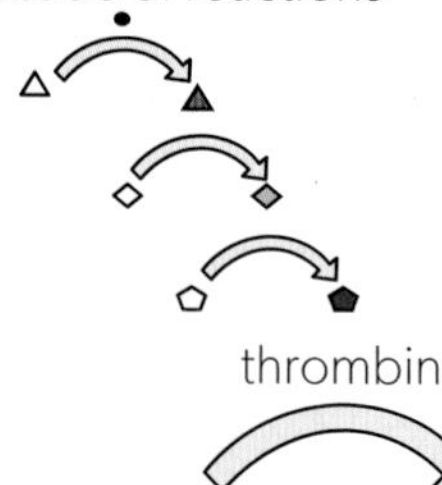
each reaction
in the cascade
produces the
catalyst for the
next reaction
thrombin
fibrinogen
fibrin
within seconds of platelets
releasing clotting factors,
thrombin produced by the
cascade is linking together
globular fibrinogen molecules
to form long strands of fibrin

▲ Figure 5 The clotting process

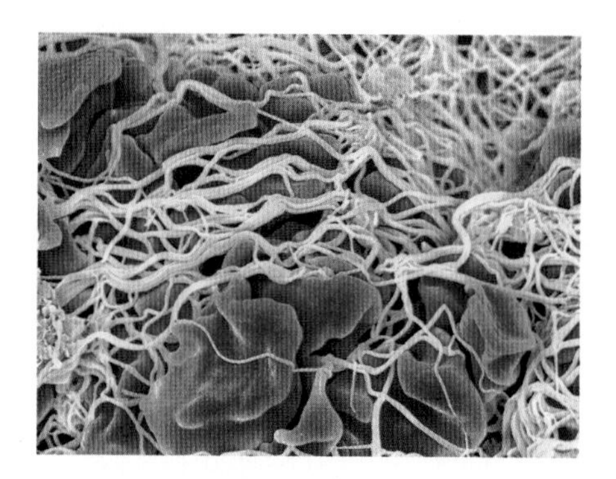

▲ Figure 6 False-colour scanning electron micrograph showing red blood cells and platelets trapped in a mesh of fibrin

C3.2.3 Sealing of cuts in skin by blood clotting

When the skin is cut, blood vessels in it are severed and start to bleed. Bleeding usually stops after a short time because of a process called clotting. Liquid blood emerging from a cut changes to a semi-solid gel. This seals the wound and prevents further loss of blood, which would lead to a decrease in blood pressure. Clotting is also important because cuts breach the barrier to infection provided by the skin. Clots prevent entry of pathogens until new tissue has grown to heal the cut.

Blood clotting involves a cascade of reactions, each of which produces a catalyst for the next reaction. As a result, blood clots very rapidly. Clotting must be strictly controlled because blood clots inside blood vessels can cause blockages.

The clotting process is initiated by blood-cell fragments called platelets. When a cut or other injury causes damage to blood vessels, platelets aggregate at the site forming a temporary plug. They then release clotting factors that trigger the clotting process.

The cascade of reactions that occurs after the release of clotting factors from platelets quickly results in the production of an enzyme called thrombin. Thrombin converts the protein fibrinogen, which is dissolved in blood plasma, into insoluble fibrin. The fibrin forms a mesh in cuts, trapping more platelets and also blood cells. The resulting clot is initially a gel, but if it is exposed to the air it dries to form a hard scab. Figure 6 shows red blood cells trapped in this mesh.

C3.2.4 Differences between the innate immune system and the adaptive immune system

The immune system protects the body against infectious diseases. There are two parts:

- The innate immune system responds to broad categories of pathogen. It does not change during an organism's life. Phagocytes are part of the innate system.
- The adaptive immune system responds in a specific way to particular pathogens. It builds up a memory of pathogens it has encountered, so it offers more effective protection against common infectious diseases. Antibody-producing lymphocytes are part of the adaptive system.

C3.2.5 Infection control by phagocytes

If microorganisms get past the physical barriers of skin and mucous membranes and enter the body, white blood cells provide the next line of defence. There are many different types of white blood cell. Some are phagocytes that squeeze out through pores in the walls of capillaries and move to sites of infection. There they engulf pathogens by endocytosis and digest them using enzymes from lysosomes. An infected wound will attract large numbers of phagocytes, resulting in the formation of a white liquid called pus.

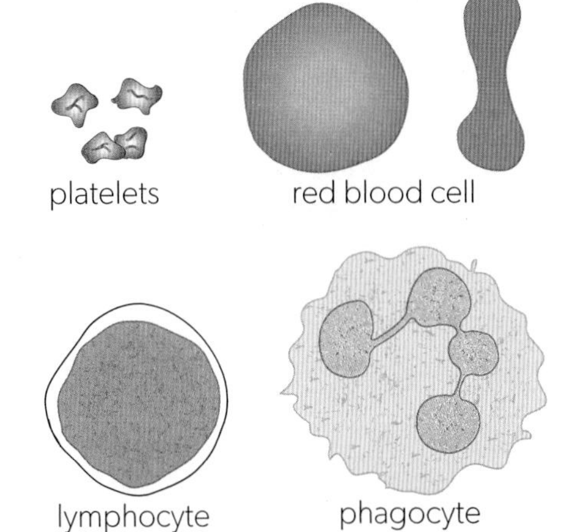

▲ Figure 7 Cells and platelets from blood. Lymphocytes and phagocytes are types of white blood cell

C3.2.6 Lymphocytes as cells in the adaptive immune system that cooperate to produce antibodies

About 25% of the white blood cells circulating in the blood are lymphocytes. These cells have a rounded nucleus and a small amount of cytoplasm. They are called lymphocytes because they also occur in the lymphatic system. This system consists of vessels that drain excess fluid from body tissues. At intervals along lymph vessels there are swollen structures called lymph nodes. There are large numbers of lymphocytes in the lymph nodes.

Lymphocytes produce antibodies—large proteins that help to destroy pathogens. Antibodies have two functional parts: a hypervariable region that recognizes and binds to a specific molecule on a pathogen and another region that helps the body to fight the pathogen, for example, by:

- making a pathogen more recognizable to phagocytes so it is more readily engulfed
- preventing viruses from docking to host cells so they cannot enter the cells.

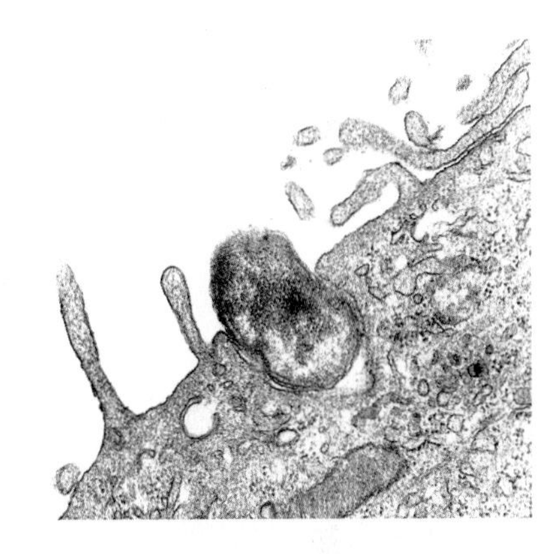

▲ Figure 8 An electron micrograph showing a cell of the pathogenic bacterium *Orientia tsutsugamushi* about to be engulfed by a phagocyte

Our bodies can become infected by many different pathogens, including new strains that have only recently evolved. The immune system as a whole can produce a vast array of different antibodies but each individual lymphocyte can produce only one type of antibody. We have only a small number of lymphocytes for producing each type of antibody. As a result, when a new pathogen infects the body, there are at first too few lymphocytes to produce enough antibodies to control the infection. When a pathogen infects the body for the first time, the lymphocytes that can produce the appropriate antibodies work together to produce one or more large clones of cells. These cells can produce antibodies in large enough quantities to control the pathogen and clear the infection.

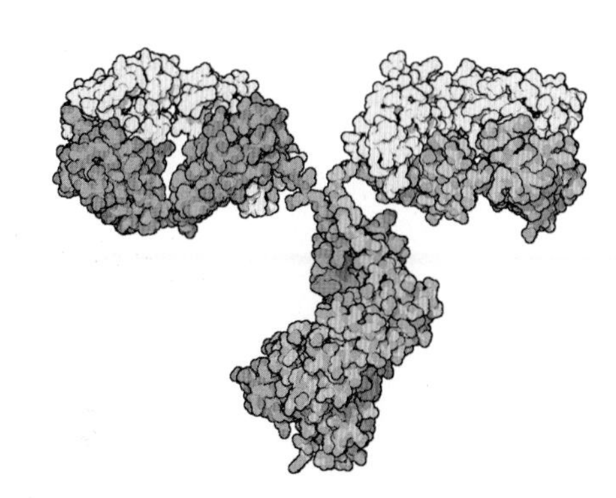

▲ Figure 9 Antibody molecules have a characteristic Y-shape, with hypervariable regions on the tips of the arms (upper right and left). The stem of the Y is the central domain, the region that destroys the pathogen

C3.2.7 Antigens as recognition molecules which trigger antibody production

Lymphocytes have to distinguish between body cells ("self-cells") and "non-self" cells such as invading pathogens. (Autoimmune diseases such as type 1 diabetes and Crohn's disease develop when lymphocytes get this wrong.) Lymphocytes recognize pathogens by differences between their molecules and those of

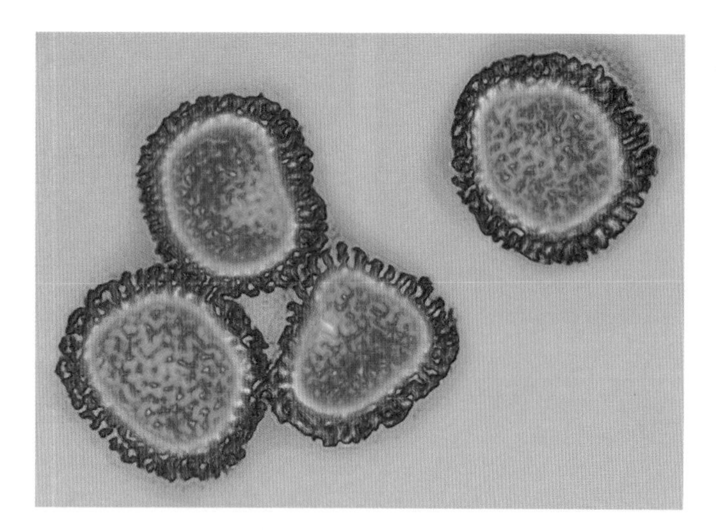

▲ Figure 10 Avian influenza viruses. In this electron micrograph of a virus in transverse section, false colour has been used to distinguish the protein coat that is recognized as antigens by the immune system (purple) from the RNA of the virus (green)

body cells. The molecules used for recognition are called antigens. They are mostly glycoproteins or other proteins and also some large polysaccharides. They are usually located on the surface of the pathogen.

The immune response to an antigen is the production of specific antibodies. Any molecule that stimulates an immune response is referred to as an antigen. Such molecules are found on the surface of cancer cells, parasites and bacteria, on pollen grains and on the coats or enveloping membranes of viruses. They may also occur on the surface of cells from another human. For example, antibodies are produced in response to molecules in transplanted organs that are of a different tissue type. If the wrong blood type is transfused into a patient then molecules on the surface of red blood cells act as antigens and trigger antibody production.

In response to an antigen, lymphocytes produce antibodies that bind to the antigen. This is similar to the binding of a ligand to a receptor, or the binding of a substrate to the active site of an enzyme: it is dependent on matching shapes and chemical properties. Protrusions on antigens match hollows on the corresponding antibody and positive charges match negative charges. However, unlike the binding of ligands to receptors, antibody to antigen binding is irreversible. Also it does not cause the antigen to change chemically, which is a property of enzymes.

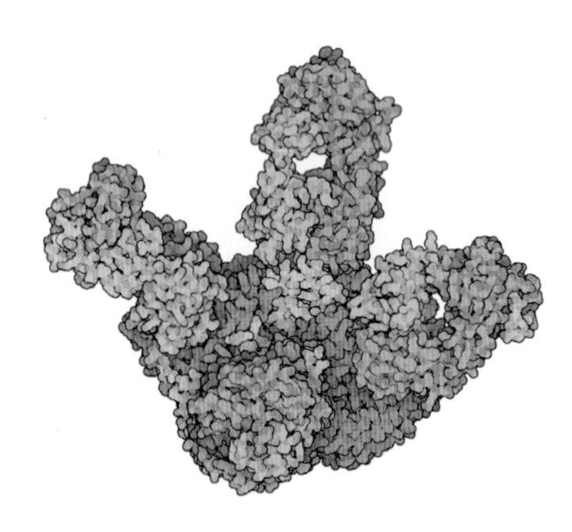

▲ Figure 11 Antibodies (blue) bound to neuraminidase (orange and red), a protein on the surface of influenza viruses that acts as an antigen. Only one arm of each antibody, with the hypervariable binding region, is shown

The part of an antibody that binds to the antigen is the hypervariable region. As the name suggests, there is immense variation in the hypervariable regions of antibodies. However, one lymphocyte only produces antibodies with one type of hypervariable region and this only binds to specific antigens. As with receptors and enzymes, there is specificity in binding.

C3.2.8 Activation of B-lymphocytes by helper T-lymphocytes

The immune system produces large amounts of the specific antibodies needed to fight an infection, without producing any of the hundreds of thousands of other types of antibodies that could be produced. This requires a series of interactions between different types of white blood cell.

Pathogens are ingested by macrophages (a type of phagocytic white blood cell). Antigens from the pathogens are then displayed in the plasma membrane of the macrophage. Cells called helper T-lymphocytes each have an antibody-like receptor protein in their plasma membranes. This can bind to antigens displayed by macrophages. There are many types of helper T-cell but only a few have receptor proteins that fit the antigen. These helper T-cells bind with and are activated by the macrophage.

The activated helper T-cells then bind to lymphocytes called B-lymphocytes (B-cells). Again, only B-cells with a specific receptor protein to which the antigen binds are selected and undergo the binding process. The helper T-cell activates the selected B-cells, both by means of the binding and by release of a signalling protein.

C3.2.9 Multiplication of activated B-lymphocytes to form clones of antibody-secreting plasma cells

When a pathogen first invades the body, development of immunity depends on the presence of B-lymphocytes capable of producing an effective antibody. These B-lymphocytes become activated, but they do not immediately start to produce the antibody. This is because there are too few of them to make significant quantities of antibody and in any case they do not yet have the organelles needed. Instead the activated B-lymphocytes divide repeatedly by mitosis to form a clone of cells that all produce the same type of antibody. These B-lymphocytes grow in size and develop an extensive endoplasmic reticulum with many ribosomes attached to it, along with a large Golgi apparatus. This allows rapid production of antibodies by protein synthesis. The cells that have grown and differentiated for antibody production are called plasma B-cells.

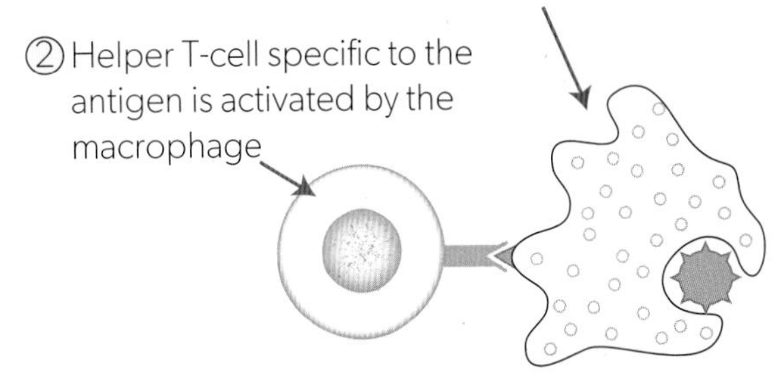
① Macrophage ingests pathogen and displays antigens from it
② Helper T-cell specific to the antigen is activated by the macrophage

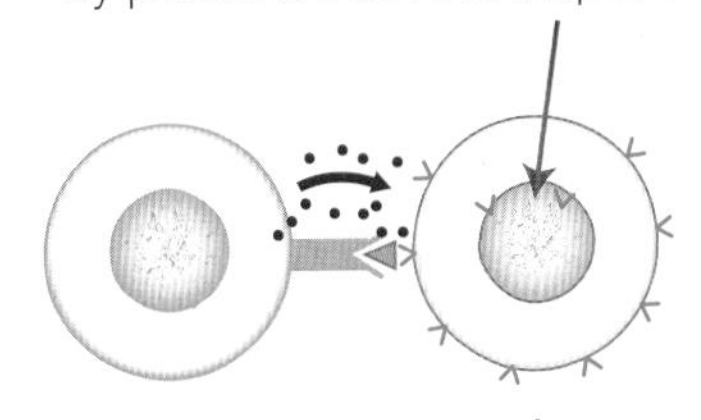
③ B-cell specific to the antigen is activated by proteins from the helper T-cell
④ B-cell divides repeatedly to produce antibody-secreting plasma cells

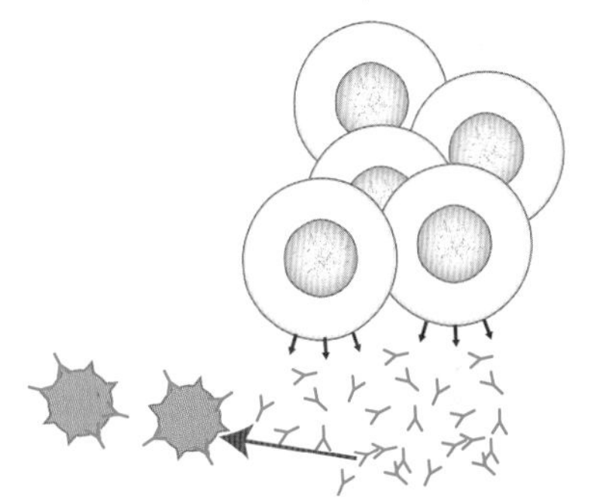
⑤ Antibodies produced by the clone of plasma cells are specific to antigens on the pathogen and help to destroy it

▲ Figure 12 Activation of lymphocytes

C3.2.10 Immunity as a consequence of retaining memory cells

Immunity is the ability to eliminate an infectious disease from the body. Antibodies can give us immunity to a disease but they only persist in the body for a few weeks or months after being secreted by plasma B-cells. The plasma cells that secrete antibodies are also gradually lost after an infection has been overcome, because the antigens associated with that infection are no longer present. However, immunity can last for much longer—in many cases, for the rest of our lives.

Most B-cells in a clone produced by mitosis become active plasma B-cells. These cells do not survive for long after fulfilling their role of rapid antibody production. A smaller number of cells in the clone do not immediately secrete antibodies but remain for a long time after the infection. These memory B-cells remain inactive unless the same pathogen infects the body again, in which case they are activated and respond very rapidly. Immunity to an infectious disease is due to having either antibodies against the pathogen, or memory cells that allow rapid production of the antibody.

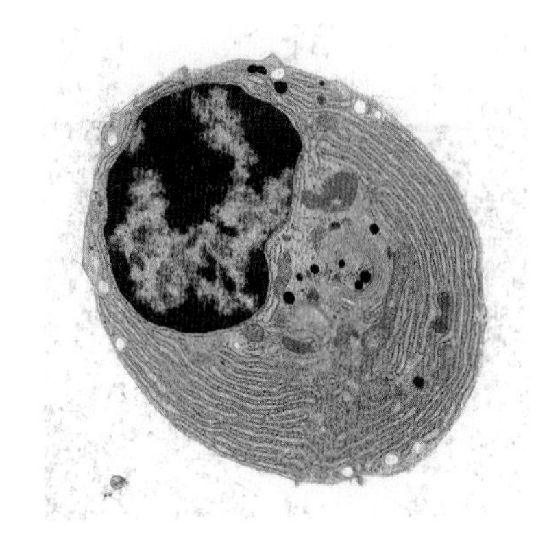

▲ Figure 13 Plasma B-cell with cytoplasm (orange) containing an unusually extensive network of rough endoplasmic reticulum (rER)

C3.2.11 Transmission of HIV in body fluids

Human immunodeficiency virus (HIV) is the cause of acquired immunodeficiency syndrome (AIDS).

The virus cannot usually survive for long outside the body and infection with HIV only occurs if blood or other body fluids pass from an infected to an uninfected person. In a person infected with HIV, there may be viruses in blood, semen,

▲ Figure 14 The red AIDS awareness ribbon is an international symbol of awareness and support for those living with HIV. It is worn on World AIDS Day each year—December. Do you know how many people in your area are affected and what can be done to support them?

vaginal fluids, rectal secretions and breast milk. These viruses may pass to an uninfected person through these actions:

- sex without a condom, during which abrasions to the mucous membranes can cause minor bleeding
- sharing of hypodermic needles by intravenous drug users
- transfusion of infected blood, or blood products such as Factor VIII
- childbirth and breastfeeding.

C3.2.12 Infection of lymphocytes by HIV with AIDS as a consequence

Production of antibodies by the immune system requires interactions between different types of lymphocyte, including helper T-cells. The human immunodeficiency virus (HIV) invades and destroys helper T-cells. This leads to a progressive loss of the capacity to produce antibodies.

In the early stages of infection, the immune system makes antibodies against HIV. If these can be detected, a person is said to be HIV-positive. The rate at which helper T-cells are destroyed by HIV varies considerably. In most HIV-positive patients who do not receive treatment, antibody production eventually becomes so ineffective that a group of opportunistic infections can strike. These are caused by pathogens which would be fought off easily by a healthy immune system. Several of the infections are so rare that they can be used as marker diseases for the latter stages of HIV infection—for example, Kaposi's sarcoma.

A collection of several diseases or conditions existing together is called a syndrome. When conditions caused by HIV are combined in a person, they have acquired immunodeficiency syndrome (AIDS).

HIV is a retrovirus so has genes made of RNA. It uses reverse transcriptase to produce DNA copies of its genes after entering a host cell. Antiretroviral drugs inhibit reverse transcriptase. Other drugs to combat HIV target the enzymes that the virus uses to insert its DNA into host cells' chromosomes or to prepare coat proteins for assembly of new virus particles. People who become HIV-positive can now be treated with a group of antiretroviral drugs, which greatly slows down or prevents damage to the immune system.

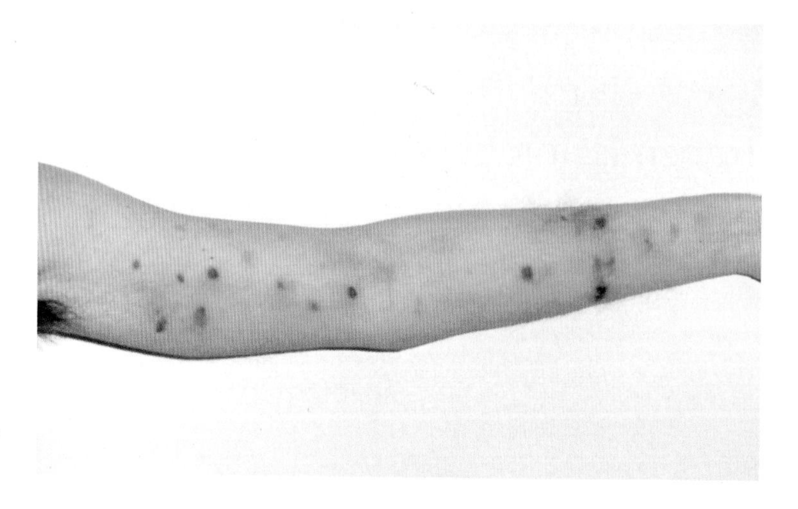

▶ Figure 15 Kaposi's sarcoma is associated with HIV/AIDS. It is an opportunistic infection, caused by a herpes virus

Data-based questions: Prevalence of HIV/AIDS

Data is available on the internet for rates of HIV/AIDS in every country. The two sample graphs in Figure 16 are for the USA and South Africa and show three measures. The scale on the y-axis is used for deaths and new infections. The number of people living with HIV is 10 times the number on the scale. In the year 2000, the population of the USA was 282 million and of South Africa was 45 million.

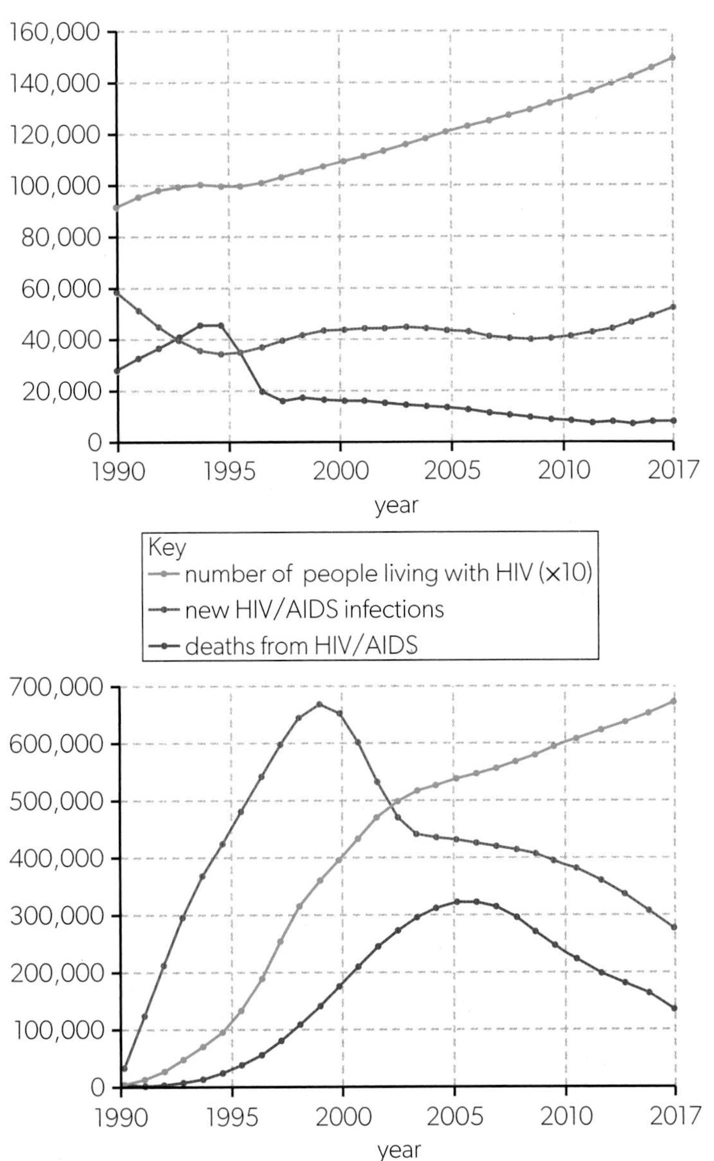

▲ Figure 16 Rates of HIV and AIDS for the USA (top) and South Africa (bottom)

1. The graphs show total numbers for the whole country.
   a. Calculate the number of new infections per 100,000 of population in the year 2000 for each country. [4]
   b. Suggest reasons for the difference between the infection rates. You may need to research this. [2]
2. a. Compare and contrast the data for deaths from HIV/AIDS in the United States and South Africa. [3]
   b. Suggest reasons for the differences. [2]
3. Suggest one explanation for:
   a. numbers of new infections rising after 2010 in the USA [2]
   b. slower increases in the numbers of people living with HIV/AIDS in South Africa after 2005. [2]

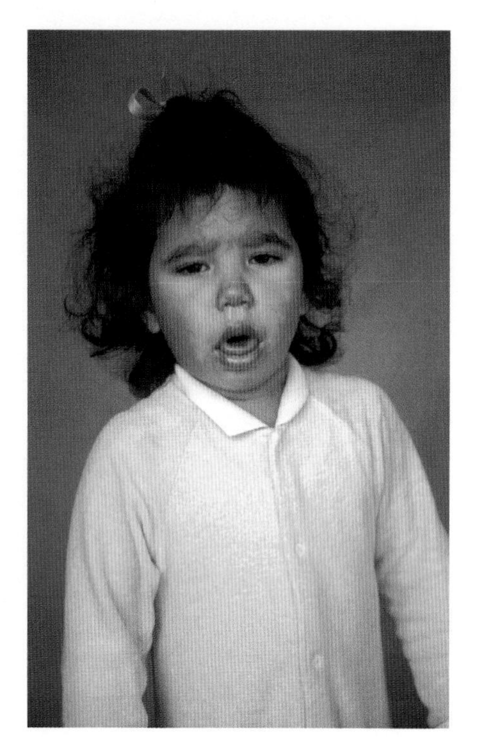

▲ Figure 17 Doctors need to be able to distinguish between viral and bacterial infections because antibiotics should only be prescribed for diseases caused by bacteria. This child probably has a common cold, caused by one of many possible viruses. Children lack immunity to most of these viruses so catch colds frequently. Unless there is a secondary bacterial infection, antibiotics should not be used

C3.2.13 Antibiotics as chemicals that block processes occurring in bacteria but not in eukaryotic cells

An antibiotic is a chemical that inhibits the growth of microorganisms. Most antibiotics are antibacterial. They block processes that occur in prokaryotes but not in eukaryotes. They can therefore be used to kill bacteria inside the body without causing harm to human cells. Antibiotics target the processes of bacterial DNA replication, transcription, translation, ribosome function and cell wall formation.

Many antibacterial antibiotics were discovered in saprotrophic fungi. These fungi compete with saprotrophic bacteria for the dead organic matter on which they both feed. By secreting antibacterial antibiotics, saprotrophic fungi inhibit the growth of their bacterial competitors. An example is penicillin. It is produced by some strains of the *Penicillium* fungus at times when nutrients are scarce and competition with bacteria would be harmful.

Viruses are non-living and can only reproduce when they are inside living cells. They use the chemical processes of a living host cell, instead of having a metabolism of their own. They do not have their own means of transcription or protein synthesis and they rely on the host cell's enzymes for ATP synthesis and other metabolic pathways. These processes cannot be targeted by drugs as the host cell would also be damaged.

All the commonly used antibiotics such as penicillin, streptomycin, chloramphenicol and tetracycline control bacterial infections but are not effective against viruses. It is inappropriate for doctors to prescribe antibiotics to treat viral infections and it contributes to the overuse of antibiotics and increases in antibiotic resistance in bacteria.

C3.2.14 Evolution of resistance to several antibiotics in strains of pathogenic bacteria

On European Antibiotic Awareness Day in 2020 this statement was released through social media by the European Centre for Disease Prevention and Control:

> Before the discovery of antibiotics, thousands of people died from bacterial diseases, such as pneumonia, or infection following surgery. Since antibiotics have been discovered and used, more and more bacteria, which were originally susceptible, have become resistant and developed numerous different means of fighting against antibiotics. Because resistance is increasing and few new antibiotics have been discovered and marketed in recent years, antibiotic resistance is now a major public health threat.
>
> Without antibiotics, we could return to the "pre-antibiotic era", when organ transplants, cancer chemotherapy, intensive care and other medical procedures would no longer be possible. Bacterial diseases would spread and could no longer be treated, causing death.
>
> Only 70 years after the introduction of antibiotics, we are facing the possibility of a future without effective drugs to treat bacterial infections.

Strains of bacteria with resistance are usually discovered soon after the introduction of an antibiotic. This is not of huge concern unless a strain develops multiple resistance, but such strains are now widespread. Methicillin-resistant *Staphylococcus aureus* (MRSA) can infect the blood or surgical wounds of hospital patients and resists all commonly used antibiotics. Another example is multidrug-resistant tuberculosis (MDR-TB). In 2020, the WHO reported about half a million cases annually, with over 200,000 deaths.

Evolution of multiple antibiotic resistance is made easier because genes can be passed from one species of bacteria to another. There is also evidence that antibiotic-resistance genes are not lost from the genomes of pathogenic bacteria rapidly when an antibiotic is no longer used.

Multiple antibiotic resistance is an avoidable problem. These measures are required:

- Doctors must prescribe antibiotics only for serious bacterial infections and for the minimum period.
- Hospital staff must maintain high standards of hygiene to prevent cross-infection.
- Farmers must avoid the use of antibiotics in animal feeds as growth stimulants.
- Pharmaceutical companies must develop new classes of antibiotic—no new types have been introduced since the 1980s.

Activity: Questions for discussion

1. Why should newly developed antibiotics be used as little as possible? Why does this make pharmaceutical companies reluctant to develop them?
2. The company Achaogen specialized in developing new antibiotics. However, despite successes, it went bankrupt in 2019. If pharmaceutical companies cannot profitably develop new antibiotics, what alternative approaches could be used?
3. Many antibiotics are modified versions of chemicals produced by soil bacteria. The number of these chemicals must be finite and every use of them to treat disease in humans reduces their future effectiveness. Use of fossil fuels causes climate change. What are the parallels between the use of antibiotics and the use of fossil fuels?
4. The World Health Organization (WHO) published this advice during Antibiotic Awareness Week 2016: "Always complete the full prescription, even if you feel better, because stopping treatment early promotes the growth of drug-resistant bacteria." There is almost no evidence base for this advice and it does not make sense in terms of natural selection. Why is this warning still given to patients?

Data-based questions: Penicillin resistance

The scatter graph in Figure 18 shows World Health Organization data for antibiotic use and penicillin resistance in *Staphylococcus pneumoniae* for 21 countries. The data was collected in 2000–2003. DDD is the defined daily dose, which is the normal quantity of antibiotic given to a patient per day.

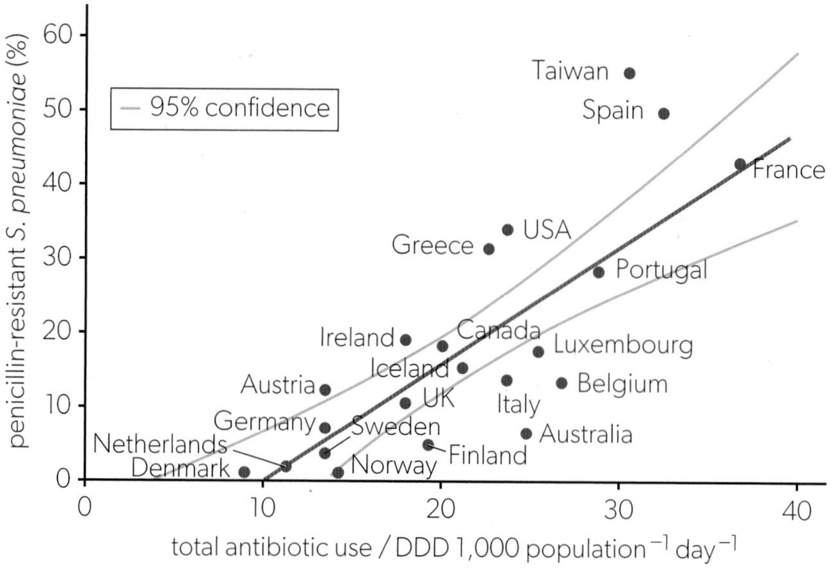

1. What are the highest and lowest levels of antibiotic resistance in any of the countries? [2]
2. a. Identify the relationship between total use of the antibiotic penicillin and penicillin resistance in *S. pneumoniae*. [1]

   b. Explain the reasons for this relationship. [2]

▲ **Figure 18** Source: WHO data 2000-2003

3. Explain what conclusion can be drawn from the blue lines on the graph showing 95% confidence limits. [2]
4. Discuss how levels of penicillin resistance could be reduced in a country. [3]

Science as a shared endeavour: Bioinformatics and chemogenomics

Computers have increased scientists' capacity to organize, store, retrieve and analyse biological data. Bioinformatics is an approach whereby multiple research groups can add information to a database, enabling other groups to query the database. One promising bioinformatics technique that has facilitated research into metabolic pathways is referred to as chemogenomics. Sometimes when a chemical binds to a target site, it can significantly alter metabolic activity. Scientists looking to develop new drugs test massive libraries of chemicals individually on a range of related organisms. For each organism, they identify a range of target sites and test a range of chemicals which are known to work on those sites. In this way, the structure of known effective antibiotics can be used to search through libraries to find other molecules with a similar structure.

C3.2.15 Zoonoses as infectious diseases that can transfer from other species to humans

Pathogens are often highly specialized, with a narrow range of hosts. For example, humans are the only known organism susceptible to the pathogens that cause syphilis, polio and measles. However, we are resistant to the virus that causes canine distemper, so that disease cannot be spread to us from dogs. Rats injected with the diphtheria toxin do not become ill, because their cells do not have the receptor that would bring the toxin into the cell. The bacterium *Mycobacterium tuberculosis* that causes most cases of tuberculosis (TB) humans does not cause disease in frogs because frogs rarely reach the 37°C temperature necessary to support the proliferation of this bacterium.

However, some pathogens can use more than one species as a host. *Mycobacterium bovis* causes tuberculosis in cattle but can also infect a wide variety of other animal species such as badgers, which live in the same area as cattle herds in Europe. Milk produced by infected cattle may contain live cells of *M. bovis* which can cause tuberculosis in humans if the milk is drunk. This is an example of a zoonosis—a disease that can be transmitted to humans from other animals in natural circumstances. Table 2 describes some other zoonotic diseases.

▼ Table 2

Zoonosis	Pathogen	Animal host from which the infection is transmitted	Mode of human infection
Tuberculosis	*Mycobacterium bovis* (a bacterium)	cattle	• drinking unpasteurized infected milk • inhaling droplets from coughing infected cows
Rabies	lyssaviruses	many mammals can be infected and transmit, but 99% of human cases are from dogs	• bite or scratch by an infected animal • eye, mouth or nose contact with saliva from an infected animal
Japanese encephalitis	Japanese encephalitis virus (JEV)	pigs or birds, via mosquitoes	by mosquito bites

Zoonoses are a current global health concern. A major factor contributing to the appearance of zoonotic diseases is humans living in close contact with livestock. For example, pigs carry the Japanese encephalitis virus which can spread to humans and cause fatal disease. Another factor is displacement of wild animals when their habitats are disrupted by the spread of human populations.

COVID-19 is thought to have originated in bats, most likely via another animal species. Because of this, it is classed as a zoonotic disease even though, since the initial transfer from animal-to-human, the spread of the disease has been human-to-human.

C3.2.16 Vaccines and immunization

Immunization involves using a vaccine to trigger immunity. Most vaccines are given by intramuscular injection (into the muscle); they can also be given by subcutaneous injection (under the skin) or by mouth (orally). Vaccines may contain any of the following active ingredients:

Active ingredient	Examples
a live but attenuated (weakened) version of the pathogen	measles, mumps and rubella vaccines
a killed form of the pathogen	rabies and influenza vaccines
subunits of the pathogen (usually proteins) that act as antigens	hepatitis B and human papillomavirus vaccines
mRNA coding for a protein that acts as an antigen, which human cells can translate to make the protein	some COVID-19 vaccines including that produced by Pfizer/BioNTech

▲ Table 3

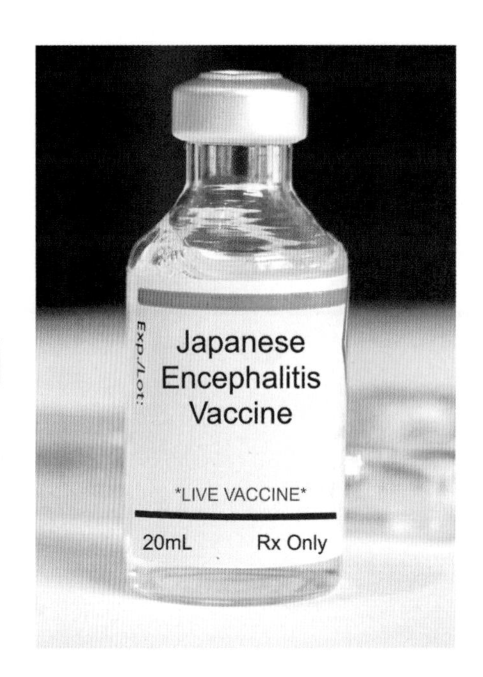

▲ Figure 19 Japanese encephalitis vaccine

▲ Figure 20 Vaccination against Japanese encephalitis in Assam, India, where there have been hundreds of deaths from this disease

All vaccines contain either antigens that allow a pathogen to be recognized by the immune system or nucleic acids from which antigens can be made. The antigens stimulate a primary immune response, by activation of T-lymphocytes and B-lymphocytes and production of plasma cells and then specific antibodies. If memory cells are also produced, long-lasting immunity develops. If a vaccine successfully triggers such immunity, the pathogenic microorganism will be destroyed by a secondary immune response if it ever enters the body.

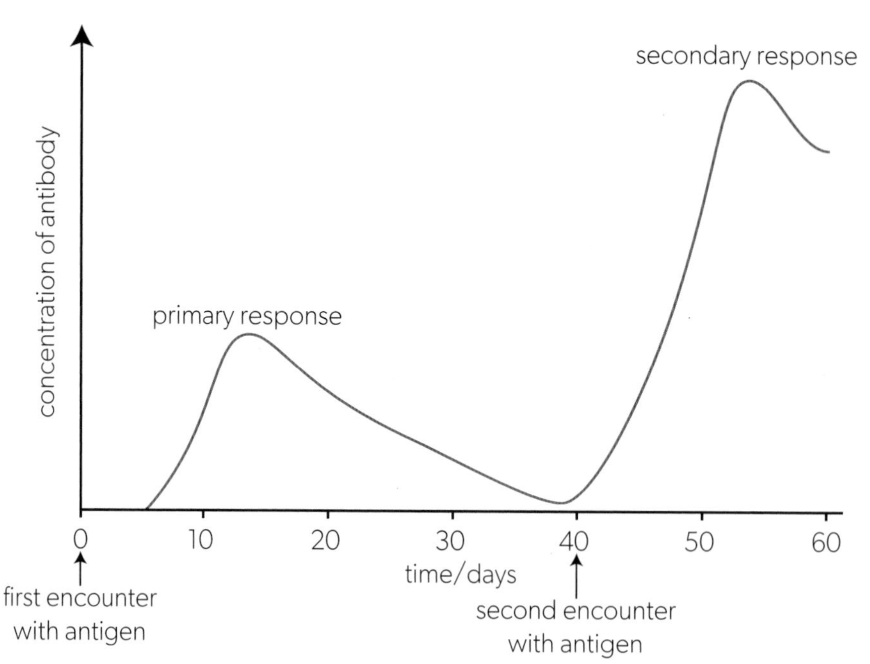

▶ Figure 21 A primary immune response happens when a pathogen first infects the body or a vaccine is administered. A secondary immune response happens after another infection or a second vaccination is given, sometimes called a "booster shot". Memory cells ensure that the second time an antigen is encountered, the body is able to respond rapidly by producing more antibodies at a faster rate

C3.2.17 Herd immunity and the prevention of epidemics

Herd immunity is achieved when a significant proportion of a population have already contracted a disease or have been vaccinated. As a result, the spread of a virus or other pathogen is impeded because it repeatedly encounters people who are already immune. When there is herd immunity, any new outbreak of the disease will decline and disappear.

Not everyone in the population has to be immune for herd immunity to develop. For this reason, epidemiologists prefer the term "herd protection". An important benefit of herd immunity is that vulnerable individuals who have compromised immune systems and cannot be vaccinated are unlikely to contract the disease.

The following formula can be used to estimate the percentage of people who must be immune for the population as a whole to be protected:

$$\left(1-\frac{1}{R}\right)\times 100\%$$

where R is the average number of people that an infected person infects. Measles is highly infectious and has an R value of 15, so $(1-1/15)\times 100\% = 93\%$ of the population must be vaccinated to reach herd immunity.

C3.2.18 Evaluation of data related to the COVID-19 pandemic

On 11 September 2021, the World Health Organization displayed this message on its website (right).

Globally, as of 4:47 pm CEST, 10 September 2021, there have been 223,022,538 confirmed cases of COVID-19, including 4,602,882 deaths, reported to WHO.

As of 6 September 2021, a total of 5,352,927,296 vaccine doses have been administered.

Table 4 shows statistics from the WHO website for a selection of countries on this day. GDP per capita (gross domestic product per head of population) is a measure of the material wealth of a country.

Country	Population	GDP per capita (International $)	Cases of COVID-19	Deaths due to COVID-19
Brazil	212,417,898	14,563	20,928,008	584,421
Canada	38,265,670	47,569	1,529,300	27,106
China	1,405,587,560	17,206	123,386	5,686
Eritrea	3,546,000	1,824	6,654	40
New Zealand	5,096,968	41,072	3,510	27
Spain	47,329,981	38,143	4,903,021	85,218
UK	66,796,807	44,288	7,132,076	133,841
USA	330,753,311	63,051	40,330,381	649,292

▲ Table 4

An evaluation is an appraisal that weighs up the strengths and limitations of evidence. The data in Table 4 can be evaluated in many different ways—and there is a lot more data available, relating to other aspects of the COVID-19 pandemic. You could use this data to answer questions such as:

1. What was the relative success of the countries in reducing COVID-19 infections during the pandemic?
2. What was the relative success of the countries in preventing deaths among COVID-19 patients?
3. To what extent was the success of countries in combating COVID-19 related to their material wealth?

To evaluate the data and answer these questions, you will need to present the data effectively and complete relevant calculations. In particular, values for percentage, percentage change and percentage difference will be useful.

Calculating percentages

To express one number as a percentage of a second number, use this formula:

$$\text{percentage} = \frac{\text{first number}}{\text{second number}} \times 100\%$$

This formula tells you to divide the first number by the second number and then multiply by 100. The % symbol is written next to the answer. If the first and second numbers have units, for example grams, the units must be the same and they cancel out, so percentages have no units.

Calculating percentage change

To calculate percentage change, use this formula:

$$\text{percentage change} = \frac{\text{final number} - \text{initial number}}{\text{initial number}} \times 100\%$$

You need to subtract the number at the start of a time period from the number at the end, then divide this difference by the number at the start. This gives a proportion, which is multiplied by 100 to find the percentage change.

Calculating percentage difference

Before calculating percentage difference, you must decide which number to use as the divisor—will the percentage difference be a percentage of number A or number B? The calculated percentage difference will not be the same. The formula given here assumes that the difference is calculated as a percentage of number A.

$$\text{percentage difference} = \frac{\text{number A} - \text{number B}}{\text{number A}} \times 100\%$$

Global impacts of science: Reporting research results

The media and scientific journals may publish ideas before they have been filtered by the scientific community. Contradictory reports may appear with equal frequency. Scientific knowledge is based on evidence but it is provisional: the "truth" becomes more certain as incorrect ideas are proved wrong. It is important for members of the public to understand this.

For example, in the early part of the COVID pandemic, there were concerns that mask wearing might lead people to touch their faces more often, increasing the risk of the spread of COVID. Over time, the balance of evidence suggested that masks were effective in preventing the spread of the virus and advice from the medical community promoted mask wearing in public.

Linking questions

1. How do animals protect themselves from threats?
   a. Outline the elements of the human non-specific immune response. (C3.2.4)
   b. Explain how vaccinations confer immunity. (C3.2.16)
   c. Explain the advantage to a population of sexual reproduction. (D3.2.1)
2. How can false-positive and false-negative results be avoided in diagnostic tests?
   a. Outline the concept of *p*-value with respect to statistical testing. (D3.2.21)
   b. Discuss the use of the chi-squared statistic to test a hypothesis. (D3.2.21)
   c. Outline the role of monoclonal antibodies in a pregnancy detection kit. (D3.1.17)

TOK

What features of knowledge have an impact on its reliability?

When a scientist carries out an experiment and gets similar results to an initial experiment then those results are said to be reliable. How are reliable results obtained? Being painstaking, meticulous and thorough are dispositions that can help.

Rosalind Franklin was an accomplished specialist in X-ray crystallography who was meticulous in her production of crystals and in photographing them under precise conditions. She is renowned for leading the team that took the photograph in Figure 1, which was an instrumental piece of data that led to the elucidation of the structure of DNA.

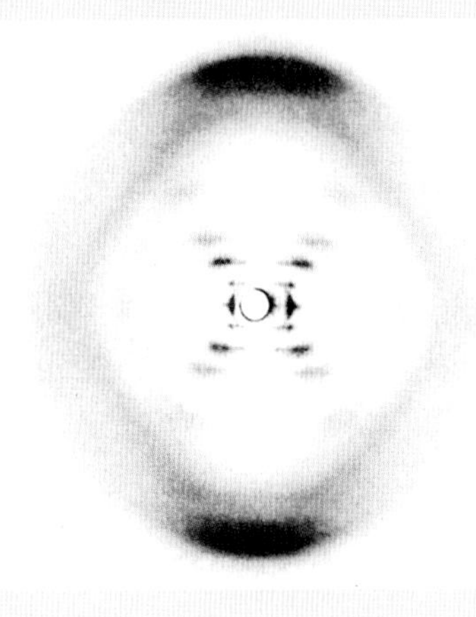

▲ Figure 1

Sometimes, theory is used to override measurement. We can use the coherence test to decide whether a result seems reasonable. In 2010, a researcher claimed that a bacterium from Mono Lake, an extremophile habitat, could substitute arsenic for phosphorus in its DNA molecules. While this discovery was announced in the media, scientists were immediately sceptical. Extraordinary claims require extraordinary evidence. Chemists were quick to point out that a backbone with arsenic would be fragile, and would be very short-lived. Despite recognition by NASA as credible research, and publication in a reputable source, the results were falsified by other research groups.

In 2008, the drug Avastin was approved by the USFDA for use in treating breast cancer, based on studies which showed a statistically significant benefit. After further studies using a larger sample size, the results no longer showed a statistically significant benefit in the treatment of breast cancer. This is an example of how sample size is important for drawing reliable conclusions.

▲ Figure 2 If ever there was going to be an environment where extraordinary biochemistry was going to be observed it would be in extreme environments such as in Mono Lake in California. The image shows alkali-loving extremophile bacteria among crystals of tellurium (upper right). This species of bacteria is haloalkaliphilic (salt and alkali-loving) and was found in the bottom sediment of Mono Lake. It can grow in oxygen-deprived conditions with a pH range of 8.5–10. It can grow on arsenate, which is water-insoluble, reducing it to arsenite, which is water-soluble and highly toxic

Peter and Rosemary Grant studied the medium ground finch, *Geospiza fortis* following a drought that affected the ecosystem of Daphne Major, an island in the Galápagos. As there was a reduction in the amount of small seeds, their main food, during the drought, those birds with beaks large enough to access bigger seeds survived to reproductive age. They noted that two generations after the drought, the mean beak size had increased. Noting that the difference was a matter of millimetres, precise tools and careful measurements were necessary to detect the approximately 5% change in beak dimensions.

▲ Figure 3 A female medium ground finch *Geospiza fortis*

End of chapter questions

1. In a study of brain organization, several factors were investigated. The relationship between the volumes of grey and white matter across mammalian species was compared.

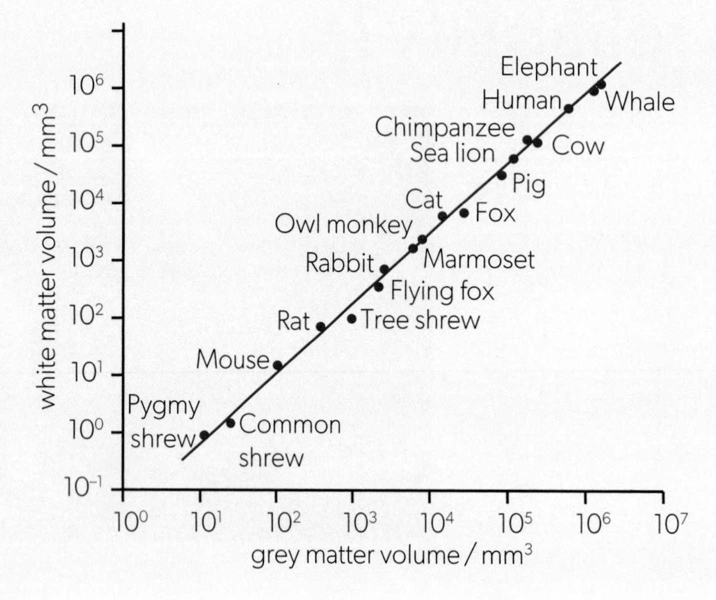

a. Describe the relationship between the volume of white matter and grey matter. [1]

b. Outline the organization of the human cerebral cortex with regard to structure and function. [3]

c. Outline **one** reason for the large energy requirement of the brain. [1]

2. During a study on exercise, researchers obtained the following data on the heart function of athletes as they increased their work rate up to the maximum. The work rate is indicated by the ventilation rate, which is the total volume of gas exchanged by the lungs in one minute.

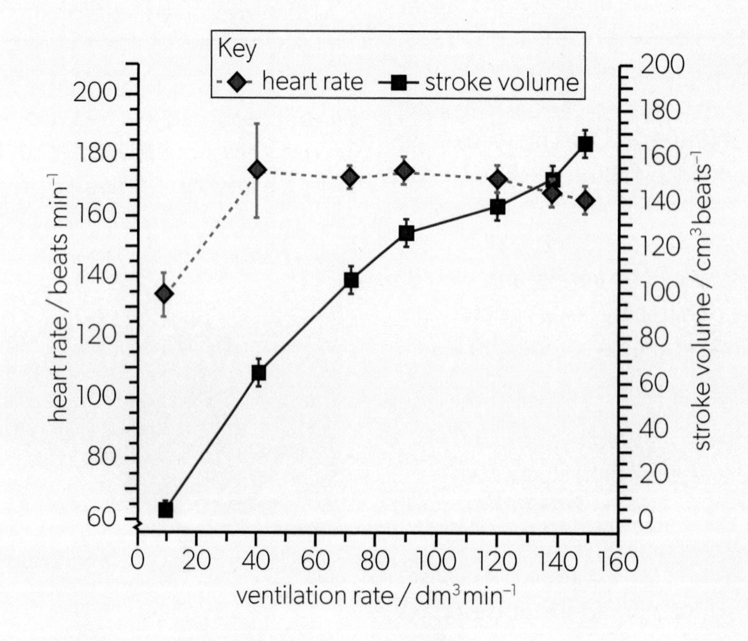

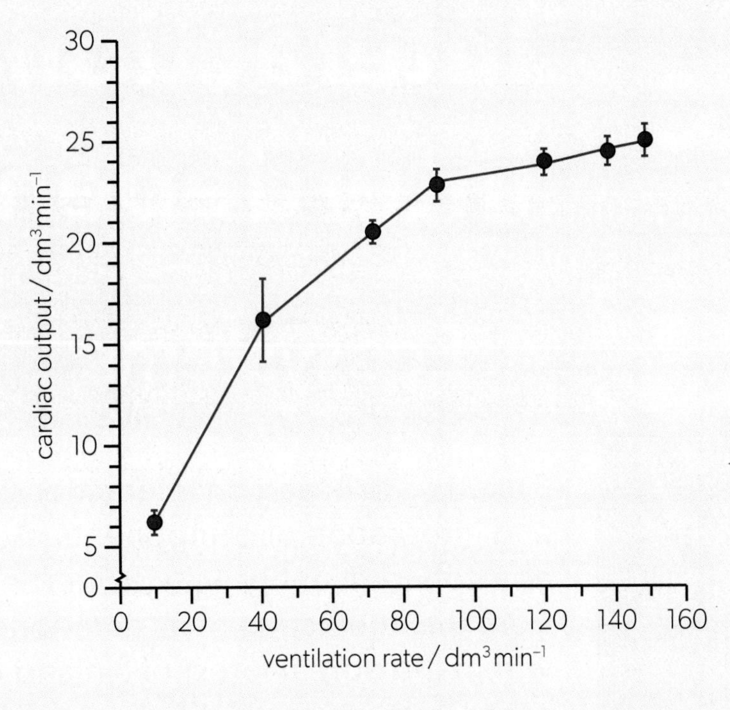

Source: I Vogiatzis et al. (2009) The Journal of Phsiology, 587 (14) pp 3665-3677

a. State the highest cardiac output, giving the units. [1]

a. Compare the trends in heart rate and cardiac output as ventilation rate increases. [2]

a. Giving one reason, deduce which ventilation rate value is likely to correspond to VO2 max, giving the units. [2]

a. Using all the data, explain how the cardiac output responds to increases in the work rate. [2]

C Interaction and interdependence

4 Ecosystems

Systems are based on interactions, interdependence and integration of components. An ecosystem functions as a coherent entity with properties that emerge from the interaction of its components. Living things within the boreal forest of Finland interact with each other through competition, predation, mutualism and commensalism. They also interact with the non-living environment. Nutrients are recycled through food chains. Photosynthesis powered by sunlight provides an endless supply of energy that supports feeding relationships. The organisms modify the environment in ways that affect other organisms.

C4.1 Populations and communities

How do interactions between organisms regulate sizes of populations in a community?

Figure 1 shows a population of European pine sawfly (*Neodiprion sertifer*) larvae feeding on a Scots pine (*Pinus sylvestris*) branch previously browsed by moose (*Alces alces*).

Organisms interact with members of their own species as well as members of other species in ways that regulate population size. Which interactions increase or maintain population size and which limit population size for the three species found in the pine forest? Why are the saw fly larva found in clusters? What are the ways in which sawfly larvae cooperate in defense? In what ways do they compete with one another?

▲ Figure 1

What interactions within a community make its populations interdependent?

The seed of red dead-nettle (*Lamium purpureum*) has a folded attachment known as an elaiosome that can be seen at the tip of the seed in Figure 2. Elaiosomes are nutritious fatty structures which are attractive to ants. The ant carries the seed to its nest and feeds the elaiosome to developing larvae. After the larvae have consumed the elaiosome, the ants take the seed to their waste disposal area, which is rich in nutrients from feces and dead ants. Which pollinators visit the flowers of the red dead-nettle? What types of herbivores consume its leaves?

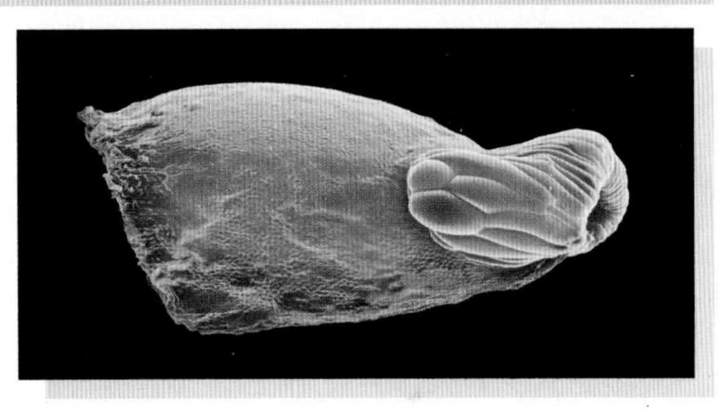

▲ Figure 2

SL and HL

C4.1.1 Populations as interacting groups of organisms of the same species living in an area
C4.1.2 Estimation of population size by random sampling
C4.1.3 Random quadrat sampling to estimate population size for sessile organisms
C4.1.4 Capture–mark–release–recapture and the Lincoln index to estimate population size for motile organisms
C4.1.5 Carrying capacity and competition for limited resources
C4.1.6 Negative feedback control of population size by density-dependent factors
C4.1.7 Population growth curves
C4.1.8 Modelling of the sigmoid population growth curve
C4.1.9 A community as all of the interacting organisms in an ecosystem
C4.1.10 Competition versus cooperation in intraspecific relationships
C4.1.11 Herbivory, predation, interspecific competition, mutualism, parasitism and pathogenicity as categories of interspecific relationship within communities
C4.1.12 Mutualism as an interspecific relationship that benefits both species
C4.1.13 Resource competition between endemic and invasive species
C4.1.14 Tests for interspecific competition
C4.1.15 Use of the chi-squared test for association between two species
C4.1.16 Predator–prey relationships as an example of density-dependent control of animal populations
C4.1.17 Top-down and bottom-up control of populations in communities
C4.1.18 Allelopathy and secretion of antibiotics

▲ Figure 3 *Chelonoidis nigra abingdonii* was a subspecies of Galápagos giant tortoise that lived on Pinta Island. The population decreased until there was only one individual left alive, nicknamed Lonesome George. He died in 2012 at the age of about 100. Is a single individual a population?

C4.1.1 Populations as interacting groups of organisms of the same species living in an area

A population is a group of individual organisms of the same species living in a given area. The members of a population normally interbreed with each other and do not interbreed with individuals in populations of other species. In addition to interbreeding, there are often many other types of interaction between individuals in a population, such as competition for food or cooperation to avoid predation.

The number of individuals in a population can be anything from a few up to billions. There may be one population of a species or many. Populations of a species are often separated by a geographical barrier, for example the sea that separates two islands.

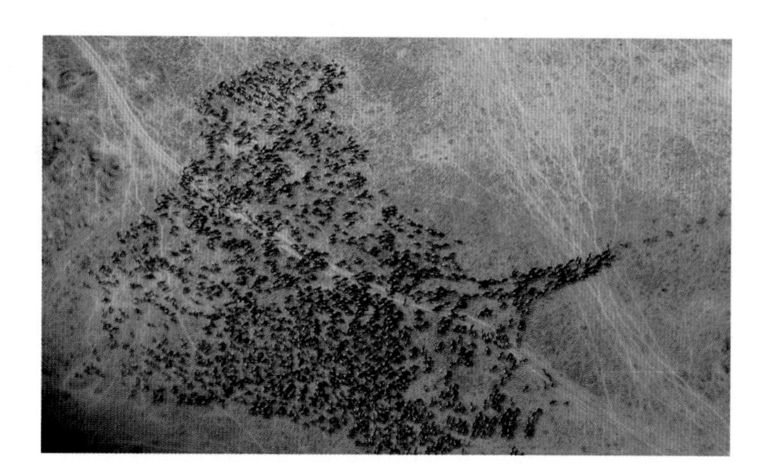

▶ Figure 4 Populations are often spread out but some animal populations form herds. This photo shows a herd of migrating wildebeest in the Masai Mara, Kenya. They are following a small group of zebras. In a population, emergent properties develop as a result of interactions between individuals. Predator–prey interactions mostly happen around the edges. Individuals within the herd can minimize their chance of predation by moving. What factors have caused the shape of this wildebeest herd to develop?

C4.1.2 Estimation of population size by random sampling

The size of a population is the total number of individuals in it. It is usually impossible to count every individual in a population. Animals use camouflage or other methods to avoid being seen and even if they are visible, they move. Populations may be spread over vast areas and contain countless individuals. Unless a population is small and easy to find, it is not usually possible to determine the exact size. Instead, it has to be estimated.

An estimate in science is not the same as a guess. It should be based on evidence. Population estimates are based on sampling. A sample is a small portion of something. One sample is unlikely to be representative of a whole population so it is better to use multiple samples. Ideally, every individual in the population should have an equal chance of being included in a sample. This can be achieved through random sampling.

A random sample is one where every member of a population has an equal chance of being selected. To select a truly random sample, you must avoid unconscious bias. Many procedures use random numbers for this purpose. There are many phone apps and websites that will generate random numbers. Table 1 shows 10 random numbers between 0 and 99, generated using an app. The random numbers will be used in different ways, depending on the type of population and the aims of the research.

39	76	49	93	85
27	12	95	35	07

▲ Table 1 Randomly selected numbers

Measurement: Sampling error

Inferential statistics is a branch of mathematics that uses data from a sample to make inferences about an entire population. It is assumed that a sample from a population represents the entire population—for example, the mean length of the leaves in a sample gives an estimate of the mean length of all the leaves on the tree.

Sampling error is the difference between a sample statistic and the equivalent value for the whole population. For example, when estimating population size, the sampling error is the difference between the estimated population size and the true size of the whole population.

Random sampling ensures that every member of a population has an equal chance of being selected and so reduces sampling error.

C4.1.3 Random quadrat sampling to estimate population size for sessile organisms

Quadrats are square sample areas, usually marked out using a quadrat frame. Quadrat sampling involves repeatedly placing a quadrat frame at random positions in a habitat and recording the numbers of organisms present each time.

The usual procedure for randomly positioning quadrats is this:

- A measuring tape is used to mark a base line along the edge of the habitat. This base line must extend all the way along the edge of the habitat.
- Random numbers are generated, using either a table or a random number generator.
- A first random number is used to determine a distance along the measuring tape. All distances along the tape must be equally likely.
- A second random number is used to determine a distance across the habitat, at right angles to the tape. All distances across the habitat must be equally likely.
- The quadrat is placed precisely at the distances determined by the two random numbers.

If this procedure is followed correctly, with a large enough number of replicates, the results will give a reliable estimate of population size. This method is only suitable for plants and other organisms that are sessile. This means they have a fixed position and do not move. Quadrat sampling is unsuitable for populations of most animals, for obvious reasons.

▶ Figure 5 Quadrat sampling of seaweed populations on a rocky shore

Mathematics: Standard deviation

Standard deviation is a measure of the variability of data. Using a calculator or computer, it is easy to find both the mean for a sample of observations and the standard deviation of the mean. The observations must be quantitative and could be counts (for example, the number of individuals in a quadrat) or measurements (such as the height of each plant in a sample).

A low standard deviation shows that there is little variation between the values in a sample; a high standard deviation shows that there is a lot of variation between values. For example, for the number of individuals per quadrat:

- a low standard deviation shows that the population is evenly spread, with a similar number of individuals in each quadrat
- a high standard deviation shows that the population is uneven, with many more individuals in some parts of the habitat than in others.

The lower the standard deviation, the more confidence you can have in any estimates based on the data. For example, if you are using quadrat counts to estimate the number of individuals in an area, you can have more confidence in the estimate if the standard deviation is low (that is, the number of individuals per quadrat is reasonably consistent).

Data-based questions: Population distributions

1. Three types of population distribution are recognized: uniform, clumped and random.

   a. Identify the types of distribution shown in Figure 6. [2]

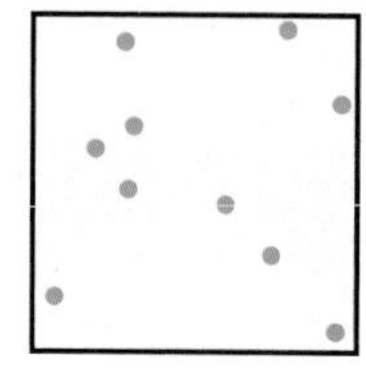

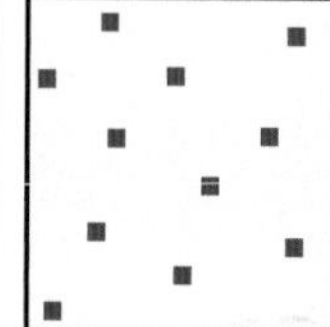

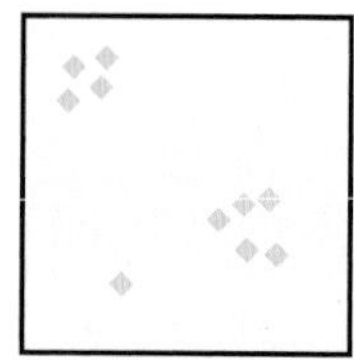

▲ Figure 6

   b. If each type of population was sampled using quadrats, which would have the highest standard deviation in mean number per quadrat? Which would have the lowest standard deviation? [2]

2. Figure 7 shows gannets nesting at Cape Kidnappers, Hawke's Bay, New Zealand.

▲ Figure 7

a. Identify the type of population distribution. [1]

b. Suggest a reason for the distribution. [1]

c. Predict whether the standard deviation of the mean would be relatively high or low for quadrat sampling of the population. [1]

3. Figures 8 and 9 show the distribution of snowdrop plants (*Galanthus nivalis*) in two populations.

a. Distinguish between the distribution of plants in the two populations. [1]

b. Snowdrops can either reproduce sexually by producing seeds or asexually by producing extra bulbs underground. Predict, with a reason, the method of reproduction used by each of the populations. [2]

▲ Figure 8

▲ Figure 9

C4.1.4 Capture–mark–release–recapture and the Lincoln index to estimate population size for motile organisms

Sessile organisms such as conifer trees or corals remain in one position. Motile organisms including basking sharks and bees move from place to place. Quadrat sampling is suitable for estimating population size for sessile organisms. However, it is not suitable for motile organisms. The capture–mark–release–recapture method and the Lincoln index can be used for organisms that are motile. Figure 10 shows the steps in this method.

1 Capture as many individuals as possible in the area occupied by the animal population, using netting, trapping or careful searching

e.g. careful searching for banded snails *(Cepaea nemoralis)*

2 Mark each individual, without making them more visible to predators.

e.g. marking the inside of the snail shell with a dot of non-toxic paint

3 Release all the marked individuals and allow them to settle back into their habitat.

The following assumptions are made about the period of time between capture and recapture:

- There is no migration into or out of the population.
- There are no deaths or births.
- Marked individuals mix back into the population and have the same chance of being captured on the second occasion as unmarked individuals.
- The marks remain visible.
- The marks do not increase the chance of predation or other threats to survival.

4 Recapture as many individuals as possible and count how many are marked and how many unmarked.

5 Calculate the estimated population size by using the Lincoln index:

$$\text{population size} = \frac{M \times N}{R}$$

where: M is the number of individuals caught and marked initially
N is the total number of individuals recaptured
R is the total number of individuals recaptured with marks.

▲ Figure 10

Applying techniques: The Lincoln index

You can apply the Lincoln index to your student population:

1. Establish a random sampling method.
2. Devise a method for marking captured students, such as placing a safety pin in a defined position on their clothing, or making a small mark with a pencil on their thumb nail.
3. Repeat the same random sampling procedure.
4. Use the Lincoln index to estimate the size of the school population.
5. Find out the actual number of students, then compare and contrast this value with your estimate.
6. Discuss how sampling error may affect your estimate.

C4.1.5 Carrying capacity and competition for limited resources

Populations take materials from their environment, such as water, oxygen or food. A larger population needs more of these resources. Resources vary in abundance but all are limited in the amount available or the rate of production. If a resource becomes scarce, the members of a population will compete for it. If a population grows too large, some individuals will be unable to obtain enough of the resource. These individuals are likely to die, reducing the population size. The maximum size of a population that an environment can support is the carrying capacity.

In practice, one resource (or a small number of resources) is likely to limit population size and so determine carrying capacity. Table 2 shows examples of resources that may limit carrying capacity in plants and animals.

Plants
• water
• light
• soil nitrogen (NO^- or NH_4^+)
Animals
• water
• space for breeding
• food, or territory for obtaining food
• dissolved oxygen

▲ Table 2

◀ Figure 11 This population of the creosote bush (*Larrea tridentata*) at Dagger Flats in the Chihuahuan Desert, Texas, is almost certainly at carrying capacity, despite large gaps between individuals. Each bush has widespread roots, so all water available in the soil is being used

C4.1.6 Negative feedback control of population size by density-dependent factors

The size of a population can rise or fall over time. If a population is successful, there may be a long-term increase. This could happen if a population fills an ecological niche that has been unoccupied. There may also be long-term decreases due to ecological changes that have negative impacts on the population. More commonly, populations fluctuate (alternately increase and decrease) but overall the population remains relatively stable over time, due to negative feedback control.

Two types of factor can increase or decrease populations.

- **Density-independent factors** have the same effect however large the size of a population. For example, seawater flooding kills all plants that are intolerant of salinity, whether the population is large or small. Forest fires kill plants and animals however many there are before the fire.
- **Density-dependent factors** have an increasing effect as the population becomes larger. They are the basis for negative feedback mechanisms because they reduce larger populations and allow smaller populations to increase.

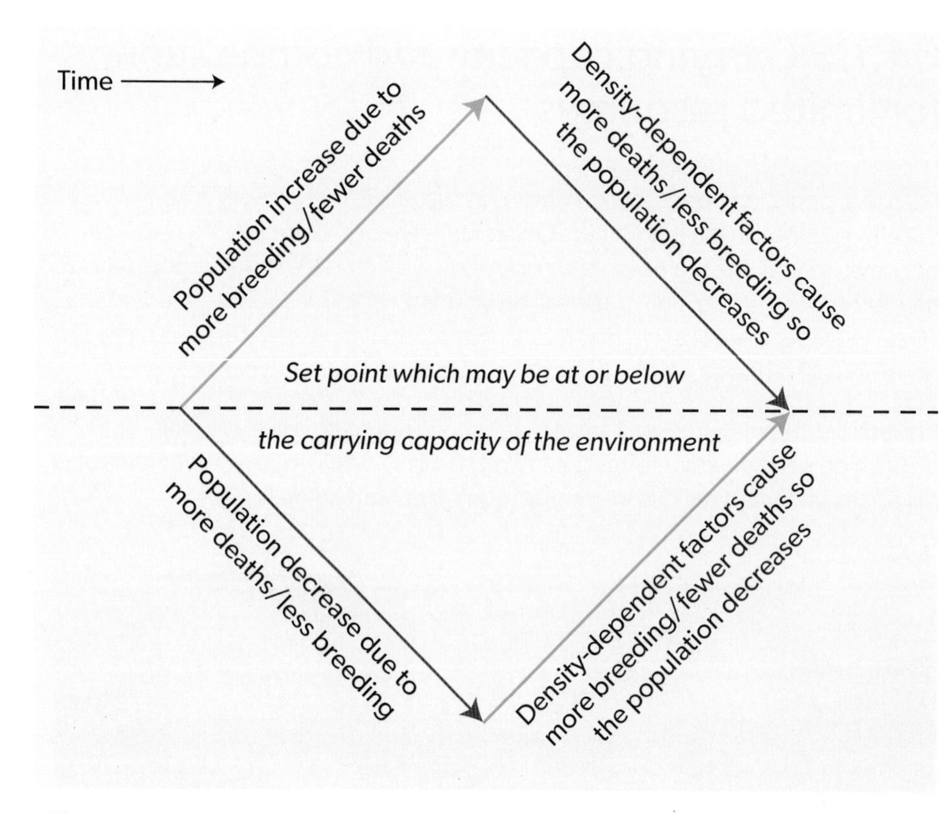
Time
Population increase due to more breeding/fewer deaths
Density-dependent factors cause more deaths/less breeding so the population decreases
Set point which may be at or below
the carrying capacity of the environment
Population decrease due to more deaths/less breeding
Density-dependent factors cause more breeding/fewer deaths so the population decreases

▶ Figure 12 Negative feedback

There are three groups of density-dependent factor:

- **Competition** for limited resources such as water, food in animals and light in plants
- **Predation**—this becomes more intense if a population of prey becomes denser and therefore easier to find, and less intense if the prey become more scarce. There are similar interactions between plants and herbivores
- **Infectious disease, parasitism and prey infestation**—these increase with population density because it is easier for pathogens, parasites and pests to spread from host to host.

C4.1.7 Population growth curves

Reproduction tends to cause exponential growth in populations. It is an example of positive feedback because breeding increases the number of individuals and a larger number of individuals can breed more. In most populations, density-dependent limiting factors lead to negative feedback effects that prevent exponential growth. If these factors are absent or ineffective, the natural tendency is for populations to go through a period of exponential growth, with numbers increasing at an accelerating rate.

Cases of exponential growth have been recorded when a species has spread into a new area and found a new ecological niche. Resources required by the population are abundant so they do not limit population growth. Initially at least there may be no predators, pathogens, parasites or pests that target the population. An example is the spread of the Eurasian collared dove (*Streptopelia decaocto*) across Europe during the middle of the 20th century. This species was originally native to parts of Asia. In Europe it has thrived in suburban areas, particularly where many people feed birds in their gardens during the winter. It is also associated with farmland where grain spilled from arable crops or livestock feeding provides food for the doves.

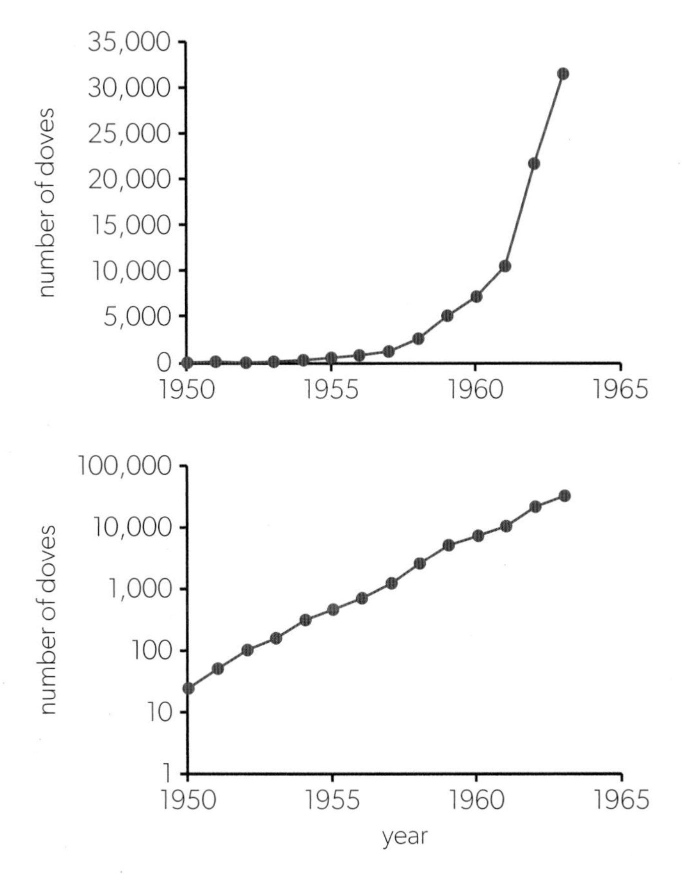

▲ Figure 13 Population growth curve for collared doves in the Netherlands, showing a characteristic J-shape. The second graph has a logarithmic *y*-axis scale and shows that the increase was very close to exponential

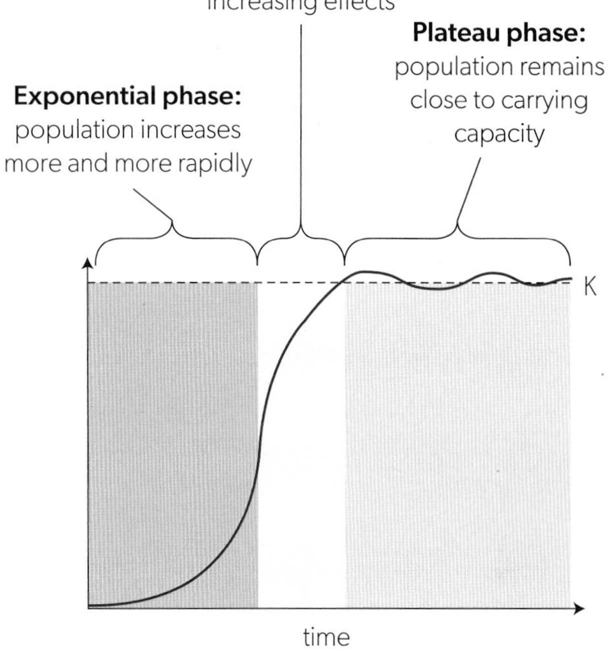

▲ Figure 14 The three phases of a sigmoid population growth curve

Population growth of the collared dove has now stopped in the UK, the Netherlands and other parts of Europe where there was exponential growth. Numbers are stable or even falling. Exponential growth of populations cannot continue indefinitely because environmental resources are limited so the carrying capacity will be reached eventually. Before that, one of the other density-dependent factors may have effects—for example, the spread of a host-specific pathogen or pest. The curve on a graph showing population size over time changes from a J-shape to an S-shape. These graph are called sigmoid curves because they are shaped like the Greek letter S, sigma.

Not all populations reach a plateau phase after exponential growth. Some show a cyclical pattern of "booms" and "busts" where the population far exceeds the carrying capacity and then drops below it, before rising again. In other cases, a population will crash after reaching its maximum. This can happen if a population produces toxins which accumulate and cause harm.

Models: Strengths and weaknesses of models

The sigmoid growth curve represents an idealized graphical model. Such models often simplify complex systems. This activity highlights the strengths and limitations of the sigmoid growth curve model.

Reindeer on St Paul Island

In 1911, the US government introduced reindeer (*Rangifer tarandus*) to the Pribilof Islands in the Bering Sea to the west of Alaska. The ecosystems on these islands are grassland and taiga, with no trees. In winter, snow covers the islands and slow-growing lichens are the main food source for herbivores. There are no natural predators of reindeer. Four males and twenty-one females were released on St Paul island. Table 3, which is available in the digital resources associated with the Course Companion, shows the numbers of reindeer killed each year by the human population and the numbers remaining after this. There are no records for years in the early 1940s because the population had been evacuated and there was a military occupation of the island.

Year	Number remaining	Number killed
1911	25	0
1912	40	0
1913	52	0
1914	75	0
1915	92	0
1916	111	0
1917	144	0
1918	155	2
1919	164	14
1920	192	22
1921	250	34
1922	190	38
1923	150	14
1924	200	13

Year	Number remaining	Number killed
1925	225	25
1926	250	10
1927	250	9
1928	315	22
1929	329	20
1930	404	17
1931	453	19
1932	485	47
1933	673	11
1934	820	14
1935	1162	23
1936	1388	37
1937	1673	80

Year	Number remaining	Number killed
1938	1943	103
1939	1800	105
1940	962	265
1941	850	326
1942	-	-
1943	-	-
1944	-	-
1945	-	56
1946	240	0
1947	250	0
1948	120	0
1949	60	0
1950	8	0

▲ Table 3

◀ **Figure 15** Image captured in 2018 using a drone, showing the St Paul island reindeer herd mustered for the annual population count

1. Display the data in the table using whatever style of graph or chart you think is most suitable. [5]
2. a. Explain the large increase in reindeer numbers between 1911 and 1938. [3]

   b. Discuss whether this increase was exponential. [2]
3. a. Between the years 1939 and 1950, the reindeer population decreased from 1800 to 8. Calculate the percentage decrease. [2]

   Various hypotheses have been proposed to account for the decrease:

   - a series of very cold winters in the early 1940s
   - hunting by humans
   - lack of availability of lichens which provide winter feed.

   b. Identify whether each of these factors is density-dependent or density-independent. [3]

   c. Evaluate each hypothesis using evidence from the population data. [3]
4. Lichens remain much scarcer on St Paul than before the introduction of reindeer, but the reindeer herd continues to offer a source of meat for the human population. Plans to introduce reindeer to parts of northern Canada where they are not endemic have been condemned. Do you agree? [2]

C4.1.8 Modelling of the sigmoid population growth curve

Sigmoid population growth in natural ecosystems can be modelled experimentally using an organism such as duckweed or yeast. A small number of organisms should be introduced initially, with abundant resources for growth and no other organisms that could limit the population. Numbers should be monitored, either by datalogging or by regular counts.

Duckweed is a stemless photosynthetic water plant that inhabits ponds and lakes. Each plant consists of a small floating frond that may have a single root on its underside. Plants can produce new fronds asexually; these fronds separate from the parent, increasing the size of the population. Duckweed can be grown on the surface of water in beakers or cups. Various experiments are possible:

- What is the carrying capacity of a given container?
- What conditions of light, nutrients or container surface area are ideal for population growth?

Yeast is a saprotrophic fungus with ovoid cells that lives on the surface of fruits and other places where sugars are available. It can reproduce asexually by budding (producing small extra cells by mitosis). Yeast is used commercially to make bread and produce alcohol by fermentation.

Figure 16 *Wolffia arrhiza* (rootless duckweed) and *Spirodela polyrhiza* (greater duckweed) growing together in a drainage ditch in Somerset, England

C4.1.9 A community as all of the interacting organisms in an ecosystem

All species depend on relationships with other species for their long-term survival. A population of one species can never live in isolation. Instead, groups of populations live together. In ecology, a group of populations living together in an area and interacting with each other is known as a community. A typical community consists of hundreds or even thousands of species living together in an ecosystem.

An important part of ecology is research into interactions between organisms in communities. These relationships are complex and varied. In some cases, the interaction between two species benefits one species and harms the other, for example, the relationship between a parasite and its host. In other cases both species benefit, as when a hummingbird feeds on nectar from a flower and helps the plant by pollinating it.

▲ Figure 17 A coral reef is a complex community with many interactions between the populations. Most corals have photosynthetic unicellular algae called zooxanthellae living inside their cells

C4.1.10 Competition versus cooperation in intraspecific relationships

An intraspecific relationship is one that exists between individuals of the same species—usually within the same population of a species. Competition and cooperation are categories of intraspecific relationship.

1. Competition

Individuals in a population are members of the same species so they share an ecological niche and are likely to require the same resources. Unless a resource is abundant, there will be competition for it. Some individuals will be more successful and gain more of the resource, helping them to survive and reproduce. As a result, there is natural selection over the generations for traits that allow individuals to compete more effectively. Figures 18–21 show examples of competition within a population.

Competition for light in plants

▲ Figure 18

Plants compete for light that is used in leaves for photosynthesis. These are leaves of wild garlic (*Allium ursinum*), which often becomes crowded in woodland, so not all leaves obtain enough light.

Competition for pollinators in flowering plants

▲ Figure 19

Flowering plants compete for pollinating insects, using bright colours, scents and nectar. Here a honey bee is dusted with pollen as it feeds on nectar in a dandelion (*Taraxacum officinale*).

Competition for food in animals

▲ Figure 20

Bohemian waxwings (*Bombycilla garrulus*) breed in northern conifer forests and migrate south to find food in winter. They compete for berries on trees and migrate further when supplies run out in an area.

Competition for breeding sites in animals

▲ Figure 21

Guillemots (*Uria aalge*) breed on ledges on sea cliffs. On the Isle of May, Scotland there is competition for the best sites which give the highest chance of breeding success.

2. Cooperation

Individuals in a population may cooperate in a variety of ways. The extent of cooperation varies, with less between plants and more in social animals such as termites. Cooperative relationships have strong advantages because all individuals benefit, whereas in competitive relationships all individuals tend to be harmed to some degree. Figures 22–25 show examples of cooperation within a population.

Communal roosting in animals

▲ Figure 22

By huddling together, birds or mammals can conserve body heat. Male emperor penguins are a well-known example. Here, fantails (*Rhipidura fuliginosa*) in New Zealand roost together in a tightly packed line.

Feeding in animals

▲ Figure 23

Hunting in groups (social predation) can increase chances of success. Here a group of chimpanzees (*Pan troglodytes*) is looking for monkeys in trees to hunt and kill, adding meat to their diet.

Defence against predation in animals

▲ Figure 24

Fish that form a tightly packed, fast-moving "bait ball" are much harder for predators to catch. Here, California sea lions (*Zalophus californianus*) are circling a bait ball of mackerel.

Parental care in animals

▲ Figure 25

Some bird species share parental care with crèches, in which one or more adults care for offspring of multiple parents. Here, one female eider duck (*Somateria mollissima*) is taking care of over 40 ducklings.

Application of techniques: Sowing density and above ground biomass

You might wish to answer the question: "What sowing density results in the maximum above ground biomass per unit area?". To explore this question, you could plant seeds or cuttings with increasing density per pot. Green onions (scallions or spring onions) or seeds of cress, radish or beans are quick-growing options.

C4.1.11 Herbivory, predation, interspecific competition, mutualism, parasitism and pathogenicity as categories of interspecific relationship within communities

Relationships between species in a community are classified by the type of interaction. Table 4 shows six categories of interspecific relationship. In the first three categories, organisms in different species encounter each other occasionally or not at all. In the other categories, species live in close association and an individual of one species will have one or more individuals of a different species either physically inside, on their outer surface or close by.

▼ Table 4

Relationship	Examples
Herbivory Primary consumers feeding on producers. The producer may or may not be killed.	• Bison grazing on grasses • Aphids feeding on phloem sap from plants • Limpets feeding on algae growing on rocky shores
Predation One consumer species (the predator) killing and eating another consumer species (the prey).	• Anteaters feeding on ants or termites • Dingoes hunting, killing and eating red kangaroos • Starfish eating oysters
Interspecific competition Two or more species using the same resource, with the amount taken by one species reducing the amount available to the other species.	• Ivy climbing up oak trees and competing for light • *Forda* and *Geoica* (gall-forming aphids) competing for phloem sap on leaves of the terebinth tree • Barnacles competing for space and food on rocky shores
Mutualism Two species living in a close association, with both species benefiting from the association.	• Nitrogen-fixing *Rhizobium* bacteria living in root nodules of plants in the Fabaceae family and exchanging materials with the plant • Mycorrhizal fungi growing into the roots of plants in the Orchidaceae family and exchanging nutrients with the orchid • Photosynthesizing zooxanthellae living in the cells of hard corals and exchanging materials with the coral
Parasitism One species (the parasite) living inside, or on the outer surface of, another species (the host) and obtaining food from them. The host is harmed and the parasite benefits.	• Ticks living on the skin of deer and feeding by sucking blood from the deer • The roundworm *Baylisascaris* living in the gut of raccoons and absorbing foods that have been digested by the raccoon • Non-photosynthesizing *Cuscuta* plants (dodders) growing on gorse and other plant hosts, absorbing foods from the host's sap
Pathogenicity One species (the pathogen) living inside another species (the host) and causing a disease in the host.	• Potato blight fungus (*Phytophthora*) infecting potato plants • Tuberculosis bacterium (*Mycobacterium tuberculosis*) infecting badgers • Myxomatosis virus infecting rabbits

Data-based questions: Barnacles

Barnacles are sessile, filter-feeding crustaceans that attach themselves to rocks on sea shores. Video clips of their method of feeding are available on the internet. Figure 26 shows a small area of rock almost covered with barnacles on the beach at St Donat's, Wales. On this beach, there were other areas of rock with the same abiotic conditions but no adult barnacles. The beach has a very large tidal range and is exposed to storms with powerful waves, especially in winter.

1. Four different species have been marked with coloured dots. Identify these species using the drawings in Figure 27. [4]
2. The magnification of the photo is ×1.2. How wide should a scale bar on the photo be to indicate a size of 20 mm? [1]
3. Discuss how the barnacles are competing with each other and whether the competition is intraspecific, interspecific, or both. [3]
4. Discuss whether there is evidence for cooperation between the barnacles. [2]

▲ Figure 26 Barnacles

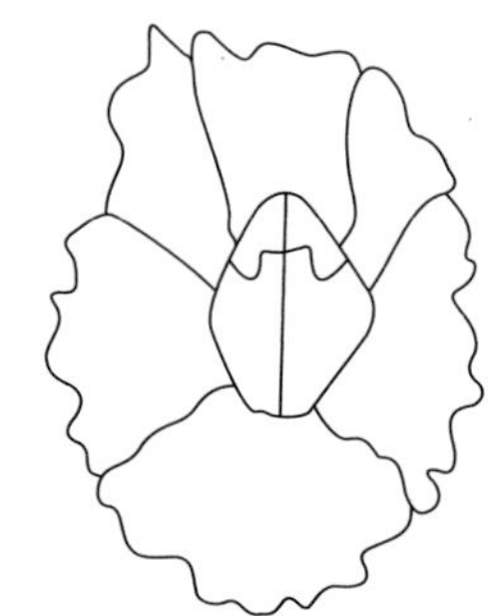

Semibalanus balanoides

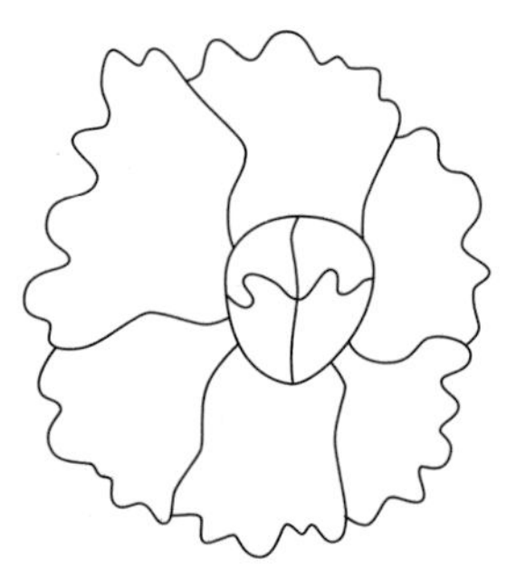

Chthamalus stellatus

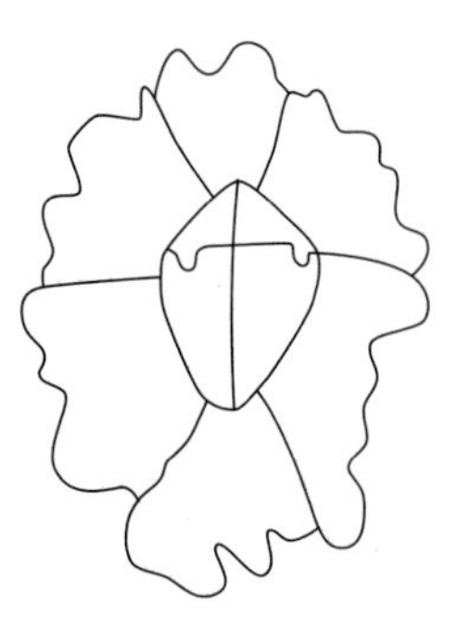

Chthamalus montagui

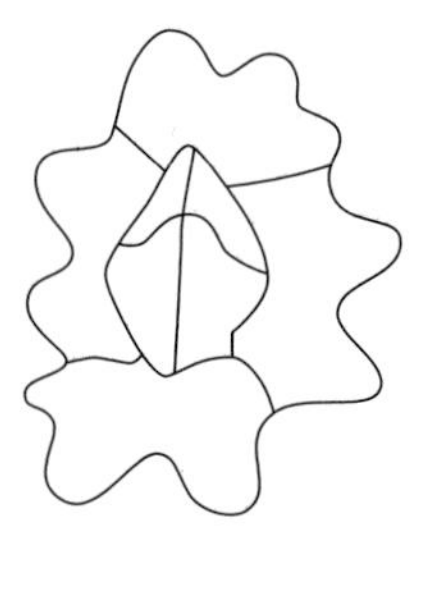

Elminius modestus

▲ Figure 27

▲ Figure 28 The sexton beetle lays its eggs on small dead birds or mammals (carrion) and then buries them. Mites that also feed on carrion use the sexton beetle to carry them from meal to meal. What relationship is this?

▲ Figure 29 Blackfly are aphids that feed on phloem sap and egest honeydew containing excess sugars. Red ants farm aphids and harvest the honeydew, placing the aphids on plants and defending them against predators. What are the relationships between plant, aphid and ant?

C4.1.12 Mutualism as an interspecific relationship that benefits both species

Mutualisms are close associations between species, where both species benefit from the relationship. In many cases, the two species are from different taxonomic kingdoms so they have different capabilities and supply different services. Three examples are described here.

Root nodules in Fabaceae

Plants need nitrogen to make amino acids and other nitrogen-containing compounds. The atmosphere is 78% nitrogen but plants cannot use this form of the element. Instead, they need a fixed form such as ammonium (NH_4^+) or nitrate (NO_3^-). Availability of fixed nitrogen is a limiting factor in many ecosystems. Some plants have developed a mutualistic relationship with nitrogen-fixing bacteria which give them an abundant supply of ammonium. For example, many plants in the Fabaceae family (peas and beans/legumes) develop a mutualistic relationship with *Rhizobium* bacteria.

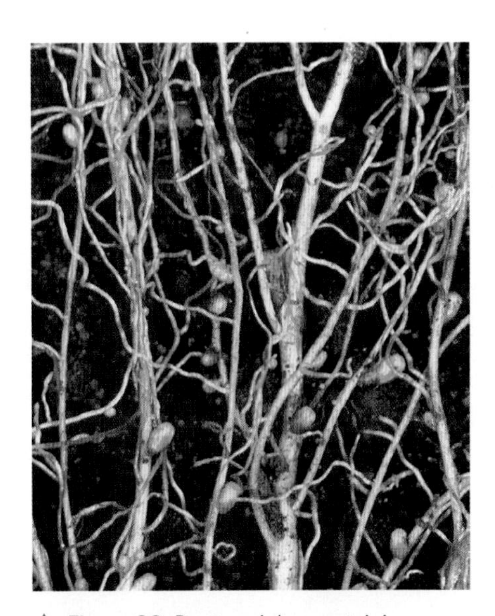

▲ Figure 30 Root nodules containing *Rhizobium* on the roots of white clover (*Trifolium repens*), a member of the Fabaceae family

The plant provides these services to *Rhizobium*	The *Rhizobium* provides these services to the plant
• Grows root nodules in which the bacteria can live with protection from consumers • Maintains low oxygen conditions inside the nodules, which suit *Rhizobium* • Supplies sugars made by photosynthesis, providing energy for the *Rhizobium*	• Absorbs nitrogen (N_2) and fixes it to produce ammonium • Supplies ammonium to the plant • Prevents nitrogen deficiency in the plant, giving it a competitive advantage over other plants

▲ Table 5

Mycorrhizae in Orchidaceae

The roots of most plants form a close association with fungi in the soil. These relationships are mutualistic and they are called mycorrhizae. Orchids are particularly dependent on mycorrhizal relationships because their seeds do not contain food reserves. At an early stage in germination, the fungal hyphae penetrate the root of an orchid seedling. They grow through cell walls but do not burst the plasma membrane of root cells; instead, the membrane forms a sheath around the hypha. Initially the fungus supplies the orchid seedling with the carbon compounds and mineral nutrients it needs for growth. Once the orchid starts to photosynthesize, there is a two-way exchange of materials between the fungus and the orchid.

▲ Figure 31 Orchids have thick unbranched roots that form mycorrhizal relationships with fungal hyphae. This orchid is growing on a tree in a tropical montane forest, together with lichens and mosses. Its roots extend for metres over the surface of the tree trunk and have green photosynthetic growing tips

The orchid provides these services for the fungus	The fungus provides these services for the orchid
• Supplies carbon compounds made by photosynthesis, including sugars	• Absorbs nitrogen and phosphorus from the soil and supplies them to the orchid; the nitrogen is mostly supplied in the form of ammonium or amino acids • Supplies fixed carbon in organic compounds, obtained from the soil by digesting dead organic matter or by parasitizing other plants • Supplies water absorbed from the soil

▲ Table 6

Zooxanthellae in hard corals

Hard corals secrete calcium carbonate to form a skeleton in which the individual animals, called polyps, can live. Coral reefs are built from these skeletons. Most reef-building hard corals contain mutualistic photosynthetic algae called zooxanthellae. These algae are absorbed from the seawater and kept alive inside coral cells.

▲ Figure 32 Polyps of a hard coral feeding on a reef in the Caribbean sea near Belize

The coral provides these services for zooxanthellae	The zooxanthellae provide these services for the coral
• Provides a safe and protected environment (the skeleton) • Grows close to the surface of the sea so the algae have a reliable source of light • Supplies carbon dioxide produced by cell respiration	• Supplies carbon compounds such as glucose and amino acids, produced by photosynthesis • Supplies oxygen produced by photosynthesis

▲ Table 7

C4.1.13 Resource competition between endemic and invasive species

Endemic species are those that occur naturally in an area. Alien species are those that were introduced by humans, deliberately or accidentally. Density-dependent factors usually regulate the population size of endemic species. However, many alien species are not effectively regulated, because the pests or predators that would control them in their native habitat are absent in their new habitat. If an alien species increases in number and spreads rapidly, it is described as invasive. Invasive species can have devastating impacts on communities.

The competitive exclusion principle predicts that two species cannot occupy the same ecological niche indefinitely. Alien species compete for resources with endemic species that have the same or overlapping requirements. To become invasive, an alien species must be successful in the competition for resources with endemic species. As a result, the endemic species may occupy a smaller realized niche, causing a decline in population, or lose its niche entirely and become extinct in an area.

This is currently a widespread phenomenon because humans transport high numbers of species to new areas, where there are no (or not enough) limiting factors to control the populations of alien species. Two examples are given in Figures 33 and 34 but local examples can be found throughout the world.

Red lionfish are endemic to coastal seas in parts of the Indo-Pacific. Because of their spectacular appearance they are sometimes kept in aquariums and have sometimes escaped. Small numbers escaped from an aquarium in southern Florida during Hurricane Andrew in 1992 and have since multiplied and spread on coral reefs in Florida and the Caribbean, helped by a lack of predators adapted to avoid the venomous spines. Red lionfish compete for prey with endemic fish species by establishing territories from which the other fish are aggressively excluded.

▲ Figure 33 Red lionfish (*Pterois volitans*)

Salvinia is a floating fern that is endemic to south-eastern Brazil. It has been grown as an ornamental plant in ponds and aquaria around the world and has escaped into the wild in many areas. It can double its population by asexual reproduction in two to three days, spreading to completely cover the surface of lakes and rivers. It eliminates endemic aquatic plants by competition for light.

▲ Figure 34 Floating ferns (*Salvinia natans*) in the Danube delta, Romania

C4.1.14 Tests for interspecific competition

If the presence or absence of more than one species is recorded in every quadrat during sampling of a habitat, it is possible to test for associations between the species. Competitive exclusion might discourage two species from growing together, so they will occur in the same quadrats less often than they would if there was no association. There are two possible hypotheses:

- H_0: two species are distributed independently (the null hypothesis)
- H_1: two species are associated

These hypotheses can be tested using a statistical procedure—the chi-squared test. This test is only valid if all the expected frequencies are 5 or larger and the sample was taken at random from the population, in this case by positioning quadrats randomly.

If the chi-squared test indicates that the null hypothesis can be rejected, this is not proof of competition. The two species could have different habitat requirements, so they tend not to occur together. Also the species might occur together more frequently than expected, rather than less frequently. This can happen if two species have the same habitat requirements but are not limited by resource competition, or if they have a relationship with each other that encourages co-location.

Stronger evidence for competition can be obtained by carrying out an experiment in a habitat—for example:

- Field manipulation could be used. One of two species could be removed from quadrats in grassland. If the other species then increases in number or biomass, you can conclude that there is interspecific competition between these species in the grassland.
- Laboratory experiments could be carried out, under carefully controlled conditions. Species could be grown together and apart to investigate whether they compete for resources.

ATL Thinking skills: Designing research questions

Wandering and wondering as the first stage of inquiry

Leaf herbivory (eating leaves) is carried out by a broad range of organisms. Take a walk through a wood and note all the different types of leaf herbivory and leaf damage you see. As you make your observations, ask yourself testable questions such as:

- Are new leaves or mature leaves preferred? Does the answer depend on the type of herbivory?
- Are there different patterns higher on the plant compared with lower on the plant?
- Are plants on the edge of the wood affected differently than those far from the edge?

Experiments versus observations

In an experimental study, researchers apply an intervention (the treatment) and investigate its effects. The experimental design may be a randomized controlled trial, in which the treatment is given to one group but not to the other control group, with the groups selected so that individuals have an equal chance of being assigned to each group. In another type of design, the treatment is applied at different levels to a series of groups.

In an observational study, researchers do not change any factors. Instead, they assess the effects of a factor of interest using methods that separate these effects from the effects of other factors.

Experiments generally provide stronger and more reliable evidence in support of a hypothesis (or against it), because cause and effect are clearer. However, some experiments would be unethical—especially where humans, other animals or natural ecosystems would be subjected to harmful treatments. Observational studies may also be more appropriate where results are gathered over many years, or where very large numbers of replicates are required.

C4.1.15 Use of the chi-squared test for association between two species

The chi-squared test has two types of use:

- testing for independence or association
- testing goodness of fit.

In both cases, a contingency table is used to compare observed and expected results. When testing for association, the expected numbers are calculated assuming independence. Tests of goodness of fit use a model such as a ratio to generate expected numbers.

Method for chi-squared test of association

1. Define the two alternative hypotheses:

   H_0: two species are distributed independently (the null hypothesis)

   H_1: two species are associated

2. Draw up a contingency table of observed frequencies, in this case, the numbers of quadrats containing or not containing the two species.

	Species A present	Species A absent	Row totals
Species B present			
Species B absent			
Column totals			

▲ Table 8

Calculate the row and column totals. Adding the row totals or the column totals should give the same grand total in the lower right cell.

3. Calculate the expected frequencies for the four species combinations, assuming independent distribution.

   Calculate each expected frequency from values in the contingency table, using this equation:

$$\text{expected frequency} = \frac{\text{row total} \times \text{column total}}{\text{grand total}}$$

Mathematics: Chi-squared test for association—worked example

Figure 35 shows an area on the summit of Caer Caradoc, a hill in Shropshire, England. The area is grazed by sheep in summer and hill walkers cross it on grassy paths. There are raised hummocks with heather (*Calluna vulgaris*) growing on them. A visual survey of this site suggested that *Rhytidiadelphus squarrosus*, a species of moss growing in this area, was associated with these heather hummocks. The presence or absence of the heather and the moss was recorded in a sample of 100 quadrats, positioned randomly.

▲ **Figure 35** Caer Caradoc summit, Shropshire, England, with purple-brown heather and green paths created by walkers

→

1. Define the two alternative hypotheses.

- H_0: heather and moss are distributed independently (the null hypothesis)
- H_1: two species are associated

2. Draw up a contingency table of observed frequencies.

   Calculate the row and column totals. Adding the row totals or the column totals should give the same grand total in the lower right cell.

	Heather present	Heather absent	Row total
Moss present	57	7	64
Moss absent	9	27	36
Column total	66	34	100

▲ Table 9

3. Calculate the expected frequencies for the four species combinations, assuming independent distribution:

   $$\text{expected frequency} = \frac{\text{row total} \times \text{column total}}{\text{grand total}}$$

	Heather present	Heather absent	Row total
Moss present	$\frac{64 \times 66}{100} = 42.2$	$\frac{64 \times 34}{100} = 21.8$	64
Moss absent	$\frac{36 \times 66}{100} = 23.8$	$\frac{36 \times 34}{100} = 12.2$	36
Column total	66	34	100

▲ Table 10

4. Calculate the number of degrees of freedom:

   $$\text{degrees of freedom} = (m-1)(n-1)$$

   where m and n are the number of rows and number of columns in the contingency table.

$$\text{Degrees of freedom} = (m-1)(n-1) = (2-1)(2-1) = 1$$

5. Find the critical region for chi-squared

The critical region (obtained from a table of chi-squared values) is 3.83 or larger.

6. Calculate chi-squared using this equation:

   $$X^2 = \sum \frac{(f_o - f_e)^2}{f_e}$$

   where f_o is the observed frequency

   f_e is the expected frequency

   Σ is the sum of.

$$\text{Chi-squared} = \frac{(57-42.2)^2}{42.2} + \frac{(7-21.8)^2}{21.8} + \frac{(9-23.8)^2}{23.8} + \frac{(27-12.2)^2}{12.2}$$

$$= 5.1905 + 10.0477 + 9.2034 + 17.9541$$

$$= 42.3957$$

7. Compare the calculated value of chi-squared with the critical region.

The calculated value of chi-squared is in the critical region, so there is evidence at the 5% level for an association between the two species and we can reject the null hypothesis H_0.

Discussion

Mosses are mostly confined to damp habitats. On this Shropshire hilltop, the moss *Rhytidiadelphus squarrosus* is associated with the heather because the heather provides shade, humidity and shelter from drying winds. Neither species can tolerate trampling on the paths created by hill walkers on this site.

Data-based questions: Chi-squared testing

In Ireland, the invasive Eastern grey squirrel occupies a similar niche to the native red squirrel, though it has a wider range of food sources. Red squirrels are endemic in Ireland. Grey squirrels were originally introduced to the woodland at Castle Forbes, County Longford in 1911, from where they spread during the 20th century. When alien grey squirrels spread into an area, the red squirrel population declines, sometimes becoming locally extinct.

The distribution map in Figure 38 shows the squirrels recorded in 10 × 10 km squares. The records were mostly produced by amateur naturalists—an example of "citizen science". Where no dot is shown, no data was recorded.

▲ **Figure 36** Red squirrel (*Sciurus vulgaris*)

▲ **Figure 37** Eastern grey squirrel (*Sciurus carolinensis*)

1. State the null and alternative hypotheses, H_0 and H_1. [2]
2. Construct a contingency table of observed values. [4]
3. Calculate the expected values, assuming no association between the species. [4]
4. Calculate the number of degrees of freedom. [2]
5. Find the critical region for chi-squared at a significance level of 5%. [2]
6. Calculate chi-squared. [4]
7. Draw conclusions about the alternative hypotheses from the calculated value of chi-squared. Explain your conclusions. [3]
8. Discuss the strength of evidence provided by this data for competition between red and grey squirrels. [4]

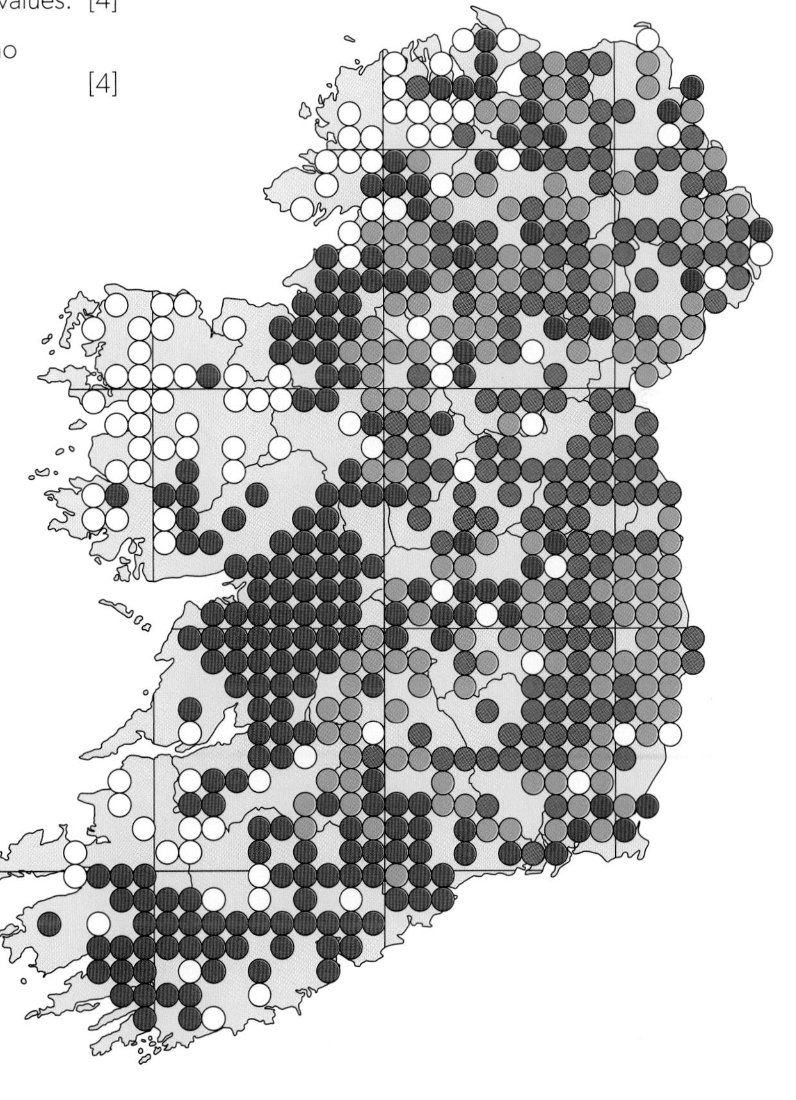

Key
- 10 km square containing both species of squirrel
- 10 km square containing only red squirrels
- 10 km square containing only grey squirrels
- 10 km square containing neither species of squirrel

▲ **Figure 38**

C4.1.16 Predator–prey relationships as an example of density-dependent control of animal populations

When a predator makes a kill and consumes its prey, the prey population is one smaller. In many communities, the population of prey does not change much overall, because new prey individuals are born at about the same rate as they are lost due to predation. Similarly, the birth and death rates of the predator are approximately equal, so the predator population remains stable.

However, there are communities where populations of prey and predators do not show this dynamic equilibrium. Instead, they undergo cyclical oscillations. There are four basic interactions in predator–prey cycles:

- A increase in prey numbers increases food availability for predators, so predator numbers rise.
- A rise in predator numbers increases predation of prey, so prey numbers fall.
- A fall in prey numbers decreases food availability for predators, so predator numbers fall.
- A fall in predator numbers decreases predation of prey, so prey numbers rise.

Oscillation due to predator–prey relationships has been investigated in part of Sweden. The graph in Figure 39 shows numbers of red fox (predator) and mountain hare (prey) that were shot in the study area. These figures show the changes over time in the relative size of the fox and hare populations, because hunters' success depends on the abundance of their target species.

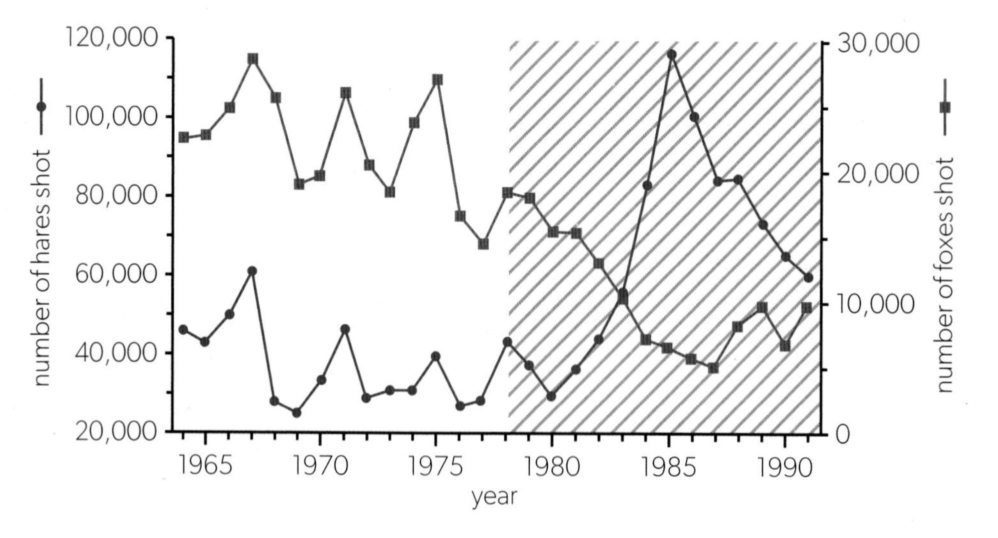

▶ Figure 39 Hunting records of red fox (*Vulpes vulpes*) and mountain hare (*Lepus timidus*) between 1964 and 1991. The shaded area was a period when there was an outbreak of sarcoptic mange in the fox population.
Source: Lindstrom *et al.* 1994. Ecology. Vol. 75. Pp 1042–1049

Four cycles of population rises and falls occurred between 1964 and 1980. After that there was an outbreak of sarcoptic mange in the fox population. This condition is due to a parasitic mite (*Sarcoptes scabiei*) which causes hair loss, skin deterioration and eventually death. In this period, the numbers of foxes declined much more than during a normal cycle and the numbers of hares rose much higher. The outbreak ended in the early 1990s and the normal fox–hare population oscillations resumed.

Cyclical oscillations in predator and prey populations are mostly seen in habitats where weather conditions vary from year to year. In northern zones, a warmer than average spring and summer causes more plant growth than normal. This provides extra food for herbivores such as mountain hares, so their numbers rise. This triggers a cycle of rising and falling predator and prey numbers in the following years.

C4.1.17 Top-down and bottom-up control of populations in communities

In many communities, interactions between trophic levels in a food chain are the basis for population control. When one population feeds on another population, there are direct interactions—for example, when a predator feeds on prey or a herbivore feeds on producers. There can also be indirect interactions between populations. For example, the population size of a predator that feeds on a herbivore could affect the population size of the producers the herbivore eats. If the number of herbivores decreases due to an increase in predator numbers, the producer population may increase.

Within a food chain, these interactions can operate in two directions:

- Top-down control acts from a higher trophic level to a lower one. For example, an increase in predator numbers will decrease the numbers of prey in lower trophic levels.
- Bottom-up control acts from a lower trophic level to a higher one. For example, a population of producers may be limited by the availability of mineral nutrients in the soil or in water.

Communities vary in whether more populations are controlled by top-down or bottom-up interactions.

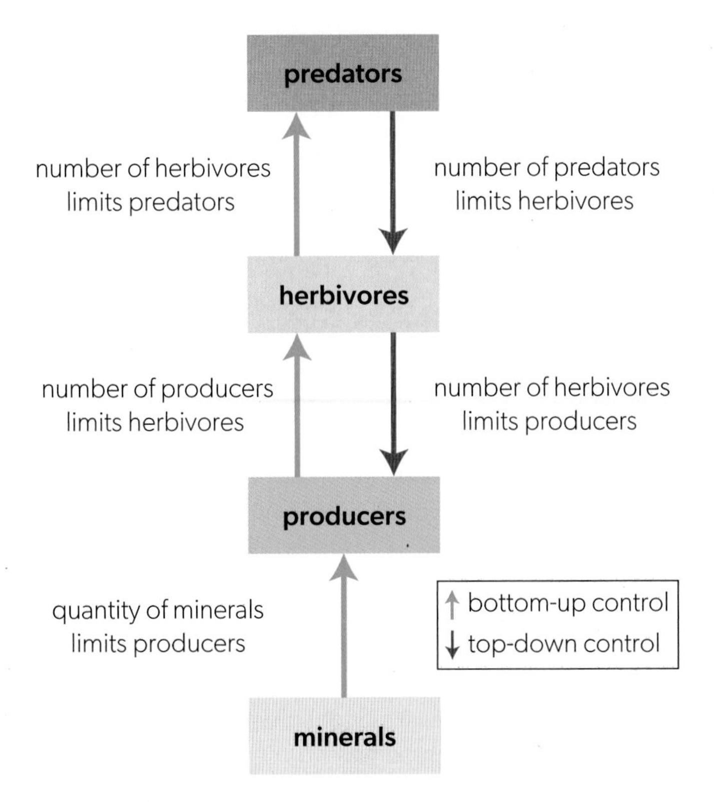

▲ Figure 40 Top-down and bottom-up population control

Top-down and bottom-up control of algae in the Shropshire meres

Over 100 small lakes called meres have formed in kettle-holes in Shropshire. These meres vary in depth. Two patterns of population control have been discovered, with implications for conservation.

Shallow meres

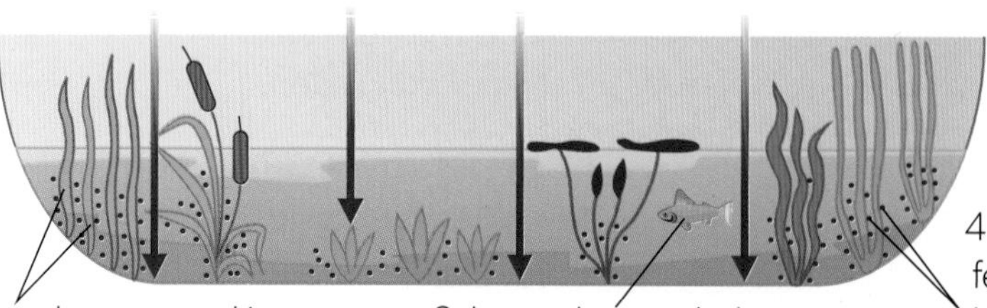

Deep meres

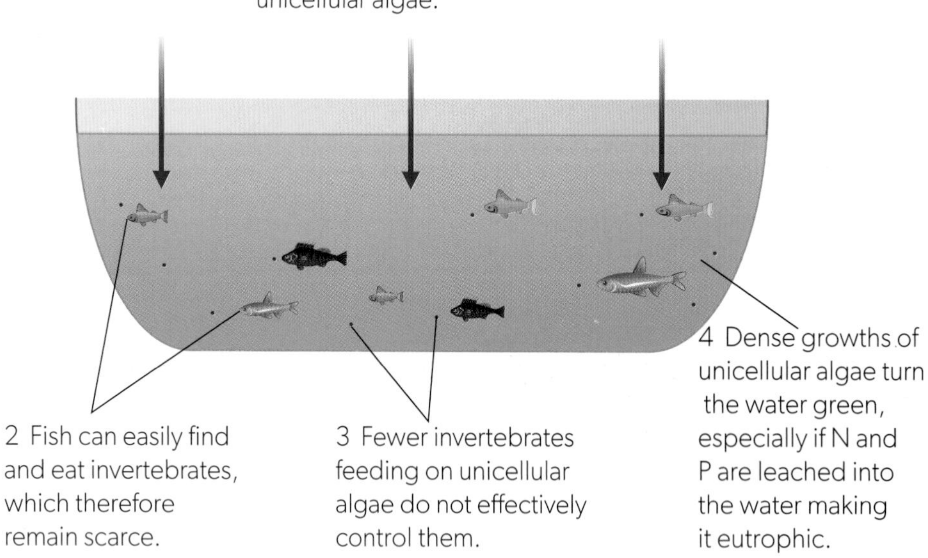

1 Sunlight cannot penetrate far in the water, due to the dense population of green unicellular algae.
2 Fish can easily find and eat invertebrates, which therefore remain scarce.
3 Fewer invertebrates feeding on unicellular algae do not effectively control them.
4 Dense growths of unicellular algae turn the water green, especially if N and P are leached into the water making it eutrophic.

▲ Figure 41 Cross-sections of shallow and deep meres

In shallow meres, water plants (hydrophytes) can root in the bed of the lake and provide shelter from predatory fish for *Daphnia* and other small crustaceans that feed on algae. These crustaceans are therefore abundant and they feed intensively on algae. As a result, the algae are scarce and the lake water remains clear. Adding extra nutrients to the lake (such as nitrogen or phosphorus) does not increase the population of algae, because it is regulated by higher trophic levels.

In deeper meres, water plants cannot root in the bed of the lake so the water is open and there is little shelter for crustaceans. They are therefore intensively predated by fish and their populations are small. This allows the algae to grow exponentially at times when nutrients are abundant; this is known as an algal bloom and turns the water of the lake bright green. Eutrophication triggers these algal blooms.

◀ Figure 42 White Mere is deep so algal numbers are under bottom-up control and algal blooms are common

C4.1.18 Allelopathy and secretion of antibiotics

Living organisms share many of the metabolic pathways required for cell growth. Substances produced by these pathways are primary metabolites, including both intermediates and end products. Secondary metabolites are substances produced by pathways that only exist in some taxonomic groups. These substances are not essential for cell growth; instead, they have a wide range of other functions. Two functional groups of secondary metabolites are antibiotics and allelopathic agents. Both of these metabolites are released into the environment, where they are toxic to other organisms and deter potential competitors.

- Antibiotics are secreted by microorganisms, mostly to kill or prevent the growth of other microorganisms.
- Allelopathic agents are secreted into the soil by plants to kill or deter the growth of neighbouring plants.

Local examples can be found throughout the world. One specific example of each is described here.

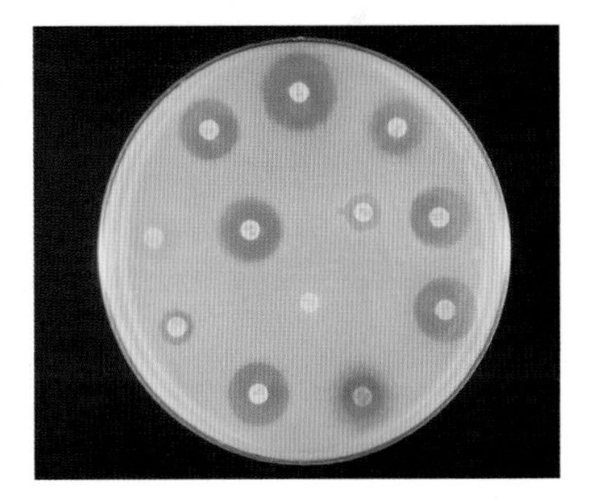

▲ Figure 43 In this antibiotic sensitivity test, a lawn of bacteria (*E. coli* and *Proteus*) has grown on the surface of nutrient agar gel. Antibiotics have diffused out from test discs, creating clear zones where they were effective in killing the bacteria

Antibiotic production by *Penicillium*

Penicillium is a genus of fungi that inhabit natural environments such as soil. They also occur on foods such as fruit, bread and cheese. They are saprotrophic fungi. They have hyphae which secrete enzymes into their environment; these enzymes digest carbohydrates, proteins and other carbon compounds. The hyphae then absorb sugars, amino acids and other products of digestion.

A risk for saprotrophs is that other organisms living in the same environment may compete for these digested foods. To reduce or prevent such competition, some species of *Penicillium* secrete the antibiotic penicillin. Penicillin interferes with cross-linking of peptidoglycan molecules in the cell walls of Gram-positive bacteria. This causes the walls to become weak so the bacteria eventually burst and die. Production of penicillin requires resources so *Penicillium* only does this when supplies of its food are scarce.

▲ Figure 44 These white spruce trees (*Picea glauca*) are growing poorly under the canopy of an Eastern black walnut tree

Allelopathy in *Ailanthus altissima*

The tree of heaven (*Ailanthus altissima*) is native to China. It has become an invasive species across North America. It releases an allelopathic chemical ailanthone, which also can be extracted from root bark and stem bark. Similarly, the Eastern black walnut tree (*Juglans nigra*) releases the chemical juglone.

Application of techniques: Assaying for allelopathy

The tree of heaven (*Ailanthus altissima*) releases ailanthone, an allelopathic chemical that can be extracted from root bark and stem bark. Eastern black walnut trees (*Juglans nigra*) release the chemical juglone. Sunflowers release a broad range of allelochemicals.

Follow these steps to test for allelopathic activity.

1. Choose a plant to test for allelopathic activity. You can test different plant parts such as bark, stems, leaves, roots, seeds or seed hulls.
2. Break down the plant material into small pieces.
3. Place the pieces in a mortar with a small quantity of sand and 25 cm^3 of water. Grind this material with a pestle until it is pulverized.
4. Filter the solution using filter paper and funnel it into a beaker or cup. Check for the possibility of allergic or other skin reactions before you do this.
5. Take ten 9 cm petri dishes and place a disc of filter paper in the bottom of each one. In five dishes, moisten the filter paper with the allelopathic plant extract; in the other five, moisten the paper with distilled water (control).
6. Obtain 100 radish or mustard seeds. Place 10 seeds in each petri dish, on the filter paper. Leave the dishes at room temperature in darkness.
7. Determine germination and root length after 0.75 (18 hours), 2, 4 and 8 days.

Linking questions

1. What are the benefits of models in studying biology?
    a. Construct a diagram of a food web using species from your local environment. (C4.2.4)
    b. Describe the fluid mosaic model of membrane structure. (B2.1.10)
    c. Outline an example of an experimental setup that is meant to model a biological system. (D4.2.4)
2. What factors can limit capacity in biological systems?
    a. Outline one example of sexual selection. (D4.1.7)
    b. Distinguish between top-down and bottom-up limiting factors. (C4.1.17)
    c. Explain the relationship between substrate concentration and reaction rate. (C1.1.8, C1.1.15)

C4.2 Transfers of energy and matter

What is the reason matter can be recycled in ecosystems but energy cannot?

Why is the role of saprotophs and detritus feeders so critical to nutrient cycling? In what ways does human activity interfere with nutrient cycles such as the carbon cycle?

The thermographic image of a polar bear (*Ursus maritimus*) shows areas of heat loss. Where does the heat come from that the bear is losing? What fraction of the energy consumed by one trophic level is available to the next level? Why is a continual supply of energy necessary for ecosystems to function?

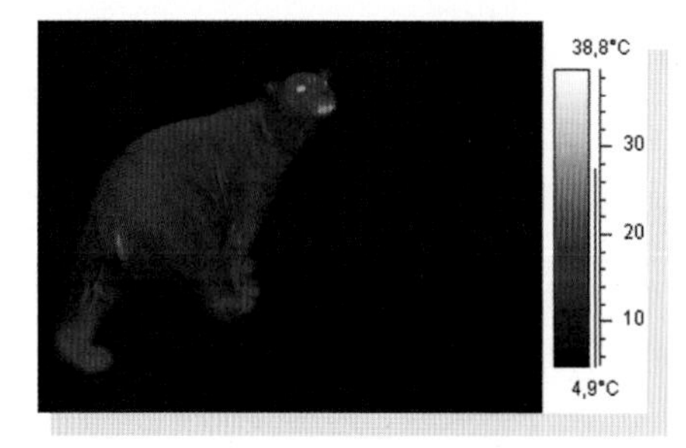

▲ Figure 1 A thermogram of a polar bear

How is the energy that is lost by each group of organisms in an ecosystem replaced?

In a famous experiment, Jan van Helmont grew a willow tree seedling and measured the amount of soil, the initial weight of the tree and the water he added. After five years the plant had gained about 74 kg. Since the amount of soil had only decreased by 57 g, he deduced that the tree's mass gain had come entirely from water. Is this the correct conclusion? Where does it come from precisely? How could you prove this experimentally?

◀ Figure 2 Life in the biosphere depends on the production of biomass by producers like the willow tree (*Salix alba*)

SL and HL

C4.2.1 Ecosystems as open systems in which both energy and matter can enter and exit
C4.2.2 Sunlight as the principal source of energy that sustains most ecosystems
C4.2.3 Flow of chemical energy through food chains
C4.2.4 Construction of food chains and food webs to represent feeding relationships in a community
C4.2.5 Supply of energy to decomposers as carbon compounds in organic matter coming from dead organisms
C4.2.6 Autotrophs as organisms that use external energy sources to synthesize carbon compounds from simple inorganic substances
C4.2.7 Use of light as the external energy source in photoautotrophs and oxidation reactions as the energy source in chemoautotrophs
C4.2.8 Heterotrophs as organisms that use carbon compounds obtained from other organisms to synthesize the carbon compounds that they require
C4.2.9 Release of energy in both autotrophs and heterotrophs by oxidation of carbon compounds in cell respiration
C4.2.10 Classification of organisms into trophic levels
C4.2.11 Construction of energy pyramids
C4.2.12 Reductions in energy availability at each successive stage in food chains due to large energy losses between trophic levels
C4.2.13 Heat loss to the environment in both autotrophs and heterotrophs due to conversion of chemical energy to heat in cell respiration
C4.2.14 Restrictions on the number of trophic levels in ecosystems due to energy losses
C4.2.15 Primary production as accumulation of carbon compounds in biomass by autotrophs
C4.2.16 Secondary production as accumulation of carbon compounds in biomass by heterotrophs
C4.2.17 Constructing carbon cycle diagrams
C4.2.18 Ecosystems as carbon sinks and carbon sources
C4.2.19 Release of carbon dioxide into the atmosphere during combustion of biomass, peat, coal, oil and natural gas
C4.2.20 Analysis of the Keeling Curve in terms of photosynthesis, respiration and combustion
C4.2.21 Dependence of aerobic respiration on atmospheric oxygen produced by photosynthesis, and of photosynthesis on atmospheric carbon dioxide produced by respiration
C4.2.22 Recycling of all chemical elements required by living organisms in ecosystems

C4.2.1 Ecosystems as open systems in which both energy and matter can enter and exit

Living organisms cannot live alone. They depend on interactions with other organisms for supplies of energy and chemical resources. They also depend on their abiotic surroundings of air, water, soil and rock. Biologists have developed the concept of ecosystems—ecological systems such as a lake or a forest. An ecosystem is composed of all the organisms in an area together with their abiotic environment.

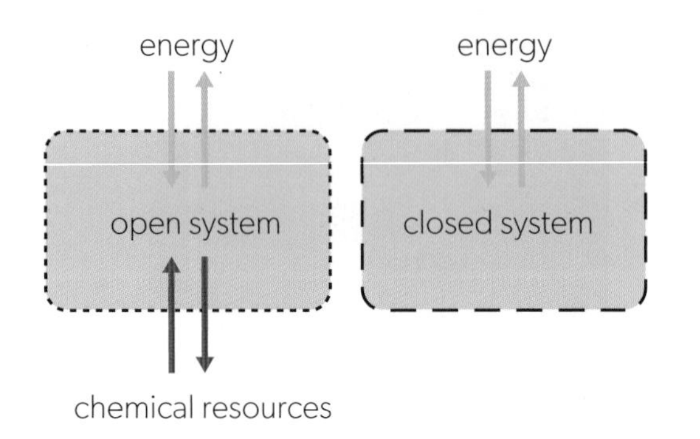

▲ Figure 3 Open and closed systems

Systems are an important concept in biology. A system is a set of interacting or interdependent components. There are two main types of system:

- open systems where resources can enter or exit, including both chemical substances and energy
- closed systems where energy can enter or exit, but chemical resources cannot be removed or replaced.

C4.2.2 Sunlight as the principal source of energy that sustains most ecosystems

The initial source of energy for most ecosystems is sunlight. Living organisms harvest this energy by photosynthesis. Three groups of organisms carry out photosynthesis: cyanobacteria, plants and eukaryotic algae including the seaweeds that grow on rocky shores. These organisms are known as producers. The energy fixed by producers in carbon compounds is available to other organisms that do not photosynthesize.

▲ Figure 4 Sunlight supplies energy to a forest ecosystem. Much of the light is absorbed by leaves in the canopy, so the forest floor is shady

The amount of energy reaching the Earth's surface in sunlight varies around the world. Also, the percentage of this light that is harvested by producers and therefore available to other organisms is greater in some ecosystems than others. For example, the intensity of sunlight is very high in the Sahara Desert but little energy is harvested because there are few producers. In redwood forests of northern California, the intensity of sunlight is far lower but much more energy becomes available to the ecosystem because producers are abundant and photosynthesis rates are high.

In marine and freshwater ecosystems, light must pass through water to reach producers. We think of water as transparent but transmission is not 100%. Photosynthesis uses light with wavelengths from 400 nm (violet) to 700 nm (red). Shorter wavelengths penetrate further in pure water, which is why the sea often appears blue. Water in marine and freshwater ecosystems contains living organisms and non-living matter, which further reduce light penetration. In open oceans there is little or no light at depths greater than 200 m. Coastal waters are often turbid (cloudy) due to suspended clay or silt and dense populations of phytoplankton, so there is little light below 50 m. Deeper ecosystems must therefore rely on other sources of energy.

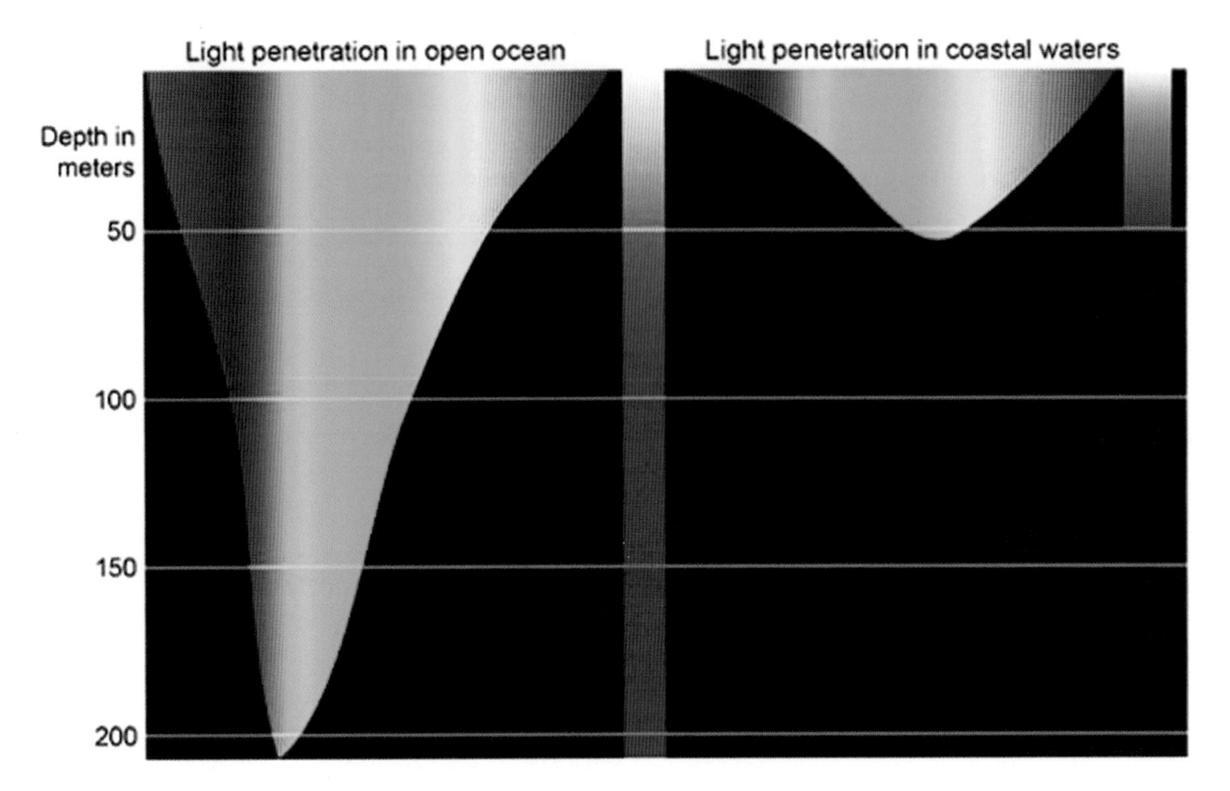

▲ Figure 5 Seawater is dark below about 200 m in open ocean and 50 m in coastal water

Ecosystems have also developed in the darkness of caves. Streams entering a cave may bring dead organic matter which provides a supply of energy; for example, dead leaves contain energy produced by photosynthesis in ecosystems outside the cave. However, some caves are isolated and do not receive inputs of energy from outside ecosystems. There are two such areas in Movile Cave, located near the Black Sea coast in Romania.

The producers in sealed caves are archaebacteria. They gain energy from chemical reactions that have methane, sulfides or other inorganic compounds as substrates. Energy from these reactions is used to synthesize carbon compounds in a type of metabolism called chemosynthesis. Microscopic invertebrates feed on biofilms of chemosynthetic archaebacteria, with other invertebrates feeding on them.

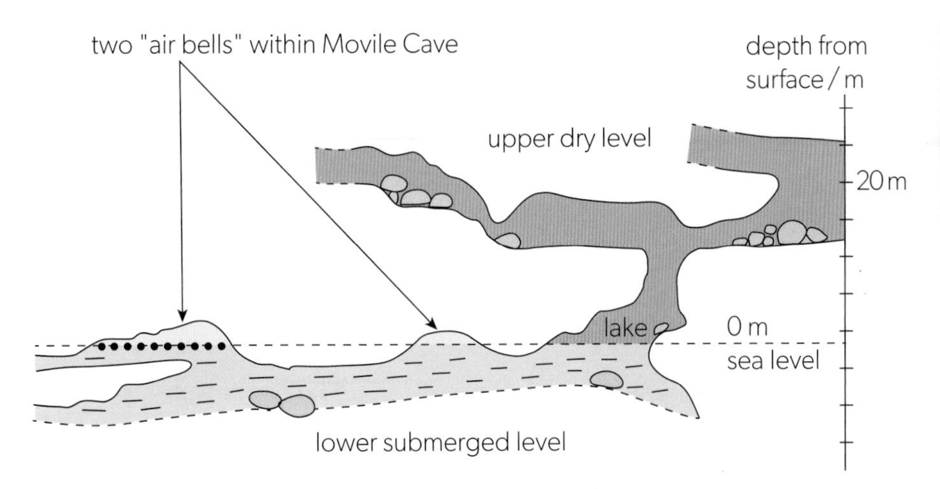

▲ Figure 6 The air bells shown in this vertical section of Movile Cave contain ecosystems based entirely on chemosynthetic producers. Are these systems closed or open?

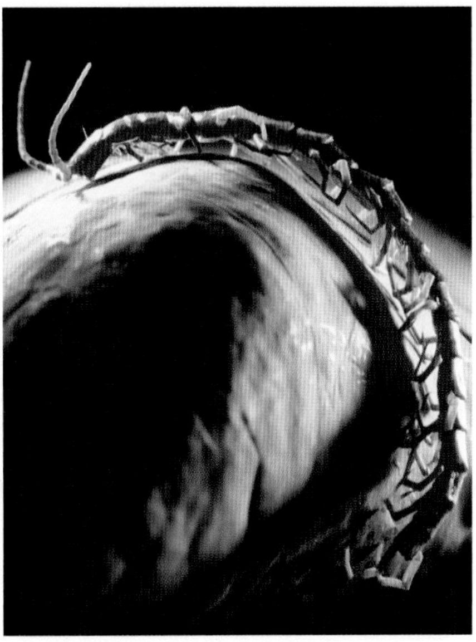

▲ Figure 7 The cave centipede (*Criptos anomalans*) is a predatory myriapod found only in Movile Cave, where life has been cut off from the outside world for the past 5.5 million years. During that time in total darkness, its eyes have been lost and oversized antennae have evolved which are used to hunt for food

Laws: The first and second laws of thermodynamics

Laws in science are generalized principles, or "rules of thumb" formulated to describe patterns observed in living organisms. They are different from theories because they do not offer explanations; instead, they simply describe phenomena. Like theories, they can be used to make predictions.

An example is the first law of thermodynamics: energy can be transferred and transformed, but it cannot be created or destroyed. Plants transform light energy into chemical energy through photosynthesis, but they do not create the energy.

Another example is the second law of thermodynamics: every energy transformation increases the entropy, or disorder, of the universe. In simpler terms, ordered forms of energy such as the chemical potential energy stored in the tissues of prey species are at least partially converted to unusable heat energy in food chains.

Hypotheses, theories and laws are all generalizations based on repeated observations of the same phenomenon. The higher the number of observations supporting a generalization, the more supported it is.

C4.2.3 Flow of chemical energy through food chains

A food chain is a sequence of organisms, each of which feeds on the previous one. There are usually between two and five organisms in a food chain. Producers are the first organisms in a food chain. The subsequent organisms are consumers.

- Most producers absorb sunlight using chlorophyll and other photosynthetic pigments. The light energy is converted to chemical energy, which is used to make carbohydrates, lipids and all the other carbon compounds that are required.
- Consumers obtain energy from the carbon compounds in the organisms on which they feed. Primary consumers feed on producers; secondary consumers feed on primary consumers; tertiary consumers feed on secondary consumers, and so on. No consumers feed on the last organism in a food chain.

The arrows in a food chain indicate the direction of energy flow.

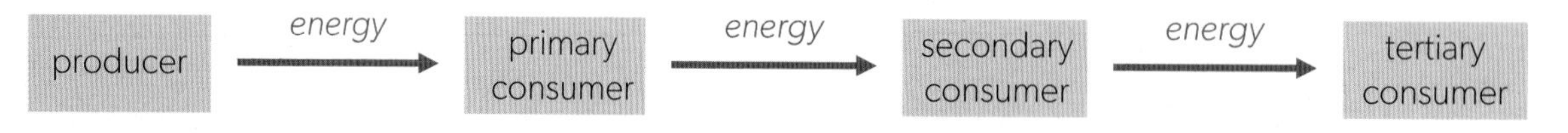

In an ecosystem, there are many specific food chains that provide organisms with a supply of energy. In the Monte Desert, which is on the foothills of the Andes in South America, leaves of a shrub with the local name of tara (*Senna arnottiana*) are eaten by guanaco (*Lama guanicoe*). Pumas (*Puma concolor*) are predators of the guanaco. They are regarded as apex predators because nothing kills or eats them, though fleas and other parasites can obtain energy from them by feeding on their blood!

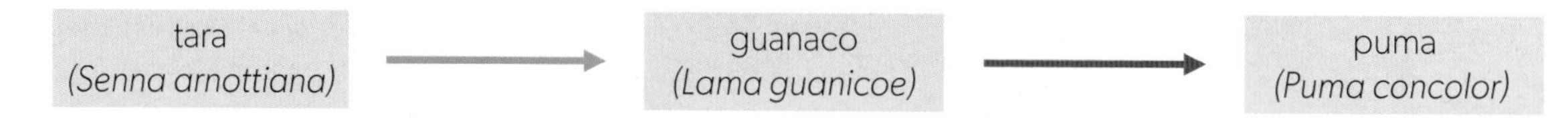

▲ Figure 8 A food chain in the Monte Desert, South America

C4.2.4 Construction of food chains and food webs to represent feeding relationships in a community

Trophic relationships within ecological communities tend to be complex and web-like. This is because many consumers feed on more than one species and are fed upon by more than one species. A food web is a model that summarizes all of the possible food chains in a community. Figure 9 shows a food web for dry thorn scrub in Argentina, showing the known feeding relationships.

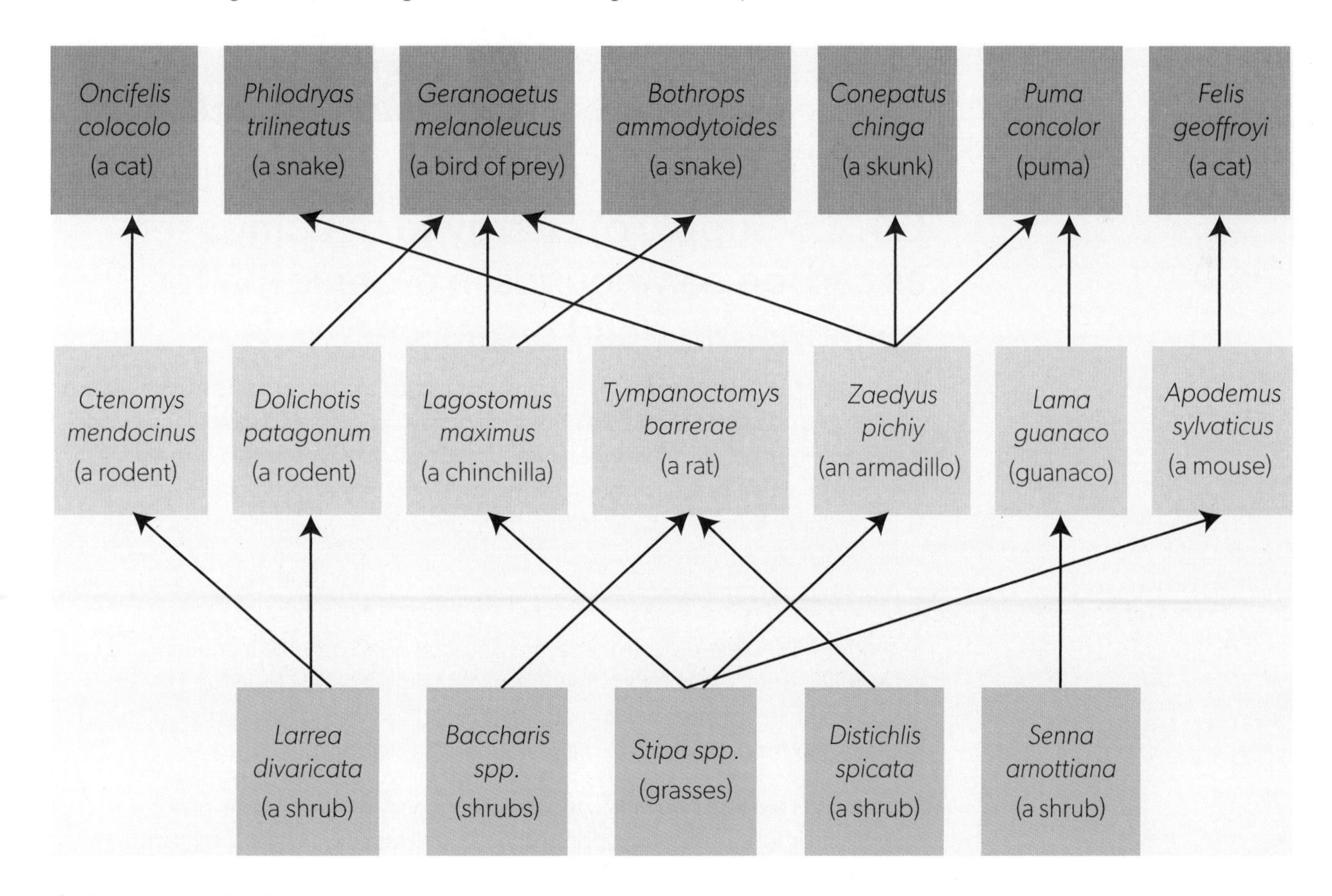

▲ Figure 9 A food web

	Species	Feeds on
1	caribou	4
2	ground squirrels	4
3	skua (jaeger)	1,4,8
4	grasses and sedges	-
5	grizzly bears	4,2
6	gulls	8
7	owls and hawks	2,8
8	voles and lemmings	4
9	weasels	2,8
10	wolves	1,2,8

▲ Table 1

When a food web is constructed, organisms at the same trophic level are often shown at the same level in the web. However, this is not always possible because some organisms feed at more than one trophic level. Table 1 shows feeding relationships for a taiga ecosystem in northern Canada. Use the data in Table 1 to construct a food web.

Application of techniques: Collecting invertebrates

Before you can construct a food web, you need to identify the organisms in an ecosystem. To do this, you can:

- use a sweep net to capture flying insects
- use a beating stick to firmly tap vegetation or foliage to dislodge insects and other organisms, while holding a beating sheet or beating net below the vegetation to catch the organisms that fall.

You can then use a smartphone app (such as *Insect This, Animal This or iNaturalist*) to identify the organisms and find out about their feeding habits. This information will allow you to construct a food web.

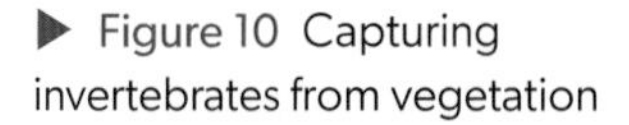

▶ Figure 10 Capturing invertebrates from vegetation

C4.2.5 Supply of energy to decomposers as carbon compounds in organic matter coming from dead organisms

The body of an organism that has died is available to other organisms as a source of energy. Body parts shed by plants and animals are also available—for example, the exoskeleton of an insect when it moults, or fallen leaves from a deciduous tree. Material that passes through animal guts undigested is egested as faeces. This is another form of organic matter that can provide energy. Some of these energy sources are eaten by animals such as earthworms and vultures. Large amounts are decomposed by saprotrophs.

Saprotrophs secrete digestive enzymes into the dead organic matter and digest it externally. They then absorb the products of digestion, including sugars and amino acids. In contrast, consumers ingest food first and digest it internally. Bacteria and fungi are the two main groups of saprotroph.

Saprotrophs are also known as decomposers, because they break down complex insoluble carbon compounds into simpler soluble ones. By doing this, they cause the gradual breakdown of solid structures. For example, a tree trunk on the forest floor will gradually soften and crumble away and fallen leaves disappear. Without the action of decomposers, dead organic matter would build up year by year and the chemical elements in it would not be recycled. Decomposers are the waste disposers and recyclers of ecosystems.

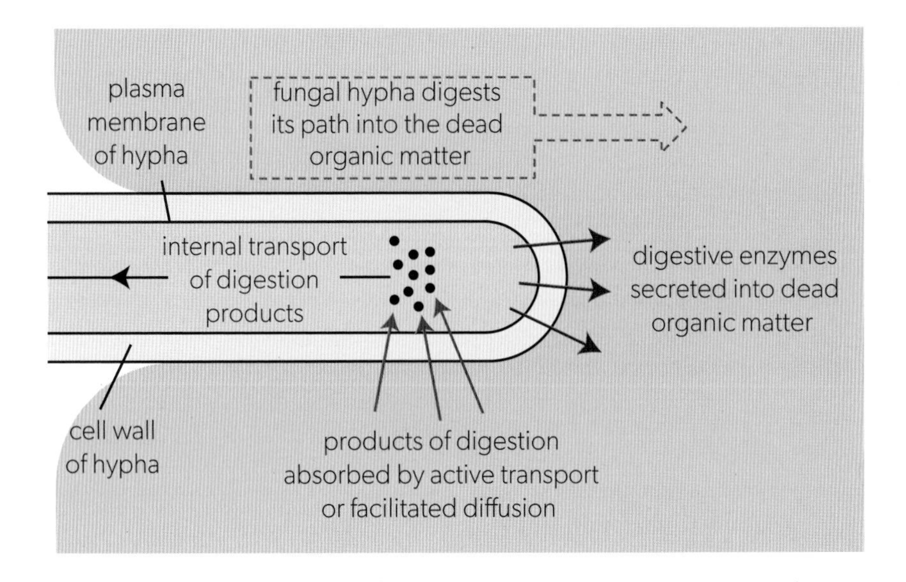

◀ Figure 11 Saprotrophic digestion by a growing fungal hypha

C4.2.6 Autotrophs as organisms that use external energy sources to synthesize carbon compounds from simple inorganic substances

All organisms have nutritional requirements. They need a variety of carbon compounds:

- amino acids for protein synthesis
- sugars for energy supply and synthesis of polysaccharides
- fatty acids for energy supply and for constructing membranes
- organic bases for synthesizing nucleic acids during DNA replication and transcription
- steroids and many other groups of carbon compounds.

Some organisms can make all of these carbon compounds themselves, using carbon dioxide (CO_2) or hydrogen carbonate (HCO_3^-) as a carbon source and nitrate, phosphate and other simple inorganic substances as sources of other elements. Organisms that do this are called autotrophs, meaning self-feeding. To carry out the anabolic reactions that build carbon compounds from simple inorganic substances, autotrophs need an external energy source. There are two possible sources of external energy: light and chemical reactions.

▲ Figure 12 Vegetables at the Chelsea Flower Show display some of the diversity produced by plants using only light energy, carbon dioxide and a few other inorganic substances

C4.2.7 Use of light as the external energy source in photoautotrophs and oxidation reactions as the energy source in chemoautotrophs

Autotrophs need an external energy source to create carbon compounds. This is because carbon in carbon dioxide or hydrogen carbonate ions is in an oxidized state and the initial reduction reactions of the Calvin cycle are endothermic. Simple carbon compounds are then linked together into larger more complex compounds by condensation reactions which are also endothermic.

Autotrophs are subdivided according to the external energy source they use:

- Photoautotrophs use light.
- Chemoautotrophs use exothermic inorganic chemical reactions.

Sunlight is used as an energy source by organisms that perform photosynthesis. Fusion reactions in the Sun generate vast amounts of energy in the form of electromagnetic radiation. A very small proportion of energy emitted by the Sun reaches the Earth, and a small proportion of that is absorbed by photosynthesis. Even so, sunlight is the main energy supply for most ecosystems. Plants, eukaryotic algae and cyanobacteria are photoautotrophs.

Inorganic chemical reactions are used as an energy source by a variety of prokaryotes, both bacteria and archaebacteria. A substrate in a reduced state—such as sulfur, hydrogen sulfide, iron, hydrogen or ammonia—is oxidized, releasing energy which is used to synthesize carbon compounds. Such organisms are autotrophic because they make their own sugars, amino acids and other carbon compounds.

Iron-oxidizing bacteria

Iron-oxidizing bacteria are chemoautotrophs. Iron sulfide (FeS_2) commonly occurs in sedimentary rocks formed in low-oxygen environments. When these rocks are exposed to air, due to mining activities or natural erosion processes, the iron sulfide reacts to produce Fe^{2+} ions, sulfate ions and sulfuric acid.

$$FeS_2 + 3\tfrac{1}{2}O_2 + H_2O \rightarrow Fe^{2+} + SO_4^{2-} + H_2SO_4$$

Iron-oxidizing bacteria then remove electrons from the Fe^{2+} ions, converting them to Fe^{3+}. The electrons carry energy released by the oxidation of iron. They are accepted by electron carriers in the plasma membrane of the iron-oxidizing bacteria. Some of the electrons are used to reduce NAD.

$$2Fe^{2+} + NAD + 2H^+ \rightarrow 2Fe^{3+} + \text{reduced NAD}$$

Some electrons are used to provide energy for proton pumping, to increase the proton gradient across the plasma membrane formed when acid is produced from iron sulfide. This proton gradient allows ATP production by chemiosmosis. These electrons are accepted by oxygen, together with hydrogen ions, to produce water.

$$2Fe^{2+} + \tfrac{1}{2}O_2 + 2H^+ \rightarrow 2Fe^{3+} + H_2O$$

Reduced NAD and ATP are used to fix carbon dioxide and produce carbon compounds by means of the Calvin cycle. *Acidithiobacillus ferrooxidans* is an example of an iron-oxidizing bacterium. It is adapted to a highly acidic environment because the initial reaction produces sulfuric acid.

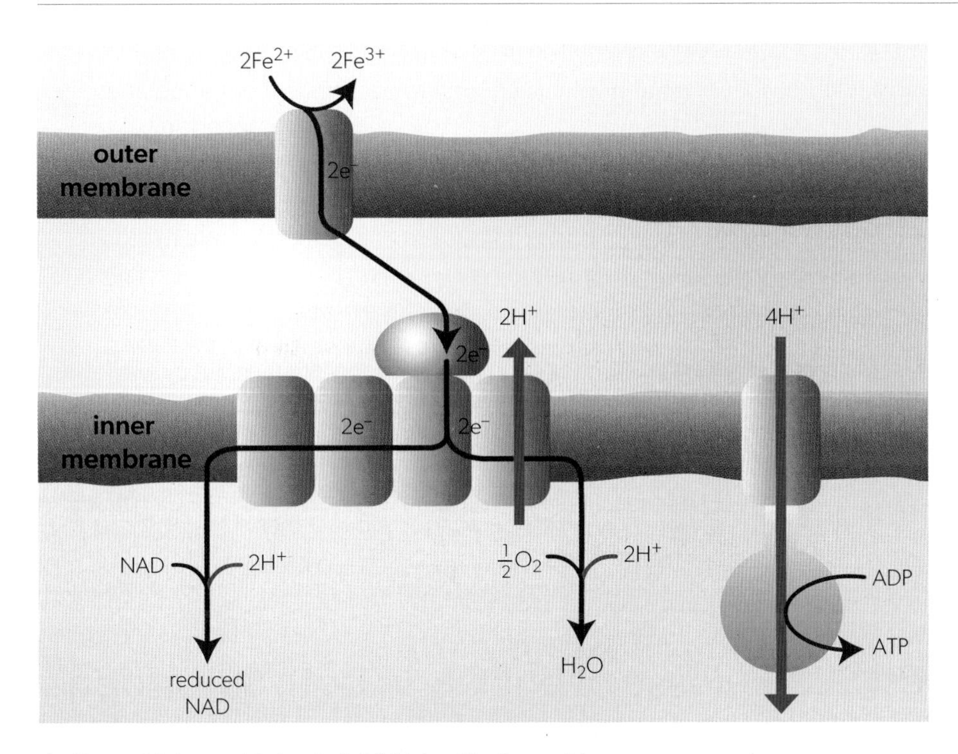

▲ Figure 13 Iron oxidation in *Acidithiobacillus ferrooxidans*

C4.2.8 Heterotrophs as organisms that use carbon compounds obtained from other organisms to synthesize the carbon compounds that they require

Many organisms fulfil their nutritional requirements by obtaining carbon compounds from other organisms. These organisms are heterotrophic, which means they feed on others. They digest carbon compounds that were part of another organism, then use the products of digestion to build the large complex carbon compounds they need. For example, guanacos digest proteins in the leaves of tara bushes, breaking them down into amino acids. Then they use these amino acids to synthesize the proteins they (the guanacos) need. The process of absorbing carbon compounds and making them part of the body is called assimilation.

◀ Figure 14 Spiders have narrow guts and they do not ingest large lumps of food. Most species inject venom and digestive enzymes into their prey to kill them and liquify their bodies. After a time, they ingest the liquid and complete the process of digestion inside their gut. The chelicerae (poison fangs) used to inject the enzymes are particularly prominent in this burgundy goliath bird-eating spider (*Theraphosa stirmi*)

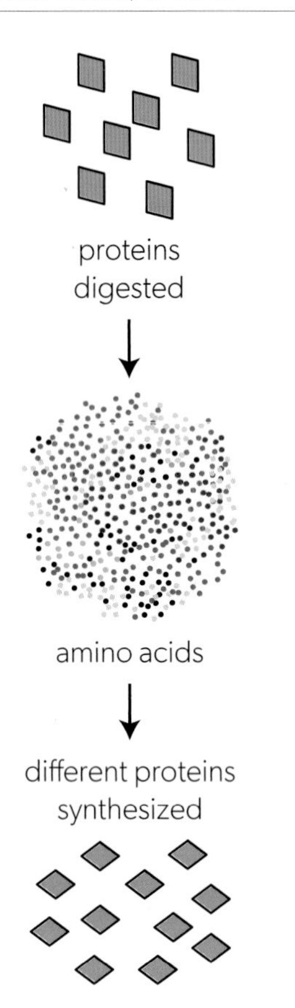

▲ Figure 15 Heterotrophs digest polymers from other organisms to produce monomers. Then they use these monomers to build their own polymers

Assimilation requires absorption of carbon compounds into cells, so the molecules must be small and soluble enough to pass across cell membranes. Proteins, polysaccharides, nucleic acids and other large compounds must be digested before they can be absorbed. Heterotrophs are subdivided according to whether they digest food internally or externally.

- Saprotrophs grow into or across the surface of food and secrete hydrolytic enzymes to digest the food externally.
- Consumers ingest their food.
- Multicellular consumers such as guanacos take food into their gut by swallowing it. Then they mix the food with enzymes from digestive glands. This is regarded as internal digestion although the food has not yet entered any cells.
- Unicellular consumers such as *Paramecium* take the food into their cells by endocytosis, then digest it inside phagocytic vacuoles. They absorb products of digestion from the vacuoles into their cytoplasm.

C4.2.9 Release of energy in both autotrophs and heterotrophs by oxidation of carbon compounds in cell respiration

All organisms require supplies of energy in the form of ATP in their cells. They use this energy for vital activities including:

- synthesizing large molecules such as DNA, RNA and proteins
- pumping molecules or ions across membranes by active transport
- moving structures within cells, for example, chromosomes or vesicles; in muscle cells, arrays of protein fibres are moved to cause muscle contraction
- maintaining constant body temperature (in birds and mammals).

In both autotrophs and heterotrophs, ATP is produced by cell respiration. Carbon compounds such as carbohydrates and lipids are oxidized to release energy and this energy is used to phosphorylate ADP, producing ATP.

C4.2.10 Classification of organisms into trophic levels

Ecologists classify organisms into groups according to how they obtain energy and carbon compounds and therefore where they are positioned in food chains. These groups are called trophic levels.

Most consumers have a varied diet and can occupy different trophic levels in different food chains. For example, red foxes (*Vulpes vulpes*) mostly eat primary consumers, making them secondary consumers. However, they also eat some leaves, roots, tubers, fruits and seeds, making them primary consumers.

▲ Figure 16

Tertiary consumers eat secondary consumers so they are fourth in food chains.

Secondary consumers eat primary consumers so they are third in food chains.

Primary consumers (herbivores) feed on producers so they are second in food chains.

Producers are autotrophic. They make their own carbon compounds using external energy sources so they are at the start of food chains.

▲ Figure 17 Trophic levels

C4.2.11 Mathematics: Construction of energy pyramids

A pyramid of energy can be used to show the amount of energy gained per year by each trophic level in an ecosystem. This is a type of bar chart with a horizontal bar for each trophic level.

- Amounts of energy are measured per unit area and per year. The units are often kilojoules per metre squared per year (kJ $m^{-2} yr^{-1}$).
- Pyramids of energy should be stepped, not triangular, with producers in the lowest bar.
- Bars should be labelled "producer", "primary consumer", "secondary consumer" and so on.
- If a suitable scale is used, the length of each bar can be proportional to the amount of energy that it shows.

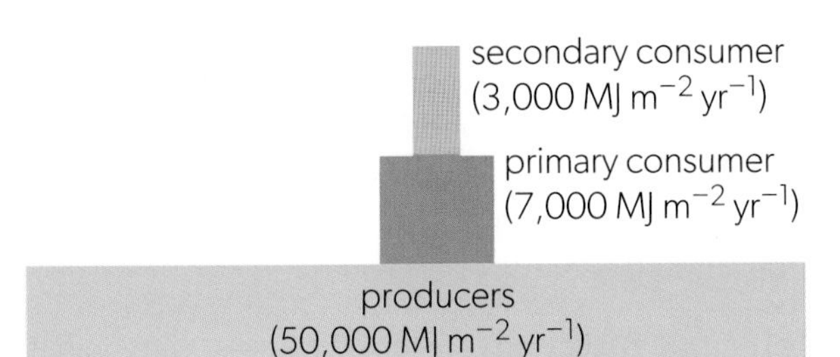

▲ Figure 18 A pyramid of energy for grassland, with the bars to scale

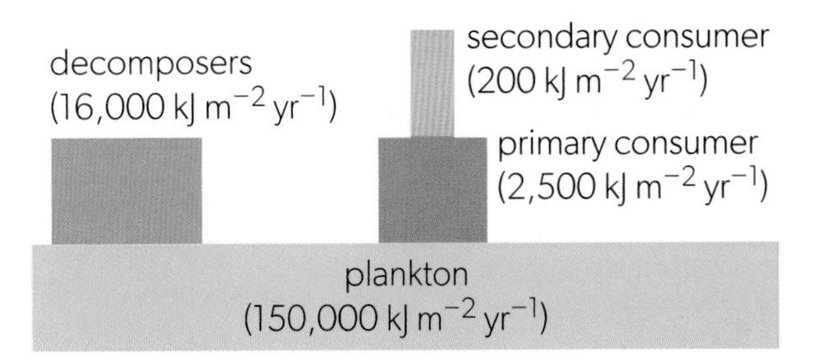

▲ Figure 19 A pyramid of energy for an aquatic ecosystem, with the bars not to scale. Choose a suitable scale and redraw this pyramid

C4.2.12 Reductions in energy availability at each successive stage in food chains due to large energy losses between trophic levels

In any ecosystem, there are large energy losses between trophic levels. As a result, there is less energy available to each successive trophic level. There are three main forms of energy loss.

1. Incomplete consumption

The organisms in a trophic level are not usually entirely consumed by organisms in the next trophic level. For example, locusts sometimes consume all the plants in an area but more usually, they only eat parts of some plants. Similarly, predators do not usually eat bones or hair from the bodies of their prey. The energy in

▲ Figure 20 Dried cow dung (faeces) is burned as part of a Hindu festival. The flames show that the faeces contains energy. This energy comes from chemical substances that the cow (primary consumer) obtained from plants (producers). These chemical substances were not digested by the cow so the energy they contain is not available to secondary consumers

dead organisms, or dead parts of organisms, passes to saprotrophs such as fungi or detritus feeders such as earthworms rather than passing to organisms in the next trophic level. This energy is lost from the food chain.

2. Incomplete digestion

Not all parts of food ingested by organisms are digested and absorbed. For example, some animals do not digest tough fibrous plant matter containing cellulose. Indigestible material is egested in faeces. Energy in faeces does not pass on along the food chain. Instead, it passes to saprotrophs or detritus feeders.

3. Cell respiration

Carbohydrates, proteins and other energy-containing substances are used as substrates in cell respiration. These substrates are oxidized to carbon dioxide and water, releasing energy. The carbon dioxide and water are waste products that cannot supply energy to the next trophic level. Only carbon compounds (foods) that are not oxidized in respiration can be passed to organisms in the next trophic level.

A smaller amount of energy flows to each successive trophic level because there are smaller amounts of carbohydrates, proteins and other energy-containing substances. It is not because these substances contain less energy per gram in higher trophic levels—they have the same energy content in every trophic level. If anything, higher trophic levels contain more energy per unit of biomass, not less. This becomes obvious if you consider the abundance of biomass in each trophic level: producers are most abundant, followed by the herbivores that eat them and then the carnivores that eat the herbivores. The relatively smaller biomass of carnivores contains a large amount of energy per gram.

It is often said that 90% of energy is lost between trophic levels, with only 10% passed on. However, there is much variation between food chains.

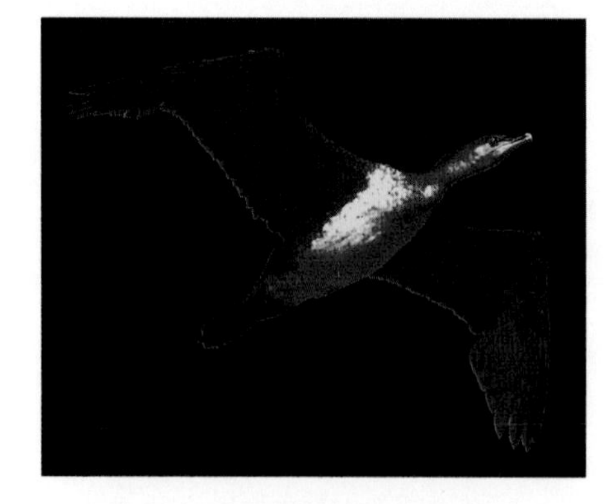

▲ Figure 21 Thermal image of a great cormorant (*Phalacrocorax carbo*) in flight. Can you explain the relatively low temperature of the wings?

C4.2.13 Heat loss to the environment in both autotrophs and heterotrophs due to conversion of chemical energy to heat in cell respiration

Energy can be converted from one form to another. Energy enters ecosystems in the form of sunlight. It is transformed into chemical energy by photosynthesis. Chemical energy flows along food chains and to decomposers. Ultimately all of this energy is transformed into heat.

All organisms—both autotrophs and heterotrophs—convert some of their chemical energy into heat. Birds and mammals sometimes increase their rate of heat generation to maintain a constant body temperature. In other organisms, heat is generated as an inevitable side-effect of activity. Cell respiration is a major source of heat in living organisms. The second law of thermodynamics states that energy transformations are never 100% efficient. This means that not all of the energy from the oxidation of carbon compounds in cell respiration is transferred to ATP. The remainder is converted to heat. Chemical energy is also converted to heat when ATP is used in cell activities. For example, muscles warm up when they contract.

Data-based questions: Finding the energy content of biomass

To study the transfer of energy in ecosystems, it is necessary to measure the amount of energy in biomass. The energy content of biomass can be determined by combustion. A known mass is burned and the heat released is used to raise the temperature of a known volume of water. The quantity of energy can be calculated from the temperature rise of the water, the amount of biomass burned and the specific heat capacity of water. The specific heat capacity of water is $4.2\,J\,g^{-1}\,°C^{-1}$, which means that 4.2 J of heat energy is needed to raise the temperature of 1 gram (or cubic centimetre) of water by 1°C.

$$\text{energy content of biomass } (J\,g^{-1}) = \frac{\text{mass of water (g)} \times \text{temperature rise of water (°C)} \times 4.2}{\text{mass of biomass (g)}}$$

1. Use the experimental results in Table 2 to calculate the energy content per gram of the cashew nut, in kilojoules. Show your working. [4]

Mass of cashew nut / g	Volume of water / cm^3	Temperature of water / °C	
		initial	final
1.6	10	20	100

▲ Table 2

2. The label on a packet of cashew nuts states that 100 g of the nuts contains 2,320 kJ.

   a. State the difference between this value for the energy content and the experimental results. [2]

   b. Suggest reasons for this difference. [4]

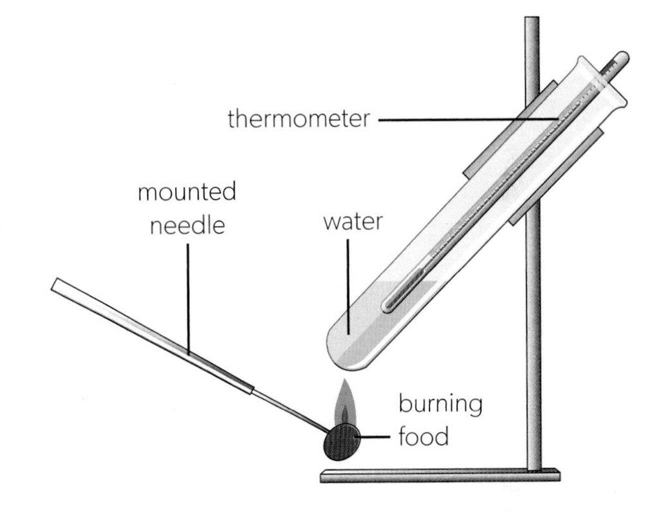

▲ **Figure 22** Simple apparatus can be used to measure the energy content of a cashew nut

Body warmth helps organisms to remain active. However, this heat is ultimately lost to the abiotic environment according to the laws of thermodynamics: heat passes from hotter to cooler bodies. This energy cannot be converted back into chemical energy or used by other trophic levels or any other organisms. For this reason, energy flows through ecosystems and cannot be recycled (unlike chemical elements). The heat may remain in an ecosystem for a while but it will eventually radiate out through the atmosphere and on into space.

C4.2.14 Restrictions on the number of trophic levels in ecosystems due to energy losses

The number of stages in a food chain varies. A favourite food of elephants in East Africa is the tree *Senegalia mellifera*. Elephants are too large to be vulnerable to predation, resulting in a two-stage food chain:

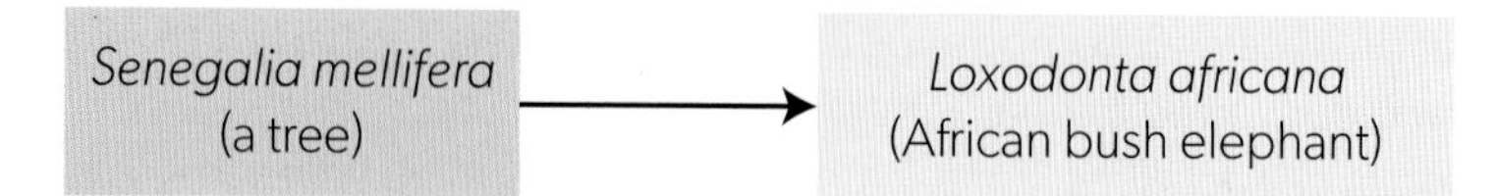

Bluefin tuna in the North Atlantic feed on species that are mostly secondary consumers or tertiary consumers, as shown in this five-stage food chain:

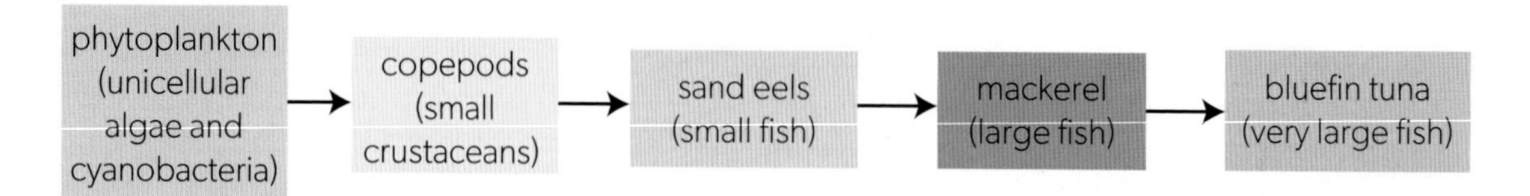

Bluefin tuna are occasionally eaten by orca (killer whales) but far more of an orca's diet comes from lower trophic levels. Orcas are apex predators—they are not eaten by other predators so they come at the end of food chains. Energy in the food eaten by apex predators has rarely passed through four or five previous trophic levels. Most of the energy has passed through only two or three levels.

We might expect food chains to be limitless, with one species being eaten by another and so on for ever. However, this does not happen. This is because so much energy is lost at each step in a food chain. There is less energy available to each successive trophic level so after only a few stages, there is not enough energy to support another trophic level. For this reason the number of trophic levels in food chains is restricted.

It is important to understand that animals in higher trophic levels do not have to eat more food to gain enough energy. Their prey contains large amounts of energy per unit mass—there just is not much prey available. A peregrine falcon for example is mainly a tertiary consumer and may need a territory of more than 100 km^2 to find enough to eat.

C4.2.15 Primary production as accumulation of carbon compounds in biomass by autotrophs

Production in ecosystems is the accumulation of carbon compounds in biomass. Biomass accumulates when living organisms grow. Reproduction can increase the number of growing organisms and thus contribute to production.

Both autotrophs and heterotrophs produce biomass by growth and reproduction. Plants and other autotrophs are primary producers because they synthesize carbon compounds from carbon dioxide and other simple substances.

- Gross primary production (GPP) is the total biomass of carbon compounds made in plants by photosynthesis.
- Net primary production is GPP minus the biomass lost due to respiration of the plant. This is the amount of biomass available to consumers.

Both GPP and NPP are generally measured over long time intervals (e.g. a year) at the ecosystem level. The units of measurement are usually grams of carbon accumulated per square metre of ecosystem per year ($g\,C\,m^{-2}\,yr^{-1}$). There are other measures, such as $g\,C\,m^{-2}\,hour^{-1}$ or tonnes $C\,hectare^{-1}\,year^{-1}$.

Biomes vary in their capacity to accumulate biomass, depending mainly on rates of photosynthesis. Satellite imaging can be used to monitor primary production.

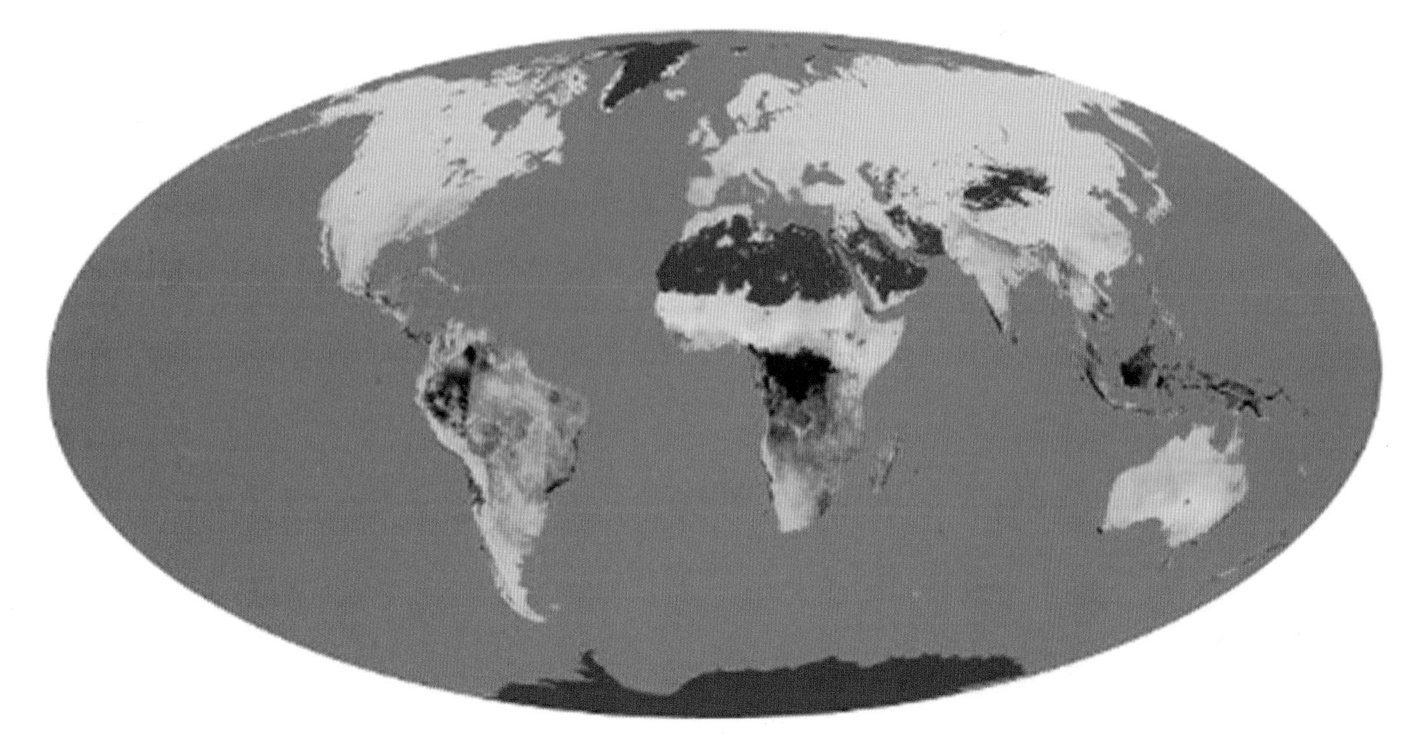

▲ Figure 23 Primary production in July 2013. This and other maps produced by NASA are available on the internet and can be analysed to find which terrestrial biomes accumulate biomass most rapidly

Data-based questions: Productivity of biomes

Table 3 shows estimates for primary production rates in the Earth's major biomes, per hectare (an area of 100 × 100 m) and for the entire Earth. The estimates are based on remotely sensed satellite data. One petagram (Pg) = 10^{15} grams.

Biomes (in decreasing order of productivity)	Primary production / g C hectare^{-1} yr^{-1}	Global primary production / Pg C yr^{-1}
Tropical forest	871–1098	13.7
Temperate forest	465–741	6.5
Boreal forest	173–238	3.2
Tropical savanna and grassland	343–393	17.7
Temperate grassland and shrubland	129–342	5.3
Desert	28–151	1.4
Tundra	80–130	1.0

▲ Table 3

1. Display the data for primary production per hectare using whatever type of chart you think is most suitable. [4]
2. Suggest reasons for the differences in primary production between the biomes. [4]
3. Display the data for global primary production using a pie chart. [4]
4. Discuss the conclusions that should be drawn from the pie chart. [3]

▼ Table 4

Product	Yield / tonnes ha^{-1} $year^{-1}$
Wheat	8.7
Beef	0.2

▲ Figure 24 Sharing plant-based (vegan) foods

C4.2.16 Secondary production as accumulation of carbon compounds in biomass by heterotrophs

Secondary production is accumulation of carbon compounds in biomass by animals and other heterotrophs. Carbon compounds such as sugars and amino acids are ingested from food and then built up into proteins and other macromolecules.

Carbon compounds are used as respiratory substrates by all organisms at every trophic level. Cell respiration results in a loss of carbon compounds and therefore a loss of biomass in every trophic level. For this reason, net production is always lower than gross production and secondary production is lower per unit area than primary production in an ecosystem. Secondary production declines with each successive trophic level from primary consumers onwards.

Differences in production are reflected in farm yields. Production of crops per unit area is always much higher than production of meat and other animal products. Typical values are shown in Table 4. This means that more humans can be fed per hectare of farmland if they eat plant products rather than meat. This is one of the reasons for increased interest in plant-based diets among environmentalists.

C4.2.17 Constructing carbon cycle diagrams

Ecologists use the terms "pool" and "flux" when describing the carbon cycle and recycling of other elements.

- A pool is a reserve of the element. It can be organic or inorganic. For example, carbon dioxide in the atmosphere is an inorganic pool of carbon. The biomass of producers in an ecosystem is an organic pool.
- Flux is the transfer of the element from one pool to another.

There are three main types of carbon flux due to living organisms in ecosystems:

- Photosynthesis—absorption of carbon dioxide from air or water and its conversion to carbon compounds
- Feeding—gaining carbon compounds from other organisms
- Respiration—release to the atmosphere of carbon dioxide produced by respiring cells.

Diagrams can be used to represent the carbon cycle. Text boxes can be used for pools and labelled arrows for fluxes. Figure 25 shows an illustrated diagram for showing processes in terrestrial ecosystems, which can be converted to a diagram of text boxes and arrows.

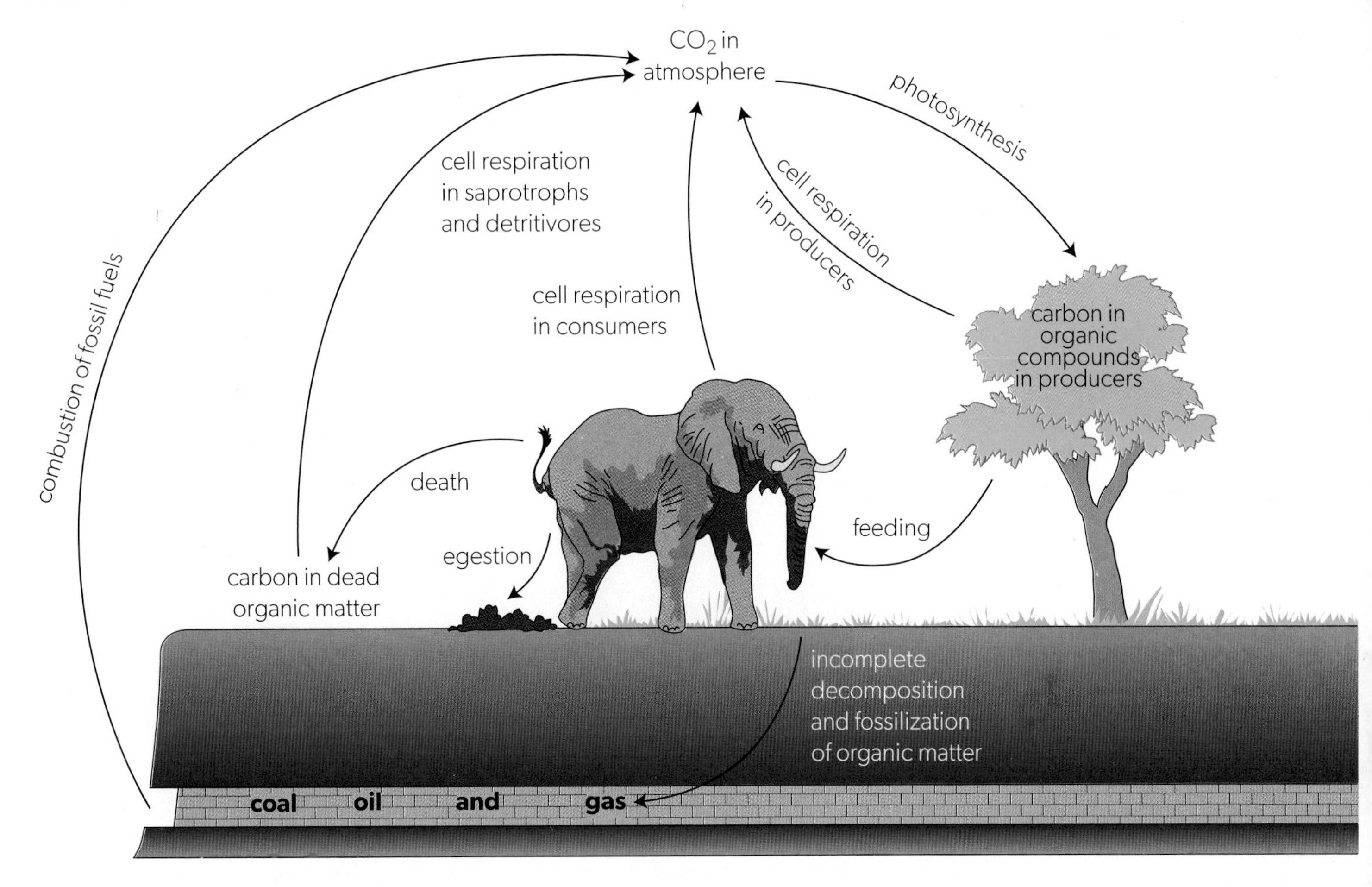

▲ Figure 25 Carbon cycle including the sequestering of carbon in fossil fuels and its release by combustion

Figure 25 only shows the carbon cycle for terrestrial ecosystems. A separate diagram could be constructed for marine or aquatic ecosystems, or a combined diagram for all ecosystems. In marine and aquatic ecosystems, the inorganic reserve of carbon is dissolved carbon dioxide and hydrogen carbonate, which is absorbed by producers and then released back into the water by various means.

C4.2.18 Ecosystems as carbon sinks and carbon sources

Ecosystems are open systems because both matter and energy can enter or exit. Carbon enters and exits mostly in the form of carbon dioxide, through the processes of photosynthesis and respiration. The rates of these two processes for an ecosystem as a whole are not necessarily equal.

- If photosynthesis exceeds respiration, there is net uptake. The ecosystem is acting as a carbon sink.
- If respiration exceeds photosynthesis, there is net release. The ecosystem is acting as a carbon source.

In most ecosystems, saprotrophs digest dead organic matter and release the carbon from it as carbon dioxide due to respiration. Conditions in some ecosystems inhibit decomposition. For example, acidic and anaerobic conditions in waterlogged habitats (bogs or swamps) prevent decomposition, so peat accumulates. During some periods in the past, this peat has turned to coal. The carbon in coal is removed from the carbon cycle for millions or even hundreds of millions of years. Removal of carbon from the cycle is known as sequestration.

Periodic fires occur naturally in some ecosystems and species in these communities are adapted to this. During a fire in a forest or other ecosystem, carbon dioxide is produced by combustion of carbon compounds in living organisms and dead organic matter. The ecosystem therefore becomes a carbon source.

▲ Figure 26 In some forests in Indonesia, up to 15 m depth of peat has accumulated over the past 5,000 years, forming a huge carbon sink. Where the forest is cleared and the peat drained for palm oil plantations, the carbon rapidly returns to the atmosphere

C4.2.19 Release of carbon dioxide into the atmosphere during combustion of biomass, peat, coal, oil and natural gas

Biomass, peat, coal, oil and natural gas are carbon sinks, because the carbon in them can remain sequestered for indefinite lengths of time. They were formed at different times in the past, by a variety of processes.

Carbon sink	When formed	How formed
natural gas and oil ▲ Figure 27	Formed over the past 550 million years or more, from the Precambrian geological period onwards	Deep burial of partially decomposed organic matter under sediments, where high temperatures caused chemical changes and produced oil and natural gas that were trapped in porous rocks

coal ▲ Figure 28	Mostly formed 325–250 million years ago, during the Pennsylvanian and Permian geological periods	Accumulation of wood and other plant matter in swamps, where it was buried under other sediments
peat ▲ Figure 29	Mostly formed over the past 10,000 years in the period since the last glaciation of the Earth	Incomplete decomposition of dead plant matter due to acidic and anaerobic conditions in waterlogged bogs and swamps
biomass ▲ Figure 30	Wood in trees has accumulated over the past few thousand years; other organic matter is more recent	Plant biomass is derived from photosynthesis, with transfer along food chains to animal biomass

▲ Table 5

When any of these materials burn in air, carbon dioxide is produced and released into the atmosphere. The scientific term for burning in air is combustion. Combustion only begins if a specific ignition temperature is reached—this is called the flash point. For coal, crude oil and natural gas, this is over 500°C. For dry wood, it is over 400°C. These temperatures are never reached during normal weather conditions on Earth. However, volcanic activity and lightning strikes can cause fires, especially when ecosystems have been desiccated (dried) by hot weather and drought.

Most coal is deeply buried so cannot burn in air. At the end of the Permian period, 250 million years ago, volcanic activity in Siberia may have led to combustion of coal deposits on a huge scale. This triggered the release of carbon dioxide, leading to extremely high temperatures on Earth due to the greenhouse effect. This caused the greatest mass extinction event of all time.

In recent years, summer wildfires have become increasingly frequent in areas of tundra located inside the Arctic Circle. Combustion of peat that accumulated over thousands of years has released large quantities of carbon dioxide. It is estimated that in 2020, nearly 250 megatonnes of CO_2 was emitted as a result of these fires—but this is far less than the 3,400 megatonnes released by humans burning fossil fuels in that year.

Humans discovered fire as much as two million years ago. This allowed them to burn carbon sinks to release energy. Until the Industrial Revolution, the rate at which this was done did not significantly change conditions on Earth. However, combustion of fossil fuels has risen rapidly since then. Today, the balance of carbon dioxide production and uptake has been radically shifted.

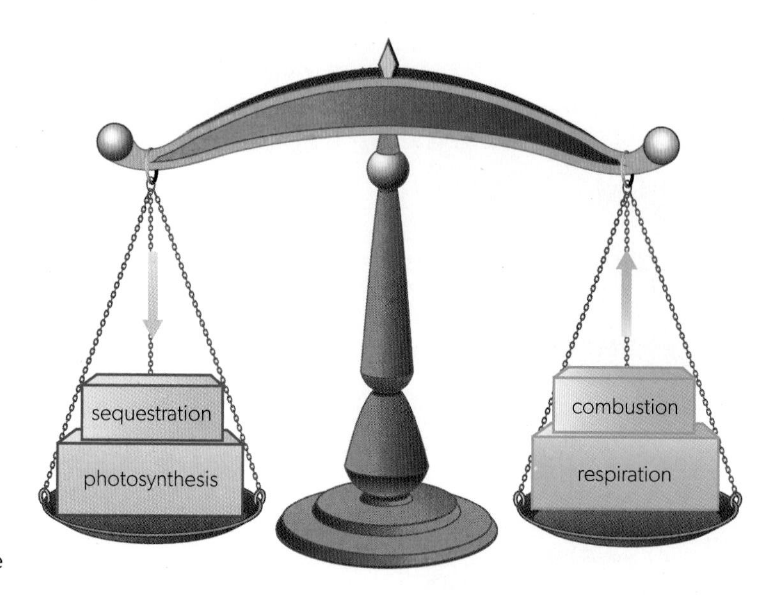

▶ Figure 31 Carbon balance

C4.2.20 Analysis of the Keeling Curve in terms of photosynthesis, respiration and combustion

Since 1959, atmospheric carbon dioxide concentrations have been measured at Mauna Loa Observatory in Hawaii. The graph plotted from these results is known as the Keeling Curve, because the collection of this data was started and continued over the decades by the American scientist Charles Keeling.

The Keeling Cuve shows two trends:

1. Annual fluctuations

Every year, the concentration of carbon dioxide increases between October and May and then falls from May to October. This is due to global imbalances in rates of carbon dioxide fixation by photosynthesis and release due to respiration. Photosynthesis is higher overall during the northern hemisphere summer, when plants over most of the Earth's land surface are in their main growth season.

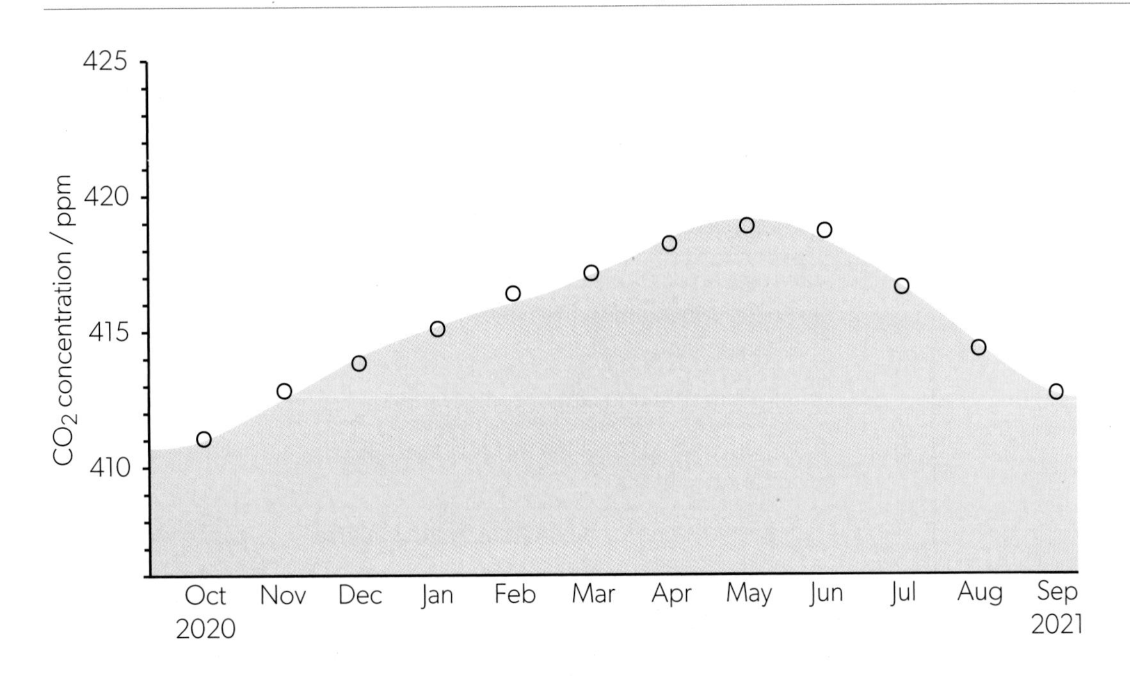

◀ Figure 32 Carbon dioxide concentration for the year ending 23 September 2021

2. Long-term trend

The graph of carbon dioxide concentrations for one year shows that the increase is not completely reversed by the decrease, so the concentration at the end of the year is higher. This is also shown by the full Keeling Curve, from 1959 onwards. This trend is largely due to burning of fossil fuels by humans, together with other anthropogenic factors such as deforestation.

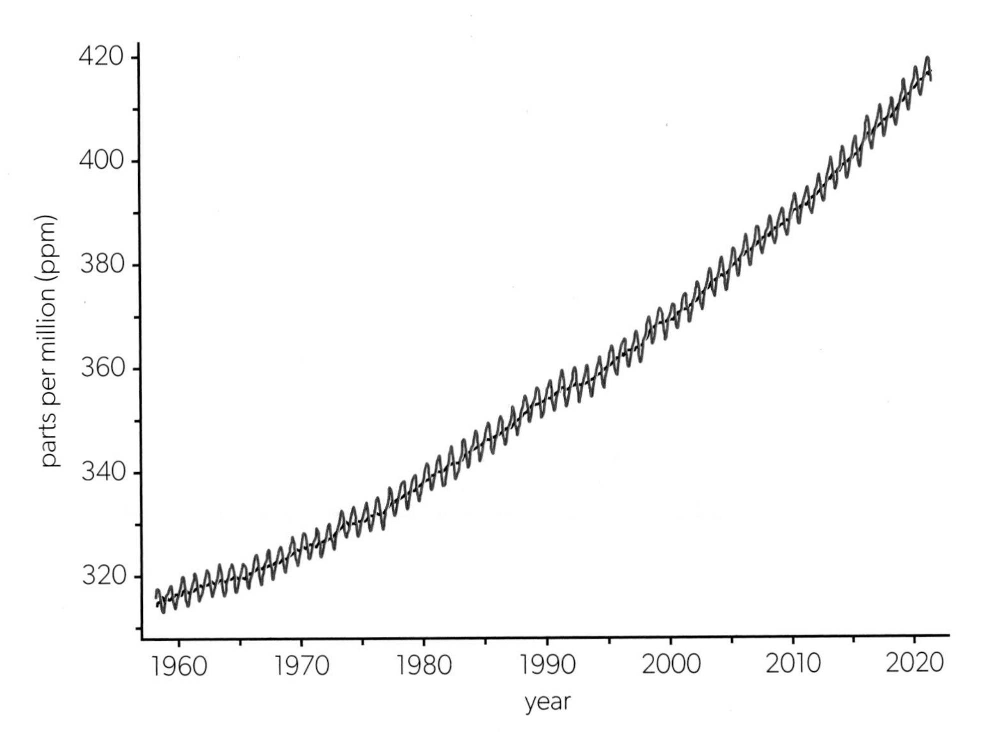

◀ Figure 33 The Keeling Curve. Source: Scripps Institution of Oceanography, NOAA Global Monitoring Laboratory

Data-based questions: Keeling Curves

Figure 34 shows the mean monthly average carbon dioxide concentrations for Mauna Loa in Hawaii, Barrow in Alaska, Samoa and Antarctica.

1. Suggest reasons for:
   a. the remote locations of the atmospheric monitoring stations, even though locating them in cities would have been more convenient and cheaper [2]
   b. multiple monitoring stations rather than only one. [2]
2. a Compare and contrast the trends shown in the data from the four monitoring stations. [3]
   b. Suggest reasons for differences in the patterns. [3]

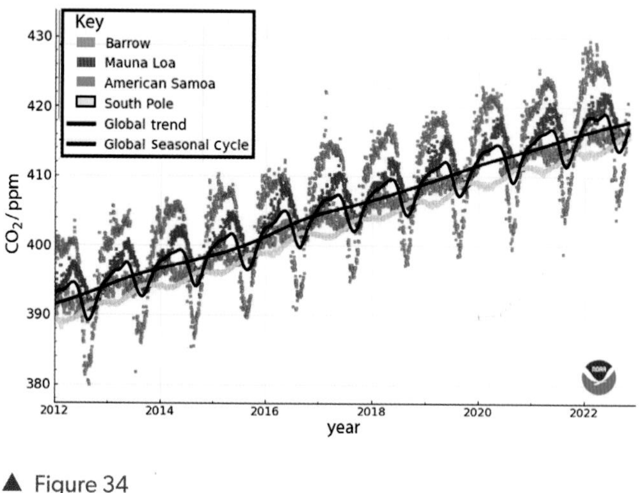

▲ Figure 34

Technology: Using databases

The relationship between atmospheric CO_2 fluctuation and degree of forestation

The database at the NOAA Global Monitoring Earth Research Laboratories contains data from over 90 atmospheric monitoring stations. The Gapminder database provides forest biomass data over time for different countries and regions. The UN REDD+ database provides information about forest cover by country.

Inquiry questions

1. One possible explanation for the differences in annual fluctuations in atmospheric CO_2 at the different monitoring stations shown in Figure 34 could be the percentage of temperate forest cover in the vicinity of the testing station. Other possibilities include differences in total land mass between the northern and southern hemispheres; seasonal temperature variations; and even volcanic activity. Gather data from the databases to test these hypotheses.
2. What other possible relationships exist between forest cover variables and atmospheric CO_2 levels, that can be tested using these or similar databases?

C4.2.21 Dependence of aerobic respiration on atmospheric oxygen produced by photosynthesis, and of photosynthesis on atmospheric carbon dioxide produced by respiration

A crucial step in photosynthesis is the release of oxygen. One oxygen atom is removed from each of two water molecules and these atoms are linked by a double covalent bond, using energy derived from light. This difficult chemical process is catalysed by the oxygen-evolving complex at the centre of photosystem II. The process developed in archaebacteria billions of years ago, and is one of the greatest triumphs of evolution.

Before the evolution of oxygenic photosynthesis there was very little oxygen in the Earth's atmosphere. The subsequent development of an oxygen-rich atmosphere allowed the evolution of aerobic respiration. Even today, oxygen only exists in the atmosphere because it is produced by photosynthesis. Respiring heterotrophs still depend on photosynthesizing autotrophs for their oxygen supply.

Photosynthesis caused the carbon dioxide concentration of the atmosphere to drop far lower than that of pre-biotic Earth. The concentration would fall further without emission of CO_2 produced by respiration in heterotrophs. CO_2 concentrations commonly limit the rate of photosynthesis. Autotrophs are therefore dependent on heterotrophs for a continued supply of carbon dioxide.

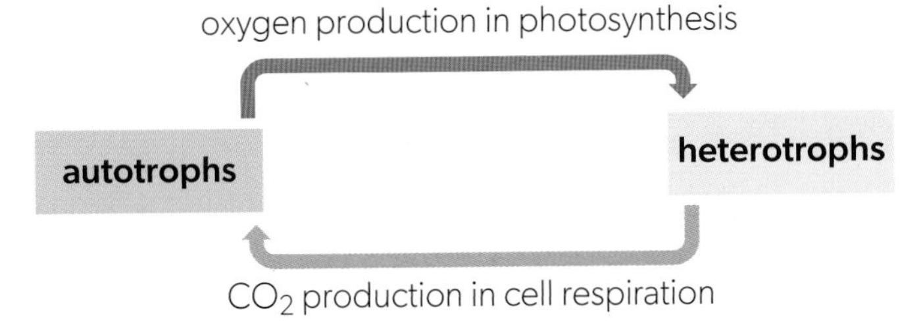

▲ Figure 35 Interdependence of heterotrophs and autotrophs

It is not possible to measure global fluxes of carbon dioxide and oxygen precisely but, as these quantities are of great interest, scientists have produced estimates for them. These estimates are based on many measurements in individual natural ecosystems, or in mesocosms. Global carbon fluxes are extremely large so estimates are in gigatonnes. One gigatonne is 10^{15} grams. It is estimated that, in terrestrial ecosystems, 120 gigatonnes of carbon are fixed by photosynthesis per year and nearly 120 gigatonnes are released by respiration.

C4.2.22 Recycling of all chemical elements required by living organisms in ecosystems

Living organisms need a supply of chemical elements:

- Carbon, hydrogen and oxygen are required to make carbohydrates, lipids and other carbon compounds on which life is based.
- Nitrogen and phosphorus are also needed to make many of these compounds.
- Approximately 15 other elements are needed by living organisms. Some of them are used in minute traces only, but they are nonetheless essential.

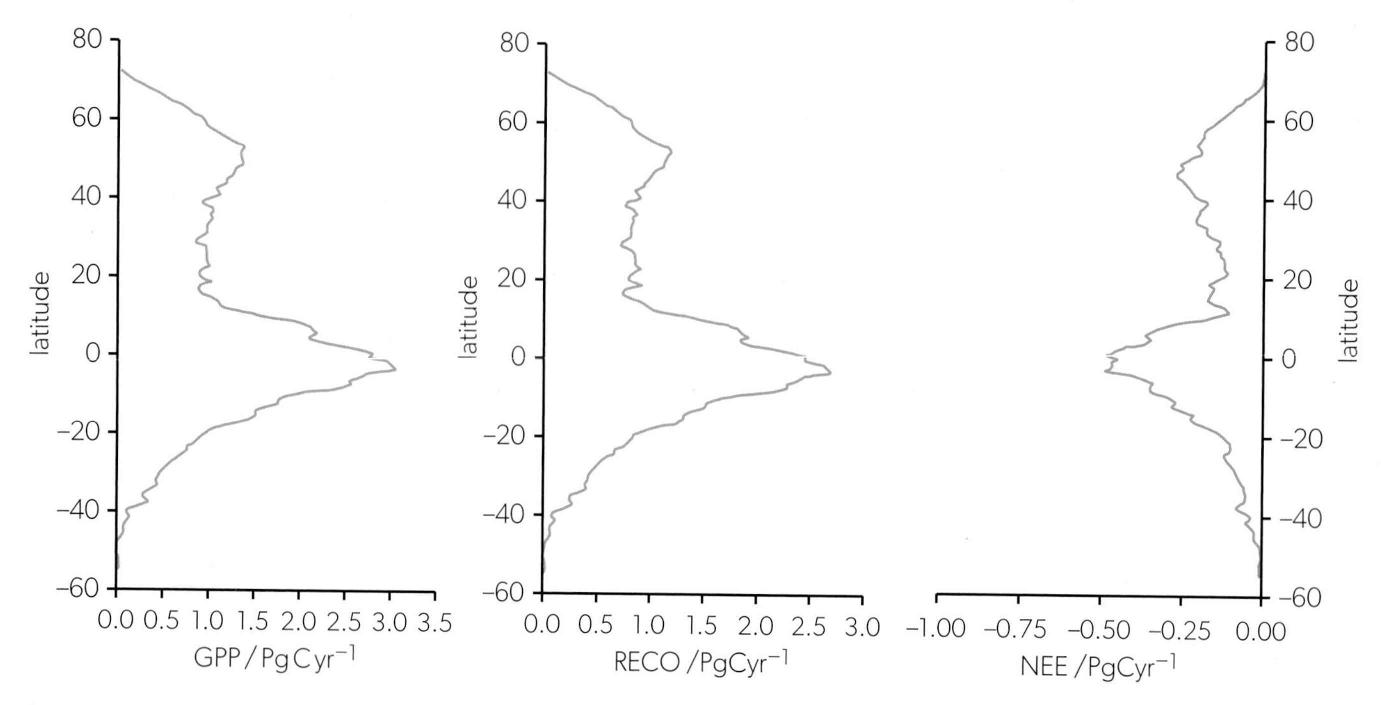

▲ Figure 36 The graphs show carbon dioxide uptake by photosynthesis (GPP), release by ecosystem respiration (RECO) and net exchange of carbon dioxide (NEE) for all terrestrial ecosystems from latitude −60° S to 80° N. Are ecosystems on land a net carbon source or sink?

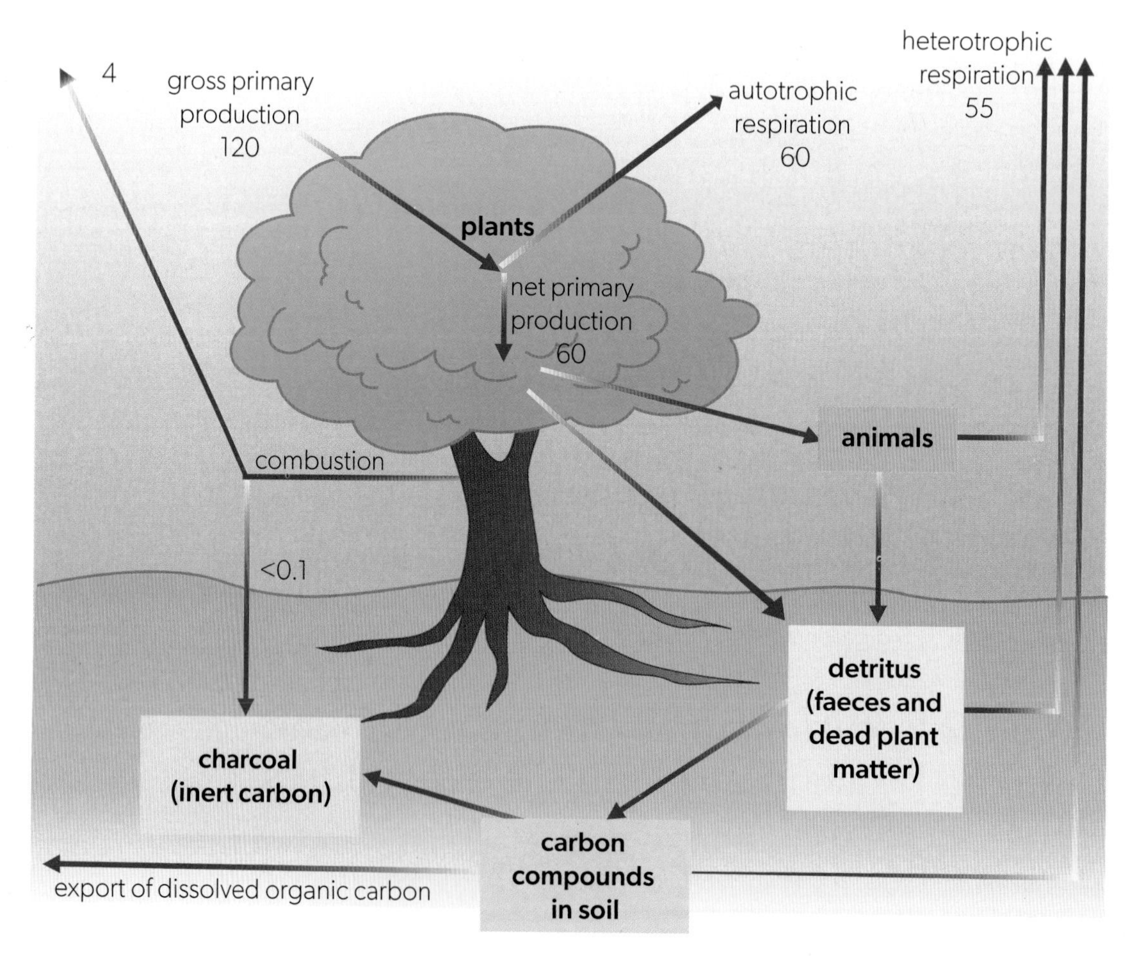

▲ Figure 37 Carbon fluxes in terrestrial ecosystems

Autotrophs obtain all of the elements they need as inorganic nutrients from the abiotic environment, including carbon and nitrogen. Heterotrophs obtain these two elements and several others as part of the carbon compounds in their food. They do however obtain other elements as inorganic nutrients from the abiotic environment, including sodium, potassium and calcium.

There are limited quantities of each chemical element on Earth. Although living organisms have been using these supplies for three billion years, they have not run out. This is because chemical elements can be endlessly recycled. Organisms absorb the elements they require as inorganic nutrients from the abiotic environment. They use these elements, then return them to the environment with the atoms unchanged.

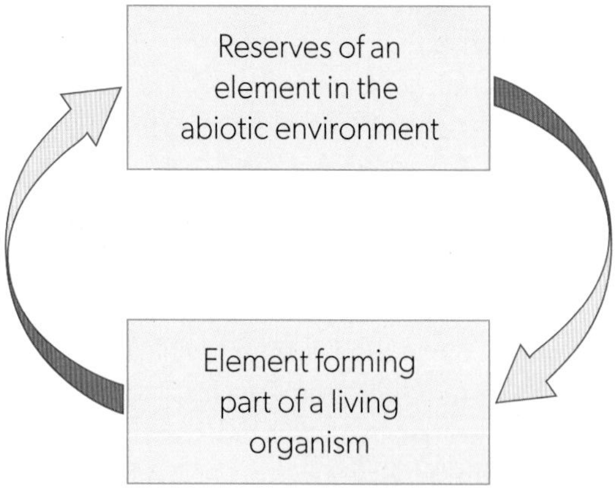

Reserves of an element in the abiotic environment

Element forming part of a living organism

▲ Figure 38 Recycling of chemical elements in ecosystems

Recycling of chemical elements is rarely as simple as shown in Figure 38. In many cases, an element is passed from organism to organism before it is released back into the abiotic environment. Decomposers play a key role because they break down the carbon compounds containing the chemical elements; for example, they release nitrogen from proteins in the form of ammonia (NH_3). The details of recycling vary from element to element. The carbon cycle is different from the nitrogen cycle, for example. Ecologists refer to these schemes collectively as nutrient cycles. The word nutrient is often ambiguous in biology; in this context, it simply means an element that an organism needs.

▲ Figure 39 Living organisms have been recycling for billions of years

Linking questions

1. What are the direct and indirect consequences of rising carbon dioxide levels in the atmosphere?
   a. Explain the pattern of atmospheric carbon dioxide concentrations illustrated by the Keeling Curve. (C4.2.20)
   b. Outline one example of a positive feedback cycle that might occur due to climate change. (D4.3.2)
   c. Outline the consequences of climate change for boreal forests. (D4.3.3)
2. How does the transformation of energy from one form to another make biological processes possible?
   a. Explain how photosynthesis can lead to an increase in plant biomass. (C4.2.15)
   b. Discuss the advantages of ATP as the energy currency of cells. (C1.2.2)
   c. Explain the concept of activation energy in relation to the activity of enzymes. (C1.1.10)

TOK

Who owns knowledge?

In 2003, a skeleton of an extinct species of human, *Homo floresiensis,* was discovered in a cave on the island of Flores, Indonesia. Since then, further remains from the same species were discovered. The fossils are property of the Indonesian state. In early 2004, the skeletal remains were removed from their repository and other scientists were not granted access. During that time, other researchers expressed concern that important scientific evidence would be available to just a small group of scientists who neither allowed access by other scientists nor published their own research. The fossils were returned with portions severely damaged and missing two leg bones. In 2005, Indonesian officials forbade access to the cave. Scientists were allowed to return to the cave in 2007, and other scientists have since been allowed to conduct research on the remains. This raises the issue of the consequences of the rare remains being seen as property rather than as a resource to be available for all scientists to research.

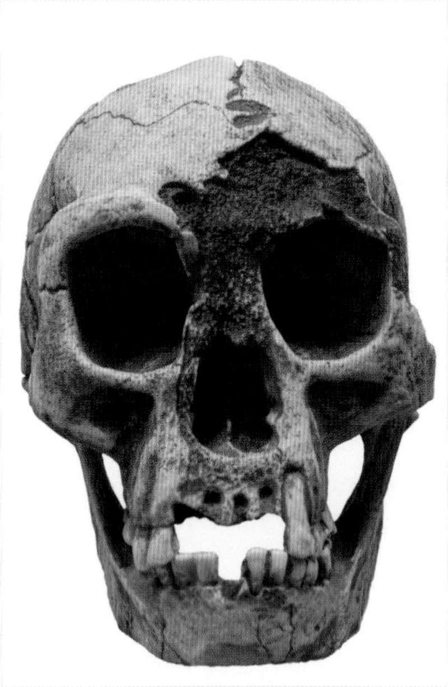

▲ Figure 1 The skeletal remains of an adult female *Homo floresiensis*

Intellectual property rights mean that indeed knowledge can be "owned". Ethical challenges arise when life-saving medicines are given intellectual property protection. Intellectual property rights allow companies to charge more than the marginal cost of production of a vaccine. This allows them to recover the costs of research and development. The challenge is that this higher price might exclude people who cannot afford the cost of the medicine.

With the COVID-19 pandemic, the genome of the COVID-19 virus was identified in China in January 2020. The virus's entire genome was published online within days. In comparison, it took three months to reach this stage with the SARS coronavirus outbreak in 2003. This open access to the genome led to the accelerated pace at which vaccines could be developed.

In 1955, Jonas Salk, the scientist credited with developing one of the first successful polio vaccines was asked: who owned the patent to the polio vaccine. "Well, the people, I would say," Salk responded. "There is no patent."

▲ Figure 2 Jonas Salk

End of chapter questions

1. White sage (*Salvia apiana*) is a native Californian shrub. Slender oat (*Avena barbata*) is a grass, originally from the Mediterranean, which was introduced to California. The map shows the distribution of the two species in relation to one another in an area near Santa Ynez, California.

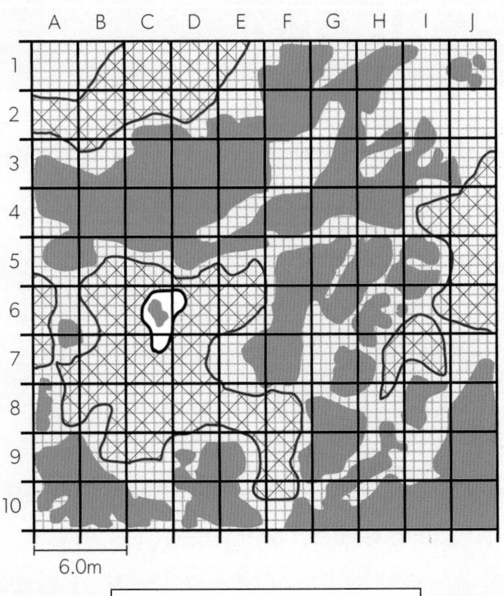

Source: http://web.csulb.edu

a. List three quadrats where both species are found. [1]

b. Using the model of the worked example in *Section C4.1.15*, carry out the chi-squared test of association for this example. [4]

c. Explain what is meant by allelopathy. [2]

d. List the kinds of interactions that can occur between plants. [3]

e. Suggest reasons why two plant species would be associated with one another; that is, found in the same locations. [2]

f. Suggest reasons why two plants species would not be found in the same locations. [2]

2. As the world's population grows, supplies of freshwater are becoming scarcer. Researchers are investigating the use of seawater to irrigate selected crops which can be fed to livestock.

The biomass yield of two freshwater-irrigated plants often used for livestock forage, alfalfa (*Medicago sativa*) and Sudan grass (*Sorghum sudanense*), were compared with those of salt-tolerant crops irrigated by seawater, saltbush (*Atriplex spp.*) and sea blite (*Sueda maritima*).

The results are shown in the bar chart below.

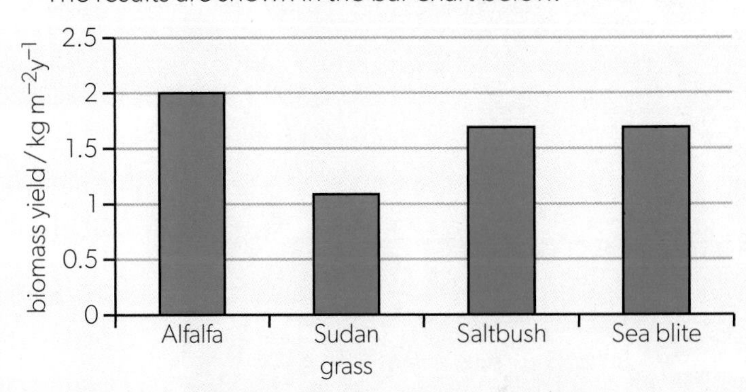

Sheep were raised on a normal diet (control sheep) and compared with sheep fed on a normal diet supplemented with salt-tolerant plants. The results are shown in the bar chart below.

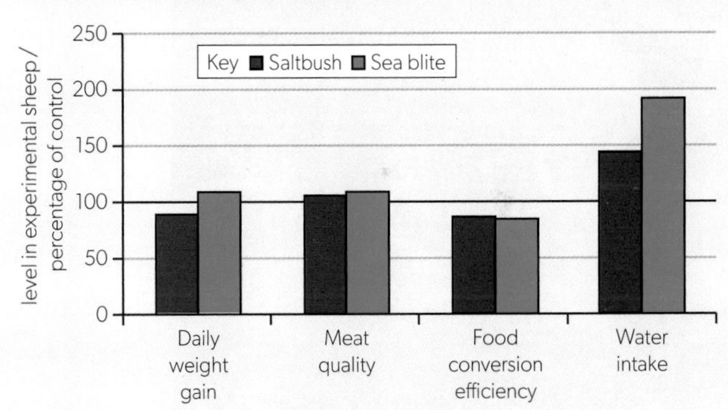

Source: E Glen et al., Scientific America (1998) pp 56-61

a. Compare the biomass yield of crops irrigated with seawater and freshwater. [2]

b. Compare the daily weight gain and water intake in sheep fed on saltbush with sheep fed on sea blite. [2]

c. Discuss, using only the data provided, the advantages and disadvantages of using crops irrigated by seawater to feed sheep. [3]

d. Outline what is meant by food conversion efficiency. What is the usual value of feed conversion efficiency between producers and herbivores? [3]

e. List variables that can affect the productivity of an area of land. [3]

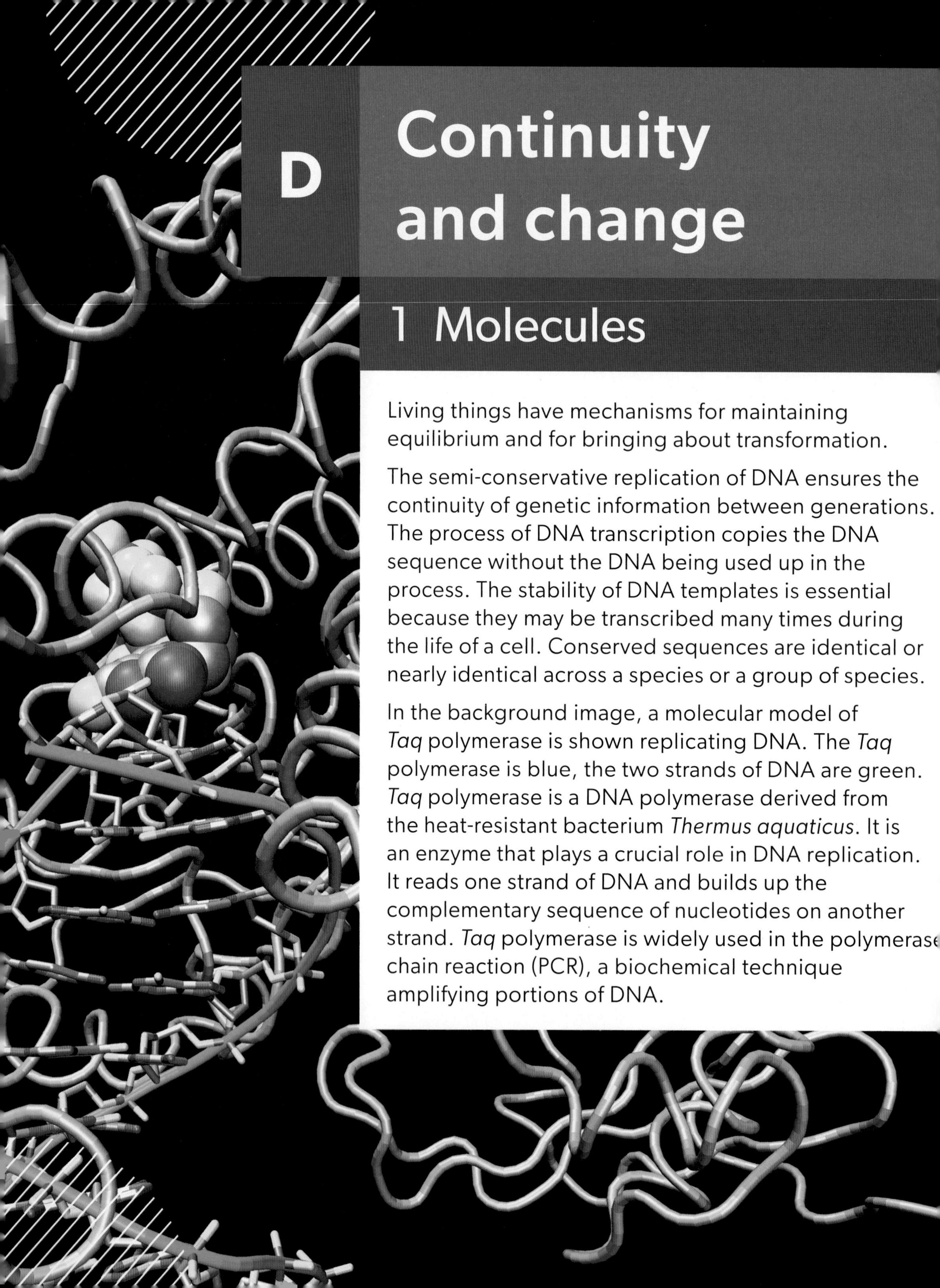

D Continuity and change

1 Molecules

Living things have mechanisms for maintaining equilibrium and for bringing about transformation.

The semi-conservative replication of DNA ensures the continuity of genetic information between generations. The process of DNA transcription copies the DNA sequence without the DNA being used up in the process. The stability of DNA templates is essential because they may be transcribed many times during the life of a cell. Conserved sequences are identical or nearly identical across a species or a group of species.

In the background image, a molecular model of *Taq* polymerase is shown replicating DNA. The *Taq* polymerase is blue, the two strands of DNA are green. *Taq* polymerase is a DNA polymerase derived from the heat-resistant bacterium *Thermus aquaticus*. It is an enzyme that plays a crucial role in DNA replication. It reads one strand of DNA and builds up the complementary sequence of nucleotides on another strand. *Taq* polymerase is widely used in the polymerase chain reaction (PCR), a biochemical technique amplifying portions of DNA.

D1.1 DNA replication

How is new DNA produced?

To prepare for cellular reproduction, DNA and organelles must be duplicated. Each of the chromosomes in Figure 1 contains two identical chromatids that have been produced by the process of DNA replication from a single-stranded chromosome. To prepare for cellular reproduction, DNA and organelles must be duplicated. What is the mechanism by which DNA duplication occurs? What is complementary base pairing and semi-conservative replication? What roles do they play in ensuring that the parent and daughter DNA are identical?

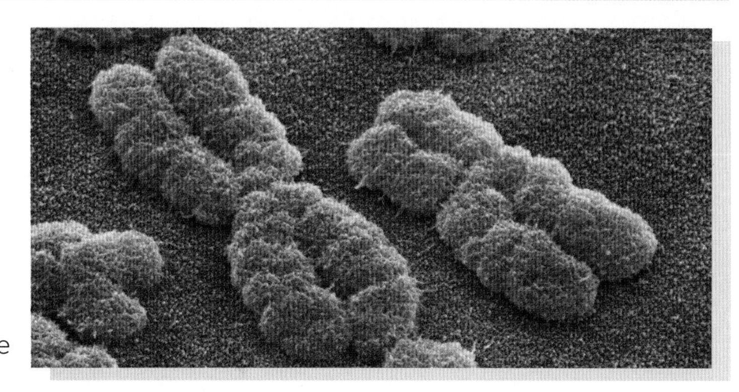

▲ Figure 1

How has knowledge of DNA replication enabled applications in biotechnology?

How is biotechnology unique in terms of human ingenuity? What biotechnology processes and products have been developed for human benefit?

Thermophilic bacteria thrive at high temperatures. For example, *Thermus aquaticus* (*Taq*) thrives in hot springs in Yellowstone National Park, Wyoming (Figure 2). How does the thermal cycling of PCR depend on the fact that the optimum temperature for *Taq* polymerase is 75°C? Bacteria use restriction enzymes to defend themselves against viruses by cutting the viral DNA. What are some of the ways that this adaptation has been utiltized in biotechnology applications?

▲ Figure 2

SL and HL	AHL only
D1.1.1 DNA replication as production of exact copies of DNA with identical base sequences D1.1.2 Semi-conservative nature of DNA replication and role of complementary base pairing D1.1.3 Role of helicase and DNA polymerase in DNA replication D1.1.4 Polymerase chain reaction and gel electrophoresis as tools for amplifying and separating DNA D1.1.5 Applications of polymerase chain reaction and gel electrophoresis	D1.1.6 Directionality of DNA polymerases D1.1.7 Differences between replication on the leading strand and the lagging strand D1.1.8 Functions of DNA primase, DNA polymerase I, DNA polymerase III and DNA ligase in replication D1.1.9 DNA proofreading

D1.1.1 DNA replication as production of exact copies of DNA with identical base sequences

A replica is an exact copy of something. DNA replication is the production of new strands of DNA with base sequences identical to existing strands. The structure of DNA makes it suited to being replicated repeatedly, without any limit on how many times this is done. DNA has been replicated since the origin of life billions of years ago, allowing the continuity of life down vast numbers of generations.

DNA replication is required for two biological processes.

- Reproduction—offspring need copies of the base sequences of their parents, so parents must replicate their DNA to reproduce.
- Growth and tissue replacement in multicellular organisms—each cell in a plant, animal or other multicellular organism needs a full set of the organism's base sequences. This means that before a cell divides into two daughter cells, it must replicate all of its DNA. Cell division is needed for growth and to replace cells in tissues where they have been lost—for example, cells worn away from the skin surface.

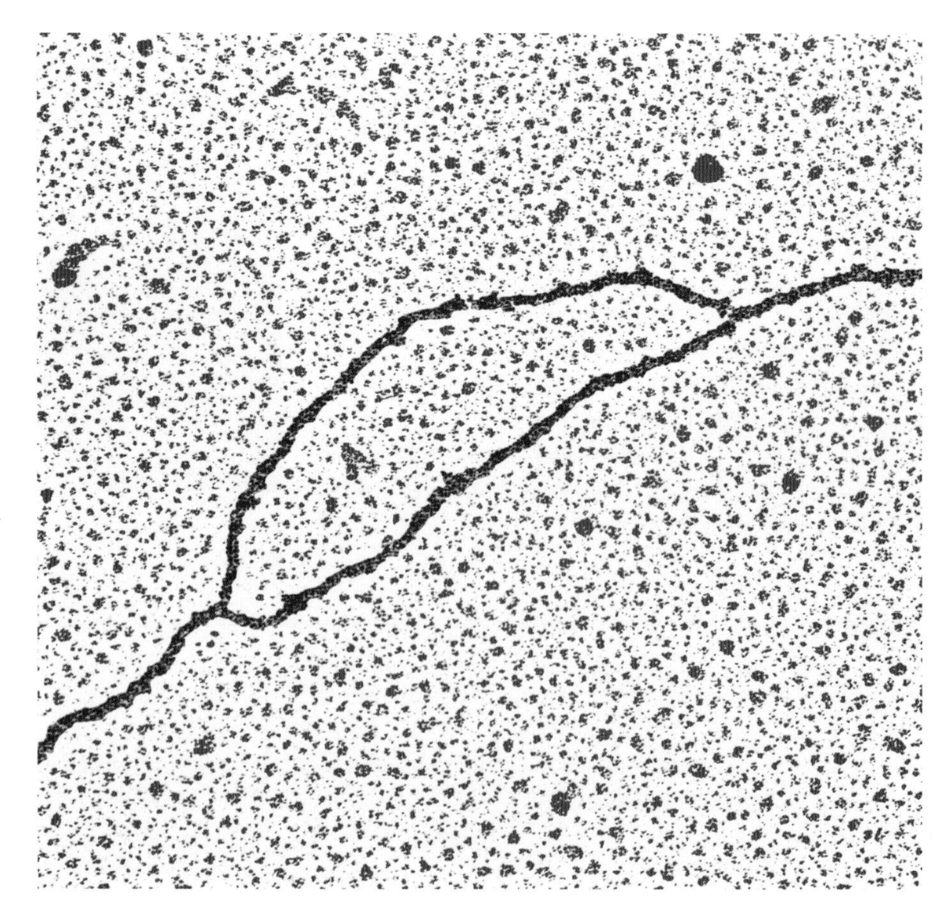

▲ Figure 3 An electron micrograph image of DNA replication—unreplicated DNA is visible to the left and right with a replicated section of the original strand in the centre

Data-based questions: The Messelsohn–Stahl experiment

Meselson and Stahl performed a classic biology experiment in the late 1950s to test three competing models of DNA replication.

- **Conservative**—a new double helix is made containing no part of the parent double helix, which is conserved.
- **Semi-conservative**—half of the parental DNA molecule is conserved in each of the two daughter molecules.
- **Dispersive**—parental DNA segments are distributed throughout both the daughter DNA molecules.

Meselson and Stahl used *Escherichia coli* bacteria which synthesize DNA nucleotides from nitrogen compounds in their food. Nitrogen has two isotopes with different atomic masses: ^{14}N and ^{15}N. At the start of the experiment, a rapidly growing population of *E. coli* was transferred from a growth medium containing only ^{15}N compounds to a growth medium with only ^{14}N compounds. DNA samples were centrifuged at high speed in a salt density gradient.

1. a. The density of the DNA band at generation 0 is 1.724 and the density of the dark band of DNA at generation 4.1 is 1.710. Estimate the density of the DNA band at generation 1.0. [1]

   b. Describe the nitrogen composition of the DNA band in the *E. coli* at generation 1.0. [1]

2. Explain the pattern shown in generation 3.0. [3]

3. This experiment was designed to demonstrate whether replication was semi-conservative or conservative. If replication was conservative, both strands of DNA of a replicated molecule would remain together after replication.

   a. Explain the evidence from the results for DNA replication being semi-conservative. [3]

   b. Predict the results after one generation if DNA were conservative. [2]

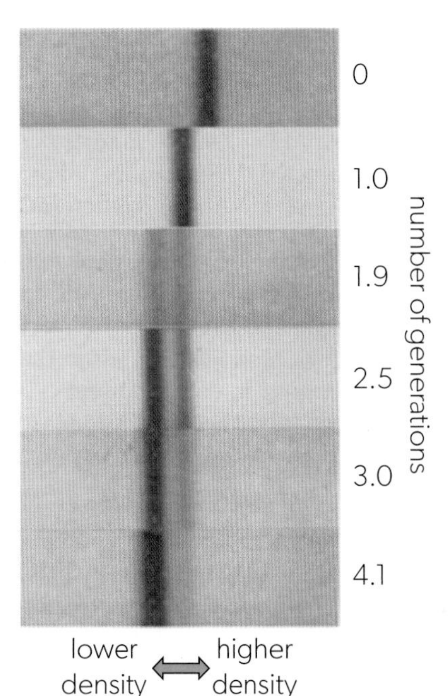

◀ **Figure 4** DNA molecules of the same density appear as a band in UV absorption photographs. The pattern of bands after different numbers of generation times is shown

D1.1.2 Semi-conservative nature of DNA replication and role of complementary base pairing

When DNA is replicated, the two strands of the double helix must separate. Both original strands are used as templates to guide the polymerization of a new strand. The new strands are formed by adding nucleotides one by one and linking them together. This happens progressively along a DNA molecule. The site at which the copying is actively occurring is a replication fork. When replication is complete, there are two DNA molecules, both composed of an original strand and a newly synthesized strand. For this reason, DNA replication is referred to as semi-conservative (Figure 5).

The base sequence on the template strand determines the base sequence on the new strand. Only a nucleotide carrying a base that is complementary to the next base on the template strand can successfully be added to the new strand. This is

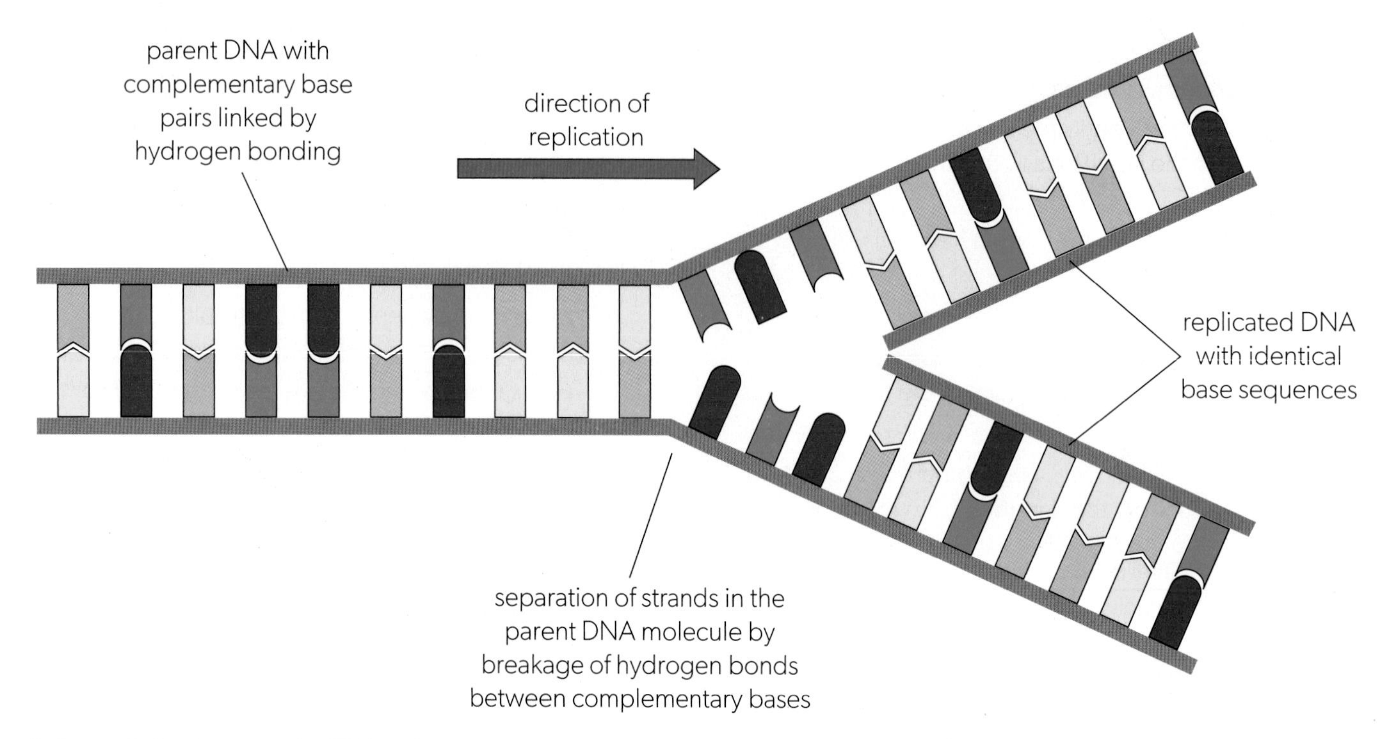

▲ Figure 5 Semi-conservative DNA replication

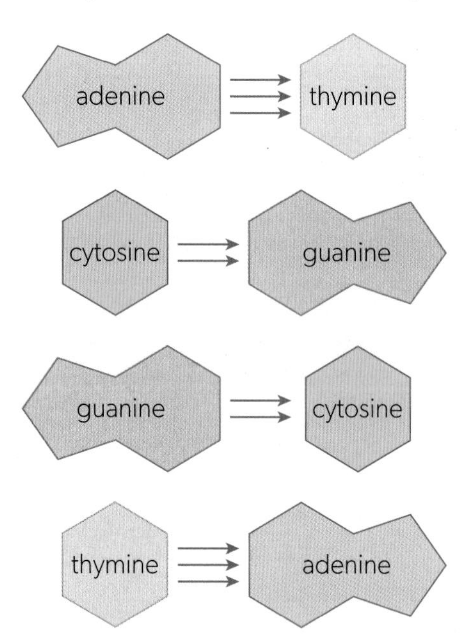

▲ Figure 6 Each of the bases in DNA will pair with only one other base. The blue arrows represent the hydrogen bonds that are the basis of this complementary base pairing

because complementary bases form hydrogen bonds with each other, stabilizing the structure. If a nucleotide with the wrong base started to be inserted, hydrogen bonding between bases would not occur and the nucleotide would be rejected. Adenine pairs with thymine, not cytosine or guanine, and cytosine only pairs with guanine (Figure 6).

The rule that one base always pairs with another is called complementary base pairing. It ensures that the two DNA molecules resulting from DNA replication are identical in their base sequences to the parent molecule. Complementary base pairing ensures a high degree of accuracy when new strands are assembled on a template strand. It also makes it possible to check the base sequence that has been assembled, recognize any mispairing, then cut out and replace the incorrect nucleotides. These processes together ensure that only about 1 in 10 billion bases is incorrect when DNA is replicated.

A diploid human cell has DNA with approximately 6 billion base pairs, so on average there are only 0.6 errors when all the DNA is replicated prior to mitosis or meiosis. This astonishing level of accuracy explains the genetic continuity that exists between generations.

D1.1.3 Role of helicase and DNA polymerase in DNA replication

DNA replication is a multi-stage process that is carried out by an assemblage of functional subunits called a replisome. Helicase and DNA polymerase are two types of protein that are essential parts of replisomes.

Helicase

Helicase is a ring-shaped protein that separates the two strands of a DNA molecule so that they can each act as a template for the formation of a new strand. As it moves along a DNA molecule, helicase breaks hydrogen bonds between bases, allowing one strand to be pulled through the hole in helicase's ring and the other to pass to the side of it. This action causes the DNA double helix to become uncoiled. The splitting into two strands is analogous to undoing a zip-fastener, so helicase's roles are often referred to as unwinding and unzipping DNA. The unwinding of the DNA double helix and separation into two strands causes tensions in the molecule which could cause supercoiling. To prevent this, the tensions are relieved by parts of the replisome.

▲ Figure 7 This black and white string is a double helix, like DNA. When the two stands were separated, they twisted into new double helices

DNA polymerase

DNA polymerase assembles new strands of DNA, using the two original strands as templates. The replisome contains separate DNA polymerases for each strand. The DNA polymerases move along the template strands, adding one nucleotide at a time. Free nucleotides with each of the four possible bases are available in the area where DNA is being replicated. Each time a nucleotide is added, only one of the four types of nucleotide has the base that can pair with the base at the position reached on the template strand. DNA polymerase brings nucleotides into the position where hydrogen bonds could form. However, if hydrogen bonds do not form and a complementary base pair is not formed, the nucleotide breaks away again.

Once a nucleotide with the correct base is in position and hydrogen bonds have formed between the two bases, DNA polymerase links the nucleotide to the end of the new strand. This is done by making a covalent bond between the phosphate group of the free nucleotide and the sugar of the nucleotide at the end of the

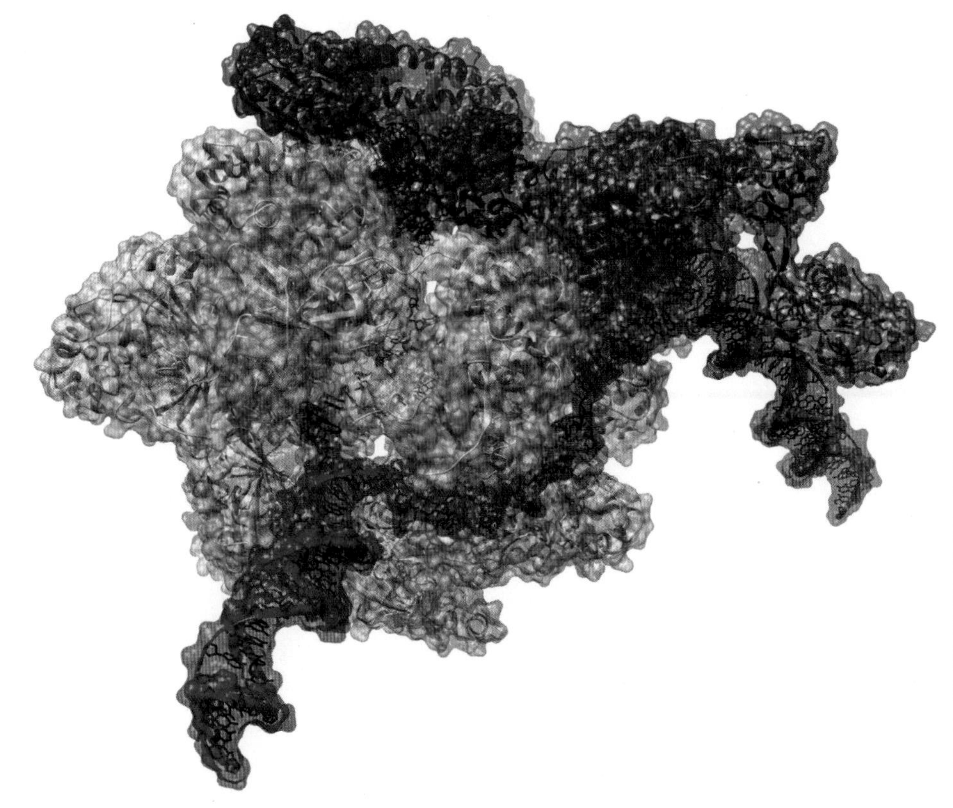

◀ Figure 8 This is the replisome of the T7 bacteriophage. The DNA double helix (red) is splitting, with one strand passing through the hole in the centre of helicase (pale blue) and the other strand passing to the right. New strands are then assembled by DNA polymerases (dark blue). One of the daughter molecules is visible on the right. The other is hidden behind the DNA polymerase assembling it on the left

▲ Figure 9 This PCR thermocycler can process 96 samples simultaneously. The wells holding samples are in metal blocks that conduct heat rapidly, allowing increases or decreases in temperature to be completed in a few seconds

new strand.

D1.1.4 Polymerase chain reaction and gel electrophoresis as tools for amplifying and separating DNA

The polymerase chain reaction, generally called PCR, is an automated method of DNA replication. PCR machines follow a cycle of steps repeatedly, doubling the quantity of DNA with each cycle. Only a very small quantity of DNA is required at the start—in theory, a single molecule. The process of producing more DNA with a specific base sequence is known as DNA amplification. The steps in a PCR cycle are triggered by changes of temperature, so a PCR machine is known as a thermal cycler or thermocycler (Figure 9). A typical cycle is shown in Figure 10.

▼ Figure 10 A cycle of PCR can be completed in less than two minutes. Thirty cycles, which amplify the DNA by a factor of a billion, take less than an hour

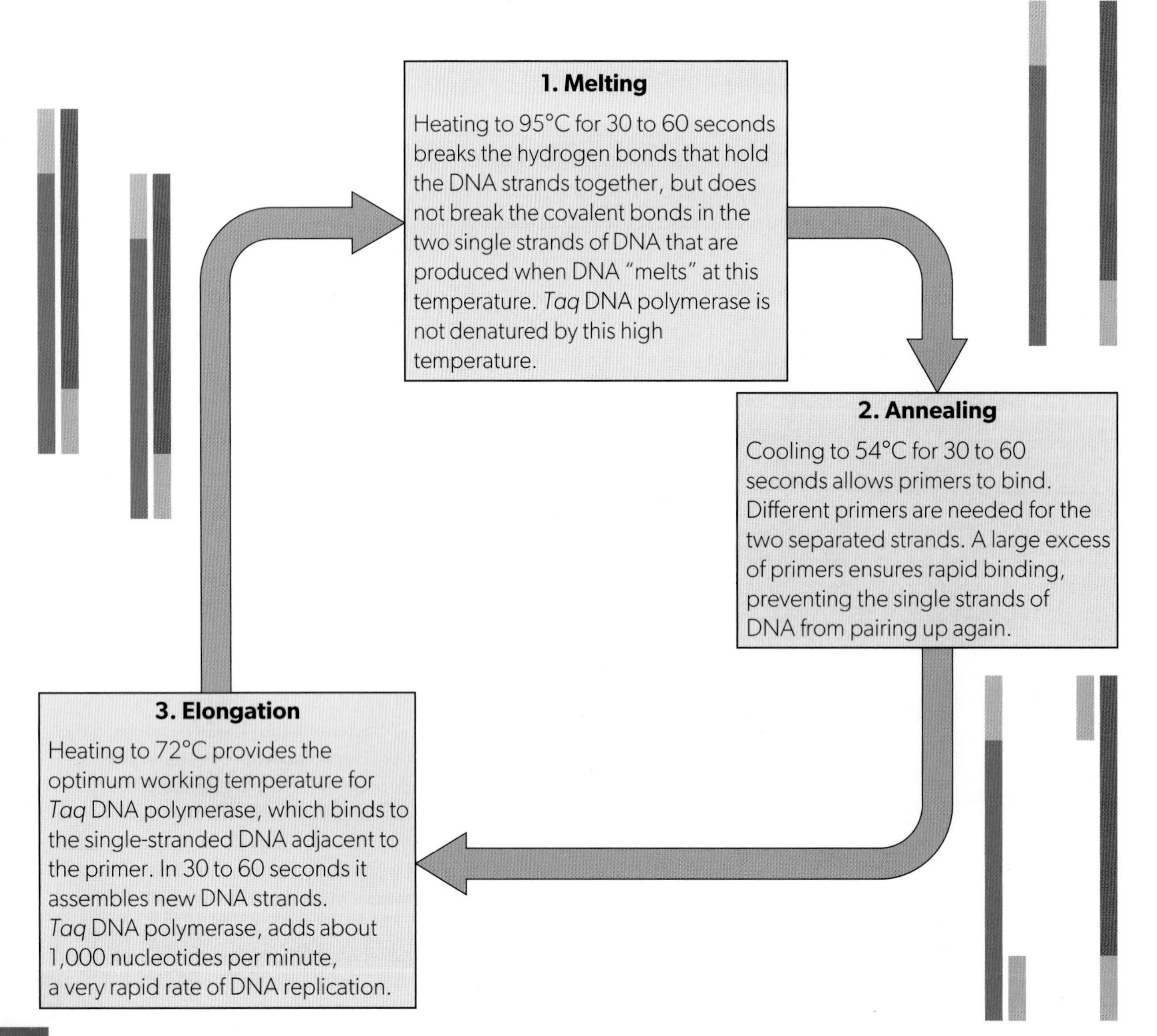

1. Melting
Heating to 95°C for 30 to 60 seconds breaks the hydrogen bonds that hold the DNA strands together, but does not break the covalent bonds in the two single strands of DNA that are produced when DNA "melts" at this temperature. *Taq* DNA polymerase is not denatured by this high temperature.
2. Annealing
Cooling to 54°C for 30 to 60 seconds allows primers to bind. Different primers are needed for the two separated strands. A large excess of primers ensures rapid binding, preventing the single strands of DNA from pairing up again.
3. Elongation
Heating to 72°C provides the optimum working temperature for *Taq* DNA polymerase, which binds to the single-stranded DNA adjacent to the primer. In 30 to 60 seconds it assembles new DNA strands. *Taq* DNA polymerase, adds about 1,000 nucleotides per minute, a very rapid rate of DNA replication.

Mixtures of DNA base sequences can be loaded into the PCR machine, but only selected sequences are amplified. The optimal length of DNA for amplification is around 100 base pairs. Primers added at the start of cycling select which sequences are copied. A primer is a short single strand of DNA designed to bind to the DNA at the point where the selected base sequence for replication begins. The ideal length for primers is 18 to 30 nucleotides.

Taq DNA polymerase and DNA nucleotides with each of the four bases are needed for PCR. *Taq* DNA polymerase evolved in the bacterium *Thermus aquaticus*, which lives in hot springs, including those of Yellowstone National Park. The temperatures of these springs can be as high as 80°C. Enzymes in most organisms would rapidly denature, but those of *T. aquaticus* are adapted to the high temperatures.

With the help of *Taq* DNA polymerase, PCR allows the production of huge numbers of copies of a selected base sequence in a very short time. Thermal cyclers can amplify multiple samples simultaneously in arrays of wells—over a hundred in larger machines.

▲ Figure 11 Samples stained with blue dye are being loaded into wells in a gel lying in an electrophoresis tank. The negative electrode is on the right

It is sometimes necessary to separate DNA molecules by length. This can be done using gel electrophoresis. The separation is done in a sheet of gel that is about 3–4 mm thick. In the gel, close to one end are rectangular holes called wells. The gel is placed in a shallow tank with electrodes at both ends. An electrolyte solution is poured over the gel to cover it. One DNA sample is pipetted into each well.

A voltage is applied across the electrodes to create an electric field. This makes charged molecules move through the gel. DNA has negative charges due to phosphates in the sugar–phosphate chains, so it moves towards the positive anode. The gel is therefore orientated in the electrophoresis tank with the wells close to the negative cathode. The gel consists of a mesh of filaments that resists the movement of molecules. Small molecules move faster than large ones, so they move further in a given time. Gel electrophoresis can therefore be used to separate fragments of DNA according to their length.

When enough time has elapsed for the smallest DNA molecules to have nearly reached the anode end of the gel, the voltage is switched off and the gel removed from the tank. It is treated with a dye to make DNA visible. The DNA has moved from the wells towards the anode in a series of parallel lanes. DNA molecules of the same length will have moved the same distance, so they form a visible band in a lane. The number of bands in a lane indicates how many different lengths of DNA were in the sample. The right-hand lane is usually used to create a "ladder". This is done by placing a mixture of DNA fragments of known length in the right-hand well. The ladder is used to estimate the lengths of bands in other lanes on the gel.

D1.1.5 Applications of polymerase chain reaction and gel electrophoresis

The technique of PCR combined with gel electrophoresis has many applications. Testing for coronaviruses and paternity testing are two examples.

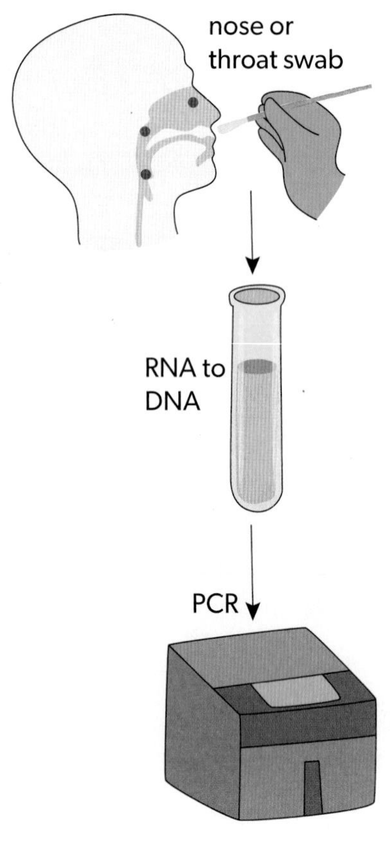

▲ Figure 12 PCR testing for coronavirus

Testing for coronaviruses

- A swab is taken in the nose or throat. Virus particles and viral RNA are rinsed off in a saline solution to produce a liquid sample.
- RNA in the sample is converted to DNA using the enzyme reverse transcriptase.
- PCR is used to amplify specific viral base sequences that are markers of the strain of coronavirus being tested for. About 35 cycles of PCR are used.
- As PCR progresses fluorescent markers are attached to any DNA produced. The level of fluorescence is monitored. If it rises above a target level, the result of the test is positive.

The PCR test for coronaviruses has advantages and disadvantages compared with other methods of testing.

Advantages

- It is very sensitive—one molecule of viral RNA is amplified to produce about 35 billion molecules of DNA, so miniscule quantities in a sample can be detected.
- It is very specific—primers can be designed so only one strain of the virus is detected.

Disadvantages

- It requires materials and equipment that are relatively expensive and are usually only available in testing laboratories.
- Because of the time taken for thermal cycling, results may not be immediately available.

DNA profiling for paternity testing

The technique of DNA profiling distinguishes between individuals using base sequences known as short tandem repeats. These are sequences of between two and seven bases that are repeated consecutively. The number of repeats of each of these sequences varies considerably between individuals. For example, at one locus on human chromosome 5, the sequence TAGA appears. It is repeated between 6 and 15 times. On chromosome 18, the sequence AGAA appears; it is repeated between 7 and 27 times.

Thirteen or more tandem repeats are widely used in DNA profiling. It is very unlikely that two humans would have the same number of repeats of each of these short tandem repeats, so individuals can be distinguished with a high degree of confidence.

These are the stages required to produce a DNA profile.

- A sample of DNA is obtained, either from a known individual or from material containing human DNA of unknown origin.
- Selected tandem repeats are copied by PCR—at least 13; some protocols copy more than this.
- The DNA produced by PCR is separated according to length of fragment and therefore number of repeats using gel electrophoresis.
- A pattern of bands of DNA is produced on the gel that is always the same with DNA taken from one individual. This is the individual's DNA profile. It is likely to be unique to that individual.

For paternity testing, DNA profiles of the child, the child's mother and the man who might be the father are compared. If any bands in the child's profile do not occur in the profile of either the mother or the man, another person must be the father. There are various reasons for paternity investigations being requested.

- Men sometimes claim that they are not the father of a child to avoid having to pay the mother to raise the child.
- Women who have had multiple partners may wish to identify the biological father of a child.
- A child may wish to prove that a deceased man was their father to show that they are the heir.

Using logarithmic scales: Analysis of a DNA profile

Logarithms are an alternative way to express an exponent. For example:

$$\log 1000 = \log 10^3 = 3 \qquad \log 100 = \log 10^2 = 2$$

In biology, very large changes in a variable are easier to represent graphically if logarithms are used. Figure 13 shows DNA fragments separated using gel electrophoresis. The fragments vary in size from 100 bp (base pairs) up to 5,000 bp. The two outside columns of the gel represent ladders—that is, mixtures of DNA fragments of known size. These were used to obtain the data in Table 1 and create the plot shown in Figure 14. The centre columns shown in Figure 13 are the unknowns.

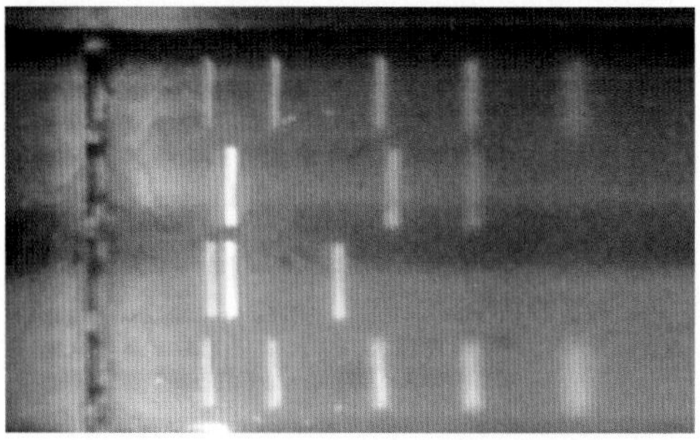

▲ **Figure 13** Gel electrophoresis. The outside columns represent ladders of known length. The two inside columns represent samples of unknown length

Known ladder fragment size / bp	5,000	2,000	850	400	100
Distance moved / mm	58	96	150	200	250

▲ Table 1

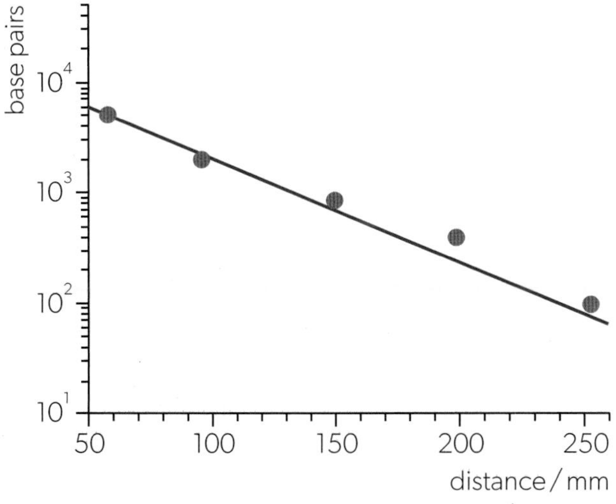

▲ **Figure 14** Distance moved as a function of fragment size in gel electrophoresis. Notice that the *y*-axis scale on this graph goes up in powers of 10. This is a logarithmic scale

Distance moved / mm	
Lane 2	**Lane 3**
60	70
70	160
130	200

▲ Table 2

The distances moved by the DNA fragments in the centre columns are shown in Table 2. The length of these fragments can be determined using the graph in Figure 14.

Data-based questions: Analysis of DNA profiles using D1S80

One commonly studied DNA locus is a variable number tandem repeat (VNTR) named D1S80. D1S80 is located on human chromosome 1. This locus is composed of repeating units of 16-nucleotide segments of DNA. The number of repeats varies from one individual to another with 29 known alleles ranging from 15 repeats to 41 repeats.

Figure 15 shows a DNA profile. The outside lanes show ladders representing multiples of 123 bp. The origin is at the top.

1. Identify the lengths of the fragments represented by each of the bands in the ladder.
2. Using a ruler, measure the distance between the origin and the bands in the ladder.
3. Measure the distance travelled by each band from the origin for individuals 1–6.
4. Using the standard curve, estimate the lengths of the bands in each individual.
5. Estimate the number of repeats represented by each band.
6. It is unclear whether individual 6 has two different copies of the same allele or different alleles. Suggest what could be done to further resolve the genotype of this individual.

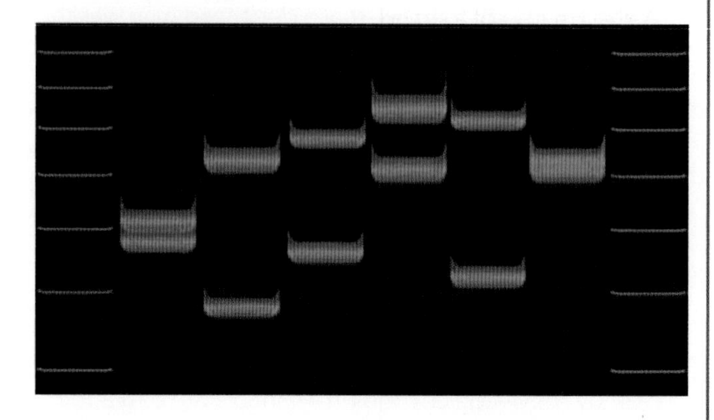

▲ **Figure 15** Photo of DNA profiles of six individuals

Measurement: The role of repeats in the justification of knowledge claims

Scientists make observations and form hypotheses based on the observations. In general, the reliability of any knowledge claim formed on the basis of observation is enhanced by increasing the number of observations or measurements in an experiment or test. In the case of DNA profiling, increasing the number of markers used reduces the probability of a false match.

For example, scientists took samples of DNA from Cape parrots (*Poicephalus robustus*) that had come from different known populations. They then conducted a blind test to determine how many microsatellite loci were necessary to correctly link them to their population.

1. Determine the percentage of correct assignments when 9 loci were used.
2. State the maximum percentage of correct assignments.
3. What was the lowest number of loci required to achieve this maximum accuracy?

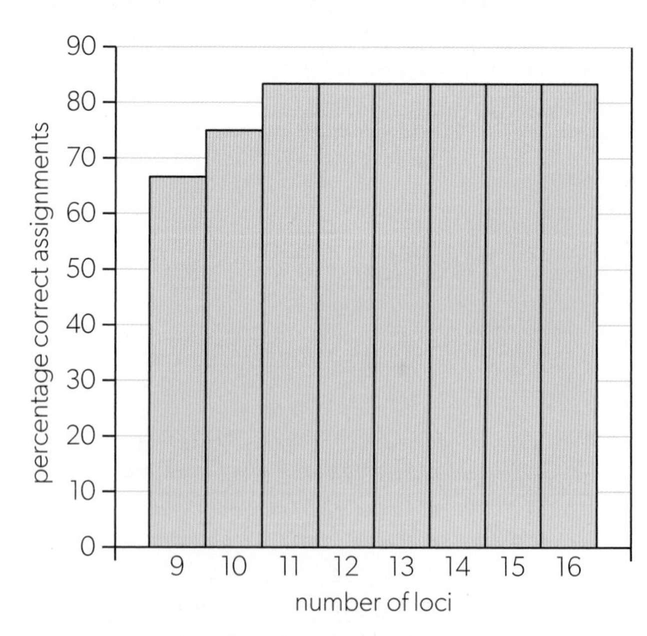

▲ **Figure 16** Graph showing the relationship between percentage correct assignments and number of loci used

D1.1.6 Directionality of DNA polymerases

Nucleotides are linked to form strands of DNA or RNA with covalent bonds between the sugar of one nucleotide and the phosphate of the next. All the nucleotides in a strand have two of these covalent bonds, apart from the nucleotides at either end. Terminal nucleotides either have a sugar or a phosphate group that has not been used to form a covalent bond with another nucleotide.

- One terminus of a strand is the 3′ end because the site available for a link to another nucleotide is the third carbon of the sugar in the terminal nucleotide.
- The other terminus of the strand is the 5′ end because the available site is the phosphate group of the nucleotide and that is linked to the fifth carbon atom of the sugar.

However many nucleotides are linked together into a strand, there will always be a 5′ end and a 3′ end.

an OH group of the phosphate could form a bond with another nucleotide. The phosphate is linked to C5 of the pentose sugar
phosphate
base
the OH group on C3 of the pentose sugar could form a bond with another nucleotide
pentose sugar

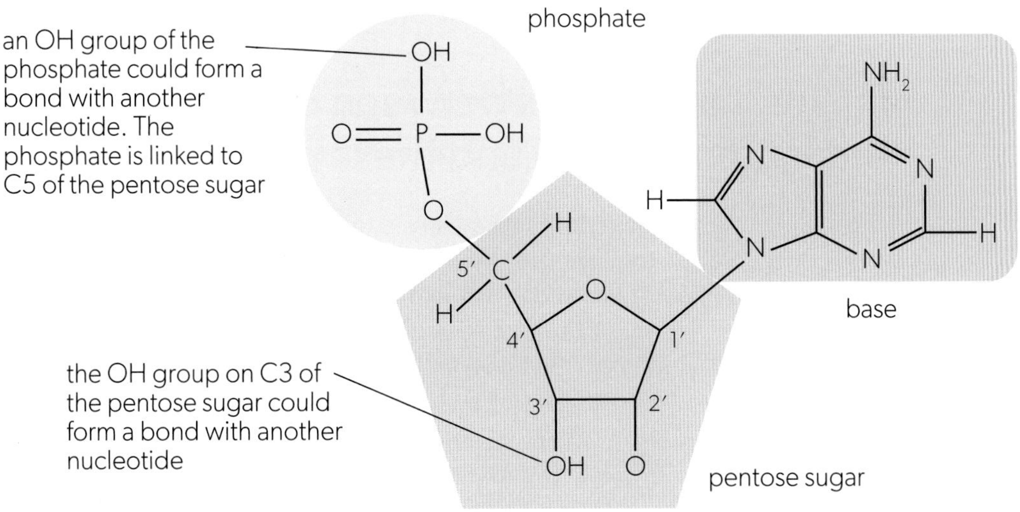

▲ Figure 17 Nucleotide structure

DNA polymerase assembles new strands of DNA by linking together a strand of nucleotides with bases complementary to those of the template strand. The links made by these enzymes are covalent bonds between the deoxyribose sugar of one nucleotide and the phosphate group of the other. In theory, DNA polymerases could either link the sugar of a free nucleotide to the phosphate of the nucleotide at the end of the growing strand or phosphate of a free nucleotide to the sugar of the nucleotide at the end of the growing strand. In practice, the direction in which DNA polymerase works is always 5′ to 3′—the 5′ phosphate of a free nucleotide is linked to the 3′ end of the growing strand.

DNA polymerases add nucleotides to a chain or polymer of nucleotides being assembled on a template strand, but they cannot initiate the process. There has to be a 3′ terminal to which a free nucleotide can be added. The details of how replication is initiated and the roles of different types of DNA replication are described later.

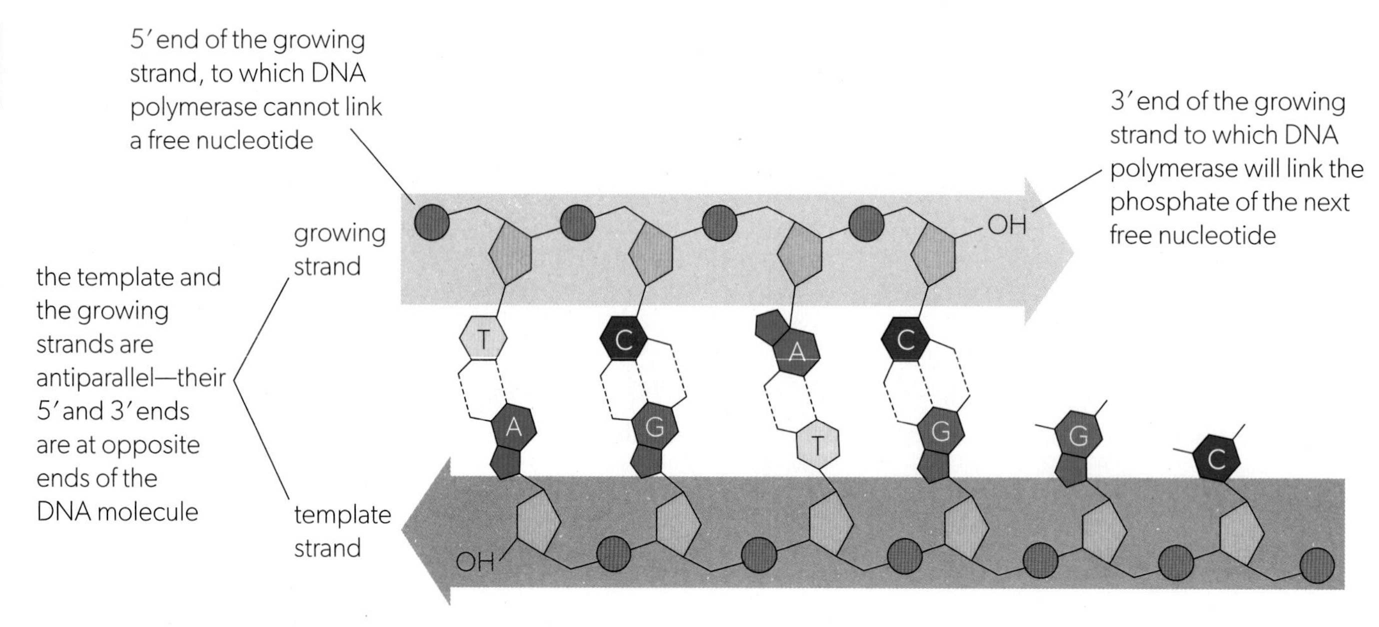

▲ Figure 18 Progress of DNA replication

D1.1.7 Differences between replication on the leading strand and the lagging strand

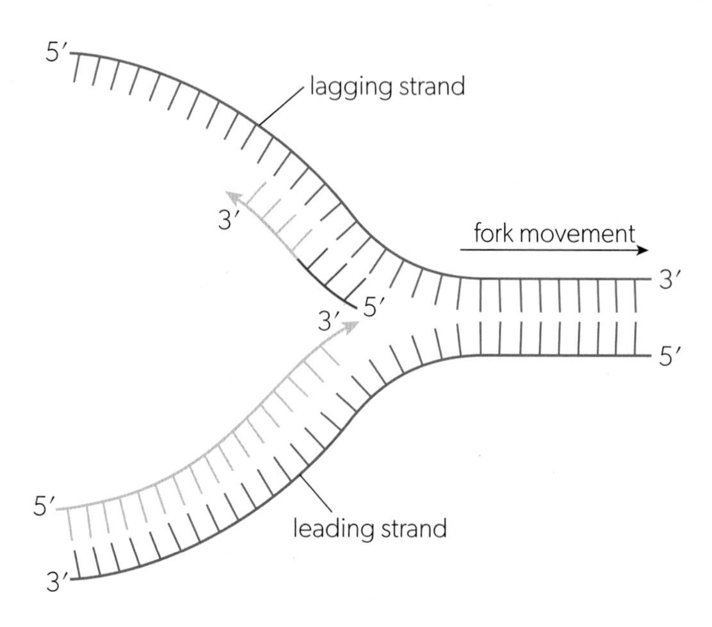

▲ Figure 19 Leading and lagging strands in DNA replication

DNA polymerase can only assemble new strands in a 5′ to 3′ direction. The two strands formed when helicase separates the DNA are antiparallel—their 5′ to 3′ directions are opposite. DNA polymerase using these strands as templates for assembling new strands moves in opposite directions—always 5′ to 3′.

- On one strand, DNA polymerase adds nucleotides moving towards the replication fork. Replication is continuous on this strand so it is completed more quickly. This is called the leading strand.
- On the other strand, DNA polymerase adds nucleotides moving away from the replication fork. They must be added in a series of lengths as the replication fork exposes more of the template strand and DNA can bind again. These short lengths of new DNA strand are called Okazaki fragments. Having to restart replication by DNA polymerase repeatedly makes the process relatively slow, This is called the lagging strand.

D1.1.8 Functions of DNA primase, DNA polymerase I, DNA polymerase III and DNA ligase in replication

Many different proteins have a role in DNA replication besides helicase. The functions of two DNA polymerases and two other enzymes are described here.

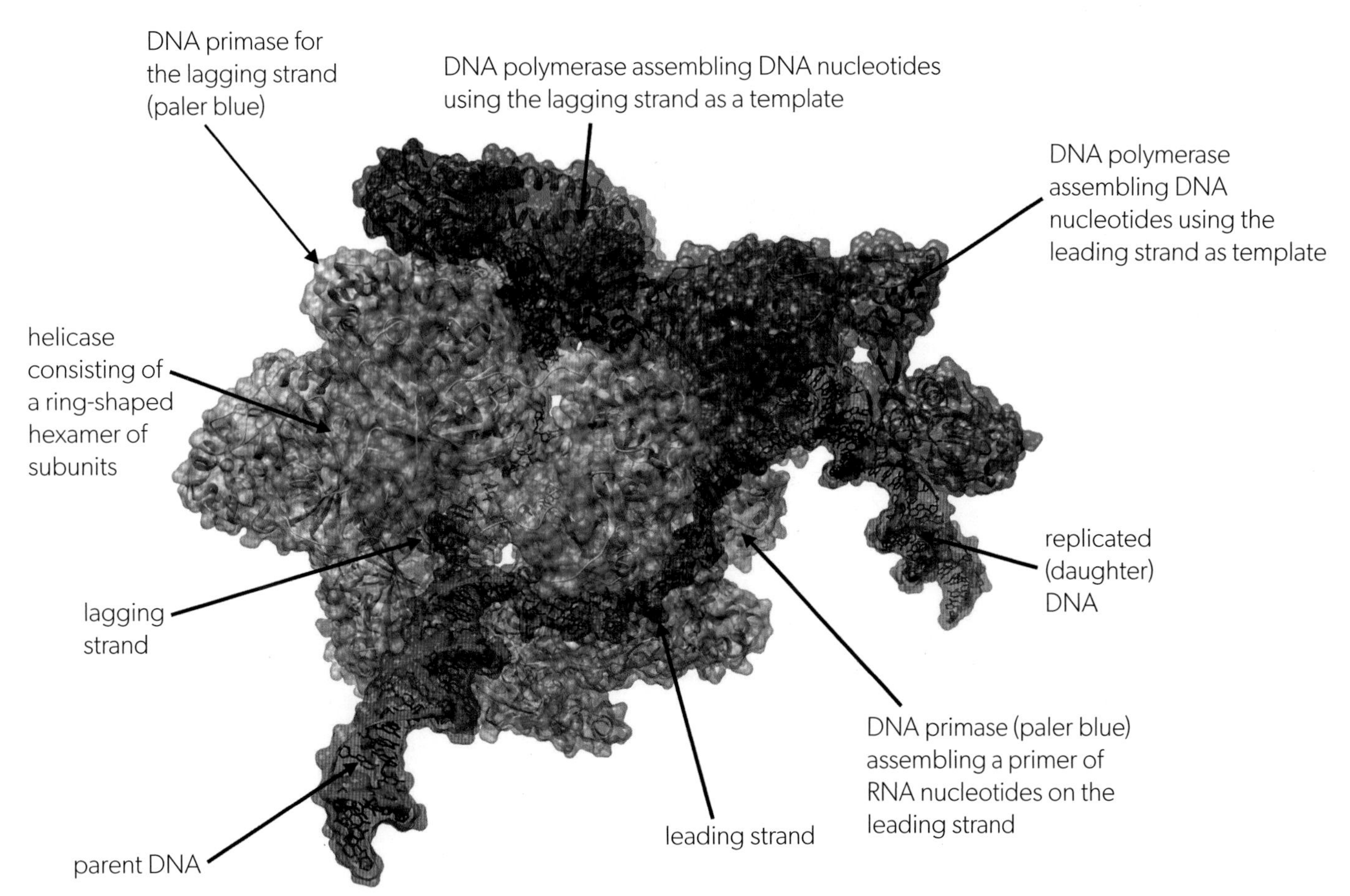

▲ Figure 20 Functional parts of the replisome of bacteriophage 17. A replisome is the multiprotein complex, including two DNA polymerase molecules that synthesize new strands of DNA on the leading and lagging strands

DNA primase is a type of RNA polymerase. During DNA replication it assembles a chain of about 10 RNA nucleotides on the template strand, to provide a site where DNA polymerase III can bind and start adding nucleotides to the 3′ end of a DNA strand. On the leading strand an RNA primer is only needed once, but primers are inserted every 100–200 nucleotides along the lagging strand.

DNA polymerase III is the principal polymerase in DNA replication (Figure 21). It binds to the template strand on the 3′ side of an RNA primer and assembles a chain of DNA nucleotides with bases complementary to those on the template strand. The nucleotides are added in a 5′ to 3′ direction, continuing until the end of the template strand or another RNA primer is reached. DNA polymerase III also proofreads each nucleotide after it has been added to the chain. This is described in *Section D1.1.9*.

DNA polymerase I is both an exonuclease because it can break bonds between nucleotides and a polymerase because it can link nucleotides. On the lagging strand, DNA polymerase I binds to the junction between the 3′ end of an Okazaki fragment and the 5′ start of a primer. It moves along the primer, removing RNA nucleotides and replacing them with DNA nucleotides. It detaches when it reaches the next Okazaki fragment. As DNA polymerases cannot make 3′ to 5′ links between nucleotides, DNA polymerase leaves a gap in the chain of nucleotides that make up the new DNA strand where a sugar–phosphate bond between two nucleotides is missing.

DNA ligase connects the gap in the chain nucleotides left by DNA polymerase I by making a sugar–phosphate bond between two adjacent nucleotides. Ligation is joining things together.

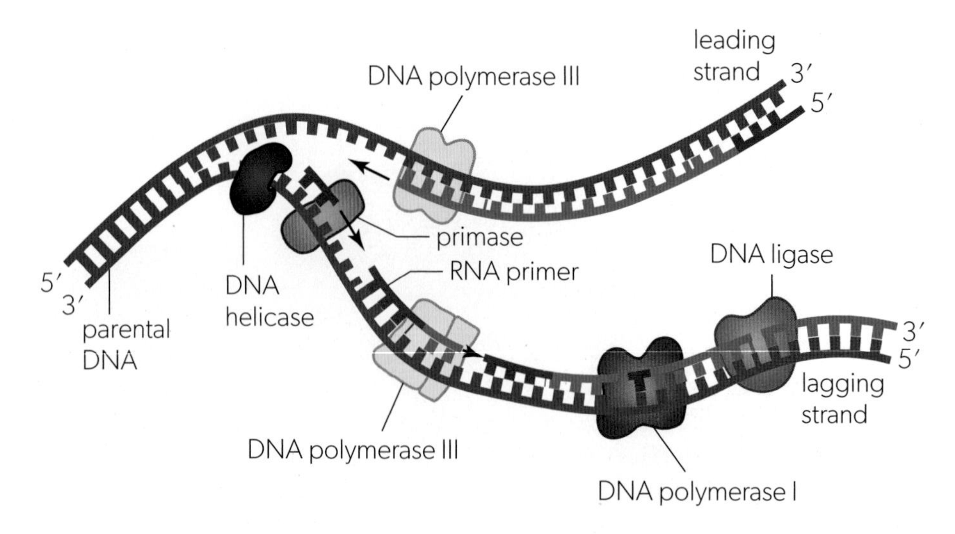

▲ Figure 21 Action of DNA replication

Data-based Questions: Discovery of Okazaki fragments

Tuneko and Reiji Okazaki performed experiments in the 1960s that led to the discovery of short DNA strands that were subsequently named Okazaki fragments. Samples of bacteria were supplied with radioactive nucleotides for a series of times (10, 30 or 60 seconds). This is called the "pulse". It was followed by adding a large excess of non-radioactive nucleotides for a longer period of time. This is the "chase". DNA was extracted from the bacteria, split into single strands by heat and then centrifuged to separate the molecules by size. The amount of radioactivity was measured at different depths in the centrifuge tube to assess the lengths of DNA strands assembled during the pulse. The closer to the top of the centrifuge tube, the shorter the DNA strand.

1. The graph shows DNA peaks at a distance of 0.5 units from the top of the tube at 10 and 30 seconds, but not at 60 seconds. These peaks represent Okazaki fragments. Using your understanding of DNA replication, explain the reasons for the changes in the numbers of Okazaki fragments with increasing lengths of pulse. [3]
2. a. Compare the data for the samples that were pulsed for 10 seconds and 30 seconds. [2]

   b. Explain the evidence for leading and lagging strands provided by the sample pulsed for 30 seconds. [3]
3. Explain the evidence for DNA ligase activity provided by the sample pulsed for 60 seconds. [3]

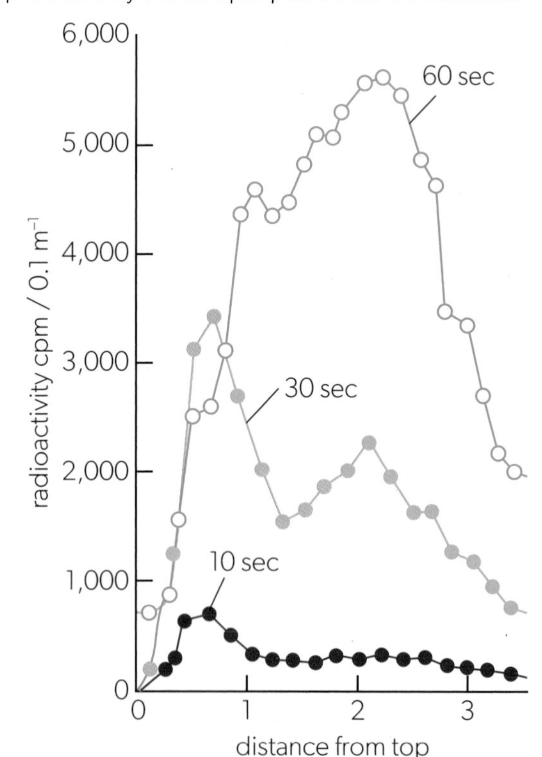

▲ Figure 22 Graph showing Okazaki fragment lengths for three time-lengths of pulse

D1.1.9 DNA proofreading

DNA polymerases replicate DNA with great fidelity. Where errors are made they are usually corrected, preventing mutations. There are multiple methods for doing this, the earliest of which is called "proofreading" and happens during replication immediately after a nucleotide with a mismatched base has been added to the growing chain.

The details of proofreading vary, but in prokaryotes it is done by DNA polymerase III. When DNA polymerase recognizes a base mismatch between the last nucleotide it has added to the growing chain and the base on the template strand, it excises the incorrect nucleotide, moves back along the template strand by one nucleotide and inserts a nucleotide with the correct base.

Linking questions

1. How is genetic continuity ensured between generations?
    a. Outline the concept of stabilizing natural selection. (D4.1.12)
    b. Explain how semi-conservative replication ensures continuity. (D1.1.2)
    c. Outline how negative feedback regulates the amount of a product created. (D3.3.2)
2. Which biological mechanisms rely on directionality?
    a. Distinguish between the leading and lagging strand in DNA replication. (D1.1.7)
    b. Outline one example of animal migration. (B3.3.9)
    c. Explain the process of the translocation of sugars. (B3.2.17)

D1.2 Protein synthesis

How does a cell produce a sequence of amino acids from a sequence of DNA bases?

What are the differences and similarities between DNA, RNA and proteins in terms of information storage?
How are the sequences of DNA and RNA decoded to enable protein translation?

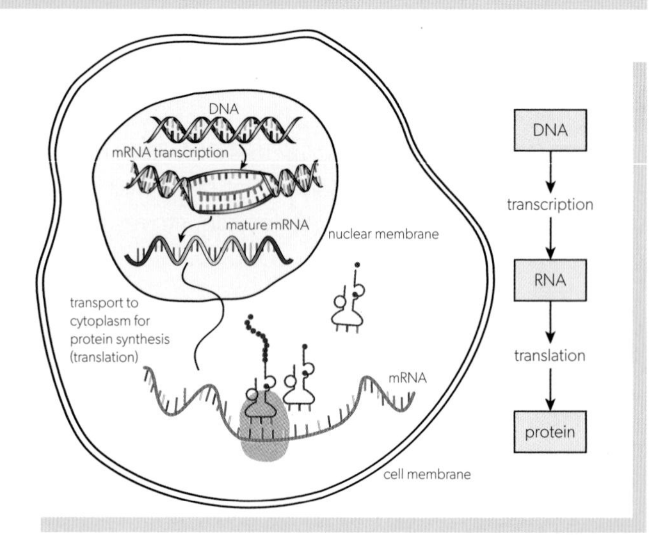

▶ Figure 1 An overview of protein synthesis

How is the reliability of protein synthesis ensured?

When events are reported in national newspapers using a different language, to what extent do transformations from the original meaning occur? Who is involved in the translation process and what effect do any transformations have on readers?

This is analogous with the translation of hereditary information in the form of mRNA to protein sequences. How is the reliability of mRNA translation ensured? Do errors in protein translation occur? What does it mean to say the genetic code is degenerate? Why do mutations not always result in a change in protein sequence?

▲ Figure 2

SL and HL	AHL only
D1.2.1 Transcription as the synthesis of RNA using a DNA template D1.2.2 Role of hydrogen bonding and complementary base pairing in transcription D1.2.3 Stability of DNA templates D1.2.4 Transcription as a process required for the expression of genes D1.2.5 Translation as the synthesis of polypeptides from mRNA D1.2.6 Roles of mRNA, ribosomes and tRNA in translation D1.2.7 Complementary base pairing between tRNA and mRNA D1.2.8 Features of the genetic code D1.2.9 Using the genetic code expressed as a table of mRNA codons D1.2.10 Stepwise movement of the ribosome along mRNA and linkage of amino acids by peptide bonding to the growing polypeptide chain D1.2.11 Mutations that change protein structure	D1.2.12 Directionality of transcription and translation D1.2.13 Initiation of transcription at the promoter D1.2.14 Non-coding sequences in DNA do not code for polypeptides D1.2.15 Post-transcriptional modification in eukaryotic cells D1.2.16 Alternative splicing of exons to produce variants of a protein from a single gene D1.2.17 Initiation of translation D1.2.18 Modification of polypeptides into their functional state D1.2.19 Recycling of amino acids by proteasomes

D1.2.1 Transcription as the synthesis of RNA using a DNA template

Transcription is the synthesis of RNA, using DNA as a template. Because RNA is single-stranded, transcription only occurs along one of the two strands of DNA. Genes can be transcribed repeatedly to provide as many copies of a base sequence as needed.

The enzyme RNA polymerase has multiple roles in transcription:

- binding to a site on the DNA at the start of the gene that is being transcribed
- unwinding the DNA double helix and separating it into two single strands (template and coding strands)
- moving along the template strand
- positioning RNA nucleotides on the template strand with bases complementary to those of the template
- linking the RNA nucleotides by covalent sugar–phosphate bonds to form a continuous strand of RNA
- detaching the assembled RNA from the template strand and allowing the DNA double helix to reform.

Transcription stops when a sequence is reached that indicates the end of the gene. The completed RNA molecule is then released. Because the RNA has a base sequence complementary to the template strand, its sequence is identical to the sense strand of the DNA, apart from one difference: uracil replaces thymine. There is no thymine in RNA. During transcription uracil not thymine pairs with adenine on the template strand.

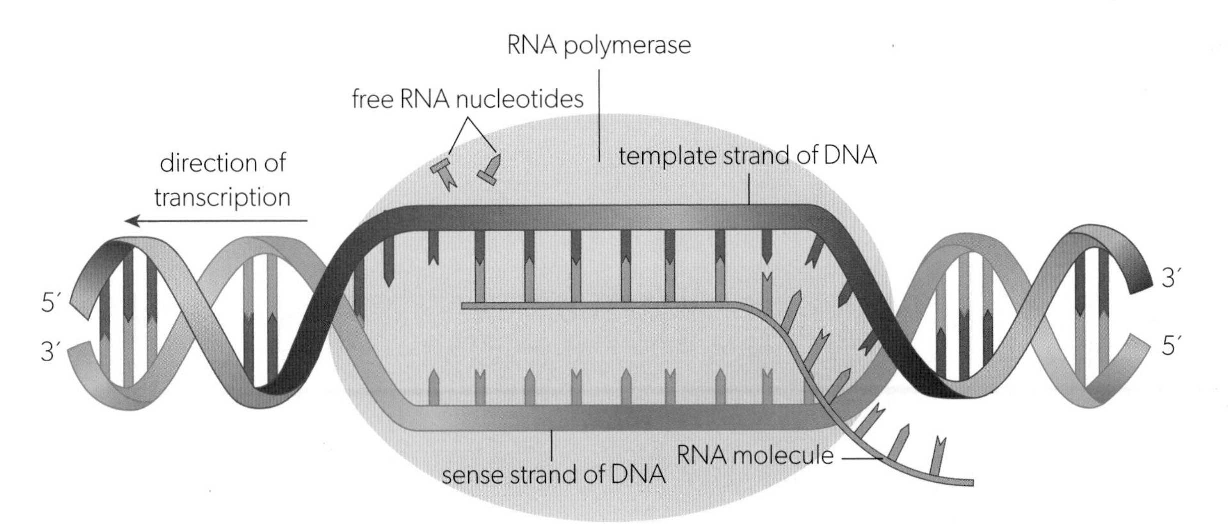

▲ Figure 3 RNA polymerase moves along the template strand, transcribing it nucleotide by nucleotide. In this diagram, RNA polymerase is moving towards the left

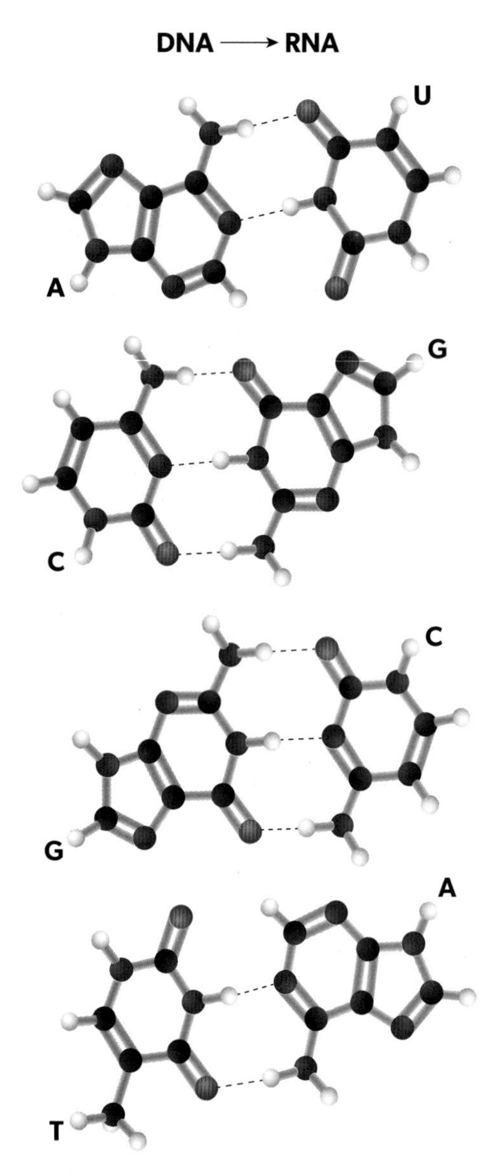

▲ Figure 4 Complementary base pairing during transcription between DNA (left) and RNA (right)

▲ Figure 6 This stability of DNA templates is particularly important in body cells (somatic cells) which in some cases live throughout an individual's life

D1.2.2 Role of hydrogen bonding and complementary base pairing in transcription

The copying of base sequences during transcription depends on complementary base pairing. Each nucleotide added to the growing RNA strand by RNA polymerase must have a base that is complementary to the corresponding base on the template DNA strand. Pairs of bases are complementary because they form hydrogen bonds with each other but not with other bases. Cytosine and guanine will only pair with each other. Thymine on the template strand will only pair with adenine on the RNA strand. Adenine on the template strand only pairs with uracil on the RNA strand.

Transcription is used to copy the base sequence of one of the two strands in a DNA molecule. The DNA strand with the base sequence to be copied into RNA is called the sense strand (or the coding strand). The other strand, which has a complementary base sequence to the sense strand, is called the template strand (or the antisense strand). Transcription of this strand results in a strand of RNA with the same base sequence as the sense strand of DNA except that uracil is replaced by thymine.

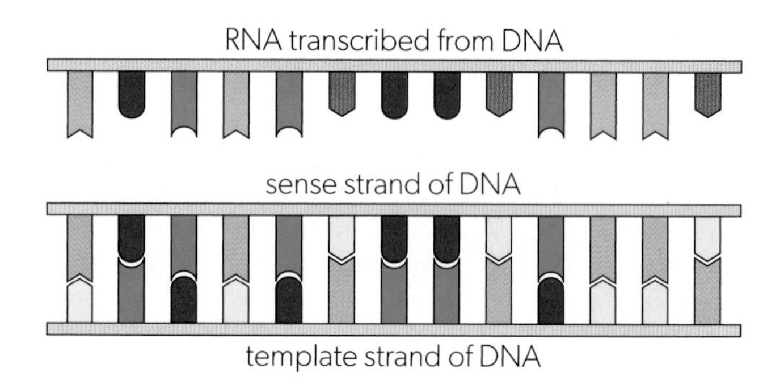

▲ Figure 5 Part of the DNA strand of a gene is represented in this diagram with the RNA strand that transcribed from it. Can you deduce which colour represents each of the five bases?

D1.2.3 Stability of DNA templates

When RNA splits DNA into single strands and uses the template strand to guide transcription, there should be no changes to the base sequence of the DNA. After transcription the two DNA strands pair up again with each base linked by hydrogen bonds to its complementary base on the opposite strand. The two strands are only parted for a short time as RNA polymerase moves along the gene, so the bases are only briefly vulnerable to chemical changes that would cause mutation.

The stability of DNA templates is essential because they may be transcribed many times during the life of a cell. If mutations were common, then frequently used templates would accumulate mutations and the RNA copies would contain more and more errors. Proteins translated from these copies would have increasing numbers of amino acid substitutions, which is very likely to make them function less well.

D1.2.4 Transcription as a process required for the expression of genes

Gene expression is the process by which information carried by a gene has observable effects on an organism. The sequence of the bases in genes does not determine the observable characteristics in an organism. The function of most genes is to specify the sequence of amino acids in a particular polypeptide. It is the proteins produced that directly or indirectly determine the observable characteristics of an individual. Two processes are needed to produce a specific polypeptide using the base sequence of a gene: transcription and translation.

▲ Figure 7 A blue pigment gives this budgerigar (*Melopsittacus undulatus*) a blue chest and tail feathers. Feathers on the head lack the blue pigment and are white

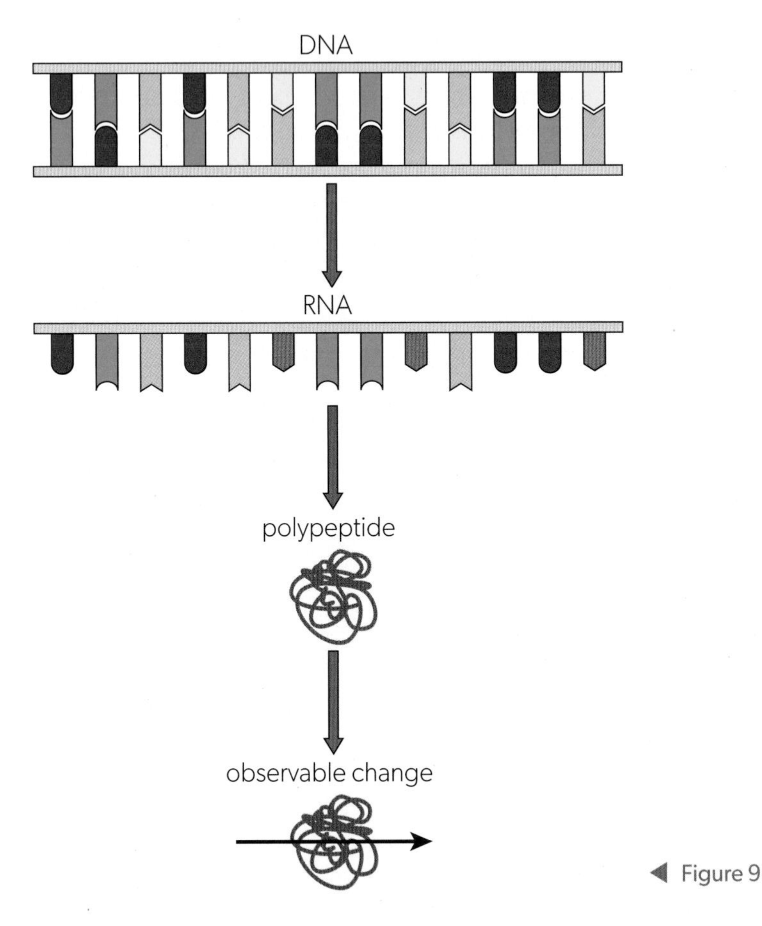

◀ Figure 9

▲ Figure 8 A gene has been expressed in this budgerigar that causes the enzyme polyketide synthase to be synthesized. The enzyme catalyses production of a yellow pigment. This changes white feathers to yellow and blue feathers to green

Transcription is the first stage in gene expression and the key stage at which it can be switched on or off. Only some genes are switched on in a cell at any particular time. Some genes may never be switched on during the life of a cell. For example, the gene for making insulin is only expressed in pancreatic β cells. In all other cells this gene is never normally transcribed. There are also genes that are always expressed because the proteins they code for are always required. These are the housekeeping genes with functions such as cell respiration.

The full range of RNA types made in a cell is its transcriptome. Within an individual, different cells or tissue types have different transcriptomes. Over time, the transcriptome changes as the activity of the cell changes.

D1.2.5 Translation as the synthesis of polypeptides from mRNA

▲ Figure 10 It is easy to confuse transcription and translation. Transcription originally meant copying a written text, as in this medieval image. Translation meant moving to a different place or changing something from one form to another

To make a specific polypeptide, amino acids must be linked together in the correct sequence. A typical polypeptide is a chain of hundreds of amino acids; each amino acid could be any one of the 20 amino acids. The information needed to make a polypeptide is held in the base sequence of an RNA molecule, copied from a gene by transcription. RNA holds the information in the form of the genetic code. To make an amino acid sequence from a base sequence this code is translated. The process of polypeptide synthesis is therefore called translation.

Translation happens in the cytoplasm. In eukaryotes, RNA is produced in the nucleus and then passes out to cytoplasm via nuclear pores. Distances are shorter in prokaryotes but transcription and translation happen in different parts of the cell. The RNA with the information for making a polypeptide travels from one place to another within a cell, so is known as messenger RNA (mRNA). Translation in biology is synthesis of polypeptides from mRNA.

ATL Thinking skills: Evaluating metaphors

Metaphors are useful to scientists because they help to make sense of phenomena, support conceptual thinking and communicate findings to both scientists and non-scientists. On the other hand, metaphors can also constrain scientific reasoning, contribute to misunderstandings and reinforce stereotypes and messages that undermine the goal of correct conceptual understanding.

1. a. Outline the ways in which the following terms are based on metaphors.

   i. Cells
   ii. Transcription
   iii. Translation
   iv. Genes as blueprints
   v. Junk DNA
   vi. (the nucleus as) Control centre
   vii. mRNA
   viii. Degeneracy

   b. Discuss the implied differences between "genes as blueprints" and "genes as recipes".

   c. To what extent can we argue that metaphors are useful and to what extent do they constrain our thinking?

D1.2.6 Roles of mRNA, ribosomes and tRNA in translation

Three components work together to synthesize polypeptides by translation.

- mRNA has a site to which a ribosome can bind and a sequence of codons that specifies the amino acid sequence of the polypeptide, with a start and a stop codon to indicate where translation should begin and end. One mRNA molecule can be translated many times but is broken down if it becomes damaged or if more copies of the polypeptide it codes for are not required.

- Transfer RNA (tRNA) translates the base sequence of mRNA into the amino acid sequence of a polypeptide. To do this tRNA molecules have an anticodon at one end consisting of three bases and at the other end an attachment point for the amino acid corresponding to the anticodon. Each type of tRNA molecule has a distinctive shape that is recognized by a dedicated activating enzyme, which attaches the correct amino acid onto the tRNA.
- Ribosomes are complex structures consisting of a small and a large subunit. The small subunit has a binding site for mRNA and the large subunit has three binding sites for tRNA. The large subunit also has a catalytic site that makes peptide bonds between amino acids, to assemble the polypeptide.

▼ Figure 11 The two subunits of the ribosome (blue and green) are shown in this image, with tRNAs (yellow) bound on each of the three binding sites. A small part of the much larger mRNA molecule (red) is shown with its codons linked to the anticodons of the tRNAs. In this diagram the ribosome is moving from right to left along the mRNA

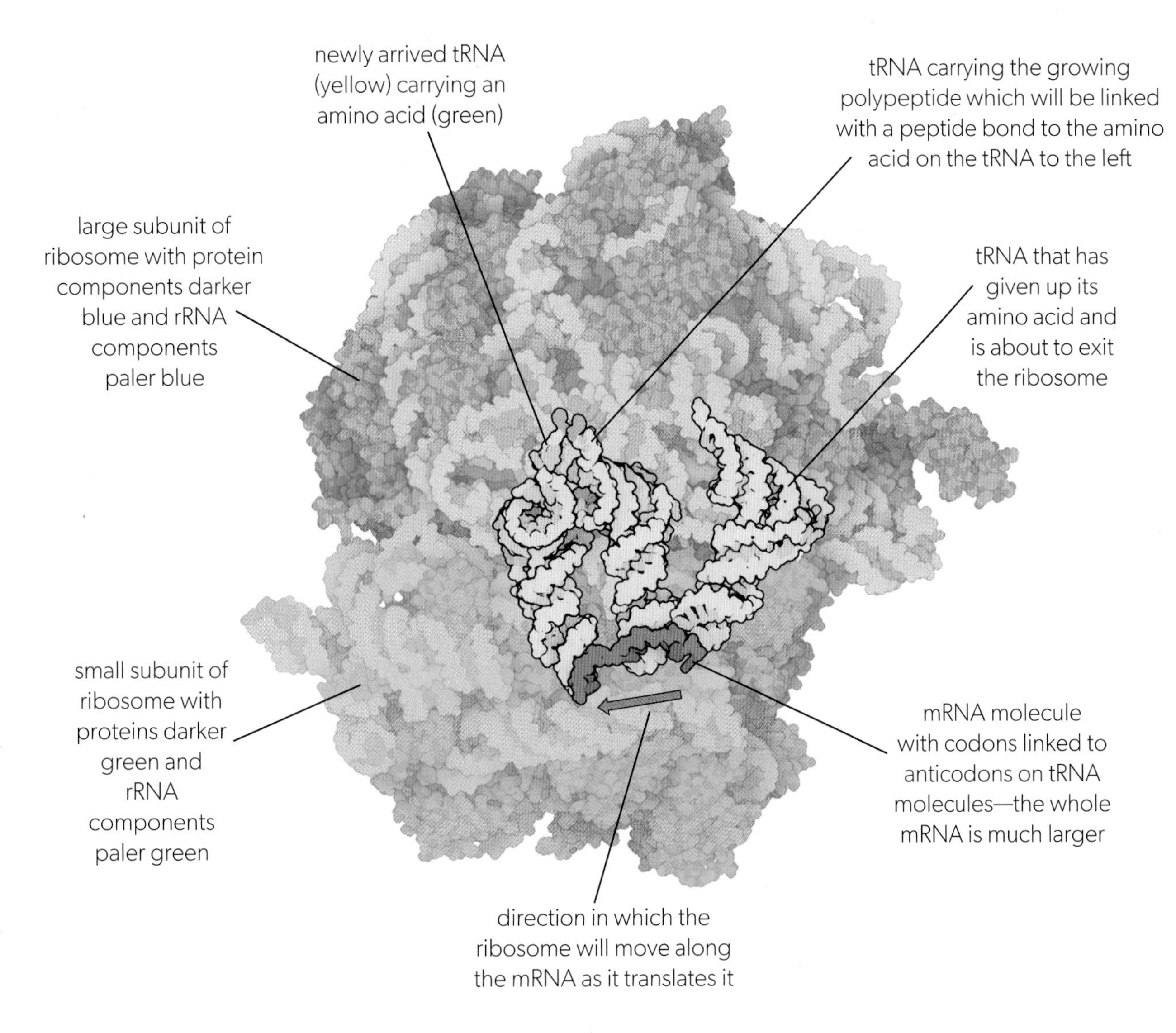

D1.2.7 Complementary base pairing between the tRNA and mRNA

The faithfulness of translation depends on complementary base pairing. The three bases of an anticodon on a tRNA must be complementary to the three bases of the next codon on mRNA for the tRNA to be able to bind to the ribosome and deliver its amino acid. Adenine and uracil are complementary and also cytosine and guanine.

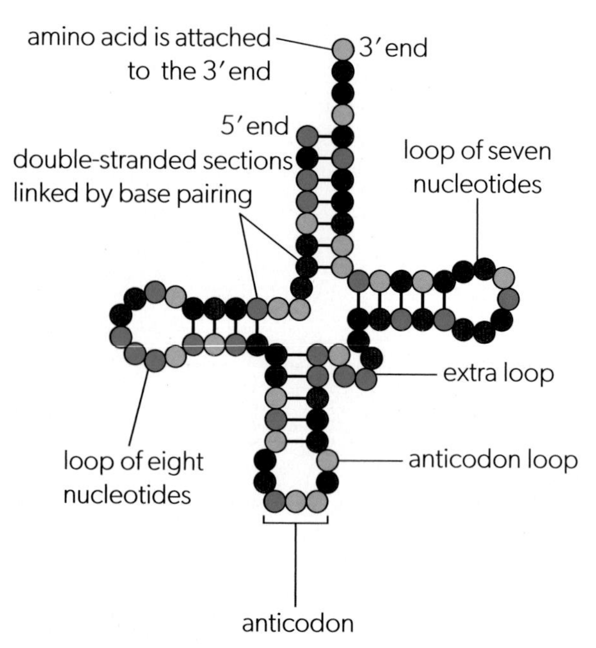

▶ Figure 12 There are double-stranded regions in tRNA molecules which result in "hairpin" folding and an overall arrangement of nucleotides that is likened to a cloverleaf. The anticodon and the position for an amino acid to attach are at opposite ends. In this diagram, nucleotides with adenine as the base are shown in blue, cytosine in red, guanine in green and uracil in purple

D1.2.8 Features of the genetic code

DNA is singularly well suited to data storage because it can hold long sequences of bases (A, C, G and T) which can be arranged in any order. The sequences can be accurately copied. The data commonly stored in DNA base sequences is the amino acid sequences of polypeptides. It is stored in a coded form. There are four different bases and twenty amino acids, so one base cannot code for one amino acid. There are 16 combinations of two bases, which is still too few to code for all 20 amino acids. Living organisms therefore use a triplet code, with groups of three bases coding for an amino acid. There are 64 combinations of three bases.

A sequence of three bases on the mRNA is called a codon. All but three of these codons cause a specific amino acid to be added to a growing polypeptide. Table 1 shows all 64 codons. The left-hand column indicates the first base in a codon, the second base is indicated by the row of bases at the top of the table, and the right-hand column indicates the third base of the codon. The coloured cells of the table tell you what amino acid is coded for. So, for example, the codon AUG codes for the amino acid methionine (often abbreviated to Met) and also acts as the start codon in translation. The codon CAU codes for histidine. There are also three stop codons that end the process of translation.

Two notable features of the genetic code are degeneracy and universality. The code is said to be degenerate because different codons can code for the same amino acid. For example, the codons GUU and GUC both code for the amino acid valine. The code is universal because it is used by all living organisms and viruses, with only very minor changes in some cases.

First base ↓	Second base in codon				Third base ↓
	U	C	A	G	
U	phenylalanine	serine	tyrosine	cysteine	U
	phenylalanine	serine	tyrosine	cysteine	C
	leucine	serine	stop	stop	A
	leucine	serine	stop	tryptophan	G
C	leucine	proline	histidine	arginine	U
	leucine	proline	histidine	arginine	C
	leucine	proline	glycine	arginine	A
	leucine	proline	glycine	arginine	G
A	isoleucine	threonine	asparagine	serine	U
	isoleucine	threonine	asparagine	serine	C
	isoleucine	threonine	lysine	arginine	A
	methionine/start	threonine	lysine	arginine	G
G	valine	alanine	aspartic acid	glycine	U
	valine	alanine	aspartic acid	glycine	C
	valine	alanine	glutamic acid	glycine	A
	valine	alanine	glutamic acid	glycine	G

▲ Table 1 The genetic code: what is translated from each of the 64 possible codons on mRNA

D1.2.9 Using the genetic code expressed as a table of mRNA codons

There is no need to try to memorize the genetic code. However, if a table showing it is available, you should know how to make various deductions.

1. Which codons correspond to an amino acid?

   Three letters are used to indicate each amino acid in the table of the genetic code. Each of the 20 amino acids has between one and six codons. Read off the three letters of each codon for the amino acid. For example, the amino acid tryptophan has one codon which is UGG.

2. What amino acid sequence would be translated from a sequence of codons in a strand of mRNA?

   The first three bases in the mRNA sequence are the codon for the first amino acid, the next three bases are the codon for the second amino acid and so on. Look down the left-hand side of the table to find the first base of a codon, across the top of the table to find the second base and down the right-hand side to find the third base. For example, GCA codes for the amino acid alanine.

3. What base sequence in DNA would be transcribed to give the base sequence of a strand of mRNA?

   A strand of mRNA is produced by transcribing the template or antisense strand of the DNA. This therefore has a base sequence complementary to the mRNA. For example, the codon AUG in mRNA is transcribed from the base sequence TAC on the template or antisense strand of the DNA. A longer example is that the base sequence GUACGUACG in mRNA is transcribed from CATGCATGC on the template strand of DNA. Note that adenine pairs with thymine in DNA but with uracil in RNA.

Activity: Interpreting the genetic code

1. Using the genetic code (Table 1), deduce the codons for:
   a. methionine
   b. tyrosine
   c. arginine [3]
2. Deduce the amino acid sequences that correspond to these mRNA sequences:
   a. ACG
   b. CACGGG
   c. CGCGCGAGG [3]
3. If mRNA contains the base sequence CUCAUCGAAUAACCC:
   a. deduce the amino acid sequence of the polypeptide translated from the mRNA [2]
   b. deduce the base sequence of the template (antisense) strand transcribed to produce the mRNA. [2]

D1.2.10 Stepwise movement of the ribosome along mRNA and linkage of amino acids by peptide bonding to the growing polypeptide chain

Translation of an mRNA molecule is done by a repeating cycle of steps. Each cycle results in the addition of one amino acid to the growing polypeptide chain. Once in each cycle the ribosome moves three bases along the mRNA, which is one codon. The cycle is shown in Figure 13. The process can also be considered by following what happens to one tRNA molecule during the process.

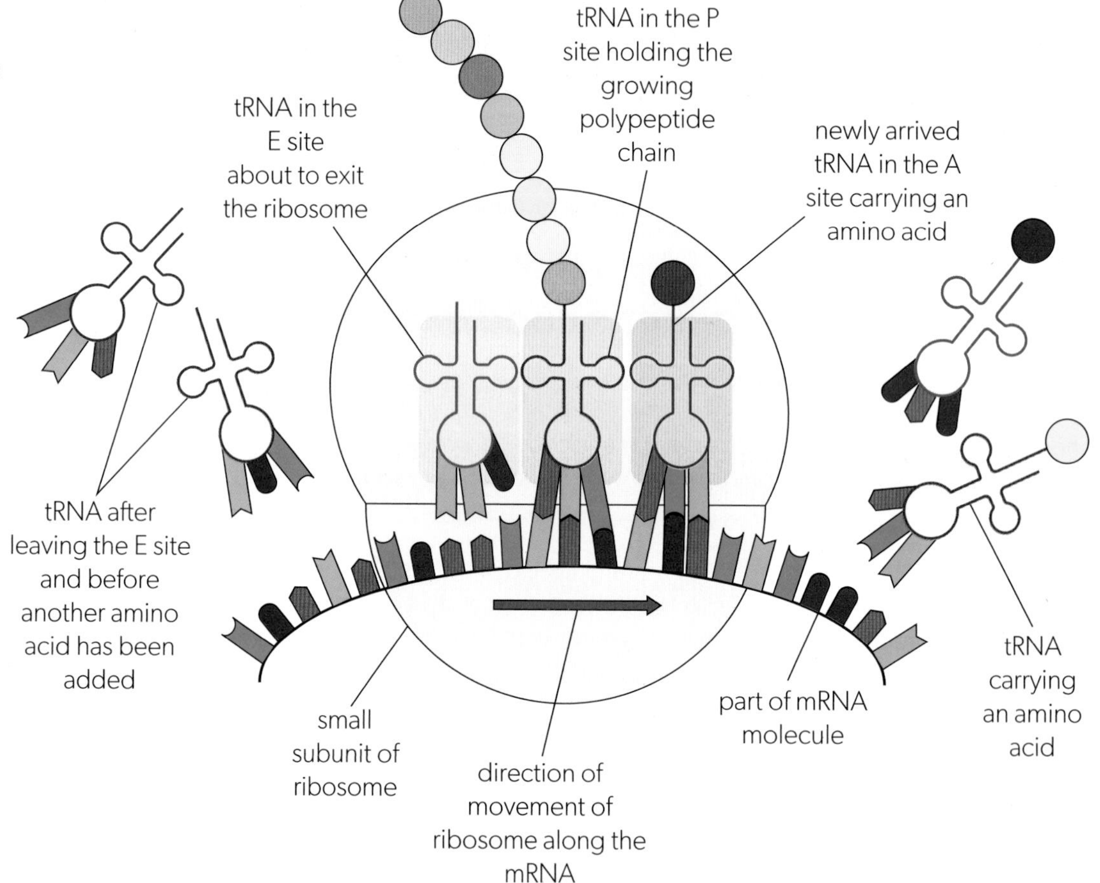

▶ Figure 13 The anticodon of the tRNAs in the P site and the A site are linked by complementary base pairing to codons on mRNA

1. An activating enzyme with an active site that fits the tRNA binds to it and attaches the specific amino acid corresponding to the anticodon of the tRNA.
2. The tRNA carrying a single, attached amino acid binds to the A (amino acyl) site on the ribosome, with its anticodon linked by complementary base pairing to the next codon on mRNA.
3. The single amino acid on the tRNA is linked to the end of the growing polypeptide by formation of a peptide bond. The tRNA is now holding the whole of the growing polypeptide.
4. The tRNA moves from the A to the P (peptidyl) site as the ribosome moves along the mRNA by one codon. The anticodon of the tRNA is still paired with the codon on the mRNA.

5. The polypeptide held by the tRNA is transferred to another tRNA that has arrived at the A site.
6. The tRNA moves from the P to the E (exit) site as the ribosome moves along the mRNA by one more codon. This causes the anticodon of the tRNA to separate from the codon on the mRNA and the tRNA to separate from the ribosome.
7. The cycle starts again with an amino acid being linked to a tRNA.

D1.2.11 Mutations that change protein structure

A gene mutation is a change to the base sequence of a gene. Even a mutation as small as a single base substitution changes a codon to a different one. The changed codon may code for a different amino acid especially if the first or second base of the triplet has been changed. A base substitution mutation therefore usually results in one amino acid changed in the polypeptide produced by translation. This might not affect the structure of the protein much, but it might affect it radically.

Sickle cell disease, the commonest inherited condition in the world, is an example of radical change due to a single base substitution. The mutated gene codes for the beta-globin polypeptide in haemoglobin. The symbol for this gene is Hb. Most humans have the allele HbA. If a base substitution mutation converts the sixth codon of the gene from GAG to GTG, a new allele is formed (form of the gene), called HbS. The mutation can be inherited by offspring if it occurs in a cell of the ovary or testis that later develops into an egg or sperm.

When the HbS allele is transcribed, the mRNA produced has GUG as its sixth codon instead of GAG, and when this mRNA is translated, the sixth amino acid in the polypeptide is valine instead of glutamic acid. This change affects the haemoglobin molecules when they are in tissues with low oxygen concentrations. In these circumstances, the haemoglobin molecules link together into chains and form bundles that are rigid enough to distort the red blood cells into a sickle shape.

These sickle cells become trapped in blood capillaries, blocking them and reducing blood flow. This causes damage to tissues. When sickle cells return to high oxygen conditions in the lung, the haemoglobin bundles break up and the cells return to their normal shape. These changes occur time after time, as the red blood cells circulate. Both the haemoglobin and the plasma membrane are damaged and the life of a red blood cell can be shortened to as little as 4 days. The body cannot replace the damaged red blood cells at a rapid enough rate and anaemia therefore develops.

So, a small change to a gene can have very harmful consequences for individuals that inherit the gene. It is not known how often this mutation has occurred but in some parts of the world the HbS allele is remarkably common. In parts of East Africa, up to 5% of newborn babies have two copies of the allele and develop severe anaemia. Another 35% have one copy so make both normal haemoglobin and the mutant form. These individuals develop mild anaemia.

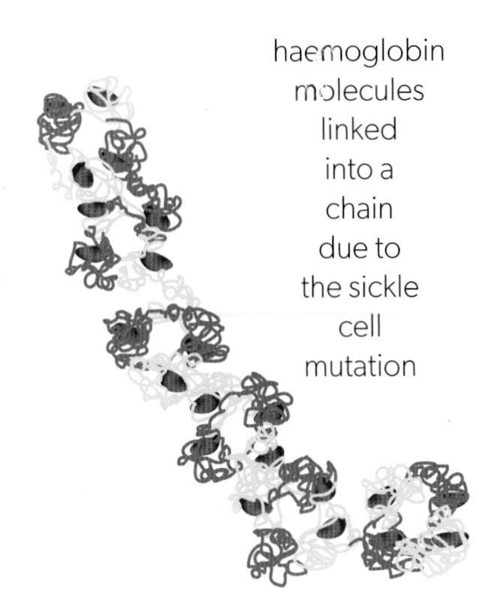

▲ Figure 14 Haemoglobin is a globular protein composed of four polypeptides. The sickle cell mutation causes the protein molecules to link together in chains

D1.2.12 Directionality of transcription and translation

A single strand of DNA or RNA has directionality because the two ends are different (Figure 15). This directionality and the specificity of enzymes, directs that both transcription and translation are unidirectional processes.

- In transcription, each RNA nucleotide is added to the growing strand by linking the phosphate group of the free nucleotide onto the ribose sugar at the end of the growing strand. Transcription is therefore always 5′ to 3′ in direction.
- In translation, the ribosome moves along the mRNA molecule towards its 3′ end. Translation is therefore also always 5′ to 3′.

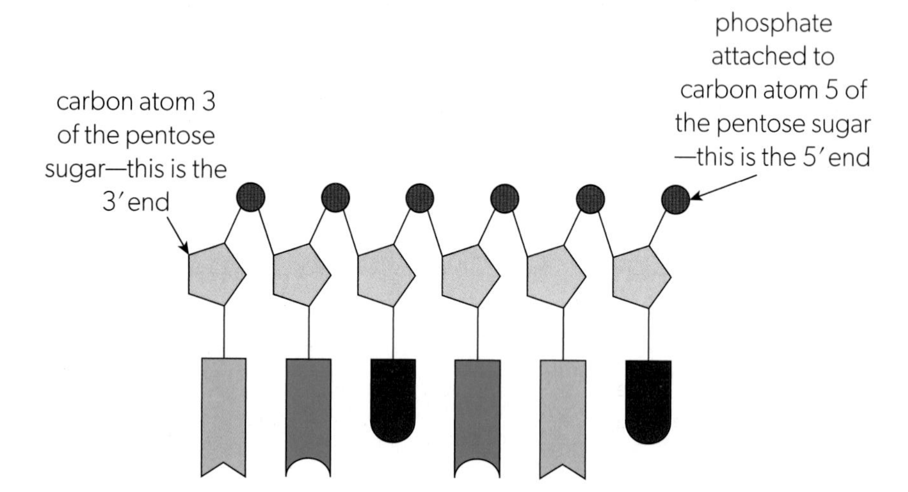

▲ Figure 15 The backbone of a single strand of DNA or RNA; the strand is a chain of alternating pentose sugars and phosphate. One end of the chain ends with a phosphate group and the other ends with a pentose sugar. These are known as the 5′ (five-prime) and 3′ (three-prime) ends respectively

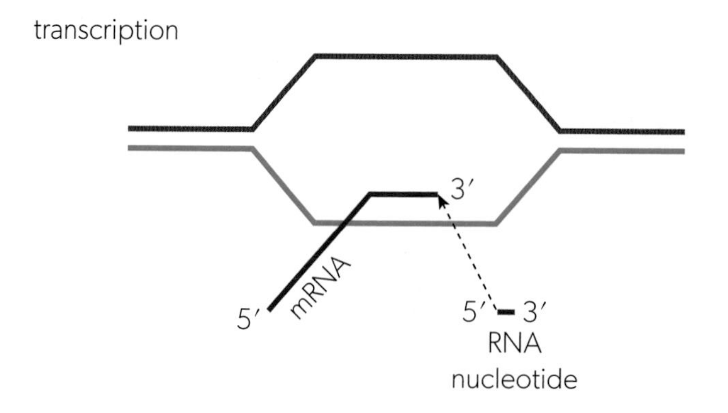

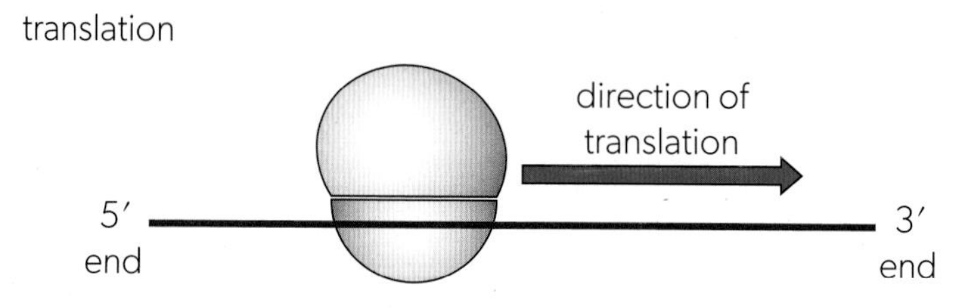

▲ Figure 16 Diagrammatic representations of transcription and translation

D1.2.13 Initiation of transcription at the promoter

A promoter is a section of the DNA that initiates gene transcription. A promoter is typically between 100 and 1,000 bases long and is located adjacent to genes on chromosomes (Figure 17). The base sequence of a promoter allows RNA polymerase to bind along with a variety of proteins that act as transcription factors.

Repressor sequences within the promoter allow transcription factors to bind that prevent binding of RNA polymerase, so transcription of a gene is prevented. Activator sequences within the promoter allow transcription factors to bind that encourage binding of RNA polymerase, leading to transcription.

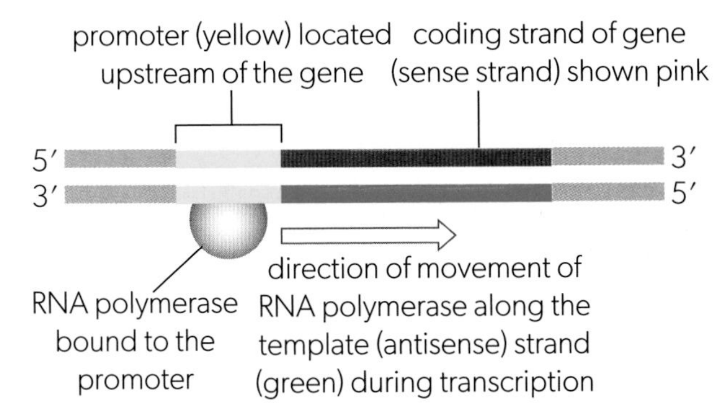

▲ Figure 17 Promoters are upstream of the gene. This means that they are between the gene and the 5′ end of the sense strand of the chromosome. RNA polymerase binds to the promoter and starts to transcribe the template (antisense) strand moving from one end of the gene to the other and making a complete RNA copy

Cells can change their pattern of gene expression depending on what transcription factors are produced to bind with the promotors. Some groups of genes have very similar promoters, so they are all expressed together even if they are located on different chromosomes. Some transcription factors only bind to the promoter after combining with another molecule. For example, the receptor for testosterone acts as a transcription factor. If testosterone binds to it, the receptor can bind to many promoters of genes with roles in male sexual development and activity.

D1.2.14 Non-coding sequences in DNA do not code for polypeptides

Many genes have the coding sequence needed for synthesizing a polypeptide. There are also base sequences in genomes that do not code for polypeptides, but have other functions. Five examples of base sequences that have a function other than coding for a polypeptide are listed here:

- base sequences that can be transcribed to produce tRNA
- base sequences that can be transcribed to produce ribosomal RNA (rRNA)

- base sequences used in the regulation of gene expression such as promoters (the binding sites for RNA polymerase), enhancers and silencers
- telomeres—the structures that form the ends of chromosomes in eukaryotes
- introns—base sequences that are edited out of mRNA during post-transcriptional modification

Data-based questions: Determining an open reading frame

Once the sequence of bases in a piece of DNA has been determined, a researcher may want to locate a gene. To do this, computers search through the sequence looking for open reading frames. An open reading frame is a series of bases uninterrupted by stop codons that could therefore code for a protein. The stop codons are UGA, UAA and UAG.

1. State the number of codons in the genetic code.
2. Determine the fraction of codons that are stop codons in the genetic code.
3. Look at Figure 18. The sequence of codons could start with the first, second or third base. These three starting positions correspond to three different reading frames: RF1, RF2 and RF3. Determine which of these is a possible open reading frame.

DNA 3′	A T T A A C T A T A A A G A C T A C A G A G A G G G C T A G T A C
mRNA 5′ RF1	U A A U U G A U A U U U C U G A U G U C U C U C C C G A U C A U G
RF2	A A U U G A U A U U U C U G A U G U C U C U C C C G A U C A U G
RF3	A U U G A U A U U U C U G A U G U C U C U C C C G A U C A U G

▲ Figure 18 A DNA sequence and three possible mRNA reading frames

Practising methodologies: Finding an open reading frame

Alcanivorax borkumensis is a rod-shaped bacterium that uses oil as an energy source. It is relatively uncommon but quickly dominates the marine microbial ecosystem after an oil spill. Scientists sequenced the genome of this bacterium in an effort to identify the genetic aspects of its oil-digesting ability. The entire genome can be accessed from the database GenBank.

Visit GenBank and search by genome to locate the genome of this organism. Click on FASTA to identify the organism's GI number. It is listed in the title (GI number 110832861). View the genome.

Go to the open reading frame finder (http://www.ncbi.nlm.nih.gov/projects/gorf/). Perhaps working as a class, divide the genome into 2,000 bp pieces. Enter the GI number and specify the range of bases that you are going to search. Share information with classmates about the open reading frames identified.

As an additional exercise, visit the Swiss Institute of Bioinformatics and look for the tool "Expasy". This allows you to paste in a DNA sequence and it will output the likely protein sequences.

D1.2.15 Post-transcriptional modification in eukaryotic cells

In prokaryotes, RNA can be translated as soon as it has been transcribed. In eukaryotes, newly produced RNA is changed into mature mRNA before transcription. These changes happen before the RNA leaves the nucleus. This modification explains the need for a nuclear membrane in eukaryotes but not prokaryotes—modification of RNA must be completed before mRNA encounters ribosomes in the cytoplasm.

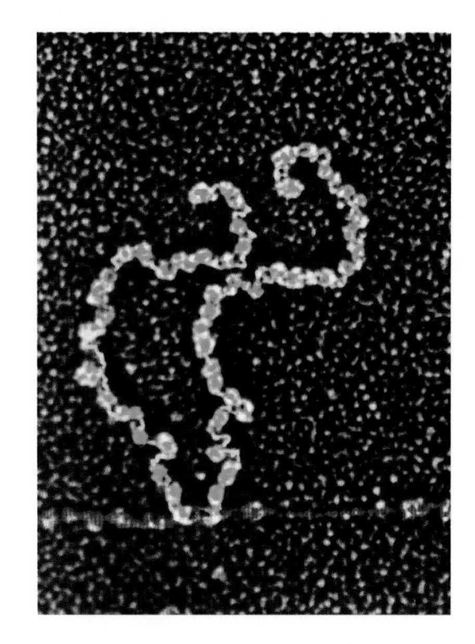

▲ Figure 19 Both DNA transcription and translation are visible in this electron micrograph of the bacterium *Escherichia coli*. DNA is coloured pink, mRNA strands are coloured green and ribosomes with polypeptides being produced by translation are coloured cyan. Transcription and translation would never be physically coupled in this way in eukaryotes

The separation of transcription and translation into separate compartments in eukaryotic cells allows for significant post-transcriptional modification to occur before the mature mRNA exits the nucleus. RNA is modified by adding extra nucleotides to give each end of mRNA a special structure.

- Five-prime caps—a modified nucleotide is added to the 5′ end of the RNA. This nucleotide has three phosphate groups instead of one, so it is similar to ATP. Its base is guanine, with an extra methyl group added.
- Poly (A) tails—between 100 and 200 adenine nucleotides are added to the 3′ end of the RNA. Translation stops before the ribosome reaches the poly (A) tail.

The 5′ cap and the poly (A) tail both stabilize the ends of the mRNA by protecting them from digestion by nuclease enzymes. They also encourage further modification of the original transcript.

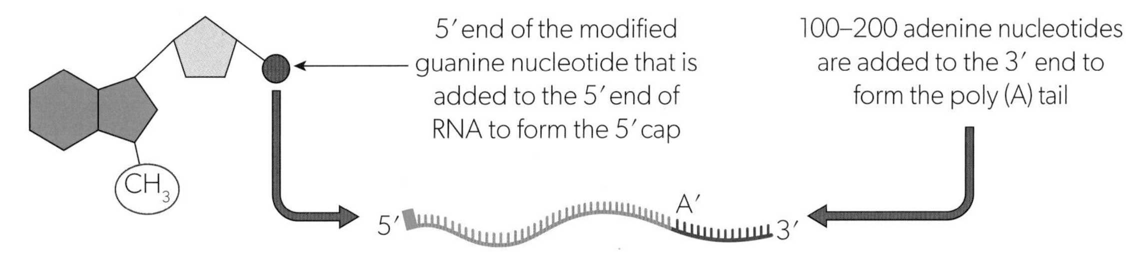

▲ Figure 20 Addition of 5′ caps and poly (A) tails

Eukaryotic genes that code for polypeptides contain two types of sequence:

- exons—coding sequences that are expressed by translation into the amino acid sequence
- introns—intervening sequences that are not expressed.

Introns are mostly between 20 and 200 nucleotides long. The mean number of introns per gene in humans is 7.8 and the mean number of exons is 8.8.

RNA transcribed from genes coding for polypeptides contains alternating exons and introns. The introns are removed and digested into single nucleotides. The remaining exons are spliced together (Figure 21), to produce an uninterrupted base sequence for translation. The mRNA is then mature and can be exported from the nucleus for translation by ribosomes in the cytoplasm.

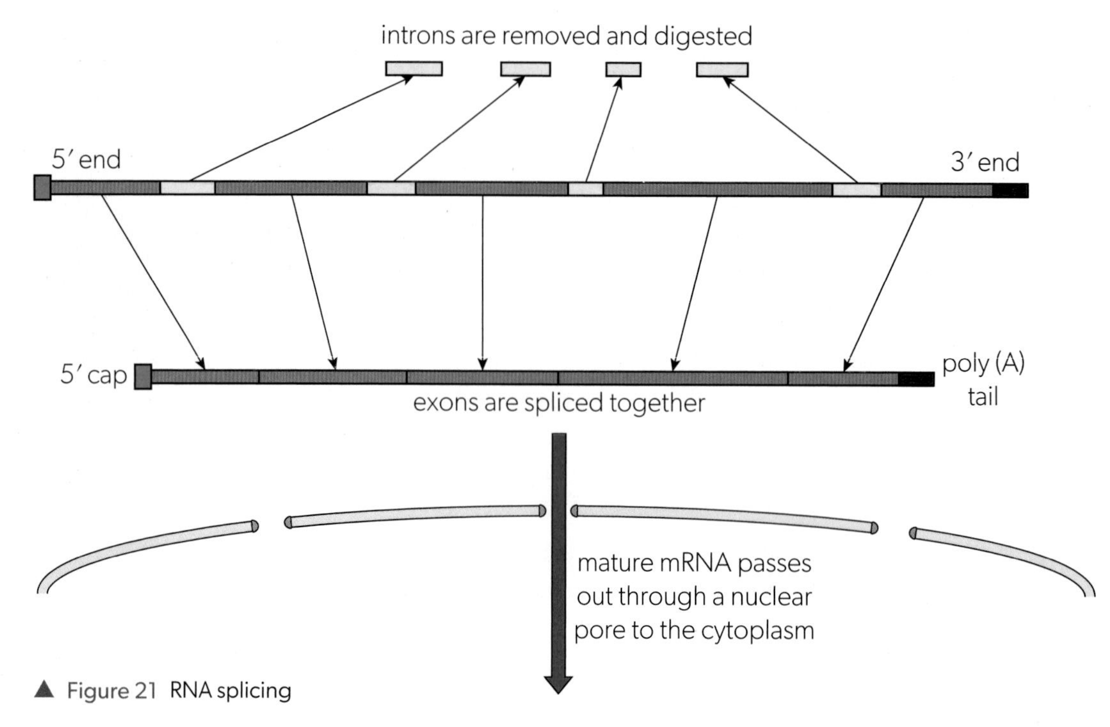

▲ Figure 21 RNA splicing

D1.2.16 Alternative splicing of exons to produce variants of a protein from a single gene

Alternative splicing is a type of change that allows several variants of a polypeptide to be produced from one gene. The primary transcript is the same in each case, but there are differences in the post-transcriptional modification. The most frequent change is exon skipping, where a sequence of bases in the mRNA is edited out to produce some variants of the polypeptide but remains in the mRNA to produce other variants. Many other methods of alternative splicing have been discovered.

It is useful to splice exons together in different ways because it allows polypeptides to be produced with changes to form and function without having to duplicate a gene in the genome. The troponin T gene expressed in heart muscle is an example of alternative splicing.

Alternative splicing of TNNT2—a gene for making a polypeptide in heart muscle cells

Troponin is a muscle protein that regulates muscle contraction and relaxation. It is attached to the actin filaments in muscle and can cause another protein (tropomyosin) to move, exposing binding sites for myosin filaments to bind. This results in muscle contraction.

Troponin consists of three subunits: troponin C, troponin T and troponin I, each playing a different role in muscle contraction. Troponin T (TnT) is the subunit that binds to tropomyosin. While there is only one version of troponin C and troponin I, there are four different forms of troponin T. Which form is present changes over the course of human development. TnT1 and TnT2 are found in the foetal heart and TnT3 is found in the normal adult human heart. TnT4 is found in a diseased heart. TNNT is the gene that codes for troponin T in heart muscle. The human TNNT gene has 17 exons and therefore 16 introns. Exons 4 and 5 are alternatively spliced to produce four different versions of troponin C:

- TnT1: all exons present in the mature mRNA
- TnT2: exon 5 present but not exon 4
- TnT3: exon 4 present but not exon 5
- TnT4: neither exon 4 nor exon 5 present.

Exons 4 and 5 both make troponin C more sensitive to calcium ions.

In mammals, the protein tropomyosin is encoded by a gene that has 11 exons. Tropomyosin pre-mRNA is spliced differently in different tissues resulting in five different forms of the protein. For example, in skeletal muscle, exon 2 is missing from the mRNA; in smooth muscle, exons 3 and 10 are missing from the mRNA.

In fruit flies, the Dscam protein is involved in guiding growing nerve cells to their targets. Research has shown that there are potentially 38,000 different mRNAs possible based on the number of different introns in the gene that could be spliced alternatively.

More broadly, alternative splicing increases the diversity of the proteome—all of the proteins expressed in an organism.

D1.2.17 Initiation of translation

Translation of mRNA starts with a series of special steps and then proceeds with the repeating cycle of steps described in *Section D1.2.10*. Here are the steps used to initiate translation.

- An activating enzyme attaches the amino acid methionine to an initiator tRNA that has the anticodon UAC.
- The initiator tRNA binds to the small subunit of the ribosome. This produces a ternary complex—a structure composed of three parts which in this case are the small subunit of the ribosome, the initiator tRNA and the amino acid.
- The ternary complex binds to the 5′ end of the mRNA and slides along it, using the UAC anticodon to scan for the start codon AUG.
- The small subunit stops moving along the mRNA, with the UAC anticodon linked to the AUG codon by hydrogen bonds.

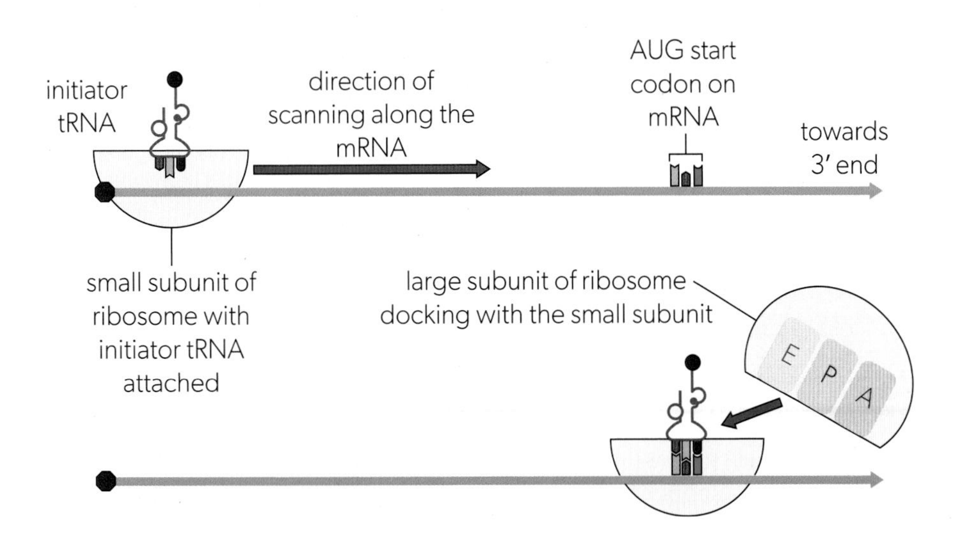

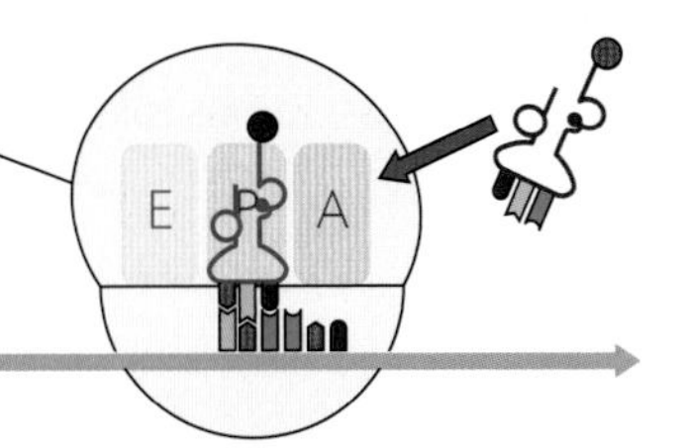

◀ Figure 22 Initiation is then followed by elongation—the repeating cycle of steps described in *Section D1.2.10*—until a stop codon is reached. The polypeptide and tRNAs are then released and the large and small subunits of the ribosome separate

- The large subunit of the ribosome binds to the small subunit, with the initiator codon on the P site.
- A tRNA with an anticodon complementary to the next codon on the mRNA binds to the A site, with the codon linked to the anticodon by hydrogen bonds.
- A peptide bond is formed between the amino acids carried by the initiator tRNA on the P site and the tRNA on the A site. The tRNA on the A site holds the dipeptide formed.
- The ribosome moves in a 5′ to 3′ direction along the mRNA by three nucleotides (one codon) causing the initiator tRNA to move to the E site and exit the ribosome.

D1.2.18 Modification of polypeptides into their functional state

Modification of polypeptides after translation turns them into functional proteins. These are the main types of change.

- Removal of the amino acid methionine from the 5′ end of the polypeptide, where it was placed during initiation of translation.
- Changes to the side chains of amino acids in the polypeptide. For example, phosphorylation or addition of carbohydrate molecules, can produce over 100 chemically different amino acids in proteins.
- Folding the polypeptide and making intramolecular interactions to stabilize the tertiary structure. For example, disulfide bonds are formed between cysteines that come close together due to folding.
- Conversion of propeptides into mature peptides by removal of part of the polypeptide chain.
- Combining two or more polypeptides into the quaternary structure proteins.
- Combining non-polypeptide components into the quaternary structure of conjugated proteins.

Modification of preproinsulin to insulin

- The insulin gene (INS) on chromosome 11 is transcribed in pancreatic beta cells to produce mRNA copies.
- The mRNA is translated by ribosomes on the rough endoplasmic reticulum. The polypeptide produced is preproinsulin and has 110 amino acids.
- Preproinsulin passes into the lumen of the rough endoplasmic reticulum (RER). Here a sequence of 24 amino acids (called the signal peptide) is removed from the N-terminal by a protease enzyme which recognizes three adjacent alanines in the amino acid sequence. A chain of 86 amino acids remains which is called proinsulin.
- Proinsulin becomes folded with three disulfide bonds stabilizing the tertiary structure.

- Peptide bonds are broken at two positions in proinsulin by proteases which recognize pairs of positively charged amino acids (lysine and arginine). This causes a sequence of 33 amino acids to be excised and leaves two separate chains: the A-chain with 21 amino acids and the B-chain, initially with 32 amino acids. These two chains are held together by the disulfide bonds made earlier.
- A pair of amino acids is removed from the C-terminal of the B-chain by a protease, to yield mature insulin with a total of 51 amino acids.

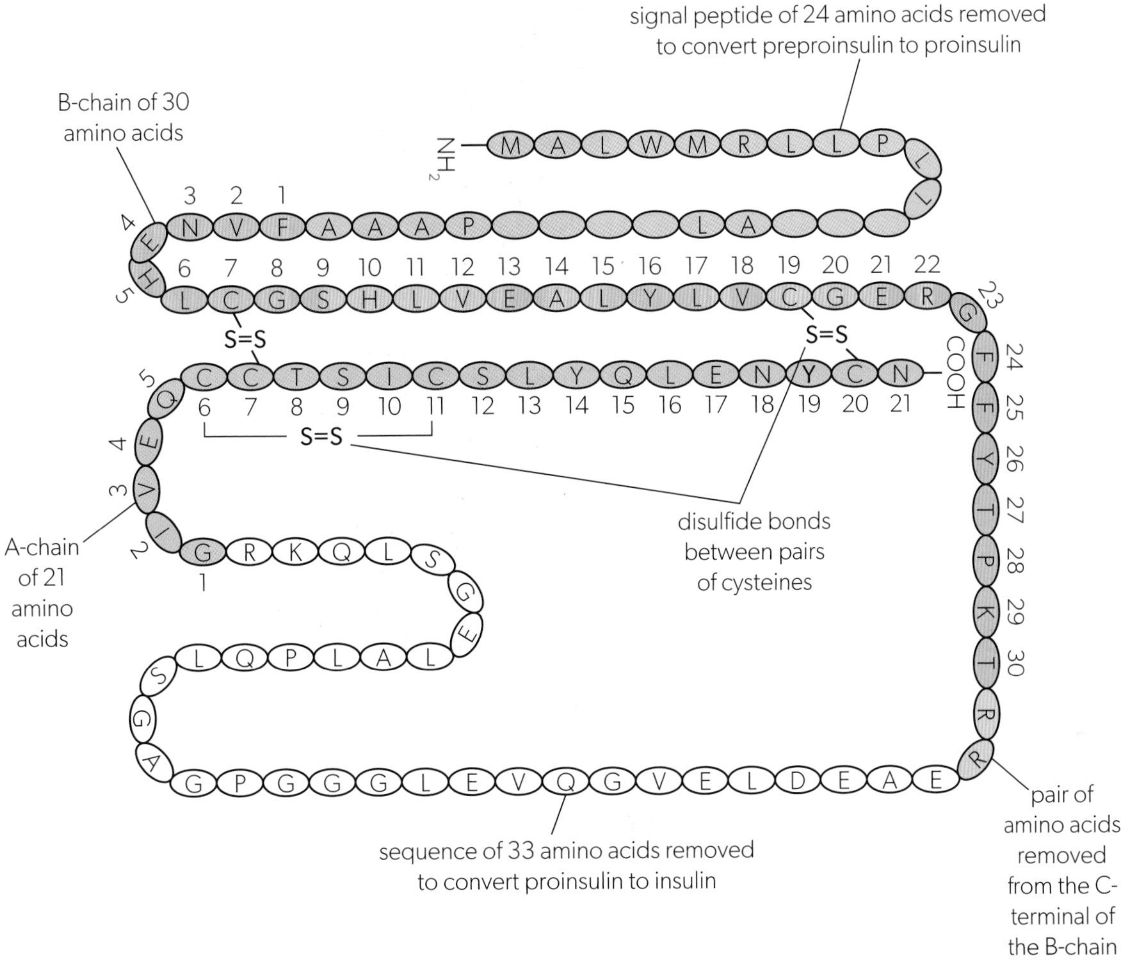

▲ Figure 23 Changes made to the sequence of amino acids in preproinsulin

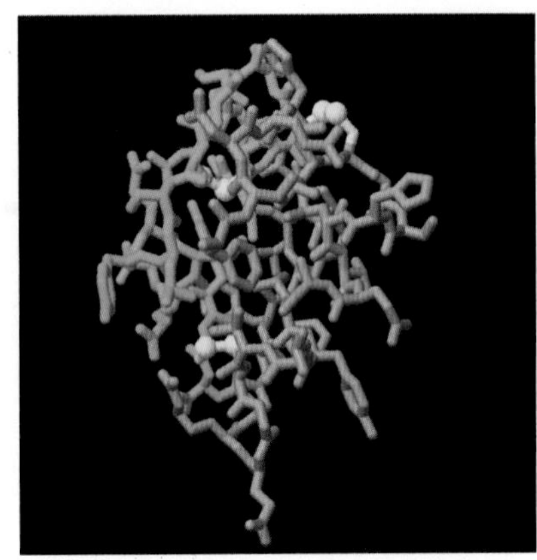

▲ Figure 24 In this image of mature insulin, the A-chain is green and the B-chain is blue. There are two disulfide bonds (yellow) between the two chains and one within the A-chain

D1.2.19 Recycling of amino acids by proteasomes

Most of the proteins produced by translation have a relatively short life in the cell. There are many reasons for this.

- The cell's activities may change, and a protein is no longer needed. An example of this is when a cell moves on from one stage in the cell cycle to the next one.
- The structure of many proteins is quite easily changed by free radicals or other reactive chemicals. If this happens the protein can no longer carry out its function.
- The protein could become misfolded or denatured in other ways.

Proteins that no longer have a function or are unable to carry out their function are broken down by proteasomes. The structure of a proteasome is shown in Figure 25.

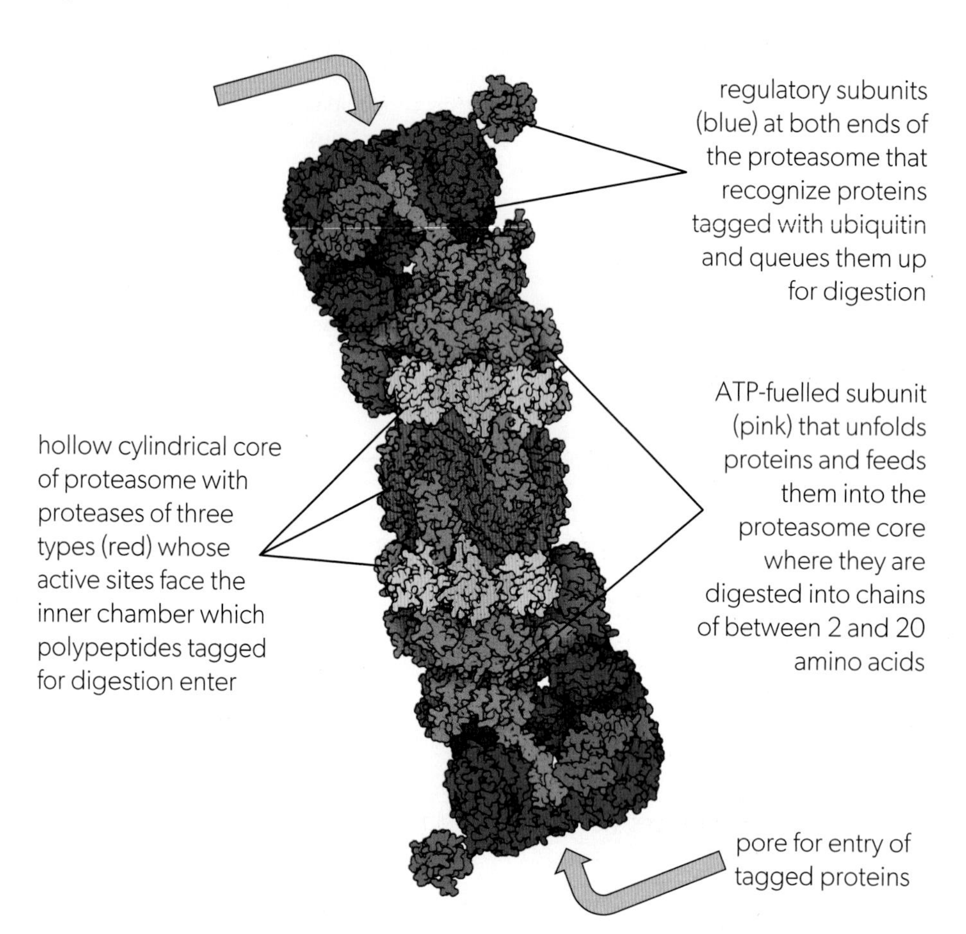
regulatory subunits (blue) at both ends of the proteasome that recognize proteins tagged with ubiquitin and queues them up for digestion
ATP-fuelled subunit (pink) that unfolds proteins and feeds them into the proteasome core where they are digested into chains of between 2 and 20 amino acids
hollow cylindrical core of proteasome with proteases of three types (red) whose active sites face the inner chamber which polypeptides tagged for digestion enter
pore for entry of tagged proteins

▲ Figure 25 The proteasome is about 15 nm long and 1.1 nm wide. The interior chamber where proteins are digested is about 5 nm wide, but the entrances at either end are about 1.3 nm wide so proteins have to be unfolded to enter

Proteins identified for degradation are tagged with a chain of small proteins called ubiquitin. This acts as a signal for proteasomes that these proteins should be digested. Ubiquitinated proteins are recognized by the regulatory subunits at either end of the proteasome. They are then unfolded and fed into the central chamber by a subunit that uses ATP as a source of energy. Active sites of multiple proteases face the central chamber and break proteins down into short chains of amino acids which pass out of the proteasome. Further digestion of these oligopeptides in the cytoplasm yields amino acids, which can be used for synthesis of new proteins. The proteasome thus removes functionless and damaged proteins from the cell, but also recycles amino acids.

Data-based questions: Proteasomes

Proteasomes are a group of proteins that destroy unwanted proteins in the cell, thereby maintaining cell health. However, chemical inhibitors can block the action of proteasomes, resulting in a build-up of unwanted proteins with negative consequences for the cell.

Proteasomes are needed to protect mitochondria from unwanted cell proteins. Mean oxygen consumption is a measure of mitochondrial activity. Mean oxygen consumption was recorded in untreated control rats, rats treated for 1 week and rats treated for 3 weeks with proteasome inhibitor.

1. Outline how proteasomes work. [2]
2. Describe the effect of the two treatments with proteasome inhibitor on mean oxygen consumption. [1]
3. Explain how mean oxygen consumption is related to energy consumption in cells. [2]
4. Proteasome inhibitors are used to target vital cell pathways in the tumour cells of cancer patients. Using the data provided in all the graphs, evaluate the risks of proteasome inhibitor treatment for cancer patients. [2]

Based on *J. Proteome Res.* (2011) 10, 12, 5275–5284
Publication Date: November 3, 2011
https://doi.org/10.1021/pr101183k
Copyright © 2011 American Chemical Society

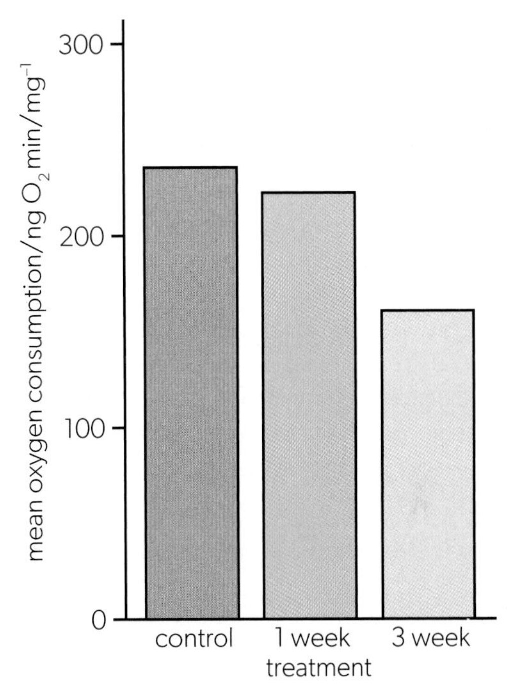

▲ **Figure 26** Bar chart showing mean oxygen consumption in three groups of rats

Linking questions

1. How does the diversity of proteins produced contribute to the functioning of the cell?
   a. Explain the relationship between dominant and recessive alleles, amino acid sequence and functional and non-functional protein variations. (D3.2.5)
   b. List five functions of membrane proteins. (B2.1.4)
   c. Compare and contrast the synthesis of hydrophobic and hydrophilic proteins. (B2.2.7)
2. What biological processes depend on hydrogen bonding?
   a. Outline the role of hydrogen bonding in the processes of transcription and translation. (D1.2.2/D1.2.7)
   b. Explain the relationship between hydrogen bonding and the movement of water in xylem. (B3.2.7)
   c. Outline the role of hydrogen bonding in the secondary and tertiary structure of proteins. (B1.2.10)

D1.3 Mutation and gene editing

How do gene mutations occur?

Antennapedia is a gene first discovered in fruit flies. It controls the formation of legs during development. Loss-of-function mutations in the regulatory region of this gene result in the development of antennae where legs are meant to be. Gain-of-function alleles result in legs where antennae should be. How is this possible? Gene mutation can be caused by errors in DNA replication or failure of repair mechanisms. In addition, mutagens are agents that can cause mutations. Chemical mutagens are called carcinogens and examples of mutagenic forms of radiation include ultraviolet radiation and radioactive isotopes.

◀ Figure 1

What are the consequences of gene mutation?

Sometimes mutations result in no discernible change. Sometimes a mutation can cause the polypeptide to have a new function.

Wrinkled pea seeds are a trait caused by a gene mutation that results in a defective, inactive starch branching enzyme. This results in an increase in sucrose concentration in the developing embryos. This in turn causes increased uptake of water during development and thereby increases the cell size. During seed maturation in these mutant seeds, a greater loss of water occurs from these larger cells. As a result, the wrinkled seed phenotype develops.

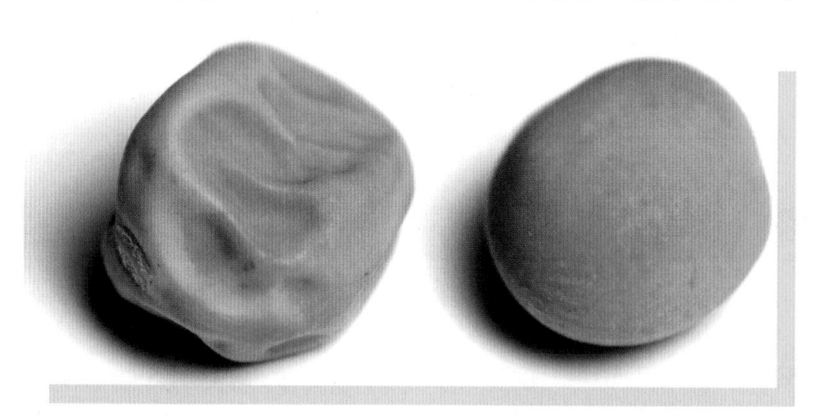

▲ Figure 2

SL and HL	AHL only
D1.3.1 Gene mutations as structural changes to genes at the molecular level D1.3.2 Consequences of base substitutions D1.3.3 Consequences of insertions and deletions D1.3.4 Causes of gene mutation D1.3.5 Randomness in mutation D1.3.6 Consequences of mutation in germ cells and somatic cells D1.3.7 Mutation as a source of genetic variation	D1.3.8 Gene knockout as a technique for investigating the function of a gene by changing it to make it inoperative D1.3.9 Use of the CRISPR sequences and the enzyme Cas9 in gene editing D1.3.10 Hypotheses to account for conserved or highly conserved sequences in genes

D1.3.1 Gene mutations as structural changes to genes at the molecular level

The DNA from which genes are made is very stable and its base sequences can be copied with great accuracy. Cells have methods of correcting errors in DNA replication, but changes do sometimes occur in DNA molecules. If a change occurs to the base sequence of a gene, it is a gene mutation. Gene mutations are random and should be distinguished from deliberate changes made by molecular biologists when editing genes.

There are three main types of gene mutation.

- **Substitution**—one base in the coding sequence of a gene is replaced by a different base. For example, adenine present at a particular point in the base sequence could be substituted by cytosine, guanine or thymine. Base substitutions can happen by chemical changes to bases or by mispairing during DNA replication. The commonest example of mispairing is G to T. If this is not corrected before replication next occurs, the result is a G to A or T to C base substitution on the coding strand.
- **Insertion**—a nucleotide is inserted, so there is an extra base in the sequence of the gene. This is a more major change as it requires a break to be made in the sugar–phosphate backbone of the DNA molecule.
- **Deletion**—a nucleotide is removed, so there is one base less in the sequence of the gene. This requires two breaks in the sugar–phosphate backbone.

Multiple insertions and deletions can occur where two or more consecutive nucleotides are added or removed.

▲ Figure 3 Eleven players are allowed in a soccer team during a match. Substitutions of one player for another are allowed. The referee may send a player off for a foul or misconduct but does not allow extra players. In a very rare example, Moscow Dynamo fielded 12 players for the second half in a match against Glasgow Rangers in 1945. In soccer, deletions are common but insertions are extremely rare!

Activity: How common are mutations?

Recent research into mutations involved finding the base sequence of all genes in parents and their offspring. It showed that there was one base mutation per 1.2×10^8 bases. Calculate how many new alleles a child is likely to have as a result of mutations in their parents. Assume that there are 25,000 human genes and these genes are 2,000 bases long on average.

D1.3.2 Consequences of base substitutions

The consequences of base substitution mutations are mostly neutral or deleterious and in some cases they are lethal.

- In the non-coding DNA between genes on chromosomes, base substitutions are unlikely to have any effect. Only changes to the coding sequences of genes can affect the amino acid sequences of polypeptides.
- Same-sense mutations are base substitutions that change one codon for an amino acid into another codon for the same amino acid. They are possible because of the redundancy of the genetic code. For example, a change from AGC to AGT still codes for the amino acid serine. Same-sense mutations do not affect the phenotype, though they may make it possible for a second mutation to change the codon into one for a different amino acid.
- Nonsense mutations change a codon that codes for an amino acid into a stop codon (ATT, ATC or ACT). Translation is therefore terminated before a polypeptide has been completed. Usually, the resulting protein does not function properly. The effects of this depend on what the protein's function is.
- Mis-sense mutations alter one amino acid in the sequence of amino acids in a polypeptide. They may not have much effect if the new amino acid has a similar structure and chemical properties to the original one (a synonymous substitution) or if it is positioned in part of a protein that is not critical in terms of function. But mis-sense mutations can also have severe and even lethal effects. Many genetic diseases are due to mis-sense mutations, for example sickle cell disease.

▲ Figure 4 Abraham Lincoln's features resemble Marfan syndrome, caused by a mutation to the fibrillin-1 (*FBN1*) gene on chromosome 15. However, recent evidence suggests that he suffered from a different genetic disease, multiple endocrine neoplasia (MEN2B). This condition is due a mutation in the *RET* gene on chromosome 10

A very small proportion of mis-sense mutations improve the functioning of a protein and increase an individual's chances of survival. Beneficial mutations are massively outnumbered by deleterious mutations but they are arguably more significant because of evolution and adaptation. When a base substitution mutation happens in one individual in a population and is inherited by offspring, a new allele of one gene has been produced and the genetic diversity of the population increases very slightly. In evolutionary terms, this is beneficial for the population as a whole.

When the DNA from individual humans is sequenced, large numbers of base substitutions are found that have happened at some time in the past. These are known as single-nucleotide polymorphisms, frequently abbreviated to SNPs and pronounced "snips". SNPs can occur in noncoding regions of DNA. The presence of some SNPs is associated with certain diseases. These correlations allow scientists to look for SNPs to determine an individual's genetic predisposition to develop a disease.

D1.3.3 Consequences of insertions and deletions

Insertions and deletions are less likely than substitutions to have beneficial consequences.

- Major insertions or deletions of nucleotides in a gene almost always result in the coded-for polypeptide ceasing to function.
- Minor insertions and deletions of one or two nucleotides can also result in total loss of function in a polypeptide. This is because they are frameshift mutations. They change the reading frame for every codon from the mutation onwards in the direction of transcription and translation. This is illustrated in Figure 5.

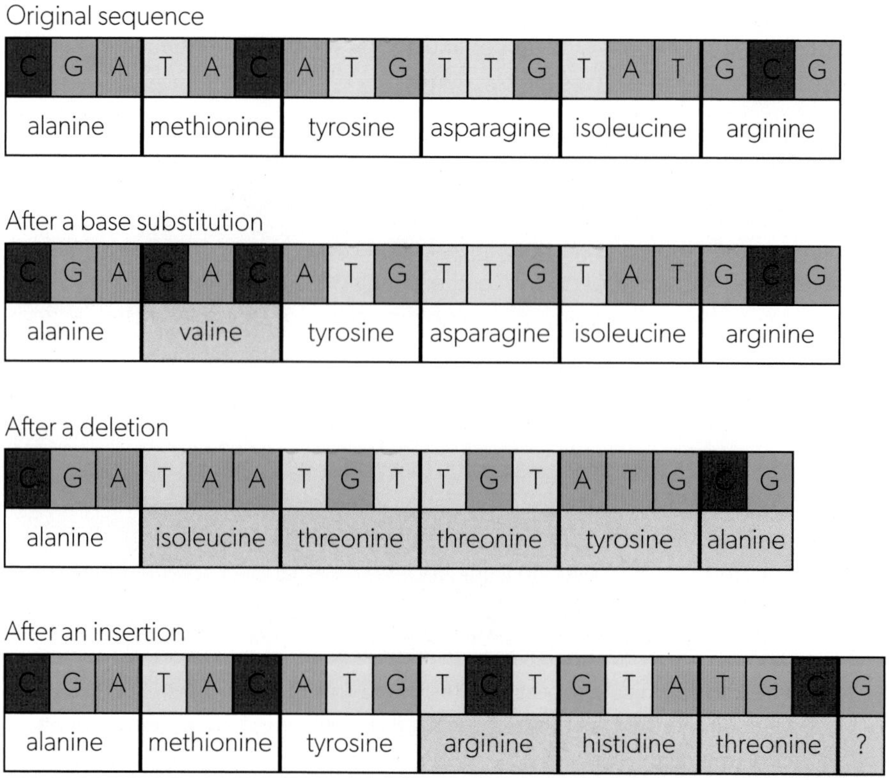

Original sequence

C G A	T A C	A T G	T T G	T A T	G C G
alanine	methionine	tyrosine	asparagine	isoleucine	arginine

After a base substitution

C G A	C A C	A T G	T T G	T A T	G C G
alanine	valine	tyrosine	asparagine	isoleucine	arginine

After a deletion

C G A	T A A	T G T	T G T	A T G	C G
alanine	isoleucine	threonine	threonine	tyrosine	alanine

After an insertion

C G A	T A C	A T G	T C T	G T A	T G C	G
alanine	methionine	tyrosine	arginine	histidine	threonine	?

▶ Figure 5 Deletions and insertions cause more changes than substitutions in the amino acid sequence translated from the base sequence of a gene, as this example of a small part of a gene shows

- Insertions and deletions of a multiple of three nucleotides are not frameshift mutations but can still have severe consequences because there will be one or more amino acids more or less in the polypeptide expressed from a mutated gene. This may cause radical changes to the structure of the protein and therefore affect whether or not it can carry out its functions.

BRCA1—an example of gene mutation

The *BRCA1* gene (pronounced "bracker-one") codes for the BRCA1 protein in humans. *BRCA1* is referred to as a tumour suppressor gene, but its actual function is DNA repair. It can mend double strand breaks and help to correct mismatches in base pairing. If this gene mutates and the BRCA1 protein cannot carry out its function, the consequence is an increased risk of other mutations due to the lack of DNA repair. In body cells, the most obvious sign of this is an increased risk of tumour formation and cancer, particularly breast, ovarian and prostate cancer.

Over 20,000 variants (alleles) of *BRCA1* have been identified, including base substitutions, deletions and insertions. These can be viewed online in the BRCA Exchange database. The numbering system for variants is explained there, using one example each of a substitution, deletion and insertion.

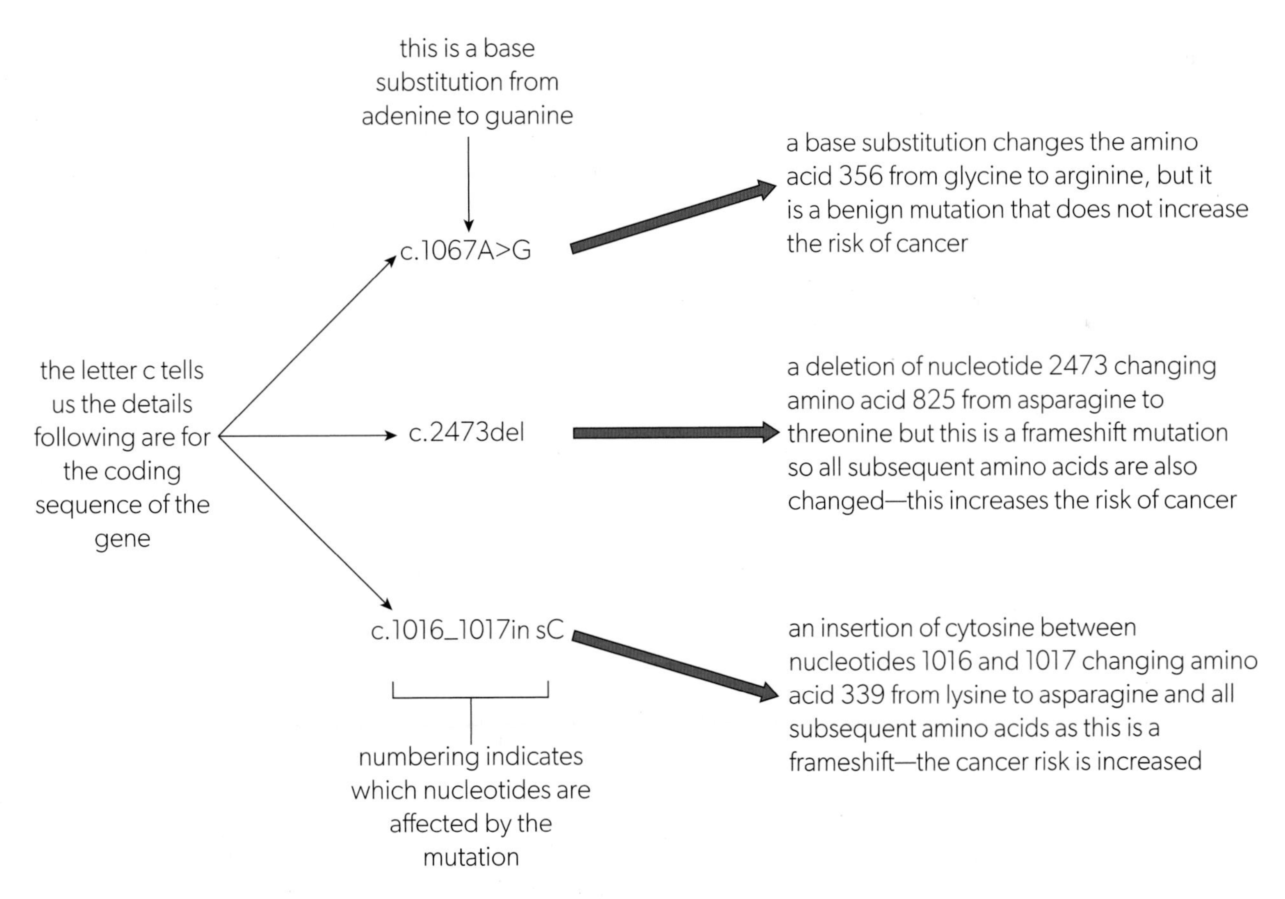

▲ Figure 6 Not all variants of the *BRCA1* gene increase the risk of cancer

Same-sense base substitutions are benign, as are some substitutions that change the amino acid in the most important functional domains in the BRCA1 protein. Of the variants that are non-benign, there is a range of increased risk of cancer but some of the variants increase the lifetime risk of breast cancer in women by as much as 80%.

Data-based questions: *BRCA1* mutations

BRCA1 is a tumour suppressor gene. There are over 500 variants of the gene that increase the risk of breast and ovarian cancer in women and the risk of some other cancers in both sexes. Most of these variants are extremely rare in human populations. Table 1 shows the frequency of types of base substitution mutation among breast cancer patients in five groups among the US population. The NAJ+H White group were non-Ashkenazi Jewish non-Hispanic Whites. Splice mutations affect the editing out of introns, which are base sequences in genes that are transcribed but not translated.

Source: Bougie O, Weberpals, JI. Clinical Considerations of BRCA1- and BRCA2-Mutation Carriers: A Review. International Journal of Surgical Oncology, vol. 2011, Article ID 374012, 2011. https://doi.org/10.1155/2011/374012

Mutation type	Number (%)				
	Hispanics (n = 21)	African Americans (n = 8)	Asian Americans (n = 3)	Ashkenazi Jewish (n = 8)	NAJ + H White (n = 14)
Frameshift	15 (71)	2 (25)	0	8 (100)	5 (36)
Mis-sense	3 (14)	3 (38)	0	0	2 (14)
Nonsense	3 (14)	1 (13)	3 (100)	0	4 (29)
Splice	0	2 (25)	0	0	3 (21)

▲ Table 1 Frequency of types of base substitution mutation among five groups of breast cancer patients in the US

1. More frameshift mutations of the *BRCA1* gene were found in this research than other types of mutation. Explain the reasons for the high frequency of frameshift mutations in patients with breast cancer. [4]
2. a. 15% of the breast cancer patients in the combined groups had a mis-sense mutation. Calculate the percentage that had a nonsense mutation. [2]
   b. Mutations are random changes to the base sequence. There are far more possible mis-sense mutations than nonsense mutations. Suggest reasons for the relatively high percentage of patients with nonsense mutations. [3]
3. a. Compare and contrast the data for African American patients with the data for Ashkenazi Jewish patients. [3]
   b. Suggest reasons for the differences. [3]

Using a database

The Online Mendelian Inheritance in Man (OMIM) Database provides information on many mutations found in humans. The Allele Frequency Database (ALFRED) provides information about how these mutations are distributed across populations around the world. An allele is a variety of a gene. A mutation causes a new allele to be formed.

An interesting human mutation is the CCR5-Δ32 mutation. The CCR5-Δ32 allele is notable for its recent origin, unexpectedly high frequency, and distinct geographic distribution. Together, these features suggest that the CCR5-Δ32 allele:

- arose from a single mutation
- was historically subject to positive selection.

The mutation is responsible for the two types of resistance to HIV. CCR5-Δ32 hampers the ability of HIV to infect immune cells. This is because the mutation causes the CCR5 receptor (normally on the outside of cells) to be reduced in size and to no longer sit outside the cell.

1. Use the OMIM database to determine which chromosome is affected by CCR5-Δ32 and what is its location on the chromosome.
2. Determine how many variants of the mutation there are.
3. Use the database to identify geographic regions with relatively high frequencies of the allele.
4. One hypothesis for the distribution of the allele is that it originated from Viking populations that migrated into Europe. Use the ALFRED to contrast the distribution of the allele in populations in Northern Europe and Southern Europe.
5. Another hypothesis was that the CCR5-Δ32 mutation enabled survival from the plague (infection by the bacterium *Yersinia pestis*). Compare the areas where there is a high frequency of the allele with areas that were significantly affected by historical outbreaks of the plague.

D1.3.4 Causes of gene mutation

Gene mutations can happen at any time, but the chance is normally very low as DNA is resistant to chemical change. There is an increased risk during DNA replication, when base-pairing errors are sometimes made and not corrected by DNA repair. The frequency of mutation is increased by external agents known as mutagens. There are two types of mutagen: radiation and chemicals.

1. Radiation increases the mutation rate if it has enough energy to cause chemical changes in DNA. Gamma rays, X-rays and alpha particles from radioactive elements such as radon are mutagenic. Short-wave ultraviolet radiation in sunlight is also mutagenic.

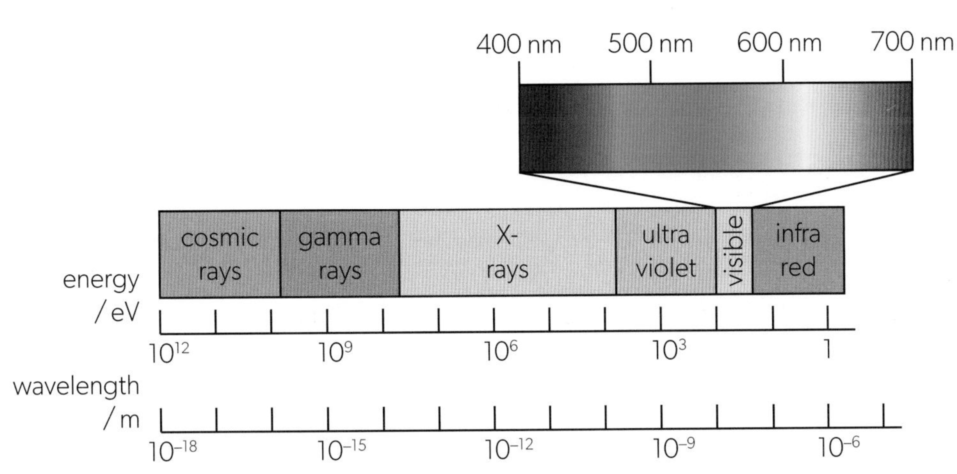

◀ Figure 7 The energy of electromagnetic radiation is inversely proportional to its wavelength

2. Some chemical substances cause chemical changes in DNA, so are mutagenic. Examples are polycyclic aromatic hydrocarbons and nitrosamines found in tobacco smoke. Mustard gas is also mutagenic and was used as a chemical weapon in the First World War.

▲ Figure 8 The mainstream smoke drawn through the filter tip of a cigarette is an aerosol containing about 10^{10} particulates per millilitre and 4,800 chemical compounds, at least 60 of which are mutagenic

D1.3.5 Randomness in mutation

Mutations are described as random changes. Whether or not anything can be truly random is a debatable question, but mutations certainly are unpredictable and cannot be directed by living organisms to achieve an intended outcome. The consequences of a mutation have no influence on the probability that this mutation will or will not occur. No natural mechanism for changing a particular base with the intention of making a beneficial change to a trait has been discovered or is likely ever to be discovered in living organisms. At most, some organisms seem to have limited control over their overall mutation rate.

Mutations can occur anywhere in the base sequences of a genome although some bases have a higher probability of mutating than others. This is because a mutation is a chemical change, and some chemical changes happen more easily than others. The position of a base within the genome also affects the chance of mutation. This is because the differences in how coding and non-coding DNA sequences are used affects the likelihood of mutation.

▲ Figure 9 Is the outcome of shaking five dice in a dice cup random?

A consequence of randomness is that a mutation is unlikely to be beneficial. Genes have developed by evolution over long periods of time, in some cases over hundreds of millions of years. So, a random change will usually be neutral or harmful.

To the best of our knowledge, there is no mechanism for testing out mutations acquired during the lifetime of an organism so that only the beneficial ones are passed to offspring. If a mutation happens in a somatic cell (body cell) its effects may be tested when the gene is expressed, but the mutation is eliminated when the individual dies. The situation is reversed for mutations in germ-line cells—for example, cells in the testes and ovaries of humans. These mutations can be passed to offspring in gametes but are not tested by gene expression in body cells. Traits that are due to mutations acquired during an individual's lifetime and that prove to be beneficial cannot be inherited by offspring. This helps to explain how evolutionary change occurs.

D1.3.6 Consequences of mutation in germ cells and somatic cells

The consequences of mutation depend on whether it occurs in a germ cell or a somatic cell. Germ cells give rise to gametes, so genes in germ cells can be passed to offspring. A new allele, produced by mutation in a germ cell can therefore be inherited. In rare cases a new allele may confer an advantage on offspring, but in far more cases the mutation causes a genetic disease. It is therefore particularly important to minimize the number of mutations in gamete-producing cells within the gonads (ovaries and testes).

Mutations in somatic cells (body cells) are eliminated when the individual dies, so they mostly have limited consequences. A cell may die, but it can generally be replaced. The consequences of mutations in one group of genes can be much greater. These are the genes that have roles in control of the cell cycle and cell division. They are known as proto-oncogenes because mutations can change them into oncogenes, which are cancer-causing genes. Cancer is due to loss of control of the cell cycle, resulting in uncontrolled cell division and therefore tumour formation. The cell cycle, cell division and cancer are described in *Topic D2.1*.

D1.3.7 Mutation as a source of genetic variation

An allele is a variant of a gene, differing in one or more bases from other alleles of the gene. Mutation changes the base sequence of a gene, so it changes one allele into another. Mutation increases the number of different alleles of genes in a population, so it increases genetic variation. Meiosis and sexual reproduction can increase variation by generating new combinations of alleles, but mutation is the original source of all genetic variation.

Most mutations are either harmful or neutral for an individual organism but nonetheless mutation is needed in all species. This is because natural selection requires genetic variation, so species cannot evolve without it. Especially during times of rapid environmental change, populations must adapt to new conditions.

Of the many alleles produced by mutation, a small proportion will confer characteristics that are needed for this adaptation. Without mutation, genetic variation in a population would decline over time, making natural selection impossible. This inevitably leads to extinction when the environment changes and the population cannot adapt and evolve.

D1.3.8 Gene knockout as a technique for investigating the function of a gene by changing it to make it inoperative

It is possible to predict which base sequences in a genome are genes by characteristic base sequence patterns. These are known as open reading frames (ORFs). RNA transcribed from them has the start codon AUG, a sequence of triplets coding for amino acids, followed by a stop codon. The function of genes may not be immediately obvious when they are discovered. Gene knockout is a technique for discovering their function.

▲ Figure 10 Collections of knockout mice are maintained in laboratories

Gene knockout has been carried in different model organisms. Mice are the model organism most intensively researched by this method. These are the stages of gene knockout in mice.

1. DNA is prepared with a base sequence that allows it to be inserted into the genome of embryonic mouse cells as a replacement for a target gene, which is therefore deleted.
2. Cells where this procedure has been successful are selected and are grown into adult mice. Instead of having two copies of the target gene, these adults will have only one copy—the other copy has been deleted.
3. Males and females with only one copy of the target gene are mated. 25% of their offspring can be expected to have no copies of the target gene—they are knockout mice.
4. The phenotype of these knockout mice is investigated to find out which traits have been changed by deletion of the target gene.

In mice, several thousand knockout strains have been developed, each lacking one gene. These strains have been widely used in research. For example, Piezo2-knockout mice were found to urinate less frequently than normal individuals. The *PIEZO2* gene has been found to code for a mechanosensitive ion channel that acts as a pressure sensor in the bladder. Humans who naturally lack a functional *PIEZO2* gene also show impaired bladder control.

In a few model organisms such as yeast (*Saccharomyces cerevisiae*) there are now libraries of knockout organisms for all genes in the genome.

D1.3.9 Use of the CRISPR sequences and the enzyme Cas9 in gene editing

CRISPR–Cas9 gene editing is based on a natural system that exists in many species of prokaryotes. The two main elements are CRISPR regions within the genome and the enzyme Cas9.

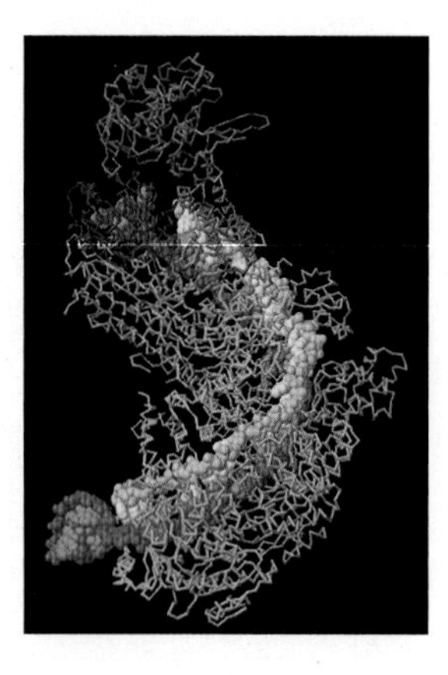

▲ Figure 11 The Cas9 enzyme is blue in this protein backbone model, the guide RNA that binds to it is red and DNA from a virus yellow. Using the guide RNA, Cas9 has recognized a target sequence in the viral DNA and will make a two-strand break in it

CRISPR

CRISPR is an acronym for "clustered, regularly interspaced, short palindromic repeats". The meaning of this complicated name can be teased out.

- *Repeats*—the same base sequence occurs several times in the genome.
- *Short*—the number of base pairs in the repeat is between 23 and 47.
- *Clustered*—the repeats are grouped in one part of the genome.
- *Regularly interspersed*—the repeats are separated by other base sequences of a similar length, called spacers. Each spacer is unique and there can be as few as 2 or over 120 in a CRISPR locus. They are derived from viruses that infect prokaryotes and allow viral DNA sequences to be recognized if reinfection occurs.
- *Palindromic*—each repeated sequence has parts which read the same backwards as forwards (dyad symmetry). It is not truly palindromic but the name CRISPR containing the letter P has stuck.

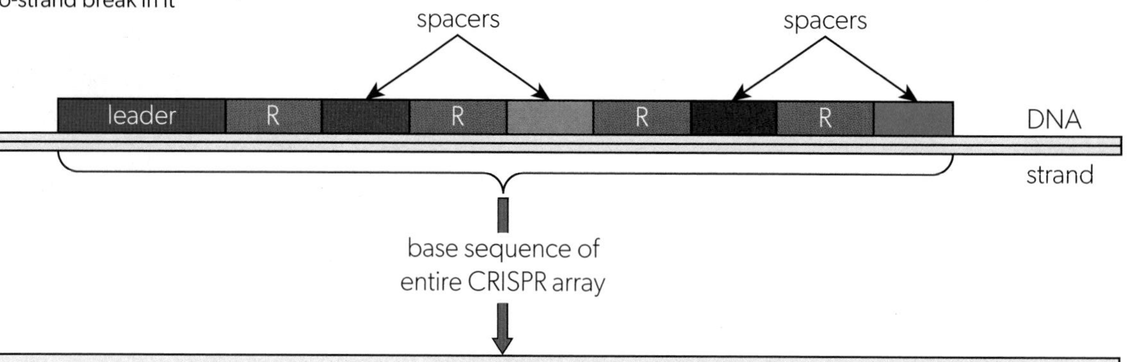

TGATTTTATATCCTGCTTACCGAGGGTTAAAAAAAATCATTTTTT
ACCCTTTGGCGAAAGAATTATTTTACAAACAGTCTGTTACCCGT
ATTATCTTACTGTTCACTGCCGCACAGGCAGCTTAGAAACCTGA
TACAATCATCCTATTTGTCCTATCCAGAAGTTCACTGCCGCACAG
GCAGCTTAGAAAGACAAGAACCGAATCTTTCGCCGTGCCGTAAA
GTTCACTGCCGCACAGGCAGCTTAGAAACCGAAATCATCAGATG
TAATTAAGATTTTTGCTGTTCACTGCCGCACAGGCAGCTTAGAA
AAGACTGATGCAAGATGGCGGTATGCGTACAGAGTTCACTGCCG
CACAGGCAGCTTATAGA

▲ Figure 12 This example of a CRISPR repeat array shows the cluster of five repeats (R), interspersed with four spacers, plus a leader sequence that is rich in AT base pairs

Prokaryotes make guide RNA by transcribing one repeat and an adjacent spacer. The repeat is used for binding to Cas9 and the spacer searches for viral base sequences in DNA.

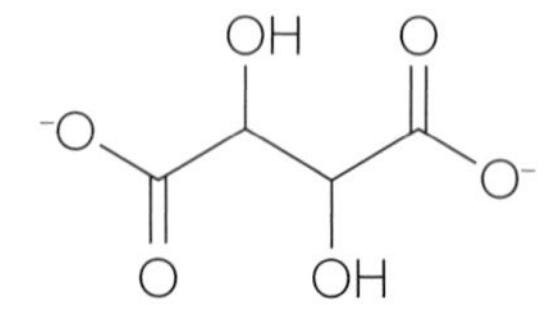

▲ Figure 13 Tartrate is formed by dissociation of tartaric acid. A molecule has been detartrated if tartrate is removed from it. What is unusual about the word "detartrated"?

Cas9

The enzyme Cas9 can find specific DNA base sequences in the genome, using the guide RNA (gRNA) that is bound to it. Guide RNA is made by transcribing one spacer and one repeat from a CRISPR array. The spacer forms a variable 17–20 base sequence at the 5′ end of the gRNA. It is complementary to the target DNA that is being searched for. The repeat forms other parts of the gRNA, which are partly double-stranded, generating loops and a distinctive molecular shape that promotes binding to Cas9.

Cas9 moves along DNA molecules, uncoiling them and bringing the DNA adjacent to the variable base sequence of the gRNA. Within its structure, Cas9 contains two endonucleases. If the target sequence is recognized, the endonucleases cut one sugar–phosphate bond in each of the DNA strands. A double-strand break is therefore made in the target DNA. This is how prokaryotes recognize and destroy foreign DNA within their cells, especially viral DNA.

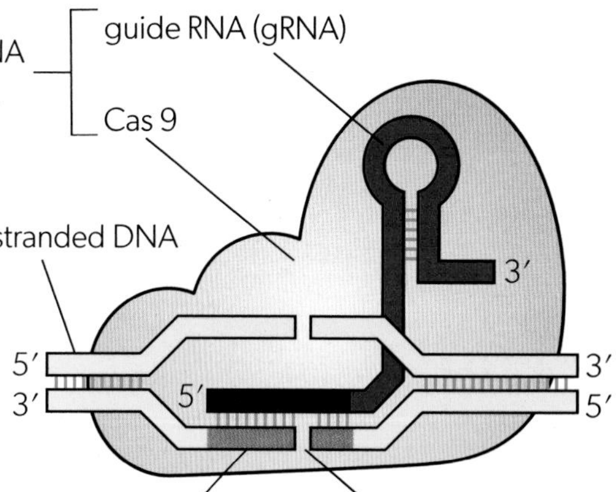

▲ Figure 14 The nucleases have made a double-stranded break in the target sequence in this Cas9–CRISPR complex

Using Cas9–CRISPR in gene editing

Gene editing has long been an aspiration of molecular biologists. If it could be done reliably, specific changes could be made to organisms such as crop plants. It would also allow genetic diseases to be eliminated from human populations. Gene editing requires a method for finding a base sequence in the genome that is the target and replacing it with the desired sequence. This is known as search and replace. The Cas9–CRISPR system can be used to find a target sequence and molecular biologists have modified it so that it can also make changes to the sequence. Prime editing is one approach that shows promise.

Prime editing guide RNA (pegRNA) is prepared. At its 5′ end it has the usual guide sequence transcribed from the repeat in a CRISPR array, together with the base sequences that allow binding to Cas9. Two extra sequences are added immediately adjacent to each other at the 3′ end of the peg RNA:

1. a primer binding site
2. an RNA copy of the desired replacement base sequence for the gene being edited, called the RT template.

A version of Cas9 is made with two modifications.

1. A reverse transcriptase enzyme is attached to it that can assemble a strand of DNA nucleotides complementary to the base sequence of the RT template.
2. An endonuclease in Cas9 is inactivated so it makes a nick in one of the DNA strands, not both.

The modified Cas9 and pegRNA are introduced into cells where they form a complex and begin the process of gene editing.

- Cas9 moves along the DNA molecule, searching for the target sequence. This is identified using the guide sequence at the 5′ end of the pegRNA.
- Where the target sequence is recognized on one of the two strands, a nick is made in the complementary sequence on the other strand. This creates 3′ and 5′ ends in that strand.
- The two DNA strands are separated, with the guide RNA linked to the target sequence by base pairing.
- The primer binding site of the pegRNA binds by base pairing to the other DNA strand, on the 3′ side of the nick. This creates a site at which reverse transcription can be initiated.
- Reverse transcriptase adds DNA nucleotides, one at a time, to extend the DNA strand from the 3′ end created when the DNA was nicked. The nucleotides are added in the usual 5′ to 3′ direction, using the RT template to determine the base sequence. This is the new genetic information that will replace the sequence edited out.

- The two strands of DNA pair up again. The sequence assembled by reverse transcriptase displaces the sequence that it is replacing, which becomes a single-stranded flap.
- Nucleotides of the flap of single-stranded DNA are removed—this is how the original base sequence is edited out.
- There will be some mispairing of bases where double-stranded DNA has reformed, because of differences between the replacement and original base sequences. DNA repair enzymes correct the mispairing. One of the two strands then has the replacement sequence and the other has sequence complementary to it.

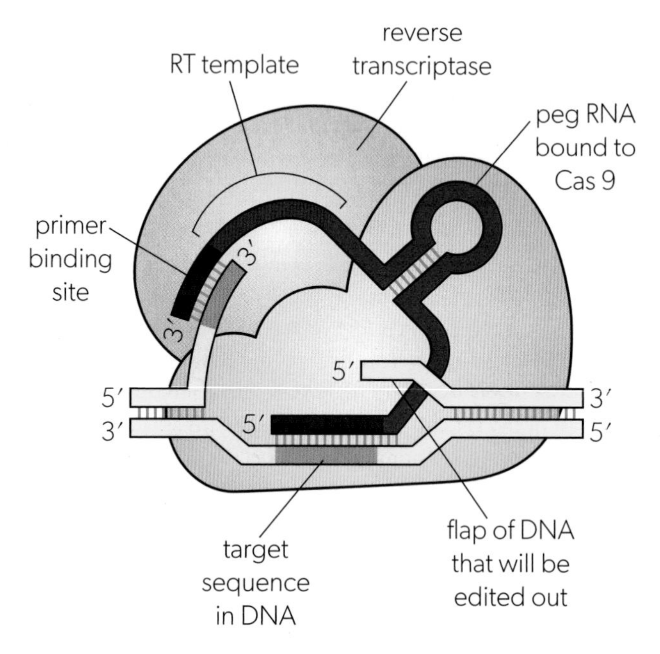

▲ Figure 15 Prime editing is shown at the stage where reverse transcriptase starts to add DNA nucleotides to the 3′ end made by the single strand break. The nucleotides will be linked by complementary base-pairing to the RT template

Science as a shared endeavour: Uneven regulations impacting scientific activity

Scientific researchers in different jurisdictions are subject to different regulations. The consequences of this for the scientific process could be wide-ranging. Because gene editing has the potential to result in harm that could have global impact, there is a need for an international effort to harmonize regulation of the application of genome editing technologies such as CRISPR.

As of August 2022, no internationally agreed regulatory framework for gene editing exists. Each country is in the process of evaluating the extent to which current regulations are adequate for research, product development, and other applications related to gene editing. Most countries are struggling to assess whether or not gene editing is different from classical genetic engineering. The debate revolves around the question: *Should gene editing be regulated differently from genetically modified organisms?* For example, in 2018, the European Court of Justice (ECJ) rejected a request for a regulatory exemption and ruled that organisms developed through gene editing are categorized as genetically modified organisms (GMOs) and are therefore subject to the same regulations as transgenic organisms.

A further debate exists about human germline editing. This is when gene editing occurs at the embryonic stage or even the gamete stage. Germline gene editing is permitted for research in Japan, whereas in Canada it is strictly prohibited, even for research purposes.

ATL Thinking skills: Evaluating websites using the ABCD framework

Gene-editing technology is a polarizing issue and it is common to find websites discussing the technology from a biased perspective.

An important question to be addressed when you are evaluating websites is: *Am I confident that this is a credible, authoritative website that can be used to justify statements in essays?*

To answer this question, use the ABCD framework. This involves evaluating the site for authority, bias, currency and documentation.

Authority

What is the name of the creator/author of the website? If you cannot locate the author, find another source that allows you to evaluate the credentials of the writer.

Is the author of the website reliable and qualified to write about the topic? If you cannot determine the credentials of the creator, find another source.

Is the website provided by an organization or a corporation? What does this organization or corporation do? You can determine this by clicking on a link that says, "About us".

What kind of website is this? Is it a blog, an advertisement, an interview? Is this website appropriate for research for school? What is the domain of the website? Typically .edu, .gov and .org sites are preferable to .com websites. Making this determination will still require you to use your judgement and to think critically about the source.

Bias

Is there any reason to doubt the information this website provides? Does this website provide objective information? Does it provide a balanced perspective?

What kind of website is this? Is it a commercial advertisement, a political statement, a parody or a news site?

What is the purpose of this website? Is its purpose to inform you, to entertain you, to persuade you, to fool you or to sell you something? How many advertisements are on the page?

Currency and content

When was this website created and when was it last updated? Do you need current information? Is this website outdated? If provided, the copyright date or date of the last update may be located at the bottom of the page.

If this website links out to other websites, do the links work? Is this website relevant to your inquiry? Is it written at your level or is it too advanced? Was it intended for a secondary school audience? Can you paraphrase the information with ease?

Documentation

Has the author provided links to the sources used to create the content on the website? Do the citations allow the reader to verify the information provided by the website? Do the citations on the site provide enough information that they can be partially evaluated without opening them? The citations should contain enough information to at least allow you to do an A, B, C evaluation.

Questions

1. Find a website that is presenting information about gene editing and carry out an ABCD evaluation. Some examples are:
    a. The Genetic Literacy Project
    b. National Association for Sustainable Agriculture (Australia)
    c. Island Conservation (Preventing Extinctions)
2. IB History students are accustomed to the OPVLC framework: a method for evaluating sources that acknowledges that a source with apparent limitations can still be of value. Give an example, with reasons, of one source of information on gene editing that has limitations but that is still of value.

D1.3.10 Hypotheses to account for conserved or highly conserved sequences in genes

Conserved sequences are identical or nearly identical across a species or a group of species. Highly conserved sequences are identical or similar over long periods of evolution, so may occur across a wider range of species.

Many conserved sequences are in protein-coding elements of the genome. They also occur in elements that are transcribed to make ribosomal or transfer RNA and in sequences used to regulate gene expression. These are all elements in the genome with an obvious function. If their functions remain unchanged, there might be little or no change in the base sequences over long periods of evolutionary time. When speciation occurs, both resulting species could inherit these sequences and even when groups diverge in many of their traits, there might be little or no change in these genes and gene regulators. This explains how some sequences are found in all mammals or even in all vertebrates.

Non-coding elements are also conserved in the genomes of many organisms. The functions of these elements are largely unknown but their conservation, sometimes across a wide range of organisms, suggests that they do have functions. But there is an alternative hypothesis that is based on the finding that mutation rates are not identical throughout the genome. The conserved non-coding elements might be in regions of the genome where mutation rates are low.

The *HACNS1* gene has 546 base pairs and is highly conserved across a wide range of birds and mammals. This implies that it has remained largely unchanged for hundreds of millions of years. However, it shows a much faster rate of change in human evolution from primate ancestors. Figure 16 shows sequence alignment of a section of the gene comparing the human genome to six other genomes.

```
Human       GC AG CC TT GG GT TC CGC AA ATA GG GCA CCC ACA GTA ACACG TGT GGCGC CG ACCC OGC C GTGCGCA AT CG GGG CT TT ATAC
Chimpanzee  A.....G...........T....................T.......A......A.......T.A..TA................AT.....................G
Rhesus      A.....G...........T....................T.......A......A.......T.A..TA................AT.....................G
Mouse       ......GT.......C..T....................T.......A.....CA.......T.A..TA................AT...................C.G
Rat         ......GT..........T....................T.......A.....CA.......T.A..TA................AT...................C.G
Dog         ......GT..........T....................T.G.....A......A.........A..TA......T.........AT..................C..G
Chicken     .GG...G...........TA...................T.......A......A.......T.A..TA......T.........AT.........A..T........G
```

▲ Figure 16 There are 16 human-specific base substitutions in *HACNS1* in humans. Thirteen of those substitutions occur in the section shown here and are shown in red. In humans, this gene has a role in wrist and thumb development

Data-based questions: Investigation of susceptibility to malaria infection of different haemoglobin types

Haemoglobin E is common in South-East Asia. It is formed by a single point mutation of the gene for normal haemoglobin A. An experiment was carried out to investigate the possible protective role of haemoglobin E against malarial parasites (*Plasmodium falciparum*). Malarial parasites were added to erythrocytes with homozygous haemoglobin A, erythrocytes with homozygous haemoglobin E, and erythrocytes with heterozygous haemoglobin AE. The susceptibility to infection gives an idea of how easily the malarial parasite can infect the cells (the higher the value, the more easily cells are infected).

1. a. State the median value of susceptibility to infection of the AA genotype. [1]

   b. Identify the greatest susceptibility to infection for heterozygous erythrocytes. [1]

   c. State the interquartile range for the EE genotype. [1]

   d. State an example of an outlier for the AA genotype. [1]

2. Compare and contrast the data shown in the graph for heterozygous AE with homozygous EE erythrocytes. [2]

3. Using the results of this experiment, suggest how both alleles of the gene for haemoglobin could be maintained in a human population. [3]

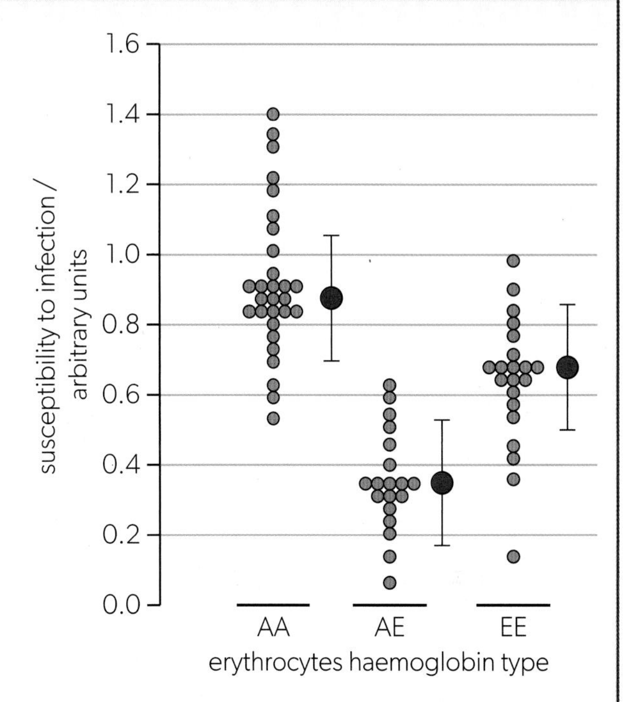

▲ **Figure 17** The error bars span the 1st quartile and the 3rd quartile and the red dot indicates the median value of the susceptibility to infection by malaria of erythrocytes with different haemoglobin types. AA: homozygous normal haemoglobin A; AE: heterozygous normal haemoglobin A and abnormal haemoglobin E; EE: homozygous abnormal haemoglobin E

Linking questions

1. How can natural selection lead both to a reduction in variation and an increase in biological diversity?

   a. List the conditions that are necessary for a population to remain in genetic equilibrium. (D4.1.14)

   b. Explain the effect of mutation and natural selection on variation in a population. (D4.1.2)

   c. Discuss the relationship between adaptive radiation and niche. (A4.1.9)

2. How does variation in subunit composition of polymers contribute to function?

   a. Explain the relationship between the primary structure of a protein and its tertiary structure. (B1.2.10)

   b. Compare and contrast the function of starch and glycogen. (B1.1.5)

   c. Explain how mutations can lead to changes in the outcome of gene expression. (D1.2.11)

TOK

What features of knowledge have an impact on its reliability?

The reliability of a knowledge claim has to do with the extent to which different observers are consistent in coming to the same conclusion when considering the evidence.

Reliability is impacted by the strength of the methods used. An experimental procedure is viewed as reliable when it gives the same or similar result under the same conditions each time it is performed. When first beginning data collection in an experiment, it is reasonable to form a preliminary generalization or hypothesis and then test it further. The greater the degree to which the hypothesis is supported by observations, the more confident we become in the hypothesis and the more reliable it becomes in predicting the outcomes of further experiments.

Chromosomes contain regions in which short sequences of DNA bases are repeated 5, 10, 15, 20 or more times. These are called short tandem repeats (STRs). These repeats are found in several different loci within the genome. The different numbers of repeats at a particular locus are called alleles. For one locus, there might be a few alleles in a population or there might be many. Two individuals can have the same allele of a short STR, but across all chromosomes, the combination of STR alleles is unique to an individual. Forensic scientists can use these markers to identify individuals. Reliability is enhanced by increasing the number of measurements in an experiment or test. In DNA profiling, increasing the number of markers used reduces the probability of a false match. In 1994, the forensic tests that were used could give a false positive in 1 in 400 cases. Combined with other evidence, this is reasonably conclusive. Since then, the number of markers used has risen to 20. This results in a false positive match occurring in only 1 in 2×10^{17} cases.

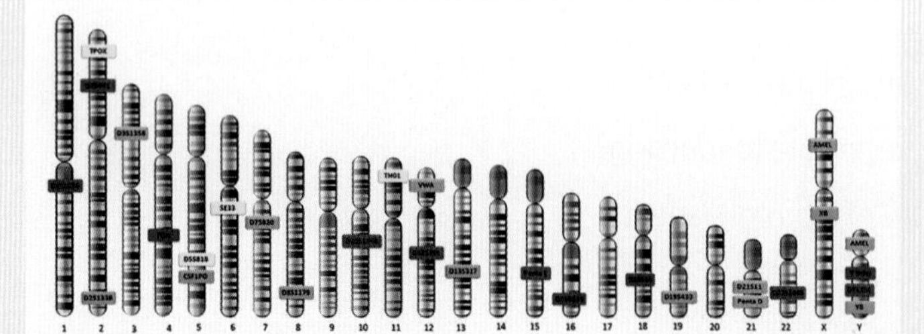

▲ Figure 1 This image shows the chromosome location of 27 STR markers. Of these, 20 are used by the US FBI. The other 7 markers are those recommended by the European Network of Forensic Science Institutes

A reliable method is one that produces a high ratio of true to false beliefs. Using similar technology to forensic scientists, commercial DNA profiling services are available at low cost and involve sending tissue samples through the mail. Some of these tests can yield information about potential future health and disease risk. Unfortunately, an untrained person can misinterpret risk which can lead to anxiety. Consulting health experts can increase the reliability of risk assessments in this case.

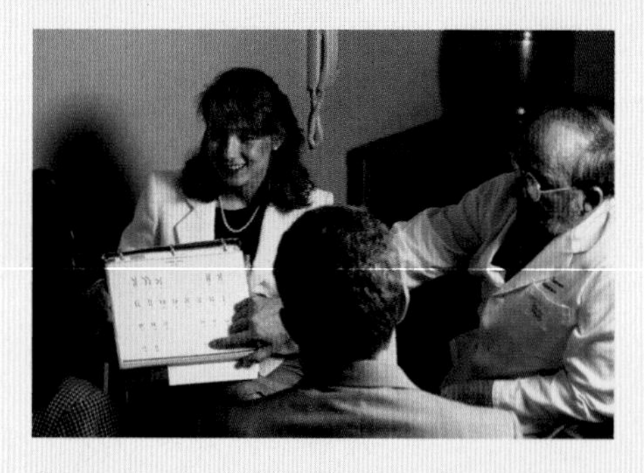

▲ Figure 2 A doctor and a genetic counsellor talking with potential parents. Genetic counselling is advice about the risks of passing on genetic ailments. Couples in certain risk groups who wish to have children are offered genetic counselling

Some knowledge claims concern debatable questions that are influenced by value judgements. How can the knowledge claims concerning debatable questions become more consistent between individuals?

Some potential uses of CRISPR raise ethical issues that must be addressed before implementation. But scientists across the world are subject to different regulatory systems. For example, in some countries human gene editing is banned by laws that could lead to large fines and prison sentences, while in others it is restricted by guidelines rather than laws. Yet the decision to allow gene editing modifications in one country could have an impact on another.

For this reason, there is an international effort to harmonize regulation of the application of genome editing technologies such as CRISPR. But that might be more difficult than it sounds. For example, if the criterion for proceeding were the precautionary principle, then agreement may be likely because the burden of proof of "no potential harm" would be the responsibility of scientists. But if the criterion for proceeding were the promotion of economic innovation, then agreement may be much less likely. Consistency of judgement when debatable questions are being considered is improved if criteria for judgement are agreed first.

End of chapter questions

1. The diagram shows the summary of events occurring during translation.

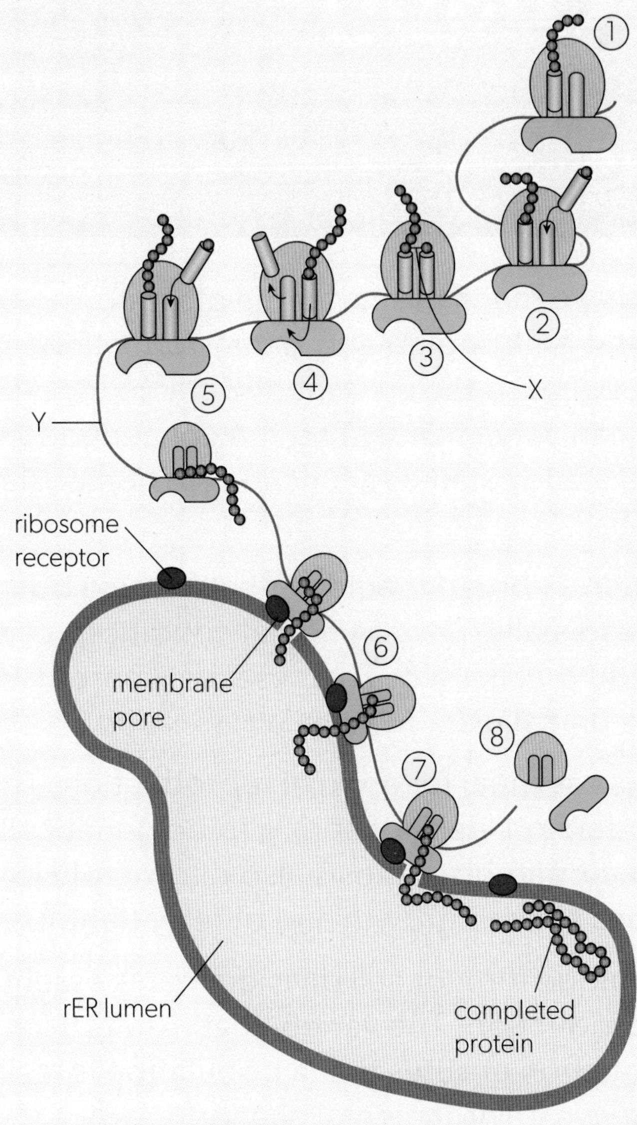

   a. State the name of the bond labelled X. [1]

   b. State the name of the molecule labelled Y. [1]

   c. Outline the processes occurring in stages 4 and 5. [2]

   d. Explain the importance of polysomes in protein synthesis. [2]

2. The ozone layer in the atmosphere is an effective barrier to ultraviolet A (UVA) and ultraviolet B (UVB) light. UVA and UVB light penetrate the surface waters of the ocean and cause damage to marine organisms. The most extensive destruction of the ozone layer occurs over Antarctica and the Southern Ocean.

   One of the major effects of UV light on the cell is damage to the DNA molecule. UV-induced damage results in adjacent thymine bases in the same strand bonding to each other (to form thymine dimers) instead of bonding to the complementary DNA strand. This results in random bases being incorporated into the new DNA strand during replication. The effect of UV light on DNA in plankton at various depths was investigated on calm and windy days. The results are shown in the table.

	DNA damage / number of thymine dimers per 10^6 bases	
Depth / m	Calm day	Windy day
0	175	75
5	175	90
9	190	125
15	100	75
20	75	–
30	25	–

   a. Outline the effect of depth on DNA damage on a calm day. [2]

   b. Compare the amount of DNA damage on windy days with DNA damage on calm days. [1]

   c. Suggest a reason for the effect of wind on DNA damage. [1]

   d. Discuss the relationship between DNA damage and cancer in humans. [3]

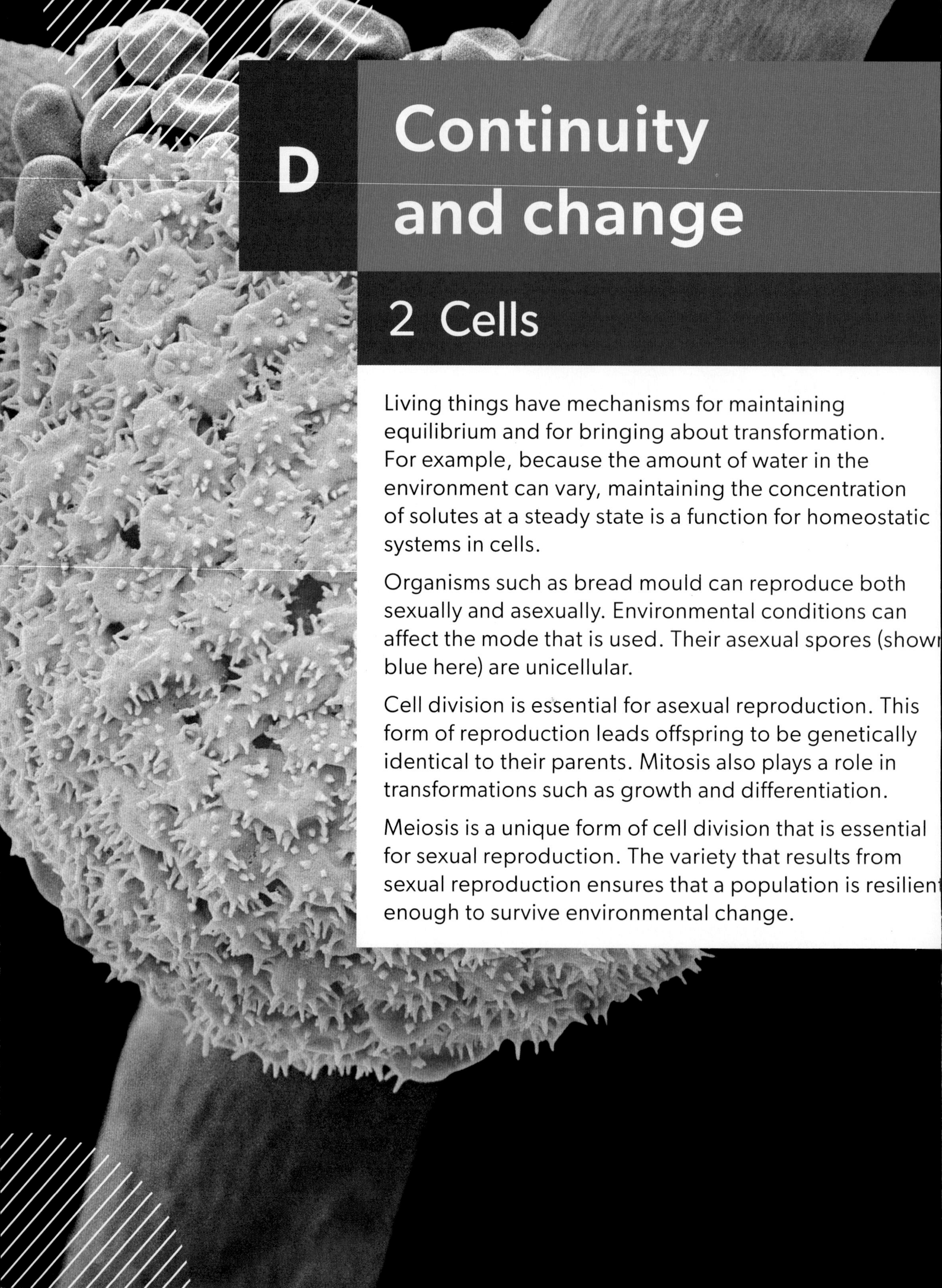

D Continuity and change

2 Cells

Living things have mechanisms for maintaining equilibrium and for bringing about transformation. For example, because the amount of water in the environment can vary, maintaining the concentration of solutes at a steady state is a function for homeostatic systems in cells.

Organisms such as bread mould can reproduce both sexually and asexually. Environmental conditions can affect the mode that is used. Their asexual spores (shown blue here) are unicellular.

Cell division is essential for asexual reproduction. This form of reproduction leads offspring to be genetically identical to their parents. Mitosis also plays a role in transformations such as growth and differentiation.

Meiosis is a unique form of cell division that is essential for sexual reproduction. The variety that results from sexual reproduction ensures that a population is resilient enough to survive environmental change.

D2.1 Cell and nuclear division

How can large numbers of genetically identical cells be produced?

Quaking aspens (Figure 1) can reproduce sexually, however, it is much more common for them to reproduce asexually by sending up new stems from a single root system. All of the stems and their single root system is called a clone. While members of the clone appear to be separate trees, they are all genetically identical.

The trees of the same clone will change leaf colour at the same time in the cooler months. What are the reasons new cells are created? Why is it essential that newly produced cells have the same genetic material as the rest of the cells in the organism?

▲ Figure 1 The trees of the same clone will change leaf colour at the same time in the cooler months and produce flowers and new leaves at the same time in warmer months

How do eukaryotes produce genetically varied cells that can develop into gametes?

In this electron micrograph, chromosomes can be seen in anaphase (I) of meiosis. Meiosis is a special form of cell division unique to the production of sperm and eggs. Why must haploid cells be produced? Why does the chromatin have to condense to form chromosomes as shown in Figure 2? How is it that meiosis can result in a very large variety of gametes? When is sexual reproduction an advantage over asexual reproduction?

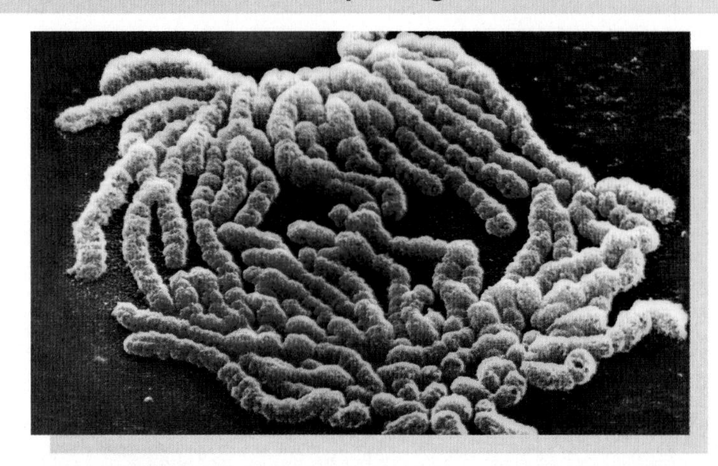

▶ Figure 2 Chromosomes in anaphase (I) of meiosis

SL and HL	AHL only
D2.1.1 Generation of new cells in living organisms by cell division D2.1.2 Cytokinesis as splitting of cytoplasm in a parent cell between daughter cells D2.1.3 Equal and unequal cytokinesis D2.1.4 Roles of mitosis and meiosis in eukaryotes D2.1.5 DNA replication as a prerequisite for both mitosis and meiosis D2.1.6 Condensation and movement of chromosomes as shared features of mitosis and meiosis D2.1.7 Phases of mitosis D2.1.8 Identification of phases of mitosis D2.1.9 Meiosis as a reduction division D2.1.10 Down syndrome and non-disjunction D2.1.11 Meiosis as a source of variation	D2.1.12 Cell proliferation for growth, cell replacement and tissue repair D2.1.13 Phases of the cell cycle D2.1.14 Cell growth during interphase D2.1.15 Control of the cell cycle using cyclins D2.1.16 Consequences of mutations in genes that control the cell cycle D2.1.17 Differences between tumours in rates of cell division and growth and in the capacity for metastasis and invasion of neighbouring tissue

D2.1.1 Generation of new cells in living organisms by cell division

All organisms need to produce new cells, for growth, maintenance and reproduction. They do this by cell division. One cell divides into two. The cell that divides is called the mother cell and those produced from it are daughter cells. The mother cell disappears as an entity in the process, unlike reproduction by animal parents.

There is strong evidence for the theory that new cells are only ever produced by division of a pre-existing cell.

The implications of this theory are profound. If we consider the trillions of cells in our bodies, each one was formed when a pre-existing cell divided in two. We can trace this back to the original cell—the zygote that was the start of our individual lives, produced by the fusion of a sperm and an egg.

Sperm and egg cells were produced by cell division in our parents. The origins of all cells in our parents' bodies goes back to the zygote from which they developed and then on through all previous generations of human ancestors. If we accept that humans evolved from pre-existing ancestral species, we can trace the origins of cells back through hundreds of millions of years to the earliest cells on Earth. This means there is a continuity of life from its beginnings to the cells in our bodies today.

▲ Figure 3 Attempts are being made by BaSyC, a research group in the Netherlands, to build a new living cell from individual lifeless components. This flying-cell artwork represents the challenges of the still unachieved endeavour

D2.1.2 Cytokinesis as splitting of cytoplasm in a parent cell between daughter cells

In cytokinesis, the cytoplasm of a cell is divided between two daughter cells. It is a part of cell division, along with nuclear division by mitosis or meiosis. The process of cytokinesis can begin as soon as chromosomes have separated and are far enough apart to ensure that none of them ends up in the wrong cell. All the cytoplasm and its contents of the mother cell are shared out between the daughter cells. Plant and animal cells carry out cytokinesis differently.

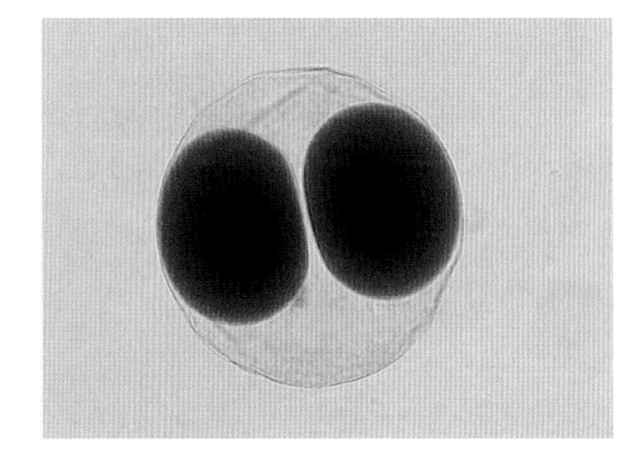

▲ Figure 4 A starfish zygote has just divided for the first time, to produce a two-cell embryo

In animal cells, the plasma membrane is pulled inwards around the equator of the cell to form a cleavage furrow. This is accomplished using a ring of contractile proteins immediately inside the plasma membrane, usually at the equator. The proteins are actin and myosin and are similar to those that cause contraction in muscle. When the cleavage furrow reaches the centre, the cell is pinched apart into two daughter cells.

In plant cells, microtubules are built into a scaffold straddling the equator, which is used to assemble a layer of vesicles. The vesicles fuse together to form plate-shaped structures. With the fusion of more vesicles, two complete layers of membrane are formed across the whole of the equator of the cell. They become the plasma membranes of the two daughter cells adjacent to the new dividing walls. They are connected to the existing plasma membranes at the sides of the cell, completing the division of the cytoplasm.

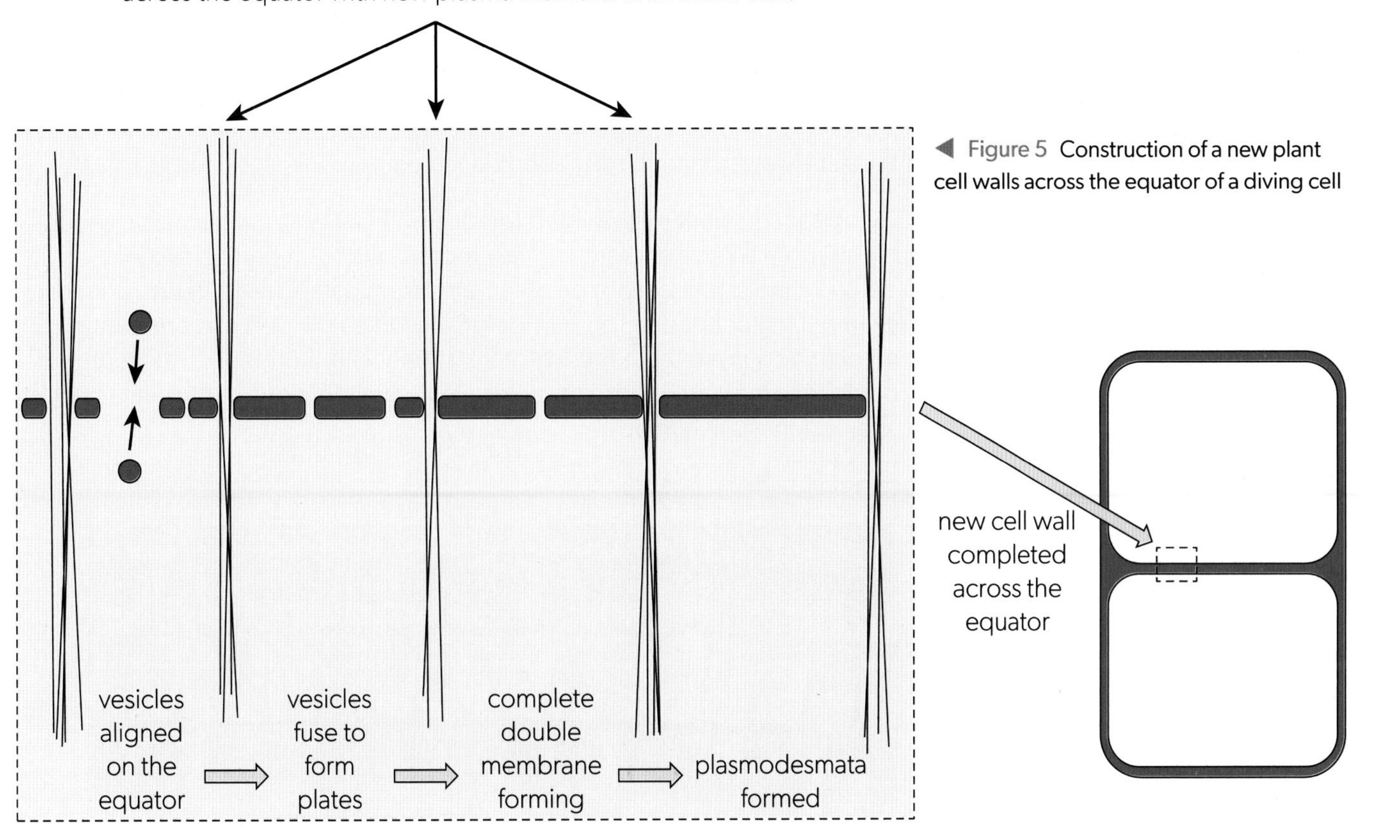
bundles of microtubules scaffold the formation of a new cell wall across the equator with new plasma membrane on either side
vesicles aligned on the equator
vesicles fuse to form plates
complete double membrane forming
plasmodesmata formed
new cell wall completed across the equator

◀ Figure 5 Construction of a new plant cell walls across the equator of a diving cell

The next stage in plants is for pectins and other substances to be brought in vesicles and deposited by exocytosis between the two new membranes. This forms the middle lamella that will link the new cell walls. Both daughter cells then bring cellulose to the equator and deposit it by exocytosis adjacent to the middle lamella. As a result, each cell builds its own cell wall across the equator.

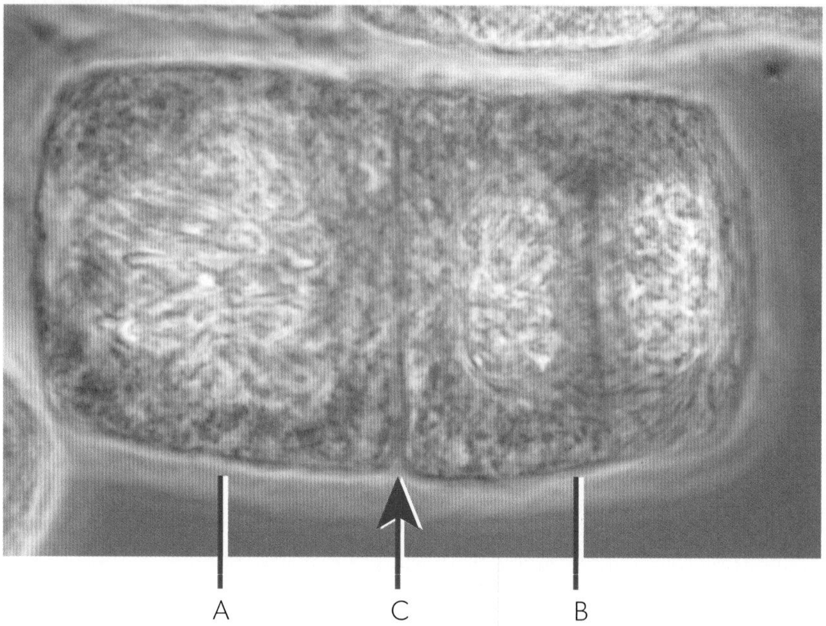

▶ Figure 6 These cells are from a growing onion root.

- Cell A is large enough to divide but is in the early stages of mitosis so cannot yet divide
- Cell B has nearly completed mitosis and is already constructing a wall across the equator, with the microtubule scaffolding visible
- Label C points to where the previous cell division occurred to produce cells A and B from a mother cell

A C B

D2.1.3 Equal and unequal cytokinesis

In many cases, cytokinesis divides the cytoplasm of the mother cell into equal halves. This happens in a growing root tip. Root growth is due to enlargement and division of cells arranged in columns. The cells in a column all differentiate in the same way, so cytoplasm is apportioned equally when they divide.

Cytoplasm is sometimes divided unequally. Small cells produced by unequal division can survive and grow if they receive a nucleus and at least one of each organelle that cannot be assembled from components in the cell. For example, mitochondria can only be produced by division of a pre-existing mitochondrion, so there must be at least one mitochondrion in a daughter cell for it to be viable. Two examples of unequal division are budding in yeast and oogenesis in humans.

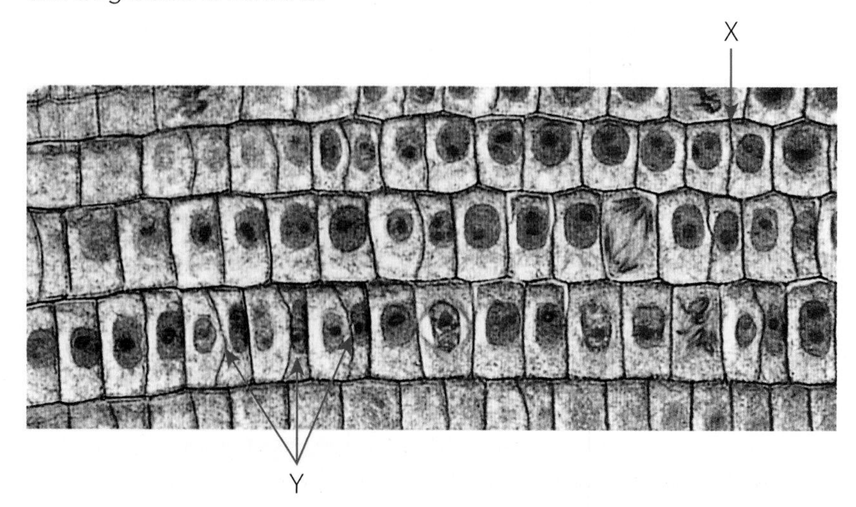

▶ Figure 7 Nuclei in these onion root cells have been stained red. New cell walls have mostly divided cytoplasm equally, as at X; but in some cases, the division was unequal, as at Y

◀ Figure 8 Occasionally a plant cell divides and its chloroplasts all pass into one of the two daughter cells. The other cell with no chloroplasts can never regain them and it produces more cells without chloroplasts when it divides. These leaves were collected from one plant and show areas of cells without chloroplasts

Budding in yeast

Yeast cells reproduce asexually in a process called budding. The nucleus divides by mitosis. A small outgrowth of the mother cell is formed. It receives one of the nuclei, but only a small share of the cytoplasm. A dividing wall is constructed, separating the two cells. The small cell then splits away, leaving a scar where it was attached to the larger cell. Yeast cells carry out this budding process repeatedly and do not have to double in size between each division.

▲ Figure 9 There are many different species of yeast. These are *Komagataella phaffii* which can use methanol as a carbon source if glucose is not available. Several cells are budding

Oogenesis in humans

The production of both sperm and eggs in humans starts with two divisions of a mother cell. During sperm production, the cytoplasm is divided equally in the first and second divisions, resulting in four, equally sized small cells, each of which develops into a mature sperm.

Whereas large numbers of sperm are needed in humans, usually only one egg cell (oocyte) is produced at a time, with enough stored food to sustain the developing embryo. There is therefore unequal division of cytoplasm during oogenesis. The first division produces one large cell with nearly all the cytoplasm and a small polar body which does not develop further. Only the large cell carries out the second division, with unequal division of the cytoplasm again resulting in one large cell and one very small polar body. The large cell develops into a mature oocyte that is ready for fertilization.

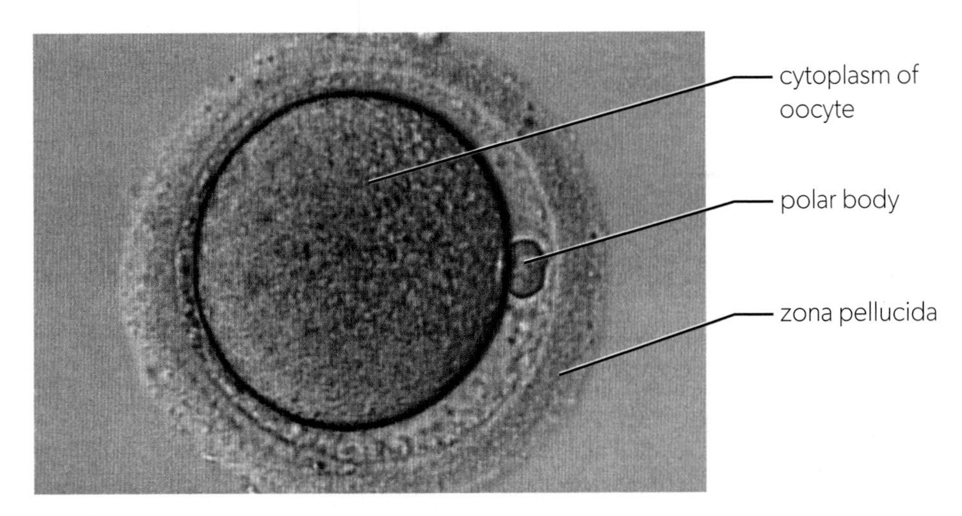

▲ Figure 10 The protective jelly coat (zona pellucida) around this human oocyte is visible. The dense cytoplasm of the oocyte contains food stores. To the right is a tiny polar body that has received a nucleus but very little cytoplasm because it has no role and will not survive

D2.1.4 Roles of mitosis and meiosis in eukaryotes

If a cell divides without first undergoing nuclear division, one daughter cell has the nucleus and the other one would be anucleate (without a nucleus). Anucleate cells cannot synthesize polypeptides, so they cannot grow or maintain themselves. They have limited lifespans. For example, red blood cells, which have no nucleus, survive for about 120 days. To produce extra nuclei before cell division, cells undergo either mitosis or meiosis. These two types of nuclear division have different roles, so most organisms use both during their life cycle.

division of a cell with only one nucleus produces one nucleated and one anucleate cell

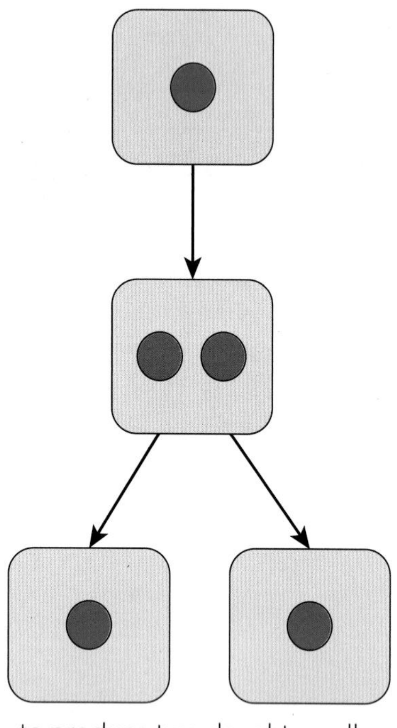

to produce two daughter cells, each with a nucleus, the mother cell's nucleus must first be divided

▲ Figure 11 Nucleated and anucleate cells produced from cell division

Mitosis—continuity
Mitosis is used to produce genetically identical cells. 2*n* is the diploid number of chromosomes. In humans, $n = 23$.

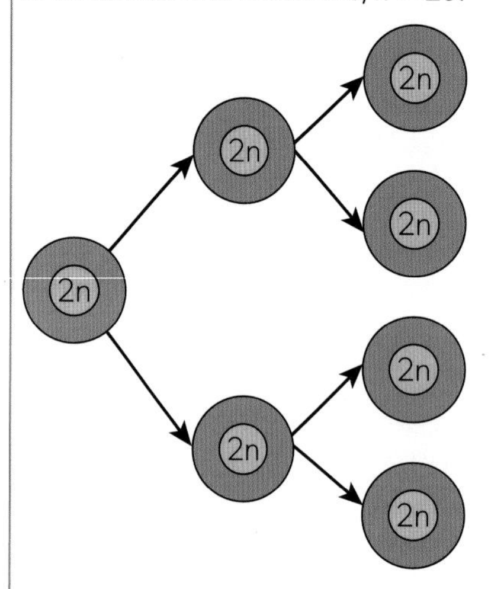

- Cells produced using mitosis have the same number of chromosomes as the parent cell, so the chromosome number is maintained.
- Cells produced by mitosis have the same genes as the parent cell, so mitosis maintains the genome. This ensures that every cell in a multicellular organism has all the genes that it needs. It also ensures that the cells in an individual are genetically identical, preventing problems such as tissue rejection.
- Mitosis allows a successful genome to be inherited without changes by offspring in asexual reproduction.

Meiosis—change
Meiosis is used to halve the chromosome number from diploid (2*n*) to haploid (*n*) and to generate genetic diversity

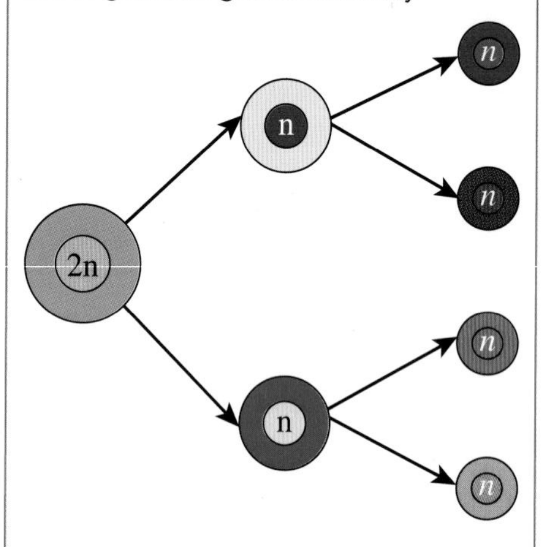

- Cells produced using meiosis have half as many chromosomes as the parent cell. Division of a nucleus with two sets of chromosomes (diploid) results in nuclei with only one set (haploid). This is essential to produce haploid gametes from diploid germ cells in sexual life cycles.
- Pairs of genes in a diploid mother cell are dealt randomly to daughter cells, so there are an almost limitless numbers of possible combinations. Meiosis therefore generates variation and genetic diversity, allowing evolution by natural selection.

D2.1.5 DNA replication as a prerequisite for both mitosis and meiosis

A cell that is preparing for nuclear division by mitosis or meiosis replicates all the DNA. This ensures that each daughter cell produced receives a full complement of genes, allowing it to perform any function required. An earlier hypothesis, now falsified, was that a single centromere held the chromatids together until anaphase, when it divided, allowing the chromatids to separate.

Before replication, the DNA within the nucleus exists as long single molecules called chromosomes. After replication, there are pairs of identical DNA molecules. These identical DNA molecules are still considered to be part of the same chromosome and they are held together by loops of a protein complex, called cohesin. The cohesin loops are not cut until the start of anaphase during mitosis or meiosis.

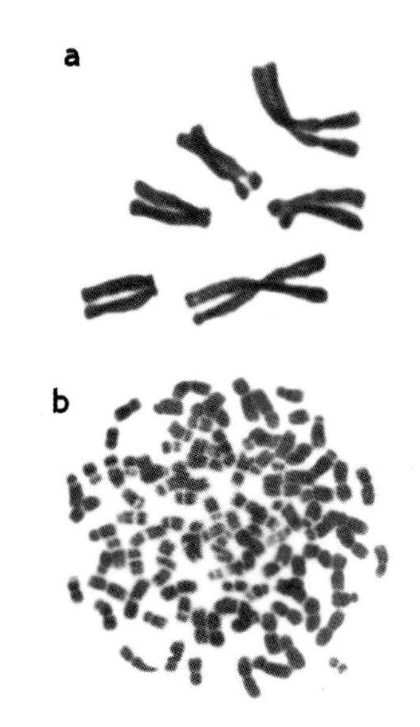
a
b

▲ Figure 13 The number and structure of chromosomes in a species can be studied by staining cells in mitosis with a pigment that binds to DNA and then by bursting the cells on a microscope slide so the chromosomes spread out. The Indian muntjac, *Muntiacus muntjak* (top) has the smallest number of chromosomes per body cell among mammals ($2n = 6$) and the Viscacha rat, *Tympanoctomys barrerae* (bottom) has the largest number ($2n = 102$)

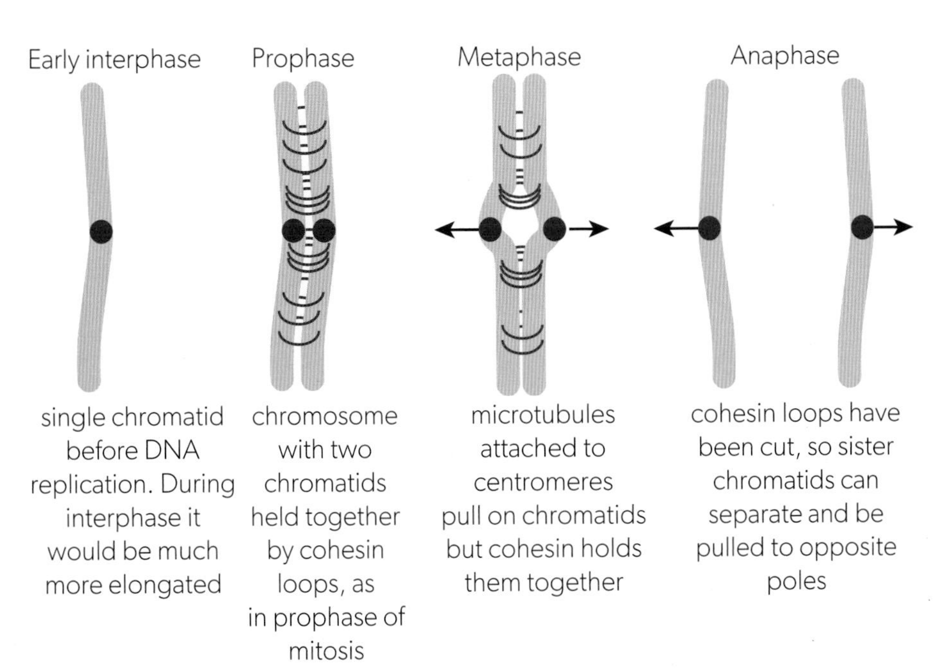
Early interphase
Prophase
Metaphase
Anaphase
single chromatid before DNA replication. During interphase it would be much more elongated
chromosome with two chromatids held together by cohesin loops, as in prophase of mitosis
microtubules attached to centromeres pull on chromatids but cohesin holds them together
cohesin loops have been cut, so sister chromatids can separate and be pulled to opposite poles

▲ Figure 12 Chromosomes, chromatids and cohesion loops

When DNA is in an elongated state, chromosomes are too narrow to be seen with a light microscope. They gradually become shorter and fatter during the early stages of mitosis or meiosis and are then visible. Eventually each chromosome can be seen to have two strands, called chromatids. Each chromatid contains a single very long DNA molecule, produced by DNA replication from an original molecule. The two strands in a chromosome are therefore known as sister chromatids and they are genetically identical. Strands on different chromosomes are non-sister chromatids and do not usually have identical genes. Figure 13 shows chromosomes consisting of pairs of sister chromosomes in two species of mammal.

D2.1.6 Condensation and movement of chromosomes as shared features of mitosis and meiosis

During mitosis and meiosis, chromosomes are moved to opposite poles of the cell, so they can become part of separate nuclei (Figure 14). The DNA molecules in these chromosomes are immensely long. For example, the average length in human chromosomes is more than 50,000 μm and the nucleus is less than 5 μm wide. To separate and move molecules as elongated as this without knots, tangles or breaks they must be packaged into much shorter structures. This condensation of chromosomes and their subsequent movement is therefore an essential feature of mitosis and meiosis.

in mitosis and the 2nd division of meiosis sister chromatids separate and are moved to opposite poles
in the 1st division of meiosis homologous chromosomes (each with two sister chromatids) are moved to opposite poles
opposite poles of cell
equator

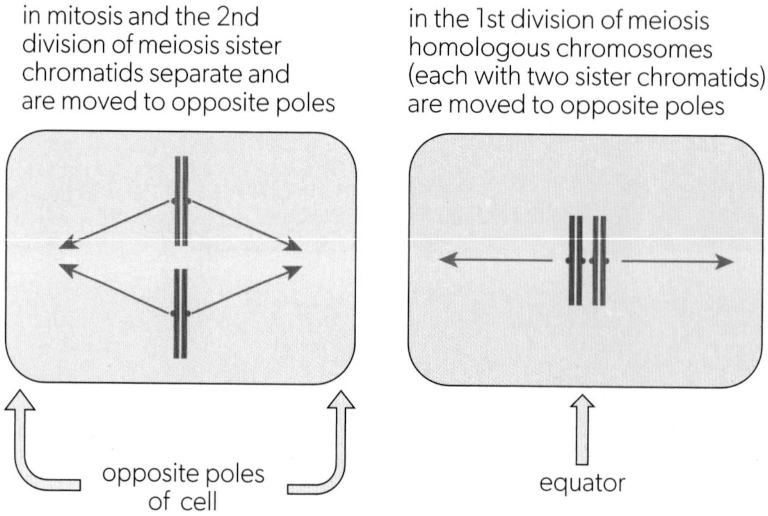

▲ Figure 14 Movement of chromosomes during mitosis and meiosis

Chromosomes are condensed by being made shorter. An initial shortening is carried out by wrapping the double helix of DNA around histone proteins to form nucleosomes, and linking the nucleosomes together. There are several more stages to condense the chromosomes but they are not yet fully understood—this is an active research field.

Chromosomes are moved using microtubules. A microtubule is a hollow cylinder of tubulin proteins that can be rapidly assembled or disassembled. During interphase, microtubules serve a variety of functions including acting as a cytoskeleton. Some of these microtubules are disassembled in the early stages of mitosis and are reassembled by microtubule organizing centres (MTOCs) at the poles of the cell, which link tubulin molecules together. Microtubules are assembled that reach the equator of the cell, forming a spindle-shaped array. At the same time, protein structures called kinetochores are assembled on the centromere of each chromatid. Some of the growing microtubules link up with these kinetochores and some attach to other microtubules from the opposite pole.

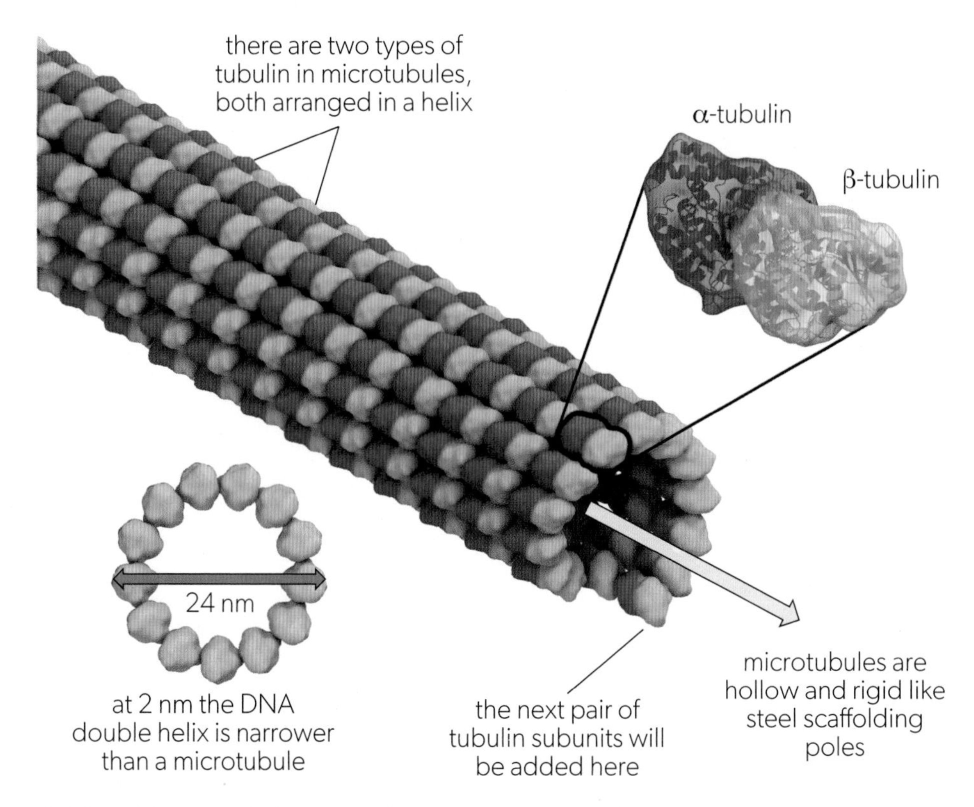
there are two types of tubulin in microtubules, both arranged in a helix
α-tubulin
β-tubulin
24 nm
at 2 nm the DNA double helix is narrower than a microtubule
the next pair of tubulin subunits will be added here
microtubules are hollow and rigid like steel scaffolding poles

▲ Figure 15 Microtubules can be changed in length by adding tubulins at one end, or removing them from the other end

The kinetochore acts as a microtubule motor by removing tubulin subunits from the attached ends of the microtubules. This shortens the microtubules linking the kinetochores to the poles, putting them under tension. Initially, in mitosis the chromatids do not move because loops of cohesin hold them together. As soon as the cohesin has been cut, shortening of the spindle microtubules by the kinetochores causes sister chromatids to move to opposite poles. In meiosis, homologous chromosomes are initially held together by knot-like structures called chiasmata, but when these have slid to the ends of the chromosomes, movement to opposite poles can begin.

D2.1.7 Phases of mitosis

Mitosis requires a precisely choreographed sequence of actions. These are usually considered as four phases.

- **Prophase**—the starting phase with condensation of chromosomes (pro = before).
- **Metaphase**—the phase after condensation with chromosomes released from the nucleus (meta = after).
- **Anaphase**—a brief phase during which the chromosomes are moved up to poles from the equator (ana = up).
- **Telophase**—the final phase in which nuclei reform and chromosomes decondense (telos = finally).

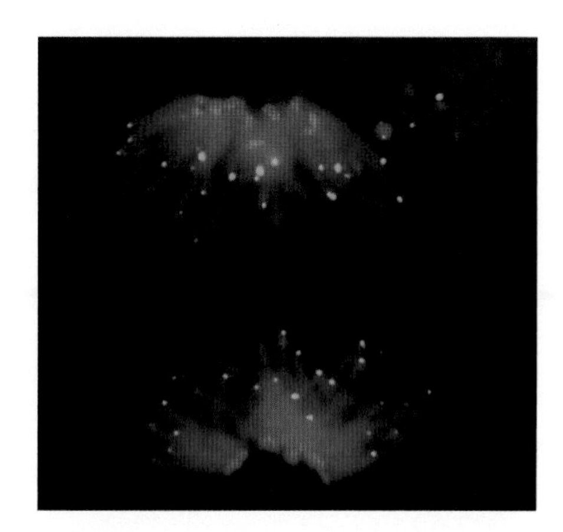

▲ Figure 16 Fluorescent stains have been used to reveal the position in anaphase of DNA (blue), kinetochores (red) and telomeres which form the ends of the chromosomes (green)

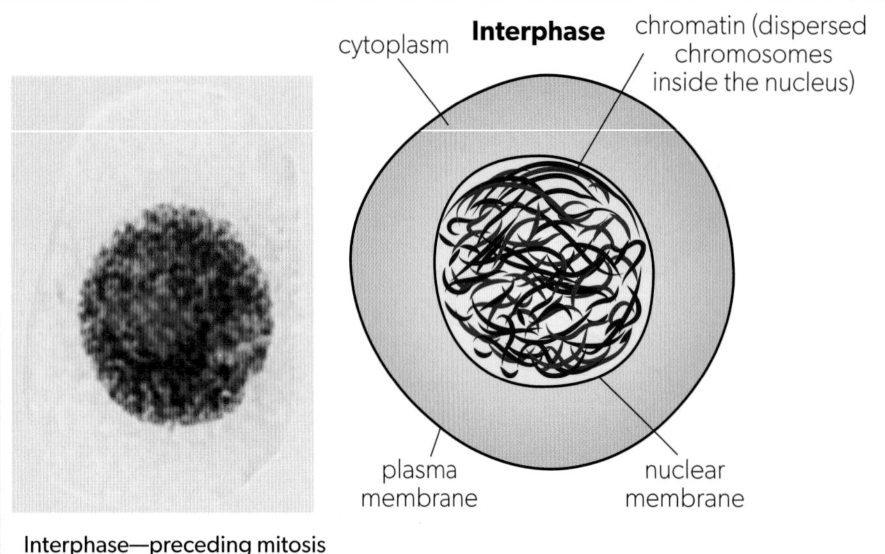

Interphase—preceding mitosis

- The chromosomes are dispersed through the nucleus so are not individually discernible.
- To prepare for mitosis, all of the DNA is replicated and each chromosome then consists of two very elongated chromatids containing identical DNA.

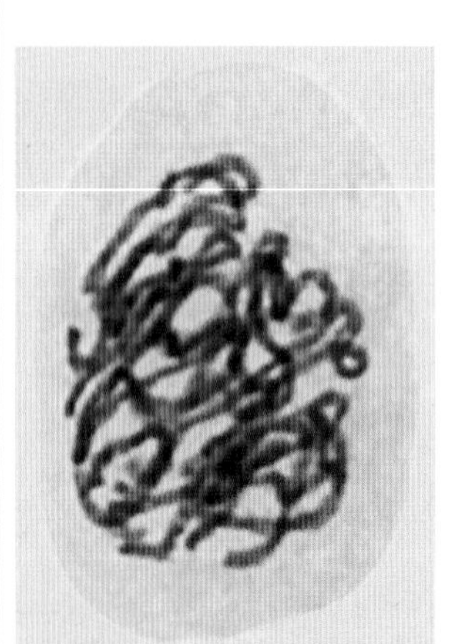

Prophase

sister chromatids held together by loops of cohesin
kinetochore attached to the centromere of the chromatid
microtubules organizing centre (MTOC)
spindle microtubules

Prophase

- The chromosomes condense by packing the DNA tightly into thicker, shorter structures. This is a protracted process that continues throughout prophase.
- Towards the end of prophase microtubules grow from structures at the poles of the cell called microtubule organizing centres (MTOCs) to form a spindle-shaped array linking the poles of the cell.
- At the end of prophase the nuclear membrane breaks down.

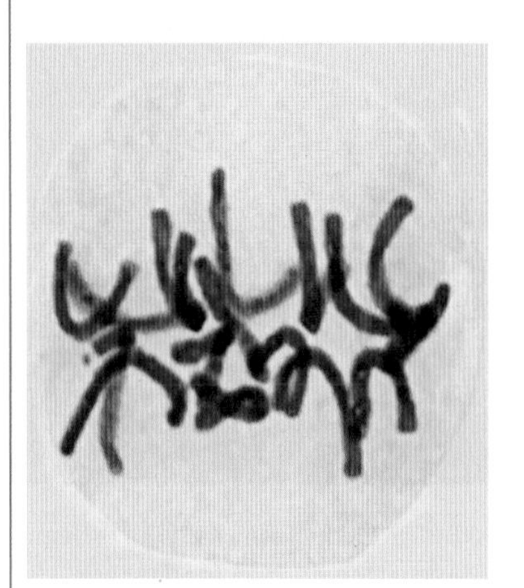

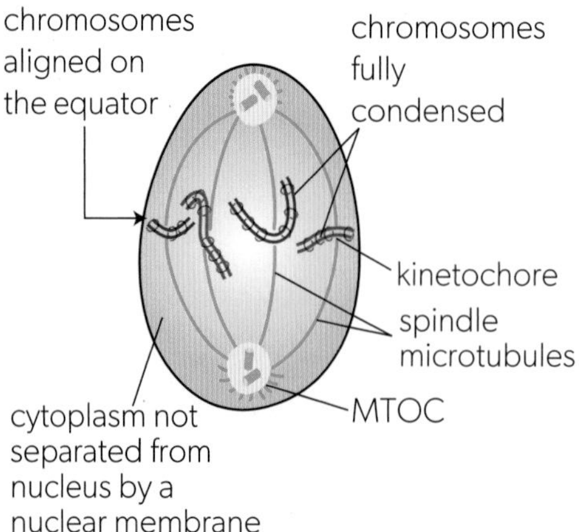

Metaphase

- Microtubules growing from the poles attach to the centromere of each chromatid. Sister chromatids within each chromosome become attached to opposite poles.
- The spindle microtubules are put under tension to test whether the attachment is correct. If the attachment is correct, the chromosomes cannot yet be pulled to either pole due to cohesin loops.
- At the end of metaphase, the chromosomes are aligned on the equator of the cell.

Anaphase

kinetochore removes tubulin subunits to shorten spindle microtubule and pull chromosome to the pole

genetically identical chromosomes (formerly sister chromatids) moving to opposite poles

chromosome arms trail behind in the viscous cytoplasm

Anaphase

- Cohesin loops that have held the sister chromatids together are now cut, so the chromatids become separate chromosomes.
- Microtubules link each chromosome to one of the poles. A kinetochore shortens the microtubules, pulling the chromosome to the pole.
- At the end of anaphase all the chromosomes have reached the poles but have not started to decondense.

Telophase

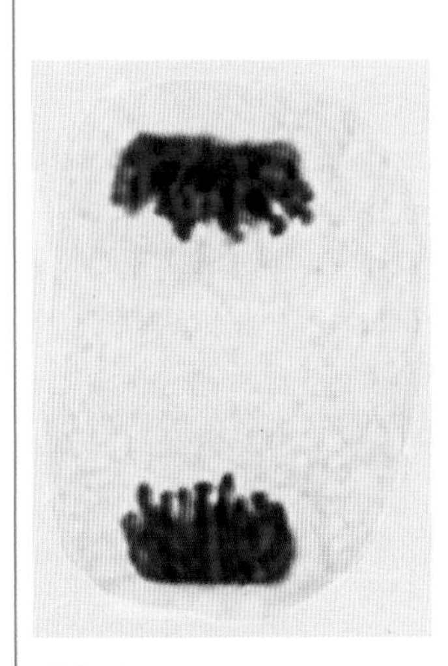

cytokinesis has started with a furrow where the plasma membrane is pulled in around the equator

chromosomes are now enclosed inside a nuclear membrane

chromosomes are decondensing

Telophase

- At each pole the chromosomes are pulled into a tight group near the MTOC and a nuclear membrane reforms around them.
- The chromosomes decondense and spread out to form dispersed chromatin inside the nucleus.
- By this stage of mitosis the cell is usually already dividing its cytoplasm and the two daughter cells produced by this enter interphase again.

Interphase

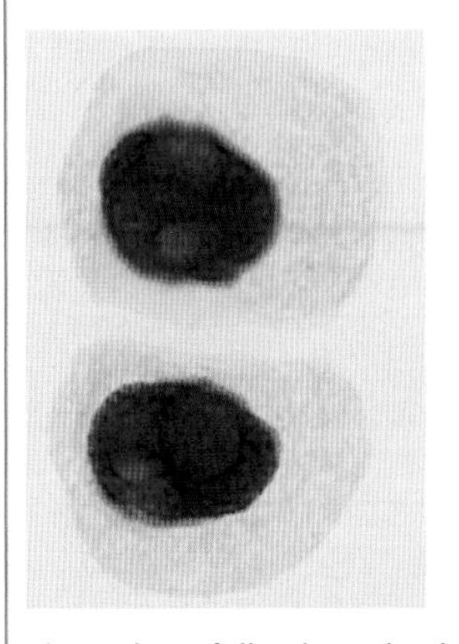

chromosomes each consisting of a single DNA molecule are dispersed throughout the nucleus

cytokinesis completed with the cytoplasm and plasma membrane pinched apart

cytoplasm is very active

Interphase following mitosis

- Genes in the decondensed chromosomes can be transcribed and the mRNA translated to synthesize proteins that the cell needs.
- The cell grows usually doubling in size before the next mitosis.

D2.1.8 Identification of phases of mitosis

Preparing and viewing a stained slide: Preparing a root tip squash

1. Suspend the base of a clove of garlic in water using toothpicks for 3 days or more, until roots grow. Cut off roots and fix them in 99 parts of 70% ethanol to 1 part of pure ethanoic acid for 24 hours.
2. When you are ready to make observations, immerse the roots in 0.1 mol dm^{-3} hydrochloric acid at 40°C for five minutes. Eye protection needed! Then rinse off the acid. This acid treatment loosens the cell walls in the root tissue.
3. Cut off about 20 mm to 30 mm from the end of two or three of the roots and immerse these root tips in 1% toluidine blue stain. Skin protection needed! The stain binds to DNA making chromosomes more visible.
4. Rinse the stain off the root tips and gently squash them under a cover slip. Observe the slide using the low power objective followed by the medium power objective.

Data-based questions: Identifying phases of mitosis

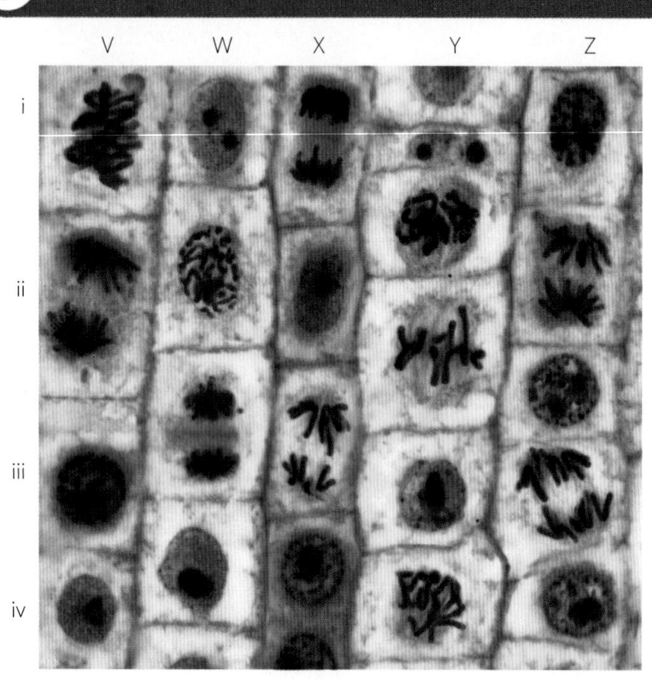

▲ **Figure 17** This micrograph shows cells in the tip of an onion root, magnified ×600.

Some of the cells are in mitosis and others are in interphase. The columns of cells are indicated by letters (V to Z) and the cells within the columns by numbering from the top (i, ii, etc.).

1. a. There are five cells in the right-hand column, Zi to Zv. Identify which of these cells are in interphase. [1]

   b. Identify the phase for each cell in column Z that is dividing its nucleus by mitosis. [2]
2. Identify the phase for each of cells Vi, Vii, Wi, Wii, Wiii, Xi, Xiii. [7]
3. There are five whole cells in column Y (Yi to Yv). Identify which, if any, of these cells is in:

   a. prophase

   b. metaphase. [3]
4. Identify two cells in which cytokinesis has begun. [2]
5. Cell Yi (the uppermost whole cell in column Y) has a different appearance from others in the micrograph. Suggest what process caused this cell to have height in the column smaller than its width.
6. Cell Wi has two dark structures in its nucleus. They are sites where ribosomes are being assembled. Suggest reasons why cell Wi needs more ribosomes. [2]
7. Identify, with a reason, which cell in the micrograph probably entered interphase most recently. [2]

D2.1.9 Meiosis as a reduction division

The nucleus of a cell contains the chromosomes. Each chromosome is a very long DNA molecule with some associated proteins. Along a chromosome is a sequence of genes. An average human chromosome has about a thousand genes in a linear sequence. When the DNA in a chromosome is replicated, the gene sequence does not change. This means the sequence of genes shows continuity through the generations in a species, potentially over millions of years. There are major disadvantages to changes in gene sequence, and such changes are only common during the evolution of new species. Chromosomes with the same sequence of genes as each other are homologous.

Although the sequence of genes on a chromosome is resistant to change, the base sequence of individual genes on a chromosome can change by mutation, resulting in new alleles. These alleles can be reshuffled during meiosis to produce new combinations. This is called recombination and together with gene mutation it explains how chromosomes can be homologous but not identical.

A species has a characteristic number of types of homologous chromosome. This is known as the haploid number and is given the symbol *n*. Body cells in most plants and animals contain two homologous chromosomes of each type, so there are 2*n* chromosomes. This is the diploid number. The terms haploid and diploid are important in cell biology and genetics.

- **Haploid** (*n*)—a nucleus, cell or organism with a single set of chromosomes, which are all non-homologous
- **Diploid** (2*n*)—a nucleus, cell or organism with two sets of chromosomes and therefore homologous pairs of chromosomes.

Diploid cells are produced by sexual reproduction. The key event is the fusion of gametes; for example, a sperm and an egg uniting to form a single cell. Gametes are haploid, so the zygote produced when male and female gametes fuse is diploid. Body cells in most plants and animals are produced from the zygote by many cycles of mitosis and cell division, so are all diploid. This continuity in the genome and chromosome number of an individual persists until sexual reproduction. If diploid body cells were to become gametes, the resulting offspring would be tetraploid (4*n*). Tetraploid gametes would result in octoploid offspring. To prevent the number of chromosomes doubling with each generation in a species, gametes must be haploid. In a sexual life cycle, there must therefore be a reduction division in which the chromosome number is halved from diploid to haploid. This counteracts the doubling effect of the male and female gametes fusing. Meiosis is this reduction division (Figure 18).

In most organisms, normal body cells are all diploid; meiosis happens during the process of gametogenesis. There are some other patterns. For example, mosses carry out meiosis earlier in the life cycle and the main moss plant is haploid. Meiosis must occur at some stage in a sexual life cycle to avoid increases in the chromosome number. Figure 18 diagrammatically compares mitosis and meiosis.

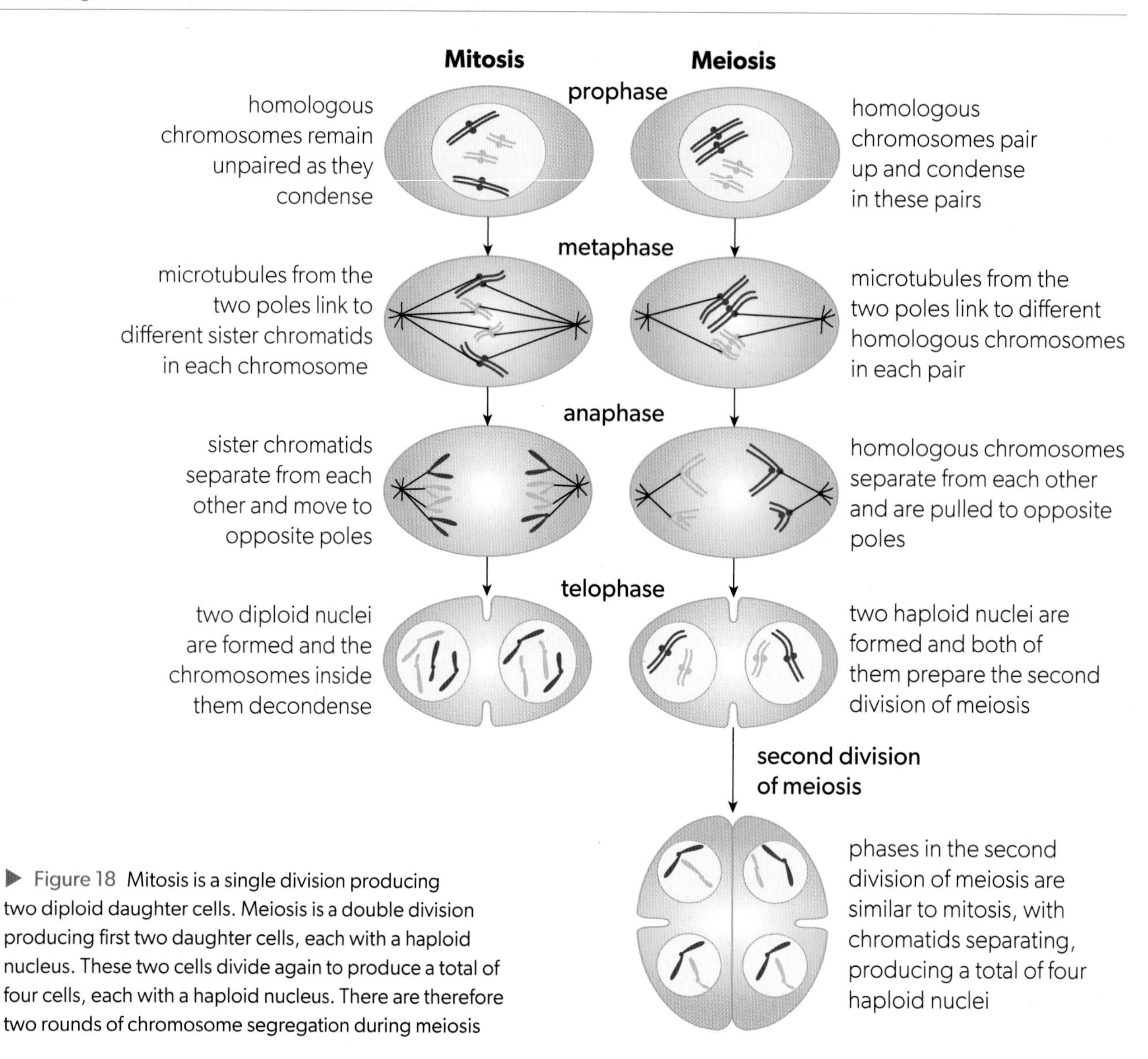
Mitosis
Meiosis
prophase
homologous chromosomes remain unpaired as they condense
homologous chromosomes pair up and condense in these pairs
metaphase
microtubules from the two poles link to different sister chromatids in each chromosome
microtubules from the two poles link to different homologous chromosomes in each pair
anaphase
sister chromatids separate from each other and move to opposite poles
homologous chromosomes separate from each other and are pulled to opposite poles
telophase
two diploid nuclei are formed and the chromosomes inside them decondense
two haploid nuclei are formed and both of them prepare the second division of meiosis
second division of meiosis
phases in the second division of meiosis are similar to mitosis, with chromatids separating, producing a total of four haploid nuclei

▶ Figure 18 Mitosis is a single division producing two diploid daughter cells. Meiosis is a double division producing first two daughter cells, each with a haploid nucleus. These two cells divide again to produce a total of four cells, each with a haploid nucleus. There are therefore two rounds of chromosome segregation during meiosis

ATL Communication skills: Responding appropriately to command terms

The command term "compare and contrast" requires that students consider the similarities and differences of a particular phenomenon in a variety of contexts. Two separate descriptions do not meet the requirements of this command term.

For example, if asked to compare and contrast the behaviour of homologous chromosomes in mitosis and meiosis, a student should write in this style: "In meiosis I, homologous pairs form a bivalent and line up along the equator, whereas in mitosis, the homologous chromosomes behave independently when they line up along the equator." In this case, the comparative term "whereas" is essential. Furthermore, responding to this command term encourages the inclusion of at least one similarity and at least one difference.

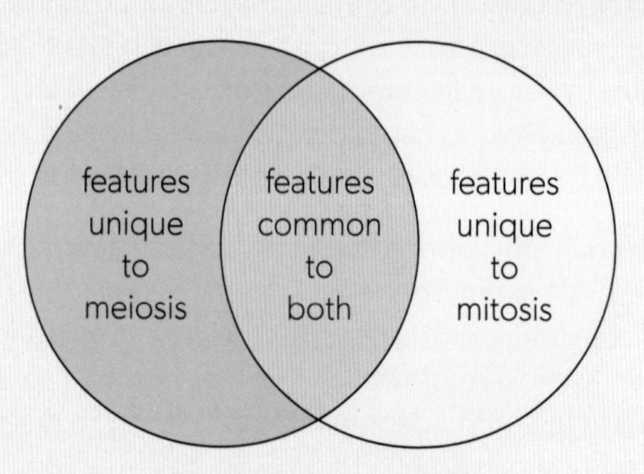
features unique to meiosis
features common to both
features unique to mitosis

▲ Figure 19 One strategy is to use a Venn diagram like this to show similarities and differences.

Data-based questions: Life cycles

▼ Figure 20 The life cycles of humans and mosses; n represents the haploid number of chromosomes and $2n$ represents the diploid number. Sporophytes of mosses grow on the main moss plant and consist of a stalk and a capsule in which spores are produced

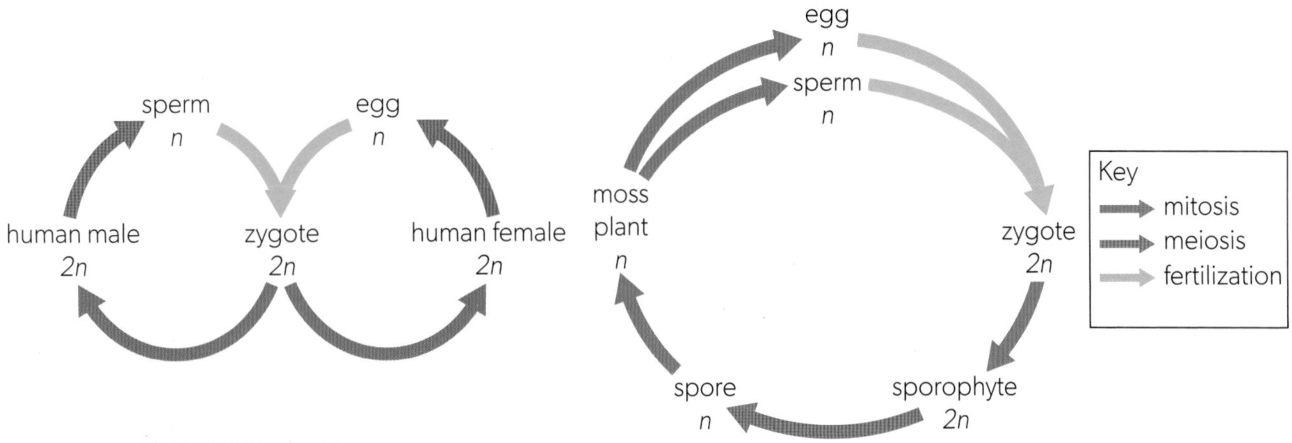

sperm n
egg n
human male $2n$
zygote $2n$
human female $2n$

egg n
sperm n
moss plant n
zygote $2n$
spore n
sporophyte $2n$

Key
mitosis
meiosis
fertilization

◀ Figure 21 Cells in the green cushion of this moss are haploid. The stalks and spore capsules growing out from it are diploid sporophytes, so genetically they are each new individuals. The moss is *Leptostomum inclinans*, growing on a tree in Waimarama Sanctuary near Nelson in New Zealand

1. Compare the life cycle of a moss and of a human by giving five similarities. [5]
2. Distinguish between the life cycles of a moss and a human by giving five differences. [5]

D2.1.10 Down syndrome and non-disjunction

If the chromosomes all separate correctly, four haploid cells are produced by meiosis in a diploid cell.

- Homologous chromosomes separate from each other and move to opposite poles in Anaphase I, halving the chromosome number
- Chromatids in each chromosome separate from each other and move to opposite poles in Anaphase II.

Errors may occur. For example, a pair of homologous chromosomes might move to the same pole in Anaphase I, or both chromatids of one chromosome might move to the same pole in Anaphase II. This is called non-disjunction and the consequence is cells with one chromosome extra or missing.

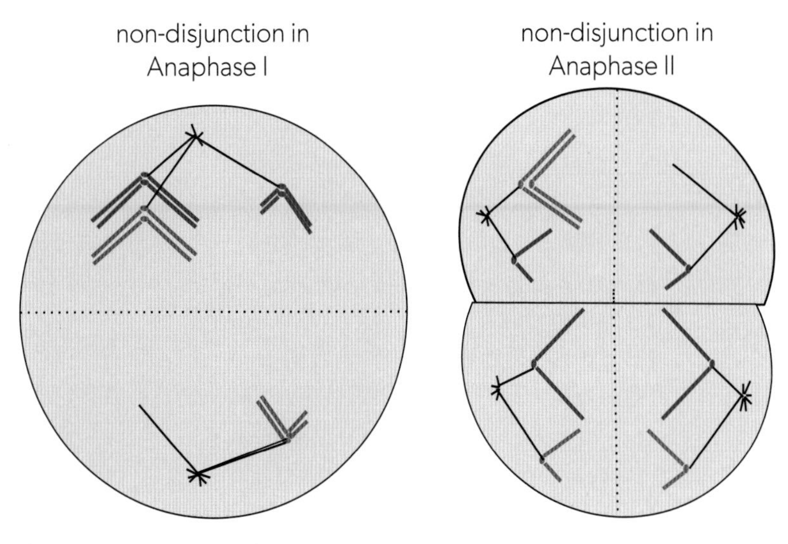

▲ Figure 22 Failure of chromosomes to separate is called non-disjunction

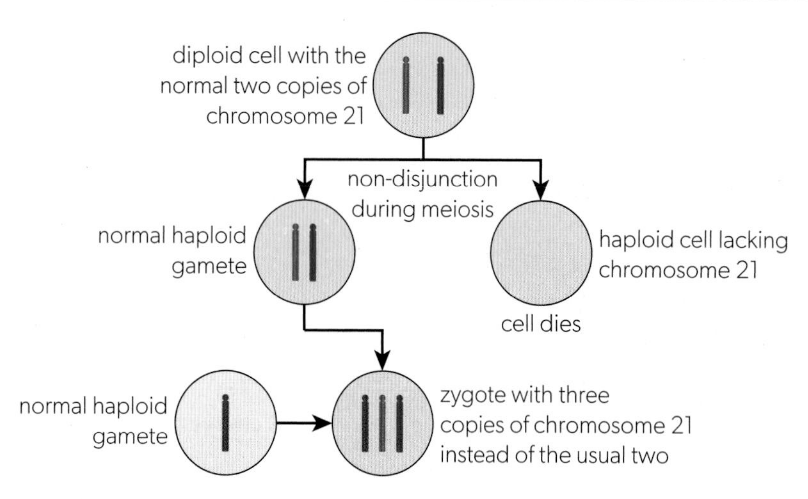

▲ Figure 23 How non-disjunction can lead to Down syndrome. Chromosomes 1 to 20, 22, X and Y are not shown

Non-disjunction can happen with any of chromosomes. In almost all cases, a missing chromosome quickly leads to the death of a haploid cell, because genes coding for essential polypeptides are lacking. Having an extra chromosome is also a lethal condition in most cases. It might lead to the death of a haploid cell produced by meiosis or the death of the zygote or embryo produced by fusion of gametes.

In a few cases an extra chromosome is not lethal. For example, Down syndrome is due to a non-disjunction event that results in a person having three copies of chromosome number 21. While individuals vary, some of the features of the syndrome are hearing loss, heart and vision disorders.

▲ Figure 24 Child with Down syndrome

Non-disjunction can also result in the birth of babies with abnormal numbers of sex chromosomes. Klinefelter's syndrome is caused by having the sex chromosomes XXY. Turner's syndrome is caused by having only one sex chromosome, an X.

Data-based questions: Parental age and non-disjunction

Studies show that the age of parents influences the chances of non-disjunction. Data in Figure 25 shows the relationship between maternal age and the incidence of Down syndrome and other conditions caused by chromosome abnormalities.

1. Outline the relationship between maternal age and the incidence of chromosomal abnormalities in live births. [2]
2. a. For 40-year-old mothers, determine the probability that their baby will have Down syndrome. [1]

   b. Using the data in Figure 25, calculate the probability that a mother aged 40 will give birth to a child with a chromosomal abnormality other than Down syndrome. [2]
3. Only a small number of possible chromosomal abnormalities are ever found among live births. More than half of all cases are Down syndrome. Suggest reasons for these trends. [3]
4. Discuss the risks and benefits of parents not having children until they are relatively old. [2]

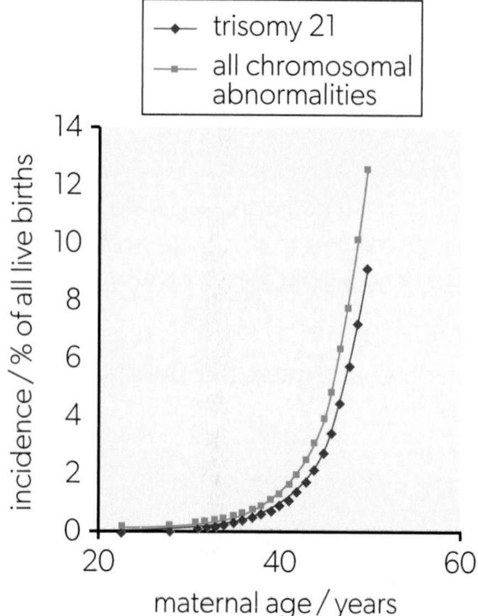

▲ Figure 25 The incidence of Down syndrome and other chromosomal abnormalities as a function of maternal age

D2.1.11 Meiosis as a source of variation

Meiosis generates genetic diversity in two ways: random orientation of bivalents (homologous chromosomes that have paired up) and crossing over. At the start of meiosis there are pairs of homologous chromosomes, one originally inherited from the mother and one from the father. These chromosomes are homologous because they have the same sequence of genes as each other but they are not identical because the alleles of some of the genes will be different.

Crossing over

Homologous chromosomes pair up at an early stage of meiosis. Because DNA replication has already occurred, each chromosome consists of two chromatids, so there are four DNA molecules associated in each pair of homologous chromosomes. A pair of homologous chromosomes is called a bivalent and the pairing process is called synapsis.

Crossing over takes place while the chromatids are still very elongated. Two non-sister chromatids are brought together at the same point along their gene sequences. The two strands of their DNA double helices are cut, one at a time, and are rejoined with the equivalent strand in the other chromatid. This results in a mutual exchange of DNA and therefore genes between the two chromatids.

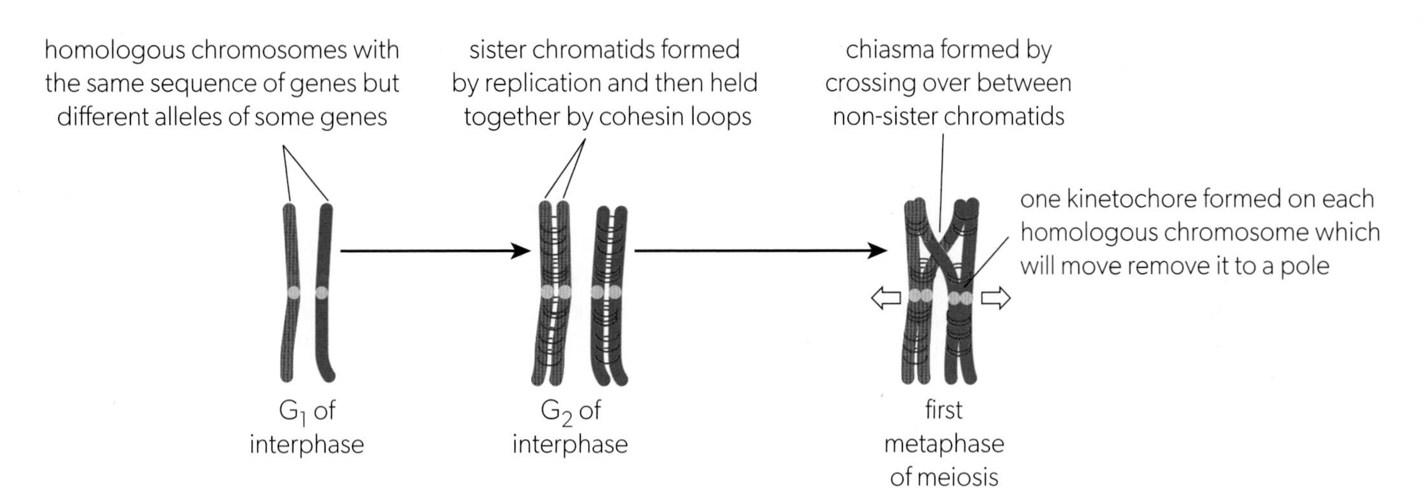

▲ Figure 26 Chiasma formation as a consequence of crossing over

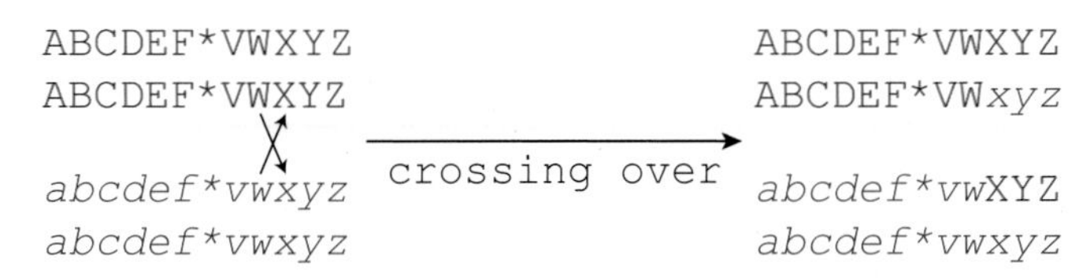

▲ Figure 27 Recombination of alleles by crossing over

As non-sister chromatids are homologous but not identical, some alleles of the exchanged genes are likely to be different. Chromatids with new combinations of alleles are therefore produced. The consequences of a crossing over can be illustrated using upper- and lowercase letters for alleles of genes.

Crossing over occurs at random positions anywhere along the chromosomes. At least one crossover occurs in each bivalent, but there is often more than one.

Random orientation of bivalents

Each homologous pair of chromosomes forms a bivalent on the equator of the cell during first metaphase of meiosis. Within each bivalent, the homologous chromosomes become attached to spindle microtubules from different poles. Which chromosome of a pair attaches to which pole is decided by chance, because orientation of bivalents at this stage is random. There is a 50% chance of each of the two possible outcomes. Although random, the orientation of bivalents has consequences, because the two homologous chromosomes have different alleles of some genes. Whether one particular allele is inherited by an individual depends only on which way a bivalent happened to be facing when spindle microtubules were attached.

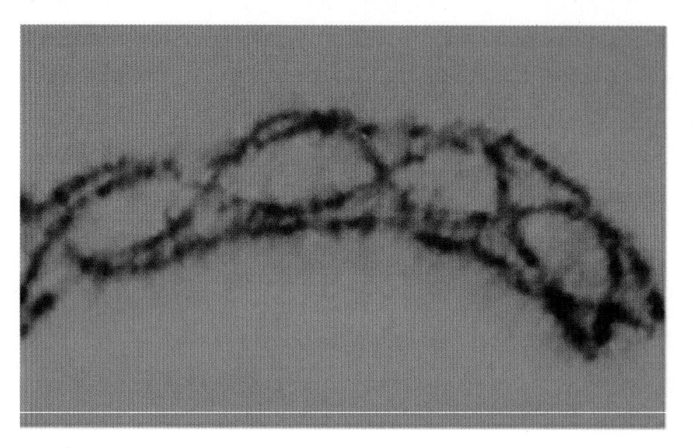

▶ Figure 28 Five chiasmata are visible in this bivalent, showing that crossing over can occur more than once

The position of one bivalent does not affect other bivalents—their orientation is independent. The number of possible combinations of chromosomes that a haploid cell produced by meiosis could contain as a consequence of the random orientation of bivalents is therefore 2^n where n is the number of bivalents (and also the haploid number for the species). With two pairs of bivalents, as in the diagrams, there are four possible combinations, but most species have more chromosomes than this, so more possible outcomes. In one human individual, random orientation of bivalents in meiosis can generate 2^{23} possible outcomes—more than 8 million. Combined with crossing over this gives almost limitless numbers of possible combinations of alleles in cells produced by meiosis.

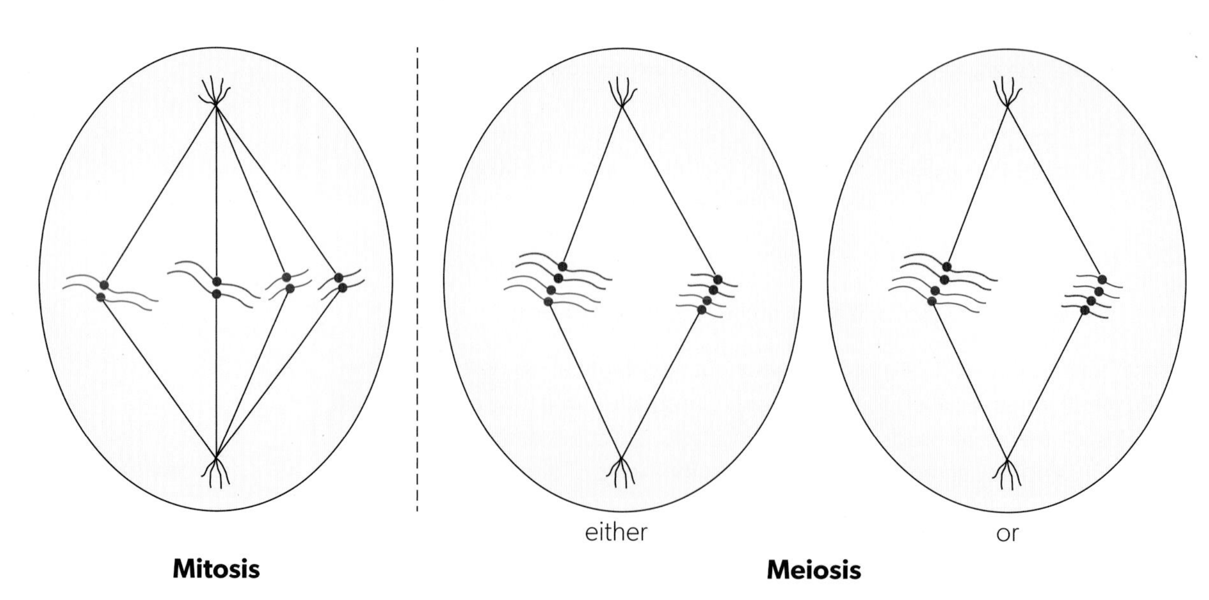

▲ Figure 29 Comparison of chromosome attachment to spindle microtubules

The first division of meiosis

Prophase I • Homologous chromosomes form parallel pairs (synapsis) • Crossing over occurs between non-sister chromatids in each bivalent • Chromatids (four per bivalent) shorten and thicken (condensation)	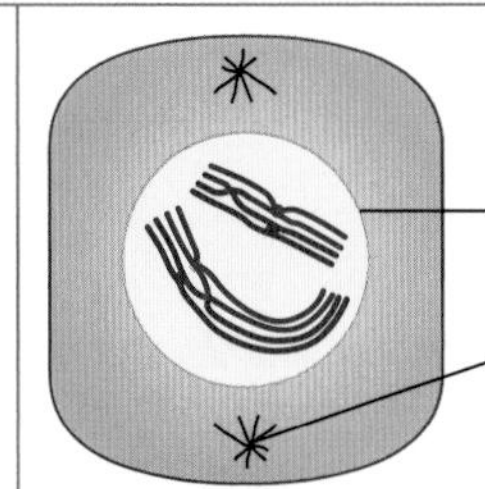	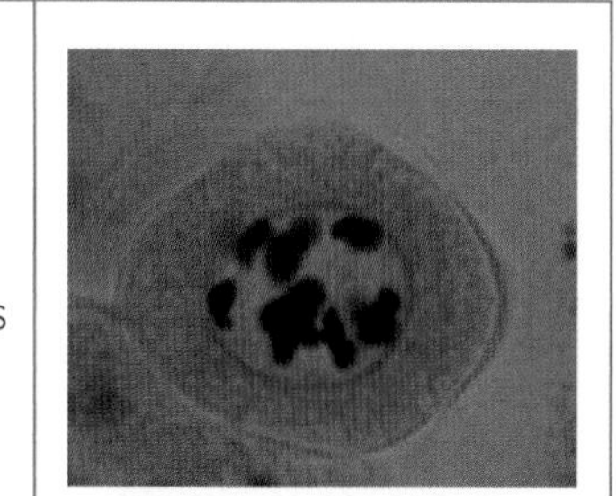
Metaphase I • Nuclear membrane disperses and homologous chromosomes in each bivalent become attached to spindle microtubules from opposite poles • Bivalents spread out in the equator • Pairing of homologous chromosome ends but chiasmata prevent separation	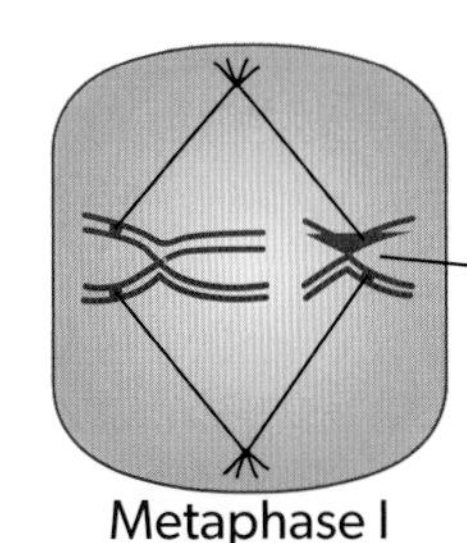	
Anaphase I • Kinetochores shorten the spindle microtubules, pulling chromosomes towards the poles • Chiasmata slide to the end of chromosomes • Pairs of homologous chromosomes separate and move to opposite poles	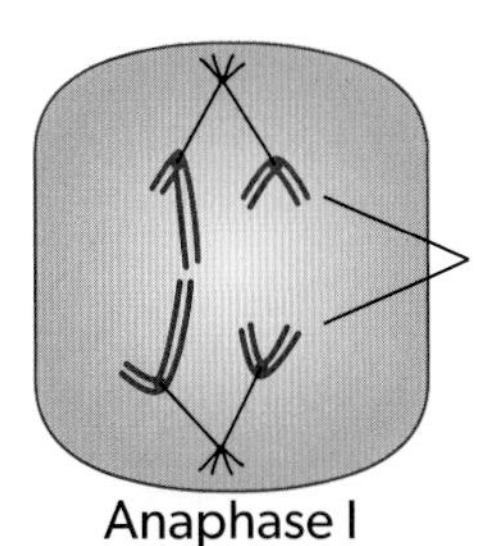	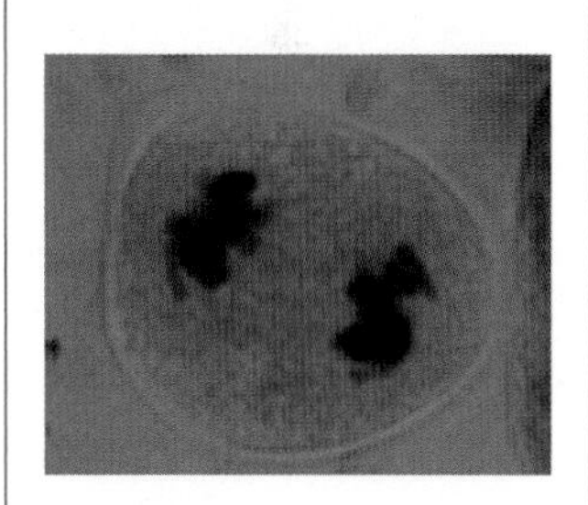
Telophase I • A nuclear membrane is assembled around the chromosomes at each pole of the cell • Chromosomes decondense, but in some species only partially • Cytokinesis divides the cytoplasm, resulting in two haploid cells		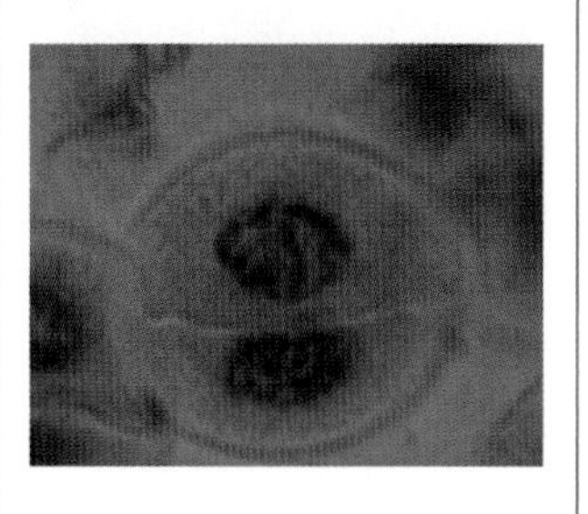

The second division of meiosis

Prophase II • DNA replication does not occur between the first and second divisions because the chromosomes each have two chromatids already. Because of this, and because there has been crossing over, the two chromatids in each chromosome are not genetically identical, unlike mitosis • Chromatids are condensed 	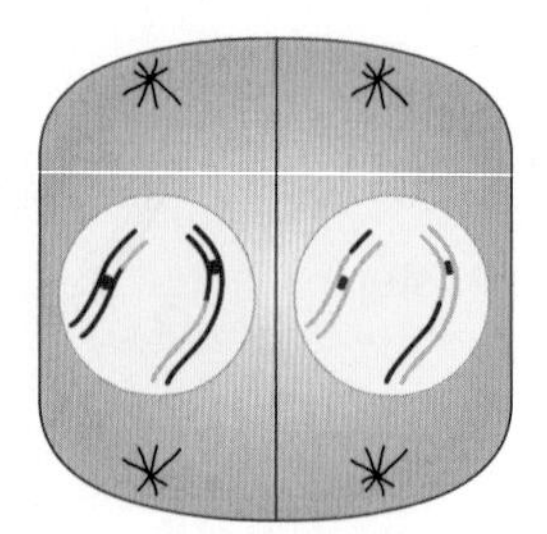 Prophase II	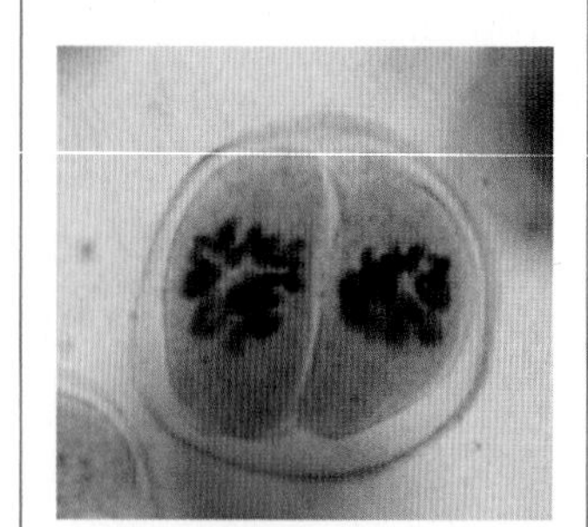
Metaphase II • Nuclear membranes disperse and the chromatids of each chromosome become attached to spindle microtubules from opposite poles • Chromosomes spread out on the equator • Cohesin loops connecting chromatids are cut allowing them to start to separate from each other	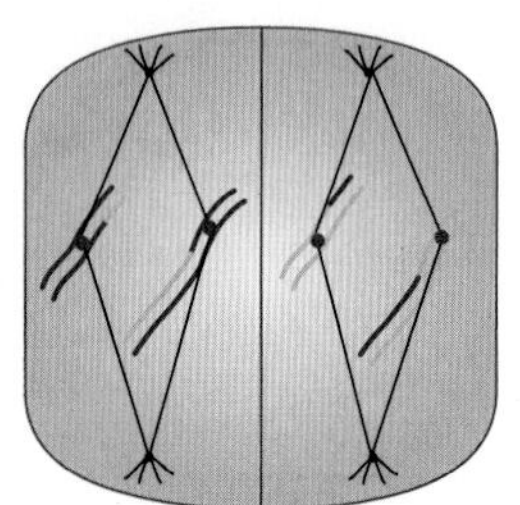 Metaphase II	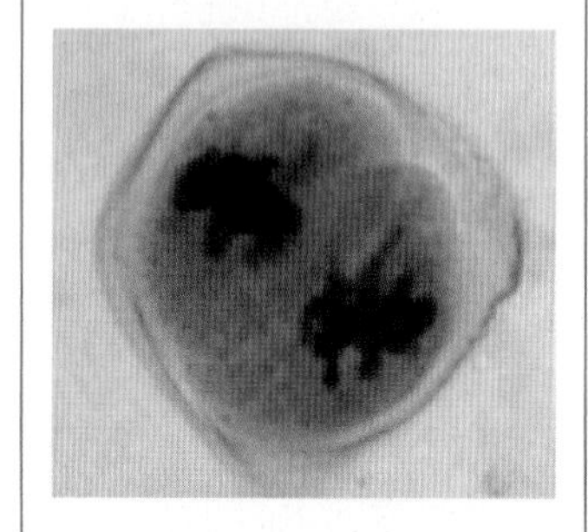
Anaphase II • Kinetochores on the centromere of each chromatid shorten the spindle microtubules, pulling the chromatids to opposite poles • The chromatids are considered to be chromosomes once they have separated • By the end of anaphase these chromosomes have reached the poles	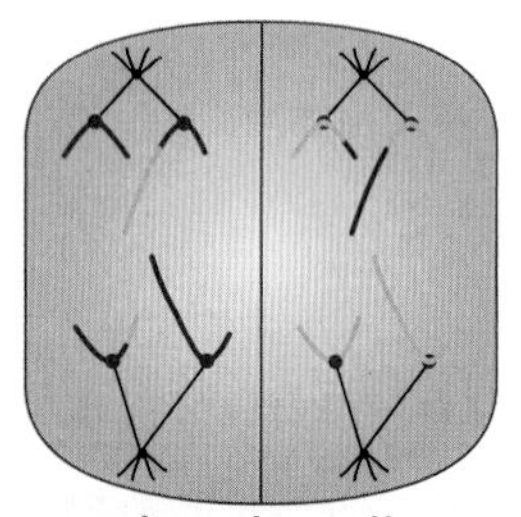 Anaphase II	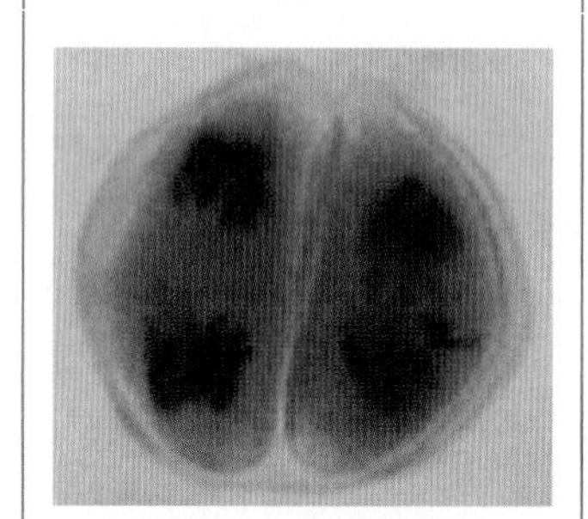
Telophase II • Nuclear membranes are assembled around the chromosomes at each pole • Decondensation spreads the chromosomes out throughout each of the four nuclei • Cytokinesis occurs, dividing both of the cells produced by the first division of meiosis, so there are now four haploid cells	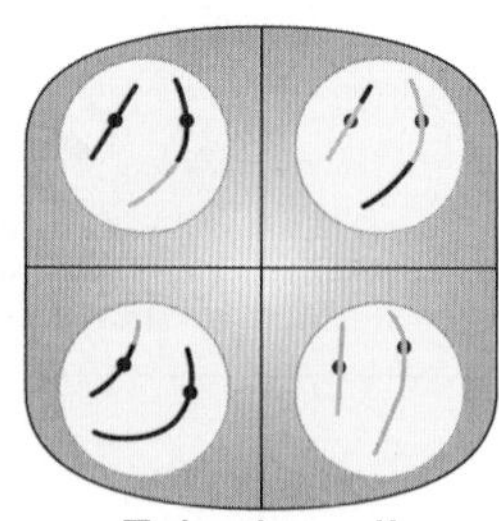 Telophase II	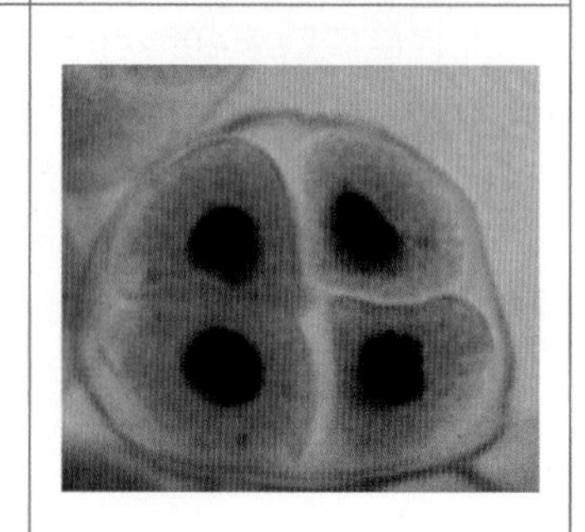

AHL

D2.1.12 Cell proliferation for growth, cell replacement and tissue repair

Cell proliferation is a rapid increase in the number of cells. It happens when cell division happens at a faster rate than cell death. Cell proliferation is needed in multicellular organisms for growth, cell replacement and tissue repair. Mitosis ensures continuity of the genome during cell proliferation in an animal, plant or other multicellular eukaryote.

Growth

In most animals, there are embryonic and juvenile phases of growth, after which adult size and form is reached and growth stops. Initially, there is cell proliferation throughout an animal embryo. In some parts of the body, cells continue to proliferate during juvenile phases of growth. For example, growth plates near the end of bones are active in childhood and adolescence, with cell divisions contributing to bone growth and rapid increase in height.

Cell proliferation in plants is confined to growth regions called meristems. Apical meristems are found at the tips of stems and roots. Some of the daughter cells formed by division remain in an apical meristem and continue to divide. Cells formed at the margin of the meristem cease divisions and instead enlarge and differentiate for a specific function. The root apical meristem generates cells for lengthening the root. The shoot apical meristem is more complex, producing cells for extension growth of the stem and for developing leaves or flowers.

dome of cells at centre of apical meristem

youngest developing leaf

developing bud

region of stem growth

▲ Figure 30 Structure of a shoot apical meristem that is producing cells for stem growth and leaf development

Cell replacement

If cells are lost from a tissue, more cells must be produced to replace them. This happens in the epidermis of the skin, where cells are abraded and replaced throughout life. Cell division happens in the basal layer of the epidermis. Cells produced here are displaced towards the skin surface by continued cell division. During this transit, cells produce large amounts of the tough protein keratin, which is hydrophobic and causes the cells to dry out. By the time the cells reach the skin surface they are flattened and dead. Friction causes them to be rubbed off after they have waterproofed and protected the body for a while.

It is debatable whether cell replacement is a true example of cell proliferation because cell division does not happen at a faster rate than cell death, but cell proliferation happens in one part of the epidermis and cell death in another.

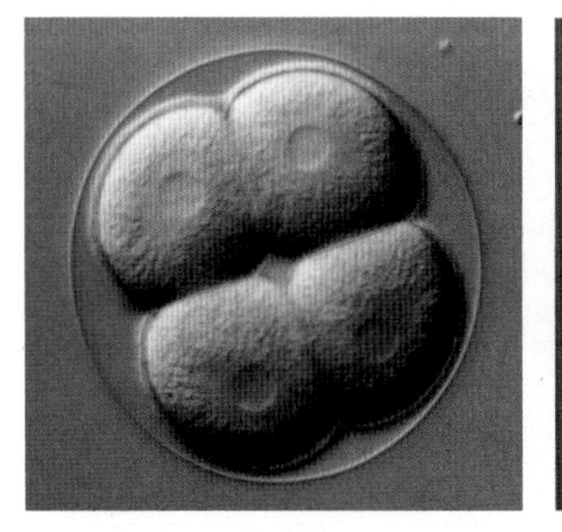

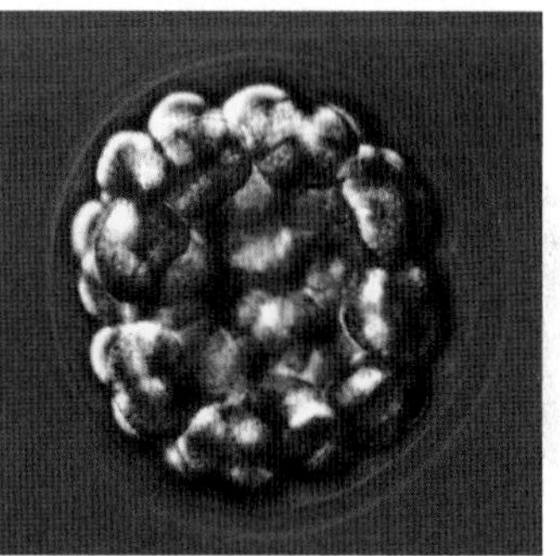

▲ Figure 31 Cell proliferation in a sea urchin begins with a large zygote, which divides repeatedly, passing through the four-cell stage (left) and the blastula (right) which is a hollow ball of cells. All the cells in the blastula continue to divide through the following embryonic stages

Tissue repair

Wounds to many parts of the body can heal, using cell division to replace lost cells. This depends on the presence of undifferentiated stem cells that can divide and then differentiate. Numbers of these stem cells vary, so some tissues are more able to repair themselves than others.

Skin is particularly effective at undertaking tissue repair after a cut or other wound. If the basal cells in the epidermis are still present, they can regenerate damaged outer layers of skin in a few days. Stem cells in the dermis can repair deeper damage, though this takes longer. If damage is so severe that there are no surviving stem cells, skin grafts may be needed.

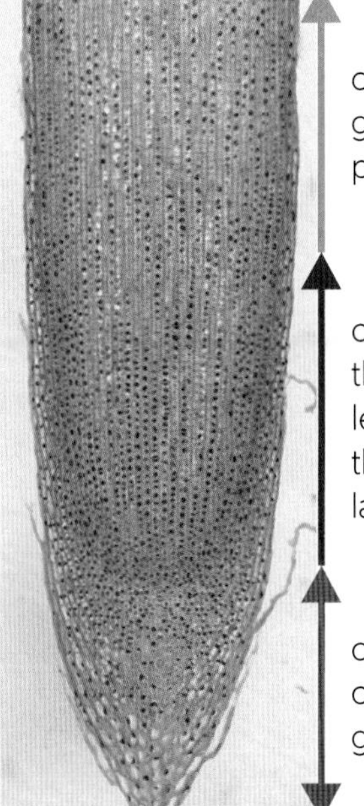

cells in this region stop dividing and then grow larger and differentiate so they can perform a specific role in the root

cell proliferation in this region produces the cells needed for the root to grow in length. The cells remain small because they divide as soon as they are large enough

cell proliferation in the root cap replaces cells that are rubbed off as extension growth pushes the root through the soil

▲ Figure 32 Longitudinal section through an onion root apical meristem

AHL

D2.1.13 Phases of the cell cycle

Cell proliferation is achieved by cells following the cell cycle. There are two main phases in this repeating sequence of processes.

- **Mitosis** is the process that divides the nucleus, after which cytokinesis divides the cell as a whole. Mitosis is subdivided into four stages: prophase, metaphase, anaphase and telophase (*Section D2.1.7*).
- **Interphase** is the period between one mitosis and the next. In interphase DNA is replicated, mitochondria divide and other cell components are synthesized. Interphase is subdivided into three stages: G_1 (Gap 1), S phase and G_2 (Gap 2).

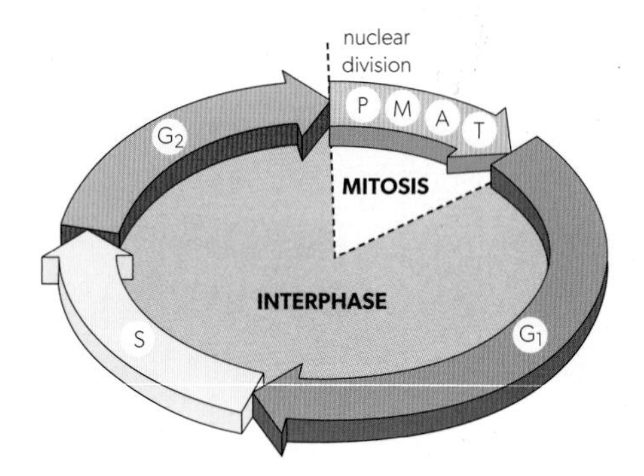

▲ Figure 33 Sequence of phases in the cell cycle

G_1—the phase after mitosis and before DNA replication when each chromosome is a single DNA molecule. This is an active growth phase.

S phase—in which all DNA in the nucleus is replicated. This results in identical pairs of DNA molecules held together by cohesin loops. The loops remain until the end of metaphase in mitosis, ensuring that the two molecules can be dispatched to different daughter cells.

G_2—the phase after all DNA in the nucleus has been replicated so each chromosome consists of two DNA molecules or chromatids. Growth may resume during this phase while the cell is preparing for mitosis.

Instead of moving on to mitosis or G_1 again, cells produced by mitosis and cytokinesis may leave the cell cycle and enter G_0 (Gap zero). In G_0, cells grow and differentiate for a specific role, but do not divide again. Of the two cells produced by one turn of the cell cycle, one or both may leave the cycle and enter G_0.

Data-based questions: DNA content per nucleus

The amount of DNA present in each cell nucleus was measured in many cells taken from two different cultures of human bone marrow (Figure 34).

1. For each group (I, II and III) in Sample B, deduce which stage of the cell cycle (S, G_1 or G_2) the cells could be in. [3]
2. Estimate the approximate amount of DNA per nucleus in these human cell types:
   a. bone marrow cells in prophase of mitosis [1]
   b. bone marrow in telophase of mitosis. [1]
3. Explain how many chromosomes there are likely to be in cells in:
   a. Sample A [1]
   b. Groups I and II in Sample B. [2]
4. Explain the mass of DNA that human egg and sperm cells are likely to contain. [2]

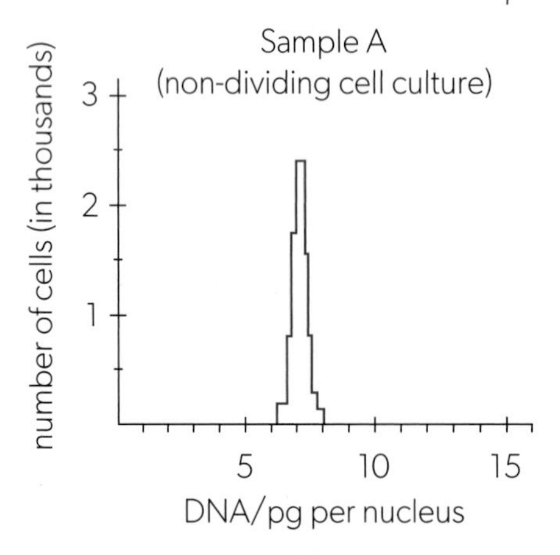

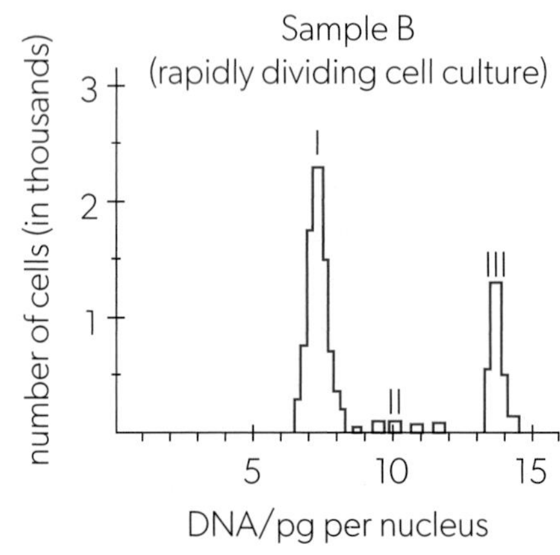

▶ Figure 34 Graphs of DNA per nucleus

D2.1.14 Cell growth during interphase

Interphase is the state of a cell when not undergoing mitosis or meiosis. DNA in regions lacking genes needed for the cell's activities, remains condensed as heterochromatin. The rest of the DNA becomes decondensed and dispersed in the nucleus as chromatin. DNA in a decondensed state can be transcribed, allowing protein synthesis by translation of mRNA.

Cells following the cell cycle typically need to double in size during interphase. DNA is doubled by replication. The volume of cytoplasm increases, so molecules within it such as enzymes must be synthesized. The area of membrane within cells is increased, requiring extra phospholipids, membrane proteins and cholesterol.

The numbers of most types of organelle must be increased. Mitochondria and chloroplasts contain DNA, so can only be propagated by division, following DNA replication and organelle growth. Other membrane-bound organelles such as Golgi bodies bud off from existing ones. Non-membrane-bound organelles are assembled de novo. Ribosomes, for example, are assembled in a region of the nucleus called the nucleolus.

Interphase is an active phase in the life of most cells, with many metabolic reactions occurring. Some of these, such as the reactions of cell respiration, also occur during mitosis and meiosis, but the biosynthesis of protein and other molecules needed for growth is a hallmark of interphase.

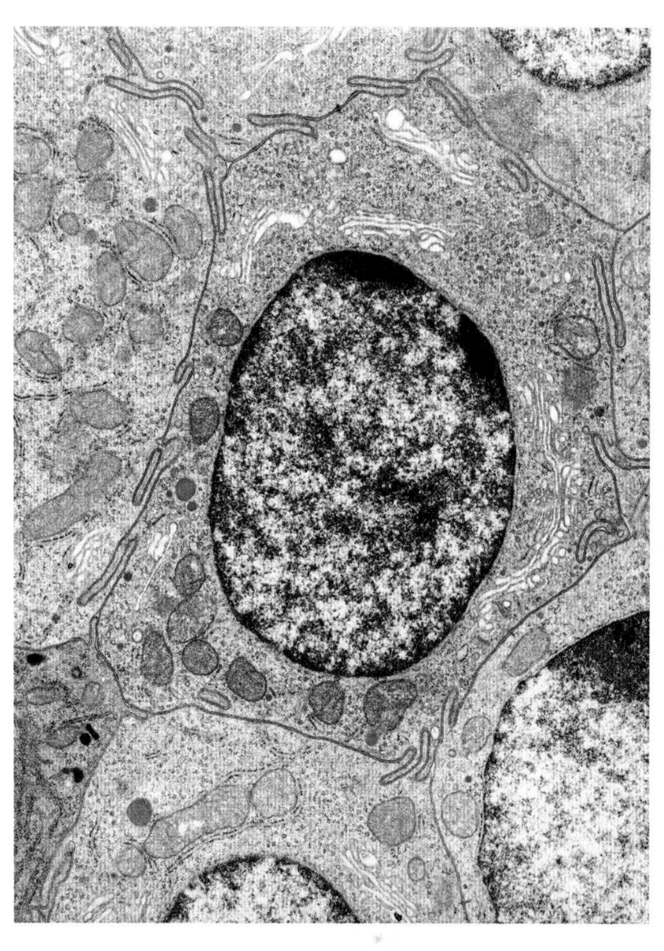

▲ Figure 35 Electron micrograph of a cell in interphase from the intestine wall, with false colour showing the nucleus (dark blue), Golgi bodies (pink) and mitochondria (purple). There are regions of heterochromatin near the margins of the nucleus

D2.1.15 Control of the cell cycle using cyclins

The cell cycle requires a coordinated sequence of changes to happen in a cell. There are checkpoints in the cycle to hold cells until it is appropriate for them to progress to the next phase. These checkpoints are also used to ensure that cells stop dividing when there has been enough cell proliferation in a tissue.

A group of proteins called cyclins coordinates the sequence of changes during the cell cycle. Cyclins activate enzymes called cyclin-dependent kinases, which bind phosphate to other proteins, activating them. In this way, each type of cyclin activates a group of proteins that carry out the actions required during a specific phase of the cycle. This ensures that all actions are performed at the correct time and that the cell only moves past checkpoints in the cycle when it benefits the whole organism.

There are four main types of cyclin in human cells. Figure 36 shows how the levels of them rise and fall. Unless these cyclins reach a threshold concentration, a cell does not progress to the next stage of the cell cycle.

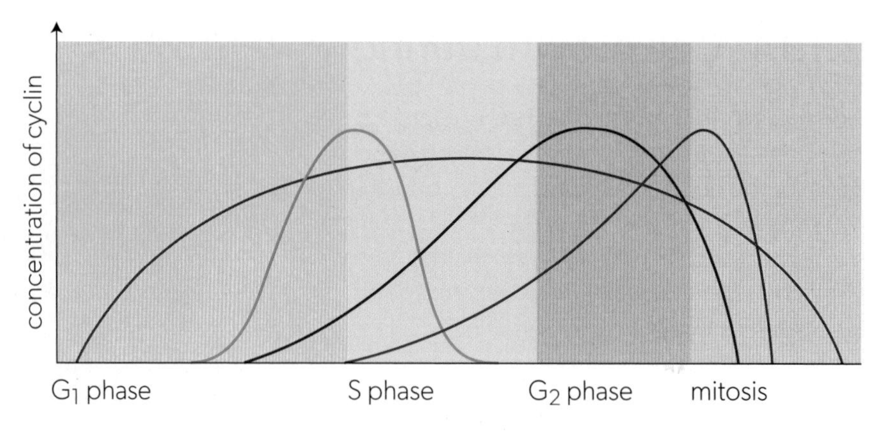

- Cyclin D triggers cells to move from G_0 to G_1 and from G_1 into S phase.
- Cyclin E prepares the cell for DNA replication in S phase.
- Cyclin A activates DNA replication inside the nucleus in S phase.
- Cyclin B promotes the assembly of the mitotic spindle and other tasks in the cytoplasm to prepare for mitosis.

▶ Figure 36 Cyclins and the cell cycle

D2.1.16 Consequences of mutations in genes that control the cell cycle

Control of the cell cycle ensures that tissues have enough cells, but not too many. Sometimes control is lost in an individual cell because of mutations to its genes. When this cell divides, its daughter cells inherit the loss of control, so also divide. The result is a group of cells that increases in number exponentially without the normal controls. This is how a tumour originates.

Mutagens increase the chance of tumour formation. There are two main classes of mutagen that do this.

- Mutagenic chemicals—the International Agency for Research on Cancer lists over 50 substances or groups of substances that are "definitely carcinogenic".
- All forms of high-energy radiation such as X-rays and ultraviolet, but not visible light or other forms of low-energy radiation.

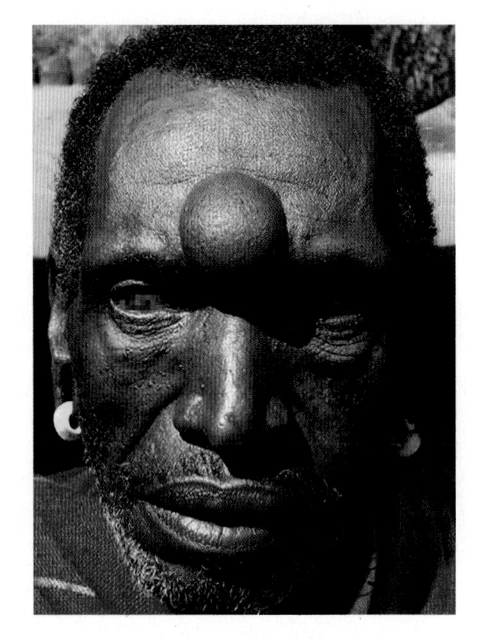

▲ Figure 37 The tumour on this patient's forehead was large but benign and was removed easily without any further treatment being needed

There are two groups of genes that can change a normal cell into a tumour cell if they mutate.

Proto-oncogenes have a variety of roles in the cell. Some regulate expression of genes concerned with cell proliferation. Others have roles in secondary messenger systems that control the cell cycle. A third group is concerned with growth factors or receptors for them.

Proto-oncogenes can mutate into oncogenes, which actively promote cell proliferation and are genetically dominant. So, only one of a pair of proto-oncogenes in a diploid cell has to mutate for an oncogene to be formed, increasing the chance of a tumour. Usually, a mis-sense mutation changes one of the amino acids in the polypeptide coded for by the gene, making it super-active.

Tumour-suppressor genes prevent cell proliferation. Some of them function as brakes at checkpoints in the cell cycle. Others are needed for DNA repair to correct errors in replication and thus reduce mutation rates. A third group has roles in programmed cell death (apoptosis) within cells where there has been irreparable DNA damage.

Mutations to tumour-suppressor genes increase the risk of tumour formation if the gene product no longer functions properly. Base substitution mutations that change a triplet of bases coding for an amino acid into a stop codon are very likely to cause loss of function, because they result in production of truncated polypeptides. However, such mutations are recessive because if one of a pair of tumour-suppressor genes in a cell is unmutated, some functioning polypeptides are produced and the risk of tumour formation is not increased.

Any cell in the body can become a tumour cell, but mutation to one gene is not usually enough. As many as 10 mutations must be present together in a single cell for some types of tumour to be formed. The chance of all these mutations happening in a single cell is extremely small. However, there are vast numbers of cells in the body and humans have long lifetimes, with mutation possible at any time. The chance of tumour formation is also increased by a mechanism that has parallels with evolution by natural selection. Mutated genes are passed to daughter cells when a cell divides and each mutation that increases the rate of cell division will result in a larger pool of cells in which further mutations needed for tumour formation could occur.

Most people will therefore develop tumours during their life, but the majority are harmless or treatable. In all cases, the variables that affect outcomes most are tumour type, promptness of detection and effectiveness of medical treatment.

▲ Figure 38 This large growth on the trunk of a beech tree is a plant tumour, caused by the pathogen *Agrobacterium tumefaciens*. Although plants sometimes develop tumours, they do not develop cancer. What hypothesis could account for this?

D2.1.17 Differences between tumours in rates of cell division and growth and in the capacity for metastasis and invasion of neighbouring tissue

When a tumour cell has been formed it divides repeatedly to form two, then four, then eight cells and so on. This group of cells is the primary tumour. Often the cells in a primary tumour adhere to each other so they remain as a single mass. Such tumours are unlikely to cause much harm and are classified as benign. They should not be thought of as cancer.

In other cases, cell-to-cell adhesion is not adequate to prevent cells becoming detached from the tumour. The cells may invade neighbouring tissues or, if a transport route such as blood or lymph is available, they may move elsewhere in the body. The spread of tumour cells from one part of the body to another is known as metastasis. If conditions are suitable, the cells that have metastasized continue dividing and develop into secondary tumours. Usually, multiple secondary tumours develop, so the consequences are greater than with a single primary tumour. Tumours that spread in this way are classified as malignant and cause what is commonly known as cancer. Certain tissues are more likely to produce malignant tumours, especially in breasts, ovaries, testes and the thyroid gland where there is hormonal stimulation of cell division.

A diagnosis of cancer is always worrying for a patient, but effective treatments have been developed for many types of tumour. Also, the rate of cell division and growth in tumours is variable and in some cases is very slow.

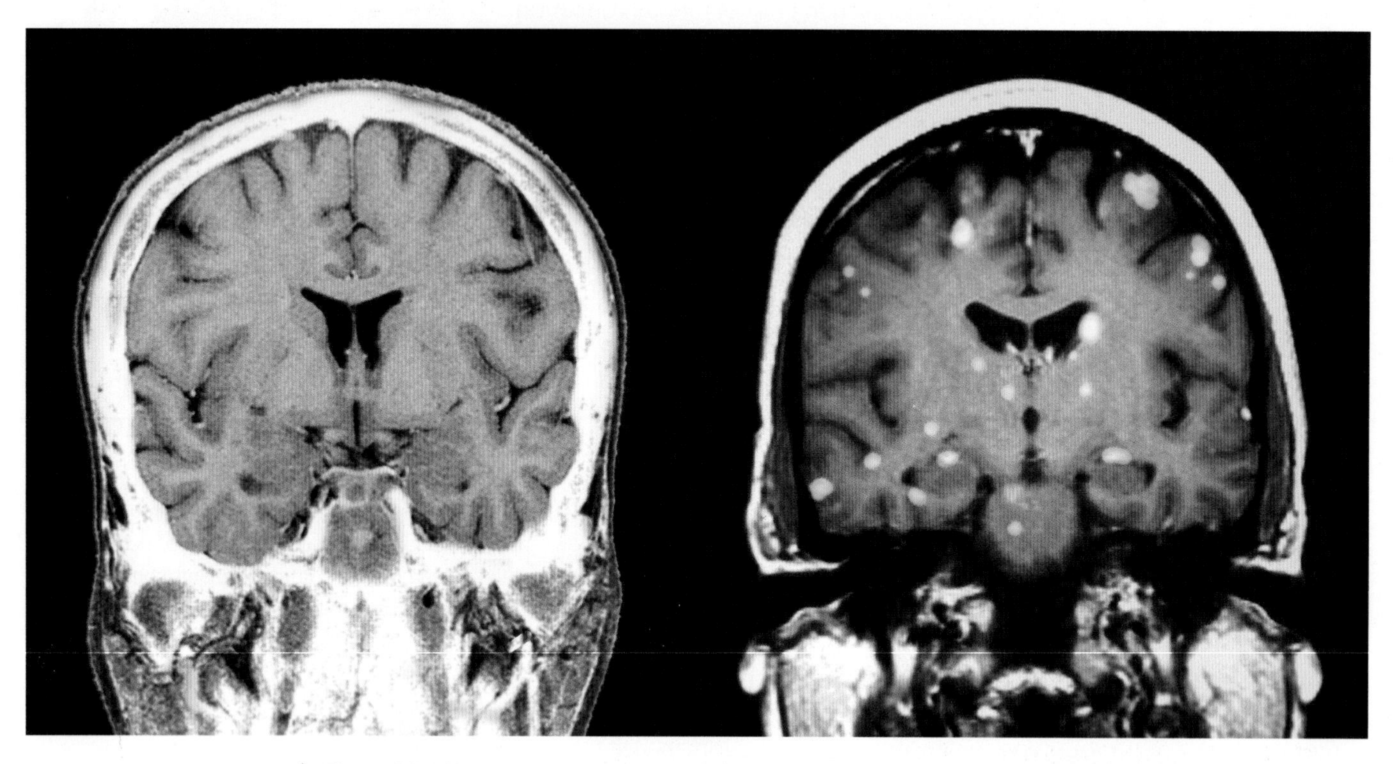

▲ Figure 39 MRI scans are used to check for the spread of cancer. The left scan shows normal brain tissue (purple). The right scan shows multiple secondary tumours due to metastasis of a primary tumour in the breast

Interpretation of a micrograph: Determination of a mitotic index

The mitotic index is the ratio of the number of cells in mitosis in a tissue to the total number of observed cells. It can be calculated using this equation:

$$\text{mitotic index} = \frac{\text{number of cells in mitosis}}{\text{total number of cells}}$$

Figure 40 is a micrograph of cells from a tumour in breast tissue. The mitotic index for this tumour can be calculated if the total number of cells in the micrograph is counted and the number of cells involved in cell division in the same area is counted.

To find the mitotic index of the part of a root tip where cells are proliferating rapidly, follow these instructions.

- Obtain a prepared slide of an onion or garlic root tip. Find and examine the meristematic region (that is, a region of rapid cell division).
- Create a tally chart. Classify each of about a hundred cells in this region as being either in interphase or in any of the stages of mitosis.
- Use this data to calculate the mitotic index.

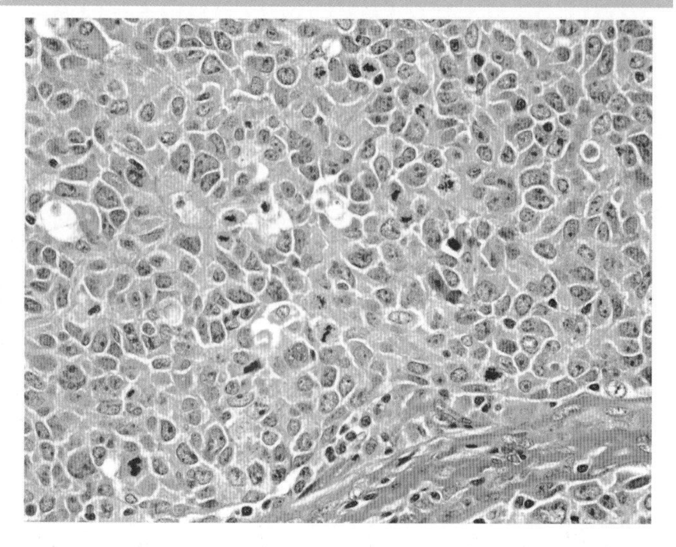

▲ Figure 40 Cells in a breast tissue tumour. What is the mitotic index for this tumour?

Data-based questions: Cell proliferation in hepatocellular carcinoma

Hepatocellular carcinoma (HCC) is the commonest type of cancer originating in the liver. Medical researchers collected data from 282 patients with HCC and calculated the mitotic index (MI) for their tumours after surgical removal. Four or fewer mitoses per high power field of view was classified as low MI and more than four as high MI. The table shows data from this research.

Variable		Low MI	High MI		Low MI	High MI
Age / years	≤55	75	88	>55	80	39
Sex	Female	22	26	Male	133	101
Tumour size/mm	≤50	106	70	>50	49	57
Invasion of small blood vessels	No	99	31	Yes	56	96
Intrahepatic metastasis	No	132	86	Yes	23	41
Early recurrence (≤ 2 years)	No	75	53	Yes	68	74
Liver cirrhosis	No	87	53	Yes	68	74
Hepatitis B or C	No	23	15	Hepatitis B	111	107
				Hepatitis C	21	5

1. Choose one of the variables on the table and test for an association with mitotic index, using the chi-squared test. Liaise with other students to ensure that as many of the variables as possible are analysed.
2. Suggest reasons to account for the association, or lack of association, between mitotic index and the variable that you chose.
3. Discuss with fellow students the conclusions for each of the variables that has been analysed.

Data-based questions: The principles of chemotherapy

Chemotherapy is a treatment for cancer that involves the use of powerful drugs to destroy cancer cells by interfering with mitosis in these cells. Many of the side-effects of chemotherapy are related to the damage caused to normal cells as well as cancer cells. The graph depicts a generalized model of the impact of chemotherapy on the number of normal and cancer cells.

1. State the effect of chemotherapy on the number of cancer cells and normal cells. [1]
2. Outline what happens in the interval between doses for both types of cells. [2]
3. Distinguish between the rate of recovery of normal cells and cancer cells after a bout of chemotherapy. [1]
4. Suggest, with a reason, one specific side-effect of chemotherapy. [2]

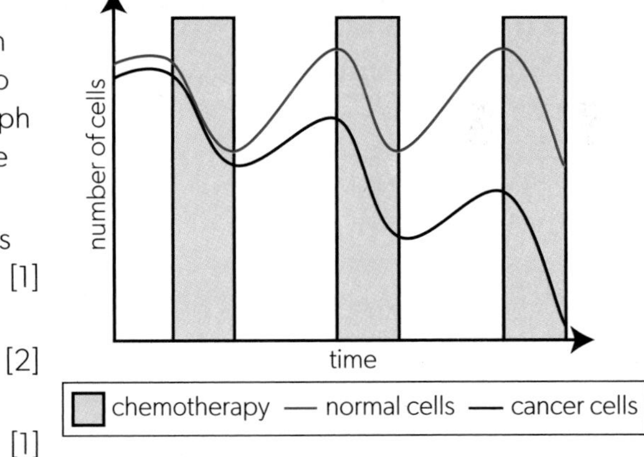

▲ **Figure 41** Effect of chemotherapy on numbers of normal cells and cancer cells

Linking questions

1. What processes support the growth of organisms?
   a. Explain the role of the central vacuole in plant growth. (A2.2.8)
   b. Outline the role of photosynthesis in plant biomass production. (C4.2.15)
   c. Describe the role of mitosis and cytokinesis in the production of new cells. (D2.1.4)
2. How does the variation produced by sexual reproduction contribute to evolution?
   a. Explain how meiosis increases variation in a population. (D2.1.11)
   b. Describe the process of evolution by natural selection. (A4.1.1)
   c. Outline the process of adaptive radiation. (A4.1.9)

D2.2 Gene expression

How is gene expression changed in a cell?

Identical twins Scott (on the right) and Mark Kelly have both served as NASA test pilots and astronauts. Scott was the commander on board the International Space Station (ISS) for a year from March 2015 to March 2016. During this period, his twin brother Mark was on Earth as the "control subject". The study showed that Scott's year in space changed gene expression compared with his twin. How is gene expression influenced by the environment? How is gene expression regulated to ensure that the correct proteins are produced at the right time? How are genes switched on and off?

▲ Figure 1

How can patterns of gene expression be conserved through inheritance?

In zoos, lions and tigers have been cross-bred. Depending on whether the male is a tiger (the cross is known as a tigon) or a lion (the cross is known as a liger), the outcome is different. Why does this happen? What is the role of epigenetic tags in regulating gene expression? Most epigenetic modifications are erased in the production of sperm and egg, but some are not. Thus, the environment experienced by a parent might have an impact on gene expression of the next generation. Is the animal in Figure 2 a liger or a tigon?

▲ Figure 2

AHL only

D2.2.1 Gene expression as the mechanism by which information in genes has effects on the phenotype
D2.2.2 Regulation of transcription by proteins that bind to specific base sequences in DNA
D2.2.3 Control of the degradation of mRNA as a means of regulating translation
D2.2.4 Epigenesis as the development of patterns of differentiation in the cells of a multicellular organism
D2.2.5 Differences between the genome, transcriptome and proteome of individual cells
D2.2.6 Methylation of the promoter and histones in nucleosomes as examples of epigenetic tags
D2.2.7 Epigenetic inheritance through heritable changes to gene expression
D2.2.8 Examples of environmental effects on gene expression in cells and organisms
D2.2.9 Consequences of removal of most but not all epigenetic tags from the ovum and sperm
D2.2.10 Monozygotic twin studies
D2.2.11 External factors impacting the pattern of gene expression

D2.2.1 Gene expression as the mechanism by which information in genes has effects on the phenotype

The genotype of an organism is all its genetic information. The phenotype is all aspects of the structures and functions of the organism—the totality of its characteristics. Gene expression is the process of turning the genotype into the phenotype. It happens in a series of stages.

- Transcription—genes within the genotype are selected and are used in production of RNA. The DNA base of a gene determines the base sequence of an RNA molecule, through complementary base pairing. RNA produced by transcribing protein-coding genes is messenger RNA (mRNA) and it is transported to ribosomes in the cytoplasm.
- Translation—mRNA is used in the production of polypeptides. The base sequence of an mRNA molecule is translated into the amino acid sequence of a polypeptide. Translation happens on ribosomes. Proteins consist of one polypeptide or more linked together, sometimes with additional non-polypeptide components.
- Protein function—the protein may form part of the structure of the organism, but many have other functions, so their effects are not visible in the outward appearance of the organism. However, the effects are observable and are therefore part of the phenotype. Many proteins act as enzymes, so have their effect on phenotype by catalysing a reaction. For example, the ability to digest lactose is due to the protein lactase.

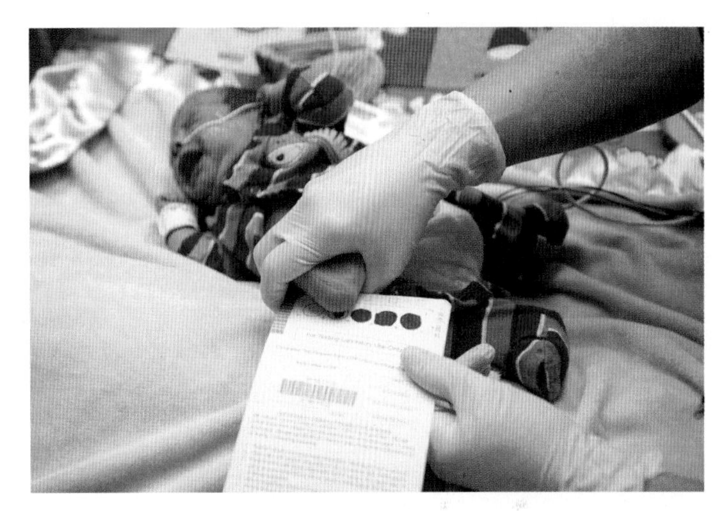

▲ Figure 3 A blood sample is taken from the heel of a newborn infant to determine if the child has phenylketonuria (PKU). This disease is due to a change in the gene that codes for the protein phenylalanine hydroxylase such that the gene does not produce a functional protein

Gene expression is not necessarily just an on or off process. It can be assessed by measuring the quantities of proteins or other gene products.

D2.2.2 Regulation of transcription by proteins that bind to specific base sequences in DNA

Gene expression requires selective transcription of RNA. The enzyme RNA polymerase binds to a sequence close to the gene, known as the promoter. Regulation of transcription is similar in many ways in prokaryotes and eukaryotes. A common feature in eukaryotes is the TATA box. This is a short sequence of bases (TATAAAA) that is part of the promoter. A general transcription factor binds to it, indicating to RNA polymerase where transcription should be initiated.

Every gene in an organism has a similar promoter but only some genes are selected for transcription at any stage in the life of a cell. Selectivity of transcription is based on DNA-binding proteins, which can locate and reversibly attach to specific DNA base sequences. The DNA double helix has two grooves where there proteins can bind to specific sequences—a wider "major groove" and a narrower "minor groove". In most cases, this happens in the major groove of the DNA molecule where differences between the bases are more pronounced.

DNA-binding proteins that regulate transcription of a gene are called transcription factors. Their binding sites lie upstream of the gene. The base sequences of the binding sites are very variable, and in some cases unique. This enables regulation of single genes or specific groups of genes. In contrast, the promoter sequence where RNA polymerase binds is shared.

Transcription factors can act as activators or repressors and turn genes "on" or "off" by binding to regulatory sequences on the DNA called enhancers and silencers.

- Enhancers are sequences to which activators bind and increase the rate of transcription.
- Silencers are sequences to which repressors bind and decrease the rate of transcription.

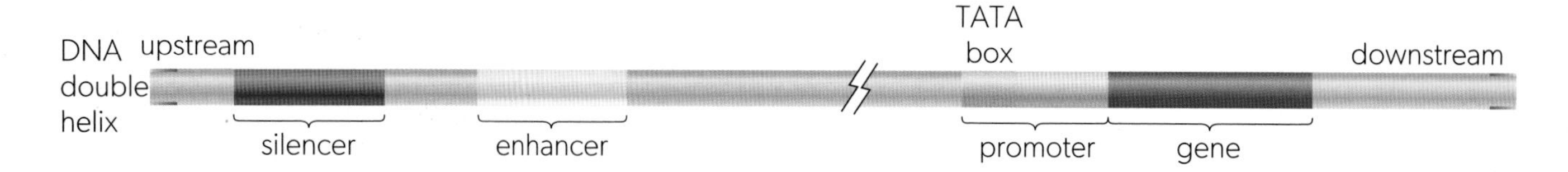

▲ Figure 4 The image shows a simple model of the kinds of sequence involved in the regulation of transcription in eukaryotes. "Repressor" proteins bind to the silencer, decreasing the rate of transcription of the gene. "Activator" proteins bind to the enhancer and increase the rate of transcription. The TATA binding protein binds to the TATA box as part of transcription initiation

D2.2.3 Control of the degradation of mRNA as a means of regulating translation

Regulation of gene expression after transcription occurs at a number of levels. One mechanism is control of the degradation of mRNA. After transcription is complete, a long sequence of adenine nucleotides is added to the 3′ end of the mRNA molecule. This is known as a poly-A tail. It increases stability, but is shortened over time. The shorter the poly-A tail, the less likely the mRNA is to be translated and the more likely that it will be degraded by a nuclease enzyme.

▲ Figure 5 The length of the poly-A tail influences how long mRNA molecules survive in the cell

The rate of shortening of the poly-A tail varies by a factor of over a thousand. In some types of mRNA, 30 nucleotides are removed per minute; in others, only one or two per hour are removed. Given an average tail length of 200 nucleotides, this implies that mRNA molecules might persist between 5 minutes and a week.

Regulating how long an mRNA molecule persists influences how much protein product can be made using it. Some proteins are produced briefly, such as those that control the cell cycle. Others such as the milk protein casein must be produced continuously for long periods of time. This necessitates regulation of mRNA degradation.

Hormones and other chemical signals can influence mRNA degradation rates. For example, the production of vitellogenin, a precursor of egg yolk, is under hormonal control in egg-laying animals. When oestrogen levels are high, the mRNA that guides the production of vitellogenin persists 3.5 times longer than when oestrogen levels are very low. Oestrogen not only affects the stability of the mRNA but also the rate of transcription.

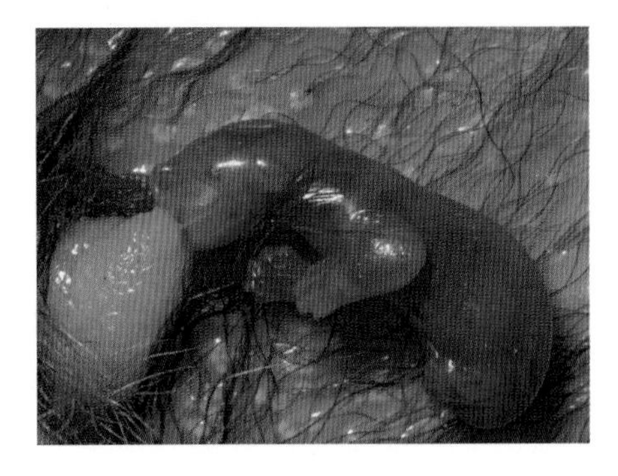

▲ Figure 6 Marsupial embryos are born at a very early stage and continue their development latched onto a teat in their mother's pouch. This common Brushtail Possum embryo (*Trichosurus vulpecula*) is less than 30 mm long

D2.2.4 Epigenesis as the development of patterns of differentiation in the cells of a multicellular organism

Epigenesis is the concept that a multicellular organism such as a plant or animal develops from undifferentiated cells. Structures and functions appear during development that were not present at the start of the organism's life.

Differentiation of cells is achieved by the activation of certain genes and the silencing of others. Chemical modifications to DNA and to the proteins associated with DNA determine which genes are activated or deactivated. These modifications are known as epigenetic tags. Epigenetic modification does not influence the genotype of an organism, just the phenotype. DNA base sequences are not altered during epigenesis.

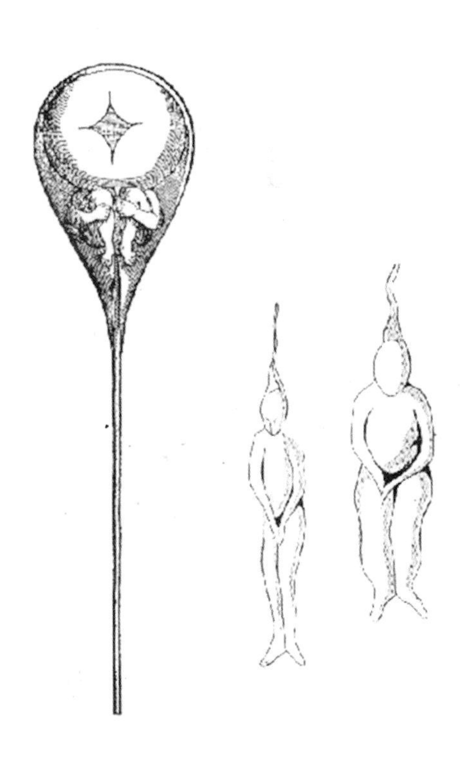

▲ Figure 7 These images of tiny people, drawn by Nicolas Hartsoeker in 1695, were predictions of what would be found inside sperm. They are based on the hypothesis that individuals begin life in a preformed state. This hypothesis has been falsified and replaced by epigenesis—the theory that the environment affects gene expression

D2.2.5 Differences between the genome, transcriptome and proteome of individual cells

A genome is the whole of the genetic information of a cell. It includes coding and non-coding sequences.

A transcriptome is the entire set of mRNAs transcribed in a cell. No cell transcribes all its genes at once, so its transcriptome is less than an RNA copy of each protein-coding gene. The transcriptome varies over time within any cell and between cells within an organism.

A proteome is the entire set of proteins produced by a cell. It is based on the transcriptome because proteins are synthesized from the base sequences of mRNA, but the quantities of each protein in a cell are not directly proportional to the quantity of the corresponding mRNAs. The number of molecules of polypeptide translated from each mRNA molecule is regulated as part of the control of gene expression. The pattern of gene expression within a cell determines how that cell differentiates.

D2.2.6 Methylation of the promoter and histones in nucleosomes as examples of epigenetic tags

Methylation of DNA is the substitution of a hydrogen with a methyl group ($-CH_3$) in a base. Methylation of cytosine frequently occurs, converting it to methylcytosine, but adenine can also be methylated. The pattern of methylation changes during the lifetime of a cell and is affected by environmental factors. It does not alter the genome of the cell, because the methyl group is in a position where it does not affect complementary base pairing.

The methyl group is an epigenetic tag and has multiple roles in the regulation of gene expression. Methylation of cytosine in the promoter of a gene tends to repress transcription of the gene and therefore its expression. The promoter is a base sequence that is located near a gene. It is the binding site of RNA polymerase.

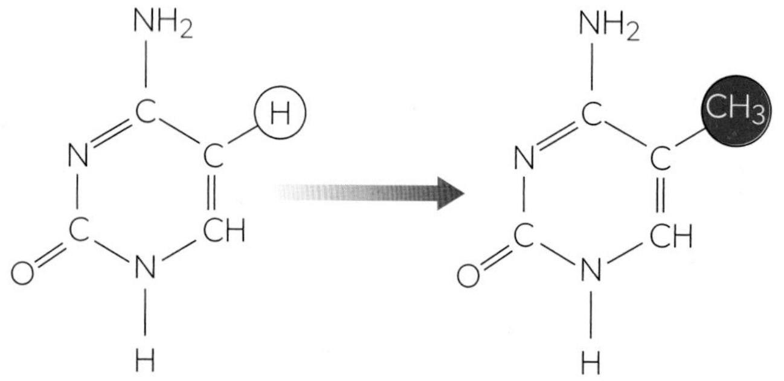

▲ Figure 8 Methylation of cytosine

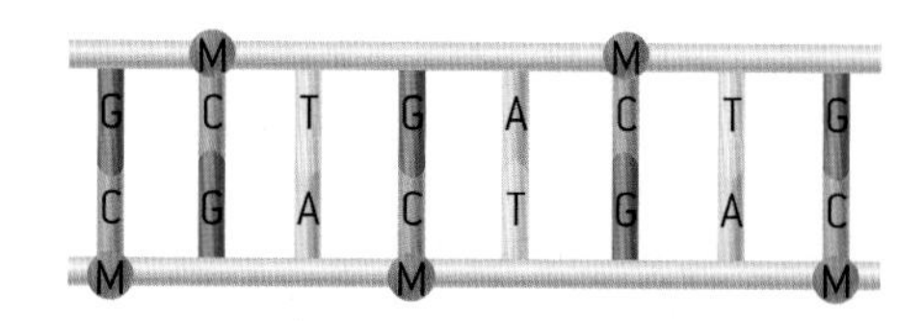

▲ Figure 9 Cytosine is the most common methylated base

DNA is combined with histone proteins to form nucleosomes in eukaryotes and nearly all archaea. A nucleosome is composed of eight histones with the DNA molecule wound round them twice. The histones have long tails extending outward from the compact nucleosome. These tails interact with neighbouring nucleosomes resulting in tight bonding during condensation of chromosomes. So, histones play a role in the tight packing of DNA. They also play a role in the regulation of gene expression.

Different types of modification can occur to amino acids in the tails of histones including the addition of methyl, acetyl or phosphate groups. These are all examples of epigenetic tags. Methylation of amino acids in histone tails can influence transcription of a gene by either decreasing or increasing the accessibility to transcription factors.

Data-based questions: Changes in methylation pattern with age in identical twins

One study compared the methylation patterns of 3-year-old identical twins with 50-year-old identical twins. Methylation patterns were dyed red on one chromosome for one twin and dyed green for the other twin on the same chromosome. Chromosome pairs in each set of twins were digitally superimposed. The result would be a yellow colour if the patterns were the same. Differences in patterns on the two chromosomes results in mixed patterns of green and red patches. This was done for four of the twenty-three chromosome pairs in the genome.

1. Explain the reason for yellow coloration if the methylation pattern is the same in the two twins. [3]
2. Identify the chromosome with the least changes as twins age. [1]
3. Identify the chromosomes with the most changes as twins age. [1]
4. Explain how these differences could arise. [3]
5. Predict, with a reason, whether identical twins will become more or less similar to each other in their characteristics as they grow older. [2]

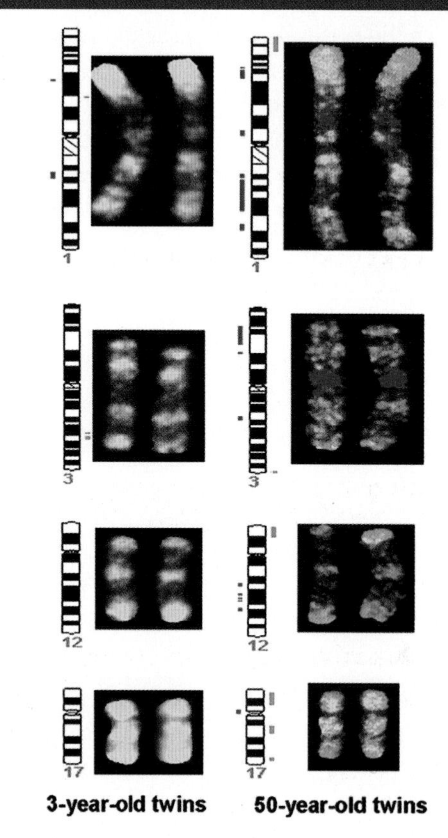

▲ Figure 10 Methylation patterns in four chromosomes of sets of identical twins aged 3 years and 50 years

Molecular visualization: The Research Collaboratory for Structural Bioinformatics (RCSB)

The RCSB interactive website contains a database of 3-dimensional images of molecules in a file type known as .pdb (protein data bank). Visit the database at https://www.rcsb.org/. Search for the nucleosome files: 1AOI and 1KX5.

1. 3-D visualization of the molecule is possible directly on the website or the file can be downloaded. Turn on the 3-D viewer—structure. Select the viewer type as "JSMol".
2. Rotate the molecule to see the association of DNA and protein. Choose different styles of presentation. Choose "amino acids" from the colour option.
3. Note that each protein has a tail that extends out from the core.
4. Note also the approximately 150 bp of DNA wrapped nearly twice around the octamer core.
5. Note the N-terminal tail that projects from the histone core for each protein. Chemical modification of this tail is involved in regulating gene expression.

AHL

6. By moving the mouse over different residues, the type of amino acid can be determined. Visualize the positively charged amino acids on the nucleosome core such as lysine. Suggest how they play a role in the association of the protein core with the negatively charged DNA. By referring to the chemical reaction shown, suggest how the reversible addition of acetyl groups would impact the interaction between DNA and histones and thus affect gene expression.
7. A range of visualization tools are available either by right clicking or pressing the control button, depending on the operating system.

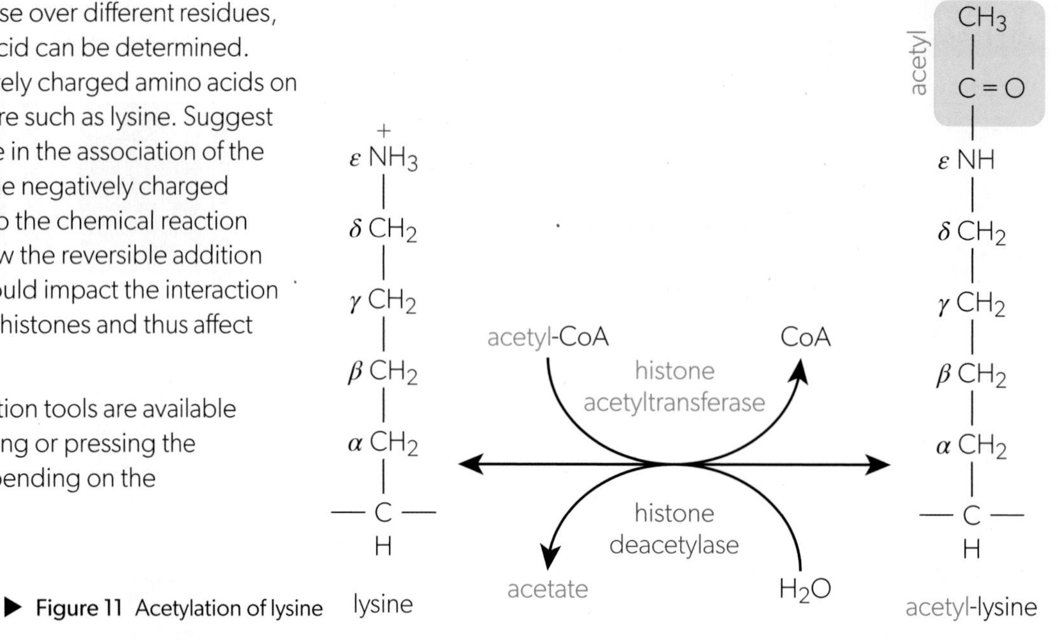

▶ **Figure 11** Acetylation of lysine

ATL Communication skills: Increasing the efficiency of browser-based research

Internet browsers often return large volumes of poorly related search results, with commercial websites often dominating the top search results. Searches can be refined by specifying the type of file. For example, to search for a nucleosome pdb file, type in the search: file type:pdb nucleosome. A similar strategy will work for searching for pdf files; for example, file type:.pdf nucleosome.

The domain type can be specified by filtering by site. URLs ending in .com are often linked to commercial operations. Contrast the search results from entering search terms in two different ways:

- site:edu genetic testing
- genetic testing.

D2.2.7 Epigenetic inheritance through heritable changes to gene expression

Chemical modifications of chromatin such as methylation of bases in DNA and acetylation or methylation of amino acids in the tails of histones are called epigenetic tags. They have an impact on gene expression and so affect the visible characteristics of an individual. The sum of all epigenetic tags in a cell or organism is the epigenome.

Each type of cell in a multicellular organism has its own distinctive patterns of epigenetic tags, so that it produces the proteins needed to perform its functions. When cells in a tissue divide by mitosis, the epigenetic pattern can be passed on to daughter cells. As a result, the new cells are differentiated for the same functions. Also, any effects of the environment on the epigenome can be conserved.

▲ Figure 12 The callipyge phenotype results in the development of much more rump muscle in sheep. It is the result of inheritance of epigenetic tags

Haploid gametes are formed by meiosis from diploid cells that have a characteristic pattern of epigenetic tags, including those due to environmental influences encountered during the organism's lifetime. Some of the pattern is passed on to the haploid cells but it has been assumed that this epigenome will be entirely erased before gametes fuse to produce a zygote. The process of "wiping" the epigenome is called "reprogramming".

There is now evidence that a few epigenetic tags may not be removed, so they remain in the zygote and are thus passed from parent to offspring. This is known as transgenerational epigenetic inheritance. It opens the possibility that the environment encountered by one generation could have an impact on gene expression in the next generation. It should be stressed that this does not amount to the environment directing specific and heritable changes in genes. Even if an epigenetic tag is passed from parent to offspring, it can easily be reversed, unlike mutations in DNA, which are changes to the base sequence and are likely to be permanent.

Data-based questions: Epigenetic inheritance

Poor nutrition of a woman during pregnancy has been associated with a variety of metabolic disorders later in the life of her offspring. During the Second World War, the normally well-fed population of Holland suffered famine over a relatively short and precisely defined period. Glucose tolerance was tested in 50–55-year-old adults who had been affected by the famine as a foetus during specific phases of pregnancy. High glucose levels in blood plasma indicate poor glucose tolerance.

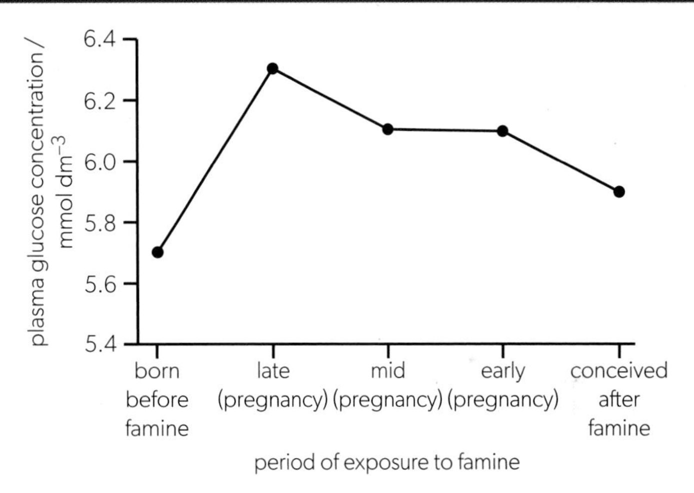

▲ Figure 13 Effect of exposure to famine at various stages of pregnancy

1. Identify the period of exposure to famine that produces the greatest decrease in glucose tolerance. [1]
2. Calculate the percentage change in plasma glucose concentration after exposure to famine from early to late pregnancy. [1]
3. Suggest a reason why glucose tolerance did not return to normal in people conceived after the famine. [1]
4. Discuss these results with respect to the concept of epigenetic inheritance. [3]

ATL Communication skills: Responding appropriately to command terms

The command term "discuss" requires you to give an account, where possible, of a range of arguments for and against a hypothesis. You should consider if there are alternative explanations other than the hypothesis under consideration. You can also indicate where any uncertainty might lie.

For example, in question 4 above, you are asked to discuss the results with respect to epigenetic inheritance. The command term signals that the answer is debatable.

The results are consistent with an impact on gene expression later in life due to famine exposure in utero. However, this could be correlation rather than cause and effect. For example, were there social factors present in the population exposed to famine that persisted into later life? There is no indication of the sample size. There is also no indication of the variability of the data which would affect conclusions about whether or not any differences were statistically significant.

D2.2.8 Examples of environmental effects on gene expression in cells and organisms

Air pollution directly damages tissue, but there is also strong evidence that it can alter the epigenome. Exposure to air pollution such as particulate matter, nitrous oxides, ozone and polyaromatic hydrocarbons has been found to decrease DNA methylation across the genome in people of all ages. One area of impact is in greater expression of proteins that regulate the immune system. These changes might be the cause of increased risk of heart disease and incidence of inflammation and diseases such as asthma.

Air pollution is particularly damaging during pregnancy because it causes changes to methylation in specific genes that have impacts during both gestation and the early years of life.

Data-based questions: Phenotypic plasticity

Phenotypic plasticity is the expression of different phenotypes from the same genome in response to changes in the environment. A clone of the New Zealand freshwater snail, *Potamopyrgus antipodarum*, has become invasive in North America and Europe. It occupies a wide range of lakes and rivers and over this habitat range shows considerable plasticity in the phenotype of its shell. Data in Figure 14 shows how differences in two characteristics of the shell are correlated with habitat.

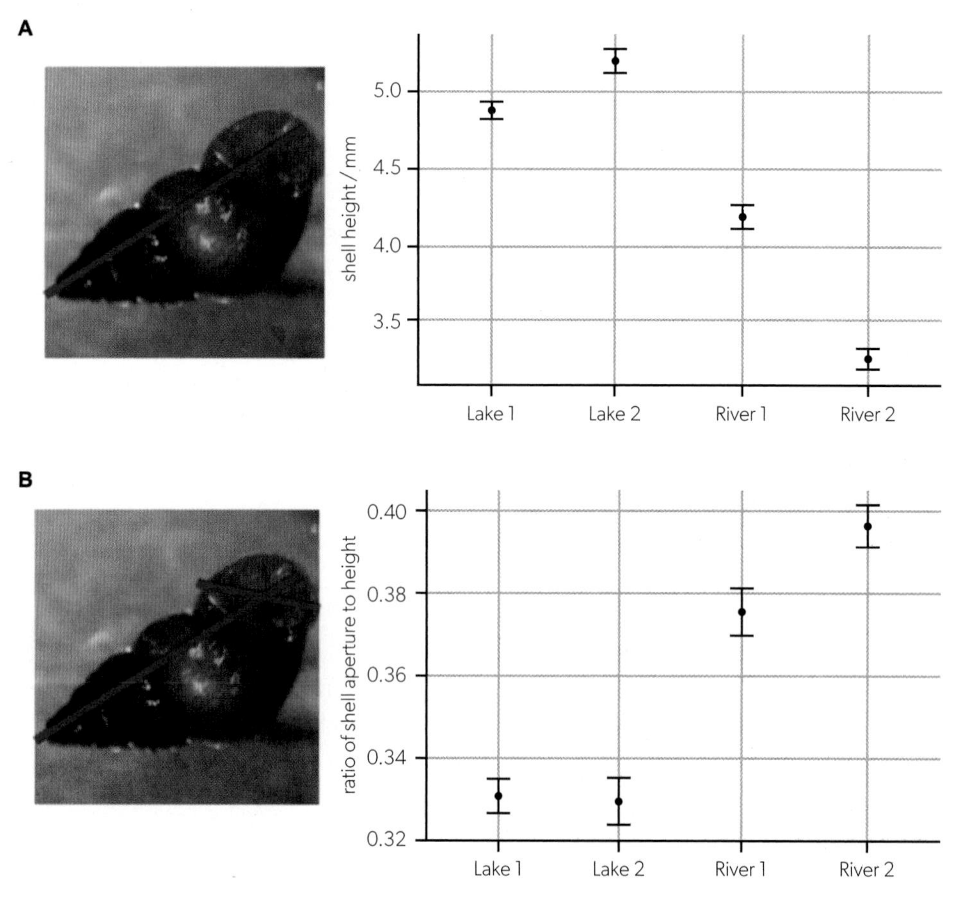

▲ **Figure 14** Shell height and aperture size in populations of *Potamopyrgus antipodarum*

1. Distinguish between the length of the snail shells in the lakes and rivers. [1]
2. Distinguish between the relative size of the aperture (opening) of the shell in lakes and rivers. [1]
3. Significant genome-wide DNA methylation differences were found. The differences between lake and river were 10× greater than the differences between replicate sites of the same habitat. Explain these results. [2]
4. Because the snails reproduce entirely asexually, genetic differences between snails from the two habitats are small. Explain how such significant phenotypic differences can result from the same genotype. [1]

D2.2.9 Consequences of removal of most but not all epigenetic tags from the ovum and sperm

During the production of sperm and eggs in humans, approximately 99% of epigenetic tags are removed. Some tags persist into the next generation. This can have an impact on the phenotype of the organism. This type of non-Mendelian inheritance is sometimes known as genomic imprinting. The epigenetic tags are "imprinted" in the gametes of the parents and can be maintained through mitotic cell divisions in the somatic cells of the offspring.

In Mendelian inheritance, if the allele inherited from one parent is recessive and does not produce a gene product, the allele from the other parent may be dominant and thus result in a functional gene product. However, in the case of genomic imprinting, the allele of a gene inherited from one parent is silenced and the allele from the other parent is expressed, so even though two copies of the gene are present, it is as if the individual were haploid. Alleles and mutations that might normally be recessive are expressed if the dominant allele is silenced by imprinting.

Activity: Ligers and tigons

In a lion–tiger cross, if the male parent is a lion, the offspring is a liger and it is often larger than both its parents. If the male parent is a tiger, then the offspring is known as a tigon and it is often smaller than both of its parents. Explain this observation.

Consider the example of a mother who is homozygous recessive for a condition and the father is homozygous dominant. Genotypically, the offspring would be heterozygous and we would expect it to show the phenotype associated with the dominant allele. However, if the allele inherited from the father is silenced, the recessive condition will appear in the offspring. One condition that is caused in this way is called Angelman syndrome.

Lions and tigers are closely related so that they have been known to interbreed in zoos. The outcome of the cross differs depending on whether the mother or the father is a lion. Female lions carry litters from multiple fathers. There is an advantage to the father to sire the largest sized cub of the litter. There is an advantage to the mother to sire the largest number of cubs. Paternal imprinting favours the production of larger offspring, and maternal imprinting favours smaller offspring. Normally, these competing priorities balance out when both parents are lions. Tiger litters are sired by a single father, so there is no competition that favours imprinting.

D2.2.10 Monozygotic twin studies

Twins can arise when two eggs are released during ovulation and both are fertilized. Twins of this type are dizygotic or fraternal. Such twins typically have about 50% of their genetic information in common.

Approximately 1 in 250 pregnancies results in the formation of monozygotic or identical twins. This occurs when the cells of a very early embryo become separated and each develops into a separate individual. Studies involving monozygotic twins are undertaken to study the relative impact of genes and the environment on phenotype.

Data-based questions: Identical twin studies

Twin studies have been used to identify the relative influence of genetic factors and environmental factors in the onset of disease (Figure 15). Identical twins have 100% of the same DNA while fraternal twins have approximately 50% of the same DNA.

1. Determine the percentage of identical twins where both have diabetes. [2]
2. Explain why a higher percentage of identical twins sharing a trait suggests that a genetic component contributes to the onset of the trait. [3]
3. With reference to any four conditions, discuss the relative role of the environment and genetics in the onset of the condition. [3]

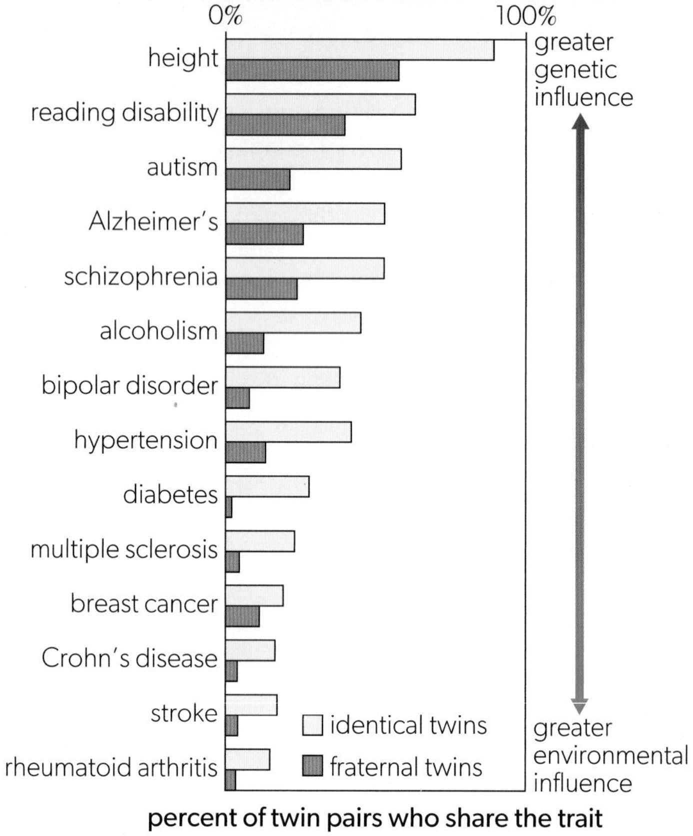

▲ **Figure 15** Results of twin studies with both identical and fraternal twins

D2.2.11 External factors impacting the pattern of gene expression

Variations in environmental factors can have an impact in gene expression. For example, the genes responsible for the absorption and metabolism of lactose in the bacterium *Escherichia coli* are expressed when lactose is present and not expressed when lactose is absent. In this case, the breakdown of lactose results in regulation of gene expression by negative feedback. A repressor protein is deactivated when lactose is present. Once the lactose has been broken down, the repressor protein is no longer deactivated and blocks genes that cause lactose metabolism when they are expressed.

When lactose is not in the environment, a repressor blocks transcription.

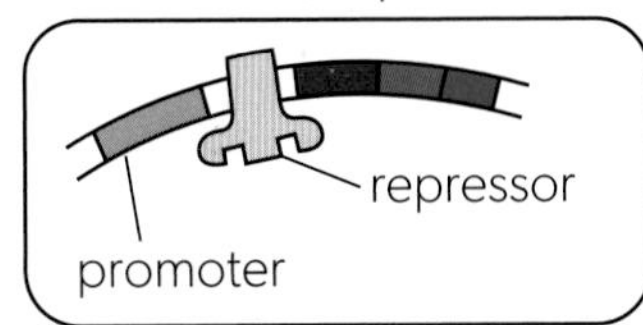

When lactose is present in the environment, the repressor is deactivated and genes involved in lactose use are transcribed.

○ = glucose ▲ = galactose

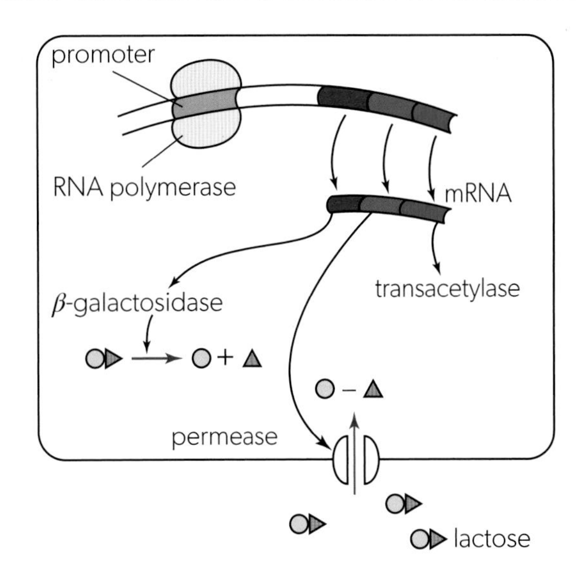

▲ Figure 16 Control of gene expression in lac operon

Each cell in a multicellular eukaryotic organism expresses only a fraction of its genes. The regulation of eukaryotic gene expression is a critical part of cellular differentiation and development. This is seen in the passage of an insect through its life cycle stages or in development of a human embryo.

Hormones can affect gene expression. For example, oestrogen passes through the phospholipid bilayer of cells in the endometrium of the uterus and binds to an oestrogen receptor within the nucleus. With oestrogen bound to it, the receptor can bind to the promoters of target genes causing their expression.

The gene that codes for production of the progesterone receptor is an example of a target gene for the oestrogen receptor. Thus, oestrogen makes the endometrium more responsive to progesterone during the luteal phase of the uterine cycle (see *Section D3.1.5*).

Linking questions

1. What mechanisms are there for inhibition in biological systems?
   a. Distinguish between competitive and non-competitive enzyme inhibition. (C1.1.15)
   b. Other than discussing enzymes, describe one example of end-product inhibition. (C2.1.14)
   c. Compare allelopathy and the secretion of antibiotics. (C4.1.18)
2. In what ways does the environment stimulate diversification?
   a. Explain the ways in which meiosis leads to increased variation in offspring. (D2.1.11)
   b. Discuss the role of adaptive radiation as a source of biodiversity. (A4.1.9)
   c. Outline the role of gradients on gene expression within an early-stage embryo. (B2.3.1)

D2.3 Water potential

What factors affect the movement of water into or out of cells?

Living organisms do not actively pump water molecules around but can cause them to move by changing solute concentrations. Squirting cucumbers (*Ecballium elaterium*) raise solute concentrations in tissue surrounding the seeds in its fruits. This causes water to enter, increasing the hydrostatic pressure to as much as 10 times atmospheric pressure and thus storing up potential energy. Eventually the end of the cucumber bursts and the seeds are ejected at high velocity. Slow-motion video clips are the best way to view this phenomenon.

▲ Figure 1 Squirting cucumber

How do plant and animal cells differ in their regulation of water movement?

Cells in animals and other organisms without a cell wall are in danger of bursting if pressure inside the cell rises above that outside. *Paramecium* continuously uses a pair of contractile vacuoles to discharge water. To cause the vacuoles to fill, positively charged protons are pumped in. This draws in negatively charged chloride ions. Together, these solutes draw in water. The vacuole swells and then expels its contents from the cell. *Paramecium* has to expend more energy in pumping the solutes back in from the freshwater environment but the alternative is death by bursting.

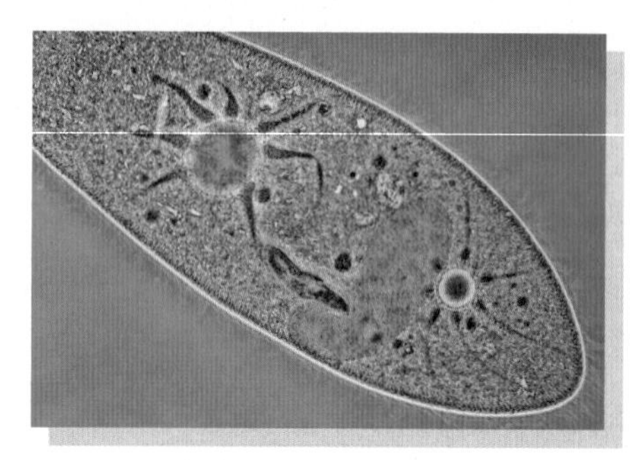

▲ Figure 2 *Paramecium* gets rid of water via contractile vacuoles

SL and HL	AHL only
D2.3.1 Solvation with water as the solvent D2.3.2 Water movement from less concentrated to more concentrated solutions D2.3.3 Water movement by osmosis into or out of cells D2.3.4 Changes due to water movement in plant tissue bathed in hypotonic and those bathed in hypertonic solutions D2.3.5 Effects of water movement on cells that lack a cell wall D2.3.6 Effects of water movement on cells with a cell wall D2.3.7 Medical applications of isotonic solutions	D2.3.8 Water potential as the potential energy of water per unit volume D2.3.9 Movement of water from higher to lower water potential D2.3.10 Contributions of solute potential and pressure potential to the water potential of cells with walls D2.3.11 Water potential and water movements in plant tissue

D2.3.1 Solvation with water as the solvent

Solvation is the combination of a solvent with the molecules or ions of a solute. The properties of water as a solvent are introduced in *Section A1.1.5*. They depend on water's polarity, with a partial negative charge at the oxygen pole and partial positive charge at the hydrogen pole of the molecule.

- Polar solutes dissolve due to attraction between the partial positive and negative charges on water molecules and solute molecules.
- Positively charged ions are attracted to the partial negative oxygen pole of water.
- Negatively charged ions are attracted to the partial positive hydrogen pole of water.

Because of these attractions, water molecules form shells around many types of ion and charged molecule, preventing them from precipitating by clumping together. Cytoplasm is a complex mixture of dissolved substances in which the chemical reactions of metabolism occur.

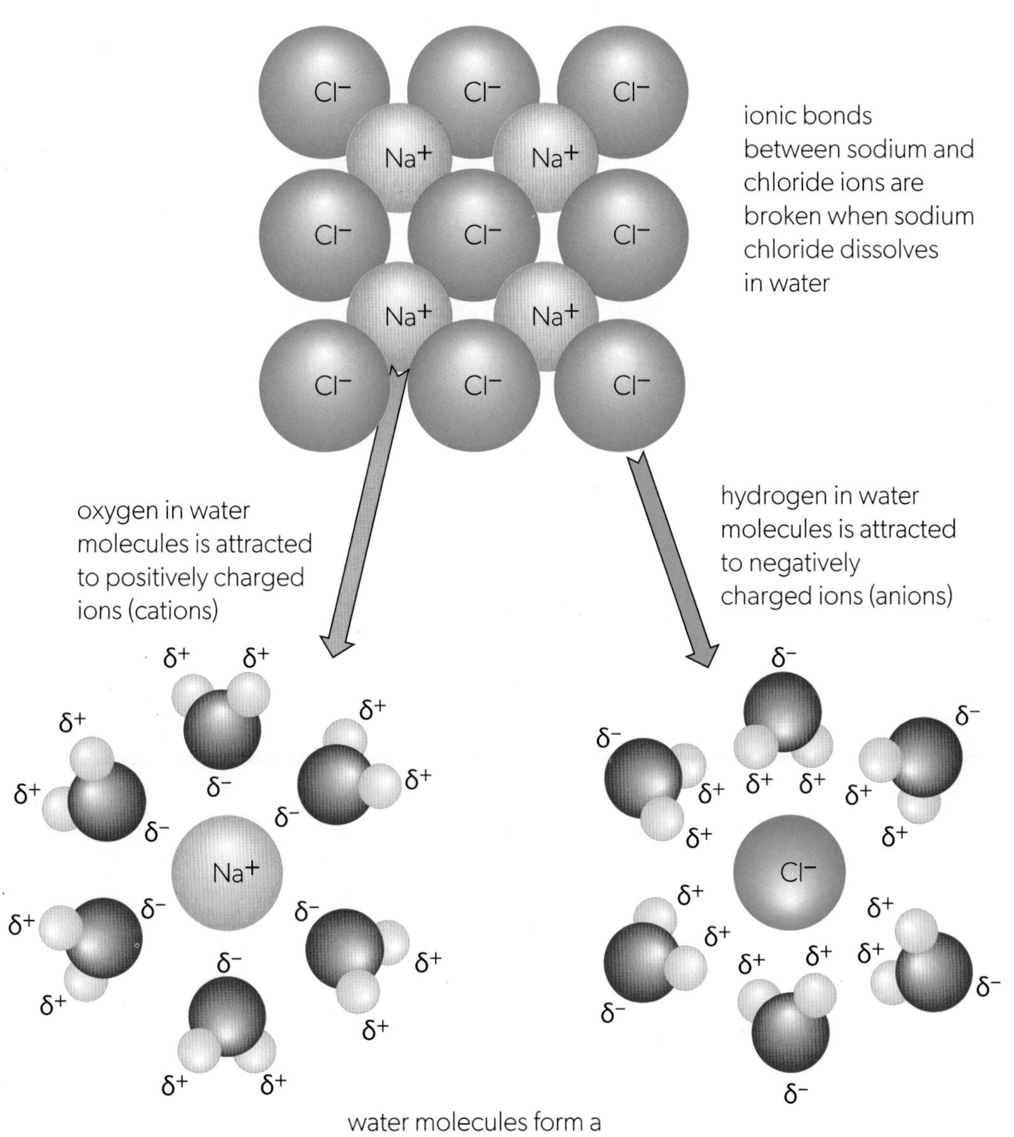
Cl⁻
Na⁺
ionic bonds between sodium and chloride ions are broken when sodium chloride dissolves in water
oxygen in water molecules is attracted to positively charged ions (cations)
hydrogen in water molecules is attracted to negatively charged ions (anions)
δ+
δ−
water molecules form a three-dimensional shell around ions with greater attraction in total than ionic bonds in a crystal

◀ Figure 3 Sodium chloride crystals are a regular array of positively and negatively charged ions which attract water molecules to surround each ion, causing solvation

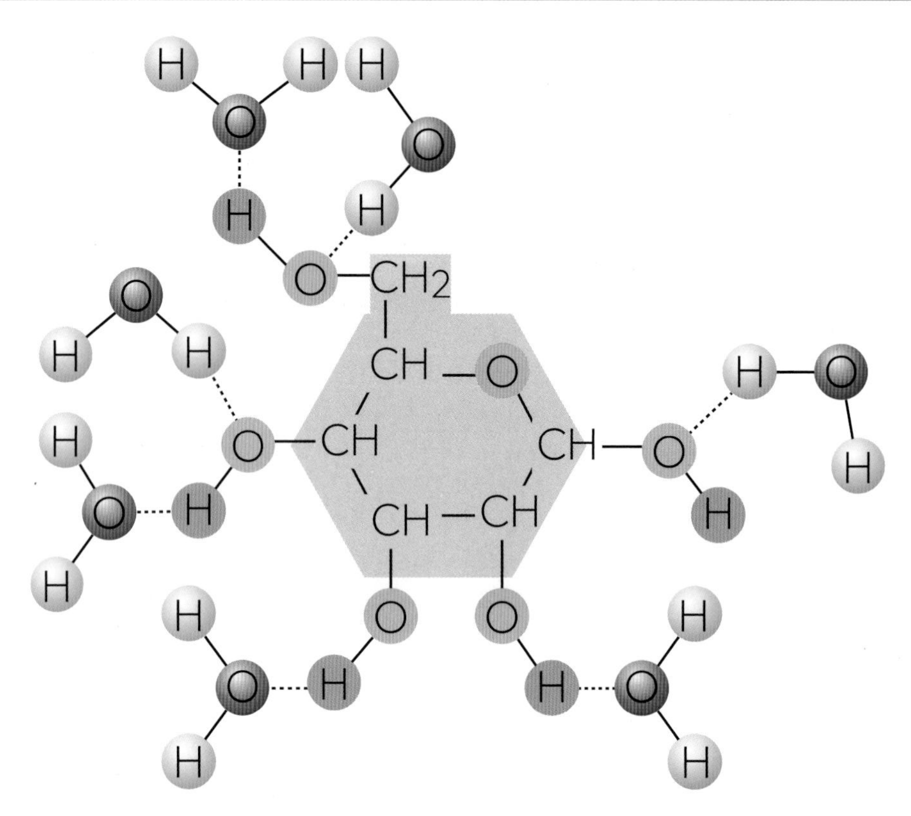

▲ Figure 4 Glucose has polar hydroxyl groups (–OH) with oxygen having a partial negative charge and hydrogen a partial positive charge. Hydrogen bonds (········) form between these hydroxyl groups and water, causing solvation

▲ Figure 5 Honey is a concentrated solution of fructose and glucose. Attractions between these sugars and the water they are dissolved in makes the honey very viscous, so it flows slowly

D2.3.2 Water movement from less concentrated to more concentrated solutions

The particles in a liquid can move, but because of intermolecular attractions they do not separate from each other completely unless the liquid is changing into a gas. As water molecules change position, hydrogen bonds are repeatedly broken and formed. However, at any given time, many bonds exist so there is a strong overall attraction between the water molecules.

Intermolecular attractions between solutes and water are even stronger, which explains how solutions form. These attractions restrict the movement of water molecules, so solutions are more viscous than pure water. Solute–water attractions influence the movement of water molecules between solutions. If water can move between two solutions, there is always movement in both directions. But more water molecules will move from the less concentrated solution to the more concentrated than in the other direction. There is therefore a net movement of water from the less concentrated to the more concentrated solution.

Osmosis is a net movement of water across a membrane due to the attractions between solutes and water. Solutes are osmotically active if intermolecular attractions form between them and water. Sodium, potassium and chloride ions and glucose are all osmotically active.

The following three terms are useful when describing the relative tendency of water to move between solutions due to the concentrations of osmotically active solutes.

- There is a net movement of water from a **hypotonic** solution to a **hypertonic** solution because the hypertonic solution has a higher concentration of osmotically active solutes.
- There is no net movement of water between two **isotonic** solutions because there is no difference between the concentrations of osmotically active solutes, so equal numbers of water molecules move between them. This is known as dynamic equilibrium.

Concentration is the amount of solute per unit volume of solution. The volume is measured in cubic metres or decimetres (m^3 or dm^3). Litres are still sometimes used as the volume unit ($1 L = 1 dm^3$). Negative indices indicate "per unit volume" (m^{-3} or dm^{-3}). The amount of solute is measured in moles, so a sodium chloride solution has a concentration of 0.5 moles dm^{-3} if a litre of solution contains half a mole of NaCl.

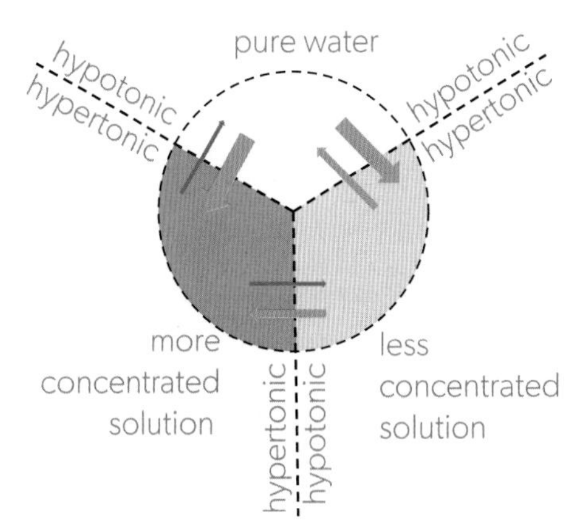

▲ Figure 6 The thickness of the arrows indicates relative amounts of water movement. What are the net movements between the three liquids?

Data-based questions: Solutes in fruits

Concentrations of solutes / moles dm^{-3}			
Sweet cherry	Sour cherry	Grape	Plum
Glucose 431.8	Glucose 287.5	Fructose 398.9	Sucrose 215.0
Fructose 393.5	Fructose 237.6	Glucose 388.3	Glucose 131.0
Sorbitol 76.9	Malic acid 161.1	Potassium 51.2	Malic acid 91.0
Malic acid 70.1	Sorbitol 86.7	Malic acid 33.7	Potassium 58.8
Potassium 56.3	Potassium 65.2	Tartaric acid 27.6	Sorbitol 59.8
Other solutes 21.3	Other solutes 31.9	Other solutes 27.3	Other solutes 67.4
Total 1050	Total 870	Total 927	Total 623

▲ Table 1 Concentrations of the five solutes that are most concentrated in each of four fruits. Concentrations of other and all solutes are also shown

1. Compare and contrast the concentrations of sugars in the four fruits. [3]
2. Explain the difference in taste between sweet and sour cherries. [2]
3. Predict the effects of placing:
   a. tissue from sweet and sour cherries in a 1.0 mole dm^{-3} glucose solution. [3]
   b. cells from a plum in grape juice. [2]

▲ Figure 7 The photo shows transverse sections of a carrot that was cut exactly in half lengthways. One half was then bathed in tap water and the other in concentrated sugar solution for 48 hours. What is the maximum radius of each of the two halves? Which half had been bathed in tap water? The carrot has two regions—outer cortex and inner stele containing xylem and phloem. Were volume changes equal in these regions?

D2.3.3 Water movement by osmosis into or out of cells

All living cells have a plasma membrane that separates the cytoplasm inside the cell from extracellular fluids. The membrane of most cells is very permeable to water, but much less permeable to solutes. A difference in solute concentration between the solutions inside and outside a cell therefore results in movement of water across the membrane rather than a movement of solutes. The net movement of water from a less concentrated to a more concentrated solution across a membrane is osmosis. Plasma membranes are semi-permeable because they are not freely permeable to all particles.

Osmosis is a passive movement. Cells can change how rapidly osmosis occurs by changing the permeability to water of their plasma membrane. They can also change the direction of movement, but only by raising or lowering the concentration of osmotically active solutes inside the cell. This is because water always moves from a hypotonic to a hypertonic solution. Root cells absorb water from the soil for example, because the cytoplasm of the cells is hypertonic compared with water in the soil.

Data-based questions: Osmosis in plant tissues

If samples of plant tissue are bathed in salt or sugar solutions for a short time, any increase or decrease in mass is due almost entirely to water entering or leaving the cells by osmosis. Figure 8 shows the percentage mass change of four tissues when they were bathed in salt solutions of different concentrations.

1. a. State whether water moved into or out of the tissues at 0.0 mol dm^{-3} sodium chloride solution. [1]

   b. State whether water moved into or out of the tissues at 1.0 mol dm^{-3} sodium chloride solution. [2]

2. Deduce which tissue had the lowest solute concentration in its cytoplasm. Include how you reached your conclusion in your answer. [2]
3. Suggest reasons for the differences in solute concentration between the tissues. [3]
4. Explain the reasons for using percentage mass change rather than the actual mass change in grams in this type of experiment. [2]

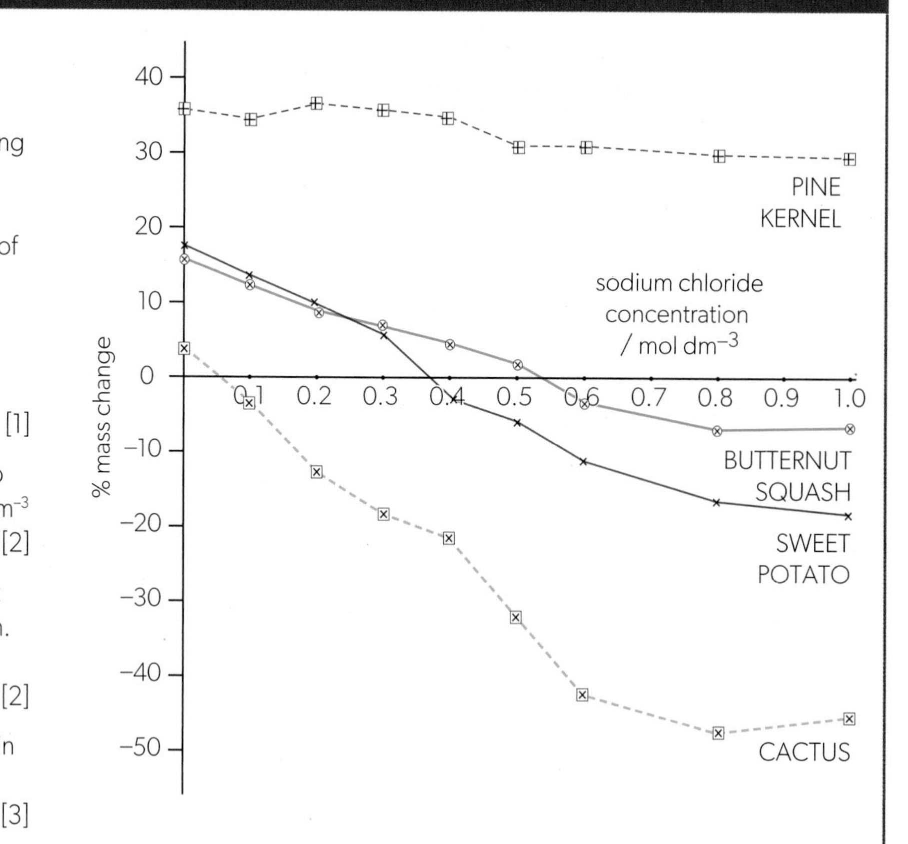

▲ Figure 8 Mass changes in plant tissues bathed in salt solutions

D2.3.4 Changes due to water movement in plant tissue bathed in hypotonic and those bathed in hypertonic solutions

The experiment in the data-based question can be repeated using potato tubers or any other plant tissue that is homogeneous and tough enough to be handled without disintegrating.

Discuss with a partner or group how you could do the following things.

1. Dilute a 1 mol dm^{-3} sodium chloride solution to obtain the concentrations shown on the graph.
2. Obtain samples of a plant tissue that are similar enough to each other to give comparable results.
3. Ensure that the surface of the tissue samples is dry when finding their mass at both the start and end of the experiment.
4. Ensure that all variables are kept constant, apart from salt concentration of the bathing solution.
5. Leave the tissue in the solutions for long enough to get a significant mass change but not so long that another factor affects the mass—for example, decomposition!

ATL Communication skills: Presenting data appropriately

When recording data in a grid, first create a table. Ensure that both rows and columns have headings. The headings should include units after a forward slash if the data is measured. An uncertainty figure should be included that allows the reader to determine how precise the measuring tool is. There should be uniform precision in the data. Sometimes, this involves standardizing the number of place values by clicking on the format "number" tab in the spreadsheet software and standardizing the precision. Uncertainty figures should not be more precise than the data. When processing the data, ensure that averages are accompanied by a measure of the variability of the data such as standard error, standard deviation or a range value.

Samples of carrot were cut with dimensions 80 × 5 × 5 mm. Five carrot samples were bathed in each of four concentrations of sodium chloride solution. After 10 hours, the samples were removed from the solutions and surface dried. Each sample was then held horizontally. Some of the samples drooped, because they had become flaccid (floppy). The amount of droop was measured as an angle. Create an appropriate data table using the results below.

Results

In 0.0 mol dm^{-3}, the angles of droop were all 0°.
In 0.33 mol dm^{-3}, angles were 37°, 32°, 27°, 22° and 39°.
In 0.66 mol dm^{-3}, angles were 46°, 41°, 55°, 42° and 48°.
In 1.0 mol dm^{-3}, angles were 64°, 62°, 63°, 72° and 75°.

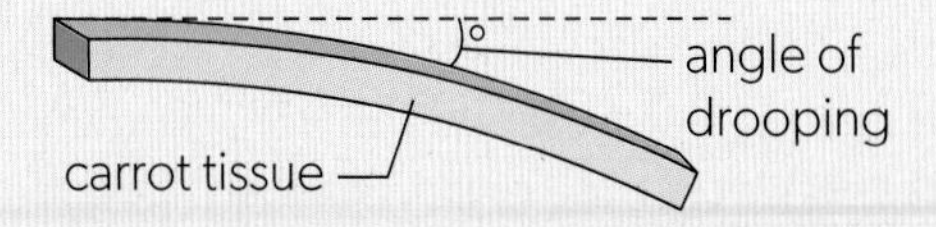

▲ Figure 9 Measuring the angle of droop

Figure 9 gives one idea for measuring changes in plant tissue due to water movement by osmosis.

Mathematical tools: Standard deviation and standard error of the mean

A sample is a small, representative portion of something. Sampling is taking a sample with a method that ensures it is representative of the whole population. Most biological samples show variation. The range of variation can be plotted as a frequency distribution. One type of frequency distribution is so common that it was named the **normal distribution**. It is also sometimes called the "bell curve" because it is bell-shaped (Figure 10). It is symmetrical around the mean. Values tend to be concentrated near the mean and decrease in frequency as the distance from the mean increases.

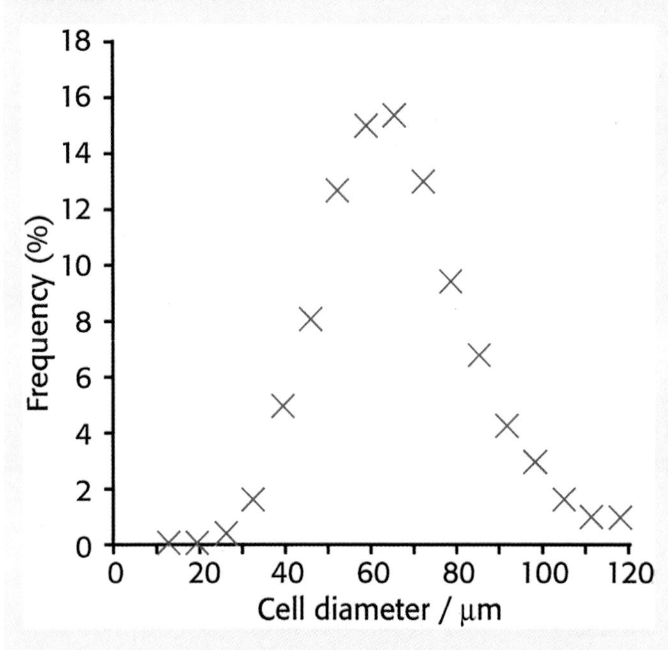

▲ Figure 10 Frequency distribution of cell diameter in a sample of root cells. This is an example of a normal distribution or bell curve

Standard deviation is a measure of the range of variation from the mean (average). A large standard deviation indicates that data is very variable. With the normal distribution, 68% of all measurements fall within one standard deviation of the mean and 95% fall within two standard deviations of the mean. Standard deviation is not affected by sample size.

A sample mean is unlikely to be precisely the same as the true mean of a whole population—it is an estimate of the true mean. The sample mean is likely to be close to the true mean if the standard deviation is low and the sample size is high. **Standard error** is a measure of how reliably the mean of a sample estimates the mean of the whole population. It is found by dividing the sample standard deviation by the square root of the sample size.

Example: Diameter of xylem vessels in a root.

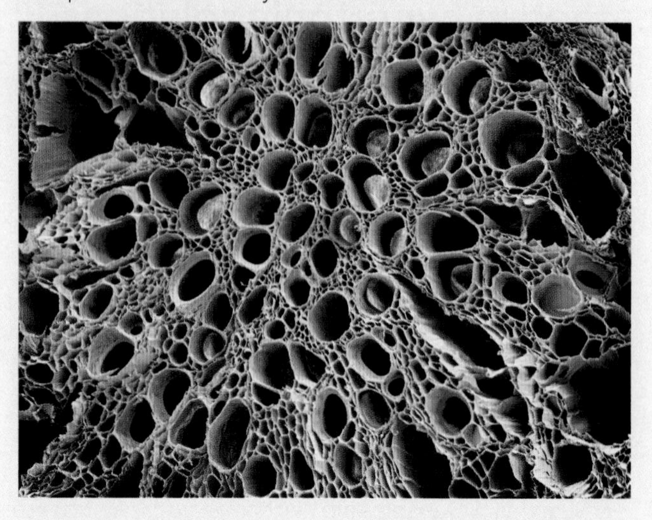

▲ Figure 11 Transverse section of the centre of a dicot root. The xylem vessels are relatively wide, with an open lumen

The data below represents the longest dimension visible for the 42 individual xylem vessels measured from an image similar to Figure 11. The measurements are in µm.

43 30 45 33 32 35 40 32 26 38 37 36 34 27
28 34 34 29 24 15 30 29 34 34 38 16 22 37
29 34 28 35 28 9 8 40 33 26 10 51 36 32

The data was entered into a graphic calculator spreadsheet in a column entitled "width".

The menus chosen were Statistics → Stat Calculations → One-variable statistics

The following data were recorded:

Sample mean cell width $\bar{x} = 3.1$

Sample standard deviation $Sx = 0.91$

$$\text{Standard error} = \frac{s.d.}{\sqrt{n}} = \frac{0.909}{\sqrt{42}} = 0.14$$

In the following two examples, what is being illustrated?

a. 68% of the sample cell width values are in the range of 3.1 ± 0.91.

b. There is a 68% probability that the population mean cell width is 3.1 ± 0.14.

ATL Self-management skills: Making accurate quantitative measurements

An ideal experiment gives results that have only one reasonable interpretation. Conclusions can be drawn from the results without any doubts or uncertainties. In most experiments, there are some doubts and uncertainties but if the design of an experiment is rigorous, these can be minimized. The experiment then provides strong evidence for or against a hypothesis.

This checklist can be used when designing an experiment.

- Results should be quantitative (if possible) because these give stronger evidence than descriptive results.
- Measurements should be as accurate as possible, using the most appropriate and best quality meters or other apparatus.
- Repeats are needed because biological samples are variable which inevitably affects results no matter how accurately quantitative measurements are taken.
- All factors that might affect the results of the experiment must be controlled so that only the factor under investigation is allowed to vary and all other factors are kept constant.

After doing an experiment, the design can be evaluated using this checklist. The evaluation might lead to improvements in the design that would make the experiment more rigorous. If you have done an osmosis experiment in which samples of plant tissue are bathed in solutions of varying solute concentration, you can evaluate its design. If you did repeats for each concentration of solution, and the results were very similar to each other, your results were probably reliable.

D2.3.5 Effects of water movement on cells that lack a cell wall

Animal cells have a plasma membrane but no cell wall. There are also unicellular eukaryotes that lack a cell wall, for example *Amoeba proteus*. Properties of membranes and cell walls are compared in Table 2.

Property	Plasma membranes of plants and animals	Cell walls of plants
Main constituent	phospholipids	cellulose
Thickness	thin—5 nm or less	much thicker—250 nm to 5 μm or more
State	liquid, allowing changes of position such as formation of vesicles by budding and diffusion of molecules in the membrane	solid, so changes of position are limited and constituent molecules of a wall do not diffuse
Tensile strength	very low—easily torn	high—stronger than steel
Permeability	semi-permeable, with some solutes scarcely able to pass through	freely permeable unless impregnated with a waterproof material (for example, cutin, lignin or suberin)

▲ Table 2 The properties of cell walls and membranes

If an animal cell is bathed in a hypotonic solution, water enters the cell by osmosis making it swell. Because it lacks the support that a wall would provide, the cell easily bursts. This can be demonstrated by placing a small droplet of blood in pure water and then examining it using a microscope. The blood cells swell up to form a spherical shape and then burst, leaving ruptured plasma membranes called red cell ghosts.

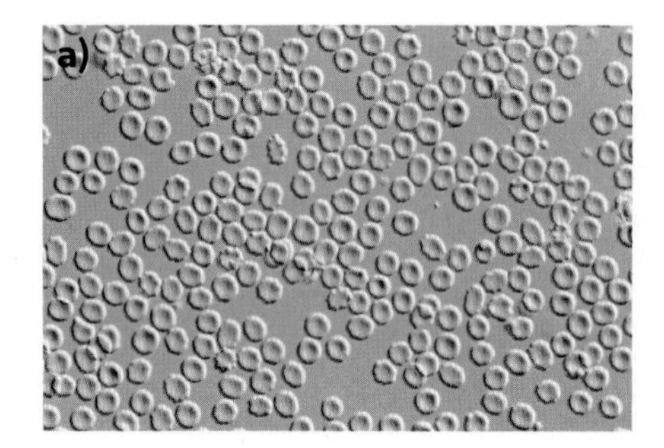

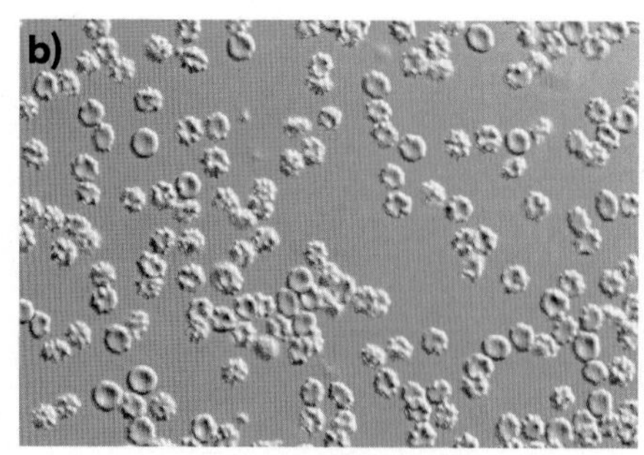

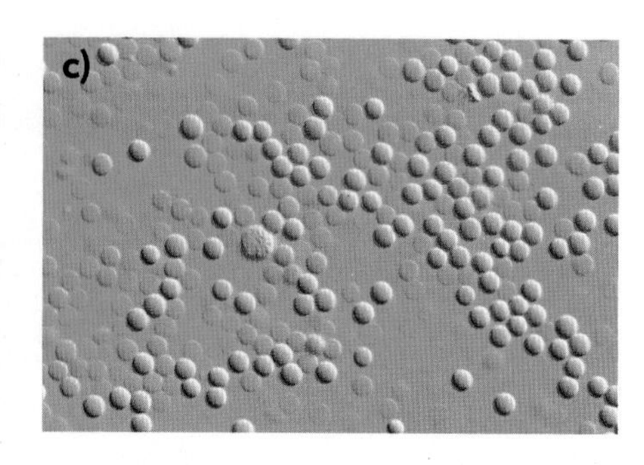

▲ Figure 12 Blood cells that have been bathed in saline solutions: (a) isotonic, (b) hypertonic, (c) hypotonic

If an animal cell is bathed in a hypertonic solution, water leaves the cell by osmosis, so the cytoplasm shrinks in volume. The area of plasma membrane does not change, so the cell develops indentations, which are sometimes called crenations.

Data-based questions: Analysing osmometer results

Osmosis can be demonstrated using a bag made from semi-permeable dialysis tubing, with a vertical glass tube connected to it. This apparatus was set up with pure water in the beaker and 1.0 mol dm^{-3} sucrose solution inside the bag plus a drop of food colouring to make the solution visible. Table 3 shows the results of measuring the height of the solution in the tube over two hours.

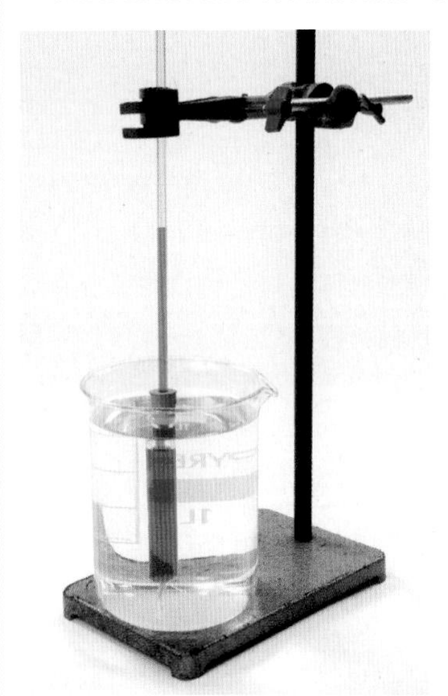

▲ Figure 13

1. Explain the change in height of the solution in the tube during the first 10 minutes. [3]
2. a. Describe the trends in the results over the course of the two-hour period. [3]
   b. Explain these trends. [3]
3. Predict, with reasons, the results of the experiment if:
   a. 0.5 mol dm^{-3} sucrose solution is placed inside the bag instead of 1.0 mol dm^{-3} [3]
   b. 2.0 mol dm^{-3} sodium chloride is placed in the beaker instead of pure water. [3]

Time elapsed / min	Height / mm
10	5
20	155
30	315
40	450
50	510
60	680
70	770
80	850
90	915
100	975
110	1025
120	1065
130	1105
140	1135
150	1165
160	1180
170	1190
180	1195

▲ Table 3 The height of the solution in the vertical glass tube compared with the initial height recorded every 10 minutes for two hours

D2.3.6 Effects of water movement on cells with a cell wall

High pressures due to entry of water by osmosis can build up inside plant cells because the cell wall is strong enough to prevent bursting. A cell that has become pressurized in this way is said to be turgid. This means it is swollen. This is the normal state of healthy plant cells. Turgid plant tissue can provide support because of its strength under compression. The stems and leaves of non-woody plants resist gravity in this way.

If plant cells lose water, the pressure of the cytoplasm decreases. If it drops to atmospheric pressure, the plasma membrane no longer pushes against the cell wall and the cell is not turgid. Further water loss causes plant cells to become flaccid, meaning limp or floppy. Leaves and stems bend downwards. This is called wilting and is seen in plants that have lost water by transpiration in hot weather or droughts. The plant usually avoids further water loss by closing stomata.

Water can also be removed from plant cells by bathing them in a hypertonic solution. The cell wall is permeable to water and does not move. This means that as the volume of cytoplasm decreases, the plasma membrane eventually pulls away from the cell wall. This is called plasmolysis. It is a damaging process and usually causes the death of the cell. Plasmolysis sometimes happens naturally, for example when seawater floods terrestrial ecosystems as a result of high tides or a tsunami.

▲ Figure 14 John Dunlop, seen here riding a bicycle, invented the pneumatic tyre. The inflatable inner tube pushes outwards on an unexpandable outer tyre, providing compressive strength in a similar way to the plasma membrane pushing on a plant cell wall

D2.3.7 Medical applications of isotonic solutions

Hypertonic solutions can damage human cells by dehydrating them. Hypotonic solutions can cause human cells to swell and burst. If extracellular fluid is isotonic, water molecules pass in and out through the plasma membrane at the same rate, so cells remain healthy. It is therefore important for isotonic solutions to be used during medical procedures. Usually, an isotonic sodium chloride solution is used (Figure 15), which is called "normal saline". It contains 9 g of NaCl per cubic decimetre of solution, which is a molarity of 0.154 mol dm^{-3}.

Normal saline is used in many medical procedures. It can be:

- safely introduced to a patient's blood system via an intravenous drip
- used to rinse wounds and skin abrasions
- used to keep areas of damaged skin moistened prior to skin grafts
- used as the basis for eye drops
- frozen to the consistency of slush for cooling hearts, kidneys and other donor organs (Figure 16) to be transported to the hospital where the transplant operation is to be done.

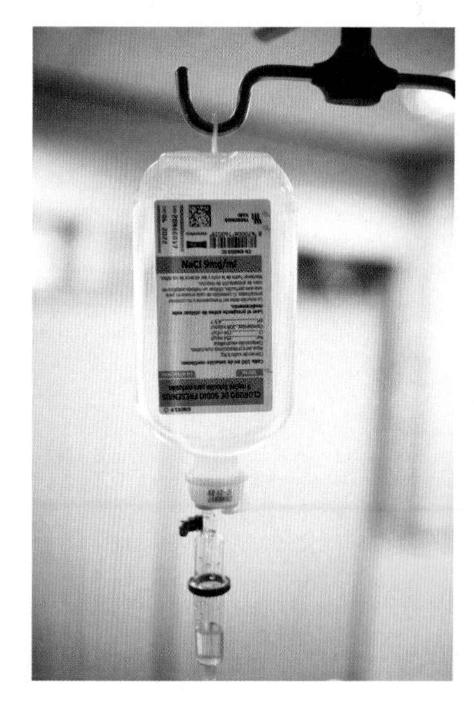

▲ Figure 15 A bag of sterile 9 mg ml^{-1} sodium chloride solution is isotonic to human blood cells so can be used as an intravenous infusion drip. What is the reason for hanging the bag 1 m higher than the patient's head? What would happen if pure water was infused?

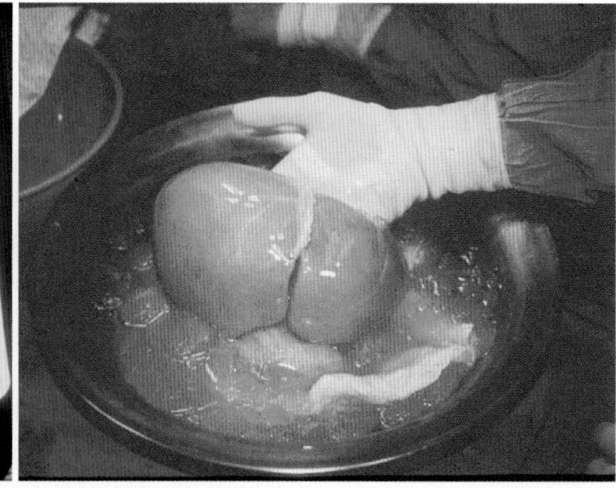

▲ Figure 16 A donor liver bathed in isotonic preserving fluid containing nutrients has been placed in a plastic bag, which is put inside a second bag containing isotonic saline, which is packed in yet another plastic bag and then kept cool in a plastic bowl surrounded by isotonic slush in an insulated transport case

AHL

D2.3.8 Water potential as the potential energy of water per unit volume

The concept of water potential helps to understand movement of water in living systems, especially plants. It is a measure of the potential energy per unit volume. The symbol for water potential is Ψ (the Greek letter *psi*) and the units for measurement are kilopascals (kPa) or megapascals (MPa). The absolute quantity of potential energy cannot be determined, so all values are relative. Pure water at standard atmospheric pressure and 20°C has been assigned a water potential of zero.

Many factors influence water potential, but in living systems only two contributors vary in such a way that they need to be considered: solute concentrations and hydrostatic pressure.

- Rises or falls in hydrostatic pressure change the potential energy of water. The higher the pressure, the more potential energy water has.
- When solutes dissolve, the potential energy of water is reduced. The higher the solute concentration, the less potential energy water has.

▲ Figure 17 Cells from a red onion floating on pure water absorb molecules from it until their water potential equals that of the water ($\Psi = 0$ kPa). There is then no more net movement, so the water potential of the plant cells cannot rise above 0 kPa

D2.3.9 Movement of water from higher to lower water potential

Water potential allows us to predict the direction in which there will be net movement of water molecules. Water moves from a higher to a lower water potential because this minimizes its potential energy. In a similar way rocks roll down hills, and atoms form chemical bonds with each other with minimization of potential energy as the motivator.

The range of water potentials found in cells is rather unusual—the maximum value is zero. Cells either have a potential of 0 kPa or a negative value, for example −200 kPa in a leaf cell. Lower water potentials are therefore more negative. Water will move from a cell with a water potential of −200 kPa to one with −300 kPa, for example.

D2.3.10 Contributions of solute potential and pressure potential to the water potential of cells with walls

Many factors can influence water potential but in living systems only two contributors vary in such a way that they need to be considered: solute potential (Ψ_s sometimes called osmotic potential) and pressure potential (Ψ_p).

Water potential is the sum of solute potential and pressure potential.

water potential (Ψ_w) = solute potential (Ψ_s) + pressure potential (Ψ_p)

The potential energy of water changes if solutes dissolve in it—this component is solute potential (Ψ_s). When solutes dissolve, the potential energy of water is reduced. With no solutes dissolved, the solute potential is zero. Because it is impossible for water to hold less than no solutes, the only possible solute potentials are zero or negative.

Rises or falls in hydrostatic pressure also change the potential energy of water—this component is the pressure potential (Ψ_p). The higher the pressure, the more potential energy water has. Pressure potential can be negative or positive because it can be greater or less than atmospheric pressure.

▼ **Figure 18** **Possible ranges of solute potential and pressure potential**

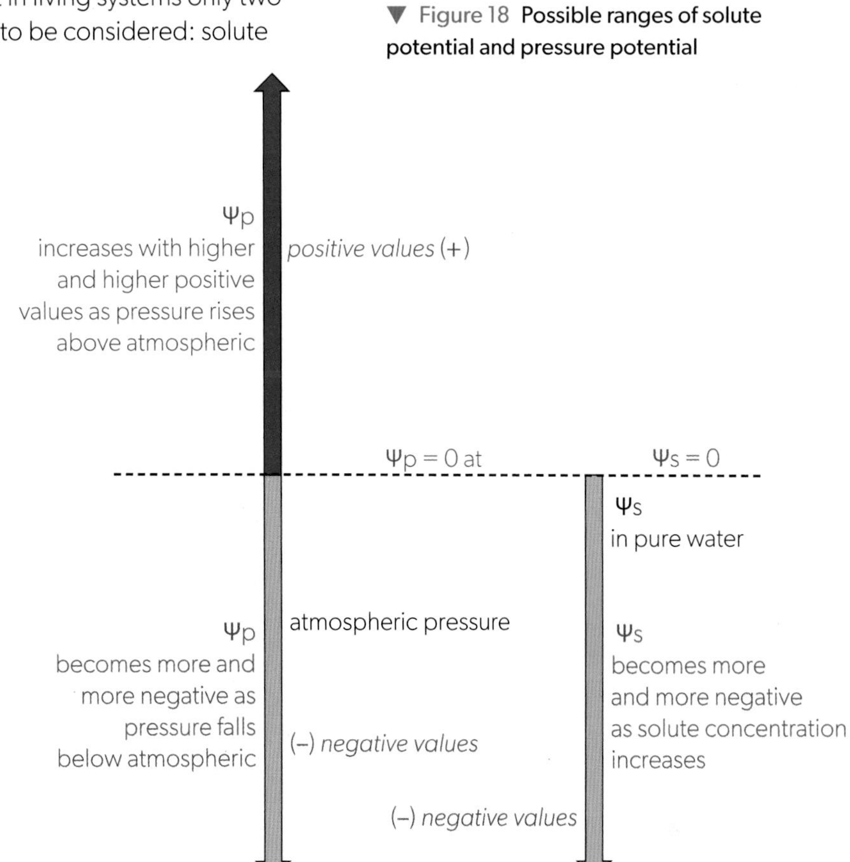
Ψp
increases with higher and higher positive values as pressure rises above atmospheric
positive values (+)
Ψp = 0 at
atmospheric pressure
Ψs = 0
Ψs
in pure water
Ψp
becomes more and more negative as pressure falls below atmospheric
(–) *negative values*
Ψs
becomes more and more negative as solute concentration increases
(–) *negative values*

D2.3.11 Water potential and water movements in plant tissue

The effects of bathing plant tissue in hypotonic or hypertonic solutions can be explained in terms of solute and pressure potentials.

Bathing in hypotonic solutions

At atmospheric pressure, $\Psi_p = 0$ so the water potential and solute potentials of the bathing solution are equal and negative. For example, a sodium chloride solution of 1.0 mol dm^{-3} has a solute and water potential of −4,540 kPa. Assuming that the water potential of the plant tissue is initially lower (more negative) there will be a net movement of water from the solution to the plant cells. This will raise the water potential of the plant tissue by making the solute potential less negative and the pressure potential more positive. When the tissue's water potential equals that of the solution, net movement of water stops.

The more dilute the hypotonic solution, the higher the water potential reached in the plant tissue. If the bathing solution is pure water in which the water potential is zero, the plant tissue will reach that level. The hydrostatic pressure of the cells cannot rise higher, so they are fully turgid. At a water potential of zero, the solute potential counterbalances the pressure potential—they are numerically equal but opposite in sign.

Bathing in hypertonic solutions

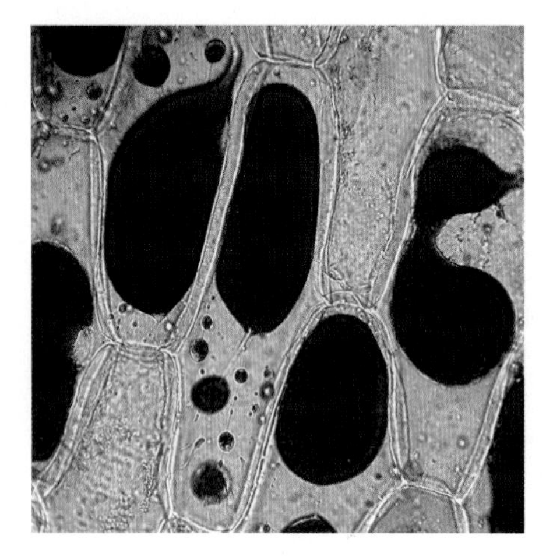

▲ Figure 19 Micrograph of red onion cells, placed in a hypertonic salt solution

The solute potential of the bathing solution is more negative than that of the tissue. The pressure potential of the bathing solution is zero, but pressure potential in the plant cells is likely to be initially above zero due to pressure above atmospheric. Both solute and pressure potentials therefore give the cells a higher water potential than the bathing solution and there will be a net movement of water out of the tissue.

Water loss from the tissue reduces the pressure inside the cells. When it drops to atmospheric pressure, the pressures inside the cells and in the bathing solution are equal and the cells are flaccid. The solute concentration of the cells will have risen due to water loss. If it is still higher than in the bathing solution, water continues to leave the plant cells, reducing the volume of cytoplasm. This causes the plasma membrane and cytoplasm to detach from the cell wall and the cells become plasmolysed. When the solute concentration of the cytoplasm equals that of the bathing solution, there is no more net movement of water because the solute potential and pressure potential of the cells and the bathing solution are equal and therefore their water potentials are equal.

Data-based questions: Water potentials in plant cells

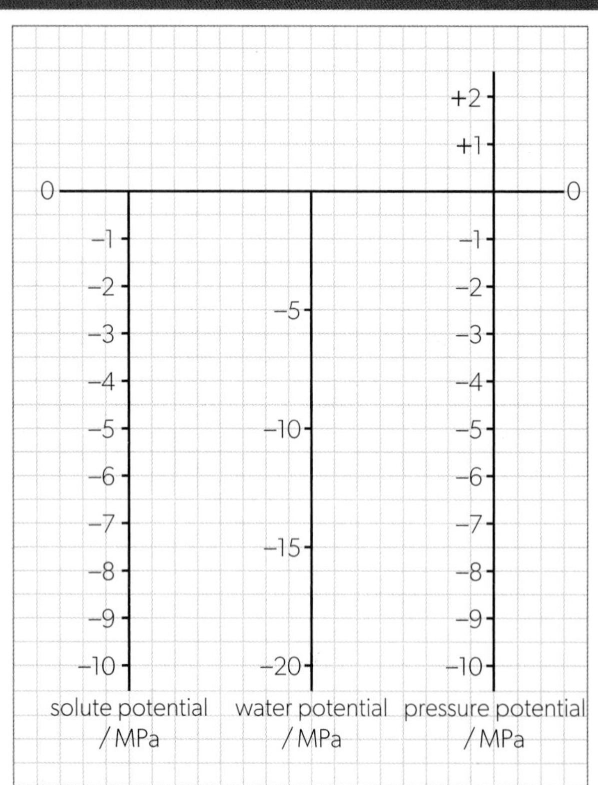

▲ Figure 20 Nomogram with maximum normal ranges of solute, water and pressure potentials in plants

If two of the potentials shown in the graph are known, the third can be determined using the nomogram.

1. Use the nomogram to determine the water potential of a plant cell with:
   a. solute potential −2 MPa and pressure potential +1 MPa
   b. solute potential 0 MPa and pressure potential −10 MPa
   c. solute potential −8 MPa and pressure potential +2 MPa
   d. Plant cells do not normally have a water potential greater than 0 MPa. Deduce the combinations of solute and pressure potentials that do not occur in plant cells. [2]
2. According to a simple understanding of osmosis, water moves from a hypotonic to a hypertonic solution.
   a. The water potential in xylem sap of oak roots at midday was −2.20 MPa. The pressure potential of the xylem sap was −2.15 MPa. Determine the solute potential of the sap inside the xylem vessels. [1]

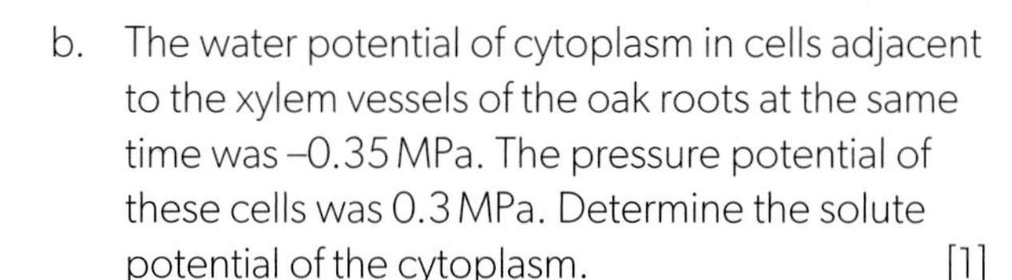

b. The water potential of cytoplasm in cells adjacent to the xylem vessels of the oak roots at the same time was −0.35 MPa. The pressure potential of these cells was 0.3 MPa. Determine the solute potential of the cytoplasm. [1]

c. Discuss the movement of water between the xylem vessels in the root and the cells surrounding them. [3]

3. Table 4 shows typical potentials in herbs (non-woody plants) at night and in deciduous trees before leaves have opened in warmer months.

Potentials	Soil around root	Root cells around xylem	Inside xylem vessels
Ψ_s / MPa	−0.25	−0.65	−0.75
Ψ_p / MPa	0.00	+0.30	+0.10
Ψ_w / MPa	−0.25	−0.35	−0.65

▲ Table 4 Water, solute and pressure potentials in some plants

a. Explain the higher pressure potential in the root cells than in the soil surrounding the root. [5]

b. At night, there is no photosynthesis in the leaf and water loss to the atmosphere is minimal. Using the data in Table 4, discuss whether or not there is a flow of sap in the xylem vessels up to the leaf. [5]

4. a. Table 5 shows potentials in cells in the root of a maize plant and in the base and tip of a leaf. Cells in the tip of the leaf had completed their growth but cells at the base were still growing. Determine the missing potentials in the table using the nomogram or the water potential equation ($\Psi w = \Psi s + \Psi p$). [3]

Potentials	Root	Leaf base	Leaf tip
Ψ_s / MPa	−0.67	−1.04	
Ψ_p / MPa	+0.23		+0.72
Ψ_w / MPa		−0.69	−0.57

▲ Table 5

b. Using the data in Table 5, explain the causes of water movement between cells in the root and the tip of the leaf. [3]

c. Suggest a hypothesis to account for growth in the base of the leaf but not the tip, based on differences in potentials between cells in these regions. [4]

Linking questions

1. What variables influence the direction of movement of materials in tissues?

   a. Explain the variables that ensure the one way flow of blood in the circulatory system. (B3.2.5)

   b. Outline two examples of where a concentration gradient plays a role in a biological process. (C1.2.4)

   c. Describe how the resting membrane potential is established in nerves. (C2.2.2)

2. What are the implications of solubility differences between chemical substances for living organisms?

   a. Explain the role that the hydrophobic interior of cell membranes plays in the movement of materials across membranes. (A1.1.5)

   b. Explain the advantage to a plant of storing carbohydrates as starch rather than as sugars. (B1.1.5)

   c. Explain the significance of the solubility differences between fibrin and fibrinogen. (C3.2.3)

TOK

How important are material tools in the production or acquisition of knowledge?

Scientific tools can extend the range of human perception beyond the range that our senses can normally detect. Electron microscopes extend the range of resolution further.

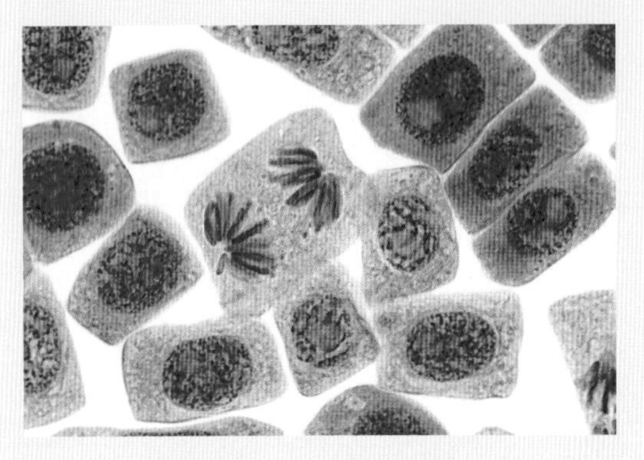

▲ Figure 1 Our eyes can distinguish between two objects that are 0.1 mm apart. A light microscope enables us to distinguish objects that are 0.0002 mm apart

New tools can open the possibility of what can be known. An example is the use of radioactive isotopes in biological research. As the United States Atomic Energy Commission (AEC) report put it in 1948, radioisotopes enabled "a new mode of perception".

Carbon-14 is an isotope of carbon that is radioactive. Radioactively labelled carbon within carbon dioxide can be fixed by plants during photosynthesis. It will release radiation that can be detected using either film or radiation detectors. As the carbon is metabolized, it will be found in different molecules within the plant. In other words, both the formation and movement of radioactive molecules can be traced.

▲ Figure 2 Melvin Calvin made use of this property of carbon-14 in working out the sequence of carbon intermediates in the light-independent reactions of photosynthesis

A consideration with respect to the use of tools is what ethical issues might be raised. For example, aphid stylets in combination with radioactivity labelled molecules have been used to study the movement of sugars within plants. This raises certain ethical questions.

Aphids penetrate plant tissues to reach the phloem (p in the first picture in Figure 3) using mouth parts called stylets (st in the first picture). If the aphid is anaesthetized and the stylet severed (process about to occur in the middle set of pictures), phloem will continue to flow out of the stylet (lower pictures). Both the rate of flow and the composition of the sap can be analysed. The closer the stylet is to the sink (the place where sugars in the sap are stored), the slower the rate at which the phloem sap will come out.

▲ Figure 3 An experiment with aphids that raises ethical issues

You would be guided to avoid this type of experiment because of the harm that would come to the aphid and the use of radioisotopes that would persist in the environment after the experiment.

End of chapter questions

1. During the development of multicellular organisms, cells differentiate into specific cell lines. A study was carried out on the early stages of differentiation in cells from mouse embryos that were grown in cultures. Two differentiated cell lines were studied, one of inner embryonic tissue (endodermal cells) and the other of external embryonic tissue (nerve cells) after 48 and 96 hours of incubation in cell cultures. A culture of undifferentiated cells was used as a control group. Cell population growth was measured by changes in cell density in all three cell lines.

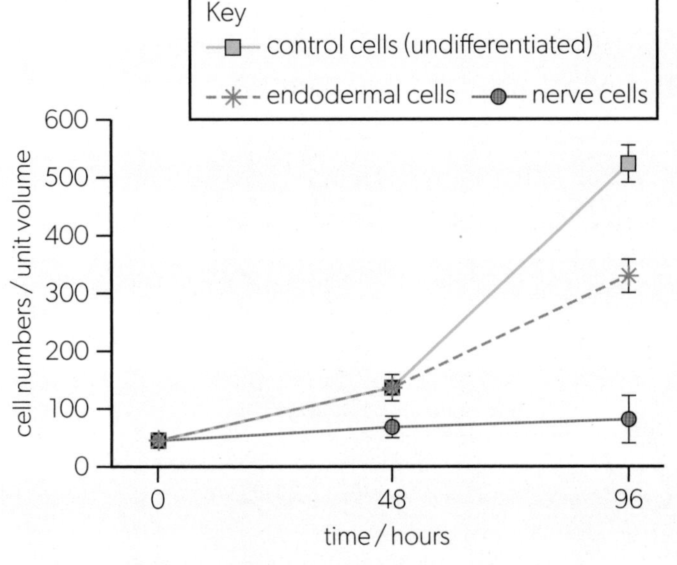

a. Distinguish between the changes in cell numbers in the three cell lines that occur during the 96-hour period. [2]

b. Using the data in the graph, deduce the relationship between cell differentiation and population growth. [1]

There are three phases of interphase in the cell cycle:

- the growth phase with the synthesis of RNA and proteins (G_1)
- the phase of DNA replication (S)
- and the pre-mitotic phase of rapid growth (G_2).

The graph shows the percentage distribution of the three cell lines in the different stages of interphase after 96 hours of incubation.

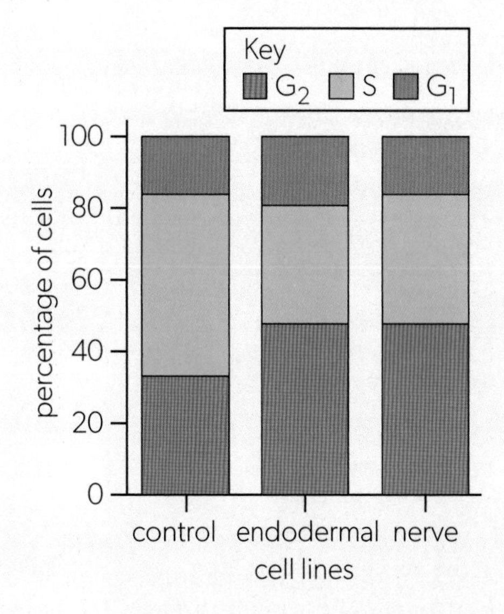

Source: Bryja, V. et al. Cell Proliferation; 41(6):875-893 doi 10.1111/j.1365-2184.2008.00556.x/epdf.

c. Compare and contrast the percentage of control and nerve cells in each of the three phases after 96 hours of incubation. [2]

d. Using the data of both graphs, deduce the relationship between the percentage of cells in each cell cycle phase and the population growth. [2]

e. Interphase is followed by mitosis. State the final product of the mitotic cell cycle. [1]

2. Identify the stage of mitosis shown in diagrams A to D. [2]

A

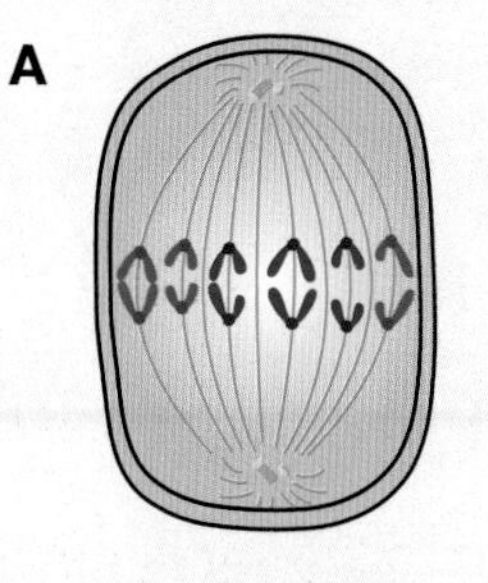

B

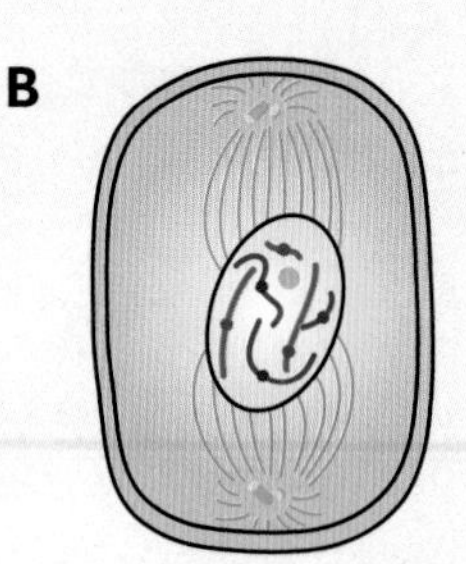

C

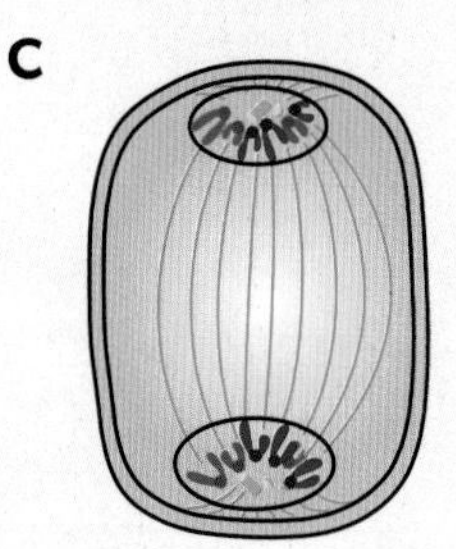

D

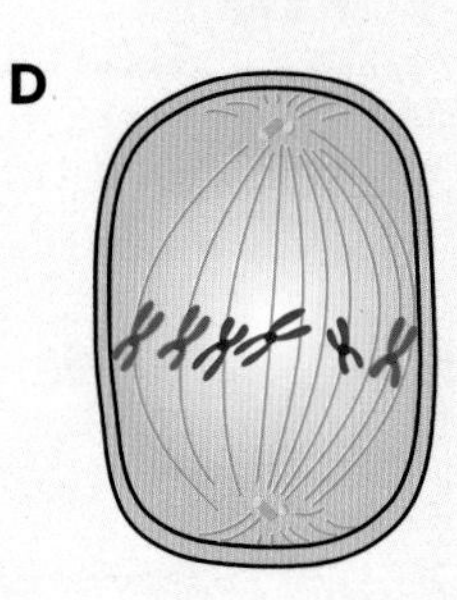

Source: Adds J et al. 1996 Cell biology and Genetics, Nelson Advanced Modular Science p71

D Continuity and change

3 Organisms

All organisms are capable of reproduction, maintenance, growth and development, and response to stimuli. Each of these phenomena is related to the theme of continuity and change. Living things have mechanisms for maintaining equilibrium when environmental conditions fluctuate. Variety among offspring resulting from sexual reproduction introduces resilience in a population to confront changing environmental conditions. Environmental change is a driver of evolution by natural selection.

D3.1 Reproduction

How does asexual or sexual reproduction exemplify themes of change or continuity?

Some organisms, such as black bread mould, can switch reproductive strategies between sexual and asexual reproduction. When conditions are ideal and stable, asexual reproduction is favoured. What is a possible explanation for this? What is the relative advantage of sexual reproduction? Figure 1 shows sexual reproduction in black bread mould (*Rhizopus stolonifer*). Usually, the mould reproduces asexually but environmental triggers can lead it to switch reproductive strategies. What might these factors be?

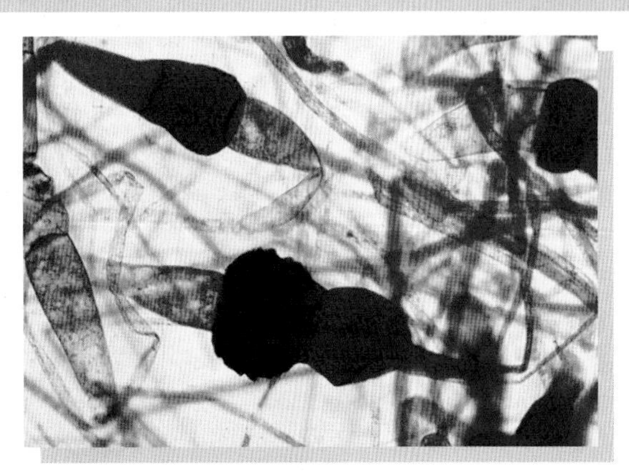

◀ Figure 1

What changes within organisms are required for reproduction?

Figure 2 shows a developing flower bud of the snapdragon, *Antirrhinum majus*. The outermost tissues (green) will develop into five sepals; a leaf-like protective covering for the bud. The red tissues are the future petals of the flower, and the central yellow region will develop sexual organs—the stamens and style. What processes trigger the ends of shoots to produce flowers and not leaves? What environmental conditions favour this switch?

In humans, puberty brings physical changes to a child's body. What causes these changes? Once the child fully matures into an adult, what ongoing biological processes occur to prepare the person for reproduction?

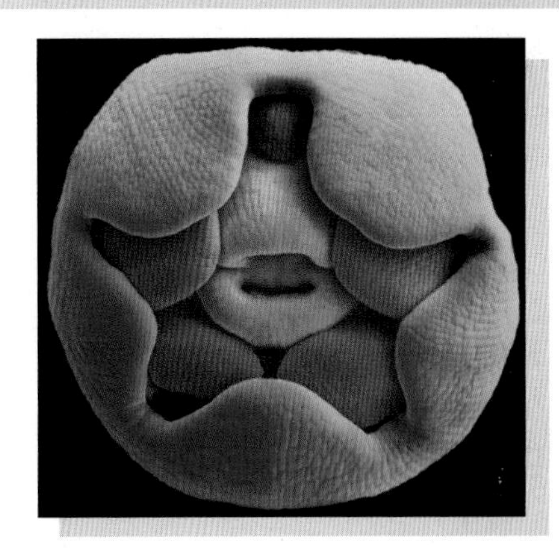

▲ Figure 2

SL and HL	AHL only
D3.1.1 Differences between sexual and asexual reproduction D3.1.2 Role of meiosis and fusion of gametes in the sexual life cycle D3.1.3 Differences between male and female sexes in sexual reproduction D3.1.4 Anatomy of the human male and female reproductive systems D3.1.5 Changes during the ovarian and uterine cycles and their hormonal regulation D3.1.6 Fertilization in humans D3.1.7 Use of hormones in in vitro fertilization (IVF) treatment D3.1.8 Sexual reproduction in flowering plants D3.1.9 Features of an insect-pollinated flower D3.1.10 Methods of promoting cross-pollination D3.1.11 Self-incompatibility mechanisms to increase genetic variation within a species D3.1.12 Dispersal and germination of seeds	D3.1.13 Control of the developmental changes of puberty by gonadotropin-releasing hormone and steroid sex hormones D3.1.14 Spermatogenesis and oogenesis in humans D3.1.15 Mechanisms to prevent polyspermy D3.1.16 Development of a blastocyst and implantation in the endometrium D3.1.17 Pregnancy testing by detection of human chorionic gonadotropin secretion D3.1.18 Role of the placenta in foetal development inside the uterus D3.1.19 Hormonal control of pregnancy and childbirth D3.1.20 Hormone replacement therapy and the risk of coronary heart disease

D3.1.1 Differences between sexual and asexual reproduction

A feature of living organisms is that they generate more members of their own species. This process is reproduction. It is achieved either sexually or asexually. Sexual reproduction brings about change whereas asexual reproduction brings about continuity. Table 1 explores this difference more fully for eukaryotes.

▼ Table 1 Reproduction in eukaryotes

Asexual reproduction	Sexual reproduction
▶ Figure 3 The adult female rose-grain aphid (*Metopolophium dirhodum*) is accompanied by 25 genetically identical daughters, produced without fertilization. The adult is giving birth to a 27th member of the clone. The black aphid below is a different species	▶ Figure 4 Fertilization is internal in insects such as this species of marsh fly (*Tetanocera elata*)
one parent	two parents, one male and one female
mitosis is used throughout an asexual life cycle	meiosis is used once per generation in a sexual life cycle
offspring are genetically identical to each other and to the parent	offspring are genetically different from each other and from their parents
existing gene combinations are maintained down the generations	new gene combinations are produced in each generation
no genetic variation is generated	genetic variation is generated
organisms adapted to an unchanging environment produce offspring that are also adapted	offspring may be better adapted than their parents if the environment is changing

D3.1.2 Role of meiosis and fusion of gametes in the sexual life cycle

Sexual life cycles in eukaryotic organisms must include two processes with opposite effects: meiosis and fusion of gametes. The fusion of a male and female gamete, also called fertilization, is the instant when a new individual is formed. The gametes usually come from two different parents, so novel combinations of genes can be generated.

Fertilization doubles the number of chromosomes each time it occurs. It would cause a doubling of the chromosome number every generation, if the number was not also halved at some stage in the life cycle. This halving of chromosome number happens during meiosis. Parental combinations of genes are broken up, allowing new combinations to form when gametes fuse. Meiosis can happen at any stage during a sexual life cycle, but in animals it happens during the process of creating the gametes. Gametes are haploid cells whereas body cells are diploid and have two copies of most genes.

Meiosis is a complex process and how it originated is not yet understood. What is clear is that evolution of meiosis was a critical step in the origin of eukaryotes. Without meiosis, there cannot be fusion of gametes and the sexual life cycle of eukaryotes could not occur.

▲ Figure 5 Wind shakes pollen off the anthers of grass flowers. The grains of pollen contain male gametes. Some of the grains are blown by chance to stigmata of flowers of the same species. This results in fertilization of the female gamete

D3.1.3 Differences between male and female sexes in sexual reproduction

The gametes that fuse in many types of fungi are outwardly identical. This is known as isogamy. Among eukaryotes, there is a trend for the evolution of two distinct types of gamete. All plants and animals are anisogamous, with different male and female gametes. It is easy to decide which is which (Table 2).

	Male (♂)	Female (♀)
Motility	male gametes travel to the female	female gametes are sessile
Size	smaller	larger
Food reserves	less—only enough for the gamete	more—enough for development of an embryo
Numbers produced	more—often very large numbers	few—sometimes only one

▲ Table 2 Anisogamy in eukaryotes

Activity: *Bangiomorpha* and the origins of sex

The first known eukaryote and first known multicellular organism is *Bangiomorpha pubescens*. Fossils of this red alga were discovered in 1,200-million-year-old rocks from northern Canada. It is the first organism known to produce two different types of gamete—a larger sessile female gamete and a smaller motile male gamete.

Bangiomorpha is therefore the first organism known to reproduce sexually. It seems unlikely that eukaryotic cell structure, multicellularity and sexual reproduction evolved simultaneously. What is the most likely sequence for these landmarks in evolution?

D3.1.4 Anatomy of the human male and female reproductive systems

Males and females have different roles in sexual reproduction, so they have different reproductive systems. The anatomy of the human male system is shown in Figure 6 and the anatomy of the human female system is shown in Figure 7.

Table 3 and Table 4 indicate functions that should be included when diagrams of male and female systems are annotated.

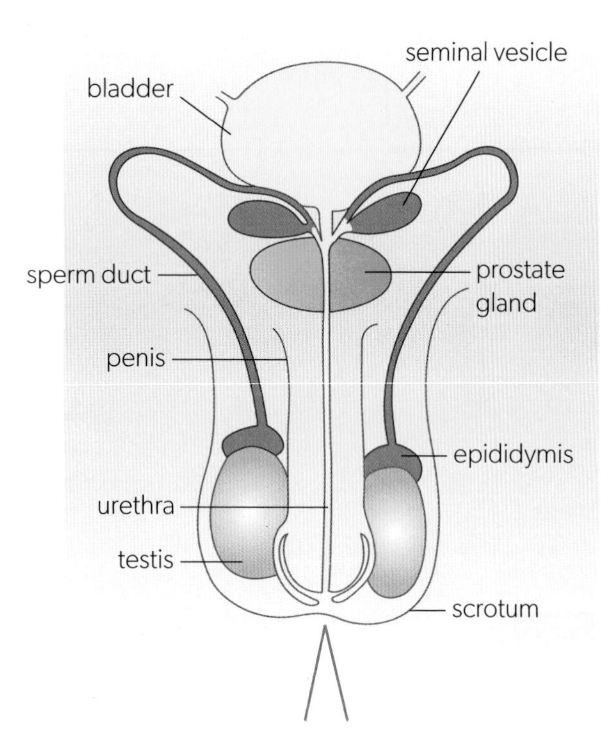

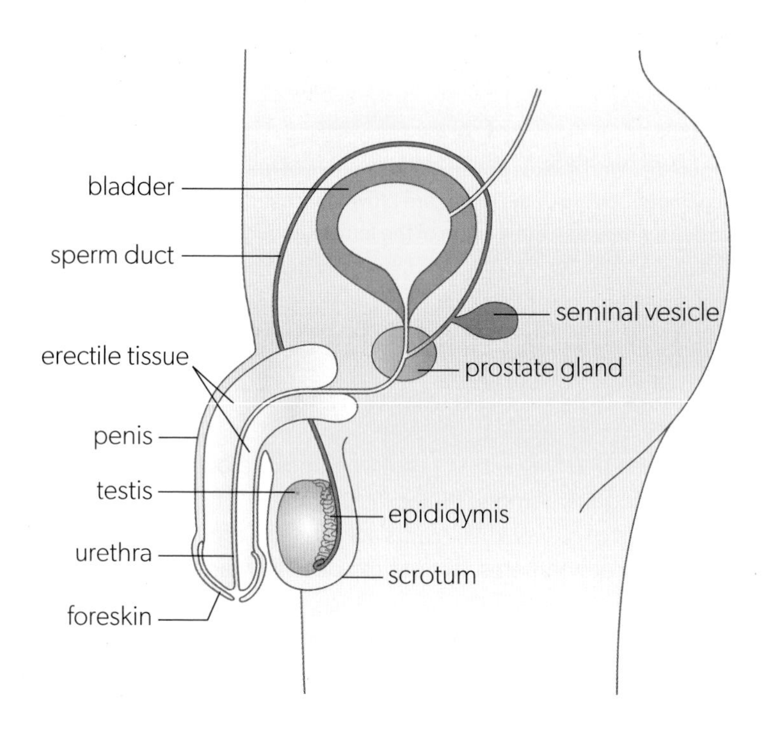

▲ Figure 6 Male reproductive system in front and side view

Structure	Function
testis	produces sperm and testosterone
scrotum	holds testes at a lower than core body temperature
epididymis	stores sperm until ejaculation
sperm duct	transfers sperm during ejaculation
seminal vesicle and prostate gland	secrete fluid containing alkali, proteins and fructose that is added to sperm to make semen
urethra	transfers semen during ejaculation and urine during urination
penis	penetrates the vagina for ejaculation of semen near the cervix

▲ Table 3 Functions of structures of the male reproductive system in humans

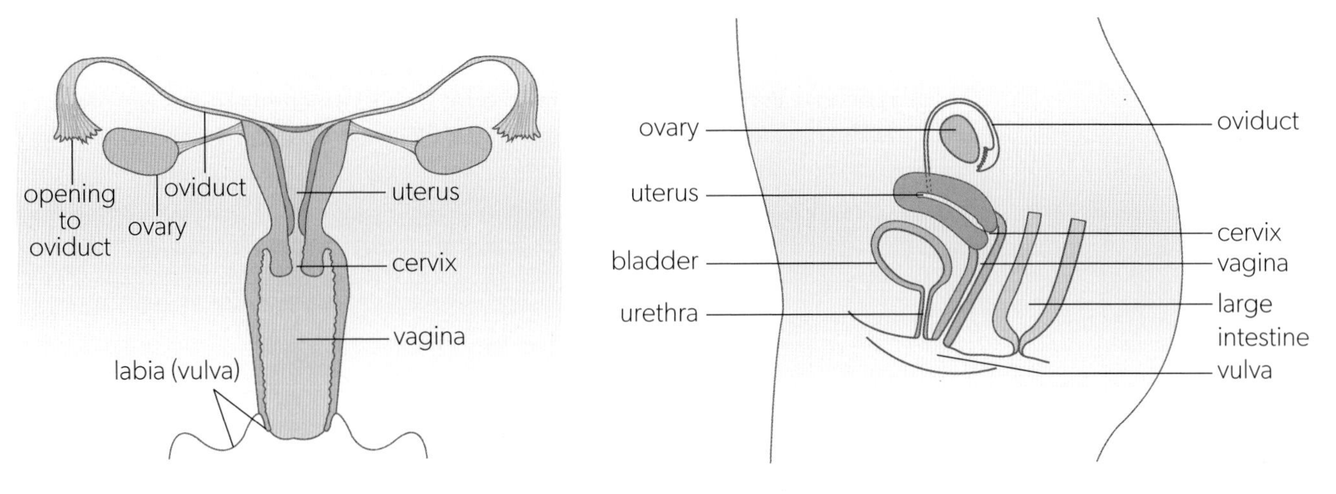

▲ Figure 7 Female reproductive system in front and side view

Structure	Function
ovary	produces eggs, oestradiol and progesterone
oviduct	collects eggs at ovulation, provides a site for fertilization then moves the embryo to the uterus
uterus	provides for the needs of the embryo and then foetus during pregnancy
cervix	protects the foetus during pregnancy and then dilates to provide a birth canal
vagina	stimulates penis to cause ejaculation and provides a birth canal
vulva	protects internal parts of the female reproductive system

▲ Table 4 Functions of structures of the female reproductive system in humans

D3.1.5 Changes during the ovarian and uterine cycles and their hormonal regulation

The menstrual cycle (Figure 8) occurs in most women from puberty until the menopause, stopping only during pregnancies or ill-health. It consists of the uterine cycle and the ovarian cycle together. Each time the menstrual cycle occurs, it provides the chance of a pregnancy. The cycles of changes in the uterus and in the ovary are both regulated by hormones.

The first half of the ovarian cycle is called the follicular phase because a group of follicles develops in the ovary. In each follicle, an egg is stimulated to grow. The most developed follicle breaks open, releasing its egg into the oviduct, while any other follicles degenerate. Release of the egg is called ovulation. Ovulation usually occurs on about Day 14 of the menstrual cycle. The second half of the ovarian cycle is called the luteal phase because the wall of the follicle that released an egg develops into a body called the corpus luteum. If fertilization does not occur, the corpus luteum breaks down and the ovary returns to the follicular phase.

The uterine cycle is the changes that occur to the endometrium (the lining of the uterus) during each menstrual cycle. The endometrium becomes thickened and more richly supplied with blood during the luteal phase, in preparation for implantation of an embryo. If there is no embryo, the thickening starts to break down towards the end of the luteal phase and material from it is then shed during menstruation. The days when this happens are usually referred to as a woman's "period". The start of a period is an obvious event so is counted as Day 1 of a menstrual cycle, when the ovaries have returned to the follicular phase. Once menstruation has ended, usually after 4 to 7 days, the lining of the uterus is repaired and starts to thicken.

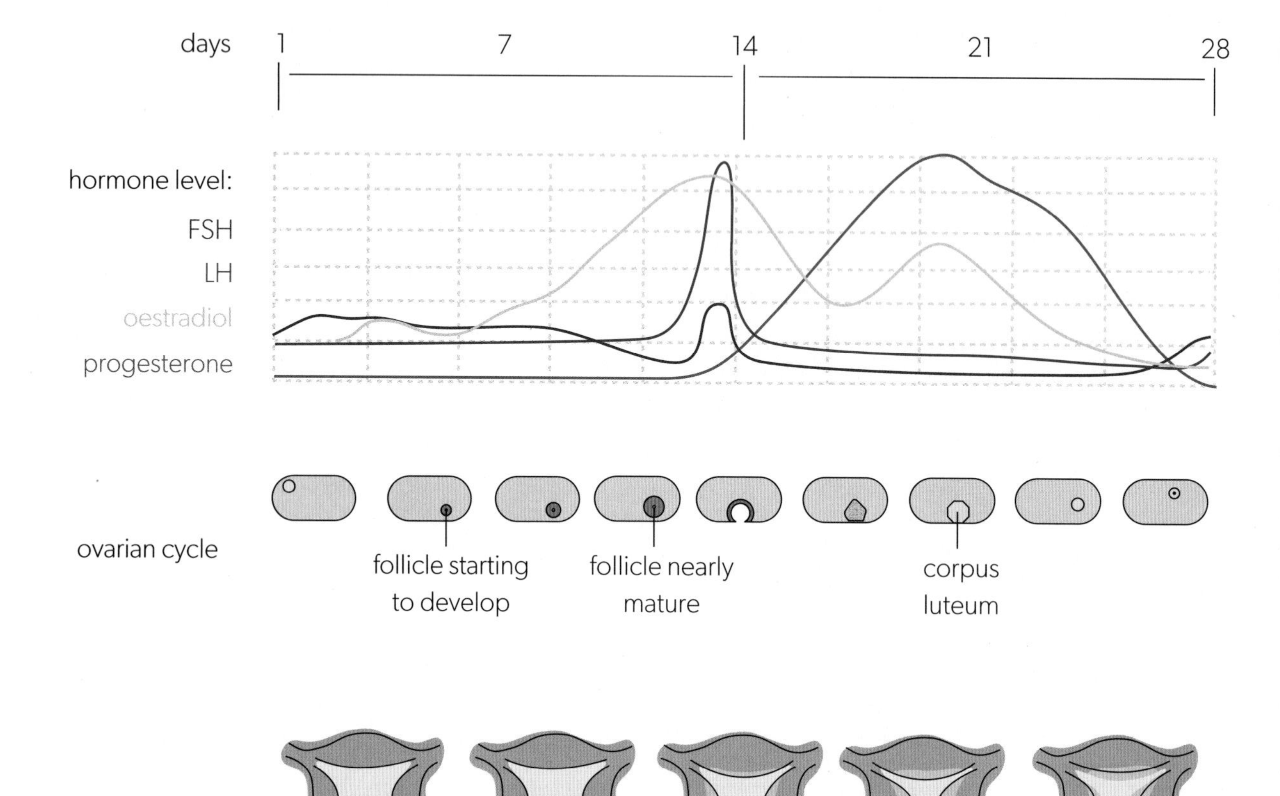

▲ **Figure 8** The menstrual cycle

The four hormones in Figure 8 regulate the menstrual cycle by negative and positive feedback. Follicle-stimulating hormone (FSH) and luteinizing hormone (LH) are protein hormones produced by the pituitary gland. They bind to FSH and LH receptors in the membranes of follicle cells. Oestradiol and progesterone are ovarian steroid hormones, produced by the wall of the follicle and the corpus luteum that develops from it. They influence gene expression and therefore development in the uterus and many other parts of the female body.

- FSH rises to a peak towards the end of the menstrual cycle and stimulates the development of follicles, each containing an oocyte and follicular fluid. FSH also stimulates secretion of oestradiol by the follicle wall.
- Oestradiol rises to a peak towards the end of the follicular phase. It stimulates the repair and thickening of the endometrium after menstruation and an increase in FSH receptors that make the follicles more receptive to FSH, boosting oestradiol production (positive feedback). When it reaches high levels, oestradiol inhibits the secretion of FSH (negative feedback) and stimulates LH secretion.
- LH rises to a sudden sharp peak towards the end of the follicular phase. It stimulates the completion of meiosis in the oocyte and partial digestion of

the follicle wall allowing it to burst open at ovulation. LH also promotes the post-ovulation development of the wall of the follicle into the corpus luteum. The corpus luteum secretes oestradiol (positive feedback) and progesterone.

- Progesterone levels rise at the start of the luteal phase, reach a peak and then drop back to a low level by the end of this phase. Progesterone promotes the thickening and maintenance of the endometrium. It also inhibits FSH and LH secretion by the pituitary gland (negative feedback).

Use of a spreadsheet: Changes in the concentrations of FSH

Daily morning blood samples were taken from five women starting on the first day of their menstrual cycle (determined by the beginning of their menstrual bleed = menses) and continued to the first day of their next menstrual bleed. The results are shown in Table 5.

Graphical presentations of the variation in concentration of FSH during the menstrual cycle show the peak in concentration occurring either in the follicular phase or during ovulation.

1. Determine the range in the length of the menstrual cycle among the five women.
2. Find two images which show the highest levels occurring at different times. Identify which is the more common representation.
3. Graph the changes in FSH over time for each woman.
4. Identify the day on which FSH peaks for each woman.
5. Suggest what is occurring on this day.

Day since menses onset	Subject 1 FSH UI dm^{-1}	Subject 2 FSH UI dm^{-1}	Subject 3 FSH UI dm^{-1}	Subject 4 FSH UI dm^{-1}	Subject 5 FSH UI dm^{-1}
1	6.73	6.66	2.69	5.49	6.6
2	6.74	8.01	4.37	5.28	6.39
3	6.33	8.49	4.25	5.94	7.49
4	6.46	7.25	4.19	6.07	7.54
5	5.41	7.62	5.33	5.09	6.49
6	5.48	7.38	4.45	5.28	5.79
7	4.3	6.71	4.7	5.25	5.19
8	3.68	4.81	5.53	4.38	4.45
9	3.58	4.29	6.21	4.6	4.21
10	3.95	4.4	6.2	3.84	4.04
11	7.01	4.65	6.89	3.27	4.62
12	13.32	3.8	7.03	2.84	9.55
13	4.74	3.79	7.2	3.55	9.3
14	4.92	4.4	5.79	16.99	3.8
15	3.7	13.92	5.54	12.16	3.07
16	3.81	8.83	5.12	10.45	3.84
17	2.04	5.73	4.36	8.32	2.68
18	1.98	5.83	3.47	7.43	2.6
19	2.33	4.13	4.24	5.55	2.35
20	2.12	3.34	15.62	5.31	2.9
21	0.84	2.95	6.68	3.99	1.9
22	1.68	2.02	3.81	4.13	3.26
23	0.5	1.3	2.24	3.58	3.14
24	0.85	1.29	2.22	3.17	2.78
25	1.12	1.37	1.6	2.27	4.4
26	3.49	1.85	1.73	1.35	5.57
27	4.92	2.45	0.96	1.37	NA
28	NA	3.2	1.14	1.54	NA
29	NA	4.32	0.91	1.39	NA
30	NA	NA	1.11	2.02	NA
31	NA	NA	1.36	4	NA
32	NA	NA	1.26	NA	NA
33	NA	NA	1.7	NA	NA

▲ Table 5 Concentration of FSH during menstrual cycles of five women. Source of data: https://www.saburchill.com/IBbiology/ICT/dataprocessing/025.html

Data-based questions: The female athelete triad

The female athlete triad is a syndrome consisting of three interrelated disorders that can affect female athletes: osteoporosis, disordered eating and menstrual disorders.

Osteoporosis is reduced bone mineral density. It can be caused by a diet low in calcium, vitamin D or energy, or by low oestradiol levels. Figure 9 shows the bone mineral density in two parts of the femur for female runners who had different numbers of menstrual cycles per year. The t-score is the number of standard deviations above or below mean peak bone mass for young women.

1. a. Outline the relationship between the number of menstrual cycles per year and bone density. [3]
   b. Compare and contrast the results for the neck of the femur with the results for the trochanter. [3]
2. Explain the reasons for some of the runners having:
   a. higher bone density than the mean [2]
   b. lower bone density than the mean. [4]
3. a. Suggest reasons for female athletes having few or no menstrual cycles. [2]
   b. Suggest one reason for eating disorders and low body weight in female athletes. [1]

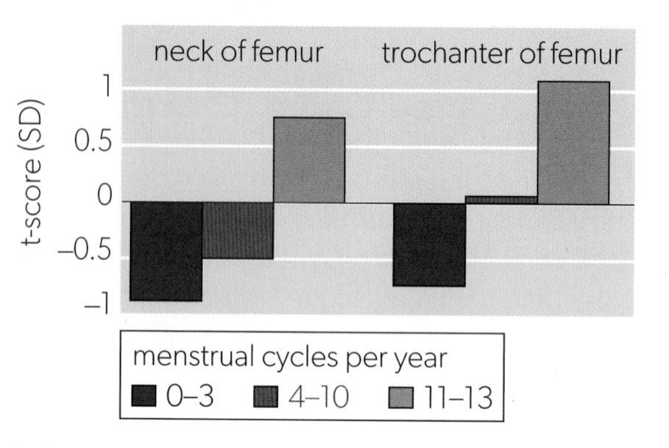

▲ Figure 9 Bone mass in women grouped according to menstrual cycles

D3.1.6 Fertilization in humans

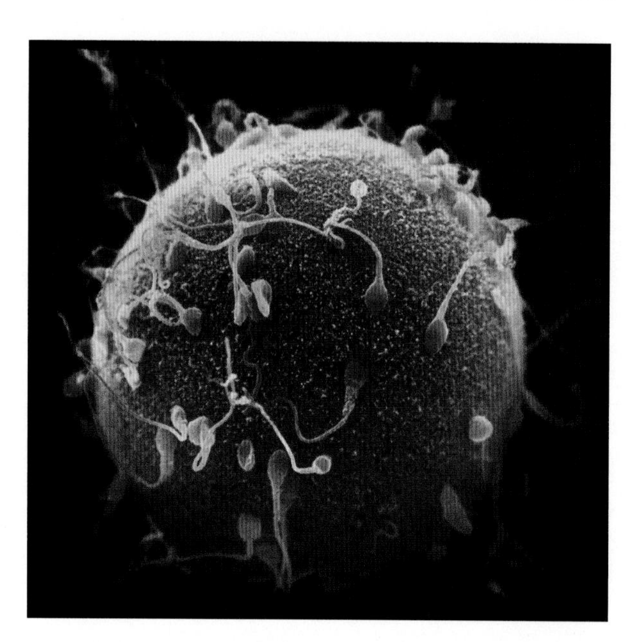

▲ Figure 10 Scanning electron micrograph of multiple sperm on the surface of an egg

Fertilization is the fusion of a sperm with an egg to form a zygote. The plasma membranes of sperm have receptors that detect chemicals released by the egg. This enables directional swimming of sperm towards the egg. The egg is surrounded by a cloud of follicle cells and a layer of glycoproteins. The sperm pushes between the cells and digests its way through the glycoproteins to reach the plasma membrane of the egg cell. The sperm's plasma membrane has proteins that bind to the egg cell's plasma membrane. The first sperm that manages to penetrate the zona pellucida binds and the membranes of sperm and egg fuse together. The sperm nucleus then enters the egg cell. This is the moment of fertilization.

Immediately after fertilization, the layer of glycoprotein around the egg hardens to prevent entry of more sperm. This ensures that a diploid zygote is produced, rather than an unviable cell with more than two sets of chromosomes.

The sperm tail either does not penetrate the egg during fertilization or is broken down inside the zygote. Sperm mitochondria may also penetrate, but they are usually all destroyed. The nuclei from the sperm and egg remain separate until the zygote's first mitosis. Then both nuclear membranes break down, releasing 23 chromosomes from each nucleus. These chromosomes, half from the mother and half from the father, participate jointly in mitosis using the same spindle of microtubules. Two genetically identical nuclei are produced, each with 46 chromosomes. Figure 11 shows the stages in the fertilization of a human egg.

D3.1.7 Use of hormones in vitro fertilization (IVF) treatment

The natural method of fertilization in humans is in vivo, meaning that it occurs inside the living tissues of the body. Fertilization can also happen outside the body in carefully controlled laboratory conditions. This is called in vitro fertilization, usually abbreviated to IVF. The procedure has been used extensively to overcome fertility problems in either the male or female parent.

There are several different protocols for IVF, but the first stage is usually down-regulation. Every day for about two weeks, the woman has an injection or nasal spray containing a drug to stop the pituitary gland secreting FSH or LH. Secretion of oestradiol and progesterone therefore also stops, suspending the woman's normal menstrual cycle and allowing doctors to control the timing and amount of egg production.

Intramuscular injections of FSH are then given daily for 7 to 12 days. This stimulates follicles to develop. The aim is to generate a much higher FSH concentration than during a normal menstrual cycle. As a consequence, far more follicles than usual develop. Between eight and fifteen follicles is ideal, each containing an egg. When the follicles are 18 mm in diameter, they are stimulated to mature by an injection of human chorionic gonadotropin (hCG). This hormone is normally secreted by a human embryo when it is about a week old, as a signal to its mother that it is alive and in need of sustenance from the endometrium.

In IVF, egg collection is performed 34 to 35 hours after the hCG injection. This is a minor surgical procedure that takes about 20 minutes. A micropipette mounted on an ultrasound scanner is passed through the uterus wall to draw the eggs out of the follicles.

Each egg is mixed with 50,000 to 100,000 sperm cells in sterile conditions in a shallow dish, which is then incubated at 37°C until the next day. If fertilization is successful, one or more embryos are placed in the uterus when they are about 48 hours old. Because the woman has not gone through a normal menstrual cycle, extra progesterone is usually given as a tablet placed in the vagina, to ensure that the uterus lining is maintained. If the embryos implant and continue to grow, the pregnancy that follows is no different from a pregnancy that began by natural conception.

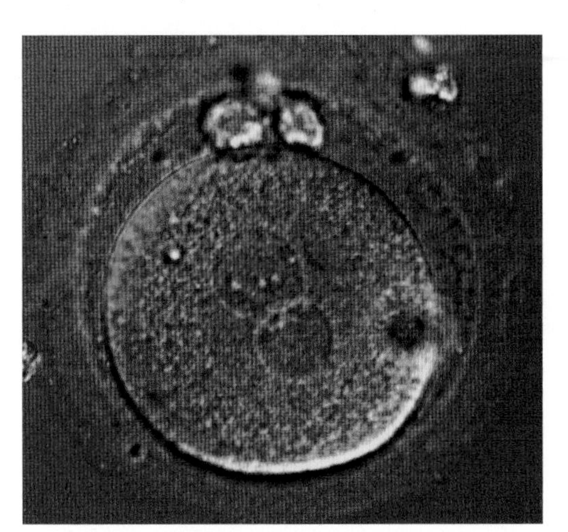

◀ Figure 12 IVF allows us to see the earliest stages of human life. This micrograph shows a zygote formed by fertilization. The nuclei of the egg and sperm are visible in the centre of the zygote. There is a protective layer of gel around the zygote, generated from the zona pellucida by the cortical reaction. Between this layer and the zygote are two small cells, called polar bodies. They are superfluous by-products of meiosis and eventually break down

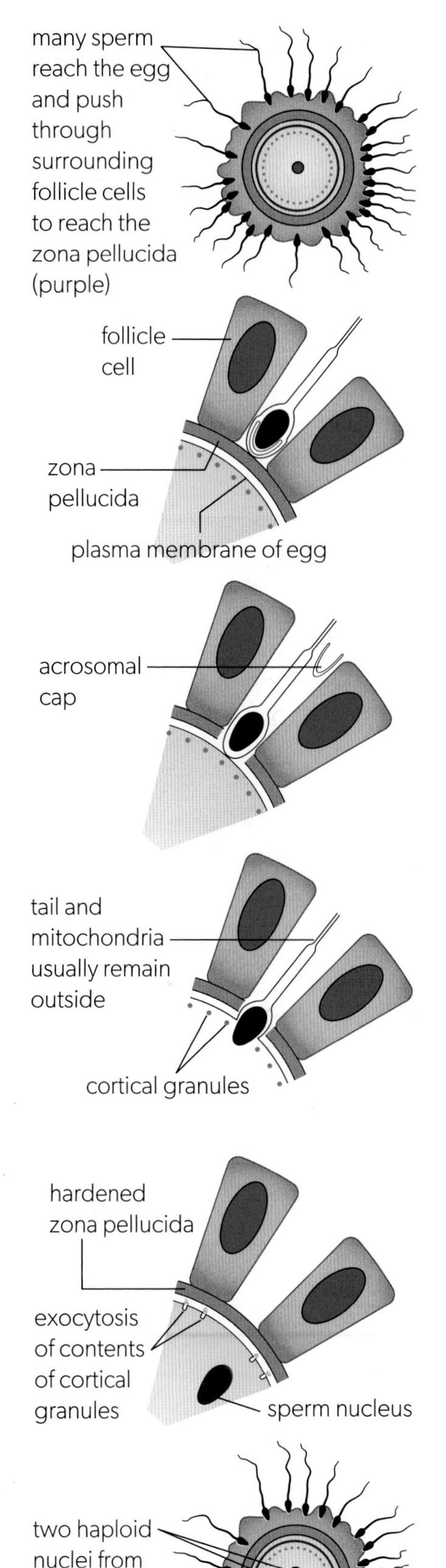

▲ Figure 11 Stages of fertilization

ATL Thinking skills: Evaluating and defending ethical positions

IVF and stem cell research has led to a growth in the global market for oocytes. In an article in the *Observer* entitled "The Cruel Cost of the Human Egg Trade" (Sunday 30 April 2006), there is a description of how transnational IVF clinics broker sales. These are usually between poor, female sellers and wealthy purchasers, often in separate countries. This is beyond the borders of national regulation, so there is limited scrutiny of the process and little, if any, protection of sellers. Academic Catherine Waldby argues that the sale of oocytes could be understood as a kind of labour and therefore it is possible to improve the protection of women who sell their eggs.

Author Ellie Houghtaling wrote a record of her experience as a donor in the article "I sold my eggs for an Ivy League education – but was it worth it?" (*The Guardian* Sunday 7 November 2021).

A systematic approach to an ethical analysis could involve considering motives, consequences or Natural Law. Natural Law is a theory that all people have inherent rights, that can be deduced through reason.

Data-based questions: Age and IVF

Use the data in Table 6 to answer the following questions.

Age of mother	Percentage of pregnancies per IVF cycle according to the number of embryos transferred					
	1 embryo	2 embryos		3 embryos		
	single	single	twins	single	twins	triplets
<30	10.4	20.1	9.0	17.5	3.6	0.4
30–34	13.4	21.8	7.9	18.2	7.8	0.6
35–39	19.1	19.1	5.0	17.4	5.6	0.6
>39	4.1	12.5	3.5	12.7	1.7	0.1

▲ Table 6 Age of mother and success rate of IVF

1. Outline the relationship between the age of the mother and the success rate of IVF. [3]
2. Outline the relationship between the number of embryos transferred and the chance of having a baby as a result of IVF. [3]
3. Discuss how many embryos fertility centres should be allowed to transfer. [4]

D3.1.8 Sexual reproduction in flowering plants

Flowering plants are a prominent part of most terrestrial ecosystems. Their flowers are the part of the plant that is used for sexual reproduction. Meiosis, gamete production and fertilization all happen inside flowers. As in mammals, a zygote produced by fertilization is retained by the female parent and supplied with food as it grows and develops into an embryo. In flowering plants, the embryo develops inside a seed which is dispersed from the female parent when fully formed.

Stamens are the male parts of a flower. They have an anther, supported by a stalk called the filament. Diploid cells inside the anther divide by meiosis to produce four haploid cells. Each of these develops into a pollen grain. The nucleus inside a pollen grain divides by mitosis, to produce three haploid nuclei. Two of these are male gametes. Genes in the third nucleus are expressed during pollen development and fertilization. Pollen grains develop a thickened wall, often with distinctive patterns on the exterior.

Carpels are the female parts of a flower. They have an ovary and a stigma, where pollen is received. Connecting the stigma to the ovary is the style, which is variable in length and usually has a hollow centre. The ovary contains one or more ovules. The ovules are ovoid in shape and multicellular. One cell in the centre of the ovule grows particularly large and then divides by meiosis to produce four haploid nuclei. One of these divides three times by mitosis to produce eight haploid nuclei, one of which is the female gamete or egg. The others assist in fertilization and subsequent embryo development.

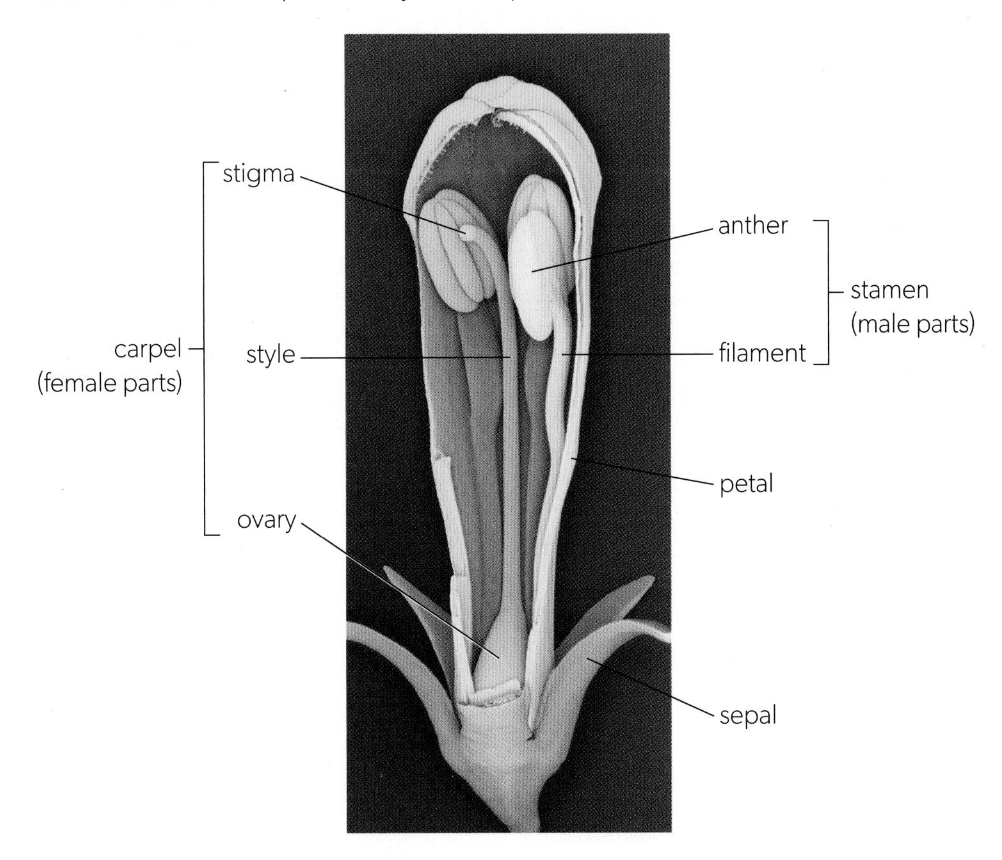

▲ **Figure 13** Flower of *Phygelius aequalis* with some parts removed to show internal structure. The flower is hermaphrodite because there are both male and female parts

Successful sexual reproduction in flowering plants depends on pollination, fertilization and embryo development.

- Pollination is the transfer of pollen from an anther to a stigma. The pollen is usually moved by wind or by animals.
- Fertilization happens inside an ovule, in the ovary. From each pollen grain on the stigma, a tube grows down the style to the ovary. The pollen tube carries male gametes near its growing tip. Once inside the ovary, the pollen tube grows towards one ovule and into one end of it. When the pollen tube reaches the centre of the ovule where the female gamete (egg) is located, the male gametes are released and fertilization occurs.
- The product of fertilization is a zygote, which develops into an embryo with an embryo root, an embryo shoot, and either one or two embryo leaves (cotyledons).

Activity: Flower structures

How many flower parts can you recognize in each of these photos?

▲ Figure 14 *Stylophorum diphyllum*

▲ Figure 15 *Corymbia ptychocarpa*

▲ Figure 16 *Passiflora caerulea*

The type of reproduction that happens inside flowers is sexual because there is meiosis and gamete production, followed by fertilization. Many flowers are hermaphrodite because they have both male and female parts, so can act as both a male parent and a female parent. Less commonly, some flowers have either male or female parts, rather than both. The male and female flowers may be on the same plant or on separate plants, but as male gametes fertilize female gametes, reproduction using flowers is sexual.

Data-based questions: Animal and wind pollination

Almost all species of flowering plant use animals or wind for pollination. Figure 17 shows results of analysing many different studies of the relative proportions of these two strategies. The data points are the mean percentage (±SD) of animal-pollinated plant species in terrestrial ecosystems at different latitudes and for all studies used in the analysis.

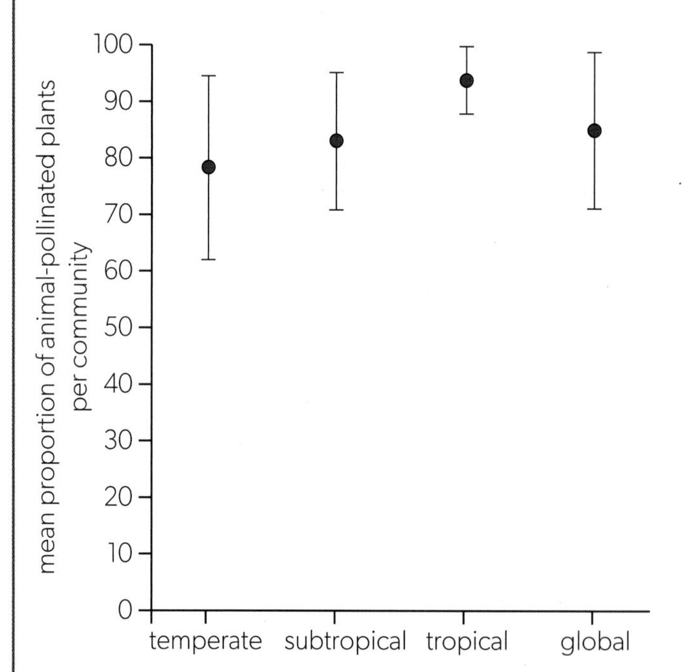

▲ Figure 17 Animal-pollinated plants in different ecosystems

1. State whether wind or animals are used by more plant species for pollination. [1]
2. a. Deduce the relationship between latitude and percentage of animal-pollinated species. [1]

   b. Suggest reasons for the relationship. [5]
3. Plant species adapted to grow in the Arctic or high in mountains do not seem to follow the trend shown in the graph, with a high proportion of animal-pollinated flowers that are very showy, such as gentians. Suggest reasons for this. [3]

D3.1.9 Features of an insect-pollinated flower

Insects were probably the pollinators of the earliest flowering plants and they are still the commonest and most diverse agents of pollination. Bees, hoverflies, wasps, beetles, butterflies and moths are the main groups. The structure, colour and scent of insect-pollinated flowers varies depending on the species of insect and the strategy for pollination. Nevertheless, there are some common features.

- Flowers have large, brightly coloured petals that advertise the flower, act as a landing stage and guide the insect's movements to the anther or stigma.
- Scent is secreted from the petals to advertise the flower.
- Pollen grains are large and spiky so they stick to insects and are attractive as a protein-rich food.
- Stigma is large and sticky to collect pollen from visiting insects.
- Glands called nectaries secrete a sugar solution (nectar) that is attractive to insects as an energy source.
- Nectaries are positioned deep inside the flower so insects can only reach them by brushing past the anthers and stigma.

Activity: Drawing flower structure

The structure of a flower is usually represented by drawing a half-view diagram. This allows all types of organ in the flower to be shown and also ovules inside the ovary. Figure 18 shows two examples of half-views.

Find a suitable specimen of an insect-pollinated flower, preferably with large floral organs. Remove all the sepals, petals and stamens and half of the ovary from one side of the flower. Draw the half flower that you now have.

Annotate your drawing with the names and functions of the flower parts.

- Petals help insects to find the flower
- Sepals protect the floral organs during development
- Anthers produce pollen containing the male gametes
- Filaments hold the anthers in a position where they are likely to brush pollen onto visiting insects
- Stigma captures pollen from insects
- Style positions the stigma where insects brush past and also guides the growing pollen tube to the ovary
- Ovary holds the ovules until fertilization and then develops into a seed-containing fruit
- Ovules hold the female gamete and develop into a seed after fertilization

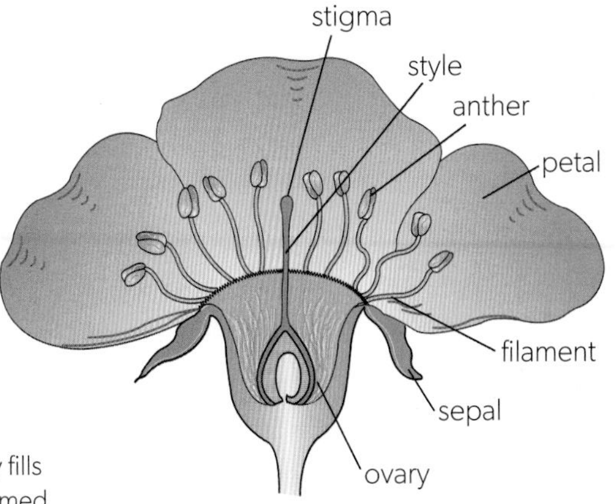
stigma
style
anther
petal
filament
sepal
ovary

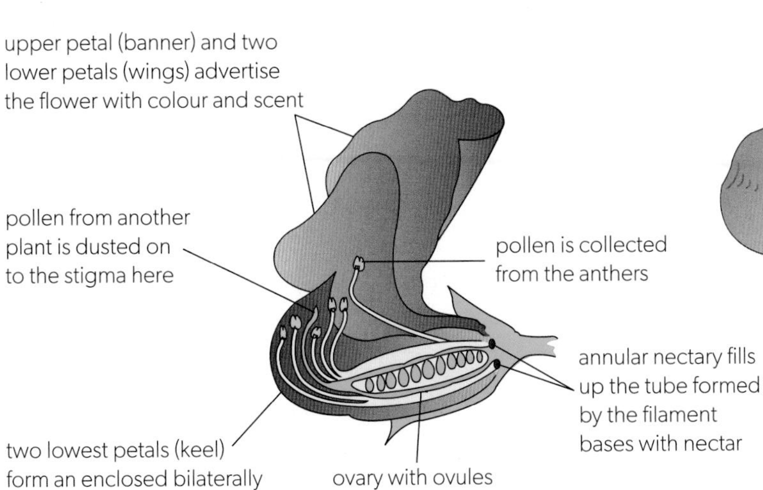
upper petal (banner) and two lower petals (wings) advertise the flower with colour and scent
pollen from another plant is dusted on to the stigma here
pollen is collected from the anthers
annular nectary fills up the tube formed by the filament bases with nectar
two lowest petals (keel) form an enclosed bilaterally symmetric chamber which bees must enter to reach pollen and nectar
ovary with ovules attached to the placenta on the upper side

◀▲ **Figure 18** Half-view drawings of two flowers: *Prunus domestica* (above) with names of structures and *Pisum sativum* (left) with functions

ATL Communication skills: Strategies for creating biological diagrams

- There should be no gaps in the lines when drawing closed shapes. For example, when drawing the pistil, the ovary and style should be shown as continuous.
- Labelling lines or arrows should end at the edge of a structure being labelled or inside it.
- The size of the diagram should be proportional to the complexity of what needs to be shown. It is difficult to distinguish detail or identify what a label is pointing to in a diagram that is unnecessarily small.
- There should be correct positioning of structures in relation to each other. For example, positioning of the seminal vesicles and the prostate gland in a diagram of human male reproductive anatomy.
- Connections between structures should be clearly and correctly shown. For example, between the filament and the anther in a diagram of a flower or the vas deferens in relation to the urethra in male reproductive anatomy.
- There should be correct proportions in relation to other structures. For example, the ovary should not be too large in relation to the uterus in a drawing of human female reproductive anatomy.

D3.1.10 Methods of promoting cross-pollination

Cross-pollination is transfer of pollen from an anther in a flower on one plant to a stigma of a flower on another plant. It leads to fusion of male and female gametes from different plants, so promotes genetic variation and therefore evolution. This is essential at times of environmental change. Cross-pollination also promotes "hybrid vigour", a phenomenon that is well-known to farmers and gardeners. The offspring of crosses between genetically unrelated plants tend to be healthy and grow strongly. This may be due to having different alleles of many genes rather than to being homozygous.

Many plants are hermaphrodite: they produce both pollen containing male gametes and ovules containing female gametes. This makes self-pollination possible—transfer of pollen from an anther to a stigma on the same plant. This is an extreme example of inbreeding, where closely related individuals reproduce.

There are taboos or laws to prevent members of most human societies from marrying a close relative (inbreeding). This is because it has long been known that such marriages tend to have higher rates of miscarriage and disease. Inbreeding increases the chance of a rare recessive allele in one ancestor being inherited twice by an individual, thus causing a genetic disorder. "Inbreeding depression" is the general trend for premature death, failure to thrive and infertility among offspring.

Mechanisms have evolved to reduce or prevent inbreeding in many plant species. Some of these mechanisms make self-pollination less likely or impossible (Table 7 and Figures 19 and 20).

Strategy	Example
adaptations to facilitate transfer of pollen from one plant to another by an outside agent	*Phyllostachys bambusoides* (madake) by wind *Vanilla planifolia* (vanilla orchid) by *Eulaema* bees
separation of anthers and stigmas/styles/ovaries in separate male and female flowers on the same plant	*Zea mays* (corn/maize) *Betula papyrifera* (paper birch)
separation of anthers and stigmas/styles/ovaries in separate male and female flowers on different plants	*Ginkgo biloba* (ginkgo tree) *Urtica dioica* (stinging nettle)
anthers and stigmas maturing at different times (protandry = anthers first, protogyny = stigma first)	*Nelumbo nucifera* (sacred lotus) protogynous *Digitalis purpurea* (foxglove) protandrous

▲ Table 7 Plant strategies to prevent self-pollination

▲ Figure 19 Vanilla orchids are native to forests in Central America but did not produce vanilla pods unless hand-pollinated when cultivated in Madagascar. What is the reason for this?

▲ Figure 20 The bamboo *Phyllostachys bambusoides* can grow to 20 m and synchronizes its flowering, with gaps of many decades between these mass flowering times. What is the benefit of this?

D3.1.11 Self-incompatibility mechanisms to increase genetic variation within a species

Despite features of flowering that reduce the chance of self-pollination, the stigma of a hermaphrodite plant occasionally receives pollen from that plant's own stamens. In many plants, this pollen fails to germinate or the pollen tube stops growing before it reaches the ovary. This prevents inbreeding due to a single individual acting as both male and female parent. It is called self-incompatibility and is the converse of the immune system, where there is rejection of non-self proteins or cells.

Self-incompatibility has evolved more than once and there are different mechanisms, but they all have a genetic basis with alternative alleles of one or more genes. Plants with the same self-incompatibility alleles cannot successfully pollinate each other.

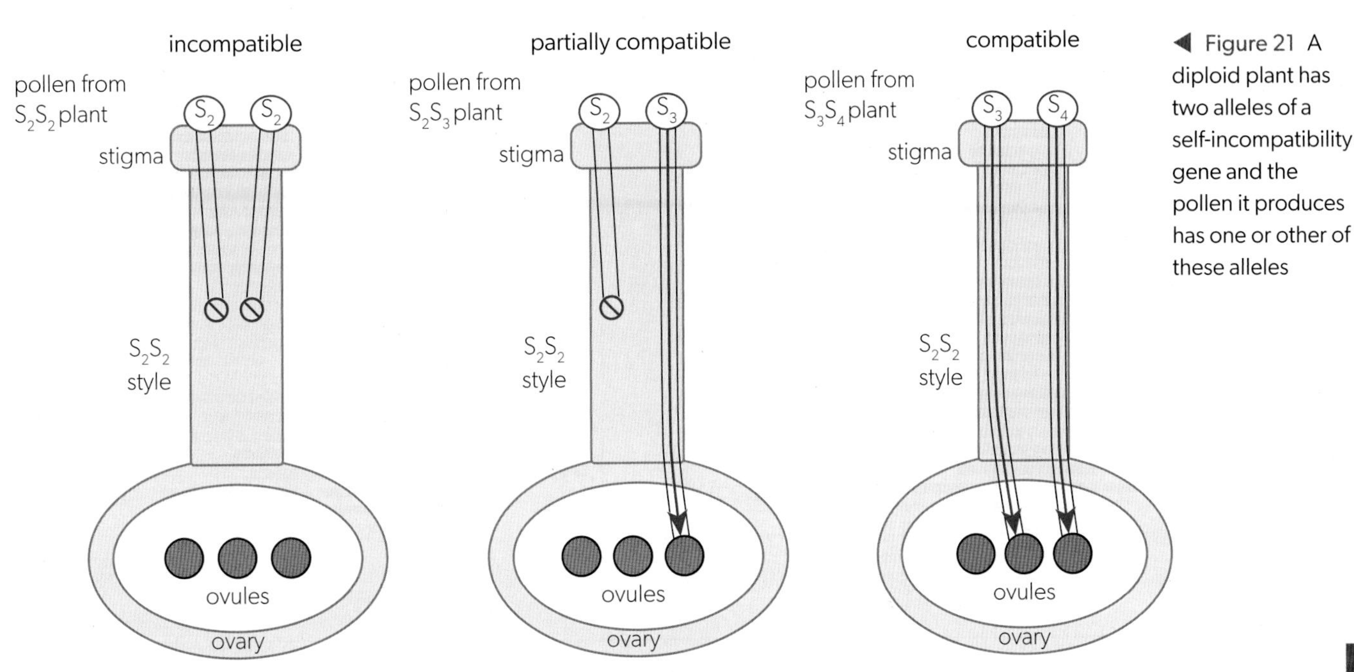

◀ Figure 21 A diploid plant has two alleles of a self-incompatibility gene and the pollen it produces has one or other of these alleles

There are consequences of self-incompatibility for fruit farmers. Planting a single variety may result in little or no fruit, but if there are two varieties in an orchard that have different alleles of self-incompatibility genes, abundant crops are produced.

Activity: Apple pollination

Jonagold is a triploid variety of apple with three self-incompatibility alleles: S_2, S_3 and S_9. How effective are the following three varieties likely to be as pollinators for Jonagold?

- Braeburn S_9 S_{24}
- Golden Delicious S_2 S_3
- Rome Beauty S_{20} S_{24}

◀ Figure 22 Jonagold apples; the S-gene in apples is explored more fully in Section D3.2.8

▲ Figure 23 Bees act as pollinators of many apple trees and many other food crops

Data-based questions: Pollination in Manchurian walnut trees

Juglans mandshurica is a wind-pollinated tree native to China and Korea. Each tree produces pollen in groups of male flowers and ovules in separate groups of female flowers. The ovules develop into walnuts after fertilization. The period of pollen release and receptiveness of stigmata to pollen was investigated in 25 trees in two successive years.

▼ Figure 24 Pollen release and stigma receptiveness in 25 Manchurian walnut trees

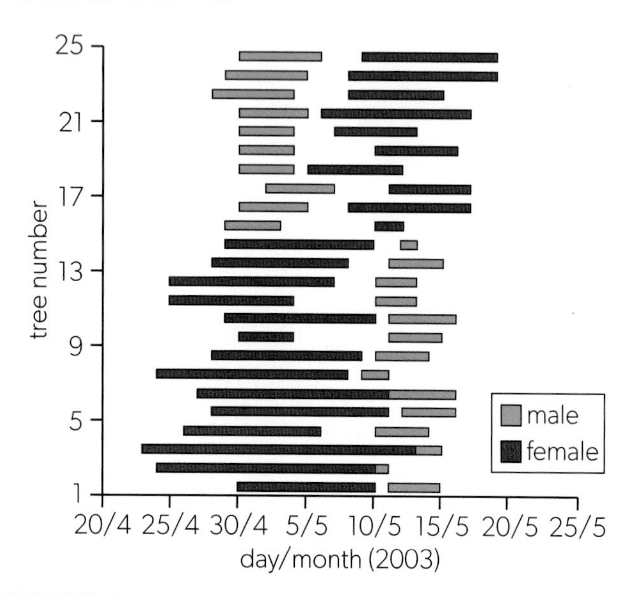

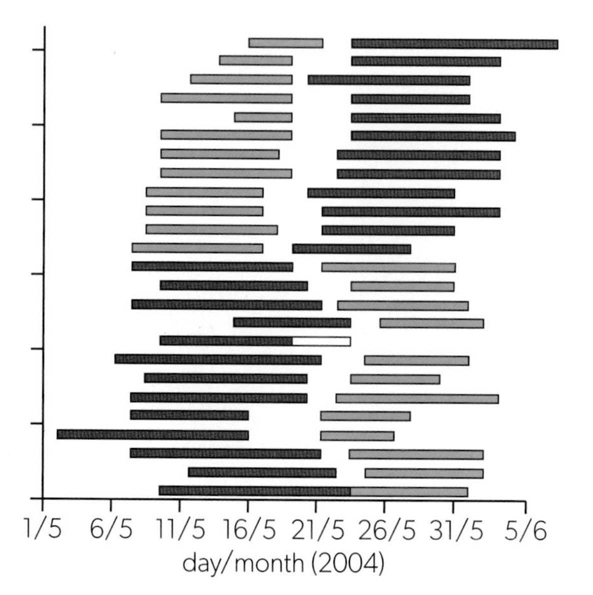

1. Explain three conclusions that can be drawn from the observations in the chart for 2003. [3]
2. Compare the results for 2003 and 2004. [2]
3. Explain the advantage of male and female parts of flowers becoming mature at different times on an individual tree. [3]
4. Explain the advantage to a species of not being entirely protandrous or protogynous. [2]

D3.1.12 Dispersal and germination of seeds

Seeds cannot move themselves, but nonetheless they often travel long distances from the parent plant. This is called seed dispersal and it reduces competition between offspring and parent. It also helps to spread the species. The type of dispersal depends on the structure of the fruit:

- dry and explosive
- fleshy and attractive for animals to eat
- feathery or winged to catch the wind
- covered in hooks that catch onto the coats of animals.

Seed dispersal and pollination are separate processes in the sexual life cycle of plants. These processes are compared and contrasted in Table 8 on the next page.

Designing methods: Seed dispersal

Seed dispersal is important for the survival of a population of plants as it minimizes competition between parent plants and their progeny. Trees use different mechanisms for seed dispersal that can be investigated.

- Some tree seeds are dispersed by water. For example, alder and willow trees often grow near bodies of water. Contrast how long an alder seed can remain floating before sinking with seeds from trees that are dispersed by other mechanisms.
- Some seeds are dispersed by explosive force such as the exploding cucumber (*Ecballium elaterium*) and jewelweed (*Impatiens capensis*). To what extent does placing gorse seed pods (*Ulex europaeus*) in a container with desiccant accelerate seed "popping"?
- Samaras are assemblages of seeds with "wings" found in maple, sycamore, elm and ash trees.

Comparisons of dispersal distances of single- and double-bladed samaras can be made. How does drop height affect the distance that the samara travels? The results of a student experiment are shown below.

1. State the dependent and independent variable in this experiment.
2. State whether the data is continuous or discrete.
3. Discuss the control of variables that would be necessary in this experiment.
4. Discuss additional trials that would be necessary to complete an investigation of the correlation between drop height and distance travelled of maple keys.

50 maple seeds, which are wind dispersed, were dropped one at a time from two different heights, 0.54 m and 10.8 m respectively. The histograms below show the distribution of the distance the maple seeds travelled.

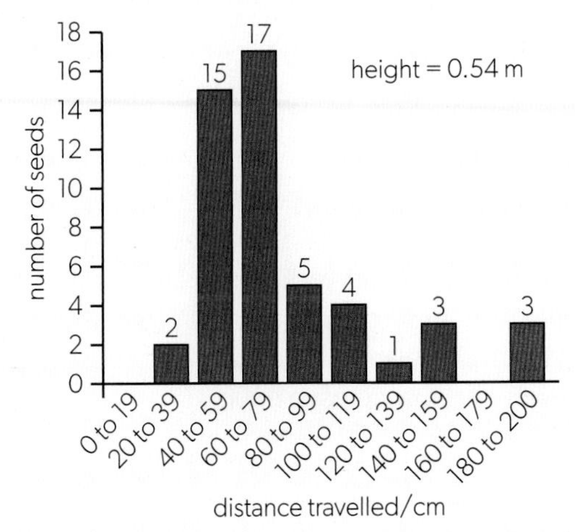

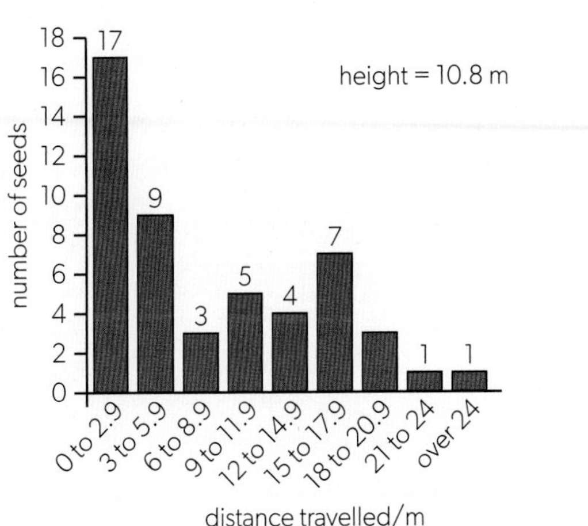

▲ **Figure 25**

▲ Figure 26 This reed bunting (*Emberiza schoeniclus*) is feeding on seeds of reedmace (*Typha latifolia*). The seeds are initially in a tight mass but as the bird pecks, hairs attached to each seed cause expansion to form a cottony fluff, from which individual seeds are blown away

	Pollination	**Seed dispersal**
What is transferred?	pollen	seeds
From where to where?	anther to stigma	from the female parent to a germination site
What are the commonest methods of movement?	wind or animals	wind, animals or explosion

▲ Table 8 Similarities and differences between seed dispersal and pollination

Activity: Making use of wind and/or animals

Dandelions are animal-pollinated and wind-dispersed whereas shagbark hickory trees are wind-pollinated and animal-dispersed. Find some local examples to complete this table. Could an animal act as both the pollinator and seed disperser for a plant?

		Agent of pollination	
		wind	**animals**
Agent of seed dispersal	**wind**		
	animals		

D3.1.13 Control of the developmental changes of puberty by gonadotropin-releasing hormone and steroid sex hormones

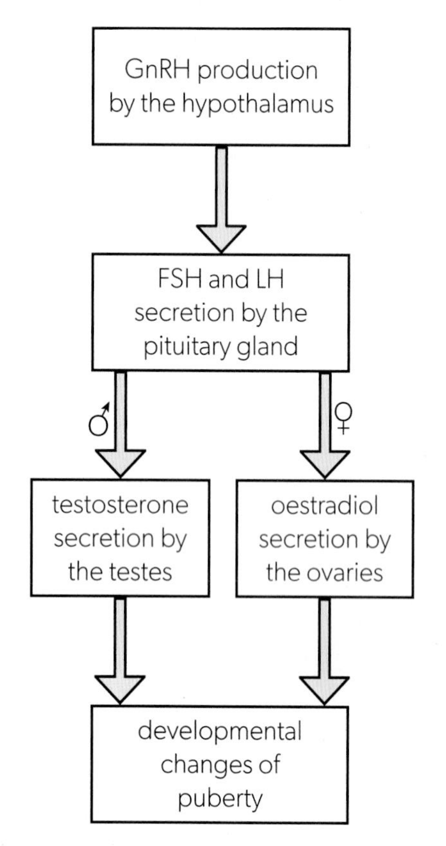

▲ Figure 27 Hormones in puberty

Puberty is the sequence of developmental changes that form the transition between childhood and sexual maturity. Onset and progress through puberty is controlled by the brain via hormone secretion. The key to this is gonadotropin-releasing hormone (GnRH), which is a peptide of 10 amino acids. A few hundred neurons in the hypothalamus synthesize this hormone and secrete it into a blood vessel that carries it directly to the pituitary gland.

Secretion of GnRH starts in a foetus about 10 weeks after fertilization. It continues throughout pregnancy and after childbirth, stopping when a baby is 4 to 6 months old. GnRH secretion only resumes when the brain determines that the time for puberty has arrived, usually at some stage during teenage years. Secretion then continues through puberty and adulthood. The hormone is released from the hypothalamus in pulses, so the concentration peaks and drops to a minimal level at least once each hour.

GnRH stimulates secretion of FSH and LH by the pituitary gland. Lower frequency pulses of GnRH stimulate FSH secretion and higher frequency pulses stimulate LH secretion. FSH and LH are gonadotropins because they cause changes in the gonads—the testes and ovaries.

In males, FSH stimulates testis growth and LH stimulates testosterone secretion by Leydig cells in the testes. FSH and LH both have roles in stimulating sperm production. Testosterone causes the development of secondary sexual characteristics in boys during puberty such as enlargement of the penis, growth of pubic hair and deepening of the voice due to growth of the larynx.

In females, FSH stimulates the development of follicles in the ovary. The wall of a follicle secretes oestradiol. LH stimulates the development of the follicle wall into the corpus luteum after ovulation. The corpus luteum secretes both oestradiol and progesterone. Oestradiol causes the development of secondary sexual characteristics in females during puberty such as enlargement of the uterus, development of the breasts and growth of pubic and underarm hair. Progesterone stimulates developmental changes in the breasts (mammary glands) to prepare for milk secretion (lactation).

D3.1.14 Spermatogenesis and oogenesis in humans

Gametogenesis is the production of gametes. Male gametes are spermatozoa, almost always abbreviated to sperm. Female gametes are oocytes, which are usually called eggs. Production of sperm in males (spermatogenesis) and eggs in females (oogenesis) happens in the same four stages: mitosis, cell growth, two divisions of meiosis and then differentiation. There are significant differences in how these stages are carried out.

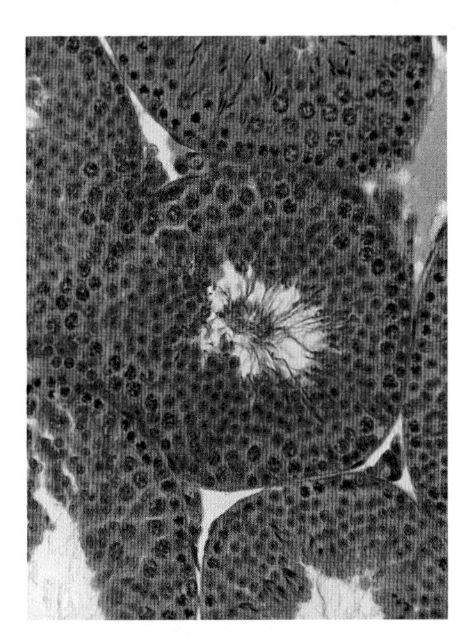

▲ Figure 28 Seminiferous tubules in transverse section, with sperm tails visible in the centre

Spermatogenesis happens in the testes, which are a coiled mass of seminiferous tubules, with small gaps or interstices between the tubules. These interstitial gaps are filled with testosterone-secreting Leydig cells. The outer layer of cells in seminiferous tubules is a germinal epithelium. Mitosis occurs continuously in the germinal epithelium, generating vast numbers of cells that are gradually displaced inwards towards the fluid-filled centre of the seminiferous tubule. As the cells migrate inwards they grow and then divide by meiosis to produce four haploid cells. The haploid cells differentiate into sperm by growing a tail and reducing their cytoplasm to a minimum. Also in the wall of the tubule are large nurse cells, called Sertoli cells.

Oogenesis happens in the ovaries, with the first stages completed before birth. The outer layer of cells in the ovaries of a female foetus is germinal epithelium. Mitosis occurs in this layer during foetal development. The cells produced migrate inwards to distribute themselves through the cortex of the ovary. When a foetus is 4 or 5 months old, these cells grow and start to divide by meiosis. By the seventh month, they are still in the first division of meiosis and are surrounded by a single layer of cells. The cell that has started meiosis, together with the surrounding follicle cells, is called a primary follicle. There are about 400,000 primary follicles in the ovaries at birth. They do not undergo further development until after puberty and no more primary follicles are ever produced. At the start of each menstrual cycle a small batch of primary follicles is stimulated to develop by FSH. Usually only one goes on to become a mature follicle, containing an egg.

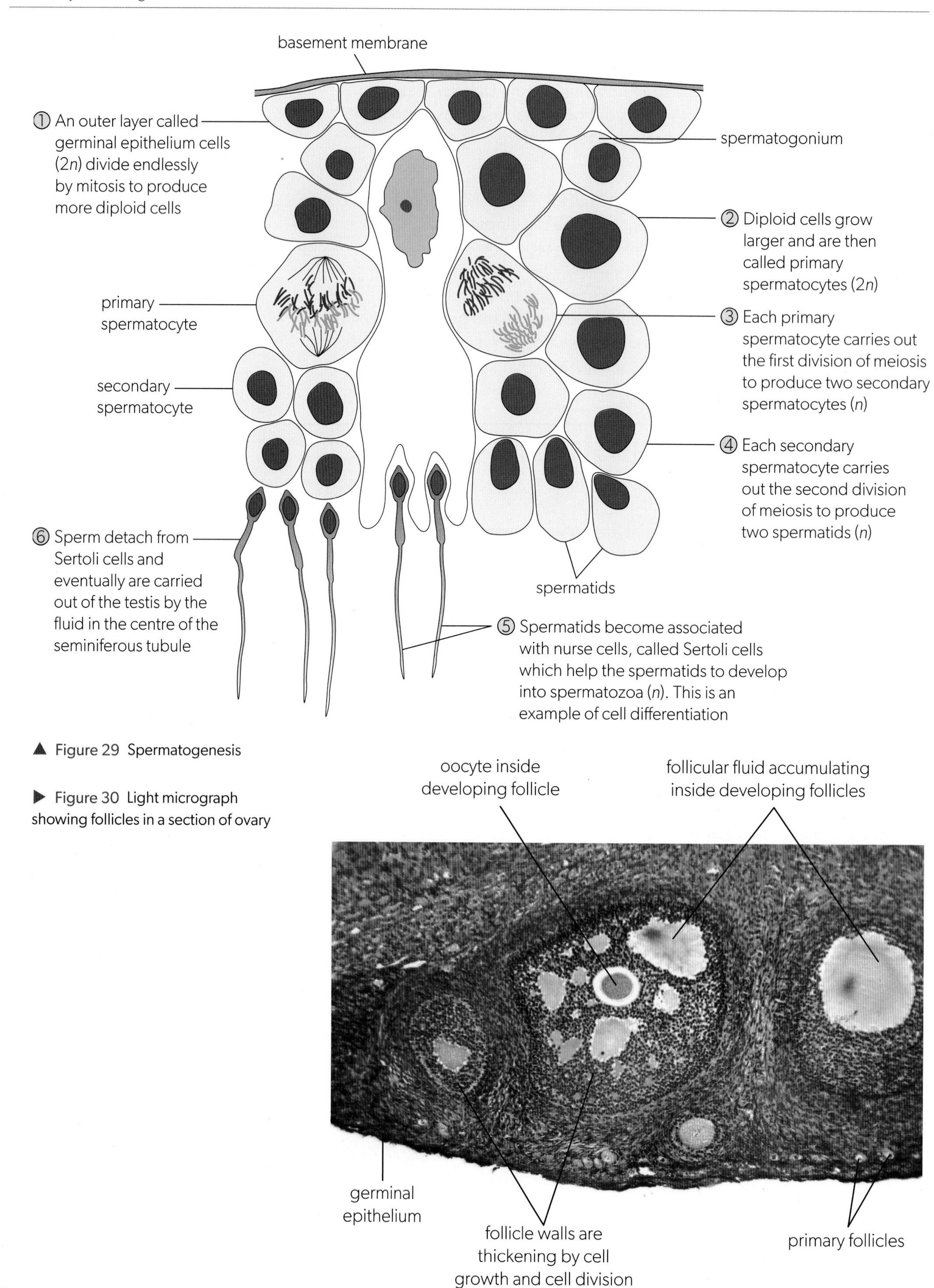

basement membrane
① An outer layer called germinal epithelium cells (2n) divide endlessly by mitosis to produce more diploid cells
spermatogonium
② Diploid cells grow larger and are then called primary spermatocytes (2n)
primary spermatocyte
③ Each primary spermatocyte carries out the first division of meiosis to produce two secondary spermatocytes (n)
secondary spermatocyte
④ Each secondary spermatocyte carries out the second division of meiosis to produce two spermatids (n)
⑥ Sperm detach from Sertoli cells and eventually are carried out of the testis by the fluid in the centre of the seminiferous tubule
spermatids
⑤ Spermatids become associated with nurse cells, called Sertoli cells which help the spermatids to develop into spermatozoa (n). This is an example of cell differentiation
oocyte inside developing follicle
follicular fluid accumulating inside developing follicles
germinal epithelium
follicle walls are thickening by cell growth and cell division
primary follicles

▲ Figure 29 Spermatogenesis

▶ Figure 30 Light micrograph showing follicles in a section of ovary

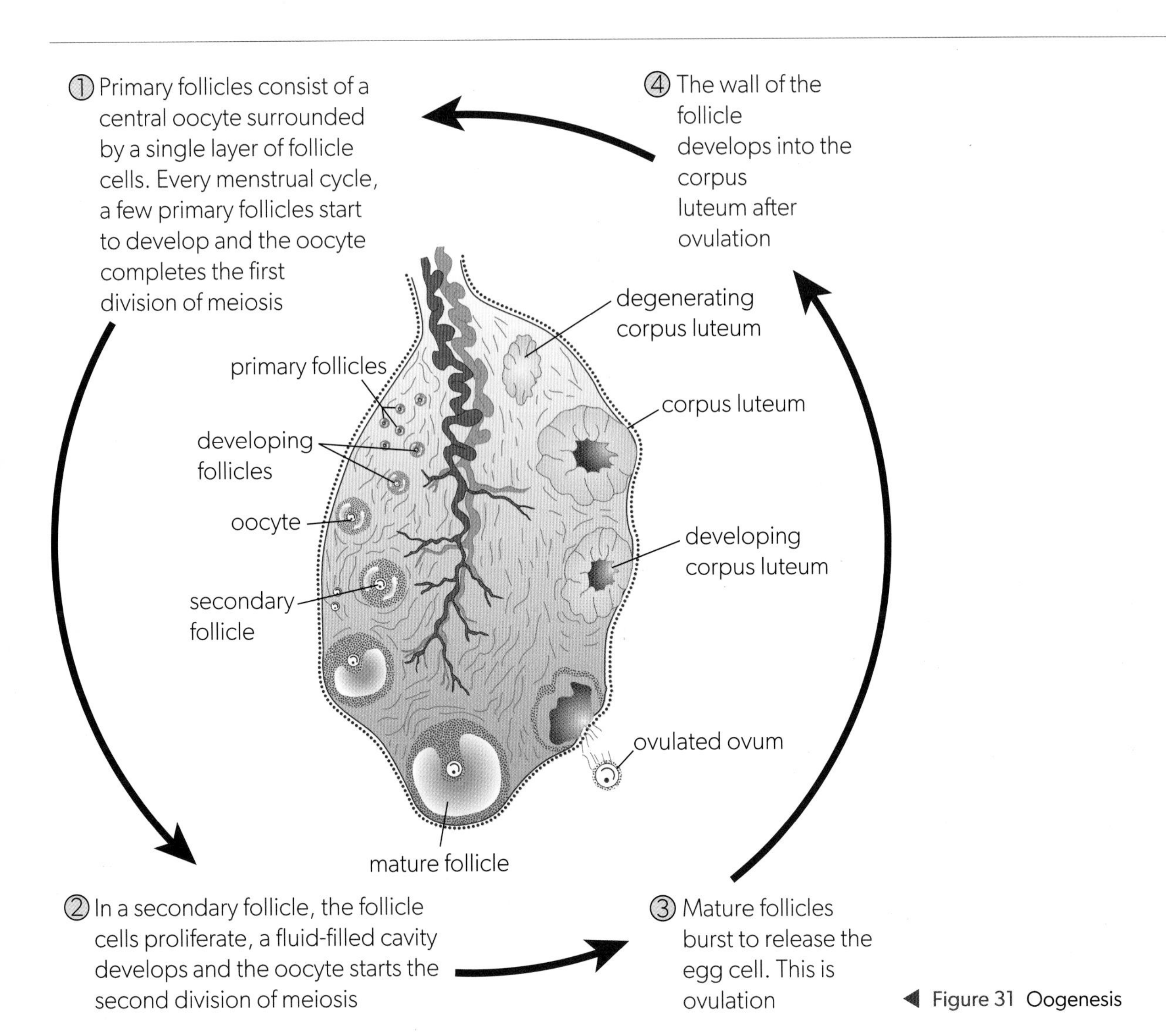
① Primary follicles consist of a central oocyte surrounded by a single layer of follicle cells. Every menstrual cycle, a few primary follicles start to develop and the oocyte completes the first division of meiosis
④ The wall of the follicle develops into the corpus luteum after ovulation
degenerating corpus luteum
primary follicles
corpus luteum
developing follicles
oocyte
developing corpus luteum
secondary follicle
ovulated ovum
mature follicle
② In a secondary follicle, the follicle cells proliferate, a fluid-filled cavity develops and the oocyte starts the second division of meiosis
③ Mature follicles burst to release the egg cell. This is ovulation

◀ Figure 31 Oogenesis

	Number of gametes from one meiosis	Rate of gamete production	Timing of release	Timing of production and release of gametes	Amount of cytoplasm
Spermatogenesis	4	millions per day	during ejaculation	produced from puberty onwards, taking about 75 days, and released during ejaculation	sperm have little cytoplasm, helping them to swim faster
Oogenesis	1	1 per month usually	at ovulation, on about Day 14 of the menstrual cycle	production initiated in female foetuses and completed during the menstrual cycle, with release on about Day 14, from puberty onwards until menopause	eggs contain more cytoplasm than any other human cell to provide food reserves for the embryo

▲ Table 9 Summary of the differences between spermatogenesis and oogenesis

AHL

D3.1.15 Mechanisms to prevent polyspermy

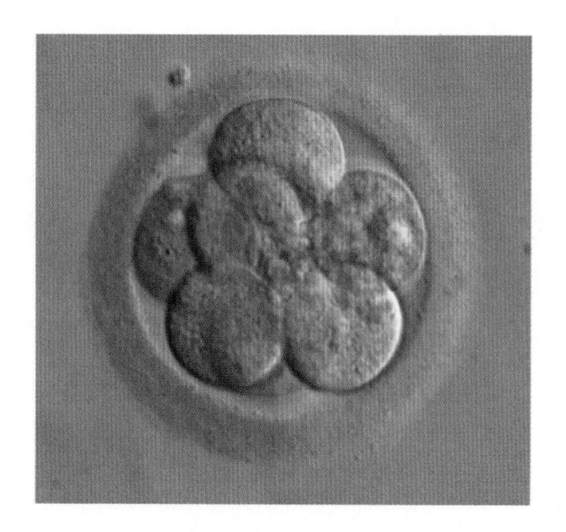

▲ Figure 32 For the first week of its life, a human embryo is protected by the coat of glycoproteins that was toughened immediately after fertilization. Here the embryo is 3 days old and has eight cells

If two sperm were to penetrate an egg at the precisely the same moment, a triploid zygote would be produced but it would soon die. In humans, only zygotes produced by fusion of one sperm with an egg are viable. Fusion of more than one sperm with an egg is polyspermy. Two processes help to make it very infrequent: the acrosome reaction and the cortical reaction.

The acrosome reaction

The acrosome is a large membrane-bound sac of enzymes in the head of a sperm. The zona pellucida is a coat of glycoproteins that surrounds and protects the egg. Sperm bind to specific glycoproteins in the zona pellucida. This triggers release of the contents of their acrosome. Enzymes from the acrosome start to digest the glycoproteins, weakening the zona pellucida. As a result, the beating of the sperm's tail is able to generate enough force for it to push through to reach the plasma membrane of the egg.

The cortical reaction

Cortical granules are the thousands of enzyme-containing vesicles located near the egg cell's plasma membrane. The nucleus of the first sperm to penetrate the zona pellucida enters the egg. This fertilizes and activates the egg. An early effect of activation is that the cortical granules move to the plasma membrane of the egg cell and release their contents by exocytosis. The released enzymes cause a general toughening of the zona pellucida making it very difficult for any more sperm to penetrate it. The enzymes also change specific glycoproteins in the zona pellucida to which sperm bind, so this can no longer happen.

D3.1.16 Development of a blastocyst and implantation in the endometrium

The zygote produced by fertilization has two haploid nuclei, one derived from the sperm and the other from the egg. Replication of DNA in each of these nuclei starts about 6 hours after fertilization and takes 3 hours to complete. The two nuclei undergo major changes in the pattern of gene expression and 30 hours after fertilization they divide jointly by mitosis, producing two genetically identical diploid nuclei, which separate when the cell divides.

During the following days the embryo undergoes rapid rounds of the cell cycle, with the number of cells doubling about every 18 hours. Because an egg cell has a large amount of cytoplasm, the early rounds of the cell cycle can happen without any cell growth between divisions, so cell size decreases. Initially the embryo is a solid ball of cells. When it is 6 or 7 days old it has changed into a hollow ball, due to unequal cell divisions and cell migration. This embryonic stage is called the blastocyst.

At 7 days old, the blastocyst has about 250 cells and is approximately 200 μm in diameter. It will have migrated to the uterus from the site of fertilization in an oviduct. This movement is due to days of gentle wafting by cilia in the oviduct wall. The toughened zona pellucida which has surrounded and protected the embryo now breaks down.

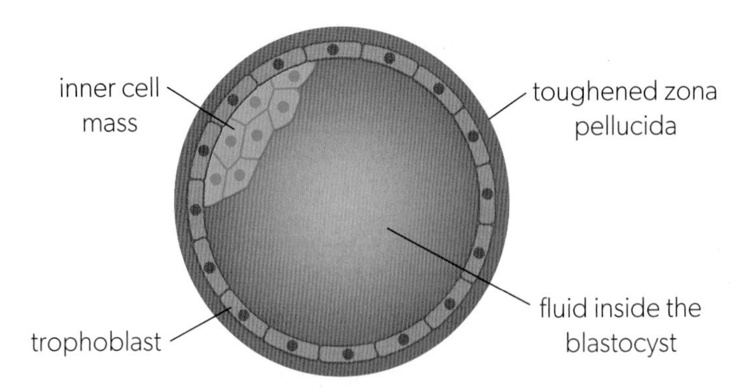

▲ Figure 33 Diagram of a blastocyst

The blastocyst has used up the reserves of the egg cell and needs an external supply of food. It obtains this by attaching itself to the endometrium (uterus lining) in a process called implantation. The outer cell layer of the blastocyst develops finger-like projections that grow into the endometrium. They exchange materials with the mother's blood, including absorbing foods and oxygen. The inner cell mass of the blastocyst grows and develops into a human body. By 8 weeks it has started to form bone tissue and is now called a foetus rather than an embryo. The new individual is recognizably human and will soon be visibly either male or female.

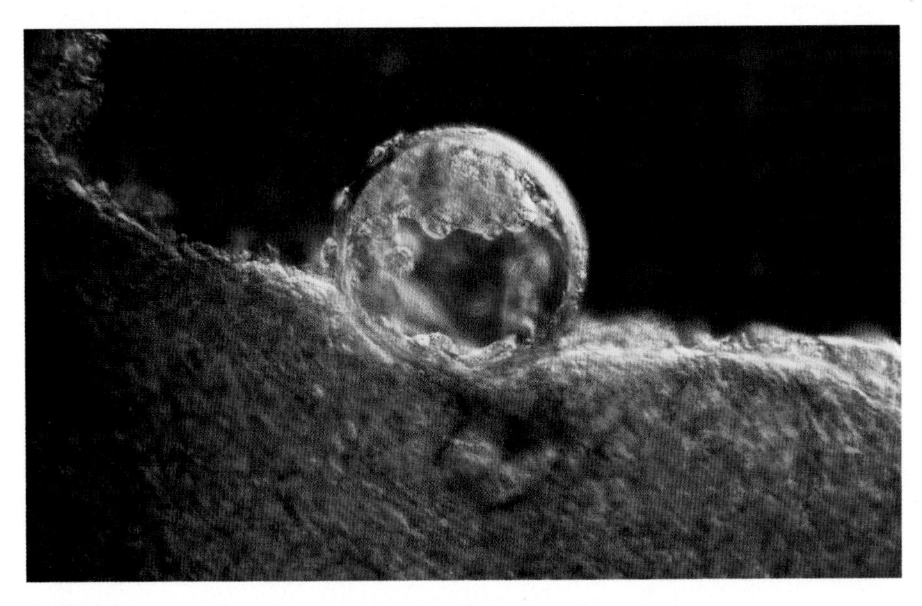

▲ Figure 34 A blastocyst has reached the uterus and is becoming implanted

D3.1.17 Pregnancy testing by detection of human chorionic gonadotropin secretion

The hormone human chorionic gonadotropin (hCG) is a medium-sized protein. It is produced by the embryo's trophoblast cells, from the blastocyst stage onwards. Continuity of hCG production is essential to maintain pregnancy and avoid miscarriage. In the first 10 or so weeks hCG stimulates the corpus luteum to develop and to secrete progesterone. This hormone helps to support the pregnancy by preventing degeneration of the uterus lining.

After 8 to 12 weeks of pregnancy the placenta starts to secrete progesterone in response to the hormone hCG. The corpus luteum stops hormone production and breaks down. Trophoblast cells in the placenta continue to secrete hCG throughout the remainder of the pregnancy, stimulating progesterone secretion by placenta cells and thereby maintenance of the endometrium.

Pregnancy tests are based on detection of hCG in a pregnant woman's urine (Figure 35).

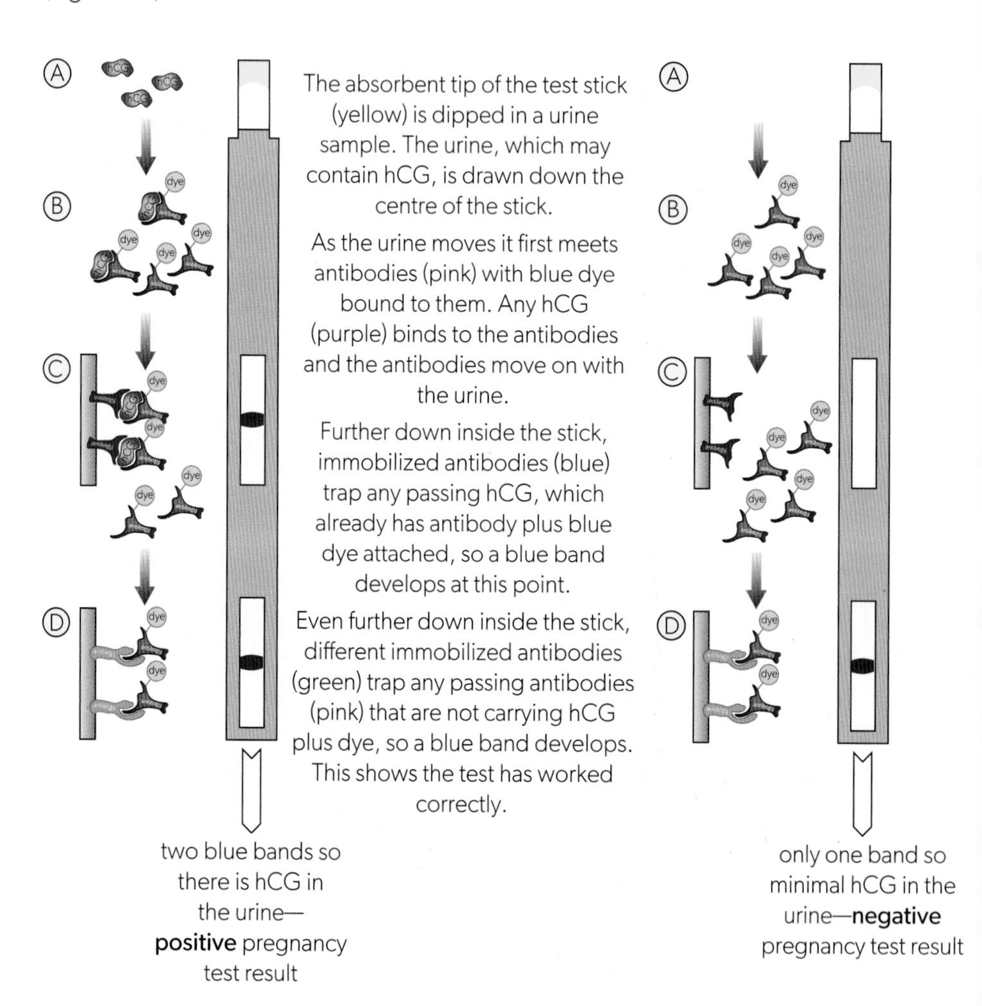

A
B
C
D
dye
The absorbent tip of the test stick (yellow) is dipped in a urine sample. The urine, which may contain hCG, is drawn down the centre of the stick.
As the urine moves it first meets antibodies (pink) with blue dye bound to them. Any hCG (purple) binds to the antibodies and the antibodies move on with the urine.
Further down inside the stick, immobilized antibodies (blue) trap any passing hCG, which already has antibody plus blue dye attached, so a blue band develops at this point.
Even further down inside the stick, different immobilized antibodies (green) trap any passing antibodies (pink) that are not carrying hCG plus dye, so a blue band develops. This shows the test has worked correctly.
two blue bands so there is hCG in the urine—**positive** pregnancy test result
only one band so minimal hCG in the urine—**negative** pregnancy test result

▲ Figure 35 How pregnancy tests using monoclonal antibodies work

Activity: Explaining pregnancy testing

Explain the answers to these questions to a classmate.

1. How does a blue band develop at point C of the test stick if the woman is pregnant?
2. Why does no blue band develop at point C if the woman is not pregnant?
3. Sometimes there is a faint blue band at point C. What hypothesis can you suggest to explain this in terms of the concentration of hCG and the stage of pregnancy?
4. What are the reasons for using immobilized monoclonal antibodies at point D, even though they do not indicate whether a woman is pregnant or not?
5. Could a test be devised that generates two red bands in the test stick if a woman is pregnant, or one blue and one red band?
6. Could a "morning-after" pregnancy test be developed?

D3.1.18 Role of the placenta in foetal development inside the uterus

Humans are placental mammals. There are two other groups of mammals: the monotremes lay eggs and the marsupials give birth to relatively undeveloped offspring that develop inside a pouch. By the stage when a marsupial would be born, a human foetus has developed a relatively complex placenta and so can remain in the uterus for months longer. The placenta is needed because the body-surface-area-to-volume ratio becomes smaller as the foetus grows larger.

The placenta is made of foetal tissues, in intimate contact with maternal tissues in the uterus wall. The foetus also develops membranes that form the amniotic sac. This contains amniotic fluid, which supports and protects the developing foetus. The basic functional unit of the placenta is a finger-like piece of foetal tissue called a placental villus. These villi increase in number during pregnancy to cope with the increasing demands of the foetus for the exchange of materials with the mother.

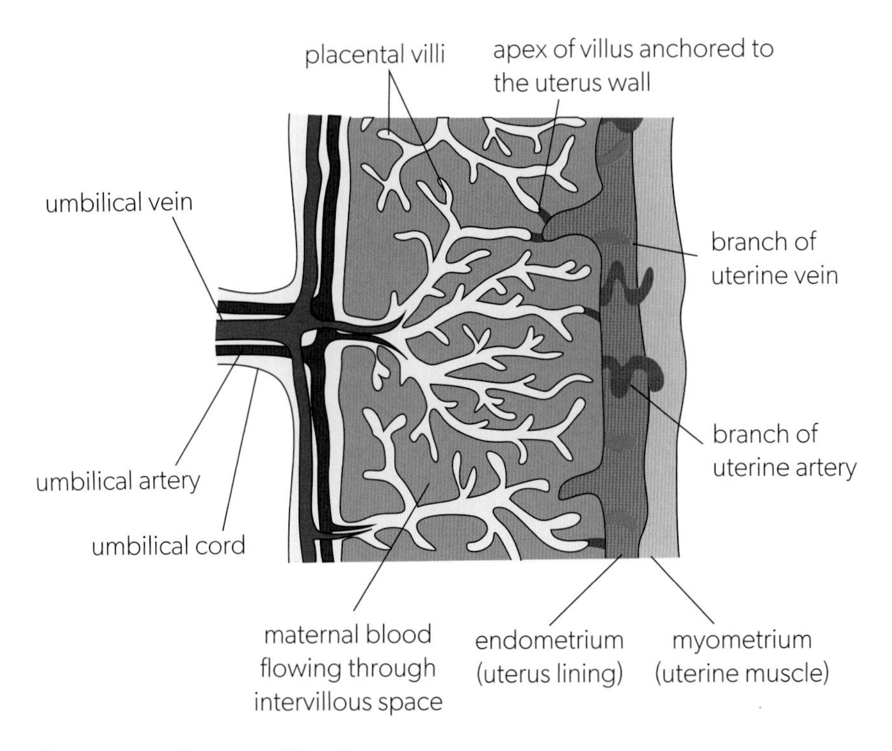

▲ Figure 36 Structure of the placenta

Maternal blood flows in spaces around the villi (intervillous spaces). This is a very unusual type of circulation as elsewhere blood is almost always retained in blood vessels. Foetal blood circulates in blood capillaries that are close to the surface of each villus. The distance between foetal and maternal blood is therefore very small—as little as 5 μm. The cells that separate maternal and foetal blood form the placental barrier. This must be selectively permeable, allowing some substances to pass, but not others.

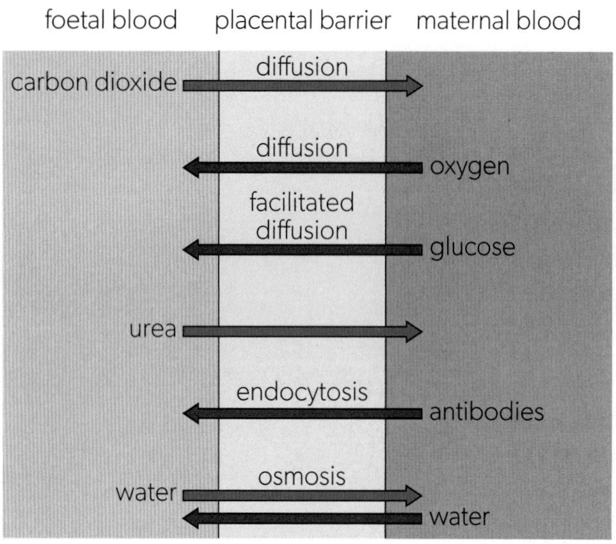

foetal blood | placental barrier | maternal blood

carbon dioxide — diffusion →
← diffusion — oxygen
← facilitated diffusion — glucose
urea →
← endocytosis — antibodies
water — osmosis →
← water

▲ Figure 37 Exchange processes in the placenta

Glucose, amino acids, oxygen and all other substances required by the foetus pass from maternal to foetal blood. Carbon dioxide, urea and other waste products pass from foetal to maternal blood. Diffusion, facilitated diffusion, osmosis and endocytosis are all used in these exchanges for specific substances.

D3.1.19 Hormonal control of pregnancy and childbirth

By about the ninth week of pregnancy, the placenta has started to secrete oestradiol and progesterone in large enough quantities to sustain the pregnancy, so the corpus luteum is no longer needed for this role. There is a danger of miscarriage at this stage of pregnancy if the switchover fails.

During pregnancy, progesterone inhibits secretion of oxytocin by the pituitary gland and inhibits contractions of the muscular outer wall of the uterus—the myometrium. At the end of pregnancy, hormones produced by the foetus signal to the placenta to stop secreting progesterone so oxytocin starts to be secreted.

Oxytocin stimulates contractions of the muscle fibres in the myometrium. These contractions are detected by stretch receptors, which signal to the pituitary gland to increase oxytocin secretion. Increased oxytocin makes the contractions more frequent and more vigorous, causing more oxytocin secretion. This is an example of positive feedback—a very unusual control system in human physiology. In this case, it has the advantage of causing a gradual increase in the myometrial contractions, allowing the baby to be born with the minimum intensity of contraction.

Relaxation of muscle fibres in the cervix causes the cervix to dilate. Uterine contraction then bursts the amniotic sac and the amniotic fluid passes out. Further uterine contractions, usually over hours rather than minutes, finally push the baby out through the cervix and vagina. The umbilical cord is broken and the baby takes its first breath and achieves physiological independence from its mother.

D3.1.20 Hormone replacement therapy and the risk of coronary heart disease

Hormone replacement therapy (HRT) is a treatment that is used to relieve menopausal symptoms. It supplements levels of the hormones oestrogen and progesterone that would naturally decrease as a woman approaches the menopause. HRT can help relieve most symptoms, such as hot flushes, night sweats, mood swings, vaginal dryness and reduced sex drive. Many menopausal symptoms pass after a few years, but they can be unpleasant and taking HRT can offer relief for many women. It can also help prevent weakening of the bones (osteoporosis), which is more common after the menopause.

It has been suggested that HRT might increase the risk of coronary heart disease (CHD). The causes and consequences of CHD are explained in *Section B3.2.6*.

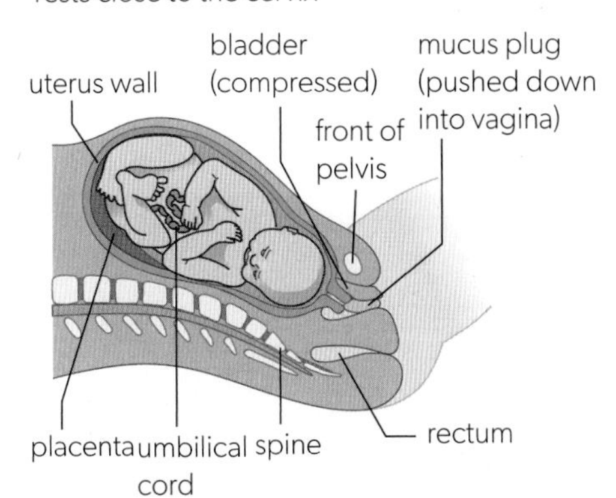

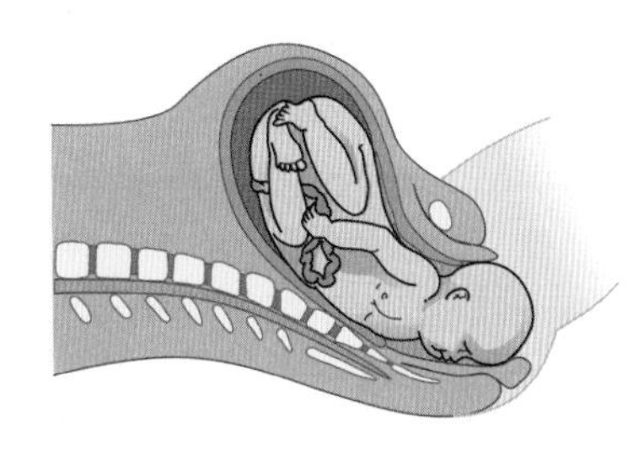

③ Baby is pushed out of mother's body

Figure 38 Stages in childbirth

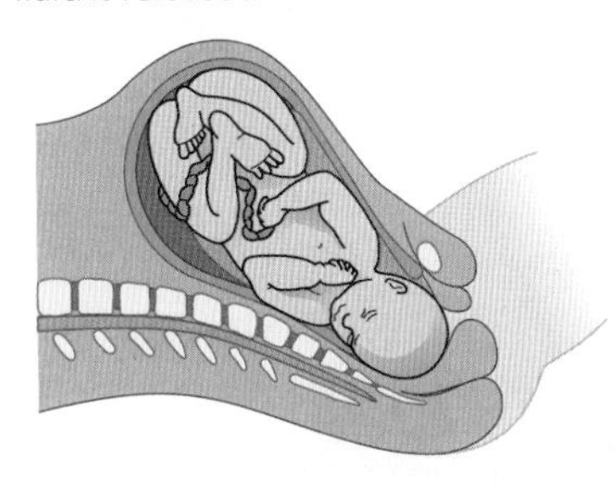

② Baby passes into vagina and amniotic fluid is released

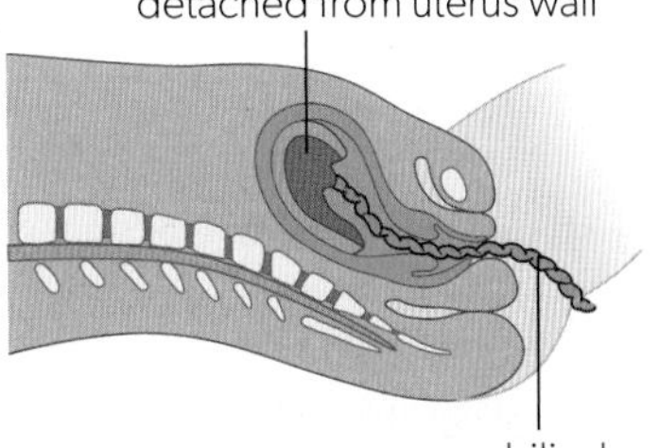

Correlations may be based on a causal relationship, but correlation does not prove causation

In early epidemiological studies, it was argued that women undergoing hormone replacement therapy (HRT) had reduced incidence of coronary heart disease (CHD) and this was deemed to be a cause-and-effect relationship. Later randomized controlled trials showed that use of HRT led to a small increase in the risk of CHD. The initial fallacy of a causal relationship between HRT and decreased incidence of CHD was coincidental because HRT patients come from groups with a higher socioeconomic status (Marks and Shinberg, 1998) and it is this that is thought to be causative of decreased CHD (Schultz et al., 2018).

Marks NF, Shinberg DS. (1998). Socioeconomic status differences in hormone therapy. *American Journal of Epidemiology*. *148(6)*, 581–593.

Schultz WM, et al. (2018). Socioeconomic status and cardiovascular outcomes: challenges and interventions. *Circulation*, *137(20)*, 2166–2178.

Linking questions

1. How can interspecific relationships assist in the reproductive strategies of living organisms?
   a. Distinguish between seed dispersal and pollination. (D3.1.12)
   b. With reference to an example of either seed dispersal or pollination, outline the concept of mutualism. (C4.1.12)
   c. Draw a representative animal-pollinated flower. (D3.1.9)
2. What are the roles of barriers in living systems?
   a. Explain the mechanisms that prevent polyspermy. (D3.1.15)
   b. Explain the mechanisms that prevent interspecific hybridization. (A4.1.10)
   c. Explain how the placenta prevents the mixing of maternal and foetal blood. (D3.1.18)

D3.2 Inheritance

What patterns of inheritance exist in plants and animals?

One type of inheritance pattern is called dominant and recessive. What are examples of this pattern?

"Sex-linked" inheritance involves different inheritance patterns in males and females. What are some examples of this? Sickle cell anaemia is a hereditary disorder, where red blood cells develop an abnormal sickle shape. This change in shape decreases the cell's flexibility and results in life- threatening complications. Because of a mutation in the haemoglobin gene individuals with one copy of the mutant gene produce a mixture of both normal and abnormal haemoglobin-containing red blood cells. Which type of inheritance pattern does this illustrate?

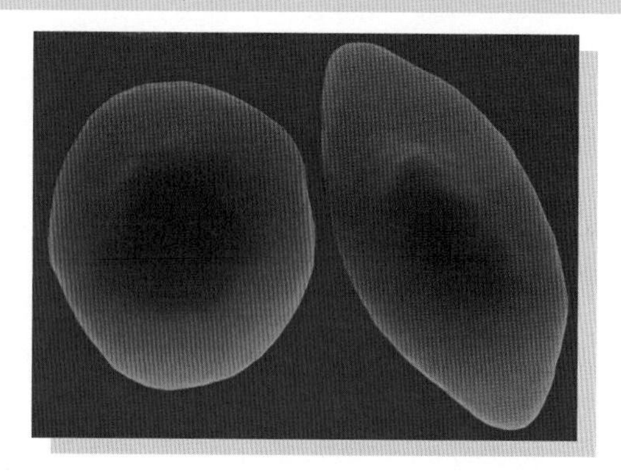

▲ Figure 1 Normal and sickle-shaped red blood cells

What is the molecular basis of inheritance patterns?

The expression of genes involves transcription of genes to produce mRNA which is then translated into a protein product. How does this relate to the dominance and recessive inheritance pattern? Albinism is a condition in which a gene for pigmentation is not expressed. Thus, it is a recessive condition. How does this relate to gene expression? Albinism is a relatively uncommon condition in animals. Are there examples where a dominant trait is relatively uncommon?

◀ Figure 2

SL and HL	AHL only
D3.2.1 Production of haploid gametes in parents and their fusion to form a diploid zygote as the means of inheritance D3.2.2 Methods for conducting genetic crosses in flowering plants D3.2.3 Genotype as the combination of alleles inherited by an organism D3.2.4 Phenotype as the observable traits of an organism resulting from genotype and environmental factors D3.2.5 Effects of dominant and recessive alleles on phenotype D3.2.6 Phenotypic plasticity as the capacity to develop traits suited to the environment experienced by an organism, by varying patterns of gene expression D3.2.7 Phenylketonuria as an example of a human disease due to a recessive allele D3.2.8 Single-nucleotide polymorphisms and multiple alleles in gene pools D3.2.9 ABO blood groups as an example of multiple alleles D3.2.10 Incomplete dominance and codominance D3.2.11 Sex determination in humans and inheritance of genes on sex chromosomes D3.2.12 Haemophilia as an example of a sex-linked genetic disorder D3.2.13 Pedigree charts to deduce patterns of inheritance of genetic disorders D3.2.14 Continuous variation due to polygenic inheritance and/or environmental factors D3.2.15 Box-and-whisker plots to represent data for a continuous variable such as student height	D3.2.16 Segregation and independent assortment of unlinked genes in meiosis D3.2.17 Punnett grids for predicting genotypic and phenotypic ratios in dihybrid crosses involving pairs of unlinked autosomal genes D3.2.18 Loci of human genes and their polypeptide products D3.2.19 Autosomal gene linkage D3.2.20 Recombinants in crosses involving two linked or unlinked genes D3.2.21 Use of a chi-squared test on data from dihybrid crosses

D3.2.1 Production of haploid gametes in parents and their fusion to form a diploid zygote as the means of inheritance

In a sexual life cycle, parents pass genes to offspring in gametes. This is the basis of inheritance. Gametes are haploid—they contain one chromosome of each type, whether they are male or female. This means male and female parents make an equal genetic contribution to their offspring.

When male and female gametes fuse, their nuclei join together, doubling the chromosome number. Thus, the nucleus of the zygote contains two chromosomes of each type—it is diploid. To produce a gamete, parents halve the chromosome number of their body cells from diploid to haploid. This is achieved by meiosis. During meiosis, a diploid nucleus divides twice to produce four haploid nuclei. A diploid nucleus contains two copies of each gene, but haploid nuclei contain only one.

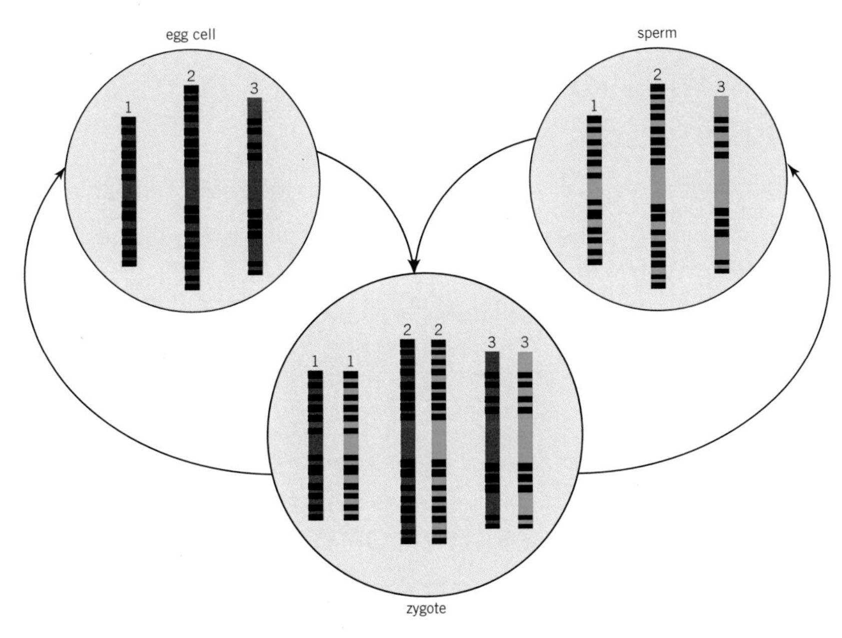
egg cell
1
2
3
sperm
1
2
3
1 1
2 2
3 3
zygote

◀ Figure 3 The sexual life cycle of the yellow fever mosquito (*Aedes aegypti*) is shown here. It has six chromosomes in its body cells and three in its egg and sperm. The human life cycle is the same, but with 46 and 23 chromosomes

D3.2.2 Methods for conducting genetic crosses in flowering plants

Patterns of inheritance can be investigated by crossing varieties of pea or other flowering plants. The method is essentially the same with all species. Pollen is transferred from the anthers of one plant (the male parent) to the stigmas of another plant (the female parent). A paint brush can be used to perform the cross by transfer of pollen or an anther with pollen can be dabbed directly onto the stigma.

Other sources of pollen must be prevented from reaching the stigma. This can be achieved by cutting off all the anthers of the same flower before their pollen becomes mature. The flower must then be enclosed in a paper bag, to prevent insects or wind transferring pollen to the stigma from other flowers on the same or other plants. Transfer of pollen from an anther on a plant to a stigma on the same plant is self-pollination and it results in self-fertilization, rather than a cross between two varieties.

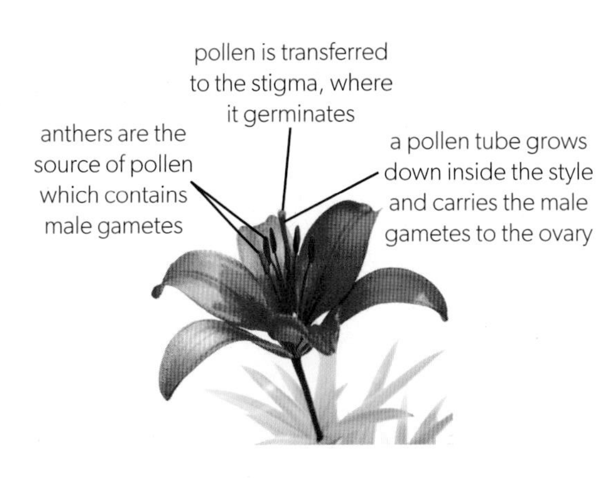
pollen is transferred to the stigma, where it germinates
anthers are the source of pollen which contains male gametes
a pollen tube grows down inside the style and carries the male gametes to the ovary

▲ Figure 4 Flower structure and pollination

If the pollen germinates and the male gametes are carried down to the ovary in a pollen tube, they can fuse with female gametes inside ovules, to form zygotes. Each zygote develops into an embryo inside a seed. The parents are known to geneticists as the P generation and the offspring inside the seeds as the F_1 (first filial) generation. Offspring of the F_1 are the F_2 (second filial) generation.

Gregor Mendel's crosses with pea plants were the first to reveal the basic pattern of inheritance. For example, he crossed tall plants with dwarf pea plants and white-flowered plants with purple-flowered plants. He counted the number of each variety in the offspring and calculated ratios. A table called a Punnett grid can be used to analyse the results of genetic crosses such as Mendel's. There are examples on the following pages. Quantifying results and analysing them mathematically is the practice today but was unusual for biologists in the middle of the 19th century.

Applying techniques: Pollination experiment

Rapidly cycling *Brassica rapa* plants can produce seeds after pollination in about 3 weeks. *B. rapa* are self-incompatible; that is, they have mechanisms which prevent self-fertilization.

A number of experiments can be undertaken.

- Are any mutant varieties of rapidly cycling *Brassica rapa* self-compatible?
- Can the incompatibility be overcome? For example, by removing the stigma and pollinating directly on to the cut surface of the style.
- Most vegetable varieties will self-pollinate. Can this be verified?

Wild plants often are not self-compatible though some are. For example, cleistogamous violets can be tested for self-compatibility.

D3.2.3 Genotype as the combination of alleles inherited by an organism

Different versions of a gene can exist. They are called **alleles** of the gene. Alleles may differ by as little as one base in the base sequence of the gene or by large sections. New alleles of a gene are generated by mutation. Humans and other diploid organisms have two alleles of most genes, one inherited from each parent. There could be two copies of one allele, or two different alleles. For example, for a gene with the alleles *D* and *d*, an individual could have *DD*, *dd* or *Dd*. Combinations of alleles such as these are known as **genotypes**.

If a parent's genotype is *DD*, all gametes produced by them will contain a single copy of allele *D*. Similarly, all the gametes produced by a parent with the genotype *dd* will have one allele *d*. Individuals with the genotypes *DD* and *dd* are **homozygous** because all the gametes they produce have the same allele of this gene.

If a parent's genotype is *Dd*, 50% of their gametes will have allele *D* and 50% will have allele *d*. Individuals with the genotype *Dd* are **heterozygous**, because they produce gametes with different alleles of the gene.

D3.2.4 Phenotype as the observable traits of an organism resulting from genotype and environmental factors

The phenotype of an organism is its observable traits or characteristics. Phenotype includes structural characteristics such as whether hair is curly or straight and functional traits such as the ability to distinguish red and green colours. "Observable" means that the trait is visible or detectable with tests (Table 1). Most phenotypic traits are due to the interaction between the genotype of an organism and the environment in which it exists, but there are some determined solely by genotype and some solely by environmental factors (Figure 5).

▲ Figure 5 Phenotypic traits determined by environment are not heritable

Genotype only	Environment interacting with genotype	Environment only
▲ Figure 6 ABO blood groups	▲ Figure 7 Skin colour	▲ Figure 8 Languages
• eye colour—brown or blue/grey • haemophilia—blood slow to clot • ability to smell β-ionone—an odorant in violets	• height in humans • autism—a personality trait • diabetes—failure to regulate blood glucose concentration	• scars due to surgery or wounds • river blindness (onchocerciasis) • body art such as tattoos and piercings

▲ Table 1 Determining factors of some phenotypic traits

D3.2.5 Effects of dominant and recessive alleles on phenotype

Gregor Mendel is regarded by most biologists as the father of genetics. He used pea plants (*Pisum sativum*) for his research. He chose clear characteristics such as red or white flower colour that can easily be followed from one generation to the next. Pea plants can be crossed to produce hybrids between two varieties or they can be allowed to self-pollinate.

Mendel investigated the inheritance of seven traits which each had two forms — for example, height (tall or dwarf) and seed shape (smooth or wrinkled). He observed that if allowed to self-pollinate, a pea plant's offspring had the same traits as its parent. Mendel developed many different "pure-breeding" varieties that produced offspring with the same traits, generation after generation, if they were self-pollinated. We now know that this is because the varieties were homozygous for all their genes.

Mendel then used cross-pollination to produce hybrid offspring of pairs of varieties. For example, he crossed a tall variety of pea with a dwarf variety. He also crossed a wrinkled-seeded with a smooth-seeded variety. We might expect the offspring to be intermediate between the two parents. This would be called blending inheritance. It is not what Mendel found.

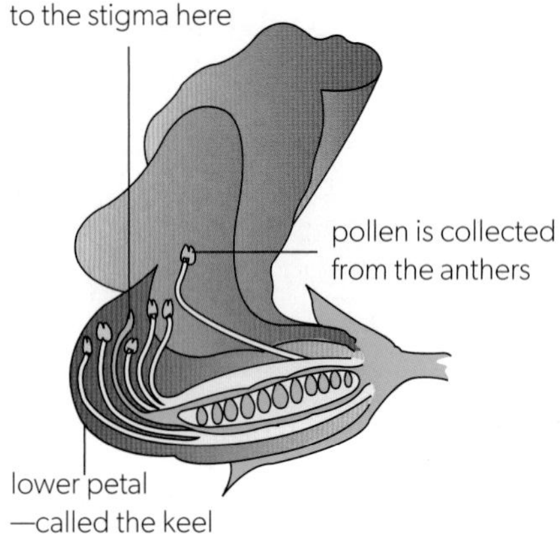

▲ Figure 9 Cross and self-pollination in pea plants

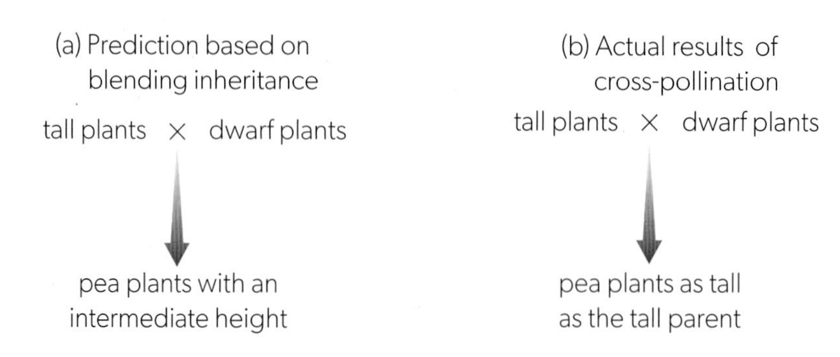

▲ Figure 10 Prediction based on blending (a), actual results (b) inheritance

In each of Mendel's crosses between different varieties of pea plant, the offspring were not intermediate in character and instead had the character of one of the parents, not the other. For example, in a cross between a tall pea plant and a dwarf pea plant, all the offspring (F_1 generation) were tall. Mendel allowed the F_1 pea plants to self-pollinate. In the offspring (F_2 generation) there were again no intermediates, but the parental character that had disappeared in the F_1 generation reappeared in a quarter of the pea plants. In the cross between tall and dwarf pea plants, there was a 3:1 ratio between tall and dwarf plants in the F_2 generation.

Mendel used algebra to explain this pattern of inheritance. His algebraic symbols are now known to correspond with the two alleles of a gene that influences height in pea plants. The alleles have the symbols *T* and *t*.

- A pure-breeding tall pea plant has two copies of the allele *T*, so it is homozygous with the genotype *TT*.
- A pure-breeding dwarf pea plant has two copies of the allele *t*, so it is homozygous with the genotype *tt*.
- Parents pass on one allele of the gene to offspring, so all the offspring of a cross between tall and dwarf plants have the heterozygous genotype *Tt*.
- When *T* and *t* are combined in one individual, it is the allele *T* that determines the height. We say that the allele for tallness is **dominant**.
- The other allele, which does not have an effect if the dominant allele is present, is **recessive**.
- When the heterozygous F_1 plants reproduce, they pass on either *T* or *t* to their offspring. 50% of offspring inherit *T* and 50% inherit *t*.
- With self-pollination there are four possible outcomes. These can be shown using a Punnett grid. Because *T* is the dominant allele, three of the outcomes result in a tall plant and one in a dwarf.

female gametes \ male gametes	T	t
T	TT tall	Tt tall
t	tT tall	tt dwarf

▲ Figure 11 A Punnett grid

In most cases, genes code for polypeptides, and recessive alleles of genes are produced by mutation from dominant alleles. A mutation alters the base sequence of a gene and therefore the amino acid sequence of a polypeptide. If this change causes the polypeptide to function less efficiently or not function at all, there could be an effect on the phenotype. This could be a mild effect or a more serious trait such as a genetic disease.

However, if one copy of the original unmutated allele of the gene is present, the polypeptide coded for by the gene can still be produced in its fully functional form and there may be no change to the phenotype. If the polypeptide is an enzyme, only a small quantity may be enough to catalyse the reaction at a fast enough rate for the phenotype to be unaffected. This is not the only possible cause of dominance and recessiveness but it is probably the commonest.

Monohybrid crosses only involve one character, for example the height of a pea plant, so they involve only one gene. Most crosses start with two pure-breeding parents. This means that the parents have two of the same allele, not two different alleles. Each parent therefore produces just one type of gamete, containing one copy of the allele.

Their offspring are also identical, although they have two different alleles. The offspring obtained by crossing the parents are called F_1 hybrids or the F_1 generation.

The F_1 hybrids have two different alleles of the gene, so they can each produce two types of gamete. If two F_1 hybrids are crossed together, or if an F_1 plant is allowed to self-pollinate, there are four possible outcomes. This can be shown using a 2×2 table, called a Punnett grid after the geneticist who first used this type of table. The offspring of a cross between two F_1 plants are called the F_2 generation.

To make a Punnett grid as clear as possible the gametes should be labelled and both the alleles and the character of the four possible outcomes should be shown on the grid. It is also useful to give an overall ratio below the Punnett grid.

D3.2.6 Phenotypic plasticity as the capacity to develop traits suited to the environment experienced by an organism, by varying patterns of gene expression

Organisms can respond to their environment by varying their patterns of gene expression and therefore their traits. This is a form of adaptation, but it is reversible because genes have only been switched on or off, not changed into new alleles. It is known as phenotypic plasticity and is particularly useful if the environment a population inhabits is heterogeneous rather than uniform.

For example, a person with pale skin may become darker-skinned if there is an increase in the intensity of sunlight or more time is spent with skin exposed to the sunlight. A change in gene expression results in increased synthesis of the black pigment melanin in the skin. If the sunlight stimulus diminishes, gene expression reverts to its former pattern and the skin gradually becomes paler again, as the melanin concentration reduces.

In some cases, phenotypic plasticity is a switch between two or more alternative forms and/or it cannot be reversed during the lifetime of the individual.

Activity: Plasticity in peas

Plants show phenotypic plasticity during their growth after germination.

You will need:

- dried peas—you can get them a food store or a garden seed company
- cardboard tubes—for example from toilet rolls
- plastic wrap
- compost.

1. Cover the sides and base of the cardboard tubes with the two or three layers of plastic wrap
2. Sow four dry pea seeds in compost in each tube at different depths:
   a. 20 mm below the top of the compost in two of the tubes
   b. halfway down in one tube
   c. 20 mm from the bottom of one tube.
3. Put each tube into a container that will support it and allow you to water the compost. A mug or a tumbler would be suitable. Water the compost thoroughly.
4. Put one of the tubes with shallowly sown seeds, in its container, into a dark cupboard
5. Put the other three tubes on a sunny window sill or other light place.
6. See what effect the different environments have on the development of the pea seedlings.

▲ **Figure 12** Pumpkin seedlings of the same variety but grown in the light and in the dark

parental genotypes: *Aa* × *Aa*

gametes: *A*, *a*, *A*, *a*

AA not carrier; *Aa*, *aA* carrier; *aa*

AA, *Aa*, *aA*: do not develop the disease

aa: disease developed

▲ **Figure 13** Inheritance of a genetic disease

D3.2.7 Phenylketonuria as an example of a human disease due to a recessive allele

A genetic disease is an illness that is due to a gene. Most genetic diseases are caused by a recessive allele of a gene. The disease therefore only develops in individuals with two copies of the recessive allele. A person with one recessive allele and one dominant allele will not show symptoms of the disease but can pass on the disease-causing recessive allele to their offspring. These individuals are called carriers.

Genetic diseases caused by a recessive allele usually appear unexpectedly. Both parents of a child with the disease must be carriers, but as they do not show symptoms of the disease they are usually unaware of this. The probability of two such parents having a child with the disease is 1 in 4 or 25%.

Phenylketonuria (PKU) is an example of a disease due to a recessive allele. The allele is produced by mutation of the gene coding for the enzyme phenylalanine hydroxylase. This enzyme converts phenylalanine into tyrosine. Both amino acids are used by cells to make proteins. The PKU allele is produced by mutation and is recessive because a carrier with one PKU allele can still produce the functioning enzyme because they have a normal allele. A person with two recessive alleles of the PKU gene cannot produce any functioning enzyme. Phenylalanine therefore accumulates in the body and tyrosine deficiency is likely to develop. In excess, phenylalanine impairs brain development, leading to intellectual disability and mental disorders. This can be prevented by screening for PKU at birth and giving affected children a diet low in phenylalanine.

The gene for phenylalanine hydroxylase is located on chromosome 12 which is an autosome—a non-sex chromosome. Boys and girls therefore have the same 25% chance of being born with PKU if both of their parents are carriers of the recessive allele.

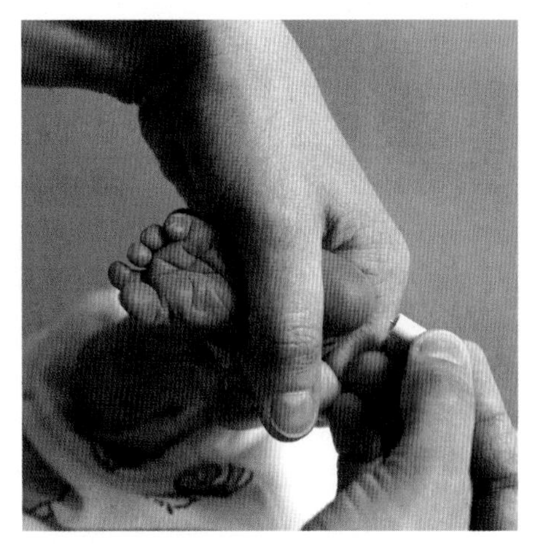

▲ Figure 14 Blood is taken from a newborn baby so phenylalanine concentration can be measured as a test for PKU

D3.2.8 Single-nucleotide polymorphisms and multiple alleles in gene pools

A gene pool is all the genes of all the individuals in a sexually reproducing population. Every new individual gets a selection of genes from the gene pool. Evolution is changes in the gene pool over time.

A gene is a length of DNA, with a base sequence that can be hundreds or thousands of bases long. The different alleles of a gene have slight variations in base sequence. Usually only one or a very small number of bases are different, for example adenine might be present at a particular position in one allele and cytosine at that position in another allele.

Positions in a gene where different bases can be present are called single-nucleotide polymorphisms (abbreviated to SNPs and pronounced snips). Even within one gene, there can be many different positions with SNPs. There can therefore be many different alleles of a gene in the gene pool. This is called multiple alleles. Each individual receives a maximum of two different alleles from the gene pool.

S-gene in apples—an example of multiple alleles

The S-gene is part of a system that prevents self-pollination and reduces the chance of inbreeding. Thirty-two different S-alleles have been discovered in the apple gene pool, numbered S_1, S_2, S_3 and so on. Each diploid apple tree has two copies of the S-gene and pollen grains produced by the tree each contain one of these. When pollen germinates on the stigma of a flower on another tree, it is rejected if any of that tree's S-gene alleles are the same as the pollen.

For example, the apple varieties Jazz and Golden Delicious both have the genotype S_2S_3 so 50% of their pollen grains will carry the S_2 allele and 50% will carry S_3. All the pollen of Jazz is rejected if transferred to flowers on a Golden Delicious tree and vice versa. The variety Gala has genotype S_2S_5 so pollen from Jazz or Golden Delicious carrying S_2 is rejected, but 50% carries S_5 and can successfully fertilize Gala, so fruit should develop. Cox's Orange Pippin has the genotype S_5S_9 so all the pollen from Golden Delicious and Jazz can achieve fertilization and an abundant crop of apples should develop.

▲ Figure 15 Apple fruits only develop after pollination if the pollen has a different S-allele from those in the tree's genotype.

D3.2.9 ABO blood groups as an example of multiple alleles

The ABO blood group system in humans is an example of multiple alleles. It is of great medical importance. Before blood is transfused, it is vital to find out the blood group of the patient and ensure that it is matched to the blood to be given to that patient. Unless this is done, there may be complications due to coagulation of red blood cells. One gene determines the ABO blood group of a person. There are three alleles of the gene: I_B I_A and i. There are therefore six possible genotypes, but only four phenotypes (Table 2).

Genotype	I_AI_A	I_Ai	I_BI_B	I_Bi	ii	I_AI_B
Phenotype	Group A	Group A	Group B	Group B	Group O	Group AB

▲ Table 2 Human blood groups and their genotypes

From the pattern of genotypes and blood groups we can deduce that allele I_A is dominant over allele i and that allele I_B is dominant over allele i. However, neither I_A nor I_B is dominant over the other allele and a person with the genotype I_AI_B has a blood group called AB. This is an example of **codominance** of alleles and is described more fully in the next section.

The reasons for two blood group alleles being codominant and the other allele being recessive are as follows.

- All three alleles cause the production of a glycoprotein in the membrane of red blood cells.
- I_A alters the glycoprotein by addition of acetylgalactosamine. This altered glycoprotein is absent from people who do not have the allele I_A so if exposed to it they make anti-A antibodies.
- I_B alters the glycoprotein by addition of galactose. This altered glycoprotein is not present in people who do not have the allele I_B so if exposed to it they make anti-B antibodies. The genotype I_AI_B causes the glycoprotein to be altered by addition of both acetylgalactosamine and galactose. As a consequence, neither anti-A nor anti-B antibodies are produced. This genotype therefore gives a different phenotype to I_AI_A and I_BI_B so the alleles I_A and I_B are codominant.
- The allele i is recessive because it does not cause the basic glycoprotein to be modified, but if either I_A or I_B are also present they do cause modification, so I_Ai gives the same phenotype as I_AI_A and I_Bi gives the same as I_BI_B.

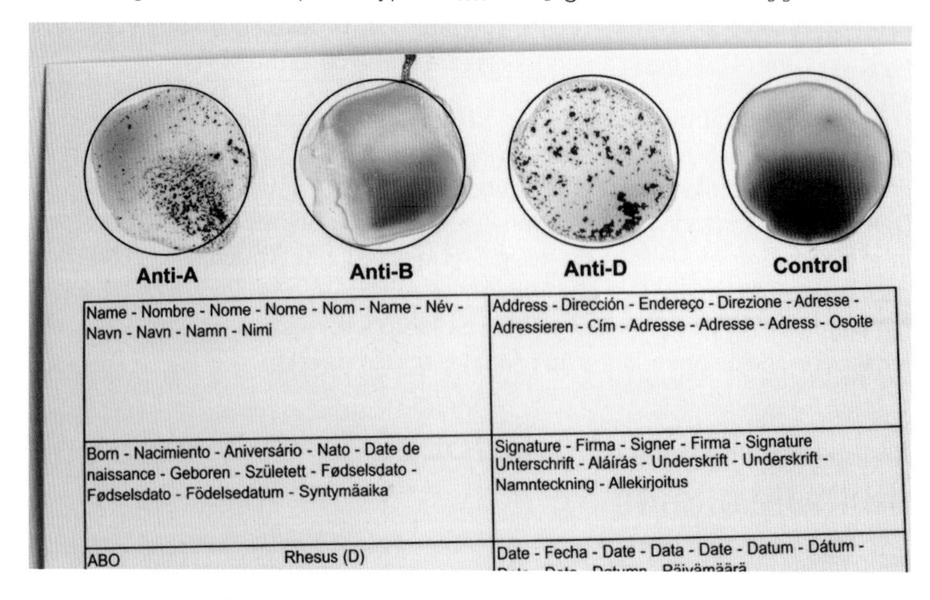

▲ Figure 16 Clotting with anti-A but not anti-B antibodies on this test card indicates that the blood is Type A. Anti-D antibodies test whether the blood is Rhesus-positive or Rhesus-negative. This blood is Rhesus-positive

Activity: Donating blood

Donating blood is an altruistic act that many citizens choose to do. In most states of the USA, you must be at least 17 years old (or 16 if have your parents' permission). Your blood must be free of any blood-borne diseases such as hepatitis or HIV. It is tested to find out your blood group, so it is only transfused into people with a compatible group. You can therefore discover your blood group by becoming a donor.

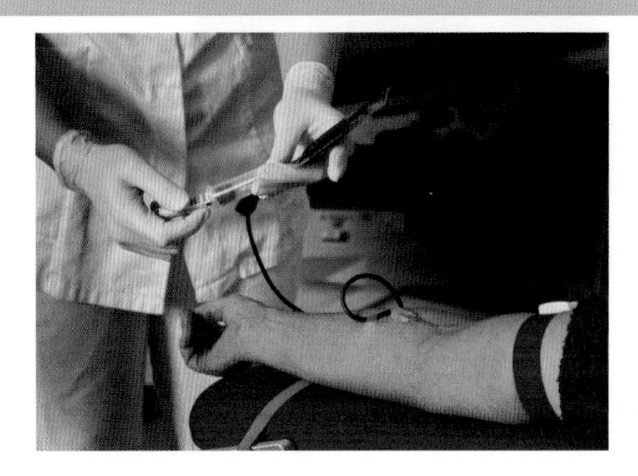

▲ Figure 17 Small samples of blood are taken from a donor for testing before the main volume of about 450 ml is taken

D3.2.10 Incomplete dominance and codominance

In each of Mendel's crosses between varieties of pea plant, one of the alleles was dominant and the other was recessive. However, some genes have pairs of alleles where neither allele is fully dominant over the other.

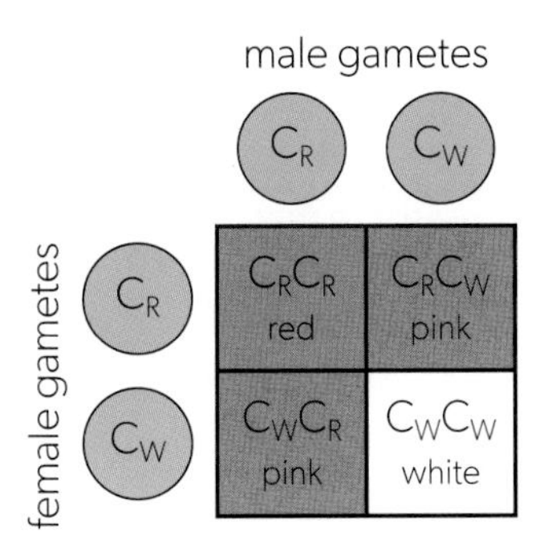

▲ Figure 18

▲ Figure 19 Red, white and pink flowers of *Mirabilis jalapa*

The ABO blood group is an example. Blood group AB is a dual phenotype and is not intermediate between group A and group B. This pattern of gene expression is called codominance.

A different pattern is seen in some other cases, such as flower colour in *Mirabilis jalapa*. If a red-flowered plant is crossed with a white-flowered plant, the offspring have pink flowers. There are two alleles of the gene for flower colour: C_R is the allele for red flowers and C_W is the allele for white flowers. A plant with the genotype C_RC_W has pink flowers, which is an intermediate phenotype (Figure 19). This pattern is known as **incomplete dominance**.

Cells in the petals of *Mirabilis jalapa* produce no red pigment if they have the genotype C_WC_W, some red pigment with the genotype C_RC_W and more with C_RC_R. The expected ratio of flower colours from a cross between two pink-flowered plants can be determined using a Punnett grid. Similarly, the differences in pigmentation of Icelandic horses can be explained in terms of the amount of pigment produced in hairs forming the coat (Figure 20).

▲ Figure 20 The alleles of a gene for coat colour in Icelandic horses show incomplete dominance

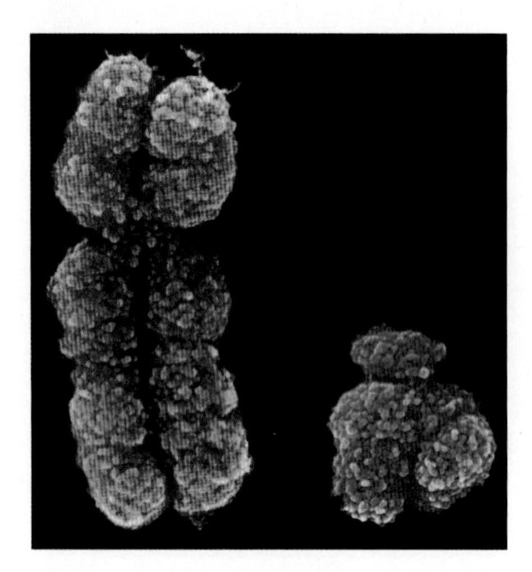

▲ Figure 21 X and Y chromosomes have different base sequences apart from two small regions called PAR1 and PAR2 which are at the tips of the X and Y chromosomes. These homologous regions are needed for pairing of the sex chromosomes during meiosis in males. What are the consequences of the sex chromosomes not separating during meiosis in males?

D3.2.11 Sex determination in humans and inheritance of genes on sex chromosomes

Sex is determined in humans by the 23rd pair of chromosomes. There are two types of sex chromosome.

- The X chromosome is relatively large and has its centromere near the middle. It contains about 900 genes, many of which are essential in both males and females. All humans must therefore have at least one X chromosome.
- The Y chromosome is much smaller and has its centromere near one end. It has about 55 genes, many of which are unique to the Y chromosome and are not needed for female development.

Females typically have a pair of X chromosomes whereas males have one X and one Y chromosome. The other 22 pairs of chromosomes are autosomes (non-sex chromosomes) and do not affect whether a foetus develops as a male or female.

Because adult females have two X chromosomes, female gametes typically contain one X chromosome. All offspring inherit an X chromosome from their mother. Adult males pass on either their X or their Y chromosome in sperm. Daughters inherit their father's X chromosome and sons inherit his Y chromosome. The sex of the offspring of a human is therefore determined by the sperm.

Sex chromosome abnormalities are relatively common in humans. An individual with two X chromosomes and one Y develops as a boy with Klinefelter's syndrome. This demonstrates that it is the presence of a Y chromosome rather than the absence of a second X chromosome that causes male development. An individual with one X and no other sex chromosome develops as a girl with Turner's syndrome, showing that it is the absence of a Y chromosome, not the presence of a second X chromosome that causes female development.

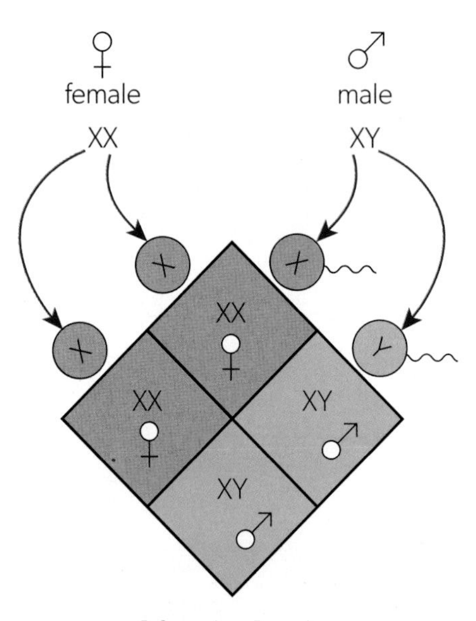

▲ Figure 22 Sex determination

A pair of gonads starts to develop in human embryos when they are about 5 weeks old. Initially they could become either testes or ovaries.

Male—One key gene on the Y chromosome called SRY causes the embryonic gonads to develop into testes. This gene is therefore known as the testis determining factor (TDF). The developing testes start to secrete testosterone, the hormone that causes development of other male organs, so a foetus with one X and one Y chromosome has a male reproductive system and will restart secretion of male sex hormone at puberty.

Female—In an embryo without a Y chromosome and therefore no TDF gene, the fall-back position is that the gonads develop into ovaries. The embryonic ovaries start to secrete oestradiol, which causes a female reproductive system to develop. Female sex hormones will be secreted from puberty onwards.

An inheritance pattern called sex linkage was discovered by American geneticist Thomas Morgan in the fruit fly, *Drosophila*. This insect is about 4 mm long and completes its life cycle in two weeks, allowing crossing experiments to be done quickly with large numbers of flies. Most crosses in *Drosophila* do not show sex linkage.

For example, these reciprocal crosses give the same results:

normal-winged ♂ × vestigial-winged ♀

vestigial-winged ♂ × normal-winged ♀

These crosses between red-eyed and white-eyed flies gave different results:

♂ white × ♀ red → all red-eyed offspring

♂ red × ♀ white → ♀ red and ♂ white

Sex determination in *Drosophila* is similar to humans, with XX females and XY males. Morgan deduced that sex linkage of eye colour could therefore be due to a gene located only on the X chromosome, with a dominant allele for red eyes and recessive allele for white. The pattern of inheritance can be shown using Punnett grids. In crosses involving sex linkage, the alleles should always be shown as a superscript letter on a letter X to represent the X chromosome. The Y chromosome should also be shown although it does not carry an allele of the gene.

It is important not to confuse sex linkage and gene linkage. Gene linkage is caused by two or more genes being located close together on a chromosome, so alleles of them tend to be inherited together rather than independently. It is possible for two genes to be both linked and sex-linked if they are located close together on the X chromosome.

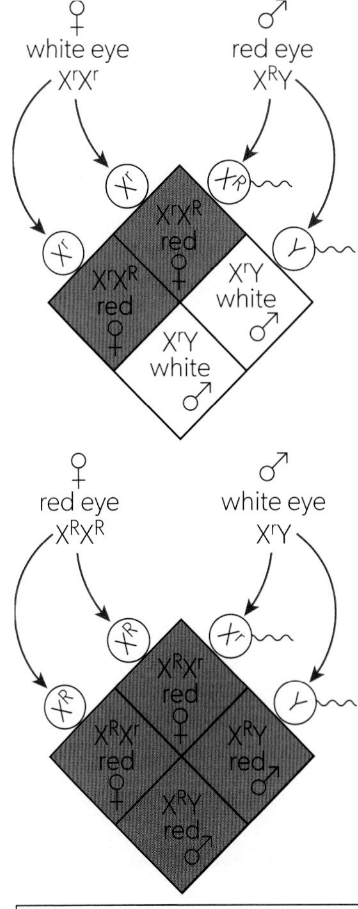

▲ Figure 23 Reciprocal sex-linkage crosses

D3.2.12 Haemophilia as an example of a sex-linked genetic disorder

Many examples of sex-linked genetic disorders have been discovered in humans. Most of these are due to recessive alleles of genes. Sex-linked genes are almost always located on the X chromosome because there are very few genes on the Y chromosome. As males only have one copy of genes on the X chromosome, they have the disorder if this one copy is the recessive allele. Females are much less likely to be affected because as long as one of their two X chromosomes carries the dominant allele, they are unaffected.

Haemophilia is an example of this pattern of sex-linked inheritance. People with this disorder, usually males, either lack or have a defective form of Factor VIII. This protein is a clotting factor that normally circulates in the blood. Cuts and other wounds bleed for much longer than normal in people with haemophilia and internal bleeding is common. Life expectancy is about 10 years if haemophilia is untreated. Treatment is by infusing Factor VIII, purified from the blood of donors.

The gene for Factor VIII is located on the X chromosome. The allele that causes haemophilia is usually considered to be recessive. The frequency of the allele is about 1 in 10,000. This is therefore the frequency of the disease in boys. Females can be carriers of the haemophilia allele, but in most cases do not have symptoms. They always develop the disease in the rare cases where both of their X chromosomes carry the allele. The frequency in girls theoretically is $(1/10{,}000)^2 = 1$ in 100,000,000. In practice, there have been even fewer cases of girls with haemophilia due to lack of Factor VIII than this. One reason is that the father would have to be haemophiliac and decide to risk passing on the condition to his children.

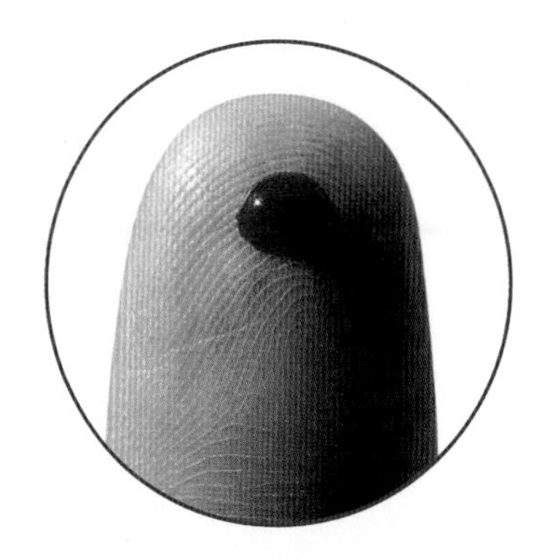

▲ Figure 24 Blood should stop flowing quickly from a pricked finger but in haemophiliacs bleeding continues for much longer as blood does not clot properly

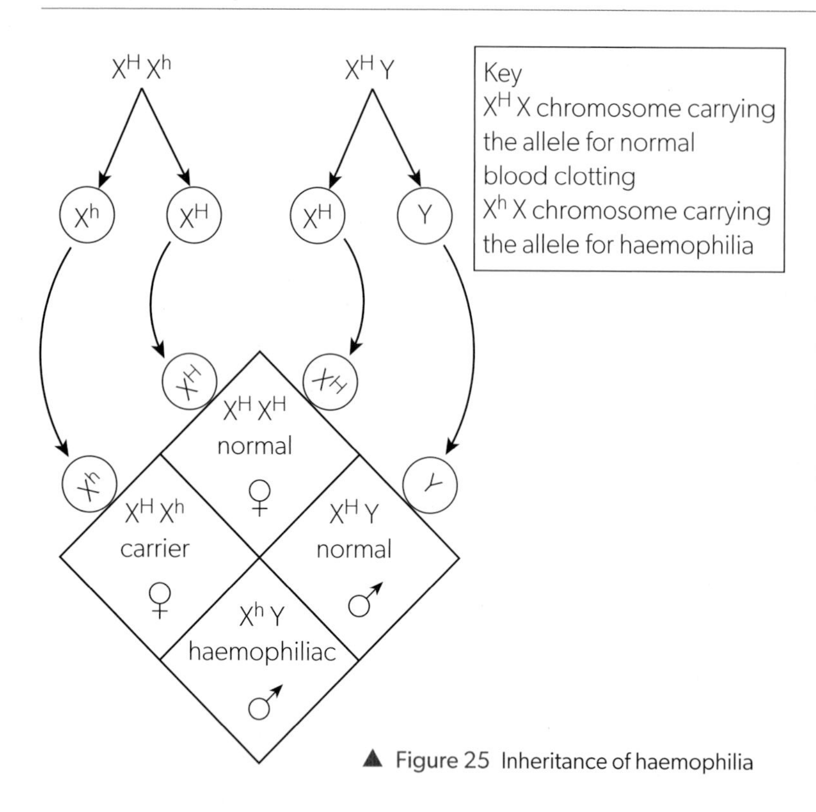

Key
X^H X chromosome carrying the allele for normal blood clotting
X^h X chromosome carrying the allele for haemophilia

▲ Figure 25 Inheritance of haemophilia

▲ Figure 26 A person with red-green colour-blindness cannot clearly distinguish between the colours of the flowers and the leaves. Red–green colour blindness is a sex-linked disorder due clearly distinguish to a recessive allele. In the Syed population in North India, the frequency in males is 11.4%. What is the expected frequency in females?

Data-based questions: Deducing genotypes from pedigree charts

This pedigree chart shows five generations of a family affected by a genetic disease.

1. Explain, using evidence from the pedigree chart, whether the condition is due to a recessive or a dominant allele. [3]
2. Explain what the probability is of the individuals in generation V having:
   a. two copies of a recessive allele [1]
   b. one recessive and one dominant allele [1]
   c. two copies of the dominant allele. [1]
3. Deduce, with reasons, the possible alleles of:
   a. 1 in generation III [1]
   b. 13 in generation II. [1]
4. Suggest two examples of genetic diseases that would fit this inheritance pattern. [2]

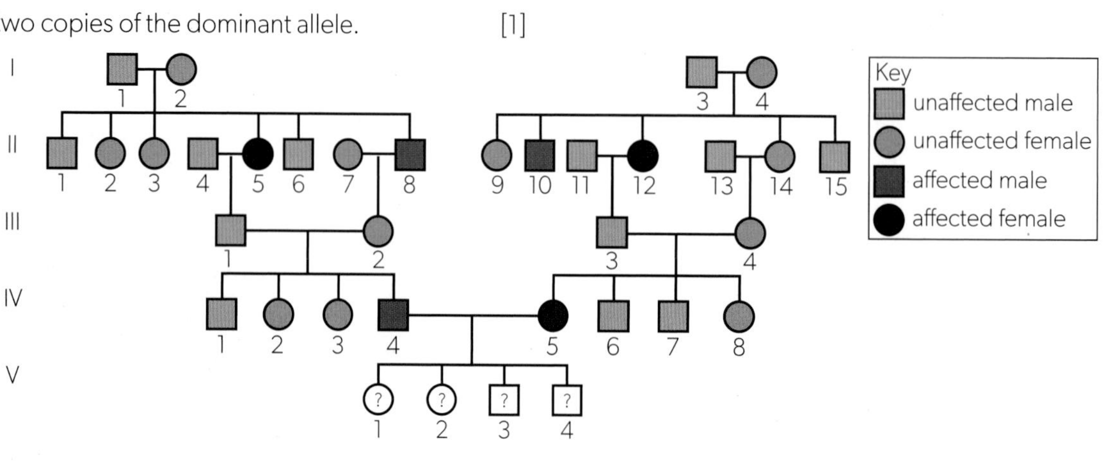

▲ Figure 27

D3.2.13 Pedigree charts to deduce patterns of inheritance of genetic disorders

Medical research has already identified more than 4,000 genetic disorders and no doubt more remain to be discovered. Given this large number, it might seem surprising that most of us do not suffer from any of these conditions. The reason is that most genetic disorders are caused by very rare recessive alleles. The chance of inheriting one allele for any specific disease is small but to develop the disorder two alleles must be inherited, one from each parent, and the chance of this is extremely small.

It is now possible to sequence the genome of an individual human cheaply and quickly. Many such sequences are now available and can be compared. This research has revealed that a typical individual carries at least five rare recessive alleles that could cause a genetic disorder, and possibly many more. But a couple can only produce a child with a genetic disease due to one of these recessive alleles if both parents of the child have the same rare allele. This is far more likely in children of a marriage between two close relatives. The same allele is inherited twice from a grandparent or earlier ancestor. Marriage between siblings, first cousins or other close relatives has been prohibited in many societies from before the time of genetic research. This is probably because it had been observed that inbreeding raised the chance of disorders in offspring.

It is not possible to investigate the inheritance of genetic disorders in humans by carrying out genetic crosses. Pedigree charts (family tree diagrams) can be used instead to deduce the pattern of inheritance. These are the usual conventions for constructing pedigree charts:

- males are shown as squares and females are shown as circles
- squares and circles are shaded or crosshatched to indicate that an individual is affected by the disease
- parents and siblings are linked with horizontal lines and parents and their offspring with vertical lines
- Roman numerals indicate generations and Arabic numbers are used for individuals in each generation.

Patterns and trends

Scientists draw general conclusions by inductive reasoning when they base a theory on observations of some but not all cases. For example, Mendel is likely to have noticed a pattern that pure-breeding plants for the seven traits he studied produced 100% of one phenotype in the F_1 generation and when the F_1 generation was self-crossed he observed a pattern in the F_2 generation that he generalized as a 3:1 ratio. Observations lead to pattern recognition and the formation of a generalization. This involves inductive reasoning. Testing a generalization involves deductive reasoning.

A pattern of inheritance may be deduced from parts of a pedigree chart and this theory may then allow genotypes of specific individuals in the pedigree to be deduced. You should be able to distinguish between inductive and deductive reasoning.

Example 1: Albinism in humans

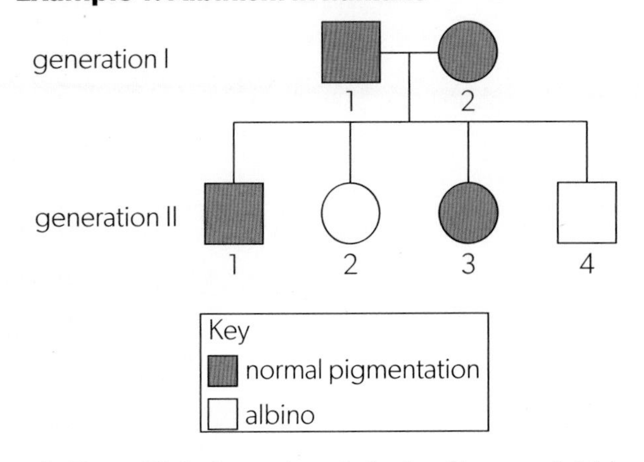

▲ **Figure 28** Pedigree chart of a family with cases of albinism

Deductions

- Two of the children are albino and yet the parents both have normal pigmentation. This suggests that albinism is caused by a recessive allele (*m*) and normal pigmentation by a dominant allele (*M*).
- There are both daughters and sons with albinism which suggests that the condition is not sex-linked. Both males and females are albino only if they have two copies of the recessive albinism allele (*mm*).
- The albino children must have inherited an allele for albinism from both parents.
- Both parents must also have one allele for normal pigmentation because they are not albino. The parents therefore have the genotype *Mm*.
- The chance of a child of these parents having albinism is 1 in 4 (25%). However, we can only be sure that 25% will be albino if the parents have a very large number of children. The actual ratio of 1 in 2 is not unexpected and does not show that our deductions about the inheritance of albinism are incorrect.

Example 2: Vitamin D-resistant rickets

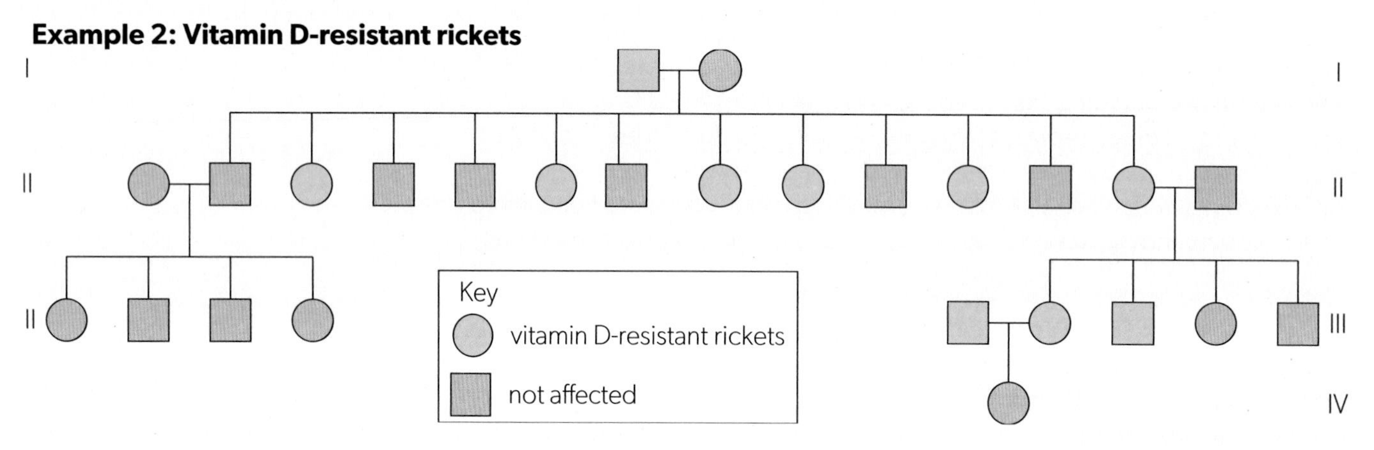

▲ **Figure 29** Pedigree chart of a family with cases of vitamin D-resistant rickets

Deductions

- Two unaffected parents only have unaffected children. Two affected parents have an unaffected child. This suggests that the disease is caused by a dominant allele.
- In generation II, the offspring of generation parents I are all affected daughters and unaffected sons. This suggests sex linkage, although the number of offspring is too small to be sure of the inheritance pattern.
- If vitamin D-resistant rickets is caused by a dominant X-linked allele, daughters of the father in generation I would inherit his X chromosome carrying the dominant allele, so all of his daughters would have the disease. The data in the pedigree shows this and so supports the theory.
- Similarly, if vitamin D-resistant rickets is caused by a dominant X-linked allele, the mother with the disease in generation II would have one X chromosome carrying the dominant allele for the disease and one with the recessive allele. All her offspring would have a 50% chance of inheriting this X chromosome and of having the disease. The data in the pedigree fits this and so supports the theory.

D3.2.14 Continuous variation due to polygenic inheritance and/or environmental factors

Variation is one of the defining features of life. Two individual organisms that are in the same species have fundamental features in common but are also likely to differ in many traits. There are two types of variation: continuous and discrete. Skin colour is an example of a continuous variation and ABO blood groups are an example of a discrete variation. Table 3 shows differences between continuous and discrete variation.

Continuous variation	Discrete variation
A continuous range of types is possible, with no distinct categories so there are many possible phenotypes. ▲ Figure 30 The spectrum of visible light is an example of a continuum, with any wavelength from 400 nm to 700 nm possible	Separate categories of variant, with no intermediates between them, so there are few possible phenotypes. α β γ δ ε ζ η θ ι κ λ μ ν ξ ο π ρ ς σ τ υ φ χ ψ ω ▲ Figure 31 The Greek alphabet is an example of disjunction, with 25 letters and no intermediates
The trait is influenced by multiple genes if there is a genetic cause.	The trait is influenced by just one or at most a few genes if there is a genetic cause.
Environmental factors may influence the trait.	Environmental factors do not usually influence the trait.
Examples: tree height, body mass of animals, human wrist circumference and skin colour	Examples: ABO blood groups, number of eggs laid by birds and left/right-handed snail shells

▲ Table 3 Comparison of continuous and discrete variation

The seven traits in peas that Mendel investigated are all influenced by single genes with no environmental effects, so every individual can be unambiguously assigned to a discrete class. For example, every pea plant was either tall or dwarf, with no intermediate heights. The gene concerned is now known to be *Le*, which codes for GA 3-oxidase, an enzyme required for production of the growth hormone gibberellin. There is one difference in the base sequence of the dominant and recessive alleles of the gene, resulting in one difference in the amino acid sequence of GA 3-oxidase. The polypeptide translated from the recessive allele is inactive as an enzyme, so growth hormone is not produced. This means stem growth is very restricted and the pea plant is dwarf. With the dominant allele, active GA 3-oxidase is produced and therefore so is growth hormone, which causes rapid stem growth and a tall pea plant.

Skin colour in humans is an example of continuous variation. It is partly due to the environment—sunlight stimulates the production of the black pigment melanin in the skin. This increases protection against harmful ultraviolet wavelengths in sunlight, which damage the skin and can cause cancer. Pale-coloured skin becomes darker in the days following increased exposure to sunlight. In people with dark skin, larger amounts of melanin are already produced without the need for sunlight as a stimulus. This is not an all-or-nothing effect. Excluding the effects of sunlight, the amount of melanin in human skin varies continuously. This is due to the influence of multiple genes, so is an example of polygenic inheritance.

Skin colour is an example of evolution in humans. Our common ancestors living in Africa almost certainly had dark skin as an adaptation to intense sunlight and not enough body hair to give protection against ultraviolet rays. When some humans spread out of Africa, those that migrated north evolved to have paler skins, because the sunlight was less intense and some UV penetration was needed to avoid vitamin D deficiency.

◀ Figure 32 Can you explain the skin coloration of this skier? Is environment or inheritance the cause?

▲ Figure 33 Skin pigmentation in humans is a continuous variable

Continuous and discrete variables

Discrete data is countable, such as the number of broadleaf weeds in a field or the frequency of blood groups (Figure 34). Continuous data is measurable and can take any value for example temperature, weight or length of corn cobs (Figure 35).

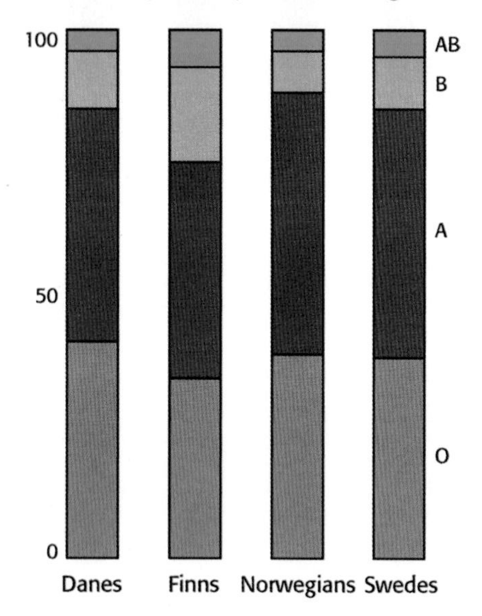

▲ Figure 34 Frequency of blood groups in four populations in northern Europe

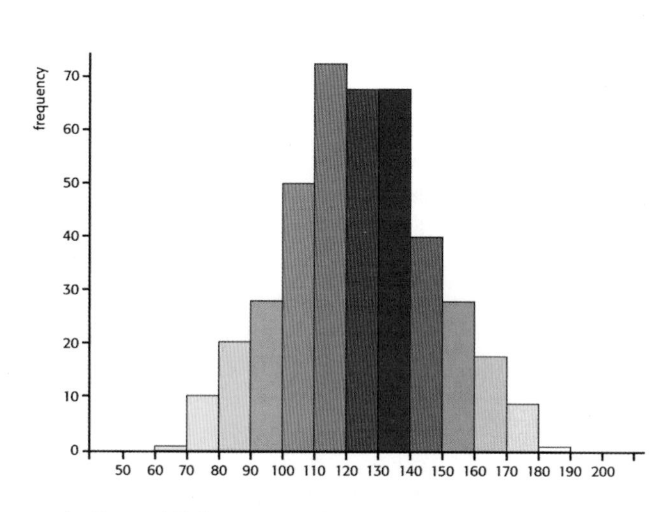

▲ Figure 35 Frequency of classes of corn cob length

Mean, median and mode are statistics known as measures of central tendency. A measure of central tendency is a single value that is meant to be representative of a set of data.

The **mean**, also known as the average, is equal to the sum of all the values in the data set divided by the number of values in the data set. The average is the most common measure of central tendency and can be used with both continuous and discrete data.

The **median** is found by choosing the middle value with the data arranged in order of increasing value. If there is an even number of values, then the median is the average of the two values in the middle. The median is a useful statistic if the data is skewed so that it does not follow a normal distribution. It is also a better choice than the average when there are outliers.

The **mode** is the most common value. The mode is most useful as a measure of central tendency when the data is nominal or categorical data, such as blood type or species name, for which a mathematical average or a median value based on ordering cannot be calculated.

Questions

1. A lawn was divided into quadrats. The number of broadleaf weeds was counted in 30 quadrats.

   22 15 6 15 33 13 25 12 15 8

   16 24 5 13 9 17 11 6 15 12

   26 16 2 34 21 18 15 12 8 4

   a. State whether the data is discrete or continuous

   b. Find the mean, median and the mode and comment on which is the most appropriate for use in this case.

2. Fourteen individuals were instructed to hold their breath for as long as they could. The time in seconds for the subjects is organized from least to greatest.

   20 22 23 23 23 58 61 65 74 79 80 81 83 92

   a. State whether the data is discrete or continuous.

   b. Find the mean, median and the mode and comment on which is the most appropriate for use in this case.

   c. Are there data values that do not appear to fit with the rest of the data? Does this help in making a selection of the most appropriate measure of central tendency?

D3.2.15 Box-and-whisker plots to represent data for a continuous variable such as student height

A box-and-whisker plot is a useful visual representation as it immediately communicates seven pieces of information: the minimum value (excluding outliers), the lower quartile, the median, the upper quartile, the maximum value (excluding outliers) and the interquartile range. In addition, it allows for a quick determination of how variable the data is and whether it is skewed. Sometimes a graph will contain multiple box-and-whisker plots and the area of the box will be adjusted so that the area indicates the relative numbers of individuals in the sample.

The other advantage of a box-and-whisker plot is that it defines a standard for identifying values as outliers. Determine the interquartile range (IQR) by subtracting the first quartile from the third quartile. Multiply the IQR by 1.5. Add this value to the 3rd quartile to determine the cut-off value for an upper outlier. Subtract this value from the 1st quartile to determine the cut-off value for a lower outlier.

Working with box-and-whisker plots

The heights of 20 students in a DP 2 class were measured in cm.

193 183 176 163 193 163 152 160 175 184
180 173 186 153 172 180 195 176 201 177

1. Enter the data into a spreadsheet in your graphic calculator and calculate single variable statistics (1-var stats).

2. Record the min value, Q_1, the median, Q_2 and the max value.

3. Determine the interquartile range.

4. Determine an example of a height (not necessarily found in this data) that would count as an outlier.

heights of DP 2 biology students

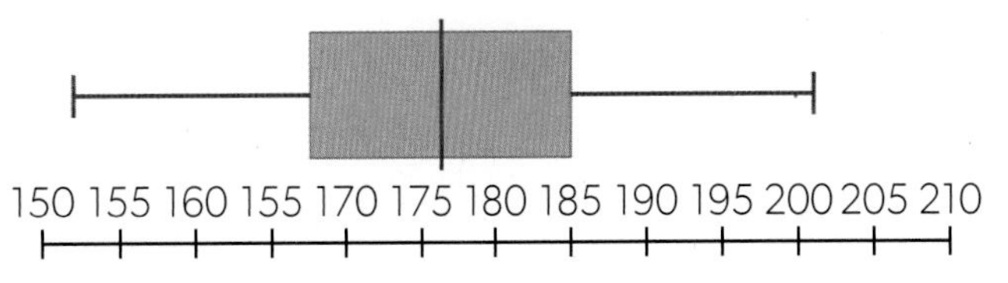

▲ **Figure 36** This graph was generated using an application found by searching for "Box and Whiskers plot generator" using an internet browser

5. Breastfed infants with rickets sometimes have seizures due to low blood calcium levels. A study was carried out to investigate the relationship between maternal blood vitamin D levels and the incidence of these infant seizures.

   a. Distinguish between the median maternal blood vitamin D in infants who have no seizures and those that do.

   b. Analyse the IQR of the two datasets.

   c. Identify one value that would be considered an upper range outlier for the "seizures" group.

   d. Determine the percentage of infants in the "seizures" group who have a maternal blood vitamin D level below 9 ng ml^{-1}.

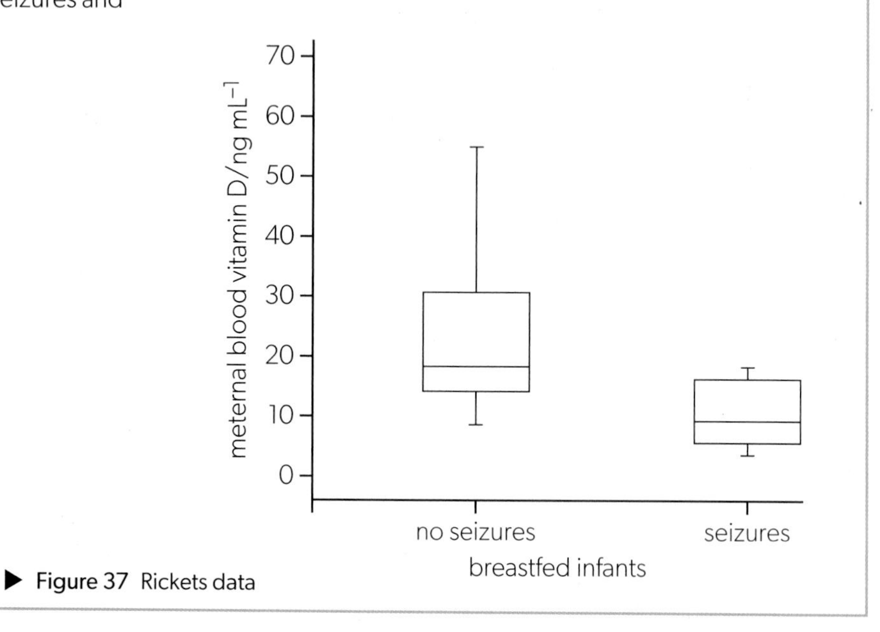

▶ Figure 37 Rickets data

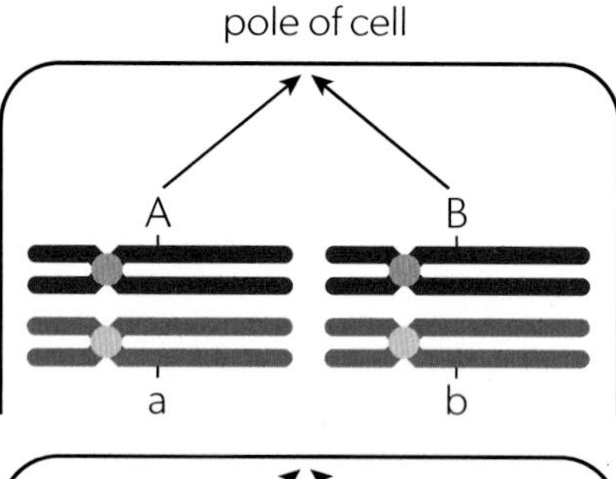

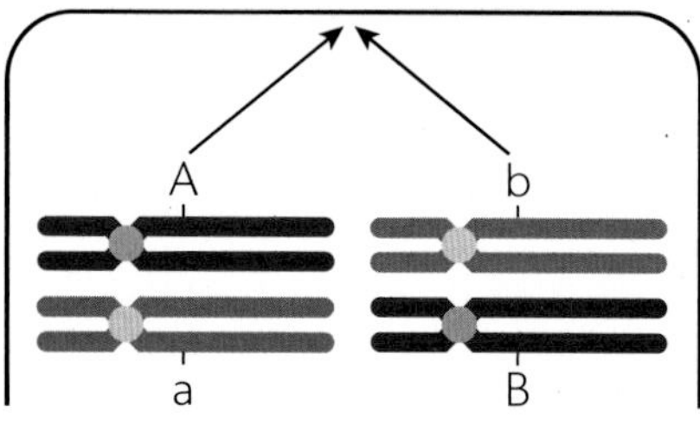

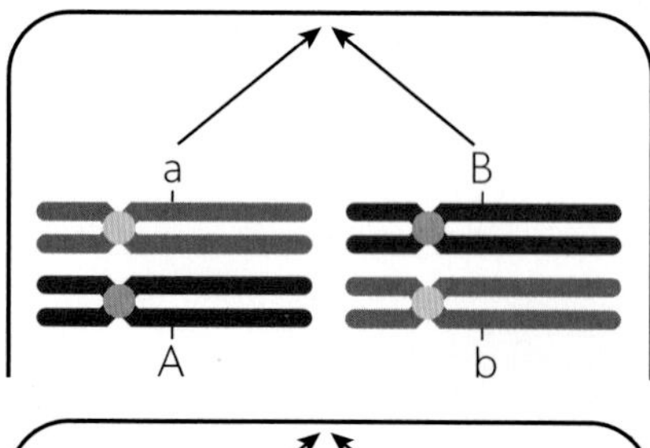

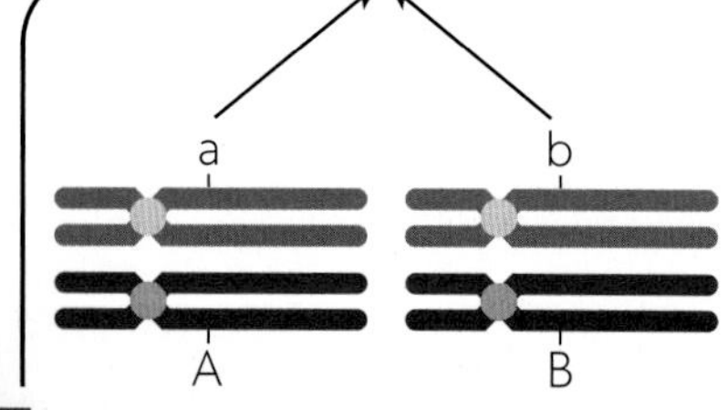

D3.2.16 Segregation and independent assortment of unlinked genes in meiosis

Segregation is separation of the alleles of a gene, so that one of the alleles is inherited but not the other. In a diploid cell, there are typically two alleles of each gene and in haploid gametes there is just one allele, so separation of alleles into different cells must occur at some stage during gamete formation. For example, in a person with the blood group AB and therefore genotype I_AI_B, the gametes produced will either contain I_A or I_B due to segregation of the alleles.

Independent assortment is the segregation of the alleles of two genes so that the outcome with each gene has no effect on the outcome with the other. The combinations of alleles that remain together after segregation are therefore random. For example, if a person has blood group AB and is a carrier of sickle cell anemia, their genotype is $I_AI_BH_AH_S$. Because of independent assortment, they will produce equal numbers of gametes with the genotypes I_AH_A, I_AH_S, I_BH_A and I_BH_S.

Segregation and independent assortment are the consequence of events in meiosis. The alleles of a gene come together within the nucleus when homologous chromosomes pair up during Prophase I of meiosis. They usually then separate during Anaphase I when the homologous chromosomes move to opposite poles. Sometimes, this separation is delayed until Anaphase II because of the effects of crossing over.

◀ Figure 38 In a cell heterozygous for two unlinked genes, meiosis can produce four different combinations of alleles. If the genes are on different chromosomes, the combinations are decided by the random orientation of bivalents (homologous pairs of chromosomes)

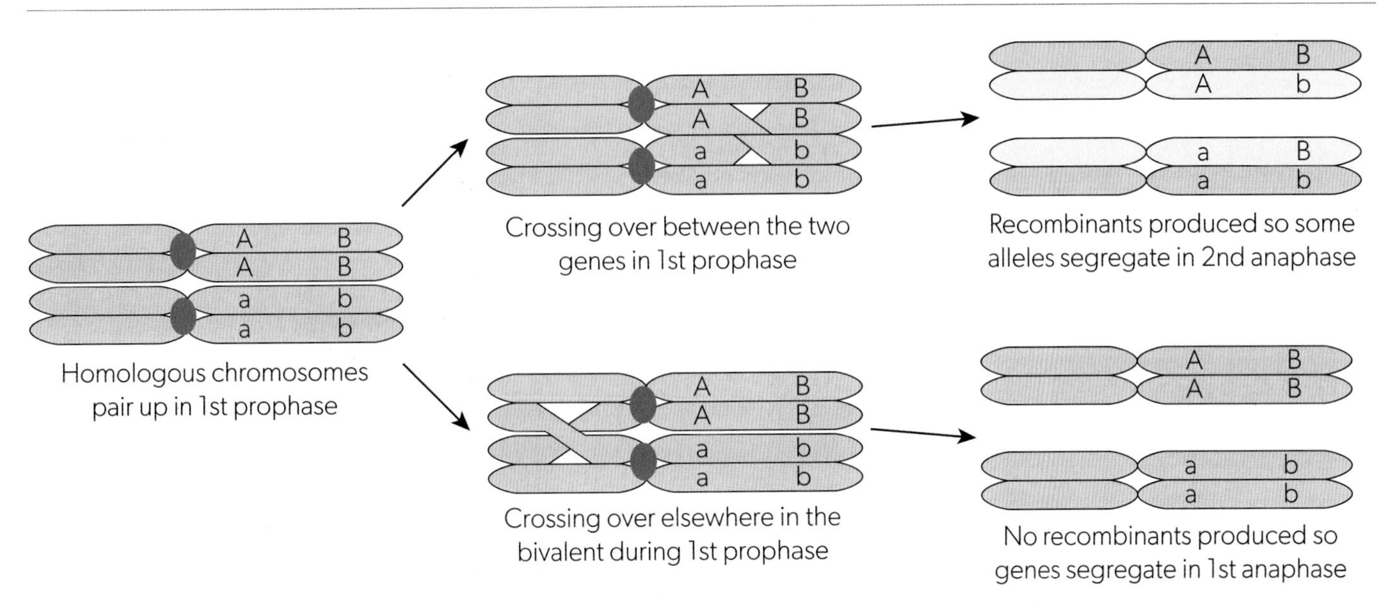

▲ Figure 39 Effects of crossing over on segregation

The pole to which each chromosome in a pair moves depends on which way the pair is facing when spindle microtubules attach to the chromosomes in Metaphase I. This is random and the consequence is that every allele has a 50% chance of being pulled to each of the two poles of the cell.

Because the orientation of each pair of chromosomes in Metaphase I is random, the way that one pair happens to be facing does not influence the orientation of other pairs of chromosomes. The alleles of one gene therefore segregate independently from the alleles of another gene. This is the basis of independent assortment.

Some genes do not assort independently because they are located close together on the same chromosome. This is gene linkage and is described later in the chapter. Independent assortment is therefore a property of genes that are unlinked, either because they are on different chromosomes or because they are far enough apart on a chromosome for them to be routinely shuffled by crossing over.

D3.2.17 Punnett grids for predicting genotypic and phenotypic ratios in dihybrid crosses involving pairs of unlinked autosomal genes

Monohybrid crosses test the inheritance of a single gene, whereas dihybrid crosses investigate the joint inheritance of two genes. Mendel performed some dihybrid crosses. For example, he crossed pure-breeding round yellow peas with pure-breeding wrinkled green peas. All the F_1 hybrids had round yellow seeds. This is not surprising, as these characters are due to dominant alleles.

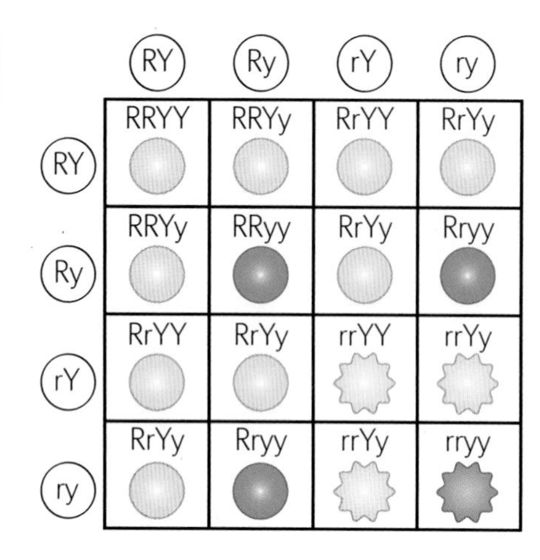

▲ Figure 40 Punnett grid for a dihybrid cross between pure-breeding yellow round and green wrinkled pea seeds. The predicted ratio of phenotypes is 9 yellow round: 3 green round: 3 yellow wrinkled: 1 green wrinkled

When Mendel allowed the F_1 plants to self-pollinate, there were four different phenotypes in the F_2 generation:

○ round yellow—one of the original parental phenotypes

● round green—a new phenotype

○ wrinkled yellow—another new phenotype

● wrinkled green—the other original parental phenotype.

If the genotype of the F_1 hybrids is *RrYy*, the gametes produced by these hybrids could contain either *R* or *r* with either *Y* or *y*. The four possible gametes are *RY*, *Ry*, *rY* and *ry*. If the inheritance of the two genes is independent (independent assortment), whether a gamete carries *R* or *r* will not affect whether it carries *Y* or *y*. The chance of a gamete containing each allele is 1 in 2, so the combined chance of containing two specific alleles is $\frac{1}{2} \times \frac{1}{2} = \frac{1}{4}$. The possible genotypes and phenotypes resulting from fusion of the four types of gamete can be shown in a 4 × 4 Punnett grid.

Constructing a Punnett grid

1. Ascertain the genotypes of the parents.
2. Deduce the genotypes of gametes that these parents can produce. Note that one copy of each gene is present in every gamete.
3. Draw a grid with one row for each genotype of male gamete and one column for each genotype of female gamete.
4. Label each row and column with the genotype of the gamete. Often these are put in a circle to make them show clearly. In a dihybrid cross there will be two alleles, one for each of the two genes.
5. Fill each cell in the grid with the genotypes from the row and column headings. Each cell then shows one of the possible combinations of alleles that can come together.
6. In each cell, indicate the phenotype resulting from the genotype. Symbols with a key can be used to do this.
7. Count the numbers of each phenotype. A tally chart may help you do this. Determine the phenotypic ratio for the offspring.

Theories and laws

Theories provide predictions and explanations. A law represents a prediction of what will reliably happen under a narrow set of conditions.

9:3:3:1 and 1:1:1:1 are ratios for dihybrid crosses that are predicted based on what has been called Mendel's Second Law which states that the alleles of one gene sort into gametes independently of the alleles of another gene. This law only operates if genes are on different chromosomes or are far apart enough on one chromosome for recombination rates to reach 50%.

Laws operate best under a narrow range of conditions. Students should recognize that there are exceptions to all biological "laws" especially if they are expanded to a greater range of conditions.

Data-based questions: Coat colour in the house mouse

In the early years of the 20th century, many crossing experiments were done in a similar way to those of Mendel. The French geneticist Lucien Cuénot used the house mouse, *Mus musculus*, to see whether the principles that Mendel had discovered in plants also operated in animals.

He crossed normal grey-coloured mice with albino mice. The hybrid mice that were produced were all grey. These grey hybrids were crossed together and produced 198 grey and 72 albino offspring.

1. Calculate the ratio between grey and albino offspring, showing your working. [2]
2. Deduce the colour of coat that is due to a recessive allele, with two reasons for your answer. [3]
3. Choose suitable symbols for the alleles for grey and albino coat and list the possible combinations of alleles of mice using your symbols, together with the coat colours associated with each combination of alleles. [3]
4. Using a Punnett grid, explain how the observed ratio of grey and albino mice was produced. [5]
5. The albino mice had red eyes in addition to white coats. Suggest how one gene can determine whether the mice had grey fur and black eyes or white fur and red eyes. [2]

Activity: Dihybrid cross practice questions

Use the following questions to develop your skill in dihybrid cross calculations.

1. A farmer has rabbits with two particular traits, each controlled by a separate gene. Brown coat colour is dominant to white. Tailed is dominant to tail-less. A brown, tailed male rabbit that is heterozygous for both genes is crossed with a white, tail-less female rabbit. A large number of offspring is produced with only two phenotypes: brown and tailed, white and tail-less. These two types are in equal numbers.
   a. State the genotypes of the male and female parents and of the gametes that are produced by each by meiosis.
   b. Predict the genotypic and phenotypic ratios of the F_2 generation. Show your working.
2. In peas, seeds can be wrinkled or smooth and stems can be tall or dwarf. A pure-breeding tall plant with smooth seeds was crossed with a pure-breeding short plant with wrinkled seeds. All the F_1 plants were tall with smooth seeds. Two of these F_1 plants were crossed and four different phenotypes were obtained in the 320 plants produced.

   How many tall plants with wrinkled seeds would you expect to find?
3. In *Drosophila* the allele for normal wings (*W*) is dominant over the allele for vestigial wings (*w*) and the allele for normal body (*G*) is dominant over the allele for ebony body (*g*). If two *Drosophila* with the genotypes *Wwgg* and *wwGg* are crossed together, what ratio of phenotypes is expected in the offspring?
4. Pure-breeding grey rabbits are crossed with pure-breeding albino rabbits. All the F_1 rabbits are grey. When these F_1 rabbits are mated with each other, the phenotypic ratio in the F_2 is 9 grey : 3 black : 4 albino. How can this ratio be explained?

D3.2.18 Loci of human genes and their polypeptide products

There are about 20,000 genes in the human genome that code for the amino acid sequence of a polypeptide. Each of these genes has a characteristic base sequence, varying somewhat between alleles, and also a locus. The locus of a gene is its specific position on one of 22 types of autosome (numbered 1 to 22) or one of the two types of sex chromosome (X and Y).

Applying technology: Using a database

In Turkey, approximately 1 in 2,500 babies is born with PKU. Assuming that both the mothers and fathers of these babies are carriers of the recessive allele, what proportion of the whole Turkish population are carriers?

This is how common the recessive PKU allele is. Is it commoner or less common than you expected for an allele causing a genetic disease?

In Australia, the frequency of the PKU allele in the population is 0.01. What proportion of newborn babies will have PKU?

Single-nucleotide polymorphisms (SNPs) are often associated with conditions like PKU. SNPs are often referred to by a reference code. The SNP rs5030858 is the most common allele associated with PKU. Use this code to search the databases OMIM and Ensembl to learn more about the population genetics of this variant.

What is meant by the "p arm" and the "q arm" of a chromosome? The protein product associated with PKU is PAH. Which chromosome locus hosts the gene for PAH?

The databases Ensembl and NCBI allow researchers to choose a chromosome and identify the genes located on that chromosome. What are five genes found on chromosome 2? What are five genes found on chromosome 1? Identify their protein products.

Chromosome 1 is longer than chromosome 2. Does it have more genes?

D3.2.19 Autosomal gene linkage

Genes that are located close to each other on the same chromosome do not assort independently and instead show gene linkage. Genes can be linked on the X chromosome, but most cases occur on autosomes (non-sex chromosomes). Gene linkage is indicated if the F_2 ratio differs significantly from the expected ratio for unlinked genes assorting independently in a dihybrid cross.

Phenotype	F2 frequencies	
	Number expected with 9:3:3:1 ratio	Number of plants actually observed
purple long	3,713	4,831
purple round	1,238	390
red long	1,238	393
red round	413	1,338

▲ Table 4 Results of a dihybrid cross showing autosomal gene linkage

The first case of autosomal gene linkage to be discovered was in the plant *Lathyrus odoratus*. A variety with purple flowers and long pollen grains was crossed with a variety with red flowers and round pollen grains. All the F_1 hybrids had purple flowers and long pollen grains. When these F_1 plants were self-pollinated, four phenotypes were observed as expected in the F_2 generation, but not in the familiar 9:3:3:1 ratio. There were more of the purple long and red round plants than expected. This is because these were the original parental combinations of alleles, which tend to be inherited together because they are physically linked as part of the DNA molecule of a chromosome.

In the cross between purple long and red round varieties of *L. odoratus*, there were some purple round and long red plants in the F_2 generation. These are known as recombinants because they show a new combination of alleles and therefore traits. They result from crossing over between the genes for flower colour and pollen shape, during Prophase I of meiosis. Without crossing over, a plant with the genotype *PpHh* could only produce gametes with genotypes *PH* and *ph*, but crossing over allows smaller numbers of recombinant *Ph* and *pH* gametes to be produced, so there are some purple round and red long plants in the F_2 generation.

In diagrams showing the inheritance of linked genes, lines should be used to represent homologous chromosomes with tick marks and letters to represent alleles. An example is shown in Figure 42. Pure-breeding spotted short-haired rabbits (*MMHH*) were mated with pure-breeding unspotted long-haired rabbits (*mmhh*). All the F_1 offspring were spotted and short-haired. These were back-crossed to unspotted long-haired rabbits. (In a backcross, an F_1 hybrid is crossed to an individual that is genetically identical to one of its parents.) In the resulting F_2 generation, 45% were spotted short, 41% were unspotted long, 7% were spotted long and 7% were unspotted short.

▲ Figure 41 Many different varieties of *L. odoratus* are grown in gardens and new varieties are developed by cross-breeding

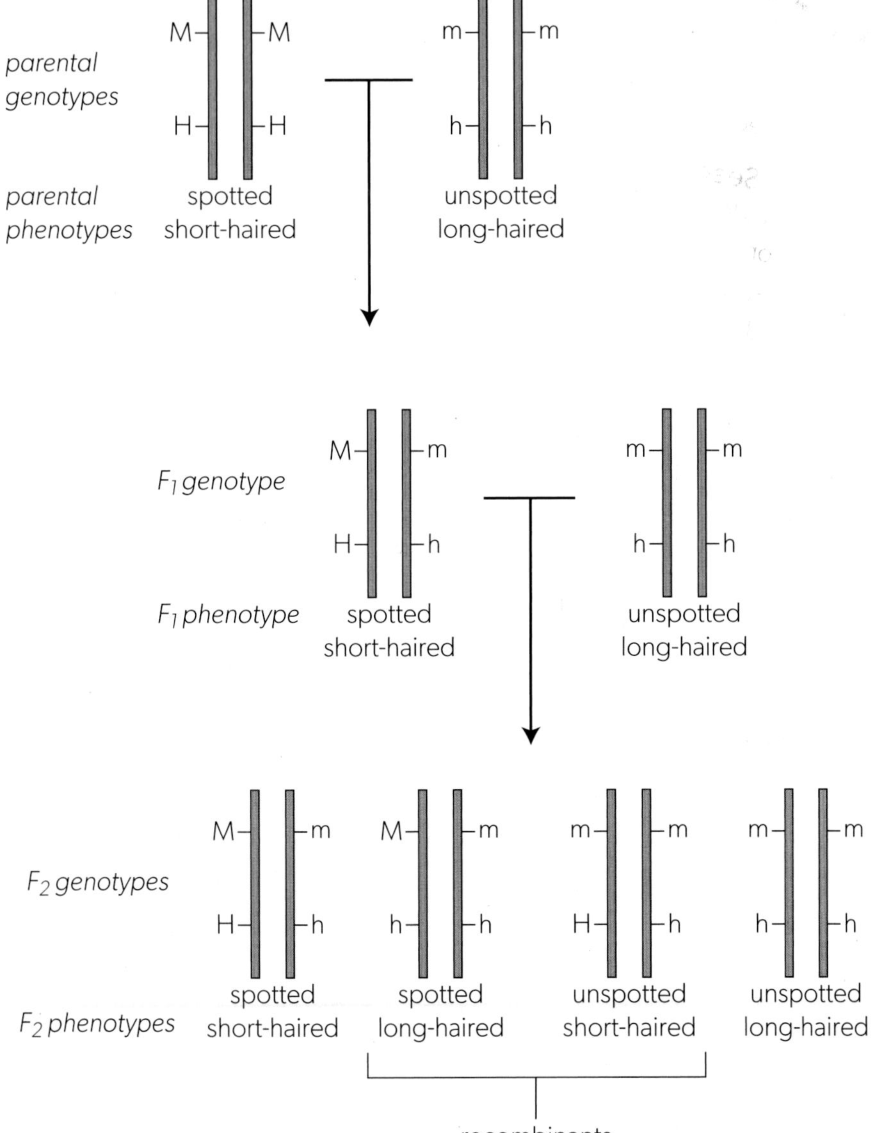

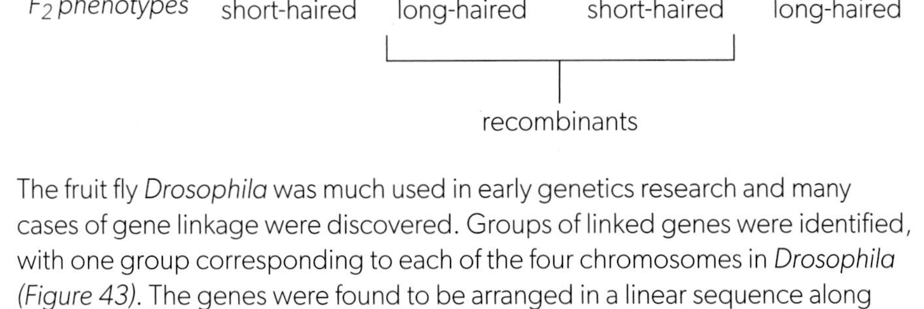

◀ Figure 42 Dihybrid cross with lines used to indicate homologous chromosomes

The fruit fly *Drosophila* was much used in early genetics research and many cases of gene linkage were discovered. Groups of linked genes were identified, with one group corresponding to each of the four chromosomes in *Drosophila (Figure 43)*. The genes were found to be arranged in a linear sequence along the chromosomes, with each particular gene found in a specific position on one chromosome, called the locus of a gene. It has since been discovered that all the genes on a chromosome are part of one DNA molecule.

AHL

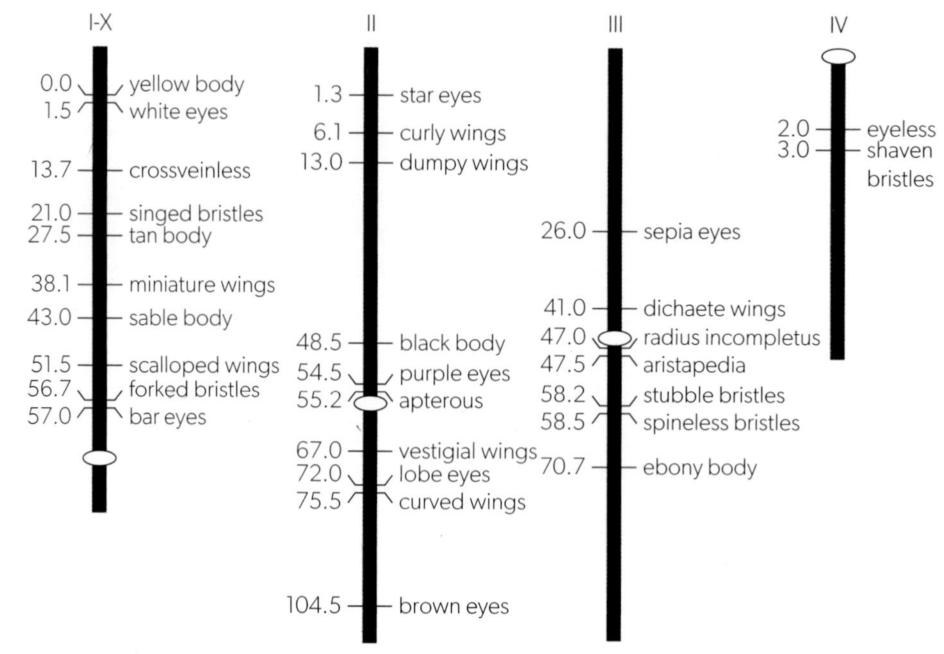

▶ Figure 43 Frequency of recombinants in dihybrid crosses showed how far apart pairs of genes are and the sequence of genes on each chromosome. This allowed maps of genes on chromosomes to be produced. Genome sequencing has superseded this cumbersome method of finding the location of genes

D3.2.20 Recombinants in crosses involving two linked or unlinked genes

A recombinant is an individual with a different combination of alleles and therefore traits from either parent.

Recombinants are the result of the process of genetic recombination, which happens during meiosis. Random orientation of bivalents results in production of new combinations of unlinked genes by independent assortment. Crossing over produces new combinations of linked genes.

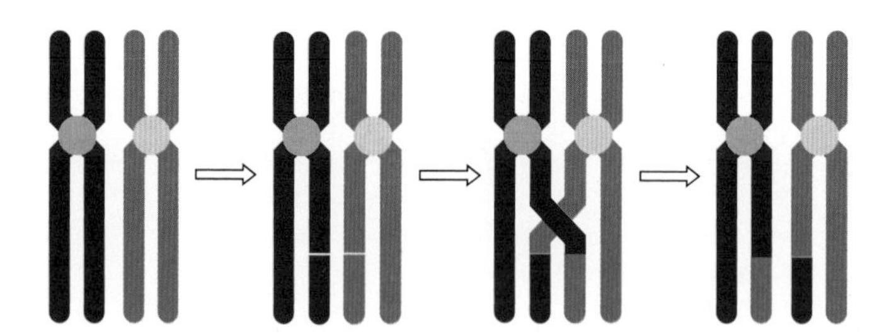

▲ Figure 44 Recombination of linked genes by crossing over

The frequency of recombination between two genes can be measured by crossing individuals that are heterozygous for both genes to individuals that are homozygous recessive for both genes.

Example 1: Seed shape and colour in peas

- The alleles for round seed (*R*) and yellow seed (*Y*) are dominant over the alleles for wrinkled seed (*r*) and green seed (*y*).
- Pure-breeding round yellow plants (*RRYY*) were crossed with pure-breeding wrinkled green plants (*rryy*). The parents produce gametes with the genotypes *RY* and *ry*. All the F_1 offspring were round yellow (*RrYy*).

- The heterozygous *RrYy* plants were backcrossed to the wrinkled green parental strain (rryy).
- In the F_2, there were 55 round yellow, 51 round green, 49 wrinkled yellow and 52 wrinkled green. This approximates to a 1:1:1:1 ratio and indicates that the two genes are unlinked.
- The recombinants are the gametes with the genotypes *Ry* and *rY* and the round green pea plants with genotype *Rryy* and wrinkled yellow pea plants with genotype *rrYy*.

gametes produced by homozygous recessive \ gametes produced by F_1 hybrids	RY	Ry	rY	ry
ry	RrYy round yellow	Rryy round green	rrYy wrinkled yellow	rryy wrinkled green

▲ Figure 45

Example 2: Seed colour and shape in corn

Corn cobs are often used for showing inheritance patterns. All the seeds on a cob have the same female parent, and with careful pollination they can also have the same male parent. A variety with coloured shrunken seeds was crossed with a variety with white non-shrunken seeds.

▲ Figure 46 The seeds on these corn cobs are all coloured and non-shrunken

- The F_1 seeds were all coloured and non-shrunken. From this we deduce that the alleles for coloured and non-shrunken are dominant and the alleles for white and shrunken are recessive. The F_1 genotype is *CcNn*.
- The F_1 plants grown from these seeds were test crossed with a homozygous recessive variety with white non-shrunken seeds with genotype ccnn.
- If the genes are unlinked, independent assortment will give a 1:1:1:1 ratio in the F_2 generation.
- The actual frequencies were:
  - coloured non-shrunken 638 (*CcNn*)
  - coloured shrunken 21,379 (*Ccnn*)
  - white non-shrunken 21,096 (*ccNn*)
  - white shrunken 672 (*ccnn*).
- This indicates that the genes for seed colour and shape are linked so the genotypes of the parents and the F_1 hybrids should be shown on a vertical line (Figure 47).
- When meiosis occurs in the F_1 plants, chromosomes with the combinations *CN* and *Cn* can be produced by crossing over. Gametes with these chromosomes are recombinants.

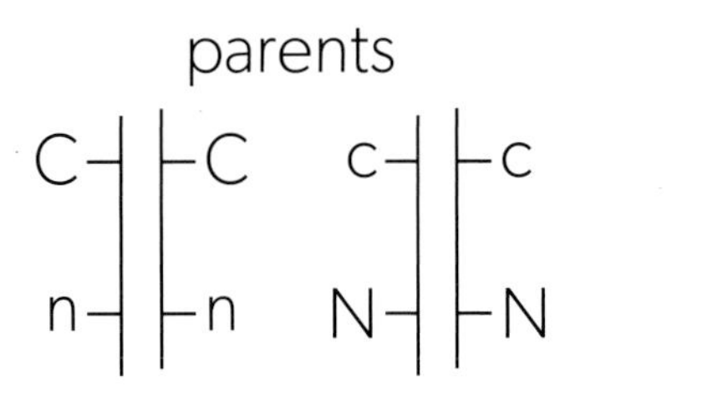

▲ Figure 47

The recombinants in the F_2 generation are the coloured non-shrunken with genotype *CcNn* and white shrunken with genotype *ccnn*.

	C N	C n	c N	c n
c n	C c / N n	C c / n n	c c / N n	c c / n n
Frequency	638	21,379	21,096	672
F_2 phenotypes	coloured non-shrunken	coloured shrunken	white non-shrunken	white shrunken

▲ Table 5 Recombination in corn

D3.2.21 Use of a chi-squared test on data from dihybrid crosses

The chi-squared test is a commonly used statistical hypothesis test. One of its uses is to test "goodness of fit", which is how well predictions from a biological or statistical model fit the observed values. In genetics, chi-squared testing can assess if observed ratios correspond with predicted Mendelian ratios such as 3:1, 1:1, 9:3:3:1 and 1:1:1:1.

There are two possible hypotheses.

- H_0: the traits fit the predicted ratio. This is the null hypothesis.
- H_1: the traits do not fit the predicted ratio. This is the alternative hypothesis.

Method for chi-squared test

1. State the null and alternative hypotheses.
2. Draw up a contingency table of observed frequencies, which are the numbers of individuals of each phenotype.
3. Calculate the expected proportion of individuals with each phenotype, using the ratio for genes that are unlinked, which in most dihybrid crosses is 9:3:3:1 or 1:1:1:1. Multiply the total number of individuals by this proportion to calculate the expected number of individuals with each phenotype.
4. Determine the number of degrees of freedom, which is one fewer than the total number of classes $(4-1)=3$ degrees of freedom.
5. Find the critical region for chi-squared from a table of chi-squared values, using the degrees of freedom that you have calculated and a significance level (p) of 0.05 (5%). The critical region is any value of chi-squared larger than the value in the table.
6. Calculate chi-squared using this equation:

$$\chi^2 = \sum \frac{(\text{obs} - \text{exp})^2}{\text{exp}}$$

7. Compare the calculated value of chi-squared with the critical region.
   - If the calculated value is not in the critical region, because it is equal to or below the value obtained from the table of chi-squared values, H_0 is not rejected. There is no evidence at the 5% level of a difference between the predicted and observed ratios.
   - If the calculated value is in the critical region, we can reject the hypothesis H_0. There is evidence at the 5% level of a difference between the predicted and observed ratios. That is, there is significant evidence at the 5% level that these Mendelian ratios do not describe the proportions of each phenotype in this case.

Example of the use of the chi-squared test: testing for gene linkage

In 1901, Bateson reported one of the first post-Mendelian studies of a cross involving two traits. White leghorn chickens with large "single" combs, were crossed with Indian game fowl with dark feathers and small "pea" combs. All the F_1 were white with pea combs, and the ratio of F_2 phenotypes involving 190 offspring was: 111 white pea, 37 white single, 34 dark pea and 8 dark single. The expected ratio is 9:3:3:1. The observed ratio was different. Were the differences between observed and expected due to sampling error or were the differences statistically significant, suggesting that the traits do not assort independently? This can be tested using the chi-squared test.

The first stage is to draw up a contingency table and calculate the expected frequencies.

	white	dark
pea	white pea (9/16) × 190 = 106.9	dark pea (3/16) × 190 = 35.6
single	white single (3/16) × 190 = 35.6	dark single (1/16) × 190 = 11.9

◀ Table 6

The number of degrees of freedom (for four classes) is (4 − 1) = 3 degrees of freedom.

The critical value is found in a table of chi-squared values.

Critical values of the χ^2 distribution										
df \ p	0.995	0.975	0.9	0.5	0.1	0.05	0.025	0.01	0.005	df
1	0.000	0.000	0.016	0.455	2.706	3.841	5.024	6.635	7.879	1
2	0.010	0.051	0.211	1.386	4.605	5.991	7.378	9.210	10.597	2
3	0.072	0.216	0.584	2.366	6.251	7.815	9.348	11.345	12.838	3

◀ Table 7

At the 0.05 level of significance, the critical value is 7.815.

We now calculate chi-squared:

$$\chi^2 = \sum \frac{(\text{obs} - \text{exp})^2}{\text{exp}}$$

$$= \frac{(111 - 106.9)^2}{106.9} + \frac{(37 - 35.6)^2}{35.6} + \frac{(34 - 35.6)^2}{35.6} + \frac{(8 - 11.9)^2}{11.9}$$

$$= 1.56$$

The calculated value for chi-squared is outside the critical region ($p > 0.05$) so we reject the alternative hypothesis and accept the null hypothesis. There is no significant evidence that the ratio 9:3:3:1 is an unsuitable model. This suggests that there is independent assortment and the two genes are unlinked.

Data-based questions: Using the chi-squared test

Warren and Hutt (1936) test-crossed a double heterozygote for two pairs of alleles in hens. One pair of alleles was for the presence (*Cr*) or absence (*cr*) of a crest; the other was for white (*I*) or non-white (*i*) plumage.

For their F_2 cross, there was a total of 754 offspring.

- 337 were white, crested
- 337 were non-white, non-crested
- 34 were non-white, crested
- 46 were white, non-crested.

1. Construct a contingency table of observed values. [4]
2. Calculate the expected values, assuming independent assortment. [4]
3. Determine the number of degrees of freedom, showing your reasoning. [2]
4. Find the critical region for chi-squared at a significance level of 5%. [2]
5. Calculate chi-squared. [4]
6. State the two hypotheses, H_0 and H_1 and the conclusions that should be drawn about them based on the calculated value for chi-squared. [4]

Models: Using samples to model populations

It is often difficult or impossible to study a phenomenon in a whole population. Biologists often choose a sample from the population to be representative of the whole population. To ensure that the population is representative, every member of the population must have an equal chance of being selected. If the population is representative, then insights on the population as a whole can be inferred from observations on the sample. The larger the sample, the more representative the sample will be. This is the reason that you are encouraged to take replicates or repeated measurements in our experimental work.

In the worked examples of the chi-squared test being applied, the F_2 generation is a sample.

Data-based questions: Flower colour and stem length in peas

Mendel used pure-breeding varieties of pea plant in an experiment. He crossed red-flowered dwarf pea plants with white-flowered dwarf plants and obtained red-flowered dwarf F_1 hybrids. He also crossed white-flowered tall plants with white-flowered dwarf plants and obtained white-flowered tall F_1 hybrids. He then crossed the red dwarf and white tall F_1 hybrids together. The numbers Mendel observed in the F_2 generation are shown in Table 8.

Phenotype	Number
red-flowered tall	47
red-flowered dwarf	38
white-flowered tall	40
white-flowered dwarf	41

▲ Table 8

1. Using the letters *E*/*e* and *H*/*h* for the alleles of the genes for flower colour and stem height, deduce the genotypes of all parent plants and the two types of F_1 hybrid. [6]
2. Deduce all the possible genotypes of gametes produced by the two types of F_1 hybrid. [4]
3. Deduce the genotypes of the F_2 generation. [4]
4. State the expected phenotypic ratio in the F_2 generation. [1]
5. Use a chi-squared test to assess whether the observed frequencies fit the expected ratio. [8]
6. Discuss which F_2 plants were recombinants. [2]

Linking questions

1. What are the principles of effective sampling in biological research?
    a. Describe the Lincoln index method of population size determination. (C4.1.4)
    b. Explain the concept of *p*-value in statistical testing. (D3.2.21)
    c. Explain the concept of random sampling with reference to the use of quadrats. (C4.1.3)
2. What biological processes involve doubling and halving?
    a. Glycolysis means "glucose splitting". Outline the process of glycolysis. (C1.2.8)
    b. Explain the relationship between meiosis and independent assortment. (D3.2.16)
    c. Outline the process that results in the doubling of DNA during the S phase of the cell cycle. (D2.1.13)

D3.3 Homeostasis

How are constant internal conditions maintained in humans?

A set point is a level of a variable that an organism seeks to maintain at a constant level. One example is body temperature. What are some other examples of biological set points? How does the body detect deviations from the set point in each of these examples? What mechanisms does it use to return the level of the variable to the set point? What is the temperature difference between this ice-swimmer's internal and external environment?

▲ Figure 1

What are the benefits to organisms of maintaining constant internal conditions?

Thermograms show the skin's temperature by recording its emission of infrared radiation. This heat radiation is displayed with each temperature in a different colour. The different coloration of the two sets of hands shown in Figure 2 has been caused by one individual smoking which in turn leads to the narrowing of blood vessels. What is the adaptive advantage of vasoconstriction? Camels have a series of physiological adaptations that allow them to withstand long periods of time without any external source of water and still maintain their blood solute concentration within homeostatic limits. What are some of these adaptations?

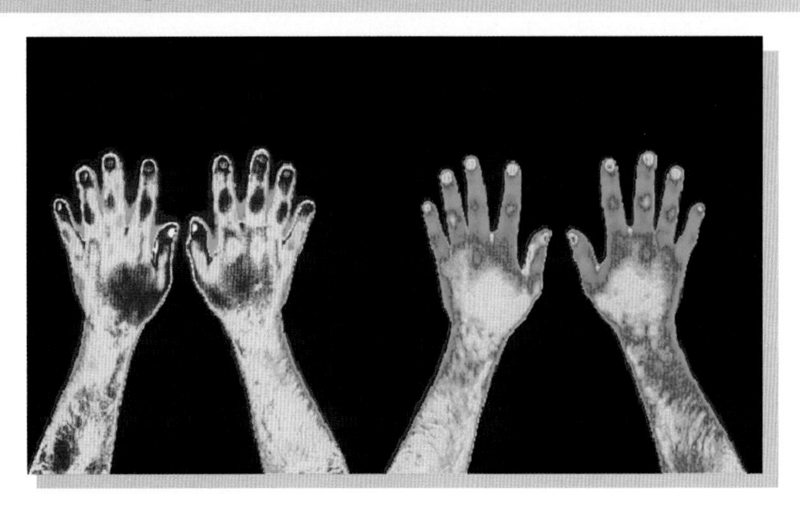

▲ Figure 2

SL and HL	AHL only
D3.3.1 Homeostasis as maintenance of the internal environment of an organism D3.3.2 Negative feedback loops in homeostasis D3.3.3 Regulation of blood glucose as an example of the role of hormones in homeostasis D3.3.4 Physiological changes that form the basis of type 1 and type 2 diabetes D3.3.5 Thermoregulation as an example of negative feedback control D3.3.6 Thermoregulation mechanisms in humans	D3.3.7 Role of the kidney in osmoregulation and excretion D3.3.8 Role of the glomerulus, Bowman's capsule and proximal convoluted tubule in excretion D3.3.9 Role of the loop of Henle D3.3.10 Osmoregulation by water reabsorption in the collecting ducts D3.3.11 Changes in blood supply to organs in response to changes in activity

D3.3.1 Homeostasis as maintenance of the internal environment of an organism

The environment for a cell inside a multicellular organism is the immediate surroundings outside the plasma membrane. For a plant cell, this is the cell wall and the fluid held in it. For an animal cell, there may be an extracellular matrix of collagen, elastin and other materials, and there is always tissue fluid filling other gaps between cells. Blood is unusual in that the volume of the tissue fluid, in this case plasma, is very large and separates the cells completely; blood is a liquid tissue.

One of the advantages of being multicellular is that the internal environment between cells can be regulated, with variables kept as close to optimal as possible. This is the process of homeostasis. Blood glucose concentration, blood osmotic concentration, blood pH and core body temperature are examples of variables that are kept relatively constant as a part of homeostasis.

D3.3.2 Negative feedback loops in homeostasis

Feedback control uses information about the outcome of a process to make decisions about the future of that process. There are two types of feedback control: positive and negative.

- Positive feedback increases the gap between the original and the new level.
- Negative feedback decreases the gap, so the original level is restored. Any rises above or falls below the original level can be reversed, keeping the variable close to the chosen set point. Negative feedback can thus be used to achieve balance.

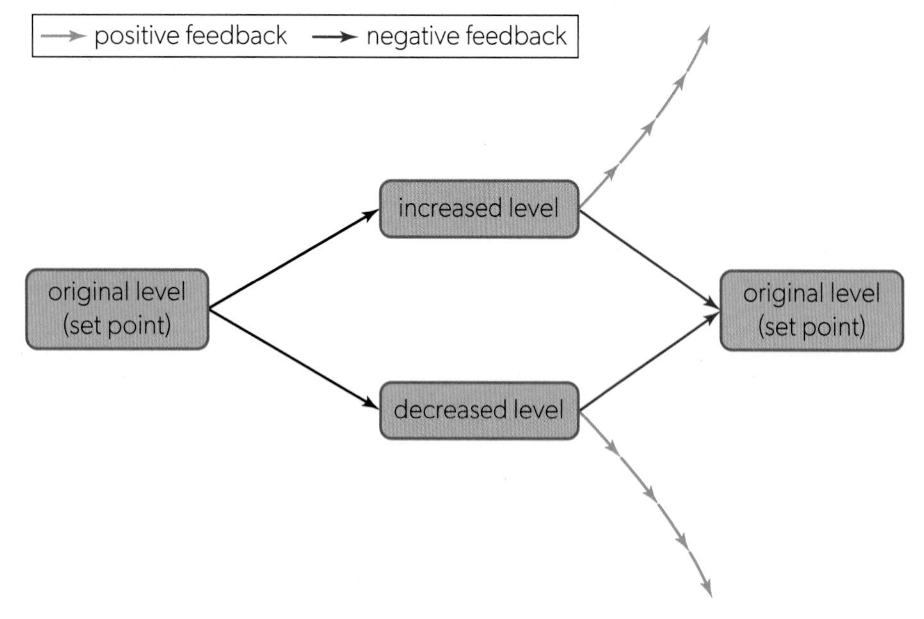

▲ Figure 3 Positive and negative feedback

Negative feedback mechanisms form the basis of the homeostatic control systems that are used to keep internal conditions in the body within narrow limits. The disadvantage is that large amounts of energy are used to maintain homeostasis. For many multicellular organisms, this is outweighed by the advantage that body cells are kept in ideal and stable conditions, despite fluctuations in the outside environment. This allows extreme and hostile environments to be inhabited so there are very few parts of Earth where life is totally absent.

Positive feedback is not common in the human body. It is used, for example, to progressively increase the force of muscle contractions during childbirth. It sometimes also occurs when normal homeostatic processes have failed. Positive feedback promotes change rather than stability so it is unsuitable for homeostasis.

ATL Communication skills: Using digital media for communicating information

Presentation software such as Google Slides contains templates for creating flowcharts that can also be used as concept maps. In Google Slides, this can be done using the menu Insert→Diagram. In PowerPoint, use the menu Insert→Smart Art. In Google docs, the menu Insert→Shape can be used to recreate the diagram shown in Figure 3 on the previous page.

Try creating a diagram to illustrate regulation of one of the following:

- osmolarity of blood
- body temperature
- blood glucose levels
- blood pH.

D3.3.3 Regulation of blood glucose as an example of the role of hormones in homeostasis

Glucose is supplied by blood to cells throughout the body. Blood glucose concentrations are kept within narrow limits (see Figure 4), by balancing the amount of glucose removed from the blood with the amount that is added. The concentration cannot be kept precisely constant. This is because the entire volume of blood in the body contains only about 5 g of glucose and body processes can add or remove it rapidly.

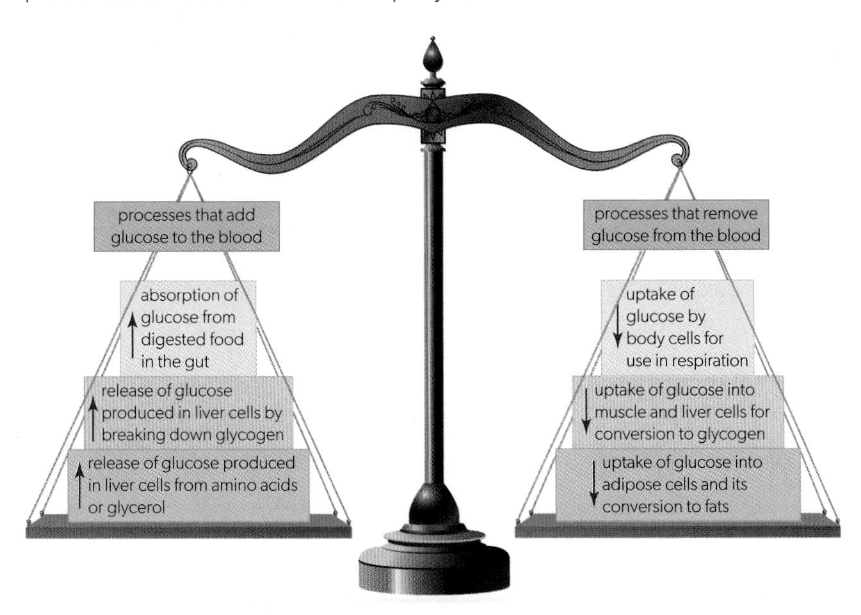

▲ **Figure 5** Balance is achieved when the quantities of glucose added to the blood and removed from it are equal

The set point for blood glucose concentration is about 5 mmol L^{-1} (dm^{-3}). If the concentration deviates substantially from this set point, homeostatic mechanisms mediated by insulin and glucagon are initiated (Figure 5). These hormones are both secreted by cells in the pancreas, which is effectively two glands in one organ. Most of the pancreas is exocrine glandular tissue that secretes digestive enzymes into ducts leading to the small intestine. There are also small regions of endocrine tissue called islets of Langerhans dotted through the pancreas. These secrete hormones directly into the bloodstream. There are two cell types in the islets of Langerhans and they secrete different hormones.

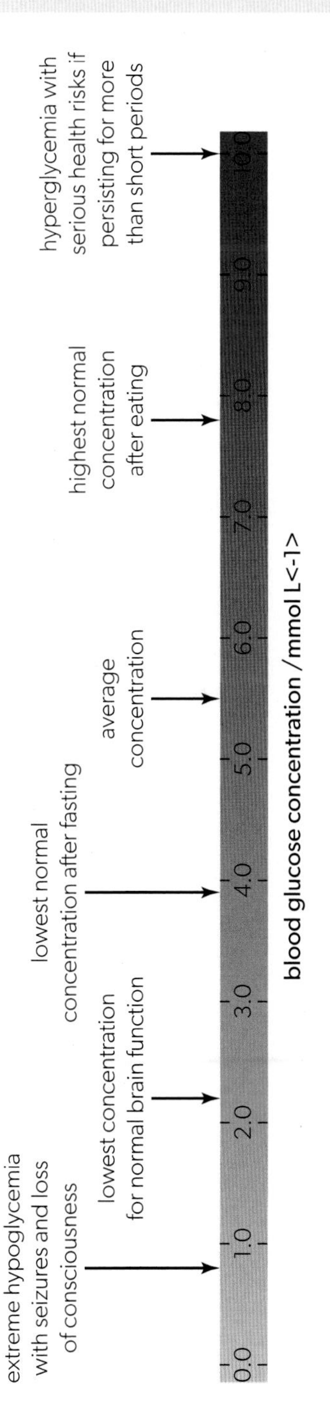

▲ **Figure 4** Blood glucose cannot be kept constant but is normally maintained within a narrow range

- Alpha cells (α cells) synthesize and secrete glucagon if the blood glucose level falls below the set point. This hormone stimulates breakdown of glycogen into glucose in liver cells and its release into the blood, increasing the concentration.
- Beta cells (β cells) synthesize insulin and secrete it when the blood glucose concentration rises above the set point. This hormone stimulates uptake of glucose by many target cells in multiple tissues. These cells then use glucose in cell respiration instead of fats. Skeletal muscle and liver are particularly active in glucose uptake and in these tissues, insulin stimulates the conversion of glucose to glycogen.

Insulin therefore reduces blood glucose concentration. Like most hormones, insulin is broken down by the cells it acts upon, so its secretion must be ongoing. Secretion of insulin begins within minutes of eating and may continue for several hours after a meal.

D3.3.4 Physiological changes that form the basis of type 1 and type 2 diabetes

Diabetes is the condition in which a person has consistently elevated blood glucose levels even during prolonged fasting. This leads to the presence of glucose in the urine. Continuously elevated glucose causes damage to tissues, particularly their proteins. It also impairs water reabsorption during urine production in the kidney, resulting in greater loss of water in urine and therefore dehydration. If a person needs to urinate more frequently, is constantly thirsty, feels tired and craves sugary drinks, they should test for glucose in the urine to check if they have developed diabetes.

There are two main types of this disease.

- Type 1 diabetes (also called early-onset diabetes) is characterized by an inability to produce sufficient insulin. It is an autoimmune disease arising from the destruction of beta cells in the islets of Langerhans by the body's own immune system. In children and young people, the more severe and obvious symptoms of the disease usually start rather suddenly. The causes of this and other autoimmune diseases are still being researched.
- Type 2 diabetes (sometimes called late-onset diabetes) is characterized by an inability to process or respond to insulin because of a deficiency of insulin receptors or glucose transporters on target cells. Onset is slow and the disease may go unnoticed for many years. Until the last few decades, this form of diabetes was very rare in people under 50 and common only in the over 65s. The causes of this form of diabetes are not well understood but the main risk factors are sugary or fatty diets, prolonged obesity due to habitual overeating and lack of exercise, together with genetic factors that affect energy metabolism.

The treatment of the two types of diabetes is different.

- Type 1 diabetes is treated by testing the blood glucose concentration regularly and injecting insulin when it is too high or likely to become too high. Injections are often done before a meal to prevent a peak of blood glucose as the food is digested and absorbed. Timing is very important because insulin molecules do not last long in the blood. Better treatments are being developed using implanted devices that can release exogenous insulin into the blood as and when it is necessary. A permanent cure may be achievable by coaxing stem cells to become fully functional replacement beta cells.

- Type 2 diabetes is treated by adjusting the diet to reduce the peaks and troughs of blood glucose. Small amounts of food should be eaten frequently rather than infrequent large meals. Foods with high sugar content should be avoided. Starchy food should only be eaten if it has a low glycaemic index, indicating that it is digested slowly. High-fibre foods should be included to slow the digestion of other foods. Strenuous exercise and weight loss are beneficial because they improve insulin uptake and action.

Data-based questions: The glucose tolerance test

The glucose tolerance test is a method used to diagnose diabetes. In this test, the patient drinks a concentrated glucose solution. The blood glucose concentration is monitored to determine the length of time required for excess glucose to be cleared from the blood. When displayed on a graph, the results are known as a blood glucose response curve.

1. With reference to Figure 6, distinguish between the person with diabetes and the person with normal glucose metabolism with respect to:
   a. the concentration of glucose at time zero (before the consumption of the glucose drink) [1]
   b. the length of time required to return to the level at time zero [1]
   c. the maximum glucose level reached [1]
   d. the time before glucose levels start to fall. [1]
2. A fasting blood sugar test measures blood glucose after an overnight fast (not eating). Predict, with a reason, whether a person with diabetes would have lower or higher than normal blood glucose in a fasting blood glucose test. [1]

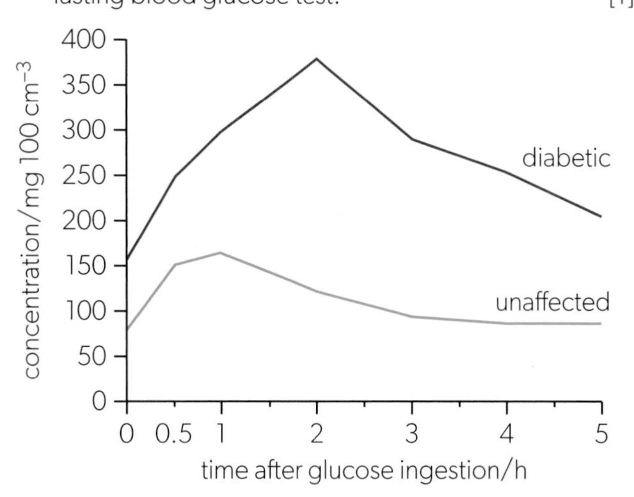

▲ **Figure 6** Glucose tolerance test results (blood glucose response curves) for a person with diabetes and an unaffected person

Activity: Foods for people living with type 2 diabetes

- Discuss which of the foods in Figure 7 are suitable for a person with type 2 diabetes.
- They should be foods with a low glycaemic index.
- For any food, the glycaemic index is the percentage of the area under the blood glucose response curve as a percentage of the area produced by ingesting the same quantity of glucose.
- Glycaemic index therefore ranges from 0 to 100.

▶ **Figure 7** Which of these foodstuffs have a low glycaemic index?

D3.3.5 Thermoregulation as an example of negative feedback control

Thermoregulation is control of core body temperature to keep it close to a set point. The set point may be different at different times of the day or year, but temperature stability is always maintained despite fluctuations in external temperature. Birds and mammals are thermoregulators, with a set point between 36°C and 42°C, depending on the species. Humans have a set point close to 37°C except during infections, when a raised temperature (fever) helps the immune system to eliminate pathogens.

Negative feedback is the basis of thermoregulation. Some processes within the body cause heat generation and others cause loss of heat. If body temperature falls below the set point or is likely to do so, heat generation can be increased, and heat loss can be reduced. The opposite happens when body temperature rises above the set point.

Body temperature is monitored by thermoreceptors, which are the free nerve endings of specialized sensory neurons. There are two types:

- cold thermoreceptors are stimulated by low temperatures
- warm thermoreceptors are stimulated by high temperatures.

Peripheral thermoreceptors are located in the skin, where body temperature is much influenced by external temperatures. These receptors can be used to anticipate rates of heat loss from the body. Central thermoreceptors are located in the core of the body, including in the hypothalamus.

The hypothalamus is a relatively small part of the brain that has some major regulatory roles. It is the integrating centre for thermoregulation. The hypothalamus monitors body temperature using sensory inputs from both peripheral and central thermoreceptors. It initiates responses if hypothermia or hyperthermia is developing to bring body temperature back to the set point.

Heat is generated by metabolism in cells. The metabolic rate can be increased or decreased to raise or lower the amount of heat generated. To increase metabolic rate, the hypothalamus secretes thyrotropin releasing hormone (TRH), a tripeptide which passes directly to the pituitary gland. In response, the pituitary gland releases thyroid stimulating hormone (TSH). This hormone is a glycoprotein that stimulates thyroxin secretion by the thyroid gland in the neck. Thyroxin increases the metabolic rate of cells. All cells respond but the most metabolically active such as liver, muscle and brain are the main targets. In this way, the hypothalamus can vary the generation of metabolic heat by increasing or decreasing secretion of TRH.

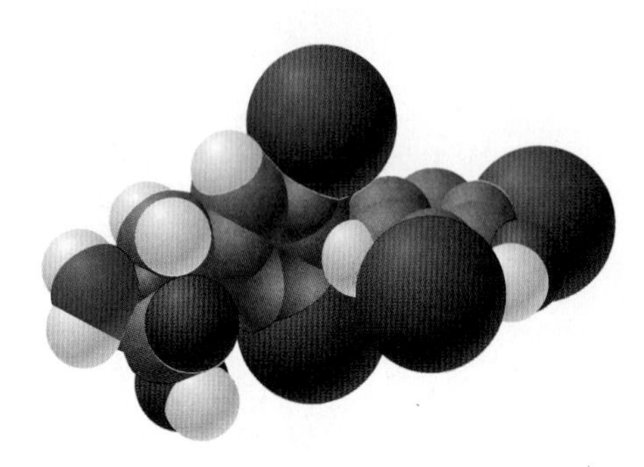

▲ Figure 8 Structure of thyroxin with iodine atoms shown purple

Some tissues have a special role as effectors of temperature change. Muscle, for example, generates heat when it contracts. Subcutaneous adipose tissue, which is adjacent to the skin, acts as an insulator and reduces heat loss. Brown adipose tissue can generate heat at a rapid rate. This is important in small mammals and new-born babies, which are particularly prone to heat loss. The thermoregulatory roles of muscle and adipose tissue are described more fully in the next section.

D3.3.6 Thermoregulation mechanisms in humans

Both birds and mammals regulate their body temperature using physiological means and behavioural changes. Table 1 details responses in humans.

Responses to cold	Responses to heat
Vasoconstriction Arterioles are branches of arteries that supply blood to part of an organ. In the walls of arterioles there are rings of muscle. When this circular muscle contracts, the circumference of the arteriole is reduced and the lumen along which blood flows is narrowed. Less blood then flows to the region supplied by the arteriole. This is vasoconstriction. When the body needs to reduce heat loss, there is vasoconstriction of arterioles supplying the skin. Less blood flows to capillaries in the skin and it cools below core body temperature. With reduced temperature difference between the skin and the external environment, less heat is lost from the body.	*Vasodilation* When the body is overheated, circular muscle cells in the walls of arterioles supplying the skin relax, so the arterioles widen. This is vasodilation. More blood flows to the skin, which warms up to core temperature. The increased temperature difference between the skin and the external environment causes more heat to be lost from the body. In a pale-skinned person, vasodilation makes the skin look pinker because there is more blood flowing through it, whereas vasoconstriction makes it look paler. It is important to realize that blood vessels in the skin do not move nearer to the skin surface or further away from it to regulate heat loss—only the amount of blood flowing through capillaries in the skin can be varied.
Shivering When muscles contract to cause movement, a side-effect is the generation of heat. Sometimes many small, involuntary muscle contractions and relaxations are carried out at a rapid rate solely to generate heat. This is known as shivering.	*Sweating* Sweat is secreted by glands in the skin and then passes through narrow ducts to the skin surface, where water in the sweat evaporates. Solutes in the sweat, especially ions such as sodium, are left on the skin surface and can sometimes be detected by their salty taste. Water has a high latent heat of vaporization so its evaporation causes significant cooling. Blood flowing through the skin loses heat and can then cool other parts of the body.
Uncoupled respiration Brown adipose tissue is a modified version of the white adipose tissue that is used for fat storage. The brown colour is due to the cells containing less fat and more mitochondria. These mitochondria oxidize fat by normal metabolic pathways but whereas the oxidation reactions are normally coupled to ATP production, in brown adipose tissue all the energy released by the oxidation is transformed into heat and no ATP is produced. This is known as uncoupled respiration. During childhood, the amount of brown adipose tissue decreases, but even in adulthood some is retained to generate heat and help prevent hypothermia. ▶ Figure 9 Infrared image of brown adipose tissue in a baby	
Hair erection In mammals with a thick coat, the air between the hairs acts as a thermal insulator. Erector muscles can move the hairs to make the coat thicker and the insulating effect greater. During human evolution, the amount of hair over most of the body has been reduced to a few short hairs. The erector muscles can still make the hairs stand up, but they do not trap air well enough to insulate the body. This ineffectual response to cold is also known as goose-bumps.	▲ Figure 10 Conditions in a sauna cause both vasodilation and sweating Sweat secretion is controlled by the hypothalamus. If the body is overheated, the hypothalamus stimulates the sweat glands to secrete up to two litres of sweat per hour. Usually, no sweat is secreted if body temperature is below the set point, but adrenaline can cause sweat secretion in anticipation of a period of intense activity that will cause overheating.

▲ Table 1 Comparison of responses to cold and heat in humans

▲ **Figure 11** How do soccer players stay warm wearing light clothing in snowy weather? How do the spectators at the match keep warm?

Designing and testing models: Modelling heat loss

Flat-bottomed tubes filled with water at 37°C can be used to model soccer spectators. How rapidly does the water cool to room temperature? Thickness of clothing and the proximity to other people could also be modelled to investigate how soccer supporters stay warm at matches in cold weather.

A data-logging thermometer can assess changes in temperature. Insulation types could be investigated such as fur or fat obtained from a butcher.

Goose-bumps occur when a person is exposed to cold temperatures. This is an adaptation in furry mammals for capturing air and using it as an insulator. How can air as an insulator be evaluated?

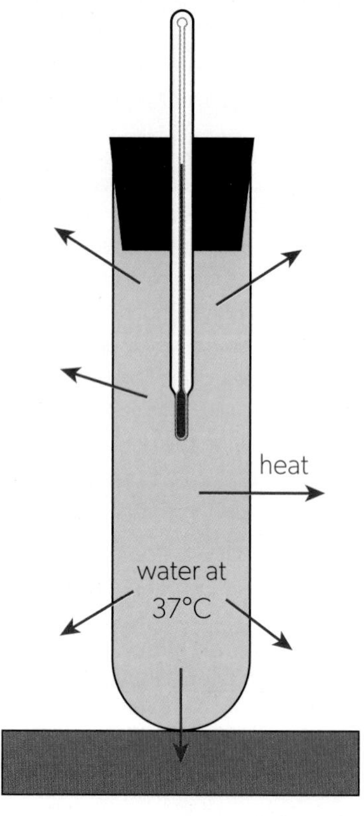

▲ **Figure 12** A model for heat loss experiments

very pale urine when the body is overhydrated, so the kidneys produce very dilute urine

dark urine when the body is dehydrated, so the kidneys produce concentrated urine

other colours of urine are often due to coloured substances absorbed from food or more rarely to drugs or specific diseases

▲ **Figure 13** Colours of urine and what they mean

D3.3.7 Role of the kidney in osmoregulation and excretion

The kidney has the twin roles of osmoregulation and excretion. It achieves these roles by filtering out about 20% of the water and solutes from the blood plasma and then selectively absorbing the substances that the body requires.

Osmoregulation is keeping the osmotic concentration of body fluids within narrow limits; it is part of homeostasis. Osmotic concentration is the overall concentration of the solutes in a fluid that can affect the movement of water by osmosis. Osmoregulation therefore allows the water content and pressure of cells to be controlled. The kidney carries out osmoregulation by varying the relative amounts of water and salts that are removed from the body in urine.

Excretion is removal of the toxic waste products of metabolism from the body. For example, nitrogen-containing compounds are produced when excess amino acids are broken down. They become toxic if they accumulate in the body so they must be excreted. In mammals, the main nitrogenous waste product is urea.

The kidneys also remove substances passively absorbed from food in the gut that are not used by the body—for example, many drugs and pigments from food. These substances may colour the urine unexpectedly (Figure 13).

The basic functional unit of the kidney is the nephron. This is a tube with a wall consisting of one layer of cells. This wall is the last layer of cells that substances cross to leave the body—it is an epithelium. There are several different parts of the nephron, which have different functions and structures (Figure 14).

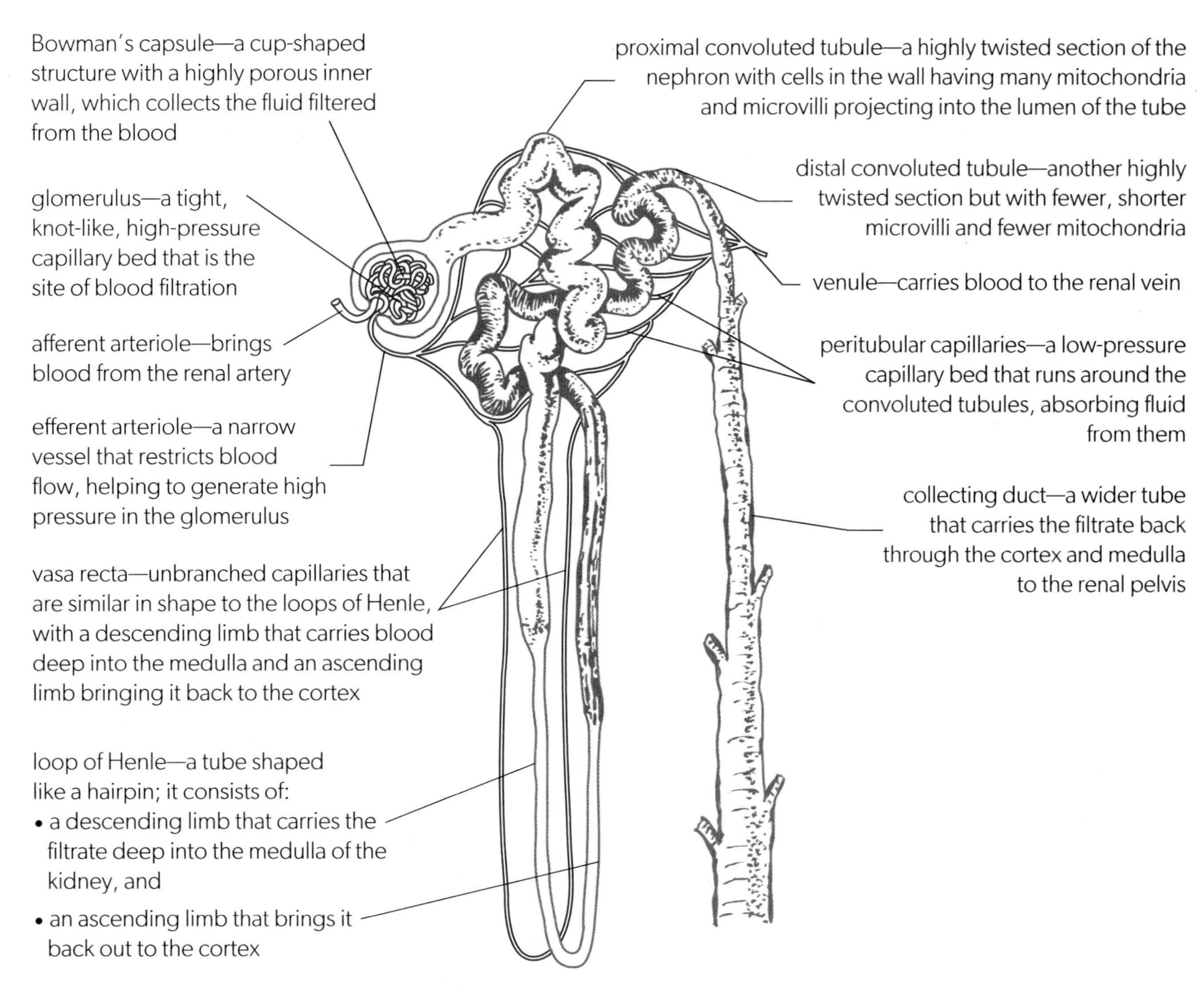
Bowman's capsule—a cup-shaped structure with a highly porous inner wall, which collects the fluid filtered from the blood
glomerulus—a tight, knot-like, high-pressure capillary bed that is the site of blood filtration
afferent arteriole—brings blood from the renal artery
efferent arteriole—a narrow vessel that restricts blood flow, helping to generate high pressure in the glomerulus
vasa recta—unbranched capillaries that are similar in shape to the loops of Henle, with a descending limb that carries blood deep into the medulla and an ascending limb bringing it back to the cortex
loop of Henle—a tube shaped like a hairpin; it consists of:
• a descending limb that carries the filtrate deep into the medulla of the kidney, and
• an ascending limb that brings it back out to the cortex
proximal convoluted tubule—a highly twisted section of the nephron with cells in the wall having many mitochondria and microvilli projecting into the lumen of the tube
distal convoluted tubule—another highly twisted section but with fewer, shorter microvilli and fewer mitochondria
venule—carries blood to the renal vein
peritubular capillaries—a low-pressure capillary bed that runs around the convoluted tubules, absorbing fluid from them
collecting duct—a wider tube that carries the filtrate back through the cortex and medulla to the renal pelvis

▲ Figure 14 The nephron and associated blood vessels

AHL

Solute	Content in a litre (dm^3) of blood plasma	Content in a litre (dm^3) of glomerular filtrate
Na^+ ions / mol	151	144
Cl^- ions / mol	110	114
glucose / mol	5	5
urea / mol	5	5
proteins / mg	740	3.5

▲ Table 2

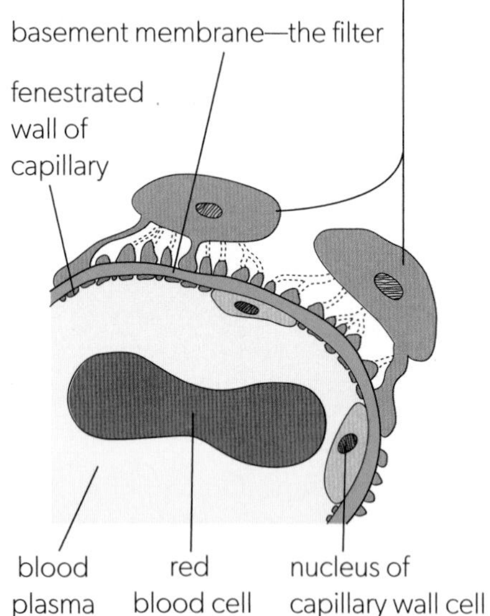

▲ Figure 15 Structure of the filter unit of the kidney, consisting of the basement membrane (green) and the gaps between the foot processes of the podocytes

D3.3.8 Role of the glomerulus, Bowman's capsule and proximal convoluted tubule in excretion

Ultrafiltration is the first stage in the production of urine. It is carried out by structures consisting of a glomerulus inside a Bowman's capsule. There are about a million of these structures located in each kidney cortex—the outer part of the kidney.

The glomerulus is a ball-shaped network of blood capillaries. Blood flows into the glomerulus through an afferent arteriole and leaves the glomerulus through an efferent arteriole. It is an arteriole rather than a venule because it carries the blood on to other capillaries in the kidney, rather than to a vein.

Fluid is filtered out through the walls of all blood capillaries, to produce tissue fluid, most of which then passes back into the capillaries. In the glomerulus, a much larger proportion of the blood plasma is filtered out—about 100 times more. This is because the blood pressure is particularly high and the capillary wall is particularly permeable. There are two factors causing the high pressure of blood in glomerular capillaries:

- the efferent arteriole being narrower than the afferent arteriole
- the contorted route that blood must follow to pass through the glomerulus.

The high permeability of the capillary wall is due to the presence of pores called fenestrations. All capillaries have pores that allow fluid to escape without having to pass through the wall cells, but the fenestrations in glomerular capillaries are unusually wide and numerous. They have a diameter of about 100 nm.

The fluid forced out of the blood plasma is called glomerular filtrate. The composition of blood plasma and filtrate is shown in Table 2. The data in the table shows that most solutes are filtered out from the blood plasma but almost all proteins are retained in the capillaries of the glomerulus. This is separation of particles differing in size by only a few nanometres and so is called ultrafiltration. All particles with a relative molecular mass below 65,000 atomic mass units can pass through. Permeability to larger molecules depends on their shape and charge. Almost all proteins are retained in the blood along with all the blood cells.

The filter unit has two layers and allows only small and medium-sized molecules to pass out of the blood.

1. The first layer is the basement membrane that covers and supports the wall of the capillaries. It is a non-cellular gel, made of negatively charged glycoproteins that are cross-linked to form a mesh. It prevents plasma proteins from being filtered out, due to their size and negative charges.
2. The second layer is the inner wall of Bowman's capsule. It consists of cells with branching outgrowths that wrap around the capillaries of the glomerulus. The cells are called podocytes and the branches are called foot processes. Very narrow gaps between adjacent foot-processes help prevent small molecules from being filtered out of blood in the glomerulus.

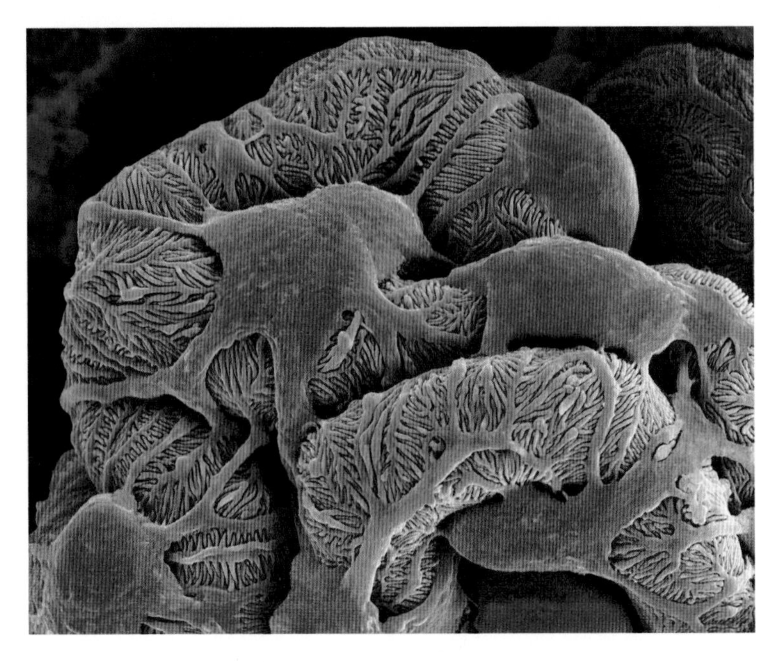

◀ Figure 16 Scanning electron micrograph of part of a glomerulus. The podocyte cell bodies (orange) have outgrowths with narrow branches (pink) that wrap around and completely hide the capillaries. Filtrate escapes from the capillaries in the narrow slits between these branches. The capillaries have a diameter of about 8 μm

The total volume of glomerular filtrate produced by the kidneys per day is huge—about 180 litres. This is several times the total volume of fluid in the body and it contains nearly 1.5 kg of salt and 5.5 kg of glucose. As the volume of urine produced per day is only about 1.5 litres and it contains no glucose and far less than 1.5 kg of salt, almost all of the filtrate must be reabsorbed into the blood. This happens in tubular structures called nephrons. Each glomerulus has an associated nephron.

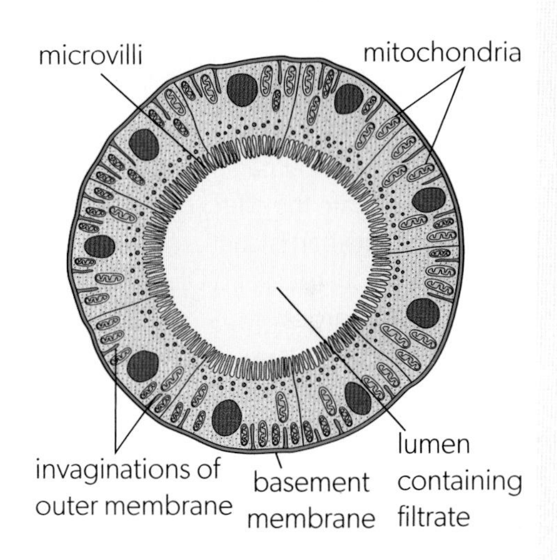

▲ Figure 17 Transverse section of the proximal convoluted tubule

The first part of the nephron is the Bowman's capsule. This is a cup-shaped structure. The outer wall of the Bowman's capsule is an impermeable layer of cells that ensures all the glomerular filtrate flows into the next part of the nephron—the proximal convoluted tubule. The convolutions increase the length of this part of the nephron, so the filtrate takes longer to flow through it, allowing more reabsorption of substances in the filtrate than the body needs. Indeed, the bulk of selective reabsorption has been accomplished by the end of the proximal tubule, with all glucose and amino acids and 80% of the water, sodium and other mineral ions returned to the blood in capillaries adjacent to the tubule.

The methods used to reabsorb four key substances are as follows.

Sodium ions—are moved by active transport from filtrate to space outside the tubule. They then pass to the peritubular capillaries. Pump proteins are located in the outer membrane of tubule cells.

Chloride ions—are attracted from filtrate to space outside the tubule because of a charge gradient set up by active transport of sodium ions.

Glucose—is moved by cotransporter proteins in the outer membrane of tubule cells. These proteins allow sodium ions to move down their concentration gradient from outside the tubule into tubule cells and at the same time glucose to move against its concentration gradient in the opposite direction. This reduces the glucose concentration of the tubule cells, causing glucose diffusion from the filtrate to the cells. The same process is used to reabsorb amino acids from the filtrate.

Water—active transport of solutes from the filtrate to the fluid outside the tubule creates a solute concentration gradient and causes water to be reabsorbed from filtrate by osmosis.

Activity: Structure of the proximal convoluted tubule

Study Figures 18 and 19 and then explain how the structure of the proximal convoluted tubule cell is adapted to carry out selective reabsorption.

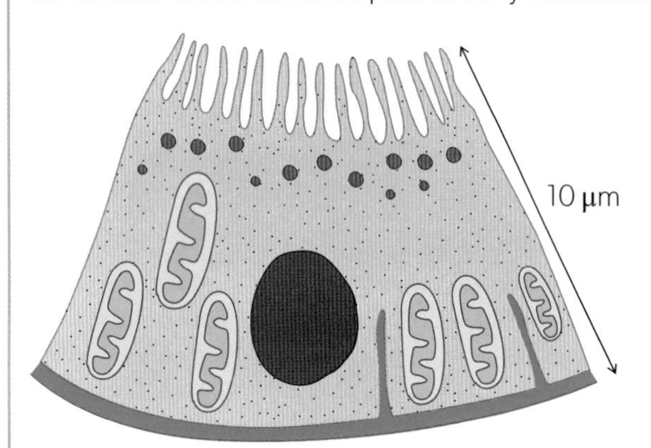

▲ **Figure 18** Structure of a cell from the wall of the proximal convoluted tubule

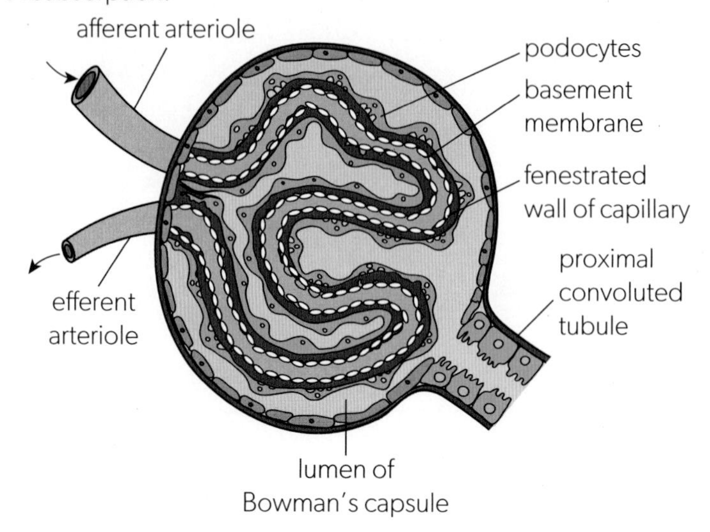

▲ **Figure 19** Relationship between glomerulus and Bowman's capsule

Data-based questions: Ultrafiltration of charged and uncharged dextrans

Dextrans are polymers of sucrose. Different sizes of dextran polymer can be synthesized, allowing their use to investigate the effect of particle size on ultrafiltration. Neutral dextran is uncharged, dextran sulfate has many negative charges, and DEAE is dextran with many positive charges.

Figure 20 shows the relationship between particle size and the permeability of the filter unit of rat glomeruli. Animal experiments like this can help us to understand how the kidney works and can be done without causing suffering to the animals.

1. State the relationship between the size of particles and the permeability to them of the filter unit of the glomerulus. [1]
2. a. Compare the permeability of the filter unit to the three types of dextran. [3]

   b. Explain these differences in permeability. [3]
3. One of the main plasma proteins is albumin, which is negatively charged and has a particle size of approximately 4.4 nm. Using the data in the graph, explain the diagnosis that is made if albumin is detected in a rat's urine. [3]

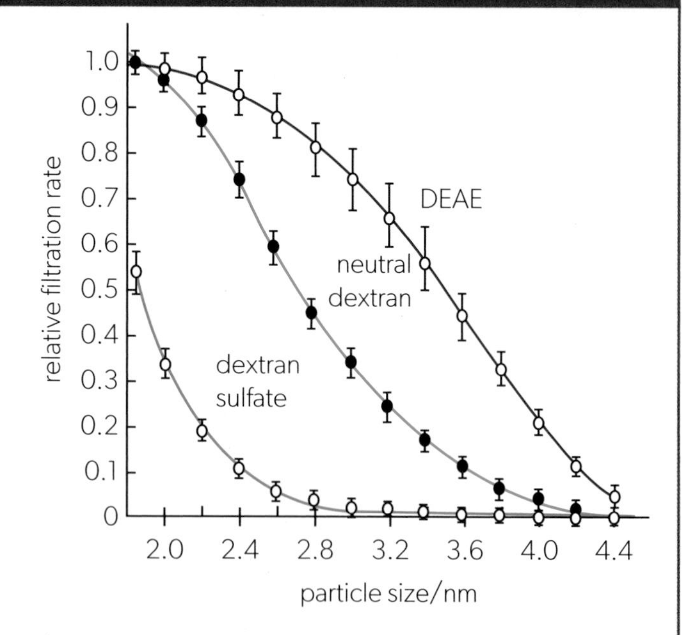

▲ **Figure 20** Relationship between particle size of dextrans and filtration rate

D3.3.9 Role of the loop of Henle

The kidney has two main regions—an outer cortex and an inner medulla. The cortex contains all the glomeruli and Bowman's capsules and the proximal and distal convoluted tubules. The medulla contains loops of Henle and collecting ducts. The cortex has the same osmotic concentration as most other tissues of the body—about 300 milliosmoles. In the medulla, there is a concentration gradient, starting at 300 mOsm near the cortex and rising to 1,200 mOsm near the centre of the kidney. This gradient can be used to extract water from the filtrate as it flows through the medulla in the collecting duct. This has the effect of producing urine with a higher osmotic concentration than normal body fluids (hypertonic urine).

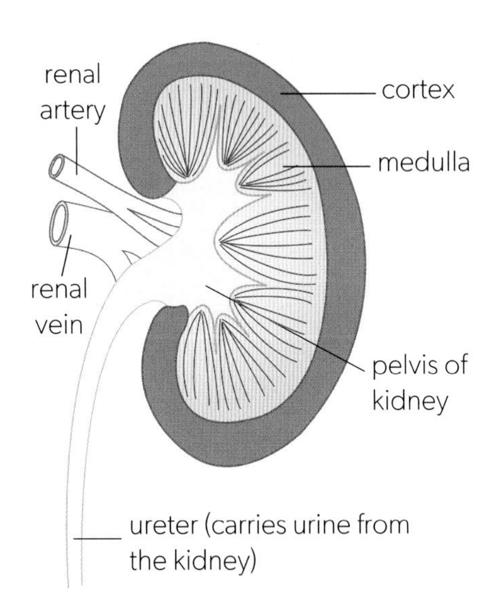

▲ Figure 21 Structure of the kidney

The role of the loop of Henle is to establish and maintain the osmotic concentration gradient in the medulla and very high osmotic concentration in the centre of the kidney. Filtrate enters the loop of Henle from the proximal convoluted tube and first flows towards the centre of the kidney in the descending limb. It then does a U-turn and flows back to the medulla in the ascending limb.

The energy needed to create the osmotic concentration gradient is expended by wall cells in the ascending limb. Sodium ions are pumped out of the filtrate to the fluid between the cells in the medulla—called the interstitial fluid. The wall of the ascending limb is unusual in that it is impermeable to water, so water is retained in the filtrate, even though the interstitial fluid is now hypertonic relative to the filtrate (that is, it has a higher solute concentration).

The pump proteins that transfer sodium ions out of the filtrate can create a gradient of up to 200 mOsm, so an interstitial concentration of 500 mOsm is achievable. The cells in the wall of the descending limb are permeable to water but are impermeable to sodium ions. As filtrate flows down the descending limb, the increased solute concentration of interstitial fluid in the medulla causes water to be drawn out of the filtrate until it reaches the same solute concentration as the interstitial fluid. If this was 500 mOsm, then filtrate entering the ascending limb would be at this concentration and the sodium pumps could raise the interstitial fluid to 700 mOsm. Fluid passing down the descending limb would therefore reach 700 mOsm, and the sodium pumps in the ascending limb could cause a further 200 mOsm rise. The interstitial fluid concentration can therefore rise further and further, until a maximum is reached. In humans, the maximum is 1,200 mOsm.

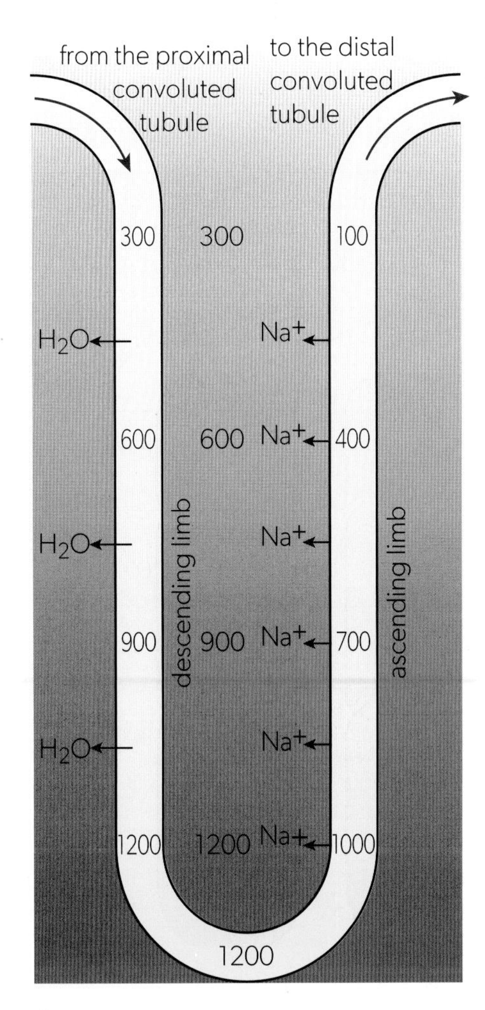

▲ Figure 22 Solute concentrations in the loop of Henle (in mOsm)

This system for raising solute concentration is an example of a countercurrent multiplier system. It is a countercurrent system because of the flows of fluid in opposite directions. It is a countercurrent multiplier because it causes a steeper gradient of solute concentration to develop in the medulla than would be possible with a concurrent system. There is also a countercurrent system in the vasa recta. This prevents the blood flowing through this vessel from diluting the solute concentration of the medulla, while still allowing the vasa recta to carry away the water removed from filtrate in the descending limb, together with some sodium ions.

As filtrate flows through the loop of Henle, proportionately more solutes than water are reabsorbed. The filtrate flowing into the distal convoluted tubule therefore has a solute concentration lower than that of normal body fluids—it is hypotonic.

Data-based questions: Medulla thickness and urine concentration

Table 3 shows the relative medullary thickness (RMT) and maximum solute concentration (MSC) of the urine in mOsm for 14 species of mammal. RMT is a measure of the thickness of the medulla in relation to the overall size of the kidney. All the species shown in the table with binomials are desert rodents.

1. Discuss the relationship between maximum solute concentration of urine and the habitat of the mammal. [3]
2. Plot a scattergraph of the data in the table, either by hand or using computer software. [7]
3. a. Using the scattergraph that you have plotted, state the relationship between RMT and the maximum solute concentration of the urine. [1]

   b. Suggest how the thickness of the medulla could affect the maximum solute concentration of the urine. [4]

Species	RMT	MSC / mOsm
beaver	1.3	517
pig	1.6	1,076
human	3.0	1,399
dog	4.3	2,465
cat	4.8	3,122
rat	5.8	2,465
Octomys mimax	6.1	2,071
Dipodomys deserti	8.5	5,597
Jaculus jaculus	9.3	6,459
Tympanoctomys barrerae	9.4	7,080
Psammomys obesus	10.7	4,952
Eligmodontia typus	11.4	8,612
Calomys mus	12.3	8,773
Salinomys delicatus	14.0	7,440

▲ Table 3

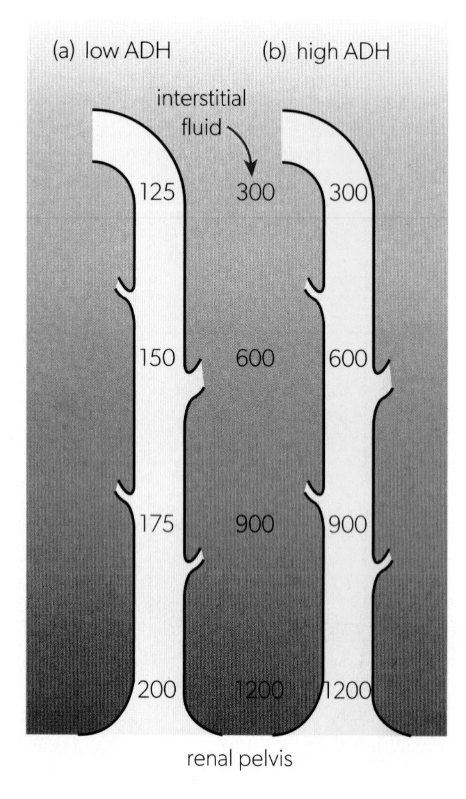

▲ Figure 23 Solute concentrations of filtrate in the collecting duct

D3.3.10 Osmoregulation by water reabsorption in the collecting duct

The kidneys can both increase and decrease the osmotic concentration of blood to keep body fluids at the normal level of about 300 mOsm. To increase the osmotic concentration, much water is removed from the blood by producing large volumes of hypotonic urine. To decrease the osmotic concentration of blood, little water is removed by producing small volumes of hypertonic urine. This is achieved by varying the permeability to water of plasma membranes in cells that form the walls of the distal convoluted tubule and collecting duct. Water permeability depends on the numbers of aquaporins in a plasma membrane. Aquaporins are channel proteins that allow water to pass through. If there are large numbers of aquaporins in a membrane the permeability to water is high. In the distal convoluted tubule, aquaporins can be removed from the membranes so permeability can become unusually low.

The filtrate that enters the distal convoluted tubule is hypotonic, with an osmotic potential of about 100 mOsm. If the solute concentration of the blood is too low, the permeability to water of plasma membranes in the distal convoluted tubule and collecting duct cells is kept low. Relatively little water is reabsorbed from the filtrate flowing past. High flow rates are maintained, further reducing the amount of water reabsorbed. This is how a large volume of urine can be produced with a low solute concentration.

The osmotic concentration of the blood is monitored in the hypothalamus. If it becomes too high, the hypothalamus responds by causing the pituitary gland to secrete antidiuretic hormone (ADH). This hormone causes more aquaporins to be moved to the plasma membranes of cells in the distal convoluted tubule and collecting duct, increasing their water permeability. As the filtrate flows along the distal convoluted tubule, water is reabsorbed by osmosis until the filtrate reaches the osmotic concentration of the surrounding cortex tissue—300 mOsm. The filtrate flows on into the collecting duct. Water is reabsorbed along the entire length because of the gradient of osmotic concentration in the medulla. So much water is reabsorbed that the flow rate of the filtrate becomes slow, allowing more time for reabsorption. An osmotic concentration of 1,200 mOsm can be achieved, which is four times the concentration of normal body fluids.

At the end of the collecting duct, the filtrate has become urine and it is discharged into the renal pelvis. This is a spongy tissue, from which the urine drains via the ureter to the bladder.

Activity: Genetic disease affecting ADH receptors

ADH binds to a receptor protein in the membranes of nephron cells in the distal convoluted tubule and collecting duct. There is a genetic disease in which this receptor protein is altered and ADH fails to bind to it.

1. Predict the health problems that would result from this disease, which can be serious enough to cause death if appropriate measures are not taken. [4]
2. Children can develop this genetic disease even if their parents show no symptoms. Explain the conclusion that you draw from this observation. [2]
3. Boys are affected far more frequently than girls. Explain the conclusion that you draw about the inheritance of the disease. [2]
4. Suggest measures that could be used to treat the disease. [2]

Data-based questions: ADH release and feelings of thirst

The plasma solute concentration, plasma antidiuretic hormone (ADH) concentration, and feelings of thirst were tested in a group of volunteers. Figure 24 shows the relationship between intensity of thirst and plasma solute concentration. Figure 25 shows the relationship between ADH and plasma solute concentration.

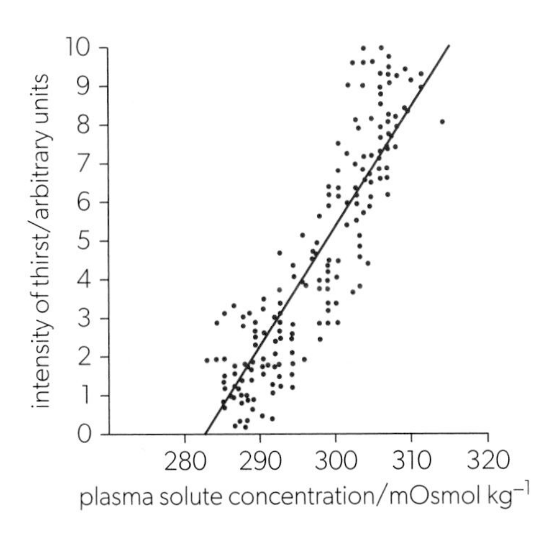

▲ Figure 24 Graph of thirst against solute plasma concentration

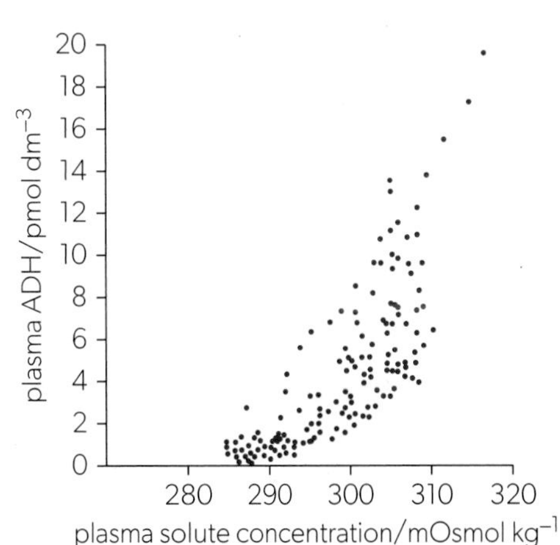

▲ Figure 25 Graph of ADH against solute plasma concentration

1. a. Predict ADH concentration in a person with a plasma solute concentration of 300 mOsmol kg^{-1} [1]
   b. Compare and contrast trends in the data shown by the two graphs. [3]
2. Outline what would happen to plasma solute concentration and ADH concentration if a person drank enough water to satisfy their thirst. [4]
3. State two reasons why a person's plasma solute concentration may increase. [2]

AHL

D3.3.11 Changes in blood supply to organs in response to changes in activity

Blood is pumped out of the left ventricle into the aorta, which then divides repeatedly to supply all the organs of the body apart from the lungs. The lungs are supplied with deoxygenated blood by the pulmonary arteries. The cardiac output of the left side of the heart is divided between the various organs, but even at its maximum is not enough to supply every organ at a maximal rate. Adjustments to the distribution of blood are therefore necessary.

Supply of blood to an organ is increased or decreased using rings of circular muscle in the walls of the arterioles serving the organ. Contraction of these rings of muscle causes vasoconstriction (narrowing of the lumen) and restricts blood flow. In some organs, shunt vessels are used to direct blood directly from arterioles to venules. Relaxation of circular muscle in arterioles causes vasodilation (widening of the lumen) and increases blood flow to the tissues served. So, using circular muscle in arterioles, the distribution of blood is adjusted to suit changes in activity.

Three contrasting patterns of blood distribution are shown in Table 4, but there are many others. For example, the pattern differs between REM and non-REM sleep.

Tidal volume during ...	...vigorous physical exercise: 25 $dm^3 min^{-1}$	...wakeful rest: 5 $dm^3 min^{-1}$	...sleep: 4 $dm^3 min^{-1}$
Skeletal muscle	greatly increased supply—to provide maximum amounts of glucose and oxygen	moderate supply—even sitting or standing requires some contraction of muscles	reduced supply—minimal muscle contraction when lying prone
Digestive system	reduced supply—digestion can be paused during relatively brief periods of vigorous exercise	variable supply—increased when there is food in the gut and reduced flow during fasting	variable supply—increased when there is food in the gut and reduced flow during fasting
Kidneys	reduced supply—causing a reduction in glomerular filtration rate	maximal supply—about 20% of cardiac output	reduced supply—avoiding the need for urination during the night
Brain	increased supply—to supply more oxygen and glucose as the brain is particularly active	moderate supply—brain tissue relies on continuous supply of oxygen and glucose	increased supply—possibly to increase the rate of removal of toxins from brain tissue

▲ Table 4 Patterns of blood distribution during exercise, rest and sleep

Data-based questions: Blood supply in piglets

Distribution of blood was measured in piglets when they were awake and when they were asleep. Cardiac output did not differ much, but the amounts of blood directed to some organs were significantly different. Figure 26 shows the percentage change in blood flow to 12 organs between piglets when asleep and awake. Negative percentages indicate that flow was lower to an organ when the piglets were asleep.

1. What is the formula for calculating percentage change? [2]
2. Suggest reasons for the change in blood flow to:
   a. skeletal muscles [1]
   b. diaphragm and intercostal muscles [2]
   c. adrenal glands [1]
   d. left ventricle [1]
   e. brain [1]
3. In adult humans, blood flow to the kidneys is reduced during sleep, but in piglets it is unchanged. Suggest reasons for the difference. [2]

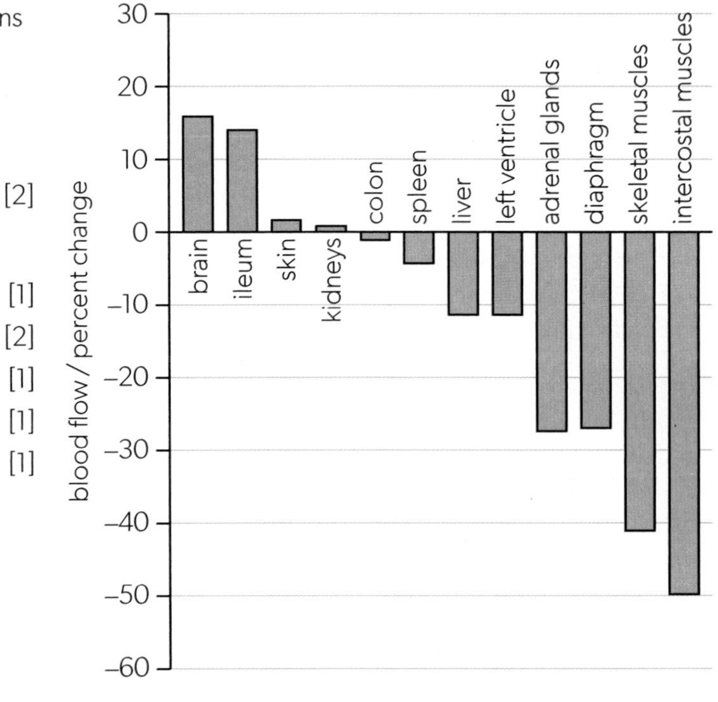

▲ Figure 26 Bar chart of blood flow to various organs in piglets

Linking questions

1. For what reasons do organisms need to distribute materials and energy?
   a. Describe the role of vasodilation and vasoconstriction in moderating the movement of heat between a human and its environment. (D3.3.6)
   b. Explain one example of a solute that is moved by active transport. (B3.2.17)
   c. Describe the movement of carbon dioxide from the point of production in a cell through to expiration to the outside air. (B3.3.3)
2. What biological systems are sensitive to temperature changes?
   a. Outline the relationship between temperature and the rate of photosynthesis. (C1.3.7)
   b. Explain the relationship between enzyme activity and temperature. (C1.1.8)
   c. Explain the behavioural and physiological adaptations that humans possess to maintain body temperature. (D3.3.6)

TOK

To what extent is certainty possible?

Scientific reasoning is often inductive. Following observations, a scientist may propose a generalization known as a hypothesis, which they can test. If the hypothesis is supported by a range of further observations, the generalization can obtain the status of a theory.

Certainty is the firm conviction that something is the case. The "problem of induction" refers to the challenge that it is not certain that everything observed to date fully predicts what will be observed in the future. Consider the following quote from Bertrand Russell, The Problems of Philosophy:

"Domestic animals expect food when they see the person who usually feeds them. We know that all these rather crude expectations of uniformity are liable to be misleading. The man who has fed the chicken every day throughout its life at last wrings its neck instead, showing that more refined views as to the uniformity of nature would have been useful to the chicken."

▲ Figure 1 Feeding chickens

How can we be certain of any conclusion that is arrived at inductively? Does science only deal in information which is "probably true"?

Scientists can gain certainty in some areas through a process called falsification. What is not the case can be ruled out with certainty. For example, Figure 2 can be used to establish with certainty that not all peppered moths are light-coloured.

Through experimentation, it can be established that the black variant has greater evolutionary fitness in polluted areas, meaning that it has a higher probability of its genes being represented in the next generation in a polluted area.

▲ Figure 2 Peppered moths

Using the pedigree chart in Figure 3, it is possible to falsify with certainty that the trait is dominant. If it were dominant, the appearance of the phenotype in individual II-3 would not be possible because the trait does not appear in either parent. The data supports the idea that the trait might be X-linked because in this pedigree it only appears in males. If it is X-linked, it will passed on through mothers as carriers. This is observed for all three males. That the trait is X-linked is treated as a pragmatic truth because it is true "probabilistically". If either individual III-5 or IV-1 had an unaffected daughter who bore an affected son, this would further support the hypothesis of X-linkage. It would not prove it with absolute certainty, but it would be unreasonable to conclude otherwise. Thus, the problem of induction is solved by the certainty of falsification and the willingness to be reasonable in the face of probability.

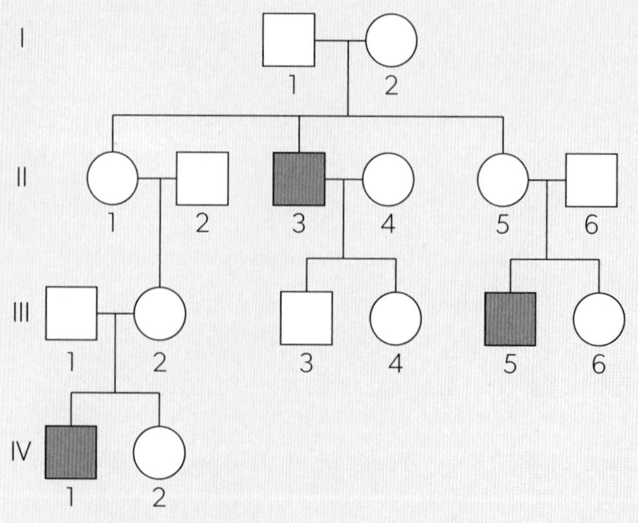

▲ Figure 3 Pedigree chart

End of chapter questions

1. Three species of closely related spiders live in a forest in eastern Austria. They build similar webs to catch similar sizes of prey in them. The box-plot below shows the height above ground level of webs constructed by the three species in each month from March to October. In months when no measurements are shown, the spiders were too young to build webs.

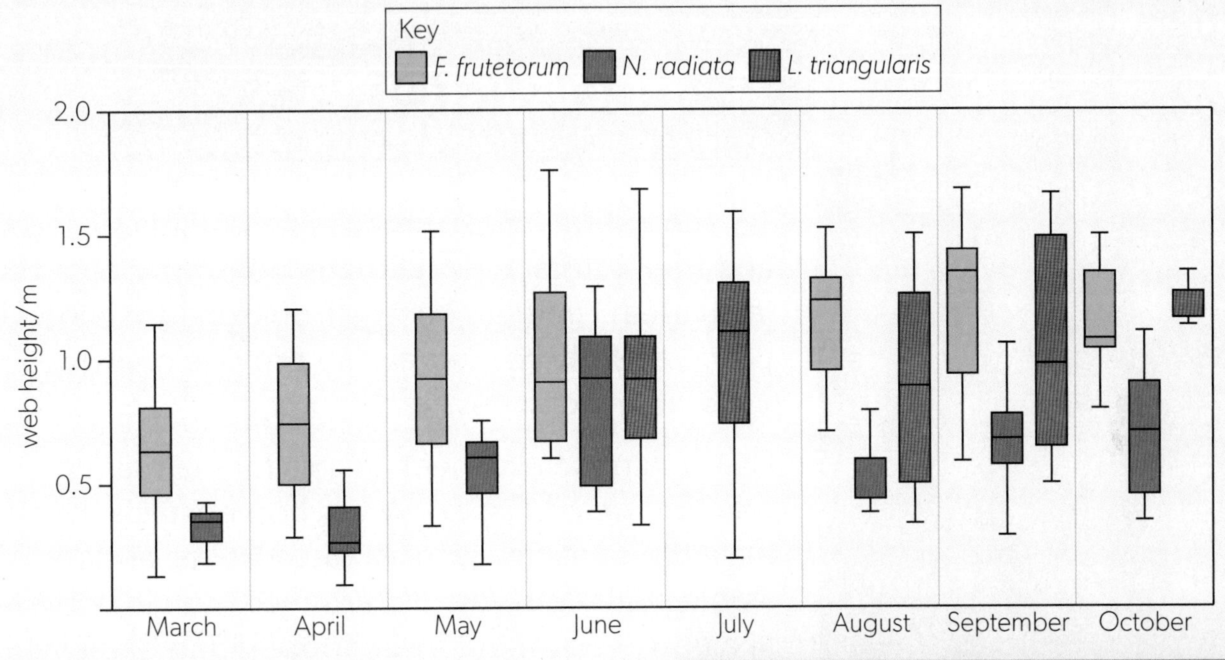

   a. State in which month the median web height of the three spiders is most similar. [1]
   b. Identify which species, in which month, builds webs over the greatest range of heights. [1]
   c. Suggest one reason for this. [1]
   d. Determine the interquartile range for *L. triangularis* in the month of September. [1]
   e. For *L. triangularis* in the month of September, give an example of a high web-height that would be considered an outlier. [2]

2. The diploid chromosome number of the pea plant which Mendel used in his experiments is 14.
   a. How many linkage groups are there in peas? [1]
   b. The table shows the chromosome location of the genes that affect each of Mendel's character differences.
      i. State one pair of character differences you expect to be unlinked. [1]
      ii. State one pair of character differences you expect to show autosomal linkage. [1]

Character difference investigated in Mendel's experiments	Chromosome on which gene is located
tall stem/dwarf stem	4
round/wrinkled seed yellow /green cotyledons	7 1
purple/white flowers full pods/constricted pods	1 4
green/yellow unripe pods flowers on stem/at stem tip	5 4

   c. If pure-breeding plants with full green pods were crossed with pure-breeding plants with constricted yellow pods and the F_1 plants were allowed to self-pollinate, what ratio of phenotypes would you expect in the F_2 generation? Use a genetic diagram to work our your answer. [5]
   d. When diploid crosses are performed with plants differing in cotyledon colour and flower colour, they give results suggesting that the two genes involved are unlinked. The two genes are located far apart from each other on chromosome number 1. Explain how this can cause them to behave as though they are in different linkage groups. [2]

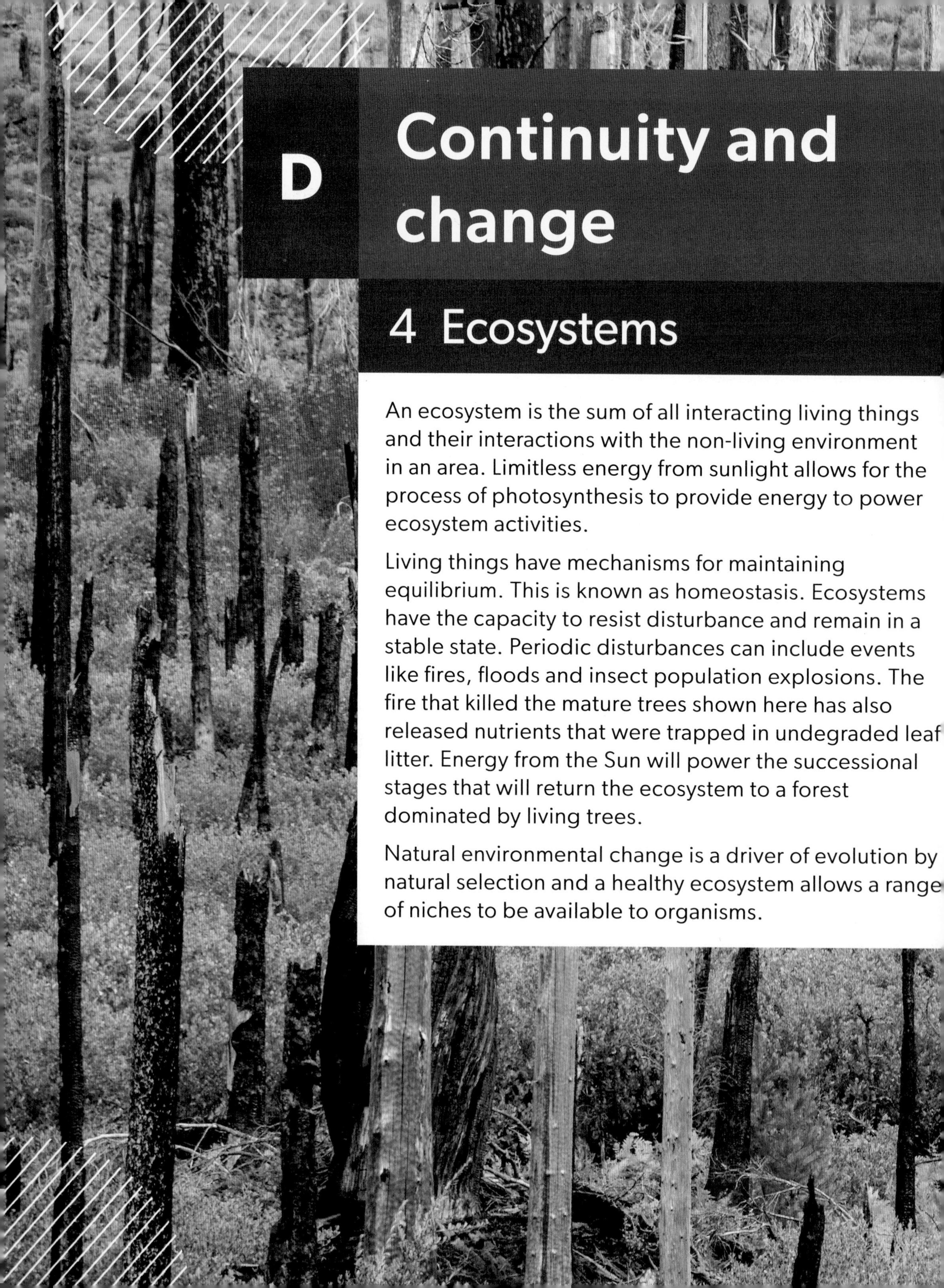

D Continuity and change

4 Ecosystems

An ecosystem is the sum of all interacting living things and their interactions with the non-living environment in an area. Limitless energy from sunlight allows for the process of photosynthesis to provide energy to power ecosystem activities.

Living things have mechanisms for maintaining equilibrium. This is known as homeostasis. Ecosystems have the capacity to resist disturbance and remain in a stable state. Periodic disturbances can include events like fires, floods and insect population explosions. The fire that killed the mature trees shown here has also released nutrients that were trapped in undegraded leaf litter. Energy from the Sun will power the successional stages that will return the ecosystem to a forest dominated by living trees.

Natural environmental change is a driver of evolution by natural selection and a healthy ecosystem allows a range of niches to be available to organisms.

D4.1 Natural selection

What processes can cause changes in allele frequencies within a population?

These two forms of peppered moth (*Biston betularia*) are a famous example of natural selection in action. The light form has recessive alleles of a gene affecting wing colour and the dark form has a dominant allele. The dark form is conspicuous when the moths are roosting on lichen-covered birch trees. Experiments have shown that it has a 10% higher chance of being predated by birds. Originally very rare, it became common in parts of Europe after industrial pollution killed lichens and stained the bark of trees black. Suggest, with a reason, which variety is more common in Britain today?

▲ Figure 1 Dark peppered moth on lichen-covered bark

What is the role of reproduction in the process of natural selection?

What role does sexual reproduction play in generating this variation? Why is variation important? Sexual selection leads to non-random mating. What consequence would this have on allele frequency? Green sea turtles (*Chelonia mydas*) reproduce sexually, with females coming ashore to lay over 100 eggs. The females each return to the same beach every two or three years, so a successful individual will lay thousands of eggs during her lifetime. But few will survive the 20 to 50 years needed to reach reproductive age. Males and females do not pair for life, increasing the genetic diversity of eggs laid by a female.

▶ Figure 2 Female green sea turtles coming ashore to lay their eggs

SL and HL	AHL only
D4.1.1 Natural selection as the mechanism driving evolutionary change D4.1.2 Roles of mutation and sexual reproduction in generating the variation on which natural selection acts D4.1.3 Overproduction of offspring and competition for resources as factors that promote natural selection D4.1.4 Abiotic factors as selection pressures D4.1.5 Differences between individuals in adaptation, survival and reproduction as the basis for natural selection D4.1.6 Requirement that traits are heritable for evolutionary change to occur D4.1.7 Sexual selection as a selection pressure in animal species D4.1.8 Modelling of sexual and natural selection based on experimental control of selection pressures	D4.1.9 Concept of the gene pool D4.1.10 Allele frequencies of geographically isolated populations D4.1.11 Changes in allele frequency in the gene pool as a consequence of natural selection between individuals according to differences in their heritable traits D4.1.12 Differences between directional, disruptive and stabilizing selection D4.1.13 Hardy–Weinberg equation and calculations of allele or genotype frequencies D4.1.14 Hardy–Weinberg conditions that must be maintained for a population to be in genetic equilibrium D4.1.15 Artificial selection by deliberate choice of traits

D4.1.1 Natural selection as the mechanism driving evolutionary change

The theory of evolution by natural selection can be explained with the following series of statements.

- Organisms produce more offspring than the environment can support.
- Among these offspring, there is variation.
- Some variants are better suited to the environment and have a higher chance of surviving to reproductive age. Less fit (less well adapted) variants have a higher risk of mortality from predation or other factors. This is sometimes called survival of the fittest.
- The features that aid survival are disproportionately inherited by successful offspring. These features therefore increase in frequency in the population. The heritable features of the population have changed, so it has evolved.

While evolution can occur rapidly in response to sudden changes in the environment, the process is often gradual. Evolution by natural selection, over the billions of years that life has existed, has resulted in the immense biodiversity now present on Earth.

Activity: Adaptations of birds' beaks

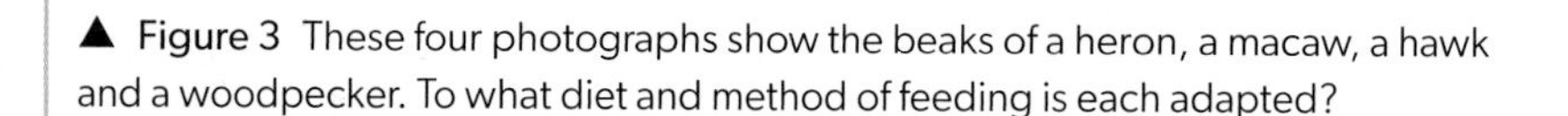

▲ **Figure 3** These four photographs show the beaks of a heron, a macaw, a hawk and a woodpecker. To what diet and method of feeding is each adapted?

Theories: Lamarckism and Paradigm shifts

In his influential book of 1962, *The Structure of Scientific Revolutions*, Thomas Kuhn adopted the word "paradigm" for a framework used to interpret information in science. A paradigm shift occurs when observations accumulate that are more easily explained by a new, "revolutionary" paradigm than the existing one.

In Darwin's time, it was widely understood that species evolved, but the mechanism was not clear. A paradigm that existed at the time was that organisms acquire advantageous traits during their lifetime and that these traits are inherited by offspring. For example, if a giraffe stretches its neck to reach leaves high in trees, its offspring will have a longer neck, even before they feed on high leaves.

Observations made by Charles Darwin led him to develop an alternative and revolutionary paradigm—evolution by natural selection. Darwin made many diverse observations during his voyage around the world on HMS *Beagle*. He developed his theory in the late 1830s and then worked to accumulate evidence for it. Darwin published his great work, *On the Origin of Species*, in 1859. In this book of nearly 500 pages, he explains evolution by natural selection and presents evidence for it collected over the previous 20 to 30 years.

The theory of evolution by inheritance of acquired characteristics is now known as Lamarckism after the French naturalist Jean-Baptiste Lamarck who proposed it in 1809. Even within Darwin's *On the Origin of Species*, the influence of Lamarckism is evident: "there can be little doubt that use in our domestic animals strengthens and enlarged certain parts, and disuse diminished them; and that such modifications are inherited… The great and inherited development of the udders in cows and goats in countries where they are habitually milked, in comparison with the state of these organs in other countries, is another instance of the effect of use."

Despite this misinterpretation, Darwin's work refuted Lamarckism and therefore caused a paradigm shift. Weismann's germ plasm theory, published in 1892, established that there is no obvious means for hereditary information to pass between somatic cells and sex cells, so acquired characteristics cannot be inherited.

ATL Research skills: Strategies for effectively reading from a textbook

Textbooks are unique in terms of their purpose and the style of writing. To get the most out of reading a textbook, employ the following strategies.

1. **Preview:** Scan through the assigned section. Read all subheadings, image captions and figure headings. This will make you curious as well as giving you a sense of the contents.
2. **Set a purpose:** Be clear on the purpose of your reading. Is it to prepare in advance for what you will learn in class or is it to understand what was covered in class?
3. **Convert headings to questions before reading:** For example, for *Section D4.1.2* convert the heading to a question: What is the role of mutation and sexual reproduction in generating the variation on which natural selection acts? When you have finished reading the section, attempt to write the answer from memory to test your understanding.

D4.1.2 Roles of mutation and sexual reproduction in generating the variation on which natural selection acts

One of the observations on which Darwin based the theory of evolution by natural selection is variation. Typical populations vary in many respects. Variation in human populations is obvious in traits such as height, skin colour and blood group. In other species, the variation may not be so immediately obvious but careful observation shows that it is there. Natural selection can only occur if there is variation within a population and some variants are removed while others survive and reproduce. The causes of variation in populations are well understood.

1. Mutation is the original source of variation. New alleles are produced by mutation, which enlarges the gene pool of a population.
2. Meiosis produces new combinations of alleles, by breaking up the existing combination in the diploid cells of a parent to produce haploid cells with only one allele of each gene. Because of crossing over and the independent orientation of bivalents, every cell produced by meiosis in an individual is likely to carry a different combination of alleles.
3. Sexual reproduction involves the fusion of male and female gametes. The gametes usually come from different parents, so offspring have a combination of alleles from two individuals. This allows mutations that occurred in different individuals to be brought together.

In species that do not carry out sexual reproduction, the only source of variation is mutation. It is generally assumed that such species will not generate enough variation to be able to evolve quickly enough for survival during times of environmental change.

▲ Figure 4 The breeding rate of pairs of southern ground hornbills is a low as 0.3 offspring per year

D4.1.3 Overproduction of offspring and competition for resources as factors that promote natural selection

Living organisms vary in the number of offspring they produce. An example of a species with a relatively slow breeding rate is the southern ground hornbill, *Bucorvus leadbeateri*. It raises one fledgling every three years on average and needs the cooperation of at least two other adults to do this. However, these birds can live for as long as 70 years so in their lifetime, a pair could theoretically raise 20 offspring. Most species have a faster breeding rate. For example, the coconut palm *Cocos nucifera*, usually produces between 20 and 60 coconuts per year. Apart from bacteria, the fastest breeding rate of all may be in the fungus *Calvatia gigantea*. It produces a huge fruiting body called a giant puffball in which there can be as many as 7 trillion spores (7,000,000,000,000).

Despite the huge variation in breeding rate, there is an overall trend for more offspring to be produced than can be supported by the available resources such as food. The population size that can be supported by the environment is known as the carrying capacity of the environment. It tends to be determined by the limiting resource that is in shortest supply. For a plant species, this is likely to be water in a desert or light in a rainforest. Darwin pointed out that overproduction of offspring results in a struggle for existence within a population. There is competition for resources and not every individual will obtain enough to allow them to survive and reproduce.

▲ Figure 5 A pride of lions (*Panthera leo*) drinking at a watering hole. Heritable characteristics that help an individual lion remain alive until reproductive age will persist and become more common in the population. Competition for access to a watering hole is a density-dependent factor

D4.1.4 Abiotic factors as selection pressures

Consider the example of a population of animals in which deaths are occurring due to shortage of food. Competition for food is acting as a selection pressure. Food is a biotic factor, because other organisms are the source of the food, not something non-living. It is density-dependent, because as population density increases, competition for food becomes more intense. The density of a population is the number of individuals per unit area of habitat. A density-dependent factor causes either birth rate to fall with rising population density or death rate to increase with population density.

Now consider the example of a population of plants experiencing a period of freezing weather that kills some of the plants but not others. Low temperature is acting as a selection pressure. It is an aspect of the physical environment and not due to other organisms, so is an abiotic factor. It is density-independent because the chance of an individual plant being killed by the freezing temperatures is the same whether the population density is high or low. Differences in tolerance to cold, rather than competition between individuals, is the basis of the selection.

There are other examples of abiotic density-independent factors causing selection pressure on plant and animal populations. They are often associated with catastrophe such as flooding, earthquakes, fire, pollution or extreme climate conditions such as heat waves or drought.

▲ Figure 7 The female elephants on the left have the "tuskless" trait. Some populations of elephants in Mozambique are under pressure from poaching. They have a 30% higher incidence of tusklessness in female elephants than similar populations that are not under poaching pressure. It is hypothesized that poaching elephants to harvest ivory has acted as a selection pressure favouring tusklessness. Elephants use their tusks for defence and, in males, for competing with one another for territory and mates. Tusks are also used for stripping bark from trees and digging for water. In different conditions, tusks can be either an advantage or a disadvantage to female elephants

▲ Figure 6 Marsh samphire (*Salicornia europaea*) grows on mudflats that are inundated by the sea tides every day, so they must tolerate high salt concentrations. Salinity is an abiotic factor that is density-independent because the chance of an individual plant being killed by high salt concentrations is the same whether the population density is high or low

D4.1.5 Differences between individuals in adaptation, survival and reproduction as the basis for natural selection

Since Darwin published his theory, numerous cases have been found of differences in adaptation that affect rates of survival and reproduction. Although not amounting to proof of Darwin's theory of evolution by natural selection, these cases provide extremely strong evidence. Some traits are adaptations to abiotic factors in the environment and others help an individual in their response to biotic factors.

The term "fitness" means how well-adapted an individual is. Something that is suited to its purpose or role is said to be fit. Fitness is therefore specific to purpose. In evolutionary biology, fitness results from having adaptations for a specific niche within an ecosystem. Different species occupy different niches so they need different adaptations.

There is variation among individuals in a sexually reproducing population. Some individuals will be fitter than others. Fitness influences whether or not an individual survives for long enough to be able to reproduce and, of those that survive, how many offspring they have. The fittest individuals tend to survive longest and have the most offspring. They therefore make the largest contribution to the gene pool of the next generation. The next generation is more like them than the less fit individuals who produced few or no offspring. This is a form of intraspecific competition (competition between members of a species) and differences in survival and reproduction are the basis of natural selection.

Data-based questions: Hunting as a selection pressure for smaller horns

A population of bighorn sheep (*Ovis canadensis*) on Ram Mountain in Alberta, Canada, has been monitored since the 1970s. Hunters can buy a licence to shoot male bighorn sheep on the mountain. The large horns of this species are very attractive to hunters, who display them as hunting trophies. Most horn growth takes place between the second and the fourth year of life in male bighorn sheep. They use their horns for fighting other males during the breeding season to try to defend groups of females and then mate with them. Younger males below the age of 6 are rarely able to compete with older males and so have fewer offspring.

The length of a male's horns is strongly influenced by its genes. It is therefore possible to predict the length of the horns of the future offspring of a male. An index of predicted horn length for the future offspring of each male on Ram Mountain has been calculated. Negative values indicate offspring with smaller horns than the mean of the population. Positive values indicate horns larger than the mean of the population.

1. Suggest one type of factor, apart from genes, that could affect the length of a male's horns. [1]

The scatter graph in Figure 8 shows the relationship between the age to which a male lived (longevity) and the predicted horn length index of the male.

2. State the most common longevity in years (the mode). [1]
3. a. Outline the relationship between longevity and the predicted horn length index of the males. [1]

   b. Suggest a reason for the relationship. [1]

Figure 9 shows the mean horn length of males on Ram Mountain, between 1975 and 2002.

4. a. Explain the change in horn length shown in the graph in terms of the longevity of the males, the number of offspring that they produce and your understanding of the process of evolution. [3]

   b. Suggest how the change in the size of horns could be reversed. [1]

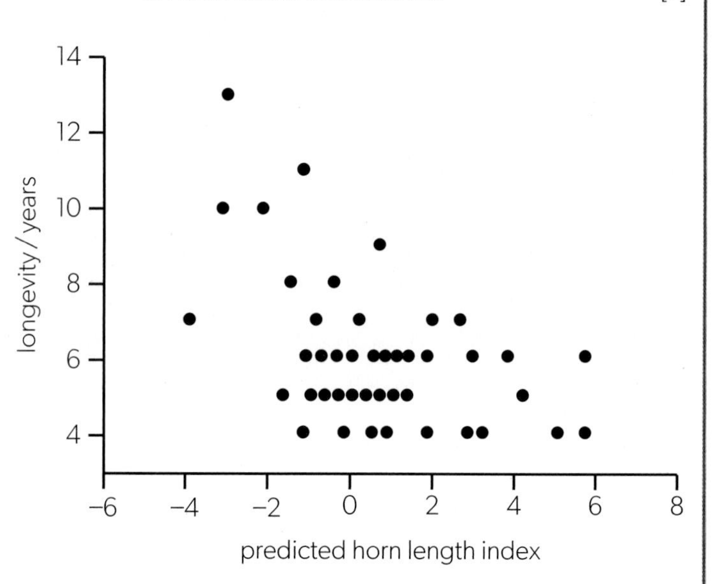

▲ **Figure 8** Scatter graph of bighorn sheep longevity and horn length

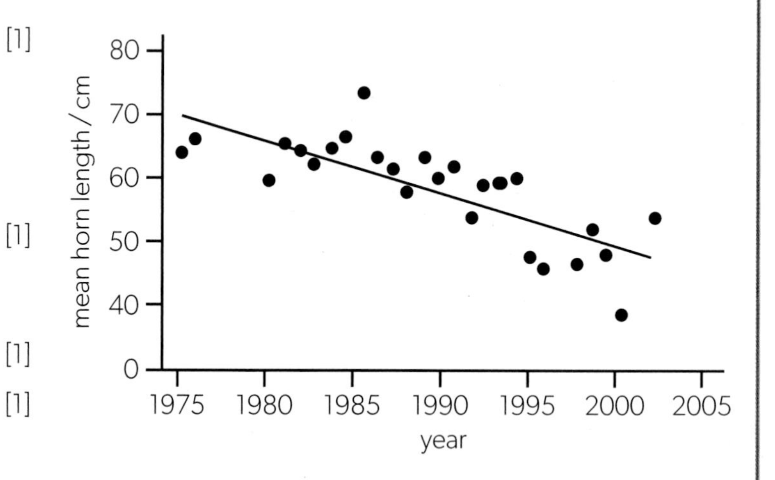

▲ **Figure 9** Mean horn length in bighorn sheep

Modelling natural selection

Make 10 or more artificial fish using modelling clay or some other malleable material. Do not make all of them fish-shaped. Drop each model in turn into a tall measuring cylinder or tube of water and time how long it takes to reach the bottom.

Discard the half of the models that were slowest. Pair up the fastest models and make intermediate shapes, to represent their offspring. Random new shapes can also be introduced to simulate mutation.

Test the new generation and repeat the elimination of the slowest and the breeding of the fastest. Does one shape gradually emerge? Describe its features.

D4.1.6 Requirement that traits are heritable for evolutionary change to occur

A common misconception about evolution is that adaptations acquired during the lifetime of an organism can be inherited by offspring. If this happened, a professional tennis player's children would be born with stronger bones and muscles in one arm, either the left or right, depending on which arm their parent used to hold the racket.

Living organisms do develop traits during their lifetime. If these are caused by random mutations, they are not adaptive and are not heritable because base sequences of genes are unaltered. Acquired traits may be caused by the environment—for example, the loss of a leg in a spider. They may also be caused by interaction between genes and environment. For example, a Himalayan rabbit kit (baby) that is exposed to as little as 20 minutes of cold develops darker fur. Normally, black fur only grows at the extremities of the body—the ears and the feet. Coldness changes gene expression and causes skin cells responsible for hair growth to produce melanin. No change in the base sequences of genes has occurred. Furthermore, the skin cells are not used for gamete production so their genes cannot be passed on to offspring. The environment has not caused specific changes to the base sequences of genes in the sex cells, so the acquired trait is not heritable. It cannot contribute to evolutionary change.

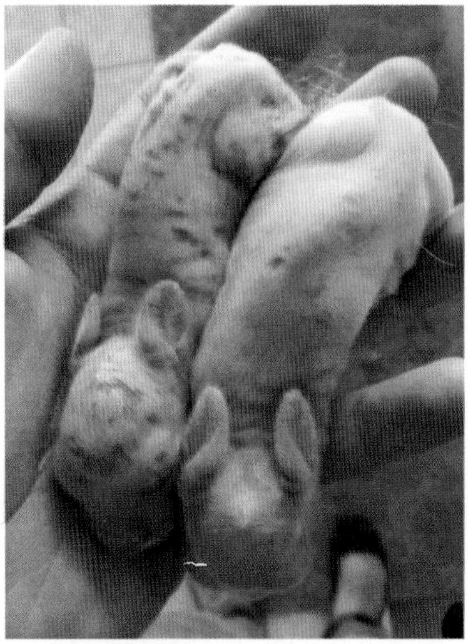

▲ Figure 10 The environment can affect the phenotype of an individual. What advantage would there be for a Himalayan rabbit to have black fur when the temperature is cold?

To a limited extent, changes in gene expression caused by the environment can be passed onto offspring. This is achieved using epigenetic tags, which are chemical markers, reversibly added by a cell to its chromosomes, which establish a pattern of gene expression. A small proportion of these tags is passed on in gametes to offspring, but it is only a pattern of gene expression that has been passed on, not differences in the base sequence of genes. This is explored more in *Topic D2.2*.

D4.1.7 Sexual selection as a selection pressure in animal species

An animal that is well-adapted enough to survive until reproductive age will usually find a mate and reproduce sexually with them. Only half of the genes in a well-adapted animal's offspring are from them. The other half are from the mate. If the mate is also well adapted, the offspring are likely to survive and reproduce, so the genes persist in the population and will contribute to evolution. If the mate is badly adapted, the offspring are unlikely to survive and reproduce, so

▲ Figure 11 Alleles from two parents come together when they have a child. Choosing a mate, means a potential parent is choosing which genes their own will mix with

▲ Figure 12 A male bird of paradise of the species *Paradisea minor*, native to Papua New Guinea

▲ Figure 13 The naturalist Robert Guppy discovered the fish now known as guppies. This illustration dates from 1903 and is by Robert's son Plantagenet

the animal's genes do not persist in the longer term. There is therefore a benefit from assessing accurately whether or not a potential mate is well-adapted. This is known as sexual selection and in many animal species it happens during a process of courtship.

Courtship behaviour is very diverse but in all cases a binary choice must be made based on the overall fitness of the potential mate. In some species, the criteria for selection are obvious, for example the ability of a male to overpower other males by fighting. In other species, the rationale is less obvious. Some animals have anatomical features that seem to the human eye to be excessive—for example, the tail feathers of the male peacock. Other animals have behaviour patterns that seem bizarre. The plumage and courtship displays of a male bird of paradise are examples of these types of exaggerated trait.

There are about 40 species of birds of paradise living on New Guinea and nearby islands. The males have very showy plumage with bright coloration and elongated or elaborate tail feathers that are of no use in flying. The females, which build the nest, incubate the eggs and rear the young, are relatively drab. Males in many of the species have a complicated and eye-catching courtship dance that they use to try to attract females. In some species, the males gather at a site called a lek and females select a mate from among the males displaying. The coloured plumage and courtship dances of birds help to avoid interspecific hybridization by allowing females to determine if a male belongs to their species. But this could be achieved in much more subtle ways than those used by birds so biologists have long speculated on the reasons for exaggerated traits.

Darwin explained such traits in terms of mate selection—females prefer to mate with males that have exaggerated traits. The reason may be that these traits indicate overall health. For example:

- if a male has enough energy to grow and maintain the elaborate plumage and repeatedly to carry out vigorous courtship displays, it indicates that the male must have fed efficiently
- if a male can survive in the rainforest with the encumbrance of its tail feathers and with bright plumage that makes it highly visible to predators, it is probably well-adapted in other ways.

This male bird is therefore a good mate to choose.

Males with showier plumage and more spectacular courtship dances have produced more offspring and continued these traits. Natural selection has therefore caused these traits to become exaggerated.

D4.1.8 Modelling of sexual and natural selection based on experimental control of selection pressures

John Endler and his colleagues conducted controlled experiments to determine the effect of predation on the physical traits of guppies (*Poecilia reticulata*) in Trinidad and Tobago. Endler found that coloration gives an advantage to males, as females prefer colorful males for mating. Coloration can be a disadvantage in shallow water, however, as it makes the guppies more visible to predatory fish. In the parts of streams where there were fewer predators, males were more colourful. Males in locations where there were more predators tended to be less colourful.

Analysing data: John Endler's simulation of natural selection leading to evolution

Endler carried out a series of experiments, starting with a population of fish that had, on average, approximately 10 spots per fish. For 6 months he kept these fish in Pond Q where there were no predators. He then transferred some of the fish into Pond R containing killifish (*Rivulus hartii*), a weak predator. Others were transferred into Pond S containing pike-cichlid fish (*Crenicichla alta*), a strong predator. Figure 14 shows the changes in the mean number of spots.

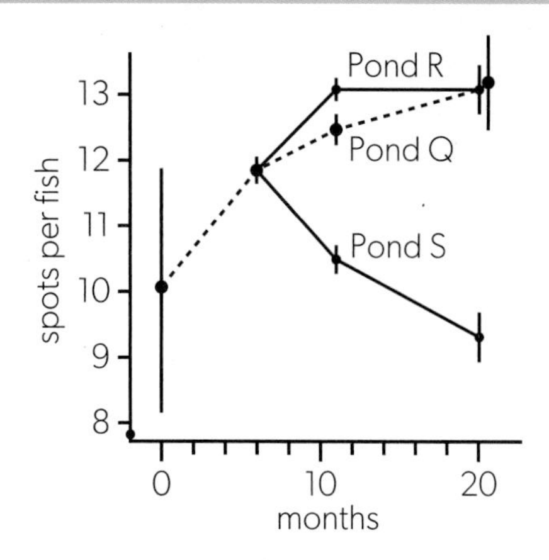

▲ Figure 14 Changes in numbers of spots

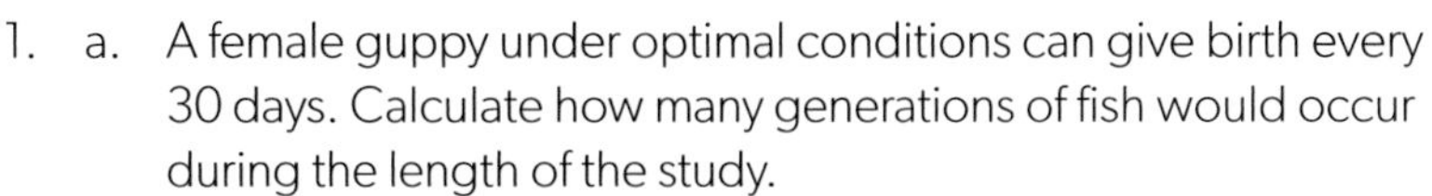

1. a. A female guppy under optimal conditions can give birth every 30 days. Calculate how many generations of fish would occur during the length of the study. [2]

   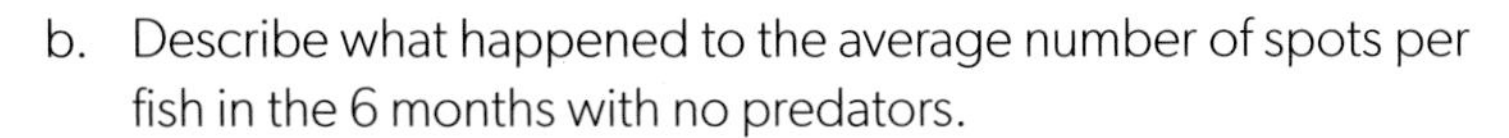

   b. Describe what happened to the average number of spots per fish in the 6 months with no predators. [2]

2. a. Distinguish between the effects of the two species of predatory fish. [2]

   b. Explain the results. [3]

Endler then transferred a population of males from a stream where pike-cichlid were present (stream A) into stream where there were killifish (stream B). After 2 years, he measured the area of spots on the guppies in stream B and in another stream where there were both guppies and killifish. Figure 15 shows the mean results.

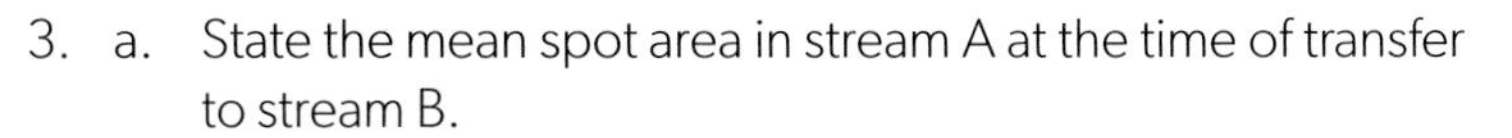

3. a. State the mean spot area in stream A at the time of transfer to stream B. [1]

   b. Determine the change in mean area in the population over the 2 years after transfer. [1]

   c. Explain the change in mean area using the concept of natural selection. [3]

4. Predict what would happen if members of the transplanted population were returned to the source stream. [1]

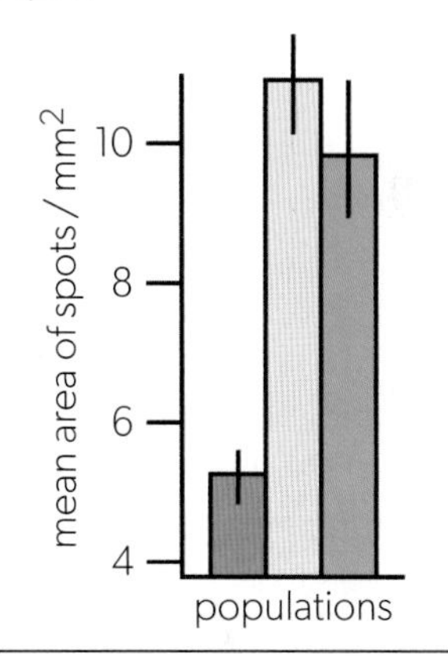

▶ Figure 15 Differences in area of spots

D4.1.9 Concept of the gene pool

A population is a group of organisms of the same species living in the same area. Individuals are added to a population when members of that population reproduce sexually with each other. A new individual inherits some of the genes that are present in the population. The combination of an individual's inherited genes depends on which individuals are the parents and on the random events that happen during meiosis in the parents. This leads to the concept of the gene pool. A gene pool is all the genes and their different alleles that are present in a population.

Individuals that reproduce contribute to the gene pool of the next generation. With natural selection, some individuals are more likely to contribute than others. Fitness can be measured in terms of contribution to the gene pool. Genetic equilibrium exists when all members of a population have an equal chance of contributing to the future gene pool.

▲ Figure 16 What are the similarities between a child's ball pit and a gene pool?

According to the biological species concept, a species is a group of organisms that can interbreed to produce fertile offspring. Most species consist of multiple populations. These may be reproductively isolated from each other due to geographical separation, in which case individuals in the different populations can potentially interbreed with each other, but do not actually do so. In this case, the separated populations have their own gene pools. It is therefore possible for multiple gene pools to exist in one species.

D4.1.10 Allele frequencies of geographically isolated populations

A common allele in one population may be much rarer in another. In humans, there are differences in allele frequency between ethnic groups and between geographical regions. The database housed at Yale University, AlFred, is searchable and allows visual comparison of allele frequencies in different populations.

Using a database to explore allele frequencies

This map is generated using AlFred and Google Maps and shows the geographic distribution of the alcohol dehydrogenase alleles ADH 1B *1 (yellow) and ADH 1B *2 (blue). Alcohol dehydrogenase breaks down alcohol (ethanol) in the liver. The ADH 1B *2 allele is associated with a more rapid metabolism of alcohol. The ADH 1B *1 allele is more common in alcoholics and in heavy drinkers than in moderate drinkers.

1. Outline the distribution of the *2 allele.
2. One hypothesis is that the *2 allele arose in the southwestern corner of modern China in the region where rice cultivation first arose. Suggest what might be the connection between early rice cultivation and alcohol consumption.
3. RS numbers are an agreed naming convention used by researchers to identify single nucleotide polymorphisms (SNP). The RS number for the *1 / *2 SNP is rs2 066 701. Use this as a keyword search term in AlFred.

   a. What is the chromosomal location of ADH 1B? Click on the "Locus name" to identify this.

   b. Click on the table view. Identify three populations with very high incidence of the *2 allele (T). Click on their "SA" number to find out more about these populations. Identify three populations with very high incidence of the *1 allele (C). Click on their "SA" number to find out more about the characteristics of these populations.

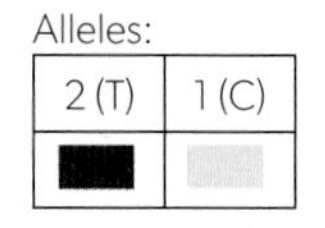

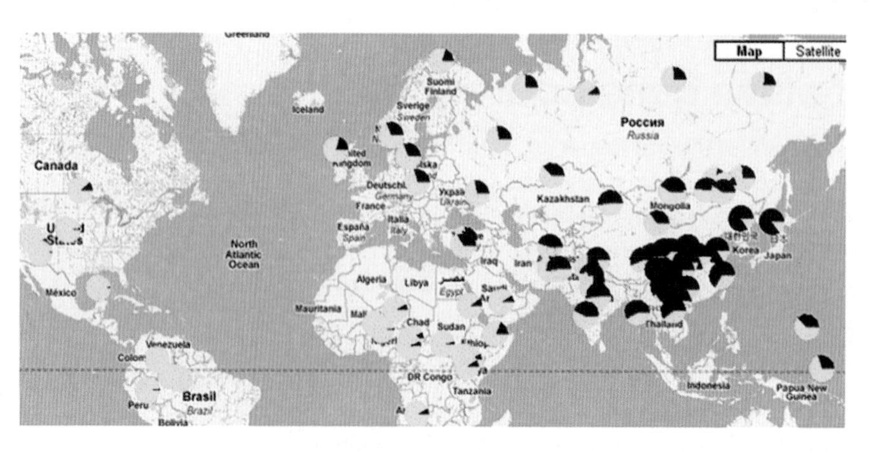

▲ Figure 17 Worldwide distribution of alcohol dehydrogenase alleles

D4.1.11 Changes in allele frequency in the gene pool as a consequence of natural selection between individuals according to differences in their heritable traits

Darwin established the concept of evolution by natural selection, which requires heritable variation. However, he did not solve the conundrum of how inheritance works. This was achieved by other biologists. Gregor Mendel laid the foundations of genetics with his crossing experiments using peas. His work was published in 1866 and established the principle of discrete inheritance, with what later became known as "the gene" as the unit of heredity. In 1892, August Weismann's germ plasm theory was published. He concluded that acquired characteristics cannot be inherited because there is no obvious mechanism for hereditary information to pass between somatic cells in the body and sex cells.

The synthesis of Darwin's theory of evolution by natural selection with Mendelian genetics and Weismann's germ plasm theory is known as Neo-Darwinism. It explains the basis of variation between individuals, how natural selection acts on it and how it is inherited. Evolution can be described as a change in the frequency of alleles in the gene pool. If an individual is better adapted to its environment than its competitors, it will tend to have more offspring. Therefore, its genotype will increase in frequency in the next generation relative to other genotypes.

Activity: Phenotype and genotypes in codominance

In the cross depicted in Figure 18, the flower colours of *Mirabilis jalapa* plants are shown in three generations of a cross experiment. The genotype C^RC^R yields red flowers, the genotype C^WC^W yields white flowers and because the alleles are codominant, the genotype C^RC^W yields pink flowers.

- In the first generation, 50% of the population are red and 50% are white.
- In the second generation, 100% of the flowers are pink.
- In the third generation, there are 50% pink, 25% white and 25% red.

Show that the allele frequency is 50% C^R and 50% C^W in each of the three generations.

In this case, the two allele frequencies did not change so there was no evolution, even though the frequencies of the phenotypes in the three generations differed. What would happen if pollinating insects preferentially visited white *Mirabilis jalapa* flowers?

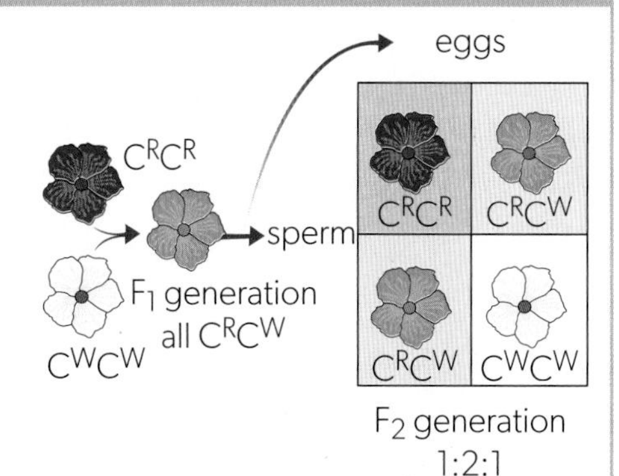

▲ **Figure 18** A change in phenotypic frequency between generations does not necessarily indicate that evolution is occurring

D4.1.12 Differences between directional, disruptive and stabilizing selection

The fitness of a genotype or phenotype determines how likely it is to be found in the next generation. Selection pressures are environmental factors that favour certain phenotypes more and others less. There are three patterns of natural selection: stabilizing selection, disruptive selection, and directional selection.

▲ Figure 19 A clutch is the number of eggs laid by a female and incubated together. Small clutch sizes may result in none of the offspring surviving into the next generation. Very large clutch sizes may result in higher mortality as the parent cannot provide adequate resources for all the offspring. This means that an intermediate clutch size is favoured

Stabilizing selection

Selection pressures act to remove extreme varieties. For example, average birth weights of human babies are favoured over low birth weight or high birth weight.

Typical clutch size in blackbirds (*Turdus merula*) is between three and five.

Disruptive selection

Selection pressures act to remove intermediate varieties, favouring the extremes. In coho salmon (*Oncorhynchus kisutch*), Figure 20, some males reach maturity as much as 50% earlier than and only reach 30% of the body size of other males in the population. Success in spawning (breeding) depends on the male releasing sperm in close proximity to the egg-laying female. Small and large males employ different strategies to gain access to females. The small-sized males, called jacks, are specialized at "sneaking". The large-sized males are specialized at fighting and coercing females to spawn. In contrast, intermediate-sized males are at a competitive disadvantage to both jacks and large males because they are more targeted for fights which they lose and are more likely to be prevented from sneaking.

Directional selection

The population changes as one extreme of a range of variation is better adapted. The body size of house mice (*Mus musculus*) introduced to Gough Island in the South Atlantic during the 19th century became about twice that of mainland populations. Given the rapid increase in body size, there must have been strong directional selection. The mouse developed a behaviour of biting ground birds and feeding on their blood (Figure 21).

▲ Figure 20 Coho salmon in the act of spawning

▲ Figure 21 House mouse on Gough island attacking an endangered Tristan albatross chick

Data-based questions: Stabilizing selection

Researchers carried out a study on 3,760 children born in a London hospital over a period of 12 years. Data was collected on the children's mass at birth and their mortality rate. The purpose of the study was to determine how natural selection acts on mass at birth.

Figure 22 shows the frequency of babies of each mass at birth. The line superimposed on the bar chart indicates the percentage mortality rate (the children that did not survive for more than 4 weeks).

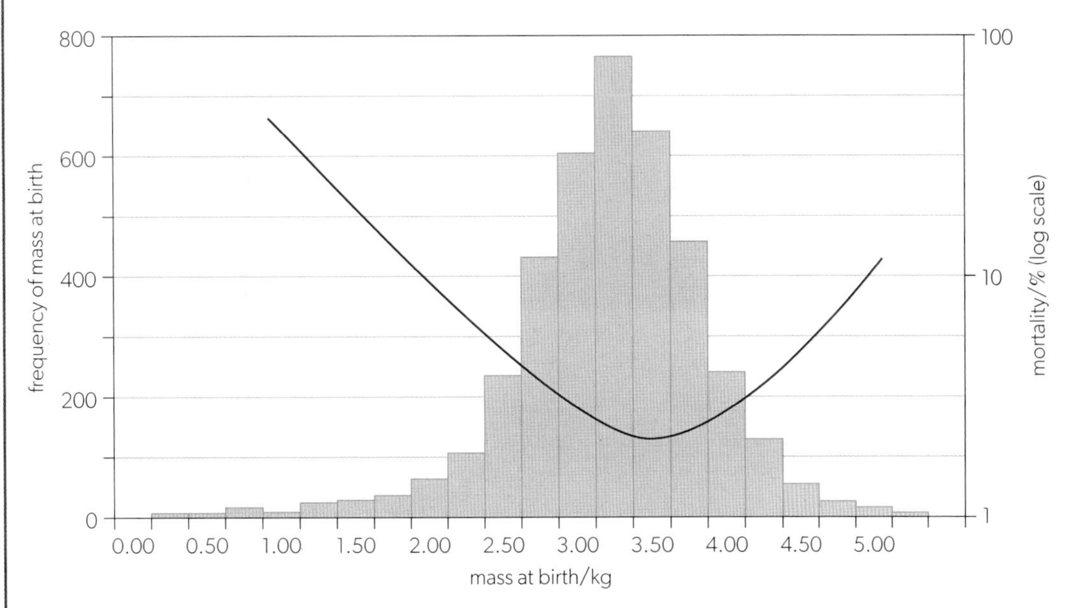

▲ Figure 22 Graph of baby birth weight and mortality rate
Source: W H Dowderswell. (1984) *Evolution, A Modern Synthesis*, page 101.

1. Identify the mode value for mass at birth. [1]
2. Identify the optimum mass at birth for survival. [1]
3. Outline the relationship between mass at birth and mortality. [3]
4. Explain how this example illustrates the pattern of natural selection called stabilizing selection. [2]

D4.1.13 Hardy–Weinberg equation and calculations of allele or genotype frequencies

The Hardy–Weinberg equation can be used to predict allele frequencies from genotype frequencies. It can also be used to predict genotypic frequencies and allele frequencies from the incidence of recessive phenotypes.

Example 1

Albinism is a lack of pigmentation in skin and hair caused by a recessive allele. Albinism occurs in North America in approximately one in 20,000 people. Let q represent the frequency of the albinism allele and p represent the frequency of the pigmented allele. To exhibit the albino phenotype, a person must be homozygous recessive. The probability of having two alleles for albinism is $q \times q$ or q^2. Thus $1/20000 = q^2$ and $q = 0.007$. The probability of p plus q must equal 1. Thus $p + 0.007 = 1$ and $p = 0.993$. The probability of the homozygous dominant genotype is $p^2 = 0.986$. Since the p and q allele can come from either the mother or the father, the probability of the heterozygous phenotype is $2pq = 2(0.993)(0.007) = 0.0139$.

		Male gametes p	Male gametes q
Female gametes	p	0.993 × 0.993 = 0.9860	0.007 × 0.993 = 0.007
Female gametes	q	0.993 × 0.007 = 0.007	0.007 × 0.007 = 0.00005

▲ Figure 23

Example 2

Tay–Sachs disease is an autosomal recessive disorder of the enzyme hexosaminidase. The disorder causes a build-up of fatty deposits in the brain. A child affected by the disease usually dies by the age of four. The frequency of Tay–Sachs disease (genotype *tt*) in a Mediterranean population is 0.0003. From this we can calculate the frequencies in the population of allele *t* and genotype *Tt*.

$q^2 = 0.0003$ so $q = 0.017$. Therefore, frequency of allele $t = 0.017$

$p = 1 - 0.017$ so $p = 0.983$. Therefore, frequency of allele $T = 0.983$

$2pq = 2(0.983)(0.017)$ so $2pq = 0.033$. Therefore, frequency of genotype $Tt = 0.033$

D4.1.14 Hardy–Weinberg conditions that must be maintained for a population to be in genetic equilibrium

According to the Hardy–Weinberg equation, if a gene has two alleles and their frequencies are p and q, the frequencies of the two homozygous genotypes will be p^2 and q^2 and the frequency of the heterozygous genotype will be $2pq$. These frequencies are only expected if the following series of conditions is met in a population.

- There is no mutation of the gene, so the alleles are not changed, and new alleles are not being generated.
- Mating is random, so no phenotype preferentially mates with another particular phenotype.
- There is no immigration or emigration that is likely to change allele frequencies.
- The population is large enough to prevent allele frequencies changing due to chance—this is called genetic drift.
- Natural selection does not favour one phenotype over another.

If the genotypic frequencies in a population do fit the predictions from the Hardy–Weinberg equation, we conclude that all the conditions are being met. And in particular, that natural selection is not causing allele frequencies to change, so evolution is not occurring. The population is said to be in genetic equilibrium.

If the genotypic frequencies diverge from the Hardy–Weinberg predictions, one or more of the conditions is not being met. If we can be sure that the first four conditions are being met, then natural selection must be favouring one phenotype over another, causing allele frequencies to change—evolution is occurring within the population.

D4.1.15 Artificial selection by deliberate choice of traits

Humans have deliberately bred and used certain animal species for thousands of years. If modern breeds of livestock are compared with the wild species that they most resemble, the differences are often huge. Consider the differences between modern egg-laying hens and the junglefowl of Southern Asia, or between Belgian Blue cattle and the aurochs of western Asia. There are many different breeds of sheep, cattle and other domesticated livestock, with much variation between breeds.

Domesticated breeds have clearly not always existed in their current form. The only credible explanation is that the change has been achieved by repeatedly selecting for and breeding the individuals with the traits most suited to human uses. This process is called artificial selection. The effectiveness of artificial selection is shown by the considerable changes that have occurred in domesticated animals over periods of time that are very short, in comparison to geological time. It shows that selection can cause evolution, but it does not prove that evolution of species has occurred naturally or that the mechanism for evolution is natural selection.

▲ Figure 24 Over the last 25,000 years many breeds of dog have been developed by artificial selection from domesticated wolves

AHL

Data-based questions: Domestication of corn

A wild grass called teosinte that grows in Central America was probably the ancestor of cultivated corn, *Zea mays*. When teosinte is grown as a crop, it gives yields of about 150 kg per hectare. This compares with a world average yield of corn of 4,100 kg per hectare at the start of the 21st century. Corn was domesticated at least 7,000 years ago. Today, lengths of corn cobs vary enormously between varieties (Table 1).

Corn variety and origin	Cob length
Teosinte—wild relative of corn	14 mm
Early primitive corn from Colombia	45 mm
Peruvian ancient corn from 500 BCE	65 mm
Imbricado—primitive corn from Colombia	90 mm
Silver Queen—modern sweetcorn	170 mm

▲ Table 1 Lengths of corn cob of a range of varieties

1. Calculate the percentage difference in length between teosinte and Silver Queen. [2]
2. Calculate the percentage difference in yield between teosinte and world average yields of corn. [2]
3. Suggest factors apart from cob length, selected for by farmers. [3]
4. Explain why improvement slows down over generations of selection. [3]

▲ Figure 25 Cobs of *Zea mays*. On the left is teosinte, the wild species that is the ancestor of cultivated corn. On the right is a modern cultivated variety. Between are older varieties. Teosinte was first cultivated in southwestern Mexico over 8,700 years ago

Linking questions

1. How do intraspecific interactions differ from interspecific interactions?
   a. Outline the concept of mutualism using herbivory and pollination as examples. (C4.1.12)
   b. With respect to a named species outline examples of intraspecific examples of competition and cooperation. (C4.1.10)
   c. Explain the role of natural selection in evolution. (D4.1.5)
2. What mechanisms minimize competition?
   a. With respect to a named species compare and contrast realized and fundamental niche. (B4.2.12)
   b. Outline the competitive exclusion principle. (B4.2.12)
   c. Explain adaptive advantage of seed dispersal as a reproductive strategy. (D3.1.12)

D4.2 Stability and change

What features of ecosystems allow stability over unlimited time periods?

Stable ecosystems are ones where the community structure is not prone to change. For an ecosystem to remain steady over time, certain requirements must be present. What are these requirements?

The water scorpion and the aquatic crustacean in Figure 1 are species found only in the Movile Cave, Romania, where life has been cut off from the outside world for the past 5.5 million years. They have evolved in total darkness over that time and have lost their eyes as a result. What challenges are presented to researchers who want to study the cave?

▲ Figure 1 Invertebrates in Movile Cave

What changes caused by humans threaten the stability of ecosystems?

Hencott Pool in Shropshire, England, was formed 10,000 years ago at the end of the last glaciation. Until recently it was a small lake with clear water and both plants and animals adapted to low concentrations of nutrients (oligotrophic). Leaching of nutrients from surrounding agricultural land has triggered an ecological succession. Larger and larger plants have colonized, adding organic matter to the water and reducing the depth until alder trees (*Alnus glutinosa*) can grow. Common duckweed (*Lemna minor*) grows over the water surface, blocking sunlight from any remaining water plants, but plants adapted to oligotrophic water have already gone.

▲ Figure 2 Alder trees and common duckweed on Hencott Pool

SL and HL	AHL only
D4.2.1 Stability as a property of natural ecosystems D4.2.2 Requirements for stability in ecosystems D4.2.3 Deforestation of Amazon rainforest as an example of a possible tipping point in ecosystem stability D4.2.4 Use of a model to investigate the effect of variables on ecosystem stability D4.2.5 Role of keystone species in the stability of ecosystems D4.2.6 Assessing sustainability of resource harvesting from natural ecosystems D4.2.7 Factors affecting the sustainability of agriculture D4.2.8 Eutrophication of aquatic and marine ecosystems due to leaching D4.2.9 Biomagnification of pollutants in natural ecosystems D4.2.10 Effects of microplastic and macroplastic pollution of the oceans D4.2.11 Restoration of natural processes in ecosystems by rewilding	D4.2.12 Ecological succession and its causes D4.2.13 Changes occurring during primary succession D4.2.14 Cyclical succession in ecosystems D4.2.15 Climax communities and arrested succession

D4.2.1 Stability as a property of natural ecosystems

Stability of an ecosystem means that it can persist indefinitely because of the mechanisms operating within it. There are examples of ecosystems that have persisted for long periods of time:

- The Daintree Rainforest in Northern Australia is estimated to have existed for 180 million years (Figure 3). Due to its age, it contains species from ancient plant families that evolved before flowering plants as well as families that have more members known only from the fossil record than are alive today.
- The Borneo Lowland Rainforest has existed for about 140 million years. It originally covered much of the island of Borneo, but there have been big losses in recent decades.
- The Namib desert in Southern Africa is relatively intact and stable. Thick fogs along the coast provide enough moisture for a number of distinctive, highly adapted animal species to survive. It is estimated to be 55–80 million years old.

The mechanisms that sustain ecosystems are fragile and easily disrupted, so ecosystems are not always stable. Even apparently minor perturbations might cause change.

Activity: Resistance and resilience

Stable ecosystems have stability in both their community structure and their functions.

- Stability is the absence of change.
- Community stability is due to either the absence of disturbance or community resistance to disturbance.
- Resistance is the ability of a community or ecosystem to remain unchanged despite disturbance.
- Resilience is the ability of an ecosystem to rebound from change.

Identify which graph in Figure 4 represents each of the following responses to ecosystem disturbance:

I. high resistance and high resilience

II. high resistance and low resilience

III. low resistance and high resilience

IV. low resistance and low resilience

▲ **Figure 3** Mangrove trees on the edge of the Daintree Rainforest in the north of Queensland, Australia, which is the oldest continuously forested rainforest area on the planet

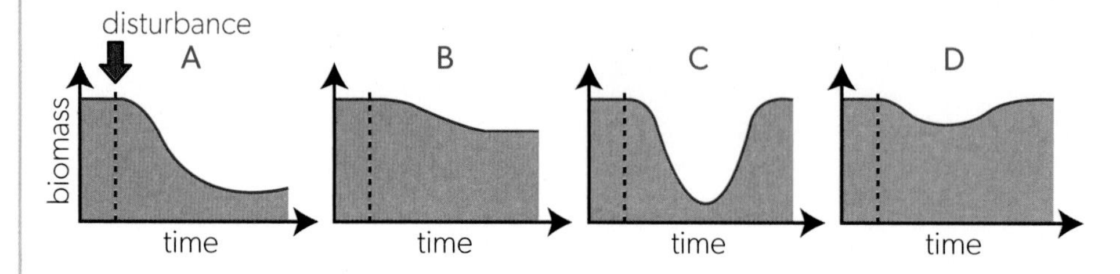

▲ **Figure 4** Graphs of responses to ecosystem disturbance

D4.2.2 Requirements for stability in ecosystems

The structural and functional stability of ecosystems depends on key conditions being met.

- There must be a steady supply of energy.
- Nutrient cycling should have no leakages which result in nutrients leaving the system.
- Individual species, especially keystone species, must have high genetic diversity so populations can survive environmental selection pressures.

Disruptions can interfere with sustainability. Several types of environmental change could act as disruptions. For example:

- harvesting and removal of materials from the environment could disrupt nutrient cycles
- erosion could result in loss of nutrients
- eutrophication is nutrient enrichment of a body of water—it can disrupt ecosystem stability by causing population imbalances
- selective removal of species by epidemics or poaching could disrupt ecosystem structure, particularly if a keystone species is removed.

If the genetic diversity of a population drops below a certain threshold, it becomes less resilient. High biodiversity tends to be associated with stable ecosystems because there will be a robust community structure.

The relationship between the climate of an area and the type of ecosystem that develops is described in *Topic B4.1*. High rainfall allows forests to develop, whereas moderate rainfall only allows grasslands. Prolonged changes in the amount of precipitation can cause ecosystem disruption. This is a concern because of climate change. There are fears that a combination of disruptions could lead to a tipping point being reached in the Amazon rainforest such that rainfall decreases to the point where grassland replaces forest.

D4.2.3 Deforestation of Amazon rainforest as an example of a possible tipping point in ecosystem stability

Given enough disturbance, systems can cross an ecological threshold that is difficult to reverse. This is known as a tipping point. If a tipping point is reached, there is abrupt change in an ecosystem rather than gradual change. The vast Amazon rainforest may seem impregnable, but to positive feedback mechanisms could endanger it. Decreases in forest cover (for example, caused by logging) reduce the total amount of transpiration from plants. This leads to a decrease in cloud formation, which reduces rainfall. Lower rainfall results in further loss of forest. Periods of drought increase the incidence of fires, causing more forest loss and further reductions in transpiration and rainfall. In combination, these changes could result in a tipping point being reached and an abrupt change from stable forest to a stable grassland ecosystem over large parts of the Amazon rainforest.

Calculating percentage change

To calculate percentage change, use the following formula:

$$\frac{\textit{final amount} - \textit{initial amount}}{\textit{initial amount}} \times 100\%$$

The data in Table 1 shows the area of primary Amazon forest in three Brazilian states in 2001 and again in 2020. The calculation below shows the percentage change for Maranhão. Complete the table for the other two states.

$$\%\text{ change in Maranhão} = \frac{2{,}483{,}153 - 3{,}185{,}732}{3{,}185{,}732} \times 100\%$$

$$= -22.1\%$$

State	Area of rainforest / hectares		% change
	In 2001	In 2020	
Maranhão	3,185,732	2,483,153	−22.1%
Rondônia	15,649,578	12,470,563	
Pará	92,225,896	83,576,973	

▲ **Table 1** Primary Amazon forest in Maranhão, Rondônia and Pará

Extracting data from a database

The interactive website at *Global Forest Watch* allows you to detect deforestation in real time with updates every week. You can track deforestation and reforestation in your country, or you can "adopt" a region in another country to monitor the state of the forests. For example, Figure 5 shows the loss of forest cover in western New Providence in The Bahamas over a 20-year period.

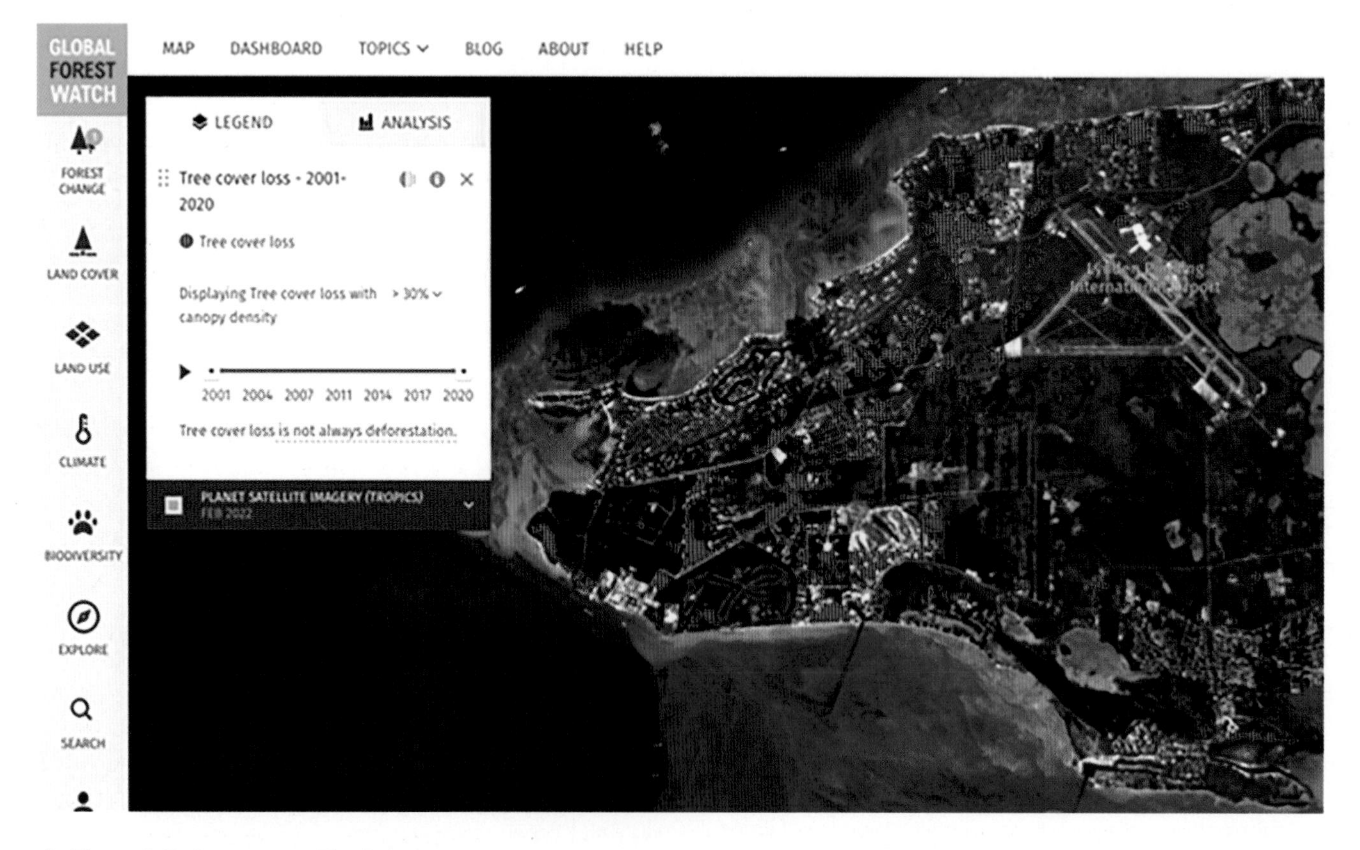

▲ **Figure 5** Deforestation in The Bahamas

Figure 6 shows changes in the percentage of land covered by forest in seven countries. In this study, forest is defined as land under natural or planted stands of trees of at least 5 metres in situ, whether productive or not, and excludes tree stands in agricultural production systems.

- Which countries show a gain in land covered by forest?
- Which countries show a decline in land covered by forest?
- Of those countries where forest has been lost, which show a declining rate of loss?

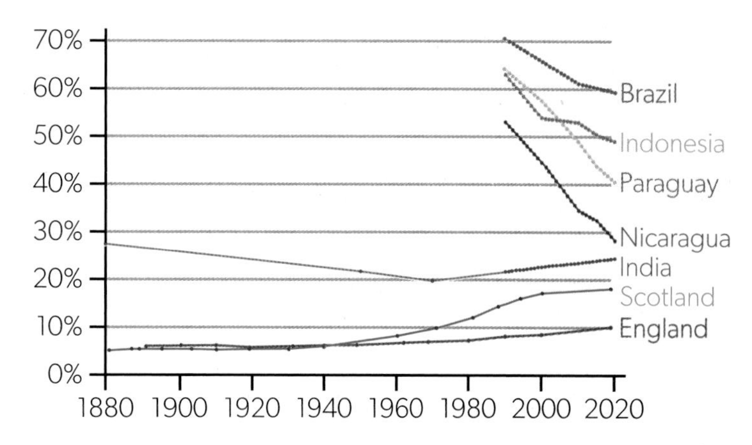

▲ **Figure 6** This graph is extracted from the website *Our World in Data* showing the land covered by forest in countries over time selected from a menu

D4.2.4 Use of a model to investigate the effect of variables on ecosystem stability

Mesocosms are small experimental areas that are set up as ecological experiments. Fenced-off enclosures in grassland or a forest could be used to model terrestrial mesocosms; tanks set up in the laboratory can be used as models of aquatic ecosystems. Ecological experiments can be done in replicated mesocosms, to find out the effects of varying one or more conditions. For example, tanks could be set up with and without fish, to investigate their effects on aquatic ecosystems. Another possible use of mesocosms is to test what types of ecosystems are sustainable. This involves sealing up a community of organisms together with air and soil or water inside a container.

▲ Figure 7 A sealed terrarium is an example of a mesocosm

Consider these questions before setting up either aquatic or terrestrial mesocosms.

- Large glass jars are ideal but transparent plastic containers could also be used. Should the sides of the container be transparent or opaque?
- Which of these groups of organisms must be included to make up a sustainable community: autotrophs, consumers, saprotrophs and detritivores?
- Once a mesocosm is sealed, no more oxygen will be able to enter. How can we ensure that the oxygen supply is sufficient for all the organisms in the mesocosm?

Impacts of science

When creating and maintaining mesocosms in the laboratory, care must be taken to meet the standards of the IB ethical experiment policy. The essential question is: Is it possible to prevent any organism from suffering as a result of it being placed in the mesocosm?

Here are two strategies that can be followed to ensure that the spirit of the policy is complied with.

- Consult with your teacher when setting up the mesocosm to collaboratively plan to minimize suffering of organisms.
- Keep all abiotic factors within the tolerance limits of the organisms in the mesocosm.

▲ Figure 8 Ochre sea star *Pisaster ochraceus*

D4.2.5 Role of keystone species in the stability of ecosystems

A keystone species is one whose activity has a disproportionate effect on the structure of an ecological community. Species diversity decreases if the keystone species is lost.

Robert Paine developed the concept of keystone species after investigating the impact of the ochre sea star *Pisaster ochraceus* on marine ecosystems. He artificially removed *Pisaster* from an 8 × 2 m area of seashore at Mukkaw Bay in Washington state and left an adjacent area unaltered as a control. The community in the control area did not change, with high diversity of animal species and benthic algae (seaweeds). In the experimental area, a species of barnacle (*Balanus glandula*) initially spread to occupy about 70% of the area. In the following year, the barnacle was crowded out by a goose-necked barnacle (*Mitella polymerus*) and a bivalve mollusc (*Mytilus californianus*). Finally, the *Mytilus* became overwhelmingly dominant and the seaweeds disappeared apart from one species that can grow on mollusc shells. Other animal species were crowded out or lost their food source, so they either died or migrated. These changes are explained by *Pisaster* being an important predator of *Mytilus*, so preventing the mollusc from becoming over-dominant. In this way, *Pisaster* maintains high species diversity.

D4.2.6 Assessing sustainability of resource harvesting from natural ecosystems

Some current human uses of resources are unsustainable (cannot continue indefinitely). Human use of fossil fuels is an example of an unsustainable activity. Supplies of fossil fuels are finite; they are not being renewed and therefore their use cannot carry on indefinitely. Natural ecosystems can teach us how to live in a sustainable way, so that our children and grandchildren will be able to live as well as we do. There are three requirements for sustainability in ecosystems.

- **Nutrient availability**—nutrients can be recycled indefinitely and if this occurs there should never be a lack of the chemical elements on which life is based. Saprotrophic bacteria and fungi (decomposers) play important roles in recycling carbon, nitrogen and other nutrients.
- **Detoxification of waste products**—the waste products of one species are usually exploited as a resource by another species. For example, ammonium ions released by decomposers are absorbed and used as an energy source by *Nitrosomonas* bacteria in the soil. Ammonium is potentially toxic but because of the action of these bacteria it does not accumulate.
- **Energy availability**—energy cannot be recycled, so sustainability depends on a continued energy supply. Most energy is supplied to ecosystems as light from the Sun. The importance of this supply can be illustrated by the consequences of the eruption of Mount Tambora in 1815. Dust in the atmosphere reduced the intensity of sunlight for some months afterwards, causing crop failures globally and deaths due to starvation. This was only a temporary phenomenon, however, and energy supplies to ecosystems in the form of sunlight will continue for billions of years.

Sustainable harvesting of plants

Brazil nuts are harvested from *Bertholletia excelsa* trees in the Amazon rainforest, which can grow to 50 m and live for a thousand years. Logging is threatening these trees in some areas. Sustainable harvesting depends on leaving some nuts to germinate and grow into new trees. In areas of intense harvesting, there are few or no young trees, so harvesting is unsustainable.

Sustainable harvesting of fish

Cod (*Gadus morhua*) inhabit the cold waters of the north Atlantic. There was once a very high density of cod on the Grand Banks off Newfoundland, but overfishing led to a total collapse of this population in the early 1990s. The cod population has still not recovered, demonstrating the importance of sustainability in harvesting of fish.

Because fish are an open-access resource on the high seas, the incentives for conservation are limited. An important component of fishing sustainably is clear data about populations. The concept of maximum sustainable yield is related to the sigmoid growth curve. At point M on the graph in Figure 9, the population is growing at its maximum rate. As long as fish are not harvested at a faster rate than this, stocks should not decline and fishing could continue indefinitely.

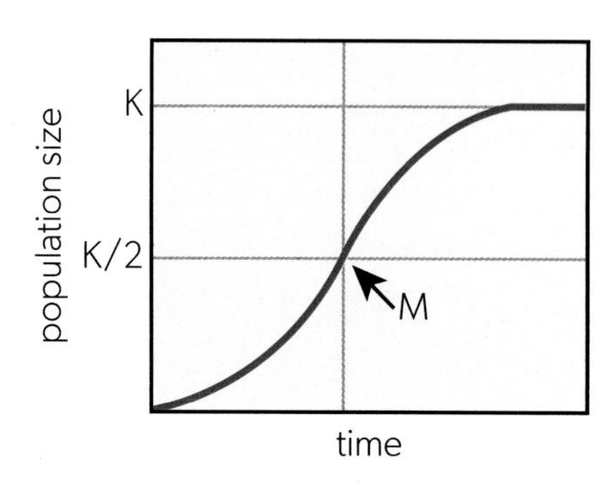

▲ Figure 9 Sigmoid population growth curve. K is the carrying capacity. Population growth is maximal at M

If there were no fishing, the yield would be zero. If there were a very high intensity of fishing, the population would be depleted and lost entirely, so again there would be no yield. The maximum rate of harvest shown on the curve in Figure 10 corresponds to point M on the sigmoid growth curve in Figure 9.

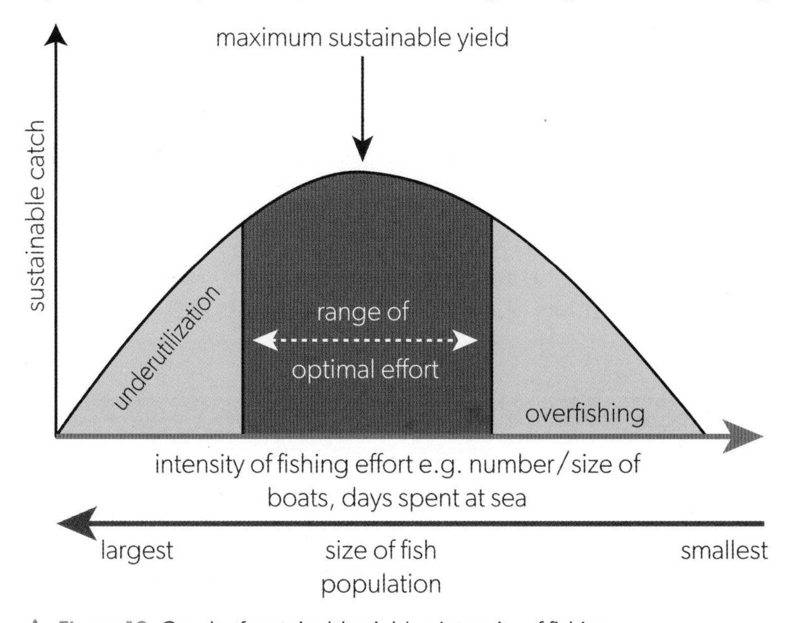

▲ Figure 10 Graph of sustainable yield vs intensity of fishing

▲ Figure 11 Atlantic cod grows to lengths up to 2 metres and can live for 25 years. It is a predator of crustacea, molluscs and smaller fish. Its prey are in several different trophic levels, with an mean trophic level of 4.4

Cod stocks in the North Sea between Britain and Norway dropped by more than 80% in the latter part of the 20th century. The measures agreed internationally to reverse this trend were:

- exclusion zones in nursery areas to allow undisturbed breeding
- increasing the size of holes in nets to allow smaller cod to escape
- reducing the size of the fishing fleet by decommissioning some boats
- monitoring the numbers and ages of cod in the North Sea population and, based on this information, setting quotas for each boat, to limit the overall quantity of cod caught per year.

Since 2005, the biomass of adult cod in the North Sea has tripled. Current quantities of cod caught per year are allowing continued increases, so cod fishing in the North Sea is sustainable.

D4.2.7 Factors affecting the sustainability of agriculture

Intensive agriculture is characterized by high levels of input factors and high output (yields). Many aspects of intensive farming overuse finite resources and thus are not sustainable indefinitely.

- Tillage is preparing soil for a crop. Soils are loosened by plowing, harrowing and other cultivations. Soil structure can become degraded and both wind and water can cause soil erosion at a much faster rate than new soil is being generated. Tropical forests that are cleared for agriculture often have soils which quickly degrade and cannot sustain high-yield agriculture for long periods of time.
- Nutrient depletion of soils is caused by removal of harvested crops and by leaching. As water drains through depleted soils, nitrogen fertilizers tend to leach quickly into streams and rivers where the nitrogen causes eutrophication. Nutrient depletion necessitates repeated application of chemical fertilizer. Phosphate and most other minerals are mined from non-renewable rock deposits. The manufacture of nitrogen fertilizer from fossil fuels requires energy.
- Cultivating large areas of a single variety of crop plant (monoculture), with the same crop grown year after year, encourages pests and weeds to become increasingly problematic. Intensive agriculture responds to this with applications of pesticide and herbicide. This can cause pollution problems, especially with persistent chemicals such as DDT. Resistance to an agrochemical tends to evolve in the pest or weed if the chemical is repeatedly used. Manufacture of agrochemicals requires an energy source, which currently is supplied by fossil fuels.
- Mechanical tillage requires power, most of which comes from diesel oil used in tractors. Energy is also required for heating glasshouses and animal housing. The carbon footprint of agriculture is high. It makes a significant contribution to climate change.

Data-based questions: Environmental impacts of human diets

A feed conversion ratio is the mass of food consumed by animals per unit mass of meat or fish produced by them as they grow (Table 2).

Meat/fish production	Feed conversion ratio
salmon	1.2
beef	8.8
chicken	1.9

▲ **Table 2 Some typical feed conversion ratios**

1. Calculate how much food will be used to produce 1 kg of:
   a. salmon [1]
   b. beef [1]
2. Suggest a reason for the differences in feed conversion ratios between salmon and chickens. [1]
3. Discuss which of these animal-based foods is most sustainable. [2]

Greenhouse gas emissions from food production shown in Figure 12 are measured in kilograms of carbon dioxide equivalents per kilogram of food product. So, methane and other non-carbon dioxide greenhouse gases are included and weighted by their relative warming impact.

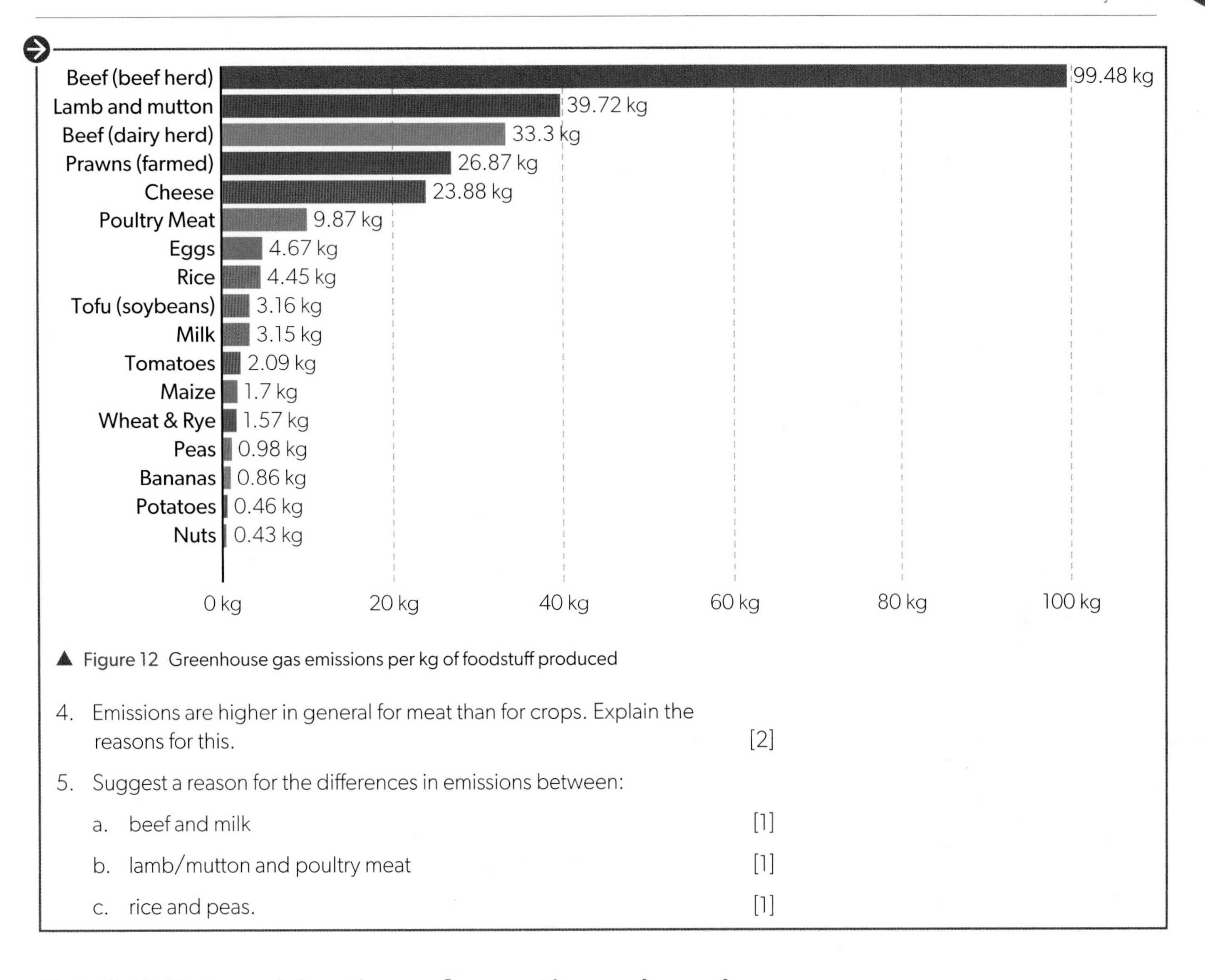

▲ Figure 12 Greenhouse gas emissions per kg of foodstuff produced

4. Emissions are higher in general for meat than for crops. Explain the reasons for this. [2]
5. Suggest a reason for the differences in emissions between:
   a. beef and milk [1]
   b. lamb/mutton and poultry meat [1]
   c. rice and peas. [1]

D4.2.8 Eutrophication of aquatic and marine ecosystems due to leaching

When rain falls on agricultural land, water-soluble nutrients such as phosphates and nitrates that have been added to crops dissolve in the water that drains through the soil. The resulting groundwater carrying nutrients enters watercourses and pollutes lakes, rivers and eventually the sea. Nutrients from manure and urine of livestock can also contribute to nutrient enrichment of bodies of water, especially if not stored correctly or if spread on land before rainfall.

▼ Figure 13 Algal bloom at Colemere in Shropshire, England

Nutrient enrichment of water is known as eutrophication. Nitrate and phosphate in particular favour the growth of algae, leading to algal blooms. During such a bloom, unicellular algae and photosynthetic bacteria increase in number unsustainably and then float to the surface. Algal blooms block light to the plants below. The algal bloom dies as do the shaded plants below the water surface. They are then decomposed by saprotrophic bacteria, which are aerobic and consume oxygen from the water. There is therefore a greatly increased biochemical oxygen demand (BOD).

With a dense algal bloom, the BOD is high and the water becomes anoxic, causing fish deaths. Eutrophication can also be caused by release of untreated sewage.

D4.2.9 Biomagnification of pollutants in natural ecosystems

Natural ecosystems in many parts of the world have been polluted with human-made chemicals. The chemicals may be at very low concentrations, yet some organisms, particularly at the ends of food chains, are found to contain concentrations great enough to be lethal. This can be explained by the associated processes of bioaccumulation and biomagnification. DDT (dichlorodiphenyltrichloroethane) caused catastrophic falls in the populations of peregrine falcons, ospreys and otters in the 1950s and 1960s.

Bioaccumulation is an increase in the concentration of a toxin in body tissues during an animal's life. It is a particular problem with toxins that are fat-soluble and therefore not easily excreted. For example, organic compounds containing mercury such as methyl mercury are more likely to accumulate in droplets of fat in adipose tissue than metallic mercury.

Biomagnification is an increase in the concentration of a chemical substance at each successive trophic level in a food chain. Predators tend to accumulate higher concentrations of a toxin than their prey. This is because the predator consumes large quantities of prey during its lifetime and bioaccumulates the toxins that they contain. Some organisms have greater concentrations of body lipids and so the accumulation rate varies across trophic levels. Sometimes, the toxin can be taken up directly from the abiotic environment rather than entering through the food chain.

▲ Figure 14 The mouse that this Eurasian kestrel (*Falco tinnunculus*) has caught should not contain DDT, but could have done in the 1950s and 1960s. Populations of kestrels and other birds of prey declined considerably in that period because DDT causing thinning of eggshells and the failure of most pairs to rear young

Data-based questions: Biomagnification

The process of biomagnification differs between aquatic and terrestrial food webs and between marine mammals and marine gill-breathing animals. The graphs in Figure 15 show biomagnification of a polychlorinated biphenyl (PCB-153) and an organochloride (β-hexachlorocyclohexane). Certain chemicals that are moderately lipid-soluble but can still dissolve in water can be eliminated into water by gill-breathing organisms but are not eliminated into the air by lung-breathing organisms.

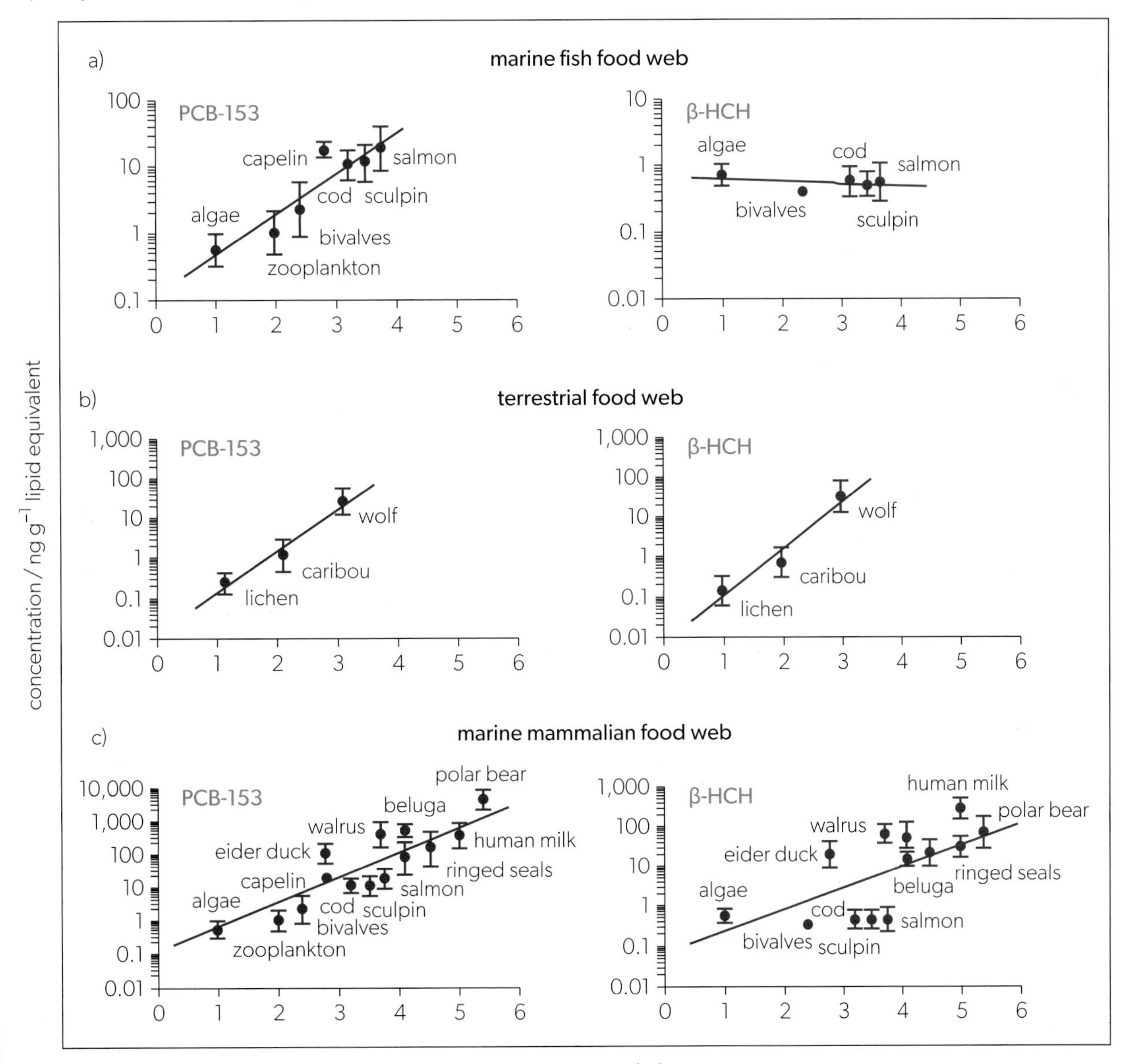

▲ Figure 15 Biomagnification in various food webs

1. Determine the trophic level of sculpin. [1]
2. Explain how it is possible to have a non-whole number trophic level. [2]
3. Outline the relationship between PCB concentration and trophic level in the terrestrial food web. [2]
4. Deduce the food web where β-HCH does not biomagnify. [2]
5. Compare the concentration of β-HCH at the third trophic level in the terrestrial food web and in the marine mammalian food web. [2]
6. Explain the differences in biomagnification of β-HCH in the terrestrial food web and the marine mammalian food web. [3]

ATL Thinking skills: Evaluating and defending ethical positions

DDT is an insecticide that was widely used in the mid-20th century, first to control disease vectors such as ticks and mosquitoes during and after the Second World War, then as an agricultural insecticide.

▲ Figure 16 A soldier demonstrating the DDT spraying equipment used to control lice which carry the typhus bacterium

DDT was made famous by the writing of Rachel Carson. Her book *Silent Spring* alerted the world to the effects that pesticides were having through bioaccumulation and biomagnification. As a consequence, DDT was banned for agricultural use under the terms of the Stockholm Convention on persistent organic chemicals, though residual indoor spraying to control mosquitoes was permitted. This use remains controversial, though it still occurs.

The World Health Organization continued to endorse the use of DDT for this purpose. Where the use of DDT was discontinued for malaria vector control in an area, rates of malaria climbed. Substitute strategies to spraying DDT were attempted but were not as successful. Many countries that had banned the use of DDT completely reapproved its use for residual indoor spraying. Concerned scientists argued that DDT may have a variety of human health effects, including reduced fertility, genital birth defects, cancer and damage to developing brains. Its metabolite, DDE, can block male hormones. The pesticide accumulates in body fat, and in breast milk, and evidence that it persists in the environment for decades is strong. With its strong record in reducing malaria cases and the WHO endorsement, the use of DDT increased globally. Pressure is mounting for governments and intergovernmental organizations to rely on DDT only as a last resort and the call is growing for the development of an alternative form of malaria vector control.

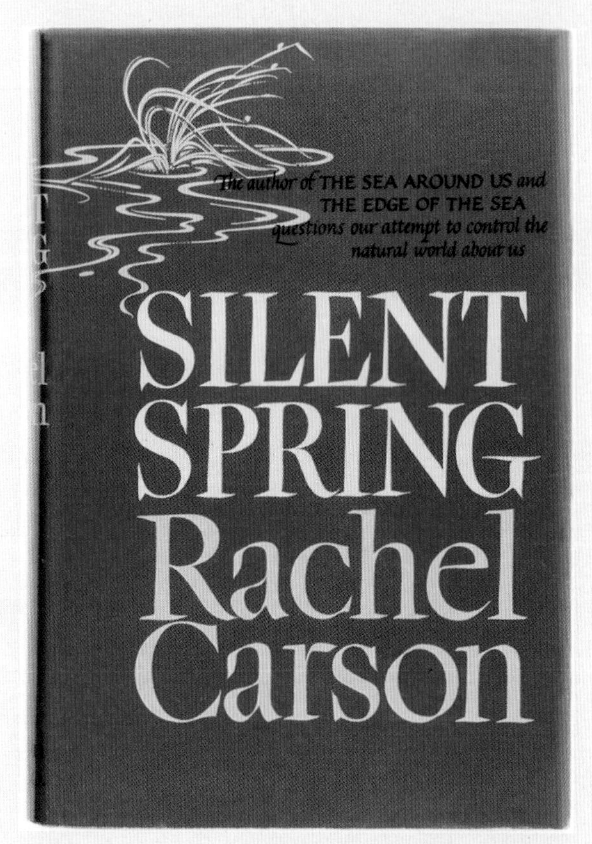

▲ Figure 17 *Silent Spring*, published in 1962

1. Discuss the competing values with respect to this issue.
2. Where are the continued "hotspots" of malaria? Discuss why this is the case.
3. What is the current status of programmes to limit the spread of malaria? How important is the continued use of DDT?

D4.2.10 Effects of microplastic and macroplastic pollution of the oceans

Plastic is a broad term that describes a range of polymers widely used in humanufacturing. Packaging and other items made from plastic are single-use in many cases and although some plastic is recycled, much is discarded every year—over 350 million tonnes in 2020. The oceans now contain huge quantities of plastic waste. It is accumulating because it is either non-biodegradable or degrades very slowly.

Degradation of plastic waste at sea releases toxic carbon compounds into the ocean that can bioaccumulate and biomagnify. Carbon compounds can also accumulate on the surface of plastic or be absorbed into it, with toxic effects for any organisms that then ingest the plastic.

Macroplastic is large visible debris including nets, ropes, drink bottles, caps and grocery bags. Marine wildlife such as seabirds and turtles can become entangled in nets and ropes and ingest plastic bags and other debris because they are mistaken for prey. Physical and chemical degradation of macroplastic in the oceans results in microplastic fragments that are less visible but far more numerous. There are estimated to be more than 50 trillion microplastic particles in the oceans and they have been found in every marine ecosystem investigated. They are also found in animal tissues. Microplastic particles are smaller than 5 mm in diameter. They may be globular or fibrous in shape. Their effects are still being investigated.

▲ Figure 18 Captain Charles Moore holding a jar of microplastics he trawled from a recent trip through the North Pacific Gyre

Data-based questions: Laysan albatrosses

The Laysan albatross (*Phoebastria immutabilis*) sometimes ingests plastic. A bolus is a pellet made of materials that the albatross cannot digest. The bolus is brought back up from the stomach to the mouth and then ejected. Figure 19 shows the mass of indigestible material in the boluses of two albatrosses.

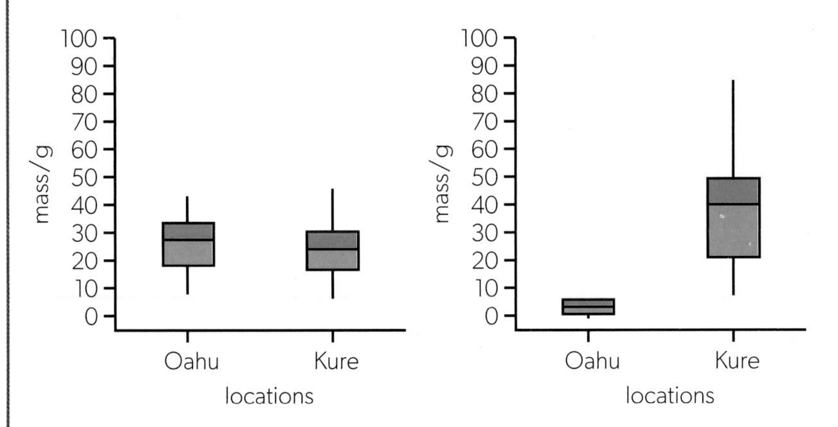

▲ Figure 19 Graphs of indigestible natural material such as bones and octopus beaks (left graph) and plastics (right graph) in the boluses of two birds at different locations

1. Suggest one reason for albatrosses ingesting indigestible plastic. [1]
2. Suggest a reason for the difference in ingested plastic in the diets of the albatrosses at the two locations. [2]
3. Outline the origin of microplastic debris in the marine environment. [2]
4. Using microplastics as an example, outline the concept of biomagnification. [2]

Science as a shared endeavour: Reporting of scientific discoveries

Much of science is an investigation of concepts and matter beyond everyday experience of the world—for example, the nature and behaviour of electromagnetic radiation or the build-up of invisible gases in the atmosphere. This makes it difficult for scientists to convince the general public that such phenomena do exist. This is especially true when the consequences of accepting their existence run counter to value systems or entrenched beliefs or when science discovers an unexpected public impact of seemingly private actions. For example, sulfates and nitrates released when coal is burned cause acid rain, so choices over which fuel to use are not just a private decision but a matter of public concern. Similarly, the discovery of the health effects of second-hand smoke turned a private decision to smoke cigarettes into a public issue and led to restrictions on smokers in public spaces.

As far back as 1971, marine biologist Ed Carpenter noticed peculiar, white specks floating amidst the mats of brown sargassum seaweed in the Sargasso Sea. These were identified as fragments of plastic. Another area where plastic was accumulating due to ocean currents was discovered in the north Pacific and described in the scientific literature in 1988. Public attention was only raised by dubbing it "the Great Pacific Garbage Patch", and by a transpacific journey in 2008 of a raft made out of plastic bottles (Figure 20). The resulting popular media coverage of the effects of plastic pollution on marine life changed public perception on the issue. However, there is concern that the term "garbage patch" underrepresents the scale of the problem. The formation and spread of microplastics throughout marine environments is harming ecosystems far more widely than accumulations in a handful of ocean gyres.

▲ **Figure 20** In 2008, Marcus Eriksen spent 88 days crossing 4,000 kilometres of ocean between California and Hawaii on a raft built from 15,000 plastic bottles wrapped in fishing nets

1. To what extent do scientists have a responsibility to educate the public about their research results?
2. For which lines of research is this most important?
3. Discuss the benefits and limitations to the public of simplified representations of research findings.

D4.2.11 Restoration of natural processes in ecosystems by rewilding

Natural ecosystems have been degraded by human actions and relatively little of the Earth's surface remains as pristine natural habitat. Among the consequences are:

- loss of biodiversity
- rapid rates of species extinction
- loss of ecosystem services such as carbon sequestration and climate regulation, flood protection, prevention of soil erosion and purification of air and water.

In many parts of the world, efforts are being made to encourage natural ecosystems to return. One approach to this is rewilding.

The essential principle of rewilding is that there should be as little intervention by humans as possible because natural processes restore habitats more effectively than humans. A first step is to stop human activities such as agriculture, logging and other forms of resource harvesting. That may be all that is required for forest to recolonize abandoned farmland or for marine ecosystems to re-establish in fishing exclusion zones. More commonly, there is a need for specific interventions to undo past human actions. Here are some examples of interventions that can speed up rewilding:

- distributing seeds of plants that should be components of the ecosystem but no natural seed source is present
- reintroduction of apex predators and other keystone species
- re-establishment of connectivity where natural ecosystems have become fragmented
- control of invasive alien species.

The Hinewai Reserve on the Banks Peninsula in New Zealand is an example of successful ecological restoration. A total of 1,250 hectares of farmland has been allowed to return to native forest. Alien mammals such as goats, brushtail possums and deer are rigorously controlled but apart from that there has been minimal interference. Alien plants are mostly tolerated because in time they are mostly eliminated by competition from better-adapted native species. Gorse (*Ulex europaeus*) is invasive on pasture land in much of New Zealand but at Hinewai it provides nurse canopies for saplings of native trees, and then dies out because of shading from the native trees that it helped to establish.

The speed of regeneration of natural ecosystems at Hinewai has been remarkable, but threats remain with increasing incidence of both droughts and floods after extreme rainfall. This can be attributed to climate change due to human activities, which threatens natural ecosystems throughout the world.

Activity: Explaining food web effects

The last wolf was killed in Yellowstone National Park in 1926, but this species was reintroduced in 1995. Wolves act as keystone species, with a disproportionate effect on the structure of the community.

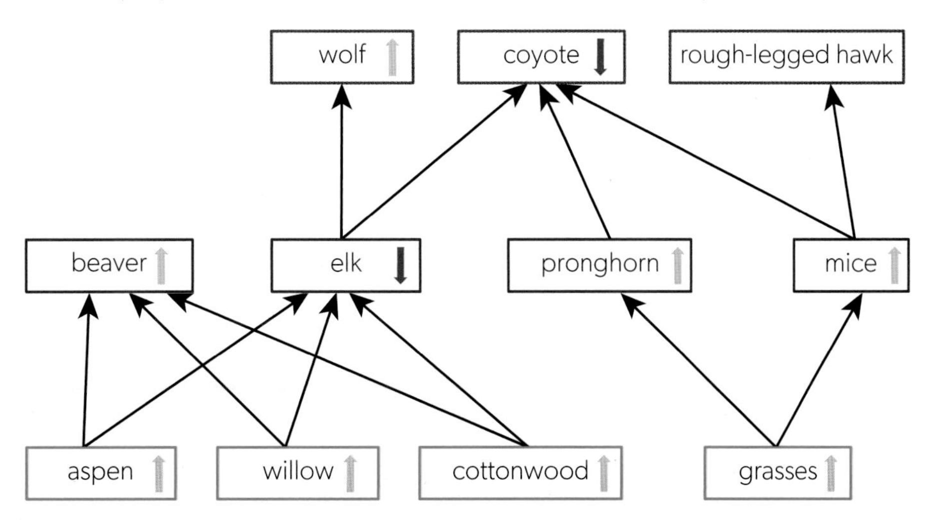

▲ **Figure 21** This food web model shows changes in populations resulting from the reintroduction of wolves. Coloured arrows pointing upwards indicate population increases and those pointing down indicate decreases

1. Define the term "apex predator".
2. Explain any four changes observed in the food web.
3. Predict the impact of the wolf reintroduction on the population of the rough-legged hawk.
4. Discuss whether the wolf should be considered to be an apex predator.

D4.2.12 Ecological succession and its causes

Ecological successions are sequences of changes that progressively transform ecosystems. Both the species composition of the community and diverse factors in the abiotic environment change over time (Figure 22).

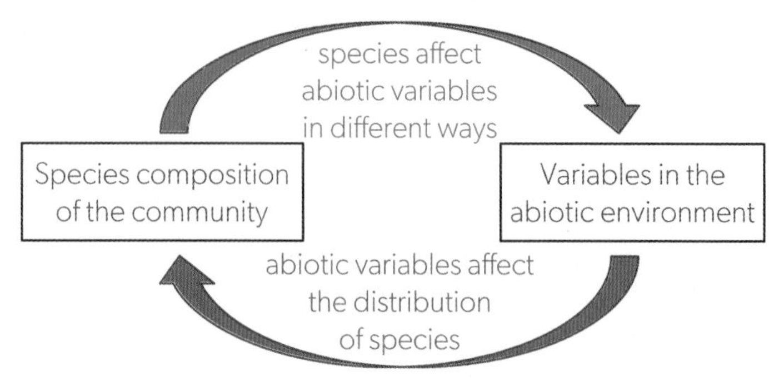

▶ Figure 22 Reciprocal interactions between living organisms and the abiotic environment cause changes that transform ecosytems

Consider an area of grassland that is being colonized by shrubs or trees. As the trees grow, light intensity at ground level diminishes and lower temperatures and higher humidity result. Leaf litter from the trees increases nutrient concentration, water infiltration and capacity for water retention of the soil. This means plant species adapted to conditions in grassland will be lost from the community and others adapted to conditions in forest will join it. There will also be changes in the animal species because of interdependency between specific species of animals and plants.

Changes in an ecosystem often trigger other changes, so one ecosystem replaces another in a series—this is an ecological succession. A stable and persistent ecosystem may then develop that does not undergo further significant change—this is the climax community. Alternatively, a recurring cycle may become established, with none of the ecosystems in the cycle persisting indefinitely.

This raises the question of what initiates an ecological succession. In some cases, the trigger is abiotic. An avalanche in mountains may sweep away forest and underlying soil, creating a bare substratum on which a new succession begins. In other cases, the trigger is biotic. For example, beavers may colonize a river and cause areas of flooding, leading to a series of successional changes that change open water into a swamp.

Activity: Changes due to succession

▲ Figure 23 The left photo was taken in 1985, the right photo 17 years later in 2002. The sign in the picture indicates that the area was covered in the ice of a retreating glacier in 1920

1. Identify changes visible in the photos that have occurred between 1985 and 2002.
2. Infer changes that will have occurred to biotic and abiotic variables during this time.

D4.2.13 Changes occurring during primary succession

There are two categories of succession: primary and secondary.

Primary succession begins in an environment where living organisms are largely or completely absent. On land, primary succession happens on bare rock or on deposits of clay, silt, sand or larger rock fragments, but no soil (Figure 24). Only organisms such as bacteria, lichens and mosses can colonize these inorganic substrata. Early colonizers generate small amounts of soil, allowing herbs with roots to start to colonize. As deeper soil develops, successively larger plants colonize—tall herbs, shrubs and then trees in most areas. Animal populations change with the plant populations, as do populations of decomposers.

These are the general principles of primary succession:

- species diversity increases as more species join the community than are eliminated
- primary production increases as larger plants colonize and there is more photosynthesis per unit area
- food webs become more complex
- nutrient cycling increases as animals and plants generate more dead organic matter.

▲ Figure 24 Primary succession as the Skaftafellsjökull glacier retreats on Iceland, leaving lifeless deposits of glacial drift—gravel, sand and clay

Data based questions: Succession in Glacier Bay

In 1794, Captain George Vancouver visited the area now known as Glacier Bay, Alaska. He made detailed notes regarding the position of the glaciers. This has allowed researchers to determine the time since the start of primary succession as the glaciers retreated.

The first species to colonize the bare rock are bacteria, lichens and moss. Mountain avens (*Dryas drummondii*) is a flowering shrub that dominates after the moss stage.

Deciduous alder trees (*Alnus sinuata*) colonize next followed by the most stable ecosystem which is a spruce and hemlock forest.

1. a. Outline the changes in mean stem diameter with time. [2]

   b. Explain the change in mean stem diameter. [2]

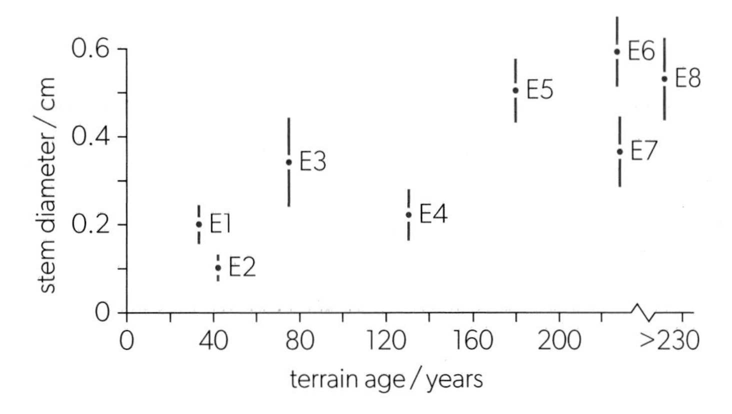

▲ Figure 25 Mean stem diameter and range of diameter of plants as a function of time since the tongue of the glacier covered the area at eight sites (E1–E8)

AHL

2. a. Outline the changes in the number of species (species richness). [2]
   b. Outline the changes in the relative numbers of species types (species evenness). [2]
3. a. Outline the changes in soil properties that are seen. [6]
   b. Deduce the stage where the greatest changes in soil properties are observed. [2]

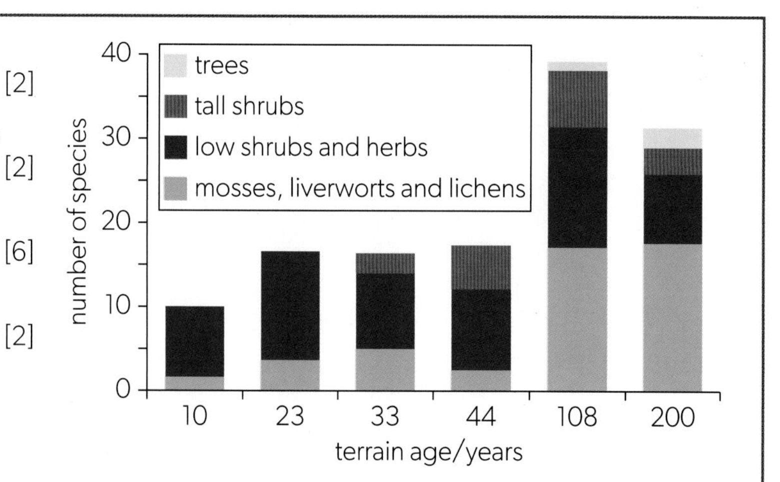

▲ **Figure 26** Number of species found in Glacier Bay, Alaska as a function of time since the glacier covered the area

▼ **Figure 27** Changes in the properties of soil as the dominant plant species change

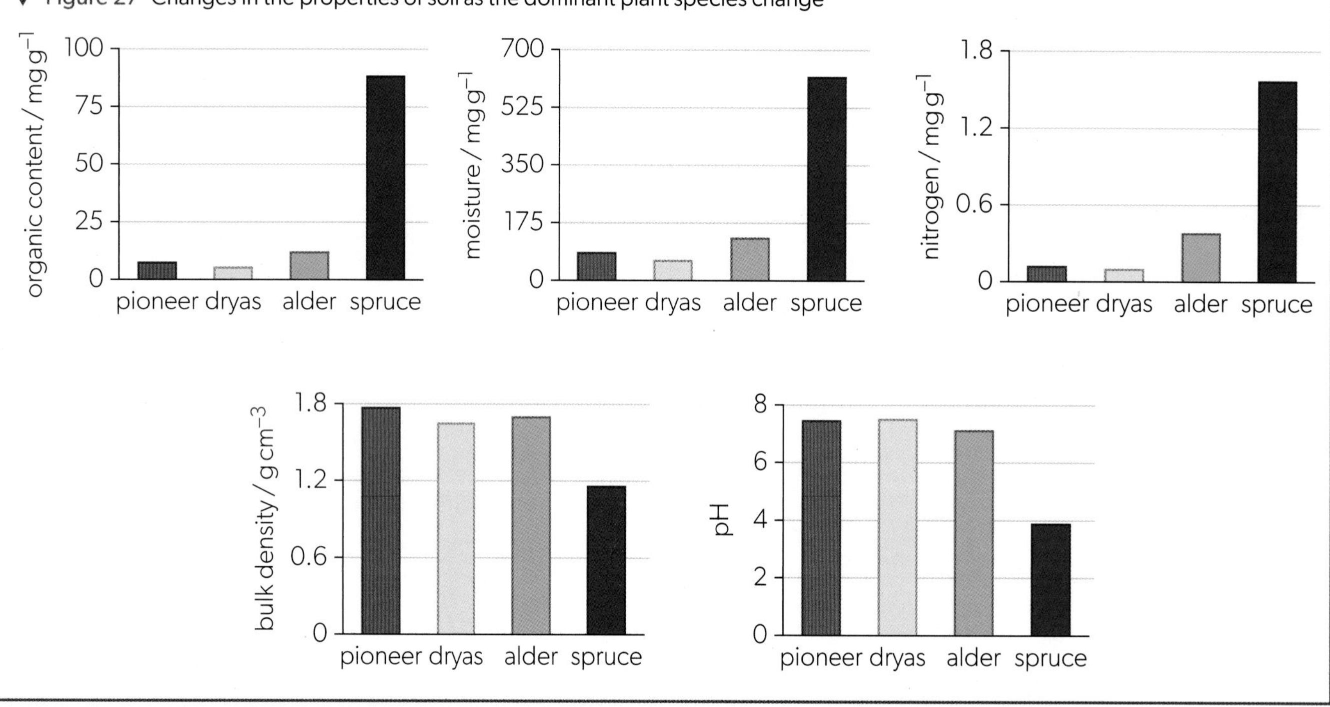

Designing investigations: Investigation into the effect of an environmental disturbance on an ecosystem

There may be opportunities to investigate secondary succession in your local area. Abandoned fields, wooded areas with disused roads and fields recovering from fire are all examples of sites where succession can be studied. Possible variables that can be studied include:

- species diversity
- stem density
- above ground biomass
- leaf area index
- volume of leaf litter
- water cycle variables including infiltration rates and run-off rates
- soil variables including soil structure, soul moisture, soil nutrient levels and compaction levels
- light levels
- bulk soil density.

D4.2.14 Cyclical succession in ecosystems

Some ecosystems are characterized by cycles of change rather than a stable climax. In cyclical succession, species replace each other over time repeatedly, even without the stimulus of large-scale disturbance. For example, forests dominated by oak in north-west Europe cannot regenerate, because oak is intolerant of shade, so they are succeeded either by forest dominated by other tree species or by grassland, depending on the intensity of browsing by deer. Oak seedlings can grow strongly in grassland, eventually shading out grasses and regenerating closed-canopy oak woodland.

Cyclical succession also occurs when a community is changed by recurring events. An example of this is the repeated fire cycles of the coastal chaparral ecosystem of California. The climax community in this type of ecosystem is dominated by deciduous shrubs and evergreen trees. Fires are the major source of disturbance and typically happen every 10 to 15 years, but sometimes at longer intervals.

On the North Island of New Zealand, a cyclic pattern of succession is observed on the rocky ocean shore. First, the honeycomb barnacles (*Chamaesipho columna*) settle on bare rock. Then crustose algae (*Ralfsia confusa*) colonize. Black mussels (*Xenostrobus pulex*) settle on top of the barnacles and crustose algae, forming a dense carpet that smothers the barnacles and algae underneath. Finally, the mussels detach, bare rock becomes available again and the cycle restarts.

▲ Figure 28 Blooming heather flowers (*Calluna*) in *Cladonia* lichen

In Scottish heathland, the dwarf *Calluna* heather is the dominant shrub. It loses its vigour as it ages and is invaded by the lichen, *Cladonia*. The lichen mat dies in time to leave bare ground, which is then invaded by bearberry (*Arctostaphylos sp.*) which is eventually outgrown by *Calluna* again.

D4.2.15 Climax communities and arrested succession

As succession proceeds, the pace of change tends to slow and eventually a climax community persists until disturbance causes further change. The nature of the climax community depends upon environmental circumstances. In the Glacier Bay ecosystem considered earlier (*Section D4.2.13*), forest dominated by western hemlock trees (*Tsuga heterophylla*) develops on well-drained, steep slopes. In poorly drained soil on shallow slopes, the climax community is muskeg swamp.

Where human activity alters the natural processes, there may be a deflected or arrested succession. If an alternative stable community develops due to human action, it is known as a plagioclimax. Two human activities that can cause a plagioclimax are grazing by farm livestock and drainage of wetlands.

Grazing

Pastoral farming is rearing cattle, sheep or other animals on pasture. The livestock are protected from predation so higher population densities can persist than in natural ecosystems. The animals feed by grazing. Grasses and many herbs can tolerate grazing, but tree and shrub seedlings are killed. This means areas that would develop into forest with lower densities of grazing animals remain as grassland.

AHL

▲ Figure 29 Mixed deciduous forest dominated by oak could develop on these Shropshire hills in England, but this is prevented by agricultural subsidy payments that allow sheep grazing to continue

Grassland ecosystems may be maintained naturally by large grazing animals in some parts of the world. The population sizes of these animals are not regulated by predation and become high enough to prevent growth of trees that are adapted to the climatic conditions. Examples are bison on the Great Plains of North America and elephants in parts of Africa.

Drainage of wetlands

Swamps are ecosystems that develop on waterlogged sites. Some plants and animals are adapted to the anaerobic conditions in the soil and they are able to thrive. Saprotrophic fungi are inhibited by the lack of oxygen, so incompletely decomposed plant matter tends to accumulate, forming peat. Considerable depths of peat can develop, increasing the waterlogging and creating acid conditions. In temperate areas, these ecosystems are known as peat bogs. Trees cannot grow on them and the dominant plants are often *Sphagnum* mosses. The large areas of peat bog in northern temperate and boreal areas store enormous quantities of carbon, helping to reduce atmospheric carbon dioxide concentrations.

Drains are sometimes constructed to remove water from wetlands and aerate the soils, making it possible to grow crops. The drains may be open ditches or porous pipes laid underground. As the soil dries out, there are major changes to the ecosystem and plants adapted to drier conditions colonize. If the land is not cultivated, trees may be able to colonize areas that were previously waterlogged. If the drains are not maintained and a site becomes waterlogged again, the changes can be reversed—the trees die due to waterlogging and peat bog re-establishes.

▲ Figure 30 Birch trees died on this swamp in Cheshire, England, after it was rewetted following a period of drainage

Linking questions

1. What is the distinction between artificial and natural processes?
   a. Discuss the strengths and limitations of using mesocosms to model ecosystems. (D4.2.4)
   b. Compare and contrast selective breeding with evolution due to nature selection. (A4.1.3)
   c. Outline the process of IVF. (D3.1.7)
2. Over what timescales do things change in different biological systems?
   a. Discuss the concept of change with respect to stable climax ecosystems. (D4.2.15)
   b. Outline the changes in atmospheric carbon dioxide since the mid-20th century. (C4.2.20)
   c. Outline the approaches used to estimate dates of the first living cells and the last universal common ancestor. (A2.1.7)

D4.3 Climate change

What are the drivers of climate change?

Greenhouse gases are the principal driver of anthropogenic climate change. What are the sources of these gases? Are there other processes contributed to climate change? The Cerrado is an ecoregion of savanna forest in Brazil that is being lost to expansion of farming. Trees in this region have roots which reach deep underground as an adaptation for surviving droughts and fires. This underground system serves as a significant carbon sink. Deforestation in the Cerrado is releasing underground stores of carbon at the same time as it is undermining the capacity of the system to store more carbon.

▲ Figure 1 Deforestation in the Cerrado, Brazil

What are the impacts of climate change on ecosystems?

Climate change will necessarily affect species distributions. It will allow some species to invade new territory while leading others to more marginal territories. Scottish mountain hares rely on camouflage to escape predators. They shed a dark coat in summer to a white coat in winter to maintain camouflage against snow-covered landscapes. This change is triggered by day length changes. Due to climate change, the duration of snow cover is decreasing. This is leading to a "mismatch" in seasonal camouflage that increases their risk of predation.

▲ Figure 2 A Scottish mountain hare in winter

SL and HL	AHL only
D4.3.1 Anthropogenic causes of climate change D4.3.2 Positive feedback cycles in global warming D4.3.3 Change from net carbon accumulation to net loss in boreal forests as an example of a tipping point D4.3.4 Melting of landfast ice and sea ice as examples of polar habitat change D4.3.5 Changes in ocean currents altering the timing and extent of nutrient upwelling D4.3.6 Poleward and upslope range shifts of temperate species D4.3.7 Threats to coral reefs as an example of potential ecosystem collapse D4.3.8 Afforestation, forest regeneration and restoration of peat-forming wetlands as approaches to carbon sequestration	D4.3.9 Phenology as research into the timing of biological events D4.3.10 Disruption to the synchrony of phenological events by climate change D4.3.11 Increases to the number of insect life cycles within a year due to climate change D4.3.12 Evolution as a consequence of climate change

D4.3.1 Anthropogenic causes of climate change

Short-wave radiation such as ultraviolet light from the Sun can penetrate the atmosphere. While some of it is reflected back into space, most of it is absorbed by the Earth and re-emitted as long-wave radiation, including thermal radiation. This is a phenomenon that is similar to what happens in a greenhouse or in a car parked in an open parking lot with the windows closed.

The Earth is kept much warmer than it otherwise would be by gases in the atmosphere that retain heat. It has been estimated that without these gases, the average temperature on Earth would be below zero Celsius. Methane and carbon dioxide are two of the most significant greenhouse gases and human activity is leading to the release of large amounts of them. Gases released due to human causes are known as anthropogenic sources of climate change.

Carbon dioxide is released into the atmosphere by cell respiration in living organisms and by combustion of biomass and fossil fuels. It is removed from the atmosphere by photosynthesis and by dissolving in the oceans. Methane is emitted from marshes and other waterlogged habitats but also from landfill sites where organic wastes have been dumped. It is released during extraction of fossil fuels and from melting permafrost in polar regions.

The two most abundant gases in the Earth's atmosphere, oxygen and nitrogen, are not greenhouse gases because they do not absorb longer-wave radiation. All the greenhouse gases together make up less than 1% of the atmosphere.

Weather consists of temperature, humidity, precipitation, visibility, wind and atmospheric pressure. Heat energy has an impact on all of these variables. Consistent and persistent increases in the heat content of the atmosphere will influence long-term weather patterns or climate.

For example, water vapour is formed by evaporation from the oceans and transpiration in plants. Both processes are accelerated by global heating. Water vapour is removed from the atmosphere by rainfall and snow. Water continues to retain heat after it condenses to form droplets of liquid water in clouds. Water, including atmospheric water, can serve as a heat sink. The water in the atmosphere absorbs heat energy and radiates it back to the Earth's surface as well as reflecting the heat energy back to Earth. This explains why the temperature drops much more quickly at night in areas with clear skies than in areas with cloud cover.

D4.3.2 Positive feedback cycles in global warming

Positive feedback is when the end product of a process results in the amplification of the process that created it. Global heating is associated with a number of positive feedback cycles.

Snow and ice reflect solar radiation. The amount of radiation they reflect into space is known as albedo. Light-coloured matter such as snow or ice has a high albedo. Dark-coloured matter such as the open ocean or dark forested areas when snow melts have a low albedo.

As polar and sea ice melt, open water becomes exposed; the water is darker and absorbs radiation more than white ice. The result is more infrared radiation is emitted at the poles, which in turn can further accelerate polar ice cap melting.

Permafrost is ground that remains frozen all year round. Some of the ground contains frozen detritus. As global heating accelerates, permafrost melts and the waterlogged detritus begins to decay, releasing methane which is a powerful greenhouse gas. This will, in turn, further accelerate global heating.

The solubility of gas in water decreases with temperature. As carbon dioxide levels in the atmosphere rise, the average global temperature also rises. The temperature of the oceans rises, releasing more carbon dioxide into the atmosphere, thus raising the global temperature further.

▲ Figure 3 Methane is bubbling up from the thawed permafrost at the bottom of the lake through the ice at its surface

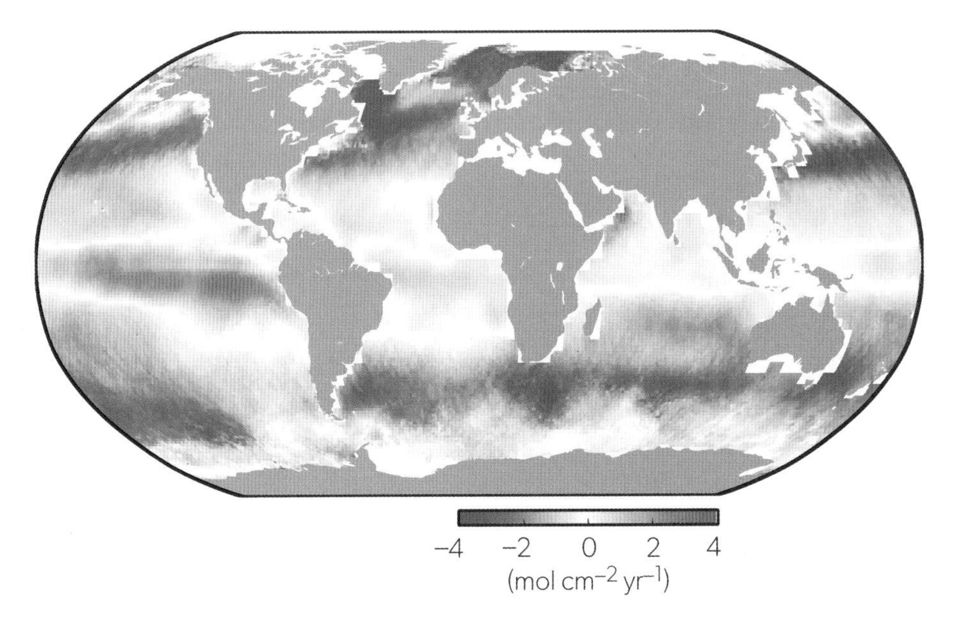

▲ Figure 4 Net flow rates of carbon from the ocean into the atmosphere. Negative values (blue) indicate that the region has a net uptake of carbon and the red indicates a net release of carbon. There tends to be a net release where the water is warmer

Frozen hydrated methane can form a cap over a store of methane, keeping it from escaping into the ocean and then into the atmosphere. As the ocean warms, the increased temperature of the seawater around some hydrate caps might dissolve them. This would cause a sudden release of large volumes of previously trapped methane.

Climate change has resulted in increases in droughts and forest fires. Increased temperatures lead to drier, more fire-prone conditions. When forests are on fire, there are more emissions from the burning vegetation. There is then reduced carbon capture by the destroyed or damaged forest, leading to amplification of the entire cycle.

▲ Figure 5 An icy cap of hydrated methane seals off a deposit of methane beneath the ocean floor

D4.3.3 Change from net carbon accumulation to net loss in boreal forests as an example of a tipping point

When the sum of environmental changes and positive feedback cycles overwhelm the resilience of an ecosystem, a tipping point is reached whereby the ecosystem is converted from one stable form to another. Cold temperatures in the boreal forest have an overall reducing effect on the rate of cellular respiration so the rate of decomposition of detritus is usually slower than the rate at which it collects. Photosynthesis by boreal forests captures carbon dioxide in tree biomass. Warmer, drier summers resulting from climate change have led to a string of droughts that predispose the forest to fires. These fires release the carbon dioxide that has been stored in centuries of detritus at the same time as undermining the capacity of the forest to store carbon. The widespread fires might lead to a tipping point where the boreal forests transition from being a carbon sink to being a carbon source. After the fires, deciduous forests composed of species such as aspen usually arise with trees that are spaced farther apart.

D4.3.4 Melting of landfast ice and sea ice as examples of polar habitat change

▲ Figure 6 A group of emperor penguins (*Aptenodytes forsteri*) walking over landfast ice at the emperor penguin colony at Snow Hill Island in the Weddell sea in Antarctica

Antarctic landfast ice is distinct from pack ice in that it does not move with wind or ocean currents but is "fastened" to the shore. The emperor penguin, *Aptenodytes forsteri*, hunts from the landfast ice edge into the open ocean. Climate variability means that the formation and break-up of the ice makes the penguins' habitat variable and unreliable. Similarly, the penguins' food sources occupy the margins of the landfast ice and the ocean. They are also impacted by the variability of the landfast ice. As nesting site distance to the ocean varies, errors in situating nesting sites can impact fledgling survival. The penguins depend on stable landfast ice for breeding success. For example, an early landfast ice break-up off Ross Island in 2018 resulted in high mortality of penguin chicks. Landfast ice provides a relatively flat surface for emperor penguins that are too large and bulky to climb over rocks or broken sea ice to raise their young.

▲ Figure 7 Approximately 1,500 walrus have hauled out of the water onto shore to rest

Walruses are large marine mammals whose habitat includes pack ice where they give birth. They are benthic feeders, which means they feed on the bottom of the ocean in shallow waters. Difficulties arise for walrus in ice-free environments. Walruses use land-based haul-out areas to rest. However, females often use sea ice for hauling out to avoid the risk of their pups being trampled by the larger males. Sea ice also expands access to a broader range of feeding sites. In the past, walruses could use nearby ice edges for resting. Land-bound walruses must now make feeding trips from coastal haul outs to areas of high prey abundance that are far from shore. Walruses lose more heat in the water than when out of the water, so they expend more energy on these trips on thermoregulation.

D4.3.5 Changes in ocean currents altering the timing and extent of nutrient upwelling

The ocean is stratified. Warmer, less salty water is less dense and floats on top of denser, colder, saltier water. Mixing between layers occurs as heat slowly seeps deeper into the ocean by the action of current, winds and tides. But the greater the difference in density between the layers, the slower and more difficult the mixing. When there is reduced current, winds, tides and movement between strata, the ocean is said to be more stable.

A warming climate increases ocean stability by making the surface ocean less dense. This happens as the water is first warmed, which expands its volume. Then melting ice adds freshwater into the ocean, decreasing the salinity of surface water. Atmospheric warming increases ocean stratification. There is evidence that the ocean has become significantly more stratified due to climate change, inhibiting the transport of heat, oxygen and carbon dioxide from the surface into the deeper ocean.

Stratification is another example of a positive feedback cycle. Increase in stratification further drives global warming. Warmer water on the surface can absorb less carbon dioxide from the atmosphere. This increases the atmospheric concentration of carbon dioxide and in turn further warms the Earth's surface, including the upper layer of the ocean. Warmer water can absorb less oxygen, and the oxygen that is absorbed cannot mix as easily with the cooler deeper ocean waters, making it difficult for marine life to thrive. Warmer ocean water also leads to increased bleaching of coral reefs.

The most active biological zone in the ocean is the surface and the top approximately 100 metres. This is because sunlight can only penetrate this far into the ocean. Carbon fixed by photosynthesis enters marine food chains. Carbon containing detritus and faeces can be circulated to the deep ocean by gravity, ocean mixing and the movement of organisms between the surface and deep waters.

Deep ocean currents that move toward continental shelves are forced upward leading to upwelling of nutrients, which contribute to nutrient cycling of the productive surface biological communities decreasing ocean primary production and energy flow through marine food chains. Climate anomalies such as El Niño decrease these upwelling events.

D4.3.6 Poleward and upslope range shifts of temperate species

Species whose habitat is on mountains are referred to as montane. Climate change is leading to warmer temperatures at each elevation. The montane species migrate upslope tracking their optimal climate. Those species that live at the highest elevations obviously cannot move higher to find a more suitable temperature. Competitive exclusion results in a need to escape competition and this often drives species to seek marginal niches as new competitors arrive.

Many climate change models predict species extinction and a disproportionate number of threatened species occupy the higher elevation habitats of mountains. The range of temperature in tropical regions is narrower than in temperate regions, including mountain habitats. It is therefore predicted that tropical montane species will be more sensitive to temperature changes than temperate montane species because they are less adapted to temperature variability.

Activity: Upslope migration of bird species

In the 1960s, researchers measured the upper altitude limit of montane bird species in two locations in Papua New Guinea: Mt. Karimui and Karkar Island. Forty-four years later, the studies were conducted again to determine if climate change had resulted in upslope migration. Figure 8 shows the results.

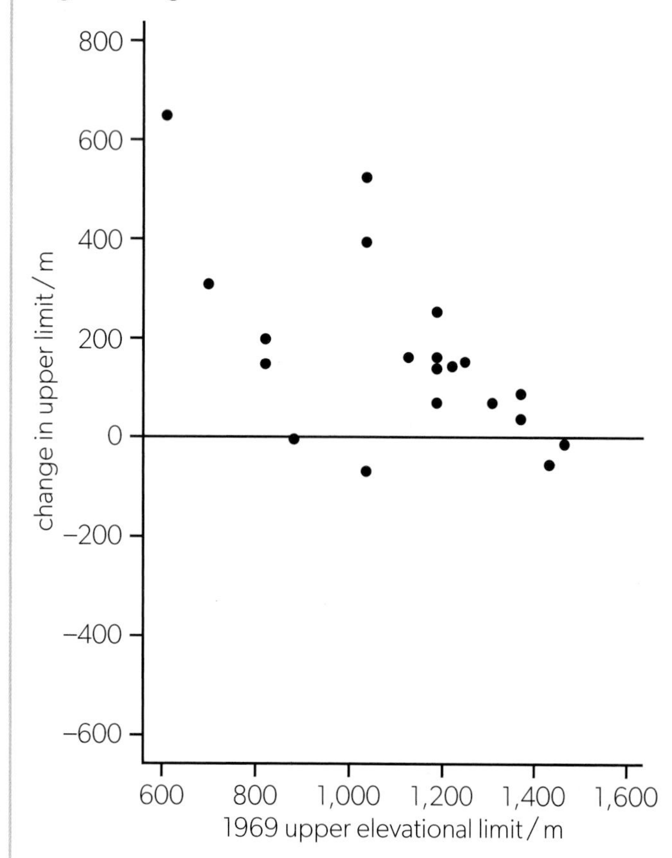

▲ **Figure 8** Comparison of the average upper elevation of 20 bird species in 1969 to the average change in upper elevation 44 years later

1. Define:
    a. montane
    b. upslope migration
2. a. Determine the new average elevation of the species which had been located at 600 m in 1969.
    b. State the relationship between upper elevation limit in 1969 and change in the upper limit more than 40 years later.
    c. Suggest why there was little change in the average elevation of species found at 425 m in 1969.
3. Outline the competitive exclusion principle.
4. a. The crested satinbird (*Cnemophilus macgregorii*) is currently restricted to the summit of Mt. Karimui. Explain why a further 1°C temperature increase would be likely to lead to the extirpation (driving out) of the population on Mt. Karimui.
    b. Suggest why populations of *C. macgregorii* might continue to exist on other mountains if there were a further 1°C temperature increase.

▲ **Figure 9** The crested satinbird (*C. macgregorii*) is currently found at the summit of Mt. Karimui in Papua New Guinea

For many species, global warming is leading to shifts of the latitude range where populations are found. These range changes are not due to migration but to changes in the relative levels of deaths and colonizations at the southern and northern boundaries of their range. In the northern hemisphere, an average northward range shift could occur when there is net extinction of populations at the southern boundary or colonization of new latitudes at the northern boundary. In a sample of 35 European butterflies, more than 60% of species have ranges that have shifted to the north, some by as much as 240 km during the 20th century.

Using a database to track changes in annual mean latitude

The Christmas Bird Count is an annual bird watching census event that is driven by volunteers across the western hemisphere. It first occurred in 1900 and still continues. The historical data from the event is available in a database at https://netapp.audubon.org/cbcobservation/. It is possible to determine the annual mean latitude of bird observations and trends can be detected. An example is the Carolina wren (*Thryothorus ludovicianus*).

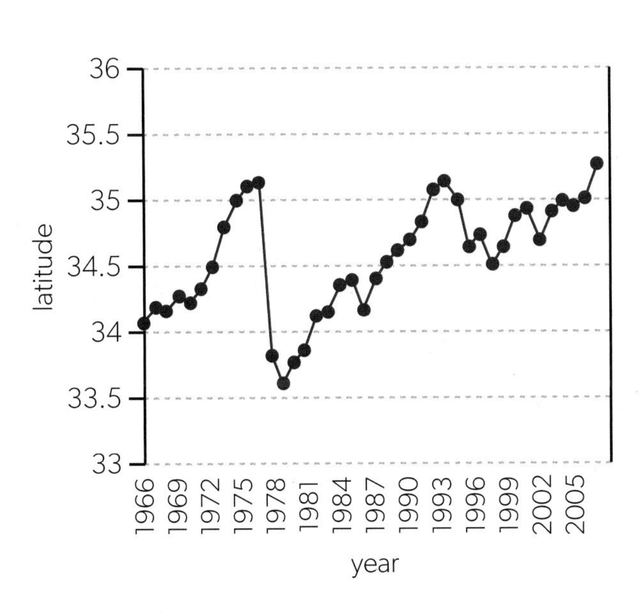

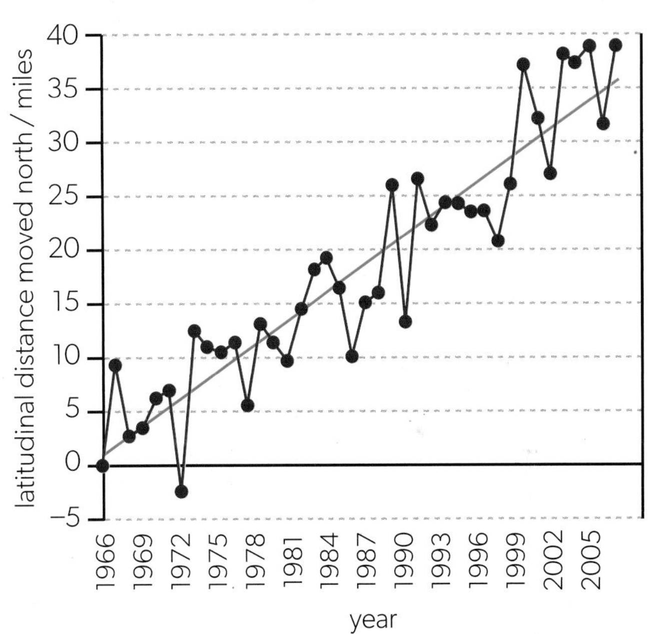

▲ **Figure 10** The graph on the left represents trends in the mean latitude of observation for the Carolina wren. The graph on the right shows the average amount of change in the latitudinal centre of abundance (in miles) of 305 widespread bird species in North America

D4.3.7 Threats to coral reefs as an example of potential ecosystem collapse

In addition to its contribution to global warming, emissions of carbon dioxide are having effects on the oceans. Over 500 billion tonnes of carbon dioxide released by humans since the start of the industrial revolution have dissolved in the oceans. The pH of surface layers of the Earth's oceans is estimated to have been 8.179 in the late 18th century when there had been little industrialization. Measurements in the mid-1990s showed that it had fallen to 8.104 and in 2014, levels had reached approximately 8.069. This seemingly small change actually represents 30% acidification. Ocean acidification will become more severe if the carbon dioxide concentration in the atmosphere continues to rise.

Marine animals such as reef-building corals that deposit calcium carbonate in their skeletons need to absorb carbonate ions from seawater. The concentration of carbonate ions in seawater is low, because they are not very soluble. Dissolved carbon dioxide makes the carbonate concentration even lower as a result of interrelated chemical reactions. Carbon dioxide reacts with water to form carbonic acid, which dissociates into hydrogen and hydrogencarbonate ions. Hydrogen ions react with dissolved carbonate ions, reducing their concentration.

$$CO_2 + H_2O \rightarrow H_2CO_3 \rightarrow H^+ + HCO_3^-$$

$$H^+ + CO_3^{2-} \rightarrow HCO_3^-$$

If carbonate ion concentrations drop, it is more difficult for reef-building corals to absorb them to make their skeletons. Also, if seawater ceases to be a saturated solution of carbonate ions, existing calcium carbonate tends to dissolve, so existing skeletons of reef-building corals are also threatened.

▲ **Figure 11** The colour in the coral is lost due to the expulsion of symbiotic unicellular algae called zooxanthellae that live within its tissues. In this case, the zooxanthellae have been expelled due to high water temperature. Some of the coral here appears brown because they still contain the zooxanthellae, which is needed to provide the coral with nutrients and thus maintain its colour

Hard corals live in a mutualistic association with photosynthetic algae known as zooxanthellae. The algae benefit by having a substrate to attach to which allows them to access the sunlight penetrating into the water. The corals get the benefit of the carbohydrates produced by the algae. When the ocean water surrounding the corals becomes too warm, the coral ejects the zooxanthellae leading to a loss of colour of the system giving the coral a bleached appearance.

Coral plays a strong role in structuring or "engineering" the community around it. Loss of coral polyps would have a disproportionate impact on the ecosystem of which they are a part.

Activity: Estimating ocean pH

Using Figure 12, determine the equation for the regression line. Use the formula $y = mx + b$. Use the y-value for 1988 as the value of b. Divide the change in y-values by the change in x-values for two different years. This will yield the value of i.

Use the regression equation to predict the year in which it is estimated that the pH of the ocean would reach 8.0.

Suggest a reason why a positive feedback cycle might result from ocean acidification.

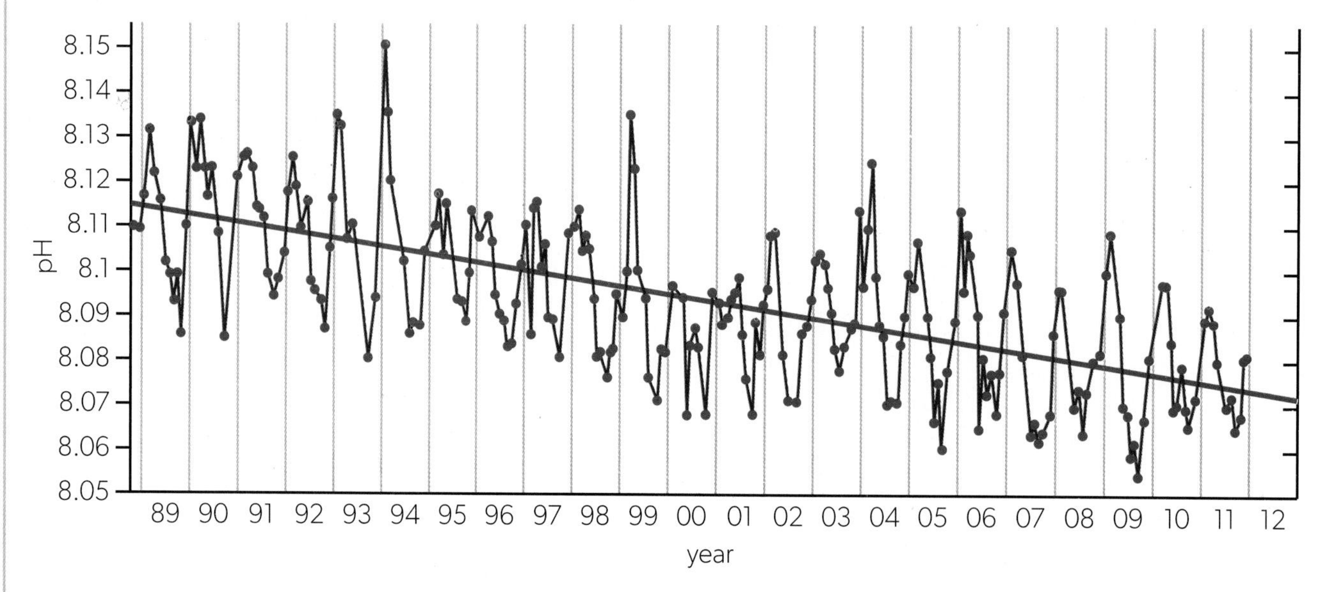

▲ **Figure 12** Graph showing ocean pH

D4.3.8 Afforestation, forest regeneration and restoration of peat-forming wetlands as approaches to carbon sequestration

Carbon sequestration involves the capture of carbon and its storage. This is a case where understanding of science is being used to address an issue of global concern.

Carbon sequestration occurs naturally through both geological processes and biological processes. Biological processes that capture and store carbon include carbon fixation in photosynthesis; growth in biomass storage in vegetation and wood and undigested detritus that becomes buried. Storage also occurs by uptake of carbon by aquatic organisms that build shells, and by the settling of detritus from the ocean surface into the deep ocean.

Afforestation involves planting trees in areas where they currently do not exist. As an approach to achieve national commitments to address climate change, a number of countries have committed to achieve numerical targets for tree planting. For example, the government of Canada has committed to plant 2 billion trees and the government of The Bahamas has committed to planting 1 million trees. Since 1990, the European Union has paid farmers to afforest their land. The Great Green Wall is a cooperative initiative between 19 north African countries to plant trees to reverse desertification.

Forest regeneration or reforestation is the restocking of forests that have been depleted, usually through forestry practices and often in the form of clearcutting. Following forest harvest, tree planters will enter the area, planting seedlings. The replacement trees are often a monoculture of commercially important trees.

Peat is partially decayed organic matter that comes from unique waterlogged ecosystems such as bogs, muskegs and moors. It is commercially useful as a source of fuel, domestic heating, fertilizer and gardening material. Peat formation naturally occurs in waterlogged soils in temperate and boreal zones and also develops very rapidly in some tropical ecosystems. Globally, wetlands such as these constitute the world's largest carbon sink. Burning in drained peatlands is a fire risk because the fires persist for a long time and are difficult to extinguish. The Global Peatlands Initiative seeks to protect, maintain, conserve and restore peatlands. Restoring peatlands requires the restoration of water levels, blocking drainage and re-establishing native species such as sphagnum moss.

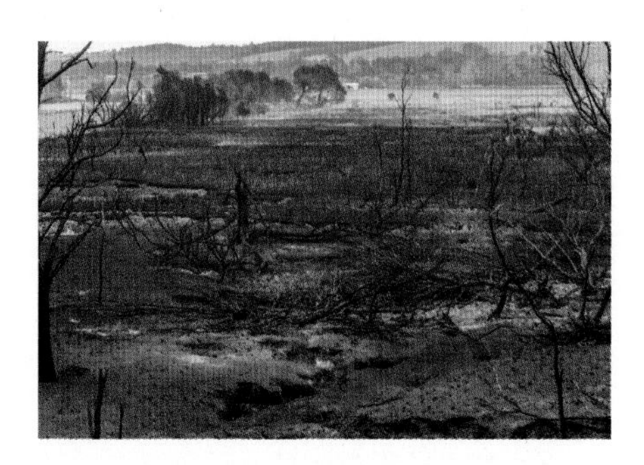

▲ Figure 13 The smoke from underground peat continues to rise into the air in Sarsfield, Australia, days after a bushfire. Dried peat is a fire hazard because it contains large amounts of stored energy

Global impacts of science: Evaluating the role of science

There is active scientific debate over whether afforestation and reforestation are the best approaches to carbon sequestration.

Sometimes, reforestation involves planting fast growing non-native species plantations or monocultures of commercially important species. It is difficult and expensive to plant anything other than a monoculture. But monocultures result in a risk of disease and a reduction in diversity. Tree planting after clear cutting will not restore the fully integrated ecosystem with all its original emergent properties.

Some concerns over afforestation and reforestation schemes include planting trees in areas where the level of rainfall does not support their survival. Fragmented forests have edge effects that prevent the full recovery of the ecosystem. A preferred approach is to plant at the edge of existing forests to extend the size of the core undisturbed area. Dense forest has a dark canopy which could exacerbate the albedo effect. Areas that are reforested or afforested are not available for other uses such as settlement and agriculture. Mitigating climate change is urgent but the carbon sequestering benefits of forests will occur in the future.

AHL

D4.3.9 Phenology as research into the timing of biological events

Phenologists are biologists who study the timing of seasonal activities in animals and plants, such as the opening of tree leaves and the laying of eggs by birds. Data such as these can provide evidence of climate changes, including global warming.

Photoperiod is the length of time that an organism is exposed to light during a 24-hour period. It is also called "day length". It varies in length as the seasons change, especially towards the poles of Earth, but is stable from year to year. It is unaffected by global warming and climate change.

Day length is an important environmental signal contributing to the timing of major developmental transitions in plants. For example, flowering in some plants occurs when day length is short. Examples of such plants are chrysanthemums and poinsettias, both important decorative flowering plants sold in the autumn and early winter. Many crop plants only flower when day length is longer; examples include lettuce, spinach, beets and potatoes.

For deciduous trees, reduction in day length in the autumn leads to the cessation of growth just prior to the onset of winter. A visible indicator of growth cessation is the formation of a bud that encloses the end of branches where the apical meristem is. This is known as bud set. In some species, bud set is determined entirely by day length. In others, the seasonal end of growth occurs in response to decreasing temperature alone or in combination with decreasing daylength. The emergence of new leaves at the start of each growing season is known as budburst. For a few species, budburst is affected by day length, but for the most part, temperature determines budburst.

Birds migrate from areas of low resources to areas of high resources. Two of the most important resources are food and nesting sites (*Section D4.3.11*). The primary physiological cue for migration is the change in day length.

Activity: To what extent are plant life cycle processes affected by temperature?

Scientists conducted a study of the timing of various life cycle events as a function of temperature.

1. Explain what is meant by R^2, making reference to the chart in Figure 14.
2. Deduce, with a reason, a life cycle event that is likely to be strongly affected by photoperiod.
3. Deduce, with a reason, an event that is likely to be strongly influenced by temperature.
4. Define "producer".
5. Explain the role of producers in food chains.
6. By referring to Figure 14, suggest an impact of climate change on deciduous forests.

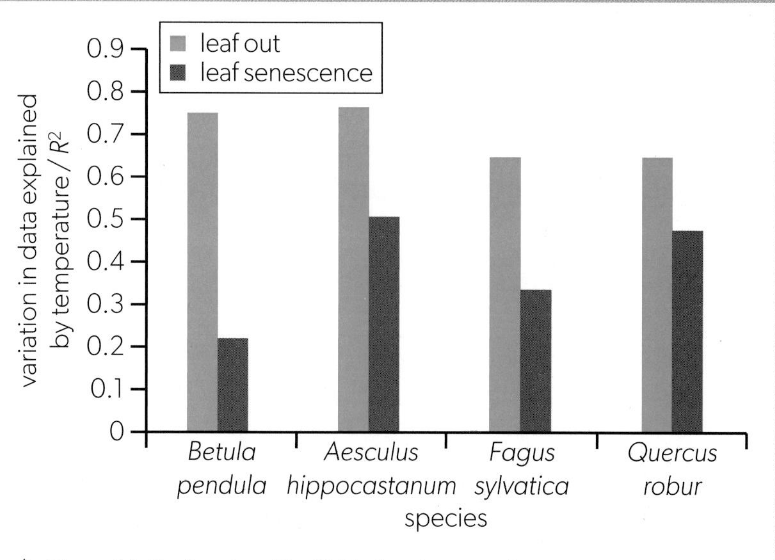

▲ **Figure 14** Budburst and leaf fall in four tree species

Data-based questions: Phenology

The date in spring when new leaves open on horse chestnut trees (*Aesculus hippocastaneum*) has been recorded in Germany since 1951. Figure 15 shows the difference between each year's date of leaf opening and the mean date of leaf opening between 1970 and 2000. Negative values indicate that leaf opening was earlier than the mean. The graph also shows the difference between each year's mean temperature during March and April and the overall mean temperature for these 2 months. The data for temperature was obtained from the records of 35 German climate stations.

1. Identify the year in which:
   a. the leaves opened earliest [1]
   b. mean temperatures in March and April were at their lowest. [1]
2. Use the data in the graph to deduce the following:
   a. the relationship between the temperatures in March and April and the date of opening of leaves on horse chestnut trees [1]
   b. whether there is evidence of global warming towards the end of the 20th century. [2]

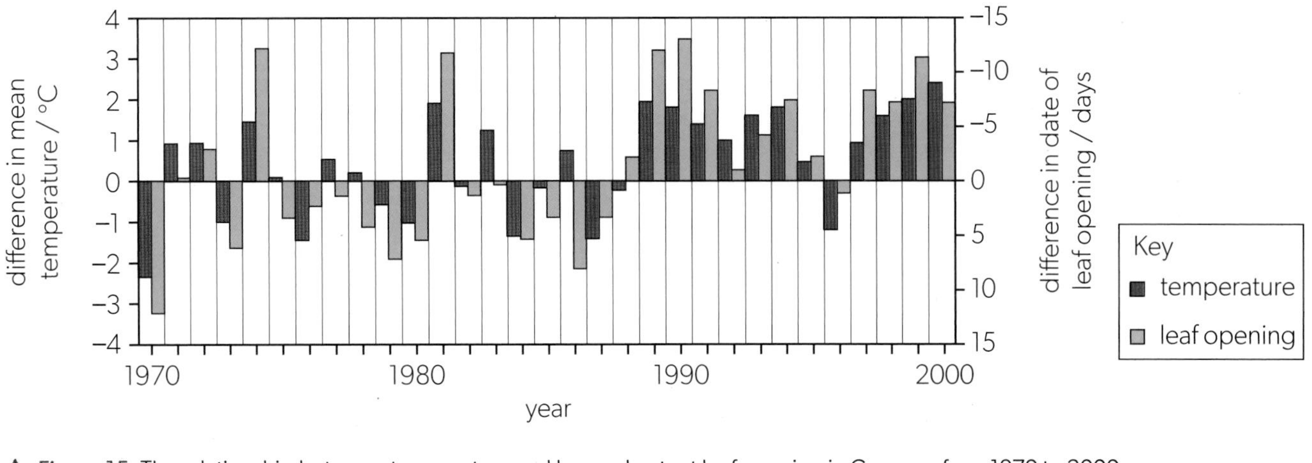

▲ **Figure 15** The relationship between temperature and horse chestnut leaf opening in Germany from 1970 to 2000

D4.3.10 Disruption to the synchrony of phenological events by climate change

Many bird species have evolved in such a way that their nesting cycle is synchronized to the period of maximum food availability. In temperate ecosystems, species of birds that migrate predominate. Migration is most often triggered by changes in daylight hours, known as photoperiod. This is a variable which is not affected by climate change. However, availability of food sources such as seeds, fruits and insects is based on events in biological life cycles that are affected by temperature. This means these events are therefore affected by climate change. Plant phenology is particularly affected by climate change. Most commonly, migratory birds arrive after quantities of food sources have peaked.

Birds like great tits (*Parus major*) rely on caterpillars as a source of food to feed their young. A 20-year study demonstrated that peak caterpillar biomass has been occurring earlier due to warmer weather in spring. The reproductive timing of great tits has not changed in alignment. The result is a mismatch that has resulted in fewer *Parus major* chicks. For those who do survive, there is a lower mean mass per chick.

A number of species interactions may become mismatched due to phenological conditions for example plant–herbivore, predator–prey and pollinator–plant interactions.

Evidence suggests that caribou populations in Greenland are declining due to a mismatch between caribou food requirements and plant availability. Caribou forage on a range of arctic plants such as the Arctic mouse-ear (*Cerastium arcticum*). Caribou time their spring migration to coincide with the emergence of food plants essential for nursing calves. They move about the landscape as the plants they graze develop at different times. Evidence suggests that climate change has led to a mismatch between plant development and caribou migration so caribou are less able to meet their nutritional needs by moving around the landscape.

D4.3.11 Increases to the number of insect life cycles within a year due to climate change

Warmer temperatures favour some species of pest—for example, the spruce bark beetle, *Dendroctonus rufipennis*. It is a native North American bark beetle that usually develops in weakened trees. Tree health increases host resistance to beetle attack: healthy spruce trees can successfully resist moderate numbers of beetle attacks. When climate events stress entire populations of trees, spruce bark beetle reproductive success is greatly increased. For example, drought, brought on by warmer temperatures and low rainfall stresses the trees.

Spruce bark beetles generally require 2 years to complete their life cycle. But when temperatures are warmer than normal, a generation can be completed in 1 year. In 2-year cycles, development is not synchronized and a portion of the beetle population emerges and attacks trees each year in low numbers. In 1-year cycles, development is synchronized and the beetles emerge in large numbers to attack trees. Important outbreaks of beetle numbers have occurred in Alaska, Colorado, British Columbia and Utah. Warm temperatures provide a double threat to the spruce trees: they can become susceptible to beetle attack and the beetles can attack in synchronized greater numbers.

▲ Figure 16 A stand of trees on Monarch Mountain, Colorado, killed by the spruce bark beetle (*Dendroctonus rufipennis*)

Data-based questions: Spruce bark beetle, climate and tree stress

1. a. Identify the two periods when the drought index remained high for 3 or more years (Figure 17). [2]
   b. Outline the trends in mean annual temperature between 1970 and 2000. [?]
   c. i. Distinguish between the beetle outbreaks in the 1970s and 1990s. [2]
      ii. Suggest reasons for the differences between the outbreaks. [2]
   d. Predict rates of destruction of spruce trees in the future, with reasons for your answer. [4]

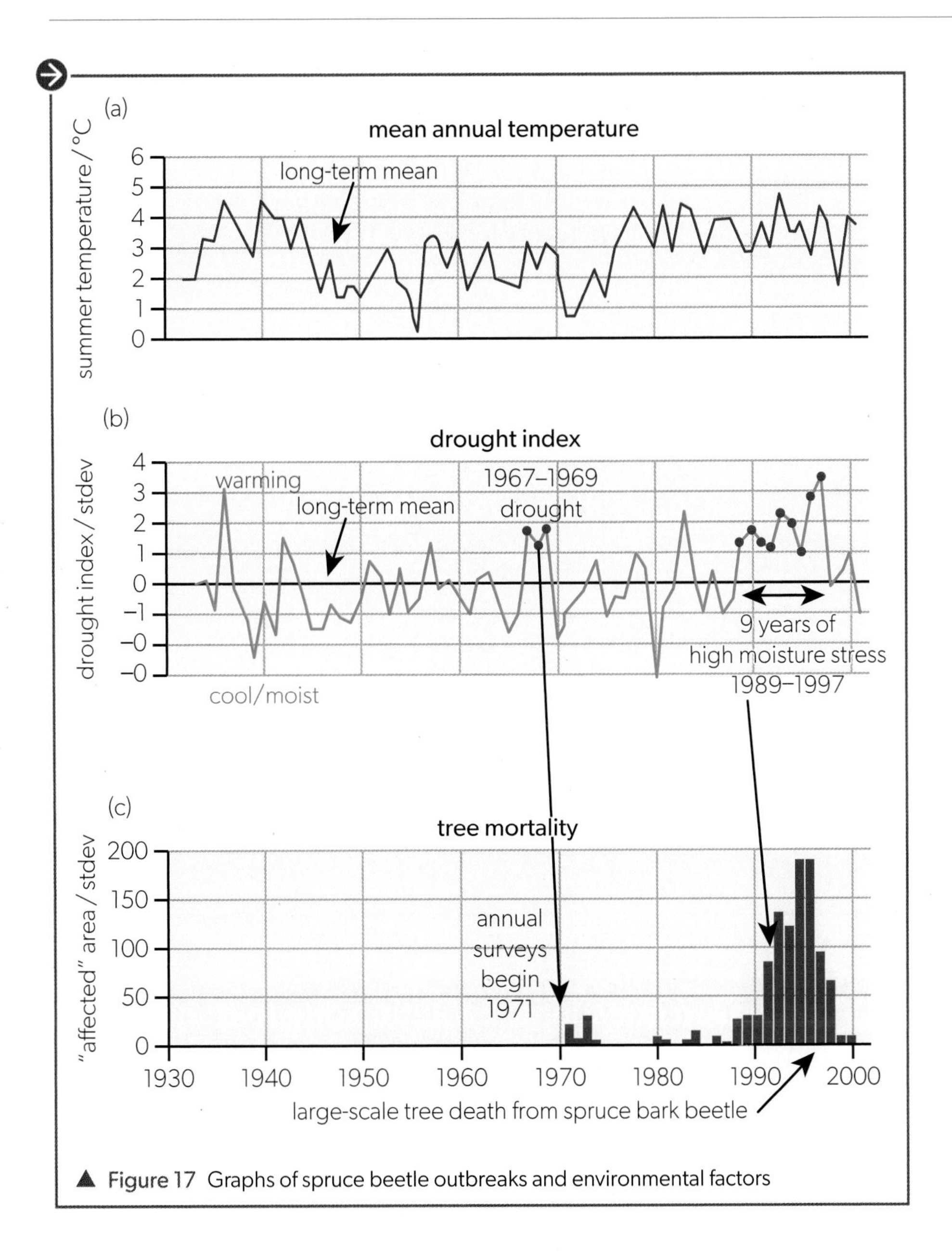

▲ **Figure 17** Graphs of spruce beetle outbreaks and environmental factors

D4.3.12 Evolution as a consequence of climate change

Data-based questions: Evolution related to climate change

Tawny owls, *Strix aluco*, vary in colour, ranging from pale grey to brown (Figure 18).

Plumage coloration in tawny owls is a heritable trait, consistent with a simple Mendelian pattern of brown (dark) dominance over grey (pale). Increasingly mild winters over the last 30 years in Finland have reduced the average snow cover. Figure 19 shows the correlation between that and the frequency of the brown variant over the 30-year period.

grey

brown

▲ **Figure 18** Tawny owl colour variants

AHL

1. a. Over the 30-year period of the study, outline:
   i. the trend in average annual snow depth [2]
   ii. the trend in the proportion of the population that has brown feathers. [1]

   b. Suggest reasons for the trend in snow depth. [2]

2. Define the term "reproductive fitness". [1]

3. Using the theory of evolution by natural selection suggest reasons why the brown variant is becoming more common and why it might have been less common before 1980. [3]

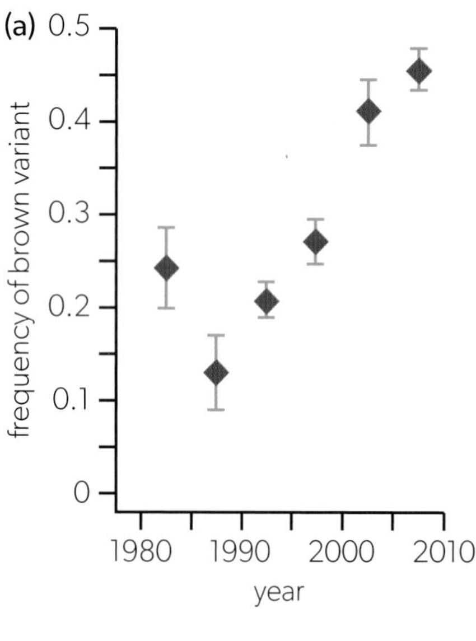

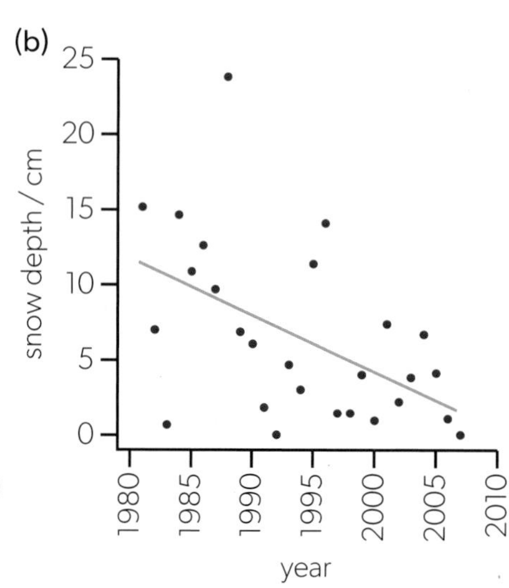

▲ **Figure 19** Snow cover and Tawny owl colour over 30 years

Data-based questions: Evolution due to climate change

In the Netherlands there are two variants of the two-spot ladybird beetle, *Adalia bipunctata*: black with red spots (melanic form) and red with black spots (non-melanic form).

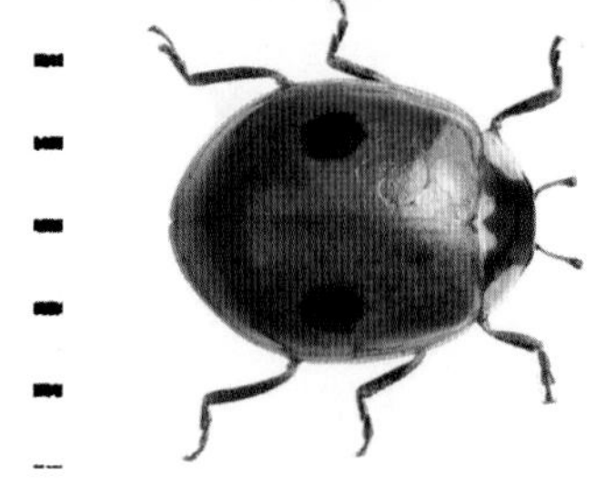

▲ **Figure 20** Two-spot ladybird

In 1980, researchers noted that the frequency of the melanic form increased along a transect away from the sea coast with the seaside having 17.8% of the melanic form and 100 km inland having an observed incidence of 56% of the melanic form (Figure 21). Changes in the difference between the median value for percentage melanics for the five most westerly localities near the coast and that for the five most easterly localities inland are shown in the graph.

1. State the difference in incidence of melanic forms in 1990 between the coastal samples and the inland samples. [1]
2. Describe the trend in the graph. [2]
3. Suggest, with a reason, whether the melanic form or the non-melanic form would absorb more light energy. [4]

AHL

4. Using the theory of evolution by natural selection suggest reasons why the non-melanic form is becoming more common inland and why it might have been less common before 1980. [3]

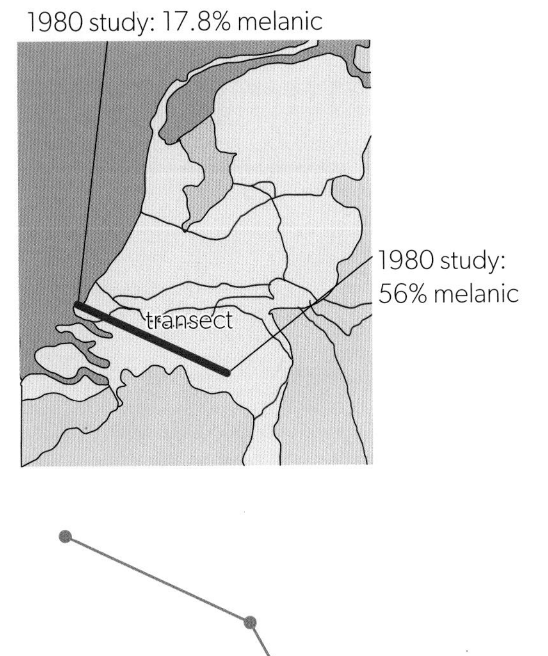

▲ **Figure 21** Map and graph showing results of 1980 study

Linking questions

1. What are the impacts of climate change at each level of biological organization?
   a. Explain reasons why rising ocean temperatures could lead to oxygen depletion. (C1.2.5)
   b. Outline the implications of lower pH for ocean life due to increases in atmospheric carbon dioxide. (D4.3.7)
   c. Outline one example of a change in the timing of a life cycle of an organism due to climate change. (D4.3.11)
2. What processes determine the distribution of organisms on Earth?
   a. Outline the relationship between temperature changes of a mountain habitat and species distribution. (D4.3.6)
   b. Explain why the top strata of the ocean is the most biologically productive referring both to photosynthesis and upwelling of nutrients. (D4.3.5)
   c. Describe the climate conditions that favour the formation of forest. (B4.1.6)

TOK

What challenges are raised by the communication of scientific knowledge?

In Darwin's time, it was widely understood that species evolved, but the mechanism was not clear. Darwin's theory of evolution by natural selection provided a convincing mechanism. Today it is argued that Darwinism replaced the theory of Lamarckism. However, in Darwin's own early writing, he refers to the inheritance of acquired characteristics in his theory of pangenesis. Darwin wrote about the theory of pangenesis in his book *Animals and Plants Under Domestication* (Figure 1), published 7 years after he published *On the Origin of Species* in which he proposed the theory of natural selection. The theory of pangenesis was based on the idea that a characteristic formed by body cells would, in response to the environment, either end up being used or disused. The cells would produce "gemmules" which travelled around the body, eventually ending up in germ cells (sperm and eggs). These gemmules were said to be microscopic particles that contained information about the required characteristics of their parent cell, and once they reached germ cells, they could pass on to the next generation the newly acquired characteristics of the parents.

THE VARIATION

OF

ANIMALS AND PLANTS

UNDER DOMESTICATION.

BY CHARLES DARWIN, M.A., F.R.S., &c.

IN TWO VOLUMES.—VOL. I.

WITH ILLUSTRATIONS.

LONDON:
JOHN MURRAY, ALBEMARLE STREET.
1868.

The right of Translation is reserved.

▲ Figure 1 The cover of *Animals and Plants Under Domestication*

The challenge of interpreting writing from the past is to attempt to understand the perspective of the time. Modern understanding of evolution by natural selection is informed by our knowledge of the mechanisms of inheritance that scientists understand today that was not available to Darwin.

Scientists often have a deep understanding of the causes of issues of global concern. They can appropriately influence the actions of citizens if they provide clear information about their research findings.

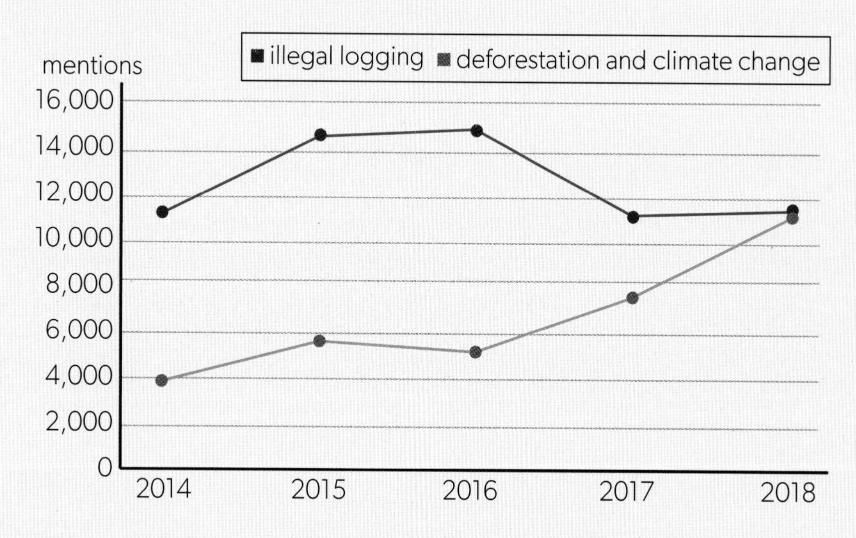

◄ Figure 2 Graph showing changes in media mentions of deforestation and climate change

However, popular media coverage of science has a publication bias toward stories that describe something novel, especially if it can capture the public's attention. One study analysed the shift in media reporting on illegal logging toward stories about concerns about deforestation and climate change. Figure 2 shows that between 2014 and 2016 there were more than twice the number of articles expressing concern about illegal logging than climate change and deforestation. However, over a 2-year period, both issues became reported with equal frequency. There are some concerns that afforestation and reforestation efforts focused on forests as carbon sinks can detract from rewilding efforts which acknowledge the complex task of redressing human impacts on ecosystems. Scientists therefore should be aware of false impressions of the natural world perpetuated by the popular media.

The linguistic relativity hypothesis argues that the words we use have an impact on our perception of the world. The orange roughy, *Hoplostethus atlanticus*, is a fish that was formerly known as "slimehead" (Figure 3). It was renamed through a US National Marine Fisheries Service programme in the 1970s. This programme identified species recommended for renaming to make them marketable. For similar reasons, the Patagonian toothfish (*Dissostichus eleginoides*) was renamed the Chilean sea bass, even though it is a member of the cod family and is mostly fished in the waters off Antarctica. In both cases, demand for both fish went up once. Unfortunately, the end result is that both have been fished unsustainably.

The strategy of choosing a term that can affect perception has helped popularize the issue of plastic in the ocean. The term "Great Pacific Garbage Patch" has captured the attention of the public in a way that brings awareness of the problem, even if the term itself leads to a misunderstanding of the scope of the problem.

Synonyms are words that sound different but have the same meaning. Sometimes synonyms have different connotations. Euphemisms are terms that have pleasant connotations while dysphemisms have negative or harsh connotations. Anyone who has ever had to deal with a toddler in cold weather is aware of their predisposition to have "elfluviate, primarily mucus, flow from their nostrils". That would be a neutral way to describe it. Dysphemistic language would be to describe the child as a "snotty nosed kid". Euphemistically, you might say to the child, "come closer while I wipe the snootle from your nose".

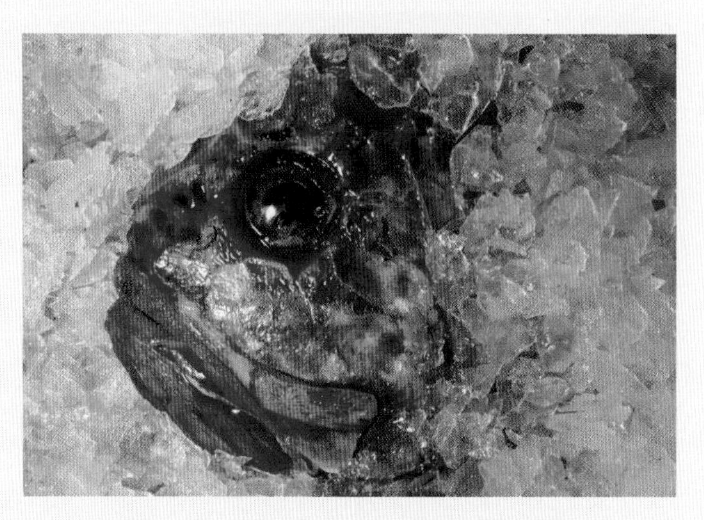

▲ Figure 3 The orange roughy, *Hoplostethus atlanticus*, packed in ice on board a trawler. *H. atlanticus* lives at depths of 80–1,000 metres. It is packed in ice to preserve it until the trawler reaches port. The fish is considered a great delicacy in some countries. It is a notably long-lived fish—some of the caught specimens are over 100 years old. The slimehead family have a canal system of lateral lines containing "slime" (mucus). This forms part of a sensory organ that allows them to detect pressure changes in their environment that might be caused by other fish and predators

Activity: Analysing changes in the frequency of term use

The search engine Google has a tool called Ngram viewer. Entering a term or phrase into the search box returns a graph showing changes in the frequency in the use of the term over time. Searches can go back to the 15th century. Carefully narrowing the date range allows for a determination of when the combination of terms is first detected in print. Entering multiple terms separated by a comma allows for a comparison of the frequency.

How has the frequency of the terms greenhouse effect, climate change, global warming, global heating, climate crisis and climate emergency changed with time?

Which terms are most common currently?

Is the language being used to discuss the consequences of the anthropogenic additions of greenhouse gases to the atmosphere becoming increasingly precise? Increasingly emotive?

End of chapter questions

1. *Drosophila subobscura* (shown in photograph below) is a species of fruit fly native to Europe. The sample on the left is from Spain, latitude 39°N, and the one on the right is from Denmark, latitude 56°N. The species was introduced into both South America and North America approximately 20 years ago. The graph below shows the wing size in arbitrary units of *D. subobscura* at different latitudes in the three locations.

Source: GW Gilchrist, et al (2004), Evolution, **58** (4), pp 768-780

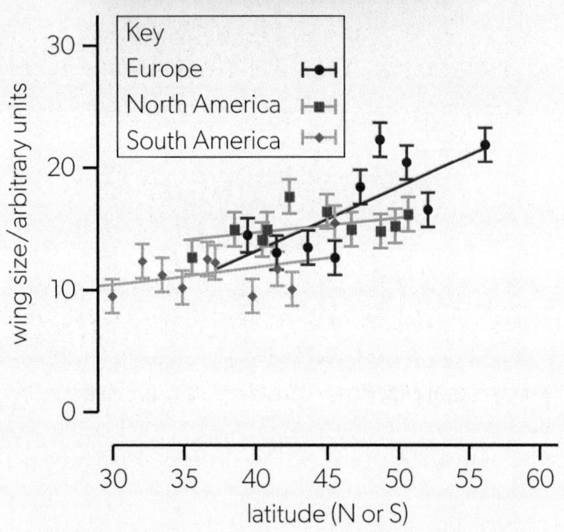

a. Identify the relationship between wing size and latitude shown by *D. subobscura* in Europe. [1]

b. Compare the data for wing size of *D. subobscura* in North and South America with wing size in Europe. [2]

c. Suggest one reason for the differences. [1]

d. Predict, with a reason, what might happen to *D. subobscura* in the future as a result of its introduction to new areas. [2]

2. Primary production is directly related to the amount of photosynthesis that occurs in a cubic metre of water. In the waters around Bermuda (32°N) in the Atlantic Ocean, microscopic phytoplankton are the producers. They use trace nutrients from seawater in their metabolism. These nutrients are a limiting factor in total population size. A dense phytoplankton population makes the water cloudy. The data shows primary production per day for each month for the year 2000 at different water depths.

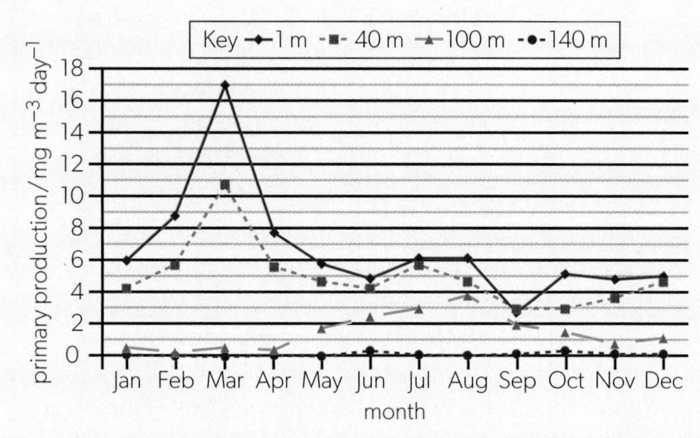

Source: DataStreme Ocean (2004) Copyright American Meteorological Society

a. State the month when total photosynthesis was greatest. [1]

b. Identify, with a reason, the water depth that receives no light. [1]

c. In the upper 40 m there is a drop in photosynthesis from March to June. This is probably due to lack of nutrients, reducing the population density. Suggest, with a reason other than nutrient levels, what might have increased photosynthesis at 100 m from April to August. [1]

d. Compare production in March with production in September. [3]

3. Antibiotics are sometimes given orally to poultry to prevent disease that may lead to reduced growth. Antibiotic resistance of bacteria from turkeys and chickens bred for meat and from egg laying hens was measured.

Excrement was collected and *Escherichia coli* bacteria were isolated. These bacteria were tested for resistance to a range of antibiotics and the results are shown below.

Number of antibiotics to which *E. coli* are resistant	**Turkeys *n* = 43**	**Chickens *n* = 45**	**Egg laying hens *n* = 20**
0	7	9	13
1	8	5	3
2	7	7	0
3	2	7	3
4	5	7	1
≥5	14	10	0

a. Calculate the percentage risk of bacteria becoming resistant to more than five kinds of antibiotics in turkeys and egg laying hens. [1]

b. Compare the incidence of drug resistance in bacteria from chickens and egg laying hens. [2]

c. Discuss the hypothesis that giving antibiotics increases antibiotic resistance in poultry bacteria. [2]

d. Suggest how antibiotic-resistant bacteria are passed from animals to humans. [1]

4. Metals such as zinc, nickel and copper are toxic to most plants. However, some terrestrial plants can store quite a lot of these metal ions in their tissues. These plants are called hyperaccumulators and could be valuable in reducing the levels of such metal ions in the soil. Some species of *Alyssum* are known to be hyperaccumulators. Two of these *Alyssum* species were grown in nutrient solutions with different concentrations of nickel ions. As a control, each species was grown in nutrient solution which contained no nickel. The following chart shows the biomass production for each species.

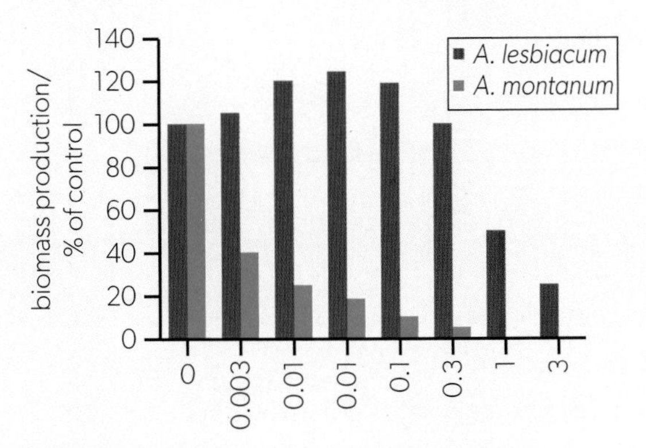

Source: Kramer U et al, (1996) Nature 379 page 635

a. Identify the nickel concentration at which the biomass production is equal to the control in *A. lesbiacum*. [1]

b. Compare the effect of nickel concentration on the growth of both species of *Alyssum*. [13]

c. Suggest why a nutrient solution was used instead of soil. [1]

The graph below shows the percentage of nickel in the dry biomass of the shoots and roots of these plants.

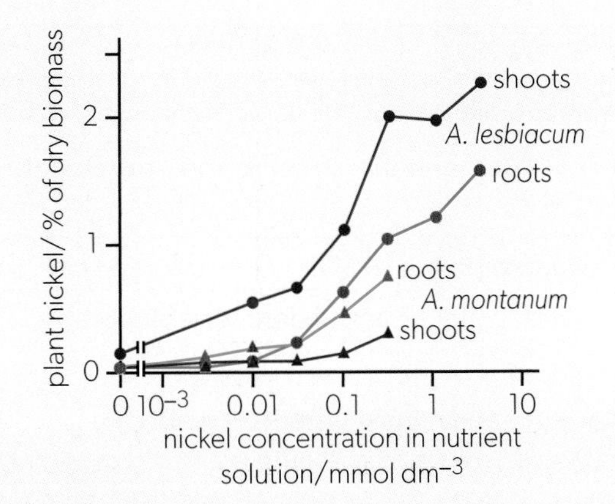

Source: Kramer U et al, (1996) Nature 379 page 635

d. Calculate the change in the percentage of nickel in the dry biomass of *A. lesbiacum* roots when the nickel concentration is increased from 0.1 to 1.0 mmol dm^{-3}. [1]

e. Compare the percentage dry biomass of nickel in the roots and shoots between the two species. [2]

f. Suggest a reason for the difference in the percentage dry biomass of nickel in roots and shoots between *A. montanum* and *A. lesbiacum*. [2]

g. Predict, with an explanation, which species would be most useful in decontaminating soils containing high levels of nickel. [2]

5. A long-term study examined the effect of ocean climate on the foraging (hunting) success of northern elephant seals (*Mirounga angustirostris*). Throughout their pregnancy, female northern elephant seals capture fish for food, which is converted to food reserves for milk production. Seal pups feed solely on their mother's milk so a pup's mass correlates positively with the mother's success at foraging. The Pacific Ocean fluctuates between extended warmer and colder periods. These bring with them changes identified broadly as either an "anchovy regime" (ocean temperatures cooler than normal, increased nutrient supply, high catches of anchovies) or a "sardine regime" (ocean temperatures warmer than normal, overall low productivity, low anchovy but high sardine catch). The graph below shows the relationship between weaning mass of northern elephant seal pups and anchovy and sardine regimes.

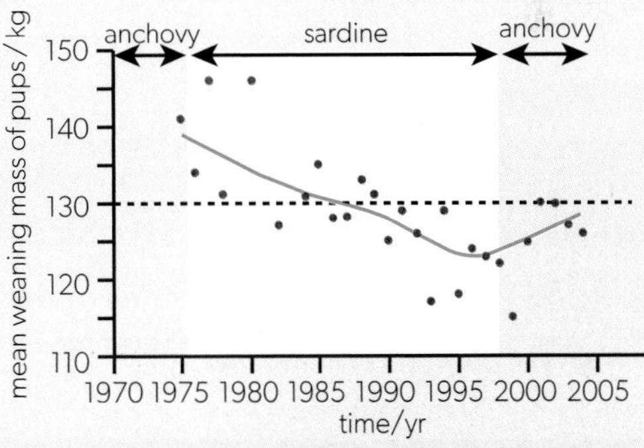

Source: Le Boeuf. BMC Biol 3, 9 (2005). https://doi.org/10.1186/1741-7007-3-9

a. Calculate the greatest difference in weaning masses during the sardine regime. [1]

b. Describe the relationship between mean weaning mass and the sardine and anchovy regimes from 1975 to 2004. [1]

c. Discuss what type of ocean water temperatures result in more successful foraging by female northern elephant seals. [3]

d. Suggest a possible effect of climate change on the breeding success of northern elephant seals. [2]

Internal assessment: The scientific investigation

▲ **Figure 1** An ideal process for coming up with a testable research question is to spend some time "wandering and wondering" observing the natural world

Throughout the IB biology syllabus, the application of skills is intended to expose you to a range of experimental and mathematical techniques as well as some suggestions for how technology can support inquiry. A subset of these skills is designated as "Practicing techniques". This is intended to introduce you to a range of protocols that can be modified and combined to carry out investigations of your own design. A culminating experience as an IB biology student is an open-ended inquiry called the Scientific investigation.

The Scientific Investigation involves production of a written report that mirrors the components and style of a professional scientist's research publication. It involves a three-stage inquiry process:

- exploring and designing
- collecting and processing data
- concluding and evaluating.

The written report of your investigation will be assessed by your teacher using four criteria:

- Research design
- Data analysis
- Conclusion
- Evaluation

Samples of your teacher's marking will be submitted to an IB-appointed moderator to ensure fair application of the criteria.

Research design

The research design criterion assesses the extent to which you effectively communicate:

- the purpose of the study
- your methodology choices
- a description of the intended execution of the methodology.

This assessment criterion has three components.

1. **The research question is described within a specific and appropriate context.**

   You should include the following points.

   - Identification of the dependent and independent variables or the variables which appear to be correlated.
   - A description of the academic and natural systems in which the research question is embedded. You should provide some context so that a reader

ATL Communication skills: Responding correctly to the language in a rubric

The research design criterion includes these expectations:

- specific and appropriate context
- relevant and sufficient data
- a methodology description which enables the investigation to be reproduced.

What evidence should be found within your report to ensure that it is accurately described by these descriptions?

who is less familiar with your topic can establish the importance of your study. So, include background theory of direct relevance that provides context for your research question. This should be accompanied by correctly formatted citations. If web-based publications are cited, then the URL and date of access must be included.

- A statement about what the scientific community currently has established and what gap is filled by your experiment.

2. **Methodological considerations associated with collecting relevant and sufficient data to answer the research question are explained.**

   You should ensure the following.

   - The methods for measuring the dependent and independent variables are justified.
   - If relevant, justification for the selection of a database or a model system is given.
   - An explanation is provided of the process of collecting sample data.
   - Justification is given for decisions about the scope, quantity and quality of measurements. This should include specification of the number of repetitions and the range, interval or values of the independent variable.
   - Decisions about the precision of measurements are explained.
   - Identification of control variables and the choice of method of their control is given. This might involve the inclusion of additional control experiments.
   - There is recognition and planning for any relevant safety, ethical or environmental issues.

3. **The description of the methodology for collecting or selecting data allows for the investigation to be reproduced.**

   - This means that a reader readily understands how the methodology was implemented and could repeat the investigation based on the method description.
   - If an organism is part of the investigation the scientific name should be included in the research question, written using the correct combination of capitals and italics.
   - You should clearly indicate additional variables that could influence the dependent variable if they are not held constant. Aim to address the most significant three or four variables that need to be controlled. When describing the method, additional controls could be discussed. In some cases, variables are not controllable—for example, in the case of fieldwork. In this case, these variables should be monitored.
   - In designing the method, you must ensure as far as possible that change in the dependent variable is solely due to changes in the independent variable. You must change only one thing and attempt to hold everything else constant.

ATL Applications of skills: Designing and explaining a valid methodology

In a group of your classmates, choose one of the following three research questions and collaboratively create a "Research design" section.

- Does an infusion of burned plant matter stimulate germination and post-germination growth of *Eucalyptus pilularis* seeds compared with deionized water? (NB wildfires in New South Wales were extinguished by torrential rainfall.)
- To what extent does proximity to the Clifton Pier power station correlate with percentage algae cover on the ocean bed? (NB Clifton Pier burns oil for electricity production. Tanker delivery often leads to oil slicks.)
- To what extent does body composition, as measured by bioelectric impedance on a bathroom scale, correlate with the rate of dive tank gas consumption in a 30-minute scuba dive?

- When describing your method, you should imagine a student who is perhaps two years younger than you attempting to follow your method. Have you written down all the steps? A photograph of the apparatus could be included. The description of the method needs to be concise as the entire report must not exceed 3000 words.
- Because of their complexity and variability, biological systems require replicate observations and multiple samples of material. As a rule, the lower limit is five measurements or a sample size of five. The number of replicates will vary within limits of the time available for an investigation, but you should aim for sample sizes larger than five if possible.

Data analysis

The data analysis criterion assesses the extent to which your report provides evidence that you have recorded, processed and presented data according to conventions. The data needs to be presented in ways that are clear and relevant to the research question.

This assessment criterion has three aspects.

1. **Communication of the recording and processing of the data is both clear and precise.**
    - Clear communication means that the method of processing can be understood easily.
    - Precise communication refers to following certain conventions correctly. The relevant conventions are those relating to the annotation of graphs and tables, the use of metric units, decimal places, uncertainty figures and consistent use of significant figures.
    - Qualitative observations can be organized into a table.
2. **Recording and processing of data shows evidence of an appropriate consideration of uncertainties.**
    - The precision of the measuring device should be indicated at the head of the column containing measurements. Usually, the precision of the device is indicated on the literature that is packed with the device, otherwise the minimum uncertainty should be stated.
3. **Processing of data relevant to addressing the research question is carried out appropriately and accurately**
    - Throughout the syllabus, a range of mathematical tools are introduced. In addition, the two IB mathematics courses (AA and AI) both include learning about mathematical tools that can be used to demonstrateunderstanding of the uncertainties in both raw data and processed data. These tools include standard deviation, standard error, measures of central tendency, box-and-whiskers plots, percentage change, error bars, R^2 and r. Mathematical standards should be used when rejecting outliers.
    - The chi-squared test is a versatile hypothesis testing tool. Consider using the *t*-test learned in your mathematics class or statistical tests learned in courses like geography.

Data-based questions: The relationship between average body length and population density in aquarium fish

Two aquaria were set up which were identical in size and shape. Ten goldfish (*Carassius auratus*) were placed into one aquarium. Forty goldfish were placed into the other. The fish were taken from the same population and assigned to the two aquaria randomly. The same amount of food and oxygen were supplied per fish to both tanks. The lengths of the fish were measured after 2 weeks.

Length of fish / mm (±1 mm)				
10 fish per aquarium	**40 fish per aquarium**			
50	61	52	52	54
41	50	59	61	60
59	45	54	63	56
52	56	57	53	51
49	57	60	60	57
58	53	54	59	56
56	52	44	46	53
47	45	51	58	51
66	60	54	44	60
50	59	51	55	58
$\bar{x}$ = 52.8 mm SD = 7.1	$\bar{x}$ = 54.5 mm SD = 5.0			

1. Of the two treatment groups, identify with a reason which data set was the most variable. [1]
2. Explain why the minimum uncertainty of a ruler is 1 mm. [2]
3. Construct a bar graph of the results that uses the values of SD as error bars. [2]
4. State possible null and alternate hypotheses. [2]
5. Examine the graph. Suggest with a reason, which hypothesis appears to be true. [1]
6. Explain the ways in which the raw and processed data provides evidence that the student has engaged with uncertainties. [3]
7. Explain how a statistical test can be carried out:
   a. to test whether the mean lengths of the two samples are different [2]
   b. to test whether the frequency distributions of the two samples are the same or different. [2]

ATL Application of skills: Designing ethical investigations

IB biology students are directed to be ethical in the execution of their experiments. Here are some guiding questions to consider when designing a research method.

- If human subjects are to be used, are they willing participants? Are they old enough to consent? Will they have full access to the conclusions drawn from the experiment?
- If animals are to be used, can they be captured from the local environment and released again unharmed? Are the treatments they are exposed to within the range of what they would experience in the wild?
- Can this investigation be conducted in a way that has lower negative impact on the environment?

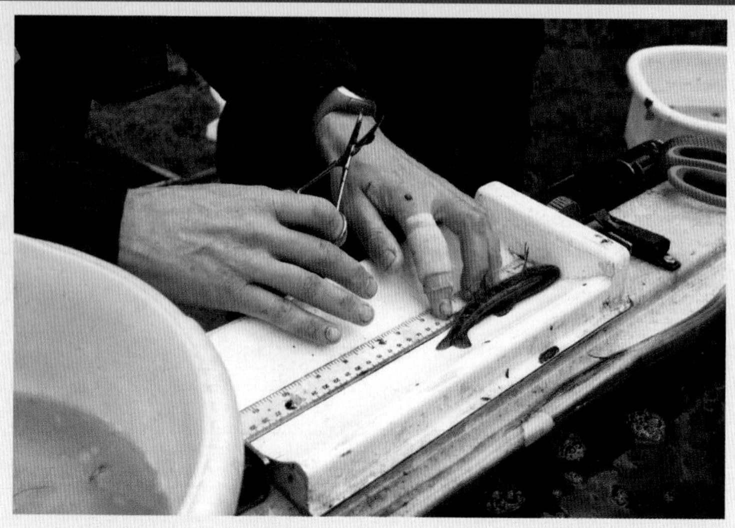

▲ **Figure 2** Protocol for the goldfish experiment

Study the protocol in Figure 2. Evaluate the goldfish experiment using the ethical guidelines published by the IB.

Conclusion

This criterion assesses the extent to which you have successfully related the experimental results to your research question and the research context.

There are two aspects to this criterion.

1. **A conclusion is justified that is relevant to the research question and fully consistent with the analysis presented.**

   A conclusion that is fully consistent requires the interpretation of processed data including acknowledgement of uncertainties.

2. **A conclusion is justified through relevant comparison to the accepted scientific context.**

   Scientific context refers to information that could come from published materials, published values, course notes, textbooks or other outside sources.

The conclusion should state the most important outcome of the experiment. You should indicate the extent to which you have succeeded in addressing the need for the study articulated in the research design section of your report. Rather than restating what has been covered in the analysis, focus on what you have established and what your findings mean. You should avoid being too definitive in your conclusions. Correlations are easier to claim than cause-and-effect relationships. It is important to say that the data supports a particular conclusion rather than proves it. The conclusion section should consistently reference the research question as well as any hypotheses. In addition, the conclusion section should thoroughly reference the relevant biology theory.

Activity: Justifying conclusions

Changes in the partial pressure of CO_2 at floor level in a classroom were measured over a 45-minute period using a data logger. Time zero represents the time when students entered the classroom.

Time / (±1)	pCO_2 at floor level / ppm (±100 ppm)
0	505
600	515
900	569
1800	620
2700	700

Note the lack of precision in the carbon dioxide sensor. Discuss whether the data supports the conclusion that the pCO_2 rose over the duration of the class.

Evaluation

The evaluation criterion assesses the extent to which your report provides evidence of evaluation of the methodology, discusses the implications of any errors on results and suggests improvements.

This assessment criterion has two aspects.

1. The report explains the relative impact of specific methodological weaknesses or limitations.
2. Realistic improvements to the investigation, that are relevant to the identified weaknesses or limitations, are explained.

Weaknesses could relate to issues relating to failing to control variables, systematic errors or random errors.

Two guiding questions for your evaluation could be:

- What makes me lack confidence in my conclusion?
- What modifications could I make to increase confidence in my conclusion?

Your evaluation should cover the design aspects of your experiment as well as performance aspects. When discussing errors, remember to include their impact on the result. For example, a systematic error could cause measurements to be larger than their true value. Random errors could decrease the reliability of data, making it more variable and leading to a decreased likelihood that significant statistical results will be obtained.

Index

p223: Panther Media GmbH / Alamy Stock Photo; **p225:** November 2005, David Goodsell, doi:10.2210/rcsb_pdb/mom_2005_11; **p227:** Lebendkulturen.de/Shutterstock; **p229(t):** Avalon.red / Alamy Stock Photo; **p229(b):** Gerry Morgan / Alamy Stock Photo; **p230:** coramueller/123RF; **p231(t):** guruxox / Alamy Stock Photo; **p231(b):** Prof. Oscar L. Miller/Science Photo Library; **p232:** Barfooz at the English Wikipedia., CC BY-SA 3.0/Wikimedia Commons; **p233(t):** CNRI / Science Photo Library; **p233(ml):** Barry Juniper; **p233(mr):** Edgloris Marys / Alamy Stock Photo; **p233(bl):** Andrew Allott; **p233(br):** Keith R. Porter / Science Photo Library; **pp235-236:** Andrew Allott; **p237:** NicoV/LGPL/Wikimedia Commons; **p238:** April 2007, Graham T. Johnson, David Goodsell; doi:10.2210/rcsb_pdb/mom_2007_4; **p240(t):** RooM the Agency / Alamy Stock Photo; **p240(bl):** Keith R. Porter / Science Photo Library; **p240(br):** Thomas Deerinck, NCMIR / Science Photo Library; **p241:** Caitlin Sedwick, CC BY 2.5/Wikimedia Commons; **p242:** Berkshire Community College Bioscience Image Library, CC0/Wikimedia Commons; **p244:** BSIP SA / Alamy Stock Photo; **p246:** Eye of Science / Science Photo Library; **p247:** Don W. Fawcett / Science Photo Library; **p248(t):** Michael Abbey/ Science Photo Library; **p248(b):** Science Photo Library / Alamy Stock Photo; **p249:** agefotostock / Alamy Stock Photo; **p252(t):** Obstetrics & Gynaecology / Science Photo Library; **p252(b):** Everett Collection/Shutterstock; **p254:** SCIEPRO/Getty Images; **p255(t):** Johner Images / Alamy Stock Photo; **p255(b):** Andrew Allott; **p256:** Iva Dimova/Shutterstock; **p259(t):** Science Photo Library / Alamy Stock Photo; **p259(b):** Mikael Häggström, M.D., CC0/Wikimedia Commons; **p261:** Mike Cerillo, CC BY-SA 2.0/Wikimedia Commons; **p262(t):** Power and Syred / Science Photo Library; **p262(m):** Nizsalóczki Zsolt, CC BY-SA 4.0/Wikimedia Commons; **p262(b):** James Mendelssohn; **pp263-264:** Cambridge University Press; **p265(t):** Andrew Allott; **p265(b):** Science Photo Library; **p266:** Charles McClean / Alamy Stock Photo; **p270(t):** ssuaphotos/Shutterstock; **p270(b):** NASA Goddard Space Flight Center from Greenbelt, MD, USA/Wikimedia Commons; **p272:** blickwinkel / Alamy Stock Photo; **p274(l):** Sean Locke Photography/Shutterstock; **p274(r):** LeventeGyori/ Shutterstock; **p275(t):** Mark Boulton / Alamy Stock Photo; **p275(b):** PhotoAlto / Alamy Stock Photo; **p276(t):** Andrew Allott; **p276(b):** Sebastian Kaulitzki/Shutterstock; **p280(t):** Dr Keith Wheeler / Science Photo Library; **p280(b):** Power and Syred / Science Photo Library; **p281:** Dr Keith Wheeler / Science Photo Library; **p282:** Jubal Harshaw/Shutterstock; **p289:** blickwinkel / Alamy Stock Photo; **p291:** Andrew Allott; **p292:** Dr. Richard Kessel and Dr. Gene Shih / Science Photo Library; **p293(t):** dpa picture alliance / Alamy Stock Photo; **p293(b):** Göran Gustafson / Alamy Stock Photo; **p294(t):** Bob Burgess / Alamy Stock Photo; **p294(b):** Minden Pictures / Alamy Stock Photo; **p295(t):** Ekaterina P. Lamber, Pascale Guicheney, and Nikos Pinotsis, CC BY 4.0/Wikimedia Commons; **p295(b):** Nourddin11, CC BY-SA 4.0/Wikimedia Commons; **p297:** Mediscan / Alamy Stock Photo; **p299(l):** Mauro Fermariello / Science Photo Library; **p299(r):** Wagner Souza e Silva/Museum of Veterinary Anatomy FMVZ USP, CC BY-SA 4.0/Wikimedia Commons; **p301:** Voxymoron; **p302(t):** John Warburton-Lee Photography / Alamy Stock Photo; **p302(m):** NSP-RF / Alamy Stock Photo; **p302(bl):** Papilio / Alamy Stock Photo; **p302(br):** Sekar B/Shutterstock; **p305(l):** Gerd Guenther / Science Photo Library; **p305(r):** Dr Keith Wheeler / Science Photo Library; **p308:** Maximillian cabinet/ Shutterstock; **p309(t):** Imagebroker Alamy Stock Photo; **p309(b):** Google Maps; **p310:** Richard Greenway; **p311(t):** Bob Gibbons / Alamy Stock Photo; **p311(m):** Scenics & Science / Alamy Stock Photo; **p311(b):** robertharding / Alamy Stock Photo; **p312(t):** Annina Kaisa Johanna Niskanen,Pekka Niittynen,Juha Aalto,Henry Väre,Miska Luoto; https://doi.org/10.1111/ddi.12889; **p312(b):** Suzanne Long / Alamy Stock Photo; **p314:** Radharc Images / Alamy Stock Photo; **p316(t):** NOAA; **p316(b):** Kevin Lino/NMFS/PIFSC/ESD/NOAA; **p318:** Sémhur/NASA/Wikimedia Commons; **p319(t):** WClarke, CC BY-SA 3.0/Wikimedia Commons; **p319(m):** Darylnovak, CC BY-SA 3.0/Wikimedia Commons; **p319(b):** Hemis / Alamy Stock Photo; **p320:** Rolf Nussbaumer Photography / Alamy Stock Photo; **p321(t):** Nature Picture Library / Alamy Stock Photo; **p321(b):** Andrew Allott; **p322:** Mykola Swarnyk, CC BY-SA 3.0/Wikimedia Commons; **p323:** Drazhnikova, CC BY-SA 4.0/Wikimedia Commons; **p325(tl):** Oxford University Press; **p325(tr):** Digital Stock/Corbis; **p325(bl):** Eric Grave / Science Photo Library; **p325(br):** SAG Culture Collection, T. Darienko 2017, CC BY-SA 4.0; **p326:** Heritage Image Partnership Ltd / Alamy Stock Photo; **p328(t):** Ivanpavlisko/Shutterstock; **p328(m):** Heritage Image Partnership Ltd / Alamy Stock Photo; **p328(b):** PRISMA ARCHIVO / Alamy Stock Photo; **p329(l):** Papilio / Alamy Stock Photo; **p329(r):** imageBROKER / Alamy Stock Photo; **p330(tl):** Dariusz Kowalczyk, CC BY-SA 4.0/Wikimedia Commons; **p330(tr):** Jeremy R. Rolfe, Date taken: 6 February 2010, Licence: CC BY; **p330(b):** aroid from San Luis Obispo, CA, USA, CC BY 2.0/Wikimedia Commons; **p331(tl):** Uwe Schmidt, CC BY-SA 4.0/Wikimedia Commons; **p331(tr):** Stephen Barlow; **p331(ml):** NickEvansKZN/Shutterstock; **p331(mr):** Quartl, CC BY-SA 3.0/Wikimedia Commons; **p331(bl):** bierchen/Shutterstock; **p331(br):** Alexander Vasenin, CC BY-SA 3.0/Wikimedia Commons; **p332:** Dorling Kindersley ltd / Alamy Stock Photo; **p336(t):** Equinox Graphics / Science Photo Library; **p336(m):** Biophoto Associates / Science Photo Library; **p336(b):** Millard H. Sharp / Science Photo Library; **p338:** Vittorio Ricci - Italy/Moment/Getty Images; **p339(t):** Darwin Dale / Science Photo Library; **p339(b):** Vsevolod Zviryk / Science Photo Library; **p340:** Andrew Allott; **p342(t):** Tony Cordoza / Alamy Stock Photo; **p342(bl):** Andrew Allott; **p342(br):** Andrew Allott; **p343(t):** Kenneth Eward/Biografx/Science Photo Library; **p343(b):** Areid3/Wikimedia Commons; **p354:** Science Photo Library / Alamy Stock Photo; **p355(t):** Steve Taylor ARPS / Alamy Stock Photo; **p355(b):** Jose Calvo / Science Photo Library; **p360:** St Mary's Hospital Medical School / Science Photo Library; **p361(t):** NOAA Okeanos Explorer Program, Gulf of Mexico 2012 Expedition; **p361(b):** Dennis Kunkel Microscopy / Science Photo Library; **p362:** Lisa F. Young/Shutterstock; **pp364-365:** Andrew Allott; **p366:** Photodisc/Getty Images; **p373:** robert cicchetti / Alamy Stock Photo; **p374(t):** Andrew Allott; **p374(m):** SpatzPhoto / Alamy Stock Photo; **p374(b):** sandy young / Alamy Stock Photo; **p377(t):** CNRI/Science Photo Library; **p377(b):** Maurice Savage / Alamy Stock Photo; **p378:** Dr David Furness, Keele University / Science Photo Library; **p380(t):** Medical Research Council of the United Kingdom; **p380(b):** Palo_ok/Shutterstock; **p381:** Rick & Nora Bowers / Alamy Stock Photo; **p382:** Igor Kovalchuk/Shutterstock; **p383:** Eric Isselee/Shutterstock; **p384(t):** Science Photo Library / Alamy Stock Photo; **p384(b):** Colin Harris / era-images / Alamy Stock Photo; **p385(t):** Corel Corporation 1994; **p385(b):** Kostyantyn Ivanyshen/Alamy Stock Photo; **p386:** Sask Photography / Alamy Stock Photo; **pp387-388:** Andrew Allott; **p389(t):** William Allott; **p389(b):** malenki, CC BY 3.0/Wikimedia Commons; **p394(t):** Thanet Earth; **p394(b):** Birmingham Institute of Forest Research (BIFoR); **p396:** January 2010, David Goodsell, doi:10.2210/rcsb_pdb/mom_2010_1; **p397:** K. M. Zielinska-Dabkowska/Asensetek Lighting Passport Pro Standard Spectrometer/Springer Nature; **p398(l):** Andrew Allott; **p398(r):** Courtesy of Plant Delights Nursery, Inc.; **p399:** Adwo / Alamy Stock Photo; **p402:** Andrew Allott; **p406:** Andrew Allott; **pp408-409:** Cambridge University Press; **p410(l):** The University of Chicago Press; **p410(r):** Charles A. Pasternak; Foreword. Biosci Rep 1 December 1991; 11 (6): 293–294. doi: https://doi.org/10.1007/BF01130210; **p412:** Steve Gschmeissner / Science Photo Library; **p413(t):** Juan Gaertner / Science Photo Library; **p413(b):** Science Photo Library; **p414:** PDB ID: 3qak_jmol, DOI Citation: Xu, F., Wu, H., Katritch, V., Han, G.W., Cherezov, V., Stevens, R., GPCR Network (GPCR) doi: 10.2210/pdb3QAK/pdb; **p415:** wildestanimal/Shutterstock; **p417:** Thomas Lukassek / Alamy Stock Photo; **p419(t):** ZUMA Press, Inc. / Alamy Stock Photo; **p419(m):** RCSB; **p419(b):** RCSB; **p422:** RCSB; **p425:** Juan Gaertner/Shutterstock; **p427:** lev dolgachov/123RF; **p428(t):** Steve Gschmeissner / Science Photo Library; **p428(b):** Don Fawcett / Science Photo Library; **p433:** Rustam Shanov/Shutterstock; **p434:** Steve Gschmeissner / Science Photo Library; **p437:** Dennis Kunkel Microscopy / Science Photo Library; **p440:** iLUXimage / Alamy Stock Photo; **p442(t):** Thomas Deerinck, NCMIR / Science Photo Library; **p442(b):** Onelia Pena/Shutterstock; **p444(t):** Taras Kushnir/Shutterstock; **p444(b):** The Oxfordshire Chilli Garden / Alamy Stock Photo; **p446:** S1001/ Shutterstock; **p447(l):** Darwin Dale / Science Photo Library; **p447(r):** Nature's Faces / Science Source / Science Photo Library; **p450:** Sebastian Kaulitzki/Science Photo Library/Getty Images; **p451(t):** Denis-Huot / Nature Picture Library / Science Photo Library; **p451(b):** Asklepios Medical Atlas/Science Photo Library; **p454(t):** Jose Calvo / Alamy Stock Photo; **p454(m):** Science Photo Library / Alamy Stock Photo; **p455:** Horizon International Images / Alamy Stock Photo; **p456(t):** Jose Calvo / Science Photo Library; **p456(b):** Photodisc/Getty Images; **p457(t):** Alvin Telser / Science Photo Library; **p457(b):** Living Art Enterprises / Science Photo Library; **p458:** Oxford University Press; **p460:** Science Photo Library / Alamy Stock Photo; **p463(t):** Science Photo Library / Alamy Stock Photo; **p463(b):** Sebastian Kaulitzki/Shutterstock; **p464(t):** ArtisticCaptures/iStock/Getty Images; **p464(b):** Okano, K., Kaczmarzyk, J.R., Dave, N. et al. Sleep quality, duration, and consistency are associated with better academic performance in college students. npj Sci. Learn. 4, 16 (2019). https://doi.org/10.1038/s41539-019-0055-z / npj Science of Learning under http://creativecommons.org/licenses/by/4.0; **p465:** PCN Photography / Alamy Stock Photo; **p471:** Andrew Allott; **p473(t):** Scenics & Science / Alamy Stock Photo; **p473(b):** Grant Heilman Photography / Alamy Stock Photo; **p475:** Andrew Allott; **p478:** Biophoto Associates / Science Photo Library; **p479-480:** Andrew Allott; **p481(t):** Lennart Nilsson, Boehringer Ingelheim International GMBH, TT / Science Photo Library; **p481(b):** Science History Images / Alamy Stock Photo; **p482:** Roger Hutchings / Alamy Stock Photo; **p483:** Andrew Allott; **p484:** Science Photo Library / Alamy Stock Photo; **p485(t):** Science History Images / Alamy Stock Photo; **p485(b):** September 2001, David Goodsell, doi:10.2210/rcsb_pdb/mom_2001_9; **p486(t):** James Cavallini / Science Photo Library; **p486(b):** May 2009, David Goodsell, doi:10.2210/rcsb_pdb/mom_2009_5; **p487:** Steve Gschmeissner/Science Photo Library; **p488(t):** Photodisc/ Getty Images; **p488(b):** RGB Ventures / SuperStock / Alamy Stock Photo; **p490:** Angela Hampton Picture Library / Alamy Stock Photo; **p493:** Samara Heisz / Alamy Stock Photo; **p494:** Pacific Press Media Production Corp. / Alamy Stock Photo; **p497(t):** Shaun Baesman / US Geological Survey / Science Photo Library; **p497(m):** Science History Images / Alamy Stock Photo; **p497(b):** Stubblefield Photography/Shutterstock; **p500:** Uri Golman / Nature Picture Library / Science Photo Library; **p501(t):** photowind/Shutterstock; **p501(b):** Dr Jeremy Burgess / Science Photo Library; **p502(t):** David Fleetham / Alamy Stock Photo; **p502(b):** T. R. Shankar Raman, CC BY 3.0/Wikimedia Commons; **p503:** Andrew Allott; **p504:** Mark Twells; **p505(l):** A Garden / Alamy Stock Photo; **p505(r):** Jonathan Need / Alamy Stock Photo; **p507:** FLPA / Alamy Stock Photo; **p510:** A. Lestenkof, Ecosystem Conservation Office, Aleut Community of St. Paul Island; **p511:** Mags Cousins; **p512:** Radius Images / Alamy Stock Photo; **p513(t):** Stephen Barlow; **p513(bl):** Stephen Barlow; **p513(br):** Mike Read / Alamy Stock Photo; **p514(tl):** Stephen Jaquiery; **p514(tr):** Minden Pictures / Alamy Stock Photo; **p514(bl):** Jaime Leonardo Gonzalez Salazar / Alamy Stock Photo; **p514(br):** High Arctic Institute; **p516:** James Mendelssohn; **p517(t):** Stephen Barlow; **p517(b):** Dr Jeremy Burgess/Science Photo Library; **p518(t):** David Menzies; **p518(b):** Carlos Villoch - MagicSea.com / Alamy Stock Photo; **p519(l):** Helmut Corneli / Alamy Stock Photo; **p519(r):** imageBROKER / Alamy Stock Photo; **p521:** Andrew Allott; **p523(l):** Duncan Shaw / Science Photo Library; **p523(r):** John Devries/Science Photo Library; **p527(t):** Andrew Allott; **p527(m):** CDC Public Health Image Library; **p527(b):** SB_Johnny, CC BY-SA 3.0, via Wikimedia Commons; **p529(t):** T-SERVICE / Science Photo Library; **p529(b):** Geoff Kidd/Science Photo Library; **p530:** Andrew Allott; **p531(t):** Kyle Carothers, NOAA-OE; **p531(b):** Patrick Landmann / Science Photo Library; **p533(l):** Jean Louis; **p533(m):** © chrisstockphotography / Alamy Stock Photo; **p533(r):** Octavio Campos Salles / Alamy Stock Photo; **p534:** Anton Sorokin / Alamy Stock Photo; **p535:** Andrew Allott; **p537:** Biosphoto / Alamy Stock Photo; **p539:** NPS Photo/ Kent Miller, Public Domain/Wikimedia Commons; **p540(t):** Dinodia Photos / Alamy Stock Photo; **p540(b):** Maximilian Buzun / Alamy Stock Photo; **p543:** NASA Earth Observations; **p544:** Photographee.eu/Shutterstock; **p546(t):** Mongabay; **p546(b):** hiroshi teshigawara/Shutterstock; **p547(t):** Inna Limanskaya/Shutterstock; **p547(m):** Madlen/Shutterstock; **p547(b):** YIFANG NIE/ Shutterstock; **p550:** Data/Image provided by NOAA Global Monitoring Laboratory, Boulder, Colorado, USA (https://esrl.noaa.gov/); **p553:** Photographers Choice/Getty Images; **p554(l):** Dima Moroz/Shutterstock; **p554(r):** NYPL / Science Source / Science Photo Library; **p556:** Dr Tim Evans / Science Photo Library; **p557(t):** Power and Syred / Science Photo Library; **p557(b):** © Brian W. Schaller / License: CC BY-NC-SA 4.0 / Wikimedia Commons; **p559:** The Book Worm / Alamy Stock Photo; **p561(t):** Andrew Allott; **p561(b):** Arek Kulczyk, Laboratory of DNA Replication and Repair, Institute for Quantitative Biomedicine, Rutgers University; **p562:** Thermo Fisher Scientific Life Technologies Ltd; **p563:** David H. Valle / Alamy Stock Photo; **p565:** Andrew Allott; **p566:** Arek Kulczyk, Laboratory of DNA Replication and Repair, Institute for Quantitative Biomedicine, Rutgers University; **p558:** Professor Oscar Miller / Science Photo Library; **p569:** Andrew Allott; **p572:** GeorgiD / Alamy Stock Photo; **p574:** Elizaveta Galitckaia / Alamy Stock Photo; **p575(t):** Design Pics/ Corey Hochachka/Getty Images; **p575(b):** Eric Isselee/Shutterstock; **p576:** Album / British Library / Alamy Stock Photo; **p577:** January 2010, David Goodsell, doi:10.2210/rcsb_pdb/mom_2010_1; **p585:** Dr Elena Kiseleva / Science Photo Library; **p589:** February 2001, David Goodsell, doi:10.2210/rcsb_pdb/mom_2001_2; **p590:** October 2013, David Goodsell, doi:10.2210/rcsb_pdb/mom_2013_10; **p592(t):** Eye of Science / Science Photo Library; **p592(b):** Martin Shields / Science Photo Library; **p593(t):** ANL/Shutterstock; **p593(b):** Dennis Kitchen Studio, Inc./Oxford University Press; **p597:** Konstantin Sutyagin/Shutterstock; **p598:** 54613/Shutterstock; **p599:** fotografixx/ iStock/Getty Images; **p600:** January 2015, David Goodsell, doi:10.2210/rcsb_pdb/mom_2015_1; **p606(t):** Will & Deni Mcintyre / Science Photo Library; **p606(b):** Genetek Biopharma GmbH; **p608:** Steve Gschmeissner / Science Photo Library; **p609(t):** Jim West / Alamy Stock Photo; **p609(b):** Adrian T Sumner / Science Photo Library; **p610:** Heleen van Rooijen; **p611:** Ed Reschke/Stone/Getty Images; **p612(t):** Catfaster, CC BY-SA 4.0 / Wikimedia Commons; **p612(b):** Berkshire Community College Bioscience Image Library, CC0 / Wikimedia Commons; **p613(t):** Andrew Allott; **p613(m):** Science Photo Library / Alamy Stock Photo; **p613(b):** Phanie / Alamy Stock Photo; **p615:** Graphodatsky et al., CC BY 2.0 / Wikimedia Commons; **p617(t):** Thomas Splettstoesser (www.scistyle.com)/Wikimedia Commons; **p617(b):** Andrew Allott; **p618-620:** Steve Gschmeissner / Science Photo Library; **p623:** Bill Malcom; **p624:** Tomasz Markowski/Dreamstime; **pp626-628:** Andrew Allott; **p629(t):** Gerard Peaucellier, ISM / Science Photo Library; **p629(b):** Science History Images / Alamy Stock Photo; **p631:** Steve Gschmeissner / Science Photo Library; **p632:** Samburu Trust; **p633:** Andy Gordon; **p634(t):**